HETEROLYTIC BOND DISSOCIATION ENERGIES

$$A:B \longrightarrow A^+ + :B^- \qquad \Delta H = \text{Heterolytic bond dissociation energy of } A:B \; (A^+ - :B^-)$$

H—H	401 1678			CH₃—H	313 1310
H—F	370 1548			CH₃—F	256 1071
H—Cl	334 1397			CH₃—Cl	227 950
H—Br	324 1356			CH₃—Br	219 916
H—I	315 1318			CH₃—I	212 887
H—OH	390 1632			CH₃—OH	274 1146

CH₃—Cl 227 950	CH₃—Br 219 916	CH₃—I 212 887	CH₃—OH 274 1146
C₂H₅—Cl 191 799	C₂H₅—Br 184 770	C₂H₅—I 176 736	C₂H₅—OH 242 1013
n-C₃H₇—Cl 185 774	n-C₃H₇—Br 178 745	n-C₃H₇—I 171 715	n-C₃H₇—OH 235 983
i-C₃H₇—Cl 170 711	i-C₃H₇—Br 164 686	i-C₃H₇—I 156 653	i-C₃H₇—OH 222 929
t-C₄H₉—Cl 157 657	t-C₄H₉—Br 149 623	t-C₄H₉—I 140 586	t-C₄H₉—OH 208 870
H₂C=CH—Cl 207 866	H₂C=CH—Br 200 837	H₂C=CH—I 194 812	
H₂C=CHCH₂—Cl 173 724	H₂C=CHCH₂—Br 165 690	H₂C=CHCH₂—I 159 665	H₂C=CHCH₂—OH 223 933
C₆H₅—Cl 219 916	C₆H₅—Br 210 879	C₆H₅—I 202 845	C₆H₅—OH 275 1151
C₆H₅CH₂—Cl 166 695	C₆H₅CH₂—Br 157 657	C₆H₅CH₂—I 149 623	C₆H₅CH₂—OH 215 900

* Values in blue represent kJ/mol

CHARACTERISTIC INFRARED ABSORPTION FREQUENCIES[a]

Bond	Compound Type	Frequency range, cm⁻¹	Reference
C—H	Alkanes	2850–2960	Sec. 16.18
		1350–1470	
C—H	Alkenes	3020–3080 (m)	Sec. 16.18
		675–1000	
C—H	Aromatic rings	3000–3100 (m)	Sec. 16.18
		675–870	
C—H	Alkynes	3300	Sec. 16.18
C=C	Alkenes	1640–1680 (v)	Sec. 16.18
C≡C	Alkynes	2100–2260 (v)	Sec. 16.18
C⩵C	Aromatic rings	1500, 1600 (v)	Sec. 16.18
C—O	Alcohols, ethers, carboxylic acids, esters	1080–1300	Sec. 18.11
			Sec. 19.18
			Sec. 23.22
			Sec. 24.25
C=O	Aldehydes, ketones, carboxylic acids, esters	1690–1760	Sec. 21.16
			Sec. 23.22
			Sec. 24.25
O—H	Monomeric alcohols, phenols	3610–3640 (v)	Sec. 18.11
			Sec. 28.14
	Hydrogen-bonded alcohols, phenols	3200–3600 (broad)	Sec. 18.11
			Sec. 28.14
	Carboxylic acids	2500–3000 (broad)	Sec. 23.22
N—H	Amines	3300–3500 (m)	Sec. 27.21
C—N	Amines	1180–1360	Sec. 27.21
C≡N	Nitriles	2210–2260 (v)	
—NO₂	Nitro compounds	1515–1560	
		1345–1385	

[a] All bands strong unless marked: *m*, moderate; *v* variable.

Organic Chemistry

Organic Chemistry

Fifth Edition

Robert Thornton Morrison

Robert Neilson Boyd

New York University

Allyn and Bacon, Inc.

Boston   London   Sydney   Toronto

Editorial-Production Service: Christine Sharrock, Omega Scientific
Photographer: Michael Freeman
Production editor: Elaine Ober
Manufacturing buyer: Ellen Glisker
Cover administrator: Linda Dickinson
Cover designer: Design Ad Cetera

Library of Congress Cataloging-in-Publication Data

Morrison, Robert Thornton
    Organic chemistry.

    Bibliography: p. 1403
    Includes index.
    1. Chemistry, Organic.   I.   Boyd, Robert Neilson.
II.   Title.
QD251.2.M67   1987   547   87–1003
ISBN 0–205–08453–2
ISBN (International) 0–205–08452–4

        Printed in the United States of America.
        10 9 8 7 6 5 4 3 2   91 90 89 88 87

Contents

4   Stereochemistry *I.*   **Stereoisomers**

Part Two

Special Topics

29   Aryl Halides   *Nucleophilic Aromatic Substitution*

Part Three

Biomolecules

Preface

In preparing this fifth edition, we have reorganized and rewritten, added new material and deleted old, clarified difficult points and updated our coverage. And, to help the students understand and to excite their interest, we have for the first time used *color*.

In organizing the new edition we had two chief aims. The first was to present organic chemistry at a rate at which students can absorb it. To accomplish this, we have moved much material to later places in the book. Although we *use* alcohols early, as before, we defer the systematic discussion of their complicated chemistry to Chapters 17 and 18. Alicyclic compounds are deferred until Chapter 12, and most of the discussion of polymerization and of rearrangements now appears as separate chapters in Part Two, Special Topics.

Our second aim was to present certain topics of growing importance in a unified way, and in a manner that emphasizes their significance to organic chemistry. To this end, we have created a number of new chapters within Part One, the Fundamentals. These chapters contain, rewritten and reorganized, much that was formerly in the chapters on nucleophilic substitution, alkenes, and alcohols.

The science of organic chemistry rests on a single premise: that chemical behavior is determined by molecular structure. Our basic approach to reactivity is to consider energy differences between reactants and transition states. We do this by examining—mentally and, by use of models, physically—the structures involved. But what we mean by "molecular structure" is constantly expanding, and our interpretation of chemical behavior must reflect this. *It was primarily to present these newer aspects of molecular structure that the book was reorganized and new chapters were created.*

In solution, all participants in a chemical reaction are solvated: the reactants and the products—and the transition state. Our examination of these *must* include any solvent molecules that help make up the structures and help determine their stabilities. And so, in Chapter 6, using as our examples the nucleophilic substitution reactions the students have just studied, we show how reactivity—and, with it, the course of reaction—is affected by the **solvent**. We show just how enormous solvent effects can be: that the presence of a solvent can speed up—or slow down—a reaction by a factor of 10^{20}; that a change from one solvent to another can bring about a million-fold change in reaction rate.

At the same time, in Chapter 6 the students are becoming acquainted with **secondary bonding**. They learn that these forces—ion–dipole, dipole–dipole, van der Waals—are involved in much more than solvent effects. They learn that, acting not only between different molecules but between different parts of the same molecule, secondary bonding plays a key role in determining the *shapes* of large molecules like proteins and DNA, shapes that determine, in turn, their biological properties.

It is becoming increasingly clear that any examination of molecular structure must be *three-dimensional.* To emphasize this, and to help guide the students through this complex area of organic chemistry, we introduce the principles of **stereochemistry** in three stages. In Chapter 4, we present the fundamentals of stereoisomerism. In Chapter 9, we deal with the concepts of *stereoselectivity* and *stereospecificity.* We show how stereochemistry helps us to understand reaction mechanisms; how this understanding can be used to control the stereochemical outcome of a reaction; and why we want to exercise this control — because the stereospecificity of biological reactions demands an equal stereoselectivity in the synthesis of drugs and hormones and pheromones.

And then, in Chapter 22, the students find that what they have learned about stereoselectivity and stereospecificity applies not only to stereochemically different molecules, but also to stereochemically different *parts of the same molecule.* They find that portions of a molecule may be stereochemically equivalent or non-equivalent, and that they must be able to distinguish between these if they are to understand subjects as widely different as NMR spectroscopy and biological oxidation and reduction. They must learn the concepts of *enantiotopic and diastereotopic ligands and faces.*

In Chapter 20, we show that three-dimensional chemistry goes far beyond what is generally thought of as stereochemistry. Up to this point, the students have learned something of the effects on reactivity of polar factors, steric factors, and the solvent. But there is another structural feature to be considered: the spatial relationship among reacting atoms and molecules. *Being in the right place,* it turns out, can be the most powerful factor of all in determining how fast a reaction goes — and what product it yields.

In this chapter we take up reactions from quite different areas, reactions seemingly quite dissimilar but having one quality in common: prior to reaction, the reactants are *brought together* and *held* in exactly the right positions for reaction to occur. They may be held by secondary bonding to an enzyme molecule; they may be held in a coordination sphere of a transition metal; they may even be two functional groups in a single molecule. Now, once they have been brought together, the substrate and the reagent are — if only temporarily — *parts of the same molecule.* And when they react, they enjoy a very great advantage over ordinary, separated reactants. The result is reaction with an enormously enhanced rate, reaction with a special stereochemistry.

The factor that makes all this possible we call **symphoria:** the bringing together of reactants into the proper spatial relationship. In Chapter 20 we introduce the concept with a set of reactions in which we can most readily *see* and *measure* symphoric effects: reactions involving neighboring group effects, where the bringing together requires nothing more than rotation about carbon–carbon bonds. Then we examine catalysis by transition metal complexes: basically the same kind of process, except that here the reactants are held, not by carbon, but by a transition metal. And, as with classical neighboring group effects, there are both rate enhancement — without the catalyst, reaction does not occur at all — and profound stereochemical consequences. Finally, we discuss catalysis by enzymes, and point out the striking similarity to the action of transition metals. An enzyme is much more complicated than a metal complex, and it binds the substrate and reagent by different forces. But fundamentally its function is the same: to bring together the reactants so that they are *near* each other and *in the right positions.*

We have greatly expanded our coverage of CMR spectroscopy, treating it in the same detailed fashion as we have proton NMR. To help the students learn, we have, as with other topics, leaned heavily on the working of problems: problems that call for the analysis of some 40 CMR spectra, problems that use CMR spectra to complement other kinds of spectra.

Finally, we have added **color** to our book. We have tried to do this thoughtfully and purposefully: not just to make the book attractive—although it *does*—but *to help the students learn.* We have used color in equations and in graphs and diagrams: to draw attention to changes that are taking place; to clarify mechanisms; to identify the chain-carrying particles in a chain reaction; to label structural units so that they can be followed through a series of reactions. We have, to the extent that it was feasible, been systematic: leaving groups are generally shown in red, for example, and nucleophiles in blue—and so are the bonds that represent the electron pairs they are taking away or bringing up. And, to bring home the importance of three-dimensional chemistry, we have included about 170 photographs of molecular models: to let the students *see* the shapes of the molecules they are dealing with, and to add reality to the formulas they write; and, we hope, to give them some sense of the *beauty*—as objects and as mental creations—of the structures that are the basis of organic chemistry.

ROBERT THORNTON MORRISON

ROBERT NEILSON BOYD

Acknowledgments

Our thanks to Sadtler Research Laboratories for the infrared and CMR spectra labeled "Sadtler", to the Infrared Data Committee of Japan for the infrared spectra labeled "IRDC", and to the following people for permission to reproduce material: Professor George A. Olah, Figure 5.6; Cornell University, Figure 5.1; the editors of The Journal of the American Chemical Society, Figures 5.6, 16.17, and 16.18; Walt Disney Productions, Figure 12.20; the Computer Graphics Laboratory, University of California, San Francisco for the photograph on page 1379; and, especially, Irving Geis for Figure 41.1. Our thanks also to Ealing Corporation for permitting us to visit their offices to photograph the molecular models in Figures 40.5, 40.6, and 41.7.

Our thanks to Michael Freeman for his splendid photographs, and for the pleasure of watching him at work. Our warm thanks to Christine Sharrock of Omega Scientific, who shepherded the book through production, from manuscript to finished pages, and proved at all times a valiant comrade-at-arms.

And, finally, our warm and totally inadequate thanks to Beverly Smith, who cheerfully took garbled dictation, rough scrawls, and crude sketches, and from these, as always, prepared an accurate and beautiful manuscript.

R.T.M.
R.N.B.

Part One

The Fundamentals

1

Structure and Properties

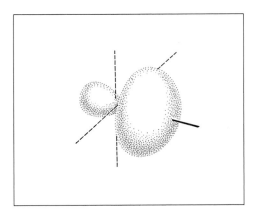

1.1 Organic chemistry

Organic chemistry is the chemistry of the **compounds of carbon**.

The misleading name "organic" is a relic of the days when chemical compounds were divided into two classes, inorganic and organic, depending upon where they had come from. Inorganic compounds were those obtained from minerals; organic compounds were those obtained from vegetable or animal sources, that is, from material produced by living organisms. Indeed, until about 1850 many chemists believed that organic compounds *must* have their origin in living organisms, and consequently could never be synthesized from inorganic material.

These compounds from organic sources had this in common: they all contained the element carbon. Even after it had become clear that these compounds did not have to come from living sources but could be made in the laboratory, it was convenient to keep the name *organic* to describe them and compounds like them. The division between inorganic and organic compounds has been retained to this day.

Today, although many compounds of carbon are still most conveniently isolated from plant and animal sources, most of them are synthesized. They are sometimes synthesized from inorganic substances like carbonates or cyanides, but more often from other organic compounds. There are two large reservoirs of organic material from which simple organic compounds can be obtained: *petroleum* and *coal*. (Both of these are "organic" in the old sense, being products of the decay of plants and animals.) These simple compounds are used as building blocks from which larger and more complicated compounds can be made.

1

We recognize petroleum and coal as the *fossil fuels*, laid down over millenia and non-renewable. They—particularly petroleum—are being consumed at an alarming rate to meet our constantly increasing demands for power. Today, less than ten percent of the petroleum used goes into making chemicals; most of it is simply burned to supply energy. There *are*, fortunately, alternative sources of power—solar, geothermal, nuclear energy—but where are we to find an alternative reservoir of organic raw material? Eventually, of course, we shall have to go to the place where the fossil fuels originally came from—the *biomass*—but this time directly, without the intervening millenia. The biomass is renewable and, used properly, can last as long on this planet as we can. In the meantime, it has been suggested, petroleum is too valuable to burn.

What is so special about the compounds of carbon that they should be separated from compounds of all the other hundred-odd elements of the Periodic Table? In part, at least, the answer seems to be this: there are so very many compounds of carbon, and their molecules can be so large and complex.

The number of compounds that contain carbon is many times greater than the number of compounds that do not contain carbon. These organic compounds have been divided into families, which generally have no counterparts among the inorganic compounds.

Organic molecules containing thousands of atoms are known, and the arrangement of atoms in even relatively small molecules can be very complicated. One of the major problems in organic chemistry is to find out how the atoms are arranged in molecules, that is, to determine the structures of compounds.

There are many ways in which these complicated molecules can break apart, or rearrange themselves, to form new molecules; there are many ways in which atoms can be added to these molecules, or new atoms substituted for old ones. Much of organic chemistry is devoted to finding out what these reactions are, how they take place, and how they can be used to synthesize compounds we want.

What is so special about carbon that it should form so many compounds? The answer to this question came to August Kekulé in 1854 during a London bus ride.

> "One fine summer evening, I was returning by the last omnibus, 'outside' as usual, through the deserted streets of the metropolis, which are at other times so full of life. I fell into a reverie and lo! the atoms were gambolling before my eyes . . . . I saw how, frequently, two smaller atoms united to form a pair, how a larger one embraced two smaller ones; how still larger ones kept hold of three or even four of the smaller; whilst the whole kept whirling in a giddy dance. I saw how the larger ones formed a chain . . . . I spent part of the night putting on paper at least sketches of these dream forms."—August Kekulé, 1890.

Carbon atoms can attach themselves to one another to an extent not possible for atoms of any other element. Carbon atoms can form chains thousands of atoms long, or rings of all sizes; the chains and rings can have branches and cross-links. To the carbon atoms of these chains and rings there are attached other atoms, chiefly hydrogen, but also fluorine, chlorine, bromine, iodine, oxygen, nitrogen, sulfur, phosphorus, and many others. (Look, for example, at cellulose on page 1339, chlorophyll on page 1207, and oxytocin on page 1357.)

Each different arrangement of atoms corresponds to a different compound, and each compound has its own characteristic set of chemical and physical properties. It is not surprising that nearly ten million compounds of carbon are known today and that this number is growing by half a million a year. It is not surprising that the study of their chemistry is a special field.

Organic chemistry is a field of immense importance to technology: it is the chemistry of dyes and drugs, paper and ink, paints and plastics, gasoline and rubber tires; it is the chemistry of the food we eat and the clothing we wear.

Organic chemistry is fundamental to biology and medicine. Aside from water, living organisms are made up chiefly of organic compounds; the molecules of "molecular biology" are organic molecules. Biology, on the molecular level, *is* organic chemistry.

1.2   The structural theory

"Organic chemistry nowadays almost drives me mad. To me it appears like a primeval tropical forest full of the most remarkable things, a dreadful endless jungle into which one does not dare enter for there seems to be no way out."—Friedrich Wöhler, 1835.

How can we even begin to study a subject of such enormous complexity? Is organic chemistry today as Wöhler saw it a century and a half ago? The jungle is still there—largely unexplored—and in it are more remarkable things than Wöhler ever dreamed of. But, so long as we do not wander too far too fast, we can enter without fear of losing our way, for we have a chart: the **structural theory**.

The structural theory is the basis upon which millions of facts about hundreds of thousands of individual compounds have been brought together and arranged in a systematic way. It is the basis upon which these facts can best be accounted for and understood.

The structural theory is the framework of ideas about how atoms are put together to make molecules. The structural theory has to do with the order in which atoms are attached to each other, and with the electrons that hold them together. It has to do with the shapes and sizes of the molecules that these atoms form, and with the way that electrons are distributed over them.

A molecule is often represented by a picture or a model—sometimes by several pictures or several models. The atomic nuclei are represented by letters or plastic balls, and the electrons that join them by lines or dots or plastic pegs. These crude pictures and models are useful to us only if we understand what they are intended to mean. Interpreted in terms of the structural theory, they tell us a good deal about the compound whose molecules they represent: how to go about making it; what physical properties to expect of it—melting point, boiling point, specific gravity, the kind of solvents the compound will dissolve in, even whether it will be colored or not; what kind of chemical behavior to expect—the kind of reagents the compound will react with and the kind of products that will be formed, whether it will react rapidly or slowly. We would know all this about a compound that we had never encountered before, simply on the basis of its structural formula and what we understand its structural formula to mean.

1.3   The chemical bond before 1926

Any consideration of the structure of molecules must begin with a discussion of *chemical bonds*, the forces that hold atoms together in a molecule.

We shall discuss chemical bonds first in terms of the theory as it had developed prior to 1926, and then in terms of the theory of today. The introduction of quantum mechanics in 1926 caused a tremendous change in ideas about how molecules are formed. For convenience, the older, simpler language and pictorial representations

are often still used, although the words and pictures are given a modern interpretation.

In 1916 two kinds of chemical bond were described: the *ionic bond* by Walther
Kossel (in Germany) and the *covalent bond* by G. N. Lewis (of the University of
California). Both Kossel and Lewis based their ideas on the following concept of
the atom.

A positively charged nucleus is surrounded by electrons arranged in concentric
shells or energy levels. There is a maximum number of electrons that can be
accommodated in each shell: two in the first shell, eight in the second shell, eight
or eighteen in the third shell, and so on. The greatest stability is reached when the
outer shell is full, as in the noble gases. Both ionic and covalent bonds arise from
the tendency of atoms to attain this stable configuration of electrons.

The **ionic bond** results from **transfer of electrons**, as, for example, in the
formation of lithium fluoride. A lithium atom has two electrons in its inner shell

$$\text{(Li) 2 1} \xrightarrow{\text{loss of } e^-} \text{(Li) 2}^{\oplus} \qquad \text{Li} \longrightarrow \text{Li}^+ + e^-$$

$$\text{(F) 2 7} \xrightarrow{\text{gain of } e^-} \text{(F) 2 8}^{\ominus} \qquad \text{F} + e^- \longrightarrow \text{F}^-$$

and one electron in its outer or valence shell; the loss of one electron would leave
lithium with a full outer shell of two electrons. A fluorine atom has two electrons
in its inner shell and seven electrons in its valence shell; the gain of one electron
would give fluorine a full outer shell of eight. Lithium fluoride is formed by the
transfer of one electron from lithium to fluorine; lithium now bears a positive
charge and fluorine bears a negative charge. The electrostatic attraction between
the oppositely charged ions is called an ionic bond. Such ionic bonds are typical of
the salts formed by combination of the metallic elements (electropositive elements)
on the far left side of the Periodic Table with the non-metallic elements (electronegative elements) on the far right side.

The **covalent bond** results from **sharing of electrons**, as, for example, in the
formation of the hydrogen molecule. Each hydrogen atom has a single electron;
by sharing a pair of electrons, both hydrogens can complete their shells of two.
Two fluorine atoms, each with seven electrons in the valence shell, can complete
their octets by sharing a pair of electrons. In a similar way we can visualize the
formation of HF, H_2O, NH_3, CH_4, and CF_4. Here, too, the bonding force is
electrostatic attraction: this time between each electron and *both* nuclei.

$$\text{H} \cdot + \cdot \text{H} \longrightarrow \text{H} : \text{H}$$

$$: \ddot{\text{F}} \cdot + \cdot \ddot{\text{F}} : \longrightarrow : \ddot{\text{F}} : \ddot{\text{F}} :$$

$$\text{H} \cdot + \cdot \ddot{\text{F}} : \longrightarrow \text{H} : \ddot{\text{F}} :$$

$$2\text{H} \cdot + \cdot \ddot{\text{O}} : \longrightarrow \text{H} : \overset{\text{H}}{\underset{\cdot\cdot}{\ddot{\text{O}}}} :$$

$$3\text{H}\cdot + \cdot\ddot{\text{N}}: \longrightarrow \text{H}:\overset{\displaystyle \text{H}}{\underset{\displaystyle \text{H}}{\ddot{\text{N}}}}:$$

$$4\text{H}\cdot + \cdot\dot{\text{C}}\cdot \longrightarrow \text{H}:\overset{\displaystyle \text{H}}{\underset{\displaystyle \text{H}}{\ddot{\text{C}}}}:\text{H}$$

$$4:\ddot{\text{F}}\cdot + \cdot\dot{\text{C}}\cdot \longrightarrow \overset{\displaystyle :\ddot{\text{F}}:}{\underset{\displaystyle :\ddot{\text{F}}:}{:\ddot{\text{F}}:\ddot{\text{C}}:\ddot{\text{F}}:}}$$

The covalent bond is typical of the compounds of carbon; it is the bond of chief importance in the study of organic chemistry.

Problem 1.1   Which of the following would you expect to be ionic, and which non-ionic? Give a simple electronic structure for each, showing only valence shell electrons.

(a) KBr	(c) NF_3	(e) $CaSO_4$	(g) PH_3
(b) H_2S	(d) $CHCl_3$	(f) NH_4Cl	(h) CH_3OH

Problem 1.2   Give a likely simple electronic structure for each of the following, assuming them to be completely covalent. Assume that every atom (except hydrogen, of course) has a complete octet, and that two atoms may share more than one pair of electrons.

(a) H_2O_2	(c) $HONO_2$	(e) HCN	(g) H_2CO_3
(b) N_2	(d) NO_3^-	(f) CO_2	(h) C_2H_6

1.4   Quantum mechanics

In 1926 there emerged the theory known as *quantum mechanics*, developed, in the form most useful to chemists, by Erwin Schrödinger (of the University of Zurich). He worked out mathematical expressions to describe the motion of an electron in terms of its energy. These mathematical expressions are called *wave equations*, since they are based upon the concept that electrons show properties not only of particles but also of waves.

A wave equation has a series of solutions, called *wave functions*, each corresponding to a different energy level for the electron. For all but the simplest of systems, doing the mathematics is so time-consuming that at present—and super-high-speed computers will some day change this—only approximate solutions can be obtained. Even so, quantum mechanics gives answers agreeing so well with the facts that it is accepted today as the most fruitful approach to an understanding of atomic and molecular structure.

"Wave mechanics has shown us what is going on, and at the deepest possible level ... it has taken the concepts of the experimental chemist—the imaginative perception that came to those who had lived in their laboratories and allowed their minds to dwell creatively upon the facts that they had found—and it has shown how they all fit together; how, if you wish, they all have one single rationale; and how this hidden relationship to each other can be brought out."—C. A. Coulson, London, 1951.

1.5   Atomic orbitals

A wave equation cannot tell us exactly where an electron is at any particular moment, or how fast it is moving; it does not permit us to plot a precise orbit about the nucleus. Instead, it tells us the *probability* of finding the electron at any particular place.

The region in space where an electron is likely to be found is called an **orbital**. There are different kinds of orbitals, which have different sizes and different shapes, and which are disposed about the nucleus in specific ways. The particular kind of orbital that an electron occupies depends upon the energy of the electron. It is the shapes of these orbitals and their disposition with respect to each other that we are particularly interested in, since these determine—or, more precisely, can conveniently be *thought of* as determining—the arrangement in space of the atoms of a molecule, and even help determine its chemical behavior.

It is convenient to picture an electron as being smeared out to form a cloud. We might think of this cloud as a sort of blurred photograph of the rapidly moving electron. The shape of the cloud is the shape of the orbital. The cloud is not uniform, but is densest in those regions where the probability of finding the electron is highest, that is, in those regions where the average negative charge, or *electron density*, is greatest.

Let us see what the shapes of some of the atomic orbitals are. The orbital at the lowest energy level is called the 1s orbital. It is a sphere with its center at the nucleus of the atom, as represented in Fig. 1.1. An orbital has no definite boundary

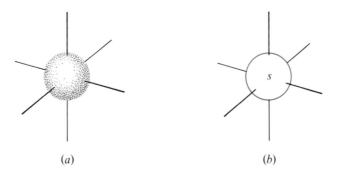

(a)   (b)

Figure 1.1   Atomic orbitals: *s* orbital. The nucleus is at the center.

since there is a probability, although a very small one, of finding the electron essentially separated from the atom—or even on some other atom! However, the probability decreases very rapidly beyond a certain distance from the nucleus, so that the distribution of charge is fairly well represented by the electron cloud in Fig. 1.1*a*. For simplicity, we may even represent an orbital as in Fig. 1.1*b*, where the solid line encloses the region where the electron spends most (say 95%) of its time.

At the next higher energy level there is the 2s orbital. This, too, is a sphere with its center at the atomic nucleus. It is—*naturally*—larger than the 1s orbital: the higher energy (lower stability) is due to the greater average distance between electron and nucleus, with the resulting decrease in electrostatic attraction. (Consider the work that must be done—the energy put into the system—to move an electron away from the oppositely charged nucleus.)

Next there are three orbitals of equal energy called $2p$ orbitals, shown in Fig. 1.2. Each $2p$ orbital is dumbbell-shaped. It consists of two lobes with the atomic nucleus lying between them. The axis of each $2p$ orbital is perpendicular to the axes of the other two. They are differentiated by the names $2p_x$, $2p_y$, and $2p_z$, where the x, y, and z refer to the corresponding axes.

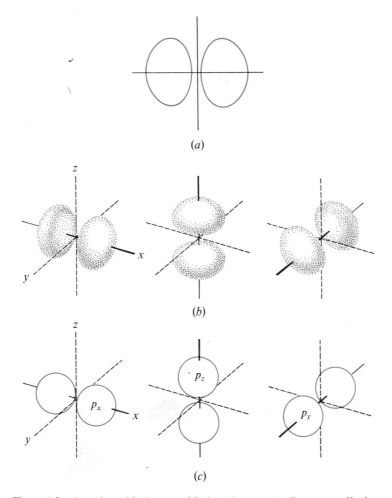

Figure 1.2   Atomic orbitals: p orbitals. Axes mutually perpendicular. (a) Cross-section showing the two lobes of a single orbital. (b) Approximate shape as pairs of distorted ellipsoids. (c) Representation as pairs of not-quite-touching spheres.

1.6   Electronic configuration. Pauli exclusion principle

There are a number of "rules" that determine the way in which the electrons of an atom may be distributed, that is, that determine the *electronic configuration* of an atom.

The most fundamental of these rules is the **Pauli exclusion principle**: *only two electrons can occupy any atomic orbital, and to do so these two must have* **opposite** *spins*. These electrons of opposite spins are said to be *paired*. *Electrons of* **like** *spin tend to get as far from each other as possible*. This tendency is the most important of all the factors that determine the shapes and properties of molecules.

The exclusion principle, advanced in 1925 by Wolfgang Pauli, Jr. (of the Institute for Theoretical Physics, Hamburg, Germany), has been called the cornerstone of chemistry.

The first ten elements of the Periodic Table have the electronic configurations shown in Table 1.1. We see that an orbital becomes occupied only if the orbitals

Table 1.1   ELECTRONIC CONFIGURATIONS

	1s	2s	2p		
H	(·)				
He	(··)				
Li	(··)	(·)	( )	( )	( )
Be	(··)	(··)	( )	( )	( )
B	(··)	(··)	(·)	( )	( )
C	(··)	(··)	(·)	(·)	( )
N	(··)	(··)	(·)	(·)	(·)
O	(··)	(··)	(··)	(·)	(·)
F	(··)	(··)	(··)	(··)	(·)
Ne	(··)	(··)	(··)	(··)	(··)

of lower energy are filled (e.g., 2s after 1s, 2p after 2s). We see that an orbital is not occupied by a pair of electrons until other orbitals of equal energy are each occupied by one electron (e.g., the 2p orbitals). The 1s electrons make up the first shell of two, and the 2s and 2p electrons make up the second shell of eight. For elements beyond the first ten, there is a third shell containing a 3s orbital, 3p orbitals, and so on.

Problem 1.3   (a) Show the electronic configurations for the next eight elements in the Periodic Table (from sodium through argon). (b) What relationship is there between electronic configuration and periodic family? (c) Between electronic configuration and chemical properties of the elements?

1.7   Molecular orbitals

In molecules, as in isolated atoms, electrons occupy orbitals, and in accordance with much the same "rules". These *molecular orbitals* are considered to be centered about many nuclei, perhaps covering the entire molecule; the distribution of nuclei and electrons is simply the one that results in the most stable molecule.

To make the enormously complicated mathematics more workable, two simplifying assumptions are commonly made: (a) that each pair of electrons is essentially localized near just two nuclei, and (b) that the shapes of these localized molecular orbitals and their disposition with respect to each other are related in a simple way to the shapes and disposition of atomic orbitals in the component atoms.

The idea of localized molecular orbitals—or what we might call *bond orbitals*— is evidently not a bad one, since mathematically this method of approximation is successful with most (although *not all*) molecules. Furthermore, this idea closely parallels the chemist's classical concept of a bond as a force acting between two atoms and pretty much independent of the rest of the molecule; it can hardly be accidental that this concept has worked amazingly well for a hundred years. Significantly, the exceptional molecules for which classical formulas do not work are just those for which the localized molecular orbital approach does not work either. (Even these cases, we shall find, can be handled by a rather simple adaptation of classical formulas, an adaptation which again parallels a method of mathematical approximation.)

The second assumption, of a relationship between atomic and molecular orbitals, is a highly reasonable one, as discussed in the following section. It has proven so useful that, when necessary, atomic orbitals of certain kinds have been *invented* just so that the assumption can be retained.

1.8 The covalent bond

Now let us consider the formation of a molecule. For convenience we shall picture this as happening by the coming together of the individual atoms, although most molecules are not actually made this way. We make physical models of molecules out of wooden or plastic balls that represent the various atoms; the location of holes or snap fasteners tells us how to put them together. In the same way, we shall make *mental* models of molecules out of mental atoms; the location of atomic orbitals—some of them imaginary—will tell us how to put these together.

For a covalent bond to form, two atoms must be located so that an orbital of one *overlaps* an orbital of the other; each orbital must contain a single electron. When this happens, the two atomic orbitals merge to form a single *bond orbital* which is occupied by both electrons. The two electrons that occupy a bond orbital must have opposite spins, that is, must be paired. Each electron has available to it the entire bond orbital, and thus may be considered to "belong to" both atomic nuclei.

This arrangement of electrons and nuclei contains less energy—that is, is more stable—than the arrangement in the isolated atoms; as a result, formation of a bond is accompanied by evolution of energy. The amount of energy (per mole) that is given off when a bond is formed (or the amount that must be put in to break the bond) is called the *bond dissociation energy*. For a given pair of atoms, the greater the overlap of atomic orbitals, the stronger the bond.

What gives the covalent bond its strength? It is the increase in electrostatic attraction. In the isolated atoms, each electron is attracted by—and attracts—one positive nucleus; in the molecule, each electron is attracted by *two* positive nuclei.

It is the concept of "overlap" that provides the mental bridge between atomic orbitals and bond orbitals. Overlap of atomic orbitals means that the bond orbital occupies much of the same region in space that was occupied by *both* atomic orbitals. Consequently, an electron from one atom can, to a considerable extent, remain in its original, favorable location with respect to "its" nucleus, and at the same time occupy a similarly favorable location with respect to the second nucleus; the same holds, of course, for the other electron.

The principle of *maximum overlap*, first stated in 1931 by Linus Pauling (at the California Institute of Technology), has been ranked only slightly below the exclusion principle in importance to the understanding of molecular structure.

As our first example, let us consider the formation of the hydrogen molecule, H_2, from two hydrogen atoms. Each hydrogen atom has one electron, which occupies the $1s$ orbital. As we have seen, this $1s$ orbital is a sphere with its center at the atomic nucleus. For a bond to form, the two nuclei must be brought closely enough together for overlap of the atomic orbitals to occur (Fig. 1.3). For hydrogen, the system is most stable when the distance between the nuclei is 0.74 Å; this distance is called the **bond length**. At this distance the stabilizing effect of overlap is exactly balanced by repulsion between the similarly charged nuclei. The resulting hydrogen molecule contains 104 kcal/mol less energy than the hydrogen atoms from which it was made. We say that the hydrogen–hydrogen bond has a length of 0.74 Å and a strength of 104 kcal.

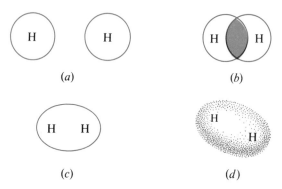

Figure 1.3 Bond formation: H_2 molecule. (*a*) Separate *s* orbitals. (*b*) Overlap of *s* orbitals. (*c*) and (*d*) The σ bond orbital.

This bond orbital has roughly the shape we would expect from the merging of two *s* orbitals. As shown in Fig. 1.3, it is sausage-shaped, with its long axis lying along the line joining the nuclei. It is cylindrically symmetrical about this long axis; that is, a slice of the sausage is circular. Bond orbitals having this shape are called σ *orbitals* (*sigma orbitals*) and the bonds are called σ *bonds*. We may visualize the hydrogen molecule as two nuclei embedded in a single sausage-shaped electron cloud. This cloud is densest in the region between the two nuclei, where the negative charge is attracted most strongly by the two positive charges.

The size of the hydrogen molecule—as measured, say, by the volume inside the 95% probability surface—is considerably *smaller* than that of a single hydrogen atom. Although surprising at first, this shrinking of the electron cloud is actually what would be expected. It is the powerful attraction of the electrons by *two* nuclei that gives the molecule greater stability than the isolated hydrogen atoms; this must mean that the electrons are held tighter, *closer*, than in the atoms.

Next, let us consider the formation of the fluorine molecule, F_2, from two fluorine atoms. As we can see from our table of electronic configurations (Table 1.1), a fluorine atom has two electrons in the $1s$ orbital, two electrons in the $2s$ orbital, and two electrons in each of two $2p$ orbitals. In the third $2p$ orbital there is a single electron which is unpaired and available for bond formation. Overlap of this p orbital with a similar p orbital of another fluorine atom permits electrons to pair and the bond to form (Fig. 1.4). The electronic charge is concentrated between the two nuclei, so that the back lobe of each of the overlapping orbitals shrinks to

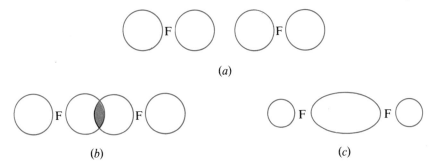

(a)

(b)   (c)

Figure 1.4   Bond formation: F_2 molecule. (*a*) Separate *p* orbitals. (*b*) Overlap of *p* orbitals. (*c*) The σ bond orbital.

a comparatively small size. Although formed by overlap of atomic orbitals of a different kind, the fluorine–fluorine bond has the same general shape as the hydrogen–hydrogen bond, being cylindrically symmetrical about a line joining the nuclei; it, too, is given the designation of σ bond. The fluorine–fluorine bond has a length of 1.42 Å and a strength of about 38 kcal.

As the examples show, a covalent bond results from the overlap of two atomic orbitals to form a bond orbital occupied by a pair of electrons. *Each kind of covalent bond has a characteristic length and strength.*

1.9   Hybrid orbitals: *sp*

Let us next consider beryllium chloride, $BeCl_2$.

Beryllium (Table 1.1) has no unpaired electrons. How are we to account for its combining with two chlorine atoms? Bond formation is an energy-releasing

	$1s$	$2s$	$2p$		
Be	⊙	⊙	○	○	○

(stabilizing) process, and the tendency is to form bonds—and as many as possible—even if this results in bond orbitals that bear little resemblance to the atomic orbitals we have talked about. If our method of mental molecule-building is to be applied here, it must be modified. We must invent an imaginary kind of beryllium atom, one that is about to become bonded to two chlorine atoms.

To arrive at this divalent beryllium atom, let us do a little electronic book-keeping. First, we "promote" one of the 2s electrons to an empty p orbital:

One electron promoted: two unpaired electrons

This provides two unpaired electrons, which are needed for bonding to two chlorine atoms. We might now expect beryllium to form one bond of one kind, using the p orbital, and one bond of another kind, using the s orbital. Again, this is contrary to fact: the two bonds in beryllium chloride are known to be equivalent.

Next, then, we *hybridize* the orbitals. Various combinations of one s orbital

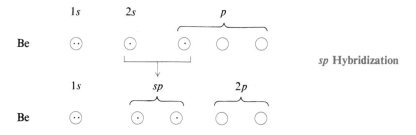

sp Hybridization

and one p orbital are taken mathematically, and the mixed (*hybrid*) orbitals with the greatest degree of *directional character* are found (Fig. 1.5). The more an atomic orbital is concentrated in the direction of the bond, the greater the overlap and the

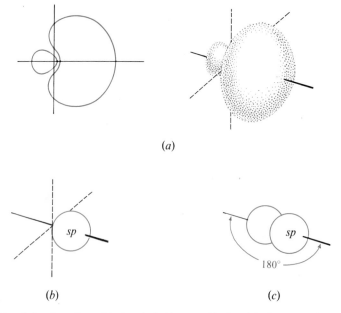

(a)

(b)   (c)

Figure 1.5 Atomic orbitals: hybrid *sp* orbitals. (*a*) Cross-section and approximate shape of a single orbital. Strongly directed along one axis. (*b*) Representation as a sphere, with the small back lobe omitted. (*c*) Two orbitals, with axes lying along a straight line.

stronger the bond it can form. Three highly significant results emerge from the calculations: (a) the "best" hybrid orbital is much more strongly directed than either the *s* or *p* orbital; (b) the two best orbitals are exactly equivalent to each other; and (c) these orbitals point in exactly opposite directions—*the arrangement that permits them to get as far away from each other as possible* (remember the Pauli exclusion principle). The angle between the orbitals is thus 180°.

These particular hybrid orbitals are called *sp* orbitals, since they are considered to arise from the mixing of *one s* orbital and *one p* orbital. They have the shape shown in Fig. 1.5*a*; for convenience we shall neglect the small back lobe and represent the front lobe as a sphere.

Using this *sp-hybridized* beryllium, let us construct beryllium chloride. An extremely important concept emerges here: **bond angle**. For maximum overlap between the *sp* orbitals of beryllium and the *p* orbitals of the chlorines, the two chlorine nuclei must lie along the axes of the *sp* orbitals; that is, they must be located on exactly opposite sides of the beryllium atom (Fig. 1.6). The angle between the beryllium–chlorine bonds must therefore be 180°.

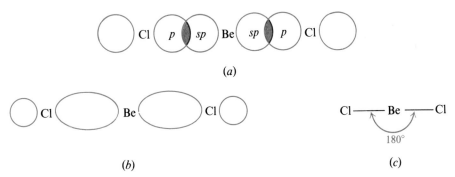

Figure 1.6   Bond formation: $BeCl_2$ molecule. (*a*) Overlap of *sp* and *p* orbitals. (*b*) The σ bond orbitals. (*c*) Shape of the molecule.

Experiment has shown that, as calculated, beryllium chloride is a *linear molecule*, all three atoms lying along a single straight line.

There is nothing magical about the increase in directional character that accompanies hybridization. The two lobes of the *p* orbital are of opposite *phase* (Sec. 33.2); combination with an *s* orbital amounts to *addition* on one side of the nucleus, but *subtraction* on the other.

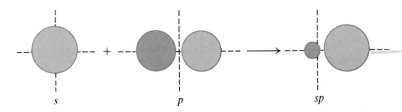

If you are curious about phase and its effect on bonding, read Secs. 33.1–33.4, which you can understand at this point.

1.10   Hybrid orbitals: *sp²*

Next, let us look at boron trifluoride, BF_3. Boron (Table 1.1) has only one unpaired electron, which occupies a 2*p* orbital. For three bonds we need three unpaired electrons, and so we promote one of the $2s$ electrons to a $2p$ orbital:

If, now, we are to "make" the most stable molecule possible, we must "make" the strongest bonds possible; for these we must provide the most strongly directed atomic orbitals that we can. Again, hybridization provides such orbitals: three hybrid orbitals, exactly equivalent to each other. Each one has the shape shown in

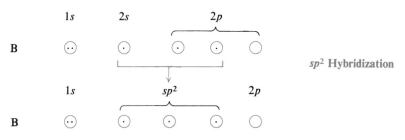

Fig. 1.7; as before, we shall neglect the small back lobe and represent the front lobe as a sphere.

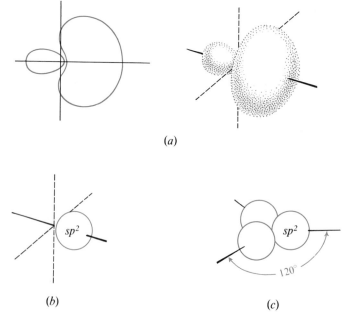

Figure 1.7 Atomic orbitals: hybrid sp^2 orbitals. (*a*) Cross-section and approximate shape of a single orbital. It is strongly directed along one axis. (*b*) Representation as a sphere, with the small back lobe omitted. (*c*) Three orbitals, with the axes directed toward the corners of an equilateral triangle.

These hybrid orbitals are called *sp²* orbitals, since they are considered to arise from the mixing of *one* s orbital and *two* p orbitals. They lie in a plane, which includes the atomic nucleus, and are directed to the corners of an equilateral triangle; the angle between any two orbitals is thus 120°. Again we see the geometry that permits the orbitals to be as far apart as possible: here, a *trigonal* (three-cornered) arrangement.

When we arrange the atoms for maximum overlap of each of the *sp²* orbitals of boron with a *p* orbital of fluorine, we obtain the structure shown in Fig. 1.8: a *flat* molecule, with the boron atom at the center of a triangle and the three fluorine atoms at the corners. Every bond angle is 120°.

Figure 1.8 BF_3 molecule.

Experiment has shown that boron trifluoride has exactly this flat, symmetrical structure calculated by quantum mechanics.

1.11   Hybrid orbitals: *sp³*

Now, let us turn to one of the simplest of organic molecules, *methane*, CH_4.

Carbon (Table 1.1) has an unpaired electron in each of the two *p* orbitals, and on this basis might be expected to form a compound CH_2. (It *does*, but CH_2 is a

highly reactive molecule whose properties center about the need to provide carbon with two more bonds.) Again, we see the tendency to form as many bonds as possible: in this case, to combine with *four* hydrogen atoms.

To provide four unpaired electrons, we promote one of the *2s* electrons to the empty *p* orbital:

Once more the most strongly directed orbitals are hybrid orbitals: this time, *sp³* orbitals, from the mixing of *one* s orbital and *three* p orbitals. Each one has the

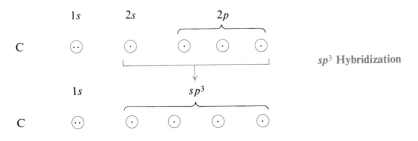

shape shown in Fig. 1.9; as with *sp* and *sp*^{2} orbitals, we shall neglect the small back lobe and represent the front lobe as a sphere.

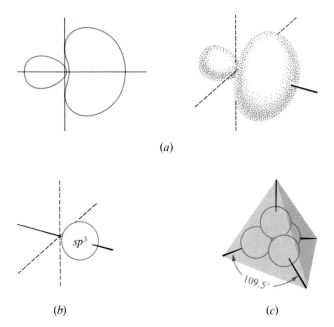

(a)

(b)   (c)

Figure 1.9   Atomic orbitals: hybrid *sp*^{3} orbitals. (*a*) Cross-section and approximate shape of a single orbital. It is strongly directed along one axis. (*b*) Representation as a sphere, with the small back lobe omitted. (*c*) Four orbitals, with the axes directed toward the corners of a tetrahedron.

Now, how are *sp*^{3} orbitals arranged in space? The answer is no surprise to us: in the way that lets them get as far away from each other as possible. They are directed to the corners of a regular tetrahedron. The angle between any two orbitals is the tetrahedral angle 109.5° (Fig. 1.9). Just as mutual repulsion among orbitals gives two linear bonds or three trigonal bonds, so it gives four tetrahedral bonds.

Overlap of each of the *sp*^{3} orbitals of carbon with an *s* orbital of hydrogen results in methane: carbon at the center of a regular tetrahedron, and the four hydrogens at the corners (Fig. 1.10).

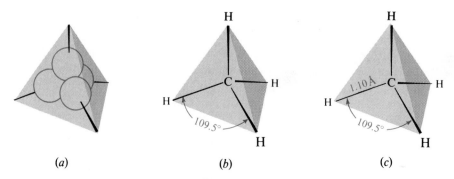

(a)   (b)   (c)

Figure 1.10   Bond formation: CH$_4$ molecule. (*a*) Tetrahedral *sp*^{3} orbitals. (*b*) Predicted shape: H nuclei located for maximum overlap. (*c*) Shape and size.

Experimentally, methane has been found to have the highly symmetrical tetrahedral structure we have assembled. Each carbon–hydrogen bond has exactly the same length, 1.10 Å; the angle between any pair of bonds is the tetrahedral angle 109.5°. It takes 104 kcal/mol to break one of the bonds of methane.

Thus, in these last three sections, we have seen that there are associated with covalent bonds not only characteristic bond lengths and bond dissociation energies but also characteristic bond *angles*. These bond angles can be conveniently related to the arrangement of atomic orbitals—including hybrid orbitals—involved in bond formation; they ultimately go back to the Pauli exclusion principle and the tendency for unpaired electrons to get as far from each other as possible.

Unlike the ionic bond, which is equally strong in all directions, *the covalent bond is a directed bond*. We can begin to see why the chemistry of the covalent bond is so much concerned with molecular size and shape.

Since compounds of carbon are held together chiefly by covalent bonds, organic chemistry, too, is much concerned with molecular size and shape. To help us in our study, we should make frequent use of molecular models. Figure 1.11 shows methane as represented by three different kinds of models: stick-and-ball, framework, and space-filling. These last are made to scale, and reflect accurately not only bond angles but also relative lengths of bonds and sizes of atoms.

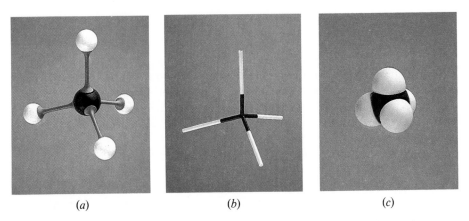

(a)   (b)   (c)

Figure 1.11   Models of methane molecule. (*a*) Stick-and-ball (Allyn & Bacon). (*b*) Framework (Prentice-Hall). (*c*) Space-filling (Corey–Pauling–Koltun, CPK); 1.25 cm equals 1.00 Å.

1.12   Unshared pairs of electrons

Two familiar compounds, ammonia (NH_3) and water (H_2O), show how *unshared pairs of electrons* can affect molecular structure.

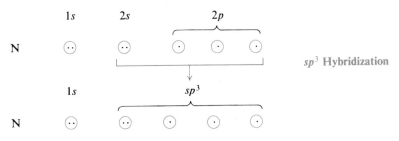

In ammonia, nitrogen resembles the carbon of methane. Nitrogen is sp^3-hybridized, but (Table 1.1) has only three unpaired electrons; they occupy three of the sp^3 orbitals. Overlap of each of these orbitals with the s orbital of a hydrogen atom results in ammonia (Fig. 1.12). The fourth sp^3 orbital of nitrogen contains a pair of electrons.

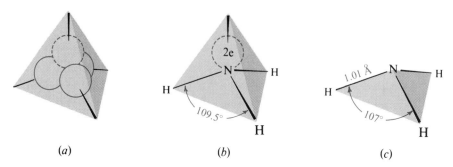

(a)   (b)   (c)

Figure 1.12   Bond formation: NH_3 molecule. (a) Tetrahedral sp^3 orbitals. (b) Predicted shape, showing the unshared pair: H nuclei located for maximum overlap. (c) Shape and size.

If there is to be maximum overlap and hence maximum bond strength, the hydrogen nuclei must be located at three corners of a tetrahedron; the fourth corner is occupied by an unshared pair of electrons. Considering only atomic nuclei, we would expect ammonia to be shaped like a pyramid with nitrogen at the apex and hydrogen at the corners of a triangular base. Each bond angle should be the tetrahedral angle 109.5°.

Experimentally, ammonia is found to have the pyramidal shape calculated by quantum mechanics. The bond angles are 107°, slightly smaller than the predicted value; it has been suggested that the unshared pair of electrons occupies more space than any of the hydrogen atoms, and hence tends to compress the bond angles slightly. The nitrogen–hydrogen bond length is 1.01 Å; it takes 103 kcal/mol to break one of the bonds of ammonia.

The sp^3 orbital occupied by the unshared pair of electrons is a region of high electron density. This region is a source of electrons for electron-seeking atoms and molecules, and thus gives ammonia its basic properties (Sec. 1.22).

There are two other conceivable electronic configurations for ammonia, but neither fits the facts.

(a) Since nitrogen is bonded to three other atoms, we might have pictured it as using sp^2 orbitals, as boron does in boron trifluoride. But ammonia is *not* a flat molecule, and so we must reject this possibility. It is the unshared pair of electrons on nitrogen that makes the difference between NH_3 and BF_3; these electrons need to stay away from those in the carbon–hydrogen bonds, and the tetrahedral shape makes this possible.

(b) We might have pictured nitrogen as simply using the p orbitals for overlap, since they would provide the necessary three unpaired electrons. But this would give bond angles of 90°—remember, the p orbitals are at right angles to each other—in contrast to the observed angles of 107°. More importantly, the unshared pair would be buried in an s orbital, and there is evidence from dipole moments (Sec. 1.16) that this is not so. Evidently the stability gained by using the highly directed sp^3 orbitals for bond formation more than makes up for raising the unshared pair from an s orbital to the higher-energy sp^3 orbital.

One further fact about ammonia: spectroscopy reveals that the molecule undergoes *inversion*, that is, turns inside-out (Fig. 1.13). There is an energy barrier

Figure 1.13   Inversion of ammonia.

of only 6 kcal/mol between one pyramidal arrangement and the other, equivalent one. This energy is provided by molecular collisions, and even at room temperature the fraction of collisions hard enough to do the job is so large that a rapid transformation between pyramidal arrangements occurs.

Compare ammonia with methane, which does *not* undergo inversion. The unshared pair plays the role of a carbon–hydrogen bond in determining the most stable shape of the molecule, tetrahedral. But, unlike a carbon–hydrogen bond, the unshared pair cannot maintain a *particular* tetrahedral arrangement; the pair points now in one direction, and the next instant in the opposite direction.

Finally, let us consider water, H_2O. The situation is similar to that for ammonia, except that oxygen has only two unpaired electrons, and hence it bonds

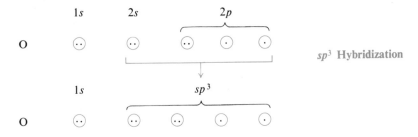

with only two hydrogen atoms, which occupy two corners of a tetrahedron. The other two corners of the tetrahedron are occupied by unshared pairs of electrons (Fig. 1.14).

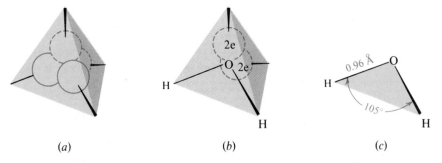

Figure 1.14   Bond formation: H_2O molecule. (*a*) Tetrahedral sp^3 orbitals. (*b*) Predicted shape, showing the unshared pairs: H nuclei located for maximum overlap. (*c*) Shape and size.

As actually measured, the H—O—H angle is 105°, smaller than the calculated tetrahedral angle, and even smaller than the angle in ammonia. Here there are two bulky unshared pairs of electrons compressing the bond angles. The oxygen–hydrogen bond length is 0.96 Å; it takes 118 kcal/mol to break one of the bonds of water.

If we examine Fig. 1.15 we can see the fundamental similarity in shape of the methane, ammonia, and water molecules: a similarity that, by the approach we have used, stems from a similarity in bonding.

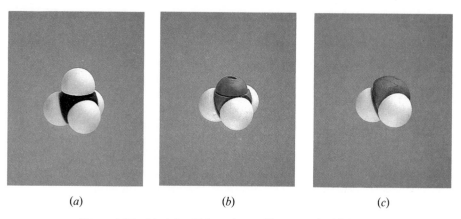

(a)   (b)   (c)

Figure 1.15   Models of (a) methane, (b) ammonia, (c) water.

Because of the unshared pairs of electrons on oxygen, water is basic, although less strongly so than ammonia (Sec. 1.22).

Problem 1.4   Predict the shape of each of the following molecules, and tell how you arrived at your prediction: (a) the ammonium ion, NH_4^+; (b) the hydronium ion, H_3O^+; (c) methyl alcohol, CH_3OH; (d) methylamine, CH_3NH_2.

1.13   Intramolecular forces

We must remember that the particular method of mentally building molecules that we are learning to use is artificial: it is a purely intellectual process involving imaginary overlap of imaginary orbitals. There are other, equally artificial ways that use different mental or physical models. Our method is the one that so far has seemed to work out best for the organic chemist. Our kit of mental atomic models will contain just three "kinds" of carbon: *tetrahedral* (sp^3-hybridized), *trigonal* (sp^2-hybridized), and *digonal* (sp-hybridized). By use of this kit, we shall find, one can do an amazingly good job of building hundreds of thousands of organic molecules.

But, however we arrive at it, we see the actual structure of a molecule to be the net result of a combination of *repulsive* and *attractive* forces, which are related to *charge* and *electron spin*.

(a) *Repulsive forces*. Electrons tend to stay as far apart as possible because they have the same charge and also, if they are unpaired, because they have the same spin (Pauli exclusion principle). The like-charged atomic nuclei, too, repel each other.

(b) *Attractive forces*. Electrons are attracted by atomic nuclei—as are the nuclei by the electrons—because of their opposite charge, and hence tend to occupy the region between two nuclei. Opposite spin *permits* (although, in itself, probably does not actually *encourage*) two electrons to occupy the same region.

In methane, for example, the four hydrogen nuclei are as widely separated as they can be. The distribution of the eight bonding electrons is such that each one occupies the desirable region near two nuclei—the bond orbital—and yet, except for its partner, is as far as possible from the other electrons. We can picture each electron accepting—perhaps reluctantly because of their similar charges—one orbital-mate of opposite spin, but staying as far as possible from all other electrons and even, as it wanders within the loose confines of its orbital, doing its best to avoid the vicinity of its restless partner.

1.14   Bond dissociation energy. Homolysis and heterolysis

We have seen that energy is liberated when atoms combine to form a molecule. For a molecule to break into atoms, an equivalent amount of energy must be consumed. *The amount of energy consumed or liberated when a bond is broken or formed is known as the* **bond dissociation energy, D**. It is characteristic of the particular bond. Table 1.2 lists bond dissociation energies that have been measured for a number of bonds. As can be seen, they vary widely, from weak bonds like I—I (36 kcal/mol) to very strong bonds like H—F (136 kcal/mol). Although the accepted values may change as experimental methods improve, certain trends are clear.

Table 1.2   HOMOLYTIC BOND DISSOCIATION ENERGIES, KCAL/MOL

$A:B \longrightarrow A\cdot + \cdot B$ ΔH = Homolytic bond dissociation energy or $D(A—B)$

H—H 104		CH_3—H 104	
H—F 136	F—F 38	CH_3—F 108	
H—Cl 103	Cl—Cl 58	CH_3—Cl 84	
H—Br 88	Br—Br 46	CH_3—Br 70	
H—I 71	I—I 36	CH_3—I 56	

CH_3—H 104	CH_3—CH_3 88	CH_3—Cl 84	CH_3—Br 70
C_2H_5—H 98	C_2H_5—CH_3 85	C_2H_5—Cl 81	C_2H_5—Br 69
n-C_3H_7—H 98	n-C_3H_7—CH_3 85	n-C_3H_7—Cl 82	n-C_3H_7—Br 69
i-C_3H_7—H 95	i-C_3H_7—CH_3 84	i-C_3H_7—Cl 81	i-C_3H_7—Br 68
t-C_4H_9—H 92	t-C_4H_9—CH_3 80	t-C_4H_9—Cl 79	t-C_4H_9—Br 63
H_2C=CH—H 108	H_2C=CH—CH_3 92	H_2C=CH—Cl 84	
H_2C=CHCH$_2$—H 88	H_2C=CHCH$_2$—CH_3 72	H_2C=CHCH$_2$—Cl 60	H_2C=CHCH$_2$—Br 47
C_6H_5—H 110	C_6H_5—CH_3 93	C_6H_5—Cl 86	C_6H_5—Br 72
$C_6H_5CH_2$—H 85	$C_6H_5CH_2$—CH_3 70	$C_6H_5CH_2$—Cl 68	$C_6H_5CH_2$—Br 51

We must not confuse *bond dissociation energy* (D) with another measure of bond strength called *bond energy* (E). If one begins with methane, for example, and breaks, successively, four carbon–hydrogen bonds, one finds four different bond dissociation energies:

$$CH_4 \longrightarrow CH_3 + H\cdot \qquad D(CH_3—H) = 104 \text{ kcal/mol}$$

$$CH_3 \longrightarrow CH_2 + H\cdot \qquad D(CH_2—H) = 106$$

$$CH_2 \longrightarrow CH + H\cdot \qquad D(CH—H) = 106$$

$$CH \longrightarrow C + H\cdot \qquad D(C—H) = 81$$

The carbon–hydrogen bond energy in methane, $E(C-H)$, on the other hand, is a single average value:

$$CH_4 \longrightarrow C + 4H \cdot \quad \Delta H = 397 \text{ kcal/mol}, \quad E(C-H) = 397/4 = 99 \text{ kcal/mol}$$

We shall generally find bond dissociation energies more useful for our purposes.

So far, we have spoken of breaking a molecule into two atoms or into an atom and a group of atoms. Thus, of the two electrons making up the covalent bond, one goes to each fragment; such bond-breaking is called *homolysis*. We shall also encounter reactions involving bond-breaking of a different kind: *heterolysis*, in which both bonding electrons go to the same fragment.

$$A \!:\! B \longrightarrow A \cdot + B \cdot \qquad \text{Homolysis: } \textit{one electron to each fragment}$$

$$A \!:\! B \longrightarrow A + \!:\! B \qquad \text{Heterolysis: } \textit{both electrons to one fragment}$$

(These words are taken from the Greek: *homo*, the same, and *hetero*, different; and *lysis*, a loosing. To a chemist *lysis* means "cleavage" as in, for example, *hydrolysis*, "cleavage by water".)

The bond dissociation energies given in Table 1.2 are for homolysis, and are therefore *homolytic* bond dissociation energies. But bond dissociation energies have also been measured for heterolysis; some of these *heterolytic* bond dissociation energies are given in Table 1.3.

Table 1.3   HETEROLYTIC BOND DISSOCIATION ENERGIES, KCAL/MOL

$A \!:\! B \longrightarrow A^+ + \!:\! B^- \qquad \Delta H = $ Heterolytic bond dissociation energy or $D(A^+ - B^-)$

H—H	401	CH$_3$—H	313
H—F	370	CH$_3$—F	256
H—Cl	334	CH$_3$—Cl	227
H—Br	324	CH$_3$—Br	219
H—I	315	CH$_3$—I	212
H—OH	390	CH$_3$—OH	274

CH$_3$—Cl 227	CH$_3$—Br 219	CH$_3$—I 212	CH$_3$—OH 274
C$_2$H$_5$—Cl 191	C$_2$H$_5$—Br 184	C$_2$H$_5$—I 176	C$_2$H$_5$—OH 242
n-C$_3$H$_7$—Cl 185	n-C$_3$H$_7$—Br 178	n-C$_3$H$_7$—I 171	n-C$_3$H$_7$—OH 235
i-C$_3$H$_7$—Cl 170	i-C$_3$H$_7$—Br 164	i-C$_3$H$_7$—I 156	i-C$_3$H$_7$—OH 222
t-C$_4$H$_9$—Cl 157	t-C$_4$H$_9$—Br 149	t-C$_4$H$_9$—I 140	t-C$_4$H$_9$—OH 208
H$_2$C=CH—Cl 207	H$_2$C=CH—Br 200	H$_2$C=CH—I 194	
H$_2$C=CHCH$_2$—Cl 173	H$_2$C=CHCH$_2$—Br 165	H$_2$C=CHCH$_2$—I 159	H$_2$C=CHCH$_2$—OH 223
C$_6$H$_5$—Cl 219	C$_6$H$_5$—Br 210	C$_6$H$_5$—I 202	C$_6$H$_5$—OH 275
C$_6$H$_5$CH$_2$—Cl 166	C$_6$H$_5$CH$_2$—Br 157	C$_6$H$_5$CH$_2$—I 149	C$_6$H$_5$CH$_2$—OH 215

If we examine these values, we see that they are considerably bigger than those in Table 1.2. Simple heterolysis of a neutral molecule yields, of course, a positive ion and a negative ion. Separation of these oppositely charged particles takes a great deal of energy: 100 kcal/mol or so *more* than separation of neutral particles. In the gas phase, therefore, bond dissociation generally takes place by the easier route, homolysis. In an ionizing solvent (Sec. 6.5), on the other hand, heterolysis is the preferred kind of cleavage.

1.15   Polarity of bonds

Besides the properties already described, certain covalent bonds have another property: **polarity**. Two atoms joined by a covalent bond share electrons; their nuclei are held by the same electron cloud. But in most cases the two nuclei do not share the electrons equally; the electron cloud is denser about one atom than the other. One end of the bond is thus relatively negative and the other end is relatively positive; that is, there is a *negative pole* and a *positive pole*. Such a bond is said to be a **polar bond**, or to *possess polarity*.

We can indicate polarity by using the symbols δ_+ and δ_-, which indicate *partial* + and − charges. (We say "delta plus" and "delta minus".) For example:

$$
\begin{array}{ccc}
\overset{\delta_+ \; \delta_-}{\text{H—F}} &
\overset{\overset{\delta_-}{\text{O}}}{\underset{\delta_+ \;\; \delta_+}{\text{H} \qquad \text{H}}} &
\overset{\overset{\delta_-}{\text{N}}}{\underset{\delta_+ \;\; \underset{\delta_+}{\text{H}} \;\; \delta_+}{\text{H} \quad \text{H} \quad \text{H}}}
\end{array}
$$

Polar bonds

We can expect a covalent bond to be polar if it joins atoms that differ in their tendency to attract electrons, that is, atoms that differ in *electronegativity*. Furthermore, the greater the difference in electronegativity, the more polar the bond will be.

The most electronegative elements are those located in the upper right-hand corner of the Periodic Table. Of the elements we are likely to encounter in organic chemistry, fluorine has the highest electronegativity, then oxygen, then nitrogen and chlorine, then bromine, and finally carbon. Hydrogen does not differ very much from carbon in electronegativity; it is not certain whether it is more or less electronegative.

Electronegativity $F > O > Cl, N > Br > C, H$

Bond polarities are intimately concerned with both physical and chemical properties. The polarity of bonds can lead to polarity of molecules, and thus profoundly affect melting point, boiling point, and solubility. The polarity of a bond determines the kind of reaction that can take place at that bond, and even affects reactivity at nearby bonds.

1.16   Polarity of molecules

A molecule is polar if the center of negative charge does not coincide with the center of positive charge. Such a molecule constitutes a *dipole*: two equal and opposite charges separated in space. A dipole is often symbolized by $\rightarrowtail$, where the arrow points from positive to negative. The molecule possesses a dipole moment, μ, which is equal to the magnitude of the charge, e, multiplied by the distance, d, between the centers of charge:

$$ \mu \;\; = \;\; e \;\; \times \;\; d $$

$$
\begin{array}{ccc}
\text{in} & \text{in} & \text{in} \\
\text{debye} & \text{e.s.u.} & \text{cm} \\
\text{units, D} & &
\end{array}
$$

In a way that cannot be gone into here, it is possible to measure the dipole moments of molecules; some of the values obtained are listed in Table 1.4. We shall be interested in the values of dipole moments as indications of the relative polarities of different molecules.

Table 1.4 DIPOLE MOMENTS, D

H_2	0	HF	1.75	CH_4	0
O_2	0	H_2O	1.84	CH_3Cl	1.86
N_2	0	NH_3	1.46	CCl_4	0
Cl_2	0	NF_3	0.24	CO_2	0
Br_2	0	BF_3	0		

It is the *fact* that some molecules are polar which has given rise to the *speculation* that some bonds are polar. We have taken up bond polarity first simply because it is convenient to consider that the polarity of a molecule is a composite of the polarities of the individual bonds.

Molecules like H_2, O_2, N_2, Cl_2, and Br_2 have zero dipole moments, that is, are non-polar. The two identical atoms of each of these molecules have, of course, the same electronegativity and share electrons equally; e is zero and hence μ is zero, too.

A molecule like hydrogen fluoride has the large dipole moment of 1.75 D. Although hydrogen fluoride is a small molecule, the very high electronegative fluorine pulls the electrons strongly; although d is small, e is large, and hence μ is large, too.

Methane and carbon tetrachloride, CCl_4, have zero dipole moments. We certainly would expect the individual bonds—of carbon tetrachloride at least—to be polar; because of the very symmetrical tetrahedral arrangement, however, they exactly cancel each other out (Fig. 1.16). In methyl chloride, CH_3Cl, the polarity

Figure 1.16 Dipole moments of some molecules. Polarity of bonds and of molecules.

of the carbon–chlorine bond is not canceled, however, and methyl chloride has a dipole moment of 1.86 D. Thus the polarity of a molecule depends not only upon the polarity of its individual bonds but also upon the way the bonds are directed, that is, upon the shape of the molecule.

Ammonia has a dipole moment of 1.46 D. This could be accounted for as a net dipole moment (a *vector sum*) resulting from the three individual bond moments, and would be in the direction shown in the diagram. In a similar way, we could account for water's dipole moment of 1.84 D.

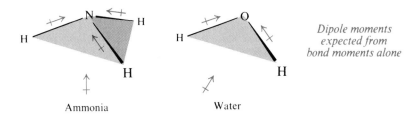

Dipole moments expected from bond moments alone

Ammonia   Water

Now, what kind of dipole moment would we expect for nitrogen trifluoride, NF$_3$, which, like ammonia, is pyramidal? Fluorine is the most electronegative element of all and should certainly pull electrons strongly from nitrogen; the N—F bonds should be highly polar, and their vector sum should be large— far larger than for ammonia with its modestly polar N—H bonds.

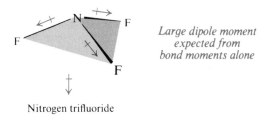

Large dipole moment expected from bond moments alone

Nitrogen trifluoride

What are the facts? Nitrogen trifluoride has a dipole moment of only 0.24 D. It is not larger than the moment for ammonia, but rather is *much smaller*.

How are we to account for this? We have forgotten the *unshared pair of electrons*. In NF$_3$ (as in NH$_3$) this pair occupies an sp^3 orbital and must contribute a dipole moment in the direction opposite to that of the net moment of the N—F bonds (Fig. 1.17); these opposing moments are evidently of about the same size,

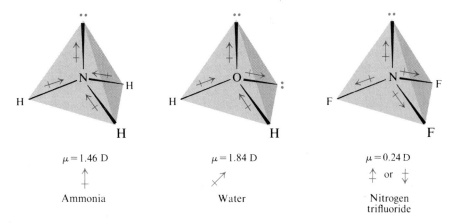

$\mu = 1.46$ D $\mu = 1.84$ D $\mu = 0.24$ D

Ammonia   Water   Nitrogen trifluoride

Figure 1.17   Dipole moments of some molecules. Contribution from unshared pairs. In NF$_3$, the moment due to the unshared pair opposes the vector sum of the bond moments.

and the result is a small moment, in which direction we cannot say. In ammonia the observed moment is probably due chiefly to the unshared pair, augmented by the sum of the bond moments. In a similar way, unshared pairs of electrons must contribute to the dipole moment of water and, indeed, of any molecules in which they appear.

Dipole moments can give valuable information about the structure of molecules. For example, any structure for carbon tetrachloride that would result in a polar molecule can be ruled out on the basis of dipole moment alone. The evidence of dipole moment thus supports the tetrahedral structure for carbon tetrachloride. (However, it does not prove this structure, since there are other conceivable structures that would also result in a non-polar molecule.)

Problem 1.5 Which of the following conceivable structures of CCl_4 would also have a zero dipole moment? (a) Carbon at the center of a square with a chlorine at each corner. (b) Carbon at the apex of a pyramid with a chlorine at each corner of a square base.

Problem 1.6 Suggest a shape for the CO_2 molecule that would account for its zero dipole moment.

Problem 1.7 In Sec. 1.12 we rejected two conceivable electronic configurations for ammonia. (a) If nitrogen were sp^2-hybridized, what dipole moment would you expect for ammonia? What *is* the dipole moment of ammonia? (b) If nitrogen used p orbitals for bonding, how would you expect the dipole moments of ammonia and nitrogen trifluoride to compare? How *do* they compare?

The dipole moments of most compounds have never been measured. For these substances we must predict polarity from structure. From our knowledge of electronegativity, we can estimate the polarity of bonds; from our knowledge of bond angles, we can then estimate the polarity of molecules, taking into account any unshared pairs of electrons.

1.17 Structure and physical properties

We have just discussed one physical property of compounds: dipole moment. Other physical properties—like melting point, boiling point, or solubility in a particular solvent—are also of concern to us. The physical properties of a new compound give valuable clues about its structure. Conversely, the structure of a compound often tells us what physical properties to expect of it.

In attempting to synthesize a new compound, for example, we must plan a series of reactions to convert a compound that we have into the compound that we want. In addition, we must work out a method of separating our product from all the other compounds making up the reaction mixture: unconsumed reactants, solvent, catalyst, by-products. Usually the *isolation* and *purification* of a product take much more time and effort than the actual making of it. The feasibility of isolating the product by distillation depends upon its boiling point and the boiling points of the contaminants; isolation by recrystallization depends upon its solubility in various solvents and the solubility of the contaminants. Success in the laboratory often depends upon making a good prediction of physical properties from structure. Organic compounds are *real* substances—not just collections of letters written on a piece of paper—and we must learn how to handle them.

We have seen that there are two extreme kinds of chemical bonds: ionic bonds, formed by the transfer of electrons, and covalent bonds, formed by the sharing of electrons. The physical properties of a compound depend largely upon which kind of bonds hold its atoms together in the molecule.

START

1.18   Melting point

In a crystalline solid the particles acting as structural units—ions or molecules—are arranged in some very regular, symmetrical way; there is a geometric pattern repeated over and over within a crystal.

Melting is the change from the highly ordered arrangement of particles in the crystalline lattice to the more random arrangement that characterizes a liquid (see Figs. 1.18 and 1.19). Melting occurs when a temperature is reached at which the thermal energy of the particles is great enough to overcome the intracrystalline forces that hold them in position.

An **ionic compound** forms crystals in which the structural units are *ions*. Solid sodium chloride, for example, is made up of positive sodium ions and negative chloride ions alternating in a very regular way. Surrounding each positive ion and

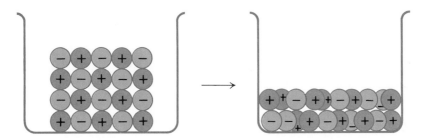

Figure 1.18   Melting of an ionic crystal. The units are ions.

equidistant from it are six negative ions: one on each side of it, one above and one below, one in front and one in back. Each negative ion is surrounded in a similar way by six positive ions. There is nothing that we can properly call a *molecule* of sodium chloride. A particular sodium ion does not "belong" to any one chloride ion; it is equally attracted to six chloride ions. The crystal is an extremely strong, rigid structure, since the electrostatic forces holding each ion in position are powerful. These powerful *interionic* forces are overcome only at a very high temperature; sodium chloride has a melting point of 801 °C.

Crystals of other ionic compounds resemble crystals of sodium chloride in having an ionic lattice, although the exact geometric arrangement may be different. As a result, these other ionic compounds, too, have high melting points. Many molecules contain both ionic and covalent bonds. Potassium nitrate, KNO_3, for example, is made up of K^+ ions and NO_3^- ions; the oxygen and nitrogen atoms of the NO_3^- ion are held to each other by covalent bonds. The physical properties of compounds like these are largely determined by the ionic bonds; potassium nitrate has very much the same sort of physical properties as sodium chloride.

A **non-ionic compound**, one whose atoms are held to each other entirely by covalent bonds, forms crystals in which the structural units are *molecules*. It is the forces holding these molecules to each other that must be overcome for melting to occur. In general, these *intermolecular* forces are very weak compared with the

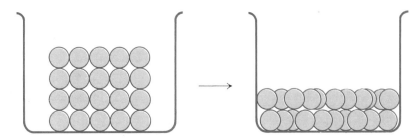

Figure 1.19   Melting of a non-ionic crystal. The units are molecules.

forces holding ions to each other. To melt sodium chloride we must supply enough energy to break ionic bonds between Na^+ and Cl^-. To melt methane, CH_4, we do not need to supply enough energy to break covalent bonds between carbon and hydrogen; we need only supply enough energy to break CH_4 molecules away from each other. In contrast to sodium chloride, methane melts at $-183\ °C$.

1.19   Intermolecular forces

What kinds of forces hold neutral molecules to each other? Like interionic forces, these forces seem to be electrostatic in nature, involving attraction of positive charge for negative charge. There are two kinds of intermolecular forces: *dipole–dipole interactions* and *van der Waals forces*.

Dipole–dipole interaction is the attraction of the positive end of one polar molecule for the negative end of another polar molecule. In hydrogen chloride, for example, the relatively positive hydrogen of one molecule is attracted to the relatively negative chlorine of another:

As a result of dipole–dipole interaction, polar molecules are generally held to each other more strongly than are non-polar molecules of comparable molecular weight; this difference in strength of intermolecular forces is reflected in the physical properties of the compounds concerned.

An especially strong kind of dipole–dipole attraction is **hydrogen bonding**, in which *a hydrogen atom serves as a bridge between two electronegative atoms, holding one by a covalent bond and the other by purely electrostatic forces*. When hydrogen is attached to a highly electronegative atom, the electron cloud is greatly distorted toward the electronegative atom, exposing the hydrogen nucleus. The strong positive charge of the thinly shielded hydrogen nucleus is strongly attracted by the negative charge of the electronegative atom of a second molecule. This attraction has a strength of about 5 kcal/mol, and is thus much weaker than the covalent bond—about 50–100 kcal/mol—that holds it to the first electronegative atom. It is, however, much stronger than other dipole–dipole attractions. Hydrogen bonding is generally indicated in formulas by a broken line:

$$
\begin{array}{cccccc}
 & & \overset{\displaystyle H}{\underset{\displaystyle H}{|}} & \overset{\displaystyle H}{\underset{\displaystyle H}{|}} & \overset{\displaystyle H}{\underset{\displaystyle H}{|}} & \overset{\displaystyle H}{|} \\
H\!-\!F\cdots H\!-\!F & & H\!-\!O\cdots H\!-\!O & H\!-\!N\cdots H\!-\!N & H\!-\!N\cdots H\!-\!O
\end{array}
$$

For hydrogen bonding to be important, both electronegative atoms must come from the group: **F, O, N**. Only hydrogen bonded to one of these three elements is positive enough, and only these three elements are negative enough, for the necessary attraction to exist. These three elements owe their special effectiveness to the concentrated negative charge on their small atoms.

There must be forces between the molecules of a non-polar compound, since even such compounds can solidify. Such attractions are called **van der Waals forces**. The existence of these forces is accounted for by quantum mechanics. We can roughly visualize them arising in the following way. The average distribution of charge about, say, a methane molecule is symmetrical, so that there is no net dipole moment. However, the electrons move about, so that at any instant the distribution will probably be distorted, and a small dipole will exist. This momentary dipole will affect the electron distribution in a second methane molecule nearby. The negative end of the dipole tends to repel electrons, and the positive end tends to attract electrons; the dipole thus *induces* an oppositely oriented dipole in the neighboring molecule:

Although the momentary dipoles and induced dipoles are constantly changing, the net result is attraction between the two molecules.

These van der Waals forces have a very short range; they act only between the portions of different molecules that are in close contact, that is, between the surfaces of molecules. As we shall see, the relationship between the strength of van der Waals forces and the surface areas of molecules (Sec. 3.12) will help us to understand the effect of molecular size and shape on physical properties.

With respect to other atoms to which it is not bonded—whether in another molecule or in another part of the same molecule—every atom has an effective "size", called its *van der Waals radius*. As two non-bonded atoms are brought together the attraction between them steadily increases, and reaches a maximum when they are just "touching"—that is to say, when the distance between the nuclei is equal to the sum of the van der Waals radii. Now, if the atoms are forced still closer together, van der Waals attraction is very rapidly replaced by van der Waals *repulsion*. Thus, non-bonded atoms welcome each other's touch, but strongly resist crowding.

We shall find both attractive and repulsive van der Waals forces important to our understanding of molecular structure.

In Chapter 6, we shall discuss in detail all of these intermolecular forces—these kinds of *secondary bonding*.

1.20   Boiling point

Although the particles in a liquid are arranged less regularly and are freer to move about than in a crystal, each particle is attracted by a number of other particles. Boiling involves the breaking away from the liquid of individual molecules or pairs of oppositely charged ions (see Figs. 1.20 and 1.21). This occurs when a temperature is reached at which the thermal energy of the particles is great enough to overcome the cohesive forces that hold them in the liquid.

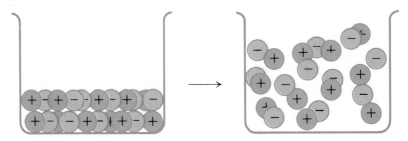

Figure 1.20   Boiling of an ionic liquid. The units are ions and ion pairs.

In the liquid state the unit of an ionic compound is again the ion. Each ion is still held strongly by a number of oppositely charged ions. Again there is nothing we could properly call a molecule. A great deal of energy is required for a pair of oppositely charged ions to break away from the liquid; boiling occurs only at a very high temperature. The boiling point of sodium chloride, for example, is 1413 °C. In the gaseous state we have an *ion pair*, which can be considered a sodium chloride molecule.

In the liquid state the unit of a non-ionic compound is again the molecule. The weak intermolecular forces here—dipole–dipole interactions and van der

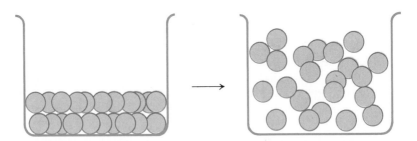

Figure 1.21   Boiling of a non-ionic liquid. The units are molecules.

Waals forces—are more readily overcome than the strong interionic forces of ionic compounds, and boiling occurs at a very much lower temperature. Non-polar methane boils at −161.5 °C, and even polar hydrogen chloride boils at only −85 °C.

Liquids whose molecules are held together by hydrogen bonds are called *associated liquids*. Breaking these hydrogen bonds takes considerable energy, and so an associated liquid has a boiling point that is abnormally high for a compound of its molecular weight and dipole moment. Hydrogen fluoride, for example, boils

100 degrees higher than the heavier, non-associated hydrogen chloride; water boils 160 degrees higher than hydrogen sulfide.

There are organic compounds, too, that contain hydrogen bonded to oxygen or nitrogen, and here, too, hydrogen bonding occurs. Let us take, for example, methane and replace one of its hydrogens with a hydroxyl group, —OH. The resulting compound, CH_3OH, is *methanol*, the smallest member of the *alcohol* family. Structurally, it resembles not only methane, but also water:

$$
\begin{array}{ccccc}
& H & & & H \\
& | & & & | \\
H-C-H & & H-O-H & & H-C-O-H \\
& | & & & | \\
& H & & & H
\end{array}
$$

$$
\text{Methane} \qquad \text{Water} \qquad \text{Methanol}
$$

Like water, it is an associated liquid with a boiling point "abnormally" high for a compound of its size and polarity.

$$
\begin{array}{c}
\qquad\qquad\quad CH_3 \\
\qquad\qquad\quad | \\
CH_3-O\cdots H-O \\
\quad\;\; | \\
\quad\;\; H
\end{array}
$$

The bigger the molecules, the stronger the van der Waals forces. Other things being equal—polarity, hydrogen bonding—boiling point rises with increasing molecular size. Boiling points of organic compounds range upward from that of tiny, non-polar methane, but we seldom encounter boiling points much above 350 °C; at higher temperatures, covalent bonds *within* the molecules start to break, and decomposition competes with boiling. It is to lower the boiling point and thus minimize decomposition that distillation of organic compounds is often carried out under reduced pressure.

Problem 1.8 Which of the following organic compounds would you predict to be *associated* liquids? Draw structures to show the hydrogen bonding you would expect. (a) CH_3OCH_3; (b) CH_3F; (c) CH_3Cl; (d) CH_3NH_2; (e) $(CH_3)_2NH$; (f) $(CH_3)_3N$.

1.21 Solubility

When a solid or liquid dissolves, the structural units—ions or molecules—become separated from each other, and the spaces in between become occupied by solvent molecules. In dissolution, as in melting and boiling, energy must be supplied to overcome the interionic or intermolecular forces. Where does the necessary energy come from? The energy required to break the bonds between solute particles is supplied by the formation of bonds between the solute particles and the solvent molecules: the old attractive forces are replaced by new ones.

Now, what are these bonds that are formed between solute and solvent? Let us consider first the case of **ionic solutes**.

A great deal of energy is necessary to overcome the powerful electrostatic forces holding together an ionic lattice. Only water or other highly polar solvents are able to dissolve ionic compounds appreciably. What kinds of bonds are formed between ions and a polar solvent? By definition, a polar molecule has a positive end and a negative end. Consequently, there is electrostatic attraction between a positive ion and the negative end of the solvent molecule, and between a negative ion and the positive end of the solvent molecule. These attractions are called **ion–dipole** bonds. Each ion–dipole bond is relatively weak, but in the aggregate they supply enough energy to overcome the interionic forces in the crystal. In solution each ion is surrounded by a cluster of solvent molecules, and is said to be *solvated*; if the solvent happens to be water, the ion is said to be *hydrated*. In solution, as in the solid and liquid states, the unit of a substance like sodium chloride is the ion, although in this case it is a solvated ion (see Fig. 1.22).

Figure 1.22   Ion–dipole interactions: a solvated cation and anion.

To dissolve ionic compounds a solvent must also have a high *dielectric constant*, that is, have high insulating properties to lower the attraction between oppositely charged ions once they are solvated.

Water owes its superiority as a solvent for ionic substances not only to its polarity and its high dielectric constant but to another factor as well: it contains the —OH group and thus can form hydrogen bonds. Water solvates both cations and anions: cations, at its negative pole (its unshared electrons, essentially); anions, through hydrogen bonding.

Now let us turn to the dissolution of **non-ionic solutes**.

The solubility characteristics of non-ionic compounds are determined chiefly by their polarity. Non-polar or weakly polar compounds dissolve in non-polar or weakly polar solvents; highly polar compounds dissolve in highly polar solvents. "Like dissolves like" is an extremely useful rule of thumb. Methane dissolves in carbon tetrachloride because the forces holding methane molecules to each other and carbon tetrachloride molecules to each other—van der Waals interactions— are replaced by very similar forces holding methane molecules to carbon tetra- chloride molecules.

Neither methane nor carbon tetrachloride is readily soluble in water. The highly polar water molecules are held to each other by very strong dipole–dipole interactions—hydrogen bonds; there could be only very weak attractive forces between water molecules on the one hand and the non-polar methane or carbon tetrachloride molecules on the other.

In contrast, the highly polar organic compound methanol, CH_3OH, is quite soluble in water. Hydrogen bonds between water and methanol molecules can readily replace the very similar hydrogen bonds between different methanol molecules and different water molecules.

An understanding of the nature of solutions is fundamental to an understan- ding of organic chemistry. Most organic reactions are carried out in solution and, it is becoming increasingly clear, the solvent does much more than simply bring different molecules together so that they can react with each other. The solvent is *involved* in the reactions that take place in it: just how much it is involved, and in what ways, is only now being realized. In Chapter 6, when we know a little more about organic reactions and how they take place, we shall return to this subject—which we have barely touched upon here—and examine in detail the role played by the solvent.

1.22   Acids and bases

Turning from physical to chemical properties, let us review briefly one familiar topic that is fundamental to the understanding of organic chemistry: acidity and basicity.

The terms *acid* and *base* have been defined in a number of ways, each definition corresponding to a particular way of looking at the properties of acidity and basicity. We shall find it useful to look at acids and bases from two of these viewpoints; the one we select will depend upon the problem at hand.

According to the **Lowry–Brønsted** definition, *an acid is a substance that gives up a proton,* and *a base is a substance that accepts a proton.* When sulfuric acid dissolves in water, the acid H_2SO_4 gives up a proton (hydrogen nucleus) to the base H_2O to form the new acid H_3O^+ and the new base HSO_4^-. When hydrogen chloride reacts with ammonia, the acid HCl gives up a proton to the base NH_3 to form the new acid NH_4^+ and the new base Cl^-.

$$H_2SO_4 \;+\; H_2O \;\rightleftharpoons\; H_3O^+ \;+\; HSO_4^-$$

Stronger acid   Stronger base   Weaker acid   Weaker base

$$HCl \;+\; NH_3 \;\rightleftharpoons\; NH_4^+ \;+\; Cl^-$$

Stronger acid   Stronger base   Weaker acid   Weaker base

According to the Lowry–Brønsted definition, the strength of an acid depends upon its tendency to give up a proton, and the strength of a base depends upon its tendency to accept a proton. Sulfuric acid and hydrogen chloride are strong acids since they tend to give up a proton very readily; conversely, bisulfate ion, HSO_4^-, and chloride ion must necessarily be weak bases since they have little tendency to hold on to protons. In each of the reactions just described, the equilibrium favors the formation of the weaker acid and the weaker base.

If aqueous H_2SO_4 is mixed with aqueous NaOH, the acid H_3O^+ (hydronium ion) gives up a proton to the base OH^- to form the new acid H_2O and the new base H_2O. When aqueous NH_4Cl is mixed with aqueous NaOH, the acid NH_4^+

$$H_3O^+ \;+\; OH^- \;\rightleftharpoons\; H_2O \;+\; H_2O$$

Stronger acid   Stronger base   Weaker acid   Weaker base

$$NH_4^+ \;+\; OH^- \;\rightleftharpoons\; H_2O \;+\; NH_3$$

Stronger acid   Stronger base   Weaker acid   Weaker base (ammonium ion) gives up a proton to the base OH^- to form the new acid H_2O and the new base NH_3. In each case the strong base, hydroxide ion, has accepted a proton to form the weak acid H_2O. If we arrange these acids in the order shown, we must necessarily arrange the corresponding (conjugate) bases in the opposite order.

Acid strength $\begin{array}{c} H_2SO_4 \\ HCl \end{array} > H_3O^+ > NH_4{}^+ > H_2O$

Base strength $\begin{array}{c} HSO_4{}^- \\ Cl^- \end{array} < H_2O < NH_3 < OH^-$

Like water, many organic compounds that contain oxygen can act as bases and accept protons; ethyl alcohol and diethyl ether, for example, form the *oxonium ions* I and II. For convenience, we shall often refer to a structure like I as a *protonated alcohol* and a structure like II as a *protonated ether*.

$$C_2H_5\overset{..}{\underset{..}{O}}H + H_2SO_4 \; \rightleftharpoons \; C_2H_5\overset{\overset{\oplus}{}}{\underset{..}{O}}\underset{\underset{H}{|}}{H} + HSO_4{}^-$$

Ethyl alcohol

I

An oxonium ion
Protonated ethyl alcohol

$$(C_2H_5)_2\overset{..}{\underset{..}{O}}: + HCl \; \rightleftharpoons \; (C_2H_5)_2\overset{\overset{\oplus}{}}{\underset{..}{O}}:H + Cl^-$$

Diethyl ether

II

An oxonium ion
Protonated diethyl ether

According to the **Lewis** definition, *a base is a substance that can furnish an electron pair to form a covalent bond*, and *an acid is a substance that can take up an electron pair to form a covalent bond*. Thus **an acid is an electron-pair acceptor** and **a base is an electron-pair donor**. This is the most fundamental of the acid–base concepts, and the most general; it includes all the other concepts.

A proton is an acid because it is deficient in electrons, and needs an electron pair to complete its valence shell. Hydroxide ion, ammonia, and water are bases because they contain electron pairs available for sharing. In boron trifluoride, BF_3, boron has only six electrons in its outer shell and hence tends to accept another pair to complete its octet. Boron trifluoride is an acid and combines with such bases as ammonia or diethyl ether.

$$\underset{\underset{F}{|}}{\overset{\overset{F}{|}}{F}}{-}B + :NH_3 \; \rightleftharpoons \; \underset{\underset{F}{|}}{\overset{\overset{F}{|}}{F}}{-}\overset{\ominus}{B}:\overset{\oplus}{N}H_3$$

     Acid   Base

$$\underset{\underset{F}{|}}{\overset{\overset{F}{|}}{F}}{-}B + :\overset{..}{\underset{..}{O}}(C_2H_5)_2 \; \rightleftharpoons \; \underset{\underset{F}{|}}{\overset{\overset{F}{|}}{F}}{-}\overset{\ominus}{B}:\overset{\oplus}{O}(C_2H_5)_2$$

     Acid   Base

Aluminum chloride, $AlCl_3$, is an acid, and for the same reason. In stannic chloride, $SnCl_4$, tin has a complete octet, but can accept additional pairs of electrons (e.g., in $SnCl_6{}^{2-}$) and hence it is an acid, too.

We write a formal negative charge on boron in these formulas because it has one more electron—half-interest in the pair shared with nitrogen or oxygen—than is balanced by the nuclear charge; correspondingly, nitrogen or oxygen is shown with a formal positive charge.

We shall find the Lewis concept of acidity and basicity fundamental to our understanding of organic chemistry. To make it clear that we are talking about this kind of acid or base, we shall often use the expression *Lewis acid* (or *Lewis base*), or sometimes *acid* (or *base*) *in the Lewis sense*.

Chemical properties, like physical properties, depend upon molecular structure. Just what features in a molecule's structure tell us what to expect about its acidity or basicity? We can try to answer this question in a general way now, although we shall return to it many times later.

To be acidic in the Lowry–Brønsted sense, a molecule must, of course, contain hydrogen. The degree of acidity is determined largely by the kind of atom that holds the hydrogen and, in particular, by that atom's ability to accommodate the electron pair left behind by the departing hydrogen ion. This ability to accommodate the electron pair seems to depend upon several factors, including (a) the atom's *electronegativity*, and (b) its *size*. Thus, within a given row of the Periodic Table, acidity increases as electronegativity increases:

Acidity

$$H—CH_3 < H—NH_2 < H—OH < H—F$$
$$H—SH < H—Cl$$

And within a given family, acidity increases as the size increases:

Acidity

$$H—F < \ H—Cl < H—Br < H—I$$
$$H—OH < H—SH < H—SeH$$

Among organic compounds, we can expect appreciable Lowry–Brønsted acidity from those containing O—H, N—H, and S—H groups.

To be acidic in the Lewis sense, a molecule must be electron-deficient; in particular, we would look for an atom bearing only a sextet of electrons.

Problem 1.9 Predict the relative acidity of: (a) methyl alcohol (CH_3OH) and methylamine (CH_3NH_2); (b) methyl alcohol (CH_3OH) and methanethiol (CH_3SH); (c) H_3O^+ and $NH_4{}^+$.

Problem 1.10 Which is the stronger acid of each pair: (a) H_3O^+ or H_2O; (b) $NH_4{}^+$ or NH_3; (c) H_2S or HS^-; (d) H_2O or OH^-? (e) What relationship is there between *charge* and acidity?

To be basic in either the Lowry–Brønsted or the Lewis sense, a molecule must have an electron pair available for sharing. The availability of these unshared electrons is determined largely by the atom that holds them: its electronegativity, its size, its charge. The operation of these factors here is necessarily opposite to what we observed for acidity; the better an atom accommodates the electron pair, the less available the pair is for sharing.

Problem 1.11 Arrange the members of each group in order of basicity:
(a) F^-, OH^-, NH_2^-, CH_3^-; (b) HF, H_2O, NH_3; (c) Cl^-, SH^-; (d) F^-, Cl^-, Br^-, I^-;
(e) OH^-, SH^-, SeH^-.

Problem 1.12 Predict the relative basicity of methyl fluoride (CH_3F), methyl alcohol (CH_3OH), and methylamine (CH_3NH_2).

Problem 1.13 Arrange the members of each group in order of basicity:
(a) H_3O^+, H_2O, OH^-; (b) NH_3, NH_2^-; (c) H_2S, HS^-, S^{2-}. (d) What relationship is there between charge and basicity?

1.23 Isomerism

Before we start our systematic study of the different kinds of organic compounds, let us look at one further concept which illustrates especially well the fundamental importance of molecular structure: the concept of **isomerism**.

The compound *ethyl alcohol* is a liquid boiling at 78 °C. Analysis (by the methods described later, Sec. 2.27) shows that it contains carbon, hydrogen, and oxygen in the proportions 2C:6H:1O. Measurement of its mass spectrum shows that it has a molecular weight of 46. The molecular formula of ethyl alcohol must therefore be C_2H_6O. Ethyl alcohol is a quite reactive compound. For example, if a piece of sodium metal is dropped into a test tube containing ethyl alcohol, there is a vigorous bubbling and the sodium metal is consumed; hydrogen gas is evolved and there is left behind a compound of formula C_2H_5ONa. Ethyl alcohol reacts with hydriodic acid to form water and a compound of formula C_2H_5I.

The compound *dimethyl ether* is a gas with a boiling point of -24 °C. It is clearly a different substance from ethyl alcohol, differing not only in its physical properties but also in its chemical properties. It does not react at all with sodium metal. Like ethyl alcohol, it reacts with hydriodic acid, but it yields a compound of formula CH_3I. Analysis of dimethyl ether shows that it contains carbon, hydrogen, and oxygen in the same proportion as ethyl alcohol, 2C:6H:1O. It has the same molecular weight as ethyl alcohol, 46. We conclude that it has the same molecular formula C_2H_6O.

Here we have two substances, ethyl alcohol and dimethyl ether, which have the same molecular formula, C_2H_6O, and yet quite clearly are different compounds. How can we account for the existence of these two compounds? The answer is: *they differ in molecular structure*. Ethyl alcohol has the structure represented by I, and dimethyl ether the structure represented by II. As we shall see, the differences in physical and chemical properties of these two compounds can readily be accounted for on the basis of the difference in structure.

        H H   H   H
        | | | |
   H—C—C—O—H   H—C—O—C—H
        | | | |
        H H   H   H
           I   II
    Ethyl alcohol   Dimethyl ether

Different compounds that have the same molecular formula are called **isomers** (Greek: *isos*, equal; *meros*, part). They contain the same numbers of the same kinds of atoms, but the atoms are attached to one another in different ways. Isomers are different compounds because they have different molecular structures.

This difference in molecular structure gives rise to a difference in properties; it is the difference in properties which tells us that we are dealing with different compounds. In some cases, the difference in structure—and hence the difference in properties—is so marked that the isomers are assigned to different chemical families, as, for example, ethyl *alcohol* and dimethyl *ether*. In other cases the difference in structure is so subtle that it can be described only in terms of three-dimensional models. Other kinds of isomerism fall between these two extremes.

PROBLEMS

1. Which of the following would you expect to be ionic, and which non-ionic? Give a simple electronic structure (Sec. 1.3) for each, showing only valence shell electrons.

(a) $MgCl_2$ (c) ICl   (e) $KClO_4$ (g) $BaSO_4$
(b) CH_2Cl_2 (d) NaOCl   (f) $SiCl_4$ (h) CH_3NH_2

2. Give a likely simple electronic structure (Sec. 1.3) for each of the following, assuming them to be completely covalent. Assume that every atom (except hydrogen, of course) has a complete octet, and that two atoms may share more than one pair of electrons.

(a) N_2H_4 (d) $COCl_2$ (g) $CO_3{}^{2-}$ (j) CH_2O
(b) H_2SO_4 (e) HONO   (h) C_2H_4 (k) CH_2O_2
(c) $HSO_4{}^-$ (f) $NO_2{}^-$ (i) C_2H_2 (l) C_3H_8

3. What shape would you expect each of the following to have?

(a) $(CH_3)_3B$ (e) the amide ion, $NH_2{}^-$
(b) the methyl anion, $CH_3:{}^-$ (f) dimethyl ether
(c) the methyl cation, $CH_3{}^+$ (g) the fluoroborate ion, $BF_4{}^-$
(d) H_2S (h) $(CH_3)_3N$

4. In many complex ions, e.g., $Co(NH_3)_6{}^{3+}$, the bonds to the central atom can be pictured as utilizing six equivalent sp^3d^2 (or d^2sp^3) hybrid orbitals. On the basis of maximum separation of orbitals, what geometry would you expect these complexes to have?

5. Indicate the direction of the dipole moment, *if any*, that you would expect for each of the following:

(a) HBr   (d) CH_2Cl_2 (g) dimethyl ether
(b) ICl   (e) $CHCl_3$ (h) $(CH_3)_3N$
(c) I_2 (f) CH_3OH (i) CF_2Cl_2

About Working Problems

Working problems is a necessary part of your work for two reasons: it will guide your study in the right direction, and, after you have studied a particular chapter, it will show whether or not you have reached your destination.

You should work all the problems that you can; you should get help with the ones you cannot work yourself. The first problems in each set are easy, but provide the drill in drawing formulas, naming compounds, and using reactions that even the best student needs. The later problems in each set are the kind encountered by practicing chemists, and test your ability to *use* what you have learned.

You can check your answers to many of the problems in the answer section in the back of the book, and by use of the index.

6. (a) Although HCl (1.27 Å) is a longer molecule than HF (0.92 Å), it has a *smaller* dipole moment (1.03 debye compared to 1.75 debye). How do you account for this fact? (b) The dipole moment of CH_3F is 1.847 debye, and of CD_3F, 1.858 debye. (D is ^2H, deuterium.) Compared with the C—H bond, what is the direction of the C—D dipole?

7. What do the differences in properties between lithium acetylacetonate (m.p. very high, insoluble in chloroform) and beryllium acetylacetonate (m.p. 108 °C, b.p. 270 °C, soluble in chloroform) suggest about their structures?

8. *n*-Butyl alcohol (b.p. 118 °C) has a much higher boiling point than its isomer diethyl ether (b.p. 35 °C), yet both compounds show the same solubility (8 g per 100 g) in water.

hydrogen bonding

```
      H  H  H  H                           H  H      H  H
      |  |  |  |                           |  |      |  |
   H—C—C—C—C—O—H                        H—C—C—O—C—C—H
      |  |  |  |                           |  |      |  |
      H  H  H  H                           H  H      H  H
```
       n-Butyl alcohol   Diethyl ether

How do you account for these facts?

9. Rewrite the following equations to show the Lowry–Brønsted acids and bases actually involved. Label each as stronger or weaker, as in Sec. 1.22.

(a) $HCl(aq) + NaHCO_3(aq) \rightleftarrows H_2CO_3 + NaCl$

(b) $NaOH(aq) + NaHCO_3(aq) \rightleftarrows Na_2CO_3 + H_2O$

(c) $NH_3(aq) + HNO_3(aq) \rightleftarrows NH_4NO_3(aq)$

(d) $NaCN(aq) \rightleftarrows HCN(aq) + NaOH(aq)$

(e) $NaH + H_2O \longrightarrow H_2 + NaOH$

(f) $CaC_2 + H_2O \longrightarrow Ca(OH)_2 + C_2H_2$
    Calcium   Acetylene
    carbide

10. What is the Lowry–Brønsted acid in (a) HCl dissolved in water; (b) HCl (un-ionized) dissolved in benzene? (c) Which solution is the more strongly acidic?

11. Account for the fact that nearly every organic compound containing oxygen dissolves in cold concentrated sulfuric acid to yield a solution from which the compound can be recovered by dilution with water.

12. How might you account for the following orders of acidity? Be as specific as you can.

$$HClO_4 > HClO_2 > HClO \quad \text{and} \quad H_2SO_4 > H_2SO_3$$

13. For each of the following molecular formulas, draw structures like those in Sec. 1.23 (a line for each shared pair of electrons) for all the isomers you can think of. Assume that every atom (except hydrogen) has a complete octet, and that two atoms may share more than one pair of electrons.

    (a) C_2H_7N (c) C_4H_{10} (e) C_3H_8O
    (b) C_3H_8 (d) C_3H_7Cl (f) C_2H_4O

14. In ordinary distillation, a liquid is placed in a flask and heated, at ordinary or reduced pressure, until distillation is complete. In the modification called *flash distillation*, the liquid is dripped into a heated flask at the same rate that it distills out, so that there is little liquid in the flask at any time. What advantage might flash distillation have, and under what conditions might you use it?

2

Methane

Energy of Activation. Transition State

2.1 Hydrocarbons

Certain organic compounds contain only two elements, hydrogen and carbon, and hence are known as **hydrocarbons**. On the basis of structure, hydrocarbons are divided into two main classes, **aliphatic** and **aromatic**. Aliphatic hydrocarbons are further divided into families: alkanes, alkenes, alkynes, and their cyclic analogs (cycloalkanes, etc.).

Hydrocarbons $C_x H_y$

Aliphatic *straight chained*   Aromatic *rings based on benzene*

Alkanes *Saturated*   Alkenes *double bond*   Alkynes *triple bond*   Cyclic aliphatic

unsaturated

The simplest member of the alkane family and, indeed, one of the simplest of all organic compounds is **methane**, CH_4. We shall study this single compound at some length, since most of what we learn about it can be carried over with minor modifications to any alkane.

2.2   Structure of methane

As we discussed in the previous chapter (Sec. 1.11), each of the four hydrogen atoms is bonded to the carbon atom by a covalent bond, that is, by the sharing of a pair of electrons. When carbon is bonded to four other atoms, its bonding orbitals (sp^3 orbitals, formed by the mixing of one s and three p orbitals) are directed to the corners of a tetrahedron (Fig. 2.1a). This tetrahedral arrangement is the one that permits the orbitals to be as far apart as possible. For each of these orbitals to overlap most effectively the spherical s orbital of a hydrogen atom, and thus to form the strongest bond, each hydrogen nucleus must be located at a corner of this tetrahedron (Fig. 2.1b).

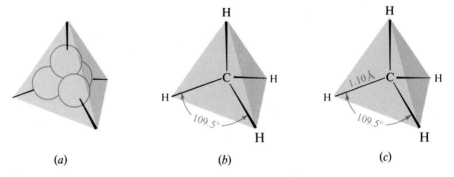

(a) (b) (c)

Figure 2.1   Methane molecule. (a) Tetrahedral sp^3 orbitals. (b) Predicted shape: H nuclei located for maximum overlap. (c) Shape and size.

The tetrahedral structure of methane has been verified by electron diffraction (Fig. 2.1c), which shows beyond question the arrangement of atoms in such simple molecules. Later on, we shall examine some of the evidence that led chemists to accept this tetrahedral structure long before quantum mechanics or electron diffraction was known.

We shall ordinarily write methane with a dash to represent each pair of electrons shared by carbon and hydrogen (I). To focus our attention on individual electrons, we may sometimes indicate a pair of electrons by a pair of dots (II). Finally, when we wish to represent the actual shape of the molecule, we shall use a simple three-dimensional formula like III or IV.

H | H—C—H | H	H ․․ H:C:H ․․ H	H ⋮ H—C—H ⋮ H	H | C H⁄ | ⁀H H
I	II	III	IV

In three-dimensional formulas of this kind, a solid wedge represents a bond coming toward us out of the plane of the paper; a broken wedge, a bond going away from us behind the plane of the paper; and an ordinary line, a bond lying in the plane of the paper. Thus formulas III and IV represent methane as in Fig. 2.2a and Fig. 2.2b, respectively.

(a)   (b)

Figure 2.2 Wedge formulas of the methane molecule oriented in different ways.

2.3 Physical properties

As we discussed in the previous chapter (Sec. 1.18), the unit of such a non-ionic compound, whether solid, liquid, or gas, is the molecule. Because the methane molecule is highly symmetrical, the polarities of the individual carbon–hydrogen bonds cancel out; as a result, the molecule itself is non-polar.

Attraction between such non-polar molecules is limited to van der Waals forces; for such small molecules, these attractive forces must be tiny compared with the enormous forces between, say, sodium and chloride ions. It is not surprising, then, that these attractive forces are easily overcome by thermal energy, so that melting and boiling occur at very low temperatures: m.p. $-183\,°C$, b.p. $-161.5\,°C$. (Compare these values with the corresponding ones for sodium chloride: m.p. $801\,°C$, b.p. $1413\,°C$.) As a consequence, methane is a gas at ordinary temperatures.

Methane is colorless and, when liquefied, is less dense than water (relative density 0.4). In agreement with the rule of thumb that "like dissolves like", it is only slightly soluble in water, but very soluble in organic liquids such as gasoline, ether, and alcohol. In its physical properties methane sets the pattern for the other members of the alkane family.

2.4 Source

Methane is an end product of the anaerobic ("without air") decay of plants, that is, of the breakdown of certain very complicated molecules. As such, it is the major constituent (up to 97%) of **natural gas**. It is the dangerous *firedamp* of the coal mine, and can be seen as *marsh gas* bubbling to the surface of swamps.

If methane is wanted in very pure form, it can be separated from the other constituents of natural gas (mostly other alkanes) by fractional distillation. Most of it, of course, is consumed as fuel without purification.

According to one theory, the origins of life go back to a primitive earth surrounded by an atmosphere of methane, water, ammonia, and hydrogen. Energy—radiation from the sun, lightning discharges—broke these simple molecules into reactive fragments (free radicals, Sec. 2.12); these combined to form larger molecules which eventually yielded the enormously complicated organic compounds that make up living organisms. (Detection of organic molecules in space has even led to the speculation that "organic seeds for life could have existed in interstellar clouds".)

Evidence that this *could* have happened was found in 1953 by the Nobel Prize winner Harold C. Urey and his student Stanley Miller at the University of Chicago. They showed that an electric discharge converts a mixture of methane, water, ammonia, and hydrogen into a large number of organic compounds, including amino acids, the building blocks from which proteins, the "stuff of life" (Chap. 40), are made. (It is perhaps appropriate that we begin this study of organic chemistry with methane and its conversion into free radicals.)

The methane generated in the final decay of a once-living organism may well be the very substance from which—in the final analysis—the organism was derived. "... *earth to earth, ashes to ashes, dust to dust....*"

2.5 Reactions

In its chemical properties as in its physical properties, methane sets the pattern for the alkane family (Sec. 3.18). Typically, it reacts only with highly reactive substances—or under very vigorous conditions, which, as we shall see, amounts to the same thing. At this point we shall take up only its oxidation: by oxygen, by halogens, and even by water.

REACTIONS OF METHANE

1. Oxidation

$$CH_4 + 2O_2 \xrightarrow{\text{flame}} CO_2 + 2H_2O + \text{heat (213 kcal/mol)} \quad \textit{Combustion}$$

$$6CH_4 + O_2 \xrightarrow{1500\,°C} 2HC\equiv CH + 2CO + 10H_2 \quad \text{Discussed in Sec. 11.5.}$$
Acetylene

$$CH_4 + H_2O \xrightarrow[\text{Ni}]{850\,°C} CO + 3H_2$$

2. Halogenation

$$CH_4 \xrightarrow{X_2} CH_3X \xrightarrow{X_2} CH_2X_2 \xrightarrow{X_2} CHX_3 \xrightarrow{X_2} CX_4$$
(each with $+ HX$) *Heat or light required*

Reactivity of X_2 $F_2 > Cl_2 > Br_2 (> I_2)$
Unreactive ∎

2.6 Oxidation. Heat of combustion

Combustion to carbon dioxide and water is characteristic of organic compounds; under special conditions it is used to determine their content of carbon and hydrogen (Sec. 2.27).

Combustion of methane is the principal reaction taking place during the burning of natural gas. It is hardly necessary to emphasize its importance in the areas where natural gas is available; the important product is not carbon dioxide or water but *heat*.

Burning of hydrocarbons takes place only at high temperatures, as provided, for example, by a flame or a spark. Once started, however, the reaction gives off heat which is often sufficient to maintain the high temperature and to permit burning to continue. *The quantity of heat evolved when one mole of a hydrocarbon is burned to carbon dioxide and water is called the* **heat of combustion**; for methane its value is 213 kcal.

Through controlled *partial* oxidation of methane and the high-temperature catalytic reaction with water, methane is an increasingly important source of products other than heat: of hydrogen, used in the manufacture of ammonia; of mixtures of carbon monoxide and hydrogen, used in the manufacture of *methanol* and other alcohols; and of *acetylene* (Sec. 11.5), itself the starting point of large-scale production of many organic compounds.

Oxidation by halogens is of particular interest to us—partly because we know more about it than the other reactions of methane—and, in one way or another, is the topic of discussion throughout the remainder of this chapter.

2.7 Chlorination: a substitution reaction

Under the influence of ultraviolet light or at a temperature of 250–400 °C a mixture of the two gases, methane and chlorine, reacts vigorously to yield hydrogen chloride and a compound of formula CH_3Cl. We say that methane has undergone **chlorination**, and we call the product, CH_3Cl, *chloromethane* or *methyl chloride* ($CH_3 = $ **methyl**).

Chlorination is a typical example of a broad class of organic reactions known as **substitution**. A chlorine atom has been substituted for a hydrogen atom of methane, and the hydrogen atom thus replaced is found combined with a second atom of chlorine.

$$
\underset{\text{Methane}}{H-\overset{\displaystyle H}{\underset{\displaystyle H}{C}}-H} + \underset{\text{Chlorine}}{Cl-Cl} \xrightarrow{\text{light or heat}} \underset{\substack{\text{Methyl chloride}\\\text{(Chloromethane)}}}{H-\overset{\displaystyle H}{\underset{\displaystyle H}{C}}-Cl} + \underset{\substack{\text{Hydrogen}\\\text{chloride}}}{H-Cl}
$$

The methyl chloride can itself undergo further substitution to form more hydrogen chloride and the compound CH_2Cl_2, *dichloromethane* or *methylene chloride* ($CH_2 = $ **methylene**).

$$
H-\overset{\displaystyle H}{\underset{\displaystyle H}{C}}-Cl + Cl-Cl \xrightarrow{\text{light or heat}} \underset{\substack{\text{Methylene chloride}\\\text{(Dichloromethane)}}}{H-\overset{\displaystyle H}{\underset{\displaystyle Cl}{C}}-Cl} + H-Cl
$$

In a similar way, chlorination may continue to yield $CHCl_3$, *trichloromethane* or *chloroform*, and CCl_4, *tetrachloromethane* or *carbon tetrachloride*. Carbon tetrachloride was once widely used as a non-flammable cleaning agent and as the fluid in certain fire extinguishers, but has been largely replaced by other materials.

$$CH_4 \xrightarrow{Cl_2} \begin{array}{c} HCl \\ + \\ CH_3Cl \end{array} \xrightarrow{Cl_2} \begin{array}{c} HCl \\ + \\ CH_2Cl_2 \end{array} \xrightarrow{Cl_2} \begin{array}{c} HCl \\ + \\ CHCl_3 \end{array} \xrightarrow{Cl_2} \begin{array}{c} HCl \\ + \\ CCl_4 \end{array}$$

| Methane | Methyl chloride | Methylene chloride | Chloroform | Carbon tetrachloride |

Heat or light required

2.8 Control of chlorination

Chlorination of methane may yield any one of four organic products, depending upon the stage to which the reaction is carried. Can we control this reaction so that methyl chloride is the principal organic product? That is, can we limit the reaction to the first stage, *mono*chlorination?

We might at first expect—naïvely, as it turns out—to accomplish this by providing only one mole of chlorine for each mole of methane. But let us see what happens if we do so. At the beginning of the reaction there is only methane for the chlorine to react with, and consequently only the first stage of chlorination takes place. This reaction, however, yields methyl chloride, so that as the reaction proceeds methane disappears and methyl chloride takes its place.

As the proportion of methyl chloride grows, it competes with the methane for the available chlorine. By the time the concentration of methyl chloride exceeds that of methane, chlorine is more likely to attack methyl chloride than methane, and the second stage of chlorination becomes more important than the first. A large amount of methylene chloride is formed, which in a similar way is chlorinated to chloroform and this, in turn, is chlorinated to carbon tetrachloride. When we finally work up the reaction product, we find that it is a mixture of all four chlorinated methanes together with some unreacted methane.

The reaction may, however, be limited almost entirely to monochlorination if we use a large excess of methane. In this case, even at the very end of the reaction unreacted methane greatly exceeds methyl chloride. Chlorine is more likely to attack methane than methyl chloride, and thus the first stage of chlorination is the principal reaction.

Because of the great difference in their boiling points, it is easy to separate the excess methane (b.p. $-161.5\ °C$) from the methyl chloride (b.p. $-24\ °C$) so that the methane can be mixed with more chlorine and put through the process again. While there is a low **conversion** of methane into methyl chloride in each cycle, the **yield** of methyl chloride based on the chlorine consumed is quite high.

The use of a large excess of one reactant is a common device of the organic chemist who wishes to limit reaction to only one of a number of reactive sites in the molecule of that reactant.

2.9 Reaction with other halogens: halogenation

Methane reacts with bromine, again at high temperatures or under the influence of ultraviolet light, to yield the corresponding bromomethanes: methyl bromide, methylene bromide, bromoform, and carbon tetrabromide.

$$CH_4 \xrightarrow{Br_2} \begin{matrix} HBr \\ + \\ CH_3Br \end{matrix} \xrightarrow{Br_2} \begin{matrix} HBr \\ + \\ CH_2Br_2 \end{matrix} \xrightarrow{Br_2} \begin{matrix} HBr \\ + \\ CHBr_3 \end{matrix} \xrightarrow{Br_2} \begin{matrix} HBr \\ + \\ CBr_4 \end{matrix}$$ *Heat or light required*

Methane   Methyl   Methylene   Bromoform   Carbon
               bromide   bromide   tetrabromide

Bromination takes place somewhat less readily than chlorination.

Methane does not react with iodine at all. With fluorine it reacts so vigorously that, even in the dark and at room temperature, the reaction must be carefully controlled: the reactants, diluted with an inert gas, are mixed at low pressure.

We can, therefore, arrange the halogens in order of reactivity.

Reactivity of halogens $F_2 > Cl_2 > Br_2 \, (> I_2)$

This same order of reactivity holds for the reaction of the halogens with other alkanes and, indeed, with most other organic compounds. The spread of reactivities is so great that only chlorination and bromination proceed at such rates as to be generally useful.

2.10   Relative reactivity

Throughout our study of organic chemistry, we shall constantly be interested in *relative reactivities*. We shall compare the reactivities of various reagents toward the same organic compound, the reactivities of different organic compounds toward the same reagent, and even the reactivities of different sites in an organic molecule toward the same reagent.

It should be understood that when we compare reactivities we compare rates of reaction. When we say that chlorine is *more reactive* than bromine toward methane, we mean that under the same conditions (same concentration, same temperature, etc.) chlorine reacts with methane *faster* than does bromine. From another point of view, we mean that the bromine reaction must be carried out under more vigorous conditions (higher concentration or higher temperature) if it is to take place as fast as the chlorine reaction. When we say that methane and iodine do not react at all, we mean that the reaction is too slow to be significant.

We shall want to know not only what these relative reactivities are but also, whenever possible, how to account for them. To see what factors cause one reaction to be faster than another, we shall take up in more detail this matter of the different reactivities of the halogens toward methane. Before we can do this, however, we must understand a little more about the reaction itself.

2.11   Reaction mechanisms

It is important for us to know not only *what* happens in a chemical reaction but also *how* it happens, that is, to know not only the *facts* but also the *theory*.

For example, we know that methane and chlorine under the influence of heat or light form methyl chloride and hydrogen chloride. Just how is a molecule of methane converted into a molecule of methyl chloride? Does this transformation involve more than one step, and, if so, what are these steps? Just what is the function of heat or light?

The answer to questions like these, that is, *the detailed, step-by-step description of a chemical reaction, is called a* **mechanism**. It is only a hypothesis; it is advanced to account for the facts. As more facts are discovered, the mechanism must also account for them, or else be modified so that it does account for them; it may even be necessary to discard a mechanism and to propose a new one.

It would be difficult to say that a mechanism had ever been *proved*. If, however, a mechanism accounts satisfactorily for a wide variety of facts; if we make predictions based upon this mechanism and find these predictions borne out; if the mechanism is consistent with mechanisms for other, related reactions; then the mechanism is said to be *well established*, and it becomes part of the theory of organic chemistry.

Why are we interested in the mechanisms of reactions? As an important part of the theory of organic chemistry, they help make up the framework on which we hang the facts we learn. An understanding of mechanisms will help us to see a pattern in the complicated and confusing multitude of organic reactions. We shall find that many apparently unrelated reactions proceed by the same or similar mechanisms, so that most of what we have already learned about one reaction may be applied directly to many new ones.

By knowing how a reaction takes place, we can make changes in the experimental conditions—not by trial and error, but logically—that will improve the yield of the product we want, or that will even alter the course of the reaction completely and give us an entirely different product. As our understanding of reactions grows, so does our power to control them.

2.12   Mechanism of chlorination. Free radicals

It will be worthwhile to examine the mechanism of chlorination of methane in some detail. The same mechanism holds for bromination as well as chlorination, and for other alkanes as well as methane; it even holds for many compounds that, while not alkanes, contain alkane-like portions in their molecules. Closely related mechanisms are involved in oxidation (combustion) and other reactions of alkanes. More important, this mechanism illustrates certain general principles that can be carried over to a wide range of chemical reactions. Finally, by studying the evidence that supports the mechanism, we can learn something of how a chemist finds out what goes on during a chemical reaction.

Among the facts that must be accounted for are these:

(a) Methane and chlorine do not react in the dark at room temperature.

(b) Reaction takes place readily, however, in the dark at temperatures over 250 °C, or

(c) under the influence of ultraviolet light at room temperature.

(d) The wavelength of light that induces chlorination is that known independently to cause dissociation of chlorine molecules.

(e) In the light-induced reaction, many (several thousand) molecules of methyl chloride are obtained for each photon of light that is absorbed by the system.

(f) The presence of a small amount of oxygen slows down the reaction for a period of time, after which the reaction proceeds normally; the length of this period depends upon how much oxygen is present.

(We shall see further evidence for the mechanism in Secs. 2.21 and 4.28.)

The mechanism that accounts for these facts most satisfactorily, and hence is generally accepted, is shown in the following equations:

(1) $$Cl_2 \xrightarrow{\text{heat or light}} 2Cl\cdot$$

(2) $$Cl\cdot + CH_4 \longrightarrow HCl + CH_3\cdot$$

(3) $$CH_3\cdot + Cl_2 \longrightarrow CH_3Cl + Cl\cdot$$

then (2), (3), (2), (3), etc.

The first step is the breaking of a chlorine molecule into two chlorine atoms. Like the breaking of any bond, this requires energy, the *bond dissociation energy*, and in Table 1.2 (p. 21) we find that in this case the value is 58 kcal/mol. The energy is supplied as either heat or light.

$$\text{energy} + :\overset{..}{\underset{..}{Cl}}:\overset{..}{\underset{..}{Cl}}: \longrightarrow :\overset{..}{\underset{..}{Cl}}\cdot + \cdot\overset{..}{\underset{..}{Cl}}:$$

The chlorine molecule undergoes *homolysis* (Sec. 1.14): that is, cleavage of the chlorine–chlorine bond takes place in a symmetrical way, so that each atom retains one electron of the pair that formed the covalent bond. This **odd electron** is not *paired* as are all the other electrons of the chlorine atom; that is, it does not have a partner of opposite spin (Sec. 1.6). *An atom or group of atoms possessing an odd (unpaired) electron is called a* **free radical**. In writing the symbol for a free radical, we generally include a dot to represent the odd electron just as we include a plus or minus sign in the symbol of an ion.

Once formed, what is a chlorine atom most likely to do? Like most free radicals, it is extremely reactive because of its tendency to gain an additional electron and thus have a complete octet; from another point of view, energy was supplied to each chlorine atom during the cleavage of the chlorine molecule, and this energy-rich particle tends strongly to lose energy by the formation of a new chemical bond.

To form a new chemical bond, that is, to react, the chlorine atom must collide with some other molecule or atom. What is it most likely to collide with? Obviously, it is most likely to collide with the particles that are present in the highest concentration: chlorine molecules and methane molecules. Collision with another chlorine atom is quite unlikely simply because there are very few of these reactive, short-lived particles around at any time. Of the likely collisions, that with a chlorine molecule causes no net change; reaction may occur, but it can result only in the exchange of one chlorine atom for another:

$$:\overset{..}{\underset{..}{Cl}}\cdot + :\overset{..}{\underset{..}{Cl}}:\overset{..}{\underset{..}{Cl}}: \longrightarrow :\overset{..}{\underset{..}{Cl}}:\overset{..}{\underset{..}{Cl}}: + :\overset{..}{\underset{..}{Cl}}\cdot \qquad \textit{Collision probable but not productive}$$

Collision of a chlorine atom with a methane molecule is both *probable* and *productive*. The chlorine atom abstracts a hydrogen atom, with one electron, to form a molecule of hydrogen chloride:

$$\begin{array}{c} H \\ H:\overset{..}{\underset{}{C}}:H + \cdot\overset{..}{\underset{..}{Cl}}: \\ H \end{array} \longrightarrow H:\overset{..}{\underset{..}{Cl}}: + \begin{array}{c} H \\ H:\overset{}{\underset{}{C}}\cdot \\ H \end{array} \qquad \textit{Collision probable and productive}$$

　Methane　　　　　　　　　　　　　Methyl radical

Now the methyl group is left with an odd, unpaired electron; the carbon atom has only seven electrons in its valence shell. One free radical, the chlorine atom, has been consumed, and a new one, the methyl radical, $CH_3\cdot$, has been formed in its place. This is step (2) in the mechanism.

Now, what is this methyl radical most likely to do? Like the chlorine atom, it is extremely reactive, and for the same reason: the tendency to complete its octet, to lose energy by forming a new bond. Again, collisions with chlorine molecules or methane molecules are the probable ones, not collisions with the relatively scarce chlorine atoms or methyl radicals. But collision with a methane molecule could at most result only in the exchange of one methyl radical for another:

$$
\begin{matrix}
\text{H} & & \text{H} & & & \text{H} & & \text{H} \\
\text{H}:\overset{..}{\underset{..}{\text{C}}}:\text{H} & + & \cdot\overset{..}{\underset{..}{\text{C}}}:\text{H} & \longrightarrow & & \text{H}:\overset{..}{\underset{..}{\text{C}}}\cdot & + & \text{H}:\overset{..}{\underset{..}{\text{C}}}:\text{H} \\
\text{H} & & \text{H} & & & \text{H} & & \text{H}
\end{matrix}
\qquad \textit{Collision probable but not productive}
$$

The collision of a methyl radical with a chlorine molecule is, then, the important one. The methyl radical abstracts a chlorine atom, with one of the bonding electrons, to form a molecule of methyl chloride:

$$
\begin{matrix}
\text{H} & & & & & \text{H} \\
\text{H}:\overset{..}{\text{C}}\cdot & + & :\overset{..}{\underset{..}{\text{Cl}}}:\overset{..}{\underset{..}{\text{Cl}}}: & \longrightarrow & & \text{H}:\overset{..}{\text{C}}:\overset{..}{\underset{..}{\text{Cl}}}: & + & :\overset{..}{\underset{..}{\text{Cl}}}\cdot \\
\text{H} & & & & & \text{H}
\end{matrix}
\qquad \textit{Collision probable and productive}
$$

Methyl   Methyl chloride
radical

The other product is a chlorine atom. This is step (3) in the mechanism.

Here again the consumption of one reactive particle has been accompanied by the formation of another. The new chlorine atom attacks methane to form a methyl radical, which attacks a chlorine molecule to form a chlorine atom, and so the sequence is repeated over and over. Each step produces not only a new reactive particle but also a molecule of product: methyl chloride or hydrogen chloride.

This process cannot, however, go on forever. As we saw earlier, union of two short-lived, relatively scarce particles is not likely; but every so often it does happen, and when it does, this particular sequence of reactions stops. Reactive particles are consumed but not generated.

$$
:\overset{..}{\underset{..}{\text{Cl}}}\cdot \; + \; \cdot\overset{..}{\underset{..}{\text{Cl}}}: \; \longrightarrow \; :\overset{..}{\underset{..}{\text{Cl}}}:\overset{..}{\underset{..}{\text{Cl}}}:
$$

$$
CH_3\cdot \; + \; \cdot CH_3 \; \longrightarrow \; CH_3:CH_3
$$

$$
CH_3\cdot \; + \; \cdot\overset{..}{\underset{..}{\text{Cl}}}: \; \longrightarrow \; CH_3:\overset{..}{\underset{..}{\text{Cl}}}:
$$

It is clear, then, how the mechanism accounts for facts (a), (b), (c), (d), and (e) on page 46: either light or heat is required to cleave the chlorine molecule and form the initial chlorine atoms; once formed, each atom may eventually bring about the formation of many molecules of methyl chloride.

2.13 Chain reactions

The chlorination of methane is an example of a **chain reaction**, *a reaction that involves a series of steps, each of which generates a reactive substance that brings about* *the next step.* While chain reactions may vary widely in their details, they all have certain fundamental characteristics in common.

(1) Cl_2 $\xrightarrow{\text{heat or light}}$ $2Cl\cdot$ Chain-initiating step

(2) $Cl\cdot + CH_4 \longrightarrow HCl + CH_3\cdot$ ⎫
                                                              ⎬ Chain-propagating steps
(3) $CH_3\cdot + Cl_2 \longrightarrow CH_3Cl + Cl\cdot$ ⎭

then (2), (3), (2), (3), etc., *until finally*:

(4) $Cl\cdot + \cdot Cl \longrightarrow Cl_2$ ⎫
    or   ⎪
(5) $CH_3\cdot + \cdot CH_3 \longrightarrow CH_3CH_3$ ⎬ Chain-terminating steps
    or   ⎪
(6) $CH_3\cdot + \cdot Cl \longrightarrow CH_3Cl$ ⎭

First in the chain of reactions is a **chain-initiating step**, in which energy is absorbed and a reactive particle generated; in the present reaction it is the cleavage of chlorine into atoms (step 1).

There are one or more **chain-propagating steps**, each of which consumes a reactive particle and generates another; here they are the reaction of chlorine atoms with methane (step 2), and of methyl radicals with chlorine (step 3).

Finally, there are **chain-terminating steps**, in which reactive particles are consumed but not generated; in the chlorination of methane these would involve the union of two of the reactive particles, or the capture of one of them by the walls of the reaction vessel.

Under one set of conditions, about 10 000 molecules of methyl chloride are formed for every quantum (photon) of light absorbed. Each photon cleaves one chlorine molecule to form two chlorine atoms, each of which starts a chain. On the average, each chain consists of 5000 repetitions of the chain-propagating cycle before it is finally stopped.

2.14 Inhibitors

Finally, how does the mechanism of chlorination account for fact (f), that a small amount of oxygen slows down the reaction for a period of time, which depends upon the amount of oxygen, after which the reaction proceeds normally?

Oxygen is believed to react with a methyl radical to form a new free radical:

$$CH_3\cdot + O_2 \longrightarrow CH_3-O-O\cdot$$

The $CH_3OO\cdot$ radical is much less reactive than the $CH_3\cdot$ radical, and can do little to continue the chain. By combining with a methyl radical, one oxygen molecule breaks a chain, and thus prevents the formation of thousands of molecules of methyl chloride; this, of course, slows down the reaction tremendously. After all the oxygen molecules present have combined with methyl radicals, the reaction is free to proceed at its normal rate.

A substance that slows down or stops a reaction even though present in small amount is called an **inhibitor**. *The period of time during which inhibition lasts, and after which the reaction proceeds normally, is called the inhibition period.* Inhibition by a relatively small amount of an added material is quite characteristic of chain reactions of any type, and is often one of the clues that first leads us to suspect that we are dealing with a chain reaction. It is hard to see how else a few molecules could prevent the reaction of so many. (We shall frequently encounter the use of oxygen to inhibit free-radical reactions.)

2.15 Heat of reaction

In our consideration of the chlorination of methane, we have so far been concerned chiefly with the particles involved—molecules and atoms—and the changes that they undergo. As with any reaction, however, it is important to consider also the energy changes involved, since these changes determine to a large extent how fast the reaction will go, and, in fact, whether it will take place at all.

By using the values of homolytic bond dissociation energies given in Table 1.2 (p. 21), we can calculate the energy changes that take place in a great number of reactions. In the conversion of methane into methyl chloride, two bonds are broken, CH_3—H and Cl—Cl, consuming $104 + 58$, or a total of 162 kcal/mol. At the same time two new bonds are formed, CH_3—Cl and H—Cl, liberating $84 + 103$, or a total of 187 kcal/mol. The result is the liberation of 25 kcal of heat for every mole of methane that is converted into methyl chloride; this is, then, an **exothermic reaction**. (This calculation, we note, does not depend on our knowing the mechanism of the reaction.)

$$CH_3\text{—H} + Cl\text{—Cl} \longrightarrow CH_3\text{—Cl} + H\text{—Cl}$$

$$\underset{162}{\underline{104 \qquad 58}} \qquad \underset{187}{\underline{84 \qquad 103}} \qquad \Delta H = -25 \text{ kcal}$$

When heat is liberated, the heat content (enthalpy), H, of the molecules themselves must decrease; the change in heat content, ΔH, is therefore given a negative sign. (In the case of an endothermic reaction, where heat is absorbed, the increase in heat content of the molecules is indicated by a positive ΔH.)

Problem 2.1 Calculate ΔH for the corresponding reaction of methane with: (a) bromine, (b) iodine, (c) fluorine.

The value of –25 kcal that we have just calculated is the *net* ΔH for the overall reaction. A more useful picture of the reaction is given by the ΔH values of the individual steps. These are calculated below:

$$(1) \qquad\qquad \underset{(58)}{Cl\text{—Cl}} \longrightarrow 2Cl\cdot \qquad\qquad \Delta H = +58 \text{ kcal}$$

$$(2) \qquad Cl\cdot + \underset{(104)}{CH_3\text{—H}} \longrightarrow CH_3\cdot + \underset{(103)}{H\text{—Cl}} \qquad \Delta H = +1$$

$$(3) \qquad CH_3\cdot + \underset{(58)}{Cl\text{—Cl}} \longrightarrow \underset{(84)}{CH_3\text{—Cl}} + Cl\cdot \qquad \Delta H = -26$$

It is clear why this reaction, even though exothermic, occurs only at a high temperature (in the absence of light). The chain-initiating step, without which reaction cannot occur, is highly **endothermic**, and takes place (at a significant rate) only at a high temperature. Once the chlorine atoms are formed, the two chain-propagating steps—one only slightly endothermic, and the other exothermic—occur readily many times before the chain is broken. The difficult cleavage of chlorine is the barrier that must be surmounted before the subsequent easy steps can be taken.

Problem 2.2   Calculate ΔH for the corresponding steps in the reaction of methane with: (a) bromine, (b) iodine, (c) fluorine.

We have assumed so far that exothermic reactions proceed readily, that is, are reasonably fast at ordinary temperatures, whereas endothermic reactions proceed with difficulty, that is, are slow except at very high temperatures. This assumed relationship between ΔH and rate of reaction is a useful rule of thumb when other information is not available; it is *not*, however, a *necessary* relationship, and there are many exceptions to the rule. We shall go on, then, to a discussion of another energy quantity, the *energy of activation*, which is related in a more exact way to rate of reaction.

2.16   Energy of activation

To see what actually happens during a chemical reaction, let us look more closely at a specific example, the attack of chlorine atoms on methane:

$$\text{Cl}\cdot + \text{CH}_3\!-\!\text{H} \longrightarrow \text{H}\!-\!\text{Cl} + \text{CH}_3\cdot \qquad \Delta H = +1 \text{ kcal} \quad E_{\text{act}} = 4 \text{ kcal}$$
$$\phantom{\text{Cl}\cdot + \text{CH}_3\!-\!} (104) (103)$$

This reaction is comparatively simple: it occurs in the gas phase, and is thus not complicated by the presence of a solvent; it involves the interaction of a single atom and the simplest of organic molecules. Yet from it we can learn certain principles that apply to any reaction.

Just what must happen if this reaction is to occur? First of all, a chlorine atom and a methane molecule must **collide**. Since chemical forces are of extremely short range, a hydrogen–chlorine bond can form only when the atoms are in close contact.

Next, to be *effective*, the collision must provide a certain *minimum amount of energy*. Formation of the H—Cl bond liberates 103 kcal/mol; breaking the CH_3—H bond requires 104 kcal/mol. We might have expected that only 1 kcal/mol additional energy would be needed for reaction to occur; however, this is not so. Bond-breaking and bond-making evidently are not perfectly synchronized, and the energy liberated by the one process is not completely available for the other. Experiment has shown that, if reaction is to occur, an additional 4 kcal/mol of energy must be supplied.

The minimum amount of energy that must be provided by a collision for reaction to occur is called the **energy of activation**, E_{act}. Its source is the kinetic energy of the moving particles. Most collisions provide less than this minimum quantity and are fruitless, the original particles simply bouncing apart. Only solid collisions between particles one or both of which are moving unusually fast are energetic enough to bring about reaction. In the present example, at 275 °C, only about one collision in 40 is sufficiently energetic.

Finally, in addition to being sufficiently energetic, the collisions must occur when the particles are properly **oriented**. At the instant of collision, the methane molecule must be turned in such a way as to present a hydrogen atom to the full force of the impact. In the present example, only about one collision in eight is properly oriented.

In general, then, *a chemical reaction requires collisions of sufficient energy* (E_{act}) *and of proper orientation*. There is an energy of activation for nearly every reaction where bonds are broken, even for exothermic reactions, in which bond-making liberates more energy than is consumed by bond-breaking.

The attack of bromine atoms on methane is more highly endothermic, with a ΔH of $+16$ kcal.

$$\text{Br·} + \text{CH}_3\text{—H} \longrightarrow \text{H—Br} + \text{CH}_3\text{·} \qquad \Delta H = +16 \text{ kcal} \quad E_{act} = 18 \text{ kcal}$$
$$\quad\quad\quad (104) \qquad\qquad\qquad (88)$$

Breaking the CH_3—H bond, as before, requires 104 kcal/mol, of which only 88 kcal is provided by formation of the H—Br bond. It is evident that, even if this 88 kcal were completely available for bond-breaking, at least an additional 16 kcal/mol would have to be supplied by the collision. In other words, the E_{act} of an endothermic reaction must be at least as large as the ΔH. As is generally true, the E_{act} of the present reaction (18 kcal) is actually somewhat larger than the ΔH.

2.17 Progress of reaction: energy changes

These energy relationships can be seen more clearly in diagrams like Figs. 2.3 and 2.4. Progress of reaction is represented by horizontal movement from reactants on the left to products on the right. Potential energy (that is, all energy except kinetic) at any stage of reaction is indicated by the height of the curve.

Let us follow the course of reaction in Fig. 2.3. We start in a potential energy valley with a methane molecule and a chlorine atom. These particles are moving,

Figure 2.3 Potential energy changes during the progress of reaction: the methane–chlorine atom reaction.

and hence possess kinetic energy in addition to the potential energy shown. The exact amount of kinetic energy varies with the particular pair of particles, since some move faster than others. They collide, and kinetic energy is converted into potential energy. With this increase in potential energy, reaction begins, and we move up the energy hill. If enough kinetic energy is converted, we reach the top of the hill and start down the far side.

During the descent, potential energy is converted back into kinetic energy, until we reach the level of the products. The products contain a little more potential energy than did the reactants, and we find ourselves in a slightly higher valley than the one we left. With this net increase in potential energy there must be a corresponding decrease in kinetic energy. The new particles break apart, and since they are moving more slowly than the particles from which they were formed, we observe a drop in temperature. Heat will be *taken up* from the surroundings.

In the bromine reaction, shown in Fig. 2.4, we climb a much higher hill and end up in a much higher valley. The increase in potential energy—and the corresponding decrease in kinetic energy—is much larger than in the chlorine reaction; more heat will be taken up from the surroundings.

$E_{act} = 18$ kcal

$\Delta H = +16$ kcal

$CH_3 \cdot + HBr$

$CH_4 + Br \cdot$

$CH_4 + Br \cdot \longrightarrow CH_3 \cdot + HBr$

Progress of reaction $\longrightarrow$

Figure 2.4   Potential energy changes during the progress of reaction: the methane–bromine atom reaction.

An exothermic reaction follows much the same course. (Take, for example, the reverse of the bromine reaction; that is, read from right to left in Fig. 2.4.) In this case, however, the products contain less potential energy than did the reactants so that we end up in a lower valley than the one we left. Since this time the new particles contain more kinetic energy than the particles from which they were formed, and hence move faster, we observe a rise in temperature. Heat will be *given off* to the surroundings.

In any reaction there are many collisions that provide too little energy for us to reach the top of the hill. These collisions are fruitless, and we slide back to our original valley. Many collisions provide sufficient energy, but take place when the molecules are improperly oriented. We then climb an energy hill, but we are off the road; we may climb very high without finding the pass that leads over into the next valley.

The difference in level between the two valleys is, of course, the ΔH; the difference in level between the reactant valley and the top of the hill is the E_{act}. We are concerned only with these differences, and not with the absolute height at any stage of the reaction. We are not even concerned with the relative levels of the reactant valleys in the chlorine and bromine reactions. We need only to know that in the chlorine reaction we climb a hill 4 kcal high and end up in a valley 1 kcal higher than our starting point; and that in the bromine reaction we climb a hill 18 kcal high and end up in a valley 16 kcal higher than our starting point.

As we shall see, it is the height of the hill, the E_{act}, that determines the rate of reaction, and not the difference in level of the two valleys, ΔH. In going to a lower valley, the hill might be very high, but *could* be very low—or even non-existent. In climbing to a higher valley, however, the hill can be no lower than the valley to which we are going; that is to say, *in an endothermic reaction the E_{act} must be at least as large as the ΔH*.

An energy diagram of the sort shown in Figs. 2.3 and 2.4 is particularly useful because it tells us not only about the reaction we are considering, but also about the reverse reaction. Let us move from right to left in Fig. 2.3, for example. We see that the reaction

$$CH_3\cdot + H{-}Cl \longrightarrow CH_3{-}H + Cl\cdot \quad \Delta H = -1 \quad E_{act} = 3$$
$$\phantom{CH_3\cdot + H{-}}(103)(104)$$

has an energy of activation of 3 kcal, since in this case we climb the hill from the higher valley. This is, of course, an exothermic reaction with a ΔH of -1 kcal.

In the same way we can see from Fig. 2.4 that the reaction

$$CH_3\cdot + H{-}Br \longrightarrow CH_3{-}H + Br\cdot \quad \Delta H = -16 \quad E_{act} = 2$$
$$\phantom{CH_3\cdot + H{-}}(88)(104)$$

has an energy of activation of 2 kcal, and is exothermic with a ΔH of -16 kcal. (We notice that, even though exothermic, these last two reactions have energies of activation.)

Reactions like the cleavage of chlorine into atoms

$$Cl{-}Cl \longrightarrow Cl\cdot + \cdot Cl \quad \Delta H = +58 \quad E_{act} = 58$$
$$\phantom{Cl{-}}(58)$$

fall into a special category: a bond is broken but no bonds are formed. The reverse of this reaction, the union of chlorine atoms, involves no bond-breaking and hence

$$Cl\cdot + \cdot Cl \longrightarrow Cl{-}Cl \quad \Delta H = -58 \quad E_{act} = 0$$
$$(58)$$

would be expected to take place very easily, in fact, with no energy of activation at all. This is considered to be generally true for reactions involving the union of two free radicals.

If there is no hill to climb in going from chlorine atoms to a chlorine molecule, but simply a slope to descend, the cleavage of a chlorine molecule must involve simply the ascent of a slope as shown in Fig. 2.5. The E_{act} for the cleavage of a chlorine molecule, then, must equal the ΔH, that is, 58 kcal. This equality of E_{act} and ΔH is believed to hold generally for reactions in which molecules dissociate into radicals.

Figure 2.5   Potential energy changes during the progress of reaction: simple dissociation.

2.18   Rate of reaction

A chemical reaction is the result of collisions of sufficient energy and proper orientation. The rate of reaction, therefore, must be the rate at which these effective collisions occur, the number of effective collisions, let us say, that occur during each second within each liter of reaction space. We can then express the rate as the product of three factors. (The number expressing the probability that a collision will have the proper orientation is commonly called the **probability factor**.) Anything that affects any one of these factors affects the rate of reaction.

number of effective collisions per L per sec	=	total number of collisions per L per sec	×	fraction of collisions that have sufficient energy	×	fraction of collisions that have proper orientation
rate	=	collision frequency	×	energy factor	×	probability factor (orientation factor)

The **collision frequency** depends upon (a) how closely the particles are crowded together, that is, concentration or pressure; (b) how large they are; and (c) how fast they are moving, which in turn depends upon their weight and the temperature.

We can change the concentration and temperature, and thus change the rate. We are familiar with the fact that an increase in concentration causes an increase in rate; it does so, of course, by increasing the collision frequency. A rise in temperature increases the collision frequency; as we shall see, it also increases the energy factor, and this latter effect is so great that the effect of temperature on collision frequency is by comparison unimportant.

The size and weight of the particles are characteristic of each reaction and cannot be changed. Although they vary widely from reaction to reaction, this variation does not affect the collision frequency greatly. A heavier weight makes the particle move more slowly at a given temperature, and hence tends to decrease the collision frequency. A heavier particle is, however, generally a larger particle, and the larger size tends to increase the collision frequency. These two factors thus tend to cancel out.

The **probability factor** depends upon the geometry of the particles and the kind of reaction that is taking place. For closely related reactions it does not vary widely.

Kinetic energy of the moving molecules is not the only source of the energy needed for reaction; energy can also be provided, for example, from vibrations among the various atoms within the molecule. Thus the probability factor has to do not only with what atoms in the molecule suffer the collision, but also with the alignment of the other atoms in the molecule at the time of collision.

By far the most important factor determining rate is the **energy factor**: the fraction of collisions that are sufficiently energetic. This factor depends upon the temperature, which we can control, and upon the energy of activation, which is characteristic of each reaction.

At a given temperature the molecules of a particular compound have an average velocity and hence an average kinetic energy that is characteristic of this system; in fact, the temperature is a measure of this average kinetic energy. But the individual molecules do not all travel with the same velocity, some moving faster than the average and some slower. The distribution of kinetic energy is shown in Fig. 2.6 by the familiar bell-shaped curve that describes the distribution

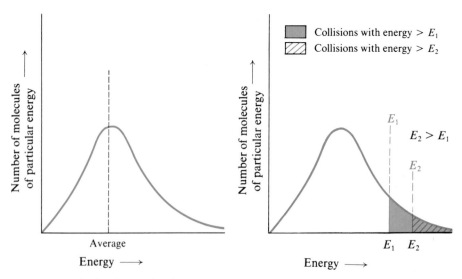

Figure 2.6   Distribution of kinetic energy among molecules.

Figure 2.7   Distribution of kinetic energy among collisions.

among individuals of so many qualities, for example, height, intelligence, income, or even life expectancy. The number of molecules with a particular kinetic energy is greatest for an energy near the average and decreases as the energy becomes larger or smaller than the average.

The distribution of collision energies, as we might expect, is described by a similar curve, Fig. 2.7. Let us indicate collisions of a particular energy, E_{act}, by a vertical line. The number of collisions with energy equal to or greater than E_{act} is indicated by the shaded area under the curve to the right of the vertical line. The fraction of the total number of collisions that have this minimum energy, E_{act}, is then the fraction of the total area that is shaded. It is evident that *the greater the value of E_{act}, the smaller the fraction of collisions that possess that energy.*

The exact relationship between energy of activation and fraction of collisions with that energy is:

$$e^{-E_{act}/RT} = \text{fraction of collisions with energy greater than } E_{act}$$

where

$$E_{act} = \text{energy of activation in cal } (not \text{ kcal})$$
$$e = 2.718 \text{ (base of natural logarithms)}$$
$$R = 1.986 \text{ (gas constant)}$$
$$T = \text{absolute temperature.}$$

Using P for the probability factor and Z for the collision frequency, we arrive at the rate equation:

$$\text{rate} = PZe^{-E_{act}/RT}$$

This exponential relationship is important to us in that it indicates that a small difference in E_{act} has a large effect on the fraction of sufficiently energetic collisions, and hence on the rate of reaction. For example, at 275 °C, out of every million collisions, 10 000 provide sufficient energy if $E_{act} = 5$ kcal, 100 provide sufficient energy if $E_{act} = 10$ kcal, and only one provides sufficient energy if $E_{act} = 15$ kcal. This means that (all other things being equal) a reaction with $E_{act} = 5$ kcal will go 100 times as fast as one with $E_{act} = 10$ kcal, and 10 000 times as fast as one with $E_{act} = 15$ kcal.

We have so far considered a system held at a given temperature. A rise in temperature, of course, increases the average kinetic energy and average velocities, and hence shifts the entire curve to the right, as shown in Fig. 2.8 (on the next page). For a given energy of activation, then, a rise in temperature increases the fraction of sufficiently energetic collisions, and hence increases the rate, as we already know.

The exponential relationship again leads to a large change in rate, this time for a small change in temperature. For example, a rise from 250 to 300 °C, which is only a 10% increase in absolute temperature, increases the rate by 50% if $E_{act} = 5$ kcal, doubles the rate if $E_{act} = 10$ kcal, and trebles the rate if $E_{act} = 15$ kcal. As this example shows, the greater the E_{act}, the greater the effect of a given change in temperature; this follows from the $e^{-E_{act}/RT}$ relationship. Indeed, it is from the relationship between rate and temperature that the E_{act} of a reaction is determined: the rate is measured at different temperatures, and from the results E_{act} is calculated.

We have examined the factors that determine rate of reaction. What we have learned may be used in many ways. To speed up a particular reaction, for example,

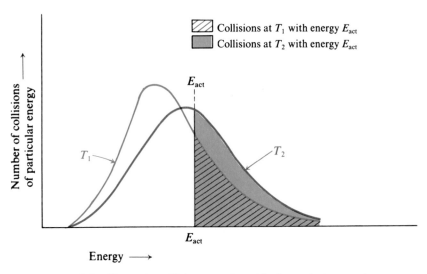

Figure 2.8 Change in collision energies with change in temperature.

we know that we might raise the temperature, or increase the concentration of reactants, or even (in ways that we shall take up later) lower the E_{act}.

Of immediate interest, however, is the matter of relative reactivities. Let us see, therefore, how our knowledge of reaction rates can help us to account for the fact that one reaction proceeds faster than another, even though conditions for the two reactions are identical.

2.19 Relative rates of reaction

We have seen that the rate of a reaction can be expressed as a product of three factors:

rate = collision frequency × energy factor × probability factor

Two reactions could proceed at different rates because of differences in any or all of these factors. To account for a difference in rate, we must first see in which of these factors the difference lies.

As an example, let us compare the reactivities of chlorine and bromine atoms toward methane; that is, let us compare the rates, under the same conditions, of the two reactions:

$$Cl\cdot \; + \; CH_3—H \;\longrightarrow\; H—Cl \; + \; CH_3\cdot \qquad \Delta H = +1 \qquad E_{act} = 4$$

$$Br\cdot \; + \; CH_3—H \;\longrightarrow\; H—Br \; + \; CH_3\cdot \qquad \Delta H = +16 \qquad E_{act} = 18$$

Since temperature and concentration must be the same for the two reactions if we are to compare them under the same conditions, any difference in **collision frequency** would have to arise from differences in particle weight or size. A bromine atom is heavier than a chlorine atom, and it is also larger; as we have seen, the effects of these two properties tend to cancel out. In actuality, the collision frequencies differ by only a few percent. It is generally true that for the same temperature and concentration, two closely related reactions differ but little in collision frequency. A difference in collision frequency therefore cannot be the cause of a large difference in reactivity.

The nature of the **probability factor** is very poorly understood. Since our two reactions are quite similar, however, we might expect them to have similar probability factors. Experiment has shown this to be true: whether chlorine or bromine atoms are involved, about one in every eight collisions with methane has the proper orientation for reaction. In general, where closely related reactions are concerned, we may assume that a difference in probability factor is *not likely* to be the cause of a large difference in reactivity.

We are left with a consideration of the **energy factor**. At a given temperature, the fraction of collisions that possess the amount of energy required for reaction depends upon how large that amount is, that is, depends upon the E_{act}. In our example E_{act} is 4 kcal for the chlorine reaction, 18 kcal for the bromine reaction. As we have seen, a difference of this size in the E_{act} causes an enormous difference in the energy factor, and hence in the rate. At 275 °C, of every 15 million collisions, 375 000 are sufficiently energetic when chlorine atoms are involved, and only *one* when bromine atoms are involved. Because of the difference in E_{act} alone, then, chlorine atoms are 375 000 times as reactive as bromine atoms toward methane.

As we encounter, again and again, differences in reactivity, we shall in general attribute them to differences in E_{act}. In many cases we shall be able to account for these differences in E_{act} on the basis of differences in molecular structure; *it must be understood that we are justified in doing this only when the reactions being compared are so closely related that differences in collision frequency and in probability factor are comparatively insignificant.*

2.20   Relative reactivities of halogens toward methane

With this background, let us return to the reaction between methane and the various halogens, and see if we can account for the order of reactivity given before, $F_2 > Cl_2 > Br_2 > I_2$, and in particular for the fact that iodine does not react at all.

From the table of bond dissociation energies (Table 1.2, p. 21) we can calculate for each of the four halogens the ΔH for each of the three steps of halogenation. Since E_{act} has been measured for only a few of these reactions, let us see what tentative conclusions we can reach using only ΔH.

		X =	F	Cl	Br	I
(1)	$X_2 \longrightarrow 2X\cdot$	$\Delta H =$	+38	+58	+46	+36
(2)	$X\cdot + CH_4 \longrightarrow HX + CH_3\cdot$		−32	+1	+16	+33
(3)	$CH_3\cdot + X_2 \longrightarrow CH_3X + X\cdot$		−70	−26	−24	−20

Since step (1) involves simply dissociation of molecules into atoms, we may quite confidently assume (Sec. 2.17 and Fig. 2.5) that ΔH in this case is equal to E_{act}. Chlorine has the largest E_{act}, and should dissociate most slowly; iodine has the smallest E_{act}, and should dissociate most rapidly. Yet this does not agree with the observed order of reactivity. Thus, except possibly for fluorine, dissociation of the halogen into atoms cannot be the step that determines the observed reactivities.

Step (3), attack of methyl radicals on halogen, is exothermic for all four halogens, and for chlorine, bromine, and iodine it has very nearly the same ΔH. For these reactions, E_{act} *could* be very small, and does indeed seem to be so; probably only a fraction of a kilocalorie. Even iodine has been found to react readily with methyl radicals generated in another way, for example, by the heating of tetramethyllead. In fact, iodine is sometimes employed as a free-radical "trap" or "scavenger" in the study of reaction mechanisms. The third step, then, cannot be the cause of the observed relative reactivities.

This leaves step (2), abstraction of hydrogen from methane by a halogen atom. Here we see a wide spread of ΔH values, from the highly exothermic reaction with the fluorine atom to the highly endothermic reaction with the iodine atom. The endothermic bromine atom reaction must have an E_{act} of at least 16 kcal; as we have seen, it is actually 18 kcal. The slightly endothermic chlorine atom reaction could have a very small E_{act}; it is actually 4 kcal. At a given temperature, then, the fraction of collisions of sufficient energy is much larger for methane and chlorine atoms than for methane and bromine atoms. To be specific, at 275 °C the fraction is about 1 in 40 for chlorine and 1 in 15 million for bromine.

A bromine atom, on the average, collides with many methane molecules before it succeeds in abstracting hydrogen; a chlorine atom collides with relatively few. During its longer search for the proper methane molecule, a bromine atom is more likely to encounter another scarce particle—a second halogen atom or a methyl radical—or be captured by the vessel wall; the chains should therefore be much shorter than in chlorination. Experiment has shown this to be so: where the average chain length is several thousand for chlorination, it is less than 100 for bromination. Even though bromine atoms are formed more rapidly than chlorine atoms at a given temperature because of the lower E_{act} of step (1), overall bromination is slower than chlorination because of the shorter chain length.

For the endothermic reaction of an iodine atom with methane, E_{act} can be no less than 33 kcal, and is probably somewhat larger. Even for this minimum value of 33 kcal, an iodine atom must collide with an enormous number of methane molecules (10^{13} or ten million million at 275 °C) before reaction is likely to occur. Virtually no iodine atoms last this long, but instead recombine to form iodine molecules; the reaction therefore proceeds at a negligible rate. Iodine atoms are easy to form; it is their inability to abstract hydrogen from methane that prevents iodination from occurring.

We cannot predict the E_{act} for the highly exothermic attack of fluorine atoms on methane, but we would certainly not expect it to be any larger than for the attack of chlorine atoms on methane. It appears actually to be smaller (about 1 kcal), thus permitting even longer chains. Because of the surprising weakness of the fluorine–fluorine bond, fluorine atoms should be formed faster than chlorine atoms; thus there should be not only longer chains in fluorination but also *more* chains. The overall reaction is extremely exothermic, with a ΔH of -102 kcal, and the difficulty of removing this heat is one cause of the difficulty of control of fluorination.

Of the two chain-propagating steps, then, step (2) is more difficult than step (3) (see Fig. 2.9). Once formed, methyl radicals react easily with any of the halogens; it is how fast methyl radicals are formed that limits the rate of overall reaction. Fluorination is fast because fluorine atoms rapidly abstract hydrogen atoms from methane; E_{act} is only 1 kcal. Iodination does not take place because iodine atoms find it virtually impossible to abstract hydrogen from methane; E_{act} is more than 33 kcal.

Figure 2.9   Potential energy changes during the progress of reaction:
chlorination of methane. The formation of the radical is the difficult step.

Values of E_{act} for step (2), we notice, parallel the values of ΔH. Since the same
bond, CH_3—H, is being broken in every case, the differences in ΔH reflect
differences in bond dissociation energy among the various hydrogen–halogen
bonds. Ultimately, it appears, the reactivity of a halogen toward methane depends
upon the strength of the bond which that halogen forms with hydrogen.

One further point requires clarification. We have said that an E_{act} of 33 kcal
is too great for the reaction between iodine atoms and methane to proceed at a
significant rate; yet the initial step in each of these halogenations requires an even
greater E_{act}. The difference is this: since halogenation is a chain reaction, dissociation of each molecule of halogen gives rise ultimately to many molecules of
methyl halide; hence, even though dissociation is very slow, the overall reaction
can be fast. The attack of iodine atoms on methane, however, is a chain-propagating
step and if it is slow the entire reaction must be slow; under these circumstances
chain-terminating steps (e.g., union of two iodine atoms) become so important
that effectively there is *no* chain.

2.21   An alternative mechanism for halogenation

In the preceding section we were concerned with the relative reactivities of
the various halogens toward methane. In the next chapter we shall change our
viewpoint, and look at the relative reactivities of various alkanes—or various
positions in one alkane—toward a given halogen. All this helps make up an
important part of our study of organic chemistry: how variations in structure lead
to variations in reactivity. But there is an even more fundamental point to consider:
how a particular type of structure leads to a particular type of reaction in the first
place. How is it, not that one halogen or one alkane reacts faster or slower than
another, but that any halogen and any alkane react together *in the way they do*?

To answer this question, let us take the chlorination of methane as an example, and examine it closely. The chain-propagating steps in our mechanism are (2a) and (3a).

(2a) $Cl\cdot + CH_4 \longrightarrow HCl + CH_3\cdot$

(3a) $CH_3\cdot + Cl_2 \longrightarrow CH_3Cl + Cl\cdot$

But consider, instead, the sequence (2b) and (3b), which represent an alternative mechanism.

(2b) $Cl\cdot + CH_4 \longrightarrow CH_3Cl + H\cdot$

(3b) $H\cdot + Cl_2 \longrightarrow HCl + Cl\cdot$

On the face of it, this mechanism would certainly seem to be worth considering; indeed, until 1940 it was just as consistent with the evidence as the now-accepted mechanism was. But this alternative mechanism is *not* the one by which chlorination takes place, and in Sec. 4.28 we shall see direct evidence against it.

Our question, then, becomes this: how is it that chlorination goes by steps (2a) and (3a) rather than by (2b) and (3b)? The crux of the matter lies in step (2). It is here that the two reaction paths divide; what happens in (2) determines the entire course of reaction. *If* (2b) were to occur, (3b) would inevitably follow; (3b) is a known reaction that takes place readily in a different system. But (2b) does *not* occur.

So we have narrowed our inquiry still further. We now ask: how is it that (2a) takes place rather than (2b)? In both reactions the chlorine atom attacks a methane

$$Cl\cdot + CH_4 \begin{cases} \longrightarrow HCl + CH_3\cdot \\ \xrightarrow{\quad\quad} CH_3Cl + H\cdot \quad \text{\textit{Does not happen}} \end{cases}$$

molecule. It can either attach itself to hydrogen and expel a methyl radical, or attach itself to carbon and expel a hydrogen atom. There is thus *competition* between the two reactions, and the faster reaction wins. If step (2a) predominates, it can only mean that (2a) *goes faster than* (2b).

How do we account for this? Ideally we would like to know the E_{act} values for the competing reactions. But E_{act} cannot, of course, be measured for (2b), since this reaction does not take place. So, as we did in Sec. 2.20, let us see what we can do using values of ΔH; these we can calculate—for real or imaginary reactions—from the homolytic bond dissociation energies of Table 1.2 (p. 21). For (2a) ΔH is $+1$ kcal; E_{act} could therefore be as small as 1 kcal and, as we have seen, is actually 4 kcal.

(2a) $Cl\cdot + CH_3{-}H \longrightarrow H{-}Cl + CH_3\cdot \quad \Delta H = +1 \text{ kcal} \quad E_{act} = 4 \text{ kcal}$
              (104)   (103)

For (2b), ΔH is $+20$ kcal; E_{act} must therefore be *at least* 20 kcal—and is probably considerably larger.

(2b) Cl· + CH$_3$—H $\longrightarrow$ CH$_3$—Cl + H· $\Delta H = +20$ kcal $E_{act} = $ *at least* 20 kcal
        (104)   (84)

The fraction of collisions providing 4 kcal or more is enormously larger than the fraction providing 20 kcal: at 275 °C, for example, *2.5 million times larger*! Just on the basis of this *minimum* estimate of the difference in E_{act}, we see that (2a) must proceed so much faster than (2b) that, in effect, (2a) is the *only* reaction that takes place.

The point is *not* that 20 kcal is in itself too high a barrier for reaction to occur; after all, the attack of Br· on methane has an E_{act} of 18 kcal, and it occurs. The point here is that a reaction with an E_{act} of 20 kcal *cannot compete successfully* with a reaction whose E_{act} is only 4 kcal. When a chlorine atom collides with a methane molecule, the collision is overwhelmingly more likely to provide enough energy for (2a) than for (2b). And so (2a) is what happens.

Finally, let us see what structural feature makes (2a) the easier of the two reactions. Both reactions involve breaking of a carbon–hydrogen bond. The difference lies in which bond is being formed, hydrogen–chlorine or carbon–chlorine. Breaking the carbon–hydrogen bond requires 104 kcal/mol—a great deal of energy. A small fraction of this is supplied by collisions. But most of it comes from the concerted formation of another bond: hydrogen–chlorine in the case of (2a), or carbon–chlorine in the case of (2b). The hydrogen–chlorine bond is a strong one (103 kcal) and its formation can supply nearly all the needed energy. But the carbon–chlorine bond is weaker (only 84 kcal) and, even if all this were available to help break the carbon–hydrogen bond, 20 kcal more would have to be supplied by collisions. The course of this reaction is, then, ultimately determined by the fact that the hydrogen–chlorine bond is stronger than the carbon–chlorine bond.

Examination of the bond dissociation energies of Table 1.2 shows that what we have just described is part of a pattern: each halogen forms a stronger bond to hydrogen than to carbon—not only carbon in methane but carbon in other alkanes as well. The result is that, whatever the halogen and whatever the alkane, halogenation follows a mechanism that is analogous to (2a) and (3a) and *not* to (2b) and (3b).

Again we have encountered relative rates of reaction, this time determining the most fundamental aspect of chemical behavior: *what type of reaction takes place*. Whenever different kinds of molecules are mixed together, there will, in principle, be more than one way in which they can react. There will be a competition between different reaction paths—very often, as we shall find, a closer competition than the one we have just used as our example. And, by and large, what the molecules actually do is *what is easiest for them*. As we encounter such cases of competition, we shall try to see what factors tend to favor one path or the other; we shall even try to see what we can do to make the path that we prefer be the easier one for the reaction to follow.

Problem 2.3 Account in a quantitative way for the fact that the first step in the thermal chlorination of methane is

Cl$_2$ $\xrightarrow{\text{heat}}$ 2Cl·   and not   CH$_4$ $\xrightarrow{\text{heat}}$ CH$_3$· + H·

What structural feature ultimately determines the nature of the chain-initiating step?

2.22   Structure of the methyl radical. sp^2 Hybridization

We have spent a good part of this chapter discussing the formation and reactions of the methyl free radical $CH_3 \cdot$. Just what is this molecule like? What is its shape? How are the electrons distributed and, in particular, where is the odd electron?

These are important questions, for the answers apply not only to this simple radical but to any free radical, however complicated, that we shall encounter. The *shape*, naturally, underlies the three-dimensional chemistry—the stereochemistry— of free radicals. The *location of the odd electron* is intimately involved with the stabilization of free radicals by substituent groups.

As we did when we "made" methane (Sec. 1.11), let us start with the electronic configuration of carbon,

and, to provide more than two unpaired electrons for bonding, promote a $2s$ electron to the empty $2p$ orbital:

Like boron in boron trifluoride (Sec. 1.10), carbon here is bonded to three other atoms. Hybridization of the $2s$ orbital and two of the p orbitals provides the

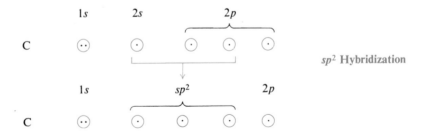

necessary orbitals: three strongly directed sp^2 orbitals which, as we saw before, lie in a plane that includes the carbon nucleus, and are directed to the corners of an equilateral triangle.

If we arrange the carbon and three hydrogens of a methyl radical to permit maximum overlap of orbitals, we obtain the structure shown in Fig. 2.10*a*. It is flat, with the carbon atom at the center of a triangle and the three hydrogen atoms at the corners. Every bond angle is 120°.

(a)   (b)

Figure 2.10   Methyl radical. (a) Only σ bonds shown. (b) The odd electron is in the p orbital above and below the plane of the σ bonds.

Now where is the odd electron? In forming the sp^2 orbitals, the carbon atom has used only two of its three p orbitals. The remaining p orbital consists of two equal lobes, one lying above and the other lying below the plane of the three sp^2 orbitals (Fig. 2.10b); it is occupied by the odd electron.

This is not the only conceivable electronic configuration for the methyl radical: an alternative treatment would lead to a pyramidal molecule like that of ammonia, except that the fourth sp^3 orbital contains the odd electron instead of an electron pair (Sec. 1.12). Quantum mechanical calculations do not offer a clear-cut decision between the two configurations. Spectroscopic studies indicate that the methyl radical is actually flat, or nearly so. Carbon is trigonal, or not far from it; the odd electron occupies a p orbital, or at least an orbital with much p character.

Compare the shapes of three molecules in which the central atom is bonded to three other atoms: (a) boron trifluoride, with no unshared electrons, trigonal; (b) ammonia, with an unshared *pair*, tetrahedral; and (c) the methyl radical, with a *single* unshared electron, trigonal or intermediate between trigonal and tetrahedral.

There is stereochemical evidence (for example, Sec. 4.28) that most other free radicals are either flat or, if pyramidal, undergo rapid *inversion* like that of the ammonia molecule (Sec. 1.12).

> **Problem 2.5**   Besides free radicals, we shall encounter two other kinds of reactive particles, carbocations (positive charge on carbon) and carbanions (negative charge on carbon). Suggest an electronic configuration, and from this predict the shape, of (a) the methyl cation, CH_3^+, and (b) the methyl anion, $CH_3:^-$.

2.23   Transition state

Clearly, the concept of E_{act} is to be our key to the understanding of chemical reactivity. To make it *useful*, we need a further concept: *transition state*.

A chemical reaction is presumably a continuous process involving a gradual transition from reactants to products. It has been found extremely helpful, however, to consider the arrangement of atoms at an intermediate stage of reaction as though it were an actual molecule. This intermediate structure is called the **transition state**; its energy content corresponds to the top of the energy hill (Fig. 2.11).

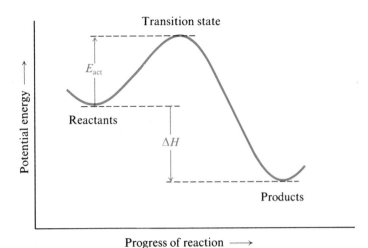

Figure 2.11 Potential energy changes during the progress of reaction: transition state at top of energy hump.

The reaction sequence is now:

$$\text{reactants} \longrightarrow \text{transition state} \longrightarrow \text{products}$$

Just as ΔH is the difference in energy content between reactants and products, so E_{act} *is the difference in energy content between reactants and transition state.*

The transition state concept is useful for this reason: we can analyze the structure of the transition state very much as though it were a molecule, and attempt to estimate its stability. Any factor that stabilizes the transition state relative to the reactants tends to lower the energy of activation; that is to say, any factor that lowers the top of the energy hill more than it lowers the reactant valley reduces the net height we must climb during reaction. Transition state stability will be the basis—whether explicit or implicit—of almost every discussion of reactivity in this book.

But the transition state is only a fleeting arrangement of atoms which, by its very nature—lying at the top of an energy hill—cannot be isolated and examined. How can we possibly know anything about its structure? Well, let us take as an example the transition state for the abstraction of hydrogen from methane by a halogen atom, and see where a little thinking will lead us.

To start with, we can certainly say this: the carbon–hydrogen bond is stretched but not entirely broken, and the hydrogen–halogen bond has started to form but is not yet complete. This condition could be represented as

$$\underset{\text{Reactants}}{H-\overset{\displaystyle H}{\underset{\displaystyle H}{C}}-H + \cdot X} \longrightarrow \underset{\text{Transition state}}{\left[H-\overset{\displaystyle H}{\underset{\displaystyle H}{\overset{\delta .}{C}}}\text{----}\overset{\delta .}{H}\text{----}X \right]} \longrightarrow \underset{\text{Products}}{H-\overset{\displaystyle H}{\underset{\displaystyle H}{C}}\cdot + H-X}$$

where the dashed lines indicate partly broken or partly formed bonds.

Now, what can we say about the shape of the methyl group in this transition state? In the reactant, where methyl holds the hydrogen, carbon is tetrahedral (sp^3-hybridized); in the product, where methyl has lost the hydrogen, carbon is trigonal (sp^2-hybridized). In the transition state, where the carbon–hydrogen bond is partly broken, hybridization of carbon is somewhere between sp^3 and sp^2. The methyl group is partly but not completely flattened; bond angles are greater than 109.5° but less than 120°.

Reactant	Transition state	Product
Tetrahedral	*Becoming trigonal*	*Trigonal*

Finally, where is the odd electron? It is on chlorine in the reactants, on the methyl group in the products, and divided between the two in the transition state. (Each atom's share is represented by the symbol $\delta\cdot$.) The methyl group *partly* carries the odd electron it will have in the product, and to this extent has taken on some of the character of the free radical it will become.

Thus, in a straightforward way, we have drawn a picture of the transition state that shows the bond-making and bond-breaking, the spatial arrangement of the atoms, and the distribution of the electrons.

(This particular transition state is intermediate between reactants and products not only in the time sequence but also in structure. Not *all* transition states are intermediate in structure: as shown on page 184, reactant and product in S_N2 reactions are tetrahedral, whereas the transition state contains pentavalent carbon.)

In Sec. 2.18, we looked at the matter of reaction rates from the standpoint of the *collision theory*. An alternative, more generally useful approach is the *transition state* (or *thermodynamic*) *theory* of reaction rates. An equilibrium is considered to exist between the reactants and the transition state, and this is handled in the same way as true equilibria of reversible reactions (Sec. 23.11). Energy of activation (E_{act}) and probability factor are replaced by, respectively, *heat (enthalpy) of activation* ($\Delta H^{\ddagger}$) and *entropy of activation* ($\Delta S^{\ddagger}$), which together make up *free energy of activation* ($\Delta G^{\ddagger}$).

$$\Delta G^{\ddagger} = \Delta H^{\ddagger} - T\Delta S^{\ddagger}$$

The smaller (the less positive) the $\Delta H^{\ddagger}$ and the larger (the more positive) the $\Delta S^{\ddagger}$, the smaller $\Delta G^{\ddagger}$ will be, and the faster the reaction.

Entropy corresponds, roughly, to the randomness of a system; equilibrium tends to favor the side in which fewer restrictions are placed on the atoms and molecules. Entropy of activation, then, is a measure of the relative randomness of reactants and transition state; the fewer the restrictions that are placed on the arrangement of atoms in the transition state—relative to the reactants—the faster the reaction will go. We can see, in a general way, how probability factor and entropy of activation measure much the same thing. A low probability factor means that a rather special orientation of atoms is required on collision. In the other language, an unfavorable (low) entropy of activation means that rather severe restrictions are placed on the positions of atoms in the transition state.

2.24   Reactivity and development of the transition state

For the abstraction of hydrogen from methane by a halogen atom, we have just seen that the transition state differs from the reactants—and this difference is, of course, what we are looking for—chiefly in being like the products. This is generally true for reactions in which free radicals (or, for that matter, carbocations or carbanions) are formed.

But just *how much* does this particular transition state resemble the products? How far have bond-breaking and bond-making gone? How flat has the methyl group become, and to what extent does it carry the odd electron?

Surprisingly, we can answer even questions like these, at least in a relative way. *In a set of similar reactions, the higher the E_{act}, the later the transition state is reached in the reaction process.* Of the theoretical considerations underlying this postulate, we shall mention only this: the difference in electronic distribution that we call a difference in structure corresponds to a difference in energy; the greater the difference in structure, the greater the difference in energy. If E_{act} is high, the transition state differs greatly from the reactants in energy and, presumably, also in electronic structure; if E_{act} is low, the transition state differs little from the reactants in energy and, presumably, also in electronic structure (see Fig. 2.12).

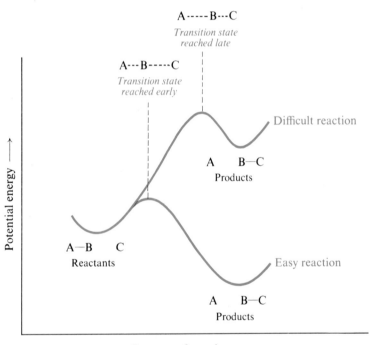

Figure 2.12 Potential energy changes during the progress of reaction: reactivity and development of the transition state. Difficult reaction: transition state reached late, resembles products. Easy reaction: transition state reached early, resembles reactants.

Practically, this postulate has been found extremely useful in the interpretation of experimental results; among other things, as we shall see, it enables us to account for the relationship between reactivity and selectivity (Sec. 3.28).

Abstraction of hydrogen by the highly reactive chlorine atom has a low E_{act}. According to the postulate, then, the transition state is reached before the reaction has proceeded very far, and when the carbon–hydrogen bond is only slightly stretched. Atoms and electrons are still distributed much as they were in the reactants; carbon is still nearly tetrahedral. The methyl group has developed little free-radical character.

Abstraction of hydrogen by the less reactive bromine atom, in contrast, has a very high E_{act}. The transition state is reached only after reaction is well along toward completion and when the carbon–hydrogen bond is more nearly broken. The geometry and electron distribution have begun to approach those of the products, and carbon may well be almost trigonal. The methyl group has developed much free-radical character.

Thus, *in the attack by a reagent of high reactivity, the transition state tends to resemble the reactants; in the attack by a reagent of low reactivity, the transition state tends to resemble the products.*

2.25   Molecular formula: its fundamental importance

In this chapter we have been concerned with the structure of methane: the way in which atoms are put together to form a molecule of methane. But first we had to know what kinds of atoms these are and how many of them make up the molecule; we had to know that methane is CH_4. Before we can assign a structural formula to a compound, we must first know its molecular formula.

Much of the chapter has been spent in discussing the substitution of chlorine for the hydrogen of methane. But first we had to know that there *is* substitution, that each step of the reaction yields a product that contains one less hydrogen atom and one more chlorine atom than the reactant; we had to know that CH_4 is converted successively into CH_3Cl, CH_2Cl_2, $CHCl_3$, and CCl_4. Before we can discuss the reactions of an organic compound, we must first know the molecular formulas of the products.

Let us review a little of what we know about the assigning of a molecular formula to a compound. We must carry out:

(a) a *qualitative elemental analysis*, to find out what kinds of atoms are present in the molecule;

(b) a *quantitative elemental analysis*, to find out the relative numbers of the different kinds of atoms, that is, to determine the *empirical formula*;

(c) a *molecular weight determination*, which (combined with the empirical formula) shows the actual numbers of the different kinds of atoms, that is, gives us the *molecular formula*.

Most of this should be familiar to the student from previous courses in chemistry. What we shall concentrate on here will be the application of these principles to organic analysis.

2.26   Qualitative elemental analysis

The presence of carbon or hydrogen in a compound is detected by **combustion**: heating with copper oxide, which converts carbon into carbon dioxide and hydrogen into water. (*Problem*: How could each of these products be identified?)

$$(C,H) + CuO \xrightarrow{\text{heat}} Cu + CO_2 + H_2O$$

Covalently bonded halogen, nitrogen, and sulfur must be converted into inorganic ions, which can then be detected in already familiar ways. This conversion is accomplished in either of two ways: (a) through **sodium fusion**, treatment with hot molten sodium metal,

$$(C,H,X,N,S) + Na \xrightarrow{\text{heat}} Na^+X^- + Na^+CN^- + Na^+S^{2-}Na^+$$

or (b) through **Schöniger oxidation** by oxygen gas.

$$(C,H,X,N,S) + O_2 \xrightarrow{\text{NaOH}} Na^+X^- + Na^+NO_2^- + Na^+SO_3^{2-}Na^+$$

(A simpler method of detecting halogen in *some* organic compounds is discussed in Sec. 5.26.)

By these methods, we could show, for example, that methane contains carbon and hydrogen, or that methyl chloride contains carbon, hydrogen, and chlorine.

Further tests would show the absence of any other element in these compounds, except possibly oxygen, for which there is no simple chemical test; presence or absence of oxygen would be shown by a quantitative analysis.

Problem 2.6 (a) How would you detect halide ion as a product of sodium fusion or oxidation? (b) If sulfur and/or nitrogen is also present in an organic molecule, this test cannot be carried out on a sodium fusion mixture until it has been acidified and boiled. Why is this so?

Problem 2.7 Only carbon and hydrogen were detected by a qualitative elemental analysis of the compound ethyl alcohol; quantitative analysis gave 52.1% carbon and 13.1% hydrogen. (a) Why would it be assumed that ethyl alcohol contains oxygen? (b) What percentage of oxygen would be assumed?

2.27 Quantitative elemental analysis: carbon, hydrogen, and halogen

Knowing what elements make up a compound, we must next determine the proportions in which they are present. To do this, we carry out very much the same analysis as before, only this time on a quantitative basis. To find out the relative amounts of carbon and hydrogen in methane, for example, we would completely oxidize a measured amount of methane and weigh the carbon dioxide and water formed.

In a quantitative combustion, a weighed sample of the organic compound is passed through a *combustion train*: a tube packed with copper oxide heated to 600–800 °C, followed by a tube containing a drying agent (usually Dehydrite, magnesium perchlorate) and a tube containing a strong base (usually Ascarite, sodium hydroxide on asbestos). The water formed is absorbed by the drying agent, and the carbon dioxide is absorbed by the base; the increase in weight of each tube gives the weight of product formed.

For example, we might find that a sample of methane weighing 9.67 mg produced 26.53 mg of CO_2 and 21.56 mg of H_2O. Now, only the fraction $C/CO_2 = 12.01/44.01$ of the carbon dioxide is carbon, and only the fraction $2H/H_2O = 2.016/18.02$ of the water is hydrogen. Therefore

wt. C = 26.53 × 12.01/44.01	wt. H = 21.56 × 2.016/18.02
wt. C (in sample) = 7.24 mg	wt. H (in sample) = 2.41 mg

and the percentage composition is

%C = (7.24/9.67) × 100   %H = (2.41/9.67) × 100

%C (in sample) = 74.9   %H (in sample) = 24.9

Since the total of carbon and hydrogen is 100%, within the limits of error of the analysis, oxygen (or any other element) must be absent.

In quantitative, as in qualitative, analysis, covalently bonded halogen must be converted into halide ion. The organic compound is heated either (a) in a bomb with sodium peroxide or (b) in a sealed tube with nitric acid (*Carius method*). The halide ion thus formed is converted into silver halide, which can be weighed.

Problem 2.8   When 7.36 mg of methyl chloride was heated in a bomb with sodium peroxide, the chloride ion liberated yielded 20.68 mg of silver chloride. (a) What percentage of chlorine is indicated by this analysis? (b) What percentage of chlorine would be expected from a compound of formula CH_3Cl? (c) What weight of silver chloride would you expect from 7.36 mg of methylene chloride? (d) Of chloroform? (e) Of carbon tetrachloride?

(We shall take up other quantitative analytical methods when we need them: nitrogen and sulfur analysis, Sec. 13.12; neutralization equivalent, Sec. 23.21; saponification equivalent, Sec. 24.24.)

2.28   Empirical formula

Knowing the percentage composition of a compound, we can now calculate the **empirical formula**: *the simplest formula that shows the relative numbers of the different kinds of atoms in a molecule.* For example, in 100 g (taken for convenience) of methane there are 74.9 g of carbon and 24.9 g of hydrogen, according to our quantitative analysis. Dividing each quantity by the proper atomic weight gives the number of moles of each element.

$$C: \frac{74.9}{12.01} = 6.24 \text{ moles}$$

$$H: \frac{24.9}{1.008} = 24.7 \text{ moles}$$

Since a mole of one element contains the same number of atoms as a mole of any other element, we now know the relative number of carbon and hydrogen atoms in methane: $C_{6.24}H_{24.7}$. Conversion to smallest whole numbers gives the empirical formula CH_4 for methane.

C : 6.24/6.24 = 1

H : 24.7/6.24 = 3.96, approximately 4

Problem 2.9   Calculate the percentage composition and then the empirical formula for each of the following compounds: (a) Combustion of a 3.02-mg sample of a compound gave 8.86 mg of carbon dioxide and 5.43 mg of water. (b) Combustion of an 8.23-mg sample of a compound gave 9.62 mg of carbon dioxide and 3.94 mg of water. Analysis of a 5.32-mg sample of the same compound by the Carius method gave 13.49 mg of silver chloride.

2.29   Molecular weight. Molecular formula

At this stage we know what kinds of atoms make up the molecule we are studying, and in what ratio they are present. This knowledge is summarized in the empirical formula.

But this is not enough. On the basis of just the empirical formula, a molecule of methane, for example, might contain one carbon and four hydrogens, or two carbons and eight hydrogens, or *any* multiple of CH_4. We still have to find the **molecular formula**: *the formula that shows the actual number of each kind of atom in a molecule.*

To find the molecular formula, we must determine the molecular weight: today, almost certainly by mass spectrometry, which gives an exact value (Sec. 16.2). *Ethane*, for example, has an empirical formula of CH_3. A molecular weight of 30 is found, indicating that, of the possible molecular formulas, C_2H_6 must be the correct one.

Problem 2.10   Quantitative elemental analysis shows that the empirical formula of a compound is CH. The molecular weight is found to be 78. What is the molecular formula?

Problem 2.11   Combustion of a 5.17-mg sample of a compound gives 10.32 mg of carbon dioxide and 4.23 mg of water. The molecular weight is 88. What is the molecular formula of the compound?

PROBLEMS

1. Calculate the percentage composition of X, Y, and Z from the following analytical data:

	wt. sample	wt. CO_2	wt. H_2O	wt. AgCl
X	4.37 mg	15.02 mg	2.48 mg	—
Y	5.95 mg	13.97 mg	2.39 mg	7.55 mg
Z	4.02 mg	9.14 mg	3.71 mg	—

2. What is the percentage composition of:

(a) C_3H_7Cl (c) $C_4H_8O_2$ (e) CH_4ON_2
(b) C_2H_6O (d) $C_6H_8O_2N_2S$ (f) C_6H_8NCl

3. What is the empirical formula of an organic compound whose percentage composition is:

(a) 85.6% C, 14.4% H   (d) 29.8% C, 6.3% H, 44.0% Cl
(b) 92.2% C, 7.8% H   (e) 48.7% C, 13.6% H, 37.8% N
(c) 40.0% C, 6.7% H   (f) 25.2% C, 2.8% H, 49.6% Cl

(*Note*: remember that oxygen often is not determined directly.)

4. A qualitative analysis of *papaverine*, one of the alkaloids in opium, showed carbon, hydrogen, and nitrogen. A quantitative analysis gave 70.8% carbon, 6.2% hydrogen, and 4.1% nitrogen. Calculate the empirical formula of papaverine.

5. *Methyl orange*, an acid–base indicator, is the sodium salt of an acid that contains carbon, hydrogen, nitrogen, sulfur, and oxygen. Quantitative analysis gave 51.4% carbon, 4.3% hydrogen, 12.8% nitrogen, 9.8% sulfur, and 7.0% sodium. What is the empirical formula of methyl orange?

6. Combustion of 6.51 mg of a compound gave 20.47 mg of carbon dioxide and 8.36 mg of water. The molecular weight was found to be 84. Calculate: (a) percentage composition; (b) empirical formula; and (c) molecular formula of the compound.

7. A liquid of molecular weight 60 was found to contain 40.0% carbon and 6.7% hydrogen. What is the molecular formula of the compound?

8. A gas of the same empirical formula as the compound in Problem 7 has a molecular weight of 30. What is its molecular formula?

9. *Indigo*, an important dyestuff, gave an analysis of 73.3% carbon, 3.8% hydrogen, and 10.7% nitrogen. Molecular weight determination gave a value of 262. What is the molecular formula of indigo?

10. The hormone *insulin* contains 3.4% sulfur. (a) What is the minimum molecular weight of insulin? (b) The actual molecular weight is 5734; how many sulfur atoms are probably present per molecule?

11. Calculate ΔH for:

(a)–(d) $H_2 + X_2 \longrightarrow 2HX$, where $X = F, Cl, Br, I$
(e) $C_2H_6 + Br_2 \longrightarrow C_2H_5Br + HBr$
(f) $C_6H_5CH_3 + Br_2 \longrightarrow C_6H_5CH_2Br + HBr$
(g) $H_2C=CHCH_3 + Br_2 \longrightarrow H_2C=CHCH_2Br + HBr$
(h) Reactions (e), (f), and (g) proceed by the same free-radical mechanism as halogenation of methane. Calculate ΔH for each step in these three reactions.

12. (a) Free methyl radicals react with methane as follows: $E_a = +13$

$$(i) \quad CH_3 \cdot + CH_4 \longrightarrow CH_4\text{---}CH_3 \cdot$$

On the basis of the bond strengths involved, show why the above reaction takes place rather than the following:

$$(ii) \quad CH_3 \cdot + CH_4 \longrightarrow CH_3\text{---}CH_3 + H \cdot$$

(b) Reaction (i) has an E_{act} of 13 kcal. In Sec. 2.12 it was listed as probable (but unproductive) on grounds of collision probability. In actuality, how probable is reaction (i) in, say, a 50:50 mixture of CH_4 and Cl_2? (*Hint*: See Secs. 2.20 and 2.18.)

13. Bromination of methane is slowed down by addition of fairly large amounts of HBr. (a) Suggest a possible explanation for this. (*Hint*: See Sec. 2.17.) (b) Account for the fact that HCl does not have a similar effect upon chlorination. (c) Any reaction tends to slow down as reactants are used up and their concentrations decrease. How do you account for the fact that bromination of methane slows down to an unusually great extent, more than, say, chlorination of methane?

14. A mixture of H_2 and Cl_2 does not react in the dark at room temperature. At high temperatures or under the influence of light (of a wavelength absorbed by chlorine) a violent reaction occurs and HCl is formed. The photochemical reaction yields as many as a million molecules of HCl for each photon absorbed. The presence of a small amount of oxygen slows down the reaction markedly. (a) Outline a possible mechanism to account for these facts. (b) Account for the fact that a mixture of H_2 and I_2 does not behave in the same way. (Hydrogen iodide is actually formed, but by an entirely different mechanism.)

15. A stream of tetramethyllead vapor, $(CH_3)_4Pb$, was passed through a quartz tube which was heated at one spot; a mirror of metallic lead was deposited at the hot point, and the gas escaping from the tube was found to be chiefly ethane. The tube was next heated upstream of the lead mirror while more tetramethyllead was passed through; a new mirror appeared at the hot point, the old mirror disappeared, and the gas escaping from the tube was now found to be chiefly tetramethyllead. Experiments like this, done by Fritz Paneth at the University of Berlin, were considered the first good evidence for the existence of short-lived free radicals like methyl. (a) Show how these experimental results can be accounted for in terms of intermediate free radicals. (b) The farther upstream the tube was heated, the more slowly the old mirror disappeared. Account for this.

16. When a small amount (0.02%) of tetraethyllead, $(C_2H_5)_4Pb$, is added to a mixture of methane and chlorine, chlorination takes place at only 140 °C instead of the usual minimum of 250 °C. In light of Problem 15, show how this fact strengthens the mechanism of Sec. 2.12.

3

Alkanes

Free-Radical Substitution

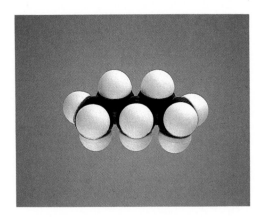

3.1 Classification by structure: the family

The basis of organic chemistry, we have said, is the structural theory. We separate all organic compounds into a number of families on the basis of structure. Having done this, we find that we have at the same time classified the compounds as to their physical and chemical properties. A particular set of properties is thus characteristic of a particular kind of structure.

Within a family there are variations in properties. All members of the family may, for example, react with a particular reagent, but some may react more readily than others. Within a single compound there may be variations in properties, one part of a molecule being more reactive than another part. These variations in properties correspond to variations in structure.

As we take up each family of organic compounds, we shall first see what structure and properties are characteristic of the family. Next we shall see how structure and properties vary within the family. We shall not simply memorize these facts, but, whenever possible, shall try to understand properties in terms of structure, and to understand variations in properties in terms of variations in structure.

Having studied methane in some detail, let us now look at the more complicated members of the alkane family. These hydrocarbons have been assigned to the same family as methane on the basis of their structure, and on the whole their properties follow the pattern laid down by methane. However, certain new points will arise simply because of the greater size and complexity of these compounds.

3.2 Structure of ethane

Next in size after methane is **ethane**, C_2H_6. If we connect the atoms of this molecule by covalent bonds, following the rule of one bond (one pair of electrons) for each hydrogen and four bonds (four pairs of electrons) for each carbon, we arrive at the structure

$$
\begin{array}{cc}
\text{H H} & \text{H H} \\
\ddot{\text{H}}:\ddot{\text{C}}:\ddot{\text{C}}:\text{H} & \text{H—C—C—H} \\
\text{H H} & \text{H H}
\end{array}
$$

Ethane

Each carbon is bonded to three hydrogens and to the other carbon.

Since each carbon atom is bonded to four other atoms, its bonding orbitals (sp^3 orbitals) are directed toward the corners of a tetrahedron. As in the case of methane, the carbon–hydrogen bonds result from overlap of these sp^3 orbitals with the s orbitals of the hydrogens. The carbon–carbon bond arises from overlap of two sp^3 orbitals.

The carbon–hydrogen and carbon–carbon bonds have the same general electron distribution, being cylindrically symmetrical about a line joining the atomic nuclei (see Fig. 3.1); because of this similarity in shape, the bonds are given the same name, σ *bonds* (*sigma bonds*).

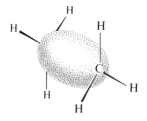

Figure 3.1 Ethane molecule. Carbon–carbon single bond: σ bond.

Figure 3.2 Ethane molecule: shape and size.

In ethane, then, the bond angles and carbon–hydrogen bond lengths should be very much the same as in methane, that is, about 109.5° and about 1.10 Å, respectively. Electron diffraction and spectroscopic studies have verified this structure in all respects, giving (Fig. 3.2) the following measurements for the molecule: bond angles, 109.5°; C—H length, 1.10 Å; C—C length, 1.53 Å. Similar studies have shown that, with only slight variations, these values are quite characteristic of carbon–hydrogen and carbon–carbon bonds and of carbon bond angles in alkanes.

3.3   Free rotation about the carbon–carbon single bond. Conformations. Torsional strain

This particular set of bond angles and bond lengths still does not limit us to a single arrangement of atoms for the ethane molecule, since the relationship between the hydrogens of one carbon and the hydrogens of the other carbon is not specified. If we examine models of ethane (Fig. 3.3), we find that we could have an arrangement like I in which the hydrogens exactly oppose each other, an arrangement like II in which the hydrogens are perfectly staggered, or an infinity of intermediate arrangements. Which of these is the actual structure of ethane? The answer is: *all of them.*

I
Eclipsed conformation

II
Staggered conformation

Figure 3.3   Models of the ethane molecule in the eclipsed and staggered conformations.

We have seen that the σ bond joining the carbon atoms is cylindrically symmetrical about a line joining the two carbon nuclei; overlap and hence bond strength should be the same for all these possible arrangements. If the various arrangements do not differ in energy, then the molecule is not restricted to any one of them, but can change freely from one to another. Since the change from one to another involves rotation about the carbon–carbon bond, we describe this freedom to change by saying that *there is free rotation about the carbon–carbon single bond.*

Different arrangements of atoms that can be converted into one another by rotation about single bonds are called **conformations**. Arrangement I is called the *eclipsed conformation*; arrangement II is called the *staggered conformation*. (The infinity of intermediate conformations are called *skew conformations*.)

To represent such conformations, we shall often use two kinds of three-dimensional formulas: *andiron formulas* (Fig. 3.4);

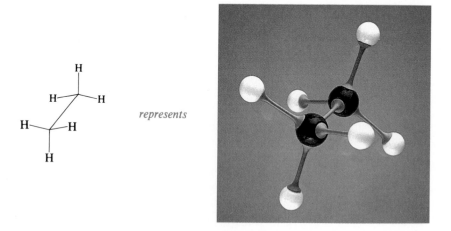

represents

Eclipsed conformation

represents

Staggered conformation

Figure 3.4 Andiron formulas of ethane in the eclipsed and staggered conformations.

and *Newman projections* (Fig. 3.5), named for M. S. Newman, of the Ohio State University, who first proposed their use.

represents

Eclipsed conformation

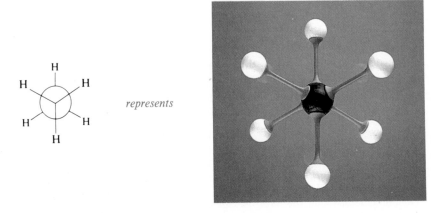

represents

Staggered conformation

Figure 3.5   Newman projections of ethane in the eclipsed and staggered conformations.

The picture is not yet complete. Certain physical properties show that rotation is *not quite free*: there is an energy barrier of about 3 kcal/mol. The potential energy of the molecule is at a minimum for the staggered conformation, increases with rotation, and reaches a maximum at the eclipsed conformation (Fig. 3.6). Most ethane molecules, naturally, exist in the most stable, staggered conformation; or, put differently, any molecule spends most of its time in the most stable conformation.

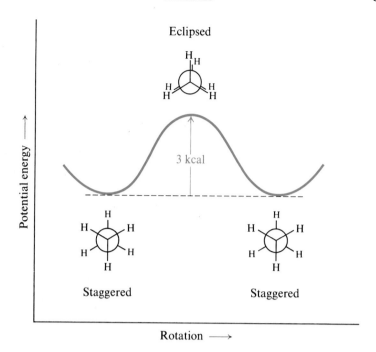

Figure 3.6 Potential energy changes during rotation about the carbon–carbon single bond of ethane.

How free are ethane molecules to rotate from one staggered arrangement to another? The 3-kcal barrier is not a very high one; even at room temperature the fraction of collisions with sufficient energy is large enough that a rapid inter-conversion between staggered arrangements occurs. For most practical purposes, we may still consider that the carbon–carbon single bond permits free rotation.

The nature of the rotational barrier in ethane is not understood or—what is not exactly the same thing—is not readily explained. It is too high to be due merely to van der Waals forces (Sec. 1.19): although thrown closer together in the eclipsed conformation than in the staggered conformation, the hydrogens on opposite carbons are not big enough for this to cause appreciable crowding (see Fig. 3.3). The barrier is considered to arise in some way from interaction among the electron clouds of the carbon–hydrogen bonds. Quantum mechanical calculations show that the barrier should exist, and so perhaps "lack of understanding" amounts to difficulty in paraphrasing the mathematics in physical terms. Like the bond orbitals in methane, the two sets of orbitals in ethane tend to be as far apart as possible—to be *staggered*.

The energy required to rotate the ethane molecule about the carbon–carbon bond is called *torsional energy*. We speak of the relative instability of the eclipsed conformation—or any of the intermediate skew conformations—as being due to *torsional strain*.

As the hydrogens of ethane are replaced by other atoms or groups of atoms, other factors affecting the relative stability of conformations appear: van der Waals forces, dipole–dipole interactions, hydrogen bonding. But the tendency for the bond orbitals on adjacent carbons to be staggered remains, and any rotation away from the staggered conformation is accompanied by torsional strain.

3.4   Propane and the butanes

The next member of the alkane family is **propane**, C_3H_8. Again following the rule of one bond per hydrogen and four bonds per carbon, we arrive at structure I.

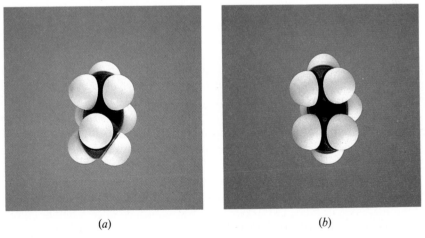

I

Propane

Here, rotation can occur about two carbon–carbon bonds, and again is essentially free. Although the methyl group is considerably larger than hydrogen, the rotational barrier (3.3 kcal/mol) is only a little higher than for ethane. Evidently there is still not very much crowding in the eclipsed conformation, and the rotational barrier is due chiefly to the same factor as the barrier in ethane: *torsional strain*. (See Fig. 3.7.)

(a)   (b)

Figure 3.7   Models of the propane molecule in (*a*) an eclipsed conformation and (*b*) a staggered conformation. There is little crowding in either conformation.

When we consider **butane**, C_4H_{10}, we find that there are two possible structures, II and III. Structure II has a four-carbon chain and III has a three-carbon chain with a one-carbon branch. There can be no doubt that these represent

$$
\begin{array}{cccc}
 & \text{H} & \text{H} & \text{H} & \text{H} \\
 & | & | & | & | \\
\text{H}- & \text{C}- & \text{C}- & \text{C}- & \text{C}-\text{H} \\
 & | & | & | & | \\
 & \text{H} & \text{H} & \text{H} & \text{H}
\end{array}
$$

II

n-Butane

$$
\begin{array}{ccc}
\text{H} & \text{H} & \text{H} \\
| & | & | \\
\text{H}-\text{C} & \text{C} & \text{C}-\text{H} \\
| & | & | \\
\text{H} & \text{H}-\text{C}-\text{H} & \text{H} \\
 & | & \\
 & \text{H} &
\end{array}
$$

III

Isobutane

different structures, since no amount of moving, twisting, or rotating about carbon–carbon bonds will cause these structures to coincide. We can see that in the *straight-chain* structure (II) each carbon possesses at least two hydrogens, whereas in the *branched-chain* structure (III) one carbon possesses only a single hydrogen; or we may notice that in the branched-chain structure (III) one carbon is bonded to three other carbons, whereas in the straight-chain structure (II) no carbon is bonded to more than two other carbons.

In agreement with this prediction, we find that two compounds of the same formula, C_4H_{10}, have been isolated. There can be no doubt that these two substances are different compounds, since they show definite differences in their physical and chemical properties (see Table 3.1); for example one boils at 0 °C and the other at -12 °C. By definition, they are *isomers* (Sec. 1.23).

Table 3.1 PHYSICAL CONSTANTS OF THE ISOMERIC BUTANES

	n-Butane	Isobutane
B.p.	0 °C	-12 °C
M.p.	-138 °C	-159 °C
Relative density at -20 °C	0.622	0.604
Solubility in 100 mL alcohol	1813 mL	1320 mL

Two compounds of formula C_4H_{10} are known and we have drawn two structures to represent them. The next question is: which structure represents which compound? For the answer we turn to the evidence of **isomer number**. Like methane, the butanes can be chlorinated; the chlorination can be allowed to proceed until there are two chlorine atoms per molecule. From the butane of b.p. 0 °C, *six* isomeric products of formula $C_4H_8Cl_2$ are obtained; from the butane of b.p. -12 °C, only *three*. We find that we can draw just six dichloro-butanes containing a straight chain of carbon atoms, and just three containing a branched chain. Therefore, the butane of b.p. 0 °C must have the straight chain, and the butane of b.p. -12 °C must have the branched chain. To distinguish between these two isomers, the straight-chain structure is called **n-butane** (spoken "normal butane") and the branched-chain structure is called **isobutane**.

Problem 3.1 Draw the structures of all possible dichloro derivatives of: (a) *n*-butane; (b) isobutane.

Problem 3.2 Could we assign structures to the isomeric butanes on the basis of the number of isomeric *mono*chloro derivatives?

3.5   Conformations of *n*-butane. Van der Waals repulsion

Let us look more closely at the *n*-butane molecule and the conformations in which it exists. Focusing our attention on the middle C—C bond, we see a molecule similar to ethane, but with a methyl group replacing one hydrogen on each carbon. As with ethane, staggered conformations have lower torsional energies and hence are more stable than eclipsed conformations. But, due to the presence of the methyl groups, two new points are encountered here: first, there are several *different* staggered conformations; and second, a factor besides torsional strain comes into play to affect conformational stabilities.

There is the *anti* conformation, I, in which the methyl groups are as far apart as they can be (dihedral angle 180°). There are two *gauche* conformations, II and III, in which the methyl groups are only 60° apart. (Conformations II and III are mirror images of each other, and are of the same stability; nevertheless, they *are* different. Make models and convince yourself that this is so.)

Figure 3.8   Potential energy changes during rotation about the C(2)–C(3) bond of *n*-butane.

The *anti* conformation, it has been found, is more stable (by 0.8 kcal/mol) than the *gauche* (Fig. 3.8). Both are free of torsional strain. But in a *gauche* conformation, the methyl groups are crowded together, that is, are thrown together closer than the sum of their van der Waals radii; under these conditions, van der Waals forces are *repulsive* (Sec. 1.19) and raise the energy of the conformation. We say that there is *van der Waals repulsion* (or *steric repulsion*) between the methyl groups, and that the molecule is less stable because of *van der Waals strain* (or *steric strain*). We can see this crowding quite clearly in scale models (Fig. 3.9).

(a)   (b)

Figure 3.9   Models of *n*-butane in (*a*) an *anti* conformation and (*b*) a gauche conformation. Note the crowding between the methyl groups in (*b*).

Van der Waals strain can affect not only the relative stabilities of various staggered conformations, but also the heights of the barriers between them. The energy maximum reached when two methyl groups swing past each other—rather than past hydrogens—is the highest rotational barrier of all, and has been estimated at 4.4–6.1 kcal/mol. Even so, it is low enough that—at ordinary temperatures, at least—the energy of molecular collisions causes rapid rotation; a given molecule exists now in a *gauche* conformation, and the next instant in the *anti* conformation.

We shall return to the relationships among conformations like these of *n*-butane in Sec. 4.20.

Problem 3.3   Both calculations and experimental evidence indicate that the dihedral angle between the methyl groups in the *gauche* conformation of *n*-butane is actually somewhat *larger* than 60°. How would you account for this?

Problem 3.4   Considering only rotation about the bond shown, draw a potential energy *vs.* rotation curve like Fig. 3.8 for:

(a) $(CH_3)_2CH$—$CH(CH_3)_2$; (b) $(CH_3)_2CH$—CH_2CH_3 ; (c) $(CH_3)_3C$—$C(CH_3)_3$.

(d) Compare the heights of the various energy barriers with each other and with those in Fig. 3.8.

3.6   Higher alkanes. The homologous series

If we examine the molecular formulas of the alkanes we have so far considered, we see that butane contains one carbon and two hydrogens more than propane, which in turn contains one carbon and two hydrogens more than ethane, and so on. *A series of compounds in which each member differs from the next member by a constant amount is called a* **homologous series**, *and the members of the series are called* **homologs**. The family of alkanes forms such a homologous series, the constant difference between successive members being CH_2. We also notice that in each of these alkanes the number of hydrogen atoms equals two more than twice the number of carbon atoms, so that we may write as a *general formula* for members of this series, C_nH_{2n+2}. As we shall see later, other homologous series have their own characteristic general formulas.

In agreement with this general formula, we find that the next alkane, *pentane*, has the formula C_5H_{12}, followed by *hexane*, C_6H_{14}, *heptane*, C_7H_{16}, and so on. We would expect that, as the number of atoms increases, so does the number of possible arrangements of those atoms. As we go up the series of alkanes, we find that this is true: the number of isomers of successive homologs increases at a surprising rate. There are 3 isomeric pentanes, 5 hexanes, 9 heptanes, and 75 decanes (C_{10}); for the twenty-carbon icosane, there are 366 319 possible isomeric structures! The carbon skeletons of the isomeric pentanes and hexanes are shown below.

C—C—C—C—C

n-Pentane
b.p. 36 °C

C—C—C—C
|
C

Isopentane
b.p. 28 °C

C
|
C—C—C
|
C

Neopentane
b.p. 9.5 °C

Pentanes

C—C—C—C—C—C

b.p. 69 °C

C—C—C—C—C
|
C

b.p. 60 °C

C—C—C—C—C
|
C

b.p. 63 °C

Hexanes

C
|
C—C—C—C
|
C

b.p. 50 °C

C—C—C—C
| |
C   C

b.p. 58 °C

It is important to practice drawing the possible isomeric structures that correspond to a single molecular formula. In doing this, a set of molecular models is especially helpful since it will show that many structures which appear to be different when drawn on paper are actually identical.

Problem 3.5 Draw the structures of: (a) the nine isomeric heptanes (C_7H_{16}); (b) the eight chloropentanes ($C_5H_{11}Cl$); (c) the nine dibromobutanes ($C_4H_8Br_2$).

3.7 Nomenclature

We have seen that the names *methane, ethane, propane, butane,* and *pentane* are used for alkanes containing respectively one, two, three, four, and five carbon atoms. Table 3.2 gives the names of many larger alkanes. Except for the first four members of the family, the name is simply derived from the Greek (or Latin) prefix for the particular number of carbons in the alkane; thus **pent**ane for five, **hex**ane for six, **hept**ane for seven, **oct**ane for eight, and so on.

Table 3.2   NAMES OF ALKANES

CH_4	methane	C_9H_{20}	nonane
C_2H_6	ethane	$C_{10}H_{22}$	decane
C_3H_8	propane	$C_{11}H_{24}$	undecane
C_4H_{10}	butane	$C_{12}H_{26}$	dodecane
C_5H_{12}	pentane	$C_{14}H_{30}$	tetradecane
C_6H_{14}	hexane	$C_{16}H_{34}$	hexadecane
C_7H_{16}	heptane	$C_{18}H_{38}$	octadecane
C_8H_{18}	octane	$C_{20}H_{42}$	icosane

You should certainly memorize the names of at least the first ten alkanes. Having done this, you will have at the same time essentially learned the names of the first ten alkenes, alkynes, alcohols, etc., since the names of many families of compounds are closely related. Compare, for example, the names *propane, propene,* and *propyne* for the three-carbon alkane, alkene, and alkyne.

But nearly every alkane can have a number of isomeric structures, and there must be an unambiguous name for each of these isomers. The butanes and pentanes are distinguished by the use of prefixes: *n*-butane and **iso**butane; *n*-pentane, **iso**pentane, and **neo**pentane. But there are 5 hexanes, 9 heptanes, and 75 decanes; it would be difficult to devise, and even more difficult to remember, a different prefix for each of these isomers. It is obvious that some systematic method of naming is needed.

As organic chemistry has developed, several different methods have been devised to name the members of nearly every class of organic compounds; each method was devised when the previously used system had been found inadequate for the growing number of increasingly complex organic compounds. Unfortunately for us, perhaps, several systems have survived and are in current use. Even if we are content ourselves to use only one system, we still have to understand the names used by other chemists; hence it is necessary for us to learn more than one system of nomenclature. But before we can do this, we must first learn the names of certain organic groups.

3.8   Alkyl groups

In our study of inorganic chemistry, we found it useful to have names for certain groups of atoms that compose only part of a molecule and yet appear many times as a unit. For example, NH_4^+ is called *ammonium*; NO_3^-, *nitrate*; SO_3^{2-}, *sulfite*; and so on.

In a similar way names are given to certain groups that constantly appear as structural units of organic molecules. We have seen that chloromethane, CH_3Cl, is also known as *methyl chloride*. The CH_3 group is called **methyl** wherever it appears, CH_3Br being *methyl* bromide; CH_3I, *methyl* iodide; and CH_3OH, *methyl* alcohol. In an analogous way, the C_2H_5 group is **ethyl**; C_3H_7, **propyl**; C_4H_9, **butyl**; and so on.

These groups are named simply by dropping *-ane* from the name of the corresponding alkane and replacing it by *-yl*. They are known collectively as **alkyl**

groups. The general formula for an alkyl group is C_nH_{2n+1}, since it contains one less hydrogen than the parent alkane, C_nH_{2n+2}.

Among the alkyl groups we again encounter the problem of isomerism. There is only one methyl chloride or ethyl chloride, and correspondingly only one methyl group or ethyl group. We can see, however, that there are two propyl chlorides, I and II, and hence that there must be two propyl groups. These groups both contain

$$
\begin{array}{cc}
\begin{array}{ccc} H & H & H \\ | & | & | \\ H-C-C-C-Cl \\ | & | & | \\ H & H & H \end{array}
&
\begin{array}{ccc} H & H & H \\ | & | & | \\ H-C-C-C-H \\ | & | & | \\ H & Cl & H \end{array}
\\
I & II \\
\textit{n}\text{-Propyl chloride} & \text{Isopropyl chloride}
\end{array}
$$

the propane chain, but differ in the point of attachment of the chlorine; they are called **_n_-propyl** and **isopropyl.** We can distinguish the two chlorides by the names

$$
\begin{array}{cc}
CH_3CH_2CH_2- & CH_3CHCH_3 \\
\textit{n}\text{-Propyl} & | \\
 & \text{Isopropyl}
\end{array}
$$

n-propyl chloride and _isopropyl chloride_; we distinguish the two propyl bromides, iodides, alcohols, and so on in the same way.

We find that there are four butyl groups, two derived from the straight-chain _n_-butane, and two derived from the branched-chain isobutane. These are given the designations **n-** (_normal_), **sec-** (_secondary_), **iso-**, and **tert-** (_tertiary_), as shown below. Again the difference between _n_-butyl and _sec_-butyl and between isobutyl and _tert_-butyl lies in the point of attachment of the alkyl group to the rest of the molecule.

$$
\begin{array}{cc}
CH_3CH_2CH_2CH_2- & CH_3CH_2CHCH_3 \\
\textit{n}\text{-Butyl} & | \\
 & \textit{sec}\text{-Butyl}
\end{array}
$$

$$
\begin{array}{cc}
\begin{array}{c} CH_3 \\ \diagdown \\ \quad\quad CHCH_2- \\ \diagup \\ CH_3 \end{array}
&
\begin{array}{c} CH_3 \\ | \\ CH_3-C- \\ | \\ CH_3 \end{array}
\\
\text{Isobutyl} & \textit{tert}\text{-Butyl}
\end{array}
$$

Beyond butyl the number of isomeric groups derived from each alkane becomes so great that it is impracticable to designate them all by various prefixes. Even though limited, this system is so useful for the small groups just described that it is widely used; a student must therefore memorize these names and learn to recognize these groups at a glance in whatever way they happen to be represented.

However large the group concerned, one of its many possible arrangements can still be designated by this simple system. The prefix _n-_ is used to designate any alkyl group in which all carbons form a single continuous chain and in which the point of attachment is the very end carbon. For example:

$$
\begin{array}{cc}
CH_3CH_2CH_2CH_2CH_2Cl & CH_3(CH_2)_4CH_2Cl \\
\textit{n}\text{-Pentyl chloride} & \textit{n}\text{-Hexyl chloride}
\end{array}
$$

The prefix *iso-* is used to designate any alkyl group (of six carbons or fewer) that has a single one-carbon branch on the next-to-last carbon of a chain and has the point of attachment at the opposite end of the chain. For example:

$$CH_3 \diagdown CHCH_2CH_2Cl \diagup CH_3 \qquad\qquad CH_3 \diagdown CH(CH_2)_2CH_2Cl \diagup CH_3$$

Isopentyl chloride   Isohexyl chloride

If the branching occurs at any other position, or if the point of attachment is at any other position, this name does not apply.

Now that we have learned the names of certain alkyl groups, let us return to the original problem: the naming of alkanes.

3.9   Common names of alkanes

As we have seen, the prefixes *n-*, *iso-*, and *neo-* are adequate to differentiate the various butanes and pentanes, but beyond this point an impracticable number of prefixes would be required. However, the prefix *n-* has been retained for any alkane, no matter how large, in which all carbons form a continuous chain with no branching:

$$CH_3CH_2CH_2CH_2CH_3 \qquad\qquad CH_3(CH_2)_4CH_3$$

n-Pentane   *n*-Hexane

An *isoalkane* is a compound of six carbons or fewer in which all carbons except one form a continuous chain and that one carbon is attached to the next-to-end carbon:

$$CH_3 \diagdown CHCH_2CH_3 \diagup CH_3 \qquad\qquad CH_3 \diagdown CH(CH_2)_2CH_3 \diagup CH_3$$

Isopentane   Isohexane

In naming any other of the higher alkanes, we make use of the IUPAC system, outlined in the following section.

(It is sometimes convenient to name alkanes as derivatives of methane; see, for example, I on page 137.

3.10   IUPAC names of alkanes

To devise a system of nomenclature that could be used for even the most complicated compounds, various committees and commissions representing the chemists of the world have met periodically since 1892. In its present modification, the system so devised is known as the **IUPAC system** (International Union of Pure and Applied Chemistry). Since this system follows much the same pattern for all families of organic compounds, we shall consider it in some detail as applied to the alkanes.

Essentially the rules of the IUPAC system are:

1. Select as the parent structure the longest continuous chain, and then consider the compound to have been derived from this structure by the replacement of hydrogen by various alkyl groups. Isobutane (I) can be considered to arise from propane by the replacement of a hydrogen atom by a methyl group, and thus may be named *methylpropane*.

CH$_3$CHCH$_3$ CH$_3$CH$_2$CH$_2$CHCH$_3$ CH$_3$CH$_2$CHCH$_2$CH$_3$

| | |

CH$_3$ CH$_3$ CH$_3$

I   II   III

Methylpropane   2-Methylpentane   3-Methylpentane

(Isobutane)

2. Where necessary, as in the isomeric methylpentanes (II and III), indicate by a number the carbon to which the alkyl group is attached.

3. In numbering the parent carbon chain, start at whichever end results in the use of the lowest numbers; thus II is called *2-methylpentane* rather than 4-methylpentane.

4. If the same alkyl group occurs more than once as a side chain, indicate this by the prefix *di-*, *tri-*, *tetra-*, etc., to show how many of these alkyl groups there are, and indicate by various numbers the positions of *each* group, as in *2,2,4-trimethyl-pentane* (IV).

CH$_3$ CH$_3$

CH$_3$CHCH$_2$CCH$_3$ CH$_2$

CH$_3$ CH$_3$ CH$_3$CH$_2$CH$_2$CH—CH—C—CH$_2$CH$_3$

CH   CH$_3$ CH$_2$

CH$_3$ CH$_3$ CH$_3$

IV   V

2,2,4-Trimethylpentane   3,3-Diethyl-5-isopropyl-4-methyloctane

5. If there are several different alkyl groups attached to the parent chain, name them in alphabetical order; as in *3,3-diethyl-5-isopropyl-4-methyloctane* (V). (Note that isopropyl comes before methyl. A dimethyl, however, would come after ethyl or diethyl.)

There are additional rules and conventions used in naming very complicated alkanes, but the five fundamental rules given above will suffice for the compounds we are likely to encounter.

The alkyl halides which appear so often in alkane chemistry are named as *haloalkanes*; that is, halogen is simply treated as a side chain. We first name the alkane as though no halogen were present, and then add *fluoro*, *chloro*, *bromo*, or *iodo*, together with any needed numbers and prefixes.

$$CH_3CH_2Cl$$

1-Chloroethane

$$CH_3CH_2CH_2Br$$

1-Bromopropane

$$CH_3CH_2\underset{\underset{Br}{|}}{C}HCH_3$$

2-Bromobutane

$$\underset{\underset{CH_3CHCH_2I}{}}{\overset{\overset{CH_3}{|}}{}}$$

1-Iodo-2-methylpropane

$$CH_3\overset{\overset{CH_3}{|}}{\underset{\underset{F}{|}}{C}}CH_3$$

2-Fluoro-2-methylpropane

$$CH_3CH_2\overset{\overset{CH_3}{|}}{\underset{\underset{Cl}{|}}{C}}\underset{\underset{Cl}{|}}{C}HCH_3$$

2,3-Dichloro-3-methylpentane

Problem 3.6   Give the IUPAC names for: (a) the isomeric hexanes shown on page 85; (b) the nine isomeric heptanes (see Problem 3.5, p. 85).

Problem 3.7   Give the IUPAC names for: (a) the eight isomeric chloropentanes; (b) the nine isomeric dibromobutanes (see Problem 3.5, p. 85).

3.11   Classes of carbon atoms and hydrogen atoms

It has been found extremely useful to classify each carbon atom of an alkane with respect to the number of other carbon atoms to which it is attached. *A* **primary** (1°) *carbon atom is attached to only one other carbon atom; a* **secondary** (2°) *is attached to two others; and a* **tertiary** (3°) *to three others.* For example:

Each hydrogen atom is similarly classified, being given the same designation of *primary, secondary*, or *tertiary* as the carbon atom to which it is attached.

We shall make constant use of these designations in our consideration of the relative reactivities of various parts of an alkane molecule.

3.12   Physical properties

The physical properties of the alkanes follow the pattern laid down by methane, and are consistent with the alkane structure. An alkane molecule is held together entirely by covalent bonds. These bonds either join two atoms of the same kind and hence are non-polar, or join two atoms that differ very little in electronegativity and hence are only slightly polar. Furthermore, these bonds are directed in a very symmetrical way, so that the slight bond polarities tend to cancel out. As a result an alkane molecule is either non-polar or very weakly polar.

As we have seen (Sec. 1.19), the forces holding non-polar molecules together (van der Waals forces) are weak and of very short range; they act only between the portions of different molecules that are in close contact, that is, between the surfaces of molecules. Within a family, therefore, we would expect that the larger the molecule—and hence the larger its surface area—the stronger the intermolecular forces.

Table 3.3 lists certain physical constants for a number of the *n*-alkanes. As we can see, the boiling points and melting points rise as the number of carbons increases. The processes of boiling and melting require overcoming the intermolecular forces of a liquid and a solid; the boiling points and melting points rise because these intermolecular forces increase as the molecules get larger.

Table 3.3 ALKANES

Name	Formula	M.p., °C	B.p., °C	Relative density (at 20 °C)
Methane	CH_4	− 183	− 162	
Ethane	CH_3CH_3	− 172	− 88.5	
Propane	$CH_3CH_2CH_3$	− 187	− 42	
n-Butane	$CH_3(CH_2)_2CH_3$	− 138	0	
n-Pentane	$CH_3(CH_2)_3CH_3$	− 130	36	0.626
n-Hexane	$CH_3(CH_2)_4CH_3$	− 95	69	.659
n-Heptane	$CH_3(CH_2)_5CH_3$	− 90.5	98	.684
n-Octane	$CH_3(CH_2)_6CH_3$	− 57	126	.703
n-Nonane	$CH_3(CH_2)_7CH_3$	− 54	151	.718
n-Decane	$CH_3(CH_2)_8CH_3$	− 30	174	.730
n-Undecane	$CH_3(CH_2)_9CH_3$	− 26	196	.740
n-Dodecane	$CH_3(CH_2)_{10}CH_3$	− 10	216	.749
n-Tridecane	$CH_3(CH_2)_{11}CH_3$	− 6	234	.757
n-Tetradecane	$CH_3(CH_2)_{12}CH_3$	5.5	252	.764
n-Pentadecane	$CH_3(CH_2)_{13}CH_3$	10	266	.769
n-Hexadecane	$CH_3(CH_2)_{14}CH_3$	18	280	.775
n-Heptadecane	$CH_3(CH_2)_{15}CH_3$	22	292	
n-Octadecane	$CH_3(CH_2)_{16}CH_3$	28	308	
n-Nonadecane	$CH_3(CH_2)_{17}CH_3$	32	320	
n-Icosane	$CH_3(CH_2)_{18}CH_3$	36		
Isobutane	$(CH_3)_2CHCH_3$	− 159	− 12	
Isopentane	$(CH_3)_2CHCH_2CH_3$	− 160	28	.620
Neopentane	$(CH_3)_4C$	− 17	9.5	
Isohexane	$(CH_3)_2CH(CH_2)_2CH_3$	− 154	60	.654
3-Methylpentane	$CH_3CH_2CH(CH_3)CH_2CH_3$	− 118	63	.676
2,2-Dimethylbutane	$(CH_3)_3CCH_2CH_3$	− 98	50	.649
2,3-Dimethylbutane	$(CH_3)_2CHCH(CH_3)_2$	− 129	58	.668

Except for the very small alkanes, *the boiling point rises 20 to 30 degrees for each carbon that is added to the chain*; we shall find that this increment of 20–30 degrees per carbon holds not only for the alkanes but also for each of the homologous series that we shall study.

The increase in melting point is not quite so regular, since the intermolecular forces in a crystal depend not only upon the size of the molecules but also upon how well they fit into a crystal lattice.

The first four *n*-alkanes are gases, but, as a result of the rise in boiling point and melting point with increasing chain length, the next thirteen (C_5–C_{17}) are liquids, and those containing 18 carbons or more are solids.

> **Problem 3.8** Using the data of Table 3.3, make a graph of: (a) b.p. *vs.* carbon number for the *n*-alkanes; (b) m.p. *vs.* carbon number; (c) density *vs.* carbon number.

There are somewhat smaller differences among the boiling points of alkanes that have the same carbon number but different structures. On pages 82 and 85 the boiling points of the isomeric butanes, pentanes, and hexanes are given. We see that in every case *a branched-chain isomer has a lower boiling point than a straight-chain isomer*, and further, that the more numerous the branches, the lower the boiling point. Thus *n*-butane has a boiling point of 0 °C and isobutane − 12 °C. *n*-Pentane has a boiling point of 36 °C, isopentane with a single branch 28 °C, and neopentane with two branches 9.5 °C. This effect of branching on boiling point is observed within all families of organic compounds. That branching should lower the boiling point is understandable: with branching the shape of the molecule tends to approach that of a sphere; and as this happens the surface area decreases, with the result that the intermolecular forces become weaker and are overcome at a lower temperature (Sec. 1.20). Compare the shapes of the isomeric pentanes, for example, as shown in Fig. 3.10.

In agreement with the rule of thumb, "like dissolves like", the alkanes are soluble in non-polar solvents such as benzene, ether, and chloroform, and are insoluble in water and other highly polar solvents. Considered themselves as solvents, the liquid alkanes dissolve compounds of low polarity and do not dissolve compounds of high polarity.

The relative density increases with size of the alkanes, but tends to level off at about 0.8; thus all alkanes are less dense than water. It is not surprising that nearly all organic compounds are less dense than water since, like the alkanes, they consist chiefly of carbon and hydrogen. In general, to be denser than water a compound must contain a heavy atom like bromine or iodine, or several atoms like chlorine.

3.13 Industrial source

The principal source of alkanes is **petroleum**, together with the accompanying **natural gas**. Decay and millions of years of geological stresses have transformed the complicated organic compounds that once made up living plants or animals into a mixture of alkanes ranging in size from one carbon to 30 or 40 carbons. Formed along with the alkanes, and particularly abundant in California petroleum, are *cycloalkanes* (Chap. 12), known to the petroleum industry as *naphthenes*.

The other fossil fuel, coal, is a potential second source of alkanes: processes are being developed to convert coal, through hydrogenation, into gasoline and fuel oil, and into synthetic gas to offset anticipated shortages of natural gas.

Natural gas contains, of course, only the more volatile alkanes, that is, those of low molecular weight; it consists chiefly of methane and progressively smaller amounts of ethane, propane, and higher alkanes. For example, a sample taken from a pipeline supplied by a large number of Pennsylvania wells contained

(a)

(b)

(c)

Figure 3.10   Molecular structure and physical properties: effect of branching. The isomeric pentanes: (a) n-pentane, b.p. 36 °C; (b) isopentane, b.p.
28 °C; (c) neopentane, b.p. 9.5 °C. Neopentane is the most highly branched
and most nearly spherical, and has the smallest surface area; intermolecular
forces are weakest, and it boils at the lowest temperature.

methane, ethane, and propane in the ratio of 12 : 2 : 1, with higher alkanes making
up only 3% of the total. The propane–butane fraction is separated from the more
volatile components by liquefaction, compressed into cylinders, and sold as *bottled
gas* in areas not served by a gas utility.

Petroleum is separated by distillation into the various fractions listed in Table
3.4; because of the relationship between boiling point and molecular weight, this
amounts to a rough separation according to carbon number. Each fraction is still
a very complicated mixture, however, since it contains alkanes of a range of carbon
numbers, and since each carbon number is represented by numerous isomers. The
use that each fraction is put to depends chiefly upon its volatility or viscosity, and
it matters very little whether it is a complicated mixture or a single pure compound.
(In gasoline, as we shall see in Sec. 3.30, the structures of the components are of
key importance.)

Table 3.4 PETROLEUM CONSTITUENTS

Fraction	Distillation temperature, °C	Carbon number
Gas	Below 20 °C	C_1–C_4
Petroleum ether	20–60 °C	C_5–C_6
Ligroin (light naphtha)	60–100 °C	C_6–C_7
Natural gasoline	40–205 °C	C_5–C_{10}, and cycloalkanes
Kerosine	175–325 °C	C_{12}–C_{18}, and aromatics
Gas oil	Above 275 °C	C_{12} and higher
Lubricating oil	Non-volatile liquids	Probably long chains attached to cyclic structures
Asphalt or petroleum coke	Non-volatile solids	Polycyclic structures

The chief use of all but the non-volatile fractions is as fuel. The gas fraction, like natural gas, is used chiefly for heating. Gasoline is used in those internal combustion engines that require a fairly volatile fuel, kerosine is used in tractor and jet engines, and gas oil is used in diesel engines. Kerosine and gas oil are also used for heating purposes, the latter being the familiar "furnace oil".

The lubricating oil fraction, especially that from Pennsylvania crude oil (*paraffin-base petroleum*), often contains large amounts of long-chain alkanes (C_{20}–C_{34}) that have fairly high melting points. If these remained in the oil, they might crystallize to waxy solids in an oil line in cold weather. To prevent this, the oil is chilled and the wax is removed by filtration. After purification this is sold as solid *paraffin wax* (m.p. 50–55 °C) or used in *petroleum jelly* (Vaseline). Asphalt is used in roofing and road building. The coke that is obtained from paraffin-base crude oil consists of complex hydrocarbons having a high carbon-to-hydrogen ratio; it is used as a fuel or in the manufacture of carbon electrodes for the electrochemical industries. Petroleum ether and ligroin are useful solvents for many organic materials of low polarity.

In addition to being used directly as just described, certain petroleum fractions are converted into other kinds of chemical compounds. Catalytic **isomerization** changes straight-chain alkanes into branched-chain ones. The **cracking** process (Sec. 3.31) converts higher alkanes into smaller alkanes and alkenes, and thus increases the gasoline yield; it can even be used for the production of "natural" gas. In addition, the alkenes thus formed are the most important raw materials for the large-scale synthesis of organic compounds. The process of **catalytic reforming** (Sec. 15.5) converts alkanes and cycloalkanes into aromatic hydrocarbons and thus provides the chief raw material for the large-scale synthesis of another broad class of compounds.

3.14 Industrial source *vs.* laboratory preparation

We shall generally divide the methods of obtaining a particular kind of organic compound into two categories: *industrial source* and *laboratory preparation*. We may contrast the two in the following way, although it must be realized that there are many exceptions to these generalizations.

An industrial source must provide large amounts of the desired material at the lowest possible cost. A laboratory preparation may be required to produce only a few hundred grams or even a few grams; cost is usually of less importance than the time of the investigator.

For many industrial purposes a mixture may be just as suitable as a pure compound; even when a single compound is required, it might be economically feasible to separate it from a mixture, particularly when the other components may also be marketed. In the laboratory a chemist nearly always wants a single pure compound. Separation of a single compound from a mixture of related substances is very time-consuming and frequently does not yield material of the required purity. Furthermore, the raw material for a particular preparation may well be the hard-won product of a previous preparation or even series of preparations, and hence one wishes to convert it as completely as possible into the desired compound. On an industrial scale, if a compound cannot be isolated from naturally occurring material, it may be synthesized along with a number of related compounds by some inexpensive reaction. In the laboratory, whenever possible, a reaction is selected that forms a single compound in high yield.

In industry it is frequently worthwhile to work out a procedure and design apparatus that may be used in the synthesis of only one member of a chemical family. In the laboratory a chemist is seldom interested in preparing the same compound over and over again, and hence makes use of methods that are applicable to many or all members of a particular family.

In our study of organic chemistry, we shall concentrate our attention on versatile laboratory preparations rather than on limited industrial methods. In learning these we may, for the sake of simplicity, use as examples the preparation of compounds that may actually never be made by the method shown. We may discuss the synthesis of ethane by the hydrogenation of ethylene, even though we can buy all the ethane we need from the petroleum industry. However, if we know how to convert ethylene into ethane, then, when the need arises, we also know how to convert 2-methyl-1-hexene into 2-methylhexane, or cholesterol into cholestanol, or, for that matter, cottonseed oil into oleomargarine.

3.15 Preparation

Each of the smaller alkanes, from methane through n-pentane and isopentane, can be obtained in pure form by fractional distillation of petroleum and natural gas; neopentane does not occur naturally. Above the pentanes the number of isomers of each homolog becomes so large and the boiling point differences become so small that it is no longer feasible to isolate individual, pure compounds; these alkanes must be synthesized by one of the methods outlined below.

In some of these equations, the symbol **R** is used to represent **any alkyl group**. This convenient device helps to summarize reactions that are typical of an entire family, and emphasizes the essential similarity of the various members.

In writing these generalized equations, however, we must not lose sight of one important point. An equation involving RCl, to take a specific example, has meaning only in terms of a reaction that we can carry out in the laboratory using a real compound, like methyl chloride or *tert*-butyl chloride. Although *typical* of alkyl halides, a reaction may differ widely in rate or yield depending upon the particular alkyl group actually concerned. We may use quite different experimental conditions for methyl chloride than for *tert*-butyl chloride; in an extreme case, a reaction that goes well for methyl chloride might go so slowly or give so many side products as to be completely useless for *tert*-butyl chloride.

PREPARATION OF ALKANES

1. **Hydrogenation of alkenes.** Discussed in Sec. 8.3.

$$C_nH_{2n} \xrightarrow{H_2 \ + \ Pt, \ Pd, \ or \ Ni} C_nH_{2n+2}$$

Alkene   Alkane

2. **Reduction of alkyl halides**

(a) **Hydrolysis of Grignard reagent.** Discussed in Sec. 3.16.

$$R-X + Mg \longrightarrow R-MgX \xrightarrow{H_2O} R-H$$

Grignard
reagent

Example:

$$CH_3CH_2CHCH_3 \xrightarrow{Mg} CH_3CH_2CHCH_3 \xrightarrow{H_2O} CH_3CH_2CHCH_3$$
$$\quad\quad | \quad\quad\quad\quad\quad\quad\quad | \quad\quad\quad\quad\quad\quad\quad\quad | $$
$$\quad\quad Br \quad\quad\quad\quad\quad\quad\quad MgBr \quad\quad\quad\quad\quad\quad\quad H$$

sec-Butyl bromide   *sec*-Butylmagnesium   *n*-Butane
bromide

(b) **Reduction by metal and acid.** Discussed in Sec. 3.15.

$$R-X + Zn + H^+ \longrightarrow R-H + Zn^{2+} + X^-$$

Example:

$$CH_3CH_2CHCH_3 \xrightarrow{Zn, \ H^+} CH_3CH_2CHCH_3$$
$$\quad\quad | \quad\quad\quad\quad\quad\quad\quad\quad\quad\quad | $$
$$\quad\quad Br \quad\quad\quad\quad\quad\quad\quad\quad\quad\quad H$$

sec-Butyl bromide   *n*-Butane

3. **Coupling of alkyl halides with organometallic compounds.** Discussed in Sec. 3.17.

$$\quad\quad\quad\quad\quad\quad\quad\quad\quad\quad\quad\quad\quad\quad R$$
$$\quad\quad\quad\quad\quad\quad\quad\quad\quad\quad\quad\quad\quad\quad | $$
$$R-X \xrightarrow{Li} R-Li \xrightarrow{CuX} R-CuLi$$

May be   Alkyllithium   Lithium
1°, 2°, 3°   dialkylcopper $\longrightarrow$ R—R′

R′X

Should be 1°

Examples:

$$\quad\quad\quad\quad\quad\quad\quad\quad\quad\quad\quad\quad\quad\quad\quad\quad CH_3CH_2$$
$$\quad\quad\quad\quad\quad\quad\quad\quad\quad\quad\quad\quad\quad\quad\quad\quad | $$
$$CH_3CH_2-Cl \xrightarrow{Li} CH_3CH_2-Li \xrightarrow{CuI} CH_3CH_2-CuLi \quad\quad CH_3(CH_2)_5CH_2Br$$

Ethyl   Ethyllithium   Lithium   *n*-Heptyl bromide
chloride   diethylcopper

$$CH_3CH_2-CH_2(CH_2)_5CH_3$$

n-Nonane

CONTINUED

—— CONTINUED ——

$$\underset{\substack{\text{CH}_3 \\ \big| \\ \text{CH}_3\text{C}\text{CH}_3 \\ \big| \\ \text{Cl}}}{} \xrightarrow{\text{Li}} \xrightarrow{\text{CuI}} t\text{-}\text{C}_4\text{H}_9\text{—CuLi}$$

tert-Butyl chloride

$$t\text{-}\text{C}_4\text{H}_9$$

$$\text{CH}_3\text{CH}_2\text{CH}_2\text{CH}_2\text{CH}_2\text{Br}$$

n-Pentyl bromide

$$\underset{\substack{\text{CH}_3 \\ \big| \\ \text{CH}_3\text{C}\text{—CH}_2\text{CH}_2\text{CH}_2\text{CH}_2\text{CH}_3 \\ \big| \\ \text{CH}_3}}{}$$

2,2-Dimethylheptane ∎

By far the most important of these methods is the hydrogenation of alkenes. When shaken under a slight pressure of hydrogen gas in the presence of a small amount of catalyst, alkenes are converted smoothly and quantitatively into alkanes of the same carbon skeleton. The method is limited only by the availability of the proper alkene. This is not a very serious limitation; as we shall see (Sec. 7.11), alkenes are readily prepared, chiefly from alcohols, which in turn can be readily synthesized (Sec. 17.8) in a wide variety of sizes and shapes.

Reduction of an alkyl halide, either via the Grignard reagent or directly with metal and acid, involves simply the replacement of a halogen atom by a hydrogen atom; the carbon skeleton remains intact. This method has about the same applicability as the previous method, since, like alkenes, alkyl halides are generally prepared from alcohols. Where either method could be used, the hydrogenation of alkenes would probably be preferred because of its simplicity and higher yield.

The coupling of alkyl halides with organometallic compounds is the only one of these methods in which carbon–carbon bonds are formed and a new, bigger carbon skeleton is generated.

3.16   The Grignard reagent: an organometallic compound

When a solution of an alkyl halide in dry ethyl ether, $(\text{C}_2\text{H}_5)_2\text{O}$, is allowed to stand over turnings of metallic magnesium, a vigorous reaction takes place: the solution turns cloudy, begins to boil, and the magnesium metal gradually disappears. The resulting solution is known as a **Grignard reagent**, after Victor Grignard (of the University of Lyons) who received the Nobel prize in 1912 for its discovery. It is one of the most useful and versatile reagents known to the organic chemist.

$$\text{CH}_3\text{I} + \text{Mg} \xrightarrow{\text{ether}} \text{CH}_3\text{MgI}$$

Methyl   Methylmagnesium
iodide   iodide

$$\text{CH}_3\text{CH}_2\text{Br} + \text{Mg} \xrightarrow{\text{ether}} \text{CH}_3\text{CH}_2\text{MgBr}$$

Ethyl bromide   Ethylmagnesium
                                    bromide

The Grignard reagent has the general formula RMgX, and the general name **alkylmagnesium halide**. The carbon–magnesium bond is covalent but highly polar, with carbon pulling electrons from electropositive magnesium; the magnesium–halogen bond is essentially ionic.

$$R : Mg^+ \; : \overset{..}{\underset{..}{X}} : \; ^-$$

Since magnesium becomes bonded to the same carbon that previously held halogen, the alkyl group remains intact during the preparation of the reagent. Thus *n*-propyl chloride yields *n*-propylmagnesium chloride, and isopropyl chloride yields isopropylmagnesium chloride.

$$\underset{\text{\textit{n}-Propyl chloride}}{CH_3CH_2CH_2Cl} + Mg \xrightarrow{\text{ether}} \underset{\text{\textit{n}-Propylmagnesium chloride}}{CH_3CH_2CH_2MgCl}$$

$$\underset{\text{Isopropyl chloride}}{CH_3CHClCH_3} + Mg \xrightarrow{\text{ether}} \underset{\text{Isopropylmagnesium chloride}}{CH_3CHMgClCH_3}$$

The Grignard reagent is the best-known member of a broad class of substances, called **organometallic** compounds, in which carbon is bonded to a metal: lithium, potassium, sodium, zinc, mercury, lead, thallium—almost any metal known. Each kind of organometallic compound has, of course, its own set of properties, and its particular uses depend on these. But, whatever the metal, it is less electronegative than carbon, and the carbon–metal bond—like the one in the Grignard reagent—is highly polar. Although the organic group is not a full-fledged *carbanion*—an anion in which carbon carries negative charge (Sec. 7.18)—it nevertheless has considerable carbanion character. As we shall see, organometallic compounds owe their enormous usefulness chiefly to one common quality: they can serve as a source from which carbon is readily transferred *with its electrons*.

$$\overset{\delta_- \quad \delta_+}{R—M}$$

The Grignard reagent is highly reactive. It reacts with numerous inorganic compounds including water, carbon dioxide, and oxygen, and with most kinds of organic compounds; in many of these cases the reaction provides the best way to make a particular class of organic compound.

The reaction with water to form an alkane is typical of the behavior of the Grignard reagent—and many of the more reactive organometallic compounds—toward acids. In view of the marked carbanion character of the alkyl group, we may consider the Grignard reagent to be the magnesium salt, RMgX, of the extremely weak acid, R—H. The reaction

$$RMgX + \underset{\substack{\text{Stronger} \\ \text{acid}}}{HOH} \longrightarrow R—H + \underset{\substack{\text{Weaker} \\ \text{acid}}}{Mg(OH)X}$$

is simply the displacement of the weaker acid, R—H, from its salt by the stronger acid, HOH.

An alkane is such a weak acid that it is displaced from the Grignard reagent by compounds that we might ordinarily consider to be very weak acids themselves, or possibly not acids at all. Any compound containing hydrogen attached to oxygen or nitrogen is tremendously more acidic than an alkane, and therefore can decompose the Grignard reagent: for example, ammonia or methyl alcohol.

$$RMgX + NH_3 \longrightarrow R\text{—}H + Mg(NH_2)X$$

Stronger   Weaker
acid   acid

$$RMgX + CH_3OH \longrightarrow R\text{—}H + Mg(OCH_3)X$$

Stronger   Weaker
acid   acid

For the preparation of an alkane, one acid is as good as another, so we naturally choose water as the most available and convenient.

Problem 3.9 (a) Which alkane would you expect to get by the action of water on *n*-propylmagnesium chloride? (b) On isopropylmagnesium chloride? (c) Answer (a) and (b) for the action of deuterium oxide ("heavy water", D_2O).

Problem 3.10 On conversion into the Grignard reagent followed by treatment with water, how many alkyl bromides would yield: (a) *n*-pentane; (b) 2-methylbutane; (c) 2,3-dimethylbutane; (d) neopentane? Draw the structures in each case.

3.17 Coupling of alkyl halides with organometallic compounds

To make an alkane of higher carbon number than the starting material requires formation of carbon–carbon bonds, most directly by the coupling together of two alkyl groups. The most versatile method of doing this is through a synthesis developed during the late 1960s by E. J. Corey and Herbert House, working independently at Harvard University and Massachusetts Institute of Technology. Coupling takes place in the reaction between a *lithium dialkylcopper*, R_2CuLi, and an alkyl halide, R'X. (R' stands for an alkyl group that may be the same as, or different from, R.)

R
|
R—CuLi   +   R'X $\longrightarrow$ R—R' + RCu + LiX

Lithium   Alkyl   Alkane
dialkylcopper halide

An alkyllithium, RLi, is prepared from an alkyl halide, RX, in much the same way as a Grignard reagent. To it is added cuprous halide, CuX, and then, finally, the second alkyl halide, R'X. Ultimately, the alkane is synthesized from the two alkyl halides, RX and R'X.

R
|
R—X $\xrightarrow{\text{Li}}$ R—Li $\xrightarrow{\text{CuX}}$ R—CuLi ┐

Alkyl   Lithium   ├→ R—R'
lithium   dialkylcopper

R'X ┘

For good yields, R′X should be a *primary* halide; the alkyl group R in the organometallic may be primary, secondary, or tertiary. For example:

$$CH_3{-}Br \xrightarrow{Li} CH_3{-}Li \xrightarrow{CuI} CH_3{-}\overset{\displaystyle CH_3}{\underset{|}{C}}uLi$$

Methyl   Methyllithium   Lithium
bromide   dimethylcopper

$$\longrightarrow CH_3{-}CH_2(CH_2)_6CH_3$$
n-Nonane

$$CH_3(CH_2)_6CH_2I$$
n-Octyl iodide

$$CH_3CH_2\overset{\displaystyle}{\underset{\underset{\displaystyle Cl}{|}}{C}}HCH_3 \xrightarrow{Li} \xrightarrow{CuI} (CH_3CH_2\underset{\underset{\displaystyle CH_3}{|}}{C}H{-})_2CuLi \qquad CH_3CH_2CH_2CH_2CH_2Br$$

sec-Butyl   *n*-Pentyl bromide
chloride

$$CH_3CH_2\underset{\underset{\displaystyle CH_3}{|}}{C}H{-}CH_2(CH_2)_3CH_3$$
3-Methyloctane

The choice of organometallic reagent is crucial. Grignard reagents or organolithium compounds, for example, couple with only a few unusually reactive organic halides. Organosodium compounds couple, but are so reactive that they couple, as they are being formed, with their parent alkyl halide; the reaction of sodium with alkyl halides (*Wurtz reaction*) is thus limited to the synthesis of symmetrical alkanes, R—R.

Organocopper compounds were long known to be particularly good at the formation of carbon–carbon bonds, but are unstable. Here, they are generated *in situ* from the organolithium, and then combine with more of it to form these relatively stable organometallics. They exist as complex aggregates but are believed to correspond roughly to $R_2Cu^-Li^+$. The anion here is an example of an *ate* complex, the negative counterpart of an *onium* complex (amm*onium*, ox*onium*).

Although the mechanism is not understood, this much is clear: the alkyl group R is transferred from copper, taking a pair of electrons with it, and becomes attached to the alkyl group R′ in place of halide ion (*nucleophilic aliphatic substitution*, Sec. 5.8).

Problem 3.11 (a) Outline two conceivable syntheses of 2-methylpentane from three-carbon compounds. (b) Which of the two would you actually use? Why?

3.18 Reactions

The alkanes are sometimes referred to by the old-fashioned name of *paraffins*. This name (Latin: *parum affinis*, not enough affinity) was given to describe what appeared to be the low reactivity of these hydrocarbons.

But reactivity depends upon the choice of reagent. If alkanes are inert toward hydrochloric and sulfuric acids, they react readily with acids like $HF{-}SbF_5$ and $FSO_3H{-}SbF_5$ ("magic acid") to yield a variety of products. If alkanes are inert toward oxidizing agents like potassium permanganate or sodium dichromate, most of this chapter is devoted to their oxidation by halogens. Certain yeasts feed happily on alkanes to produce proteins—certainly a chemical reaction. As Professor

M. S. Kharasch (p. 309) used to put it, consider the "inertness" of a room containing natural gas, air, and a lighted match.

Still, on a comparative basis, reactivity *is* limited. "Magic acid" is, after all, one of the strongest acids known; halogenation requires heat or light; combustion needs a flame or spark to get it started.

Much of the chemistry of alkanes involves free-radical chain reactions, which take place under vigorous conditions and usually yield mixtures of products. A reactive particle—typically an atom or free radical—is needed to begin the attack on an alkane molecule. It is the generation of this reactive particle that requires the vigorous conditions: the dissociation of a halogen molecule into atoms, for example, or even (as in pyrolysis) dissociation of the alkane molecule itself.

In its attack, the reactive particle abstracts hydrogen from the alkane; the alkane itself is thus converted into a reactive particle which continues the reaction sequence, that is, carries on the chain. But an alkane molecule contains many hydrogen atoms and the particular product eventually obtained depends upon *which* of these hydrogen atoms is abstracted. Although an attacking particle may show a certain selectivity, it can abstract a hydrogen from any part of the molecule, and thus bring about the formation of many isomeric products.

REACTIONS OF ALKANES

1. **Halogenation.** Discussed in Secs. 3.19–3.22.

$$\underset{\displaystyle |}{\overset{\displaystyle |}{-C}}-H + X_2 \xrightarrow{\text{250–400 °C, or light}} \underset{\displaystyle |}{\overset{\displaystyle |}{-C}}-X + HX$$

Usually a
mixture

Reactivity X_2: $Cl_2 > Br_2$

                    H: $3° > 2° > 1° > CH_3-H$

Example:

$$\underset{\text{Isobutane}}{CH_3-\underset{\displaystyle \underset{CH_3}{|}}{CH}-CH_3} \xrightarrow[\text{250–400 °C}]{Cl_2} \underset{\text{Isobutyl chloride}}{CH_3-\underset{\displaystyle \underset{CH_3}{|}}{CH}-CH_2Cl} + \underset{\textit{tert}\text{-Butyl chloride}}{CH_3-\underset{\displaystyle \underset{Cl}{|}}{\overset{\displaystyle \overset{CH_3}{|}}{C}}-CH_3}$$

2. **Combustion.** Discussed in Sec. 3.30.

$$C_nH_{2n+2} + \text{excess } O_2 \xrightarrow{\text{flame}} nCO_2 + (n+1)H_2O \qquad \Delta H = \text{heat of combustion}$$

Example:

$$n\text{-}C_5H_{12} + 8O_2 \xrightarrow{\text{flame}} 5CO_2 + 6H_2O \qquad \Delta H = -845 \text{ kcal}$$

3. **Pyrolysis (cracking).** Discussed in Sec. 3.31.

$$\text{alkane} \xrightarrow[\text{without catalysis}]{\text{400–600 °C; with or}} H_2 + \text{smaller alkanes} + \text{alkenes} \qquad \blacksquare$$

3.19   Halogenation

As we might expect, halogenation of the higher alkanes is essentially the same as the halogenation of methane. It can be complicated, however, by the formation of mixtures of isomers.

Under the influence of ultraviolet light, or at 250–400 °C, chlorine or bromine converts alkanes into chloroalkanes (alkyl chlorides) or bromoalkanes (alkyl bromides); an equivalent amount of hydrogen chloride or hydrogen bromide is formed at the same time. When diluted with an inert gas, and in an apparatus designed to carry away the heat produced, fluorine has recently been found to give analogous results. As with methane, iodination does not take place at all.

Depending upon which hydrogen atom is replaced, any of a number of isomeric products can be formed from a single alkane. Ethane can yield only one haloethane; propane, *n*-butane, and isobutane can yield two isomers each; *n*-pentane can yield three isomers, and isopentane, four isomers. Experiment has shown that on halogenation an alkane yields a mixture of all possible isomeric products, indicating that all hydrogen atoms are susceptible to replacement. For example, for chlorination:

$$CH_3CH_3 \xrightarrow[\text{light, 25 °C}]{Cl_2} CH_3CH_2-Cl$$

Ethane   b.p. 13 °C
Chloroethane
Ethyl chloride

$$CH_3CH_2CH_3 \xrightarrow[\text{light, 25 °C}]{Cl_2} CH_3CH_2CH_2-Cl \quad + \quad CH_3CHCH_3$$

Propane   b.p. 47 °C | Cl

1-Chloropropane
n-Propyl chloride
45%

b.p. 36 °C
2-Chloropropane
Isopropyl chloride
55%

$$CH_3CH_2CH_2CH_3 \xrightarrow[\text{light, 25 °C}]{Cl_2} CH_3CH_2CH_2CH_2-Cl \quad + \quad CH_3CH_2CHCH_3$$

n-Butane   b.p. 78.5 °C | Cl

1-Chlorobutane
n-Butyl chloride
28%

b.p. 68 °C
2-Chlorobutane
sec-Butyl chloride
72%

$$\overset{\displaystyle CH_3}{\underset{}{CH_3CHCH_3}} \xrightarrow[\text{light, 25 °C}]{Cl_2} \overset{\displaystyle CH_3}{\underset{}{CH_3CHCH_2-Cl}} \quad + \quad \overset{\displaystyle CH_3}{\underset{\displaystyle Cl}{CH_3CCH_3}}$$

Isobutane

b.p. 69 °C
1-Chloro-
2-methylpropane
Isobutyl chloride
64%

b.p. 51 °C
2-Chloro-
2-methylpropane
tert-Butyl chloride
36%

Bromination gives the corresponding bromides but in different proportions:

$$CH_3CH_3 \xrightarrow[\text{light, 127 °C}]{Br_2} CH_3CH_2Br$$
Ethane

$$CH_3CH_2CH_3 \xrightarrow[\text{light, 127 °C}]{Br_2} CH_3CH_2CH_2Br \quad + \quad CH_3CHCH_3$$
Propane 3% $|$
                                                                          Br
                                                                        97%

$$CH_3CH_2CH_2CH_3 \xrightarrow[\text{light, 127 °C}]{Br_2} CH_3CH_2CH_2CH_2Br \quad + \quad CH_3CH_2CHCH_3$$
n-Butane 2% $|$
                                                                            Br
                                                                          98%

$$CH_3\overset{\displaystyle CH_3}{\underset{|}{CHCH_3}} \xrightarrow[\text{light, 127 °C}]{Br_2} CH_3\overset{\displaystyle CH_3}{\underset{|}{CHCH_2Br}} \quad + \quad CH_3\overset{\displaystyle CH_3}{\underset{|}{\underset{Br}{\overset{|}{CCH_3}}}}$$
Isobutane   trace

over 99%

Problem 3.12 Draw the structures of: (a) the three monochloro derivatives of n-pentane; (b) the four monochloro derivatives of isopentane.

Although both chlorination and bromination yield mixtures of isomers, the results given above show that the *relative amounts* of the various isomers differ markedly depending upon the halogen used. Chlorination gives mixtures in which no isomer greatly predominates; in bromination, by contrast, one isomer may predominate to such an extent as to be almost the only product, making up 97–99% of the total mixture. In bromination, there is a high degree of *selectivity* as to which hydrogen atoms are to be replaced. (As we shall see in Sec. 3.28, this characteristic of bromination is due to the relatively low reactivity of bromine atoms, and is an example of a general relationship between *reactivity* and *selectivity*.)

With rare exceptions, *halogenation of alkanes is not suitable for the laboratory preparation of alkyl halides*. In chlorination, any one product is necessarily formed in low yield, and is difficult to separate from its isomers, whose boiling points are seldom far from its own. Even bromination of alkanes is seldom used. As we shall see in Chapter 5, there are excellent alternative ways to make alkyl halides, conveniently and from readily available precursors.

We begin our study of organic reactions with halogenation of methane and other alkanes, not for its utility in laboratory synthesis; synthesis is only one aspect of organic chemistry. But this reaction offers an easily understood approach to principles underlying all reactions we shall study. Alkanes are simple compounds. The mechanism is well-understood, and based upon evidence that we can readily grasp. We can deal rigorously and quantitatively with the matter of relative rates and orientation, since values of E_{act} and ΔH are accurately known. The nature of the transition state is unclouded by uncertainty as to the role played by a solvent. Finally, the study of free radicals is, in itself, an important part of organic chemistry.

On an industrial scale, chlorination of alkanes is important. For many purposes—for example, use as a solvent—a mixture of isomers is just as suitable as, and much cheaper than, a pure compound. It may be even worthwhile, when necessary, to separate a mixture of isomers if each isomer can then be marketed.

3.20   Mechanism of halogenation

Halogenation of alkanes proceeds by the same mechanism as halogenation of methane:

(1) $\qquad$ $X_2 \xrightarrow[\substack{\text{or} \\ \text{ultraviolet} \\ \text{light}}]{250\text{–}400\ ^\circ C} 2X\cdot$ $\qquad$ **Chain-initiating step**

(2) $\qquad$ $X\cdot + RH \longrightarrow HX + R\cdot$

(3) $\qquad$ $R\cdot + X_2 \longrightarrow RX + X\cdot$ $\qquad$ **Chain-propagating steps**

then (2), (3), (2), (3), etc., until finally a chain is terminated (Sec. 2.13)

A halogen atom abstracts hydrogen from the alkane (RH) to form an alkyl radical (R·). The radical in turn abstracts a halogen atom from a halogen molecule to yield the alkyl halide (RX).

Which alkyl halide is obtained depends upon which alkyl radical is formed.

$$CH_4 \xrightarrow{X\cdot} CH_3\cdot \xrightarrow{X_2} CH_3\text{—}X$$

Methane   Methyl   Methyl
                 radical   halide

$$CH_3CH_3 \xrightarrow{X\cdot} CH_3CH_2\cdot \xrightarrow{X_2} CH_3CH_2\text{—}X$$

Ethane   Ethyl   Ethyl
                radical   halide

$CH_3CH_2CH_3 \xrightarrow{X\cdot}$ Propane

abstraction of 1° H $\longrightarrow CH_3CH_2CH_2\cdot \xrightarrow{X_2} CH_3CH_2CH_2\text{—}X$

n-Propyl   *n*-Propyl
radical   halide

abstraction of 2° H $\longrightarrow CH_3\overset{\cdot}{C}HCH_3 \xrightarrow{X_2} CH_3\overset{|}{C}HCH_3$
                                                                          X

Isopropyl   Isopropyl
radical   halide

This in turn depends upon the alkane and which hydrogen atom is abstracted from it. For example, *n*-propyl halide is obtained from a *n*-propyl radical, formed from propane by abstraction of a primary hydrogen; isopropyl halide is obtained from an isopropyl radical, formed by abstraction of a secondary hydrogen.

How fast an alkyl halide is formed depends upon how fast the alkyl radical is formed. Here also, as was the case with methane (Sec. 2.20), of the two chain-propagating steps, step (2) is more difficult than step (3), and hence controls the rate of overall reaction. Formation of the alkyl radical is difficult, but once formed the radical is readily converted into the alkyl halide (see Fig. 3.11).

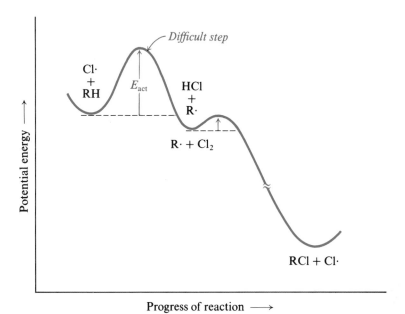

Figure 3.11   Potential energy changes during the progress of reaction: chlorination of an alkane. Formation of the radical is the rate-controlling step.

3.21   Orientation of halogenation

With this background let us turn to the problem of **orientation**; that is, let us examine the factors that determine *where* in a molecule reaction is most likely to occur. It is a problem that we shall encounter again and again, whenever we study a compound that offers more than one reactive site to attack by a reagent. It is an important problem, because orientation determines what product we obtain.

As an example let us take chlorination of propane. The relative amounts of *n*-propyl chloride and isopropyl chloride obtained depend upon the relative rates at which *n*-propyl radicals and isopropyl radicals are formed. If, say, isopropyl radicals are formed faster, then isopropyl chloride will be formed faster, and will make up a larger fraction of the product. As we can see, *n*-propyl radicals are formed by abstraction of primary hydrogens, and isopropyl radicals by abstraction of secondary hydrogens.

Thus *orientation is determined by the relative rates of competing reactions.* In this case we are comparing the rate of abstraction of primary hydrogens with the rate of abstraction of secondary hydrogens. What are the factors that determine the rates of these two reactions, and in which of these factors may the two reactions differ?

First of all, there is the collision frequency. This must be the same for the two reactions, since both involve collisions of the same particles: a propane molecule and a chlorine atom.

Next, there is the probability factor. If a primary hydrogen is to be abstracted, the propane molecule must be so oriented at the time of collision that the chlorine atom strikes a primary hydrogen; if a secondary hydrogen is to be abstracted, the propane must be so oriented that the chlorine collides with a secondary hydrogen. Since there are six primary hydrogens and only two secondary hydrogens in each molecule, we might estimate that the probability factor favors abstraction of primary hydrogens by the ratio of 6:2, or 3:1.

Considering only collision frequency and our guess about probability factors, we predict that chlorination of propane would yield *n*-propyl chloride and isopropyl chloride in the ratio of 3:1. As shown on page 102, however, the two chlorides are formed in roughly equal amounts, that is, in the ratio of about 1:1, or 3:3. The proportion of isopropyl chloride is about three times as great as predicted. Evidently, about three times as many collisions with secondary hydrogens are successful as collisions with primary hydrogens. If our assumption about the probability factor is correct, this means that E_{act} is less for abstraction of a secondary hydrogen than for abstraction of a primary hydrogen.

Chlorination of isobutane presents a similar problem. In this case, abstraction of one of the nine primary hydrogens leads to the formation of isobutyl chloride, whereas abstraction of a single tertiary hydrogen leads to the formation of *tert*-butyl chloride. We would estimate, then, that the probability factor favors

formation of isobutyl chloride by the ratio of 9:1. The experimental results given on page 102 show that the ratio is roughly 2:1, or 9:4.5. Evidently, about 4.5 times as many collisions with the tertiary hydrogen are successful as collisions with the primary hydrogens. This, in turn, probably means that E_{act} is less for abstraction of a tertiary hydrogen than for abstraction of a primary hydrogen, and, in fact, even less than for abstraction of a secondary hydrogen.

Study of the chlorination of a great many alkanes has shown that these are typical results. After allowance is made for differences in the probability factor, the rate of abstraction of hydrogen atoms is always found to follow the sequence $3° > 2° > 1°$. At room temperature, for example, the relative rates *per hydrogen atom* are 5.0:3.8:1.0. Using these values we can predict quite well the ratio of isomeric chlorination products from a given alkane. For example:

$$CH_3CH_2CH_2CH_3 \xrightarrow[\text{light, 25 °C}]{Cl_2} CH_3CH_2CH_2CH_2Cl + CH_3CH_2CHClCH_3$$

$$\quad\quad n\text{-Butane} \quad\quad\quad\quad\quad n\text{-Butyl chloride} \quad\quad sec\text{-Butyl chloride}$$

$$\frac{n\text{-butyl chloride}}{sec\text{-butyl chloride}} = \frac{\text{no. of 1° H}}{\text{no. of 2° H}} \times \frac{\text{reactivity of 1° H}}{\text{reactivity of 2° H}}$$

$$= \frac{6}{4} \times \frac{1.0}{3.8}$$

$$= \frac{6}{15.2} \quad equivalent\ to \quad \frac{28\%}{72\%}$$

In spite of these differences in reactivity, chlorination rarely yields a great preponderance of any single isomer. In nearly every alkane, as in the examples we have studied, the less reactive hydrogens are the more numerous; their lower reactivity is compensated for by a higher probability factor, with the result that appreciable amounts of every isomer are obtained.

Problem 3.13 Predict the proportions of isomeric products from chlorination at room temperature of: (a) propane; (b) isobutane; (c) 2,3-dimethylbutane; (d) n-pentane (*Note*: There are *three* isomeric products); (e) isopentane; (f) 2,2,3-trimethylbutane; (g) 2,2,4-trimethylpentane. For (a) and (b) check your calculations against the experimental values given on page 102.

The same sequence of reactivity, $3° > 2° > 1°$, is found in bromination, but with enormously larger reactivity ratios. At 127 °C, for example, the relative rates per hydrogen atom are 1600:82:1. Here, differences in reactivity are so marked as vastly to outweigh probability factors.

Problem 3.14 Answer Problem 3.13 for bromination at 127 °C.

3.22 Relative reactivities of alkanes toward halogenation

The best way to measure the relative reactivities of different compounds toward the same reagent is by the **method of competition**, since this permits an exact quantitative comparison under identical reaction conditions. Equimolar amounts of two compounds to be compared are mixed together and allowed to react with a limited amount of a particular reagent. Since there is not enough reagent for both compounds, the two compete with each other. Analysis of the reaction products shows which compound has consumed more of the reagent and hence is more reactive.

For example, if equimolar amounts of methane and ethane are allowed to react with a small amount of chlorine, about 400 times as much ethyl chloride as methyl chloride is obtained, showing that ethane is 400 times as reactive as methane. When allowance is made for the relative numbers of hydrogens in the two kinds of molecules, we see that each hydrogen of ethane is about 270 times as reactive as each hydrogen of methane.

$$CH_3Cl \xleftarrow{CH_4} Cl_2 \xrightarrow{C_2H_6} C_2H_5Cl$$

| 1 | light, 25 °C | 400 |

Problem 3.15   Because of the rather large difference in reactivity between ethane and methane, competition experiments have actually used mixtures containing more methane than ethane. If the molar ratio of methane to ethane were 10:1, what ratio of ethyl chloride to methyl chloride would you expect to obtain? What practical advantage would this experiment have over one involving a 1:1 ratio?

Data obtained from similar studies of other compounds are consistent with this simple generalization: *the reactivity of a hydrogen depends chiefly upon its class, and not upon the alkane to which it is attached.* Each primary hydrogen of propane, for example, is about as easily abstracted as each primary hydrogen in *n*-butane or isobutane; each secondary hydrogen of propane, about as easily as each secondary hydrogen of *n*-butane or *n*-pentane; and so on.

The hydrogen atoms of methane, which fall into a special class, are even less reactive than primary hydrogens, as shown by the above competition with ethane.

Problem 3.16   On chlorination, an equimolar mixture of ethane and neopentane yields neopentyl chloride and ethyl chloride in the ratio of 2.3:1. How does the reactivity of a primary hydrogen in neopentane compare with that of a primary hydrogen in ethane?

3.23   Ease of abstraction of hydrogen atoms. Energy of activation

At this stage we can summarize the effect of structure on halogenation of alkanes in the following way. The controlling step in halogenation is abstraction of hydrogen by a halogen atom:

$$R{-}H + X\cdot \longrightarrow H{-}X + R\cdot$$

The relative ease with which the different classes of hydrogen atoms are abstracted is:

Ease of abstraction $3° > 2° > 1° > CH_4$
of hydrogen atoms

This sequence applies (a) to the various hydrogens within a single alkane and hence governs **orientation** of reaction, and (b) to the hydrogens of different alkanes and hence governs **relative reactivities**.

Earlier, we concluded that these differences in ease of abstraction—like most differences in rate between closely related reactions (Sec. 2.19)—are probably due to differences in E_{act}. By study of halogenation at a series of temperatures (Sec. 2.18), the values of E_{act} listed in Table 3.5 were measured. In agreement with our tentative conclusions, the increasing rate of reaction along the series, methyl, 1°, 2°, 3°, is paralleled by a decreasing E_{act}. In chlorination the differences in E_{act}, like the differences in rate, are small; in bromination both differences are large.

Table 3.5 ENERGIES OF ACTIVATION, KCAL/MOL

	R—H + X· $\longrightarrow$ R· + H—X	
R	X = Cl	X = Br
CH_3	4	18
1°	1	13
2°	0.5	10
3°	0.1	7.5

We have seen (Sec. 2.18) that the larger the E_{act} of a reaction, the larger the increase in rate brought about by a given rise in temperature. We have just found that the differences in rate of abstraction among primary, secondary, and tertiary hydrogens are due to differences in E_{act}. We predict, therefore, that a rise in temperature should speed up abstraction of primary hydrogens (with the largest E_{act}) most, and abstraction of tertiary hydrogens (with the smallest E_{act}) least; the three classes of hydrogen should then display more nearly the same reactivity.

This leveling-out effect has indeed been observed: as the temperature is raised, the relative rates per hydrogen atom change from 5.0:3.8:1.0 toward 1:1:1. At very high temperatures virtually every collision has enough energy for abstraction of even primary hydrogens. It is generally true that *as the temperature is raised a given reagent becomes less selective in the position of its attack*; conversely, as the temperature is lowered it becomes more selective.

How can we account for the effect of structure on ease of abstraction of hydrogen atoms? Since this is a matter of E_{act}, we must look for our answer, as always, in the transition state. To do this, however, we must first shift our focus from the hydrogen atom being abstracted to the radical being formed.

3.24 Stability of free radicals

In Table 1.2 (p. 21) we find the homolytic dissociation energies of the bonds that hold hydrogen atoms to a number of groups. These are the ΔH values for the following reactions:

$$CH_3—H \longrightarrow CH_3· + H· \qquad\qquad \Delta H = 104 \text{ kcal}$$

$$CH_3CH_2—H \longrightarrow CH_3CH_2· + H· \qquad\qquad \Delta H = 98$$
$$\text{A 1° radical}$$

$$CH_3CH_2CH_2—H \longrightarrow CH_3CH_2CH_2· + H· \qquad\qquad \Delta H = 98$$
$$\text{A 1° radical}$$

$$CH_3\underset{\underset{H}{|}}{C}HCH_3 \longrightarrow \underset{\text{A 2° radical}}{CH_3\dot{C}HCH_3} + H\cdot \qquad\qquad \Delta H = 95$$

$$\underset{\underset{H}{|}}{CH_3\overset{\overset{CH_3}{|}}{C}CH_3} \longrightarrow \underset{\text{A 3° radical}}{CH_3\overset{\overset{CH_3}{|}}{\dot{C}}CH_3} + H\cdot \qquad\qquad \Delta H = 92$$

By definition, this bond dissociation energy is the amount of energy that must be supplied to convert a mole of alkane into radicals and hydrogen atoms. As we can see, the amount of energy needed to form the various classes of radicals decreases in the order: $CH_3\cdot > 1° > 2° > 3°$.

$$R\text{—}H \longrightarrow R\cdot + H\cdot \qquad \Delta H = \text{homolytic bond dissociation energy}$$

If less energy is needed to form one radical than another, it can only mean that, *relative to the alkane from which it is formed*, the one radical contains less energy than the other, this is to say, is *more stable* (see Fig. 3.12).

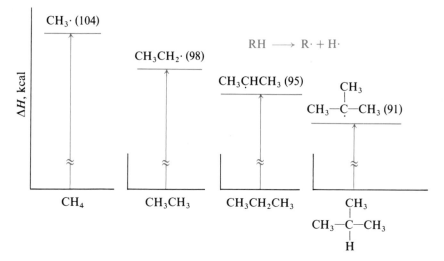

Figure 3.12   Relative stabilities of free radicals. (The plots are aligned with each other for easy comparison.)

We are not attempting to compare the absolute energy contents of, say, methyl and ethyl radicals; we are simply saying that the difference in energy between methane and methyl radicals is greater than the difference between ethane and ethyl radicals. *When we compare stabilities of free radicals, it must be understood that our standard for each radical is the alkane from which it is formed.* As we shall see, this is precisely the kind of stability that we are interested in.

Relative to the alkane from which each is formed, then, the order of stability of free radicals is:

**Stability of
free radicals** $3° > 2° > 1° > CH_3$

3.25   Ease of formation of free radicals

Let us return to the halogenation of alkanes. Orientation and reactivity, we have seen (Sec. 3.23), are governed by the relative ease with which the different classes of hydrogen atoms are abstracted. But by definition, the hydrogen being abstracted and the radical being formed belong to the same class. Abstraction of a primary hydrogen yields a primary radical, abstraction of a secondary hydrogen yields a secondary radical, and so on. For example:

$$CH_3CH_2CH_2\text{—}H + Br\cdot \longrightarrow H\text{—}Br + CH_3CH_2CH_2\cdot$$
$$\text{A 1° hydrogen} \qquad\qquad\qquad \text{A 1° radical}$$

$$CH_3\overset{|}{\underset{|}{C}}HCH_3 + Br\cdot \longrightarrow H\text{—}Br + CH_3\dot{C}HCH_3$$
$$\text{H} \qquad\qquad\qquad\qquad \text{A 2° radical}$$
$$\text{A 2° hydrogen}$$

$$\overset{CH_3}{\underset{H}{\overset{|}{CH_3\overset{|}{C}CH_3}}} + Br\cdot \longrightarrow H\text{—}Br + \overset{CH_3}{\underset{}{\overset{|}{CH_3\dot{C}CH_3}}}$$
$$\text{A 3° radical}$$
$$\text{A 3° hydrogen}$$

If the ease of abstraction of hydrogen atoms follows the sequence $3° > 2° > 1° > CH_4$, then the ease of formation of free radicals must follow the same sequence:

Ease of formation
of free radicals $3° > 2° > 1° > CH_3\cdot$

In listing free radicals in order of their ease of formation, we find that we have at the same time listed them in order of their stability. **The more stable the free radical, the more easily it is formed.**

This is an extremely useful generalization. *Radical stability seems to govern orientation and reactivity in many reactions where radicals are formed.* The addition of bromine atoms to alkenes (Sec. 8.19), for example, is a quite different sort of reaction from the one we have just studied; yet, there too, orientation and reactivity can be governed by radical stability. (Even in those cases where other factors—steric hindrance, polar effects—are significant or even dominant, it is convenient to use radical stability as a point of departure.)

3.26   Transition state for halogenation

Is it reasonable that the more stable radical should be formed more easily?

We have already seen that the differences in reactivity toward halogen atoms are due chiefly to differences in E_{act}: the more stable the radical, then, the lower the E_{act} for its formation. This, in turn, means that the more stable the radical, the more stable the transition state leading to its formation—both stabilities being measured, as they must be, against the same standard, the reactants. (*Remember*: E_{act} is the difference in energy content between reactants and transition state.)

Examination of the transition state shows that this is exactly what we would expect. As we saw before (Sec. 2.23), the hydrogen–halogen bond is partly formed and the carbon–hydrogen bond is partly broken. To the extent that the bond is

$$\begin{array}{ccc} -\overset{|}{\underset{|}{C}}-H + \cdot X & \longrightarrow & \left[-\overset{|}{\underset{|}{C}}\overset{\delta \cdot}{\cdots}H\cdots\overset{\delta \cdot}{X} \right] & \longrightarrow & -\overset{|}{\underset{|}{C}}\cdot + H\text{---}X \end{array}$$

Reactants	Transition state	Products
Halogen has odd electron	*Carbon acquiring free-radical character*	*Carbon has odd electron*

broken, the alkyl group possesses character of the free radical it will become. *Factors that tend to stabilize the free radical tend to stabilize the incipient free radical in the transition state.*

We have seen that the stabilities of free radicals follow the sequence $3° > 2° > 1° > CH_3\cdot$. A certain factor (*delocalization of the odd electron*, Sec. 10.11) causes the energy difference between isobutane and the *tert*-butyl radical, for example, to be smaller than between propane and the isopropyl radical. It is not unreasonable that this same factor should cause the energy difference between isobutane and the *incipient tert*-butyl radical in the transition state to be smaller than between propane and the *incipient* isopropyl radical in its transition state (Fig. 3.13).

Figure 3.13 Molecular structure and reactivity: free-radical substitution. The stability of the transition state parallels the stability of the radical: the more stable radical is formed faster. (The plots are aligned with each other for easy comparison.)

3.27   Orientation and reactivity

Throughout our study of organic chemistry, we shall approach the problems of orientation and reactivity in the following way.

Both problems involve comparing the rates of closely related reactions: in the case of orientation, reactions at different sites in the same compound; in the case of reactivity, reactions with different compounds. For such closely related reactions, variations in rate are due mostly to differences in E_{act}; by definition, E_{act} is the difference in energy content between reactants and transition state.

We shall examine the most likely structure for the transition state, then, to see what structural features affect its stability without at the same time affecting by an equal amount the stability of the reactants; that is, we shall look for factors that tend to increase or decrease the energy difference between reactants and transition state. Having decided what structural features affect the E_{act}, we shall compare the transition states for the reactions whose rates we wish to compare: the more stable the transition state, the faster the reaction.

In many, if not most, reactions where a free radical is formed, as in the present case, the transition state differs from the reactants chiefly in being like the product. It is reasonable, then, that the factor most affecting the E_{act} should be the *radical character* of the transition state. Hence we find that the more stable the radical the more stable the transition state leading to its formation, and the faster the radical is formed.

3.28   Reactivity and selectivity

In its attack on alkanes, the bromine atom is much more selective than the chlorine atom (with relative rate factors of $1600:82:1$ as compared with $5.0:3.8:1$). It is also much less reactive than the chlorine atom (only $1/375\,000$ as reactive toward methane, for example, as we saw in Sec. 2.19). This is just one example of a general relationship: in a set of similar reactions, the *less reactive* the reagent, the *more selective* it is in its attack.

To account for this relationship, we must recall what we learned in Sec. 2.24. In the attack by the comparatively unreactive bromine atom, the transition state is reached late in the reaction process, after the alkyl group has developed considerable radical character. In the attack by the highly reactive chlorine atom, the transition state is reached early, when the alkyl group has gained very little radical character.

Bromination

$$R-H + Br\cdot \quad \longrightarrow \quad \left[\overset{\delta.}{R}\text{---}H\text{--}\overset{\delta.}{Br}\right] \quad \longrightarrow \quad R\cdot + H-Br$$

Low reactivity;   Transition state
high selectivity
                         Reached late:
                         much radical
                         character

Chlorination

$$R-H + Cl\cdot \quad \longrightarrow \quad \left[\overset{\delta.}{R}\text{--}H\text{----}\overset{\delta.}{Cl}\right] \quad \longrightarrow \quad R\cdot + H-Cl$$

High reactivity;   Transition state
low selectivity
                         Reached early:
                         little radical
                         character

Now, by "selectivity" we mean here the differences in rate at which the various classes of free radicals are formed; a more stable free radical is formed faster, we said, because the factor that stabilizes it—delocalization of the odd electron (Sec. 10.11)—also stabilizes the incipient radical in the transition state. If this is so, then the more fully developed the radical character in the transition state, the more effective delocalization will be in stabilizing the transition state. The isopropyl radical, for example, is 3 kcal more stable than the n-propyl radical; if the radicals were *completely* formed in the transition state, the difference in E_{act} would be 3 kcal. Actually, in bromination the difference in E_{act} is 3 kcal: equal, within the limits of experimental error, to the maximum potential stabilization, indicating, as we expected, a great deal of radical character. In chlorination, by contrast, the difference in E_{act} is only 0.5 kcal, indicating only very slight radical character.

A similar situation exists for reactions of other kinds. Whatever the factor responsible for differences in stability among a set of transition states—whether it is delocalization of an odd electron, or accommodation of a positive or negative charge, or perhaps a change in crowding of the atoms—the factor will operate more effectively when the transition state is more fully developed, that is, when the reagent is less reactive.

3.29   Non-rearrangement of free radicals. Isotopic tracers

Our interpretation of orientation (Sec. 3.21) was based on an assumption that we have not yet justified: that the relative amounts of isomeric halides we find in the product reflect the relative rates at which various free radicals were formed from the alkane. From isobutane, for example, we obtain twice as much isobutyl chloride as *tert*-butyl chloride, and we assume from this that, by abstraction of hydrogen, isobutyl radicals are formed twice as fast as *tert*-butyl radicals.

Yet how do we know, in this case, that every isobutyl radical that is formed ultimately yields a molecule of isobutyl chloride? Suppose some isobutyl radicals were to change—by *rearrangement* of atoms—into *tert*-butyl radicals, which then react with chlorine to yield *tert*-butyl chloride. This supposition is not so far-fetched

as we, in our present innocence, might think; the doubt it raises is a very real one. We shall shortly see that another kind of reactive intermediate particle, the carbocation, is very prone to rearrange, with less stable ions readily changing into more stable ones (Sec. 5.23).

H. C. Brown (p. 650) and Glen Russell (now of Iowa State University) decided to test the possibility that free radicals, like carbocations, might rearrange, and chose the chlorination of isobutane as a good test case, because of the large difference in stability between *tert*-butyl and isobutyl radicals. If rearrangement of alkyl radicals can indeed take place, it should certainly happen here.

What the problem comes down to is this: does every abstraction of primary hydrogen lead to isobutyl chloride, and every abstraction of tertiary hydrogen lead to *tert*-butyl chloride? This, we might say, we could never know, because all hydrogen atoms are exactly alike. But are they? Actually, three isotopes of hydrogen exist: ^1H, *protium*, ordinary hydrogen; ^2H or D, *deuterium*, heavy hydrogen; and ^3H or T, *tritium*. Protium and deuterium are distributed in nature in the ratio of $5000:1$. (Tritium, the unstable, radioactive isotope, is present in traces, but can be made by neutron bombardment of ^6Li.) Modern methods of separation of isotopes have made very pure deuterium available, at moderate prices, in the form of deuterium oxide, D_2O, heavy water.

Brown and Russell prepared the deuterium-labeled isobutane I,

photochemically chlorinated it, and analyzed the products. The DCl:HCl ratio (determined by the mass spectrometer) was found to be equal (within experimental error) to the *tert*-butyl chloride:isobutyl chloride ratio. Clearly, every abstraction of a tertiary hydrogen (*deuterium*) gave a molecule of *tert*-butyl chloride, and every abstraction of a primary hydrogen (*protium*) gave a molecule of isobutyl chloride. *Rearrangement of the intermediate free radicals did not occur.*

All the existing evidence indicates quite strongly that, although rearrangement of free radicals occasionally happens, it is not very common and does not involve simple alkyl radicals.

Problem 3.17 (a) What results would have been obtained if some isobutyl radicals *had* rearranged to *tert*-butyl radicals? (b) Suppose that, instead of rearranging, isobutyl radicals were, in effect, converted into *tert*-butyl radicals by the reaction

What results would Brown and Russell have obtained?

Problem 3.18 Keeping in mind the availability of D_2O, suggest a way to make I from *tert*-butyl chloride. (*Hint*: See Sec. 3.16.)

The work of Brown and Russell is just one example of the way in which we can gain insight into a chemical reaction by using isotopically labeled compounds. We shall encounter many other examples in which isotopes, used either as *tracers*, as in this case, or for the detection of *isotope effects* (Sec. 7.17), give us information about reaction mechanisms that we could not get in any other way.

Besides deuterium and tritium, isotopes commonly used in organic chemistry include: ^{14}C, available as $^{14}CH_3OH$ and $Ba^{14}CO_3$; ^{18}O, as $H_2^{18}O$; ^{15}N, as $^{15}NH_3$, $^{15}NO_3^-$, and $^{15}NO_2^-$; ^{36}Cl, as chlorine or chloride; ^{131}I, as iodide.

Problem 3.19 Bromination of methane is slowed down by the addition of HBr (Problem 13, p. 73); this is attributed to the reaction

$$CH_3\cdot + HBr \longrightarrow CH_4 + Br\cdot$$

which, as the reverse of one of the chain-carrying steps, slows down bromination. How might you test whether or not this reaction actually occurs in the bromination mixture?

Problem 3.20 In Sec. 2.12 the reaction

$$Cl\cdot + Cl_2 \longrightarrow Cl_2 + Cl\cdot$$

was listed as probable but unproductive. Given ordinary chlorine (made up of ^{35}Cl and ^{37}Cl) and $^{36}Cl_2$, and a mass spectrometer, how would you go about finding out whether or not the reaction actually occurs?

3.30 Combustion

The reaction of alkanes with oxygen to form carbon dioxide, water, and—*heat* is the chief reaction occurring in the internal combustion engine; its tremendous practical importance is obvious.

The mechanism of this reaction is extremely complicated and is not yet fully understood. There seems to be no doubt, however, that it is a free-radical chain reaction. The reaction is extremely exothermic and yet requires a very high temperature, that of a flame, for its initiation. As in the case of chlorination, a great deal of energy is required for the bond-breaking that generates the initial reactive particles; once this energy barrier is surmounted, the subsequent chain-carrying steps proceed readily and with the evolution of energy.

A higher compression ratio has made the modern gasoline engine more efficient than earlier ones, but has at the same time created a new problem. Under certain conditions the smooth explosion of the fuel–air mixture in the cylinder is replaced by **knocking**, which greatly reduces the power of the engine.

The problem of knocking has been successfully met in two general ways: (a) proper selection of the hydrocarbons to be used as fuel, and (b) addition of tetraethyllead.

Experiments with pure compounds have shown that hydrocarbons of differing structures differ widely in knocking tendency. The relative antiknock tendency of a fuel is generally indicated by its **octane number**. An arbitrary scale has been set up, with *n*-heptane, which knocks very badly, being given an octane number of zero, and 2,2,4-trimethylpentane ("isooctane") being given the octane number of 100. There are available today fuels with better antiknock qualities than "isooctane".

The gasoline fraction obtained by direct distillation of petroleum (*straight-run gasoline*) is improved by addition of compounds of higher octane number; it is sometimes entirely replaced by these better fuels. Branched-chain alkanes and alkenes, and aromatic hydrocarbons generally have excellent antiknock qualities; these are produced from petroleum hydrocarbons by *catalytic cracking* (Sec. 3.31) and *catalytic reforming* (Sec. 15.5). Highly branched alkanes are synthesized from alkenes and alkanes by *alkylation* (Sec. 8.17).

In 1922 T. C. Midgley, Jr., and T. A. Boyd (of the General Motors Research Laboratory) found that the octane number of a fuel is greatly improved by addition of a small amount of tetraethyllead, $(C_2H_5)_4Pb$. Gasoline so treated is called *ethyl* gasoline or *leaded* gasoline. Nearly 50 years of research finally showed that tetraethyllead probably works by producing tiny particles of lead oxides, on whose surface certain reaction chains are broken.

In addition to carbon dioxide and water, however, the gasoline engine discharges other substances into the atmosphere, substances that are either smog-producing or downright poisonous: unburned hydrocarbons, carbon monoxide, nitrogen oxides, and, from leaded gasoline, various compounds of lead—in the United States, formerly, hundreds of tons of lead a day. Growing public concern about these pollutants has caused a minor revolution in the petroleum and auto industries. *Converters* have been developed to clean up exhaust emissions: by catalytic oxidation of hydrocarbons and carbon monoxide, and by the breaking down of nitrogen oxides into nitrogen and oxygen. But most of these oxidation catalysts contain platinum, which is poisoned by lead; there was a move to get the lead out of gasoline—not, initially, to cut down on lead pollution, but to permit converters to function. This has, in turn, brought back the problem of knocking, which is being met in two ways: (a) by lowering the compression ratio of the new automobiles being built; and (b) by increasing the octane number of gasoline through changes in hydrocarbon composition—through addition of aromatics and through increased use of isomerization (Sec. 3.13).

3.31 Pyrolysis: cracking

Decomposition of a compound by the action of heat alone is known as **pyrolysis**. This word is taken from the Greek *pyr*, fire, and *lysis*, a loosing, and hence to chemists means "cleavage by heat"; compare *hydro-lysis*, "cleavage by water".

The pyrolysis of alkanes, particularly when petroleum is concerned, is known as **cracking**. In *thermal cracking* alkanes are simply passed through a chamber heated to a high temperature. Large alkanes are converted into smaller alkanes, alkenes, and some hydrogen. This process yields predominantly ethylene (C_2H_4) together with other small molecules. In a modification called *steam cracking*, the hydrocarbon is diluted with steam, heated for a fraction of a second to 700–900 °C, and rapidly cooled. Steam cracking is of great importance in the production of hydrocarbons as chemicals, including ethylene, propylene, butadiene, isoprene, and cyclopentadiene. Another source of smaller hydrocarbons is *hydrocracking*, carried out in the presence of a catalyst and hydrogen, at high pressure and at much lower temperatures (250–450 °C).

The low-molecular-weight alkenes obtained from these cracking processes can be separated and purified, and are the most important raw materials for the large-scale synthesis of aliphatic compounds.

Most cracking, however, is directed toward the production of fuels, not chemicals, and for this *catalytic cracking* is the major process. Higher boiling petroleum fractions (typically, gas oil) are brought into contact with a finely divided silica–alumina catalyst at 450–550 °C and under slight pressure. Catalytic cracking not only increases the yield of gasoline by breaking large molecules into smaller ones, but also improves the quality of the gasoline: this process involves carbo-cations (Sec. 5.17), and yields alkanes and alkenes with the highly branched structures desirable in gasoline.

Through the process of *alkylation* (Sec. 8.17) some of the smaller alkanes and alkenes are converted into high-octane synthetic fuels.

Finally, by the process of *catalytic reforming* (Sec. 15.5) enormous quantities of the aliphatic hydrocarbons of petroleum are converted into *aromatic* hydro-carbons which are used not only as superior fuels but as the starting materials in the synthesis of most aromatic compounds (Chap. 13).

3.32 Determination of structure

One of the commonest and most important jobs in organic chemistry is to determine the structural formula of a compound just synthesized or isolated from a natural source.

The compound will fall into one of two groups, although at first we probably shall not know *which* group. It will be either (a) a previously reported compound, which we must identify, or (b) a new compound, whose structure we must prove.

If the compound has previously been encountered by some other chemist who determined its structure, then a description of its properties will be found some-where in the chemical literature, together with the evidence on which its structure was assigned. In that case, we need only to show that our compound is identical with the one previously described.

If, on the other hand, our compound is a new one that has never before been reported, then we must carry out a much more elaborate proof of structure.

Let us see—in a general way now, and in more detail later—just how we would go about this job. We are confronted by a flask filled with gas, or a few milliliters of liquid, or a tiny heap of crystals. We must find the answer to the question: *what is it?*

First, we purify the compound and determine its physical properties: melting point, boiling point, density, refractive index, and solubility in various solvents. In the laboratory today, we would measure various spectra of the compound (Chap. 16), in particular the infrared spectrum and the NMR spectrum; indeed, because of the wealth of information to be gotten in this way, spectroscopic examination might well be the first order of business after purification. From the mass spectrum we would get a very accurate molecular weight. Increasingly, structure is being determined in the most direct way possible: by x-ray analysis, which can show the precise distribution of atoms in a molecule.

We would carry out a qualitative elemental analysis to see what elements are present (Sec. 2.26). We might follow this with a quantitative analysis, and from this and the molecular weight we could calculate a molecular formula (Sec. 2.27); we would certainly do this if the compound is suspected of being a new one.

Next, we study systematically the behavior of the compound toward certain reagents. This behavior, taken with the elemental analysis, solubility properties, and spectra, generally permits us to *characterize* the compound, that is, to decide what family the unknown belongs to. We might find, for example, that the compound is an alkane, or that it is an alkene, or an aldehyde, or an ester.

Now the question is: *which* alkane is it? Or which alkene, or which aldehyde, or which ester? To find the answer, we first go to the chemical literature and look up compounds of the particular family to which our unknown belongs.

If we find one described whose physical properties are identical with those of our unknown, then the chances are good that the two compounds are identical. For confirmation, we generally convert the unknown by a chemical reaction into a new compound called a **derivative**, and show that this derivative is identical with the product derived in the same way from the previously reported compound.

If, on the other hand, we do not find a compound described whose physical properties are identical with those of our unknown, then we have a difficult job on our hands: we have a new compound, and must prove its structure. We may carry out a *degradation*: break the molecule apart, identify the fragments, and deduce what the structure must have been. To clinch any proof of structure, we attempt to *synthesize* the unknown by a method that leaves no doubt about its structure.

Problem 3.21   The final step in the proof of structure of an unknown alkane was its synthesis by the coupling of lithium di(*tert*-butyl)copper with *n*-butyl bromide. What was the alkane?

In Chapter 16, after we have become familiar with more features of organic structure, we shall see how spectroscopy fits into the general procedure outlined above.

3.33   Analysis of alkanes

An unknown compound is characterized as an alkane on the basis of negative evidence.

Upon qualitative elemental analysis, an alkane gives negative tests for all elements except carbon and hydrogen. A quantitative combustion, if one is carried out, shows the absence of oxygen; taken with a molecular weight determination, the combustion gives the molecular formula, C_nH_{2n+2}, which is that of an alkane.

An alkane is insoluble not only in water but also in dilute acid and base and in concentrated sulfuric acid. (As we shall see, most kinds of organic compounds dissolve in one or more of these solvents.)

An alkane is unreactive toward most chemical reagents. Its infrared spectrum lacks the absorption bands characteristic of groups of atoms present in other families of organic compounds (like OH, C=O, C=C, etc.).

Once the unknown has been characterized as an alkane, there remains the second half of the problem: finding out *which* alkane.

On the basis of its physical properties—boiling point, melting point, density, refractive index, and, most reliable of all, its infrared and mass spectra—it may be identified as a previously studied alkane of known structure.

If it turns out to be a new alkane, the proof of structure can be a difficult job. Combustion and molecular weight determination give its molecular formula. Clues about the arrangement of atoms are given by its infrared and NMR spectra. (For compounds like alkanes, it may be necessary to lean heavily on x-ray diffraction and mass spectrometry.)

Final proof lies in synthesis of the unknown by a method that can lead only to the particular structure assigned.

(The spectroscopic analysis of alkanes will be discussed in Chapter 16.)

PROBLEMS

1. Give the structural formula of:

(a) 2,2,3,3-tetramethylpentane
(b) 2,3-dimethylbutane
(c) 3,4,4,5-tetramethylheptane
(d) 4-ethyl-3,4-dimethylheptane
(e) 4-ethyl-2,4-dimethylheptane

(f) 2,5-dimethylhexane
(g) 3-ethyl-2-methylpentane
(h) 2,2,4-trimethylpentane
(i) 3-chloro-2-methylpentane
(j) 1,2-dibromo-2-methylpropane

2. Draw out the structural formula and give the IUPAC name of:

(a) $(CH_3)_2CHCH_2CH_2CH_3$
(b) $CH_3CBr_2CH_3$
(c) $CH_3CH_2C(CH_3)_2CH_2CH_3$
(d) $(C_2H_5)_2C(CH_3)CH_2CH_3$
(e) $CH_3CH_2CH(CH_3)CH(CH_3)CH(CH_3)_2$
(f) $CH_3CH_2CHCH_2CHCH_2CH_3$
$\ \ \ \ \underset{\displaystyle CH_3}{|}\ \ \ \underset{\displaystyle CH_2CH_2CH_3}{|}$

(g) $(CH_3)_3CCH_2C(CH_3)_3$

(h) $(CH_3)_2CCICH(CH_3)_2$
(i) $(CH_3)_2CHCH_2CH_2CH(C_2H_5)_2$
(j) $(CH_3)_2CHCH(CH_3)CH_2C(C_2H_5)_2CH_3$
(k) $(CH_3)_2CHC(C_2H_5)_2CH_2CH_2CH_3$
(l) $CH_3CH_2CHCH_2CHCHCH_3$
$\ \ \ \ \ \overset{\displaystyle CH_3}{|}\ \ \ \ \ \ \overset{\displaystyle CH_3}{|}$
$\ \ \ \ \ \ \ \ \ \ \ \ \ \ \ \ \underset{\displaystyle CH_2CH_2CH_3}{|}$

3. Pick out an alkane in Problem 1 or 2 that has: (a) no tertiary hydrogen; (b) one tertiary hydrogen; (c) two tertiary hydrogens; (d) no secondary hydrogen; (e) two secondary hydrogens; (f) half the number of secondary hydrogens as primary hydrogens.

4. Pick out an alkane (if any) in Problem 1 or 2 that contains:

(a) one isopropyl group
(b) two isopropyl groups
(c) one isobutyl group
(d) two isobutyl groups
(e) one sec-butyl group
(f) two sec-butyl groups

(g) one tert-butyl group
(h) two tert-butyl groups
(i) an isopropyl group and a sec-butyl group
(j) a tert-butyl group and an isobutyl group
(k) a methyl, an ethyl, a n-propyl, and a sec-butyl group

5. What alkane or alkanes of molecular weight 86 have: (a) two monobromo derivatives? (b) three? (c) four? (d) five? (e) How many dibromo derivatives does the alkane in (a) have? (f) Name the monobromo derivatives in (a).

6. How many mono-, di-, and trichloro derivatives are possible for cyclopentane? (Structure given in Sec. 12.2.)

7. Without referring to tables, list the following hydrocarbons in order of decreasing boiling points (i.e., highest boiling at top, lowest at bottom):

(a) 3,3-dimethylpentane
(b) n-heptane

(c) 2-methylheptane
(d) n-pentane

(e) 2-methylhexane

8. Write balanced equations, naming all organic products, for the following reactions:

(a) isobutyl bromide + Mg/ether
(b) tert-butyl bromide + Mg/ether
(c) product of (a) + H_2O
(d) product of (b) + H_2O

(e) product of (a) + D_2O
(f) sec-butyl chloride + Li, then CuI
(g) product of (f) + ethyl bromide

9. Write equations for the preparation of *n*-butane from:

(a) *n*-butyl bromide
(b) *sec*-butyl bromide
(c) ethyl chloride

(d) 1-butene, $CH_3CH_2CH=CH_2$
(e) 2-butene, $CH_3CH=CHCH_3$

10. Draw structures of all products expected from monochlorination at room temperature of:

(a) *n*-hexane
(b) isohexane

(c) 2,2,4-trimethylpentane
(d) 2,2-dimethylbutane

11. Predict the proportions of products in the preceding problem.

12. (a) Reaction of an aldehyde with a Grignard reagent is an important way of making alcohols. Why must one scrupulously dry the aldehyde before adding it to the Grignard reagent? (b) Why would one not prepare a Grignard reagent from $BrCH_2CH_2OH$?

13. On the basis of bond strengths in Table 1.2, page 21, add the following free radicals to the stability sequence of Sec. 3.24:

(a) *vinyl*, $H_2C=CH\cdot$
(b) *allyl*, $H_2C=CHCH_2\cdot$
(c) *benzyl*, $C_6H_5CH_2\cdot$

Check your answer in Sec. 15.15.

14. On the basis of your answer to Problem 13, predict how the following would fit into the sequence (Sec. 3.23) that shows ease of abstraction of hydrogen atoms:

(a) *vinylic* hydrogen, $H_2C=CH-H$
(b) *allylic* hydrogen, $H_2C=CHCH_2-H$
(c) *benzylic* hydrogen, $C_6H_5CH_2-H$

Check your answer against the facts in Secs. 10.3 and 15.14.

15. Free-radical chlorination of *either* *n*-propyl or isopropyl bromide gives 1-bromo-2-chloropropane, and of *either* isobutyl or *tert*-butyl bromide gives 1-bromo-2-chloro-2-methylpropane. What appears to be happening? Is there any pattern to this behavior?

16. (a) If a rocket were fueled with kerosine and liquid oxygen, what weight of oxygen would be required for every liter of kerosine? (Assume kerosine to have the average composition of n-$C_{14}H_{30}$.) (b) How much heat would be evolved in the combustion of one liter of kerosine? (Assume 157 kcal/mol for each $-CH_2-$ group and 186 kcal/mol for each $-CH_3$ group.) (c) If it were to become feasible to fuel a rocket with free hydrogen atoms, what weight of fuel would be required to provide the same heat as a liter of kerosine and the necessary oxygen? (Assume H_2 as the sole product.)

17. By what two quantitative methods could you show that a product isolated from the chlorination of propane was a monochloro or a dichloro derivative of propane? Tell exactly what results you would expect from each of the methods.

18. On the basis of certain evidence, including its infrared spectrum, an unknown compound of formula $C_{10}H_{22}$ is suspected of being 2,7-dimethyloctane. How could you confirm or disprove this tentatively assigned structure?

19. (a) A solution containing an unknown amount of methyl alcohol (CH_3OH) dissolved in *n*-octane is added to an excess of methylmagnesium iodide dissolved in the high-boiling solvent, *n*-butyl ether. A gas is evolved, and is collected and its volume measured: 1.04 mL (corrected to STP). What is the gas, and how is it formed? What weight of methyl alcohol was added to the Grignard reagent?

(b) A sample of 4.12 mg of an unknown alcohol, ROH, is added to methylmagnesium iodide as above; there is evolved 1.56 mL of gas (corrected to STP). What is the molecular weight of the alcohol? Suggest a possible structure or structures for the alcohol.

(c) A sample of 1.79 mg of a compound of mol. wt. about 90 gave 1.34 mL of the gas (corrected to STP). How many "active (that is, acidic) hydrogens" are there per molecule?

Assuming all these to be in —OH groups, suggest a structure for the alcohol. (This is an example of the *Zerewitinoff active hydrogen determination*.)

20. (a) *tert*-Butyl peroxide is a stable, easy-to-handle liquid that serves as a convenient source of free radicals:

$$(CH_3)_3CO—OC(CH_3)_3 \xrightarrow[\text{or light}]{130\ °C} 2(CH_3)_3CO \cdot$$

A mixture of isobutane and CCl_4 is quite stable at 130–140 °C. If a small amount of *tert*-butyl peroxide is added, a reaction occurs that yields (chiefly) *tert*-butyl chloride and chloroform. A small amount of *tert*-butyl alcohol (($CH_3)_3COH$, equivalent to the peroxide used) is also isolated. Give all steps in a likely mechanism for this reaction.

(b) When irradiated with ultraviolet light, or in the presence of a small amount of peroxides, *tert*-butyl hypochlorite, $(CH_3)_3C—O—Cl$, reacts with alkanes to form, in equimolar amounts, alkyl chlorides and *tert*-butyl alcohol. Outline all steps in a likely mechanism for this reaction.

4

Stereochemistry I. Stereoisomers

4.1 Stereochemistry and stereoisomerism

The science of organic chemistry, we said, is based on the relationship between molecular structure and properties. That part of the science which deals with structure *in three dimensions* is called **stereochemistry** (Greek: *stereos*, solid).

One aspect of stereochemistry is *stereoisomerism*. Isomers, we recall, are different compounds that have the same molecular formula. The particular kind of isomers that are different from each other *only* in the way the atoms are oriented in space (but are like one another with respect to which atoms are joined to which other atoms) are called **stereoisomers**.

Pairs of stereoisomers exist that differ so little in structure—and hence in properties—that of all the physical measurements we can make, only one, involving a special instrument and an unusual kind of light, can distinguish between them. Yet, despite this close similarity, the existence of such stereoisomers provides us with one of our most sensitive probes into mechanisms of chemical reactions; very often, one of these isomers is selected for study, not because it is different from ordinary compounds in its three-dimensional chemistry, but because it can be made to reveal what ordinary compounds hide. And, again despite their close similarity, one isomer of such a pair may serve as a nourishing food, or as an antibiotic, or as a powerful heart stimulant, and the other isomer may be useless.

We have already (Secs. 3.3 and 3.5) begun our study of the branch of stereochemistry called *conformational analysis*. In this chapter we shall, first, learn how to predict the existence of the kinds of stereoisomers called *enantiomers* and *diastereomers*, how to represent and designate their structures, and, in a general way, how their properties will compare. Then, in the latter part of the chapter, the emphasis will shift from what these isomers *are* to how they are formed, what they do, and what they can tell us. But stereochemistry permeates organic chemistry, and we shall return to it again and again throughout the rest of the book: to add to our knowledge of the fundamental concepts of stereochemistry; and simply to use it to help us understand what is going on in chemical reactions.

4.2 Isomer number and tetrahedral carbon

Let us begin our study of stereochemistry with methane and some of its simple substitution products. Any compound, however complicated, that contains carbon bonded to four other atoms can be considered to be a derivative of methane; and whatever we learn about the shape of the methane molecule can be applied to the shapes of vastly more complicated molecules.

The evidence of electron diffraction, x-ray diffraction, and spectroscopy shows that when carbon is bonded to four other atoms its bonds are directed toward the corners of a tetrahedron. But as early as 1874, years before the direct determination of molecular structure was possible, the tetrahedral carbon atom was proposed by J. H. van't Hoff (while he was still a student at the University of Utrecht) and, independently, J. A. LeBel. Their proposal was based upon the evidence of **isomer number**.

For any atom Y, *only one substance of formula* CH_3Y *has ever been found.* Chlorination of methane yields only one compound of formula CH_3Cl; bromination yields only one compound of formula CH_3Br. Similarly, only one CH_3F is known, and only one CH_3I. Indeed, the same holds true if Y represents, not just an atom, but a group of atoms (unless the group is so complicated that in itself it brings about isomerism); there is only one CH_3OH, only one CH_3COOH, only one CH_3SO_3H.

What does this suggest about the arrangement of atoms in methane? It suggests that every hydrogen atom in methane is equivalent to every other hydrogen atom, so that replacement of any one of them gives rise to the same product. If the hydrogen atoms of methane were not equivalent, then replacement of one would yield a different compound than replacement of another, and isomeric substitution products would be obtained.

In what ways can the atoms of methane be arranged so that the four hydrogen atoms are equivalent? There are three such arrangements: (a) a *planar* arrangement (I) in which carbon is at the center of a rectangle (or square) and a hydrogen atom is at each corner; (b) a *pyramidal* arrangement (II) in which carbon is at the apex of a pyramid and a hydrogen atom is at each corner of a square base; (c) a

I   II   III

tetrahedral arrangement (III) in which carbon is at the center of a tetrahedron and a hydrogen atom is at each corner.

How do we know that each of these arrangements could give rise to only one substance of formula CH_3Y? As always for problems like this, the answer lies in the use of molecular models. (Gumdrops and toothpicks can be used to make structures like I and II, for which the bond angles of ordinary molecular models are not suited.) For example, we make two identical models of I. In one model we replace, say, the upper right-hand H with a different atom Y, represented by a differently colored ball or gumdrop; in the other model we similarly replace, say, the lower right-hand H. We next see whether or not the two resulting models are *superimposable*; that is, we see whether or not, by any manipulations except bending or breaking bonds, we can make the models coincide in all their parts. If the two models are superimposable, they simply represent two molecules of the same compound; if the models are not superimposable, they represent molecules of different compounds which, since they have the same molecular formula, are by definition *isomers* (p. 37). Whichever hydrogen we replace in I (or in II or III), we get the same structure. From any arrangement other than these three, we would get more than one structure.

As far as compounds of the formula CH_3Y are concerned, the evidence of isomer number limits the structure of methane to one of these three possibilities.

Problem 4.1 How many isomers of formula CH_3Y would be possible if methane were a pyramid with a *rectangular* base? What are they? (*Hint*: If you have trouble with this question now, try it again after you have studied Sec. 4.7.)

For any atom Y and for any atom Z, only one substance of formula CH_2YZ has ever been found. Halogenation of methane, for example, yields only one compound of formula CH_2Cl_2, only one compound of formula CH_2Br_2, and only one compound of formula CH_2ClBr.

Of the three possible structures of methane, only the tetrahedral one is consistent with this evidence.

Problem 4.2 How many isomers of formula CH_2YZ would be expected from each of the following structures for methane? (a) Structure I with carbon at the center of a rectangle; (b) structure I with carbon at the center of a square; (c) structure II; (d) structure III.

Thus, only the tetrahedral structure for methane agrees with the evidence of isomer number. It is true that this is negative evidence; one might argue that isomers exist which have never been isolated or detected simply because the experimental techniques are not good enough. But, as we said before, any compound that contains carbon bonded to four other atoms can be considered to be a derivative of methane; in the preparation of hundreds of thousands of compounds of this sort, the number of isomers obtained has always been consistent with the concept of the tetrahedral carbon atom.

There is additional, positive evidence for the tetrahedral carbon atom: the finding of just the kind of isomers—*enantiomers*—that are predicted for compounds of formula CWXYZ. It was the existence of enantiomers that convinced van't

Hoff and LeBel that the carbon atom is tetrahedral. But to understand what enantiomers are, we must first learn about the property called *optical activity*.

4.3   Optical activity. Plane-polarized light

Light possesses certain properties that are best understood by considering it to be a wave phenomenon in which the vibrations occur at right angles to the direction in which the light travels. There are an infinite number of planes passing through the line of propagation, and ordinary light is vibrating in all these planes. If we consider that we are looking directly into the beam of a flashlight, Fig. 4.1 shows schematically the sort of vibrations that are taking place, all perpendicular

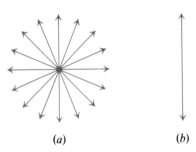

Figure 4.1   Schematic representation of (*a*) ordinary light and (*b*) plane-polarized light. The light is traveling perpendicular to the page; the vibrations are in the plane of the page.

(*a*)   (*b*)

to a line between our eye and the paper (flashlight). **Plane-polarized light** *is light whose vibrations take place in only one of these possible planes.* Ordinary light is turned into plane-polarized light by passing it through a lens made of the material known as Polaroid or more traditionally through pieces of *calcite* (a particular crystalline form of $CaCO_3$) so arranged as to constitute what is called a *Nicol prism.*

An **optically active substance** *is one that rotates the plane of polarized light.* When polarized light, vibrating in a certain plane, is passed through an optically active substance, it emerges vibrating in a different plane.

4.4   The polarimeter

How can this rotation of the plane of polarized light—this optical activity—be detected? It is both detected and measured by an instrument called the **polarimeter**, which is represented schematically in Fig. 4.2. It consists of a light source, two lenses (Polaroid or Nicol), and between the lenses a tube to hold the substance that is being examined for optical activity. These are arranged so that the light passes through one of the lenses (*polarizer*), then the tube, then the second lens (*analyzer*), and finally reaches our eye. When the tube is empty, we find that the maximum amount of light reaches our eye when the two lenses are so arranged that they pass light vibrating in the same plane. If we rotate the lens that is nearer our eye, say, we find that the light dims, and reaches a minimum when the lens is at right angles to its previous position.

Let us adjust the lenses so that a maximum amount of light is allowed to pass. (In practice, it is easier to detect a minimum than a maximum; the principle remains the same.) Now let us place the sample to be tested in the tube. If the substance does not affect the plane of polarization, light transmission is still at a maximum and the substance is said to be **optically inactive**. If, on the other hand, the substance rotates the plane of polarization, then the lens nearer our eye must

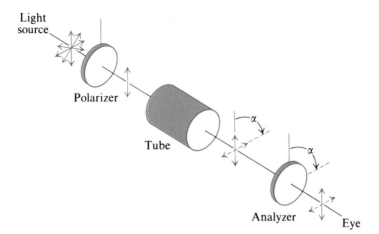

Light source

Polarizer

Tube

Analyzer

Eye

Figure 4.2   Schematic representation of a polarimeter. Solid lines: before rotation. Broken lines: after rotation. α is the angle of rotation.

be rotated to conform with this new plane if light transmission is again to be a maximum, and the substance is said to be **optically active**. If the rotation of the plane, and hence our rotation of the lens, is to the right (clockwise), the substance is **dextrorotatory** (Latin: *dexter*, right); if the rotation is to the left (counterclockwise), the substance is **levorotatory** (Latin: *laevus*, left).

We can determine not only that the substance has rotated the plane, and in which direction, but also *by how much*. The amount of rotation is simply the number of degrees that we must rotate the lens to conform with the light. The symbols + and − are used to indicate rotations to the right and to the left, respectively.

The lactic acid (p. 130) that is extracted from muscle tissue rotates light to the right, and hence is known as *dextrorotatory* lactic acid, or (+)-lactic acid. The 2-methyl-1-butanol that is obtained from fusel oil (a by-product of the fermentation of starch to ethyl alcohol) rotates light to the left, and is known as *levorotatory* 2-methyl-1-butanol, or (−)-2-methyl-1-butanol.

4.5   Specific rotation

Since optical rotation of the kind we are interested in is caused by individual molecules of the active compound, *the amount of rotation depends upon how many molecules the light encounters in passing through the tube*.

The light will encounter twice as many molecules in a tube 20 cm long as in a tube 10 cm long, and the rotation will be twice as large. If the active compound is in solution, the number of molecules encountered by the light will depend upon the concentration. For a given tube length, light will encounter twice as many molecules in a solution of 2 g per 100 mL of solvent as in a solution containing 1 g per 100 mL of solvent, and the rotation will be twice as large. When allowances are made for the length of tube and the concentration, it is found that the amount of rotation, as well as its direction, is a characteristic of each individual optically active compound.

Specific rotation is the number of degrees of rotation observed if a 1-dm (10-cm) tube is used, and the compound being examined is present to the extent of 1 g/mL. This is usually calculated from observations with tubes of other lengths and at different concentrations by means of the equation

$$[\alpha] = \frac{\alpha}{l \times d}$$

$$\text{specific rotation} = \frac{\text{observed rotation (degrees)}}{\text{length (dm)} \times \text{g/mL}}$$

where d represents density for a pure liquid or concentration for a solution.

The specific rotation is as much a property of a compound as its melting point, boiling point, density, or refractive index. Thus the specific rotation of the 2-methyl-1-butanol obtained from fusel oil is

$$[\alpha]_D{}^{20} = -5.90°$$

Here 20 is the temperature and D is the wavelength of the light used in the measurement (D line of sodium, 5893 Å).

Problem 4.3 The concentration of cholesterol dissolved in chloroform is 6.15 g per 100 mL of solution. (a) A portion of this solution in a 5-cm polarimeter tube causes an observed rotation of $-1.2°$. Calculate the specific rotation of cholesterol. (b) Predict the observed rotation if the same solution were placed in a 10-cm tube. (c) Predict the observed rotation if 10 mL of the solution were diluted to 20 mL and placed in a 5-cm tube.

Problem 4.4 A sample of a pure liquid in a 10-cm tube is placed in a polarimeter, and a reading of $+45°$ is made. How could you establish that $[\alpha]$ is really $+45°$ and not $-315°$? That it is $+45°$ and not $+405°$ or, for that matter, $+765°$?

4.6   Enantiomerism: the discovery

The optical activity we have just described was discovered in 1815 at the Collège de France by the physicist Jean-Baptiste Biot.

In 1848 at the École normale in Paris the chemist Louis Pasteur made a set of observations which led him a few years later to make a proposal that is the foundation of stereochemistry. Pasteur, then a young man, had come to the École normale from the Royal College of Besançon (where he had received his *baccalauréat ès sciences* with the rating of *médiocre* in chemistry), and had just won his *docteur ès sciences*. To gain some experience in crystallography, he was repeating another chemist's earlier work on salts of tartaric acid when he saw something that no one had noticed before: optically inactive sodium ammonium tartrate existed as a mixture of two different kinds of crystals, which were *mirror images* of each other. Using a hand lens and a pair of tweezers, he carefully and laboriously separated the mixture into two tiny piles—one of right-handed crystals and the other of left-handed crystals—much as one might separate right-handed and left-handed gloves lying jumbled together on a shop counter. Now, although the original mixture was optically inactive, each set of crystals dissolved in water was found to be *optically active*! Furthermore, the specific rotations of the two solutions were exactly *equal, but of opposite sign*; that is to say, one solution rotated plane-polarized light to the right, and the other solution an equal number of degrees to the left. In all other properties the two substances were identical.

Since the difference in optical rotation was observed *in solution*, Pasteur concluded that it was characteristic, not of the crystals, but of the *molecules*. He proposed that, like the two sets of crystals themselves, the molecules making up the crystals were *mirror images of each other*. He was proposing the existence of isomers whose structures differ only in being mirror images of each other, and whose properties differ only in the direction of rotation of polarized light.

There remained only for van't Hoff and LeBel to point out that a *tetrahedral* carbon atom would account not only for the absence of isomers of formula CH_3Y and CH_2YZ, but also for the existence of mirror-image isomers—*enantiomers*—like Pasteur's tartaric acids.

4.7 Enantiomerism and tetrahedral carbon

Let us convince ourselves that such mirror-image isomers should indeed exist. Starting with the actual, tetrahedral arrangement for methane, let us make a model of a compound CWXYZ, using a ball of a different color for each different atom or group represented as W, X, Y, and Z. Let us then imagine that we are holding this model before a mirror, and construct a second model of what its mirror image would look like. We now have two models that look like this:

mirror

which, using our wedge formulas, we can represent as:

mirror

$$\begin{array}{ccc} & X & \\ & \vdots & \\ W - &C& - Y \\ & \vdots & \\ & Z & \end{array} \qquad\qquad \begin{array}{ccc} & X & \\ & \vdots & \\ Y - &C& - W \\ & \vdots & \\ & Z & \end{array}$$

Not superimposable: isomers

As we saw before (Fig. 2.2, p. 41), a solid wedge represents a bond coming toward us out of the plane of the paper, and a broken wedge represents a bond going away from us behind the plane of the paper. (An ordinary line would represent a bond lying in the plane of the paper.)

Now, are these two models superimposable? *No*. We may twist and turn them as much as we please—so long as no bonds are broken—but, although two groups of each may coincide, the other two do not. (In our mind's eye, we can do the same thing with the wedge formulas.) The models are not superimposable, and therefore must represent two isomers of formula CWXYZ.

As predicted, mirror-image isomers do indeed exist, and thousands of instances besides the tartaric acids are known. There are, for example, two isomeric *lactic*

$$\begin{array}{cccc}
\text{COOH} & \text{COOH} & \text{C}_2\text{H}_5 & \text{C}_2\text{H}_5 \\
\text{H}-\text{C}-\text{OH} & \text{HO}-\text{C}-\text{H} & \text{HOCH}_2-\text{C}-\text{H} & \text{H}-\text{C}-\text{CH}_2\text{OH} \\
\text{CH}_3 & \text{CH}_3 & \text{CH}_3 & \text{CH}_3
\end{array}$$

Lactic acid   2-Methyl-1-butanol

acids and two *2-methyl-1-butanols*, two *chloroiodomethanesulfonic acids* and two *sec-butyl chlorides*.

$$\begin{array}{cccc}
\text{H} & \text{H} & \text{C}_2\text{H}_5 & \text{C}_2\text{H}_5 \\
\text{I}-\text{C}-\text{Cl} & \text{Cl}-\text{C}-\text{I} & \text{H}-\text{C}-\text{Cl} & \text{Cl}-\text{C}-\text{H} \\
\text{SO}_3\text{H} & \text{SO}_3\text{H} & \text{CH}_3 & \text{CH}_3
\end{array}$$

Chloroiodomethanesulfonic acid   *sec*-Butyl chloride

As we can see, the structures of each pair are mirror images; as we can easily verify by use of models, the structures of each pair are not superimposable and therefore represent isomers. (In fact, we have *already* verified this, since the models we made for CWXYZ can, of course, stand for any of these.)

At this point we do not need to know the chemistry of these compounds, or even what structure a particular collection of letters (—COOH, say, or —CH$_2$OH) stands for; we can tell when atoms or groups are the *same* or *different* from each other, and whether or not a model can be superimposed on its mirror image. Even two isotopes of the same element, like protium (ordinary hydrogen, H) and deuterium (heavy hydrogen, D) are different enough to permit detectable isomerism:

$$\begin{array}{cc}
\text{CH}_3 & \text{CH}_3 \\
\text{H}-\text{C}-\text{D} & \text{D}-\text{C}-\text{H} \\
\text{C}_6\text{H}_5 & \text{C}_6\text{H}_5
\end{array}$$

α-Deuterioethylbenzene

We must remember that *everything* (except, of course, a vampire) has a mirror image, including all molecules. Most molecules, however, are superimposable on their mirror images, as, for example, bromochloromethane, and do not show this mirror-image isomerism.

mirror

$$Cl \qquad\qquad Cl$$

$$H-C-H \qquad\qquad H-C-H$$

$$Br \qquad\qquad Br$$

Bromochloromethane
Superimposable: no isomerism

Mirror-image isomers are called *enantiomers.* Since they differ from one another only in the way the atoms are oriented in space, enantiomers belong to the general class called *stereoisomers.* Later on we shall encounter stereoisomers that are *not* mirror images of each other; these are called *diastereomers. Any two stereoisomers are thus classified either as enantiomers or as diastereomers, depending upon whether or not they are mirror images of each other.*

The non-superimposability of mirror images that brings about the existence of enantiomers also, as we shall see, gives them their optical activity, and hence enantiomers are often referred to as (one kind of) *optical isomers.* We shall make no use of the term *optical isomer,* since it is hard to define—indeed, is often used undefined—and of doubtful usefulness.

4.8   Enantiomerism and optical activity

Most compounds do not rotate the plane of polarized light. How is it that *some* do? It is not the particular chemical family that they belong to, since optically active compounds are found in all families. To see what special structural feature gives rise to optical activity, let us look more closely at what happens when polarized light is passed through a sample of a single pure compound.

When a beam of polarized light passes through an individual molecule, in nearly every instance its plane is rotated a tiny amount by interaction with the charged particles of the molecule; the direction and extent of rotation varies with the orientation of the particular molecule in the beam. For most compounds, because of the random distribution of the large number of molecules that make up even the smallest sample of a single pure compound, for every molecule that the light encounters, there is another (identical) molecule oriented *as the mirror image of the first,* which exactly cancels its effect. The net result is no rotation, that is, optical inactivity. Thus optical inactivity is not a property of individual molecules, but rather of the *random distribution of molecules that can serve as mirror images of each other.*

Optical inactivity requires, then, that one molecule of a compound act as the mirror image of another. But in the special case of CWXYZ, we have found (Sec. 4.7) a molecule whose mirror image is not just another, identical molecule, but rather a molecule of a different, isomeric compound. In a pure sample of a single enantiomer, no molecule can serve as the mirror image of another; there is no exact canceling-out of rotations, and the net result is optical activity. Thus, the same non-superimposability of mirror images that gives rise to enantiomerism also is responsible for optical activity.

4.9   Prediction of enantiomerism. Chirality

Molecules that are not superimposable on their mirror images are **chiral.**
Chirality is the necessary and sufficient condition for the existence of enan-

tiomers. That is to say: *a compound whose molecules are chiral can exist as enantiomers*; *a compound whose molecules are achiral* (without chirality) *cannot exist as enantiomers*.

When we say that a molecule and its mirror image are superimposable, we mean that if—in our mind's eye—we were to bring the image from behind the mirror where it seems to be, it could be made to coincide in all its parts with the molecule. To decide whether or not a molecule is chiral, therefore, we make a model of it and a model of its mirror image, and see if we can superimpose them. This is the safest way, since properly handled it must give us the right answer. It is the method that we should use until we have become quite familiar with the ideas involved; even then, it is the method we should use when we encounter a new type of compound.

After we have become familiar with the models themselves, we can draw wedge formulas to represent them, and *mentally* try to superimpose these. Some, we find, are not superimposable, like these:

Chloroiodomethanesulfonic acid
Not superimposable: enantiomers

These molecules are chiral, and we know that chloroiodomethanesulfonic acid can exist as enantiomers, which have the structures we have just made or drawn.

Others, we find, are superimposable, like these:

Isopropyl chloride
Superimposable: no enantiomers

These molecules are achiral, and so we know that isopropyl chloride cannot exist as enantiomers.

"I call any geometrical figure, or any group of points, *chiral*, and say it has *chirality*, if its image in a plane mirror, ideally realized, cannot be brought to coincide with itself."—Lord Kelvin, 1893.

In 1964, Cahn, Ingold, and Prelog (see p. 138) proposed that chemists use the terms "chiral" and "chirality" as defined by Kelvin. Based on the Greek word for "hand" (*cheir*), chirality means "handedness", in reference to that pair of non-superimposable mirror images we constantly have before us: our two hands. There has been widespread acceptance of Kelvin's terms, and they have largely displaced the earlier "dissymmetric" and "dissymmetry" (and the still earlier—and less accurate—"asymmetric" and "asymmetry"), although one must expect to encounter the older terms in the older chemical literature.

Whatever one calls it, it is non-superimposability-on-mirror-image that is the necessary and sufficient condition for enantiomerism; it is also a necessary—but *not* sufficient—condition for optical activity (see Sec. 4.13).

4.10   The chiral center

So far, all the chiral molecules we have talked about happen to be of the kind CWXYZ; that is, in each molecule there is a carbon (C*) that holds four different groups.

$$
\begin{array}{c}
\text{H} \\
| \\
\text{C}_2\text{H}_5-\overset{\displaystyle}{\text{C}}{}^*-\text{CH}_2\text{OH} \\
| \\
\text{CH}_3
\end{array}
\qquad
\begin{array}{c}
\text{H} \\
| \\
\text{CH}_3-\overset{\displaystyle}{\text{C}}{}^*-\text{COOH} \\
| \\
\text{OH}
\end{array}
\qquad
\begin{array}{c}
\text{H} \\
| \\
\text{C}_2\text{H}_5-\overset{\displaystyle}{\text{C}}{}^*-\text{CH}_3 \\
| \\
\text{Cl}
\end{array}
\qquad
\begin{array}{c}
\text{H} \\
| \\
\text{C}{}^*-\text{CH}_3 \\
| \\
\text{D}
\end{array}
$$

2-Methyl-1-butanol   Lactic acid   *sec*-Butyl chloride   α-Deuterioethylbenzene

A carbon atom to which four different groups are attached is a **chiral center**. (Sometimes it is called *chiral carbon,* when it is necessary to distinguish it from *chiral nitrogen, chiral phosphorus,* etc.)

Many—*but not all*—molecules that contain a chiral center are chiral. Many—*but not all*—chiral molecules contain a chiral center. There are molecules that contain chiral centers and yet are achiral (Sec. 4.18). (Such achiral molecules *always* contain *more than one* chiral center; if there is only one chiral center in a molecule, we can be certain that the molecule is chiral.) There are chiral molecules that contain no chiral centers (see, for example, Problem 6, p. 472).

The presence or absence of a chiral center is thus no criterion of chirality. However, most of the chiral molecules that we shall take up do contain chiral centers, and it will be useful for us to look for such centers; if we find a chiral center, then we should consider the *possibility* that the molecule is chiral, and hence can exist in enantiomeric forms. We shall later (Sec. 4.18) learn to recognize the kind of molecule that may be achiral in spite of the presence of chiral centers; such molecules contain more than one chiral center.

After we become familiar with the use of models and of wedge formulas, we can make use of even simpler representations of molecules containing chiral centers, which can be drawn much faster. This is a more dangerous method, however, and must be used properly to give the right answers. We simply draw a cross and attach to the four ends the four groups that are attached to the chiral center. The chiral center is understood to be located where the lines cross. Chemists have agreed that such a diagram stands for a particular structure: *the horizontal lines represent bonds coming toward us out of the plane of the paper, whereas the vertical lines represent bonds going away from us behind the plane of the paper.* That is to say:

$$
\begin{array}{c}
\text{C}_2\text{H}_5 \\
\vdots \\
\text{H}-\text{C}\rightarrow\text{Cl} \\
\vdots \\
\text{CH}_3
\end{array}
\qquad
\begin{array}{c}
\text{C}_2\text{H}_5 \\
\vdots \\
\text{Cl}-\text{C}\rightarrow\text{H} \\
\vdots \\
\text{CH}_3
\end{array}
$$

can be represented by

$$
\begin{array}{c}
\text{C}_2\text{H}_5 \\
| \\
\text{H}-\!\!\!+\!\!\!-\text{Cl} \\
| \\
\text{CH}_3
\end{array}
\qquad
\begin{array}{c}
\text{C}_2\text{H}_5 \\
| \\
\text{Cl}-\!\!\!+\!\!\!-\text{H} \\
| \\
\text{CH}_3
\end{array}
$$

In testing the superimposability of two of these flat, two-dimensional representations of three-dimensional objects, we must follow a certain procedure and obey certain rules. First, we use these representations only for molecules that contain a chiral center. Second, we draw one of them, and then draw the other as its mirror image. (Drawing these formulas *at random* can lead to some interesting but quite *wrong* conclusions about isomer numbers.) Third, in our mind's eye we may slide these formulas or rotate them end for end, *but we may not remove them from the plane of the paper*. Used with caution, this method of representation is convenient; it is not foolproof, however, and in doubtful cases models or wedge formulas should be used.

Problem 4.5   Using cross formulas, decide which of the following compounds are chiral. Check your answers by use of wedge formulas, and finally by use of models.
(a) 1-chloropentane
(b) 2-chloropentane
(c) 3-chloropentane
(d) 1-chloro-2-methylpentane
(e) 2-chloro-2-methylpentane
(f) 3-chloro-2-methylpentane
(g) 4-chloro-2-methylpentane
(h) 2-bromo-1-chlorobutane

Problem 4.6   (a) Neglecting stereoisomers for the moment, draw all isomers of formula C_3H_6DCl. (b) Decide, as in Problem 4.5, which of these are chiral.

4.11   Enantiomers

Isomers that are mirror images of each other are called **enantiomers**. The two different lactic acids whose models we made in Sec. 4.7 are enantiomers (Greek: *enantio-*, opposite). So are the two 2-methyl-1-butanols, the two *sec*-butyl chlorides, etc. How do the properties of enantiomers compare?

Enantiomers have identical physical properties, except for the direction of rotation of the plane of polarized light. The two 2-methyl-1-butanols, for example,

	(+)-2-Methyl-1-butanol	(−)-2-Methyl-1-butanol (fermentation product)
Specific rotation	+5.90°	−5.90°
Boiling point	128.9 °C	128.9 °C
Relative density	0.8193	0.8193
Refractive index	1.4107	1.4107

have identical melting points, boiling points, densities, refractive indexes, and any other physical constant one might measure, except for this: one rotates plane-polarized light to the right, the other to the left. This fact is not surprising, since the interactions of both kinds of molecule with their fellows should be the same. Only the *direction* of rotation is different; the *amount* of rotation is the same, the specific rotation of one being +5.90°, the other −5.90°. It is reasonable that these molecules, being so similar, can rotate light by the same amount. The molecules are mirror images, and so are their properties: the mirror image of a clockwise rotation is a counterclockwise rotation—and of exactly the same *magnitude*.

Enantiomers have identical chemical properties except toward optically active reagents. The two lactic acids are not only acids, but acids of exactly the same strength; that is, dissolved in water at the same concentration, both ionize to exactly the same degree. The two 2-methyl-1-butanols not only form the same products—*alkenes* on treatment with hot sulfuric acid, *alkyl bromides* on treatment with HBr, *esters* on treatment with acetic acid—but also form them at exactly the same rate. We can see why this must be so: the atoms undergoing attack in each case are influenced in their reactivity by exactly the same combination of substituents. The reagent approaching either kind of molecule encounters the same environment, except, of course, that one environment is the mirror image of the other.

(There is only one way in which enantiomers may differ in their reactions with ordinary, optically inactive reagents: *sometimes* they give products that are not identical but enantiomeric—still, of course, at exactly the same rate. As we shall see, whether or not this is the case can be highly significant, both practically and theoretically.)

In the special case of a reagent that is itself optically active, on the other hand, the influences exerted on the reagent are *not* identical in the attack on the two enantiomers, and reaction rates will be different—so different, in some cases, that reaction with one isomer does not take place at all. In biological systems, for example, such stereochemical specificity is the rule rather than the exception, since the all-important catalysts, *enzymes*, and most of the compounds they work on, are optically active. The sugar (+)-glucose plays a unique role in animal metabolism (Sec. 38.3) and is the basis of a multimillion-dollar fermentation industry (Sec. 17.6); yet (−)-glucose is neither metabolized by animals nor fermented by yeasts. When the mold *Penicillium glaucum* feeds on a mixture of enantiomeric tartaric acids, it consumes only the (+) enantiomer and leaves (−)-tartaric acid behind. The hormonal activity of (−)-adrenaline is many times that of its enantiomer; only one stereoisomer of chloromycetin is an antibiotic. (+)-Ephedrine not only has no activity as a drug, but actually interferes with the action of its enantiomer. Among amino acids, only one asparagine and one leucine are sweet, and only one glutamic acid enhances the flavor of food. It is (−)-carvone that gives oil of spearmint its characteristic odor; yet the enantiomeric (+)-carvone is the essence of caraway.

Consider, as a crude analogy, a right and left hand of equal strength (the enantiomers) hammering a nail (an optically inactive reagent) or, alternatively, inserting a right-handed screw (an optically active reagent). Hammering requires exactly corresponding sets of muscles in the two hands, and can be done at identical rates. Inserting the screw uses different sets of muscles: the right thumb pushes, for example, whereas the left thumb pulls.

Or, let us consider reactivity in the most precise way we know: by the transition state approach (Sec. 2.23).

Take first the reactions of two enantiomers with an optically inactive reagent. The reactants in both cases are of exactly the same energy: one enantiomer plus the reagent, and the other enantiomer plus the same reagent. The two transition states for the reactions are mirror images (they are enantiomeric), and hence are of exactly the same energy, too. Therefore, the energy differences between reactants and transition states—the E_{act} values—are identical, and so are the rates of reaction.

Now take the reactions of two enantiomers with an optically *active* reagent. Again the reactants are of the same energy. The two transition states, however, are *not* mirror images of each other (they are diastereomeric, Sec. 4.17), and hence are of *different* energies; the E_{act} values are different, and so are the rates of reaction.

The principle underlying all this is: enantiomers show different properties—physical or chemical—*only in a chiral medium*. Polarized light provides such a medium, and in it enantiomers differ in a physical property: direction of the rotation of the light. They may also differ in solubility in an optically active solvent, or in adsorption on an optically active surface. For enantiomers to react at different rates, the necessary chiral medium can be provided in a number of ways: by an optically active reagent; by a chiral solvent or the chiral surface of a catalyst; even—for some light-catalyzed reactions—by irradiation with circularly polarized light. For simplicity, we shall often use the term "optically active reagent" or "chiral reagent" in speaking of reaction under any of these chiral conditions. We shall use the term "optically inactive reagent" or "achiral reagent" or even "ordinary conditions" in speaking of reaction in the absence of a chiral medium.

4.12 The racemic modification

A mixture of equal parts of enantiomers is called a **racemic modification**. *A racemic modification is optically inactive*: when enantiomers are mixed together, the rotation caused by a molecule of one isomer is exactly canceled by an equal and opposite rotation caused by a molecule of its enantiomer.

The prefix $\pm$ is used to specify the racemic nature of the particular sample, as, for example, $(\pm)$-lactic acid or $(\pm)$-2-methyl-1-butanol.

It is useful to compare a racemic modification with a compound whose molecules are superimposable on their mirror images, that is, with an achiral compound. They are both optically inactive, and for exactly the same reason. Because of the random distribution of the large number of molecules, for every molecule that the light encounters there is a second molecule, a mirror image of the first, aligned just right to cancel the effect of the first one. In a racemic modification this second molecule happens to be an isomer of the first; for an achiral compound it is not an isomer, but another, identical molecule (Sec. 4.8).

(For an optically active substance uncontaminated by its enantiomer, we have seen, such cancellation of rotation cannot occur since no other molecule can serve as the mirror image of another, no matter how random the distribution.)

Problem 4.7 To confirm the statements of the three preceding paragraphs, make models of: (a) a pair of enantiomers, e.g., CHClBrI; (b) a pair of identical achiral molecules, e.g., CH_2ClBr; (c) a pair of identical chiral molecules, e.g., CHClBrI. (d) Which pairs are mirror images?

The identity of most physical properties of enantiomers has one consequence of great practical significance. They cannot be separated by ordinary methods: not by fractional distillation, because their boiling points are identical; not by fractional crystallization, because their solubilities in a given solvent are identical (unless the solvent is optically active); not by chromatography, because they are held equally strongly on a given adsorbent (unless it is optically active). The separation of a racemic modification into enantiomers—the *resolution* of a racemic modification— is therefore a special kind of job, and requires a special kind of approach (Sec. 4.27).

The first resolution was, of course, the one Pasteur carried out with his hand lens and tweezers (Sec. 4.6). But this method can almost never be used, since racemic modifications seldom form mixtures of crystals recognizable as mirror images. Indeed, even sodium ammonium tartrate does not, unless it crystallizes at a temperature below 28 °C. Thus partial credit for Pasteur's discovery has been given to the cool Parisian climate—and, of course, to the availability of tartaric acid from the winemakers of France.

The method of resolution nearly always used—one also discovered by Pasteur—involves the use of optically active reagents, and is described in Sec. 4.27.

Although popularly known chiefly for his great work in bacteriology and medicine, Pasteur was by training a chemist, and his work in chemistry alone would have earned him a position as an outstanding scientist.

4.13 Optical activity: a closer look

We have seen (Sec. 4.8) that, like enantiomerism, optical activity results from—and *only* from—chirality: the non-superimposability of certain molecules on their mirror images. Whenever we observe (molecular) optical activity, we know we are dealing with chiral molecules.

Is the reverse true? Whenever we deal with chiral molecules—with compounds that exist as enantiomers—must we always observe optical activity? *No.* We have just seen that a 50:50 mixture of enantiomers is optically inactive. Clearly, if we are to *observe* optical activity, the material we are dealing with must contain an *excess* of one enantiomer: enough of an excess that the net optical rotation can be detected by the particular polarimeter at hand.

Furthermore, this excess of one enantiomer must persist long enough for the optical activity to be measured. If the enantiomers are rapidly interconverted, then before we could measure the optical activity due to one enantiomer, it would be converted into an equilibrium mixture, which—since enantiomers are of exactly the same stability—must be a 50:50 mixture and optically inactive.

Even if all these conditions are met, the magnitude—and hence the detectability—of the optical rotation depends on the structure of the particular molecule concerned. In compound I, for example, the four groups attached to the chiral center differ only in chain length.

$$CH_3CH_2CH_2CH_2CH_2CH_2-\overset{\overset{\displaystyle CH_2CH_3}{|}}{\underset{\underset{\displaystyle CH_2CH_2CH_3}{|}}{C}}-CH_2CH_2CH_2CH_3$$

I

Ethyl-*n*-propyl-*n*-butyl-*n*-hexylmethane

It has been calculated that this compound should have the tiny specific rotation of $0.000\,01°$—far below the limits of detection by any existing polarimeter. In 1965, enantiomerically pure samples of both enantiomers of I were prepared (see Problem 19, p. 1230), and each was found to be optically inactive.

At our present level of study, the matter of speed of interconversion will give us no particular trouble. Nearly all the chiral molecules we encounter in this book lie at either of two extremes, which we shall easily recognize: (a) molecules, like those described in this chapter, which owe their chirality to chiral centers; here interconversion of enantiomers (*configurational* enantiomers) is so slow—because bonds have to be broken—that we need not concern ourselves at all about interconversion; (b) molecules whose enantiomeric forms (*conformational* enantiomers) are interconvertible simply by rotations about single bonds; here—for the compounds we shall encounter—interconversion is so fast that ordinarily we need not concern ourselves at all about the existence of the enantiomers.

4.14   Configuration

The arrangement of atoms that characterizes a particular stereoisomer is called its **configuration**.

Using the test of superimposability, we conclude, for example, that there are two stereoisomeric *sec*-butyl chlorides; their *configurations* are I and II. Let us say

$$C_2H_5 \qquad\qquad C_2H_5$$
$$H - C - Cl \qquad Cl - C - H$$
$$CH_3 \qquad\qquad CH_3$$
$$\text{I} \qquad\qquad\quad \text{II}$$

sec-Butyl chloride

that, by methods we shall take up later (Sec. 4.27), we have obtained in the laboratory samples of two compounds of formula $C_2H_5CHClCH_3$. We find that one rotates the plane of polarized light to the right, and the other to the left; we put them into two bottles, one labeled "(+)-*sec*-butyl chloride" and the other "(−)-*sec*-butyl chloride".

We have made two models to represent the two configurations of this chloride. We have isolated two isomeric compounds of the proper formula. Now the question arises, which configuration does each isomer have? Does the (+) isomer, say, have configuration I or configuration II? How do we know which structural formula, I or II, to draw on the label of each bottle? That is to say, how do we *assign configuration*?

Until 1951 the question of configuration could not be answered in an absolute sense for any optically active compound. But in that year J. M. Bijvoet—most fittingly Director of the van't Hoff Laboratory at the University of Utrecht (Sec. 4.2)—reported that, using a special kind of x-ray analysis (the method of anomalous scattering), he had determined the actual arrangement in space of the atoms of an optically active compound. The compound was a salt of (+)-tartaric acid, the same acid that—almost exactly 100 years before—had led Pasteur to his discovery of optical isomerism. Over the years prior to 1951, the relationships between the configuration of (+)-tartaric acid and the configurations of hundreds of optically active compounds had been worked out (by methods that we shall take up later, Sec. 4.24); when the configuration of (+)-tartaric acid became known, these other configurations, too, immediately became known. (In the case of the *sec*-butyl chlorides, for example, the (−) isomer is known to have configuration I, and the (+) isomer configuration II.)

4.15   Specification of configuration: *R* and *S*

Now, a further problem arises. How can we specify a particular configuration in some simpler, more convenient way than by always having to draw its picture? The most generally useful way yet suggested is the use of the prefixes *R* and *S*. According to a procedure proposed by R. S. Cahn (The Chemical Society, London), Sir Christopher Ingold (University College, London), and V. Prelog (Eidgenössiche Technische Hochschule, Zurich), two steps are involved.

Step 1. Following a set of *sequence rules* (Sec. 4.16), we assign a sequence of priority to the four atoms or groups of atoms—that is, the four *ligands*—attached to the chiral center.

In the case of CHClBrI, for example, the four atoms attached to the chiral center are all different and priority depends simply on atomic number, the atom of higher number having higher priority. Thus I, Br, Cl, H.

Bromochloroiodomethane

Step 2. We visualize the molecule oriented so that the ligand of *lowest* priority is directed *away* from us, and observe the arrangement of the remaining ligands. If, in proceeding from the ligand of highest priority to the ligand of second priority and thence to the third, our eye travels in a clockwise direction, the configuration is specified *R* (Latin: *rectus*, right); if counterclockwise, the configuration is specified *S* (Latin: *sinister*, left).

Thus, configurations I and II are viewed like this:

and are specified *R* and *S*, respectively.

A complete name for an optically active compound reveals—if they are known—both configuration and direction of rotation, as, for example, (*S*)-(+)-*sec*-butyl chloride. A racemic modification can be specified by the prefix *RS*, as, for example, (*RS*)-*sec*-butyl chloride.

(Specification of compounds containing more than one chiral center is discussed in Sec. 4.19.)

We must not, of course, confuse the direction of optical rotation of a compound—a physical property of a real substance, like melting point or boiling point—with the direction in which our eye happens to travel when we imagine a molecule held in an arbitrary manner. So far as we are concerned, unless we happen to know what has been established experimentally for a specific compound, we have no idea whether (+) or (−) rotation is associated with the *R* or the *S* configuration.

4.16 Sequence rules

For ease of reference and for convenience in reviewing, we shall set down here those sequence rules we shall have need of. You should study Rules 1 and 2 now, and Rule 3 later when the need for it arises.

Sequence Rule 1. If the four atoms attached to the chiral center are all different, priority depends on atomic number, with the atom of higher atomic number getting higher priority. If two atoms are isotopes of the same element, the atom of higher mass number has the higher priority.

For example, in chloroiodomethanesulfonic acid the sequence is I, Cl, S, H; in α-deuterioethyl bromide it is Br, C, D, H.

$$\underset{\substack{| \\ \text{I}}}{\overset{\substack{\text{Cl} \\ |}}{\text{H}-\text{C}-\text{SO}_3\text{H}}}$$

$$\underset{\substack{| \\ \text{D}}}{\overset{\substack{\text{H} \\ |}}{\text{H}_3\text{C}-\text{C}-\text{Br}}}$$

Chloroiodomethanesulfonic
acid

α-Deuterioethyl bromide

Problem 4.8   Make models and then draw both wedge formulas and cross formulas for the enantiomers of: (a) chloroiodomethanesulfonic acid and (b) α-deuterioethyl bromide. Label each as *R* or *S*.

Sequence Rule 2. If the relative priority of two groups cannot be decided by Rule 1, it shall be determined by a similar comparison of the next atoms in the groups (and so on, if necessary, working outward from the chiral center). That is to say, if two atoms attached to the chiral center are the same, we compare the atoms attached to each of these first atoms.

For example, take *sec*-butyl chloride, in which two of the atoms attached to the chiral center are themselves carbon. In CH_3 the second atoms are H, H, H;

$$\underset{\substack{| \\ \text{Cl}}}{\overset{\substack{\text{H} \\ |}}{\text{CH}_3-\text{CH}_2-\text{C}-\text{CH}_3}}$$

sec-Butyl chloride

in C_2H_5 they are C, H, H. Since carbon has a higher atomic number than hydrogen, C_2H_5 has the higher priority. A complete sequence of priority for *sec*-butyl chloride is therefore Cl, C_2H_5, CH_3, H.

In 3-chloro-2-methylpentane the C, C, H of isopropyl takes priority over the C, H, H of ethyl, and the complete sequence of priority is Cl, isopropyl, ethyl, H.

$$\underset{\substack{| \\ \text{Cl}}}{\overset{\substack{\text{CH}_3\ \text{H} \\ |\ \ \ |}}{\text{CH}_3-\text{CH}-\text{C}-\text{CH}_2-\text{CH}_3}}$$

$$\underset{\substack{| \\ \text{Cl}}}{\overset{\substack{\text{CH}_3\ \text{H} \\ |\ \ \ |}}{\text{CH}_3-\text{CH}-\text{C}-\text{CH}_2\text{Cl}}}$$

3-Chloro-2-methylpentane

1,2-Dichloro-3-methylbutane

In 1,2-dichloro-3-methylbutane the Cl, H, H of CH_2Cl takes priority over the C, C, H of isopropyl. Chlorine has a higher atomic number than carbon, and the fact that there are *two* C's and only *one* Cl does not matter. (One higher number is worth more than two—or three—of a lower number.)

Problem 4.9   Into what sequence of priority must these alkyl groups always fall: CH_3, 1°, 2°, 3°?

Sequence Rule 3. (*You should defer study of this rule until you need it.*)
Where there is a double or triple bond, both atoms are considered to be duplicated or triplicated. Thus

$$-\overset{|}{C}{=}A \quad equals \quad -\overset{|}{\underset{|}{C}}-A \quad and \quad -C{\equiv}A \quad equals \quad -\overset{A\ C}{\underset{A\ C}{C}}-A$$

For example, in glyceraldehyde the OH group has the highest priority of all,

$$H-\overset{\overset{\overset{H}{|}}{\underset{|}{C}{=}O}}{\underset{\underset{CH_2OH}{|}}{C}}-OH \qquad -\overset{H}{\underset{|}{C}}{=}O \quad equals \quad -\overset{\overset{H}{|}}{\underset{\underset{O\ C}{|}}{C}}-O$$

Glyceraldehyde

and the O, O, H of —CHO takes priority over the O, H, H of —CH_2OH. The complete sequence is then —OH, —CHO, —CH_2OH, —H.

The phenyl group, C_6H_5—, is handled as though it had one of the Kekulé structures:

benzene ring equals cyclohexadiene-like structure equals HC, CH, C, C structure

equals ... equals ...

In 1-amino-2-methyl-1-phenylpropane, for example, the C, C, C of phenyl takes

phenyl—C(H)(NH_2)—CH(CH_3)_2

priority over the C, C, H of isopropyl, but not over N, which has a higher atomic number. The entire sequence is then NH_2, C_6H_5, C_3H_7, H.

The vinyl group, $CH_2{=}CH$—, takes priority over isopropyl.

$$-CH{=}CH_2 \quad equals \quad -\overset{\overset{H\ H}{|\ |}}{\underset{\underset{C\ H}{|\ |}}{C}-C}-C \quad takes\ priority\ over \quad -\overset{\overset{H}{|}}{\underset{\underset{CH_3}{|}}{C}}-CH_3$$

Following the "senior" branch, —CH_2—C, we arrive at C in vinyl as compared with H in the —CH_2—H of isopropyl.

Problem 4.11   Draw and specify as *R* or *S* the enantiomers (if any) of:
(a) 3-chloro-1-pentene   (e) methylethyl-*n*-propylisopropylmethane
(b) 3-chloro-4-methyl-1-pentene   (f) $C_6H_5CHOHCOOH$, mandelic acid
(c) $HOOCCH_2CHOHCOOH$, malic acid   (g) $CH_3CH(NH_2)COOH$, alanine
(d) $C_6H_5CH(CH_3)NH_2$

4.17   Diastereomers

Next, we must learn what stereoisomers are possible for compounds whose molecules contain, not just one, but *more than one* chiral center. (In Chapter 38, we shall be dealing regularly with molecules that contain *five* chiral centers.)

Let us start with 2,3-dichloropentane. This compound contains two chiral

$$CH_3CH_2\overset{*}{-}CH\overset{*}{-}CH-CH_3$$
$$\underset{Cl}{|}\ \underset{Cl}{|}$$

2,3-Dichloropentane

centers, C–2 and C–3. (What four groups are attached to each of these carbon atoms?) How many stereoisomers are possible?

Using models, let us first make structure I and its mirror image II, and see if these are superimposable. We find that I and II are not superimposable, and hence

mirror

Not superimposable
Enantiomers

must be enantiomers. (As before, we may represent the structures by wedge formulas, and mentally try to superimpose these. Or, we may use the simple "cross" representations, being careful, as before (Sec. 4.10), not to remove the drawings from the plane of the paper or blackboard.)

Next, we try to interconvert I and II by rotations about carbon–carbon bonds. We find that they are not interconvertible in this way, and hence each of them is capable of retaining its identity and, if separated from its mirror image, of showing optical activity.

Are there any other stereoisomers of 2,3-dichloropentane? We can make structure III, which we find to be non-superimposable on either I or II; it is not, of

mirror

$$CH_3$$
$$H-C-Cl$$
$$H-C-Cl$$
$$C_2H_5$$

$$CH_3$$
$$Cl-C-H$$
$$Cl-C-H$$
$$C_2H_5$$

$$CH_3$$
$$H \!-\!\!\!+\!\!\!- Cl$$
$$H \!-\!\!\!+\!\!\!- Cl$$
$$C_2H_5$$
III

$$CH_3$$
$$Cl \!-\!\!\!+\!\!\!- H$$
$$Cl \!-\!\!\!+\!\!\!- H$$
$$C_2H_5$$
IV

Not superimposable
Enantiomers

course, the mirror image of either. What is the relationship between III and I? Between III and II? They are stereoisomers but not enantiomers. *Stereoisomers that are not mirror images of each other are called* **diastereomers**. Compound III is a diastereomer of I, and similarly of II.

Now, is III chiral? Using models, we make its mirror image, structure IV, and find that this is not superimposable on (or interconvertible with) III. Structures III and IV represent a second pair of enantiomers. Like III, compound IV is a diastereomer of I and of II.

How do the properties of diastereomers compare?

Diastereomers have similar chemical properties, since they are members of the same family. Their chemical properties are *not identical*, however. In the reaction of two diastereomers with a given reagent, neither the two sets of reactants nor the two transition states are mirror images, and hence—except by sheer coincidence— will not be of equal energies. The E_{act} values will be different and so will the rates of reaction.

Diastereomers have different physical properties: different melting points, boiling points, solubilities in a given solvent, densities, refractive indexes, and so on. Diastereomers differ in specific rotation; they may have the same or opposite signs of rotation, or some may be inactive.

As a result of their differences in boiling point and in solubility, they can, in principle at least, be separated from each other by fractional distillation or fractional crystallization; as a result of differences in molecular shape and polarity, they differ in adsorption, and can be separated by chromatography.

Given a mixture of all four stereoisomeric 2,3-dichloropentanes, we could separate it, by distillation, for example, into two fractions but no further. One fraction would be the racemic modification of I plus II; the other fraction would be the racemic modification of III plus IV. Further separation would require *resolution* of the racemic modifications by use of optically active reagents (Sec. 4.27).

Thus the presence of two chiral centers can lead to the existence of as many as four stereoisomers. For compounds containing three chiral centers, there could be as many as eight stereoisomers; for compounds containing four chiral centers, there could be as many as sixteen stereoisomers, and so on. The maximum number of stereoisomers that can exist is equal to 2^n, where n is the number of chiral centers. (In any case where *meso* compounds exist, as discussed in the following section, there will be fewer than this maximum number.)

The sugar (+)-glucose is by far the most important and abundant of the carbohydrates (Chap. 38). It is the compound oxidized in our cells to provide energy; it is the building block making up starch, from which our food ultimately comes, and cellulose, the framework of the plants that synthesize this starch. Glucose contains five chiral centers; this could—and does—give rise to 2^5 or 32 stereoisomers. Of these only *one*, α-D-glucose, is the unit of starch, and only one, β-D-glucose, is the unit of cellulose.

4.18 *Meso* structures

Now let us look at 2,3-dichlorobutane, which also has two chiral centers. Does this compound, too, exist in four stereoisomeric forms?

$$\overset{*}{CH_3-CH}-\overset{*}{CH}-CH_3$$
$$\quad\quad | \quad\quad |$$
$$\quad\quad Cl \quad\quad Cl$$

2,3-Dichlorobutane

Using models as before, we arrive first at the two structures V and VI. These are mirror images that are not superimposable or interconvertible; they are therefore enantiomers, and each should be capable of optical activity.

mirror

CH₃	CH₃
H—C—Cl	Cl—C—H
Cl—C—H	H—C—Cl
CH₃	CH₃

CH₃	CH₃
H——Cl	Cl——H
Cl——H	H——Cl
CH₃	CH₃
V	VI

Not superimposable

Enantiomers

Next, we make VII, which we find to be a diastereomer of V and of VI. We now have three stereoisomers; is there a fourth? *No.* If we make VIII, the mirror image of VII, we find the two to be superimposable; turned end-for-end, VII coincides in every respect with VIII. In spite of its chiral centers, VII is not chiral. It cannot exist in two enantiomeric forms, and it cannot be optically active. It is called a *meso* compound.

mirror

CH$_3$

H—C—Cl

H—C—Cl

CH$_3$

CH$_3$

Cl—C—H

Cl—C—H

CH$_3$

CH$_3$

H———Cl

H———Cl

CH$_3$

VII

CH$_3$

Cl———H

Cl———H

CH$_3$

VIII

Superimposable
A *meso* compound

A **meso compound** *is one whose molecules are superimposable on their mirror images even though they contain chiral centers.* A *meso* compound is optically inactive for the same reason as any other compound whose molecules are achiral: the rotation caused by any one molecule is canceled by an equal and opposite rotation caused by another molecule that is the mirror image of the first (Sec. 4.8).

We can often recognize a *meso* structure on sight by the fact that (in at least one of its conformations) one half of the molecule is the mirror image of the other half. This can be seen for *meso*-2,3-dichlorobutane by imagining the molecule to be

CH$_3$

H—C—Cl

H—C—Cl

CH$_3$

CH$_3$

H———Cl

H———Cl

CH$_3$

cut by a plane lying where the dotted line is drawn. The molecule has *a plane of symmetry*, and cannot be chiral. (*Caution*: If we do not see a plane of symmetry, however, this does not necessarily mean that the molecule is chiral.)

Problem 4.12 Draw stereochemical formulas for all the possible stereoisomers of the following compounds. Label pairs of enantiomers, and *meso* compounds. Tell which isomers, if separated from all other stereoisomers, will be optically active. Pick out several examples of diastereomers.

(a) 1,2-dibromopropane
(b) 3,4-dibromo-3,4-dimethylhexane
(c) 2,4-dibromopentane
(d) 2,3,4-tribromohexane

(e) 1,2,3,4-tetrabromobutane
(f) 2-bromo-3-chlorobutane
(g) 1-chloro-2-methylbutane
(h) 1,3-dichloro-2-methylbutane

4.19   Specification of configuration: more than one chiral center

Now, how do we specify the configuration of compounds which, like these, contain more than one chiral center? They present no special problem; we simply specify the configuration about *each* of the chiral centers, and by use of numbers tell which specification refers to which carbon.

Consider, for example, the 2,3-dichloropentanes (Sec. 4.17). We take each of the chiral centers, C–2 and C–3, in turn—ignoring for the moment the existence of

$$CH_3CH_2-\overset{3}{\underset{Cl}{\overset{*}{CH}}}-\overset{2}{\underset{Cl}{\overset{*}{CH}}}-CH_3$$

2,3-Dichloropentane

the other—and follow the steps of Sec. 4.15 and use the Sequence Rules of Sec. 4.16. In order of priority, the four ligands attached to C–2 are Cl, CH_3CH_2CHCl-, CH_3, H. On C–3 they are Cl, CH_3CHCl-, CH_3CH_2-, H. (Why is CH_3CHCl- "senior" to CH_3CH_2-?)

Taking in our hands—or in our mind's eye—a model of the particular stereoisomer we are interested in, we focus our attention first on C–2 (ignoring C–3), and then on C–3 (ignoring C–2). Stereoisomer I (p. 142), for example, we specify as $(2S,3S)$-2,3-dichloropentane. Similarly, II is $(2R,3R)$, III is $(2S,3R)$, and IV is $(2R,3S)$. These specifications help us to analyze the relationships among the stereoisomers. As enantiomers, I and II have opposite—that is, mirror-image—configurations about both chiral centers: $2S,3S$ and $2R,3R$. As diastereomers, I and III have opposite configurations about one chiral center, and the same configuration about the other: $2S,3S$ and $2S,3R$.

We would handle 2,3-dichlorobutane (Sec. 4.18) in exactly the same way. Here it happens that the two chiral centers occupy equivalent positions along the chain,

$$CH_3-\overset{*}{\underset{Cl}{CH}}-\overset{*}{\underset{Cl}{CH}}-CH_3$$

2,3-Dichlorobutane

and so it is not necessary to use numbers in the specifications. Enantiomers V and VI (p. 144) are specified (S,S)- and (R,R)-2,3-dichlorobutane, respectively. The *meso* isomer, VII, can, of course, be specified as either (R,S)- or (S,R)-2,3-dichlorobutane—the absence of numbers emphasizing the equivalence of the two specifications. The mirror-image relationship between the two ends of this molecule is consistent with the *opposite* designations of R and S for the two chiral centers. (Not all R,S isomers, of course, are *meso* structures—only those whose two halves are chemically equivalent.)

Problem 4.13   Give the R/S specification for each stereoisomer you drew in Problem 4.12 (p. 145).

4.20　Conformational isomers

In Sec. 3.5, we saw that there are several different staggered conformations of *n*-butane, each of which lies at the bottom of an energy valley—at an *energy minimum*—separated from the others by energy hills (see Fig. 3.8, p. 83). *Different conformations corresponding to energy minima are called* **conformational isomers**, *or* **conformers**. Since conformational isomers differ from each other only in the way their atoms are oriented in space, they, too, are stereoisomers. Like stereoisomers of any kind, a pair of conformers can either be mirror images of each other or not.

n-Butane exists as three conformational isomers, one *anti* and two *gauche* (Sec. 3.5). The *gauche* conformers, II and III, are mirror images of each other, and hence are (conformational) enantiomers. Conformers I and II (or I and III) are *not* mirror images of each other, and hence are (conformational) diastereomers.

Although the barrier to rotation in *n*-butane is a little higher than in ethane, it is still low enough that—at ordinary temperatures, at least—interconversion of conformers is easy and rapid. Equilibrium exists, and favors a higher population of the more stable *anti* conformer; the populations of the two *gauche* conformers—mirror images, and hence of exactly equal stability—are, of course, equal. Put differently, any given molecule spends the greater part of its time as the *anti* conformer, and divides the smaller part equally between the two *gauche* conformers. As a result of the rapid interconversion, these isomers cannot be separated.

Problem 4.14　Return to Problem 3.4 (p. 84) and, for each compound: (a) tell how many conformers there are, and label pairs of (conformational) enantiomers; (b) give the order of relative abundance of the various conformers.

Easy interconversion is characteristic of nearly every set of conformational isomers, and is the quality in which such isomers differ most from the kind of stereoisomers we have encountered so far in this chapter. This difference in interconvertibility is due to a difference in height of the energy barrier separating stereoisomers, which is, in turn, due to a difference in origin of the barrier. By definition, interconversion of conformational isomers involves rotation about single bonds; the rotational barrier is—in most cases—a very low one and interconversion is easy and fast. The other kind of stereoisomers, *configurational isomers*, or *inversional isomers*, differ from one another in configuration about a chiral center. Interconversion here involves the breaking of a covalent bond, for which there is a very high barrier: 50 kcal/mol or more (Sec. 1.14). Interconversion is difficult, and—unless one deliberately provides conditions to bring it about—is negligibly slow.

Interconvertibility of stereoisomers is of great practical significance because it limits their *isolability*. Hard-to-interconvert stereoisomers can be separated (with special methods, of course, for resolution of enantiomers) and studied individually; among other things, their optical activity can be measured. Easy-to-interconvert isomers cannot be separated, and single isolated isomers cannot be studied; optical activity cannot be observed, since any chiral molecules are present only as non-resolvable racemic modifications.

Our general approach to stereoisomers involves, then, two stages: first, we test the *superimposability* of possible isomeric structures, and then we test their *interconvertibility*. Both tests are best carried out with models. We make models of the two molecules and, without allowing any rotations about single bonds, we try to superimpose them: if they cannot be superimposed, they represent isomers. Next, we allow the models all possible rotations about single bonds, and repeatedly try to superimpose them: if they still cannot be superimposed, they are non-interconvertible, and represent *configurational isomers*; but if they can be superimposed after rotation, they are interconvertible and represent *conformational isomers*.

In dealing with those aspects of stereochemistry that depend on isolation of stereoisomers—isomer number or optical activity, for example, or study of the reactions of a single stereoisomer—we can ignore the existence of easy-to-inter-convert isomers, which means *most* conformational isomers. For convenience the following "ground rule" will hold for discussions and problems in this book: unless specifically indicated otherwise, *the terms "stereoisomers", "enantiomers", and "diastereomers" will refer only to configurational isomers, including geometric isomers* (Sec. 7.6), and will exclude conformational isomers. The latter will be referred to as "conformational isomers", "conformers", "conformational enantiomers", and "conformational diastereomers".

There is no sharp boundary between easy-to-interconvert and hard-to-interconvert stereoisomers. Although we can be sure that interconversion of configurational isomers will be hard, we cannot be sure that interconversion of conformational isomers will be easy. Depending upon the size and nature of substituents, the barrier to rotation about single bonds can be of any height, from the low one in ethane to one comparable to that for breaking a covalent bond. Some conformational isomers exist that are readily isolated, kept, and studied; indeed, study of such isomers (*atropisomers*) makes up a large and extremely important part of stereochemistry, one which, unfortunately, we shall not be able to take up in this beginning book. Other conformational isomers exist that can be isolated, not at ordinary temperatures, but at lower temperatures, where the average collision energy is lower. The conformational isomers that we shall encounter in this book, however, have low rotational barriers, and we may assume—until we learn otherwise—that when we classify stereoisomers as configurational or conformational, we at the same time classify them as hard to interconvert or easy to interconvert.

Problem 4.15 At low temperatures, where collision energies are small, two isomeric forms of the badly crowded $CHBr_2CHBr_2$ have been isolated by crystallization. (a) Give a formula or formulas (Newman projections) corresponding to each of the separable forms. (b) Which, if either, of the materials, as actually isolated at low temperatures, would be optically active? Explain.

4.21 Reactions involving stereoisomers

So far, our study of stereochemistry has been limited chiefly to finding out what the various kinds of stereoisomers are, how to predict their existence, and how to name and classify them. We have compared their properties, but only in a very general way.

Now let us go on from the *existence* of stereoisomers, and look at their *involvement* in chemical reactions: reactions in which stereoisomers are *formed*, and reactions in which stereoisomers are *consumed*; reactions in which the reagent is of the ordinary (i.e., optically inactive) kind and those in which the reagent is optically active.

We shall take up:

(a) the conversion of an achiral molecule into a chiral molecule, with the generation of a chiral center;

(b) reactions of chiral molecules in which bonds to the chiral center are not broken, and see how such reactions can be used to relate the configuration of one compound to that of another;

(c) reactions of the kind in (b) in which a second chiral center is generated;

(d) reactions of chiral compounds with optically active reagents.

Then we shall examine the stereochemistry of a reaction we have already studied—free-radical halogenation of alkanes—and see how stereochemistry can be used to get information about reaction mechanism. In doing this, we shall take up:

(e) a reaction of a chiral compound in which a bond to a chiral center is broken.

4.22   Generation of a chiral center. Synthesis and optical activity

One of the products of chlorination of *n*-butane is the chiral compound, *sec*-butyl chloride.

$$CH_3CH_2CH_2CH_3 \xrightarrow{\text{Cl}_2,\ \text{heat or light}} CH_3CH_2\overset{*}{-}\overset{|}{\underset{\underset{Cl}{|}}{C}H}-CH_3 + n\text{-Butyl chloride}$$

n-Butane

Achiral   *sec*-Butyl chloride

Chiral

It can exist as two enantiomers, I and II, which are specified (Sec. 4.16) as S and R, respectively.

CH$_3$ CH$_3$

H—C—Cl   Cl—C—H

C$_2$H$_5$ C$_2$H$_5$

I   II

(S)-(+)-*sec*-Butyl chloride   (R)-(−)-*sec*-Butyl chloride

Each enantiomer should, of course, be optically active. Now, if we were to put the *sec*-butyl chloride actually prepared by the chlorination of *n*-butane into a polarimeter, would it rotate the plane of polarized light? The answer is *no*, because prepared as described it would consist of the racemic modification. The next question is: *why is the racemic modification formed?*

In the first step of the reaction, a chlorine atom abstracts hydrogen to yield hydrogen chloride and a *sec*-butyl free radical. The carbon that carries the odd electron in the free radical is sp^2-hybridized (*trigonal*, Sec. 2.22), and hence a part of the molecule is *flat*, the trigonal carbon and the three atoms attached to it lying in the same plane. In the second step, the free radical abstracts chlorine from a chlorine molecule to yield *sec*-butyl chloride. But chlorine may become attached to either face of the flat radical, and, depending upon which face, yield either of two

Figure 4.3 Generation of a chiral center. Chlorine becomes attached to either face of the flat free radical, via (*a*) or (*b*), to give enantiomers, and in equal amounts.

Enantiomers

Formed in equal amounts

products: *R* or *S* (see Fig. 4.3). Since the chance of attachment to one face is exactly the same as for attachment to the other face, the enantiomers are obtained in exactly equal amounts. The product is the racemic modification.

If we were to apply the approach just illustrated to the synthesis of any compound whatsoever—and on the basis of any mechanism, correct or incorrect—we would arrive at the same conclusion: as long as neither the starting material nor the reagent (nor the environment) is optically active, we should obtain an optically inactive product. At some stage of the reaction sequence, there will be two alternative paths, one of which yields one enantiomer and the other the opposite enantiomer. The two paths will always be equivalent, and selection between them *random*. The facts agree with these predictions. **Synthesis of chiral compounds from achiral reactants always yields the racemic modification.** This is simply one aspect of the more general rule: **optically inactive reactants yield optically inactive products.**

> **Problem 4.16** Show in detail why racemic *sec*-butyl chloride would be obtained if: (a) the *sec*-butyl radical were not flat, but pyramidal; (b) chlorination did not involve a free *sec*-butyl radical at all, but proceeded by a mechanism in which a chlorine atom displaced a hydrogen atom, taking the position on the carbon atom formerly occupied by that hydrogen.

To purify the *sec*-butyl chloride obtained by chlorination of *n*-butane, we would carry out a fractional distillation. But since the enantiomeric *sec*-butyl chlorides have exactly the same boiling point, they cannot be separated, and are collected in the same distillation fraction. If recrystallization is attempted, there can again be no separation since their solubilities in every (optically inactive) solvent are identical. It is easy to see, then, that whenever a racemic modification is *formed* in a reaction, we will *isolate* (by ordinary methods) a racemic modification.

If an ordinary chemical synthesis yields a racemic modification, and if this cannot be separated by our usual methods of distillation, crystallization, etc., how do we know that the product obtained *is* a racemic modification? It is optically inactive; how do we know that it is actually made up of a mixture of two optically active substances? The separation of enantiomers (called *resolution*) can be accomplished by special methods; these involve the use of optically active reagents, and will be discussed later (Sec. 4.27).

Problem 4.17　Isopentane is allowed to undergo free-radical chlorination, and the reaction mixture is separated by careful fractional distillation. (a) How many fractions of formula $C_5H_{11}Cl$ would you expect to collect? (b) Draw structural formulas, stereochemical where pertinent, for the compounds making up each fraction. Specify each enantiomer as *R* or *S*. (c) Which, if any, of the fractions, as collected, would show optical activity? (d) Account in detail—just as was done above—for the optical activity or inactivity of each fraction.

4.23　Reactions of chiral molecules. Bond-breaking

Having made a chiral compound, *sec*-butyl chloride, let us see what happens when it, in turn, undergoes free-radical chlorination. A number of isomeric dichlorobutanes are formed, corresponding to attack at various positions in the molecule. (*Problem*: What are these isomers?)

$$\underset{\underset{\text{sec-Butyl chloride}}{\text{Cl}}}{CH_3CH_2-\overset{*}{C}H-CH_3} \xrightarrow{\text{Cl}_2,\ \text{heat or light}} \underset{\underset{\text{1,2-Dichlorobutane}}{\text{Cl}}}{CH_3CH_2-\overset{*}{C}H-CH_2Cl} + \text{other products}$$

Let us take, say, (*S*)-*sec*-butyl chloride (which, we saw in Sec. 4.22, happens to rotate light to the right), and consider only the part of the reaction that yields 1,2-dichlorobutane. Let us make a model (I) of the starting molecule, using a single ball for —C_2H_5 but a separate ball for each atom in —CH_3. Following the familiar steps of the mechanism, we remove an —H from —CH_3 and replace it with a —Cl. Since we break no bond to the chiral center in either step, the model we arrive at necessarily has configuration II, in which the spatial arrangement about

$$
\underset{\substack{| \\ (S)\text{-}sec\text{-Butyl chloride}}}{
\begin{array}{c} CH_3 \\ \vdots \\ H-C-Cl \\ \vdots \\ C_2H_5 \end{array}}
\xrightarrow{Cl\cdot}
\begin{array}{c} CH_2\cdot \\ \vdots \\ H-C-Cl \\ \vdots \\ C_2H_5 \end{array}
\xrightarrow{Cl_2}
\underset{\substack{II \\ (R)\text{-}1,2\text{-Dichlorobutane}}}{
\begin{array}{c} CH_2Cl \\ \vdots \\ H-C-Cl \\ \vdots \\ C_2H_5 \end{array}}
$$

the chiral center is unchanged—or, as we say, *configuration is retained*—with —CH_2Cl now occupying the same relative position that was previously occupied by —CH_3. It is an axiom of stereochemistry that molecules, too, behave in just this way, and that *a reaction that does not involve the breaking of a bond to a chiral center proceeds with retention of configuration about that chiral center.*

(If a bond to a chiral center *is* broken in a reaction, we can make no general statement about stereochemistry, except that configuration *can* be—and more than likely *will* be—changed. As is discussed in Sec. 4.28, just what happens depends on the mechanism of the particular reaction.)

Problem 4.18 We carry out free-radical chlorination of (*S*)-*sec*-butyl chloride, and by fractional distillation isolate the various isomeric products. (a) Draw stereochemical formulas of the 1,2-, 2,2-, and 1,3-dichlorobutanes obtained in this way. Give each enantiomer its proper *R* or *S* specification. (b) Which of these fractions, as isolated, will be optically active, and which will be optically inactive?

Now, let us see how the axiom about bond-breaking is applied in relating the configuration of one chiral compound to that of another.

4.24 Reactions of chiral molecules. Relating configurations

We learned (Sec. 4.14) that the configuration of a particular enantiomer can be determined directly by a special kind of x-ray diffraction, which was first applied in 1951 by Bijvoet to (+)-tartaric acid. But the procedure is difficult and time-consuming, and can be applied only to certain compounds. In spite of this limitation, however, the configurations of thousands of other compounds are now known, since they had already been related by chemical methods to (+)-tartaric acid. Most of these relationships were established by application of the axiom given above; that is, *the configurational relationship between two optically active compounds can be determined by converting one into the other by reactions that do not involve breaking of a bond to a chiral center.*

Let us take as an example (−)-2-methyl-1-butanol (the enantiomer found in fusel oil) and accept, for the moment, that it has configuration III, which we would specify *S*. We treat this alcohol with hydrogen chloride and obtain the alkyl chloride, 1-chloro-2-methylbutane. Without knowing the mechanism of this reac-

$$
\begin{array}{cc}
\text{CH}_3 & \text{CH}_3 \\
| & | \\
\text{H}-\text{C}-\text{CH}_2\!+\!\text{OH} \xrightarrow{\text{HCl}} & \text{H}-\text{C}-\text{CH}_2-\text{Cl} \\
| & | \\
\text{C}_2\text{H}_5 & \text{C}_2\text{H}_5 \\
\text{III} & \text{IV} \\
(S)\text{-}(-)\text{-2-Methyl-1-butanol} & (S)\text{-}(+)\text{-1-Chloro-2-methylbutane}
\end{array}
$$

tion, we can see that the carbon–oxygen bond is the one that is broken. *No bond to the chiral center is broken*, and therefore configuration is retained, with —CH₂Cl occupying the same relative position in the product that was occupied by —CH₂OH in the reactant. We put the chloride into a tube, place this tube in a polarimeter, and find that the plane of polarized light is rotated to the right; that is, the product is (+)-1-chloro-2-methylbutane. Since (−)-2-methyl-1-butanol has configuration III, (+)-1-chloro-2-methylbutane must have configuration IV.

Or, we oxidize (−)-2-methyl-1-butanol with potassium permanganate, obtain the acid 2-methylbutanoic acid, and find that this rotates light to the right.

Again, no bond to the chiral center is broken, and we assign configuration V to (+)-2-methylbutanoic acid.

$$\begin{array}{ccc}
\text{CH}_3 & & \text{CH}_3 \\
\mid & & \mid \\
\text{H}\!-\!\text{C}\!-\!\text{CH}_2\text{OH} & \xrightarrow{\text{KMnO}_4} & \text{H}\!-\!\text{C}\!-\!\text{COOH} \\
\mid & & \mid \\
\text{C}_2\text{H}_5 & & \text{C}_2\text{H}_5 \\
\text{III} & & \text{V}
\end{array}$$

(S)-(−)-2-Methyl-1-butanol   (S)-(+)-2-Methylbutanoic acid

We can nearly always tell whether or not a bond to a chiral center is broken by simple inspection of the formulas of the reactant and product, as we have done in these cases, and without a knowledge of the reaction mechanism. We must be aware of the possibility, however, that a bond may break and re-form during the course of a reaction without this being evident on the surface. This kind of thing does not happen at random, but in certain specific situations which an organic chemist learns to recognize. Indeed, stereochemistry plays a leading role in this learning process: one of the best ways to detect hidden bond-breaking is so to design the experiment that, if such breaking occurs, it must involve a chiral center.

But how do we know in the first place that (−)-2-methyl-1-butanol has configuration III? Its configuration was related in this same manner to that of another compound, and that one to the configuration of still another, and so on, going back ultimately to (+)-tartaric acid and Bijvoet's x-ray analysis.

We say that the (−)-2-methyl-1-butanol, the (+)-chloride, and the (+)-acid have *similar* (or the *same*) configurations. The enantiomers of these compounds, the (+)-alcohol, (−)-chloride, and (−)-acid, form another set of compounds with similar configurations. The (−)-alcohol and, for example, the (−)-chloride are said to have *opposite* configurations. As we shall find, we are usually more interested in knowing whether two compounds have similar or opposite configurations than in knowing what the actual configuration of either compound actually is. That is to say, we are more interested in *relative* configurations than in *absolute* configurations.

In this set of compounds with similar configurations, we notice that two are dextrorotatory and the third is levorotatory. The sign of rotation is important as a means of keeping track of a particular isomer—just as we might use boiling point or refractive index to tell us whether we have *n*-butane or isobutane, *now that their structures have been assigned*—but the fact that two compounds happen to have the same sign or opposite sign of rotation means little; they may or may not have similar configurations.

The three compounds all happen to be specified as S, but this is simply because —CH$_2$Cl and —COOH happen to have the same relative priority as —CH$_2$OH. If we were to replace the chlorine with deuterium (*Problem*: How could this be done?), the product would be specified R, yet obviously it would have the same configuration as the alcohol, halide, and acid. Indeed, looking back to *sec*-butyl chloride and 1,2-dichlorobutane, we see that the similar configurations I and II *are* specified differently, one S and the other R; here, a group (—CH$_3$) that has a lower priority than —C$_2$H$_5$ is converted into a group (—CH$_2$Cl) that has a higher priority. We cannot tell whether two compounds have the same or opposite configurations by simply looking at the letters used to specify their configurations; we must work out and compare the absolute configurations indicated by those letters.

Problem 4.19 Which of the following reactions could safely be used to relate configurations?

(a) $(+)$-$C_6H_5CH(OH)CH_3 + PBr_3 \longrightarrow C_6H_5CHBrCH_3$

(b) $(+)$-$CH_3CH_2CHClCH_3 + C_6H_6 + AlCl_3 \longrightarrow C_6H_5CH(CH_3)CH_2CH_3$

(c) $(-)$-$C_6H_5CH(OC_2H_5)CH_2OH + HBr \longrightarrow C_6H_5CH(OC_2H_5)CH_2Br$

(d) $(+)$-$CH_3CH(OH)CH_2Br + NaCN \longrightarrow CH_3CH(OH)CH_2CN$

(e) $(+)$-$CH_3CH_2C-^{18}OCH(CH_3)C_2H_5 + OH^- \longrightarrow CH_3CH_2COO^-$
$\overset{\displaystyle \|}{}$
$O + CH_3CH_2CH^{18}OHCH_3$

(f) $(-)$-$CH_3CH_2CHBrCH_3 + C_2H_5O^- Na^+ \longrightarrow C_2H_5-O-CH(CH_3)CH_2CH_3$

(g) $(+)$-$CH_3CH_2CHOHCH_3 \xrightarrow{\text{Na}} CH_3CH_2CH(ONa)CH_3 \xrightarrow{C_2H_5Br}$
$ C_2H_5-O-CH(CH_3)CH_2CH_3$

Problem 4.20 What general conclusion must you draw from each of the following observations? (a) After standing in an aqueous acidic solution, optically active $CH_3CH_2CHOHCH_3$ is found to have lost its optical activity. (b) After standing in solution with potassium iodide, optically active $n\text{-}C_6H_{13}CHICH_3$ is found to have lost its optical activity. (c) Can you suggest experiments to test your conclusions? (See Sec. 3.29.)

4.25   Optical purity

Reactions in which bonds to chiral centers are not broken can be used to get one more highly important kind of information: the specific rotations of optically pure compounds. For example, the 2-methyl-1-butanol obtained from fusel oil (which happens to have specific rotation $-5.90°$) is *optically pure*—like most chiral compounds from biological sources—that is, it consists entirely of the one enantiomer, and contains none of its mirror image. When this material is treated with hydrogen chloride, the 1-chloro-2-methylbutane obtained is found to have specific rotation of $+1.67°$. Since no bond to the chiral center is broken, every molecule of alcohol with configuration III is converted into a molecule of chloride with configuration IV; since the alcohol was optically pure, the chloride of specific rotation $+1.67°$ is also optically pure. Once this *maximum rotation* has been established, anyone can determine the optical purity of a sample of 1-chloro-2-methylbutane in a few moments by simply measuring its specific rotation.

If a sample of the chloride has a rotation of $+0.835°$, that is, 50% of the maximum, we say that it is *50% optically pure*. We consider the components of the mixture to be $(+)$ isomer and $(\pm)$ isomer (not $(+)$ isomer and $(-)$ isomer). (*Problem*: What are the percentages of $(+)$ isomer and $(-)$ isomer in this sample?)

Problem 4.21 Predict the specific rotation of the chloride obtained by treatment with hydrogen chloride of 2-methyl-1-butanol of specific rotation $+3.54°$.

4.26   Reactions of chiral molecules. Generation of a second chiral center

Let us return to the reaction we used as our example in Sec. 4.23, free-radical chlorination of *sec*-butyl chloride, but this time focus our attention on one of the other products, one in which a second chiral center is generated: 2,3-dichlorobutane. This compound, we have seen (Sec. 4.18), exists as three stereoisomers, *meso* and a pair of enantiomers.

$$\underset{\text{sec-Butyl chloride}}{CH_3CH_2-\overset{*}{\underset{|}{\underset{Cl}{CH}}}-CH_3} \xrightarrow{\text{Cl}_2,\text{ heat or light}} \underset{\text{2,3-Dichlorobutane}}{CH_3-\overset{*}{\underset{|}{\underset{Cl}{CH}}}-\overset{*}{\underset{|}{\underset{Cl}{CH}}}-CH_3} + \text{ other products}$$

Let us suppose that we take optically active *sec*-butyl chloride (the S isomer, say), carry out the chlorination, and by fractional distillation separate the 2,3-dichlorobutanes from all the other products (the 1,2 isomer, 2,2 isomer, etc.). Which stereoisomers can we expect to have?

Figure 4.4 Generation of a second chiral center. The configuration at the original chiral center is unchanged. Chlorine becomes attached via (*a*) or (*b*) to give diastereomers, and in unequal amounts.

Diastereomers
Formed in unequal amounts

Figure 4.4 shows the course of reaction. Three important points are illustrated:

(1) Since no bond to the original chiral center, C–2, is broken, its configuration is retained in all the products. This must apply in all cases where a second chiral center is generated.

(2) There are two possible configurations about the new chiral center, C–3, and both of these appear; in this particular case, they result from attacks (*a*) and (*b*) on opposite sides of the flat portion of the free radical, giving the diastereomeric *S,S* and *R,S* (or *meso*) products.

In some reactions, both configurations may not actually be generated, but we must always consider the *possibility* that they will be; this is our point of departure. (We shall return to this point in Sec. 9.1.)

(3) The diastereomeric products will be formed in unequal amounts, in this

case because attack (*a*) and attack (*b*) are not equally likely. This must apply in all cases where diastereomeric products are formed.

In Sec. 4.22 we saw that generation of the first chiral center in a compound yields equal amounts of enantiomers, that is, yields an optically inactive racemic modification. Now we see that generation of a new chiral center in a compound that is already optically active yields an optically active product containing unequal amounts of diastereomers.

Suppose (as is actually the case) that the products from (*S*)-*sec*-butyl chloride show an *S,S*:*meso* ratio of 29:71. What would we get from chlorination of (*R*)-*sec*-butyl chloride? We would get *R,R* and *meso* products, and the *R,R*:*meso* ratio would be exactly 29:71. Whatever factor favors *meso* product over *S,S* product will favor *meso* product over *R,R* product, and to exactly the same extent.

Finally, what can we expect to get from optically inactive, racemic *sec*-butyl chloride? The *S* isomer that is present would yield *S,S* and *meso* products in the ratio of 29:71; the *R* isomer would yield *R,R* and *meso* products, and in the ratio of 29:71. Since there are exactly equal quantities of *S* and *R* reactants, the two sets of products would exactly balance each other, and we would obtain racemic and *meso* products in the ratio of 29:71. Optically inactive reactants yield optically inactive products.

One point requires further discussion. Why are the diastereomeric products formed in unequal amounts? It is because the intermediate 3-chloro-2-butyl radical in Fig. 4.4 already contains a chiral center. The free radical is chiral, and lacks the symmetry that is necessary for attack at the two faces to be equally likely. (Make a model of the radical and assure yourself that this is so.)

Problem 4.22 Each of the following reactions is carried out, and the products are separated by careful fractional distillation or recrystallization. For each reaction tell how many fractions will be collected. Draw stereochemical formulas of the compound or compounds making up each fraction, and give each its *R/S* specification. Tell whether each fraction, as collected, will show optical activity or optical inactivity.

(a) monochlorination of (*R*)-*sec*-butyl chloride at 300 °C
(b) monochlorination of racemic *sec*-butyl chloride at 300 °C
(c) monochlorination of racemic 1-chloro-2-methylbutane at 300 °C

4.27   Reactions of chiral molecules with optically active reagents. Resolution

So far in this chapter we have discussed the reactions of chiral compounds only with optically inactive reagents. Now let us turn to reactions with optically *active* reagents, and examine one of their most useful applications: **resolution of a racemic modification**, that is, *the separation of a racemic modification into enantiomers.*

We know (Sec. 4.22) that, when optically inactive reactants form a chiral compound, the product is the racemic modification. We know that the enantiomers making up a racemic modification have identical physical properties (except for direction of rotation of polarized light), and hence cannot be separated by the usual methods of fractional distillation or fractional crystallization. Yet throughout this book are frequent references to experiments carried out using optically active compounds like (+)-*sec*-butyl alcohol, (−)-2-bromooctane, (−)-α-phenylethyl chloride, $(+)$-α-phenylpropionamide. How are such optically active compounds obtained?

Some optically active compounds are obtained from natural sources, since living organisms usually produce only one enantiomer of a pair. Thus only $(-)$-2-methyl-1-butanol is formed in the yeast fermentation of starches, and only $(+)$-lactic acid, $CH_3CHOHCOOH$, in the contraction of muscles; only $(-)$-malic acid, $HOOCCH_2CHOHCOOH$, is obtained from fruit juices, and only $(-)$-quinine from the bark of the cinchona tree. Indeed, we deal with optically active substances to an extent that we may not realize. We eat optically active bread and optically active meat, live in houses, wear clothes, and read books made of optically active cellulose. The proteins that make up our muscles and other tissues, the glycogen in our liver and glucose in our blood, the enzymes and hormones that enable us to grow and that regulate our bodily processes—all these are optically active. Naturally occurring compounds are optically active because the enzymes that bring about their formation—and often the raw materials from which they are made—are themselves optically active. As to the origin of the optically active enzymes, we can only speculate.

Amino acids, the units from which proteins are made, have been reported present in meteorites, but in such tiny amounts that the speculation has been made that "what appears to be the pitter-patter of heavenly feet is probably instead the print of an earthly thumb". Part of the evidence that the amino acids found in a meteorite by Cyril Ponnamperuma (of the University of Maryland) are really extraterrestrial in origin is that they are optically *inactive*—not optically active as earthly contaminants from biological sources would be.

From these naturally occurring compounds, other optically active compounds can be made. We have already seen, for example, how $(-)$-2-methyl-1-butanol can be converted without loss of configuration into the corresponding chloride or acid (Sec. 4.24); these optically active compounds can, in turn, be converted into many others.

Most optically active compounds are obtained by the resolution of a racemic modification, that is, by a separation of a racemic modification into enantiomers. Most such resolutions are accomplished through the use of reagents that are themselves optically active; these reagents are generally obtained from natural sources.

The majority of resolutions that have been carried out depend upon the reaction of organic bases with organic acids to yield salts. Let us suppose, for example, that we have prepared the racemic acid, $(\pm)$-HA. Now, there are isolated from various plants very complicated bases called *alkaloids* (that is, *alkali-like*), among which are cocaine, morphine, strychnine, and quinine. Most alkaloids are produced by plants in only one of two possible enantiomeric forms, and hence they are optically active. Let us take one of these optically active bases, say a levorotatory one, $(-)$-B, and mix it with our racemic acid $(\pm)$-HA. The acid is present in two configurations, but the base is present in only one configuration; there will result, therefore, crystals of two different salts, $[(-)\text{-BH}^+ \ (+)\text{-A}^-]$ and $[(-)\text{-BH}^+ \ (-)\text{-A}^-]$.

What is the relationship between these two salts? They are not superimposable, since the acid portions are not superimposable. They are not mirror images, since the base portions are not mirror images. The salts are stereoisomers that are not enantiomers, and therefore are *diastereomers*.

These diastereomeric salts have, of course, different physical properties, including solubility in a given solvent. They can therefore be separated by frac-

$$\xrightarrow{\text{H}^+} \quad (+)\text{-HA} \quad + \quad (-)\text{-BH}^+$$

$$
\begin{array}{c}
(+)\text{-HA} \\
\\
(-)\text{-HA}
\end{array}
\quad + \quad (-)\text{-B} \longrightarrow
\quad
\begin{array}{c}
[(-)\text{-BH}^+ \, (+)\text{-A}^-] \\
\\
[(-)\text{-BH}^+ \, (-)\text{-A}^-]
\end{array}
$$

$$\xrightarrow{\text{H}^+} \quad (-)\text{-HA} \quad + \quad (-)\text{-BH}^+$$

Enantiomers:	Alkaloid	Diastereomers:	Resolved	Alkaloid
in a racemic modification	*base*	*separable*	enantiomers	*as a salt*

tional crystallization. Once the two salts are separated, optically active acid can be recovered from each salt by addition of strong mineral acid, which displaces the weaker organic acid. If the salt has been carefully purified by repeated crystallizations to remove all traces of its diastereomer, then the acid obtained from it is *optically pure*. Among the alkaloids commonly used for this purpose are (−)-brucine, (−)-quinine, (−)-strychnine, and (+)-cinchonine.

Resolution of organic bases is carried out by reversing the process just described: using naturally occurring optically active acids, (−)-malic acid, for example. Resolution of alcohols, which we shall find to be of special importance in synthesis, poses a special problem: since alcohols are neither appreciably basic nor acidic, they cannot be resolved by direct formation of salts. Yet they can be resolved by a rather ingenious adaptation of the method we have just described: one attaches to them an acidic "handle", which permits the formation of salts, and then when it is no longer needed can be removed.

Compounds other than organic bases, acids, or alcohols can also be resolved. Although the particular chemistry may differ from the salt formation just described, the principle remains the same: **a racemic modification is converted by an optically active reagent into a mixture of diastereomers which can then be separated.**

4.28 Reactions of chiral molecules. Mechanism of free-radical chlorination

So far, we have discussed only reactions of chiral molecules in which bonds to the chiral center are not broken. What is the stereochemistry of reactions in which the bonds to the chiral center *are* broken? The answer is: *it depends*. It depends upon the *mechanism* of the reaction that is taking place; because of this, stereochemistry can often give us information about a reaction that we cannot get in any other way.

For example, stereochemistry played an important part in establishing the mechanism that was the basis of our entire discussion of the halogenation of alkanes (Chap. 3). The chain-propagating steps of this mechanism are:

(2a) $$X \cdot \; + \; R{-}H \; \longrightarrow \; H{-}X \; + \; R \cdot$$

(3a) $$R \cdot \; + \; X_2 \; \longrightarrow \; R{-}X \; + \; X \cdot$$

Until 1940 the existing evidence was just as consistent with the following alternative steps:

(2b) $X\cdot\ +\ R{-}H\ \longrightarrow\ R{-}X\ +\ H\cdot$

(3b) $H\cdot\ +\ X_2\ \longrightarrow\ H{-}X\ +\ X\cdot$

To differentiate between these alternative mechanisms, H. C. Brown, M. S. Kharasch, and T. H. Chao, working at the University of Chicago, carried out the photochemical halogenation of optically active (S)-$(+)$-1-chloro-2-methylbutane. A number of isomeric products were, of course, formed, corresponding to attack at various positions in the molecule. (*Problem*: What were these products?) They focused their attention on just *one* of these products: 1,2-dichloro-2-methylbutane, resulting from substitution at the chiral center (C–2).

They had planned the experiment on the following basis. The two mechanisms differed as to whether or not a free alkyl radical is an intermediate. The most likely structure for such a radical, they thought, was *flat*—as, it turns out, it very probably is—and the radical would lose the original chirality. Attachment of chlorine to either face would be equally likely, so that an optically inactive, racemic product would be formed. That is to say, the reaction would take place *with racemization* (see Fig. 4.5).

Figure 4.5   Racemization through free-radical formation. Chlorine becomes attached to either face of the free radical, via (*a*) or (*b*), to give enantiomers, and in equal amounts.

For the alternative mechanism, in which chlorine would become attached to the molecule while the hydrogen was being displaced, they could make no prediction, except that formation of an optically inactive product would be highly unlikely: there was certainly no reason to expect that *back-side* attack (on the face opposite the hydrogen) would take place to exactly the same extent as *front-side* attack. (In heterolytic displacements, attack is generally back-side.)

By careful fractional distillation they separated the 1,2-dichloro-2-methyl-butane from the reaction mixture, and found it to be *optically inactive*. From this they concluded that the mechanism involving free alkyl radicals, (2a), (3a), is the correct one. This mechanism is accepted without question now; in Sec. 2.21 we saw how the relative strengths of the hydrogen–chlorine and carbon–chlorine bonds force the reaction to follow this course. Today, the work of Brown, Kharasch, and Chao is frequently referred to as evidence of the stereochemical behavior of free radicals, with the original significance of the work exactly reversed.

We can begin to see how stereochemistry provides the organic chemist with one of the most powerful tools for finding out what is going on in a chemical reaction.

Problem 4.23 This work does *not* prove that free radicals are flat. Racemization is consistent with what other structure for free radicals? Explain. (*Hint*: See Sec. 2.22.)

Problem 4.24 Altogether, the free-radical chlorination of (S)-(+)-1-chloro-2-methylbutane gave six fractions of formula $C_5H_{10}Cl_2$. Four fractions were found to be optically active, and two fractions optically inactive. Draw structural formulas for the compounds making up each fraction. Account in detail for optical activity or inactivity in each case.

PROBLEMS

1. What is meant by each of the following?

(a) optical activity
(b) dextrorotatory
(c) levorotatory
(d) specific rotation
(e) chirality
(f) chiral molecule
(g) chiral center
(h) superimposable
(i) enantiomers
(j) diastereomers

(k) *meso* compound
(l) racemic modification
(m) configuration
(n) conformations
(o) *R*
(p) *S*
(q) +
(r) −
(s) configurational isomers
(t) conformational isomers

2. (a) What is the necessary and sufficient condition for enantiomerism? (b) What is a necessary but not a sufficient condition for optical activity? (c) What conditions must be met for the observation of optical activity? (d) How can you tell from its formula whether or not a compound can exist as enantiomers? (e) What restrictions, if any, must be applied to the use of planar formulas in (d)? To the use of models in (d)? (f) Exactly how do you go about deciding whether a molecule should be specified as *R* or as *S*?

3. Compare the dextrorotatory and levorotatory forms of *sec*-butyl alcohol, $CH_3CH_2CHOHCH_3$, with respect to:

(a) boiling point
(b) melting point
(c) relative density

(d) specific rotation *this cod ?k*
(e) refractive index
(f) solubility in 100 g of water

(g) rate of reaction with HBr
(h) infrared spectrum
(i) NMR spectrum

(j) adsorption on alumina
(k) retention time in gas chromatography
(l) specification as R or S

4. Which of the following objects are chiral?

(a) nail, screw, pair of scissors, knife, spool of thread
(b) glove, shoe, sock, pullover sweater, coat sweater, scarf tied around your neck
(c) child's block, rubber ball, Pyramid of Cheops, helix (p. 1371), double helix (p. 1397)
(d) basketball, football, tennis racket, golf club, baseball bat, shotgun barrel, rifle barrel
(e) your hand, your foot, your ear, your nose, yourself

5. Assuming both your hands to be of equal strength and skill, which of the following operations could you perform with equal speed and efficiency?

(a) driving a screw, sawing a board, drilling a hole
(b) opening a door, opening a soft-drink can, opening a coffee jar, turning on the hot water
(c) signing your name, sharpening a pencil, throwing a ball, shaking hands with someone's right hand, turning to page 162

6. Draw and specify as R or S the enantiomers (if any) of:

(a) 3-bromohexane
(b) 3-chloro-3-methylpentane
(c) 1,2-dibromo-2-methylbutane
(d) 1,3-dichloropentane

(e) 3-chloro-2,2,5-trimethylhexane
(f) 1-deuterio-1-chlorobutane,
     $CH_3CH_2CH_2CHDCl$

7. (a) What is the lowest-molecular-weight alkane that is chiral? Draw stereochemical formulas of the enantiomers and specify each as R or S. (b) Is there another alkane of the same molecular weight that is also chiral? If there is, give its structure and name, and specify the enantiomers as R or S.

8. Draw stereochemical formulas for all the possible stereoisomers of the following compounds. Label pairs of enantiomers, and *meso* compounds. Tell which isomers, if separated from all other stereoisomers, will be optically active. Give one isomer of each set its R/S specification.

(a) $CH_3CHBrCHOHCH_3$
(b) $CH_3CHBrCHBrCH_2Br$
(c) $C_6H_5CH(CH_3)CH(CH_3)C_6H_5$
(d) $CH_3CH_2CH(CH_3)CH_2CH_2CH(CH_3)CH_2CH_3$
(e) $CH_3CH(C_6H_5)CHOHCH_3$
(f) $CH_3CHOHCHOHCHOHCH_2OH$

(g) $HOCH_2(CHOH)_3CH_2OH$
(h) CH_2-CHCl
     | | (*Make models.*)
   CH_2-CHCl
(i) CH_2-CHCl
     | |
   $CHCl-CH_2$

(j) *n*-butylethylmethyl-*n*-propylammonium chloride, $(RR'R''R'''N)^+Cl^-$ (see Sec. 1.12)
(k) *sec*-butylethylmethyl-*n*-propylammonium chloride

9. (a) In a study of chlorination of propane, four products (A, B, C, and D) of formula $C_3H_6Cl_2$ were isolated. What are their structures?

(b) Each was chlorinated further, and the number of trichloro products $(C_3H_5Cl_3)$ obtained from each was determined by gas chromatography. A gave one trichloro product; B gave two; and C and D each gave three. What is the structure of A? Of B? Of C and D?

(c) By another synthetic method, compound C was obtained in optically active form. Now what is the structure of C? Of D?

(d) When optically active C was chlorinated, one of the trichloropropanes obtained was optically active, and the other two were optically inactive. What is the structure of the optically active one? Of the other two?

10. Draw configurational isomers (if any) of: (a) CH_2BrCH_2Cl; (b) $CH_3CHBrCH_2Cl$. (c) For each substance of (a) and (b), draw all conformers. Label pairs of conformational enantiomers.

11. The more stable conformer of *n*-propyl chloride, $CH_3CH_2-CH_2Cl$, is the *gauche*. What does this indicate about the interaction between $-Cl$ and $-CH_3$? How do you account for this interaction? (*Hint*: See Sec. 1.19.)

12. (a) What must be the dipole moment of the *anti* conformation of 1,2-dichloro-ethane, $CH_2Cl\!-\!CH_2Cl$? (b) At 32 °C in the gas phase, the measured dipole moment of 1,2-dichloroethane is 1.12 D. What does this single fact tell you about the conformational make-up of the compound? (c) The dipole moment of a mixture of X and Y is given by the expression

$$\mu^2 = N_X\mu_X^2 + N_Y\mu_Y^2$$

where N is the mole fraction of each kind of molecule. From bond moments, it has been estimated that the *gauche* conformation of 1,2-dichloroethane should have a dipole moment of about 3.2 D. Calculate the conformational composition of 1,2-dichloroethane at 32 °C in the gas phase.

13. It is February, 1929. In a lonely cottage in Devonshire, George Harrison, a middle-aged amateur mycologist, has died shortly after eating a mushroom stew he prepared from warty caps (*Amanita rubescens*) collected in nearby Five-Acre Wood. Cause of death: poisoning by *muscarine*, an alkaloid found in the fly agaric (*Amanita muscaria*).

Muscarine
An alkaloid found in
the mushroom *Amanita muscaria*

You are Sir James Lubbock, Home Office Analyst, and you have been asked to help solve a knotty problem crucial to the investigation: whether (a) a deadly *Amanita muscaria* found its way accidentally into the mess of closely similar, but harmless, *Amanita rubescens*; or (b) a lethal dose of synthetic muscarine (filched from a London laboratory) was deliberately added to the stew pot—perhaps by the lover of beautiful Mrs. Harrison.

You have available a solution of muscarine that you isolated from left-over stew, a well-equipped (for 1929) laboratory, and ten minutes. Tell what you can do that might give a definite answer to the question: was there a fly agaric in Mr. Harrison's soup—or did a second cook, wilfully and with malice aforethought, spoil the broth?

14. Each of the following reactions is carried out, and the products are separated by careful fractional distillation or recrystallization. For each reaction tell how many fractions will be collected. Draw stereochemical formulas of the compound or compounds making up each fraction, and give each its R/S specification. Tell whether each fraction, as collected, will show optical activity or optical inactivity.

(a) *n*-pentane + Cl_2 (300 °C) $\longrightarrow$ $C_5H_{11}Cl$
(b) 1-chloropentane + Cl_2 (300 °C) $\longrightarrow$ $C_5H_{10}Cl_2$
(c) (*S*)-2-chloropentane + Cl_2 (300 °C) $\longrightarrow$ $C_5H_{10}Cl_2$
(d) (*R*)-2-chloro-2,3-dimethylpentane + Cl_2 (300 °C) $\longrightarrow$ $C_7H_{14}Cl_2$
(e) *meso*-$HOCH_2CHOHCHOHCH_2OH$ + HNO_3 $\longrightarrow$ $HOCH_2CHOHCHOHCOOH$
(f) (*S*)-3-chloro-1-butene + HCl $\longrightarrow$ 2,3-dichloro-2-methylbutane
(g) racemic $C_6H_5COCHOHC_6H_5$ + H_2/Ni, catalyst $\longrightarrow$ $C_6H_5CHOHCHOHC_6H_5$

15. Give the absolute configuration and R/S specification of compounds E–K.

(a) (*R*)-$HOCH_2CHOHCH\!=\!CH_2$ + cold alkaline $KMnO_4$ $\longrightarrow$ E (optically active) + F (optically inactive); E and F are $HOCH_2CHOHCHOHCH_2OH$
(b) (*S*)-1-chloro-2-methylbutane + Li, then + CuI $\longrightarrow$ G
(c) G + (*S*)-1-chloro-2-methylbutane $\longrightarrow$ H
(d) (*R,R*)-$HOCH_2CHOHCHOHCH_2OH$ + HBr $\longrightarrow$ I ($HOCH_2CHOHCHOHCH_2Br$)
(e) (*R*)-2-ethyl-3-methyl-1-pentene ($CH_3CH_2CH(CH_3)C(C_2H_5)\!=\!CH_2$) + H_2/Ni $\longrightarrow$ J (optically active) + K (optically inactive); both J and K are C_8H_{18}

16. An excess of the racemic acid $CH_3CHClCOOH$ is allowed to react with (S)-2-methyl-1-butanol to form the ester,

$$CH_3CHClC-OCH_2CH(CH_3)CH_2CH_3$$
$$\overset{\displaystyle \|}{\underset{\displaystyle O}{}}$$

and the reaction mixture is carefully distilled. Three fractions are obtained, each of which is optically active. Draw stereochemical formulas of the compound or compounds making up each fraction.

5

Alkyl Halides

Nucleophilic Aliphatic Substitution

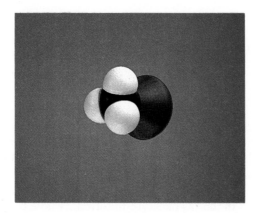

5.1 Homolytic and heterolytic chemistry

Chemistry owes its existence as a science, of course, to the chemical change: the conversion of one substance into another. Old molecules are changed into new ones, which means that old bonds must be broken, and new bonds must be formed—covalent bonds, mostly, in the case of organic chemistry.

Now, the breaking of a covalent bond, we have seen (Sec. 1.14), can take place in two fundamentally different ways, depending upon what happens to the two electrons making up the bonding pair: in homolysis, one electron goes to each fragment; in heterolysis, both electrons go to the same fragment. The nouns "homolysis" and "heterolysis" are used only in their literal sense (p. 22) to mean bond-*breaking*. But the adjectives "homolytic" and "heterolytic"—for want of better words—are used in a broader sense, to include the bond-making process as well, and so define two broad classes of organic reactions.

Thus, **homolytic reactions** are those in which the electrons of the bonding pair are taken away—or provided—*singly*. Whether bonds are being broken,

$$A:B \longrightarrow A\cdot + \cdot B$$

or formed,

$$A\cdot + \cdot B \longrightarrow A:B$$

165

or simultaneously broken and formed,

$$C\cdot + A:B \longrightarrow C:A + \cdot B$$

each of the atoms being separated takes one of the bonding electrons, and each of the atoms being joined together provides one of the bonding electrons.

Heterolytic reactions are those in which the bonding electrons are taken away—or provided—*in pairs*. Whether bonds are being broken,

$$A:B \longrightarrow A + :B$$

or formed,

$$A + :B \longrightarrow A:B$$

or simultaneously broken and formed,

$$C: + A:B \longrightarrow C:A + :B$$

one of the atoms being separated takes both bonding electrons, and one of the atoms being joined together provides both electrons.

Homolytic chemistry is thus the chemistry of the odd electron; heterolytic chemistry is the chemistry of the electron pair. Where homolytic chemistry deals with the neutral particles called free radicals, heterolytic chemistry deals with positive and negative charges, with cations and anions. Homolytic reactions are typically carried out in the gas phase, or in solvents whose principal function is to provide an inert medium in which the reacting molecules can move about. Heterolytic reactions are typically carried out in solution; and the solvents, as we shall see, exert powerful effects—just how powerful is only now being realized.

So far, the reaction we have been chiefly concerned with—free-radical substitution, as exemplified by the halogenation of alkanes—is a part of homolytic chemistry. Now let us begin our study of heterolytic chemistry. The larger part of organic chemistry is heterolytic, and it is the kind that will take up most of our time in the remainder of this book. The reaction we shall start with is, like halogenation, substitution, but of a quite different kind: heterolytic, and of the specific type called *nucleophilic aliphatic substitution*.

5.2 Relative rates of competing reactions

In our study of nucleophilic substitution, we shall enormously broaden our understanding of the principle underlying all of organic chemistry: that chemical behavior depends upon molecular structure. Before we go further, let us remind ourselves of how, in general, we approach the matter of chemical behavior.

In a reaction vessel there is a collection of molecules, banging blindly about, colliding with one another. In principle, a number of options are open to them: a number of reactions that they can conceivably undergo. Which of these reactions actually takes place—or, at least, predominates—is the one that *goes fastest*. Chemical behavior thus comes down to a matter of *relative rates of competing reactions*.

As we have seen from our study of halogenation of alkanes, relative rates of reaction thus determine

(a) *what* happens: a halogen atom, for example, attaches itself to a hydrogen of methane and abstracts the hydrogen; the alternative, attachment to carbon with expulsion of a hydrogen atom, is vastly slower and, in effect, does not take place (Sec. 2.21).

(b) *where* it happens: a halogen atom abstracts hydrogen from ethane in preference to methane; it abstracts a tertiary hydrogen in preference to a primary hydrogen (Sec. 3.23).

(c) even *whether* it happens: a chlorine atom abstracts hydrogen from an alkane; an iodine atom does not, because it recombines with another iodine atom faster (Sec. 2.20).

A chemical reaction is, then, the result of a *competition*; it is a race that is won by the fastest runner. And, we have learned, the most important factor determining how fast a reaction goes is the energy of activation. What our collection of molecules tend to do, by and large, is *what is easiest for them*. They follow the course that makes the smallest demand for energy; that is, they undergo the reaction with the smallest E_{act}.

And, finally, to help us understand E_{act}—to interpret and, sometimes, even to predict—we have our all-important intellectual tool, the transition state. The more stable the transition state relative to the reactant, then the smaller the E_{act} and the faster the reaction. It is the concept of the transition state that is our mental link between molecular structure and chemical behavior.

What we have said above is based on the premise that the products we *obtain* from a chemical reaction, and their relative proportions, reflect the relative rates at which they are initially *formed*; that is to say, once formed, a particular product sits and waits unchanged for the completion of reaction. For most of the reactions we study this premise is correct; under the conditions employed, most organic reactions are essentially irreversible, that is, they are one-way reactions.

But this is *not always* the case. Some reactions are reversible, and equilibrium exists among the various products; what we then obtain reflects, not which product is initially formed faster, but which product is eventually favored by the equilibrium. We shall see examples of this kind of behavior. Irreversibility, therefore, is not something that can be simply assumed for an organic reaction. It must be established by experiment; and only when it has been established are we justified in interpreting product composition on the basis of relative rates.

In our study of nucleophilic substitution, we shall have much to do with competition between reaction pathways: in this chapter, competition between different mechanisms for substitution itself; in later chapters, competition between nucleophilic substitution and a reaction of a quite different type, elimination. We shall be concerned with the relative rates at which these competing reactions take place, and the kinds of transition states they pass through. Most important, we shall learn about the factors that determine the stability of these transition states, factors that we shall work with throughout the rest of our study of organic chemistry.

5.3   Structure of alkyl halides. The functional group

In this introduction to nucleophilic substitution, we shall deal chiefly with a family of compounds already familiar to us, the *alkyl halides*. Alkyl halides have the general formula RX, in which R is an alkyl or substituted alkyl group.

$$R—X$$

An alkyl halide

The characteristic feature of the alkyl halide structure is the halogen atom, —X, and the characteristic reactions of an alkyl halide are those that take place at the halogen atom. *The atom or group of atoms that defines the structure of a particular family of organic compounds and, at the same time, determines their properties is called the* **functional group**.

In alkyl halides the functional group is the halogen atom. We must not forget that an alkyl halide has an alkyl group attached to this functional group; under the proper conditions, the alkyl portion of these molecules undergoes the reactions typical of alkanes. However, the reactions that are *characteristic* of the family are the ones that occur at the halogen atom.

A large part of organic chemistry is therefore the chemistry of the various functional groups. We shall learn to associate a particular set of properties with a particular group wherever we may find it. When we encounter a complicated molecule, which contains a number of different functional groups, we may expect the properties of this molecule to be roughly a composite of the properties of the various functional groups. A compound that contains both —X and —OH, for example, is both an alkyl halide and an alcohol; depending upon experimental conditions, it may undergo reactions characteristic of either kind of compound. The properties of one group may be modified, of course, by the presence of another group, and it is important for us to understand these modifications; but our point of departure is the chemistry of individual functional groups.

In this chapter, we shall take up alkyl halides in a systematic way. We shall outline their chemistry, and then concentrate on their most important reaction: nucleophilic substitution.

5.4 Classification and nomenclature of alkyl halides

We classify a carbon atom as *primary*, *secondary*, or *tertiary*, according to the number of other carbon atoms attached to it (Sec. 3.11). An alkyl halide is classified according to the kind of carbon that bears the halogen:

H	R	R
|	|	|
R—C—X	R—C—X	R—C—X
|	|	|
H	H	R
Primary	Secondary	Tertiary
(1°)	(2°)	(3°)

As members of the same family, containing the same functional group, alkyl halides of different classes tend to undergo the same kinds of reactions. They differ in rates of reaction, however, and these differences in rates may lead to other, deeper differences.

As we have seen (Secs. 3.8 and 3.10), alkyl halides can be given two kinds of names: **common names** (for the simpler halides); and **IUPAC names**, in which the compound is simply named as an alkane with a halogen attached as a side chain.

For example:

$$CH_3CH_2CH_2CH_2Br$$

n-Butyl bromide

1-Bromobutane

(1°)

$$CH_3CHCH_3$$

Cl

Isopropyl chloride

2-Chloropropane

(2°)

$$CH_3$$

$$CH_3CHCH_2Cl$$

Isobutyl chloride

1-Chloro-2-methylpropane

(1°)

$$CH_3$$

$$CH_3CH_2\text{—}C\text{—}CH_3$$

I

tert-Pentyl iodide

2-Iodo-2-methylbutane

(3°)

$$CH_3$$

$$CH_3CH_2CHCHCH_3$$

Cl

3-Chloro-2-methylpentane

(2°)

$$CH_3 \quad CH_3$$

$$CH_3CH_2CCH_2CHCH_3$$

Br

4-Bromo-2,4-dimethylhexane

(3°)

We should notice that similar names do not always mean the same classification; for example, isopropyl chloride is a secondary chloride, whereas isobutyl chloride is a primary chloride.

Problem 5.1   Label as primary, secondary, or tertiary each of the isomeric chloropentanes whose structures you drew in Problem 3.5, p. 85.

5.5   Alcohols: a brief introduction

We cannot go very far into organic chemistry of almost any kind without encountering another family of compounds: the *alcohols*. The chemistry of alkyl halides is no exception. To begin with, alcohols are the compounds from which alkyl halides—and, in fact, most substances undergoing nucleophilic substitution— are nearly always made. Next, once we have made alkyl halides and are allowing them to undergo nucleophilic substitution, we encounter alcohols again playing a variety of roles: they may be the reagents with which alkyl halides react, or the products, or the solvents in which reaction takes place. (Indeed, as we shall see, the preparation of alkyl halides involves nucleophilic substitution, too, this time with alcohols as the reactants undergoing substitution.)

We shall treat the chemistry of alcohols systematically later (Chaps. 17 and 18). But for now let us learn just enough about these compounds to enable us to work with them.

Alcohols are compounds of the general formula ROH, where R is any alkyl group and OH is the hydroxyl group. A simple alcohol can be named just by naming the alkyl group that holds the hydroxyl group and following this with the word *alcohol*. The IUPAC name is derived by replacing the terminal *-e* of the corresponding alkane by *-ol*. (This nomenclature is discussed in detail in Sec. 17.4, which you should consult whenever it seems necessary.) Like an alkyl halide, an alcohol is classified as *primary*, *secondary*, or *tertiary*, according to the kind of carbon that bears the —OH group.

CH_3CH_2OH

Ethyl alcohol

Ethanol

(1°)

CH_3CHCH_3
|
OH

Isopropyl alcohol

2-Propanol

(2°)

CH_3
|
CH_3CHCH_2OH

Isobutyl alcohol

2-Methyl-1-propanol

(1°)

CH_3
|
CH_3-C-CH_3
|
OH

tert-Butyl alcohol

2-Methyl-2-propanol

(3°)

Of the varied chemical properties of alcohols, there is just one pair that we should become acquainted with at this point: their *acidity* and *basicity*. These properties reside in the hydroxyl group, —OH, the functional group of alcohols. This group is like the hydroxyl group of water, a compound with which we are already familiar. Like water, alcohols are weak acids and weak bases—roughly, about as acidic and about as basic as water.

Like water, alcohols are acidic enough to react with active metals to liberate hydrogen gas.

$$ROH + Na \longrightarrow RONa + \tfrac{1}{2}H_2$$

An
alcohol

A weak acid

A sodium
alkoxide

A strong base

The products are called *alkoxides*: sodium ethoxide, for example, or potassium *n*-propoxide. Alkoxides are strong bases, like sodium hydroxide.

Like water, alcohols are basic enough to accept a proton from strong acids like hydrogen chloride and hydrogen sulfate, and thus bring about complete dissociation of these compounds. For example:

$$ROH + H_2SO_4 \rightleftharpoons ROH_2^+ + HSO_4^-$$

Alcohol

*Stronger
base*

Protonated
alcohol

*Weaker
base*

Problem 5.2 Write equations for the reactions you would expect to take place between:

(a) ethanol and *n*-propylmagnesium bromide
(b) sodium ethoxide (C_2H_5ONa) and aqueous sulfuric acid
(c) methanol (CH_3OH) and dry hydrogen chloride
(d) *tert*-butyl alcohol and potassium metal

5.6 Physical properties of alkyl halides

Because of their greater molecular weights, haloalkanes have considerably higher boiling points (Table 5.1) than alkanes with the same number of carbons. For a given alkyl group, the boiling point increases with increasing atomic weight of the halogen, so that a fluoride is the lowest boiling, an iodide the highest boiling.

For a given halogen, the boiling point rises with increasing carbon number; as with alkanes, the boiling point rise is 20–30 degrees for each added carbon, except for the very small homologs. As before, branching—involving either alkyl groups or the halogen itself—lowers the boiling point.

Table 5.1 Alkyl Halides

Name	Chloride		Bromide		Iodide	
	B.p., °C	Relative density at 20 °C	B.p., °C	Relative density at 20 °C	B.p., °C	Relative density at 20 °C
Methyl	−24		5		43	2.279
Ethyl	12.5		38	1.440	72	1.933
n-Propyl	47	0.890	71	1.335	102	1.747
n-Butyl	78.5	.884	102	1.276	130	1.617
n-Pentyl	108	.883	130	1.223	157	1.517
n-Hexyl	134	.882	156	1.173	180	1.441
n-Heptyl	160	.880	180		204	1.401
n-Octyl	185	.879	202		225.5	
Isopropyl	36.5	.859	60	1.310	89.5	1.705
Isobutyl	69	.875	91	1.261	120	1.605
sec-Butyl	68	.871	91	1.258	119	1.595
tert-Butyl	51	.840	73	1.222	100dec.	
Cyclohexyl	142.5	1.000	165			
Vinyl (Haloethene)	−14		16		56	
Allyl (3-Halopropene)	45	.938	71	1.398	103	
Crotyl (1-Halo-2-butene)	84				132	
Methylvinylcarbinyl (3-Halo-1-butene)	64					
Propargyl (3-Halopropyne)	65		90	1.520	115	
Benzyl	179	1.102	201		93[10]	
α-Phenylethyl	92[15]		85[10]			
β-Phenylethyl	92[20]		92[11]		127[19]	
Diphenylmethyl	173[19]		184[20]			
Triphenylmethyl	310		230[15]			
Dihalomethane	40	1.336	99	2.49	180dec.	3.325
Trihalomethane	61	1.489	151	2.89	subl.	4.008
Tetrahalomethane	77	1.595	189.5	3.42	subl.	4.32
1,1-Dihaloethane	57	1.174	110	2.056	179	2.84
1,2-Dihaloethane	84	1.257	132	2.180	dec.	2.13
Trihaloethylene	87		164	2.708		
Tetrahaloethylene	121				subl.	
Benzal halide	205		140[20]			
Benzotrihalide	221	1.38				

In spite of their modest polarity, alkyl halides are insoluble in water, probably because of their inability to form hydrogen bonds. They are soluble in the typical organic solvents of low polarity, like benzene, ether, chloroform, or ligroin.

Iodo, bromo, and polychloro compounds are more dense than water.

Alkanes and alkyl halides, then, have the physical properties we might expect of compounds of low polarity, whose molecules are held together by van der Waals forces or weak dipole–dipole attraction. They have relatively low melting points and boiling points, and are soluble in non-polar solvents and insoluble in water.

There is a further result of their low polarity: while alkanes and alkyl halides are themselves good solvents for other compounds of low polarity—each other, for example—they cannot solvate simple ions appreciably, and hence cannot dissolve inorganic salts.

5.7   Preparation of alkyl halides

In the laboratory alkyl halides are most often prepared by the methods outlined below.

───── PREPARATION OF ALKYL HALIDES ──────────────────────────

1. From alcohols. Discussed in Secs. 5.7 and 5.25.

$$R—OH \xrightarrow{\text{HX or PX}_3} R—X$$

Examples:

$$CH_3CH_2CH_2OH \xrightarrow[\substack{\text{or} \\ \text{NaBr, H}_2\text{SO}_4, \\ \text{heat}}]{\text{conc. HBr}} CH_3CH_2CH_2Br$$

n-Propyl alcohol   *n*-Propyl bromide

1-Phenylethanol   1-Bromo-1-phenylethane
α-Phenylethyl alcohol   α-Phenylethyl bromide

$$CH_3CH_2OH \xrightarrow{P + I_2} CH_3CH_2I$$

Ethyl alcohol   Ethyl iodide

2. Halogenation of certain hydrocarbons. Discussed in Secs. 3.19, 10.3, 15.13–15.14.

$$R—H \xrightarrow{X_2} R—X + HX$$

Examples:

Neopentane   Neopentyl chloride

Toluene   Benzyl bromide

3. Addition of hydrogen halides to alkenes. Discussed in Secs. 8.5–8.6.

CONTINUED

CONTINUED

4. **Addition of halogens to alkenes and alkynes**

$$-\overset{|}{C}{=}\overset{|}{C}- \quad \xrightarrow{X_2} \quad -\overset{|}{\underset{X}{C}}-\overset{|}{\underset{X}{C}}-$$

Discussed in Secs. 8.13–8.14 and 9.5–9.6.

$$-C{\equiv}C- \quad \xrightarrow{2X_2} \quad -\overset{X\ \ X}{\underset{X\ \ X}{\underset{|}{\overset{|}{C}}-\underset{|}{\overset{|}{C}}}}-$$

Discussed in Sec. 11.7.

5. **Halide exchange.** Discussed in Sec. 5.7.

$$R{-}X + I^- \quad \xrightarrow{\text{acetone}} \quad R{-}I + X^-$$

■

Alkyl halides are nearly always prepared from alcohols. Alcohols, in turn, are readily available in a wide variety of shapes and sizes. Simpler alcohols are

$$\underset{\text{An alcohol}}{R{-}OH} \quad \xrightarrow{\text{HX or PX}_3} \quad \underset{\substack{\text{An alkyl}\\\text{halide}}}{R{-}X}$$

produced commercially (Sec. 17.6); the more complicated ones are readily synthesized (Secs. 17.8 and 17.14–17.15). Although certain alcohols tend to undergo rearrangement (Sec. 5.25) during replacement of —OH by —X, this tendency can be minimized by use of phosphorus halides.

In the laboratory, alcohols are the most common starting point for the synthesis of aliphatic compounds, and one of the commonest first steps in such a synthesis is the conversion of the alcohol into an alkyl halide. Once the alkyl halide is made,

$$R{-}OH \longrightarrow R{-}X \left\{ \begin{array}{l} \longrightarrow \text{substitution products} \\ \longrightarrow \text{elimination products} \\ \longrightarrow \text{organometallic compounds} \end{array} \right.$$

the synthesis can follow any one of dozens of pathways, depending upon the reaction that the alkyl halide is allowed to undergo—and, as we shall see in the following section, there are dozens of possibilities.

Alkyl halides are *almost never* prepared by direct halogenation of alkanes. *From the standpoint of synthesis in the laboratory, an alkane is a dead-end.* Halogenation generally gives a mixture of isomers; even if, occasionally, one isomer greatly predominates—as in the bromination of isobutane, say—it is probably not the one we want. How much more practical simply to pick an alcohol that has the —OH in the proper position, and then to replace that —OH by halide!

An alkyl iodide is often prepared from the corresponding bromide or chloride by treatment with a solution of sodium iodide in acetone; the less soluble sodium bromide or sodium chloride precipitates from solution and can be removed by filtration.

5.8   Reactions of alkyl halides. Nucleophilic aliphatic substitution

When methyl bromide is treated with sodium hydroxide in a solvent that dissolves both reagents, there is obtained methanol and sodium bromide. This is a *substitution* reaction: the —OH group is substituted for —Br in the original compound.

$$\text{CH}_3:\text{Br} + :\text{OH}^- \longrightarrow \text{CH}_3:\text{OH} + :\text{Br}^-$$

       Methyl bromide   Methanol

It is clearly heterolytic: the departing halide ion takes with it the electron pair it has been sharing with carbon; hydroxide ion brings with it the electron pair needed to bind it to carbon. Carbon loses one pair of electrons and gains another pair. This is just one example of the class of reactions called *nucleophilic aliphatic substitution.*

Nucleophilic substitution is characteristic of alkyl halides. To see why this is so, we must look at the functional group of this family: halogen.

A halide ion is an extremely weak base. This is shown by its readiness to release a proton to other bases, that is, by the high acidity of the hydrogen halides. In an alkyl halide, halogen is attached to carbon; and, just as halide readily releases a proton, so it readily releases carbon—again, to other bases. These bases possess an unshared pair of electrons and are seeking a relatively positive site, that is, are seeking a nucleus with which to share their electron pair.

Basic, electron-rich reagents that tend to attack the nucleus of carbon are called **nucleophilic reagents** (from the Greek, *nucleus-loving*) or simply **nucleophiles**. When this attack results in substitution, the reaction is called **nucleophilic substitution**.

$$\text{R}:\text{W} \quad + \quad :\text{Z} \longrightarrow \text{R}:\text{Z} \quad + \quad :\text{W}^-$$

      Substrate   Nucleophile   Leaving
                                         group

The carbon compound that undergoes a particular kind of reaction—here, the compound on which substitution takes place—is called the **substrate**. In the case of nucleophilic substitution, the substrate is characterized by the presence of a **leaving group**: the group that becomes displaced from carbon and, taking the electron pair with it, departs from the molecule.

It should be understood that the nucleophile :Z can be negatively charged or neutral; the product R:Z will then be neutral or positively charged. The substrate R:W can be neutral or positively charged; the leaving group will then be negatively charged or neutral.

In the example we started with, methyl bromide is the substrate, bromide is the leaving group, and hydroxide ion is the nucleophile.

Because the weakly basic halide ion is a good leaving group, then, alkyl halides are good substrates for nucleophilic substitution. They react with a large number of nucleophilic reagents, both inorganic and organic, to yield a wide variety of important products. As we shall see, these reagents include not only negative ions like hydroxide, alkoxide, and cyanide, but also neutral bases like ammonia and water; their characteristic feature is an unshared pair of electrons.

As a synthetic tool, nucleophilic substitution is one of the three or four most useful classes of organic reactions. Nucleophilic substitution is the work-horse of organic synthesis; in its various forms, it is the reaction we shall turn to first when faced with the basic job of replacing one functional group by another. The synthesis of aliphatic compounds, we said, most often starts with alcohols. But the —OH group, we shall find, is a very poor leaving group; it is only conversion of alcohols into alkyl halides—or other compounds with good leaving groups—that opens the door to nucleophilic substitution.

A large number of nucleophilic substitutions are listed below to give an idea of the versatility of alkyl halides; many will be left to later chapters for detailed discussion.

REACTIONS OF ALKYL HALIDES _____

1. **Nucleophilic substitution**

$$R:X + :Z \longrightarrow R:Z + :X^-$$

$R:X + :OH^-$	$\longrightarrow R:OH + :X^-$	Alcohol
$+ H_2O$	$\longrightarrow R:OH$	Alcohol
$+ :OR'^-$	$\longrightarrow R:OR'$	Ether (Williamson synthesis, Sec. 19.5)
$+ {}^-:C{\equiv}CR'$	$\longrightarrow R:C{\equiv}CR'$	Alkyne (Sec. 11.13)
$+ \overset{\delta-}{R'}{-}\overset{\delta+}{M}$	$\longrightarrow R:R'$	Alkane (coupling, Sec. 3.17)
$+ :I^-$	$\longrightarrow R:I$	Alkyl iodide
$+ :CN^-$	$\longrightarrow R:CN$	Nitrile (Sec. 23.8)
$+ R'COO:^-$	$\longrightarrow R'COO:R$	Ester
$+ :NH_3$	$\longrightarrow R:NH_2$	Primary amine (Sec. 26.10)
$+ :NH_2R'$	$\longrightarrow R:NHR'$	Secondary amine (Sec. 26.13)
$+ :NHR'R''$	$\longrightarrow R:NR'R''$	Tertiary amine (Sec. 26.13)
$+ :P(C_6H_5)_3$	$\longrightarrow R:P(C_6H_5)_3{}^+X^-$	Phosphonium salt (Sec. 25.10)
$+ :SH^-$	$\longrightarrow R:SH$	Thiol (mercaptan)
$+ :SR'^-$	$\longrightarrow R:SR'$	Thioether (sulfide)

CONTINUED

CONTINUED

$$R:X + Ar{-}H + AlCl_3 \longrightarrow Ar{-}R \qquad \text{Alkylbenzene (Friedel–Crafts reaction, Sec. 15.7)}$$

$$+ [CH(COOC_2H_5)_2]^- \longrightarrow R:CH(COOC_2H_5)_2 \qquad \text{(Malonic ester synthesis, Sec. 30.2)}$$

$$+ [CH_3COCHCOOC_2H_5]^- \longrightarrow \underset{\text{R}}{CH_3COCHCOOC_2H_5} \qquad \text{(Acetoacetic ester synthesis, Sec. 30.3)}$$

2. Dehydrohalogenation: elimination. Discussed in Secs. 7.12 and 7.24.

$$-\underset{H}{\overset{|}{C}}-\underset{X}{\overset{|}{C}}- \xrightarrow{\text{base}} -\overset{|}{C}=\overset{|}{C}-$$

3. Preparation of Grignard reagent. Discussed in Secs. 3.16 and 17.14.

$$RX + Mg \xrightarrow{\text{dry ether}} RMgX$$

4. Reduction. Discussed in Sec. 3.15.

$$RX + M + H^+ \longrightarrow RH + M^+ + X^-$$

Examples:

$$(CH_3)_3CCl \xrightarrow{\text{Mg}} (CH_3)_3CMgCl \xrightarrow{\text{D}_2\text{O}} (CH_3)_3CD$$

7,7-Dibromonorcarane　　　　　　　Norcarane

With nucleophilic substitution we shall encounter many things new to us: a new reaction, of course—several new reactions, actually—and a new kind of reactive particle, the *carbocation*. To find out what is going on in these reactions, we shall use a new tool, *kinetics*, and use an old tool, *stereochemistry*, in a new way. We shall be introduced to new factors affecting reactivity—*dispersal of charge, polar factors, steric hindrance, acid catalysis*—factors that we shall work with throughout the rest of our study.

In Chapter 6, still working with nucleophilic substitution, we shall see how reactivity—and, with it, the course of reaction—is affected by the *solvent*. Then, in Chapter 20, we shall use nucleophilic substitution as the starting point for our study of *symphoria*—the bringing together of reactants into the proper spatial relationship—and see how, to an extent never before possible, we can control what happens in an organic reaction: rate, orientation, and even the stereochemical outcome.

Alkyl halides undergo not only substitution but also **elimination**, a reaction that we shall take up in Chapter 7. Both elimination and substitution are brought about by basic reagents, and hence there will always be *competition* between the two reactions. We shall be interested to see how this competition is affected by such factors as the structure of the halide and the particular nucleophilic reagent used.

Alkyl halides are the substances most commonly **converted into organometallic compounds**: compounds that contain carbon attached to a metal—magnesium (as in the Grignard reagent), lithium, copper, and a host of others. We have already met some of these compounds, and shall have a great deal to do with them as we go along. As we shall see (Sec. 11.13), conversion of alkyl halides into organo-metallic compounds changes the nature of the central carbon atom in a funda-mental way, and gives us a class of reagents with unique properties.

5.9   Nucleophilic aliphatic substitution. Nucleophiles and leaving groups

The components required for nucleophilic substitution are: *substrate, nucleo-phile*, and *solvent*. The substrate consists of two parts, *alkyl group* and *leaving group*. We shall be concerned with the alkyl group throughout much of the chapter; we

$$\underset{\text{Substrate}}{\overset{\overset{\displaystyle\textit{Alkyl}}{\overset{\displaystyle\textit{group}}{\searrow}}\quad\overset{\displaystyle\textit{Leaving}}{\overset{\displaystyle\textit{group}}{\swarrow}}}{\text{R—W}} \quad + \quad \underset{\text{Nucleophile}}{:\text{Z}} \quad\xrightarrow{\text{solvent}}\quad \text{R—Z} \quad + \quad \underset{\substack{\textit{Leaving} \\ \textit{group}}}{:\text{W}}$$

shall study the roles played by the solvent in Chapter 6. At this point let us examine the other components of these systems, nucleophiles and leaving groups.

We have already seen enough to realize that basicity plays an important part in our understanding of nucleophiles and leaving groups. Nucleophiles are char-acterized by being bases, and leaving groups are characterized by being *weak* bases. We may find a rough correlation between *degree* of basicity, on the one hand, and nucleophilic power or leaving ability, on the other: the stronger of two bases is often the more powerful nucleophile, and the weaker of two bases is often the better leaving group. But this holds true only for closely related sets of nucleophiles or sets of leaving groups: ones that, among other things, involve the same central element—oxygen, say, or nitrogen. There are many exceptions to such a correlation, and clearly basicity is only *one* of the factors involved.

We should have clear in our minds the distinction between basicity and nucleophilic power or leaving ability. All have to do with the tendency—or, in the case of leaving ability, *lack* of tendency—to share an electron pair to form a covalent bond. But there are two fundamental differences:

(a) Basicity is a matter of *equilibrium*; nucleophilic power and leaving ability are matters of *rate*. Of two bases, one is said to be the stronger because at equilibrium it holds a greater proportion of the acid. Of two nucleophiles, one is said to be the more powerful because it attacks carbon *faster*; of two leaving groups, one is said to be the better because it leaves carbon *faster*.

(b) Basicity (in the Lowry–Brønsted sense) involves interaction with a proton; nucleo-philic power and leaving ability involve interactions with carbon.

It is not surprising, then, that there is no exact parallel between basicity and these two other properties. The surprise, perhaps, is that the parallel is as good as it is.

Let us have a look at some of the **nucleophiles** we shall be working with. Many of the products formed are new to us, but at this point we need see only how the structure of a particular product is the natural result of the structure of a particular nucleophile. For now, we shall use alkyl halides as our examples of substrates.

Some nucleophiles are anions, like *hydroxide* ion;

$$R\!-\!X + :OH^- \longrightarrow R\!-\!OH + :X^-$$

Hydroxide   An alcohol
ion

the closely related *alkoxides*, methoxide, say;

$$R\!-\!X + :OCH_3^- \longrightarrow R\!-\!OCH_3 + :X^-$$

Methoxide   An ether
ion
An alkoxide
ion

cyanide (the strongly basic anion of the very weak acid, HCN);

$$R\!-\!X + :CN^- \longrightarrow R\!-\!CN + :X^-$$

Cyanide   A nitrile
ion

or even another *halide* ion which, while only weakly basic, does after all possess unshared electrons.

$$R\!-\!X + :I^- \longrightarrow R\!-\!I + :X^-$$

Iodide   Alkyl
ion   iodide

But neutral molecules, too, can possess unshared electrons, be basic, and hence act as nucleophiles. *Water*, for example, attacks an alkyl halide to yield, ultimately, an alcohol. But the oxygen of water already has two hydrogens, and when it attaches

$$R\!-\!X + :OH_2 \longrightarrow R\!-\!OH_2^+ + :X^-$$

Protonated
alcohol

itself to carbon there is formed initially, not the alcohol itself, but its conjugate acid, the protonated alcohol. This easily changes itself into the alcohol by loss of the proton.

$$R\!-\!OH_2^+ \rightleftharpoons R\!-\!OH + H^+$$

Protonated   Alcohol
alcohol

An important point arises here. For convenience we shall often show the loss or gain of a hydrogen ion, H^+. But it should be understood that we are *not* actually dealing with a naked proton, but rather with the *transfer* of a proton from one base

$$ROH_2^+ + H_2O \rightleftharpoons ROH + H_3O^+$$

Protonated alcohol	Water	Alcohol	Hydronium ion
Acid	*Base*	*Base*	*Acid*

to another. In the present case, for example, protonated alcohol is converted into alcohol by transfer of the proton to water—about as basic as the alcohol itself, and much more abundant.

Problem 5.3 What product would you expect to obtain from the reaction of *n*-propyl bromide with methanol, CH_3OH? Write equations for all steps involved.

Next, let us look at some of the **leaving groups** we shall encounter. So far we have used alkyl halides as our chief examples, and we shall continue to do this in following sections. But we should realize that these reactions take place in exactly the same ways with a variety of other substrates: compounds which, like alkyl halides, contain good leaving groups.

Of these other substrates, alkyl esters of sulfonic acids, $ArSO_2OR$, are most commonly used in place of alkyl halides: usually in the study of reaction mechanisms, but also in synthesis. Sulfonic acids, $ArSO_3H$, are related to sulfuric acid

$$ArSO_3H \rightleftharpoons ArSO_3^- + H^+$$

A sulfonic   A sulfonate
acid   anion

Strong acid   *Weak base*

and, like sulfuric, are *strong* acids. Their anions, the sulfonates, are weak bases and hence good leaving groups.

$$ArSO_2O-R + :Z \longrightarrow R-Z + ArSO_3^-$$

An alkyl   Nucleophile   A sulfonate
sulfonate   anion

*Weak base:
good leaving
group*

Most commonly used are esters of *p*-toluenesulfonic acid: the *p*-toluenesulfonates. (We shall understand the structures of these aromatic compounds later; for now we need only know that sulfonates are good leaving groups.) The name of the *p*-toluenesulfonyl group is often abbreviated to *tosyl* (Ts), and *p*-toluenesulfonates become *tosylates* (TsOR).

Tosyl or *Ts*   *Brosyl* or *Bs*   *Mesyl* or *Ms*   *Trifyl* or *Tf*

Like alkyl halides, alkyl sulfonates are readily made from alcohols. For example:

TsCl $TsOC_2H_5$

p-Toluenesulfonyl   Ethyl *p*-toluenesulfonate
chloride

Tosyl chloride   Ethyl tosylate

As we shall see in Sec. 18.5, the two syntheses differ in one very important way.

Now let us begin our study of the mechanism of nucleophilic aliphatic substitution. For generations this reaction has fascinated chemists, including many of the "greats" whose names are—or will become—familiar to us: J. A. LeBel, G. N. Lewis, T. M. Lowry; and that giant of organic chemistry, Emil Fischer, who, we shall find, opened up the two vast fields of carbohydrates and proteins.

In its various forms, nucleophilic aliphatic substitution has been the most widely studied—and most strongly disputed—area of organic chemistry. The fascination—and the argument—has lain in two related questions. The bond to the leaving group is being broken and the bond to the nucleophile is being formed. (a) What is the *timing* of these two processes? (b) Where does the energy required to break the bond to the leaving group come from?

We shall begin our study of the mechanism where the modern history of the reaction begins: with the *kinetics* of nucleophilic aliphatic substitution. But, first, what *is* kinetics?

5.10   Rate of reaction: effect of concentration. Kinetics

We have seen (Sec. 2.18) that the rate of a chemical reaction can be expressed as a product of three factors:

$$\text{rate} = \frac{\text{collision}}{\text{frequency}} \times \frac{\text{energy}}{\text{factor}} \times \frac{\text{probability}}{\text{factor}}$$

So far, we have used this relationship in comparing rates of *different* reactions: to help us understand orientation and relative reactivity, and why a particular reaction takes place at all. So that comparisons of this sort may be as fair as possible, we keep the conditions that we can control—temperature, concentration—the same. If this is done, then closely related reactions proceed at different rates chiefly because they have different energy factors, that is to say, different E_{act} values; and to account for different E_{act} values we must estimate relative stabilities of transition states.

It is also useful to study an *individual* reaction to see how its rate is affected by deliberate changes in experimental conditions. We can determine E_{act}, for example, if we measure the rate at different temperatures (Sec. 2.18). But perhaps the most valuable information about a reaction is obtained by studying the effect of *changes in concentration* on its rate.

How does a change in concentration of reactants affect the rate of a reaction at a constant temperature? An increase in concentration cannot alter the fraction of collisions that have sufficient energy, or the fraction of collisions that have the proper orientation; it can serve only to increase the total number of collisions. If more molecules are crowded into the same space, they will collide more often and the reaction will go faster. Collision frequency, and hence rate, depends in a very exact way upon concentration.

The field of chemistry that deals with rates of reaction, and in particular with dependence of rates on concentration, is called **kinetics**. Let us see what kinetics can tell us about nucleophilic aliphatic substitution.

5.11   Kinetics of nucleophilic aliphatic substitution. Second-order and first-order reactions

Let us take a specific example, the reaction of methyl bromide with sodium hydroxide to yield methanol:

$$CH_3Br + OH^- \longrightarrow CH_3OH + Br^-$$

This reaction would probably be carried out in aqueous ethanol, in which both reactants are soluble.

If the reaction results from collision between a hydroxide ion and a methyl bromide molecule, we would expect the rate to depend upon the concentration of both these reactants. If either OH^- concentration, $[OH^-]$, or CH_3Br concentration, $[CH_3Br]$, is doubled, the collision frequency should be doubled and the reaction rate doubled. If either concentration is cut in half, the collision frequency, and consequently the rate, should be halved.

This is found to be so. We say that the rate of reaction depends upon both $[OH^-]$ and $[CH_3Br]$, and we indicate this by the expression

$$\text{rate} = k[CH_3Br][OH^-]$$

If concentrations are expressed in, say, moles per liter, then k is the number which, multiplied by these concentrations, tells us how many moles of methanol are formed in each liter during each second. At a given temperature and for a given solvent, k always has the same value and is characteristic of this particular reaction; k is called the **rate constant**. For example, for the reaction between methyl bromide and hydroxide ion in a mixture of 80% ethanol and 20% water at 55 °C, the value of k is 0.0214 liters per mole per second.

What we have just seen is, of course, not surprising; we all know that an increase in concentration causes an increase in rate. But now let us look at the corresponding reaction between *tert*-butyl bromide and hydroxide ion:

$$\underset{\underset{\text{Br}}{|}}{\overset{\overset{\text{CH}_3}{|}}{CH_3-C-CH_3}} + OH^- \longrightarrow \underset{\underset{\text{OH}}{|}}{\overset{\overset{\text{CH}_3}{|}}{CH_3-C-CH_3}} + Br^-$$

As before, if we double [RBr] the rate doubles; if we cut [RBr] in half the rate is halved. But if we double $[OH^-]$, or if we cut $[OH^-]$ in half, there is no change in the rate. *The rate of reaction is independent of* $[OH^-]$.

The rate of reaction of *tert*-butyl bromide depends only upon [RBr]. This is indicated by the expression

$$\text{rate} = k[RBr]$$

For the reaction of *tert*-butyl bromide in 80% alcohol at 55 °C, the rate constant is 0.010 per second. This means that, of every mole of *tert*-butyl bromide present, 0.010 mole reacts each second, whatever the $[OH^-]$.

The methyl bromide reaction is said to follow **second-order kinetics**, since its rate is dependent upon the concentrations of *two* substances. The *tert*-butyl bromide reaction is said to follow **first-order kinetics**; its rate depends upon the concentration of only *one* substance.

5.12   Nucleophilic aliphatic substitution: duality of mechanism

By the 1930s, kinetics studies of nucleophilic substitution had been carried out with a variety of substrates, and the following had been found. Like methyl, primary substrates react by second-order kinetics. Like *tert*-butyl, other tertiary substrates react by first-order kinetics. Secondary substrates show borderline behavior: sometimes second-order, sometimes first-order; often a mixture of the two.

Besides the kinetic order, the rate studies had revealed something else about the substitution: the relative reactivities of the various substrates. Typically, at a given concentration of a nucleophile like OH^-, reactivity was found to vary something like this:

$$CH_3X > 1° > 2° < 3°$$

That is, as one proceeds along the series CH_3, $1°$, $2°$, $3°$, reactivity at first decreases, then passes through a minimum (usually at $2°$), and finally rises (see Fig. 5.1). Significantly, the minimum occurs at just the point in the series where the kinetics changes from second-order to first-order.

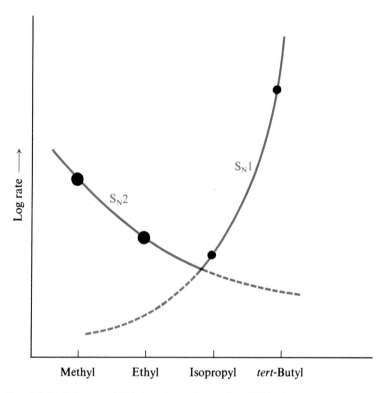

Figure 5.1   Nucleophilic aliphatic substitution: typical effect on rate caused by variations in structure of substrate, RX. The minimum in rate is attributed to the crossing of two opposing curves, that is, to a change in mechanism from S_N2 to S_N1.

In 1935, E. D. Hughes and Sir Christopher Ingold (University College, London) took these two sets of facts—kinetic order and relative reactivity—and on them built a broad theory of nucleophilic aliphatic substitution. The keystone of their theory was this: that *nucleophilic aliphatic substitution can proceed by two different mechanisms*. These mechanisms, for reasons that will become clear, they named S_N2 and S_N1. Different substrates react by different kinetic orders because they are reacting by different mechanisms: some, like methyl, by S_N2; others, like *tert*-butyl, by S_N1.

Reactivity passes through a minimum with secondary substrates because the mechanism changes at this point, from S_N2 to S_N1. The occurrence of a minimum or maximum in a property—reactivity, acidity, antibacterial activity—as one proceeds along a logical series, suggests the working of opposing factors. Here, Hughes and Ingold proposed, the factors are the opposing reactivity sequences for the two different mechanisms. As one passes along the series, reactivity by the S_N2

S_N2 increases
←————————————

$RX = CH_3X \quad 1° \quad 2° \quad 3°$

————————————→
S_N1 increases

S_N2
vs.
S_N1

mechanism decreases from CH_3 to $1°$, and at $2°$ is so low that the S_N1 reaction begins to contribute significantly; reactivity, now by S_N1, rises sharply to $3°$ (Fig. 5.1).

In the following sections, we shall see what these two mechanisms are, the facts on which they are based, and how they account for these facts. We shall see, for example, how they account for the difference in kinetic order and, in particular, for the puzzling fact that the rate of the *tert*-butyl bromide reaction is independent of $[OH^-]$. We shall see what factors are believed to be responsible for the opposing reactivity sequences for the two mechanisms. Finally, we shall see how this mechanistic pattern drawn in 1935 has stood the test of time.

5.13   The S_N2 reaction: mechanism and kinetics

The reaction between methyl bromide and hydroxide ion to yield methanol follows second-order kinetics; that is, the rate depends upon the concentrations of both reactants:

$$CH_3Br + OH^- \longrightarrow CH_3OH + Br^-$$

$$rate = k[CH_3Br][OH^-]$$

The simplest way to account for the kinetics is to assume that reaction requires a collision between a hydroxide ion and a methyl bromide molecule. On the basis of evidence we shall shortly discuss, it is known that in its attack the hydroxide ion stays as far away as possible from the bromine; that is to say, it attacks the molecule from the rear.

The reaction is believed to take place as shown in Fig. 5.2. When hydroxide ion collides with a methyl bromide molecule at the face most remote from the bromine, and when such a collision has sufficient energy, a C—OH bond forms and the C—Br bond breaks, liberating the bromide ion.

$$HO^- \quad -C-Br \quad \longrightarrow \quad HO \cdots \overset{\delta-}{C} \cdots \overset{\delta-}{Br} \quad \longrightarrow \quad HO-C \quad Br^-$$

Figure 5.2 The S_N2 reaction: complete inversion of configuration. The nucleophilic reagent attacks the back side.

The transition state can be pictured as a structure in which carbon is partially bonded to both —OH and —Br; the C—OH bond is not completely formed, the C—Br bond is not yet completely broken. Hydroxide has a diminished negative charge, since it has begun to share its electrons with carbon. Bromine has developed a partial negative charge, since it has partly removed a pair of electrons from carbon.

The —OH and —Br are located as far apart as possible; the three hydrogens and the carbon lie in a single plane, all bond angles being 120°. The C—H bonds are thus arranged like the spokes of a wheel, with the C—OH and the C—Br bonds lying along the axle.

We can see how the geometry of this transition state arises. Carbon holds the three hydrogens through overlap of three sp^2 orbitals: trigonal, and hence flat and 120° apart. The partial bonds to the leaving group and the nucleophile are formed through overlap of the remaining p orbitals: 180° apart, and perpendicular to the plane of the sp^2 orbitals.

This is the mechanism that is called S_N2: *substitution nucleophilic bimolecular*. The term *bi*molecular is used here since the rate-determining step involves collision of *two* particles.

What evidence is there that alkyl halides can react in this manner? First of all, as we have just seen, the mechanism is consistent with the kinetics of a reaction like the one between methyl bromide and hydroxide ion. In general, **an S_N2 reaction follows second-order kinetics**. Let us look at some of the other evidence.

5.14 The S_N2 reaction: stereochemistry. Inversion of configuration

Both 2-bromooctane and 2-octanol are chiral; that is, they have molecules that are not superimposable on their mirror images. Consequently, these compounds can exist as enantiomers, and can show optical activity. Optically active 2-octanol has been obtained by resolution of the racemic modification (Sec. 4.27), and from it optically active 2-bromooctane has been made.

The following configurations have been assigned (Sec. 4.24):

$$\begin{array}{cc}
C_6H_{13} & C_6H_{13} \\
| & | \\
H-C-Br & H-C-OH \\
| & | \\
CH_3 & CH_3 \\
(-)\text{-2-Bromooctane} & (-)\text{-2-Octanol} \\
[\alpha] = -39.6° & [\alpha] = -10.3°
\end{array}$$

We notice that the (−)-bromide and the (−)-alcohol have similar configurations; that is, —OH occupies the same relative position in the (−)-alcohol as —Br does in the (−)-bromide. As we know, compounds of similar configuration do not *necessarily* rotate light in the same direction; they just happen to do so in the present case. (As we also know, compounds of similar configuration are not necessarily given the same specification of *R* and *S* (Sec. 4.24); it just happens that both are *R* in this case.)

Now, when (−)-2-bromooctane is allowed to react with sodium hydroxide under conditions where second-order kinetics are followed, there is obtained (+)-2-octanol.

$$\underset{\substack{\text{(−)-2-Bromooctane} \\ [\alpha] = -39.6° \\ \text{\textit{optical purity} } 100\%}}{\overset{\displaystyle C_6H_{13}}{\underset{\displaystyle CH_3}{H\!-\!C\!\rightarrow\!Br}}} \quad \xrightarrow[\text{S}_\text{N}2]{\text{NaOH}} \quad \underset{\substack{\text{(+)-2-Octanol} \\ [\alpha] = +10.3° \\ \text{\textit{optical purity} } 100\%}}{\overset{\displaystyle C_6H_{13}}{\underset{\displaystyle CH_3}{HO\!-\!C\!\leftarrow\!H}}}$$

We see that the —OH group has not taken the position previously occupied by —Br; the alcohol obtained has a configuration *opposite* to that of the bromide. *A reaction that yields a product whose configuration is opposite to that of the reactant is said to proceed with* **inversion of configuration**.

(In this particular case, inversion of configuration happens to be accompanied by a change in specification, from *R* to *S*, but this is not always true. We cannot tell whether a reaction proceeds with inversion or retention of configuration simply by looking at the letters used to specify the reactant and product; we must work out and compare the absolute configurations indicated by those letters.)

Now the question arises: does a reaction like this proceed with *complete* inversion? That is to say, is the configuration of *every* molecule inverted? The answer is *yes*. **An S$_N$2 reaction proceeds with complete stereochemical inversion.**

To answer a question like this, we must in general know the optical purity both of the reactant that we start with and of the product that we obtain: in this case, of 2-bromooctane and 2-octanol. To know these we must, in turn, know the maximum rotation of the bromide and of the alcohol; that is, we must know the rotation of an optically pure sample of each.

Suppose, for example, that we know the rotation of optically pure 2-bromooctane to be 39.6° and that of optically pure 2-octanol to be 10.3°. If, then, a sample of optically pure bromide were found to yield optically pure alcohol, we would know that the reaction had proceeded with *complete* inversion. Or—and this is much more practicable—if a sample of a halide of rotation, say, −32.9° (83% optically pure) were found to yield alcohol of rotation +8.55° (83% optically pure), we would draw exactly the same conclusion.

In developing the ideas of S$_N$1 and S$_N$2 reactions, Hughes and Ingold studied the reaction of optically active 2-bromooctane and obtained results which led them to conclude that the S$_N$2 reaction proceeds, within limits of experimental error, with complete inversion.

The particular value that Hughes and Ingold used for the rotation of optically pure 2-bromooctane has been questioned, but the basic idea of complete inversion in S$_N$2 reactions is established beyond question: by the studies of systems other than alkyl halides and by elegant work involving radioactivity and optical activity (Problem 5.5, below).

Problem 5.4 In 1923 Henry Phillips (Battersea Polytechnic, London) reported the following experiment:

$$
\begin{array}{ccccc}
& \text{CH}_3 & & \text{CH}_3 & & \text{CH}_3 \\
& | & & | & & | \\
\text{C}_6\text{H}_5\text{CHOK} & \xleftarrow{\;\;\text{K}\;\;} & \text{C}_6\text{H}_5\text{CHOH} & \xrightarrow{\;\;\text{TsCl}\;\;} & \text{C}_6\text{H}_5\text{CHOTs}
\end{array}
$$

α-Phenylethyl alcohol
α = +33.02°

C₂H₅OTs ↓   C₂H₅OK ↓

$$
\begin{array}{ccc}
\text{CH}_3 & & \text{CH}_3 \\
| & & | \\
\text{C}_6\text{H}_5\text{CHOC}_2\text{H}_5 & & \text{C}_6\text{H}_5\text{CHOC}_2\text{H}_5
\end{array}
$$

Ethyl   Ethyl
α-phenylethyl ether   α-phenylethyl ether
α = +19.84°   α = −19.90°

(a) Account for the fact that the ethers obtained by the two routes have *opposite but equal* optical rotations. (*Hint*: See Sec. 4.24.) (b) Does it matter what the optical purity of the starting alcohol is? (c) What is the fundamental significance of this finding?

Problem 5.5 When optically active 2-iodooctane was allowed to stand in acetone solution containing Na¹³¹I (radioactive iodide), the alkyl halide was observed to lose optical activity and to exchange its ordinary iodine for radioactive iodine. The rate of each of these reactions depended on both [RI] and [I⁻], but loss of optical activity was exactly *twice* as fast as gain of radioactivity. Combining as it does kinetics and stereochemistry, this experiment, reported in 1935 by E. D. Hughes (p. 183), is considered to have established the stereochemistry of the S_N2 reaction: that each molecule undergoing substitution suffers inversion of configuration. Show exactly how this conclusion is justified. (*Hint*: Take one molecule of alkyl halide at a time, and consider what happens when it undergoes substitution.)

It was to account for inversion of configuration that back-side attack was first proposed for substitution of the S_N2 kind. As —OH becomes attached to carbon, three bonds are forced apart until they reach the planar "spoke" arrangement of the transition state; then, as bromide is expelled, they move on to a tetrahedral arrangement *opposite* to the original one. This process has often been likened to the turning-inside-out of an umbrella in a gale.

$$
\begin{array}{ccccc}
& \text{C}_6\text{H}_{13} & & \text{C}_6\text{H}_{13}\;\text{H} & & \text{C}_6\text{H}_{13} \\
& \vdots & & {}^{\delta-}\diagdown\;\diagup\;{}^{\delta-} & & \vdots \\
\text{HO}^- \;\; \text{H}\!-\!\text{C}\!-\!\text{Br} & \longrightarrow & \text{HO}\text{-}\text{-}\text{-}\text{C}\text{-}\text{-}\text{-}\text{-}\text{Br} & \longrightarrow & \text{HO}\!-\!\text{C}\!-\!\text{H} \;\; \text{Br}^- \\
& \text{CH}_3 & & \text{CH}_3 & & \text{CH}_3
\end{array}
$$

S_N2: complete inversion

The stereochemistry of the 2-bromooctane reaction indicates back-side attack in accordance with the S_N2 mechanism; studies of other optically active compounds, under conditions where the reactions follow second-order kinetics, show similar results. It is not possible to study the stereochemistry of most halides, since they are not optically active; however, there seems no reason to doubt that they, too, undergo back-side attack.

Inversion of configuration is the general rule for reactions occurring at chiral centers, being much commoner than retention of configuration. Oddly enough, it is the very prevalence of inversion that made its detection difficult. Paul Walden (at the Polytechnicum in Riga, Latvia) discovered the phenomenon of inversion in 1896 when he encountered one of the exceptional reactions in which inversion does *not* take place.

Problem 5.6 Show the absolute configuration and give the *R/S* specification and specific rotation of the 2-octanol expected from the S$_N$2 reaction of 2-bromooctane of $[\alpha] + 24.9°$.

But, besides the spatial orientation of attack, there is another feature of the S$_N$2 reaction, a feature that is even more fundamental since it defines the mechanism: reaction occurs *in a single step*, and hence bond-making and bond-breaking occur simultaneously, in a *concerted* fashion. This feature, too, is supported by the stereochemistry: not by the fact that there is inversion, but by the fact that there is *complete* inversion. Every molecule of substrate suffers the same stereochemical fate—inversion, as it happens. This specificity is completely consistent with the mechanism: the leaving group is still attached to carbon when nucleophilic attack begins, and controls the direction from which that attack occurs. (We shall appreciate the significance of this point better when we see the contrast offered by the S$_N$1 reaction.)

Actually, we have already encountered a contrasting situation: the free-radical chlorination of optically active 1-chloro-2-methylbutane (Sec. 4.28). First, hydrogen is extracted from the chiral center. Then, in a subsequent step, chlorine becomes attached to that carbon. But, with the hydrogen gone, there is nothing left to direct chlorine to a particular face of the carbon; attack occurs randomly at either face, and the racemic modification is obtained.

The S$_N$2 mechanism is supported, then, by stereochemical evidence. Indeed, the relationship between mechanism and stereochemistry is so well established that in the absence of other evidence complete inversion is taken to indicate an S$_N$2 reaction.

We see once more how stereochemistry can give us a kind of information about a reaction that we cannot get by any other means.

5.15   The S$_N$2 reaction: reactivity. Steric hindrance

Now let us turn to the matter of *reactivity* in nucleophilic aliphatic substitution, and see how it is affected by changes in the structure of the alkyl group.

According to the dual mechanism theory (Sec. 5.12), the commonly observed order of reactivity, with a minimum at 2°, is simply the composite of two opposing orders of reactivity, one for S$_N$2 and the other for S$_N$1. Clearly, a test of this hypothesis would be to carry out substitution under conditions where all members of a series—methyl through 3°—react to a significant extent by, say, second-order kinetics, and measure the second-order rate constants; then, to repeat the process, this time selecting conditions that favor first-order reaction, and measure the first-order rate constants. Let us look at results obtained in this way, first for the S$_N$2 reaction and then, in a later section, for the S$_N$1 reaction.

Direct measurement of S_N2 rates for a series of substrates gives results like the following. (DMF is *dimethylformamide*, a solvent that, as we shall find in Sec. 6.4, favors the S_N2 reaction here.)

S_N2 substitution: relative reactivity

$$R—Br + Cl^- \xrightarrow{\text{DMF}} R—Cl + Br^-$$

H	H	CH$_3$	CH$_3$
H—C—Br >	CH$_3$—C—Br >	CH$_3$—C—Br >	CH$_3$—C—Br
H	H	H	CH$_3$
Methyl	Ethyl	Isopropyl	*tert*-Butyl

Relative rate (S_N2)	Methyl	Ethyl	Isopropyl	*tert*-Butyl
	37	1.0	0.02	0.0008

As postulated, then, the reactivity of substrates in the S_N2 reaction is:

Reactivity in S_N2 $CH_3W > 1° > 2° > 3°$

How are we to account for this order of reactivity? As always to answer a question like this, we must take the specific reaction involved—here, the S_N2 reaction—and compare the structure of the reactants with the structure of the transition state. In contrast to free-radical substitution, this time the structure of the transition state is *not* intermediate between the structures of the reactants and products; this time, we cannot simply expect that factors stabilizing the product will also stabilize the transition state.

During many reactions, as we shall discover, there is a change in electron distribution such that a negative or positive charge develops in the reacting molecule; and very often reactivity depends upon how easily the molecule accommodates that charge. Accommodation of charge depends, in turn, upon the *polar effects* of substituents, that is, upon how well the substituents tend to withdraw or release electrons. First of all, then, let us examine the S_N2 reaction with regard to changes in electron distribution, again using the reaction of an alkyl halide with hydroxide ion as our example. In the transition state as we have described it (Sec. 5.13), there is a partly formed bond between carbon and hydroxide ion, and a

$$HO^- + R—X \longrightarrow \left[\overset{\delta_-}{HO}\text{----}R\text{----}\overset{\delta_-}{X} \right] \longrightarrow HO—R + X^-$$

Reactants	Transition state	Products

partly broken bond between carbon and halogen; hydroxide ion has brought electrons to carbon, and halide ion has taken electrons away. Unless one of the two processes, bond-making or bond-breaking, has gone much further than the other, carbon is not appreciably more negative or positive than it was at the start of the reaction. If this is so, it would appear that the reactivity sequence for the S_N2 reaction does *not* result from the polar effects of substituent groups.

To understand how structure does influence the rate, let us compare transition state and reactants with regard to *shape*, starting with the methyl bromide reaction. The carbon in reactant and product is tetrahedral, whereas carbon in the transition state is bonded to five atoms. As indicated before, the C—H bonds are arranged like the spokes of a wheel, with the C—OH and C—Br bonds lying along the axle (Fig. 5.3).

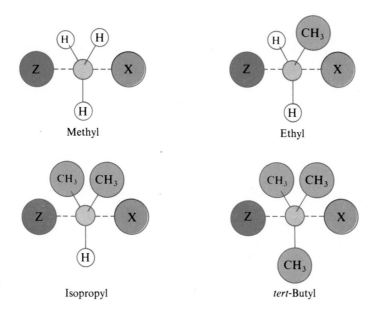

Methyl **Ethyl**

Isopropyl ***tert*-Butyl**

Figure 5.3   Molecular structure and reactivity: the steric factor in the S$_N$2 reaction. Crowding raises the energy of the transition state and slows down reaction.

What would be the effect of replacing the hydrogens successively by methyl groups? That is, how would the transition state differ as we go from methyl bromide through ethyl bromide and isopropyl bromide to *tert*-butyl bromide? As hydrogen atoms are replaced by the larger methyl groups, there is increased crowding about the carbon. As can be seen from scale models (Fig. 5.4, on the next page), the back side of the molecule, where attack must take place, becomes increasingly inaccessible.

This crowding is particularly severe in the transition state, where the methyls are thrown close to both —OH and —Br (Fig. 5.3). Non-bonded interaction raises the energy of the crowded transition state more than the energy of the roomier reactant; E_{act} is higher and the reaction is slower.

This interpretation is the one that is generally accepted today. *Differences in rate between two S$_N$2 reactions are due chiefly to* **steric factors**, and not to polar factors; that is to say, differences in rate are related to the *bulk* of the substituents and not to their effect on electron distribution. As the number of substituents attached to the carbon bearing the halogen is increased, the reactivity toward S$_N$2 substitution decreases, as measurements have shown for the series methyl, 1°, 2°, 3°.

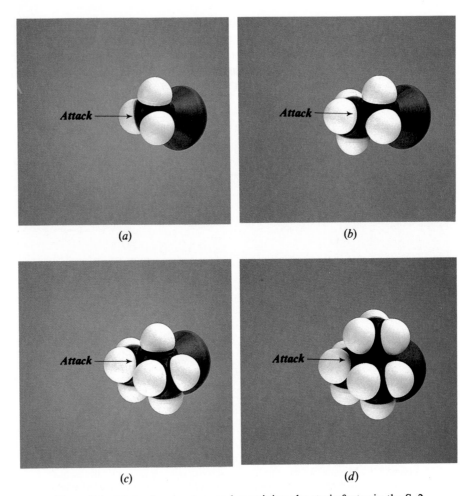

Figure 5.4 Molecular structure and reactivity: the steric factor in the S_N2 reaction. Models of alkyl bromides: (*a*) methyl, (*b*) ethyl, (*c*) isopropyl, (*d*) *tert*-butyl. As the number of substituents on the carbon bearing —Br increases, crowding at the point of nucleophilic attack increases.

That steric factors are at work here is confirmed by the relative rates of another series of substrates. This time all the substrates are primary, and hence have the

S_N2 substitution: relative reactivity

$$R\!-\!Br + Cl^- \xrightarrow{\ DMF\ } R\!-\!Cl + Br^-$$

$$CH_3\!-\!CH_2\!-\!Br > CH_3CH_2\!-\!CH_2\!-\!Br > CH_3\overset{\displaystyle CH_3}{\underset{}{CH}}\!-\!CH_2\!-\!Br > CH_3\!-\!\overset{\displaystyle CH_3}{\underset{\displaystyle CH_3}{C}}\!-\!CH_2\!-\!Br$$

	Ethyl	*n*-Propyl	Isobutyl	Neopentyl
Relative rate (S_N2)	1.0	0.69	0.33	0.000 006

same *number* of substituents—one—attached to the carbon bearing halogen. But now the *size* of the substituent is steadily increased: in ethyl bromide the substituent is methyl; in *n*-propyl bromide, ethyl; in isobutyl bromide, isopropyl; and in neopentyl bromide, *tert*-butyl. And as the size of the (single) substituent increases, so does steric hindrance to attack, and the rate falls off. (See Fig. 5.5.)

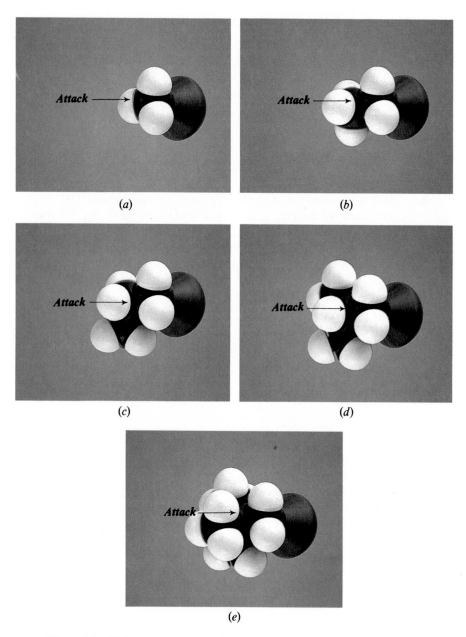

(a)

(b)

(c)

(d)

(e)

Figure 5.5   Molecular structure and reactivity: the steric factor in the S$_N$2 reaction. Models of alkyl bromides: (*a*) methyl, (*b*) ethyl, (*c*) *n*-propyl, (*d*) isobutyl, (*e*) neopentyl. As the size of the single substituent on the carbon bearing —Br increases, crowding at the point of nucleophilic attack increases.

Thus we see that the S_N2 mechanism is supported by three lines of evidence: kinetics, stereochemistry, and effect of structure on reactivity.

We shall return to the S_N2 reaction later in this chapter, but for now let us turn to another mechanism by which nucleophilic aliphatic substitution can take place.

5.16 The S_N1 reaction: mechanism and kinetics. Rate-determining step

The reaction between *tert*-butyl bromide and hydroxide ion to yield *tert*-butyl alcohol follows first-order kinetics; that is, the rate depends upon the concentration of only one reactant, *tert*-butyl bromide.

$$CH_3-\underset{\underset{Br}{|}}{\overset{\overset{CH_3}{|}}{C}}-CH_3 + OH^- \longrightarrow CH_3-\underset{\underset{OH}{|}}{\overset{\overset{CH_3}{|}}{C}}-CH_3 + Br^-$$

$$\text{rate} = k[RBr]$$

How are we to interpret the fact that the rate is independent of $[OH^-]$? If the rate of reaction does not depend upon $[OH^-]$, it can only mean that the reaction *whose rate we are measuring* does not involve OH^-.

These observations are quite consistent with the following mechanism. *tert*-Butyl bromide slowly dissociates (step 1) into a bromide ion and a cation derived

$$(1) \qquad CH_3-\underset{\underset{Br}{|}}{\overset{\overset{CH_3}{|}}{C}}-CH_3 \longrightarrow CH_3-\underset{\oplus}{\overset{\overset{CH_3}{|}}{C}}-CH_3 + Br^- \qquad \textit{Slow}$$

$$S_N1$$

$$(2) \qquad CH_3-\underset{\oplus}{\overset{\overset{CH_3}{|}}{C}}-CH_3 + OH^- \longrightarrow CH_3-\underset{\underset{OH}{|}}{\overset{\overset{CH_3}{|}}{C}}-CH_3 \qquad \textit{Fast}$$

from the *tert*-butyl group: a *carbocation*. This carbocation then combines rapidly (step 2) with a hydroxide ion to yield *tert*-butyl alcohol.

The rate of the overall reaction is determined by the slow breaking of the C—Br bond to form the carbocation; once formed, the carbocation reacts rapidly to form the product. It is step (1) whose rate we are actually measuring; this step does not involve OH^-, and its rate does not depend upon $[OH^-]$. *A single step whose rate determines the overall rate of a stepwise reaction is called a* **rate-determining step**.

It is not surprising that the rate-determining step here is the one that involves the *breaking* of a bond, an energy-demanding process. We recognize this particular bond-breaking as an example of *heterolysis*, cleavage in which both bonding electrons go to the same fragment: a process that takes even more energy (Sec. 1.14) than the homolysis that we encountered in free-radical substitution. (In Sec. 6.5, we shall find out just where all this energy comes from.)

Nor is it surprising that the combining of the carbocation with hydroxide ion is a very fast step, since it involves only the *formation* of a bond, an energy-releasing process. We recognize this combining as an acid–base reaction in the Lewis sense.

We are familiar with hydroxide ion as a strong base; as we shall see, carbocations are powerful Lewis acids.

This is the mechanism that is called **S$_N$1**: *substitution nucleophilic unimolecular*. The term *uni*molecular is used here since the rate-determining step involves only *one* molecule (the many necessary solvent molecules being disregarded).

We must not imagine that the laws of chemistry are somehow magically suspended for the second, fast step. It involves a reaction with OH$^-$ and its rate depends on [OH$^-$]. What is special here is that, even if step (2) is slowed down by a low [OH$^-$], it is still *much* faster than step (1), and any change in its rate does not affect the overall rate.

Let us see what we mean by rate-determining step in a reaction like this,

(1) $$A \; \underset{k_{-1}}{\overset{k_1}{\rightleftarrows}} \; R + B$$

(2) $$R + C \; \overset{k_2}{\longrightarrow} \; product$$

where R is a reactive intermediate (carbocation, free radical, carbanion) whose concentration is maintained at some low *steady state* throughout the reaction. The exact kinetics expression for the formation of the product is

(3) $$rate = \frac{k_1[A]}{1 + \dfrac{k_{-1}[B]}{k_2[C]}}$$

Without going into the derivation of this equation, let us see what it means.

The term $k_1[A]$ is in the numerator and the term $k_2[C]$ is in the denominator of the denominator; the bigger they are, the faster the rate. This is reasonable, since $k_1[A]$ is the rate of step (1) and $k_2[C]$ contributes to the rate of step (2). The term $k_{-1}[B]$ is in the denominator; the bigger it is, the slower the rate. This, too, is understandable, since it contributes to the rate of the reverse of step (1).

Now if $k_2[C]$ happens to be *much larger* than $k_{-1}[B]$, the term $k_{-1}[B]/k_2[C]$ is very small—insignificant relative to 1—and drops out. Under these conditions we get our familiar rate expression for first-order kinetics:

$$rate = k_1[A]$$

But if $k_2[C]$ is much larger than $k_{-1}[B]$, it must mean that *step (2) is much faster than the reverse of step (1)*. This is the *real* requirement for step (1) to be rate-determining. Does this mean that, contrary to what was said before, step (1)—in the forward direction—need not be slower than step (2)? Step (1) must still be a slow step, for otherwise the reactive intermediate would be formed faster than it could be consumed, and its concentration would build up—contrary to the nature of the reactive intermediate, and a condition different from the one for which the kinetics expression (3) holds.

What evidence is there that alkyl halides can react by the S$_N$1 mechanism? As we have just seen, the mechanism is consistent with the first-order kinetics of a reaction like the one between *tert*-butyl bromide and hydroxide ion. In general, **an S$_N$1 reaction follows first-order kinetics**. The rate of the entire reaction is determined by how fast the alkyl halide ionizes, and hence depends only upon the concentration of alkyl halide.

In following sections, we shall look at some of the other evidence. But to understand this evidence, we must know something about the intermediate that lies at the heart of the mechanism—and, indeed, at the heart of much of organic chemistry—the *carbocation*. And so, for a time, we shall find ourselves tracing two intertwining threads through the pattern of organic chemistry: concentrating alternately on the S_N1 reaction and on the fundamental chemistry of carbocations.

5.17 Carbocations

To account for the observed facts, we saw earlier, a certain mechanism was advanced for the halogenation of alkanes; central to this mechanism is the fleeting existence of free radicals, highly reactive neutral particles bearing an odd electron.

To account for a host of observations—of which the first-order kinetics described in the preceding section is just one—another kind of reactive particle has been proposed: the **carbocation**, *a group of atoms that contains a carbon atom bearing only six electrons.*

Carbocations are classified as primary, secondary, or tertiary after the carbon bearing the positive charge. They are named by use of the word *cation*. For example:

$$
\begin{array}{cccc}
\overset{\text{H}}{\underset{\text{H}}{\text{H:}\overset{..}{\text{C}}\oplus}} &
\overset{\text{H}}{\underset{\text{H}}{\text{CH}_3\text{:}\overset{..}{\text{C}}\oplus}} &
\overset{\text{H}}{\underset{\oplus}{\text{CH}_3\text{:}\overset{..}{\text{C}}\text{:CH}_3}} &
\overset{\text{CH}_3}{\underset{\oplus}{\text{CH}_3\text{:}\overset{..}{\text{C}}\text{:CH}_3}}
\end{array}
$$

Methyl cation	Ethyl cation	Isopropyl cation	*tert*-Butyl cation
	(primary, 1°)	*(secondary, 2°)*	*(tertiary, 3°)*

We must expect to encounter two other names for what we have called the *carbocation*. *Carbonium ion* is almost the only name used in the older literature; it is still very commonly used, although sometimes with a special meaning. Olah (below) has proposed that *carbenium ion* be used for the species we have described above, with the name *carbonium ion* reserved for such species as CH_5^+ (analogous to *ammonium ion*, etc.); carbenium ions and carbonium ions together would be called *carbocations*.

Like the free radical, the carbocation is an exceedingly reactive particle, and for the same reason: the tendency to complete the octet of carbon. Since it takes a *pair* of electrons to complete the octet here, the carbocation is a Lewis acid, and an extremely powerful one. Unlike the free radical, the carbocation carries a positive charge.

One kind of unusually stable carbocation (Problem 15.10, p. 557) was recognized as early as 1902 by the salt-like character of certain organic compounds. But for simple alkyl cations such direct observation should be exceedingly difficult, because of the very reactivity—and hence short life—that is attributed to them. Nevertheless, during the 1920s and 1930s, alkyl cations were proposed as intermediates in many organic reactions, and their existence was generally accepted, due largely to the work of three chemists: Hans Meerwein of Germany, "the father of modern carbonium ion chemistry"; Sir Christopher Ingold of England; and Frank Whitmore of the United States. The evidence consisted of a wide variety of observations made in studying the chemistry of alkyl halides, alcohols, alkenes, and many other kinds of organic compounds: observations that revealed a basically similar pattern of behavior most logically attributed to intermediate carbocations. A sizeable part of this book will be devoted to seeing what that pattern is.

In 1963, George Olah (now at University of Southern California) reported the *direct observation* of simple alkyl cations. Dissolved in the extremely powerful Lewis acid SbF_5, alkyl fluorides (and, later, other halides) were found to undergo ionization to form the cation, which could be studied at leisure. There was a dramatic change in the NMR spectrum (Chap. 16), from the spectrum of the alkyl fluoride to the spectrum of a molecule that contains no fluorine but instead sp^2-hybridized carbon with a very low electron density.

$$RF + SbF_5 \longrightarrow R^+ SbF_6^-$$

Figure 5.6 shows what was observed for the *tert*-butyl fluoride system: a simple spectrum but, by its very simplicity, enormously significant. Although potentially very reactive, the *tert*-butyl cation can do little in this environment except try to regain the fluoride ion—and the SbF_5 is an even stronger Lewis acid than the cation. This is an acid–base reaction; the SbF_5 (a so-called *superacid*) is an even

a (CH$_3$)$_3$CF neat
b (CH$_3$)$_3$CF in SbF$_5$

J_{HF} 20 Hz

Courtesy of *The Journal of American Chemical Society*

Figure 5.6 Proton NMR spectrum of (*a*) *tert*-butyl fluoride and (*b*) *tert*-butyl cation. In (*a*), the proton signal split into two peaks by coupling with nearby fluorine. In (*b*), a single peak, shifted far downfield; there is strong deshielding due to low electron density on the positive carbon.

stronger Lewis acid than the alkyl cation, and keeps the base it has won, the fluoride ion.

By methods like this, Olah opened the door to the study not just of the existence of organic cations of many kinds, but of intimate details of their structure.

A highly meaningful sequel to this reaction can be carried out. If, now, the solution containing $R^+SbF_6^-$ is diluted with water, there is obtained the alcohol, ROH. What we have here are, essentially, the two steps proposed for the S_N1 reaction—generation of a carbocation, and its combination with a nucleophile—

(1) $$RF + SbF_5 \longrightarrow R^+SbF_6^-$$

(2) $$R^+SbF_6^- + H_2O \longrightarrow ROH_2^+ + SbF_6^- \longrightarrow ROH + SbF_5 + HF$$

but observed as discrete processes, separated by as long a time period as we care to wait.

With only a sextet of electrons on carbon, a carbocation is an unstable, highly reactive particle. It can undergo a wide variety of reactions, as we shall see; just which one occurs depends upon the experimental conditions. But all reactions of a carbocation have a common end: *to provide a pair of electrons to complete the octet of the positively charged carbon.* In the second step of an S_N1 reaction we see perhaps the most direct way of going about this: combining with a nucleophile, a basic, electron-rich molecule.

$$\mathbf{R^+} \quad + \quad \mathbf{:Z} \quad \longrightarrow \quad \mathbf{R:Z}$$

Carbocation   Nucleophile

5.18   Structure of carbocations

In a carbocation, the electron-deficient carbon is bonded to three other atoms, and for this bonding uses sp^2 orbitals. As we have seen (Sec. 1.10), sp^2 orbitals lie in one plane, that of the carbon nucleus, and are directed toward the corners of an equilateral triangle. This part of a carbocation is therefore *flat*, the electron-deficient carbon and the three atoms attached to it lying in the same plane (Fig. 5.7a).

(a)   (b)

Figure 5.7   A carbocation. (a) Only σ bonds shown. (b) An empty *p* orbital above and below the plane of the σ bonds.

But our description of the molecule is not yet quite complete. Carbon still has a *p* orbital, with its two lobes lying above and below the plane of the σ bonds (Fig. 5.7b); in a carbocation, the *p* orbital is *empty*. Although formally empty, this *p* orbital, we shall find, is intimately involved in the chemistry of carbocations: in their stability, and in the stability of various transition states leading to their formation. This comes about through overlap of the *p* orbital with certain nearby orbitals—overlap that is made geometrically possible by the flatness of the carbocation.

There can be little doubt that carbocations actually are flat. The quantum mechanical picture of a carbocation is exactly the same as that of boron trifluoride (Sec. 1.10), a molecule whose flatness is firmly established. NMR and infrared spectra of the stabilized carbocations studied by Olah are consistent with sp^2 hybridization and flatness: in particular, infrared and Raman spectra of the *tert*-butyl cation are strikingly similar to those of trimethylboron, known to be flat.

5.19   The S_N1 reaction: stereochemistry

We shall continue with the fundamental chemistry of carbocations in Sec. 5.20, but for now let us pick up the thread of our original discussion, nucleophilic substitution, and look at an aspect that is directly related to the shape of carbocations: the stereochemistry of the S_N1 reaction. Here, as in stereochemical studies of the S$_N$2 reaction (Sec. 5.14), substitution is carried out on an optically active substrate; the product is isolated, and its configuration and optical purity are compared with those of the starting material. As before, relative configurations of reactant and product must have been assigned, and rotations of optically pure samples must be known so that optical purities can be calculated.

Such studies have been made of the reactions between several tertiary substrates and the solvent methanol, CH$_3$OH: reactions of a type most likely to proceed by S$_N$1. In each case there is obtained a product of opposite configuration

$$
\begin{array}{cc}
\overset{\displaystyle CH_3}{\underset{\displaystyle C_2H_5}{R-C-W}} + CH_3OH & \longrightarrow \quad \overset{\displaystyle CH_3}{\underset{\displaystyle C_2H_5}{R-C-OCH_3}} + W^- + H^+ \\
\textit{Optically active} & \textit{Opposite configuration;} \\
& \textit{lower optical purity}
\end{array}
$$

S$_N$1: racemization plus inversion

from the starting material, and of considerably *lower optical purity*. Optically pure substrate, for example, gives a product that is only about 50% optically pure—and in some cases much less pure than that.

Now, optically pure starting material contains only the one enantiomer, whereas the product clearly must contain both. The product is thus a mixture of the inverted compound and the racemic modification, and we say that the reaction has proceeded with inversion plus *partial racemization*.

Let us get our terms straight. Consider the case where optically pure substrate gives product of opposite configuration and 50% optical purity. Of every 100 molecules of product, 75 are formed with inversion of configuration, and 25 with retention. The 25 of retained configuration cancel the rotation of 25 of the molecules of inverted configuration, leaving an excess of 50 molecules of inverted configuration to provide the observed optical rotation: 50% of the maximum value.

One could say that the reaction proceeds with 75% inversion and 25% retention; equally accurately one could say that reaction proceeds with 50% inversion and 50% racemization. But it is the latter way that we generally use: the percentage of racemization, as we shall see, is a measure of stereochemical randomness, and the percentage of *net* inversion (or, as happens in some kinds of reactions, net retention) is a measure of stereoselectivity (Sec. 9.2).

Problem 5.7   Optically pure (*R*)-α-phenylethyl chloride (C$_6$H$_5$CHClCH$_3$) has [α] −109°; optically pure (*R*)-α-phenylethyl alcohol has [α] −42.3°. When chloride of [α] −34° is treated with dilute aqueous NaOH, there is obtained alcohol of [α] +1.7°. Calculate (a) the optical purity of reactant and of product; (b) the percentage of retention and of inversion; (c) the percentage of racemization and of retention or inversion.

How do we account for the stereochemistry observed for the S$_N$1 reaction? Let us see first why racemization occurs, and then why it is only partial and is accompanied by some net inversion.

In an S$_N$2 reaction, we saw (Sec. 5.14), the nucleophile attacks the substrate molecule itself, and the *complete* inversion observed is a direct consequence of that fact: the leaving group is still attached to carbon at the time of attack, and directs this attack on every molecule in the same way—to the back side. Now, in an S_N1 reaction the nucleophile attacks, *not* the substrate, but the intermediate, the carbocation; the leaving group has already become detached and, we might have thought, can no longer affect the spatial orientation of attack.

Let us see where this line of reasoning leads us. In the first step the optically active substrate—an alkyl halide, say—dissociates to form halide ion and the carbocation. The nucleophilic reagent, Z:, then attaches itself to the carbocation. But it may attach itself to either face of this flat ion and, depending upon which face, yield one or the other of the two enantiomeric products (see Fig. 5.8). Together, the two enantiomers constitute the racemic modification. Thus, the

Figure 5.8   The S_N1 reaction: racemization plus inversion. The nucleophilic reagent attacks both (*a*) the back side and (*b*) the front side of the carbocation. Back-side attack predominates.

racemization that accompanies these reactions is consistent with the S_N1 mechanism and the formation of an intermediate carbocation.

(So far, our discussion parallels what was said about the stereochemistry of free-radical chlorination (Sec. 4.28) where, we remember, random attack on the two faces of a free radical gives total racemization.)

Now, if attack on the two faces of the carbocation were purely random, we would expect to obtain equal amounts of the two enantiomers; that is to say, we would expect to obtain only the racemic modification. Yet, although racemization is sometimes very high—90% or more—it is seldom complete, and in general the inverted product exceeds its enantiomer. Reaction proceeds with racemization *plus some net inversion.*

How do we accommodate even this limited net inversion within the framework of the S_N1 mechanism? How do we account for the fact that attack on the carbocation is *not* purely random? Clearly, the excess of inversion is due, in some way, to the leaving group: it must still be exerting a measure of control over the stereochemistry. In the complete absence of the leaving group, the flat carbocation would lose all chirality and could not yield a product with any optical activity. (*Remember*: Synthesis of chiral compounds from achiral reactants always yields the racemic modification.)

How can the leaving group be involved? To find an answer, let us consider the process of heterolysis. As reaction proceeds, the distance between carbon and halogen steadily increases until finally the covalent bond breaks. The two oppositely charged ions are formed—but *not*, immediately, as completely *free* ions. Initially, they must be close together, close enough for electrostatic attraction to be sizeable; and so they exist—for a time—as an *ion pair* (Sec. 6.4). As first formed, the ions are in contact with each other. Then, as they diffuse apart, layer after layer of solvent intervenes until finally they are independent of each other, and we speak of "free" ions.

$$R:X \longrightarrow R^+X^- \longrightarrow R^+\|X^- \longrightarrow R^+ + X^-$$

| | Intimate ion pair | Solvent-separated ion pair | Free ions |

Now, nucleophilic attack can, conceivably, take place at any time after the heterolysis, and thus can involve any species from the initially formed ion pair to the free carbocation. Attack on the free carbocation is random, and yields the racemic modification. But attack on the ion pair is not random: the anion clings more or less closely to the front side of the carbocation and thus shields this side from attack; as a result, back-side attack is preferred. To the extent, then, that attack occurs before the ion pair has completely separated, inversion of configuration competes with racemization.

Thus the S_N1 mechanism can accommodate the fact that racemization is not complete. But the important thing—the important contrast to the S_N2 stereochemistry—is that racemization occurs at all. Unlike an S_N2 reaction, which proceeds with complete inversion, **an S_N1 reaction proceeds with racemization**.

That there are two kinds of stereochemistry supports the central idea that there are two different mechanisms. The particular form of the stereochemistry gives powerful support for the particular mechanisms proposed. Complete inversion in the S_N2 reaction supports the idea of concerted bond-breaking and bond-making, *in a single step*; occurrence of racemization in the S_N1 reaction shows that bond-breaking and bond-making occur separately, *in different steps*.

The next aspect of the S_N1 reaction that we shall take up is the matter of *reactivity*. But to understand that, we must first return to the chemistry of carbocations, and examine what will be to us their most important property: their *relative stabilities*.

5.20 Relative stabilities of carbocations

When we wished to compare stabilities of free radicals (Sec. 3.24), we made use of homolytic bond dissociation energies, since these apply to reactions in which free radicals are generated.

Now we wish to compare stabilities of carbocations, and to do this we shall follow exactly the same line of reasoning that we followed for free radicals. This time, however, we must start with the *heterolytic* bond dissociation energies in Table 1.3 (p. 22), since these apply to reactions in which carbocations are generated. In this table we find energies of the bonds that hold bromine to a number of groups.

These are the ΔH values of the following reactions:

$$CH_3-Br \longrightarrow CH_3^{\oplus} + Br^- \qquad \Delta H = 219 \text{ kcal}$$

Methyl bromide　　Methyl cation

$$CH_3CH_2-Br \longrightarrow CH_3CH_2^{\oplus} + Br^- \qquad \Delta H = 184$$

Ethyl bromide　　　　Ethyl cation
　　　　　　　　　　　A 1° cation

$$CH_3\underset{\underset{Br}{|}}{CH}CH_3 \longrightarrow CH_3\overset{\oplus}{C}HCH_3 + Br^- \qquad \Delta H = 164$$

Isopropyl bromide　　　Isopropyl cation
　　　　　　　　　　　A 2° cation

$$CH_3-\underset{\underset{Br}{|}}{\overset{\overset{CH_3}{|}}{C}}-CH_3 \longrightarrow CH_3-\underset{\oplus}{\overset{\overset{CH_3}{|}}{C}}-CH_3 + Br^- \qquad \Delta H = 149$$

tert-Butyl bromide　　　*tert*-Butyl cation
　　　　　　　　　　　　A 3° cation

By definition, this bond dissociation energy is the amount of energy that must be supplied to convert a mole of alkyl bromide into carbocations and bromide ions.

$$R-Br \longrightarrow R^+ + Br^- \qquad \Delta H = \text{heterolytic bond dissociation energy}$$

As we can see, the amount of energy needed to form the various classes of carbocations decreases in the order: $CH_3^+ > 1° > 2° > 3°$.

If less energy is needed to form one carbocation than another, it can only mean that, *relative to the alkyl bromide from which it is formed*, the one carbocation contains less energy than the other, that is to say, is *more stable* (see Fig. 5.9).

Figure 5.9 Stabilities of carbocations relative to alkyl bromides. (The plots are aligned with each other for easy comparison.)

We are not attempting to compare the absolute energy contents of, say, isopropyl and *tert*-butyl cations; we are simply saying that the difference in energy between isopropyl bromide and isopropyl cations is greater than the difference between *tert*-butyl bromide and *tert*-butyl cations. *When we compare stabilities of carbocations, it must be understood that our standard for each cation is the substrate from which it is formed.* As we shall see, this is precisely the kind of stability that we are interested in.

We used alkyl bromides for our comparison above, but we could just as well have used alkyl fluorides, chlorides, or iodides, or the corresponding alcohols. For all these compounds the bond dissociation energies in Table 1.3 show the same order of stability of carbocations. Even the sizes of the energy differences, in kcal/mol, are very nearly the same, whatever the class of parent compounds. The difference between methyl and *tert*-butyl cations, for example, relative to various substrates is: fluorides, 67 kcal; chlorides, 70 kcal; bromides, 70 kcal; iodides, 72 kcal; and alcohols, 66 kcal.

Relative to the substrate from which each is formed, then, the order of stability of carbocations is:

Stability of carbocations $\qquad$ $3° > 2° > 1° > CH_3{}^+$

We shall find that this same order of stability applies not only when carbocations are formed by heterolysis, but also when they are formed by entirely different processes.

Differences in stability between carbocations are *much* larger than between free radicals. The *tert*-butyl free radical, for example, is only 12 kcal more stable than the methyl free radical; the *tert*-butyl cation is, depending upon the substrate, 66–72 kcal more stable than the methyl cation. As we shall see, these much larger differences in stability give rise to much larger effects on reactivity.

So far in this section, our discussion has been based on bond dissociation energies, which are measured in the gas phase. But nearly all carbocation chemistry takes place in solution, and solvents, as we know, can exert powerful stabilizing effects on ionic solutes. Does the order of stability that we have arrived at hold for carbocations in solution? The answer to this question has been given most directly by measurement, in a variety of solvents, of the ΔH values for the generation of carbocations by Olah's superacid method. The values obtained reveal the same

$$RCl + SbF_5 \longrightarrow R^+SbF_5Cl^- \qquad \Delta H = \text{heat of ionization}$$

order of carbocation stability, relative to the parent substrate, as do the dissociation energies. Even the *differences* in stability, in kcal/mol, are much the same.

So now we have arrived at an order of stability of carbocations which holds for solution as well as for gas phase, and which applies to the generation of carbocations from a wide variety of substrates and, we shall see, in a wide variety of chemical reactions. As we continue our study, we shall add other kinds of carbocations to our series, and examine other kinds of reactions by which they can be generated.

Now, let us see how this order of stability can be accounted for.

5.21 Stabilization of carbocations. Accommodation of charge. Polar effects

The characteristic feature of a carbocation is, by definition, the electron-deficient carbon and the attendant positive charge. The relative stability of a carbocation is determined chiefly by how well it *accommodates* that charge.

According to the laws of electrostatics, **the stability of a charged system is increased by dispersal of the charge**. Any factor, therefore, that tends to spread out the positive charge of the electron-deficient carbon and distribute it over the rest of the ion must stabilize a carbocation.

Consider a substituent, G, attached to an electron-deficient carbon in place of a hydrogen atom. Compared with hydrogen, G may either release electrons or withdraw electrons. Such *an effect on the availability of electrons at the reaction center is called a* **polar effect**.

Carbocation stability

G releases electrons:	G withdraws electrons:
disperses charge,	*intensifies charge,*
stabilizes cation	*destabilizes cation*

An electron-releasing substituent tends to reduce the positive charge at the electron-deficient carbon; in doing this, the substituent itself becomes somewhat positive. This dispersal of the charge stabilizes the carbocation.

An electron-withdrawing substituent tends to intensify the positive charge on the electron-deficient carbon, and hence makes the carbocation less stable.

The order of stability of carbocations, we have just seen, is the following:

Stability of carbocations $3° > 2° > 1° > CH_3{}^+$

Now, by definition, the distinction among primary, secondary, and tertiary cations is the number of alkyl groups attached to the electron-deficient carbon. The facts are, then, *that the greater the number of alkyl groups, the more stable the carbocation.*

| Methyl cation | Primary cation | Secondary cation | Tertiary cation |

Electron release: *Disperses charge, stabilizes ion*

If our generalization about dispersal of charge applies in this case, alkyl groups must *release electrons* here.

Is electron release what we would have expected of alkyl groups? Ingold (p. 183) has suggested that alkyl groups, lacking strong polar tendencies of their own, can do pretty much what is demanded of them by other groups in the molecule. There is increasing evidence that this is so: alkyl groups often tend to stabilize both cations and anions, indicating electron release or electron withdrawal *on demand*. In a carbocation, electron-deficient carbon has an urgent need for electrons—it is like a different element, a very electronegative one—and it *induces* alkyl groups to release electrons to meet that need.

Now, how does a substituent exert its polar effect? Despite the vast amount of work that has been done—and is still being done—on this problem, there is no general agreement, except that at least two factors must be at work. We shall consider electron withdrawal and electron release to result from the operation of two factors: the *inductive effect* and the *resonance effect*.

The **inductive effect** depends upon the "intrinsic" tendency of a substituent to release or withdraw electrons—by definition, its electronegativity—acting either through the molecular chain or through space. The effect weakens steadily with increasing distance from the substituent. Most elements likely to be substituted for hydrogen in an organic molecule are more electronegative than hydrogen, so that most substituents exert electron-withdrawing inductive effects: for example, —F, —Cl, —Br, —I, —OH, —NH$_2$, —NO$_2$.

The **resonance effect** involves *delocalization* of electrons—typically, those called π (pi) electrons. It depends upon the overlap of certain orbitals, and therefore can only operate when the substituent is located in certain special ways relative to the charge center. By its very nature, as we shall see (Sec. 10.14), the resonance effect is a stabilizing effect, and so it amounts to electron withdrawal from a negatively charged center, and electron release to a positively charged center.

The nature of the electron release by alkyl groups is not clear. It may be an inductive effect; it may be a resonance effect (*hyperconjugation*, Sec. 10.16), electrons being provided by overlap of σ bonds with the empty p orbital of the electron-deficient carbon. It may very well be a combination of the two. When we refer to the inductive effect of alkyl groups in this book, it should be understood that this may well include a contribution from hyperconjugation.

However it arises, the polar effect of alkyl groups is not a powerful one, as such effects go. Yet it leads to very large differences in stability among the various classes of carbocations. And it is these differences that we must keep uppermost in our minds in dealing with the varied chemistry of carbocations.

5.22   The S$_N$1 reaction: reactivity. Ease of formation of carbocations

Once again let us return to nucleophilic substitution, and the matter of how the structure of the alkyl group affects reactivity. We have already seen (Sec. 5.15) that reactivity in S$_N$2 decreases along the series CH$_3$W, 1°, 2°, 3°, as postulated by Hughes and Ingold (Sec. 5.12). Now, what are the facts with regard to the other half of their duality theory: does reactivity by S$_N$1 change in the opposite direction along this same series?

Under conditions that greatly favor S$_N$1, results like the following have been obtained:

S$_N$1 substitution: relative reactivity

$$R-W + CF_3COOH \longrightarrow R-OCOCF_3 + H-W$$

	CH$_3$		CH$_3$		H		H
	|		|		|		|
CH$_3$—C—W	>	CH$_3$—C—W	>	CH$_3$—C—W	>	H—C—W
	|		|		|		|
	CH$_3$		H		H		H

	tert-Butyl	Isopropyl	Ethyl	Methyl
Relative rate (S$_N$1)	$> 10^6$	1.0	$< 10^{-4}$	$< 10^{-5}$

Thus, the postulated order of reactivity is confirmed. Also as postulated—see the sharply rising S_N1 curve of Fig. 5.1 (p. 182)—the differences in reactivity are much greater than those found for the S_N2 reaction. By S_N1, tertiary substrates are more than a million times as reactive as secondary, which in turn are at least ten thousand—and probably more than a million—times as reactive as primary.

Even these differences are believed to be underestimations. Reactivities of primary and methyl substrates are *very* much less than the maximum values indicated; it is likely that even the small rates measured for them are in large part not for S_N1, but for S_N2 with the solvent acting as nucleophile (Sec. 6.9).

The reactivity of substrates in the S_N1 reaction, then, follows the sequence:

Reactivity in S_N1 $3° > 2° > 1° > CH_3W$

Now, the rate-determining step in S_N1 is formation of the carbocation; that is to say, one substrate undergoes S_N1 faster than another because it forms a carbocation faster. Our reactivity sequence therefore leads directly to a sequence showing the relative rates of formation of carbocations:

Rate of formation of carbocations $3° > 2° > 1° > CH_3^+$

In listing carbocations in order of their rates of formation, we find we have at the same time listed them in order of their stability. **The more stable the carbocation, the faster it is formed.**

This is probably the most useful generalization about structure and reactivity that appears in this book—or, indeed, that exists in organic chemistry. Carbocations are formed from many compounds other than alkyl halides, and in reactions quite different from nucleophilic substitution. *Yet in all these reactions in which carbocations are formed, carbocation stability plays a leading role in governing reactivity and orientation.*

How can we account for the fact that the rate of formation of a carbocation depends upon its stability? As always to answer a question like this, we must take the specific reaction involved—here, the S_N1 reaction—and compare the structure of the reactants with the structure of the transition state.

In an S_N1 reaction of an alkyl halide, the carbocation is formed by heterolysis of the substrate molecule, that is, by breaking of the carbon–halogen bond. In the reactant an electron pair is shared by carbon and halogen; except for a modest polarity, these two atoms are neutral. In the products, halogen has taken away the electron pair, and carbon is left with only a sextet; halide carries a full negative charge, and the carbocation carries a full positive charge centered on carbon.

In the transition state, the C—X bond must be partly broken, halogen having partly pulled the electron pair away from carbon. Halogen has partly gained the negative charge it is to carry in the halide ion. Most important, *carbon has partly gained the positive charge it is to carry in the carbocation.*

$$R:X \longrightarrow \begin{bmatrix} \overset{\delta_+}{R}\text{----}\overset{\delta_-}{:}X \end{bmatrix} \longrightarrow R^+ + :X^-$$

Reactant	Transition state	Products
	R *has partial*	R *has full*
	positive charge	*positive charge*

Electron-releasing groups tend to disperse the partial positive charge (δ_+) developing on carbon, and in this way stabilize the transition state. Stabilization of the transition state lowers the E_{act} and permits a faster reaction (see Fig. 5.10).

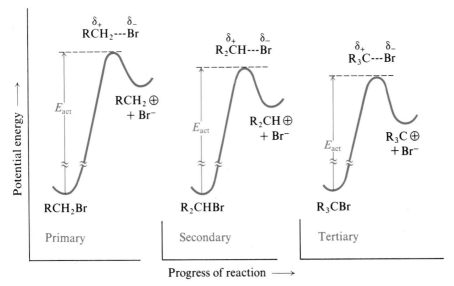

Figure 5.10   Molecular structure and reactivity. The stability of the transition state parallels the stability of the carbocation: the more stable carbocation is formed faster. (The plots are aligned with each other for easy comparison.)

Thus, to the extent that the C—X bond is broken, the alkyl group possesses character of the carbocation it is to become. The same factor, electron release, that stabilizes the carbocation also stabilizes the *incipient* carbocation in the transition state.

In 1979, Edward Arnett (Duke University) and Paul Schleyer (University of Erlangen-Nürnberg) reported "an extraordinary corroboration of the fundamental soundness of the 'carbocation theory of organic chemistry'". For a set of substrates of widely varying structures, they compared the E_{act} values of S$_N$1 reactions with the heats of ionization in superacid solutions, and found a direct *quantitative* dependence of rate of formation of carbocations on carbocation stability. The more stable the carbocation, they found, the faster it is formed.

As we encounter other reactions in which carbocations are formed, we must, for each of these reactions, examine the structure of the transition state. In most, if not all, of these reactions, we shall find that the transition state differs from the reactants chiefly in being like the product. The *carbocation character* of the transition state will be the factor most affecting E_{act}; hence, the more stable the carbocation, the more stable the transition state leading to its formation, and the faster the carbocation will be formed.

But what we have just learned here will be applied in an even more general way. We shall return again and again to the relationship between polar effects and dispersal of charge, and between dispersal of charge and stability. We shall find that these relationships will help us to understand, not only carbocation reactions of many kinds, but all reactions in which a charge—positive or negative—develops or disappears. These will include reactions as seemingly different from S$_N$1 substitution as dehydration of alcohols, addition to alkenes, and aromatic substi-tution—both electrophilic and nucleophilic—and the fundamental properties of acidity and basicity.

Differences in reactivity by S_N1, then, depend upon differences in stability among the various classes of carbocations. In the next section, we shall see how these same differences in stability lead to rather surprising behavior on the part of carbocations.

Problem 5.8   Neopentyl halides are notoriously slow in nucleophilic substitution, whatever the experimental conditions. How can you account for this?

5.23   Rearrangement of carbocations

We spoke earlier of a pattern of behavior that led to the development of the carbocation theory. The most striking feature of that pattern is the occurrence of *rearrangements*.

In nucleophilic substitution, for example, it is sometimes observed that the entering group, Z, becomes attached to a different carbon atom than the one that originally held the leaving group, X. For example:

$$CH_3CH_2CH_2-X \xrightarrow{\ :Z\ } CH_3CHCH_3$$
$$\qquad\qquad\qquad\qquad\qquad\qquad\quad |$$
$$\qquad\qquad\qquad\qquad\qquad\qquad\quad Z$$

n-Propyl substrate   Isopropyl product

$$\qquad\ CH_3 \qquad\qquad\qquad\qquad\qquad CH_3$$
$$\qquad\ \ | \qquad\qquad\qquad\qquad\qquad\quad |$$
$$CH_3-C-CH_2-X \xrightarrow{\ :Z\ } CH_3-C-CH_3$$
$$\qquad\ \ | \qquad\qquad\qquad\qquad\qquad\quad |$$
$$\qquad\ \ H \qquad\qquad\qquad\qquad\qquad\quad Z$$

Isobutyl substrate   *tert*-Butyl product

$$\qquad\ CH_3 \qquad\qquad\qquad\qquad\qquad CH_3$$
$$\qquad\ \ | \qquad\qquad\qquad\qquad\qquad\quad |$$
$$CH_3-C-CH-CH_3 \xrightarrow{\ :Z\ } CH_3-C-CH-CH_3$$
$$\qquad\ \ |\ \ | \qquad\qquad\qquad\qquad\quad\ |\ \ |$$
$$\qquad\ \ H\ \ X \qquad\qquad\qquad\qquad\quad Z\ \ H$$

3-Methyl-2-butyl substrate   2-Methyl-2-butyl product
                                            (*tert*-Pentyl product)

In each of these cases we see that, to accommodate Z in the new position, there must be a rearrangement of hydrogen atoms in the substrate. The transformation of a *n*-propyl group into an isopropyl group, for example, requires removal of one H from C–2, and attachment of one H to C–1.

Sometimes, there is even a rearrangement of the carbon skeleton:

$$\qquad\ CH_3 \qquad\qquad\qquad\qquad\qquad CH_3$$
$$\qquad\ \ | \qquad\qquad\qquad\qquad\qquad\quad |$$
$$CH_3-C-CH_2-X \xrightarrow{\ :Z\ } CH_3-C-CH_2CH_3$$
$$\qquad\ \ | \qquad\qquad\qquad\qquad\qquad\quad |$$
$$\qquad\ \ CH_3 \qquad\qquad\qquad\qquad\qquad Z$$

Neopentyl substrate   *tert*-Pentyl product

$$
\begin{array}{ccc}
\overset{\displaystyle CH_3}{\underset{\displaystyle CH_3\,X}{CH_3-\overset{|}{\underset{|}{C}}-CH-CH_3}} & \xrightarrow{\;:Z\;} & \overset{\displaystyle CH_3}{\underset{\displaystyle Z\;\;CH_3}{CH_3-\overset{|}{\underset{|}{C}}-CH-CH_3}} \\
\text{3,3-Dimethyl-2-butyl substrate} & & \text{2,3-Dimethyl-2-butyl product}
\end{array}
$$

In reactions of quite different types—elimination, addition—rearrangements are also observed, and these rearrangements are of the same pattern as those above. This similarity in behavior suggests a similarity in mechanism. However different the various mechanisms might be, they all have one feature in common: at some stage the same intermediate is formed, and it is this that undergoes the actual rearrangement. This intermediate, as was first clearly proposed in 1922 by Hans Meerwein (p. 194), is the *carbocation*.

Now, of the two mechanisms advanced for nucleophilic substitution, only S_N1 is postulated to involve an intermediate carbocation, and therefore we expect only reactions proceeding by S_N1 to be accompanied by these characteristic rearrangements. By contrast, the single step postulated for S_N2 simply provides no opportunity for such rearrangements.

This expectation is borne out by experiment, as the following example illustrates. Neopentyl substrates are particularly prone to rearrange to *tert*-pentyl products. In a solution of sodium ethoxide in ethanol, neopentyl bromide undergoes a (slow) second-order reaction to yield the unrearranged ethyl neopentyl ether. In a solution of ethanol alone, it undergoes a (slow) first-order reaction to yield ethyl *tert*-pentyl ether (and other rearrangement products).

$$
\begin{array}{c}
\overset{\displaystyle CH_3}{\underset{\displaystyle CH_3}{CH_3-\overset{|}{\underset{|}{C}}-CH_2-Br}} \\
\text{Neopentyl} \\
\text{bromide}
\end{array}
\qquad
\begin{array}{l}
\xrightarrow[S_N2]{C_2H_5O^-} \quad
\overset{\displaystyle CH_3}{\underset{\displaystyle CH_3}{CH_3-\overset{|}{\underset{|}{C}}-CH_2-OC_2H_5}} \quad \textit{No rearrangement} \\
\qquad\qquad\qquad \text{Ethyl neopentyl ether} \\[3em]
\xrightarrow[S_N1]{C_2H_5OH} \quad
\overset{\displaystyle CH_3}{\underset{\displaystyle OC_2H_5}{CH_3-\overset{|}{\underset{|}{C}}-CH_2-CH_3}} \\
\qquad\qquad\qquad \text{Ethyl }\textit{tert}\text{-pentyl ether} \\
\qquad\qquad\qquad\qquad + \\
\qquad\qquad\qquad\quad \text{alkenes} \\
\qquad\qquad\qquad\quad \text{(Sec. 7.21)}
\end{array}
\quad \Big\} \;\textit{Rearrangement}
$$

The occurrence or non-occurrence of rearrangement is a striking difference, and it provides one more piece of evidence that there *are* two mechanisms for nucleophilic substitution. In addition, rearrangement gives powerful support to the particular form of the S_N1 mechanism—the intermediacy of carbocations—by linking this mechanism to the mechanisms of those other kinds of reactions where rearrangements are observed. The correlation between rearrangement and intermediate cations is so strong that, in the absence of other information about a particular example of nucleophilic substitution, rearrangement is generally taken as evidence that reaction is by S_N1.

On this basis, then, we can account for the observed products in the following way. A *n*-propyl substrate, for example, yields the *n*-propyl cation; this rearranges to the isopropyl cation, which combines with the nucleophile to give the isopropyl product.

$$CH_3CH_2CH_2\text{—}X \longrightarrow CH_3CH_2CH_2^{\oplus} + :X^-$$

$$\underset{\textit{A 1° cation}}{CH_3CH_2CH_2^{\oplus}} \longrightarrow \underset{\textit{A 2° cation}}{CH_3\underset{\oplus}{C}HCH_3} \qquad \textit{Rearrangement}$$

$$CH_3\underset{\oplus}{C}HCH_3 + :Z \longrightarrow \underset{\underset{Z}{|}}{CH_3CHCH_3}$$

In a similar way, the isobutyl cation rearranges to the *tert*-butyl cation,

$$\underset{\textit{A 1° cation}}{\overset{\overset{\displaystyle CH_3}{|}}{\underset{\overset{\displaystyle |}{H}}{CH_3\text{—}C\text{—}CH_2^{\oplus}}}} \longrightarrow \underset{\textit{A 3° cation}}{\overset{\overset{\displaystyle CH_3}{|}}{CH_3\text{—}\underset{\oplus}{C}\text{—}CH_3}} \qquad \textit{Rearrangement}$$

the 3-methyl-2-butyl cation rearranges to the 2-methyl-2-butyl cation,

$$\underset{\textit{A 2° cation}}{\overset{\overset{\displaystyle CH_3}{|}}{\underset{\overset{\displaystyle |}{H}}{CH_3\text{—}C\text{—}\underset{\oplus}{C}H\text{—}CH_3}}} \longrightarrow \underset{\textit{A 3° cation}}{\overset{\overset{\displaystyle CH_3}{|}}{CH_3\text{—}\underset{\oplus}{C}\text{—}\overset{\overset{\displaystyle }{|}}{\underset{\overset{\displaystyle |}{H}}{CH}}\text{—}CH_3}} \qquad \textit{Rearrangement}$$

the neopentyl cation rearranges to the *tert*-pentyl cation,

$$\underset{\textit{A 1° cation}}{\overset{\overset{\displaystyle CH_3}{|}}{\underset{\overset{\displaystyle |}{CH_3}}{CH_3\text{—}C\text{—}CH_2^{\oplus}}}} \longrightarrow \underset{\textit{A 3° cation}}{\overset{\overset{\displaystyle CH_3}{|}}{CH_3\text{—}\underset{\oplus}{C}\text{—}CH_2CH_3}} \qquad \textit{Rearrangement}$$

and the 3,3-dimethyl-2-butyl cation rearranges to the 2,3-dimethyl-2-butyl cation.

$$\underset{\underset{CH_3}{|}}{\overset{\overset{CH_3}{|}}{CH_3-\underset{\oplus}{C}-CH-CH_3}} \longrightarrow \underset{\underset{CH_3}{|}}{\overset{\overset{CH_3}{|}}{CH_3-\underset{\oplus}{C}-\underset{|}{CH}-CH_3}} \qquad Rearrangement$$

A 2° cation   A 3° cation

We see that *in each case rearrangement takes place in such a way that a less stable carbocation is converted into a more stable one*: a primary into a secondary, a primary into a tertiary, or a secondary into a tertiary.

Just how does this rearrangement occur? Frank Whitmore (Pennsylvania State University) pictured rearrangement as taking place in this way: a hydrogen atom or alkyl group migrates *with a pair of electrons* from an adjacent carbon to the carbon bearing the positive charge. The carbon that loses the migrating group acquires the positive charge. A migration of hydrogen with a pair of electrons is known as a **hydride shift**; a similar migration of an alkyl group is known as an **alkyl shift**. These are just two examples of the most common kind of rearrangement, the **1,2-shifts**: *rearrangements in which the migrating group moves from one atom to the very next atom.*

A *hydride* shift

1,2-Shifts

An *alkyl* shift

We can account for rearrangements in S_N1 reactions in the following way. A carbocation is formed by loss of the leaving group from the substrate. **If a 1,2-shift of hydrogen or alkyl can form a more stable carbocation, then such a rearrangement takes place.** The new carbocation now combines with the nucleophile to yield the substitution product.

In the case of *n*-propyl cation, for example, a shift of hydrogen yields the more stable isopropyl cation; migration of a methyl group would simply form a different *n*-propyl cation.

:H migrates

$$CH_3-\underset{\underset{\oplus}{|}}{\overset{\overset{H}{|}}{C}}-CH_2-H$$

A 2° cation

$$CH_3-\underset{\underset{H}{|}}{\overset{\overset{H}{|}}{C}}-CH_2^{\oplus}$$

A 1° cation

if :CH₃ migrated

$$\underset{\underset{H}{|}}{\overset{\overset{H}{|}}{\oplus C}}-CH_2-CH_3$$

A 1° cation

In the case of the isobutyl cation, a hydride shift yields a tertiary cation, and hence is preferred over a methyl shift, which would only yield a secondary cation.

$$CH_3 \overset{\displaystyle CH_3}{\underset{\displaystyle H}{-\overset{|}{\underset{|}{C}}-}} CH_2^{\oplus}$$

A 1° cation

:H migrates →

$$CH_3 \overset{\displaystyle CH_3}{-\underset{\oplus}{\overset{|}{C}}-} CH_2-H$$

A 3° cation

if :CH₃ migrated →

$$CH_3 \underset{\displaystyle H}{-\overset{\oplus}{\underset{|}{C}}-} CH_2-CH_3$$

A 2° cation

In the case of the 3,3-dimethyl-2-butyl cation, on the other hand, a methyl shift can yield a tertiary cation and is the rearrangement that takes place.

$$CH_3 \overset{\displaystyle CH_3}{\underset{\displaystyle CH_3}{-\overset{|}{\underset{|}{C}}-}} \overset{\displaystyle H}{\underset{\displaystyle H}{\overset{|}{\underset{|}{CH}}}} \overset{}{\underset{}{\overset{\oplus}{C}}}-H$$

A 2° cation

:CH₃ migrates →

$$CH_3 \underset{\oplus}{\overset{\displaystyle CH_3}{-\overset{|}{C}-}} \underset{\displaystyle CH_3}{\overset{|}{CH}}-CH_3$$

A 3° cation

if :H migrated →

$$CH_3 \underset{\displaystyle CH_3 H}{-\overset{\displaystyle CH_3}{\overset{|}{C}}-CH-}CH_2^{\oplus}$$

A 1° cation

We can view rearrangement as an intramolecular acid–base reaction in which, as usual, the stronger acid gets the base. The base is the migrating group with its electrons (hydride or alkyl). Competing for it are two Lewis acids: the electron-deficient carbons in the alternative carbocations. In the *n*-propyl–isopropyl rearrangement, for example, C–1 is more electron-deficient and hence the stronger acid, and it ends up holding the base.

Just as the reality of carbocations has been verified, so has the reality of their rearrangement. Prepared under the superacid conditions of Olah, and studied by spectroscopy, carbocations have been *observed* to rearrange; the rates of some rearrangements have even been measured, and the E_{act} values estimated. If water is added, it combines with the rearranged cation, and the —OH appears at the new position in the molecule. Here again we are observing as discrete processes steps proposed for S_N1, this time with rearrangement: first, formation of a carbocation; then, its rearrangement into a new cation; and finally, combination of this new cation with the nucleophile.

In our short acquaintance with the carbocation, we have encountered two of its reactions. A carbocation may:

(a) combine with a nucleophile;
(b) rearrange to a more stable carbocation.

This list will grow rapidly.

In rearrangement, as in every other reaction of a carbocation, the electron-deficient carbon atom gains a pair of electrons, this time at the expense of a neighboring carbon atom, one that can better accommodate the positive charge.

Problem 5.9   When the alkene 3,3-dimethyl-1-butene is treated with hydrogen iodide, there is obtained a mixture of products:

$$CH_3-\underset{\underset{CH}{|}}{\overset{\overset{CH_3}{|}}{C}}-CH=CH_2 \quad \xrightarrow{HI} \quad CH_3-\underset{\underset{CH_3}{|}}{\overset{\overset{CH_3}{|}}{C}}-\underset{\underset{I}{|}}{CH}-CH_3 \quad + \quad CH_3-\underset{\underset{I}{|}}{\overset{\overset{CH_3}{|}}{C}}-\underset{\underset{H}{|}}{\overset{\overset{CH_3}{|}}{C}}-CH_3$$

3,3-Dimethyl-1-butene   3-Iodo-2,2-dimethylbutane   2-Iodo-2,3-dimethylbutane

What does the formation of the second product suggest to you? Propose a likely mechanism for this reaction, which is an example of *electrophilic addition*, the reaction most typical of alkenes. Check your answer in Secs. 8.9 and 8.10.

In Chapter 6, we shall look at another aspect of the S_N1 reaction, and examine the factor that makes it all possible: the solvent. For now, let us return to the place where we started, the competition between S_N1 and S_N2.

5.24 S_N2 *vs.* S_N1

We have so far described two mechanisms for nucleophilic substitution: the S_N2, characterized by

(a) second-order kinetics,
(b) complete stereochemical inversion,
(c) absence of rearrangement, and
(d) the reactivity sequence $CH_3W > 1° > 2° > 3°$;

and the **S_N1**, characterized by

(a) first-order kinetics,
(b) racemization,
(c) rearrangement, and
(d) the reactivity sequence $3° > 2° > 1° > CH_3W$.

Except for a brief discussion in Sec. 5.12, we have discussed these mechanisms as separate topics. Now let us turn to the relationship between the two. For a given substrate under a given set of conditions, which mechanism will be followed? And what, if anything, can we do to throw reaction toward one mechanism or another?

To answer these questions, let us consider just what can happen to a molecule of substrate. It can either suffer back-side attack by the nucleophile, or undergo heterolysis to form a carbocation. Whichever of these two processes goes faster

$$\begin{array}{l}
\xrightarrow{\;:Z\;} \quad [Z\text{-}\text{-}\text{-}R\text{-}\text{-}\text{-}\text{-}W] \quad \longrightarrow \quad Z\text{---}R \;+\; :W \qquad\qquad S_N2 \\[4pt]
\hspace{9.5cm} \textit{Nucleophilic attack}
\end{array}$$

R—W

$$\begin{array}{l}
\xrightarrow{\quad} \quad \left[\overset{\delta+}{R}\text{-}\text{-}\text{-}W\right] \quad \longrightarrow \quad R^{\oplus} \;+\; :W \qquad\qquad S_N1 \\[4pt]
\hspace{4.7cm} \Big|_{\xrightarrow{\;:Z\;}}\; Z\text{---}R \hspace{2.6cm} \textit{Heterolysis}
\end{array}$$

determines which mechanism predominates. (*Remember*: Heterolysis is the first—and rate-determining—step of the S_N1 mechanism.) Once again, we find, we must turn to the matter of *relative rates of competing reactions*.

Let us examine each of the components of the reaction system, and see what effect it exerts on this competition between nucleophilic attack and heterolysis.

Let us begin with the **substrate**, which consists of two parts, the alkyl group and the leaving group. The *nature of the leaving group* is, of course, vital to the very occurrence of substitution. Whichever process is taking place, nucleophilic attack or heterolysis, the bond to the leaving group is being broken; the easier it is to break this bond—that is, the better the leaving group—the faster the reaction occurs. A better leaving group thus speeds up reaction by both mechanisms; and, as it happens, it speeds up both to about the same degree. As a result, the nature of the leaving group has little effect on which mechanism, S_N2 or S_N1, is predominant.

In contrast, the *nature of the alkyl group, R, of the substrate* exerts a profound effect on which mechanism is to be followed. In R, two structural factors are at work: *steric hindrance*, which largely determines ease of back-side attack; and *ability to accommodate a positive charge*, which largely determines ease of heterolysis. As we proceed along the simple alkyl series CH_3, 1°, 2°, 3°, the group R becomes, by definition, more branched. There is a regular increase in the number of substituents on carbon: bulky, electron-releasing substituents. Steric hindrance increases; back-side attack becomes more difficult and hence slower. At the same time, ability to accommodate a positive charge increases; heterolysis becomes easier and hence faster.

$$\begin{array}{c}
\xleftarrow{\quad\; S_N2 \text{ increases} \quad\;} \\
RX = CH_3X \quad 1° \quad 2° \quad 3° \\
\xrightarrow{\quad\; S_N1 \text{ increases} \quad\;}
\end{array}
\qquad
\begin{array}{c}
S_N2 \\
vs. \\
S_N1
\end{array}$$

The result is the pattern we encountered earlier: for methyl and primary substrates, a predisposition toward S_N2; for tertiary substrates, a predisposition toward S_N1. For secondary substrates there is a tendency toward intermediate behavior: a mixture of the two mechanisms or, as we shall see in the following chapter, perhaps a mechanism with characteristics of both S_N2 and S_N1.

Despite this predisposition of a particular substrate toward a particular mechanism, we can still control the course of reaction to a considerable degree by our choice of experimental conditions. To see how this can be done, we must examine the other components of the reaction system.

Next, then, let us turn to the **nucleophile**. The key difference between the S$_N$2 and S$_N$1 mechanisms is the matter of *when* the nucleophile participates: *in* the rate-determining step of S$_N$2, but *after* the rate-determining step of S$_N$1. This difference in timing leads directly to two factors that help determine the mechanism to be followed: the *concentration of the nucleophile*, and the *nature of the nucleophile*.

The rate of S$_N$2 depends upon the concentration of the nucleophile, [:Z]; reaction, as we have seen (Sec. 5.13), is second-order.

$$\text{rate} = k[\text{RW}][:\text{Z}] \qquad \text{S}_N2$$

The rate of S$_N$1 is independent of [:Z]; reaction (Sec. 5.16) is first-order.

$$\text{rate} = k[\text{RW}] \qquad \text{S}_N1$$

An increase in [:Z] speeds up the second-order reaction but has no effect on the first-order reaction; the fraction of reaction by S$_N$2 increases. A decrease in [:Z] slows down the second-order reaction but has no effect on the first-order reaction; the fraction of reaction by S$_N$2 decreases. The net result is that, other things being equal, a *high concentration of nucleophile favors the S$_N$2 reaction*, and a *low concentration favors the S$_N$1 reaction*.

Problem 5.10   In 80% ethanol at 55°C, isopropyl bromide reacts with hydroxide ion according to the following kinetic equation, where the rate is expressed as moles per liter per second:

$$\text{rate} = 4.7 \times 10^{-5}[\text{RX}][\text{OH}^-] + 0.24 \times 10^{-5}[\text{RX}]$$

What percentage of the isopropyl bromide reacts by the S$_N$2 mechanism when [OH$^-$] is: (a) 0.001 molar, (b) 0.01 molar, (c) 0.1 molar, (d) 1.0 molar, (e) 5.0 molar?

In the same way, the rate of S$_N$2 depends upon the nature of the nucleophile: a stronger nucleophile attacks the substrate faster. The rate of S$_N$1 is independent of the nature of the nucleophile: stronger or weaker, the nucleophile waits until the carbocation is formed. The net result is that, other things being equal, a *strong nucleophile favors the S$_N$2 reaction*, and a *weak nucleophile favors the S$_N$1 reaction*.

We have already seen an illustration of this effect in Sec. 5.23. Neopentyl bromide reacts with the strong nucleophile, ethoxide, to give the unrearranged ethyl neopentyl ether: clearly an S$_N$2 reaction. It reacts with the weak nucleophile, ethanol, to give the rearranged ethyl *tert*-pentyl ether: clearly an S$_N$1 reaction.

But we have neglected the third component of the system, the one that offers the most scope for control of the reaction: the **solvent**. We shall examine the role of the solvent in detail in Chapter 6.

At the beginnning of this chapter, we said that a chemical reaction is the result of a competition: what actually happens when a particular set of reactants is mixed together under a particular set of conditions is what happens *fastest*. Here, we have been concerned with competition between different pathways for nucleophilic substitution. But in Chapter 7 we shall find still further competition: between nucleophilic substitution and an entirely different type of reaction, *elimination*. In all this, we are concerned with the factors that favor one mechanism or one type of reaction over another, and, where possible, with what we can do to control the outcome.

5.25   Reaction of alcohols with hydrogen halides. Acid catalysis

One method of making alkyl halides, we saw (Sec. 5.7), is by the reaction of alcohols with hydrogen halides. Let us look more closely at this reaction, not just as an important synthetic method, but as an example of nucleophilic substitution.

$$R\text{—OH} + HX \longrightarrow R\text{—}X + H_2O$$

Alcohol   Alkyl halide

In doing this, we shall see something completely new to us: how we can change a very poor leaving group into a very good leaving group *instantaneously*, and with no more effort than it takes to pour a solution from a bottle into a flask. What we shall see is the most important—and simplest—kind of catalytic effect known to the organic chemist: an effect that plays a key role in the chemistry of compounds of all kinds, in the test tube and in the living organism.

Alcohols react readily with hydrogen halides to yield alkyl halides and water. The reaction is carried out either by passing the dry halogen halide gas into the alcohol, or by heating the alcohol with the concentrated aqueous acid. Sometimes hydrogen bromide is generated in the presence of the alcohol by reaction between sulfuric acid and sodium bromide.

The least reactive of the hydrogen halides, HCl, generally requires the presence of zinc chloride for reaction with primary and secondary alcohols; on the other hand, the very reactive *tert*-butyl alcohol is converted to the chloride by simply being shaken with concentrated hydrochloric acid at room temperature. For example:

$$CH_3CH_2CH_2CH_2OH \xrightarrow[\substack{\text{or} \\ \text{NaBr, H}_2\text{SO}_4, \\ \text{heat}}]{\text{dry HBr}} CH_3CH_2CH_2CH_2Br$$

n-Butyl alcohol   *n*-Butyl bromide

$$CH_3CH_2CH_2OH \xrightarrow[\text{heat}]{\text{HCl + ZnCl}_2} CH_3CH_2CH_2Cl$$

n-Propyl alcohol   *n*-Propyl chloride

$$\underset{\substack{| \\ OH}}{\overset{\substack{CH_3 \\ |}}{CH_3\text{—}C\text{—}CH_3}} \xrightarrow[\text{room temp.}]{\text{conc. HCl}} \underset{\substack{| \\ Cl}}{\overset{\substack{CH_3 \\ |}}{CH_3\text{—}C\text{—}CH_3}}$$

tert-Butyl alcohol   *tert*-Butyl chloride

Let us list some of the facts that are known about the reaction between alcohols and hydrogen halides.

(a) The reaction is catalyzed by acids. Even though the aqueous hydrogen halides are themselves strong acids, the presence of additional sulfuric acid speeds up the formation of alkyl halides.

(b) Rearrangement of the alkyl group occurs, except with most primary alcohols. The alkyl group in the halide does not always have the same structure as the alkyl group in the parent alcohol. For example:

$$CH_3-\underset{\underset{CH_3}{|}}{\overset{\overset{CH_3}{|}}{C}}-\underset{\underset{OH}{|}}{\overset{\overset{H}{|}}{C}}-CH_3 \xrightarrow{\text{HCl}} CH_3-\underset{\underset{Cl}{|}}{\overset{\overset{CH_3}{|}}{C}}-\underset{\underset{CH_3}{|}}{\overset{\overset{H}{|}}{C}}-CH_3 \quad (\text{but } no \ CH_3-\underset{\underset{CH_3}{|}}{\overset{\overset{CH_3}{|}}{C}}-\underset{\underset{Cl}{|}}{\overset{\overset{H}{|}}{C}}-CH_3)$$

3,3-Dimethyl-2-butanol　　　　　　　2-Chloro-2,3-dimethylbutane

$$CH_3-\underset{\underset{CH_3}{|}}{\overset{\overset{CH_3}{|}}{C}}-CH_2-OH \xrightarrow{\text{HCl}} CH_3-\underset{\underset{Cl}{|}}{\overset{\overset{CH_3}{|}}{C}}-CH_2-CH_3$$

Neopentyl alcohol　　　　　　　　*tert*-Pentyl chloride

We see that the halogen does not always become attached to the carbon that originally held the hydroxyl (the first example); even the carbon skeleton may be different from that of the starting material (the second example).

On the other hand, as shown above for *n*-propyl and *n*-butyl alcohols, most primary alcohols give high yields of primary halides *without* rearrangement.

(c) The order of reactivity of alcohols toward HX is $3° > 2° > 1° < CH_3$. Reactivity decreases through most of the series (and this order is the basis of the *Lucas test*, Sec. 18.9), passes through a *minimum* at 1°, and rises again at CH_3.

What do the facts that we have just listed suggest to us about the mechanism of reaction between alcohols and hydrogen halides?

Catalysis by acid suggests that the protonated alcohol $ROH_2{}^+$ is involved. The occurrence of *rearrangement* suggests that carbocations are intermediates—although *not* with primary alcohols. The idea of carbocations is strongly supported by the *order of reactivity* of alcohols, which parallels the stability of carbocations—*except* for methyl.

On the basis of this evidence, we formulate the following mechanism. The

(1)　　　　　$R-OH + HX \rightleftharpoons R-OH_2{}^+ + :X^-$

(2)　　　　　　　$R-OH_2{}^+ \rightleftharpoons R^+ + H_2O:$

(3)　　　　$R^+ + :X^- \longrightarrow R-X$

S_N1:
all except methanol and most 1° alcohols

alcohol accepts (step 1) the hydrogen ion to form the protonated alcohol, which dissociates (step 2) into water and a carbocation; the carbocation then combines (step 3) with a halide ion (not necessarily the one from step 1) to form the alkyl halide.

Looking at the mechanism we have written, we recognize the reaction for what it is: *nucleophilic substitution*, with the protonated alcohol as substrate and halide ion as the nucleophile. Once the reaction type is recognized, the other pieces of evidence fall into place.

The particular set of equations written above is, of course, the S_N1 mechanism for substitution. Primary alcohols do not undergo rearrangement simply because they do not react by this mechanism. Instead, they react by the alternative S_N2 mechanism:

$$:X^- + R-OH_2{}^+ \longrightarrow \left[\overset{\delta-}{X}---R----\overset{\delta+}{O}H_2\right] \longrightarrow X-R + H_2O:$$

S_N2:
most 1° alcohols and methanol

What we see here is another example of that characteristic of nucleophilic substitution: a shift in the molecularity of reaction, in this particular case occurring between 2° and 1°. This shift is confirmed by the fact that reactivity passes through a minimum at 1° and rises again at methyl.

Let us review what is probably happening here, beginning at the methyl end of the series. The methyl substrate is least capable of heterolysis and most open to nucleophilic attack; it reacts by a full-fledged S_N2 reaction. So do primary substrates but, because of greater steric hindrance, they react less rapidly than the methyl. Secondary substrates give still more steric hindrance, but are more capable of forming carbocations. For them heterolysis is faster than nucleophilic attack by a halide ion, and the mechanism changes here to S_N1. With the change in mechanism, the rate begins to rise. Tertiary substrates, too, react by an S_N1 mechanism; they react faster than secondary substrates because of the greater dispersal of charge in the incipient carbocations.

So far, we have discussed this reaction in terms of the very useful classification of substrates as 1°, 2°, or 3°. But we must always keep in mind that it is not this classification—*as such*—that is important. It is the factors actually at work: in this reaction, *steric hindrance* to nucleophilic attack, and *dispersal of charge* in the incipient carbocation. These factors give rise—*among other things*—to the relationship between 1°, 2°, 3° and the S_N2–S_N1 competition. But they do more than that. They can make a substrate of one class act like a substrate of another class; and yet such behavior is understandable if we simply examine the structures involved. Let us look at two such examples.

As shown above, neopentyl alcohol reacts with almost complete rearrangement, showing that, although primary, it follows the carbocation mechanism. This is contrary to our generalization, but readily accounted for. Although neopentyl is a primary group, it is a very bulky one and, as we have seen (Sec. 5.15), neopentyl substrates undergo S_N2 reactions very slowly. Formation of the neopentyl cation here is slow, too, but is nevertheless much faster than the alternative bimolecular reaction.

Our second example involves 1-chloro-2-propanol. Although technically a secondary alcohol, it reacts with hydrogen halides "abnormally" slowly, and at about the rate of a primary alcohol. This time we are dealing, not with a steric

$$ClCH_2\!-\!\overset{\displaystyle |}{\underset{\displaystyle OH}{CH}}\!-\!CH_3 \;+\; HX \;\longrightarrow\; ClCH_2\!-\!\overset{\displaystyle |}{\underset{\displaystyle X}{CH}}\!-\!CH_3 \;+\; H_2O$$

      1-Chloro-2-propanol   1-Chloro-2-halopropane

effect, but with a polar effect. The rate of an S_N1 reaction, we have seen (Sec. 5.22), depends upon the stability of the carbocation being formed. Let us compare, then, the 1-chloro-2-propyl cation with a simple secondary cation, the isopropyl cation, say. Electronegative chlorine has an electron-withdrawing inductive effect. As we have seen (Sec. 5.21), this intensifies the positive charge on the electron-deficient

$$CH_3\!-\!\overset{\displaystyle \oplus}{CH}\!-\!CH_3 \qquad\qquad Cl\!\leftarrow\!CH_2\!-\!\overset{\displaystyle \oplus}{CH}\!-\!CH_3$$

       Isopropyl cation   1-Chloro-2-propyl cation
                                   Chlorine withdraws electrons:
                                       intensifies charge,
                                       destabilizes cation

carbon and makes the carbocation less stable. This same electron withdrawal destabilizes the incipient cation in the transition state, raises E_{act}, and slows down the reaction.

Now let us turn to what is to us the most important aspect of the reaction between alcohol and hydrogen halides: the *acid catalysis*. What does acid do? In the first step, it converts the alcohol into the protonated alcohol, which is the substrate actually undergoing substitution. In the absence of acid, substitution—by either mechanism—would require loss of the hydroxide ion: strongly basic, and an extremely poor leaving group. Substitution with the protonated alcohol as

$$R-OH_2\oplus \longrightarrow R\oplus + H_2O \qquad\qquad \text{Easy}$$

Weak base:
good leaving group

$$R-OH \longrightarrow R\oplus + OH^- \qquad\qquad \text{Difficult}$$

Strong base:
poor leaving group

substrate, on the other hand, involves loss of water: weakly basic, and a very good leaving group. Protonation of the alcohol involves a simple acid–base equilibrium, and takes place instantaneously on mixing of the reagents. Yet it changes a very poor leaving group into a very good one and permits reaction to occur. The evidence indicates that separation of a hydroxide ion from an alcohol almost never occurs; reactions involving cleavage of the C—O bond of an alcohol seem in nearly every case to require an acidic catalyst, the purpose of which, as here, is to form the protonated alcohol.

Thus alcohols, like alkyl halides, undergo nucleophilic substitution by both S_N2 and S_N1 mechanisms, but alcohols lean more toward the unimolecular mechanism. We can see, in a general way, why this is so. To undergo substitution an alcohol must be protonated, and this requires an acidic medium. An S_N2 reaction, we have seen, is favored by the use of a strong nucleophile, something that is quite feasible in reactions of alkyl halides. But we cannot have a strong nucleophile—a strong *base*—present in the acidic medium required for protonation of an alcohol; any base much stronger than the alcohol itself would become protonated at the expense of the alcohol. Restricted, then, to reaction with weakly basic, weakly nucleophilic reagents, alcohols react chiefly by the carbocation mechanism.

In the opening paragraph of this section it was, of course, protonation that was referred to as the most important—and simplest—catalytic effect in organic chemistry. In the presence of acid many kinds of atoms found in organic compounds are protonated to a significant degree: oxygen, nitrogen, sulfur, often even carbon. And, as we shall see in nearly every chapter of this book, this protonation exerts powerful effects on reactions of many kinds involving nearly every class of compound.

Problem 5.11 Because of the great tendency of the neopentyl cation to rearrange, neopentyl chloride cannot be prepared from the alcohol. How might neopentyl chloride be prepared?

Problem 5.12   (a) Write the steps in the reaction of an alcohol with HCl by the S_N1 mechanism. (b) What is the rate-determining step? (c) The rate of reaction depends upon the concentration of what substance? (d) The concentration of this substance depends in turn upon the concentrations of what other compounds? (e) Will the rate depend *only* on [ROH]? Does an S_N1 reaction always follow first-order kinetics?

5.26   Analysis of alkyl halides

Simple alkyl halides respond to the common characterization tests in the same manner as alkanes: they are insoluble in cold concentrated sulfuric acid; they are inert to bromine in carbon tetrachloride, to aqueous permanganate, and to chromic anhydride. They are readily distinguished from alkanes, however, by qualitative analysis (Sec. 2.26), which shows the presence of halogen.

In many cases, the presence of halogen can be detected without a sodium fusion or Schöniger oxidation. An unknown is warmed for a few minutes with alcoholic silver nitrate (the alcohol dissolves both the ionic reagent and the organic compound); halogen is indicated by formation of a precipitate that is insoluble in dilute nitric acid.

As in almost all reactions of organic halides, reactivity toward alcoholic silver nitrate follows the sequence RI > RBr > RCl. For a given halogen atom, reactivity decreases in the order $3° > 2° > 1°$, the sequence typical of carbocation formation; as we shall see, allylic halides (Sec. 10.13) and benzylic halides (Sec. 15.18) are highly reactive. Other evidence (stereochemistry, rearrangements) suggests that this reaction is of the S_N1 type. Silver ion is believed to accelerate reaction by *pulling* halide away from the alkyl group.

$$R:X + Ag^+ \longrightarrow R^+ + Ag^+X^-$$

(Vinyl and aryl halides do not react, Secs. 10.18 and 29.5.)

As mentioned earlier (Sec. 5.3), substituted alkyl halides also undergo the reactions characteristic of their other functional groups.

(Analysis of alkyl halides by spectroscopy will be discussed in Chapter 16.)

PROBLEMS

1. Give the structural formula of:

(a) 1-bromo-2,2-dimethylpropane
(b) neopentyl alcohol
(c) 2-bromo-1-propanol
(d) 2-chloro-2,3-dimethylpentane

(e) 3-methyl-2-pentanol
(f) potassium ethoxide
(g) 2,2,2-trifluoroethanol
(h) isobutyl tosylate (use Ts for tosyl)

2. Draw out the structural formula and give the IUPAC name of:

(a) $(CH_3)_2CHCH_2I$
(b) $(CH_3)_2CHCHClCH_3$
(c) $CH_3CHBrC(CH_3)_2CH_2CH_3$

(d) $(CH_3)_3COH$
(e) $CH_3CH_2CHClCHOHCH_3$
(f) $(CH_3)_2CClCBr(CH_3)_2$

3. Write equations for the preparation of *n*-propyl iodide from:

(a) *n*-propyl alcohol   (b) *n*-propyl bromide   (c) *n*-propyl tosylate

4. Give the structures and names of the chief organic products expected from the reaction (if any) of *n*-butyl bromide with:

(a) NaOH(aq)
(b) cold conc. H_2SO_4
(c) Zn, H^+
(d) Li, then CuI, ethyl bromide
(e) Mg, ether

(f) product (e) + D_2O
(g) dilute neutral $KMnO_4$
(h) NaI in acetone
(i) Br_2/CCl_4

5. Referring when necessary to the list on pages 175–176, give structures of the chief organic products expected from the reaction of *n*-butyl bromide with:

(a) NH_3
(b) $C_6H_5NH_2$
(c) NaCN

(d) $NaOC_2H_5$
(e) CH_3COOAg
(f) $NaSCH_3$

6. Give the reagents, inorganic or organic, needed to convert *n*-butyl bromide into:

(a) *n*-butyl iodide
(b) *n*-butyl chloride
(c) *n*-butyl methyl ether ($CH_3CH_2CH_2CH_2OCH_3$)
(d) *n*-butyl alcohol 4a
(e) pentanenitrile ($CH_3CH_2CH_2CH_2CN$)
(f) *n*-butylamine ($CH_3CH_2CH_2CH_2NH_2$)
(g) *n*-butylmagnesium bromide
(h) lithium di-*n*-butylcopper

7. Arrange the compounds of each set in order of reactivity toward S_N2 displacement:

(a) 2-bromo-2-methylbutane, 1-bromopentane, 2-bromopentane
(b) 1-bromo-3-methylbutane, 2-bromo-2-methylbutane, 3-bromo-2-methylbutane
(c) 1-bromobutane, 1-bromo-2,2-dimethylpropane, 1-bromo-2-methylbutane, 1-bromo-3-methylbutane

8. Arrange the compounds of each set in order of reactivity toward S_N1 displacement:

(a) the compounds of Problem 7(a)
(b) the compounds of Problem 7(b)

9. Consider, as an example, the reaction between an alkyl halide and NaOH in a mixture of water and ethanol. In a table, with one column for S_N2 and another for S_N1, compare the two mechanisms with regard to:

(a) stereochemistry
(b) kinetic order
(c) occurrence of rearrangements
(d) relative rates for CH_3X, C_2H_5X, iso-C_3H_7X, *tert*-C_4H_9X
(e) relative rates for RCl, RBr, and RI
(f) effect on rate of a rise in temperature
(g) effect on rate of doubling [RX]
(h) effect on rate of doubling [OH^-]

10. Arrange the alcohols of each set in order of reactivity toward gaseous HBr:

(a) 2-butanol, 2-methyl-1-propanol, 2-methyl-2-propanol
(b) 3-pentanol, 2-fluoro-3-pentanol, 2,2-difluoro-3-pentanol, 1-fluoro-3-pentanol

11. Account for the fact that *either* 2-pentanol *or* 3-pentanol reacts with HCl to give *both* 2-chloropentane and 3-chloropentane.

12. Starting from (R)-sec-butyl alcohol, and using any optically inactive reagents, show all steps in the synthesis of:

(a) (R)-sec-butyl ethyl ether ($CH_3CH_2CH(CH_3)OC_2H_5$)
(b) (S)-sec-butyl ethyl ether

13. Optically active sec-butyl alcohol retains its activity indefinitely in contact with aqueous base, but is rapidly converted into optically inactive (racemic) sec-butyl alcohol by dilute sulfuric acid. How do you account for these facts? Suggest a detailed mechanism or mechanisms for the racemization by dilute acid.

14. When isopropyl alcohol is heated in the presence of H_2SO_4, there can be obtained diisopropyl ether (i-Pr_2O).

$$2CH_3CHCH_3 \xrightarrow{\ H_2SO_4\ } \underset{\text{Diisopropyl ether}}{CH_3\overset{\displaystyle CH_3}{\overset{|}{C}}H-O-\overset{\displaystyle CH_3}{\overset{|}{C}}HCH_3}$$

$\underset{\text{Isopropyl alcohol}}{\overset{|}{OH}}$

(a) To what type of reaction does this belong?
(b) Show all steps in a likely mechanism or mechanisms.

15. The most important way to make *alkenes* (Chap. 7) is through *base-promoted 1,2-elimination*:

$$-\overset{|}{\underset{H}{C}}-\overset{|}{\underset{X}{C}}- \ + \ :B \longrightarrow \underset{\text{An alkene}}{\diagdown C=C \diagup} \ + \ H:B \ + \ X^-$$

(a) When 3-bromo-2,2-dimethylbutane is heated with a dilute solution of C_2H_5ONa in C_2H_5OH, or with C_2H_5OH alone, reaction follows first-order kinetics; along with substitution, there also occurs elimination, to yield alkenes I and II. What does the formation of these particular alkenes suggest to you? Propose a likely mechanism for the reaction by which they are formed.

$$\underset{I}{CH_3-\overset{\displaystyle H_3C}{\overset{|}{C}}=\overset{\displaystyle CH_3}{\overset{|}{C}}-CH_3} \qquad \underset{II}{CH_3-\overset{\displaystyle CH_3}{\overset{|}{C}}H-\overset{\displaystyle CH_3}{\overset{|}{C}}=CH_2} \qquad \underset{III}{CH_3-\overset{\displaystyle CH_3}{\overset{|}{C}}-CH=CH_2}$$

(b) When the same halide is allowed to react with a concentrated solution of C_2H_5ONa in C_2H_5OH, reaction follows second-order kinetics; again elimination accompanies substitution, this time to yield, not alkenes I and II, but alkene III. Propose a likely mechanism or mechanisms for the elimination taking place under these conditions.
(c) How do you account for the shift in mechanism between (a) and (b)?
(d) What substitution product or products would you expect in each case?

16. A liquid of boiling point 39–41 °C was insoluble in water, dilute acids or bases, or concentrated H_2SO_4. It did not react with Br_2/CCl_4 or dilute $KMnO_4$. It was subjected to sodium fusion, and the resulting solution was filtered, acidified with nitric acid, and boiled. Addition of $AgNO_3$ gave a precipitate.
(a) On the basis of Table 5.1, what compound or compounds might this have been?
(b) Several milliliters of CCl_4 was added to a portion of the acidified solution from the fusion, and the mixture was shaken with chlorine water. A violet color appeared in the CCl_4 layer. Which compound or compounds of (a) are still possible? (c) How would each of the other possibilities have responded in (b)?

17. Outline all steps in the conversion of *n*-propyl alcohol into each of the following compounds. (See the general instructions about synthesis in the box below.)

(a) *n*-propyl bromide
(b) *n*-propyl iodide
(c) sodium *n*-propoxide
(d) butanenitrile ($CH_3CH_2CH_2CN$)
(e) *n*-propylamine ($CH_3CH_2CH_2NH_2$)

(f) di-*n*-propylamine (($CH_3CH_2CH_2$)$_2$NH)
(g) di-*n*-propyl ether (($CH_3CH_2CH_2$)$_2$O)
(h) propane
(i) propane-1-*d* ($CH_3CH_2CH_2D$)
(j) *n*-hexane

18. Starting *from alcohols of four carbons or fewer*, and making use of any necessary solvents or reagents, outline a possible synthesis for each of the following compounds:

(a) 2-chloropropane
(b) ethyl tosylate
(c) potassium *tert*-butoxide
(d) propanenitrile (CH_3CH_2CN)
(e) isobutane
(f) ethyl *n*-propyl ether ($CH_3CH_2CH_2OC_2H_5$)
(g) butane-2-*d* ($CH_3CH_2CHDCH_3$)
(h) 3-methylhexane

About Synthesis

Each synthesis should be the one that gives a reasonably pure product in reasonably good yield.

It is not necessary to complete and balance each equation. Simply draw the structure of the organic compounds, and write on the arrow the necessary reagents and any critical conditions. For example:

$$CH_3CH_2OH \xrightarrow{\text{HBr}} CH_3CH_2Br \xrightarrow{\text{CN}^-} CH_3CH_2CN$$

At this stage you may be asked to make a particular compound that can readily be bought, or that might better be made by another method: the synthesis of propane in Problem 17, for example. But if you can work out a way to make propane from *n*-propyl alcohol, then, when the need arises, you will also know how to make a complicated alkane from a complicated alcohol, and, in fact, how to replace an —OH group by —H in just about any compound you encounter. Furthermore, you will have gained practice in putting together what you have learned about several different kinds of compounds.

Remember: Alkyl halides are *almost never* prepared by direct halogenation of alkanes. *From the standpoint of synthesis in the laboratory, an alkane is a dead-end.*

6

Role of the Solvent

Secondary Bonding

6.1 Role of the solvent

Most of organic chemistry, we said at the beginning of Chapter 5, is heterolytic; and heterolytic reactions, we said, are typically carried out in solution. If we are to understand organic reactions, then, we must learn something about the role played by the component of the reaction system that seldom appears in our equations, but that is nearly always present: the **solvent**.

The role played by the solvent is not a minor one. Study of heterolytic reactions in the absence of any solvent—in the gas phase—has provided a standard to show just how enormous solvent effects can be. The presence of a solvent can speed up— or slow down—a reaction by a factor of 10^{20}; a change from one solvent to another can bring about a million-fold change in reaction rate. Solvent effects can be more powerful than the effects exerted by any other factor: vastly more powerful than polar or steric effects; sometimes more powerful, even, than the symphoric effects that we shall study in Chapter 20. The solvent—the particular choice of solvent— can be the major factor determining how fast a reaction occurs, and even whether it takes place at all; it can determine which of several alternative pathways a reaction actually follows.

Clearly, a solvent is not simply a place—a kind of gymnasium—where solute molecules may gambol about and occasionally collide. The solvent is intimately *involved* in any reaction that takes place in it, and it is important for us to find out how much it is involved and in what ways.

We have already learned enough (Sec. 1.21) to realize that solute molecules and ions do not exist in solution as naked particles; they are *solvated*. Clinging to each dissolved particle is a cluster of solvent molecules, held there by bonds; it was formation of these bonds that provided the energy needed to break the bonds that held solute particles to each other. The very fact that dissolution has taken place shows this: new bonds have replaced old ones.

The science of organic chemistry rests on a simple premise: that chemical behavior is determined by molecular structure. Yet, in solution all participants in a chemical reaction are solvated: the reactants and the products—and the *transition state*. Now, our basic approach to chemical reactivity is to consider energy differences between reactants and transition states; that is, we estimate relative stabilities of these species. We do this by examining—mentally and, by use of models, physically—the structures involved; this examination *must* include any solvent clusters that help make up those structures and help determine their stabilities.

In this chapter, using as our examples the nucleophilic substitution reactions we have just studied, we shall see how reactivity—and, with it, the course of reaction—is affected by the solvent. The solvent adds a new dimension to our study of organic chemistry; if it complicates things, it at the same time adds richness. It offers us the most practical way to *control* what happens in a chemical reaction. The effect exerted by a solvent is one kind of medium effect—environmental effect—and in that sense is just the beginning of a trail that leads all the way to the ultimate organic reaction, the action of an enzyme; this (literally) vital action is possible only because the substrate is *dissolved in* the enzyme, held to it by the same kinds of forces that a solvent uses.

Let us begin our study of the role of the solvent, then, by learning more about the kinds of bonds that are broken and formed when dissolution takes place.

6.2   Secondary bonding

In our discussion of melting, boiling, and solubility in Chapter 1, we looked briefly at the kinds of forces that act between molecules, between ions, and between molecules and ions. All of these, we said, are electrostatic forces—the attraction of positive for negative. These attractive forces—these *bonds*—are the following.

(a) **Ion–ion bonds:** the attraction between the opposite charges on a cation and an anion.

(b) **Dipole–dipole bonds:** the attraction between the positive end of one polar molecule and the negative end of another polar molecule.

The most powerful of these dipole–dipole bonds is the **hydrogen bond**, in which a hydrogen atom acts as a bridge between two electronegative atoms (**F, O, N**); it is held to one—the *hydrogen-bond donor*—by a covalent bond, and to the other—the *hydrogen-bond acceptor*— by purely electrostatic attraction. For example:

The strength of a particular hydrogen bond depends upon the nature of the hydrogen-bond donor and of the hydrogen-bond acceptor. *A more acidic donor forms a stronger hydrogen bond.* The strength of a hydrogen bond depends upon how positive the hydrogen is (Sec. 1.19); acidity depends upon how well the conjugate base accommodates the electron pair left behind by the departed proton. Both of these properties are increased by the same factor: electron withdrawal in the group attached to the hydrogen. *A more basic acceptor forms a stronger hydrogen bond.* The strength of a hydrogen bond depends upon how negative the acceptor atom is—that is, how available its electrons are; and availability of electrons is what makes a molecule basic.

(c) **Van der Waals forces**: the attraction between the oppositely charged ends of momentary, induced dipoles in neighboring molecules. These forces act between all molecules, even non-polar ones.

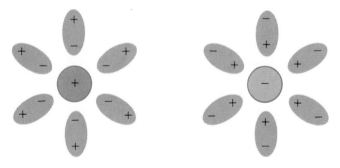

(d) **Ion–dipole bonds**: the attraction of a positive ion for the negative end of a polar solvent molecule, and of a negative ion for the positive end of a polar solvent molecule (Fig. 6.1).

Figure 6.1   Ion–dipole interaction: solvated cation and anion.

Except for full-fledged ionic bonds in an ionic crystal, these attractive forces are often called **secondary bonds**. They act *between* different molecules and ions, in contrast to the covalent bonds that act *within* a molecule or ion and hold the atoms together. Individually, each secondary bond is relatively weak. But, acting to-gether—and they work as a team—a collection of such bonds is extremely powerful; their formation, as we shall see, can supply enough energy to break a covalent bond.

If this seems surprising, remember this. To make sodium chloride boil, we must heat it to 1413 °C; at room temperature we can dissolve it in a few moments by simply stirring it into a beaker of water. Yet the interionic forces being overcome in both processes are exactly the same.

In this chapter, we shall be primarily concerned with secondary bonding as it is involved in the action of the solvent: dissolving solutes, affecting their reactivity, and even, in a very direct way, reacting with them. But secondary bonding is involved in much more than solvent effects. These same forces, acting between the long, thread-like molecules of cotton, wool, silk, and nylon, give strength that is needed in the formation of fibers (p. 1254). Even the weakest of them, van der Waals forces, acting between non-polar chains of phospholipids, are the mortar in the walls of living cells (p. 1274).

Secondary bonding exists not only between different molecules but between different parts of the same molecule. In this way it plays a key role in determining the *shapes* of large molecules like proteins and nucleic acids, shapes that determine in turn, their biological properties: the size of the "pockets" in the hemoglobin molecule, for example, just big enough to hold heme groups with their oxygen-carrying iron atoms (p. 1367); the helical shape of α-keratin and collagen molecules that makes wool and hair strong, and tendons and skin tough (p. 1364). It is secondary bonding that makes the double helix of DNA *double*—and thus permits the self-duplication of molecules that is the basis of heredity (p. 1395).

And so, our study in this chapter will have two aims: to understand better the role of the solvent; and, at the same time, to understand better the nature of secondary bonding.

6.3 Solubility: non-ionic solutes

The solubility characteristics of non-ionic solutes, we said earlier (Sec. 1.21), depend chiefly upon their polarity—and in particular their ability to form hydrogen bonds. "Like dissolves like" is our rule-of-thumb.

Let us consider the kinds of compounds we have already encountered, beginning with hydrocarbons and alkyl halides. These are non-polar or weakly polar, and dissolve in solvents of similar polarity: in hydrocarbons like ligroin or benzene; in alkyl halides like chloroform and carbon tetrachloride; in diethyl ether. The forces holding the solute molecules to each other—and the solvent molecules to each other—are readily replaced by very similar forces holding solute molecules to solvent molecules. Hydrocarbons and alkyl halides do not dissolve in water, whose molecules are highly polar and held strongly to each other by hydrogen bonds.

Next, let us turn to alcohols. Structurally, an alcohol is a composite of an alkane and water: it contains an alkane-like alkyl group and a water-like hydroxyl group.

R—H   H—OH   R—OH
An alkane   Water   An alcohol

The hydroxyl group is quite polar and, most important, contains hydrogen attached to the highly electronegative element oxygen. Through the hydroxyl group, alcohols are capable of forming hydrogen bonds: hydrogen bonds to each other, which give alcohols abnormally high boiling points (Secs. 1.20 and 17.5);

$$R-O{\cdots}H-O \overset{R}{\underset{H}{|}} \qquad R-O{\cdots}H-O \overset{H}{\underset{H}{|}}$$

hydrogen bonds to other molecules, which tend to make alcohols soluble in other hydroxyl compounds, such as water. For the smallest alcohol, methanol (CH_3OH), we have seen, the result is complete solubility in water (Sec. 1.21). Hydrogen bonds between water and methanol molecules readily replace the very similar hydrogen bonds between different methanol molecules and different water molecules.

Now, because of the very special status of water as a solvent—especially in biological systems—the terms *hydrophilic* (water loving) and *hydrophobic* (water hating) are used in reference to water solubility and water insolubility. Instead of hydrophobic, the term *lipophilic* (fat loving) is often used; this emphasizes not so much insolubility in water as solubility in non-polar solvents. Thus, methanol is hydrophilic, and alkanes and alkyl halides are lipophilic (or hydrophobic).

Since it is easier to work with a term for a positive quality than one for a negative quality, in this book we shall generally use *lipophilic*. This term is meant simply to indicate the *fact* of solubility in non-polar solvents. It may well be—as is widely held—that this solubility is chiefly due to rejection by water rather than positive acceptance by a non-polar solvent.

Next, let us consider a series of alcohols, and the effect of the alkyl group on solubility. Where the hydroxyl group is hydrophilic, the alkyl group is lipophilic. Table 6.1 gives the water solubility of a series of alcohols. For the lower members of the series, the —OH group constitutes a large portion of the molecule, and these compounds are miscible with water. But, we see, as the number of carbons increases, the solubility steadily decreases; a long chain with an —OH at one end of it is mostly hydrocarbon, and its solubility shows this. (See Fig. 6.2, on the next page.)

Table 6.1   SOLUBILITY OF ALCOHOLS IN WATER

Alcohol	Solubility, g/100 g H_2O
CH_3OH	∞
CH_3CH_2OH	∞
$CH_3CH_2CH_2OH$	∞
$CH_3CH_2CH_2CH_2OH$	7.9
$CH_3CH_2CH_2CH_2CH_2OH$	2.3
$CH_3CH_2CH_2CH_2CH_2CH_2OH$	0.6
$CH_3CH_2CH_2CH_2CH_2CH_2CH_2OH$	0.2
$CH_3CH_2CH_2CH_2CH_2CH_2CH_2CH_2OH$	0.05

Now, if a molecule is big enough—if an alcohol, say, has a chain of 16 to 20 carbons or more—hydrophilic and lipophilic parts display their individual solubility properties. The hydrophilic parts dissolve in water; the lipophilic parts dissolve in a non-polar solvent or, if there is none about, cluster together—in effect, dissolve in each other. Such dual solubility behavior gives soaps and detergents their cleansing power (Secs. 37.3 and 37.5), and controls the alignment of molecules in cell membranes (Sec 37.8); a globular protein molecule—an enzyme, say—coils up to expose its hydrophilic parts to the surrounding water and to hide its lipophilic

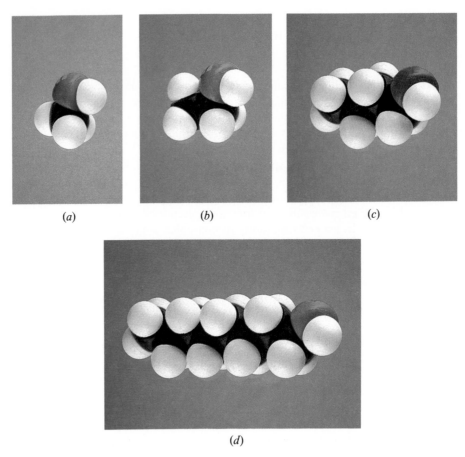

Figure 6.2 Molecular structure and physical properties: solubility. (a) Methyl alcohol, (b) ethyl alcohol, (c) n-butyl alcohol, and (d) n-octyl alcohol. As the alkyl group gets bigger, the molecule becomes increasingly alkane-like and the water solubility decreases.

parts, and in doing this takes on the particular shape needed for its characteristic biological properties (Sec. 40.11).

6.4 Solubility: ionic solutes. Protic and aprotic solvents. Ion pairs

Now let us turn to the dissolution of ionic compounds.

The forces holding together an ionic lattice are powerful, and a great deal of energy is needed to overcome them. This energy is supplied by the formation of many ion–dipole bonds between the ions and the solvent. About each ion there gathers a cluster of solvent molecules, their positive ends turned toward a negative ion, their negative ends turned toward a positive ion (Fig. 6.1, p. 225).

To dissolve ionic compounds, then, a solvent must be *highly polar*. In addition, we have seen, it must have a *high dielectric constant*; that is, it must be a good insulator, to lower the attraction between oppositely charged ions once they are solvated.

But water owes its superiority as a solvent for ionic substances only *partly* to its polarity and its high dielectric constant. There are other liquids that have very large dipole moments and high dielectric constants, and yet are very poor solvents for ionic compounds. What is needed is *solvating power*: the ability to form strong bonds to dissolved ions. Solvating power is not simply a matter of high dipole moment; it has to do with the nature of the ion–dipole bonds that are formed. To see what is meant by this we must look more closely at the structure of the solvent. Let us start with water.

Cations, we said, are attracted to the negative pole of a polar solvent. In water the negative pole is clearly on oxygen. Oxygen is highly electronegative and, most important, it has *unshared pairs of electrons*.

Furthermore, with only two tiny hydrogens attached to it, the oxygen is *well exposed*; a number of oxygen atoms in a number of water molecules can cluster closely about the cation without crowding.

Anions, we said, are attracted to the positive pole of a polar molecule. In water the positive poles are clearly on hydrogen. The ion–dipole bonds holding anions to water, we recognize, are *hydrogen bonds*.

Hydrogen bonding permits particularly strong solvation of anions. Not only is there a strong positive charge concentrated on a very small atom, hydrogen, but this hydrogen juts out from the molecule and is well exposed; the anion can be held by a number of hydrogen bonds on a number of water molecules without crowding.

Thus, water owes a large part of its special solvating power to its —OH group: it solvates cations strongly through the unshared pairs on oxygen; it solvates anions strongly through hydrogen bonding.

Methanol resembles water in having an —OH group. It is not surprising that it, too, dissolves ionic compounds. (It is, however, inferior to water. It is less polar, and the CH_3 group is bigger and causes more crowding than the second H of water.)

Solvents like water and methanol are called **protic solvents**: solvents containing hydrogen that is attached to oxygen or nitrogen and hence is acidic enough to form hydrogen bonds. Other protic solvents solvate ions in the same way that water does: *cations, through unshared pairs*; *anions, through hydrogen bonding*.

Recent years have seen the development and widespread use of **aprotic solvents**: polar solvents with moderately high dielectric constants, which do not contain acidic hydrogen. For example:

Dimethylsulfoxide   Dimethylformamide   Hexamethylphosphorotriamide
DMSO   DMF   HMPT

They dissolve ionic compounds, but in doing this their action differs in a very important way from that of protic solvents: *they cannot form hydrogen bonds to anions.*

These aprotic solvents are highly polar, with dipole moments several times as large as that of water. As indicated on the formulas, the negative pole in each of our examples is on an oxygen atom that juts out from the molecule (see Fig. 6.3). Through unshared pairs of electrons on these negatively charged, well-exposed atoms, cations are solvated very strongly.

The positive pole, on the other hand, is buried within the molecule. Through this shielded, diffuse charge, the molecule can solvate anions only very weakly. *Aprotic solvents thus dissolve ionic compounds chiefly through their solvation of cations.*

Figure 6.3 A polar aprotic solvent: hexamethylphosphorotriamide (HMPT). The molecule is shown with its negative pole, on the red oxygen atom, up, and the positive pole, on yellow phosphorus, down. As we see, oxygen juts out from the molecule, exposed and accessible; through its unshared pairs of electrons, it bonds strongly to cations. Phosphorus is buried within the molecule; its positive charge is shielded from the outside by bulky groups, and is only very weakly attracted to anions.

Now, as we have already seen for nucleophilic substitution, much of organic chemistry is concerned with reactions between non-ionic compounds (generally organic) and ionic compounds (inorganic and organic), and it is necessary to select a solvent in which both of the reagents will dissolve. Water dissolves ionic compounds very well, but it is a poor solvent for most organic compounds. Non-polar solvents—ether, chloroform, benzene—are good solvents for organic compounds, but very poor solvents for inorganic salts. Alcohols, particularly the smaller ones like methanol and ethanol, offer one way—the traditional way—out of this difficulty. Their lipophilic alkyl groups help them to dissolve non-ionic organic reagents; their hydroxyl groups permit them to dissolve ionic reagents. And so, alone or mixed with water, methanol and ethanol provide a medium in which, for example, nucleophilic aliphatic substitution has been commonly carried out.

But water and alcohols are protic solvents. Through hydrogen bonding, we have seen, such solvents solvate anions strongly; and anions, as it turns out, are usually the important half of an ionic reagent. Thus, although protic solvents dissolve the reagent and bring it into contact with the organic molecule, they at the same time stabilize the anions and lower their reactivity drastically; their basicity is weakened and, with it, the related property, nucleophilic power (Sec. 5.9).

This is where aprotic solvents come in. Through their lipophilic portions, they dissolve organic compounds. They also dissolve inorganic compounds, but they do this, as we have just seen, chiefly through their solvation of cations. Anions are left relatively unencumbered and highly reactive; they are more basic and more nucleophilic.

By use of these aprotic solvents, dramatic effects have been achieved on a wide variety of reactions. Reactions that, in protic solvents, proceed slowly at high temperatures to give low yields may be found, in an aprotic solvent, to proceed rapidly—often at room temperature—to give high yields. A change to an aprotic solvent may increase the reaction rate as much as a million-fold.

Just as solvents differ in their ability to solvate ions, so ions differ in their tendency *to be solvated*. The concentrated charge on a small, "hard" ion leads to stronger ion–dipole bonding than the diffuse charge on a larger, "soft" ion. Thus, in a given solvent, F^- is more strongly solvated than Cl^-, and Li^+ is more strongly solvated than Na^+.

There is an alternative way to view the stabilization of an ion by a solvent. According to the laws of electrostatics, we have seen (Sec. 5.21), the *stability of a charged system is increased by dispersal of charge*. Consider, for example, a solvated anion. The positive ends of the solvent molecules are turned toward the anion and partially neutralize its charge; in doing this they are themselves partially neutralized. This leaves the solvent molecules with a net negative charge; that is, the outer, negative ends are no longer quite balanced by the inner, positive ends. The negative charge originally concentrated on the anion is now distributed over the very large outer surface of the solvent cluster. This amounts to a very large dispersal of charge and, with it, an enormous stabilization of the anion. In the same way, of course, cations are stabilized by dispersal of their positive charge over the solvent cluster.

Such dispersal is more important for the stabilization of a small ion like F^- or Li^+ than for a larger ion like I^- or Rb^+, in which the charge is already dispersed over a considerable surface.

Dispersal of charge—either through solvation or within the ion itself—tends to stabilize organic cations and anions as well as inorganic ones. This concept plays a key role in our understanding of the large fraction of organic chemistry that involves such intermediate particles, as we have already begun to realize from our study of carbocations in Chapter 5.

So far in this section we have discussed the interaction of an ion only with the solvent. But there is another component of the solution to be considered. Each ion has a *counter-ion*, that is, an ion of opposite charge that is also necessarily present. In dilute aqueous solutions an inorganic ion is strongly solvated and effectively insulated from the charge of its counter-ion. But in a solvent of weaker solvating power or lower dielectric constant—in methanol, for example, or one of the aprotic solvents we have described—it feels this charge, and is attracted by it. There is a measure of ionic bonding, and the pair of oppositely charged ions is called an **ion pair**.

The strength of this ionic bonding depends upon the nature of the solvent. In solvents that solvate weakly, ionic bonding is strong; there are no solvent molecules between the pair of ions, and we speak of a *tight ion pair*. In solvents that solvate strongly, ionic bonding is weak; a layer or layers of solvent molecules may separate the pair of ions, and we speak of a *loose ion pair*.

Ion pairs—organic as well as inorganic—play an exceedingly important part in organic chemistry. An ion in solution is subject to many forces, and the stabilizing effect of a counter-ion—like that of the solvent—is one that must always be reckoned with.

Problem 6.1 Label each of the following solvents as protic or aprotic:

(a) $NH_3(l)$ (b) $SO_2(l)$ (c) CH_2Cl_2 (d) CH_3CH_2OH
    Ammonia   Sulfur dioxide   Methylene chloride   Ethanol

(e) $CH_3CH_2OCH_2CH_3$ (f) $CH_3\overset{O}{\overset{\|}{C}}$—OH   (g) $CH_3\overset{O}{\overset{\|}{C}}CH_3$ (h) $H-\overset{O}{\overset{\|}{C}}-NH_2$
       Diethyl ether   Acetic acid   Acetone   Formamide

(i) $H-\overset{O}{\overset{\|}{C}}-\overset{H}{\overset{|}{N}}-CH_3$ (j) $CH_3C{\equiv}N$ (k) $\underset{H_2C\diagdown_{O}\diagup CH_2}{H_2C-CH_2}$ (l) $\underset{H_2C\diagdown_{S}\diagup CH_2}{H_2C-CH_2}$
       N-Methylformamide   Acetonitrile

                                                            Tetrahydrofuran $\underset{O\quad O}{\diagup\diagdown}$
                                                                                    Sulfolane

Problem 6.2 Like most inorganic salts, ammonium chloride is insoluble in non-polar organic solvents. If the hydrogens of NH_4^+ are replaced by CH_3 groups, however,

$$H-\overset{\overset{H}{|}}{\underset{\underset{H}{|}}{N}}-H^+\,Cl^-$$

$$CH_3-\overset{\overset{CH_3}{|}}{\underset{\underset{CH_3}{|}}{N}}-CH_3\overset{\oplus}{}\,Cl^-$$

      Ammonium chloride   Tetramethylammonium chloride
                                A quaternary ammonium salt

the resulting salt shows appreciable solubility in these solvents. (a) How do you account for this contrast? (b) How might you increase the solubility still further?

Now let us see how what we have discussed so far comes into play in chemical reactions.

6.5 The S_N1 reaction: role of the solvent. Ion–dipole bonds

In discussing each of the reactions, S_N2 and S_N1, we accounted for differences in reactivity among various substrates on the basis of differences in the amount of energy required: one substrate reacts faster than another chiefly because of a lower E_{act}. In S_N1, for example, the difference in rate between tertiary and secondary substrates corresponds to a difference in E_{act} of about 15 kcal.

But we have not taken up a more basic matter—one that involves much larger amounts of energy. How do we account for the fact that substitution occurs *at all*, even for the most reactive substrates? By either mechanism, S_N2 or S_N1, a bond is broken between carbon and the leaving group—the carbon–halogen bond, for example, in an alkyl halide—and bond-breaking requires energy. Where does this energy come from?

For an S$_N$2 reaction, the answer is clear: most of the energy needed to break the bond to the leaving group is supplied by the making of the bond to the nucleophile. In attack by OH$^-$, say, the carbon–halogen bond is being broken, and simultaneously a carbon–oxygen bond is being formed.

But what can we say about an S$_N$1 reaction? Here, the rate-determining step is "simple" heterolysis—bond-breaking without, apparently, bond-making to balance it. In the gas phase, bond dissociation energies show, heterolysis of an alkyl halide would require a great deal of energy: 149 kcal/mol for *tert*-butyl bromide, and even more for other substrates. Yet in an S$_N$1 reaction heterolysis occurs readily at moderate temperatures with an E_{act} of only 20 to 30 kcal/mol. This leaves a difference of 130 kcal or more to be provided. Where does this very large amount of energy come from?

The answer is, once again, from bond formation: not formation of one bond, as in the S$_N$2 reaction, but formation of *many* bonds—bonds between the ions produced and the *solvent*. The ions are not generated as naked particles in the near-emptiness of the gas phase; instead, they are generated as *solvated* ions. Clustered about each ion is a group of polar solvent molecules, oriented with their negative ends toward the carbocation and their positive ends toward the anion (Fig. 6.4). Individually, each of these ion–dipole bonds is relatively weak, but altogether they provide a great deal of energy.

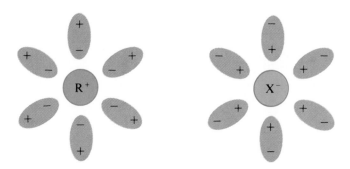

Figure 6.4   Ion–dipole interactions: a solvated carbocation and anion.

But the ions are the *products* of heterolysis. Since we are concerned here with the rate of heterolysis, we must consider not the products but the transition state, and compare *its* stability with the stability of the reactant.

The reactant has a dipole moment, and forms dipole–dipole bonds to solvent molecules. (Indeed, the solvent would have been selected partly for this purpose, since otherwise the reactant would not have dissolved in the first place.) The transition state, we have seen, has a stretched carbon–halogen bond and well-

$$R\!-\!X \longrightarrow \left[\overset{\delta+}{R}\text{----}\overset{\delta-}{X} \right] \longrightarrow R^+ + X^-$$

Reactant   Transition state   Products
More polar than reactant:
stabilized more
by solvation

developed positive and negative charges. It has a *much* greater dipole moment than the reactant, and forms *much* stronger dipole–dipole bonds to the solvent. The solvent thus stabilizes the transition state more than it does the reactant, lowers the E_{act}, and speeds up reaction (Fig. 6.5). Just as a polar solvent stabilizes the ions formed in heterolysis, so it stabilizes the *incipient* ions in the transition state leading to their formation. In an S_N1 reaction, the substrate molecule does not simply fall apart; it is *pulled* apart by the solvent molecules.

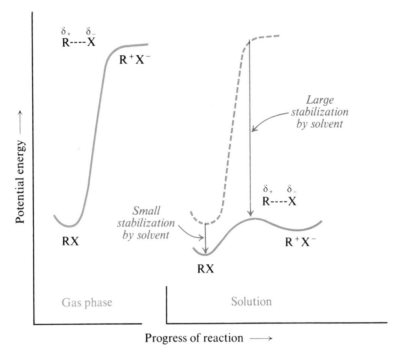

Figure 6.5   Effect of the solvent on the rate of heterolysis of an alkyl halide. The transition state is more polar than the reactant, and is more stabilized by dipole–dipole bonds.

What we have been discussing so far is the difference between heterolysis in the absence and in the presence of a solvent. Clearly the effect of the solvent is enormous: it lowers the E_{act} by 130 kcal or more, and thus allows the reaction to take place.

Now let us take the next step in our analysis and ask: what kind of solvents are best at promoting heterolysis? That is, what kind of solvents have the greatest *ionizing power*? For simplicity, let us discuss solvation of the products, the ions, on the reasonable assumption that the same factors that stabilize them also stabilize the incipient ions in the transition state. On this basis, then, the ionizing power of the solvent depends upon how well it solvates ions. In turn, the ability to solvate ions depends, *in part*, on the polarity of the solvent: other things being equal, the more polar the solvent, the stronger the ion–dipole bonds. Thus, S_N1 reactions of neutral substrates go faster in water than in ethanol; they go faster in, say, 20% ethanol (a 20:80 ethanol:water mixture) than in 80% ethanol.

But, we saw in Sec. 6.4, much more than simple polarity is involved. Cations are solvated chiefly through unshared pairs of electrons; anions are solvated chiefly through hydrogen bonding. Now, here, the cations are *carbo*cations; because of their dispersed charge, they form weaker ion–dipole bonds than smaller metal cations. In the ionization of these organic substrates, therefore, solvation of the cation is relatively weak, whatever the solvent; it is *solvation of the anion* that is particularly important. For this we want solvents capable of hydrogen bonding, that is, *protic* solvents. Thus S$_N$1 reactions proceed more rapidly in water, alcohols, and mixtures of water and alcohols than in aprotic solvents like DMF, DMSO, and HMPT.

We can go further than this. Among protic solvents, ionizing power is highest for the solvents that form the *strongest* hydrogen bonds, that is, the solvents with the most acidic hydrogens. For example, because of powerful electron withdrawal by the fluorine atoms, 2,2,2-trifluoroethanol (CF_3CH_2OH) is much more acidic than ethanol; it forms stronger hydrogen bonds to the leaving group, and is a better solvent for S$_N$1 reactions. In the same way, formic acid (HCOOH) and trifluoro-acetic acid (CF_3COOH) are excellent ionizing solvents.

What we have seen in this section, then, is how the solvent promotes heterolysis by pulling apart the substrate molecule. In Sec. 6.9 we shall see that the solvent can sometimes do more than pull—it can *push*, too.

Problem 6.3 What we have discussed in this section is heterolysis of a *neutral* substrate. Using the same approach, account for the fact that increasing the solvent polarity causes a modest *decrease* in the rate of the following S$_N$1 reaction:

$$RS(CH_3)_2{}^+ \longrightarrow (CH_3)_2S + R^+$$
$$\xrightarrow{\text{H}_2\text{O, C}_2\text{H}_5\text{OH}} ROH + ROC_2H_5$$

6.6 The S$_N$2 reaction: role of the solvent. Protic and aprotic solvents

Now let us turn to the S$_N$2 reaction, and see how it is affected by the solvent. Let us consider what is by far the most common kind of system, one in which the substrate is a neutral molecule and the nucleophile is an anion: the reaction of an alkyl halide with hydroxide ion, for example.

$$R\text{—}X + OH^- \longrightarrow R\text{—}OH + X^-$$

Let us begin as we did with S$_N$1, and see how the reaction as it is ordinarily carried out, in solution, compares with the reaction in the gas phase—that is, with no solvent at all. Once again, it is found, the solvent exerts a powerful effect—*but in the opposite direction*. Where the solvent speeds up an S$_N$1 reaction enormously, it *slows down* the S$_N$2 reaction—and by a factor as large as 10^{20}!

Now, how are we to account for this dramatic reversal? As always when dealing with an effect on rate of reaction, we must compare the reactants with the transition state; this time, we must see how each is affected by the solvent. By definition, there are *two* reactants to consider in the rate-determining step of an S$_N$2 reaction: here, the alkyl halide and the hydroxide ion. The alkyl halide, as we saw, has a dipole moment and forms weak dipole–dipole bonds to the solvent. The hydroxide ion carries a full negative charge, and forms very powerful ion–dipole bonds to the solvent. The transition state carries a full negative charge, too, but

$$HO^- + R{-}X \longrightarrow \left[\begin{matrix} {\scriptstyle\delta-} & & {\scriptstyle\delta-} \\ HO\text{-}\text{-}\text{-}R\text{-}\text{-}\text{-}X \end{matrix}\right] \longrightarrow HO{-}R + X^-$$

Reactants	Transition state	Products
Concentrated charge:	*Dispersed charge*	
stabilized more than		
transition state by solvation		

the charge here is divided between the attacking hydroxyl and the departing halide. Bonding of the solvent to this dispersed charge is much weaker than to the concentrated charge of the small hydroxide ion. The solvent thus stabilizes the reactants—specifically, the nucleophile—more than it does the transition state, raises the E_{act}, and slows down reaction (Fig. 6.6).

Figure 6.6 Effect of the solvent on the rate of attack by hydroxide ion on an alkyl halide. The nucleophile has a more concentrated charge than the transition state, and is more stabilized by ion–dipole bonds.

Solvation of the anionic nucleophile is thus the overriding factor here. By stabilizing it—relative to the transition state—the solvent *deactivates* the nucleophile. Deactivation of the nucleophile by the solvent has actually been measured, molecule by molecule. The gas-phase reaction of methyl bromide with hydroxide ions that are hydrated to varying degrees has been studied, and the following results have been obtained:

$$CH_3Br + (H_2O)_n \cdot OH^- \xrightarrow{\text{gas phase}} CH_3OH + Br^-$$

$n =$	0	1	2	3	In solution
Relative $k =$	1	0.6	.002	.0002	10^{-16}

Starting from the water-free system, we see that as the number of water molecules per hydroxide ion goes up, the rate goes steadily down; finally, in solution, the rate drops to a tiny fraction of its original value.

But the strength of solvation varies from anion to anion, and so does the deactivation it causes. Consider the reaction of methyl bromide with various halide ions.

$$CH_3Br + X^- \longrightarrow CH_3X + Br^-$$

In the gas phase, the order of reactivity of halide ions is $F^- > Cl^- > Br^- > I^-$, reflecting the strength of the C—X bond being formed. Yet in methanol solution the order of reactivity is *reversed*, and becomes $I^- > Br^- > Cl^- > F^-$.

The explanation is straightforward. The strength of solvation varies from anion to anion and, with it, the degree of deactivation. Fluoride is the smallest halide, with the most concentrated charge; as we saw (Sec. 6.4), it forms the strongest ion–dipole bonds—hydrogen bonds in methanol—and hence is the most deactivated. Iodide is the biggest of these halides, with a dispersed charge; it is solvated the least strongly and hence is deactivated the least. In methanol we are not comparing a naked fluoride ion with a naked iodide ion; we are comparing a strongly solvated fluoride ion with a weakly solvated iodide ion. Iodide reacts fastest, not—as was once thought—because of its greater intrinsic reactivity, but because it is solvated least. The solvent is an integral part of the structure of a dissolved molecule; fluoride ion in methanol is a *different reagent* from fluoride ion in the gas phase—or, for that matter, from fluoride ion in DMF. We observe two different orders of reactivity for the reaction with methyl bromide because we are dealing with two different sets of nucleophiles: unsolvated and solvated.

So far, we have been discussing the difference between an S$_N$2 reaction in the absence and in the presence of a solvent. Now, what is the effect of changing from one solvent to another?

Among similar solvents, in general, the greater the polarity, the slower the S$_N$2 reaction; stabilization by the more polar solvent is stronger for the anionic nucleophile than for the transition state, and E_{act} is increased. (Again, this is the opposite of what is observed for an S$_N$1 reaction.)

But these effects of polarity alone are not very big ones. In contrast, the effects of changing from a protic solvent to an aprotic solvent are spectacular. S$_N$2 reactions in solvents like dimethylsulfoxide (DMSO), dimethylformamide (DMF), or hexamethylphosphorotriamide (HMPT) go as much as *a million times faster* than in an alcohol or an alcohol–water mixture. Again solvation of the anion is of overriding importance: the more strongly it is solvated—relative to the transition state—the slower the reaction. The strongest solvation of anions, we know, is through hydrogen bonding—something that is possible for protic solvents but not for aprotic solvents. Aprotic solvents dissolve ionic reagents chiefly through their bonding to the cation; they leave the anion relatively free and highly reactive.

We must not forget that, implicitly at least, we are discussing effects on the anion *relative to* effects on the transition state. Justification for concentrating our attention on the anion is simply that solvation is more important here, and usually—but *not always*—so are differences in solvation.

Problem 6.4   What we have discussed in this section is the commonest kind of S$_N$2 reaction, in which an anionic nucleophile attacks a neutral substrate. Using the same approach, suggest a possible explanation for the following facts.

(a) Increasing solvent polarity causes a large increase in the rate of the S_N2 attack by ammonia on an alkyl halide.

$$RX + NH_3 \longrightarrow RNH_3^+ + X^-$$

(b) Increasing solvent polarity causes a large decrease in the rate of the S_N2 attack by hydroxide ion on trimethylsulfonium ion.

$$HO^- + (CH_3)_3S^+ \longrightarrow CH_3OH + (CH_3)_2S$$

Trimethylsulfonium   Dimethyl sulfide
ion

(c) Increasing solvent polarity causes a small decrease in the rate of the S_N2 attack by trimethylamine on trimethylsulfonium ion.

$$(CH_3)_3N + (CH_3)_3S^+ \longrightarrow CH_3N(CH_3)_3^+ + (CH_3)_2S$$

Trimethylamine   Trimethylsulfonium   Tetramethylammonium   Dimethyl sulfide
ion   ion

6.7   The S_N2 reaction: phase-transfer catalysis

In changing from a protic to an aprotic solvent, then, we have taken a step in the direction of that "ideal" S_N2 reaction medium: the gas phase, where the anion is completely unencumbered and extremely reactive. Yet even an aprotic solvent does solvate anions; it is polar, and forms ion–dipole bonds. From the standpoint of nucleophile reactivity alone, we might imagine that an ideal solvent would be one of very low polarity, like a hydrocarbon or an organic halide: benzene (C_6H_6) or methylene chloride (CH_2Cl_2), for example. But the purpose of the solvent is to bring the reactants together; the organic substrate would dissolve in such a solvent, but the ionic reagent would not. This problem—like the reagent—seems insoluble. But is it?

Take, for example, the reaction of an alkyl halide with sodium cyanide. Cyanide is a strongly basic, nucleophilic anion, and displaces halide to yield the

$$R—X + CN^- \longrightarrow R—CN + X^-$$

Alkyl halide   Cyanide ion   Alkyl cyanide   Halide ion
(A *nitrile*)

alkyl cyanide or *nitrile*. (As we shall see in Sec. 23.8, this is an important step in the synthesis of carboxylic acids.) The traditional way to carry out this reaction would be to use a solvent—protic or aprotic—that dissolves both reagents.

Consider, instead, that we have a solution of the alkyl halide in a non-polar organic solvent and a solution of sodium cyanide in water, and that we mix the two solutions together. The solvents are immiscible and form two layers—two *phases*. We can heat this mixture for a very long time, but nothing will happen. The substrate remains in the organic layer and the nucleophile remains in the water layer, and they cannot do what they must do if they are to react: they cannot *collide*.

Next, to this mixture we add a small amount of a *quaternary ammonium salt*: a compound in which the hydrogens of the ammonium ion have been replaced by alkyl groups—methyl or, even better, *n*-butyl groups. For simplicity we shall refer to this cation as *quat* (Q^+). For reasons that will become clear, the anion of this salt might well be bisulfate, HSO_4^-.

H
|
H—N—H$^+$ X$^-$
|
H

An ammonium salt

R
|
R—N—R$^{\oplus}$ X$^-$
|
R

A quaternary ammonium salt

And now, a remarkable thing happens: in the presence of a catalytic amount of this quat salt, alkyl halide and cyanide—apparently still separated, each in its own phase—react rapidly and under mild conditions to give a high yield of the nitrile.

This is an example of what Charles M. Starks (Continental Oil Company), one of the pioneers in the field, has named **phase-transfer catalysis**. Now, just how does it work? Starks has summarized the catalytic cycle as shown in Fig. 6.7. Everything hinges on the fact that the alkyl groups of the quat ion make it lipophilic, and hence capable of entering the organic phase. But it cannot go alone; to balance its positive charge it must take an anion along. This anion will occasionally be its original counter-ion, bisulfate; this weakly basic anion has virtually no nucleophilic power, and does nothing.

Q$^+$Z$^-$ + R—X $\longrightarrow$ R—Z + Q$^+$X$^-$ *Organic phase*

Substrate   Product

Q$^+$Z$^-$ + Na$^+$X$^-$ $\rightleftharpoons$ Na$^+$Z$^-$ + Q$^+$X$^-$ *Aqueous phase*

Figure 6.7   Phase-transfer catalysis. The quaternary ammonium ion (Q$^+$) is both hydrophilic and lipophilic. It shuttles back and forth between the aqueous phase and the organic phase, taking an anion with it: the nucleophile (Z$^-$) or the leaving group (X$^-$). In the organic phase, the nucleophile is virtually unsolvated and reacts rapidly with the substrate (R—X).

But most of the anions in the aqueous phase are cyanide ions (or whatever nucleophile is being used), and they are the ones most likely to be conducted into the organic phase. We now have cyanide ions in a very unlikely medium: a non-polar solvent. Their concentration there may be very low, but they are virtually unsolvated and highly reactive. Substitution rapidly takes place. The nitrile is formed and a halide ion is liberated. This halide ion is conducted into the aqueous phase by the quat ion as it makes its return trip.

And so reaction continues. The quat ion shuttles back and forth between the two phases taking anions with it: sometimes the original counter-ion; sometimes one of the displaced halide ions; and sometimes the nucleophile, cyanide ion. And when this last happens, reaction can occur. Catalysis is thus due to the *transfer* of the nucleophile from one phase to another.

There is another factor involved here. In most solvents, as we have seen (Sec. 6.4), salts exist to some extent as *ion pairs*. An ion feels the opposite charge of its counter-ion, and is attracted by it. The less polar the solvent—that is, the weaker the solvation—the stronger the ion pairing: one kind of bonding is replaced by another. This electrostatic attraction, too, tends to stabilize an anion; and in doing this it deactivates the anion as a nucleophile and as a base. And so, we might think, in going to a non-polar solvent we are simply exchanging one kind of deactivation for another.

But here we find another advantage of the quat ion as a phase-transfer catalyst. The alkyl groups that make it lipophilic are bulky groups, and they shield the anion from the positive charge on nitrogen (see Fig. 6.8). The anion is attracted much less strongly to this charge buried within the quat ion than to the concentrated charge on a metal cation. The ion pair is only a loose one, and the anion is comparatively free and very reactive.

(a)   (b)

Figure 6.8 Molecular structure and physical properties: solubility. Models of: (a) the ammonium ion, NH_4^+; (b) the tetramethylammonium ion, $(CH_3)_4N^+$. The four lipophilic methyl groups in (b) make the ion soluble in organic solvents. At the same time, they act as a shield between the positive charge on nitrogen and the negative charge on any counter-ion, and thus minimize ion pairing.

The power of phase-transfer catalysis thus lies in the fact that it minimizes the two chief deactivating forces acting on the anion: solvation and ion pairing. There are many variations of the method. There need be no aqueous phase: cyanide can be transferred into the organic phase directly from solid sodium cyanide. There need be no added organic solvent: the substrate itself, if it is a liquid, can act as solvent. The phase-transfer agent need not be ionic, but can be a neutral molecule instead; the most important of these neutral catalysts, as we shall find, are the *crown ethers*—and with the study of them we enter the fascinating area of *host–guest* relationships (Sec. 19.10).

In its various forms phase-transfer catalysis has started a revolution in the technique by which organic reactions are carried out, in the laboratory and in industry: not just nucleophilic substitution, but reactions of all kinds—elimination, addition, oxidation, reduction. It has given the organic chemist a new and powerful tool to help achieve a major aim: *control* of the organic reaction.

Problem 6.5 What is the advantage of using a quat bisulfate as catalyst instead of, say, a quat hydroxide?

6.8 S_N2 vs. S_N1: effect of the solvent

Nucleophilic substitution, we have seen, can proceed by two different mechanisms, S_N2 and S_N1, and in this chapter we have discussed the effect of the solvent on each of them. Indeed, in these solvent effects we see one more piece of evidence that there *are* two mechanisms: one more difference between the two kinds of reaction which, in sheer magnitude, is the most striking of all. We have two reactions that in the gas phase differ in rate by a factor of astronomical size: one reaction immeasurably slow, the other extremely fast. Yet the solvent speeds one up and slows the other down to such an extent that, in dealing with ordinary solution chemistry, we must actually concern ourselves with competition between the two.

In Sec. 5.24, we discussed this competition between S_N2 and S_N1. For a given substrate under a given set of conditions, we asked, which reaction path will be followed? And what, if anything, can we do to steer the reaction in one direction or the other?

The matter comes down simply to this. The mechanism followed depends upon which of two reactions the substrate undergoes faster: attack by the nucleophile, or heterolysis to form a carbocation.

$$\begin{array}{l}
\xrightarrow{\ :Z\ } \quad [Z\text{----}R\text{----}W] \ \longrightarrow \ Z\text{---}R \ + \ :W \qquad\qquad S_N2 \\
\hspace{14em} \textit{Nucleophilic attack} \\[2em]
R\text{---}W \\[2em]
\xrightarrow{\quad} \ \left[\overset{\delta+}{R}\text{----}W\right] \ \longrightarrow \ R^{\oplus} \ + \ :W \qquad\qquad S_N1 \\
\hspace{12em} \textit{Heterolysis} \\
\hspace{6em} \xrightarrow{\ :Z\ } \ Z\text{---}R
\end{array}$$

To see what factors determine these relative rates, we have already discussed the **substrate** and the **nucleophile**. Now let us turn to the third component of the reaction system, the **solvent**.

To predict the solvent effect on reaction by either mechanism, as we know, we must compare the reactants with the transition state for the particular kind of system involved. Let us consider the commonest type of nucleophilic substitution: attack by an anionic nucleophile on a neutral substrate. We examined the system in detail in Secs. 6.5–6.7, and saw that solvent effects are sharply different for reactions by the two mechanisms. Reaction by S_N1 is favored by solvents of high ionizing power, that is, by polar protic solvents that help to pull the leaving group out of the molecule. Reaction by S_N2 is favored by solvents that stabilize (and thus deactivate) the anionic nucleophile *least*: aprotic solvents or solvents of low polarity, as with phase-transfer catalysis.

It is not accidental that, to illustrate the effect of structure on reactivity in Secs. 5.15 and 5.22, we chose for S_N2 a reaction carried out in the aprotic solvent DMF, and for S_N1 a reaction carried out in the polar, strongly hydrogen-bonding (and weakly nucleophilic) solvent CF_3COOH.

(In the effect of the solvent we are really seeing, in part, a factor already discussed: the nature of the nucleophile. In an aprotic solvent or under phase-transfer conditions, we are providing a more powerful nucleophile, and this of course favors S_N2.)

Of all the components of this reaction system, we said earlier, it is the solvent that offers the most scope for control of the reaction. We are restricted in our selection of substrate and nucleophile by our desire to make a particular product. But in choosing the environment in which to carry out the reaction, we have open to us a rapidly widening range of possibilities: from strongly ionizing, weakly nucleophilic solvents at one end to aprotic solvents or phase transfer at the other.

In this section we have discussed the effect of the solvent on the competition between reactions occurring by the two mechanisms, S_N2 and S_N1. Now let us continue with the matter of competition, not between two neatly separated mechanisms, but between the factors that actually determine what happens: nucleophilic power, steric hindrance, and dispersal of charge. And here, we shall find the solvent playing a very important part—the title role, in fact.

6.9 Solvolysis. Nucleophilic assistance by the solvent

We said earlier (Sec. 5.9) that, in its various aspects, nucleophilic aliphatic substitution has been for years the most widely studied—and most strongly disputed—area of organic chemistry. The particular aspect about which most of the study—and most of the dispute—has centered is the special case in which the nucleophile is the solvent: *solvolysis*.

$$R\!-\!X + :S \xrightarrow{\quad\quad} R\!-\!S + :X^-$$
solvent

There is no added strong nucleophile and so, for many substrates, solvolysis falls into the category we have called S_N1; that is, reaction proceeds by two—or more—steps, with the intermediate formation of an organic cation. It is this intermediate that lies at the center of the problem: its nature, how it is formed, and how it reacts. In studying solvolysis we are studying all S_N1 reactions and, in many ways, all reactions involving intermediate carbocations.

Perhaps the biggest question to be answered is: just what is the role played by the solvent? Does it, at one extreme, simply cluster about the carbocation and the anion—and the transition state leading to their formation—and thus aid in heterolysis through formation of ion–dipole bonds? Or, at the other extreme, does a single solvent molecule act as a nucleophile and help push the leaving group out of the molecule? (See Fig. 6.9.)

The two extremes that we have just described correspond, of course, to our descriptions of the mechanisms S_N2 and S_N1. Can we not, then, simply resolve the problem by study of the kinetics? Does the rate depend upon the concentration of the nucleophile, or does it not? Here we encounter the special problem posed by solvolysis. The nucleophile is the solvent, and the solvent's concentration *does not change* during the course of reaction. Regardless of the role played by

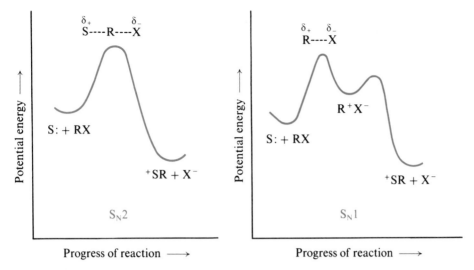

Figure 6.9   Potential energy changes during progress of reaction: solvolysis by classical S_N2 and S_N1 mechanisms. S_N2 involves a single step, with nucleophilic attack on the substrate by the solvent to yield the product directly. S_N1 involves two (or more) steps, with no nucleophilic attack by the solvent on the substrate; the intermediate is a carbocation.

the solvent in the rate-determining step, we *observe* first-order kinetics; the rate depends only upon the concentration of the substrate.

$$rate = k_{obs}[RX]$$

Reaction *could* be full-fledged S_N1. But, for all the kinetics tells us, it could equally well be full-fledged S_N2; the kinetics could be *pseudo* first-order, and the observed rate constant, k_{obs}, could actually be a true rate constant multiplied by the concentration of the solvent, that is, $k_{obs} = k[:S]$.

We cannot tell, then, whether or not the rate depends upon the concentration of the nucleophile. This fact is undoubtedly the origin of much of the interest in solvolysis: the sheer difficulty of the problem is a challenge to the organic chemist. But that is not all. Compared with anions like hydroxide, alkoxides, or cyanide, solvents are mild nucleophiles. Against their mild action other forces can compete, and this competition can be observed and measured: the force exerted by a migrating group in rearrangement, for example, or by the just-departed leaving group. Mild nucleophiles permit the formation of carbocations, and the study of these particles is always fascinating.

It seems clear that the solvent can give *nucleophilic assistance* to solvolysis. How strong this assistance is depends upon:

(a) the nucleophilic power of the solvent;
(b) how badly assistance is needed; and
(c) how accessible, sterically, carbon is to the assisting molecule.

Water, methanol, and ethanol, for example, are strongly nucleophilic—for solvents, that is; acetic acid (CH_3COOH) is weaker, and formic acid ($HCOOH$) is weaker yet. Trifluoroacetic acid (CF_3COOH), trifluoroethyl alcohol (CF_3CH_2OH), and hexafluoroisopropyl alcohol ($CF_3CHOHCF_3$) are very weak;

the highly electronegative fluorine atoms pull electrons strongly from oxygen, and thus lower its basicity and nucleophilic power.

Reactivity of tertiary substrates is found to depend little upon the nucleophilic power of the solvent and chiefly upon its ionizing power (Sec. 6.5). Formation of tertiary cations is relatively easy and needs little nucleophilic assistance; in any case, crowding would discourage such assistance. Reactivity of secondary substrates is found to depend upon both nucleophilic power and ionizing power of the solvent. Formation of secondary cations is more difficult, and needs much nucleophilic assistance. With most primary substrates, reaction is probably straightforward S_N2: a single step with solvent acting as nucleophile.

Let us concentrate, then, on secondary alkyl substrates. Just what is meant by the term *nucleophilic assistance*? First of all, it differs from the S_N2 kind of attack in this way: it leads to the formation, not of the product, but of an intermediate cation. Next, it differs from general "solvation" in this way: a single solvent molecule is involved, not a cluster. The solvent molecule attacks the substrate at the back side and, acting as a nucleophile, helps to push the leaving group out the front side. There is formed a carbocation—or, rather, something with a great deal of carbocation character. Clinging to its back side is the solvent molecule and to

$$:S \ + \ R{-}X \ \longrightarrow \ \left[S\text{----}\overset{\delta_+}{R}\text{----}\overset{\delta_-}{X} \right] \ \longrightarrow \ S\text{---}R^+X^- \ \longrightarrow \ \text{product}$$

Transition state   Intermediate
Solvent gives   *A nucleophilically*
nucleophilic assistance   *solvated carbocation*

the front side, the leaving group. Each may be bonded to carbon through overlap of a lobe of a p orbital on carbon—the empty p orbital of the classical carbocation. The geometry is similar to that of the S_N2 transition state, but this is an *intermediate*, and corresponds to an energy minimum in a progress-of-reaction plot. (Compare Fig. 6.10 with Fig. 6.9.) If the leaving group is an anion, and if the solvent is of only moderate polarity, bonding between cation and anion may be chiefly electrostatic and one speaks of an *ion pair*.

This cationic intermediate—this nucleophilically solvated carbocation—now reacts. It has open to it the wide variety of reactions that, as we shall find, carbocations may undergo. In the reaction that we are concerned with here, it combines with the solvent molecule—with formation of a full-fledged bond—to yield product. If, at the time of reaction, the leaving group is still bonded to the front side—or is still lurking there—reaction with solvent occurs at the back side. If, on the other hand, the cation has lasted long enough for the leaving group to be exchanged for a second solvent molecule—thus forming a symmetrical intermediate—reaction is equally likely at front or back. Solvolysis can occur with complete inversion or with inversion plus varying amounts of racemization.

Elegant work by Saul Winstein (University of California, Los Angeles) revealed the detailed behavior of ion pairs that are intermediates in certain cases of solvolysis: *tight* (or *intimate*) ion pairs, the cation of which is free enough to pivot about and lose configuration, and yet is held tightly enough that recombination to the covalently bonded compound is the favored process; *loose* (or *solvent-separated*) ion pairs, the cation of which is susceptible to attack by outside nucleophiles. The exact role played by ion pairs in nucleophilic substitution has been the subject of a great deal of research, and has been perhaps more hotly debated than the role of the solvent; but this is a big area of complicated chemistry, and we cannot go into it here.

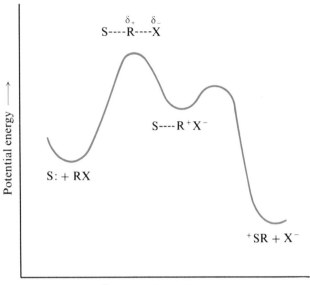

Figure 6.10   Potential energy changes during progress of reaction: solvolysis with nucleophilic assistance by the solvent. Reaction involves two (or more) steps with nucleophilic attack on the substrate by the solvent; the intermediate is a nucleophilically solvated carbocation.

It has been suggested that there is a continuous spectrum of mechanisms for solvolysis ranging from the classical S_N1 reaction at the one end to the single-step S_N2 reaction at the other. On progress-of-reaction plots, the energy minimum for the carbocation becomes shallower and shallower as we move away from the S_N1 end; at the S_N2 end the minimum has disappeared, and we have a single maximum.

In between the ends of the spectrum, there lie mechanisms involving varying degrees of nucleophilic assistance by the solvent. Paul Schleyer (p. 205), whose picture of nucleophilic assistance is essentially the one we have described, has called these mechanisms "S_N2 (intermediate)", that is, S_N2 reactions involving formation of an intermediate. Such a mechanism has characteristics of both classical mechanisms, the single-step S_N2 and the S_N1. Schleyer's terminology emphasizes the S_N2 aspect: that nucleophilic attack provides part of the driving force for the reaction. In this book, however, we shall treat the mechanism as a modification of S_N1: there is a cationic intermediate formed—one that, presumably, is capable of all that a carbocation is capable—and dispersal of the developing positive charge provides much of the driving force for reaction. We shall refer to such a reaction as one following the S_N1 mechanism with nucleophilic assistance from the solvent, and shall call the intermediate a *nucleophilically solvated carbocation* or, sometimes, an *encumbered carbocation*. The exact terminology we use is not important, so long as we understand each other. What is important is that we see here the operation of the same basic factors first recognized by Hughes and Ingold fifty years ago: nucleophilic attack, with its susceptibility to steric hindrance; and dispersal of charge, by substituents and by the solvent. What is new is a growing understanding of how important *both* these factors can be.

We must keep our sense of perspective here. We are discussing the special case of solvolysis, and most of what we say has to do only with secondary alkyl substrates. The differences in stability between the various classes of carbocations are great enough that, by and large, reactions fall into three separate groups: (a) for primary substrates, single-step S_N2; (b) for tertiary substrates, S_N1 with an intermediate that approximates our idea of a simple (solvated) carbocation; (c) for secondary substrates, a two-step reaction that is S_N1-like to the extent that there is a cationic intermediate, but one formed with nucleophilic assistance and still encumbered with nucleophile (solvent) and leaving group. Nucleophilic assistance is an important factor in determining the relative reactivities among secondary substrates, and their reactivities in various solvents—but so is the ionizing power of the solvent. And nucleophilic assistance is not so powerful a factor as the dispersal of charge that makes tertiary substrates react—without any nucleophilic assistance—more rapidly than secondary substrates.

6.10 The medium: a message

What we have said in this chapter about control of the medium in which nucleophilic substitution takes place is only the beginning. We shall see that reactions of many kinds can be carried out between reagents held in the coordination sphere of transition metals (Secs. 20.5–20.8) or residing as *guests* within cavities of large, tailor-made *host* molecules (Secs. 19.10 and 39.10). Control of the reaction medium can be used to bring about new reactions or to speed up old ones, and to achieve a degree of selectivity—in stereochemistry, and in orientation and relative reactivity—never before possible. And yet *this*, too, is only a beginning.

PROBLEMS

1. In the gas phase the heat liberated from the interaction of an ion with each successive molecule of water has been measured: the first molecule, the second, the third, etc. How do you account for the relative quantities (in kcal/mol) in each of the following examples?

(a) For the first molecule of water: H^+, 165; Li^+, 34; Na^+, 24; K^+, 18; Rb^+, 16.

(b) For Li^+, each successive molecule of water: 34, 26, 21, 16, 14, 12.

2. Bulky carbocations are sometimes described as being "self-solvated". How would you justify the use of this term? What fundamental similarity is being referred to?

3. Suggest a reason for representing hydrogen bonding in aqueous ammonia

4. Return to the table you made in Problem 9 (p. 219), and add the following entries:

(i) effect on rate of increasing the water content of the solvent
(j) effect on rate of increasing the alcohol content of the solvent
(k) replacing the alcohol–water solvent by HMPT.

5. How do you account for the fact that in the solvent DMSO the order of reactivity of halide ions with methyl bromide is $F^- > Cl^- > Br^- > I^-$, *opposite to* that observed in methanol solution?

6. The following table lists some physical properties of five compounds of about the same molecular weight. How do you account for: (a) their relative boiling points, and (b) their relative solubilities in water?

Table 6.2 STRUCTURE AND PHYSICAL PROPERTIES

Name	Structure	Dipole moment D	B.p., °C	Solubility g/100 g H_2O
n-Pentane	$CH_3CH_2CH_2CH_2CH_3$	0	36	insol.
Diethyl ether	CH_3CH_2—O—CH_2CH_3	1.18	35	8
n-Propyl chloride	$CH_3CH_2CH_2Cl$	2.10	47	insol.
n-Butyraldehyde	$CH_3CH_2CH_2CHO$	2.72	76	7
n-Butyl alcohol	$CH_3CH_2CH_2CH_2OH$	1.63	118	8

7. For each of the following second-order reactions suggest a possible explanation (or explanations) for the solvent effects given.

(a) $$^{131}I^- + CH_3I \longrightarrow CH_3{}^{131}I + I^-$$

Relative rates: in water, 1; in methanol, 16; in ethanol, 44

(b) $$(n\text{-}C_3H_7)_3N + CH_3I \longrightarrow (n\text{-}C_3H_7)_3NCH_3{}^+I^-$$

Relative rates: in n-hexane, 1; in chloroform, 13 000

(c) $$Br^- + CH_3OTs \longrightarrow CH_3Br + TsO^-$$

Relative rates: in methanol, 1; in HMPT, 10^5

8. The photograph on p. 699 shows a *crown ether*, a doughnut-shaped molecule, holding a potassium ion in its hole. (a) What forces hold the ion in this position? (b) Although a neutral molecule, the crown ether acts as a very efficient phase-transfer agent. Explain in detail how it does this.

9. The following reaction is carried out in the weakly ionizing solvent, acetone, $(CH_3)_2C=O$. (Bs is *brosyl, p*-bromobenzenesulfonyl; like tosylate, brosylate is a good leaving group.)

$$n\text{-}C_4H_9OBs + X^- \longrightarrow n\text{-}C_4H_9X + BsO^-$$

The order of reactivity of halide ions depends upon the salt that is used as their source: if Li^+X^- is used, $I^- > Br^- > Cl^-$; if $(n\text{-}C_4H_9)_4N^+X^-$ is used, $Cl^- > Br^- > I^-$. How do you account for this contrast in behavior?

10. As every chemistry student knows, concentrated sulfuric acid is a thick, viscous liquid. Can you suggest an explanation for this? (*Hint:* So is glycerol, $CH_2OHCHOHCH_2OH$.)

7

Alkenes I. Structure and Preparation

Elimination

7.1 Unsaturated hydrocarbons

In our discussion of the alkanes we mentioned briefly another family of hydrocarbons, the **alkenes**, which contain less hydrogen, carbon for carbon, than the alkanes, and which can be converted into alkanes by addition of hydrogen. The alkenes were further described as being obtained from alkanes by loss of hydrogen in the cracking process.

Since alkenes evidently contain less than the maximum quantity of hydrogen, they are referred to as **unsaturated hydrocarbons.** This unsaturation can be satisfied by reagents other than hydrogen and gives rise to the characteristic chemical properties of alkenes.

7.2 Structure of ethylene. The carbon–carbon double bond

The simplest member of the alkene family is **ethylene**, C_2H_4. In view of the ready conversion of ethylene into ethane, we can reasonably expect certain structural similarities between the two compounds.

To start, then, we connect the carbon atoms by a covalent bond, and then attach two hydrogen atoms to each carbon atom. At this stage we find that each carbon atom possesses only six electrons in its valence shell, instead of the required eight, and that the entire molecule needs an additional pair of electrons if it is to be neutral. We can solve both these problems by assuming that the carbon atoms can share two pairs of electrons. To describe this sharing of two pairs of electrons, we say that the carbon atoms are joined by a *double bond*. *The* **carbon–carbon double bond** *is the distinguishing feature of the alkene structure.*

Ethylene

Quantum mechanics gives a more detailed picture of ethylene and the carbon–carbon double bond. To form bonds with three other atoms, carbon makes use of three equivalent hybrid orbitals: sp^2 orbitals, formed by the mixing of *one s* and *two p* orbitals. As we have seen (Sec. 1.10), sp^2 orbitals lie in one plane, that of the carbon nucleus, and are directed toward the corners of an equilateral triangle; the angle between any pair of orbitals is thus 120°. This **trigonal** arrangement (Fig. 7.1)

Figure 7.1   Atomic orbitals: hybrid sp^2 orbitals. The axes are directed toward the corners of an equilateral triangle.

permits the hybrid orbitals to be as far apart as possible. Just as mutual repulsion among orbitals gives four tetrahedral bonds, so it gives three trigonal bonds.

If we arrange the two carbons and four hydrogens of ethylene to permit maximum overlap of orbitals, we obtain the structure shown in Fig. 7.2. Each

Figure 7.2   Ethylene molecule: only σ bonds are shown.

carbon atom lies at the center of a triangle, at whose corners are located the two hydrogen atoms and the other carbon atom. Every bond angle is 120°. Although distributed differently about the carbon nucleus, these bonds individually are very similar to the bonds in ethane, being cylindrically symmetrical about a line joining the nuclei, and are given the same designation: σ *bond* (*sigma bond*).

The molecule is not yet complete, however. In forming the sp^2 orbitals, each carbon atom has used only two of its three *p* orbitals. The remaining *p* orbital consists of two equal lobes, one lying above and the other lying below the plane of the three sp^2 orbitals (Fig. 7.3); it is occupied by a single electron. If the *p* orbital

Figure 7.3   Ethylene molecule: carbon–carbon double bond. Overlap of *p* orbitals gives a π bond; there is a π cloud above and below the plane.

of one carbon atom overlaps the *p* orbital of the other carbon atom, the electrons pair up and an additional bond is formed.

Because it is formed by the overlap of *p* orbitals, and to distinguish it from the differently shaped σ bonds, this bond is called a π *bond* (*pi bond*). It consists of two parts, one electron cloud that lies above the plane of the atoms, and another electron cloud that lies below. Because of lesser overlap, the π bond is weaker than the carbon–carbon σ bond. As we can see from Fig. 7.3, this overlap can occur only when all six atoms lie in the same plane. Ethylene, then, is a *flat molecule*.

The carbon–carbon "double bond" is thus made up of a strong σ bond and a weak π bond. The total bond energy of 146 kcal is greater than that of the carbon–carbon single bond of ethane (88 kcal). Since the carbon atoms are held more tightly together, the C—C distance in ethylene is less than the C—C distance in ethane; that is to say, the carbon–carbon double bond is shorter than the carbon–carbon single bond.

The σ bond in ethylene has been estimated to have a strength of about 95 kcal: stronger than the one in ethane because it is formed by overlap of *sp²* orbitals (Sec. 7.4). On this basis, we would estimate the strength of the π bond to be 51 kcal.

Figure 7.4   Ethylene molecule: shape and size.

$$H \diagdown \; \overset{1.10 \,\text{Å}}{} C \overset{1.34\,\text{Å}}{=\!=\!=} C \diagup H$$
$$117.5°$$
$$121°$$

This quantum mechanical structure of ethylene is verified by direct evidence. Electron diffraction and spectroscopic studies show ethylene (Fig. 7.4) to be a flat molecule, with bond angles very close to 120°. The C—C distance is 1.34 Å as compared with the C—C distance of 1.53 Å in ethane. (See Fig. 7.5.)

Figure 7.5   Electronic configuration and molecular shape. Model of the ethylene molecule: two views.

In addition to these direct measurements, we shall soon see that two important aspects of alkene chemistry are consistent with the quantum mechanical picture of the double bond, and are most readily understood in terms of that picture. These are (a) the concept of *hindered rotation* and the accompanying phenomenon of *geometric isomerism* (Sec. 7.6), and (b) the kind of reactivity characteristic of the carbon–carbon double bond (Sec. 8.2).

7.3   Propylene

The next member of the alkene family is **propylene**, C_3H_6. In view of its great similarity to ethylene, it seems reasonable to assume that this compound also contains a carbon–carbon double bond. Starting with two carbons joined by a double bond, and attaching the other atoms according to our rule of one bond per hydrogen and four bonds per carbon, we arrive at the structure

$$
\begin{array}{cccc}
& H & H & H \\
& | & | & | \\
H{-}C & {-}C & {=}C & {-}H \\
& | & & \\
& H & &
\end{array}
$$

Propylene

7.4   Hybridization and orbital size

The carbon–carbon double bond in alkenes is shorter than the carbon–carbon single bond in alkanes because four electrons bind more tightly than two. But, in addition, certain other bonds in alkenes are significantly shorter than their counterparts in alkanes: for example, the C—H distance is 1.103 Å in ethylene compared with 1.112 Å in ethane. To account for this and other differences in bond length, we must consider differences in hybridization of carbon.

The carbon–hydrogen bonds of ethylene are single bonds just as in, say, ethane, but they are formed by overlap of sp^2 orbitals of carbon, instead of sp^3 orbitals as in ethane. Now, compared with an sp^3 orbital, an sp^2 orbital has less p character and more s character. A p orbital extends some distance from the nucleus; an s orbital, on the other hand, lies close about the nucleus. As the s character of a hybrid orbital increases, the effective size of the orbital decreases and, with it, the length of the bond to a given second atom. Thus an sp^2–s carbon–hydrogen bond should be shorter than an sp^3–s carbon–hydrogen bond.

Benzene, in most ways a quite different kind of molecule from ethylene (Sec. 13.8), also contains sp^2–s carbon–hydrogen bonds; the C—H bond distance is 1.10 Å, almost exactly the same as in ethylene. Acetylene (Sec. 11.2) contains sp-hybridized carbon which, in view of the even greater s character of the orbitals, should form even shorter bonds than in ethylene; this expectation is correct, the sp–s bond being only 1.079 Å.

A consideration of hybridization and orbital size would lead one to expect an sp^2–sp^3 bond to be shorter than an sp^3–sp^3 bond. In agreement, the carbon–carbon single bond distance in propylene is 1.501 Å, as compared with the carbon–carbon distance of 1.534 Å in ethane. The sp–sp^3 carbon–carbon single bond in methylacetylene (Sec. 10.24) is even shorter, 1.459 Å. These differences in carbon–carbon single bond lengths are greater than the corresponding differences in carbon–hydrogen bond lengths; however, another factor (Sec. 10.23) besides the particular hybridization of carbon may be at work here.

Consideration of hybridization and orbital size helps us to understand other properties of molecules besides bond length: the relative acidities of certain hydrocarbons (Sec. 11.11), for example, and the relative basicities of certain amines (Sec. 35.11). We might reasonably expect shorter bonds to be stronger bonds, and in agreement Table 1.2 (p. 21) shows that the C—H bond dissociation energy in ethylene (108 kcal) is larger than that in ethane (98 kcal), and the C—C (single) bond dissociation energy in propylene (92 kcal) is greater than that in ethane (88 kcal). Indeed, as will be discussed in Sec. 10.24, by affecting the stability of molecules, changes in hybridization may be of more fundamental importance than has been generally recognized.

7.5 The butylenes

Going on to the **butylenes**, C_4H_8, we find that there are a number of possible arrangements. First of all, we may have a straight-chain skeleton as in *n*-butane, or a branched-chain structure as in isobutane. Next, even when we restrict ourselves to the straight-chain skeleton, we find that there are two possible arrangements that differ in position of the double bond in the chain. So far, then, we have a total of three structures; as indicated, these are given the names *1-butene, 2-butene,* and *isobutylene.*

```
    H H H H              H H H H            H      H
    | | | |              | | | |            |      |
 H—C—C—C=C—H          H—C—C=C—C—H       H—C—C=C—H
    | |                  |     |            |
    H H                  H     H            H
                                          H—C—H
                                            |
    1-Butene             2-Butene           H

                                         Isobutylene
```

How do the facts agree with the prediction of three isomeric butylenes? Experiment has shown that not three but *four* alkenes of the formula C_4H_8 exist; they have the physical properties shown in Table 7.1.

Table 7.1 PHYSICAL PROPERTIES OF THE BUTYLENES

Name	B.p., °C	M.p., °C	Relative density (at −20 °C)	Refractive index (at −12.7 °C)
Isobutylene	−7	−141	0.640	1.3727
1-Butene	−6	−195	.641	1.3711
trans-2-Butene	+1	−106	.649	1.3778
cis-2-Butene	+4	−139	.667	1.3868

On hydrogenation, the isomer of b.p. −7 °C yields isobutane; this butylene evidently contains a branched chain, and has therefore the structure we have designated isobutylene.

On hydrogenation, the other three isomers all yield the same compound, *n*-butane; they evidently have a straight-chain skeleton. In ways that we shall study later (Sec. 8.23), it is possible to break an alkene molecule apart at the double bond, and from the fragments obtained deduce the position of the double bond in the molecule. When this procedure is carried out, the isomer of b.p. −6 °C yields products indicating clearly that the double bond is at the end of the chain; this butylene has therefore the structure we have designated 1-butene. When the same procedure is carried out on the two remaining isomers, both yield the same mixture of products; these products show that the double bond is in the middle of the chain.

Judging from the products of hydrogenation and the products of cleavage, we would conclude that the butylenes of b.p. $+1\,°C$ and $+4\,°C$ *both* have the structure we have designated 2-butene. Yet the differences in boiling point, melting point, and other physical properties show clearly that they are not the same compound, that is, that they are isomers. In what way can their structures differ?

To understand the kind of isomerism that gives rise to two 2-butenes, we must examine more closely the structure of alkenes and the nature of the carbon–carbon double bond. Ethylene is a flat molecule. We have seen that this flatness is a result of the geometric arrangement of the bonding orbitals, and in particular the overlap that gives rise to the π orbital. For the same reasons, a portion of any alkene must also be flat, the two doubly bonded carbons and the four atoms attached to them lying in the same plane.

If we examine the structure of 2-butene more closely, and particularly if we use molecular models, we find that there are two quite different ways, I and II, in which the atoms can be arranged (aside from the infinite number of possibilities arising from rotation about the single bonds). In one of the structures the methyl groups lie on the same side of the molecule (I), and in the other structure they lie on opposite sides of the molecule (II).

H⋯C=C⋯H
CH₃   CH₃

I

CH₃⋯C=C⋯H
H   CH₃

II

Now the question arises: can we expect to isolate two isomeric 2-butenes corresponding to these two different structures, or are they too readily inter-converted—like, say, the conformations of *n*-butane (Sec. 3.5)?

Conversion of I into II involves rotation about the carbon–carbon double bond. The possibility of isolating isomers depends upon the energy required for this rotation. We have seen that the formation of the π bond involves overlap of the p orbitals that lie above and below the plane of the σ orbitals. To pass from one of these 2-butenes to the other, the molecule must be twisted so that the p orbitals no longer overlap; that is, the π bond must be broken (see Fig. 7.6).

Figure 7.6 Hindered rotation about the carbon–carbon double bond. Rotation would prevent overlap of the p orbitals and would break the π bond.

Breaking the π bond requires about 70 kcal of energy; at room temperature an insignificant proportion of collisions possess this necessary energy, and hence the rate of this interconversion is extremely small. Because of this 70-kcal energy barrier, then, *there is* **hindered rotation** *about the carbon–carbon double bond.* As a result of this hindered rotation, two isomeric 2-butenes can be isolated. These are, of course, the butylenes of b.p. $+1\,°C$ and b.p. $+4\,°C$.

7.6 Geometric isomerism

Since the isomeric 2-butenes differ from one another *only* in the way the atoms are oriented in space (but are like one another with respect to which atoms are attached to which other atoms), they belong to the general class we have called *stereoisomers* (Sec. 4.1). They are not, however, mirror images of each other, and hence are not enantiomers. As we have already said, *stereoisomers that are not mirror images of each other are called* **diastereomers**.

The particular kind of diastereomers that owe their existence to hindered rotation about double bonds are called **geometric isomers**. The isomeric 2-butenes, then, are diastereomers, and more specifically, geometric isomers.

We recall that the arrangement of atoms that characterizes a particular stereoisomer is called its *configuration*. The configurations of the isomeric 2-butenes are the structures I and II. These configurations are differentiated in their names by the prefixes *cis-* (Latin: on this side) and *trans-* (Latin: across), which indicate that the methyl groups are on the same side or on opposite sides of the molecule. In a way that we shall take up shortly (Sec. 7.9), the isomer of b.p. $+4\,°C$ has been assigned the *cis* configuration, and the isomer of b.p. $+1\,°C$ has been assigned the *trans* configuration.

Figure 7.7 shows scale models of the isomeric 2-butenes. In the *trans* isomer, we see, the methyl groups are well separated; in the *cis* isomer, however, the methyls are thrown closely enough together to cause crowding. From this, we might expect the *cis* isomer to be less stable than the *trans*, and, as we shall see (Sec. 8.4), it *is*.

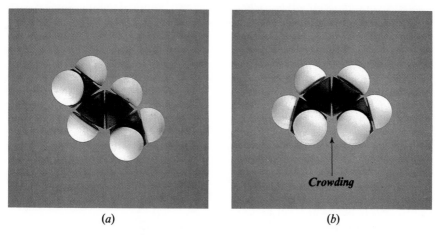

Figure 7.7 Models of the geometric isomers of 2-butene: (*a*) *trans*, (*b*) *cis*. Note the crowding between the methyl groups in the *cis* isomer.

There is hindered rotation about *any* carbon–carbon double bond, but it gives rise to geometric isomerism only if there is a certain relationship among the groups attached to the doubly bonded carbons. We can look for this isomerism by drawing the possible structures (or better yet, by constructing them from molecular models), and then seeing if these are indeed isomeric, or actually identical. On this basis we find that propylene, 1-butene, and isobutylene should not show isomerism; this

No geometric isomerism

Propylene   1-Butene   Isobutylene

conclusion agrees with the facts. Many higher alkenes may, of course, show geometric isomerism.

If we consider compounds other than hydrocarbons, we find that 1,1-dichloro- and 1,1-dibromoethene should not show isomerism, whereas the 1,2-dichloro- and 1,2-dibromoethenes should. In every case these predictions have been found correct. Isomers of the following physical properties have been isolated.

cis-	*trans-*	*cis-*	*trans-*
1,2-Dichloroethene		1,2-Dibromoethene	
b.p. 60 °C	b.p. 48 °C	b.p. 110 °C	b.p. 108 °C
m.p. −80 °C	m.p. −50 °C	m.p. −53 °C	m.p. −6 °C

As we soon conclude from our examination of these structures, geometric isomerism cannot exist if either carbon carries two identical groups. Some possible combinations are shown below.

Isomerism   *Isomerism*   *No isomerism*

The phenomenon of geometric isomerism is a general one and can be encountered in any class of compounds that contain carbon–carbon double bonds (or even double bonds of other kinds).

The prefixes *cis-* and *trans-* work very well for disubstituted ethylenes and some trisubstituted ethylenes. But how are we to specify configurations like the following?

Z	E	Z	E	Z	E
1-Bromo-1-chloropropene		2-Bromo-1-chloropropene		1-Bromo-1,2-dichloroethene	
$CH_3 > H$		$Br > CH_3$		$Cl > H$	
$Br > Cl$		$Cl > H$		$Br > Cl$	

Which groups are our reference points? Looking at each doubly bonded carbon in turn, we arrange its two atoms or groups in their Cahn–Ingold–Prelog sequence (Sec. 4.16). We then take the group of higher priority on the one carbon and the group of higher priority on the other carbon, and tell whether they are on the same side of the molecule or on opposite sides. So that it will be clear that we are using this method of specification, we use the letter **Z** to mean *on the same side*, and the letter **E** to mean *on opposite sides*. (From the German: *zusammen*, together, and *entgegen*, opposite.) The appropriate letter then becomes part of the name of such an alkene: (Z)-1-bromo-1-chloropropene, for example.

A pair of geometric isomers are, then, diastereomers. Where do they fit into the other classification scheme, the one based on how stereoisomers are interconverted (Sec. 4.20)? There are, we saw:
  (a) *configurational isomers*, interconverted by inversion (turning-inside-out) at a chiral center; and
  (b) *conformational isomers*, interconverted by rotations about single bonds.
Now, we add:
  (c) *geometric isomers*, interconverted—in principle—by rotation about a double bond.
  The operation required—rotation—is the same for interconversion of geometric and conformational isomers. But, from the very practical standpoint of *isolability*, geometric isomers are more akin to configurational isomers: interconversion requires bond-breaking—a π bond in the case of geometric isomers—and hence is always a difficult process. Conformational isomers are interconverted by the (usually) easy process of rotation about single bonds.
  For convenience, we laid down the following "ground rule" for discussions and problems in this book: unless specifically indicated otherwise, *the terms "stereoisomers", "enantiomers", and "diastereomers" will refer only to configurational isomers and geometric isomers*, and will exclude conformational isomers. The latter will be referred to as "conformational isomers", "conformers", "conformational enantiomers", and "conformational diastereomers".

In so far as chemical and physical properties are concerned, geometric isomers show the same relationship to each other as do the other diastereomers we have encountered (Sec. 4.17). They contain the same functional groups and hence show similar chemical properties. Their chemical properties are *not identical*, however, since their structures are neither identical nor mirror images; they react with the same reagents, but at different rates. (Under certain conditions—especially in biological systems—geometric isomers can vary widely in their chemical behavior, as we shall see in Chapter 9.)

As the examples above illustrate, geometric isomers have different physical properties: different melting points, boiling points, refractive indices, solubilities, densities, and so on. On the basis of these different physical properties, they can be distinguished from each other and, once the configuration of each has been determined, identified. On the basis of these differences in physical properties they can, in principle at least, be separated. (See Sec. 4.17.)

When we take up the physical properties of the alkenes (Sec. 7.9), we shall discuss one of the ways in which we can tell whether a particular substance is the *cis* or *trans* isomer, that is, one of the ways in which we *assign configuration*.

7.7　Higher alkenes

As we can see, the butylenes contain one carbon and two hydrogens more than propylene, which in turn contains one carbon and two hydrogens more than ethylene. The alkenes, therefore, form another homologous series, the increment being the same as for the alkanes: CH_2. The general formula for this family is C_nH_{2n}.

As we ascend the series of alkenes, the number of isomeric structures for each member increases even more rapidly than in the case of the alkane series; in addition to variations in the carbon skeletons, there are variations in the position of the double bond for a given skeleton, and the possibility of geometric isomerism.

Problem 7.1　Neglecting enantiomerism, draw structures of: (a) the six isomeric pentylenes (C_5H_{10}); (b) the four chloropropylenes (C_3H_5Cl); (c) the eleven chloro-butylenes (C_4H_7Cl). Specify as Z or E each geometric isomer.

7.8　Names of alkenes

Common names are seldom used except for three simple alkenes: *ethylene*, *propylene*, and *isobutylene*. The various alkenes of a given carbon number are, however, sometimes referred to collectively as the *pentylenes* (*amylenes*), *hexylenes*, *heptylenes*, and so on. (One sometimes encounters the naming of alkenes as derivatives of ethylene: as, for example, *tetramethylethylene* for $(CH_3)_2C=C(CH_3)_2$.) Most alkenes are named by the IUPAC system.

The rules of the IUPAC system are:

1. Select as the parent structure the longest continuous chain *that contains the carbon–carbon double bond*; then consider the compound to have been derived from this structure by replacement of hydrogen by various alkyl groups. The parent structure is known as *ethene, propene, butene, pentene,* and so on, depending upon the number of carbon atoms; each name is derived by changing the ending *-ane* of the corresponding alkane name to **-ene**:

$$H_2C=CH_2 \qquad CH_3-CH=CH_2 \qquad \overset{4}{C}H_3\overset{3}{C}H_2\overset{2}{C}H=\overset{1}{C}H_2 \qquad CH_3CH=CHCH_3$$

Ethene　　　　　　　Propene　　　　　　　1-Butene　　　　　　　　2-Butene
　　　　　　　　　　　　　　　　　　　　　　　　　　　　　　　　　　　　(*cis*- or *trans*-)

$$CH_3-\overset{\displaystyle CH_3}{\underset{}{C}}=CH_2 \qquad CH_3-\overset{\displaystyle CH_3}{\underset{\displaystyle CH_3}{C}}-\overset{2}{C}H=\overset{1}{C}H_2 \qquad CH_3-\overset{\displaystyle CH_3}{\underset{\displaystyle H}{C}}-\overset{3}{C}H=\overset{2}{C}H-CH_3$$

2-Methylpropene　　　　　3,3-Dimethyl-1-butene　　　　4-Methyl-2-pentene
　　　　　　　　　　　　　　　　　　　　　　　　　　　　　　　　(*cis*- or *trans*-)

2. Indicate by a number the position of the double bond in the parent chain. Although the double bond involves two carbon atoms, designate its position by the number of the *first* doubly bonded carbon encountered when numbering from the end of the chain nearest the double bond; thus *1-butene* and *2-butene*.

3. Indicate by numbers the positions of the alkyl groups attached to the parent chain.

When a geometric isomer is to be specified, a prefix is added: **cis-** or **trans-**, or **(Z)-** or **(E)-**.

An alkene containing halogen is generally named as a **haloalkene**, that is, as an alkene containing halogen as a side chain. Two unsaturated groups are so

CH_2=CH—Cl

Chloroethene

Vinyl chloride

CH_2=CH—CH_2Br

3-Bromopropene

Allyl bromide

$$\underset{\substack{CH_3 \\ }}{\overset{\substack{H}}{\diagdown}}C=C\underset{\substack{CH_2Cl}}{\overset{\substack{H}}{\diagup}}$$

(Z)-1-Chloro-2-butene

commonly encountered that they are given special names: **vinyl**, CH_2=CH—; and **allyl**, CH_2=CH—CH_2—.

Problem 7.2   Give the structural formula of:

(a) 2,3-dimethyl-2-butene

(b) 3-bromo-2-methylpropene

(c) *cis*-2-methyl-3-heptene

(d) (*E*)-2-chloro-2-butene

Problem 7.3   Referring to your answer to Problem 7.1 (p. 258), give IUPAC names for: (a) the isomeric pentylenes; (b) the isomeric chloropropenes.

7.9 Physical properties

As a class, the alkenes possess physical properties that are essentially the same as those of the alkanes. They are insoluble in water, but quite soluble in non-polar solvents like benzene, ether, chloroform, or ligroin. They are less dense than water. As we can see from Table 7.2 (on the next page), the boiling point rises with increasing carbon number; as with the alkanes, the boiling point rise is 20–30 degrees for each added carbon, except for the very small homologs. As before, branching lowers the boiling point. A comparison of Table 7.2 with Table 3.3 (p. 91) shows that the boiling point of an alkene is very nearly the same as that of the alkane with the corresponding carbon skeleton.

Like alkanes, alkenes are at most only weakly polar. Since the loosely held π electrons of the double bond are easily pulled or pushed, dipole moments are larger than for alkanes. They are still small, however: compare the dipole moments shown for propylene and 1-butene, for example, with the moment of 1.83 D for methyl chloride. The bond joining the alkyl group to the doubly bonded carbon has a small polarity, which is believed to be in the direction shown, that is, with the alkyl group releasing electrons to the doubly bonded carbon. Since this polarity is not canceled by a corresponding polarity in the opposite direction, it gives a net dipole moment to the molecule.

Table 7.2 ALKENES

Name	Formula	M.p., °C	B.p., °C	Relative density (at 20 °C)
Ethylene	CH_2=CH_2	−169	−102	
Propylene	CH_2=$CHCH_3$	−185	−48	
1-Butene	CH_2=$CHCH_2CH_3$		−6.5	
1-Pentene	CH_2=$CH(CH_2)_2CH_3$		30	0.643
1-Hexene	CH_2=$CH(CH_2)_3CH_3$	−138	63.5	.675
1-Heptene	CH_2=$CH(CH_2)_4CH_3$	−119	93	.698
1-Octene	CH_2=$CH(CH_2)_5CH_3$	−104	122.5	.716
1-Nonene	CH_2=$CH(CH_2)_6CH_3$		146	.731
1-Decene	CH_2=$CH(CH_2)_7CH_3$	−87	171	.743
cis-2-Butene	cis-CH_3CH=$CHCH_3$	−139	4	
trans-2-Butene	trans-CH_3CH=$CHCH_3$	−106	1	
Isobutylene	CH_2=$C(CH_3)_2$	−141	−7	
cis-2-Pentene	cis-CH_3CH=$CHCH_2CH_3$	−151	37	.655
trans-2-Pentene	trans-CH_3CH=$CHCH_2CH_3$		36	.647
3-Methyl-1-butene	CH_2=$CHCH(CH_3)_2$	−135	25	.648
2-Methyl-2-butene	CH_3CH=$C(CH_3)_2$	−123	39	.660
2,3-Dimethyl-2-butene	$(CH_3)_2C$=$C(CH_3)_2$	−74	73	.705

Propylene
$\mu = 0.35$ D

1-Butene
$\mu = 0.37$ D

cis-2-Butene, with two methyl groups on one side of the molecule and two hydrogens on the other, should have a small dipole moment. In trans-2-butene, on the other hand, with one methyl and one hydrogen on each side of the molecule, the bond moments should cancel out. Although the dipole moments have not been

cis-2-Butene
expect small $\longmapsto$
b.p. +4 °C
m.p. −139 °C

trans-2-Butene
expect $\mu = 0$
b.p. +1 °C
m.p. −106 °C

measured directly, a small difference in polarity is reflected in the higher boiling point of the cis isomer.

This same relationship exists for many pairs of geometric isomers. Because of its higher polarity the cis isomer is generally the higher boiling of a pair; because of its lower symmetry it fits into a crystalline lattice more poorly, and thus generally has the lower melting point.

The differences in polarity, and hence the differences in melting point and boiling point, are greater for alkenes that contain elements whose electronegativities differ widely from that of carbon. For example:

cis	trans	cis	trans	cis	trans
$\mu = 1.85$ D	$\mu = 0$	$\mu = 1.35$ D	$\mu = 0$	$\mu = 0.75$ D	$\mu = 0$
b.p. 60 °C	b.p. 48 °C	b.p. 110 °C	b.p. 108 °C	b.p. 188 °C	b.p. 192 °C
m.p. -80 °C	m.p. -50 °C	m.p. -53 °C	m.p. -6 °C	m.p. -14 °C	m.p. $+72$ °C

The relationship between configuration and boiling point or melting point is only a rule of thumb, to which there are many exceptions (for example, the boiling points of the diiodoethenes). Measurement of dipole moment, on the other hand, frequently enables us positively to designate a particular isomer as *cis* or *trans*.

Problem 7.4 (a) Indicate the direction of the net dipole moment for each of the dihaloethenes. (b) Would *cis*-2,3-dichloro-2-butene have a larger or smaller dipole moment than *cis*-1,2-dichloroethene? (c) Indicate the direction of the net dipole moment of *cis*-1,2-dibromo-1,2-dichloroethene. Will it be larger than the dipole moment of *cis*-1,2-dichloroethene? Why?

7.10 Industrial source

Petroleum and natural gas provide the alkanes that are the chief primary source of organic chemicals: the chemicals on which a vast industry is built and the chemicals we use in the laboratory. Now, alkanes themselves are ill-suited for direct conversion into a variety of other compounds: they are comparatively unreactive, and the reactions they do undergo take place more or less indiscriminately over the molecule to yield complex mixtures.

But from alkanes there are obtained, by cracking in its various forms (Sec. 3.31), certain more reactive substances: the *aromatic hydrocarbons* benzene, toluene, and the xylenes (Chap. 15); and the smaller *alkenes* ethylene, propylene, and the butylenes. From these few compounds, plus methane, most aromatic and aliphatic chemicals are ultimately made. Ethylene, for example, is the organic compound consumed in the largest amount by the chemical industry—and it ranks fifth among all compounds, following only sulfuric acid, lime, ammonia, and oxygen.

In contrast to alkanes, we shall find, alkenes are highly reactive by virtue of their functional group, the carbon–carbon double bond. (Alkanes really have *no* functional group—or, if they do, it is —H, which occurs everywhere in the molecule.) Not only do alkenes undergo a wide variety of reactions, but these reactions take place at well-defined places in the molecule: at the double bond itself, or at certain positions having a specific relationship to the double bond. The conditions under which alkenes are allowed to react on an industrial scale may, for practical, economic reasons, differ vastly from those used in the laboratory; but in the final analysis the reactions actually taking place are the same ones that we shall study in Chapter 8.

To the extent that the renewable biomass some day replaces the non-renewable fossil mass as the primary source of organic chemicals, alkenes will undoubtedly continue to play a central role. Ethylene, for example, is readily formed by dehydration of ethanol, produced by fermentation of carbohydrates.

7.11   Preparation

Alkenes containing up to four carbon atoms can be obtained in pure form from the petroleum industry. Pure samples of more complicated alkenes must be prepared by methods like those outlined below.

The introduction of a carbon–carbon double bond into a molecule containing only single bonds must necessarily involve the **elimination** of atoms or groups from two adjacent carbons:

$$-\underset{Y}{\overset{|}{C}}-\underset{Z}{\overset{|}{C}}- \longrightarrow -\overset{|}{C}=\overset{|}{C}- \qquad\qquad \text{Elimination}$$

In the cracking process already discussed, for example, the atoms eliminated are both hydrogen atoms:

$$-\underset{H}{\overset{|}{C}}-\underset{H}{\overset{|}{C}}- \xrightarrow{\text{heat}} -\overset{|}{C}=\overset{|}{C}- + H_2$$

The elimination reactions described below not only can be used to make simple alkenes, but also—and this is much more important—provide the best general ways to introduce carbon–carbon double bonds into molecules of all kinds.

_____ PREPARATION OF ALKENES _____

1. **Dehydrohalogenation of alkyl halides.** Discussed in Secs. 7.12 and 7.24.

$$-\underset{H}{\overset{|}{C}}-\underset{X}{\overset{|}{C}}- + KOH \xrightarrow{\text{alcohol}} -\overset{|}{C}=\overset{|}{C}- + KX + H_2O$$

Ease of dehydrohalogenation
of alkyl halides

$$3° > 2° > 1°$$

Examples:

$$CH_3CH_2CH_2CH_2Cl \xrightarrow{\text{KOH (alc)}} CH_3CH_2CH=CH_2$$

n-Butyl chloride   1-Butene

$$CH_3CH_2CHClCH_3 \xrightarrow{\text{KOH (alc)}} CH_3CH=CHCH_3 + CH_3CH_2CH=CH_2$$

sec-Butyl chloride   2-Butene   1-Butene

                              80%   20%

—— CONTINUED ——

_____ CONTINUED _____

2. **Dehydration of alcohols.** Discussed in Sec. 7.25.

$$-\overset{|}{\underset{H}{C}}-\overset{|}{\underset{OH}{C}}- \quad \xrightarrow{acid} \quad -\overset{|}{C}=\overset{|}{C}- + H_2O \qquad \begin{array}{c} \text{Ease of dehydration of alcohols} \\ 3° > 2° > 1° \end{array}$$

$$\begin{array}{ccc} \text{Alcohols} & & \text{Alkenes} \end{array}$$

Examples:

$$H-\overset{\overset{\displaystyle H}{|}}{\underset{\underset{\displaystyle H}{|}}{C}}-\overset{\overset{\displaystyle H}{|}}{\underset{\underset{\displaystyle OH}{|}}{C}}-H \quad \xrightarrow{acid} \quad H-\overset{\overset{\displaystyle H}{|}}{C}=\overset{\overset{\displaystyle H}{|}}{C}-H + H_2O$$

$$\begin{array}{ccc} \text{Ethyl alcohol} & & \text{Ethylene} \end{array}$$

$$CH_3CH_2CH_2CH_2OH \quad \xrightarrow{acid} \quad CH_3CH_2CH=CH_2 + CH_3CH=CHCH_3$$

$$\begin{array}{ccc} \textit{n-}\text{Butyl alcohol} & \text{1-Butene} & \text{2-Butene} \\ & & \textit{Chief product} \end{array}$$

$$CH_3CH_2-\overset{\displaystyle}{\underset{\underset{\displaystyle OH}{|}}{C}}H-CH_3 \quad \xrightarrow{acid} \quad CH_3CH=CHCH_3 + CH_3CH_2CH=CH_2$$

$$\begin{array}{ccc} \textit{sec-}\text{Butyl alcohol} & \text{2-Butene} & \text{1-Butene} \\ & \textit{Chief product} & \end{array}$$

3. **Dehalogenation of vicinal dihalides.** Discussed in Sec. 7.11.

$$-\overset{|}{\underset{X}{C}}-\overset{|}{\underset{X}{C}}- + Zn \quad \longrightarrow \quad -\overset{|}{C}=\overset{|}{C}- + ZnX_2$$

Example:

$$CH_3CHBrCHBrCH_3 \quad \xrightarrow{Zn} \quad CH_3CH=CHCH_3$$

$$\begin{array}{cc} \text{2,3-Dibromobutane} & \text{2-Butene} \end{array}$$

4. **Reduction of alkynes.** Discussed in Sec. 11.8.

The most important of these methods of preparation—since they are the most generally applicable—are the *dehydrohalogenation of alkyl halides*, promoted by base,

Dehydrohalogenation: 1,2-elimination of HX

$$-\overset{|}{\underset{H}{C}}-\overset{|}{\underset{X}{C}}- \quad + \quad :B \quad \longrightarrow \quad \overset{\diagdown}{\diagup}C{=}C\overset{\diagup}{\diagdown} \quad + \quad H{:}B \quad + \quad :X^-$$

<table>
<tr><td>Alkyl halide</td><td>Base</td><td>Alkene</td><td>Protonated base</td><td>Halide ion</td></tr>
</table>

and the *dehydration of alcohols*, catalyzed by acid.

Dehydration: 1,2-elimination of H$_2$O

$$-\overset{|}{\underset{H}{C}}-\overset{|}{\underset{OH}{C}}- \quad \xrightarrow[\text{heat}]{\text{acid}} \quad \overset{\diagdown}{\diagup}C{=}C\overset{\diagup}{\diagdown} \quad + \quad H_2O$$

<table>
<tr><td>Alcohol</td><td>Alkene</td><td>Water</td></tr>
</table>

Not surprisingly, alkyl sulfonates undergo a base-promoted elimination closely analogous to dehydrohalogenation; most of what we have to say about dehydrohalogenation applies equally well to this reaction, too.

As we have seen (Sec. 5.7), alkyl halides and sulfonates are nearly always prepared from the corresponding alcohols, and hence all these methods ultimately involve preparation from alcohols; however, base-promoted elimination generally

$$\text{alcohol} \quad \begin{cases} \xrightarrow{PX_3} \text{alkyl halide} \xrightarrow{\text{base}} \text{alkene} \\ \xrightarrow{TsCl} \text{alkyl tosylate} \xrightarrow{\text{base}} \text{alkene} \\ \xrightarrow{\text{acid}} \text{alkene} \end{cases}$$

leads to fewer complications and is often the preferred method despite the extra step in the sequence.

Both dehydrohalogenation and dehydration suffer from the disadvantage that, where the structure permits, hydrogen can be eliminated from the carbon on either side of the carbon bearing the —X or —OH; this frequently produces isomers. To make certain that the double bond appears only between a given pair of carbon atoms, we need a substrate from which only a given pair of substituents—neither of them ubiquitous hydrogen—can be eliminated, as, for example, in the *Wittig reaction* (Sec. 25.10).

Dehalogenation of vicinal (Latin: *vicinalis*, neighboring) dihalides is severely limited by the fact that these dihalides are themselves generally prepared from the alkenes. However, it is sometimes useful to convert an alkene into a dihalide while we perform some operation on another part of the molecule, and then to regenerate the alkene by treatment with zinc; this procedure is referred to as *protecting the double bond*.

Just as a carbon–carbon double bond can be generated from a carbon–carbon single bond by elimination, so it can be generated from a carbon–carbon triple bond by *addition*.

$$-C\equiv C- \; + \; YZ \; \longrightarrow \; \underset{Y \quad Z}{\overset{}{C=C}} \; \text{or} \; \underset{Y}{\overset{Z}{C=C}} \qquad \text{Addition to triple bond}$$

Alkyne

Since such addition can often be controlled to yield either the *cis* or *trans* alkene, as desired, triply bonded compounds (the *alkynes*, Chap. 11) are important intermediates in the synthesis of stereochemically pure *cis* or *trans* alkenes.

Key intermediates in the above syntheses are alcohols and alkynes. Both these kinds of compounds, we shall find, are themselves readily prepared from smaller, simpler substances. By combining the chemistry of alkenes with the chemistry of alcohols and alkynes, we shall be able to make alkenes in a wide variety of sizes and shapes.

7.12   Dehydrohalogenation of alkyl halides: 1,2-elimination

When isopropyl bromide is treated with a hot concentrated alcoholic solution of a strong base like potassium hydroxide, there is obtained propylene, potassium bromide, and water.

$$\underset{\underset{Br}{|}}{CH_3CHCH_3} \; + \; KOH \; \xrightarrow[\text{heat}]{C_2H_5OH} \; \underset{\text{Propylene}}{CH_3CH=CH_2} \; + \; KBr \; + \; H_2O$$

Isopropyl bromide

This is an example of **dehydrohalogenation**: *1,2-elimination of the elements of hydrogen halide.* Dehydrohalogenation involves loss—elimination—of the halogen atom and of a hydrogen atom from a carbon adjacent to the one losing the halogen. The reagent required is a *base*, whose function is to abstract the hydrogen as a proton. (The base :B can be neutral or negatively charged: for example, H_2O or OH^-. The conjugate acid $H:B$ will then be positively charged or neutral: for example, H_3O^+ or H_2O.)

Dehydrohalogenation: 1,2-elimination of HX

$$\underset{\underset{\text{Alkyl halide}}{H \quad X}}{-\overset{|}{\underset{|}{C}}-\overset{|}{\underset{|}{C}}-} \; + \; \underset{\text{Base}}{:B} \; \longrightarrow \; \underset{\text{Alkene}}{C=C} \; + \; \underset{\substack{\text{Protonated} \\ \text{base}}}{H:B} \; + \; \underset{\substack{\text{Halide} \\ \text{ion}}}{:X^-}$$

Now, how does such an elimination generate a double bond? Regardless of the exact mechanism, the products of reaction show that what must happen is the following. Halogen leaves the molecule as halide ion, and hence must take its electron pair along. Hydrogen is abstracted by the base as a proton, and hence must leave its electron pair behind; it is this electron pair that is available to form the second bond—the π bond—between the carbon atoms.

$$-\overset{|}{\underset{|}{C}}:\overset{\overset{X}{\vdots}}{\underset{\overset{\cdot\cdot}{H}}{C}}- \longrightarrow \; \underset{\diagup}{\overset{\diagdown}{C}}::\overset{\diagup}{\underset{\diagdown}{C}} \; + \; H:B \; + \; :X^-$$

$$\overset{\curvearrowright}{\underset{\raise2pt\hbox{$\curvearrowright$}}{:B}}$$

represented as

$$-\overset{|}{\underset{\underset{\displaystyle H}{|}}{C}}\!\!-\!\!\overset{\overset{\displaystyle X}{|}}{\underset{|}{C}}- \longrightarrow \; \underset{\diagup}{\overset{\diagdown}{C}}\!=\!\overset{\diagup}{\underset{\diagdown}{C}} \; + \; H:B \; + \; :X^-$$

$$\curvearrowleft :B$$

where arrows show the
direction of electron shift

At this point, what we have shown is intended to be only a kind of electronic book-keeping, to show the changes in bonding—the movement of electrons—that must occur. Actually, it represents a specific mechanism—the most common one, as it happens—in which the entire process takes place in a single step, with the various bond-makings and bond-breakings occurring simultaneously. But, as we shall see, there are other mechanisms to be considered: mechanisms in which the same changes in bonding occur, but with different *timing*. As in nucleophilic substitution and in many of the other reactions we shall study, this matter of timing is often the crucial point of difference between alternative mechanisms, and it is to establish the timing that much of the experimental evidence is gathered.

We have called this *1,2*-elimination: for the double bond to form, the hydrogen must come from a carbon that is adjacent to the carbon holding the halogen. Now, the carbon holding the halogen is commonly called the α-carbon (*alpha*-carbon).

$$-\overset{\beta}{\underset{|}{C}}-\overset{\overset{\displaystyle H^\beta \quad H^\beta}{|\qquad|}}{\underset{\underset{\displaystyle X}{|}}{\overset{\alpha}{C}}}\!\!-\!\!\overset{\beta}{\underset{|}{C}}-$$

Any carbon attached to the α-carbon is a β-carbon (*beta*-carbon), and its hydrogens are β-hydrogens. *Elimination, then, involves loss of a β-hydrogen.*

In some cases, dehydrohalogenation yields a single alkene,

$$\underset{\textit{n}\text{-Propyl bromide}}{CH_3CH_2CH_2Br} \xrightarrow{\ \text{KOH (alc)}\ } \underset{\text{Propylene}}{CH_3CH=CH_2} \xleftarrow{\ \text{KOH (alc)}\ } \underset{\underset{\textstyle \text{Isopropyl bromide}}{\underset{\textstyle Br}{|}}}{CH_3CHCH_3}$$

$$\underset{\textit{n}\text{-Butyl bromide}}{CH_3CH_2CH_2CH_2Br} \xrightarrow{\ \text{KOH (alc)}\ } \underset{\text{1-Butene}}{CH_3CH_2CH=CH_2}$$

and in other cases yields a mixture.

$$\underset{\underset{\textstyle \textit{sec}\text{-Butyl bromide}}{\underset{\textstyle Br}{|}}}{CH_3CH_2CHCH_3} \xrightarrow{\ \text{KOH (alc)}\ } \underset{\underset{\textstyle 81\%}{\text{2-Butene}}}{CH_3CH=CHCH_3} + \underset{\underset{\textstyle 19\%}{\text{1-Butene}}}{CH_3CH_2CH=CH_2}$$

To predict which products can be formed in a given reaction, we have only to examine the structure of the substrate. We can expect an alkene corresponding to the loss of *any* one of the β-hydrogens—*but no other alkenes*. *n*-Butyl bromide, for example, can lose hydrogen only from C–2,

$$
\begin{array}{c}
\text{H} \\
| \\
\text{CH}_3\text{CH}_2\text{—C—CH}_2 \quad \xrightarrow{\text{KOH (alc)}} \quad \text{CH}_3\text{CH}_2\text{CH=CH}_2 \\
| \ \ | \\
\text{H} \ \ \text{Br}
\end{array}
$$

n-Butyl bromide   1-Butene
                        Only product

and hence yields only 1-butene. *sec*-Butyl bromide, on the other hand, can lose hydrogen either from C–1,

$$
\begin{array}{c}
\text{H} \ \ \text{H} \\
| \ \ | \\
\text{CH}_3\text{CH}_2\text{—C—C—H} \quad \xrightarrow{\text{KOH (alc)}} \quad \text{CH}_3\text{CH}_2\text{CH=CH}_2 \\
| \ \ | \\
\text{Br} \ \ \text{H}
\end{array}
$$

sec-Butyl bromide   1-Butene
                        19%

or from C–3,

$$
\begin{array}{c}
\text{H} \ \ \text{H} \\
| \ \ | \\
\text{CH}_3\text{—C—C—CH}_3 \quad \xrightarrow{\text{KOH (alc)}} \quad \text{CH}_3\text{CH=CHCH}_3 \\
| \ \ | \\
\text{H} \ \ \text{Br}
\end{array}
$$

sec-Butyl bromide   2-Butene
                        81%

and hence yields both 1-butene and 2-butene. Where the two alkenes can be formed, 2-butene is the chief product; this fact fits into a general pattern for dehydrohalogenation which we shall discuss later (Sec. 7.20).

Problem 7.5 Give structures of all alkenes expected from dehydrohalogenation by strong base of:

    (a) 1-chloropentane   (e) 3-chloro-2-methylbutane
    (b) 2-chloropentane   (f) 2-chloro-2,3-dimethylbutane
    (c) 3-chloropentane   (g) 1-chloro-2,2-dimethylpropane
    (d) 2-chloro-2-methylbutane

Problem 7.6 What alkyl halide (*if any*) would yield each of the following pure alkenes upon dehydrohalogenation by strong base?

    (a) isobutylene   (c) 2-pentene   (e) 2-methyl-2-butene
    (b) 1-pentene   (d) 2-methyl-1-butene   (f) 3-methyl-1-butene

---

(What we have just discussed assumes that no rearrangement takes place, an assumption that is justified for dehydrohalogenation carried out under the usual conditions: in concentrated alcoholic solutions of strong base. We shall learn to recognize situations where rearrangements are likely, and to predict the elimination products in those cases, too.)

In studying dehydrohalogenation we shall learn a good deal about the entire class of reactions to which it belongs, and of which it is typical: **1,2-elimination.**

1,2-Elimination

Such elimination reactions are characterized by the following:

(a) The substrate contains a **leaving group**, an atom or group that leaves the molecule, taking its electron pair with it.

(b) In a position *beta* to the leaving group, the substrate contains an atom or group—nearly always **hydrogen**—that can be extracted by a base, leaving its electron pair behind.

(c) Reaction is brought about by action of a **base**.

Typically, the base is a strongly basic anion like hydroxide, or an alkoxide derived from an alcohol (Sec. 5.5): ethoxide, $C_2H_5O^-$; *tert*-butoxide, $(CH_3)_3CO^-$; etc. But the solvent itself, a neutral substance like an alcohol or water, sometimes serves as the base, although a considerably weaker one.

For convenience, particularly in designating solvents or reagents, one often abbreviates the names of the simpler alkyl groups: methyl, Me; ethyl, Et; *n*-propyl, *n*-Pr; isopropyl, *i*-Pr; *tert*-butyl, *t*-Bu. Thus, methanol becomes MeOH; sodium methoxide, NaOMe; methoxide ion, MeO$^-$.

In elimination, a good leaving group is a weakly basic anion or molecule, just as in nucleophilic substitution—and for exactly the same reasons. As a weak base, it readily releases a proton; as a good leaving group, it readily releases carbon (Sec. 5.9). In dehydrohalogenation the leaving group is the very weakly basic halide ion; it is not just accidental that alkyl halides are important substrates in both nucleophilic substitution and elimination.

Nor is it just accidental that the same alternatives to alkyl halides can be used in both kinds of reaction—other substrates that can release weakly basic anions. Chief among these other substrates are the *tosylates* that we encountered in Sec. 5.9, and their relatives, other sulfonates.

(The similarity of substrates in nucleophilic substitution and in elimination, coupled with the fact that both nucleophiles and bases are electron-rich reagents—indeed, are very often the *same* reagent—can lead to problems: potentially, there is always *competition* between the two reactions (Sec. 7.24).)

Now, what mechanism or mechanisms does dehydrohalogenation follow? Just by examining the structures of the reactants and products, we have arrived at certain conclusions about what happens during the reaction: the carbon–halogen and carbon–hydrogen bonds are broken; the base–hydrogen bond and the π bond are formed. But this description is still far from being a mechanism. Bonds are being broken and bonds are being formed, and we must ask the question: what is the *timing* of all these bond-breakings and bond-makings? Do they all happen at once, or one after another? And, if one after another, which is first?

So far as timing is concerned, there are a number of possibilities, each one corresponding to a different mechanism. In following sections, we shall take up the experimental evidence—the collection of *facts*—that shows which of these mechanisms elimination actually follows under a particular set of conditions. But first let us see what these various mechanisms are, so that we can then better grasp the significance of each fact as it is presented. In all this, we should realize that more important than what we learn about elimination is what we learn about the methods one uses to find out what is going on in a chemical reaction.

Problem 7.7   Starting with an alcohol in each case, outline the synthesis of isobutylene by two different routes.

7.13   Kinetics of dehydrohalogenation. Duality of mechanism

The theory of elimination reactions developed in a way remarkably similar to the way the theory of nucleophilic substitution developed (Sec. 5.12). Again it was in the mid-1930s that a broad theory of the reaction was proposed, and again it was Hughes and Ingold who proposed it. Here, too, they proposed two mechanisms differing in molecularity. Much of what we shall discuss is based on work done since their initial proposals, and by other workers. This subsequent work has led to refinements in the theory, and has given us a closer look at just what is going on; but, by and large, it has fitted remarkably well into the pattern they laid out.

Let us begin our study where Hughes and Ingold did, with the *kinetics* of elimination. As ordinarily carried out, with a concentrated solution of a strong base, dehydrohalogenation *follows second-order kinetics.* That is, the rate of alkene formation depends upon the concentration of *two* substances: alkyl halide and base. This second-order reaction is observed for all classes of alkyl halides.

$$\text{rate} = k[\text{RX}][:\text{B}]$$

Now, if one proceeds along a series of substrates, 1° to 2° to 3°, and if one reduces the strength or concentration of the base, a second kind of behavior begins to appear: *first-order kinetics.* The rate of elimination depends only upon the concentration of alkyl halide, and is independent of the concentration of base.

$$\text{rate} = k[\text{RX}]$$

In general, this first-order reaction is encountered only with secondary or tertiary substrates, and in solutions where the base is either weak or in low concentration.

To account for the two kinds of kinetics behavior, Hughes and Ingold proposed that elimination, like nucleophilic substitution, can proceed by two different mechanisms. These mechanisms, for reasons that will emerge, they named **E2** and **E1**.

7.14   The E2 mechanism

For the reaction that proceeds by second-order kinetics, Hughes and Ingold proposed the **E2 mechanism**. Reaction involves *a single step*: base pulls a proton away from carbon; simultaneously a halide ion departs and the double bond forms. Halogen takes its electron pair with it; hydrogen leaves its electron pair behind, to form the double bond. These are the electronic changes that we said must happen in dehydrohalogenation; what characterizes this particular mechanism is that they are all happening simultaneously, in a single step, via a single transition state.

E2

Bimolecular elimination

In this transition state, two bonds are being broken: C—H and C—X. Now, where does the energy for this bond-breaking come from? As usual, it comes *from bond-making*: formation of the bond between the proton and the base, and formation of the π bond. (Although weaker than a σ bond, the π bond does supply about 70 kcal/mol.)

Consider what happens as the base begins to pull the proton away from the molecule. The β-carbon, armed with the electron pair the departing proton is leaving behind, begins to form a bond to the α-carbon: a second bond, the π bond. As the π bond starts to form, the carbon–halogen bond starts to break: the π bond-making helps to supply energy for the carbon–halogen bond-breaking. Halogen is being *pushed out* in what, from the viewpoint of the α-carbon, is a kind of nucleophilic attack, not unlike an S_N2 reaction. (We shall return to this point in Sec. 9.7.)

This mechanism, we said, was proposed for second-order elimination. Second-order kinetics is, of course, exactly what must be observed for a reaction proceeding by the E2 mechanism. The rate-determining step—the *only* step—involves reaction between a molecule of alkyl halide and a molecule of base, and its rate is proportional

$$\text{rate} = k[\text{RX}][:\text{B}]$$

E2 reaction
Second-order kinetics

to the concentration of both reactants. This mechanism was named E2, that is, *elimination, bimolecular*, because in the rate-determining step two molecules undergo covalency changes.

7.15   The E1 mechanism

For the reaction proceeding by first-order kinetics, Hughes and Ingold proposed the **E1 mechanism**. In this mechanism the electronic changes—the bond-breaking and bond-making—are the same as in E2; here, however, they are taking place, not simultaneously, but one after the other. Where E2 involves a single step, E1 involves two steps. In step (1) the substrate undergoes slow heterolysis to form halide ion and a carbocation. In step (2) the carbocation rapidly loses a proton to the base and forms the alkene.

E1

Unimolecular elimination

(1)

$$\begin{array}{c}\overset{\curvearrowleft X}{\underset{|}{-C-}}\overset{|}{\underset{|}{C-}}\\ \quad\ H\end{array}\ \longrightarrow\ :X^- \ +\ \begin{array}{c}\overset{\oplus}{\underset{|}{-C-}}\overset{|}{\underset{|}{C-}}\\ \quad\ H\end{array}\qquad Slow$$

A carbocation

(2)

$$\begin{array}{c}\overset{\oplus}{\underset{|}{-C-}}\overset{|}{\underset{|}{C-}}\\ \quad\ H\\ \quad\ :B\end{array}\ \longrightarrow\ \begin{array}{c}\diagdown\quad\diagup\\ C\!=\!C\\ \diagup\quad\diagdown\end{array}\ +\ H:B\qquad Fast$$

We recognize step (1) as identical to the first step in S_N1. In the second step of S_N1 the carbocation combines with a nucleophile to yield the substitution product; in step (2) of E1 the carbocation reacts with the base to yield the elimination product.

Here, as always, the reactions of a carbocation have a common end: *they provide a pair of electrons to complete the octet of the electron-deficient carbon.* In S_N1 these electrons are an unshared pair on the nucleophile; in E1 they are the pair originally shared by the proton, and made available—through π bond formation—by departure of the proton.

On the basis of step (2), we can add another reaction to our list of Sec. 5.23.

A carbocation may:

   (a) combine with a nucleophile;
   (b) rearrange to a more stable carbocation;
   (c) eliminate a proton to form an alkene.

This list will continue to grow (Sec. 8.16).

The E1 reaction follows first-order kinetics just as an S_N1 reaction does, and for exactly the same reason. The overall rate of reaction is determined only by the slow first step. Except for the many necessary solvent molecules, this *rate-determining* step involves only substrate, and its rate depends only on the concentration of substrate. The rate of an E1 reaction is independent of base concentration

$$\text{rate} = k[\text{RX}]$$

E1 reaction
First-order kinetics

because the reaction *whose rate we are measuring* does not involve base. Again it is the rate of formation of carbocations that determines how fast a reaction goes. Once formed, the carbocations rapidly react to yield product—in this case, the alkene.

This mechanism was named E1, that is, *elimination, unimolecular*, because in the rate-determining step only one molecule, substrate, undergoes covalency change.

7.16 Evidence for the E2 mechanism. Absence of rearrangement

Now let us look at the evidence that shows which of these mechanisms is actually followed under a particular set of conditions. Let us begin with the kind of system we have already described: dehydrohalogenation by a concentrated alcoholic solution of a strong base. Under these conditions, we have seen, reaction follows second-order kinetics.

Elimination reactions that

(a) *follow second-order kinetics*

also:

(b) *are not accompanied by rearrangements*;
(c) *show a large hydrogen isotope effect*;
(d) *are not accompanied by hydrogen exchange*; and
(e) *show a large element effect*.

The mechanism that is consistent with all these facts, and is therefore generally accepted for the second-order reaction, is the E2 mechanism. It is the most important of the mechanisms that we have described, since it is the principal path followed, not only by dehydrohalogenation, but by 1,2-elimination in general. Let us take up each piece of evidence: let us see how it supports the E2 mechanism and helps to rule out the alternative mechanisms.

First, these reactions (a) *follow second-order kinetics*. As we have seen (Sec. 7.14), this is consistent with the E2 mechanism. It is *not* consistent with the E1 mechanism, which must lead to first-order kinetics.

Next, these second-order eliminations (b) *are not accompanied by rearrangements*. This fact, too, is consistent with the E2 mechanism, whose single step simply provides no opportunity for rearrangement. Like the kinetics, this fact is inconsistent with the E1 mechanism, since carbocations, as we know (Sec. 5.23), are particularly prone to rearrangement.

7.17 Evidence for the E2 mechanism. Isotope effects

Now we come to the third piece of evidence for the E2 mechanism. These second-order eliminations (c) *show a large hydrogen isotope effect*. To understand what this means, we must first learn what an isotope effect is and what, in general, it signifies.

Different isotopes of the same element have, by definition, the same electronic configuration, and hence similar chemical properties. This similarity is the basis of the isotopic tracer technique (Sec. 3.29): one isotope does pretty much what another will do, but, from its radioactivity or unusual mass, can be traced through a chemical sequence.

Yet different isotopes have, also by definition, different masses, and because of this their chemical properties are *not identical*: the same reactions can occur but at somewhat different rates (or, for reversible reactions, with different positions of equilibrium). *A difference in rate (or position of equilibrium) due to a difference in the isotope present in the reaction system is called an* **isotope effect.**

Theoretical considerations, which we cannot go into, supported by much experimental evidence, lead to the conclusion: *if a particular atom is less tightly bound in the transition state of a reaction than in the reactant, the reaction involving the heavier isotope of that atom will go more slowly*. The hydrogen isotopes have the greatest proportional differences in mass: deuterium (D) is twice as heavy as protium (H), and tritium (T) is three times as heavy. As a result, hydrogen isotope effects are the biggest, the easiest to measure, and—because of the special importance of hydrogen in organic chemistry—the most often studied. (If you doubt the importance of hydrogen, look at the structure of almost any compound in this book.)

One kind of reaction in which an atom is less tightly bound in the transition state than in the reactant is a reaction in which a bond to that atom is being broken. Isotope effects due to the breaking of a bond to the isotopic atom are called *primary isotope effects*. They are in general the biggest effects observed for a particular set of isotopes.

In this book we shall be concerned with **primary hydrogen isotope effects**, which amount to this: *a bond to protium* (*H*) *is broken faster than a bond to deuterium* (*D*). For many reactions of this kind,

(1) $\sim\text{C}-\text{H} + \text{Z} \xrightarrow{k^H} [\sim\text{C}----\text{H}----\text{Z}] \longrightarrow \sim\text{C} + \text{H}-\text{Z}$

(2) $\sim\text{C}-\text{D} + \text{Z} \xrightarrow{k^D} [\sim\text{C}----\text{D}----\text{Z}] \longrightarrow \sim\text{C} + \text{D}-\text{Z}$

in which hydrogen is abstracted as an atom, positive ion, or negative ion, deuterium isotope effects (k^H/k^D) as large as 5 to 8 (at room temperature) have been observed; that is to say, the reaction is 5 to 8 times as fast for ordinary hydrogen as for deuterium. (Tritium isotope effects, k^H/k^T, are about twice as large as deuterium isotope effects.)

These differences in rate can be measured in a variety of ways. In some cases, the rates of the two individual reactions (1) and (2) can be measured directly and the rate constants k^H and k^D compared. Often, however, it is more feasible to use our familiar method of competition (Sec. 3.22) in either of two ways.

In *intermolecular* competition, a mixture of labeled and unlabeled reactants compete for a limited amount of reagent; reactions (1) and (2) thus go on in the same mixture, and we measure the relative amounts of H—Z and D—Z produced. (Sometimes, larger amounts of the reagent Z are used, and the relative amounts of the two reactants—ordinary and labeled—left *unconsumed* are measured; the less reactive will have been used up more slowly and will predominate. The relative rates of reaction can be calculated without much difficulty.)

In *intramolecular* competition, a single reactant is used which contains several equivalent positions, some labeled and some not:

(3) $\text{H}-\text{Z} + \text{C} \longrightarrow \text{Product(D)}$

$$\begin{array}{l} \text{C}-\text{H} \\ \text{C}-\text{D} \end{array} \text{Z}$$

(4) $\text{D}-\text{Z} + \text{C}-\text{H} \longrightarrow \text{Product(H)}$

One can then measure either the relative amounts of H—Z and D—Z, or the relative amounts of the D-containing product formed by reaction (3) and the H-containing product formed by reaction (4).

Problem 7.8 (a) When excess toluene-α-*d* ($\text{C}_6\text{H}_5\text{CH}_2\text{D}$) was photochemically mono-chlorinated at 80 °C with 0.1 mol of chlorine, there were obtained 0.0212 mol DCl and 0.0868 mol HCl. What is the value of the isotope effect k^H/k^D (*per hydrogen atom, of course*)? (b) What relative amounts of DCl and HCl would you expect to get from $\text{C}_6\text{H}_5\text{CHD}_2$?

The presence—or *absence*—of an isotope effect for a particular reaction can be of enormous significance to the organic chemist. As our first example of how this concept can be used, let us return to our original topic, the evidence supporting the E2 mechanism.

Let us consider the substituted alkyl halide 2-phenylethyl bromide, $C_6H_5CH_2CH_2Br$. The *phenyl* group, $-C_6H_5$, is derived from the aromatic compound *benzene*, C_6H_6. (Phenyl is often represented by $-Ph$.) When attached to an alkyl halide, the group exerts certain effects on the reactions undergone by the halide, including elimination; we shall point out these effects as we encounter them, and discuss later how they arise (Chap. 15). But for the present we need only know that the $-C_6H_5$ group itself is inert toward the reagents that bring about elimination, and can be considered as just another substituent.

The labeled 2-phenylethyl bromide $C_6H_5CD_2CH_2Br$ was prepared. This compound, we see, contains deuterium at both β-positions, the positions from which hydrogen must be lost in elimination. The rate constant (k^D) for its dehydrobromination by sodium ethoxide was measured, and compared with the rate constant (k^H) for reaction of ordinary (unlabeled) 2-phenylethyl bromide under the same conditions. It was found that $k^H/k^D = 7$, that is, the compound containing

$$C_6H_5-\overset{\overset{\text{H}}{|}}{\underset{\overset{|}{\text{H}}\ \overset{|}{\text{Br}}}{C}}-CH_2 + C_2H_5O^- \xrightarrow{k^H} C_6H_5CH{=}CH_2 + C_2H_5O{-}H$$

$$C_6H_5-\overset{\overset{\text{D}}{|}}{\underset{\overset{|}{\text{D}}\ \overset{|}{\text{Br}}}{C}}-CH_2 + C_2H_5O^- \xrightarrow{k^D} C_6H_5CD{=}CH_2 + C_2H_5O{-}D$$

$$k^H/k^D = 7$$

protium reacts *seven times as fast* as the compound containing deuterium. An isotope effect of this size, we saw, is what we would expect for the breaking of a carbon–hydrogen bond.

Now, what is significant about the existence of an isotope effect here? Not that it shows the breaking of a β-carbon–hydrogen bond; we already know that, just from the products of reaction. What is significant is that it shows the breaking of a β-carbon–hydrogen bond *in a rate-determining step*.

This fact is, of course, consistent with the E2 mechanism: a β-carbon–hydrogen bond is broken in the only—and, hence, rate-determining—step.

To appreciate the significance of this point, consider what would be expected of a reaction proceeding by the E1 mechanism. Here, too, a β-carbon–hydrogen bond is broken, and protium would be lost faster than deuterium—but from the carbocation, in the second, fast step, whose rate has no effect on the overall rate of reaction. The rate is determined by the first step, formation of the carbocation; and this step does not involve the breaking of a carbon–hydrogen bond. (There would be an isotope effect on this first step, but it would be a secondary isotope effect, and much smaller than the primary effect observed for the E2 reaction.) The overall rate of reaction by E1, as measured by the disappearance of the substrate, would show no primary isotope effect.

7.18   Evidence for the E2 mechanism. Absence of hydrogen exchange

Facts (a), (b), and (c) are, then, exactly what we would expect for the E2 mechanism. The rate-determining step (the *only* step) involves a reaction between a molecule of alkyl halide and a molecule of base in which a carbon–hydrogen bond is broken and in which there is no opportunity for rearrangement. In particular, these three facts rule out a carbocation (E1) mechanism for second-order elimination.

So far we have considered only two mechanisms for dehydrohalogenation: one in which halogen and hydrogen leave the substrate at the same time (E2); and one in which halogen leaves first (E1). But there is a third reasonable possibility, a mechanism in which hydrogen leaves first: the *carbanion mechanism*.

Such a mechanism involves two steps. In step (1) the substrate loses a proton to the base to form a negatively charged particle, a *carbanion*. In step (2) this carbanion loses halide ion to yield the alkene.

(1)

$$-\underset{\underset{\overset{\uparrow}{:B}}{\overset{|}{H}}}{\overset{\overset{X}{|}}{C}}-\overset{\overset{X}{|}}{\underset{|}{C}}- \quad\rightleftharpoons\quad -\overset{\overset{X}{|}}{\underset{|}{C}}-\overset{|}{\underset{|}{\overset{..}{C}}{}_{\ominus}}- \quad + \quad H\!:\!B$$

A carbanion

Elimination via carbanion

Seldom observed

(2)

$$-\overset{\overset{\nearrow X}{|}}{\underset{|}{C}}-\overset{|}{\underset{\ominus}{C}}- \quad\longrightarrow\quad :X^{-} \; + \; \underset{/}{\overset{\backslash}{C}}{=}\overset{/}{\underset{\backslash}{C}}$$

Step (1) is an acid-base reaction in the Lowry–Brønsted sense, with the substrate acting as an acid. *Since the acidic proton is attached to carbon, the substrate is called a* **carbon acid**. Like most carbon acids, it is extremely weak.

The products of step (1) are the conjugate acid of the base—water, say, from the base, hydroxide ion—and the conjugate base of the carbon acid, a carbanion. *A* **carbanion** *is the conjugate base of a carbon acid.*

$$-\overset{|}{\underset{|}{C}}{-}H \; + \; :B \quad\rightleftharpoons\quad -\overset{|}{\underset{|}{C}}\!:\!\ominus \; + \; H\!:\!B$$

A carbon acid   Base   A carbanion: Conjugate
                                   conjugate base   acid
                                   of carbon acid

(Elimination via carbanions is often called E1cB, *elimination, unimolecular*, of the *conjugate base*.)

The carbanion mechanism, like the E2, is consistent with facts (a), (b), and (c). In attempts to distinguish between these two possibilities, experiments have been carried out using deuterium as a label: this time, not to test for isotope effects, but simply as a tracer, to test for *hydrogen exchange*. Let us see how this approach works.

Consider the dehydrohalogenation of 2-phenylethyl bromide, $C_6H_5CH_2CH_2Br$. (This substrate was selected because, for reasons that we shall

$$C_6H_5CH_2CH_2Br \xrightarrow[C_2H_5OH]{C_2H_5ONa} C_6H_5CH=CH_2$$

2-Phenylethyl bromide   Phenylethylene

see in Sec. 15.17, the phenyl group, C_6H_5, should strongly favor formation of carbanions.) Dehydrohalogenation was brought about by the strong base sodium ethoxide, C_2H_5ONa, in ethanol solution. Formation of carbanions would involve conversion of the base, ethoxide ion, into its conjugate acid, ethanol, which is the solvent.

$$C_6H_5-\underset{\underset{\text{H}}{|}}{CH}-\underset{\underset{\text{Br}}{|}}{CH_2} + C_2H_5O^- \rightleftharpoons C_6H_5-\underset{\underset{\ominus}{|}}{CH}-\underset{\underset{\text{Br}}{|}}{CH_2} + C_2H_5OH$$

2-Phenylethyl bromide   Ethoxide ion   Carbanion   Ethanol
Acid   *Base*   *Base*   *Acid*

Now, in the actual experiment, the substrate was ordinary (unlabeled) 2-phenylethyl bromide, and the solvent was *labeled* ethanol, C_2H_5OD. Consider what would happen if carbanions were formed—and formed reversibly. Most of them would regain hydrogen many times to regenerate starting material before eventually losing halide ion to yield alkene. And they would regain this hydrogen *from the solvent*, the conjugate acid of the base and, in fact, the only acid around

$$C_6H_5-\underset{\underset{\text{H}}{|}}{CH}-\underset{\underset{\text{Br}}{|}}{CH_2} \underset{C_2H_5OH}{\overset{C_2H_5O^-}{\rightleftharpoons}} C_6H_5-\underset{\underset{\ominus}{|}}{\overset{\cdot\cdot}{CH}}-\underset{\underset{\text{Br}}{|}}{CH_2} \underset{C_2H_5OD}{\overset{C_2H_5O^-}{\rightleftharpoons}} C_6H_5-\underset{\underset{\text{D}}{|}}{CH}-\underset{\underset{\text{Br}}{|}}{CH_2}$$

Unlabeled halide   Labeled halide
Starting material   *Exchange product*

$$C_6H_5-CH=CH_2 + Br^-$$

Elimination product

of appreciable acidity. But nearly all the molecules of solvent are C_2H_5OD, not C_2H_5OH; and so, in this reversal, the carbanion would be almost certain to gain a *deuteron*, not a proton.

Reaction was allowed to run until about half the substrate had been converted into alkene. Reaction was then interrupted, and unconsumed 2-phenylethyl bromide was recovered. Mass spectrometric analysis showed that it contained *no deuterium*. Similar experiments with other systems have given similar results. Typical second-order elimination reactions (d) *are not accompanied by hydrogen exchange*.

Fact (d) thus rules out the mechanism in which carbanions are formed reversibly. It is, of course, consistent with the E2 mechanism, which provides no opportunity for hydrogen exchange.

7.19 Evidence for the E2 mechanism. The element effect

These second-order eliminations (e) *show a large element effect*.

Let us look at the two steps of the carbanion mechanism again. The absence of hydrogen exchange discussed in the preceding section does not completely rule out such a mechanism. It simply shows that *if* carbanions are formed, they are

(1)
$$-\overset{\underset{\displaystyle |}{X}}{\underset{\displaystyle |}{C}}-\overset{\displaystyle |}{\underset{\displaystyle |}{\underset{H}{C}}}- \quad + \quad :B \quad \underset{k_{-1}}{\overset{k_1}{\rightleftharpoons}} \quad -\overset{\underset{\displaystyle |}{X}}{\underset{\displaystyle |}{C}}-\overset{\displaystyle |}{\underset{\displaystyle \ominus}{C}}- \quad + \quad H:B \qquad \textit{Slow}$$

(2)
$$-\overset{\underset{\displaystyle |}{X}}{\underset{\displaystyle |}{C}}-\overset{\displaystyle |}{\underset{\displaystyle \ominus}{C}}- \quad \overset{k_2}{\longrightarrow} \quad \overset{\diagup}{\underset{\diagdown}{C}}{=}\overset{\diagup}{\underset{\diagdown}{C}} \quad + \quad :X^- \qquad \textit{Fast}$$

$$k_2 \gg k_{-1} \qquad \textit{Carbanions formed irreversibly}$$

formed *irreversibly*: that they lose halide ions much faster than they regain protons. That is, k_2 would have to be much larger than k_{-1}.

Now, if this were so, step (1) would be rate-determining, and the rate of step (2) would have no effect on the overall rate of reaction—just as in an S_N1 or E1 reaction. Depending upon conditions, step (2) might go faster or slower, but it really would not matter; step (1) would be the bottle-neck and its rate would determine how fast elimination occurs.

So, at this point in our analysis, we are left with two possible mechanisms for second-order elimination: the E2 mechanism; and a carbanion mechanism whose rate-determining step is formation of the carbanion. To choose between these two, we must recognize this key difference: the carbon–halogen bond is being broken in the rate-determining step of E2, but not in the rate-determining step of the carbanion mechanism. The ease of breaking the carbon–halogen bond should therefore affect the rate of elimination by E2, but not the rate of elimination by the (irreversible) carbanion mechanism.

Now, to tell whether or not a particular bond is being broken in the rate-determining step, we might consider looking for an isotope effect, as was done in studying the cleavage of the carbon–hydrogen bond (Sec. 7.17). But here that would be a more difficult job. We are not dealing with loss of hydrogen, whose isotopes differ two- and three-fold in mass. We are dealing with loss of heavier elements like chlorine, whose isotopes differ by only a few percent, with correspondingly small differences in the ease with which bonds are broken.

It has been pointed out by Joseph Bunnett (University of California, Santa Cruz) that evidence on this point has existed for many years in what he has named the *element effect*.

Heterolytic bond dissociation energies (Table 1.3, p. 22) show that the strength of carbon–halogen bonds follows the sequence

Heterolytic bond dissociation energy	$R-F > R-Cl > R-Br > R-I$

In both S_N2 and S_N1 reactions the carbon–halogen bond is broken in the rate-determining step. And, as expected, reactivity in nucleophilic substitution follows the sequence,

Reactivity toward S_N2 or S_N1	$R-I > R-Br > R-Cl > R-F$

with the rate of reaction reflecting the ease of breaking the carbon–halogen bond. The differences in rate here are quite large: alkyl bromides, for example, react 25 to 50 times as fast as the corresponding alkyl chlorides. These element effects are, in fact, much larger than the isotope effects observed for the breaking of bonds to protium and deuterium—as, indeed, they *should* be, in view of the much greater differences in bond strength.

Now, in these elimination reactions the reactivity of alkyl halides follows the same sequence as for substitution,

Reactivity toward E2 $R—I > R—Br > R—Cl > R—F$

and with element effects of just about the same size: alkyl bromides react 40 to 60 times as fast as the chlorides, and—to take the full range of reactivity—alkyl iodides react more than 25 000 times as fast as the fluorides. Clearly, the rate of breaking the carbon–halogen bond *does* affect the overall rate of elimination.

Leaving-group isotope effects have, in fact, been measured for elimination reactions, and found to be of a size consistent with considerable bond-breaking in the transition state.

Thus, only the E2 mechanism fits all the facts, and is generally accepted as the principal pathway followed by 1,2-elimination.

7.20   The E2 reaction: orientation and reactivity

So far we have been concerned with the evidence that second-order dehydrohalogenation proceeds by the E2 mechanism. Now let us look at some other characteristics of this reaction.

Dehydrohalogenation, we have seen (Sec. 7.12), often yields a mixture of isomeric alkenes. In such a case, which isomer, if any, will predominate? Study of many reactions has shown that one isomer generally does predominate, and that it is possible to predict which isomer this will be—that is, to predict the *orientation* of elimination—on the basis of molecular structure.

$$CH_3CH_2CHBrCH_3 \xrightarrow{\text{KOH (alc)}} CH_3CH{=}CHCH_3 \quad + \quad CH_3CH_2CH{=}CH_2$$
$$\qquad\qquad\qquad\qquad\qquad\qquad\qquad 81\% \qquad\qquad\qquad\qquad 19\%$$

$$CH_3CH_2CH_2CHBrCH_3 \xrightarrow{\text{KOH (alc)}} CH_3CH_2CH{=}CHCH_3 \; + \; CH_3CH_2CH_2CH{=}CH_2$$
$$\qquad\qquad\qquad\qquad\qquad\qquad\qquad 71\% \qquad\qquad\qquad\qquad 29\%$$

$$\overset{\displaystyle CH_3}{\underset{\displaystyle}{CH_3CH_2\overset{|}{C}BrCH_3}} \xrightarrow{\text{KOH (alc)}} \overset{\displaystyle CH_3}{CH_3CH{=}\overset{|}{C}CH_3} \quad + \quad \overset{\displaystyle CH_3}{CH_3CH_2\overset{|}{C}{=}CH_2}$$
$$\qquad\qquad\qquad\qquad\qquad\qquad\qquad 71\% \qquad\qquad\qquad\qquad 29\%$$

Take, for example, *sec*-butyl bromide. Attack by base at any one of three β-hydrogens (those on C–1) can lead to the formation of 1-butene; attack at either of two β-hydrogens (on C–3) can lead to the formation of 2-butene. We see that 2-butene is the preferred product *despite* a probability factor of 3:2 working against its formation.

$$CH_3-\overset{\overset{\displaystyle H}{|}}{C}-\overset{\overset{\displaystyle H}{|}}{\underset{\underset{\displaystyle H}{|}}{C}}=\overset{\overset{\displaystyle H}{|}}{C}-H$$

1-Butene

$$CH_3-\overset{\overset{\displaystyle H}{|}}{\underset{\underset{\displaystyle H}{|}}{C}}-\overset{\overset{\displaystyle H}{|}}{\underset{\underset{\displaystyle Br}{|}}{C}}-\overset{\overset{\displaystyle H}{|}}{\underset{\underset{\displaystyle H}{|}}{C}}-H \quad\xrightarrow{OH^-}$$

4   3   2   1

$$CH_3-\overset{\overset{\displaystyle H}{|}}{C}=\overset{\overset{\displaystyle H}{|}}{C}-\overset{\overset{\displaystyle H}{|}}{\underset{\underset{\displaystyle H}{|}}{C}}-H$$

Preferred product

2-Butene

If we focus our attention, not on the hydrogen being lost, but on the alkene being formed, we see the following. The preferred product, 2-butene, is a *di*substituted alkene, whereas 1-butene is a *mono*substituted alkene; that is, in 2-butene there are two alkyl groups (two CH_3 groups) attached to the doubly bonded carbons, and in 1-butene there is only one alkyl group (C_2H_5).

$$CH_3-\overset{\overset{\displaystyle H}{|}}{C}=\overset{\overset{\displaystyle H}{|}}{C}-CH_3 \quad \textit{preferred over} \quad CH_3CH_2-\overset{\overset{\displaystyle H}{|}}{C}=\overset{\overset{\displaystyle H}{|}}{C}-H$$

Disubstituted alkene   Monosubstituted alkene

In the other examples we see that a disubstituted alkene is preferred over a monosubstituted alkene, and a trisubstituted alkene is preferred over a disubstituted alkene.

$$CH_3CH_2-\overset{\overset{\displaystyle H}{|}}{C}=\overset{\overset{\displaystyle H}{|}}{C}-CH_3 \quad \textit{preferred over} \quad CH_3CH_2CH_2-\overset{\overset{\displaystyle H}{|}}{C}=\overset{\overset{\displaystyle H}{|}}{C}-H$$

Disubstituted alkene   Monosubstituted alkene

$$CH_3-\overset{\overset{\displaystyle H}{|}}{C}=\overset{\overset{\displaystyle CH_3}{|}}{C}-CH_3 \quad \textit{preferred over} \quad CH_3CH_2-\overset{\overset{\displaystyle CH_3}{|}}{\underset{\underset{\displaystyle H}{|}}{C}}=\overset{}{\underset{\underset{\displaystyle H}{|}}{C}}-H$$

Trisubstituted alkene   Disubstituted alkene

These form part of a pattern first observed by the Russian chemist Alexander Saytzeff (University of Kazan), who in 1875 formulated a "rule" which can be summarized as: *in dehydrohalogenation the preferred product is the alkene that has the greater number of alkyl groups attached to the doubly bonded carbon atoms.*

Now, dehydrohalogenation is an irreversible reaction, so that once again orientation is determined by the relative rates of competing reactions. More 2-butene than 1-butene is obtained from *sec*-butyl bromide because 2-butene is formed faster than 1-butene. The alkene with the greater number of alkyl groups is the preferred product because it is formed faster than alternative alkenes. What the Saytzeff rule gives us, then, is a sequence showing the relative rates of formation of alkenes.

Ease of formation of alkenes

$$R_2C=CR_2 > R_2C=CHR > R_2C=CH_2,\ RCH=CHR > RCH=CH_2$$

In Sec. 8.4 we shall find evidence that the stability of alkenes follows exactly the same sequence.

Stability of alkenes

$$R_2C=CR_2 > R_2C=CHR > R_2C=CH_2,\ RCH=CHR > RCH=CH_2 > CH_2=CH_2$$

On this basis we can recast **Saytzeff's rule** to read: **in dehydrohalogenation, the more stable the alkene, the faster it is formed.** Predominant formation of the more stable isomer is called **Saytzeff orientation**.

In this form the rule is more generally useful, since it applies to cases where alkene stability is determined by structural features other than alkyl substituents (Secs. 10.12 and 15.19). Furthermore, this formulation leads directly to the factor actually at work.

Consider the transition state for the E2 reaction. Bonds to hydrogen and the leaving group are partly broken, and the double bond is partly formed. The transition state has thus acquired considerable *alkene character*. Factors that

Transition state:
partly formed
double bond

stabilize the alkene—alkyl groups in these cases—also stabilize the incipient alkene in the transition state. E_{act} is lowered, and the alkene is formed faster. Once again, as in the formation of free radicals and of carbocations, the product character of the transition state is a major factor in determining its stability, and hence the rate of reaction.

But the alkene character of the transition state is not the only factor at work in elimination and, as a result, orientation is not always Saytzeff. This is particularly true when substrates other than alkyl halides and alkyl sulfonates are involved. In Sec. 27.6, we shall look at another kind of orientation, *Hofmann*, and at the factors that lie behind it, too. We shall see that orientation in elimination is the net result of the working of several factors—often opposing each other—and that the Saytzeff orientation generally observed for elimination from alkyl halides and sulfonates simply reflects a transition state where one factor, alkene stability, is dominant.

Alkene stability not only determines *orientation* of dehydrohalogenation, but also is an important factor in determining the *reactivity* of an alkyl halide toward elimination, as shown below, for example, for reaction with sodium ethoxide in ethanol at 55 °C. We see that, even after we have allowed for the number of β-hydrogens, the relative rate *per hydrogen* increases as the alkene becomes more highly substituted.

Substrate $\longrightarrow$	Product	Relative rates	Relative rates *per* H
$CH_3CH_2Br \longrightarrow$	$CH_2=CH_2$	1.0	1.0
$CH_3CH_2CH_2Br \longrightarrow$	$CH_3CH=CH_2$	3.3	5.0
$CH_3CHBrCH_3 \longrightarrow$	$CH_3CH=CH_2$	9.4	4.7
$(CH_3)_3CBr \longrightarrow$	$(CH_3)_2C=CH_2$	120	40

As one proceeds along a series of alkyl halides from 1° to 2° to 3°, the structure by definition becomes more branched at the carbon carrying the halogen. This increased branching has two results: it provides a greater number of β-hydrogens for attack by base, and hence a more favorable probability factor toward elimination; and it leads to a more highly branched, more stable alkene, and hence a more stable transition state and lower E_{act}. As a result of this combination of factors, **in E2 dehydrohalogenation the order of reactivity of alkyl halides is**

Reactivity of RX toward E2 $3° > 2° > 1°$

We can, however, look deeper than this in analyzing the structures of substrates. A substrate may be of the same class as another and yet yield a more highly branched alkene; and, in general, we expect it to be more reactive. This is usually true even though the number of β-hydrogens is smaller; where the two factors oppose each other, alkene stability tends to outweigh the probability factor.

Problem 7.9   Predict the major product of each dehydrohalogenation in Problem 7.5 (p. 267).

Problem 7.10   Predict the order of reactivity toward E2 dehydrohalogenation of the following compounds: ethyl bromide, *n*-propyl bromide, isobutyl bromide, neopentyl bromide. Explain your answer in detail.

In Chapter 9, we shall examine a further aspect of the E2 reaction: its *stereochemistry* (Sec. 9.7).

7.21   Evidence for the E1 mechanism

So far our focus has been on E2, the mechanism Hughes and Ingold proposed for dehydrohalogenation with second-order kinetics. But dehydrohalogenation can also proceed by first-order kinetics, and for this reaction Hughes and Ingold proposed another mechanism, the E1.

This mechanism, we saw (Sec. 7.15), involves two steps: slow heterolysis of the alkyl halide (step 1) to form the carbocation, which then rapidly loses a proton to the base (step 2) to yield the alkene.

E1

Unimolecular elimination

(1) $-\overset{\overset{\displaystyle X}{|}}{\underset{|}{C}}-\overset{|}{\underset{|}{\underset{H}{C}}}- \longrightarrow :X^- + -\overset{\overset{\displaystyle \oplus}{|}}{C}-\overset{|}{\underset{|}{\underset{H}{C}}}-$ *Slow*

A carbocation

(2) $-\overset{\overset{\displaystyle \oplus}{|}}{C}-\overset{|}{\underset{|}{\underset{H}{C}}}- \longrightarrow \overset{}{\underset{}{C}}=\overset{}{\underset{}{C}} + H:B$ *Fast*

:B

Reaction by E1 follows first-order kinetics because step (1) is rate-determining, and the rate of this step depends only on [substrate]. The overall rate is independent of [base] because base does not enter into the reaction until the second, fast step.

$$\text{rate} = k[\text{RX}]$$

E1 reaction
First-order kinetics

But consider the situation in which elimination is brought about, not by an added base, but by the solvent itself: by ethanol, say. For all the kinetics tells us,

$$R\text{---}X \xrightarrow{\ \text{EtOH}\ } \text{alkene} + HX$$

reaction *could* be taking place by the E2 mechanism, with the solvent abstracting a proton in the single, rate-determining step. Yet the kinetics would not reveal this: the concentration of the solvent does not change throughout the reaction, and we would observe first-order kinetics—*pseudo* first-order kinetics, but indistinguishable from true first-order kinetics. (The observed rate constant, k_{obs}, would actually

$$\text{rate} = k[\text{RX}][\text{EtOH}] = k_{obs}[\text{RX}]$$

where $k_{obs} = k[\text{EtOH}]$

be a true rate constant *multiplied by* the concentration of the solvent.) The situation is strictly analogous to that in solvolysis (Sec. 6.9); in fact, it is under solvolytic conditions that the E1 mechanism is most commonly followed. Clearly, in this situation other evidence is needed.

Besides the kinetics, then, what is the evidence for the E1 mechanism? The elimination reactions that

(a) *follow first-order kinetics*

also

(b) *are not accompanied by a primary hydrogen isotope effect;*
(c) show the same *effect of structure on reactivity* as S$_N$1 reactions do; and
(d) where the structure permits, *are accompanied by rearrangement.*

Let us examine each piece of evidence.

These first-order reactions (b) *are not accompanied by a primary hydrogen isotope effect*. Such an isotope effect, we have seen (Sec. 7.17), would be expected in elimination only if the β-carbon–hydrogen bond is broken *in the rate-determining step*. It is expected in E2, with its single step; and a large, primary isotope effect is in fact observed in second-order eliminations. It is *not* expected in E1, where the proton is lost in the second, fast step (see Sec. 7.15). And it is a fact that a primary isotope effect is not observed in first-order eliminations.

Next, first-order eliminations (c) show the same *effect of structure on reactivity* as S_N1 reactions do. To understand this evidence, we need only recall something we recognized earlier (Sec. 7.15): that E1 involves *exactly the same first step* as S_N1.

Since this first step is rate-determining, it follows that the order of reactivity of alkyl halides in E1 must be the same as in S_N1. Experiment has shown that this is so.

Reactivity in E1 $3° > 2° > 1°$

In E1, as in S_N1, reactivity is determined by the rate of formation of the carbocation; and this, we have seen (Sec. 5.22), depends upon the stability of the carbocation.

Where the structure permits, these first-order eliminations (d) *are accompanied by rearrangement*. Again we turn to the fact that the first step is the same as in S_N1. Since this first step yields carbocations, it follows that E1 should be susceptible to rearrangements, and of exactly the same kind as those characteristic of S_N1 (Sec. 5.23). This, too, is confirmed by experiment. The double bond appears in places remote from the carbon that held the leaving group:

$$\underset{\substack{\\ \text{3-Methyl-2-butyl}\\ \text{tosylate}}}{\overset{\overset{\displaystyle CH_3}{|}}{CH_3CHCHCH_3}} \underset{\displaystyle OTs}{} \xrightarrow{\textit{n-}BuOH} \underset{\text{2-Methyl-2-butene}}{CH_3\overset{\overset{\displaystyle CH_3}{|}}{C}=CHCH_3} + \underset{\text{3-Methyl-1-butene}}{CH_3\overset{\overset{\displaystyle CH_3}{|}}{C}HCH=CH_2} + \underset{\text{2-Methyl-1-butene}}{CH_2=\overset{\overset{\displaystyle CH_3}{|}}{C}CH_2CH_3}$$

Sometimes the carbon skeleton is changed:

$$\underset{\text{3,3-Dimethyl-2-bromobutane}}{CH_3-\overset{\overset{\displaystyle CH_3}{|}}{\underset{\underset{\displaystyle CH_3}{|}}{C}}-\overset{}{\underset{\underset{\displaystyle Br}{|}}{C}}H-CH_3} \xrightarrow{EtOH} \underset{\text{2,3-Dimethyl-2-butene}}{CH_3-\overset{\overset{\displaystyle CH_3}{|}}{C}=\overset{}{\underset{\underset{\displaystyle CH_3}{|}}{C}}-CH_3} + \underset{\text{2,3-Dimethyl-1-butene}}{CH_3\overset{\overset{\displaystyle CH_3}{|}}{C}H-\overset{}{\underset{\underset{\displaystyle CH_3}{|}}{C}}=CH_2}$$

Alkenes are even obtained from substrates that do not contain a β-hydrogen:

$$\underset{\text{Neopentyl bromide}}{CH_3-\overset{\overset{\displaystyle CH_3}{|}}{\underset{\underset{\displaystyle CH_3}{|}}{C}}-CH_2Br} \xrightarrow{EtOH} \underset{\text{2-Methyl-2-butene}}{CH_3-\overset{\overset{\displaystyle CH_3}{|}}{C}=CHCH_3} + \underset{\text{2-Methyl-1-butene}}{CH_2=\overset{\overset{\displaystyle CH_3}{|}}{C}-CH_2CH_3}$$

In each case it is evident that if, indeed, the alkene is formed from a carbocation, *it is not the same carbocation that was initially formed from the substrate*. And, of course, it is not.

In each of these examples the initially formed carbocation can rearrange by a 1,2-shift to form a more stable carbocation. And—as we saw for S_N1 reactions (Sec. 5.23)—when this *can* happen, it *does*.

$$
\underset{\underset{\text{H}\quad\text{OTs}}{|}}{\overset{\overset{\text{CH}_3}{|}}{\text{CH}_3-\text{C}-\text{CH}-\text{CH}_3}} \xrightarrow{-\text{OTs}^-} \underset{\overset{\curvearrowleft}{\text{H}}\overset{\nearrow}{\ }\oplus}{\overset{\overset{\text{CH}_3}{|}}{\text{CH}_3-\text{C}-\text{CH}-\text{CH}_3}} \longrightarrow \underset{\underset{\text{H}}{\oplus}}{\overset{\overset{\text{CH}_3}{|}}{\text{CH}_3-\text{C}-\text{CH}-\text{CH}_3}}
$$

$$\qquad\qquad\qquad\qquad\qquad\quad (2°)\qquad\qquad\qquad\qquad (3°)$$

$$
\underset{\underset{\text{CH}_3\quad\text{Br}}{|}}{\overset{\overset{\text{CH}_3}{|}}{\text{CH}_3-\text{C}-\text{CH}-\text{CH}_3}} \xrightarrow{-\text{Br}^-} \underset{\overset{\curvearrowleft}{\text{CH}_3}\overset{\nearrow}{\ }\oplus}{\overset{\overset{\text{CH}_3}{|}}{\text{CH}_3-\text{C}-\text{CH}-\text{CH}_3}} \longrightarrow \underset{\underset{\text{CH}_3}{\oplus}}{\overset{\overset{\text{CH}_3}{|}}{\text{CH}_3-\text{C}-\text{CH}-\text{CH}_3}}
$$

$$\qquad\qquad\qquad\qquad\qquad\quad (2°)\qquad\qquad\qquad\qquad (3°)$$

$$
\underset{\underset{\text{CH}_3}{|}}{\overset{\overset{\text{CH}_3}{|}}{\text{CH}_3-\text{C}-\text{CH}_2\text{Br}}} \xrightarrow{-\text{Br}^-} \underset{\overset{\curvearrowleft}{\text{CH}_3}}{\overset{\overset{\text{CH}_3}{|}}{\text{CH}_3-\text{C}-\text{CH}_2\oplus}} \longrightarrow \underset{\underset{\oplus}{}}{\overset{\overset{\text{CH}_3}{|}}{\text{CH}_3-\text{C}-\text{CH}_2-\text{CH}_3}}
$$

$$\qquad\qquad\qquad\qquad\qquad\quad (1°)\qquad\qquad\qquad\qquad (3°)$$

It is this new carbocation that loses the proton—in a perfectly straightforward way from the β-position—to yield the "unexpected" alkenes.

$$
\underset{\underset{\oplus}{}}{\overset{\overset{\text{CH}_3}{|}}{\text{CH}_3-\text{C}-\text{CH}_2-\text{CH}_3}} \xrightarrow{-\text{H}^+} \overset{\overset{\text{CH}_3}{|}}{\text{CH}_3-\text{C}=\text{CH}-\text{CH}_3} \quad + \quad \overset{\overset{\text{CH}_3}{|}}{\text{CH}_2=\text{C}-\text{CH}_2-\text{CH}_3}
$$

Rearranged cation   *Chief product*

$$
\underset{\overset{\oplus}{\underset{\text{CH}_3}{|}}}{\overset{\overset{\text{CH}_3}{|}}{\text{CH}_3-\text{C}-\text{CH}-\text{CH}_3}} \xrightarrow{-\text{H}^+} \underset{\underset{\text{CH}_3}{|}}{\overset{\overset{\text{CH}_3}{|}}{\text{CH}_3-\text{C}=\text{C}-\text{CH}_3}} \quad + \quad \overset{\overset{\text{CH}_3}{|}}{\text{CH}_2=\text{C}-\text{CH}_2-\text{CH}_3}
$$

Rearranged cation   *Chief product*

$$
\underset{\underset{\oplus}{}}{\overset{\overset{\text{CH}_3}{|}}{\text{CH}_3-\text{C}-\text{CH}_2-\text{CH}_3}} \xrightarrow{-\text{H}^+} \overset{\overset{\text{CH}_3}{|}}{\text{CH}_3-\text{C}=\text{CH}-\text{CH}_3} \quad + \quad \overset{\overset{\text{CH}_3}{|}}{\text{CH}_2=\text{C}-\text{CH}_2-\text{CH}_3}
$$

Rearranged cation   *Chief product*

We can begin to see the pattern of rearrangements that runs through reactions of many different types, the pattern first glimpsed by Meerwein (p. 194) in 1922, and which led him to conceive of the carbocation as a reactive intermediate.

Problem 7.11   When heated with catalytic amounts of strong acids like H_2SO_4 or $HClO_4$, alcohols are converted into alkenes. The order of reactivity of alcohols is *tert*-butyl > isopropyl > ethyl. The alcohol 3,3-dimethyl-2-butanol gives 2,3-dimethyl-2-butene together with a smaller amount of 2,3-dimethyl-1-butene.

Assuming that these observations represent typical behavior (they *do*), write all steps in a possible mechanism for *dehydration of alcohols*.

7.22   The E1 reaction: orientation

Elimination by E1 shows strong Saytzeff orientation. That is to say, when more than one alkene can be formed, the more highly branched—the *more stable*—alkene is the preferred product. Thus a disubstituted alkene is preferred over a monosubstituted,

$$CH_3CH_2CH_2\underset{\underset{\displaystyle \text{2-Pentyl tosylate}}{|}}{\overset{\displaystyle |}{\underset{\text{OTs}}{C}}}HCH_3 \xrightarrow{n\text{-BuOH}} \underset{\underset{89\%}{\text{2-Pentene}}}{CH_3CH_2CH=CHCH_3} + \underset{\underset{11\%}{\text{1-Pentene}}}{CH_3CH_2CH_2CH=CH_2}$$

and a trisubstituted over a disubstituted.

$$\underset{\underset{\text{tert-Pentyl bromide}}{}}{CH_3CH_2\overset{\overset{\displaystyle CH_3}{|}}{\underset{\underset{}{|}}{\underset{Br}{C}}}CH_3} \xrightarrow{EtOH} \underset{\underset{82\%}{\text{2-Methyl-2-butene}}}{CH_3CH=\overset{\overset{\displaystyle CH_3}{|}}{C}CH_3} + \underset{\underset{18\%}{\text{2-Methyl-1-butene}}}{CH_3CH_2\overset{\overset{\displaystyle CH_3}{|}}{C}=CH_2}$$

How do we account for this kind of orientation? In other reactions that we have taken up so far, orientation and reactivity have gone hand-in-hand. Both are determined by relative rates of reaction, and *in the same step*: abstraction of a hydrogen by a chlorine atom, say, or the formation of a double bond by concerted loss of a proton and the leaving group.

But here, in E1, we find a difference. Orientation and reactivity are still determined by relative rates of reaction—but *of different steps*. How fast the substrate reacts is determined by the rate of step (1). But which alkene is produced is clearly determined by which β-proton is lost faster from the carbocation in step (2). The 2-pentyl cation, for example, can lose either a proton from C–3 to form

$$CH_3CH_2CH_2\overset{\displaystyle |}{\underset{\underset{}{|}}{\underset{OTs}{C}}}HCH_3$$

$$(1) \Big\downarrow -OTs^-$$

$$\underset{\underset{\textit{Major product}}{\text{2-Pentene}}}{CH_3CH_2CH=CHCH_3} \xleftarrow[-H^+]{(2)} \underset{\underset{\text{2-Pentyl cation}}{}}{\overset{3 \quad\; 2 \quad\; 1}{CH_3CH_2\underset{\underset{H}{|}}{C}H-\overset{\oplus}{\underset{\underset{}{}}{C}}H-\underset{\underset{H}{|}}{C}H_2}} \xrightarrow[-H^+]{(2)} \underset{\underset{\textit{Minor product}}{\text{1-Pentene}}}{CH_3CH_2CH_2CH=CH_2}$$

2-pentene or a proton from C–1 to form 1-pentene. There is a competition, and more 2-pentene is obtained because 2-pentene is formed faster.

Let us examine the transition state, then, for this product-determining step. The carbon–hydrogen bond is partly broken, and the double bond is partly formed.

Transition state:
partly formed
double bond

The transition state has acquired *alkene character*. As in E2, factors that stabilize the alkene also stabilize the incipient alkene in the transition state. E_{act} is lowered, and the alkene is formed faster.

When rearrangement occurs in E1, we still predict orientation by Saytzeff's rule. But now we must consider the loss of β-protons from the rearranged cations as well as from the cations initially formed.

Problem 7.12 When 2-methyl-3-pentyl tosylate was heated in *n*-butyl alcohol with no added base, the following alkenes were obtained in the proportions indicated: 2-methyl-2-pentene (80%), 4-methyl-2-pentene (11%), 2-methyl-1-pentene (9%). How do you account for (a) the formation of each of these products, (b) their relative proportions, and (c) the fact that the 4-methyl-2-pentene was entirely the *trans* isomer?

7.23 Elimination: E2 *vs.* E1

How can we tell which mechanism, E2 or E1, is likely to operate under a particular set of conditions?

First, let us look at the effect of the nature of the alkyl group of the substrate. As one proceeds along the sequence 1°, 2°, 3°, reactivity by both mechanisms increases, although for different reasons. Reactivity by E2 increases chiefly because of the greater stability of the more highly branched alkenes being formed. Reactivity by E1 increases because of the greater stability of the carbocations being formed in the rate-determining step. Thus, except that it is very difficult for primary substrates even to form carbocations, we can expect no abrupt shift in mechanism due simply to changes in the alkyl group.

But if we turn to the role played by the other reagent, the base, we find a striking difference between the two mechanisms: in E2, base takes part in the rate-determining step; in E1, it does not. (We have already (Sec. 5.24) encountered an analogous competition between a bimolecular (S_N2) and a unimolecular (S_N1) mechanism, and what follows will come as no surprise to us.)

The rate of E2 depends upon the *concentration* of the base; the rate of E1 does not. The rate of E2 depends upon the *nature* of the base; a stronger base pulls a proton away from the substrate faster. The rate of E1 is independent of the nature of the base; stronger or weaker, the base waits until the carbocation is formed.

For a given substrate, then, the more concentrated the base, or the stronger the base, the more E2 is favored over E1. Under the conditions typically used to bring about dehydrohalogenation—a concentrated solution of a strong base—the E2 mechanism is the path taken by elimination. In general, the E1 mechanism is encountered only with secondary or tertiary substrates, and in solutions where the base is either in low concentration or weak—typically, where the base is the solvent.

Problem 7.13 Dehydrohalogenation of isopropyl bromide, which requires several hours of refluxing in alcoholic KOH, is brought about in less than a minute at room temperature by $t\text{-BuO}^- \text{K}^+$ in DMSO. Suggest a possible explanation for this.

7.24 Elimination *vs.* substitution

The most commonly used substrates for base-promoted 1,2-elimination, we have said, are alkyl halides and alkyl sulfonates. These are, of course, the same compounds that also serve as substrates for nucleophilic substitution—and for a very good reason: both reactions require substrates with good leaving groups. Furthermore, the reagents required to bring about the two kinds of reactions, bases and nucleophiles, are similar—indeed, are very often the same reagent. Both reagents are electron-rich; bases are nucleophilic, and nucleophiles are basic. It follows, then, that there will nearly always be—in principle, at least—*competition* between substitution and elimination.

Let us consider first the bimolecular reactions, S$_N$2 and E2. Both reactions result from attack on the substrate by the reagent :Z. Acting as a nucleophile, it attacks carbon to bring about substitution; acting as a base, it attacks hydrogen to bring about elimination.

Among substrates, we have seen that the order of reactivity by E2 is 3° > 2° > 1°. In S$_N$2, we recall (Sec. 5.15), the order of reactivity is just the opposite. As one proceeds along the series 1°, 2°, 3°, then, reactivity by E2 increases, and reactivity by S$_N$2 decreases.

Primary substrates undergo elimination slowest and substitution fastest; tertiary substrates undergo elimination fastest and substitution slowest. *Where bimolecular substitution and elimination are competing reactions, the proportion of elimination increases as the structure of the substrate is changed from primary to secondary to tertiary.* Many tertiary substrates yield almost exclusively alkenes under these conditions.

Problem 7.14 Account for the fact that in bimolecular reaction with the strongly basic sodium ethoxide, C_2H_5ONa, these proportions of substitution and elimination products are obtained from the following alkyl bromides, *all primary*: ethyl, 99% substitution, 1% elimination; *n*-propyl, 91% substitution, 9% elimination; isobutyl, 40% substitution, 60% elimination.

The nature of the alkyl group is thus perhaps the major factor influencing the competition between S_N2 and E2. But there are other factors. Many nucleophiles are, like hydroxide ion, quite strong bases; and elimination competes strongly with substitution. There are some reagents, however, that are good nucleophiles but comparatively weak bases, and with these, substitution tends to be favored.

A less polar solvent tends to favor elimination; so does a higher temperature. Thus, hot alcoholic KOH is the classical reagent for dehydrohalogenation; a lower temperature and the presence of the more polar water in the solvent tend to increase the proportion of the substitution product, the alcohol.

Now let us turn to the competition between the unimolecular reactions, S_N1 and E1. Both, as we have seen, have the same first step, heterolysis to form the carbocation. It is at the second step that the reaction path forks: one branch leads to substitution; the other leads to elimination.

$$R:X \longrightarrow R^{\oplus} \xrightarrow{\quad :Z \quad} \begin{cases} \longrightarrow R:Z & \qquad S_N1 \\ \\ \longrightarrow \text{alkene} + H:Z & \qquad E1 \end{cases}$$
$$+$$
$$:X^-$$

In this second step there is attack by the nucleophilic, basic reagent :Z, which is typically the solvent. This time the attack is not on the substrate itself, but on the carbocation. Attack at carbon brings about substitution; attack at hydrogen brings about elimination.

$$\begin{array}{c} \overset{\oplus}{\underset{S \curvearrowleft}{-C-}} \overset{|}{\underset{\underset{E}{\uparrow}{H}}{C-}} \\ Z: \end{array} \qquad \text{E1 } \textit{vs. } S_N1$$

Now, the proportions of products that are ultimately obtained—how much substitution product, how much alkene—are determined by the relative rates of these alternative second steps. How fast a carbocation loses a proton, we concluded earlier (Sec. 7.22), depends upon the stability of the alkene being formed—for simple alkenes, upon how branched it is. The rate of elimination from a carbocation, therefore, follows the sequence $3° > 2° > 1°$. We would expect the rate of substitution—the combining with a nucleophile—to follow the opposite sequence, $1° > 2° > 3°$, with the least stable being the shortest-lived. (In fact, we have seen (Sec. 6.9), heterolysis of even secondary substrates generally involves nucleophilic assistance from the solvent. The cation formed has clinging to its back side a solvent molecule; nucleophilic substitution has, to a degree, already begun.)

The facts agree with our analysis: in the unimolecular reactions tertiary substrates give the highest proportion of elimination. In aqueous ethanol at 80 °C, for example, *tert*-butyl bromide gives 19% of the alkene, whereas isopropyl bromide gives only 5%.

What we are faced with, then, is this. When we want the product of a substitution reaction, elimination is a nuisance to be avoided. But we cannot always do this. With some nucleophiles, we are faced with the fact that an acceptable yield can be obtained only with primary and possibly secondary substrates, and that tertiary substrates give virtually all elimination. But when we want an alkene, elimination is what we are trying to bring about. To do this, we generally try to drive reaction toward bimolecular elimination: we use a solvent of low polarity, and a high concentration of a strong base.

Problem 7.15 Account in detail for the difference in the percentage of alkene obtained within each set of compounds on treatment with aqueous ethanol at 80 °C: (a) isopropyl bromide, 5%; *sec*-butyl bromide, 9%; (b) 2-bromopentane, 7%; 3-bromopentane, 15%; (c) *tert*-butyl bromide, 19%; *tert*-pentyl bromide, 36%.

Problem 7.16 The reaction of *tert*-butyl chloride in water to yield (chiefly) *tert*-butyl alcohol is not appreciably affected by dissolved sodium fluoride; in DMSO, however, sodium fluoride brings about rapid formation of isobutylene. How do you account for this contrast?

7.25 Dehydration of alcohols

So far, we have been dealing with the kind of 1,2-elimination that is promoted by *base*. Now let us turn to 1,2-elimination that is catalyzed by *acid*: the *dehydration of alcohols*. Despite the drastic change in reaction conditions, we shall find, dehydration is fundamentally not very different from the elimination we have already discussed.

An alcohol is converted into an alkene by **dehydration**: *elimination of a molecule of water*.

Dehydration: 1,2-elimination of H_2O

$$-\overset{|}{\underset{H}{C}}-\overset{|}{\underset{OH}{C}}- \quad \xrightarrow[\text{heat}]{\text{acid}} \quad \overset{\diagdown}{\diagup}C=C\overset{\diagup}{\diagdown} \quad + \quad H_2O$$

Alcohol   Alkene   Water

Dehydration requires the presence of an acid and the application of heat. It is generally carried out in either of two ways: (a) by heating the alcohol with sulfuric or phosphoric acid; or (b) by passing the alcohol vapor over a catalyst, commonly alumina (Al_2O_3), at high temperatures. (The alumina functions as an acid: either as a Lewis acid or, through —OH groups on its surface, as a Lowry–Brønsted acid.)

The various classes of alcohols differ widely in ease of dehydration, the order of reactivity being

Ease of dehydration of alcohols $3° > 2° > 1°$

The following examples show how these differences in reactivity affect the experimental conditions of the dehydration. (Certain tertiary alcohols are so prone to dehydration that they can be distilled only if precautions are taken to protect the system from the acid fumes present in the ordinary laboratory.)

$$CH_3CH_2OH \xrightarrow[170\ °C]{95\%\ H_2SO_4} CH_2\!\!=\!\!CH_2$$

Ethyl alcohol   Ethylene

$$CH_3CH_2CH_2CH_2OH \xrightarrow[140\ °C]{75\%\ H_2SO_4} CH_3CH\!\!=\!\!CHCH_3$$

n-Butyl alcohol   2-Butene

Chief product

$$CH_3CH_2CHOHCH_3 \xrightarrow[100\ °C]{60\%\ H_2SO_4} CH_3CH\!\!=\!\!CHCH_3$$

sec-Butyl alcohol   2-Butene

Chief product

$$CH_3\!-\!\overset{\displaystyle CH_3}{\underset{\displaystyle OH}{\overset{|}{\underset{|}{C}}}}\!-\!CH_3 \xrightarrow[85\text{–}90\ °C]{20\%\ H_2SO_4} CH_3\!-\!\overset{\displaystyle CH_3}{\overset{|}{C}}\!\!=\!\!CH_2$$

tert-Butyl alcohol   Isobutylene

For dehydration of secondary and tertiary alcohols the following mechanism is generally accepted. Step (1) is a fast acid–base reaction between the alcohol and

Dehydration

(1)
$$-\overset{|}{\underset{\underset{\textstyle H}{|}}{C}}\!-\!\overset{|}{\underset{\underset{\textstyle OH}{|}}{C}}- \;+\; H\!:\!B \;\rightleftharpoons\; -\overset{|}{\underset{\underset{\textstyle H}{|}}{C}}\!-\!\overset{|}{\underset{\underset{\textstyle OH_2^+}{|}}{C}}- \;+\; B\!:$$

(2)
$$-\overset{|}{\underset{\underset{\textstyle H}{|}}{C}}\!-\!\overset{|}{\underset{\underset{\textstyle OH_2^+}{|}}{C}}- \;\rightleftharpoons\; -\overset{|}{\underset{\underset{\textstyle H}{|}}{C}}\!-\!\overset{|}{\underset{\underset{\textstyle \oplus}{|}}{C}}- \;+\; H_2O\!:$$

(3)
$$-\overset{|}{\underset{\underset{\textstyle H}{|}}{C}}\!-\!\overset{|}{\underset{\underset{\textstyle \oplus}{|}}{C}}- \;\rightleftharpoons\; \overset{\diagdown}{\diagup}C\!\!=\!\!C\overset{\diagup}{\diagdown} \;+\; H\!:\!B$$

B:⤴

the catalyzing acid which gives the protonated alcohol and the conjugate base of the acid. In step (2) the protonated alcohol undergoes heterolysis to form the carbocation and water. In step (3) the carbocation loses a proton to the base to yield alkene.

In steps (2) and (3) of this mechanism we recognize a kind of E1 elimination with the protonated alcohol as substrate. Step (1) is simply the fast, reversible prelude that produces the actual substrate.

Let us look at the facts about dehydration and see how they are accounted for by this mechanism.

Dehydration is acid-catalyzed. Acid is needed to convert the alcohol into the protonated alcohol, which can then undergo heterolysis to lose the weakly basic water molecule. In the absence of acid, heterolysis would require loss of the strongly basic hydroxide ion: a process which, as we have seen (Sec. 5.25), is so difficult that it seldom if ever happens. Acid transforms the very poor leaving group, —OH, into the very good leaving group, $-OH_2^+$.

We spoke of dehydrohalogenation as being base-*promoted*: base is consumed by the reaction, and must be present in molar amounts. We speak of dehydration as being acid-*catalyzed*; acid is not consumed and, for the more reactive alcohols, need be present in only trace amounts. This fact is consistent with the mechanism: the acid used in step (1) is regenerated in step (3). Take, for example, dehydration in aqueous sulfuric acid. The acid H:B is the hydronium ion, H_3O^+; the conjugate base :B is water. In step (1) H_3O^+ loses a proton to form H_2O; in step (3) H_2O is the base that takes a proton from the carbocation and, in doing this, is reconverted into H_3O^+.

We notice here the fundamental similarity of dehydration to dehydrohalogenation. Once the alcohol has been protonated—and this requires an acidic medium—a base plays its customary essential role in the elimination process by abstracting a proton.

Problem 7.17   In the dehydration of *tert*-butyl alcohol by the addition of a drop of concentrated sulfuric acid to the dry alcohol, what is the principal base :B of our mechanism? What is the acid H:B? Write equations to show exactly what happens.

Dehydration is reversible. Unlike base-promoted 1,2-elimination, this elimination is reversible. As we shall soon see, acid catalyzes the hydration of alkenes to give alcohols. In agreement with this fact, each step of the mechanism is shown as reversible. Under the conditions of dehydration the alkene, being quite volatile, is generally driven from the reaction mixture, and thus equilibrium (3) is shifted to the right. As a consequence the entire reaction sequence is forced toward elimination.

Now, according to the **principle of microscopic reversibility**, a reaction and its reverse follow exactly the same path but in opposite directions. (The lowest pass across a mountain ridge from one side is also the lowest from the other side.) On this basis, dehydration of alcohols must involve exactly the same steps—but in reverse—that are involved in hydration of alkenes. Any evidence, therefore, that is gathered about the mechanism of hydration—and there is a good deal (Secs. 8.9–8.12)—adds to our understanding of the mechanism of dehydration.

Problem 7.18   In light of what you have learned so far, write a detailed mechanism for the hydration of alkenes, that is, the acid-catalyzed addition of water to alkenes to give alcohols.

The **order of reactivity of alcohols** toward dehydration, we have seen, is

Ease of dehydration of alcohols $3° > 2° > 1°$

There is evidence (some of it from the study of hydration) that the rate of dehydration depends upon both step (2), formation of the carbocation, and step (3), its loss of a proton. Tertiary alcohols undergo dehydration the most rapidly of the alcohols, because they form the most stable carbocations and then, once formed, these cations yield the most stable alkenes.

Strictly speaking, then, dehydration is not an E1 reaction of the protonated alcohol. In a true E1 elimination, the rate of reaction depends only upon the heterolysis step, since every carbocation formed goes rapidly on to product; that is, loss of a proton is much faster than regeneration of substrate. Here that is not the case: carbocations are formed reversibly from the protonated alcohol, and every so often one loses a proton to yield alkene.

Problem 7.19 *tert*-Butyl alcohol was heated with sulfuric acid in water that was enriched with the isotope ^{18}O. At intervals samples were withdrawn and analyzed for isobutylene and for labeled alcohol, t-Bu^{18}OH. The kinetics showed that formation of the labeled alcohol (that is, isotopic exchange) was 20 to 30 times as fast as alkene formation. How do you interpret these findings, and what is their significance?

Where the structure of the alkyl group permits, **rearrangement takes place**. This follows the pattern we observed for E1 dehydrohalogenation (Sec. 7.21). For example:

$$
\begin{array}{ccc}
CH_3 & CH_3 & CH_3 \\
| & | & | \\
CH_3CH_2CHCH_2OH & \xrightarrow{H^+} CH_3CH=CCH_3 & + \ CH_3CH_2C=CH_2 \\
\text{2-Methyl-1-butanol} & \text{2-Methyl-2-butene} & \text{2-Methyl-1-butene} \\
 & \textit{Chief product} &
\end{array}
$$

$$
\begin{array}{ccc}
CH_3 & CH_3 & CH_3 \\
| & | & | \\
CH_3CCHOHCH_3 & \xrightarrow{H^+} CH_3C=CCH_3 & + \ CH_2=CCHCH_3 \\
| & | & | \\
CH_3 & CH_3 & CH_3 \\
\text{3,3-Dimethyl-2-butanol} & \text{2,3-Dimethyl-2-butene} & \text{2,3-Dimethyl-1-butene} \\
 & \textit{Chief product} &
\end{array}
$$

In each case we can account for the products on the usual basis: the initially formed carbocation rearranges to a more stable carbocation. The alkenes obtained are those formed by loss of a proton from this rearranged carbocation as well as from the original one.

Problem 7.20 As was done on page 284, account in detail for all the alkenes formed in the examples shown above.

Orientation is strongly Saytzeff. Where more than one alkene can be formed, the preferred product is the more stable one. For example:

$$
\begin{array}{ccc}
CH_3 & CH_3 & CH_3 \\
| & | & | \\
CH_3CH_2CCH_3 & \xrightarrow{H^+} CH_3CH=CCH_3 & + \ CH_3CH_2C=CH_2 \\
| & \text{2-Methyl-2-butene} & \text{2-Methyl-1-butene} \\
OH & \textit{Chief product} & \\
\textit{tert}\text{-Pentyl alcohol} & &
\end{array}
$$

(In addition, look again at the examples of rearrangement given above.) This is, of course, exactly what we would expect for loss of a proton from a carbocation, as we discussed in Sec. 7.22.

Another factor comes in here. Since dehydration is reversible, the composition of the product does not necessarily reflect which alkene is formed faster but—depending upon how nearly reaction approaches equilibrium—which alkene is more stable. As we have seen, however, the more stable alkene generally *is* formed faster. On either basis, orientation is consistent with the mechanism, and the predictions we make about orientation are likely to be good ones.

Secondary and tertiary alcohols, we have said, react by this carbocation mechanism. Primary alcohols pose a special problem. As we have seen (Sec. 5.22), primary carbocations are extremely difficult to form. Yet dehydration of primary alcohols typically gives the rearrangements so characteristic of carbocation reactions. For example:

$$CH_3CH_2CH_2CH_2OH \quad \xrightarrow[\text{heat}]{H^+} \quad CH_3CH=CHCH_3$$

n-Butyl alcohol   2-Butene

Chief product

There are several possible explanations. It may be that, in the concentrated acid used to dehydrate primary alcohols, a primary cation *is* generated—heavily encumbered, but capable of rearrangement. It may be that for these substrates dehydration is an E2 reaction of the protonated alcohol. In that case, the rearranged alkenes result, not from the rearrangement of a primary cation, but from the *reversibility* of dehydration. (See Problem 8.6, p. 320.)

In dehydration we see once again the vital role played by protonation of the —OH group: to transform a very poor leaving group into a very good leaving group. In the reaction of alcohols with hydrogen halides (Sec. 5.25), this transformation makes nucleophilic substitution possible; here, it makes elimination possible.

In dehydration the protonated alcohol reacts, in most cases, by the carbocation route, as in E1; alkyl halides, on the other hand, mostly undergo E2. We encountered the same situation in nucleophilic substitution (Sec. 5.25), and the explanation here is essentially the same. To undergo dehydration an alcohol must be protonated, and therefore an acidic medium is required. For E2 elimination we need a fairly strong base to attack the substrate without waiting for it to dissociate into carbocations. But a strong base and an acidic medium are, of course, incompatible: any base much stronger than the alcohol itself would become protonated at the expense of the alcohol. Forced, then, to take place in the absence of strong base, dehydration generally follows the carbocation route. Since alcohols are the usual precursors of alkyl halides and sulfonates, all the eliminations in this chapter are, in a sense, illustrations of the same thing: transformation of —OH into a better leaving group. Conversion into an alkyl halide or sulfonate accomplishes this. So does protonation; it is simpler, but it exacts a price—we are limited in our choice of that key reagent, the base.

Problem 7.21 On treatment with strong base, quaternary ammonium ions, R_4N^+, undergo elimination. For example:

$$CH_3CHCH_3 + OH^- \xrightarrow{\text{heat}} CH_3CH=CH_2 + (CH_3)_3N + H_2O$$

|$_+N(CH_3)_3$ Propylene   Trimethylamine

Isopropyltrimethylammonium
ion

The corresponding ammonium ions RNH_3^+ do not, although the basicity of NH_3 is not very different from that of $(CH_3)_3N$. How do you account for this difference in behavior?

PROBLEMS

1. Give the structural formula of:

(a) 3,6-dimethyl-1-octene
(b) 3-chloropropene
(c) 2,4,4-trimethyl-2-pentene
(d) *trans*-3,4-dimethyl-3-hexene

(e) (*Z*)-3-chloro-4-methyl-3-hexene
(f) (*E*)-1-deuterio-2-chloropropene
(g) (*R*)-3-bromo-1-butene
(h) (*S*)-*trans*-4-methyl-2-hexene

2. Draw out the structural formula and give the IUPAC name of:

(a) isobutylene
(b) *cis*-$CH_3CH_2CH=CHCH_2CH_3$
(c) $(CH_3)_3CCH=CH_2$

(d) *trans*-$(CH_3)_2CHCH=CHCH(CH_3)_2$
(e) $(CH_3)_2CHCH_2CH=C(CH_3)_2$
(f) $(CH_3CH_2)_2C=CH_2$

3. Indicate which of the following compounds show geometric (*cis–trans*) isomerism, draw the isomeric structures, and specify each as *Z* or *E*.

(a) 1-butene
(b) 2-butene
(c) 1,1-dichloroethene
(d) 1,2-dichloroethene
(e) 2-methyl-2-butene
(f) 1-pentene

(g) 2-pentene
(h) 1-chloropropene
(i) 1-chloro-2-methyl-2-butene
(j) 4-ethyl-3-methyl-3-hexene
(k) 2,4-hexadiene
  ($CH_3CH=CHCH=CHCH_3$)

4. There are 13 isomeric hexylenes (C_6H_{12}), geometric isomerism disregarded. (a) Draw the structure and give the IUPAC name for each. (b) Indicate which ones show geometric isomerism, draw the isomeric structures, and specify each as *Z* or *E*. (c) One of the hexylenes is chiral. Which one is it? Draw structures of the enantiomers, and specify each as *R* or *S*.

5. In which of the following will *cis*-3-hexene differ from *trans*-3-hexene?

(a) b.p.
(b) m.p.
(c) adsorption on alumina
(d) infrared spectrum
(e) dipole moment
(f) refractive index

(g) rate of hydrogenation
(h) product of hydrogenation
(i) solubility in ethyl alcohol
(j) density
(k) retention time in gas chromatography

(l) Which *one* of the above would absolutely prove the configuration of each isomer?

6. Write balanced equations for preparation of propylene from:

(a) $CH_3CH_2CH_2OH$ (*n*-propyl alcohol)
(b) $CH_3CHOHCH_3$ (isopropyl alcohol)
(c) isopropyl chloride

(d) *n*-propyl tosylate (use Ts for tosyl)
(e) 1,2-dibromopropane
(f) the alkyne, $CH_3C≡CH$

7. Give structures of the products expected from dehydrohalogenation of:

(a) 1-bromohexane
(b) 2-bromohexane
(c) 1-bromo-2-methylpentane
(d) 2-bromo-2-methylpentane

(e) 3-bromo-2-methylpentane
(f) 4-bromo-2-methylpentane
(g) 1-bromo-4-methylpentane
(h) 3-bromo-2,3-dimethylpentane

8. In those cases in Problem 7 where more than one product can be formed, predict the *major* product.

9. Which alcohol of each pair would you expect to be more easily dehydrated?

(a) $CH_3CH_2CH_2CH_2CH_2OH$ or $CH_3CH_2CH_2CHOHCH_3$
(b) $(CH_3)_2C(OH)CH_2CH_3$ or $(CH_3)_2CHCHOHCH_3$
(c) $(CH_3)_2CHC(OH)(CH_3)_2$ or $(CH_3)_2CHCH(CH_3)CH_2OH$

10. Arrange the compounds of each set in order of reactivity toward dehydrohalogenation by strong base:

(a) 2-bromo-2-methylbutane, 1-bromopentane, 2-bromopentane, 3-bromopentane
(b) 1-bromo-3-methylbutane, 2-bromo-2-methylbutane, 3-bromo-2-methylbutane
(c) 1-bromobutane, 1-bromo-2,2-dimethylpropane, 1-bromo-2-methylbutane, 1-bromo-3-methylbutane

11. Outline the sequence of steps that best accounts for the following:

2,2,4-trimethyl-3-pentanol $\xrightarrow{Al_2O_3,\ heat}$ 2,4,4-trimethyl-2-pentene + 2,2,4-trimethyl-1-pentene + 2,3,4-trimethyl-2-pentene + 2,3,4-trimethyl-1-pentene + 2-isopropyl-3-methyl-1-butene + 3,3,4-trimethyl-1-pentene

12. Ethers can be made by the reaction between alkyl halides and sodium or potassium alkoxides:

$$RX\ +\ R'ONa\ \longrightarrow\ R\!-\!O\!-\!R'\ +\ NaX$$
$$\text{An ether}$$

(a) Using this method, outline all steps in two conceivable alternative routes to *tert*-butyl ethyl ether.

(b) One of these routes gives excellent yields, and the other is worthless. Which is the worthless route, and why? Write equations to show exactly what is happening.

13. When neopentyl alcohol, $(CH_3)_3CCH_2OH$, is heated with acid, it is slowly converted into an 85:15 mixture of two alkenes of formula C_5H_{10}. What are these alkenes, and how are they formed? Which one would you think is the major product, and why?

14. When 3,3-dimethyl-1-butene is treated with hydrogen chloride there is obtained a mixture of 3-chloro-2,2-dimethylbutane and 2-chloro-2,3-dimethylbutane. What does the formation of the second product suggest to you? Propose a likely mechanism for this reaction, which is an example of *electrophilic addition*, the most important class of reactions undergone by alkenes. Check your answer in Secs. 8.9 and 8.10.

8

Alkenes II. Reactions of the Carbon-Carbon Double Bond

Electrophilic and Free-Radical Addition

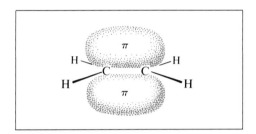

8.1 Reactions of alkenes

The characteristic feature of the alkene structure, we have said, is the carbon–carbon double bond. It is thus the *functional group* of alkenes and, as the functional group, it determines the characteristic reactions that alkenes undergo.

These reactions are of two kinds. (a) First, there are those that take place at the double bond itself and, in doing this, destroy the double bond. These reactions we shall take up in the present chapter.

(b) Next, there are the reactions that take place, not at the double bond, but at certain positions having special relationships to the double bond. Outwardly the double bond is not involved; it is found intact in the product. Yet it plays an essential, though hidden, part in the reaction: it determines how fast reaction takes place and by which mechanism—even whether it takes place at all. Reactions of this kind we shall take up in Chapter 10.

8.2 Reactions at the carbon–carbon double bond. Addition

What kind of reactions can we expect of the carbon–carbon double bond? The double bond consists of a strong σ bond and a weak π bond; we might expect, therefore, that reaction would involve breaking of this weaker bond. This expectation is correct; the typical reactions of the double bond are of the sort,

$$-\overset{|}{\underset{}{C}}=\overset{|}{\underset{}{C}}- + \text{YZ} \longrightarrow -\overset{|}{\underset{Y}{C}}-\overset{|}{\underset{Z}{C}}- \qquad \text{Addition}$$

where the π bond is broken and two strong σ bonds are formed in its place.

A reaction in which two molecules combine to yield a single molecule of product is called an **addition reaction**. The reagent is simply *added to* the substrate, in contrast to a substitution reaction where part of the reagent is *substituted for* a part of the substrate. Addition reactions are necessarily limited to compounds that contain atoms sharing more than one pair of electrons, that is, to compounds that contain multiply bonded atoms. Formally, addition is the opposite of elimination; just as elimination generates a multiple bond, so addition destroys it.

What kind of reagent can we expect to add to the carbon–carbon double bond? In our structure of the bond there is a cloud of π electrons above and below the plane of the atoms (see Fig. 8.1). These π electrons are less involved than the σ

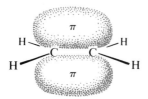

Figure 8.1 Carbon–carbon double bond: the π bond is a source of electrons.

electrons in holding together the carbon nuclei. As a result, they are themselves held less tightly. These loosely held π electrons are particularly available to a reagent that is seeking electrons. It is not surprising, then, that in many of its reactions the carbon–carbon double bond serves as a **source of electrons**: that is, it acts as a **base**. The compounds with which it reacts are those that are deficient in electrons, that is, are *acids. These acidic reagents that are seeking a pair of electrons are called* **electrophilic reagents** (Greek: electron-loving). *The typical reaction of an alkene is* **electrophilic addition**, or, in other words, addition of acidic reagents.

Reagents of another kind, *free radicals*, seek electrons—or, rather, seek *an* electron. And so we find that alkenes also undergo **free-radical addition**.

Most alkenes contain not only the carbon–carbon double bond but also alkyl groups, which have essentially the alkane structure. Besides the addition reactions characteristic of the carbon–carbon double bond, therefore, alkenes may undergo the free-radical substitution characteristic of alkanes. The most important of these addition and substitution reactions are summarized below, and will be discussed in detail in following sections: in this chapter and later chapters.

There are reagents that can add either as acids or as free radicals, and with strikingly different results; there are reagents that are capable both of adding to the double bond and of bringing about substitution. We shall see how, by our choice of conditions, we can lead these reagents along the particular reaction path—electrophilic or free-radical, addition or substitution—we want them to follow.

The alkyl groups attached to the doubly bonded carbons modify the reactions of the double bond; the double bond modifies the reactions of the alkyl groups. We shall see what these modifications are and, where possible, how they can be accounted for.

In Chapters 9 and 20, we shall take up the stereochemistry of these addition reactions, both for the practical reason of knowing what we are likely to obtain in a synthesis, and for what it can tell us about how these reactions take place.

REACTIONS OF ALKENES

Addition Reactions

$$-\overset{|}{\underset{|}{C}}=\overset{|}{\underset{|}{C}}- \ + \ YZ \ \longrightarrow \ -\overset{|}{\underset{Y}{C}}-\overset{|}{\underset{Z}{C}}-$$

1. Addition of hydrogen. Catalytic hydrogenation. Discussed in Secs. 8.3 and 20.5–20.7.

$$-\overset{|}{\underset{|}{C}}=\overset{|}{\underset{|}{C}}- \ + \ H_2 \ \xrightarrow{\text{Pt, Pd, or Ni}} \ -\overset{|}{\underset{H}{C}}-\overset{|}{\underset{H}{C}}-$$

Example:

$$CH_3CH{=}CH_2 \ \xrightarrow{\text{H}_2,\ \text{Ni}} \ CH_3CH_2CH_3$$

Propene   Propane
(Propylene)

2. Addition of halogens. Discussed in Secs. 8.13–8.14, and 9.5–9.6.

$$-\overset{|}{\underset{|}{C}}=\overset{|}{\underset{|}{C}}- \ + \ X_2 \ \longrightarrow \ -\overset{|}{\underset{X}{C}}-\overset{|}{\underset{X}{C}}- \qquad X_2 = Cl_2, Br_2$$

Example:

$$CH_3CH{=}CH_2 \ \xrightarrow{\text{Br}_2 \text{ in CCl}_4} \ CH_3CHBrCH_2Br$$

Propene   1,2-Dibromopropane
(Propylene)   (Propylene bromide)

3. Addition of hydrogen halides. Discussed in Secs. 8.5–8.6, and 8.18–8.19.

$$-\overset{|}{\underset{|}{C}}=\overset{|}{\underset{|}{C}}- \ + \ HX \ \longrightarrow \ -\overset{|}{\underset{H}{C}}-\overset{|}{\underset{X}{C}}- \qquad HX = HCl, HBr, HI$$

Examples:

$$CH_3CH{=}CH_2 \ \xrightarrow{\text{HI}} \ CH_3CHICH_3$$

Propene   2-Iodopropane
                         (Isopropyl iodide)

$$CH_3CH{=}CH_2 \quad \xrightarrow{\text{HBr}}$$

no peroxides → $CH_3CHBrCH_3$ **Markovnikov addition**
2-Bromopropane
(Isopropyl bromide)

peroxides → $CH_3CH_2CH_2Br$ **Anti-Markovnikov addition**
1-Bromopropane
(*n*-Propyl bromide)

CONTINUED

CONTINUED

4. Addition of sulfuric acid. Discussed in Sec. 8.7.

$$-\overset{|}{C}=\overset{|}{C}- \ + \ H_2SO_4 \ \longrightarrow \ -\overset{|}{\underset{H}{C}}-\overset{|}{\underset{OSO_3H}{C}}-$$

Example:

$$CH_3CH{=}CH_2 \ \xrightarrow{\text{conc. } H_2SO_4} \ CH_3\underset{OSO_3H}{CHCH_3}$$

    Propene   Isopropyl hydrogen sulfate

5. Addition of water. Hydration. Discussed in Sec. 8.8.

$$-\overset{|}{C}=\overset{|}{C}- \ + \ HOH \ \xrightarrow{H^+} \ -\overset{|}{\underset{H}{C}}-\overset{|}{\underset{OH}{C}}-$$

Example:

$$CH_3CH{=}CH_2 \ \xrightarrow{H_2O, \ H^+} \ CH_3\underset{OH}{CHCH_3}$$

    Propene   Isopropyl alcohol
                            (2-Propanol)

6. Halohydrin formation. Discussed in Sec. 8.15.

$$-\overset{|}{C}=\overset{|}{C}- \ + \ X_2 + H_2O \ \longrightarrow \ -\overset{|}{\underset{X}{C}}-\overset{|}{\underset{OH}{C}}- \ + \ HX \qquad X_2 = Cl_2, \ Br_2$$

Example:

$$CH_3CH{=}CH_2 \ \xrightarrow{Cl_2, \ H_2O} \ CH_3\underset{OH}{CH}{-}\underset{Cl}{CH_2}$$

    Propylene   Propylene chlorohydrin
    (Propene)   (1-Chloro-2-propanol)

7. Dimerization. Discussed in Sec. 8.16.

Example:

$$\underset{\text{Isobutylene}}{CH_3{-}\overset{\overset{\displaystyle CH_3}{|}}{C}{=}CH_2} + CH_3{-}\overset{\overset{\displaystyle CH_3}{|}}{C}{=}CH_2 \ \xrightarrow{acid} \ CH_3{-}\overset{\overset{\displaystyle CH_3}{|}}{\underset{\underset{\displaystyle CH_3}{|}}{C}}{-}CH{=}\overset{\overset{\displaystyle CH_3}{|}}{C}{-}CH_3$$

                                            2,4,4-Trimethyl-2-pentene

$$+ \ CH_3{-}\overset{\overset{\displaystyle CH_3}{|}}{\underset{\underset{\displaystyle CH_3}{|}}{C}}{-}CH_2{-}\overset{\overset{\displaystyle CH_3}{|}}{C}{=}CH_2$$

                                            2,4,4-Trimethyl-1-pentene

CONTINUED

CONTINUED

8. **Alkylation.** Discussed in Sec. 8.17.

$$-\overset{|}{C}=\overset{|}{C}- + R-H \xrightarrow{\text{acid}} -\overset{|}{\underset{H}{C}}-\overset{|}{\underset{R}{C}}-$$

Example:

$$\underset{\text{Isobutylene}}{CH_3-\overset{\overset{\displaystyle CH_3}{|}}{C}=CH_2} + \underset{\underset{\displaystyle CH_3}{|}}{\underset{\text{Isobutane}}{CH_3-\overset{\overset{\displaystyle CH_3}{|}}{C}-H}} \xrightarrow{H_2SO_4} \underset{\text{2,2,4-Trimethylpentane}}{CH_3\overset{\overset{\displaystyle CH_3}{|}}{C}-CH_2-\overset{\overset{\displaystyle CH_3}{|}}{\underset{\underset{\displaystyle CH_3}{|}}{C}}-CH_3}$$

9. **Oxymercuration–demercuration.** Discussed in Sec. 17.9.

$$-\overset{|}{C}=\overset{|}{C}- + H_2O + Hg(OAc)_2 \longrightarrow -\overset{|}{\underset{HO}{C}}-\overset{|}{\underset{HgOAc}{C}}- \xrightarrow{NaBH_4} -\overset{|}{\underset{HO}{C}}-\overset{|}{\underset{H}{C}}-$$

Markovnikov orientation

10. **Hydroboration–oxidation.** Discussed in Secs. 17.10–17.12.

$$-\overset{|}{C}=\overset{|}{C}- + (BH_3)_2 \longrightarrow -\overset{|}{\underset{H}{C}}-\overset{|}{\underset{B-}{C}}- \xrightarrow[OH^-]{H_2O_2} -\overset{|}{\underset{H}{C}}-\overset{|}{\underset{OH}{C}}-$$

Diborane

Anti-Markovnikov orientation

11. **Addition of free radicals.** Discussed in Secs. 8.19 and 8.20.

$$-\overset{|}{C}=\overset{|}{C}- + Y-Z \xrightarrow[\text{or light}]{\text{peroxides}} -\overset{|}{\underset{Y}{C}}-\overset{|}{\underset{Z}{C}}-$$

Example:

$$\underset{\text{1-Octene}}{n\text{-}C_6H_{13}CH=CH_2} + \underset{\text{Bromotrichloromethane}}{BrCCl_3} \xrightarrow{\text{peroxides}} n\text{-}C_6H_{13}\underset{\underset{\displaystyle Br}{|}}{CH}-CH_2-CCl_3$$

3-Bromo-1,1,1-trichlorononane

12. **Polymerization.** Discussed in Secs. 8.21, 10.30, and 36.3–36.6.

13. **Addition of carbenes.** Discussed in Secs. 12.16–12.17.

CONTINUED

CONTINUED

14. **Hydroxylation. Glycol formation.** Discussed in Secs. 8.22 and 19.13.

$$-\overset{|}{C}=\overset{|}{C}- \ + \ KMnO_4 \ \text{or} \ HCO_2OH \ \longrightarrow \ -\overset{|}{\underset{OH}{C}}-\overset{|}{\underset{OH}{C}}-$$

Example:

$$CH_3CH{=}CH_2 \ \xrightarrow{\ KMnO_4 \ \text{or} \ HCO_2OH \ } \ CH_3-\underset{OH}{CH}-\underset{OH}{CH_2}$$

Propylene
(Propene)

1,2-Propanediol
(Propylene glycol)

Substitution Reactions

15. **Halogenation. Allylic substitution.** Discussed in Secs. 10.1–10.4.

$$H-\overset{|}{\underset{|}{C}}-\overset{|}{C}=\overset{|}{C}- \ + \ X_2 \ \xrightarrow{\ heat \ } \ X-\overset{|}{\underset{|}{C}}-\overset{|}{C}=\overset{|}{C}- \qquad X_2 = Cl_2, Br_2$$

*Low
concentration*

Examples:

$$CH_3CH{=}CH_2 \ \xrightarrow{\ Cl_2, \ 600 \ °C \ } \ Cl-CH_2CH{=}CH_2$$

Propylene
(Propene)

Allyl chloride
(3-Chloro-1-propene)

Cyclohexene　　　　3-Bromocyclohexene

N-Bromosuccinimide

Cleavage Reactions

16. **Ozonolysis.** Discussed in Sec. 8.23.

$$-\overset{|}{C}=\overset{|}{C}- \ + \ O_3 \ \longrightarrow \ \overset{O}{\underset{O-O}{\diagup\diagdown}} \ \xrightarrow{\ H_2O, \ Zn \ } \ -\overset{|}{C}{=}O \ + \ O{=}\overset{|}{C}-$$

Ozone　　　　　　　Ozonide　　　　　Aldehydes and ketones

*Used to
determine
structure*

CONTINUED

—— CONTINUED ——

Examples:

$$CH_3CH_2CH=CH_2 \xrightarrow{O_3} \xrightarrow{H_2O,\ Zn} CH_3CH_2\overset{H}{\underset{|}{C}}=O + O=\overset{H}{\underset{|}{C}}H$$

1-Butene

$$CH_3-\overset{CH_3}{\underset{|}{C}}=CH_2 \xrightarrow{O_3} \xrightarrow{H_2O,\ Zn} CH_3\overset{CH_3}{\underset{|}{C}}=O + O=\overset{H}{\underset{|}{C}}H$$

Isobutylene   ∎

8.3   Hydrogenation. Heat of hydrogenation

We have already encountered hydrogenation as the most useful method for preparing alkanes (Sec. 3.15). It is not limited to the synthesis of alkanes, but is a general method for the conversion of a carbon–carbon double bond into a carbon–carbon single bond in almost any kind of compound we encounter. Using the same apparatus, the same catalyst, and very nearly the same conditions, we can convert an alkene into an alkane, an unsaturated alcohol into a saturated alcohol, or an unsaturated ester into a saturated ester. By varying the catalyst and conditions, we can *selectively* hydrogenate one multiple bond but not another in the same molecule: a carbon–carbon double bond but not a carbon–oxygen double bond; a triple bond but not a double bond; even one carbon–carbon double bond but not another. We can even, as we shall see, convert an optically inactive unsaturated compound into an optically active product!

Hydrogenation is of two general kinds: (a) *heterogeneous* (two-phase) and (b) *homogeneous* (one-phase). In both cases a catalyst brings about addition of molecular hydrogen, H_2, to the double bond.

Heterogeneous hydrogenation is the classical method, and is still widely used. The catalyst is a finely divided metal, usually platinum, palladium, or nickel. A solution of the alkene is shaken under a slight pressure of hydrogen gas in the presence of a small amount of the catalyst. Reaction takes place rapidly and smoothly and, when it is complete, the solution of the saturated product is simply filtered from the insoluble catalyst.

The much newer **homogeneous hydrogenation** offers a flexibility not possible with the old-style catalysts. Through modifications in the catalysts, hydrogenation can be carried out with unprecedented selectivity. The catalysts are organic complexes of transition metals like rhodium or iridium: *Wilkinson's catalyst*, for example (Sec. 20.5). They are soluble in organic solvents, and hydrogenation thus takes place in a single phase, the solution. An inconvenience of the method has been the difficulty of separating the catalyst from the product once reaction is over. Methods are, however, being worked out to avoid this problem: the catalyst is attached—built-in chemically—to a solid soluble polymer (giant molecule), thus permitting easy filtration at the end of the reaction. Homogeneous hydrogenation thus becomes heterogeneous; but the mode of action seems to remain the same. In Chapter 20, we shall discuss these catalysts in some detail: their structure, how they work, and in particular how they permit stereochemical control of hydrogenation and many other reactions.

Since the reaction is generally quantitative, and since the volume of hydrogen consumed can be easily measured, hydrogenation is frequently used as an analytical tool; it can, for example, tell us the number of double bonds in a compound.

$$\begin{matrix} | \ \ | \\ -\text{C}\!=\!\text{C}- \end{matrix} + \text{H}-\text{H} \longrightarrow \begin{matrix} | \ \ | \\ -\text{C}-\text{C}- \\ | \ \ | \\ \text{H} \ \ \text{H} \end{matrix} \qquad \Delta H = \text{heat of hydrogenation}$$

Hydrogenation is exothermic: the two σ bonds (C—H) being formed are, together, stronger than the σ bond (H—H) and π bond being broken. *The quantity of heat evolved when one mole of an unsaturated compound is hydrogenated is called the* **heat of hydrogenation**; it is simply ΔH of the reaction, but the minus sign is not included. The heat of hydrogenation of nearly every alkene is fairly close to a value of 30 kcal for each double bond in the compound (see Table 8.1).

Table 8.1 HEATS OF HYDROGENATION OF ALKENES

Alkene	Heat of hydrogenation, kcal/mol
Ethylene	32.8
Propylene	30.1
1-Butene	30.3
1-Pentene	30.1
1-Heptene	30.1
3-Methyl-1-butene	30.3
3,3-Dimethyl-1-butene	30.3
4,4-Dimethyl-1-pentene	29.5
cis-2-Butene	28.6
trans-2-Butene	27.6
Isobutylene	28.4
cis-2-Pentene	28.6
trans-2-Pentene	27.6
2-Methyl-1-butene	28.5
2,3-Dimethyl-1-butene	28.0
2-Methyl-2-butene	26.9
2,3-Dimethyl-2-butene	26.6

Although hydrogenation is an exothermic reaction, it proceeds at a negligible rate in the absence of a catalyst, even at elevated temperatures. The uncatalyzed reaction must have, therefore, a very large energy of activation. The function of the catalyst is to lower the energy of activation (E_{act}) so that the reaction can proceed rapidly at room temperature. The catalyst does not, of course, affect the net energy change of the overall reaction; it simply lowers the energy hill between the reactants and products (see Fig. 8.2).

A catalyst lowers E_{act} by permitting reaction to take place in a different way, that is, by a different mechanism. In this case, the reactants are adsorbed on the

Figure 8.2 Potential energy changes during progress of reaction: effect of the catalyst.

enormous surface of a finely divided solid metal, or bonded temporarily to a soluble metal ion. Reaction under these conditions is very different from the reaction that would have to take place otherwise. It is believed, for example, that the surface of a solid catalyst breaks the π bond of the alkene prior to reaction with hydrogen. In homogeneous hydrogenation, the complex metal ion breaks the hydrogen–hydrogen bond and transfers the hydrogens, one at a time, to the double bond.

Lowering the energy hill, as we can see, decreases the energy of activation of the reverse reaction as well, and thus increases the rate of *de*hydrogenation. We might expect, therefore, that the solid catalysts platinum, palladium, and nickel, under the proper conditions, should serve as dehydrogenation catalysts; this is indeed the case (Sec. 34.7). We are familiar with the fact that, although a catalyst speeds up a reaction, it does not shift the position of equilibrium; this is, of course, because it speeds up *both* the forward and reverse reactions.

Like hydrogenation, the addition of other reagents to the double bond is generally exothermic. The energy consumed by the breaking of the Y—Z and π bonds is almost always less than that liberated by formation of the C—Y and C—Z bonds.

$$-\overset{|}{C}=\overset{|}{C}- + Y-Z \longrightarrow -\overset{|}{\underset{Y}{C}}-\overset{|}{\underset{Z}{C}}- + \text{heat}$$

8.4 Heat of hydrogenation and stability of alkenes

Heats of hydrogenation can often give us valuable information about the relative stabilities of unsaturated compounds. For example, of the isomeric 2-butenes, the *cis* isomer has a heat of hydrogenation of 28.6 kcal, the *trans* isomer one of 27.6 kcal. Both reactions consume one mole of hydrogen and yield the same product, *n*-butane. Therefore, if the *trans* isomer *evolves* 1 kcal less energy than the *cis* isomer, it can only mean that it *contains* 1 kcal less energy; in other words, the *trans* isomer is *more stable* by 1 kcal than the *cis* isomer (see Fig. 8.3). In a similar way, *trans*-2-pentene (heat of hydrogenation = 27.6 kcal) must be more stable by 1.0 kcal than *cis*-2-pentene (heat of hydrogenation = 28.6 kcal).

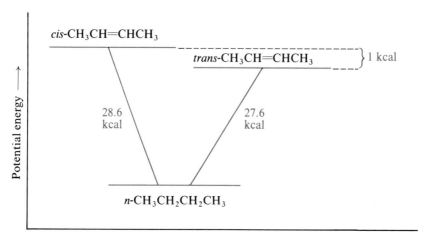

Figure 8.3 Heats of hydrogenation and stability: *cis*- and *trans*-2-butene.

Of simple dialkylethylenes, it is usually the *trans* isomer that is the more stable. The two larger substituents are located farther apart than in the *cis* isomer; there is less crowding, and less van der Waals strain (Sec. 7.6).

Heats of hydrogenation show that the stability of an alkene also depends upon the position of the double bond. The following examples are typical:

$CH_3CH_2CH=CH_2$ $CH_3CH=CHCH_3$
30.3 kcal   *cis* 28.6; *trans* 27.6

$CH_3CH_2CH_2CH=CH_2$ $CH_3CH_2CH=CHCH_3$
30.1 kcal   *cis* 28.6; *trans* 27.6

$\begin{matrix} CH_3 \\ | \\ CH_3CHCH=CH_2 \end{matrix}$ $\begin{matrix} CH_3 \\ | \\ CH_2=CCH_2CH_3 \end{matrix}$ $\begin{matrix} CH_3 \\ | \\ CH_3C=CHCH_3 \end{matrix}$
30.3 kcal   28.5   26.9

Each set of isomeric alkenes yields the same alkane. The differences in heat of hydrogenation must therefore be due to differences in stability. In each case, **the greater the number of alkyl groups attached to the doubly bonded carbon atoms, the more stable the alkene.**

Stability of alkenes

$R_2C=CR_2 > R_2C=CHR > R_2C=CH_2, RCH=CHR > RCH=CH_2 > CH_2=CH_2$

We have already seen the important role that stability of alkenes plays in orientation and reactivity in elimination reactions.

Problem 8.1   (a) Write a balanced equation for combustion of 1-butene. (b) How does this equation compare with the corresponding one for *cis*-2-butene? For *trans*-2-butene? (c) The following heats of combustion have been measured for these three butenes: 648.1, 647.1, 649.8 kcal. Which heat of combustion do you think applies to each butene? (d) Assign the following heats of combustion to 1-pentene, and *cis*- and *trans*-2-pentene: 804.3, 806.9, 805.3 kcal.

8.5   Addition of hydrogen halides. Markovnikov's rule. Regioselective reactions

An alkene is converted by hydrogen chloride, hydrogen bromide, or hydrogen iodide into the corresponding alkyl halide.

$$
\underset{\text{Alkene}}{-\overset{|}{C}=\overset{|}{C}-} \quad + \quad \underset{\text{(HX = HCl, HBr, HI)}}{HX} \quad \longrightarrow \quad \underset{\substack{\\ H \quad X \\ \text{Alkyl halide}}}{-\overset{|}{\underset{|}{C}}-\overset{|}{\underset{|}{C}}-}
$$

The reaction is frequently carried out by passing the dry gaseous hydrogen halide directly into the alkene. Sometimes the moderately polar solvent, acetic acid, which will dissolve both the polar hydrogen halide and the non-polar alkene, is used. The familiar aqueous solutions of the hydrogen halides are not generally used; in part, this is to avoid the addition of water to the alkene (Sec. 8.8).

Problem 8.2   (a) What is the acid in an aqueous solution of HBr? In dry HBr? (b) Which is the stronger acid? (c) Which can better transfer a hydrogen ion to an alkene?

In this way, ethylene is converted into ethyl halide, the hydrogen becoming attached to one doubly bonded carbon and the halogen to the other.

$$
\underset{\text{Ethylene}}{CH_2{=}CH_2} + HI \longrightarrow \underset{\text{Ethyl iodide}}{CH_3CH_2I}
$$

Propylene could yield either of two products, the *n*-propyl halide or the isopropyl halide, depending upon the orientation of addition, that is, depending upon which carbon atoms the hydrogen and halogen become attached to. Actually, only the isopropyl halide is formed.

$$
\underset{\substack{\\ H{-}I}}{CH_3{-}CH{=}CH_2} \;\not\!\!\longrightarrow\; \underset{\substack{| \quad | \\ H \quad I \\ \textit{n}\text{-Propyl iodide}}}{CH_3{-}CH{-}CH_2}
$$

$$
\underset{\substack{\\ I{-}H}}{CH_3{-}CH{=}CH_2} \;\longrightarrow\; \underset{\substack{| \quad | \\ I \quad H \\ \text{Isopropyl iodide}}}{CH_3{-}CH{-}CH_2} \qquad\qquad \textit{Actual product}
$$

In the same way, isobutylene could yield either of two products, isobutyl halide or *tert*-butyl halide; here the orientation of addition is such that only the *tert*-butyl halide is formed.

$$CH_3-\underset{\underset{H-I}{\overset{\overset{\displaystyle CH_3}{|}}{C}}}{C}=CH_2 \quad \not\longrightarrow \quad CH_3-\underset{\underset{H}{\overset{\overset{\displaystyle CH_3}{|}}{C}}}{C}-\underset{I}{\overset{}{C}}H_2$$

Isobutyl iodide

$$CH_3-\underset{\underset{I-H}{\overset{\overset{\displaystyle CH_3}{|}}{C}}}{C}=CH_2 \quad \longrightarrow \quad CH_3-\underset{\underset{I}{\overset{\overset{\displaystyle CH_3}{|}}{C}}}{C}-\underset{H}{\overset{}{C}}H_2 \qquad\qquad \textit{Actual product}$$

tert-Butyl iodide

Thus, in the addition of a reagent YZ to an alkene, orientation depends upon which doubly bonded carbon accepts Y and which accepts Z.

On examination of a large number of such additions, the Russian chemist Vladimir Markovnikov (University of Kazan) observed that, where two isomeric products are possible, one of them usually predominates. He pointed out in 1869 that the orientation of addition follows a pattern which we can summarize as: *in the addition of an acid to the carbon–carbon double bond of an alkene, the hydrogen of the acid attaches itself to the carbon that already holds the greater number of hydrogens.* This statement is generally known as **Markovnikov's rule**. Thus: "Unto everyone that hath shall be given", or "Them as has, gits".

Thus, in the addition to propylene we see that the hydrogen goes to the carbon bearing two hydrogen atoms rather than to the carbon bearing one. In the addition to isobutylene, the hydrogen goes to the carbon bearing two hydrogens rather than to the carbon bearing none.

Using Markovnikov's rule, we can correctly predict the principal product of many reactions. For example:

$$CH_3CH_2CH=CH_2 + HI \longrightarrow CH_3CH_2CHICH_3$$

1-Butene   *sec*-Butyl iodide
                            (2-Iodobutane)

$$CH_3\underset{\overset{\overset{\displaystyle CH_3}{|}}{}}{C}=CH-CH_3 + HI \longrightarrow CH_3-\underset{\underset{I}{\overset{\overset{\displaystyle CH_3}{|}}{}}}{C}-CH_2-CH_3$$

2-Methyl-2-butene

tert-Pentyl iodide
(2-Iodo-2-methylbutane)

$$CH_3CH=CHCH_3 + HI \longrightarrow CH_3CHICH_2CH_3$$

2-Butene   *sec*-Butyl iodide
                            (2-Iodobutane)

$$CH_2=CHCl + HI \longrightarrow CH_3CHICl$$

Vinyl chloride   1-Chloro-1-iodoethane
(Chloroethene)

$$CH_3CH_2CH{=}CHCH_3 + HI \longrightarrow CH_3CH_2CHICH_2CH_3 + CH_3CH_2CH_2CHICH_3$$

      2-Pentene   3-Iodopentane   2-Iodopentane

In 2-pentene each of the doubly bonded carbons holds one hydrogen, so that according to the rule we should expect neither product to predominate. Here again the prediction is essentially correct, roughly equal quantities of the two isomers actually being obtained.

The examples have involved the addition of hydrogen iodide; exactly similar results are obtained in the addition of hydrogen chloride and, except for special conditions indicated in the following section, of hydrogen bromide.

Reactions that, from the standpoint of orientation, give exclusively or nearly exclusively *one* of several possible isomeric products are called *regioselective*. (From the Latin *regio*, direction, and pronounced "reejio".)

Problem 8.3 Saytzeff actually stated his rule for orientation of elimination (Sec. 7.20) in terms, not of alkyl groups on the alkene being formed, but, like Markovnikov, of numbers of hydrogens on carbon atoms of the substrate. Suggest a wording for this original Saytzeff rule.

8.6 Addition of hydrogen bromide. Peroxide effect

Addition of hydrogen chloride and hydrogen iodide to alkenes follows Markovnikov's rule. Until 1933 the situation with respect to hydrogen bromide was exceedingly confused. It had been reported by some workers that addition of hydrogen bromide to a particular alkene yields a product in agreement with Markovnikov's rule; by others, a product in contradiction to Markovnikov's rule; and by still others, a mixture of both products. It had been variously reported that the product obtained depended upon the presence or absence of water, or of light, or of certain metallic halides; it had been reported that the product obtained depended upon the solvent used, or upon the nature of the surface of the reaction vessel.

In 1933, M. S. Kharasch and F. R. Mayo at the University of Chicago brought order to this chemical chaos by discovering that the orientation of addition of hydrogen bromide to the carbon–carbon double bond is determined solely by the presence or absence of **peroxides.**

Organic peroxides are compounds containing the —O—O— linkage. They are encountered, generally in only very small amounts, as impurities in many organic compounds, where they have been slowly formed by the action of oxygen. Certain peroxides are deliberately synthesized, and used as reagents: *tert*-butyl peroxide (p. 122) or benzoyl peroxide, $(C_6H_5COO)_2$, for example.

Kharasch and Mayo found that if one carefully excludes peroxides from the reaction system, or if one adds certain **inhibitors**—*hydroquinone* (p. 1092), for example, or *diphenylamine* (p. 933)—the addition of HBr to alkenes follows Markovnikov's rule. On the other hand, if one does not exclude peroxides, or if one deliberately puts peroxides into the reaction system, HBr adds to alkenes in exactly the reverse direction.

$$\text{CH}_3\text{CH}{=}\text{CH}_2 \xrightarrow{\text{HBr}} \begin{cases} \xrightarrow{\text{no peroxides}} \text{CH}_3\text{CHBrCH}_3 & \text{Markovnikov addition} \\ & \text{Isopropyl bromide} \\ \xrightarrow{\text{peroxides}} \text{CH}_3\text{CH}_2\text{CH}_2\text{Br} & \text{Anti-Markovnikov addition} \\ & n\text{-Propyl bromide} \end{cases}$$

Propylene

$$\begin{array}{c}\text{CH}_3 \\ | \\ \text{CH}_3{-}\text{C}{=}\text{CH}_2\end{array} \xrightarrow{\text{HBr}}$$

Isobutylene

no peroxides →

$$\begin{array}{c}\text{CH}_3 \\ | \\ \text{CH}_3{-}\text{C}{-}\text{CH}_3 \\ | \\ \text{Br}\end{array}$$ Markovnikov addition

tert-Butyl bromide

peroxides →

$$\begin{array}{c}\text{CH}_3 \\ | \\ \text{CH}_3{-}\text{C}{-}\text{CH}_2\text{Br} \\ | \\ \text{H}\end{array}$$ Anti-Markovnikov addition

Isobutyl bromide

This reversal of the orientation of addition caused by the presence of peroxides is known as the **peroxide effect**. Of the reactions we are studying, *only* the addition of hydrogen bromide shows the peroxide effect. The presence or absence of peroxides has no effect on the orientation of addition of hydrogen chloride, hydrogen iodide, sulfuric acid, water, etc. As we shall see (Secs. 8.12 and 8.18), both Markovnikov's rule and the peroxide effect can readily be accounted for in ways that are quite consistent with the chemistry we have learned so far.

8.7 Addition of sulfuric acid

Alkenes react with cold concentrated sulfuric acid to form compounds of the general formula $ROSO_3H$, known as **alkyl hydrogen sulfates**. These products are formed by addition of hydrogen to one carbon of the double bond and bisulfate ion to the other.

$$\underset{\text{Alkene}}{{-}\overset{|}{\text{C}}{=}\overset{|}{\text{C}}{-}} + \underset{\text{Sulfuric acid}}{\text{H}{-}\text{O}{-}\overset{\overset{\text{O}}{\|}}{\underset{\underset{\text{O}}{}}{\text{S}}}{-}\text{O}{-}\text{H}} \longrightarrow \underset{\text{Alkyl hydrogen sulfate}}{{-}\overset{|}{\underset{\underset{\text{H}}{|}}{\text{C}}}{-}\overset{|}{\text{C}}{-}\text{O}{-}\overset{\overset{\text{O}}{\|}}{\underset{\underset{\text{O}}{}}{\text{S}}}{-}\text{O}{-}\text{H}}$$

Like alkyl sulfonates (Sec. 5.9), these compounds are *esters*: esters of sulfuric acid, just as alkyl sulfonates are esters of sulfonic acids.

Reaction is carried out simply by bringing the reactants into contact: a gaseous alkene is bubbled into the acid, and a liquid alkene is stirred or shaken with the acid. Since alkyl hydrogen sulfates are soluble in sulfuric acid, a clear solution results. The alkyl hydrogen sulfates are deliquescent solids, and are difficult to isolate. As the examples below show, the concentration of sulfuric acid required for reaction depends upon the particular alkene involved; we shall account for this later.

$$CH_2=CH_2 \xrightarrow{\text{98\% } H_2SO_4} CH_3CH_2OSO_3H \xrightarrow{H_2O, \text{ heat}} CH_3CH_2OH + H_2SO_4$$

Ethylene   Ethyl hydrogen sulfate   Ethyl alcohol

$$CH_3CH=CH_2 \xrightarrow{\text{80\% } H_2SO_4} \underset{\underset{OSO_3H}{|}}{CH_3CHCH_3} \xrightarrow{H_2O, \text{ heat}} \underset{\underset{OH}{|}}{CH_3CHCH_3}$$

Propylene

Isopropyl hydrogen   Isopropyl alcohol
sulfate

$$\underset{\underset{Isobutylene}{}}{\overset{\overset{CH_3}{|}}{CH_3-C=CH_2}} \xrightarrow{\text{63\% } H_2SO_4} \overset{\overset{CH_3}{|}}{\underset{\underset{OSO_3H}{|}}{CH_3-C-CH_3}} \xrightarrow{H_2O, \text{ heat}} \overset{\overset{CH_3}{|}}{\underset{\underset{OH}{|}}{CH_3-C-CH_3}}$$

tert- Butyl hydrogen   *tert-*Butyl alcohol
sulfate

If the sulfuric acid solution of the alkyl hydrogen sulfate is diluted with water and heated, there is obtained an alcohol bearing the same alkyl group as the original alkyl hydrogen sulfate. The ester has been cleaved by water to form the alcohol and sulfuric acid, and is said to have been *hydrolyzed*. This sequence of reactions affords a route to the alcohols, and it is for this purpose that addition of sulfuric acid to alkenes is generally carried out. This is an excellent method for the large-scale manufacture of alcohols, since alkenes are readily obtained by the cracking of petroleum. As the examples show, *the addition of sulfuric acid follows Markovnikov's rule.* Consequently, certain alcohols cannot be obtained by this method. For example, isopropyl alcohol can be made, but not *n*-propyl alcohol; *tert*-butyl alcohol, but not isobutyl alcohol.

The fact that alkenes dissolve in cold concentrated sulfuric acid to form the alkyl hydrogen sulfates is made use of in the purification of certain other kinds of compounds. Alkanes or alkyl halides, for example, which are insoluble in sulfuric acid, can be freed from alkene impurities by washing with sulfuric acid. A gaseous alkane is bubbled through several bottles of sulfuric acid, and a liquid alkane is shaken with sulfuric acid in a separatory funnel.

Problem 8.4 (a) To what class of reactions does the hydrolysis of alkyl hydrogen sulfates probably belong? Explain. (b) Suggest a mechanism or mechanisms for hydrolysis of ethyl hydrogen sulfate; (c) for hydrolysis of *tert*-butyl hydrogen sulfate.

8.8 Addition of water. Hydration

Water adds to the more reactive alkenes in the presence of acids to yield alcohols. Since this addition, too, follows Markovnikov's rule, the alcohols are the

$$\underset{\text{Alkene}}{\overset{\overset{|}{}\overset{|}{}}{-C=C-}} + H_2O \xrightarrow{H^+} \underset{\underset{H \quad OH}{}}{\overset{\overset{|}{}\overset{|}{}}{-C-C-}}$$

Alcohol

same as those obtained by the two-step synthesis just described. Hydration of alkenes, directly or via the alkyl hydrogen sulfates, is the principal industrial source of those lower alcohols whose formation is consistent with Markovnikov's rule.

$$
\underset{\text{Isobutylene}}{CH_3-\overset{\overset{\displaystyle CH_3}{|}}{C}=CH_2} \xrightarrow{H_2O,\ H^+} CH_3-\underset{\underset{\displaystyle OH}{|}}{\overset{\overset{\displaystyle CH_3}{|}}{C}}-CH_3
$$

tert-Butyl alcohol

8.9 Electrophilic addition: mechanism

Before we take up other reactions of alkenes, let us examine the mechanism of the reactions we have discussed so far. After we have done this, we shall return to our systematic consideration of alkene reactions, prepared to understand them better in terms of these earlier reactions.

Addition of the acidic reagent, HZ, involves two steps:

(1) $\quad -C{=}C- + H{:}Z \longrightarrow -\overset{|}{\underset{H}{C}}-\overset{|}{\underset{\oplus}{C}}- + :Z \qquad HZ = HCl,\ HBr,\ HI, \qquad$ *Slow*
$$H_2SO_4,\ H_3O^+$$

(2) $\quad -\overset{|}{\underset{H}{C}}-\overset{|}{\underset{\oplus}{C}}- + \ :Z \longrightarrow -\overset{|}{\underset{H}{C}}-\overset{|}{\underset{Z}{C}}- \qquad :Z = Cl^-,\ Br^-,\ I^-, \qquad$ *Fast*
$$HSO_4^-,\ H_2O$$

Step (1) is the transfer of hydrogen ion from :Z to the alkene to form a carbocation: a transfer of a proton from one base to another.

Electrophilic addition

$$
-\underset{Z-H}{\overset{|}{C}{=}\overset{|}{C}}- \longrightarrow Z: + -\overset{|}{C}-\overset{|}{\underset{H}{\underset{\oplus}{C}}}-
$$

Step (2) is the combining of the carbocation with the base :Z.

Let us see what happens in step (1), focusing our attention on HZ and the two doubly bonded carbons of the alkene. Hydrogen is transferred as a proton—that is, *without its electrons*, which are left behind on the base :Z. To form the bond to hydrogen, carbon uses the π electrons formerly shared with the other carbon. This leaves this other carbon with only a sextet of electrons; it thus becomes the electron-deficient carbon of a carbocation.

Step (1) is the slow, difficult step, and its rate largely or entirely controls the overall rate of addition. This step involves attack by an acidic, electron-seeking reagent—that is, an *electrophilic* reagent—and hence the reaction is an example of **electrophilic addition**. The electrophile is not necessarily a Lowry–Brønsted acid transferring a proton, as shown here, but, as we shall see, can be almost any kind of electron-deficient molecule (Lewis acid).

The general mechanism can be illustrated by specific examples: addition of hydrogen chloride,

(1) $CH_3-CH=CH_2 + H:\overset{..}{\underset{..}{Cl}}: \longrightarrow CH_3-\overset{\oplus}{CH}-CH_3 + :\overset{..}{\underset{..}{Cl}}:^-$

(2) $CH_3-\overset{\oplus}{CH}-CH_3 + :\overset{..}{\underset{..}{Cl}}:^- \longrightarrow CH_3-\underset{\underset{Cl}{|}}{CH}-CH_3$

of sulfuric acid,

(1) $CH_3-CH=CH_2 + H:OSO_3H \longrightarrow CH_3-\overset{\oplus}{CH}-CH_3 + :OSO_3H^-$

(2) $CH_3-\overset{\oplus}{CH}-CH_3 + :OSO_3H^- \longrightarrow CH_3-\underset{\underset{OSO_3H}{|}}{CH}-CH_3$

and of water.

(1) $CH_3-CH=CH_2 + H:OH_2^+ \rightleftarrows CH_3-\overset{\oplus}{CH}-CH_3 + :OH_2$

(2a) $CH_3-\overset{\oplus}{CH}-CH_3 + :OH_2 \rightleftarrows CH_3-\underset{\underset{\oplus OH_2}{|}}{CH}-CH_3$

(2b) $CH_3-\underset{\underset{\oplus OH_2}{|}}{CH}-CH_3 + :OH_2 \rightleftarrows CH_3-\underset{\underset{OH}{|}}{CH}-CH_3 + H:OH_2^+$

We notice that the carbocation combines with water to form not the alcohol but the protonated alcohol; in a subsequent reaction this protonated alcohol releases a hydrogen ion to another base to form the alcohol. This sequence of reactions, we can see, is just the reverse of that proposed for the dehydration of alcohols (Sec. 7.25). In dehydration, the equilibria are shifted in favor of the alkene chiefly by the removal of the alkene from the reaction mixture by distillation: in hydration, the equilibria are shifted in favor of the alcohol partly by the high concentration of water.

Now, what is the evidence for this mechanism? It includes the following:

(a) The *rate of reaction* depends upon the concentration of both the alkene and the reagent HZ.

(b) *Reaction requires an acidic reagent.*

(c) Where the structure permits, *reaction is accompanied by rearrangements.*

(d) *The alkene does not undergo appreciable hydrogen exchange.*

In addition, the mechanism is consistent with:

(e) the *orientation* of addition; and

(f) the *relative reactivities* of alkenes.

Let us examine this evidence.

First, (a) the *rate of reaction* depends upon concentration of both the alkene and the reagent HZ. This fact is, of course, consistent with a mechanism that starts with reaction between these two reagents.

Next, (b) *reaction requires an acidic reagent*. According to the mechanism, the first step in all these reactions is the transfer of a proton to the alkene. This agrees with the fact that all these reagents except water are strong acids in the Lowry–Brønsted sense; that is, they can readily transfer protons. The exception, water, requires the presence of added strong acid for reaction to occur. An alkene is a *weak* base, and accepts protons to a significant degree only from strong acids.

Even the familiar acetic acid (CH_3COOH) is not strong enough and, alone, does not react with alkenes. Yet when acetic acid is used as the solvent for the addition of HCl, there is obtained, along with the alkyl chloride, a smaller amount of alkyl acetate (CH_3COOR). This is not formed by a substitution reaction between acetic acid and alkyl chloride already formed by addition; under the reaction conditions, the alkyl chloride is inert to acetic acid. The simplest interpretation is that strong acid converts the alkene into *something* that can react with water or with acetic acid. According to the mechanism, that "something" is the carbocation.

8.10　Electrophilic addition: rearrangements

Where the structure permits, (c) *reaction is accompanied by rearrangements*. The product sometimes contains the group Z attached to a carbon that was not doubly bonded in the substrate; sometimes the product even has a carbon skeleton different from that of the substrate.

These "unexpected" products, it turns out, are readily accounted for by rearrangements of the carbocations proposed as intermediates. These rearrangements follow exactly the same pattern that we have come to expect from our study of carbocations in S_N1 substitution (Sec. 5.23) and in E1 elimination (Secs. 7.21 and 7.25).

Addition of hydrogen chloride to 3-methyl-1-butene, for example, yields not only the expected 3-chloro-2-methylbutane, but also 2-chloro-2-methylbutane.

Since a 1,2-shift of hydrogen can convert the initially formed secondary cation into a more stable tertiary cation, such a rearrangement does occur, and much of the product is derived from this new cation.

Addition of hydrogen iodide to 3,3-dimethyl-1-butene yields not only the expected 3-iodo-2,2-dimethylbutane, but also 2-iodo-2,3-dimethylbutane:

$$
\underset{\substack{\text{3,3-Dimethyl-1-butene}}}{\overset{\displaystyle CH_3}{\underset{\displaystyle CH_3}{CH_3-\overset{|}{\underset{|}{C}}-CH=CH_2}}}
\xrightarrow{\text{HI}}
\underset{\substack{\text{A 2° cation}}}{\overset{\displaystyle CH_3}{\underset{\displaystyle CH_3}{CH_3-\overset{|}{\underset{|}{C}}-\overset{\oplus}{CH}-CH_3}}}
\xrightarrow{\text{I}^-}
\underset{\substack{\text{3-Iodo-2,2-dimethylbutane}}}{\overset{\displaystyle CH_3}{\underset{\displaystyle H_3C\quad I}{CH_3-\overset{|}{\underset{|}{C}}-CH-CH_3}}}
$$

methyl shift

$$
\underset{\substack{\text{A 3° cation}}}{\overset{\displaystyle CH_3}{\underset{\displaystyle CH_3}{CH_3-\overset{|}{\underset{|}{\overset{\oplus}{C}}}-CH-CH_3}}}
\xrightarrow{\text{I}^-}
\underset{\substack{\text{2-Iodo-2,3-dimethylbutane}}}{\overset{\displaystyle CH_3}{\underset{\displaystyle I\quad CH_3}{CH_3-\overset{|}{\underset{|}{C}}-CH-CH_3}}}
$$

Here again we see the conversion of a secondary cation into a tertiary, this time brought about by a 1,2-shift of a methyl group.

The change in carbon skeleton accompanying this last example of *addition* is identical to that accompanying two reactions of 3,3-dimethyl-2-butanol: dehydration (p. 292), an *elimination* reaction; and conversion into the chloride (p. 215), a *substitution* reaction. This is a particularly dramatic example of the kind of evidence that gave rise to the idea that these apparently unrelated reactions proceed through the same intermediate: the carbocation.

Of all the evidence supporting the mechanism we have given for electrophilic addition, the strongest single piece is the occurrence of rearrangements, since this bears directly on the heart of the mechanism: the formation of the carbocation.

8.11   Electrophilic addition: absence of hydrogen exchange

In electrophilic addition, (d) *the alkene does not undergo appreciable hydrogen exchange.* Addition of D_2O to 2-methyl-2-butene in the presence of D_3O^+ was found, as we might expect, to yield the deuterated alcohol I.

$$
\underset{\substack{\text{2-Methyl-2-butene}}}{\overset{\displaystyle CH_3}{CH_3-\overset{|}{C}=CH-CH_3}} + D_2O
\xrightarrow{D_3O^+}
\underset{\substack{\text{I}}}{\overset{\displaystyle CH_3}{\underset{\displaystyle DO\quad D}{CH_3-\overset{|}{\underset{|}{C}}-CH-CH_3}}}
$$

When the reaction was about half over, it was interrupted and the unconsumed alkene was isolated. Mass spectrometric analysis showed that it contained almost no deuterium; that is, the alkene had not undergone appreciable exchange of hydrogen or deuterium.

Now, what is the significance of this finding? (We have already encountered this technique (Sec. 7.18), and should have a suspicion of where the discussion is leading.) Consider what would happen if carbocations were formed rapidly and reversibly in step (1), and then—every so often—slowly combined with the base to complete the addition. In that case most carbocations would lose hydrogen many times to regenerate the alkene before eventually going on to product. But in the carbocation there are *two* hydrogens on C–3: the protium (H) that has been there all along, and the newly acquired deuterium (D). In reverting to alkene, the carbocation would be just as likely—*more* likely, actually (*Why?*)—to lose protium as deuterium, and thus leave deuterium in the alkene. While stewing in the solution

$$CH_3-\underset{\underset{H}{|}}{\overset{\overset{CH_3}{|}}{C}}=C-CH_3 \underset{D_2O}{\overset{(1) \ D^+}{\rightleftharpoons}} CH_3-\underset{\underset{D}{|}}{\overset{\overset{H_3C \ \ H}{|}}{C}}\overset{\oplus}{-}\underset{}{C}-CH_3 \underset{D_2O}{\rightleftharpoons} CH_3-\underset{\underset{D}{|}}{\overset{\overset{CH_3}{|}}{C}}=C-CH_3$$

Unlabeled alkene   Labeled alkene
Starting material   *Exchange product:*
                            not obtained

(2) $\big\downarrow D_2O$

$$CH_3-\underset{\underset{+OD_2}{|}}{\overset{\overset{CH_3}{|}}{C}}-CHD-CH_3 \xrightarrow{-D^+} CH_3-\underset{\underset{OD}{|}}{\overset{\overset{CH_3}{|}}{C}}-CHD-CH_3$$

until half-reaction time, unconsumed alkene would exchange much of its protium for deuterium and, on recovery, would be found to be heavily deuterated—*contrary to fact.*

What this evidence shows is that *if* carbocations are formed—and other evidence shows that they *are*—they combine with base much faster than they revert to alkene. That is to say, as the mechanism on page 312 shows, step (1) is the slow, rate-determining step. How fast addition takes place depends chiefly on how fast the carbocation is formed.

8.12   Electrophilic addition: orientation and reactivity

The mechanism is consistent with (e) the *orientation* of addition of acidic reagents, and (f) the effect of structure on the *relative reactivities* of alkenes.

Addition of hydrogen chloride to three typical alkenes is outlined below, with the two steps of the mechanism shown. In accord with Markovnikov's rule, propylene yields isopropyl chloride, isobutylene yields *tert*-butyl chloride, and 2-methyl-2-butene yields *tert*-pentyl chloride.

$$CH_3-CH=CH_2 \xrightarrow{HCl}$$
Propylene

$$\to CH_3-\overset{\oplus}{CH}-CH_3 \xrightarrow{Cl^-} CH_3-\underset{\underset{Cl}{|}}{CH}-CH_3 \quad \textit{Actual product}$$
A 2° cation
Isopropyl chloride

$$\xcancel{\to} CH_3-CH_2-CH_2\oplus$$
A 1° cation

$$CH_3-\underset{\underset{CH_3}{|}}{C}=CH_2 \xrightarrow{HCl}$$
Isobutylene

$$\to CH_3-\underset{\underset{CH_3}{|}}{\overset{\oplus}{C}}-CH_3 \xrightarrow{Cl^-} CH_3-\underset{\underset{Cl}{|}}{\overset{\overset{CH_3}{|}}{C}}-CH_3 \quad \textit{Actual product}$$
A 3° cation
tert-Butyl chloride

$$\xcancel{\to} CH_3-\underset{\underset{H}{|}}{\overset{\overset{CH_3}{|}}{C}}-CH_2\oplus$$
A 1° cation

$$\begin{array}{c} \text{CH}_3 \\ | \\ \text{CH}_3\text{—CH}_2\text{—}\overset{\oplus}{\text{C}}\text{—CH}_3 \end{array} \xrightarrow{\text{Cl}^-} \begin{array}{c} \text{CH}_3 \\ | \\ \text{CH}_3\text{—CH}_2\text{—C—CH}_3 \\ | \\ \text{Cl} \end{array}$$

A 3° cation

tert-Pentyl chloride
(2-Chloro-2-methylbutane)

Actual product

$$\begin{array}{c} \text{CH}_3 \\ | \\ \text{CH}_3\text{—CH=C—CH}_3 \end{array} \xrightarrow{\text{HCl}}$$

2-Methyl-2-butene

$$\begin{array}{c} \text{CH}_3 \\ | \\ \text{CH}_3\text{—}\overset{\oplus}{\text{CH}}\text{—C—CH}_3 \\ | \\ \text{H} \end{array}$$

A 2° cation

According to the mechanism, hydrogen from the reagent adds to one or the other of the two doubly bonded carbons to give one or the other of two possible carbocations. For example, if hydrogen goes to C–2 of propylene, there is formed the *n*-propyl cation; if it goes to C–1, there is formed the isopropyl cation. Once formed, the carbocation rapidly reacts to yield product. Which halide is obtained, then, depends upon which carbocation is formed in the first step. The fact that propylene yields isopropyl chloride rather than *n*-propyl chloride shows that the isopropyl cation is formed rather than—that is, *faster than*—the *n*-propyl cation. Thus, orientation in electrophilic addition is determined by the relative rates of two competing reactions: *formation of one carbocation or the other*.

In each of the examples given above, the product obtained shows that in the initial step a secondary cation is formed faster than a primary, or a tertiary faster than a primary, or a tertiary faster than a secondary. Examination of the orientation in many cases shows that this is a general rule: in electrophilic addition the rate of formation of carbocations follows the sequence

Rate of formation of carbocations $3° > 2° > 1° > \text{CH}_3{}^+$

In listing carbocations in order of their rate of formation from alkenes, we find that, once again (compare Sec. 5.22), we have listed them in order of their stability (Sec. 5.20).

Stability of carbocations $3° > 2° > 1° > \text{CH}_3{}^+$

We can now reword **Markovnikov's rule** as: **electrophilic addition to a carbon–carbon double bond involves the intermediate formation of the more stable carbocation.**

As with Saytzeff's rule (Sec. 7.20), this rewording gives a rule that not only is more generally applicable, but leads us to the factor actually at work.

How can we account for the fact that the rate of formation of a carbocation in electrophilic addition depends upon its stability? Once more we must compare the structure of the reactants with the structure of the transition state. In the reactants, hydrogen is attached to :Z, and the doubly bonded carbons are held to each other not only by a σ bond but also by a π bond. In the products, hydrogen is attached to one of the carbons; the π bond is broken, and the other carbon is left with only a sextet of electrons and hence a positive charge. In the transition state, the bond between hydrogen and :Z is partly broken, and the bond between hydrogen and carbon is partly formed. The π bond is partly broken, and *carbon has partly gained the positive charge it will carry in the carbocation.*

Electron-releasing groups tend to disperse the partial positive charge developing on carbon, and in this way stabilize the transition state. Stabilization of the transition state lowers E_{act} and permits a faster reaction (see Fig. 8.4). To the extent that the π bond is broken, the organic group possesses the character of the carbocation it is to become. As before, the same factor, electron release, that

Figure 8.4 Molecular structure and orientation of reaction. The stability of the transition state parallels the stability of the carbocation: the more stable carbocation is formed faster.

stabilizes the carbocation also stabilizes the *incipient* carbocation in the transition state. Once again, we find, *the more stable the carbocation, the faster it is formed.*

Thus, the rate of addition of a hydrogen ion to a double bond depends upon the stability of the carbocation being formed. As we might expect, this factor determines not only the **orientation** of addition to a single alkene, but also the **relative reactivities** of different alkenes.

Alkenes generally show the following order of reactivity toward addition of acids:

Reactivity of alkenes toward acids

$$\begin{array}{c} CH_3 \\ \diagdown \\ \diagup \\ CH_3 \end{array} C{=}CH_2 > CH_3CH{=}CHCH_3, \ CH_3CH_2CH{=}CH_2, \ CH_3CH{=}CH_2 >$$

$$CH_2{=}CH_2 > CH_2{=}CHCl$$

Isobutylene, which forms a tertiary cation, reacts faster than 2-butene, which forms a secondary cation. 1-Butene, 2-butene, and propylene, which form secondary cations, react faster than ethylene, which forms a primary cation.

$$CH_3\text{—}\overset{\overset{\displaystyle CH_3}{|}}{C}\text{=}CH_2 + H:Z \longrightarrow CH_3\text{—}\overset{\overset{\displaystyle CH_3}{|}}{\underset{\oplus}{C}}\text{—}CH_3 + :Z$$

Isobutylene

A 3° cation

$$CH_3CH\text{=}CHCH_3 + H:Z \longrightarrow CH_3CH_2\underset{\oplus}{C}HCH_3 + :Z$$

2-Butene

A 2° cation

$$CH_3CH_2CH\text{=}CH_2 + H:Z \longrightarrow CH_3CH_2\underset{\oplus}{C}HCH_3 + :Z$$

1-Butene

A 2° cation

$$CH_3CH\text{=}CH_2 + H:Z \longrightarrow CH_3\underset{\oplus}{C}HCH_3 + :Z$$

Propylene

A 2° cation

$$CH_2\text{=}CH_2 + H:Z \longrightarrow CH_3CH_2\oplus + :Z$$

Ethylene

A 1° cation

As substituents, halogens tend to attract electrons. Just as electron release by alkyl groups disperses the positive charge and stabilizes a carbocation, so electron withdrawal by halogens intensifies the positive charge and destabilizes the carbocation. We saw that this electron withdrawal slows down the formation of carbocations in heterolysis (Sec. 5.25); in the same way, it slows down formation of carbocations in electrophilic addition. Vinyl chloride, CH_2=$CHCl$, for example is *less* reactive than ethylene.

When we said that the carbocation is the heart of the mechanism of electrophilic addition, we meant not just that it is an intermediate; we meant that, as in other carbocation reactions we have studied, it is the *rate of formation of the carbocation* that determines the course of reaction.

We can begin to see what a powerful weapon we have for attacking the problems that arise in connection with a wide variety of reactions that involve carbocations. We know that the more stable the carbocation, the faster it is formed; that its stability depends upon dispersal of charge; and that dispersal of charge is determined by the electronic effect of the attached groups. We have already found that this same approach enables us to deal with such seemingly different matters as (a) the relative reactivities of substrates in S_N1 substitution; (b) the relative ease of dehydration of alcohols; (c) the relative reactivities of alkenes toward addition of acids; (d) the orientation of addition of acids to alkenes; and (e) the pattern of rearrangements that can occur in all these reactions.

Problem 8.5   What we really want as standards for the stabilities of carbocations are the compounds they are being generated from: alkenes at this point, not the alkyl halides on which the discussion of Sec. 5.20 was based. For the addition of protons to alkenes in the gas phase,

$$\text{>C=C<} \quad + \quad H^+ \quad \longrightarrow \quad -\overset{|}{\underset{|}{C}}-\overset{|}{\underset{\oplus}{C}}-$$
$$H$$

the following ΔH values have been measured: ethylene, -160.6 kcal; propylene, -180.4 kcal; isobutylene, -193.5 kcal.

(a) Using a diagram similar to Fig. 5.9 (p. 200), derive the order of stability of ethyl, isopropyl, and *tert*-butyl cations *relative to the alkene from which each is formed*. (*Caution*: Be sure to note the *sign* of ΔH.)

(b) On this basis, what is the difference in stability between the ethyl and isopropyl cations? Between the ethyl and *tert*-butyl cations?

(c) How does each of these differences compare with the corresponding differences based on alkyl bromides as standards? Alkyl chlorides? Alkyl iodides?

Problem 8.6 We have seen (Sec. 7.25) that dehydration of alcohols is reversible; its reverse is, of course, hydration of alkenes. On the basis of what you have just learned, show how dehydration of 1-butanol could give rise to 2-butene without involving rearrangement of—or even formation of—a *n*-butyl cation.

8.13   Addition of halogens

Alkenes are readily converted by chlorine or bromine into saturated compounds that contain two atoms of halogen attached to adjacent carbons; iodine generally fails to react.

$$-\overset{|}{\underset{|}{C}}=\overset{|}{\underset{|}{C}}- \quad + \quad X_2 \quad \longrightarrow \quad -\overset{|}{\underset{X}{C}}-\overset{|}{\underset{X}{C}}-$$
Alkene $(X_2 = Cl_2, Br_2)$

Vicinal dihalide

The reaction is carried out simply by mixing together the two reactants, usually in an inert solvent like carbon tetrachloride. The addition proceeds rapidly at room temperature or below, and does not require exposure to ultraviolet light; in fact, we deliberately avoid higher temperatures and undue exposure to light, as well as the presence of excess halogen, since under those conditions substitution might become an important side reaction.

This reaction is by far the best method of preparing **vicinal dihalides**. For example:

$$CH_2=CH_2 + Br_2 \xrightarrow{CCl_4} CH_2-CH_2$$
$$\overset{|}{Br} \quad \overset{|}{Br}$$

Ethene
(Ethylene)

1,2-Dibromoethane

$$CH_3CH=CH_2 + Br_2 \xrightarrow{CCl_4} CH_3-CH-CH_2$$
$$\overset{|}{Br} \quad \overset{|}{Br}$$

Propene
(Propylene)

1,2-Dibromopropane

$$
\underset{\substack{\text{2-Methylpropene}\\ \text{(Isobutylene)}}}{\underset{\substack{| \\ CH_3}}{CH_3-C=CH_2}} + Br_2 \xrightarrow{\text{CCl}_4} \underset{\substack{\text{1,2-Dibromo-2-methylpropane}}}{\underset{\substack{| | \\ Br Br}}{\underset{\substack{| \\ CH_3}}{CH_3-C-CH_2}}}
$$

Addition of bromine is extremely useful for detection of the carbon–carbon double bond. A solution of bromine in carbon tetrachloride is red; the dihalide, like the alkene, is colorless. Rapid decolorization of a bromine solution is characteristic of compounds containing the carbon–carbon double bond. (However, see Sec. 8.24.)

8.14 Mechanism of addition of halogens

The addition of halogens to alkenes, like the addition of protic acids, is believed to be electrophilic addition, and to involve two steps. Again, the first step involves the formation of a cation. But this cation, in most cases, is not a carbocation, but something new to us: a *halonium ion*. Let us see what a halonium ion is, and what evidence there is for its formation.

Let us use addition of bromine as our example. In step (1) bromine is transferred from a bromine molecule to the alkene: not to just one of the doubly bonded carbons, but to both, forming a cyclic **bromonium ion**.

(1)

A bromonium ion

(2)

Step (1) does indeed represent electrophilic addition. Bromine is transferred as *positive* bromine: that is, without a pair of electrons, which are left behind on the newly formed bromide ion. In step (2) this bromide ion, or more probably another just like it, reacts with the bromonium ion to yield the product, the dibromide.

What is being proposed here is *not* a π-complex (Sec. 13.10), in which the (acidic) bromine molecule is held by the (basic) π cloud of the alkene. Bromine is bonded by two σ bonds—one to each carbon—to form a ring. A π-complex of molecular Br_2 and alkene may, however, be a reversibly formed precursor of the bromonium ion.

The transfer of a proton from a strong acid to an alkene, while new to us, does fit into a familiar framework of acid–base reactions. But how are we to understand the transfer of positive bromine from a bromine molecule? To begin with, it *is* an acid–base reaction—although not in the Lowry–Brønsted sense. Just as alkenes are bases, so halogens are acids, of the Lewis type.

We can understand this reaction better if we change our viewpoint. An acid is an acid only in the presence of a base, and vice versa. In the same way, an electrophile is an electrophile only in the presence of a nucleophile. By definition, whatever reacts with an electrophile must be a nucleophile; and, again, vice versa.

As organic chemists we tend to speak of reactions from the standpoint of the organic molecule, the substrate. It, we say, undergoes nucleophilic substitution or electrophilic addition. But in nucleophilic substitution, the substrate is acting as an electrophile; and in electrophilic addition, the substrate is acting as nucleophile. From the standpoint of a halogen molecule, the reaction with an alkene is nucleophilic substitution. Acting as a nucleophile, the alkene attaches itself to one of the bromines and pushes the other bromine out as bromide ion. Bromide ion is the leaving group; and, as we have seen, bromide ion is a very *good* leaving group.

What are the facts upon which this mechanism is based? They are:

(a) the *effect of the structure of the alkene on reactivity*;

(b) the *effect of added nucleophiles* on the products obtained;

(c) the fact that halogens add with *complete stereoselectivity* and in the *anti* sense;

(d) the *direct observation of halonium ions* under superacid conditions; and

(e) the role played by halonium ions in *neighboring group effects*.

We shall examine each of these pieces of evidence: (a) and (b) now, and (c) and (d) in Secs. 9.5–9.6, and (e) in Secs. 20.2–20.4.

First, there is (a) the *effect of the structure of the alkene on reactivity*. Alkenes show the same order of reactivity toward halogens as toward the acids already studied: electron-releasing substituents activate an alkene, and electron-withdrawing substituents deactivate. This fact supports the idea that addition is indeed electrophilic—that the alkene is acting as an electron source, and that halogen acts as an acid.

Next, there is (b) the *effect of added nucleophiles* on the products obtained. If a halonium ion is the intermediate, and capable of reacting with halide ion, then we might expect it to react with almost any negative ion or basic molecule we care to provide. The bromonium ion formed in the reaction between ethylene and bromine, for example, should be able to react not only with bromide ion but also— if these are present—with fluoride ion, iodide ion, nitrate ion, or water.

The facts are in complete agreement with this expectation. When ethylene is bubbled into an aqueous solution of bromine and sodium chloride, there is formed not only the dibromo compound but also the bromochloro compound and the

$$CH_2=CH_2 \xrightarrow{Br_2} CH_2\overset{\overset{\displaystyle Br^{\oplus}}{\diagup\diagdown}}{-}CH_2$$

$$\xrightarrow{Br^-} CH_2Br-CH_2Br$$
1,2-Dibromoethane

$$\xrightarrow{Cl^-} CH_2Br-CH_2Cl$$
2-Bromo-1-chloroethane

$$\xrightarrow{I^-} CH_2Br-CH_2I$$
2-Bromo-1-iodoethane

$$\xrightarrow{NO_3^-} CH_2Br-CH_2ONO_2$$
2-Bromoethyl nitrate

$$\xrightarrow{H_2O} CH_2Br-CH_2\overset{\oplus}{O}H_2$$

$$\xrightarrow{-H^+} CH_2Br-CH_2OH$$
2-Bromoethanol

bromoalcohol. Aqueous sodium chloride *alone* is completely inert toward ethylene; chloride ion or water can react only after the halonium ion has been formed by the action of bromine. In a similar way bromine and aqueous sodium iodide or sodium nitrate convert ethylene into the bromoiodo compound or the bromo-nitrate, as well as the dibromo compound and the bromoalcohol. Bromine in water with no added ion yields the dibromo compound and the bromoalcohol.

Now, this elegant work certainly shows that ethylene reacts with bromine to form *something* that can react with these other nucleophiles—but it need not be a bromonium ion. On this evidence alone the intermediate cation could be the simple open carbocation $BrCH_2CH_2{}^+$. As we shall see in Secs. 9.5–9.6, it was the *stereochemistry* of the reaction that led to the concept of an intermediate bromonium ion, a concept that has since been supported by actual observation of such ions.

8.15   Halohydrin formation: addition of the elements of hypohalous acids

As we have seen (Sec. 8.14), addition of chlorine or bromine in the presence of water can yield compounds containing halogen and hydroxyl on adjacent carbon atoms. These compounds are thus chloro- or bromoalcohols. They are commonly referred to as **halohydrins**: *chlorohydrins* or *bromohydrins*. Under proper conditions they can be made the major products. For example:

$$CH_2{=}CH_2 \xrightarrow{\text{Br}_2,\ \text{H}_2\text{O}} \underset{\substack{| \quad\ | \\ \text{OH}\ \ \text{Br}}}{CH_2{-}CH_2}$$

Ethylene

2-Bromoethanol
(Ethylene bromohydrin)

$$CH_3{-}CH{=}CH_2 \xrightarrow{\text{Cl}_2,\ \text{H}_2\text{O}} \underset{\substack{| \quad\ | \\ \text{OH}\ \ \text{Cl}}}{CH_3{-}CH{-}CH_2}$$

Propylene

1-Chloro-2-propanol
(Propylene chlorohydrin)

There is evidence, of a kind we are not prepared to go into here, that these compounds are not formed by addition of preformed hypohalous acid, HOX, but by reaction of the alkene with, successively, halogen and water, as was shown in Sec. 8.14.

(1) X_2 + $\diagup C{=}C \diagdown$ $\longrightarrow$ halonium ion + X^-

A halonium ion

(2)   halonium ion + H_2O $\longrightarrow$ $-\underset{|}{C}-\underset{\underset{OH_2{}^+}{|}}{C}-$ $\xrightarrow{-H^+}$ $-\underset{|}{C}-\underset{\underset{OH}{|}}{C}-$

A halohydrin

Halogen adds (step 1) to form the halonium ion; this then reacts, in part, not with bromide ion, but with water (step 2) to yield the protonated alcohol. Whatever the mechanism, the result is addition of the elements of hypohalous acid (HO— and —X), and the reaction is often referred to in that way.

Propylene, we see above, gives the chlorohydrin in which chlorine is attached to the terminal carbon. This is typical behavior for an unsymmetrical alkene; orientation follows Markovnikov's rule, with positive halogen going to the same carbon that the hydrogen of a protic reagent would.

Now, this orientation would be perfectly understandable if the intermediate were an open carbocation: the initial addition of halogen yields the more stable carbocation—secondary, in the case of propylene. But the stereochemistry indicates that in halohydrin formation, too, the intermediate is a cyclic halonium ion. Orientation depends upon which carbon of the cation suffers nucleophilic attack by water: the terminal carbon (path *a*) or the middle carbon (path *b*). The product obtained shows that the path (*b*) is greatly favored.

$$
\underset{a}{\xrightarrow{\hspace{1cm}}}
CH_3-\underset{\overset{|}{Cl}}{CH}-CH_2OH_2{}^+
\xrightarrow{-H^+}
CH_3-\underset{\overset{|}{Cl}}{CH}-CH_2OH
$$

$$
\underset{\substack{b \qquad a \\ H_2O}}{CH_3-\underset{\overset{|}{\overset{\oplus}{Cl}}}{CH}-CH_2}
$$

$$
\xrightarrow{b}
CH_3-\underset{\overset{|}{OH_2{}^+}}{CH}-CH_2Cl
\xrightarrow{-H^+}
CH_3-\underset{\overset{|}{OH}}{CH}-CH_2Cl
$$

Actual product

As we shall see (Sec. 9.6), the stereochemistry of halohydrin formation strongly indicates that this nucleophilic attack is of the S_N2 type: cleavage of the carbon–halogen bond and formation of the carbon–oxygen bond occur in a single step. Reactivity in S_N2, we saw (Sec. 5.15), typically depends upon steric hindrance. How, then, are we to account for preferential attack at the *more hindered* carbon of the halonium ion?

In an S_N2 reaction, we said earlier, carbon loses electrons to the leaving group and gains electrons from the nucleophile, and as a result does not become appreciably positive or negative in the transition state; electronic factors are unimportant, and steric factors largely control reactivity.

But here the substrate is a halonium ion. Bonding between carbon and halogen is very weak: partly because of angle strain (Sec. 12.7) in the three-membered ring, but mostly because halogen is, after all, sharing a second pair of electrons and carrying a positive charge. And so this halogen is an *exceedingly good leaving group*. (*Remember*: The leaving group here is not a halide ion—a good leaving group itself—but a neutral halogen already attached to another carbon.)

The nucleophile, on the other hand, is a poor one: water. Although there are both partly broken and partly formed bonds in the transition state, bond-breaking has proceeded further than bond-making; the leaving group has taken electrons away to a much greater extent than the nucleophile has brought them up, and the carbon has acquired a considerable positive charge.

Transition state
*Bond-breaking exceeds
bond-making:
positive charge on carbon*

Crowding, on the other hand, is relatively unimportant, because both leaving group and nucleophile are far away. Stability of the transition state is determined chiefly, therefore, by electronic factors, not steric factors. We speak of such a reaction as having considerable S_N1 *character*. Attack occurs, not at the less hindered carbon, but *at the carbon that can best accommodate the positive charge*.

Thus, the orientation observed is what would be expected if an open carbocation were the intermediate. This kind of orientation, we shall find, is commonly observed in cases like this one, where there is a three-membered cyclic intermediate with weak bonding to the leaving group: a *mercurinium ion* (Sec. 17.9), for example, or a *protonated epoxide* (Sec. 19.13).

8.16   Addition of alkenes. Dimerization

Under proper conditions, isobutylene is converted by sulfuric or phosphoric acid into a mixture of two alkenes of molecular formula C_8H_{16}. Hydrogenation of either of these alkenes produces the same alkane, 2,2,4-trimethylpentane (Sec. 3.30). The two alkenes are isomers, then, and differ only in position of the double bond. (*Problem:* Could they, instead, be geometric isomers?) When studied by the methods discussed at the end of this chapter (Sec. 8.23), these two alkenes are found to have the structures shown:

Since the alkenes produced contain exactly twice the number of carbon and hydrogen atoms as the original isobutylene, they are known as **dimers** (*di* = two, *mer* = part) of isobutylene, and the reaction is called **dimerization**. Other alkenes undergo analogous dimerizations.

Let us see if we can devise an acceptable mechanism for this dimerization. There are a great many isomeric octenes; if our mechanism should lead us to just the two that are actually formed, this in itself would provide considerable support for the mechanism.

Since the reaction is catalyzed by acid, let us write as step (1) addition of a hydrogen ion to isobutylene to form the carbocation; the tertiary cation would, of course, be the preferred ion.

$$
\text{(1)} \qquad
\underset{\begin{array}{c}|\\ \end{array}}{CH_3}-\overset{\begin{array}{c}CH_3\\|\end{array}}{C}=CH_2 + H:B \longrightarrow CH_3-\overset{\begin{array}{c}CH_3\\|\end{array}}{\underset{\oplus}{C}}-CH_3 + :B
$$

A carbocation undergoes reactions that provide electrons to complete the octet of the positively charged carbon atom. A carbon–carbon double bond is an excellent electron source, and a carbocation might well go there in its quest for electrons. Let us write as step (2), then, addition of the *tert*-butyl cation to isobutylene; again, the orientation of addition is such as to yield the more stable

$$
\text{(2)} \qquad CH_3-\overset{CH_3}{\underset{}{C}}=CH_2 + \oplus\overset{CH_3}{\underset{CH_3}{C}}-CH_3 \longrightarrow CH_3-\overset{CH_3}{\underset{\oplus}{C}}-CH_2-\overset{CH_3}{\underset{CH_3}{C}}-CH_3
$$

tertiary cation. Step (2) brings about the union of two isobutylene units, which is, of course, necessary to account for the products.

What is this new carbocation likely to do? We might expect that it could add to another molecule of alkene and thus make an even larger molecule; under certain conditions this does indeed happen. Under the present conditions, however, we know that this reaction stops at eight-carbon compounds, and that these compounds are alkenes. Evidently, the carbocation undergoes a reaction familiar to us: loss of a hydrogen ion (step 3). Since the hydrogen ion can be lost from a carbon on either side of the positively charged carbon, two products should be possible.

$$
\text{(3)} \qquad CH_3-\overset{CH_3}{\underset{\oplus}{C}}-CH_2-\overset{CH_3}{\underset{CH_3}{C}}-CH_3 \quad :B
$$

$$
\longrightarrow H:B + CH_2=\overset{CH_3}{\underset{}{C}}-CH_2-\overset{CH_3}{\underset{CH_3}{C}}-CH_3
$$

$$
\longrightarrow H:B + CH_3-\overset{CH_3}{\underset{}{C}}=CH-\overset{CH_3}{\underset{CH_3}{C}}-CH_3
$$

We find that the products expected on the basis of our mechanism are just the ones that are actually obtained. The fact that we can make this prediction simply on the basis of the fundamental properties of carbocations as we understand them is, of course, powerful support for the entire carbocation theory.

From what we have seen here, we can add one more reaction to our list of Sec. 7.15. A **carbocation may**:

(d) add to an alkene to form a larger carbocation.

We have studied this dimerization, not for its great industrial importance— "isooctane" is made by a new, cheaper process—but for what it reveals about carbocations and alkenes. The attachment of carbocations (or carbocation-like species) to π-electron systems is a fundamental reaction type that is encountered both in ordinary organic chemistry (Secs. 15.8 and 36.5) and—in a modified form— in biogenesis, the sequence of reactions by which a compound is formed in living systems, plant or animal (Sec. 10.31 and Problem 23, p. 413).

8.17   Addition of alkanes. Alkylation

Now let us look at the industrial method that *is* used today to make the large amounts of 2,2,4-trimethylpentane ("isooctane") that are consumed as high-test gasoline (Sec. 3.30). In doing this we shall learn still more about the fundamental properties of carbocations—and something rather surprising about alkanes.

When isobutylene and isobutane are allowed to react in the presence of an acidic catalyst, they form directly 2,2,4-trimethylpentane. This reaction is, in effect, addition of an alkane to an alkene.

$$CH_3-C(CH_3)=CH_2 + H-C(CH_3)(CH_3)-CH_3 \xrightarrow{conc.\ H_2SO_4,\ or\ HF,\ 0\text{--}10\ °C} CH_3-C(CH_3)(H)-CH_2-C(CH_3)(CH_3)-CH_3$$

Isobutylene   Isobutane   2,2,4-Trimethylpentane

The commonly accepted mechanism of this **alkylation** is based on the study of many related reactions and involves in step (3) a reaction of carbocations that we have not previously encountered.

(1) $\quad CH_3-C(CH_3)=CH_2 + H:B \longrightarrow CH_3-\overset{\oplus}{C}(CH_3)-CH_3 + :B$

(2) $\quad CH_3-C(CH_3)=CH_2 + \overset{\oplus}{C}(CH_3)(CH_3)-CH_3 \longrightarrow CH_3-\overset{\oplus}{C}(CH_3)-CH_2-C(CH_3)(CH_3)-CH_3$

(3) $\quad CH_3-\overset{\oplus}{C}(CH_3)-CH_2-C(CH_3)(CH_3)-CH_3 + H-C(CH_3)(CH_3)-CH_3 \longrightarrow CH_3-C(CH_3)(H)-CH_2-C(CH_3)(CH_3)-CH_3 + \overset{\oplus}{C}(CH_3)(CH_3)-CH_3$

then (2), (3), (2), (3), etc.

The first two steps are identical with those of the dimerization reaction. In step (3) a carbocation abstracts a hydrogen atom *with its pair of electrons* (a **hydride ion**, essentially) from a molecule of alkane. This abstraction of hydride ion yields an alkane of eight carbons, and a new carbocation to continue the chain. As we might expect, abstraction occurs in the way that yields the *tert*-butyl cation rather than the less stable (1°) isobutyl cation.

This is not our first encounter with the transfer of hydride ion to an electron-deficient carbon; we saw much the same thing in the 1,2-shifts accompanying the rearrangement of carbocations (Sec. 5.23). There, transfer was *intramolecular* (within a molecule); here, it is *intermolecular* (between molecules). This reaction shows us what an extremely strong acid a carbocation is. At the same time, it illustrates something we hinted at earlier (Sec. 3.18): that the "inertness" of alkanes is greatly exaggerated. With a strong enough acid as reagent, an alkane reacts quite readily, and in a heterolytic fashion, too.

Now let us bring our list of carbocation reactions up to date. **A carbocation may:**

(a) combine with a negative ion or other basic molecule;
(b) rearrange to a more stable carbocation;
(c) eliminate a proton to form an alkene;
(d) add to an alkene to form a larger carbocation;
(e) abstract a hydride ion from an alkane.

A carbocation formed by (b) or (d) can subsequently undergo any of the reactions.

As we see, all reactions of a carbocation have a common end: *they provide a pair of electrons to complete the octet of the positively charged carbon.*

Problem 8.7   When ethylene is alkylated by isobutane in the presence of acid, there is obtained, not neohexane, $(CH_3)_3CCH_2CH_3$, but chiefly 2,3-dimethylbutane. Account in detail for the formation of this product.

8.18   Free-radical addition. Mechanism of the peroxide-initiated addition of HBr

In the absence of peroxides, hydrogen bromide adds to alkenes in agreement with Markovnikov's rule; in the presence of peroxides, the direction of addition is exactly reversed (see Sec. 8.6).

To account for this *peroxide effect*, Kharasch and Mayo proposed that addition can take place by two entirely different mechanisms: Markovnikov addition by the electrophilic mechanism that we have just discussed, and anti-Markovnikov addition by a free-radical mechanism. Peroxides initiate the free-radical reaction; in their absence (or if an inhibitor, p. 309, is added), addition follows the usual electrophilic path.

The essence of the mechanism is that hydrogen and bromine add to the double bond homolytically rather than heterolytically; the intermediate is a *free radical*

(1)   peroxides $\longrightarrow$ Rad·   ⎫
                                                                   ⎬ Chain-initiating steps
(2)   Rad· + H:Br $\longrightarrow$ Rad:H + Br·   ⎭

(3)   Br· + —C=C— $\longrightarrow$ —C—C—
                                        |
                                        Br

                                                        ⎫
                                                        ⎬ Chain-propagating steps
(4)   —C—C— + H:Br $\longrightarrow$ —C—C— + Br·   ⎭
         | | |
         Br   Br H

then (3), (4), (3), (4), etc.

rather than a carbocation. Like halogenation of alkanes, this is a chain reaction, this time involving addition rather than substitution.

Decomposition of the peroxide (step 1) to yield free radicals is a well-known reaction. The free radical thus formed abstracts hydrogen from hydrogen bromide (step 2) to form a bromine atom. In step (3) this bromine atom attaches itself to one of the doubly bonded carbons; in doing this, it uses its odd electron and *one* of the π electrons. The other carbon is left with an odd electron, and the alkene is thus converted into a free radical.

$$-\overset{|}{\underset{Br\cdot}{C}}::\overset{|}{C}- \longrightarrow -\overset{|}{\underset{Br}{C}}:\overset{|}{C}-$$ Free-radical addition

This free radical, like the free radical initially generated from the peroxide, abstracts hydrogen from hydrogen bromide (step 4). Addition is now complete, and a new bromine atom has been generated to continue the chain. As in halogenation of alkanes, every so often a reactive particle combines with another one, or is captured by the wall of the reaction vessel, and a chain is terminated.

The mechanism is well supported by the facts. The fact that a very few molecules of peroxide can change the orientation of addition of many molecules of hydrogen bromide strongly indicates a chain reaction. So, too, does the fact that a very few molecules of inhibitor can prevent this change in orientation. It is not surprising to find that these same compounds are efficient inhibitors of many other chain reactions. Although their exact mode of action is not understood, it seems clear that they break the chain, presumably by forming unreactive radicals.

We must not confuse the effects of peroxides, which may have been formed by the action of oxygen, with the effects of oxygen itself. Peroxides *initiate* free-radical reactions; oxygen *inhibits* free-radical reactions (Sec. 2.14).

The mechanism involves addition of a bromine atom to the double bond. It is supported, therefore, by the fact that anti-Markovnikov addition is caused not only by the presence of peroxides but also by irradiation with light of a wavelength known to dissociate hydrogen bromide into hydrogen and bromine atoms.

The light-catalyzed addition of hydrogen bromide to several alkenes was studied by means of ESR (electron spin resonance) spectroscopy, which not only can detect the presence of free radicals at extremely low concentrations, but also can tell something about their structure (Sec. 16.17). Organic free radicals were shown to be present at appreciable concentration, in agreement with the mechanism.

8.19 Orientation of free-radical addition. Polar factors

Now, how do we account for the fact that free-radical addition of hydrogen bromide occurs with orientation opposite to that of electrophilic addition? Let us compare the two kinds of addition to propylene.

Electrophilic addition yields isopropyl bromide because the isopropyl cation is formed faster than the *n*-propyl cation. This we have already accounted for: the isopropyl cation is the more stable cation, and the same factors that stabilize it stabilize the transition state leading to its formation (Sec. 8.12).

Electrophilic addition: *Markovnikov orientation*

$$CH_3-CH=CH_2 \xrightarrow{HBr} \begin{cases} CH_3-CH-CH_3 \overset{\oplus}{} \xrightarrow{Br^-} CH_3-CH-CH_3 \\ \quad\quad\text{A 2}^\circ\text{ cation} \quad\quad\quad\quad\;\; | \\ \quad\quad\quad\quad\quad\quad\quad\quad\quad\;\; Br \\ \text{(X) } CH_3-CH_2-CH_2\oplus \\ \quad\quad\quad\text{A 1}^\circ\text{ cation} \end{cases}$$

Propylene

Isopropyl bromide
Actual product

Free-radical addition: *Anti-Markovnikov orientation*

$$CH_3-CH=CH_2 \xrightarrow{Br\cdot} \begin{cases} CH_3-CH-CH_2Br \xrightarrow{HBr} CH_3-CH_2-CH_2Br \\ \quad\quad\text{A 2}^\circ\text{ free radical} \quad\quad\quad\quad n\text{-Propyl bromide} \\ \quad\quad\quad\quad\quad\quad\quad\quad\quad\quad\quad\quad\quad \textit{Actual product} \\ \text{(X) } CH_3-CH-CH_2\cdot \\ \quad\quad\quad\quad\quad | \\ \quad\quad\quad\quad\; Br \\ \quad\text{A 1}^\circ\text{ free radical} \end{cases}$$

Propylene

Free-radical addition yields *n*-propyl bromide because the secondary free radical is formed faster than the primary. Now, why does this happen? Study of the addition of many different free radicals to many different alkenes indicates that three factors can be involved:

(a) the stability of the free radical being formed;
(b) polar factors; and
(c) steric factors.

Let us look at each of these, using free-radical addition of HBr as our example. As always in dealing with relative rates, we must consider the transition state for the reaction, and see how its stability might be affected by each of these factors.

Let us begin with the **stability of the free radical being formed**. This is a factor with which we are already familiar (Sec. 3.26). In the transition state, the bond between bromine and one of the carbons is partly formed. The π bond is partly broken and the other carbon has partly gained the odd electron it will carry in the

$$CH_3-CH=CH_2 + Br\cdot \longrightarrow \left[CH_3-\overset{\delta\cdot}{CH}\text{---}CH_2 \atop Br_{\overset{.}{\delta}} \right] \longrightarrow CH_3-\overset{\cdot}{CH}-CH_2 \atop | \atop Br$$

Reactants   Transition state   Product

Bromine has   *Carbon has*   *Carbon has*
odd electron   *partially gained*   *odd electron*
                        odd electron

intermediate free radical. To some degree, the organic group possesses the character of the free radical it is to become. Factors that stabilize the free radical also stabilize the incipient free radical in the transition state. Thus, in our example, the secondary free radical is formed faster than the primary because it is more stable. That is one interpretation, then, and the most obvious: rate of reaction depends upon the free-radical character of the transition state.

A great many observations in other areas of free-radical chemistry (Sec. 36.4) have made it clear that reactions of free radicals can be affected—and sometimes even controlled—by **polar factors**. Although free radicals are neutral, they have certain tendencies to gain or lose electrons, and hence they partake of the character of electrophilic or nucleophilic reagents. The transition states for their reactions can be polar, with the radical moiety acquiring a partial negative or positive charge at the expense of the substrate.

Now, because of its electronegativity we would expect the bromine atom to be *electrophilic*. In the transition state, bromine holds more than its share of electrons, at the expense of the alkene. The transition state is thus a polar one, and the substrate moiety has not only free-radical character but also carbocation character.

$$CH_3-CH{=}CH_2 \ + \ Br\cdot \ \longrightarrow \ \left[CH_3-\overset{\delta+}{CH}\text{---}CH_2 \atop \right. \!\!\!\! \left. {Br}_{\delta-} \right] \ \longrightarrow \ CH_3-\overset{\cdot}{CH}-CH_2 \atop Br$$

Transition state

Bromine has negative charge;
carbon has positive charge

The stability of the transition state, and hence the rate of reaction, depends upon the ability of the substrate not only to accommodate the odd electron, but also to accommodate the partial positive charge.

The polar factor will thus favor the orientation that places the charge on the carbon that can best accommodate it. In our example, addition of Br· to C–1 is favored, since in this way positive charge develops on C–2 rather than C–1; and secondary carbocation character is more stabilizing than primary.

Finally, there is the **steric factor**. Addition of a free radical to the terminal carbon, C–1, is less hindered than addition to C–2; the transition state is less crowded (compare Sec. 5.15), and therefore more stable.

In the particular reaction we are studying here, free-radical addition of hydrogen bromide, all three factors would be expected to favor formation of the same intermediate and hence bring about the same orientation. The question is: what is the relative importance of each? Which, if any, is the controlling factor?

This is a difficult question to answer. There is little doubt that each factor exists and, in the proper system, can be dominant. Free-radical addition to conjugated dienes (Sec. 10.28) and styrenes (Sec. 15.21) is clearly controlled by the stability of the radical being formed. Orientation of addition of very bulky radicals like ·CBr$_3$ is very probably determined by steric factors. Addition of powerfully electrophilic radicals like ·CF$_3$ is subject to marked polar effects—particularly if the alkene, too, contains substituents with strongly electron-withdrawing or electron-releasing tendencies.

But each of these examples is an extreme case: a very stable radical is being formed; a very bulky or a very electrophilic radical is the reagent. What does this tell us about the addition of the bromine atom—only moderately electrophilic and not very big—to a simple alkene, with formation of the only moderately stable secondary radical? Probably, just this: that all three factors may well be at work.

Orientation in both electrophilic and free-radical addition of hydrogen bromide is determined by preferential formation of the more highly substituted particle, whether carbocation or free radical. Orientation is reversed simply because the hydrogen adds first in the electrophilic reaction, and bromine adds first in the radical reaction.

8.20 Other free-radical additions

In the years since the discovery of the peroxide effect, dozens of reagents besides HBr have been found (mostly by Kharasch) to add to alkenes in the presence of peroxides or light. Exactly analogous free-radical mechanisms are generally accepted for these reactions, too.

For the addition of carbon tetrachloride to an alkene, for example,

$$RCH{=}CH_2 + CCl_4 \xrightarrow{\text{peroxides}} RCH{-}CH_2{-}CCl_3$$
$$\underset{Cl}{|}$$

the following mechanism has been proposed:

(1) $\qquad\qquad\qquad$ peroxide $\longrightarrow$ Rad·

(2) $\qquad\qquad$ Rad· + Cl:CCl$_3$ $\longrightarrow$ Rad:Cl + ·CCl$_3$

(3) $\qquad\qquad$ ·CCl$_3$ + RCH=CH$_2$ $\longrightarrow$ RCH–CH$_2$–CCl$_3$

(4) $\quad$ RCH–CH$_2$–CCl$_3$ + Cl:CCl$_3$ $\longrightarrow$ RCH–CH$_2$–CCl$_3$ + ·CCl$_3$
$$\qquad\qquad\qquad\qquad\qquad\qquad\qquad\underset{Cl}{|}$$

then (3), (4), (3), (4), etc.

In the next section, we shall encounter another example of free-radical addition—*polymerization*—which has played a key part in the creation of this age of plastics.

Problem 8.8   In the presence of a trace of peroxide or under the influence of ultraviolet light, 1-octene reacts:
(a) with CHCl$_3$ to form 1,1,1-trichlorononane;
(b) with CHBr$_3$ to form 1,1,3-tribromononane;
(c) with CBrCl$_3$ to form 1,1,1-trichloro-3-bromononane;
(d) with H–S–CH$_2$COOH (thioglycolic acid) to yield n-C$_8$H$_{17}$–S–CH$_2$COOH;
(e) with aldehydes, R–C=O, to yield ketones, n-C$_8$H$_{17}$–C–R.
$$\underset{H}{|}\qquad\qquad\qquad\qquad\overset{\|}{O}$$
Show all steps of a likely mechanism for these reactions.

Problem 8.9   From the addition of CCl$_4$ to alkenes, RCH=CH$_2$, there is obtained not only RCHClCH$_2$CCl$_3$, but also RCHClCH$_2$–CHCH$_2$CCl$_3$. *Using only*
$$\underset{R}{|}$$
the kinds of reactions you have already encountered, suggest a mechanism for the formation of this second product.

Problem 8.10 In the dark at room temperature, a solution of chlorine in tetra-chloroethylene can be kept for long periods with no sign of reaction. When irradiated with ultraviolet light, however, the chlorine is rapidly consumed, with the formation of hexachloroethane; many molecules of product are formed for each photon of light absorbed; this reaction is slowed down markedly when oxygen is bubbled through the solution.

(a) How do you account for the absence of reaction in the dark? (b) Outline all steps in the most likely mechanism for the photochemical reaction. Show how it accounts for the facts, including the effect of oxygen.

8.21 Free-radical polymerization of alkenes

When ethylene is heated under pressure with oxygen, there is obtained a compound of high molecular weight (about 20 000), which is essentially an alkane with a very long chain. This compound is made up of many ethylene units and

$$n CH_2{=}CH_2 \xrightarrow{\text{O}_2,\ \text{heat, pressure}} {\sim}CH_2{-}CH_2{-}CH_2{-}CH_2{-}CH_2{-}CH_2{\sim}$$

$$\text{or } ({-}CH_2CH_2{-})_n$$
Polyethylene

hence is called *polyethylene*. It is familiar to most of us as the plastic material of packaging films.

The formation of polyethylene is a simple example of the process called **polymerization**: the *joining together of many small molecules to form very large molecules*. The compound composed of these very large molecules is called a **polymer** (Greek: *poly* + *meros*, many parts). The simple compounds from which polymers are made are called **monomers** (*mono*, one).

Polymerization of substituted ethylenes yields compounds whose structures contain the long chain of polyethylene, with substituents attached at more or less regular intervals. For example, vinyl chloride yields *poly(vinyl chloride)*, used to

$$n CH_2{=}CH \xrightarrow{\text{peroxides}} {\sim}CH_2{-}CH{-}CH_2{-}CH{-}CH_2{-}CH{\sim}$$

with Cl substituents on the CH groups.

Vinyl chloride

$$\text{or } \left({-}CH_2{-}CH{-} \atop \qquad\ \ Cl \right)_n$$
Poly(vinyl chloride)

make phonograph records, plastic pipe, and—when plasticized with high-boiling esters—raincoats, shower curtains, and coatings for metals and upholstery fabrics.

Many other groups (e.g., $-COOCH_3$, $-CH_3$, $-C_6H_5$) may be attached to the doubly bonded carbons. These substituted ethylenes polymerize more or less readily, and yield plastics of widely differing physical properties and uses, but the polymerization process and the structure of the polymer are basically the same as for ethylene or vinyl chloride.

Polymerization requires the presence of a small amount of an **initiator**. Among the commonest of these initiators are peroxides, which function by breaking down to form a free radical. This radical adds to a molecule of alkene, and in doing so generates another free radical. This radical adds to another molecule of alkene to generate a still larger radical, which in turn adds to another molecule of alkene, and so on. Eventually the chain is terminated by steps, such as union of two radicals, that consume but do not generate radicals.

$$\text{Peroxide} \longrightarrow \text{Rad}\cdot$$

$$\text{Rad}\cdot + \underset{\underset{\text{G}}{|}}{CH_2}{=}CH \longrightarrow \text{Rad}{-}CH_2{-}\underset{\underset{\text{G}}{|}}{CH}\cdot \quad \Bigg\} \quad \text{Chain-initiating steps}$$

$$\text{Rad}{-}CH_2{-}\underset{\underset{\text{G}}{|}}{CH}\cdot + \underset{\underset{\text{G}}{|}}{CH_2}{=}CH \longrightarrow \text{Rad}{-}CH_2{-}\underset{\underset{\text{G}}{|}}{CH}{-}CH_2{-}\underset{\underset{\text{G}}{|}}{CH}\cdot \longrightarrow \quad etc.$$

Chain-propagating step

This kind of polymerization, each step of which consumes a reactive particle and produces another, similar particle, is an example of *chain-reaction polymerization*. In Chapter 36, we shall encounter chain-reaction polymerization that takes place, not by way of free radicals, but by way of organic ions, or within the coordination sphere of a transition metal complex. We shall also encounter *step-reaction polymerization*, which involves a series of reactions each of which is essentially independent of the others.

Problem 8.11   Give the structure of the monomer from which each of the following polymers would most likely be made:
(a) Orlon (fibers, fabrics), $\sim\!\!CH_2CH(CN)CH_2CH(CN)\!\!\sim$;
(b) Saran (packaging film, seat covers), $\sim\!\!CH_2CCl_2CH_2CCl_2\!\!\sim$;
(c) Teflon (chemically resistant articles), $\sim\!\!CF_2CF_2CF_2CF_2\!\!\sim$.

Problem 8.12   Can you suggest a reason why polymerization should take place in a way ("head-to-tail") that yields a polymer with regularly alternating groups?

8.22   Hydroxylation. Formation of 1,2-diols

Certain oxidizing agents convert alkenes into **1,2-diols**: dihydroxy alcohols containing the two —OH groups on adjacent carbons. (They are also known as *glycols*.) The reaction amounts to addition of two hydroxyl groups to the double bond.

$$-\overset{|}{C}{=}\overset{|}{C}- \quad \xrightarrow[\text{or } HCO_2OH]{\text{cold alkaline } KMnO_4} \quad -\underset{\underset{OH}{|}}{C}-\underset{\underset{OH}{|}}{C}-$$

A 1,2-diol

Of the numerous oxidizing agents that bring about hydroxylation, two of the most commonly used are (a) cold alkaline potassium *permanganate* ($KMnO_4$), and (b) *peroxy acids*, such as peroxyformic acid (HCO_2OH).

Since permanganate is one of the most important oxidizing agents in organic chemistry, we should perhaps become familiar now with certain of its general characteristics. It is a powerful oxidizing agent, and conditions must be carefully selected—acidity or alkalinity, temperature, quantity of the reagent used—to avoid over-oxidation, that is, taking reaction past the oxidation stage we want. A major problem has been that of solubility: one must get the water-soluble permanganate into contact with the very often water-insoluble substrate. Yet many solvents commonly used to bring polar and non-polar reagents together—alcohols, for example—are themselves oxidized by permanganate. In recent years, this problem has been solved in part by use of phase-transfer catalysts (Sec. 6.7). Quaternary ammonium ions can carry permanganate ions from an aqueous layer into a non-aqueous layer (benzene, say, or dichloromethane) where the substrate awaits. Crown ethers (Sec. 19.10) can complex potassium ions and thus make solid $KMnO_4$ soluble in benzene; the resulting "purple benzene" is an excellent oxidizing agent.

Hydroxylation with permanganate is carried out by stirring together at room temperature the alkene and the aqueous permanganate solution: either neutral— the reaction produces OH^-—or, better, slightly alkaline. Higher yields are some-times obtained by use of "purple benzene" solutions. *Mild conditions* are the key consideration. Heat and the addition of acid are avoided, since these more vigorous conditions promote further oxidation of the diol, with cleavage of the carbon–carbon double bond (Sec. 8.23).

Hydroxylation with peroxyformic acid is carried out by allowing the alkene to stand with a mixture of hydrogen peroxide and formic acid, HCOOH, for a few hours, and then heating the product with water to hydrolyze certain intermediate compounds.

For example:

$$3CH_2{=}CH_2 + 2KMnO_4 + 4H_2O \longrightarrow 3CH_2{-}CH_2 + 2MnO_2 + 2KOH$$

Ethylene
$\qquad\qquad\qquad\qquad\qquad\qquad\qquad\qquad\quad$ OH $\quad$ OH

1,2-Ethanediol

$$CH_3{-}CH{=}CH_2 \xrightarrow{\text{HCO}_2\text{OH}} \xrightarrow{\text{H}_2\text{O}} CH_3{-}CH{-}CH_2$$

Propylene $\qquad\qquad\qquad\qquad\qquad\qquad\qquad\qquad$ OH $\quad$ OH

1,2-Propanediol

Hydroxylation of alkenes is the most important method for the synthesis of 1,2-diols, with the special feature of permitting *stereochemical control* by the choice of reagent (Sec. 19.13).

Oxidation by permanganate is the basis of a very useful analytical test known as the Baeyer test (Sec. 8.24).

8.23 Cleavage: determination of structure by degradation. Ozonolysis

So far we have discussed the addition reactions of alkenes; in Chapter 10 we shall take up their substitution reactions. But there is a third general kind of alkene reaction, **cleavage**: a reaction in which the double bond is completely broken and the alkene molecule converted into two smaller molecules.

The classical reagent for cleaving the carbon–carbon double bond is ozone. **Ozonolysis** (cleavage by ozone) is carried out in two stages: first, addition of ozone to the double bond to form an *ozonide*; and second, hydrolysis of the ozonide to yield the cleavage products.

Ozone gas is passed into a solution of the alkene in some inert solvent like carbon tetrachloride; evaporation of the solvent leaves the ozonide as a viscous oil. This unstable, explosive compound is not purified, but is treated directly with water, generally in the presence of a reducing agent.

In the cleavage products a doubly bonded oxygen is found attached to each of the originally doubly bonded carbons:

Ozonolysis

| Alkene | Molozonide | Ozonide | Cleavage products (Aldehydes and ketones) |

These compounds containing the C=O group are called *aldehydes* and *ketones*; at this point we need only know that they are compounds that can readily be identified (Sec. 21.15). The function of the reducing agent, which is frequently zinc dust, is to prevent formation of hydrogen peroxide, which would otherwise react with the aldehydes and ketones. (Aldehydes, RCHO, are often converted into acids, RCOOH, for ease of isolation.)

Knowing the number and arrangement of carbon atoms in these aldehydes and ketones, we can work back to the structure of the original alkene. For example, for three of the isomeric hexylenes:

One general approach to the determination of the structure of an unknown compound is **degradation**, the breaking down of the unknown compound into a number of smaller, more easily identifiable fragments. Ozonolysis is a typical means of degradation.

Another method of degradation that gives essentially the same information is oxidation by sodium periodate (NaIO$_4$) in the presence of *catalytic amounts* of permanganate. Periodate, we shall find, is much used for cleavage of 1,2-diols

(Secs. 18.10 and 38.6). The permanganate hydroxylates the double bond (Sec. 8.22) to give the 1,2-diol, and is itself reduced to the manganate state. The periodate then (a) cleaves the 1,2-diol and (b) oxidizes manganate back up to permanganate, and the reaction continues.

$$-\overset{|}{C}=\overset{|}{C}- \xrightarrow{\text{KMnO}_4} \left[-\overset{|}{\underset{\overset{|}{OH}}{C}}-\overset{|}{\underset{\overset{|}{OH}}{C}}- \right] \xrightarrow{\text{IO}_4^-} \text{acids, ketones, CO}_2$$

Carboxylic acids, RCOOH, are generally obtained instead of aldehydes, RCHO. A terminal $=CH_2$ group is oxidized to CO_2. For example:

$$\underset{\substack{\text{Carboxylic} \\ \text{acid}}}{\text{CH}_3\text{COOH}} + \underset{\text{Ketone}}{\overset{\overset{\displaystyle CH_3}{|}}{O=C}-CH_3} \xleftarrow[\text{NaIO}_4]{\text{KMnO}_4} \underset{\text{2-Methyl-2-butene}}{\text{CH}_3\text{CH}=\overset{\overset{\displaystyle CH_3}{|}}{C}-CH_3}$$

$$\underset{\substack{\text{Carboxylic} \\ \text{acid}}}{\text{CH}_3\text{CH}_2\text{CH}_2\text{COOH}} + \underset{\substack{\text{Carbon} \\ \text{dioxide}}}{\text{CO}_2} \xleftarrow[\text{NaIO}_4]{\text{KMnO}_4} \underset{\text{1-Pentene}}{\text{CH}_3\text{CH}_2\text{CH}_2\text{CH}=\text{CH}_2}$$

Problem 8.13 What products would you expect from each of the dimers of iso-butylene (Sec. 8.16) upon cleavage by: (a) ozonolysis; (b) $NaIO_4/KMnO_4$?

8.24 Analysis of alkenes

The functional group of an alkene is the carbon–carbon double bond. To characterize an unknown compound as an alkene, therefore, we must show that it undergoes the reactions typical of the carbon–carbon double bond. Since there are so many of these reactions, we might at first assume that this is an easy job. But let us look at the problem more closely.

First of all, which of the many reactions of alkenes do we select? Addition of hydrogen bromide, for example? Hydrogenation? Let us imagine ourselves in the laboratory, working with gases and liquids and solids, with flasks and test tubes and bottles.

We could pass dry hydrogen bromide from a tank through a test tube of an unknown liquid. But what would we see? How could we tell whether or not a reaction takes place? A colorless gas bubbles through a colorless liquid; a different colorless liquid may or may not be formed.

We could attempt to hydrogenate the unknown compound. Here, we might say, we could certainly tell whether or not reaction takes place: a drop in the hydrogen pressure would show us that addition had occurred. This is true, and hydrogenation can be a useful analytical tool. But a catalyst must be prepared, and a fairly elaborate piece of apparatus must be used; the whole operation might take hours.

Whenever possible, *we select for a characterization test a reaction that is rapidly and conveniently carried out, and that gives rise to an easily observed change.* We select a test that requires a few minutes and a few test tubes, a test in which a color appears or disappears, or bubbles of gas are evolved, or a precipitate forms or dissolves.

Experience has shown that an alkene is best characterized, then, by its property of decolorizing both a solution of bromine in carbon tetrachloride (Sec. 8.13) and a cold, dilute, neutral permanganate solution (the Baeyer test, Sec. 8.22). Both tests are easily carried out; in one, a red color disappears, and in the other, a purple color disappears and is replaced by brown manganese dioxide.

$$\underset{\substack{\text{Alkene}}}{\text{C=C}} + \underset{\substack{\text{Red}}}{\text{Br}_2/\text{CCl}_4} \longrightarrow \underset{\substack{\text{Colorless}}}{-\overset{|}{\underset{\underset{\text{Br}}{|}}{\text{C}}}-\overset{|}{\underset{\underset{\text{Br}}{|}}{\text{C}}}-}$$

$$\underset{\substack{\text{Alkene} \qquad \text{Purple}}}{\text{C=C} + \text{MnO}_4^-} \longrightarrow \underset{\substack{\text{Brown ppt.}}}{\text{MnO}_2 +} \underset{\substack{\text{Colorless}}}{-\overset{|}{\underset{\underset{\text{OH}}{|}}{\text{C}}}-\overset{|}{\underset{\underset{\text{OH}}{|}}{\text{C}}}-} \text{ or other products}$$

Granting that we have selected the best tests for the characterization of alkenes, let us go on to another question. We add bromine in carbon tetrachloride to an unknown organic compound, let us say, and the red color disappears. What does this tell us? Only that our unknown is a compound that reacts with bromine. It *may* be an alkene. But it is not enough merely to know that a particular kind of compound reacts with a given reagent; we must also know what *other* kinds of compounds also react with the reagent. In this case, the unknown may equally well be an alkyne. (It may also be any of a number of compounds that undergo rapid *substitution* by bromine; in that case, however, hydrogen bromide would be evolved and could be detected by the cloud it forms when we blow our breath over the test tube.)

In the same way, decolorization of permanganate does not prove that a compound is an alkene, but only that it contains some functional group that can be oxidized by permanganate. The compound *may* be an alkene; but it may instead be an alkyne, an aldehyde, or any of a number of easily oxidized compounds. It may even be a compound that is contaminated with an *impurity* that is oxidized; alcohols, for example, are not oxidized under these conditions, but often contain impurities that *are*. We can usually rule out this by making sure that more than a drop or two of the reagent is decolorized.

By itself, a single characterization test seldom proves that an unknown is one particular kind of compound. It may limit the number of possibilities, so that a final decision can then be made on the basis of additional tests. Or, conversely, if certain possibilities have already been eliminated, a single test may permit a final choice to be made. Thus, the bromine or permanganate test would be sufficient to differentiate an alkene from an alkane, or an alkene from an alkyl halide, or an alkene from an alcohol.

The tests most used in characterizing alkenes, then, are the following: (a) rapid decolorization of bromine in carbon tetrachloride without evolution of HBr, a test also given by alkynes; (b) decolorization of cold, dilute, neutral, aqueous permanganate solution (the Baeyer test), a test also given by alkynes and aldehydes. Also helpful is the solubility of alkenes in cold concentrated sulfuric acid, a test also given by a great many other compounds, including all those containing oxygen (they form soluble oxonium salts) and compounds that are readily sulfonated (Secs. 15.12 and 28.10). Alkanes or alkyl halides are not soluble in cold concentrated sulfuric acid. (A cyclopropane readily dissolves in concentrated sulfuric acid, but is not oxidized by permanganate.)

Of the compounds we have dealt with so far, alcohols also dissolve in sulfuric acid. Alcohols can be distinguished from alkenes, however, by the fact that alcohols give a negative test with bromine in carbon tetrachloride and a negative Baeyer test—so long as we are not misled by impurities. Primary and secondary alcohols *are* oxidized by chromic anhydride, CrO_3, in aqueous sulfuric acid: within *two seconds*, the clear orange solution turns blue-green and becomes opaque.

$$ROH + HCrO_4^- \longrightarrow \textit{Opaque, blue-green}$$
$$1° \text{ or } 2° \quad \textit{Clear,}$$
$$\textit{orange}$$

Tertiary alcohols do not give this test; nor do alkenes.

Problem 8.14 Describe simple chemical tests (if any) that would distinguish between: (a) an alkene and an alkane; (b) an alkene and an alkyl halide; (c) an alkene and a secondary alcohol; (d) an alkene, an alkane, an alkyl halide, and a secondary alcohol. Tell exactly what you would *do* and *see*.

Problem 8.15 Assuming the choice to be limited to alkane, alkene, alkyl halide, secondary alcohol, and tertiary alcohol, characterize compounds A, B, C, D, and E on the basis of the following information:

Compound	Qual. elem. anal.	H_2SO_4	Br_2/CCl_4	$KMnO_4$	CrO_3
A	----	Insoluble	−	−	−
B	----	Soluble	−	−	+
C	Cl	Insoluble	−	−	−
D	----	Soluble	+	+	−
E	----	Soluble	−	−	−

Once characterized as an alkene, an unknown may then be identified as a previously reported alkene on the basis of its physical properties, including its infrared spectrum and molecular weight. Proof of structure of a new compound is best accomplished by degradation: cleavage by ozone or periodate/permanganate, followed by identification of the fragments formed (Sec. 8.23).

(Spectroscopic analysis of alkenes will be discussed in Chapter 16, particularly in Secs. 16.18–16.19.)

Problem 8.16 Describe simple chemical tests (if any) that would distinguish between:

(a) 2-bromoethanol and 1,2-dibromoethane
(b) 4-chloro-1-butene and *n*-butyl chloride
(c) 1-hexene and 2-hexanol
(d) 1-chloro-2-methyl-2-propanol and 1,2-dichloro-2-methylpropane

Tell exactly what you would *do* and *see*.

PROBLEMS

1. Give structures and names of the products (if any) expected from reaction of isobutylene with:

(a) H_2, Ni
(b) Cl_2
(c) Br_2
(d) I_2
(e) HBr
(f) HBr (peroxides)

(g) HI
(h) HI (peroxides)
(i) H_2SO_4
(j) H_2O, H^+
(k) Br_2, H_2O
(l) Br_2 + NaCl(aq)

(m) H_2SO_4 ($\longrightarrow C_8H_{16}$)
(n) isobutane + HF
(o) cold alkaline $KMnO_4$
(p) hot $KMnO_4$
(q) HCO_2OH
(r) O_3; then Zn, H_2O

2. Which alkene of each pair would you expect to be more reactive toward addition of H_2SO_4?

(a) ethylene or propylene
(b) ethylene or vinyl bromide
(c) propylene or 2-butene
(d) 2-butene or isobutylene

(e) vinyl chloride or 1,2-dichloroethene
(f) 1-pentene or 2-methyl-1-butene
(g) ethylene or CH_2=CHCOOH
(h) propylene or 3,3,3-trifluoropropene

3. Give structures and names of the principal products expected from addition of HI to:

(a) 2-butene
(b) 2-pentene
(c) 2-methyl-1-butene
(d) 2-methyl-2-butene

(e) 3-methyl-1-butene (two products)
(f) vinyl bromide
(g) 2,3-dimethyl-1-butene
(h) 2,2,4-trimethyl-2-pentene

4. Account for the fact that addition of $CBrCl_3$ in the presence of peroxides takes place faster to 2-ethyl-1-hexene than to 1-octene.

5. (a) In methyl alcohol solution (CH_3OH), bromine adds to ethylene to yield not only ethylene bromide but also Br—CH_2CH_2—OCH_3. How can you account for this? Write equations for all steps. (b) Predict the products formed under the same conditions from propylene.

6. As an alternative to the one-step 1,2-hydride shift described in Sec. 5.23, one might instead propose—in view of the reactions we have studied in this chapter—that carbocations rearrange by a two-step mechanism, involving the intermediate formation of an alkene:

$$-\overset{|}{\underset{H}{C}}-\overset{|}{\underset{\oplus}{C}}- \longrightarrow -\overset{|}{C}=\overset{|}{C}- + H^+ \longrightarrow -\overset{|}{\underset{\oplus}{C}}-\overset{|}{\underset{H}{C}}-$$

When (by a reaction we have not yet taken up) the isobutyl cation was generated in D_2O containing D_3O^+, there was obtained *tert*-butyl alcohol containing *no* deuterium attached to carbon. How does this experiment permit one to rule out the two-step mechanism?

7. In Sec. 8.18 a mechanism was presented for free-radical addition of hydrogen bromide. Equally consistent with the evidence given there is the following alternative mechanism:

(2a)   Rad· + HBr $\longrightarrow$ Rad—Br + H·

(3a)   H· + $-\overset{|}{C}=\overset{|}{C}-$ $\longrightarrow$ $-\overset{|}{C}-\overset{|}{\underset{H}{C}}-$

(4a) $-\overset{|}{C}-\overset{|}{\underset{H}{C}}-$ + HBr $\longrightarrow$ $-\overset{|}{\underset{H}{C}}-\overset{|}{\underset{Br}{C}}-$ + H·

then (3a), (4a), (3a), (4a), etc.

(a) In steps (2a) and (4a) an alkyl radical abstracts bromine instead of hydrogen from hydrogen bromide. On the basis of homolytic bond dissociation energies (Table 1.2, p. 21), is this mechanism more or less likely than (2)–(4) in Sec. 8.18? Explain.

(b) The ESR study (Sec. 8.18) showed that the intermediate free radical from a given alkene is the *same* whether HBr or DBr (deuterium bromide) is being added to the double bond. Explain how this evidence permits a definite choice between mechanism (2a)–(4a) and mechanism (2)–(4).

8. (a) Write all steps in the free-radical addition of HBr to propylene. (b) Write all steps that would be involved in the free-radical addition of HCl to propylene.

(c) List ΔH for each reaction in (a) and (b). Assume the following homolytic bond dissociation energies: π bond, 51 kcal; $1°$ R—Br, 69 kcal; $1°$ R—Cl, 82 kcal; $2°$ R—H, 95 kcal.

(d) Suggest a possible reason why the peroxide effect is observed for HBr but not for HCl.

9. When isobutylene and chlorine are allowed to react in the dark at $0°C$ in the absence of peroxides, the principal product is not the addition product but methallyl chloride (3-chloro-2-methyl-1-propene). Bubbling oxygen through the reaction mixture produces no change.

This reaction was carried out with labeled isobutylene ([1-^{14}C]2-methyl-1-propene, $(CH_3)_2C=^{14}CH_2$), and the methallyl chloride contained was collected, purified, and subjected to ozonolysis. Formaldehyde ($H_2C=O$) and chloroacetone ($ClCH_2COCH_3$) were obtained; all (97% or more) of the radioactivity was present in the chloroacetone.

(a) Give the structure, including the position of the isotopic label, of the methallyl chloride obtained. (b) Judging from the evidence, is the reaction ionic or free-radical? (c) Using only steps with which you are already familiar, outline a mechanism that accounts for the formation of this product. (d) Can you suggest one reason why isobutylene is more prone than 1- or 2-butene to undergo this particular reaction? (e) Under similar conditions, and in the presence of oxygen, 3,3-dimethyl-1-butene yields mostly the addition product, but also a small yield of 4-chloro-2,3-dimethyl-1-butene. In light of your answer to (c) how do you account for the formation of this minor product?

10. When treated with bromine and water, allyl bromide gives chiefly (80%) the primary alcohol, $CH_2BrCHBrCH_2OH$, in contrast to propylene, which gives the secondary alcohol, $CH_3CHOHCH_2Br$. In light of the discussion of Sec. 8.15, can you suggest an explanation for this difference in orientation?

11. (a) Hydration of either 2-methyl-1-butene or 2-methyl-2-butene yields the same alcohol. Which alcohol would you expect this to be? Showing all steps in the reactions, explain your answer.

(b) Each of these alkenes *separately* was allowed to react with aqueous HNO_3. When hydration was about half over, reaction was interrupted and unconsumed alkene was recovered. In each case, *only* the original alkene was recovered; there was *none* of its isomer present.

How do you interpret this finding? What is its fundamental significance to the mechanism of electrophilic addition?

12. Give the structure of the alkene that yields on ozonolysis:

(a) $CH_3CH_2CH_2CHO$ and HCHO
(b) CH_3—CH—CHO and CH_3CHO
  |
  CH_3
(c) Only CH_3—CO—CH_3
(d) CH_3CHO and HCHO and OHC—CH_2—CHO
(e) Only OHC—$CH_2CH_2CH_2$—CHO (*Think hard.*)
(f) What would each of these alkenes yield upon cleavage by $NaIO_4/KMnO_4$?

13. Describe simple chemical tests that would distinguish between:

(a) isobutane and isobutylene
(b) 2-hexene and *tert*-butyl bromide
(c) 2-chloropentane and *n*-heptane
(d) *tert*-pentyl alcohol and 2,2-dimethylhexane
(e) *n*-propyl alcohol and allyl alcohol $(CH_2\!=\!CHCH_2OH)$
(f) *sec*-butyl alcohol and *n*-heptane
(g) 1-octene and *n*-pentyl alcohol
(h) allyl bromide and 1-hexene
(i) *tert*-butyl alcohol, *tert*-butyl chloride, and 2-hexene
(j) 2-chloroethanol, 1,2-dichloroethane, and 1,2-ethanediol
(k) *n*-pentyl alcohol, *n*-pentane, 1-pentene, and *n*-pentyl bromide

Tell exactly what you would *do* and *see*. (Qualitative elemental analysis is a simple chemical test; degradation is not.)

14. A hydrocarbon, A, adds one mole of hydrogen in the presence of a platinum catalyst to form *n*-hexane. When A is oxidized vigorously with $KMnO_4$, a single carboxylic acid, containing three carbon atoms, is isolated. Give the structure and name of A. Show your reasoning, including equations for all reactions.

15. Give the structure of the alkene you would start with, and the reagents and any special conditions necessary to convert it into each of these products:

(a) *tert*-butyl alcohol
(b) isopropyl iodide
(c) isobutyl bromide
(d) 1-chloro-2-methyl-2-butanol
(e) 2-methylpentane

16. Starting with alcohols of four carbons or fewer, outline all steps in a possible synthesis of each of the following:

(a) 1,2-dichloropropane
(b) 1,2-dichlorobutane

(c) 1,2-propanediol
(d) 1-bromo-2-methyl-2-propanol

9

Stereochemistry II.
Stereoselective and Stereospecific
Reactions

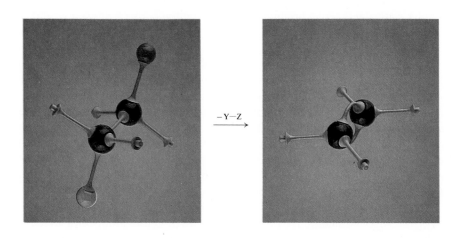

$$-Y-Z \longrightarrow$$

9.1 Organic chemistry in three dimensions

Stereochemistry, we said earlier, permeates organic chemistry; as organic chemistry has grown, so has stereochemistry. And with this growth has come something else: the increasing realization of just how important to organic chemistry stereochemistry really is.

Organic chemistry depends upon the relationship between molecular structure and properties. Our basic approach to chemical reactivity, we have seen, is to consider energy differences between reactants and transition states; that is, we examine the structures of these molecules and estimate their relative stabilities.

Now, molecules are not two-dimensional formulas existing in an imaginary Flatland. They are three-dimensional objects, and they move about, collide, and react in three-dimensional space. We cannot understand molecules or the reactions they undergo unless we understand them *in three dimensions*. And the part of chemistry that deals with molecular structure in three dimensions is, of course, stereochemistry.

As we have already begun to see, stereochemistry can give us a three-dimensional picture of a reaction: the direction of attack; the shape of a transition state. The carbon–carbon double bond is highly reactive, in part because its planar faces are open to attack. Reactivity in the S_N2 reaction is largely determined by crowding about pentavalent carbon in the transition state. And stereochemistry can often give, indirectly, other information as well: the timing of bond-breaking and bond-making; the nature of an intermediate.

But, as we shall find in this chapter, stereochemistry does not stop here. With this understanding of the mechanism comes the power to *control* the stereochemistry of the reaction: to select the proper reagent, conditions, and catalyst so that we obtain our product in just the stereochemical form we want. And we shall find that, as organic chemistry moves to fill the gap between it and biochemistry, there is a growing *need* for this control.

In Chapter 4 we discussed, in a general way, the involvement of stereoisomers in chemical reactions: reactions in which stereoisomers are formed, and reactions in which stereoisomers are consumed. Some of what we learned took the form of hard-and-fast rules: if no bond to a chiral center is broken, configuration must be retained about that center; optically inactive reactants in an achiral medium can only give optically inactive products. On the other hand, much of what we did was simply to set limits on what *can* and what *cannot* happen in certain general situations.

(a) When a chiral center or a double bond is generated in a molecule, we must *consider the possibility* that both configurations about that chiral center or double bond will result. But, in fact, both configurations are *not always* produced; some reactions are *selective* among possible stereoisomeric products, and actually yield fewer than the maximum number allowed.

(b) In general, we said, stereoisomers react similarly—even, in some cases, identically. But this is *not always* true. There are reactions in which each stereoisomeric reactant displays its own *specific* behavior, which can be dramatically different from that of its counterpart.

In this chapter, then, we shall examine the concepts of *stereoselectivity* and *stereospecificity*. We shall move on from what can or cannot happen in a reaction to what actually *does* happen. In doing this, we shall study the stereochemistry of two fundamental reaction types that we took up in the previous chapters: addition and elimination.

But, first, let us go back to the S_N2 reaction and look at its stereochemistry again: this time, not for what it tells us about the mechanism, but as an illustration of certain stereochemical concepts.

9.2 Stereoselective reactions

When (*R*)-2-bromooctane is allowed to react with hydroxide ion under S_N2 conditions, we saw (Sec. 5.14), there is obtained (*S*)-2-octanol.

$$
\underset{\substack{(R)\text{-2-Bromooctane}}}{\overset{\displaystyle C_6H_{13}}{\underset{\displaystyle CH_3}{H-\overset{|}{\underset{|}{C}}-Br}}}
\quad \xrightarrow[S_N2]{NaOH} \quad
\underset{\substack{(S)\text{-2-Octanol}\\ \textit{Only product}}}{\overset{\displaystyle C_6H_{13}}{\underset{\displaystyle CH_3}{HO-\overset{|}{\underset{|}{C}}-H}}}
$$

Although two enantiomeric 2-octanols exist, only *one* of them is obtained here. That is to say, instead of random formation of both enantiomers, there is *selective* formation of just one. Since this selectivity is stereochemical, it is called *stereo-selectivity*, and the reaction is said to be *stereoselective*.

There are stereoisomers of a kind other than enantiomers, and selectivity can be observed in their formation, too: selective formation of one or sometimes two diastereomers of a large number of possible diastereomeric products. Our definition is, then, the following. *A* **stereoselective reaction** *is a reaction that yields predominantly one enantiomer of a possible pair, or one diastereomer (or one enantiomeric pair) of several possible diastereomers.*

Stereoselective reactions can be of two types: *enantioselective*, in which selection is between enantiomeric products; and *diastereoselective*, in which selection is between diastereomeric products. The S_N2 reaction is thus enantioselective.

Stereoselectivity can be observed in various degrees, and reactions are often said to be "highly stereoselective", "moderately stereoselective", and so on. The S_N2 reaction is *completely* stereoselective.

We have already seen what this stereochemistry tells us about the mechanism of the reaction. The S_N2 reaction is stereoselective. To describe the type of stereoselectivity, we say that reaction proceeds *with inversion of configuration*; this fact is the evidence that attack occurs from the back side. The degree of stereo-selectivity is *complete*; this fact is strong evidence that reaction involves a single step, with concerted bond-making and bond-breaking.

Earlier (Sec. 4.28), we saw that free-radical chlorination of an optically active alkane gives an optically inactive product. One enantiomer is *not* selected over the other; instead, there is random formation of both enantiomers to give a racemic product. The reaction is thus completely *non-stereoselective*. This fact is evidence that the carbon–hydrogen bond to the chiral center is broken *before* the carbon–chlorine bond is formed.

Here we have, then, two reactions with quite different stereochemistry: the S_N2 reaction and free-radical chlorination. One is completely stereoselective, and the other is completely non-stereoselective. In each case the very existence or non-existence of stereoselectivity provides powerful evidence for a particular mecha-nism. In addition, for the S_N2 reaction the nature of the stereoselectivity (inversion) gives direct evidence of the orientation of attack—something that could not have been determined in any other way.

Between these two extremes of stereochemical behavior we have encountered a third kind, *partial* stereoselectivity, in the S_N1 reaction (Sec. 5.19), which again gives essential information about the reaction mechanism.

In our study of reactions of other kinds, we shall find other examples of stereoselectivity—in some cases stereoselectivity of quite different types from inversion of configuration. And we shall find other examples of non-stereoselec-tivity. Whatever the stereochemistry, it must, of course, be accounted for by a satisfactory mechanism.

But stereoselectivity in a reaction does more than give information about the mechanism: it provides a way to make a compound in just the stereoisomeric form that we want. Now let us see why we might *want* a particular stereoisomer.

9.3 Stereospecific reactions

There is still more for us to learn from the stereochemistry of the S_N2 reaction. Suppose we start, not with (*R*)-2-bromooctane, but with its enantiomer, (*S*)-2-bromooctane. Again inversion of configuration takes place, and we obtain, not

$$
\begin{array}{ccc}
\text{C}_6\text{H}_{13} & & \text{C}_6\text{H}_{13} \\
| & \xrightarrow[\text{S}_\text{N}2]{\text{NaOH}} & | \\
\text{Br}-\text{C}-\text{H} & & \text{H}-\text{C}-\text{OH} \\
| & & | \\
\text{CH}_3 & & \text{CH}_3
\end{array}
$$

(S)-2-Bromooctane (R)-2-Octanol

Only product

(S)-2-octanol as before, but its enantiomer, (R)-2-octanol. Just which product we obtain depends in a specific way on just which stereoisomer we start with. Such a reaction, in which stereochemically different reactants give stereochemically different products, is called a *stereospecific* reaction.

But the term *stereospecific* is used in a much broader sense, to indicate any kind of *discrimination on a stereochemical basis* between different reactant molecules. Our definition is, then, the following. *A* **stereospecific reaction** *is one in which stereochemically different molecules react differently.*

The S$_\text{N}$2 reaction is thus not only completely stereoselective but completely stereospecific as well.

By "stereochemically different molecules" is meant stereoisomers: enantiomers or diastereomers. To "react differently" means to show any difference whatsoever in chemical behavior. In a stereospecific reaction, stereoisomers can:

(a) yield different stereoisomers as products;

(b) react at different rates—in some cases to such an extent that, while one stereoisomer reacts readily, the other does not react at all;

(c) react by different paths to yield quite different kinds of compounds as products.

Stereospecificity toward enantiomers is called **enantiospecificity**. In reactions *with achiral reagents*, enantiomers can show only difference (a): they can yield different stereoisomers as products, as in the S$_\text{N}$2 reaction, but in all other respects they must react identically—at identical rates to yield products that are identical except for their stereochemistry.

On the other hand, in reactions *with optically active reagents*—or in a chiral medium of any sort—enantiomers may show all the differences in behavior that we have listed. We have already encountered enantiospecificity in the resolution of racemic modifications by use of optically active reagents (Sec. 4.27). There, they yielded stereochemically different products—not enantiomers, as in the S$_\text{N}$2 reaction, but diastereomers.

Biological systems, we have seen (Sec. 4.11), generally discriminate sharply between stereoisomers. The organism responds to only one enantiomer of a pair, or responds differently to the two; only one is metabolized, or serves as a hormone or drug, tastes sweet, and so on. Now, biological activity, in the final analysis, depends upon chemical reactions in the organism—in this case, reactions with one enantiomer or the other. The discrimination is the result of virtually complete enantiospecificity in these reactions. Such enantiospecificity is the rule for the countless reactions taking place in the chiral medium provided by the optically active enzymes of living organisms.

We have already accounted for this contrast in behavior toward optically inactive and optically active reagents. It stems from the fact that—whether we are comparing reactants or comparing transition states—enantiomers are of equal energy and diastereomers are of unequal energy (Sec. 4.11).

Stereospecificity toward diastereomers is called **diastereospecificity**. Diastereomers can differ in all the ways that we have listed above, whether the reagent is optically active or inactive. Indeed, as we have seen (Sec. 4.17), a difference in rate of reaction is the *rule* for diastereomers; in this respect at least, diastereomers will always react stereospecifically, although often to only a modest degree.

We have already seen (Sec. 4.17) why this must be so. Since diastereomers are neither identical nor mirror images, they are of different energies. In the reaction of two diastereomers with a given reagent, both the two sets of reactants and the two transition states are diastereomeric, and hence—except by sheer coincidence—will not be of equal energies. E_{act} values will be different and so will the rates of reaction.

In later sections of this chapter, we shall examine in detail examples of diastereospecificity in reactions we have already studied: differences in behavior between geometric isomers; and differences in behavior between configurational diastereomers, compounds containing more than one chiral center.

Reactions in biological systems are highly stereospecific not only toward enantiomers, but also toward diastereomers, including geometric isomers. This is especially evident in the action of *pheromones*, compounds produced by an organism for the purpose of communicating with other organisms of the same species: to attract members of the opposite sex; to spread an alarm; to mark the trail to food; or simply to carry the message, "Let's all get together". (This communication can span remarkable distances: the male gypsy moth receives the signal from a female a mile away!) There are, for example, four geometric isomers of 10,12-hexadecadien-1-ol; only one of these, the (10*E*,12*Z*) isomer, is the sex attractant produced by the female silk moth—and it is a billion times as attractive to the male as any of the other isomers. The male grape berry moth is attracted by (*Z*)-9-dodecen-1-yl acetate; the male European pine shoot moth is attracted by the (*E*) isomer of the same compound—yet this attraction is completely nullified by the presence of only 3% of the (*Z*) isomer. Moving the double bond over one position gives (*Z*)-8-dodecen-1-yl acetate, which is the sex attractant of the oriental fruit moth—but only if 7% of the (*E*) isomer is present; pure (*Z*) is completely inactive. (This requirement of a precise *mixture* of stereoisomers is very common.)

(10*E*,12*Z*)-10,12-Hexadecadien-1-ol
Sex attractant of silk moth

(*Z*)-9-Dodecen-1-yl acetate
Sex attractant of
grape berry moth

(*E*)-9-Dodecen-1-yl acetate
Sex attractant of
European pine shoot moth

(*Z*)-8-Dodecen-1-yl acetate
(93%)

(*E*)-8-Dodecen-1-yl acetate
(7%)

Sex attractant of oriental fruit moth

To find out how stereospecificity and the simple light-catalyzed transformation of a *cis* alkene into a *trans* alkene enables us—and all other sighted animals—to *see*, read Sec. 41.3.

9.4   Stereoselectivity *vs.* stereospecificity

Many reactions are, like the S_N2 reaction, both stereoselective and stereospecific. But this is not always true. *Some reactions are stereoselective but not stereospecific*: one particular stereoisomer is the predominant product regardless of the stereochemistry of the reactant, or regardless of whether the reactant even exists as stereoisomers.

Some reactions are stereospecific but not stereoselective. Stereoisomers may react at widely different rates, but yield the same stereoisomers as the product—or yield products that differ in ways other than in their stereochemistry. Sometimes one stereoisomer reacts readily, and another does not react at all, as in the biological reactions we have referred to.

The quality of **stereoselectivity** is concerned solely with the **products**, and their stereochemistry. Of a number of possible stereoisomeric products, the reaction *selects* one or two to be formed.

The quality of **stereospecificity** is focused on the **reactants** and their stereochemistry; it is concerned with the products, too, but only as they provide evidence of a difference in behavior between reactants. Of stereoisomeric reactants, each behaves in its own *specific* way.

The stereospecificity of biological reactions has given a powerful impetus to the development of synthetic methods that are highly stereoselective. In synthesizing a drug, for example, or a hormone, a chemist wants to use (stereoselective) reactions that produce just the correct stereoisomer, since only that stereoisomer will show (stereospecific) activity in a biological system.

For example, the sex attractants of insects have been studied extensively in recent years, with the aim—already realized in some cases—of synthesizing them to serve as bait with which to lure and entrap the female-seeking males of a species before they can mate, or to confuse them and disrupt their search. To be effective these synthetic materials must duplicate the stereochemical make-up of the natural pheromones; the stereospecificity of their action demands an equal stereoselectivity in their synthesis—*enantioselectivity* to match enantiospecificity, and *diastereoselectivity* to match diastereospecificity. And so a large part of the research in the field of pheromones—and of other biologically active substances—involves development of new, highly stereoselective ways to introduce the carbon–carbon double bond or other structural elements into a molecule: new reagents, new catalysts, new reaction media. Later on, we shall look at some of these ways.

9.5   Stereochemistry of addition of halogens to alkenes. *syn-* and *anti-*Addition

Now let us turn to the stereochemistry of *addition*, using as our example a familiar reaction: addition of halogens to alkenes. In this section we shall look at the stereochemical facts, and, in the next, see what these facts tell us about the mechanism.

Addition of bromine to 2-butene yields 2,3-dibromobutane.

$$CH_3CH{=}CHCH_3 + Br_2 \longrightarrow CH_3{-}\overset{*}{C}H{-}\overset{*}{C}H{-}CH_3$$

2-Butene

$$\begin{array}{cc} | & | \\ Br & Br \end{array}$$

2,3-Dibromobutane

Two chiral centers are generated in the reaction, and the product, we can easily show (Sec. 4.18), can exist as a pair of enantiomers (I and II) and a *meso* compound (III).

$$
\begin{array}{ccc}
\text{CH}_3 & \text{CH}_3 & \text{CH}_3 \\
\text{H}-\!\!\!-\text{Br} & \text{Br}-\!\!\!-\text{H} & \text{H}-\!\!\!-\text{Br} \\
\text{Br}-\!\!\!-\text{H} & \text{H}-\!\!\!-\text{Br} & \text{H}-\!\!\!-\text{Br} \\
\text{CH}_3 & \text{CH}_3 & \text{CH}_3 \\
\text{I} & \text{II} & \text{III} \\
(S,S) & (R,R) & \textit{Meso}
\end{array}
$$

Enantiomers

2,3-Dibromobutane

The reactants, too, exist as stereoisomers: a pair of geometric isomers, *cis* and *trans*.

$$
\begin{array}{cc}
\underset{\text{H}}{\overset{\text{H}}{\diagdown}}\text{C}\overset{\text{CH}_3}{\diagup} & \underset{\text{H}}{\overset{\text{CH}_3}{\diagdown}}\text{C}\overset{\text{H}}{\diagup} \\
\diagup \text{C} \diagdown & \diagup \text{C} \diagdown \\
\text{H}\quad\text{CH}_3 & \text{H}\quad\text{CH}_3 \\
\textit{cis} & \textit{trans}
\end{array}
$$

2-Butene

If we start with, say *cis*-2-butene, which of the stereoisomeric products do we get? A mixture of all of them? *No.* The *cis* alkene yields *only* racemic 2,3-dibromobutane, I plus II; none of the *meso* compound is obtained.

$$
\text{cis-2-Butene} \xrightarrow{\text{Br}_2}
\begin{array}{cc}
\text{CH}_3 & \text{CH}_3 \\
\text{H}-\!\!\!-\text{Br} & \text{Br}-\!\!\!-\text{H} \\
\text{Br}-\!\!\!-\text{H} & \text{H}-\!\!\!-\text{Br} \\
\text{CH}_3 & \text{CH}_3 \\
\text{I} & \text{II}
\end{array}
$$

rac-2,3-Dibromobutane

Only product

Since the reaction yields only one enantiomeric pair of several possible diastereomers, it is *stereoselective*. Since the selectivity is between diastereomeric products, the reaction is *diastereoselective*.

Now, suppose we start with *trans*-2-butene. Does this, too, yield the racemic dibromide? *No.* The *trans* alkene yields *only* meso-2,3-dibromobutane.

$$
\text{trans-2-Butene} \xrightarrow{\text{Br}_2}
\begin{array}{c}
\text{CH}_3 \\
\text{H}-\!\!\!-\text{Br} \\
\text{H}-\!\!\!-\text{Br} \\
\text{CH}_3
\end{array}
$$

meso-2,3-Dibromobutane

Only product

Just which product we obtain depends upon which stereoisomer we start with. Since stereochemically different molecules react differently—they yield stereoisomeric products—the reaction is *stereospecific*. Since this specificity is toward diastereomers, the reaction is *diastereospecific*.

Other studies have shown that these results are typical: *addition of halogens to simple alkenes is completely stereoselective and completely stereospecific.*

To describe the kinds of stereochemistry possible in addition reactions, the concepts of **syn**-*addition* and **anti**-*addition* are used. These terms are not the names of specific mechanisms, but simply indicate the stereochemical facts: that the added groups become attached to the same face (**syn**) or to opposite faces (**anti**) of the double bond (Fig. 9.1).

syn-Addition

anti-Addition

Figure 9.1   *syn*- and *anti*-Addition.

Addition of bromine to the 2-butenes involves *anti*-addition. Let us see that this is so. If we start (Fig. 9.2) with *cis*-2-butene, we can attach the bromines to opposite faces of the double bond in two different ways. Attachment as in (*a*) gives enantiomer I; attachment as in (*b*) gives enantiomer II. Since, whatever the mechanism, (*a*) and (*b*) are equally likely, we obtain the racemic modification.

Figure 9.2   *anti*-Addition to *cis*-2-butene. Attachment as in (*a*) or (*b*) is equally likely: it gives the racemic modification.

Starting with *trans*-2-butene (Fig. 9.3), we can again attach the bromines to opposite faces of the double bond in two ways: as in (*c*) or in (*d*). Whichever way we choose, we obtain III, which we recognize as the *meso* dibromide.

Figure 9.3   *anti*-Addition to *trans*-2-butene. Attachment as in (*c*) or (*d*) gives the *meso* product.

anti-Addition is the general rule for the reaction of bromine or chlorine with simple alkenes. We shall encounter other examples of stereoselective additions, some *anti* and some *syn*.

Problem 9.1   On treatment with permanganate, *cis*-2-butene yields 2,3-butanediol of m.p. 34 °C, and *trans*-2-butene yields 2,3-butanediol of m.p. 19 °C. Both diols are optically inactive. Handling as described in Sec. 4.27 converts the diol of m.p. 19 °C (but not the one of m.p. 34 °C) into two optically active fractions of equal but opposite rotation.

(a) What is the configuration of the diol of m.p. 19 °C? Of m.p. 34 °C?

(b) Assuming that these results are typical (they are), *what is the stereochemistry of hydroxylation with permanganate*?

(c) Treatment of the same alkenes with peroxy acids gives the opposite results: the diol of m.p. 19 °C from *cis*-2-butene, and the diol of m.p. 34 °C from *trans*-2-butene. *What is the stereochemistry of hydroxylation with peroxy acids*?

Now let us see what the stereochemistry of halogen addition tells us about the mechanism of this reaction.

9.6   Mechanism of addition of halogens to alkenes

We saw earlier (Sec. 8.14) that addition of halogens to alkenes is believed to proceed by two steps. In step (1) a halogen is transferred, without a pair of electrons, from a halogen molecule to the carbon–carbon double bond; there is formed a halide ion and an organic cation. In step (2) this cation reacts with a halide ion to yield the addition product.

(1)

A halonium ion

(2)

In Sec. 8.14, we listed five facts that provide evidence for this mechanism, but discussed only two of them:

(a) the *effect of the structure of the alkene on reactivity*; and

(b) the *effect of added nucleophiles on the products obtained*.

Now, it is the nature of the intermediate cation that is our chief concern here. As we have shown it, it is a *halonium ion*: a cyclic ion in which halogen is attached to both carbons and carries a positive charge. Yet, from facts (a) and (b) alone, the cation could be a simple carbocation—open, not cyclic.

In the last section we learned another fact:

(c) halogens add *with complete stereoselectivity and in the anti sense*.

What does this stereochemistry tell us about the nature of the intermediate?
Assume first that reaction proceeds via an open carbocation.

(1)

$$\ce{>C=C<} + \ce{X-X} \longrightarrow \ce{-\underset{X}{C}-\overset{\oplus}{C}-} + \ce{X-}$$

(2)

$$\ce{-\underset{X}{C}-\overset{\oplus}{C}-} + \ce{X-} \longrightarrow \ce{-\underset{X}{C}-\underset{X}{C}-}$$

Is the observed stereochemistry consistent with a mechanism involving such an
intermediate? Let us use addition of bromine to *cis*-2-butene as an example. A
positive bromine ion is transferred to, say, the top face of the alkene to form
carbocation IV. Then, a bromide ion attacks the *bottom* face of the positively

| cis-2-Butene | Cation IV | (S,S)-2,3-Dibromobutane |

charged carbon to complete the *anti*-addition; attack at this face is preferred, we
might say, because it permits the two bromines to be as far apart as possible in the
transition state. (We obtain the racemic product: the *S,S*-dibromide as shown, the
R,R-dibromide through attachment of positive bromine to the near end of the
alkene molecule.)

   But this picture of the reaction is not satisfactory, and for two reasons. First,
to account for the *complete* stereospecificity of addition, we must assume that attack
at the bottom face of the cation is not just preferred, but is the *only* line of attack:
conceivable, but—especially in view of other reactions of carbocations (Sec. 5.19)—
not likely. Then, even if we accept this exclusively bottom-side attack, we are faced
with a second problem. Rotation about the carbon–carbon bond would convert
cation IV into cation V; bottom-side attack on cation V would yield not the racemic
dibromide but the *meso* dibromide—in effect *syn*-addition, and contrary to fact.

| Cation IV | Cation V | meso-2,3-Dibromobutane |

To accommodate the stereochemical facts, then, we would have to make two assumptions about halogen addition: after the carbocation is formed, it is attacked by bromide ion (a) before rotation about the single bond can occur, and (b) exclusively from the face away from the halogen already in the cation. Neither of these assumptions is very likely; together, they make the idea of an open carbocation intermediate hard to accept.

It was to account better for the observed stereochemistry that, in 1937, I. Roberts and G. E. Kimball at Columbia University proposed the bromonium ion mechanism that we have given.

(1)

A bromonium ion

(2)

Now, how does the bromonium ion mechanism account for *anti*-addition? Using models, let us first consider addition of bromine to *cis*-2-butene (Fig. 9.4).

cis-2-Butene   VI

cis-Bromonium ion

VII

VIII

VII *and* VIII *are enantiomers*
rac-2,3-Dibromobutane

Figure 9.4   Addition of bromine to *cis*-2-butene via a cyclic bromonium ion. Opposite-side attacks (*a*) and (*b*) are equally likely, and give enantiomers in equal amounts.

In the first step, positive bromine becomes attached to either the top or bottom face of the alkene. Let us see what we would get if bromine becomes attached to the top face. When this happens, the carbon atoms of the double bond tend to become tetrahedral, and the hydrogens and methyls are displaced downward. The methyl groups are, however, still located across from each other, as they were in the alkene. In this way, bromonium ion VI is formed.

Now bromonium ion VI is attacked by bromide ion. A new carbon–bromine bond is formed, and an old carbon–bromine bond is broken. This is a familiar reaction, nucleophilic substitution; bromide ion is the nucleophile, and the positive bromine is the leaving group. As we might expect, then, attack by bromide ion is *from the back side*; on the bottom face of VI, so that the bond being formed is on the opposite side of carbon from the bond being broken. There is *inversion of configuration* about the carbon being attacked.

Attack on VI can occur by path (*a*) to yield structure VII or by path (*b*) to yield structure VIII. We recognize VII and VIII as enantiomers. Since attack by either (*a*) or (*b*) is equally likely, the enantiomers are formed in equal amounts, and thus we obtain the racemic modification. The same results are obtained if positive bromine initially becomes attached to the bottom face of *cis*-2-butene. (Show with models that this is so.)

Next, let us carry through the same operation on *trans*-2-butene (Fig. 9.5). This time, bromonium ion IX is formed. Attack on it by path (*c*) yields X; attack by (*d*) yields XI. If we simply rotate either X or XI about the carbon–carbon bond, we readily recognize the symmetry of the compound. It is *meso*-2,3-dibromo-butane; X and XI are identical. The same results are obtained if positive bromine is initially attached to the bottom face of *trans*-2-butene. (Show with models that this is so.)

Figure 9.5   Addition of bromine to *trans*-2-butene via a cyclic bromonium ion. Opposite-side attacks (*c*) and (*d*) give the same product.

Problem 9.2   (a) What is the relationship between the bromonium ions formed by attachment of positive bromine to the top and bottom faces of *trans*-2-butene? In what proportions are they formed? (b) Answer the same questions for *cis*-2-butene. (c) For *trans*-2-pentene. (d) For *cis*-2-pentene.

Problem 9.3   (a) Predict the products of addition of bromine to *trans*-2-pentene. Is attack by bromide ion by the two paths equally likely? Account for the fact that inactive material is actually obtained. (b) Do the same for *cis*-2-pentene.

The concept of a halonium ion solves both of the problems associated with an open carbocation: a halogen bridge prevents rotation about the carbon–carbon bond, and at the same time restricts attack by bromide ion exclusively to the opposite face of the intermediate. The stereochemistry of halogen addition thus not only gives powerful support for a two-step mechanism, but it shows, in a way no other evidence could, just what those two steps almost certainly are.

That such cyclic intermediates *can* give rise to *anti*-addition is demonstrated by hydroxylation with peroxy acids (Problem 9.1, p. 352): there, analogous intermediates—perfectly respectable compounds called *epoxides* (Chap. 19)—can actually be isolated and studied.

$$-\overset{|}{\underset{}{C}}-\overset{|}{\underset{}{C}}-$$
$$\diagdown\!\!\diagup$$
$$O$$

An epoxide

Cyclic halonium ions were first proposed, then, simply as the most reasonable explanation for the observed stereochemistry. Since that time, however, more direct evidence has been discovered:

(d) the *direct observation of halonium ions* under superacid conditions.

In 1967, Olah (p. 194) prepared cations whose NMR spectra indicate that they are indeed cyclic halonium ions. For example:

$$(CH_3)_2\underset{\underset{F}{|}}{C}-\underset{\underset{Br}{|}}{C}HCH_3 + SbF_5 \xrightarrow{\text{liquid } SO_2} (CH_3)_2C-CHCH_3 \quad SbF_6^-$$
$$\underset{Br\oplus}{\diagdown\!\!\diagup}$$

The idea of a bromonium or chloronium ion may appear strange to us, in contrast to the already familiar oxonium and ammonium ions. The tendency for halogen to share two pairs of electrons and acquire a positive charge, we might say, should be weak because of the high electronegativity of halogens. But the evidence—here and, as we shall see, in other connections—shows that this tendency is *appreciable*. In halogen addition we are concerned with this question: which is more stable, an open carbocation in which carbon has only a sextet of electrons, or a halonium ion in which each atom (except hydrogen, of course) has a complete octet? It is not a matter of which atom, halogen or carbon, can better accommodate a positive charge; it is a matter of completeness or incompleteness of octets.

In halonium ion formation we see one more example of what underlies all carbocation behavior: *the need to get a pair of electrons to complete the octet of the positively charged carbon.*

There are exceptions to the rule of *anti*-addition of halogens, but exceptions that are quite understandable. If the alkene contains substituents that can strongly stabilize the open carbocation—as, for example, in a *benzyl* cation (Sec. 15.17)—then addition proceeds with little or no stereoselectivity. Carbon is getting the electrons it needs, but in a different way.

Problem 9.4 Olah treated compounds of the formula $(CH_3)_2CXCF(CH_3)_2$ with SbF_5. He observed the formation of halonium ions when X = Cl, Br, or I, but an open carbocation when X = F. How do you account for the difference in behavior of the difluoro compound? (*Hint*: See Sec. 1.15.)

In Secs. 20.2–20.4, we shall examine still another piece of evidence for the existence of bromonium ions:

(e) the role played by halonium ions in *neighboring group effects*.

9.7 Stereochemistry of the E2 reaction. *syn-* and *anti-*Elimination

Next, let us look at the stereochemistry of *elimination*, using as our example another familiar reaction: dehydrohalogenation under E2 conditions.

Consider dehydrohalogenation of the alkyl halide 1-bromo-1,2-diphenylpropane. (As we saw in Sec. 7.18, phenyl, $-C_6H_5$, is an aromatic hydrocarbon group that is inert under these reaction conditions.) This compound contains two chiral

$$C_6H_5\overset{*}{C}H-\overset{*}{C}H-C_6H_5 \longrightarrow C_6H_5CH{=}C(CH_3)C_6H_5$$

$$\underset{Br}{|} \quad \underset{CH_3}{|}$$

1-Bromo-1,2-diphenylpropane   1,2-Diphenylpropene

centers, and we can easily show that it can exist as two pairs of enantiomers: I and II, called *erythro*; and III and IV, called *threo*. Each pair is diastereomeric with the other pair.

I   II   III   IV

Erythro   *Threo*

1-Bromo-1,2-diphenylpropane

The designations *erythro* and *threo* are very commonly used by organic chemists to distinguish between certain diastereomers containing two chiral carbons. They are derived from the names of diastereomeric aldoses (Carbohydrates I, Chap. 38), *erythrose* and *threose*. If we draw cross formulas for these aldoses so that the biggest groups are at top and bottom,

Erythrose   Threose

the H's and OH's lie on the two sides. In erythrose, we see, the similar substituents (the two H's, say) lie on the same side of the formula; in threose they lie on opposite sides. In the same way, in the 1-bromo-1,2-diphenylpropanes—with the large C_6H_5's at top and bottom—the H's are on the same side in the *erythro* isomers, and on opposite sides in the *threo* isomers. (*To help you remember*: In **E** (*erythro*) the horizontal bars are on the same side; in **T** (*threo*) they are on opposite sides.)

The product, too, exists as stereoisomers: a pair of geometric isomers, *Z* and *E*.

(Z)   (E)

1,2-Diphenyl-1-propene

Now, if we start with the *erythro* halide, I and II, we obtain *only* the Z alkene. If we start with the *threo* halide, III and IV, we obtain only the E alkene.

I   II

(Z)-1,2-Diphenyl-1-propene

Erythro

1-Bromo-1,2-diphenylpropane

III   IV

(E)-1,2-Diphenyl-1-propene

Threo

1-Bromo-1,2-diphenylpropane

Other studies have shown that these results are typical: *E2 elimination is both stereoselective and stereospecific.*

To describe the kinds of stereoselectivity that may be observed in elimination reactions, the concepts of ***syn*-elimination** and ***anti*-elimination** are used. These terms are not the names of specific mechanisms. They simply indicate the stereochemical facts: that the eliminated groups are lost from the same face (***syn***) or opposite faces (***anti***) of the developing double bond (Fig. 9.6).

syn-Elimination

anti-Elimination

Figure 9.6   *syn*- and *anti*-Elimination.

As this example and many others show, E2 elimination typically involves *anti*-elimination: in the transition state the hydrogen and the leaving group are located in the *anti* relationship (Sec. 3.5) as contrasted to *gauche* or *eclipsed* (Fig. 9.7).

Figure 9.7 The E2 reaction of alkyl halides: *anti*-elimination. Hydrogen and the leaving group, —X, are as far apart as possible, in the *anti* relationship.

Thus, diastereomer I (or its enantiomer, II) gives the *Z* alkene:

I

(Z)-1,2-Diphenyl-1-propene

and diastereomer III (or its enantiomer, IV) gives the *E* alkene:

III

(E)-1,2-Diphenyl-1-propene

Earlier (Sec. 7.14), we likened E2 to S$_N$2 in that halide is pushed out of the molecule by a kind of nucleophilic attack: the "nucleophile" is the β-carbon which, using the electron pair left behind by the departing proton, begins to form a bond—the π bond—to the α-carbon. On this basis, the preference for *anti*-elimination indicates that this "nucleophilic attack" takes place preferentially at the face of the α-carbon most remote from the departing halide—the familiar *back-side attack* of nucleophilic substitution.

The preference for *anti*-elimination can be very strong, as we shall see in our study of cyclic compounds (Sec. 12.15). But, as we shall also see, under certain circumstances E2 reactions can proceed by *syn*-elimination. Whatever happens, the reacting molecules, as usual, are doing *what is easiest for them*.

The stereochemistry observed for these elimination reactions is entirely consistent with the E2 mechanism. The high degree of stereoselectivity indicates a strong preference for a particular spatial relationship between the two departing groups—something that is quite understandable if they are departing simultaneously.

Problem 9.5   When treated with C_2H_5OK in C_2H_5OH, diastereomer V (and its

enantiomer) gave *cis*-2-butene without loss of deuterium and *trans*-2-butene with loss of deuterium; diastereomer VI (and its enantiomer) gave *trans*-2-butene without loss of deuterium. How do you account for these findings? What is the stereochemistry of elimination here?

9.8   A look ahead

In this chapter, we have studied the stereochemistry of two basic reaction types: addition and elimination. More important, we have added to our knowledge of the fundamentals of stereochemistry two new concepts: *stereoselectivity* and *stereospecificity*.

We have seen additional examples of how stereochemistry helps us to understand reaction mechanisms. We have begun to learn how we can use this understanding to control the stereochemistry of a reaction, and why we want to exercise this control.

But we have hardly begun our study of chemistry in three dimensions. In following chapters, we shall find out how we can induce a reaction to give us, not just the diastereomer we want, but even the enantiomer we want—directly, and without laborious resolution.

We shall find that what we have said about stereoselectivity and stereospecificity applies not only to stereochemically different molecules, but also to stereochemically different *parts of the same molecule*. Portions of a molecule may be stereochemically equivalent or non-equivalent, and we must be able to distinguish between these if we are to understand subjects as widely different as NMR spectra and biological oxidation and reduction.

And in Chapter 20, where we discuss *symphoria*, we shall find that three-dimensional chemistry goes far beyond what is generally thought of as stereochemistry. Of all the factors determining the course of an organic reaction, we shall find, the most powerful can be the relative locations of reacting atoms: being *near* each other, and *in just the right positions to react*.

PROBLEMS

1. Homogeneous hydrogenation with deuterium of the unsaturated carboxylic acid *butenedioic acid* gives the saturated acid *butanedioic acid* containing two deuterium atoms.

$$\text{HOOC—CH=CH—COOH} + \text{D}_2 \xrightarrow[\text{catalyst}]{\text{Wilkinson's}} \text{HOOCCHDCHDCOOH}$$

$\quad\quad\quad$ Butenedioic acid $\quad\quad\quad\quad\quad\quad\quad\quad\quad$ (2,3-D$_2$)Butanedioic acid

cis-Butenedioic acid yields only the *meso* product; *trans*-butenedioic acid yields only the racemic product. Assuming that these results are typical (they are), *what is the stereochemistry of homogeneous hydrogenation?*

2. In Problem 9.1 (p. 352) we saw that *hydroxylation with permanganate is syn,* and *hydroxylation with peroxy acids is anti.* Keeping in mind that epoxides are intermediates in this latter reaction (p. 356), and given the fact that reactions of epoxides are acid-catalyzed, suggest a detailed mechanism for hydroxylation with peroxy acids. Show exactly how this mechanism accounts for the observed stereochemistry and for the catalysis by acid. (Check your answer in Sec. 19.13.)

3. Addition of chlorine water to 2-butene yields not only 2,3-dichlorobutane but also the chlorohydrin, 3-chloro-2-butanol. *cis*-2-Butene gives only the (racemic) *threo* chlorohydrin, and *trans*-2-butene gives only the (racemic) *erythro* chlorohydrin.

and enantiomer $\quad\quad\quad\quad$ and enantiomer
$\quad\quad$ *Threo* $\quad\quad\quad\quad\quad\quad\quad$ *Erythro*
$\quad\quad\quad\quad$ 3-Chloro-2-butanol

(a) Assuming that these results are typical (they are), *what is the stereochemistry of halohydrin formation?*

(b) Following the pattern of Fig. 9.4 (p. 354) and Fig. 9.5 (p. 355), show all steps in the formation of the chlorohydrin from *cis*-2-butene.

(c) Do the same thing starting with *trans*-2-butene.

(d) For each of the reactions (b) and (c) identify the step that actually leads to a racemic product.

4. What stereochemistry would you expect of E1 elimination? Explain your answer in detail. (*Hint:* See Sec. 9.7.)

5. (a) Alfred Hassner (at the University of Colorado) found iodine azide, IN$_3$, to add to terminal alkenes with the orientation shown, and with complete stereoselectivity (*anti*) to the 2-butenes. Suggest a mechanism for this reaction.

$$\text{RCH=CH}_2 + \text{IN}_3 \longrightarrow \underset{\underset{\text{N}_3}{|}}{\text{RCHCH}_2\text{I}}$$

(b) In polar solvents like nitromethane, BrN$_3$ adds with the same orientation and stereoselectivity as IN$_3$. In non-polar solvents like *n*-pentene, however, orientation is reversed, and addition is non-stereoselective. In solvents of intermediate polarity like methylene chloride, mixtures of products are obtained; light or peroxides favor formation of RCHBrCH$_2$N$_3$; oxygen favors formation of RCH(N$_3$)CH$_2$Br. Account in detail for these observations.

6. Each of the following reactions is carried out, and the products are separated by careful distillation, recrystallization, or chromatography. For each reaction tell how many fractions will be collected. Draw a stereochemical formula of the compound or compounds making up each fraction. Tell whether each fraction, as collected, will be optically active or optically inactive.

(a) *trans*-2-pentene + D$_2$ (Wilkinson's catalyst) $\longrightarrow$ C$_5$H$_{10}$D$_2$
(b) (*S*)(*Z*)-3-penten-2-ol + KMnO$_4$ $\longrightarrow$ C$_5$H$_{12}$O$_3$
(c) (*S*)(*Z*)-3-penten-2-ol + HCO$_2$OH $\longrightarrow$ C$_5$H$_{12}$O$_3$
(d) racemic (*E*)-4-methyl-2-hexene + Br$_2$ $\longrightarrow$ C$_6$H$_{12}$Br$_2$
(e) (*S*)-HOCH$_2$CHOHCH=CH$_2$ + KMnO$_4$ $\longrightarrow$ C$_4$H$_{10}$O$_4$
(f) (*R*)-2-ethyl-3-methyl-1-pentene + H$_2$/Ni $\longrightarrow$ C$_8$H$_{18}$

7. The 2-butene obtained by the E2 reaction of *sec*-butyl chloride consists mostly of the *trans* isomer, with a *trans*:*cis* ratio of 6:1.

(a) Can you suggest a possible cause or causes for this (moderate) stereoselectivity? (*Hint*: As usual for irreversible reactions, compare the structures of reactants and transition states for the competing paths.)

(b) The corresponding reaction of *sec*-butyl bromide also yields more *trans*- than *cis*-2-butene, but the *trans*:*cis* ratio here is only 3:1. How do you account for the lower ratio? *Be specific.* (*Hint*: See Secs. 2.24 and 7.19.)

8. To obtain each of the following products by addition, give the structure of the unsaturated compound you would start with, and the reagent and any special conditions you would use.

(a) *erythro*-2,3-dichloropentane
(b) *meso*-3,4-hexanediol
(c) *meso*-3,4-hexanediol (from a different alkene)
(d) *threo*-3-bromo-2-butanol
(e) racemic (2,3-D$_2$)butane (CH$_3$CHDCHDCH$_3$)

9. (a) In the work described in Problem 9.5 (p. 361) the 2-butenes were obtained in the following *trans*:*cis* ratios: from the *erythro* isomer (V and its enantiomer), 0.82; from the *threo* isomer (VI and its enantiomer), 10.6; from the unlabeled *sec*-butyl bromide under the same conditions, 2.84. How do you account for these differences in the *trans*:*cis* ratio?

(b) In the same work the ratios of each 2-butene to 1-butene were measured. The *trans*-2-butene:1-butene ratios were: from unlabeled, 2.82; from *erythro*, 0.82; from *threo*, 2.82. The *cis*-2-butene:1-butene ratios were: unlabeled, 0.99; *erythro*, 0.98; *threo*, 0.27. How do you account for the differences in these ratios? Is your answer consistent with your answer to part (a)? *Be as quantitative as you can.*

10. (a) On treatment with HBr, *threo*-3-bromo-2-butanol is converted into racemic 2,3-dibromobutane, and *erythro*-3-bromo-2-butanol is converted into *meso*-2,3-dibromobutane. What appears to be the stereochemistry of the reaction? Does it proceed with inversion or retention of configuration?

and enantiomer   and enantiomer
Threo   *Erythro*
3-Bromo-2-butanol

(b) When optically active *threo*-3-bromo-2-butanol is treated with HBr, *racemic* 2,3-dibromobutane is obtained. Now what is the stereochemistry of the reaction? Can you think of a mechanism that accounts for this stereochemistry?

(c) These observations, reported in 1939 by Saul Winstein (p. 244) and Howard J. Lucas (of The California Institute of Technology), are the first of many described as "neighboring group effects". Does this term help you find an answer to (b)?

11. (a) It has been proposed that the conversion of vicinal dihalides into alkenes by the action of iodide ion can proceed by either a one-step mechanism (i) or a three-step mechanism (ii).

(i)
$$-\underset{\underset{Br}{|}}{C}-\underset{\underset{Br}{|}}{C}- \xrightarrow{\ I^- \ } -C{=}C- + IBr + Br^-$$

(ii)
$$-\underset{\underset{Br}{|}}{C}-\underset{\underset{Br}{|}}{C}- \xrightarrow{\ I^- \ } -\underset{\underset{I}{|}}{C}-\underset{\underset{Br}{|}}{C}- \xrightarrow{\ I^- \ } -\underset{\underset{I}{|}}{C}-\underset{}{C}- \longrightarrow -C{=}C-$$

Show the details, particularly the expected stereochemistry, of each step of each mechanism.

(b) The following stereochemical observations have been made:

meso-1,2-dibromo-1,2-dideuterioethane (CHDBrCHDBr) + I⁻ ⟶

only *cis*-CHD=CHD

meso-2,3-dibromobutane + I⁻ ⟶ only *trans*-2-butene
racemic 2,3-dibromobutane + I⁻ ⟶ only *cis*-2-butene

On the basis of the observed stereochemistry, which mechanism is most probably followed by each halide? Explain in detail. How do you account for the difference in behavior between the halides?

12. On treatment with the aromatic base pyridine (Sec. 35.11), racemic 1,2-dibromo-1,2-diphenylethane loses HBr to yield *trans*-1-bromo-1,2-diphenylethene; in contrast, the *meso* dibromide loses Br₂ to yield *trans*-1,2-diphenylethene. (a) Suggest a mechanism for the reaction of each stereoisomer. (b) How do you account for the difference in their behavior? (*Hint*: Phenyl is a *large* group.)

10

Conjugation and Resonance

Dienes

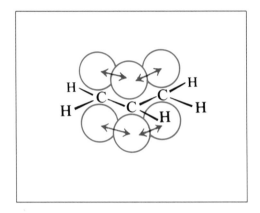

10.1 The carbon–carbon double bond as a substituent

In Chapter 8, we began our study of the chemistry of the carbon–carbon double bond. We saw the double bond as a place in the alkene molecule where reaction can occur: electrophilic or free-radical addition. But that is only part of the story. Besides providing a site for addition, the double bond exerts powerful effects on certain reactions taking place elsewhere on the molecule. Although suffering no permanent change itself, the double bond plays an essential role in determining the course of reaction. It is this part of alkene chemistry that we shall take up in this chapter: the carbon–carbon double bond, not as a functional group, but *as a substituent*.

Now, at this point in our study we have discussed several families of compounds: alkanes, alkyl halides (and related compounds), alcohols, and alkenes. We have seen some of the chemical properties that are associated with the functional group of each of these families: C—H of alkanes, —X and —OH of alkyl halides and alcohols, the carbon–carbon double bond of alkenes. This approach has led us to several of the major types of organic reactions: free-radical substitution, nucleophilic substitution, elimination, and addition. We have discussed the effects exerted on these reactions by substituents—alkyl groups, mostly: their polar effects, steric effects, and (until now unspecified) effects on the stability of free radicals and alkenes. We have looked at the inductive effect of halogens.

In this chapter we shall return to each of these families of compounds and each of these reaction types, and look at the effects exerted by a different kind of substituent: the carbon–carbon double bond. A double bond, we shall find, exerts its effect differently from an alkyl group, and, as a result, its effects are often more powerful. Most of these effects stem from the structural feature called *conjugation*: the location of the π orbital in such a way that it can overlap other orbitals in the molecule. And to implement our discussion of conjugation we shall make use of the structural theory called *resonance*.

10.2   Free-radical halogenation of alkenes: substitution *vs.* addition

Let us look at the structure of the simple alkene, propylene. It contains a carbon–carbon double bond, where the same addition reactions that are characteristic of ethylene take place. With hydrogen chloride, for example, propylene

$$CH_2=CH-CH_3$$
Propylene

undergoes electrophilic addition; with hydrogen bromide in the presence of peroxides, it undergoes free-radical addition.

$$CH_3-CH=CH_2 \; + \; HCl \; \longrightarrow \; CH_3-\underset{\underset{Cl}{|}}{C}H-\underset{\underset{H}{|}}{C}H_2$$

More readily than for ethylene

Isopropyl chloride
Only product

$$CH_3-CH=CH_2 \; + \; HBr \; \xrightarrow{\text{peroxides}} \; CH_3-\underset{\underset{H}{|}}{C}H-\underset{\underset{Br}{|}}{C}H_2$$

More readily than for ethylene

n-Propyl bromide
Only product

But propylene also contains a methyl group, and this modifies the reactions taking place at the double bond. Because of the methyl group, the electrophilic addition takes place faster than with ethylene itself, and gives exclusively isopropyl chloride. And because of the methyl group, the free-radical addition takes place faster than with ethylene, and gives exclusively *n*-propyl bromide. Thus, as a substituent, the methyl group affects the reactivity of the carbon–carbon double bond and determines the orientation of attack.

Now let us change our point of view and consider the methyl group, not as a substituent, but as the site of reaction. What kind of reactions can we expect to take place here? The methyl group has an alkane-like structure, and hence we might expect it to undergo alkane-like reactions: free-radical substitution of a halogen, for example.

$$CH_2=CH-CH_3$$

Alkene-like: Alkane-like:
site of addition   *site of substitution*

Let us consider, then, the reaction of propylene with halogens. But the propylene molecule presents *two* sites where halogen can attack, the double bond and the methyl group. Can we direct the attack to just one of these sites? The answer is yes, *by our choice of reaction conditions.*

We know that alkanes undergo substitution by halogen at high temperatures or under the influence of ultraviolet light, and generally in the gas phase: conditions that favor formation of free radicals. We know that alkenes undergo addition of halogen at low temperatures and in the absence of light, and generally in the liquid phase: conditions that favor heterolytic reactions, or at least do not aid formation of radicals.

$$-\overset{|}{C}=\overset{|}{C}-\overset{|}{\underset{H}{C}}-$$

Heterolytic
attack
Addition

Free-radical
attack
Substitution

If we wish to direct the attack of halogen to the alkyl portion of an alkene molecule, then, we choose conditions that are favorable for the free-radical reaction and unfavorable for the heterolytic reaction. Chemists of the Shell Development Company found that, at a temperature of 500–600 °C, a mixture of gaseous propylene and chlorine yields chiefly the substitution product, 3-chloro-1-propene, known as *allyl chloride* ($CH_2=CH-CH_2-$ = **allyl**). Bromine behaves similarly.

$$CH_3-CH=CH_2 \xrightarrow{Cl_2}$$
Propylene

$\xrightarrow[\text{CCl}_4 \text{ soln.}]{\text{low temp.}}$ $CH_3-\underset{Cl}{\overset{|}{C}}H-\underset{Cl}{\overset{|}{C}}H_2$ Heterolytic: *addition*

1,2-Dichloropropane
Propylene chloride

$\xrightarrow[\text{gas phase}]{500\text{–}600\,°\text{C}}$ $Cl-CH_2-CH=CH_2 + HCl$ Free-radical: *substitution*

3-Chloro-1-propene
Allyl chloride

In view of Secs. 8.18–8.19, we might wonder why a halogen atom does not add to a double bond, instead of abstracting a hydrogen atom. H. C. Brown (of Purdue University) has suggested that the halogen atom *does* add but, at high temperatures, is expelled before the second step of free-radical addition can occur.

Free-radical addition

$$CH_3-CH-CH_2-X \xrightarrow{X_2} CH_3-CH-CH_2-X + X\cdot$$
$$\qquad\qquad\qquad \text{I} \qquad\qquad\qquad\qquad \overset{|}{X}$$

$$X\cdot + CH_3-CH=CH_2$$

Free-radical substitution

$$HX + \underset{\cdot}{C}H_2-CH=CH_2 \xrightarrow{X_2} X-CH_2-CH=CH_2 + X\cdot$$
Allyl radical   Allyl halide

*Actual product at
high temperature or
low halogen concentration*

(X = Cl, Br)

Consistent with Brown's explanation is the finding that *low concentration* of halogen can be used instead of high temperature to favor substitution over (free-radical) addition. Addition of the halogen atom gives radical I, which falls apart (to regenerate the starting material) if the temperature is high or if it does not soon encounter a halogen molecule to complete the addition. The allyl radical, on the other hand, once formed, has little option but to wait for a halogen molecule, whatever the temperature or however low the halogen concentration.

Problem 10.1   (a) What would the allyl radical have to do to return to the starting material? (b) From bond dissociation energies, calculate the minimum E_{act} for this reaction.

The compound **N-bromosuccinimide (NBS)** is a reagent used *for the specific purpose of brominating alkenes at the allylic position*; NBS functions simply by providing a constant, low concentration of bromine. As each molecule of HBr is formed by the halogenation, NBS converts it into a molecule of Br_2.

$$HBr + \text{N-Bromosuccinimide} \longrightarrow Br_2 + \text{Succinimide}$$

N-Bromosuccinimide
(NBS)   Succinimide

10.3   Free-radical substitution in alkenes: orientation and reactivity

The alkyl groups of alkenes, then, undergo substitution by halogen in exactly the same manner as alkanes do. But attached to these alkyl groups there is a substituent, the double bond. Just as the alkyl groups affect the reactivity of the double bond, so the double bond affects the reactivity of the alkyl groups. Let us see what this effect is, and how it arises.

Halogenation of many alkenes has shown that: (a) hydrogens attached to doubly bonded carbons undergo very little substitution; and (b) hydrogens attached to carbons adjacent to doubly bonded carbons are particularly reactive toward substitution. Examination of reactions which involve attack not only by halogen atoms but by other free radicals as well has shown that this is a general rule: hydrogens attached to doubly bonded carbons, known as **vinylic** hydrogens, are harder to abstract than ordinary primary hydrogens; hydrogens attached to a carbon atom next to a double bond, known as **allylic** hydrogens, are even easier to abstract than tertiary hydrogens.

Vinylic hydrogen: *hard to abstract*

—C—H   **Allylic hydrogen:** *easy to abstract*

We can now expand the reactivity sequence of Sec. 3.23.

**Ease of abstraction
of hydrogen atoms**   allylic > 3° > 2° > 1° > CH_4 > vinylic

Substitution in alkenes proceeds by the same mechanism as substitution in alkanes. For example:

$$CH_2{=}CH{-}H \xrightarrow{\text{Cl·}} CH_2{=}CH· \xrightarrow{Cl_2} CH_2{=}CH{-}Cl$$

Ethylene   Vinyl radical   Vinyl chloride

$$CH_2{=}CH{-}CH_2{-}H \xrightarrow{\text{Cl·}} CH_2{=}CH{-}CH_2· \xrightarrow{Cl_2} CH_2{=}CH{-}CH_2Cl$$

Propylene   Allyl radical   Allyl chloride

Evidently the vinyl radical is formed very slowly and the allyl radical is formed very rapidly. We can now expand the sequence of Sec. 3.25.

**Ease of formation
of free radicals**   allyl > 3° > 2° > 1° > $CH_3·$ > vinyl

Are these findings in accord with our rule that *the more stable the radical, the more rapidly it is formed*? Is the slowly formed vinyl radical relatively unstable, and the rapidly formed allyl radical relatively stable?

The bond dissociation energies in Table 1.2 (p. 21) show that 108 kcal of energy is needed to form vinyl radicals from a mole of ethylene, as compared with 98 kcal for formation of ethyl radicals from ethane. Relative to the hydrocarbon from which each is formed, then, the vinyl radical contains more energy and is less stable than a methyl radical.

On the other hand, bond dissociation energies show that only 88 kcal is needed for formation of allyl radicals from propylene, as compared with 92 kcal for formation of *tert*-butyl radicals. Relative to the hydrocarbon from which each is formed, the allyl radical contains less energy and is more stable than the *tert*-butyl radical.

We can now expand the sequence of Sec. 3.24; relative to the hydrocarbon from which each is formed, the order of stability of free radicals is:

**Stability of
free radicals**   allyl > 3° > 2° > 1° > $CH_3·$ > vinyl

In some way, then, the double bond affects the stability of certain free radicals; it exerts a similar effect on the incipient radicals of the transition state, and thus affects the rate of their formation. Through these effects on rate of reaction, the double bond helps to determine both the *orientation* of free-radical substitution in an alkene, and the *relative reactivities* of different alkenes. Thus, the cyclic alkene cyclohexene is brominated almost exclusively at the allylic positions,

$$
\begin{array}{c}
\text{CH} \\
H_2C \diagup \quad \diagdown CH \\
\mid \qquad\qquad \mid \\
H_2C \diagdown \quad \diagup CH_2 \\
\text{CH}_2
\end{array}
\xrightarrow{\text{NBS}}
\begin{array}{c}
\text{CH} \\
H_2C \diagup \quad \diagdown CH \\
\mid \qquad\qquad \mid \\
H_2C \diagdown \quad \diagup CHBr \\
\text{CH}_2
\end{array}
$$

Cyclohexene   3-Bromocyclohexene

and reacts much faster than the saturated hydrocarbon cyclohexane despite a probability factor of 12:4 favoring attack on the saturated compound. (*Problem*: Why 12:4?)

As we know, free radicals are formed, not only by abstraction of hydrogen atoms, but also by addition to a double bond. Here too, we shall find, a double bond—a *second* double bond, not the one undergoing addition—can, through its effect on the stability of the incipient free radical, help to determine orientation and reactivity.

We have already seen (Sec. 7.4) a possible explanation for the low stability of vinylic radicals. Bonding of a vinylic hydrogen to carbon results from overlap with an sp^2 orbital of carbon rather than the sp^3 orbital of saturated carbon; this carbon–hydrogen bond is therefore shorter and stronger, and more energy must be supplied to break it. Relative to the hydrocarbon from which it is made, then, a vinylic radical is relatively unstable.

The high stability of allylic radicals is, as we shall see, readily accounted for by the structural theory: specifically, by the concept of resonance. But before we turn to resonance, let us look at some other characteristics of allylic radicals which, like their stability, are unusual.

10.4 Free-radical substitution in alkenes: allylic rearrangement

Since we shall use the allyl radical as our introduction to both the concept of conjugation and the theory of resonance, let us examine its structure in detail. Besides the fact that (a) *the allyl radical is especially stable*, there are other facts that must be accounted for by a satisfactory structure. Let us see what these facts are.

(b) *Free-radical substitution at allylic positions can lead to allylic rearrangement.* When 1-octene, for example, is treated with *N*-bromosuccinimide, there is obtained not only the expected 3-bromo-1-octene, but also—and in larger amounts— 1-bromo-2-octene (both *Z* and *E*). It is an allylic hydrogen on C–3 that is abstracted,

$$CH_3(CH_2)_3CH_2CH_2CH=CH_2$$
1-Octene

$$\downarrow NBS$$

$$CH_3(CH_2)_3CH_2\underset{\underset{Br}{|}}{C}HCH=CH_2 \quad + \quad CH_3(CH_2)_3CH_2CH=CH\underset{\underset{Br}{|}}{C}H_2$$

3-Bromo-1-octene   1-Bromo-2-octene
                                         Rearrangement product

but in much of the product bromine appears on C–1. Whenever the structure permits, such allylic rearrangement occurs, and according to a well-defined pattern:

$$-\overset{|}{C}=\overset{|}{C}-\overset{|}{\underset{\cdot}{C}}- \quad \xrightarrow{X_2} \quad -\overset{|}{C}=\overset{|}{C}-\overset{|}{\underset{\underset{X}{|}}{C}}- \quad + \quad -\overset{|}{\underset{\underset{X}{|}}{C}}-\overset{|}{C}=\overset{|}{C}-$$

As we see, the allylic radical reacts to give two different products: one in which halogen has become attached to the carbon that lost the hydrogen; and the other in which halogen has become attached to the carbon at the other end of the three-carbon unit—the allylic system—that we represent as C=C—C·.

Examination of the structures involved shows us that such rearrangement involves no migration of atoms or groups; only the double bond appears in a different position from the one it occupied in the reactant.

Problem 10.2 Free-radical chlorination with *tert*-butyl hypochlorite (*t*-BuOCl, Problem 20, p. 122) shows a strong preference for allylic substitution rather than addition. Whether one starts with 1-butene or 2-butene (*cis* or *trans*), such chlorination yields a mixture of the same chloroalkenes (neglecting stereoisomerism). What are these chloroalkenes likely to be, and how are they formed?

10.5 Symmetry of the allyl radical

(c) *The allyl radical is a symmetrical molecule.*

A carbon–carbon double bond, we have seen, is quite different from a carbon–carbon single bond: it is shorter and stronger; rotation about it is hindered; the doubly bonded carbons hold other atoms—hydrogens, say—by shorter, stronger bonds.

If the allyl radical actually possessed the "classical" structure that we have so far drawn for it,

it would be unsymmetrical about the central carbon atom; that is, the two ends of the molecule would be different from each other. It would contain two kinds of carbon–carbon bonds: a long single bond and a short double bond.

Now, an ESR spectrum (electron spin resonance spectrum, Sec. 16.17) reflects the structure of a free radical by what it shows about the *hydrogens* in the molecule: among other things, how many different "kinds" of hydrogen the free radical contains. It gives a signal for each hydrogen or each set of *equivalent* hydrogens—that is, each set of hydrogens in the same environment (Sec. 16.7).

Let us examine the classical structure of the allyl radical. The two vinylic hydrogens (H_a and H_b) on the terminal carbon would be non-equivalent (diastereotopic, actually), since one is *cis* and the other *trans* to —CH$_2$. The two hydrogens (H_c) of —CH$_2$ would be equivalent; because of rapid rotation about the

Allyl radical
Classical structure:
expect 4 ESR signals

carbon–carbon single bond they would be in the same *average* environment. Finally, there is the vinylic hydrogen (H_d) on the central carbon; it is different from all the others. If the allyl radical had the classical structure, then, we would expect an ESR spectrum corresponding to *four* kinds of hydrogens.

In fact, however, the ESR spectrum actually measured reveals only *three* kinds of hydrogens. Each vinylic hydrogen at one end of the molecule has an exact counterpart at the other end.

Allyl radical

Actual structure:
gives 3 ESR signals

(The two hydrogens labeled H_a are equivalent, and so are the ones labeled H_b.) The two ends of the molecule are equivalent; both carbon–carbon bonds are of exactly the same kind. The allyl radical is perfectly *symmetrical* about the central carbon.

Our classical structure of the allyl radical is clearly not satisfactory. What is required is a structure that accounts for the unusual stability of this radical, the occurrence of allylic rearrangements, and the symmetry revealed by ESR. To see what the structure is, we must turn to the theory of *resonance*.

10.6   The theory of resonance

It will be helpful first to list some of the general principles of the concept of resonance, and then to discuss these principles in terms of a specific example, the structure of the allyl radical.

(a) *Whenever a molecule can be represented by two or more structures that differ only in the arrangement of electrons—that is, by structures that have the same arrangement of atomic nuclei—there is* **resonance.** The molecule is a **hybrid** of all these structures, and cannot be represented satisfactorily by any one of them. Each of these structures is said to **contribute** to the hybrid.

(b) *When these contributing structures are of about the same stability (that is, have about the same energy content), then* **resonance is important.** The contribution of each structure to the hybrid depends upon the relative stability of that structure: the more stable structures make the larger contribution.

(c) *The resonance hybrid is more stable than any of the contributing structures.* This increase in stability is called the **resonance energy.** The more nearly equal in stability the contributing structures, the greater the resonance energy.

There can be resonance only between structures that contain the *same number of odd electrons*. We need concern ourselves about this restriction only in dealing with *di*-radicals: molecules that contain *two* unpaired electrons. There cannot be resonance between a diradical structure and a structure with all electrons paired.

10.7   The allyl radical as a resonance hybrid

In the language of the resonance theory, then, the allyl radical is a resonance hybrid of the two structures, I and II.

$$CH_2{=}CH{-}CH_2\cdot \qquad \cdot CH_2{-}CH{=}CH_2$$

I   II

This simply means that the allyl radical does not correspond to either I or II, but rather to a structure intermediate between I and II. Furthermore, since I and II are exactly equivalent, and hence have exactly the same stability, the resonance hybrid is equally related to I and to II; that is, I and II are said to make *equal contributions to the hybrid*.

This does *not* mean that the allyl radical consists of molecules half of which correspond to I and half to II, nor does it mean that an individual molecule changes back and forth between I and II. All molecules are the same; each one has a structure intermediate between I and II.

An analogy to biological hybrids that was suggested by Professor G. W. Wheland of the University of Chicago is helpful. When we refer to a mule as a hybrid of a horse and a donkey, we do not mean that some mules are horses and some mules are donkeys; nor do we mean that an individual mule is a horse part of the time and a donkey part of the time. We mean simply that a mule is an animal that is related to both a horse and a donkey, and that can be conveniently defined in terms of those familiar animals.

An analogy used by Professor John D. Roberts of the California Institute of Technology is even more apt. A medieval European traveler returns home from a journey to India, and describes a rhinoceros as a sort of cross between a dragon and a unicorn—a quite satisfactory description of a real animal in terms of two familiar but entirely imaginary animals.

It must be understood that our drawing of two structures to represent the allyl radical does not imply that either of these structures (or the molecules each would singly represent) has any existence. The two pictures are necessary because of the limitations of our rather crude methods of representing molecules. We draw two pictures because no *single* one would suffice. It is not surprising that certain molecules cannot be represented by one structure of the sort we have employed; on the contrary, the surprising fact is that the crude dot-and-dash representation used by organic chemists has worked out to the extent that it has.

The resonance theory further tells us that the allyl radical does not contain one carbon–carbon single bond and one carbon–carbon double bond (as in I or II), but rather contains two *identical* bonds, each one intermediate between a single and a double bond. This new type of bond—this **hybrid bond**—has been described as a *one-and-a-half bond*. It is said to possess one-half single-bond character and one-half double-bond character.

$$[CH_2{=}CH{-}CH_2\cdot \quad \cdot CH_2{-}CH{=}CH_2] \quad \textit{equivalent to} \quad CH_2{\cdots}CH{\cdots}CH_2$$

I   II   III

The odd electron is not localized on one carbon or the other but is *delocalized*, being equally distributed over both terminal carbons. We might represent this symmetrical hybrid molecule as in III, where the broken lines represent half bonds.

What we have arrived at is, of course, exactly the kind of highly symmetrical structure indicated by the ESR spectrum of the allyl radical.

Allylic rearrangement is a natural consequence of the hybrid character of an allylic radical. The terminal carbons of the three-carbon allylic system are exactly equivalent in the allyl radical itself, and very similar in an unsymmetrically substituted allylic radical. When halogen reacts with such a radical, it can become

attached to either of these terminal carbons. Where the structure permits, as in 1-octene for example, this attachment to either end is shown by the formation of two different products. In the case of the unsubstituted allyl radical itself, the same

product is obtained whichever end receives the halogen, and so no rearrangement is *seen*; but there can be little doubt that here, too, both carbons are subject to attack.

Problem 10.3   Actually, one *could* detect "rearrangement"—that is, attachment to either end of the allyl radical—in the chlorination of propylene. Tell how.

Problem 10.4   The nitro group, $-NO_2$, is usually represented as

Actual measurement shows that the two nitrogen–oxygen bonds of a nitro compound have exactly the same length. In nitromethane, CH_3NO_2, for example, the two nitrogen–oxygen bond lengths are each 1.21 Å, as compared with a usual length of 1.36 Å for a nitrogen–oxygen single bond and 1.18 Å for a nitrogen–oxygen double bond. What is a better representation of the $-NO_2$ group?

Problem 10.5   The carbonate ion, CO_3^{2-}, might be represented as

Actual measurement shows that all the carbon–oxygen bonds in $CaCO_3$ have the same length, 1.31 Å, as compared with a usual length of about 1.36 Å for a carbon–

oxygen single bond and about 1.23 Å for a carbon–oxygen double bond. What is a better representation of the $CO_3{}^{2-}$ ion?

Problem 10.6 The addition of $BrCCl_3$ to 1,3-butadiene in the presence of a peroxide gives a mixture of IV and V. How do you account for the formation of these two products?

$$CH_2{=}CHCHCH_2CCl_3 \qquad\qquad CH_2CH{=}CHCH_2CCl_3$$
$$\ \ \ \ \ \ \ \ \ \ \ \ | \qquad\qquad\qquad\qquad\ \ \ |$$
$$\ \ \ \ \ \ \ \ \ \ Br \qquad\qquad\qquad\qquad\ \ Br$$

$$\text{IV} \qquad\qquad\qquad\qquad\qquad \text{V}$$

10.8 Stability of the allyl radical

A further, most important outcome of the resonance theory is this: *as a resonance hybrid, the allyl radical is more stable (that is, contains less energy) than either of the contributing structures.* This additional stability possessed by the molecule is referred to as *resonance energy.* Since these particular contributing structures are exactly equivalent and hence of the same stability, we expect stabilization due to resonance to be large.

Just *how* large is the resonance energy of the allyl radical? To know the exact value, we would have to compare the actual, hybrid allyl radical with a *non-existent* radical of structure I or II—something we cannot do, experimentally. We can, however, estimate the resonance energy by comparing two reactions: dissociation of propane to form a *n*-propyl radical, and dissociation of propylene to form an allyl radical.

$$CH_3CH_2CH_3 \quad\longrightarrow\quad CH_3CH_2CH_2\cdot + H\cdot \qquad \Delta H = +98\ \text{kcal}$$
$$\text{Propane} \qquad\qquad\quad \textit{n}\text{-Propyl radical}$$

$$CH_2{=}CH{-}CH_3 \quad\longrightarrow\quad CH_2{=}CH{-}CH_2\cdot + H\cdot \qquad \Delta H = +88$$
$$\text{Propylene} \qquad\qquad \text{Allyl radical}$$

Propane, the *n*-propyl radical, and propylene are each fairly satisfactorily represented by a single structure; the allyl radical, on the other hand, is a resonance hybrid. We see that the energy difference between propylene and the allyl radical is 10 kcal/mol less $(98 - 88)$ than the energy difference between propane and the *n*-propyl radical; we attribute the lower dissociation energy entirely to resonance stabilization of the allyl radical, and estimate the resonance energy to be 10 kcal/mol.

10.9 Orbital picture of the allyl radical

To get a clearer picture of what a resonance hybrid is—and, especially, to understand how resonance stabilization arises—let us consider the bond orbitals in the allyl radical.

Since each carbon is bonded to three other atoms, it uses sp^2 orbitals (as in ethylene, Sec. 7.2). Overlap of these orbitals with each other and with the *s* orbitals of five hydrogen atoms gives the molecular skeleton shown in Fig. 10.1, with all bond angles 120°. In addition, each carbon atom has a *p* orbital which, as we know, consists of two equal lobes, one lying above and the other lying below the plane of the σ bonds; it is occupied by a single electron.

$$[CH_2\!=\!CH\!-\!CH_2\cdot \quad \cdot CH_2\!-\!CH\!=\!CH_2] \qquad equivalent\ to \qquad CH_2\text{---}CH\text{---}CH_2$$

Figure 10.1 Allyl radical. The *p* orbital of the middle carbon overlaps *p* orbitals on both sides to permit delocalization of electrons.

As in the case of ethylene, the *p* orbital of one carbon can overlap the *p* orbital of an adjacent carbon atom, permitting the electrons to pair and a bond to be formed. In this way we would arrive at either of the contributing structures, I or II, with the odd electron occupying the *p* orbital of the remaining carbon atom. But the overlap is not limited to a pair of *p* orbitals as it was in ethylene; the *p* orbital of the middle carbon atom overlaps equally well the *p* orbitals of *both* the carbon atoms to which it is bonded. The result is two continuous π electron clouds, one lying above and one lying below the plane of the atoms.

Since no more than two electrons may occupy the same orbital (Pauli exclusion principle), these π clouds are actually made up of *two* orbitals (Sec. 33.5). One of these, containing two π electrons, encompasses all three carbon atoms; the other, containing the third (odd) π electron, is divided equally between the terminal carbons.

The overlap of the *p* orbitals in both directions, and the resulting participation of each electron in two bonds, is equivalent to our earlier description of the allyl radical as a resonance hybrid of two structures. These two methods of representation, the drawing of several resonance structures and the drawing of an electron cloud, are merely our crude attempts to convey by means of pictures the idea that *a given pair of electrons may serve to bind together more than two nuclei*. It is this ability of π electrons to participate in several bonds, this **delocalization of electrons**, that results in stronger bonds and a more stable molecule. For this reason the term *delocalization energy* is frequently used instead of *resonance energy*.

The covalent bond owes its strength to the fact that an electron is attracted more strongly by two nuclei than by one. In the same way an electron is more strongly attracted by three nuclei than by two.

We saw earlier (Sec. 2.22) that the methyl radical may not be quite flat: that hybridization of carbon may be intermediate between sp^2 and sp^3. For the allyl radical, on the other hand—and for many other free radicals—flatness is clearly required to permit the overlap of *p* orbitals that leads to stabilization of the radical.

In terms of the conventional valence-bond structures we employ, it is difficult to visualize a single structure that is intermediate between the two structures, I and II. The orbital approach, on the other hand, gives us a rather clear picture of the allyl radical: the density of electrons holding the central carbon to each of the others is intermediate between that of a single bond and that of a double bond.

For generations, chemists have used the word *conjugated* to describe molecules containing alternating single and double (or triple) bonds: 1,3-butadiene, for example, or (and especially) benzene. A special name was given to compounds

$$CH_2 = CH - CH = CH_2$$

1,3-Butadiene   Benzene

with this structural feature since it was observed that they had certain special properties in common.

With the advent of the theory of resonance in the 1930s, the special properties of these conjugated molecules were attributed to interaction of the π orbitals of two or more double bonds: overlap much like what we have just described for the "double bond" of an allyl radical with the *p* orbital containing the odd electron. The meaning of the word *conjugation* became broadened to include the juxtaposition of a double bond and any π or *p* orbital—juxtaposition that permits overlap. And with *hyperconjugation*, the concept has been further broadened to include a similar juxtaposition of bonds of any kind—σ as well as π or *p*—juxtaposition, again, that permits sideways overlap.

The allyl radical is, then, a conjugated molecule. We interpret its special properties, as we shall do for other conjugated molecules, by the use of the theory of resonance. We can expect the carbon–carbon double bond to play a special role as a substituent whenever its location in a molecule creates a conjugated system: a system that, according to our interpretation, must exist as a resonance hybrid.

Problem 10.7 In the reaction described in Problem 10.2 (p. 371), the 1-chloro-2-butene obtained from *cis*-2-butene is exclusively the *cis* isomer, and the 1-chloro-2-butene obtained from *trans*-2-butene is exclusively the *trans* isomer. What does this show about the intermediate allylic radicals? How do you account for this on the basis of their structure? (*Hint*: See Sec. 7.5.)

10.10   Using the resonance theory

The great usefulness, and hence the great value, of the resonance theory lies in the fact that it retains the simple though crude type of structural representation which we have used so far in this book. Particularly helpful is the fact that the stability of a structure can often be roughly estimated from its **reasonableness**. If only one reasonable structure can be drawn for a molecule, the chances are good that this one structure adequately describes the molecule.

The criterion of reasonableness is not so vague as it might appear. The fact that a particular structure seems reasonable to us means that we have previously encountered a compound whose properties are pretty well accounted for by a structure of that type; the structure must, therefore, represent a fairly stable kind of arrangement of atoms and electrons. For example, each of the contributing structures for the allyl radical appears quite reasonable because we have encountered compounds, alkenes and free radicals, that possess the features of this structure.

There are a number of other criteria that we can use to estimate relative stabilities, and hence relative importance, of contributing structures. One of these has to do with (a) *electronegativity and location of charge.*

For example, a convenient way of indicating the polarity (*ionic character*) of the hydrogen–chlorine bond is to represent HCl as a hybrid of structures I and II. We judge that II is appreciably stable and hence makes significant contribution, because in it a negative charge is located on a highly electronegative atom, chlorine.

$$H\text{—}Cl \qquad\qquad H^+Cl^-$$
$$\text{I} \qquad\qquad\qquad \text{II}$$

On the other hand, we consider methane to be represented adequately by the single structure III.

$$
\begin{array}{c}
H \\
| \\
H\text{—}C\text{—}H \\
| \\
H
\end{array}
$$

$$\text{III}$$

Although it is possible to draw additional, ionic structures like IV and V, we judge these to be unstable since in them a negative charge is located on an atom of low

$$
\begin{array}{ccc}
H^+ & & H \\
| & & | \\
H\text{—}\overset{-}{C}\text{—}H & \qquad & H\text{—}\overset{-}{C}\ \ H^+\ \text{etc.} \\
| & & | \\
H & & H \\
\text{IV} & & \text{V}
\end{array}
$$

electronegativity, carbon. We expect IV and V to make negligible contribution to the hybrid and hence we ignore them.

In later sections we shall use certain other criteria to help us estimate stabilities of possible contributing structures: (b) *number of bonds* (Sec. 10.22); (c) *dispersal of charge* (Sec. 14.16); (d) *complete vs. incomplete octet* (Secs. 10.15 and 14.18); (e) *separation of charge* (Sec. 23.12).

Finally, we shall find certain cases where the overwhelming weight of evidence—bond lengths, dipole moments, reactivity—indicates that an accurate description of a given molecule requires contribution from structures of a sort that may appear quite unreasonable to us (Secs. 10.11 and 10.16); this simply reminds us that, after all, we know very little about the structure of molecules, and must be prepared to change our ideas of what is reasonable to conform with evidence provided by experimental facts.

In the next section, we shall encounter contributing structures that are very strange looking indeed.

Problem 10.8 *Benzene*, C_6H_6, is a flat molecule with all bond angles 120° and all carbon–carbon bonds 1.39 Å long. Its heat of hydrogenation (absorption of three moles of hydrogen) is 49.8 kcal/mol, as compared with values of 28.6 for cyclohexene (one mole of hydrogen) and 55.4 for 1,3-cyclohexadiene (two moles of hydrogen).

(a) Is benzene adequately represented by the Kekulé formula shown? (b) Suggest a better structure for benzene in both valence-bond and orbital terms. (Check your answer in Secs. 13.7–13.8.)

$$
\begin{array}{ccc}
\underset{\substack{\text{H}_2\text{C} \quad \text{CH} \\ | \qquad | \\ \text{H}_2\text{C} \quad \text{CH} \\ \text{CH}_2}}{\overset{\text{CH}_2}{}} &
\underset{\substack{\text{H}_2\text{C} \quad \text{CH} \\ | \qquad \| \\ \text{HC} \quad \text{CH} \\ \text{CH}}}{\overset{\text{CH}_2}{}} &
\end{array}
$$

Cyclohexene   1,3-Cyclohexadiene   Benzene
(Kekulé formula)

10.11   Resonance stabilization of alkyl radicals. Hyperconjugation

At this point let us look at an extension of the resonance theory which, although it does not involve a double bond, does nevertheless involve a kind of conjugation.

The relative stabilities of tertiary, secondary, and primary alkyl radicals are accounted for on exactly the same basis as the stability of the allyl radical: *delocalization of electrons*, this time through overlap between the p orbital occupied by the odd electron and a σ orbital of the alkyl group (Fig. 10.2). Through this

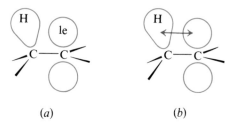

Figure 10.2   Hyperconjugation in an alkyl free radical. (a) Separate σ and p orbitals. (b) Overlapping orbitals.

overlap, individual electrons can, to an extent, help bind together three nuclei, two carbons and one hydrogen. This kind of delocalization, involving σ bond orbitals, is called **hyperconjugation.**

In resonance language, we would say that the ethyl radical, for example, is a hybrid of not only the usual structure, I, but also three additional structures, II, III,

and IV, in which a double bond joins the two carbons, and the odd electron is held by a hydrogen atom.

Individually, each of these "no-bond" resonance structures appears strange but, taken together, they mean that the carbon–hydrogen bond is something less than a single bond, that the carbon–carbon bond has some double bond character, and that the odd electron is partly accommodated by hydrogen atoms. Contribution from these unstable structures is not nearly so important as from, say, the equivalent structures for the allyl radical, and the resulting stabilization is not nearly so large. It is believed, however, to stabilize the ethyl radical to the extent of 6 kcal relative to the methyl radical ($104 - 98$, Sec. 3.24) for which such resonance is not possible.

If we extend this idea to the isopropyl radical, we find that instead of three hyperconjugation structures we now have six. (*Draw them.*) The larger number of contributing structures means more extensive delocalization of the odd electron, and hence greater stabilization of the radical. In agreement with this expectation, we find that the bond dissociation energy of the isopropyl–hydrogen bond is only 95 kcal, indicating a resonance energy of 9 kcal/mol ($104 - 95$).

For the *tert*-butyl radical there should be nine such hyperconjugation structures. (*Draw them.*) Here we find a bond dissociation energy of 92 kcal, indicating a resonance stabilization of 12 kcal/mol ($104 - 92$).

In summary, the relative stabilities of the free radicals we have studied are determined by delocalization of electrons. Delocalization takes place through overlap of the p orbital occupied by the odd electron: overlap with the π cloud of a double bond in the allyl radical, or overlap with σ bonds in alkyl radicals.

First advanced in 1939 by R. S. Mulliken at the University of Chicago, the idea of hyperconjugation, in some of its applications at least, has aroused considerable controversy (Sec. 10.24). A great deal of research has been done—and is still going on—in an effort to evaluate the importance of hyperconjugative effects.

Problem 10.9   It has been postulated that the relative stabilities of alkyl cations are determined not only by inductive effects but also by resonance stabilization. How might you account for the following order of stability of cations?

tert-butyl > isopropyl > ethyl > methyl

10.12   The allyl cation as a resonance hybrid

Let us turn to heterolytic chemistry, and see how this is affected by the presence of a double bond in the substrate molecule. Since carbocations are key intermediates in much of heterolytic chemistry, let us begin by examining the structure of the *allyl cation*:

$$CH_2=CH-CH_2{}^+$$

Allyl cation
Classical structure

Before we look at the facts, let us see what predictions we might make about this carbocation, using our newly acquired theory of resonance. We begin, as usual, by examining the structure of the molecule. We have drawn the allyl cation as I, but we could just as well have drawn its structure as II. The structures I and II, we

$$[CH_2{=}CH{-}CH_2{}^+ \quad {}^+CH_2{-}CH{=}CH_2] \quad \text{\textit{equivalent to}} \quad CH_2{\cdots}CH{\cdots}CH_2$$

$$\text{I} \qquad\qquad\qquad \text{II} \qquad\qquad\qquad\qquad\qquad\qquad \underset{\oplus}{\underbrace{\phantom{CH_2{\cdots}CH{\cdots}CH_2}}}$$

$$\text{III}$$

now immediately recognize, meet the conditions for resonance: *structures that differ only in the arrangement of electrons.*

According to the resonance theory, neither I nor II adequately represents the cation; it is, instead, a *hybrid* of I and II, and has a structure we might represent as III. Since I and II are exactly equivalent, and hence of exactly the same stability, they make equal contributions to the hybrid. Like the allyl radical, the allyl cation does not contain one carbon–carbon single bond and one carbon–carbon double bond; it contains two identical bonds, each one intermediate between a single and double bond. The positive charge is not located on either terminal carbon atom, but is spread over both.

As was true with the allyl radical, we can get a clearer picture of this molecule by examining the bond orbitals involved. In either of the contributing structures, there is an empty *p* orbital on the electron-deficient carbon. Overlap of this empty *p* orbital with the π cloud of the double bond results in delocalization of the π electrons: each of these two electrons helps to hold together all three carbon nuclei (Fig. 10.3).

$$[CH_2{=}CH{-}CH_2{}^+ \qquad {}^+CH_2{-}CH{=}CH_2] \qquad \text{\textit{equivalent to}} \qquad CH_2{\cdots}CH{\cdots}CH_2$$

$$\underset{\oplus}{\underbrace{\phantom{CH_2{\cdots}CH{\cdots}CH_2}}}$$

Figure 10.3   Allyl cation. The *p* orbital of the middle carbon overlaps *p* orbitals on both sides to permit delocalization of electrons.

Now, on the basis of the structure we have arrived at, what predictions can we make about the properties of the allyl cation? First, since I and II are exactly equivalent, we expect resonance to be *important*, and *to give rise to considerable stabilization of the molecule.*

This prediction is borne out by the facts, as Table 1.3 (p. 22) shows. The heterolytic bond dissociation energy for allyl chloride is 173 kcal, 12 kcal less than for *n*-propyl chloride, and about the same as for isopropyl chloride (170 kcal). Thus, although structure I or II is, formally, that of a primary cation, the allyl cation is about as stable as a secondary cation. We can now expand the sequence of Sec. 5.20.

Stability of
carbocations
$$3° > \begin{matrix}\text{allyl}\\2°\end{matrix} > 1° > CH_3{}^+$$

Next, we expect the allyl cation *to be symmetrical about the central carbon.* Again the facts show that this is so. The allyl cation and symmetrically substituted allylic cations have been prepared under strongly acidic conditions and studied spectroscopically. The infrared spectrum for such a cation is particularly revealing. There are not two absorption bands for carbon–carbon stretching (Sec. 16.18), one for C—C and one for C=C; instead, there is just *one*. This band appears at a frequency intermediate between those characteristic of C—C and of C=C, indicating two equivalent C=C bonds. The intensity of this band—it is the most intense infrared band observed for organic compounds—indicates a system with positive charge located on both terminal carbons.

10.13 Nucleophilic substitution in allylic substrates: S_N1. Reactivity. Allylic rearrangement

So far, our predictions about the properties of the allyl cation have been correct. Now let us see what we might expect of a reaction in which allylic cations are intermediates: nucleophilic substitution of the S_N1 kind.

Consider, for example, the solvolysis of allyl chloride,

$$CH_2=CH-CH_2-Cl \xrightarrow{C_2H_5OH} CH_2=CH-CH_2-OC_2H_5$$

Allyl chloride   Allyl ethyl ether

or the reaction of allyl alcohol with a hydrogen halide.

$$CH_2=CH-CH_2-OH \xrightarrow{HBr} CH_2=CH-CH_2-Br$$

Allyl alcohol   Allyl bromide

As their names indicate, these are *allylic substrates*, since the leaving group is attached to a carbon *next to* a doubly bonded carbon. What is the effect of this double bond on the way these substrates react? Will they behave differently from, say, their saturated analogs, *n*-propyl chloride and *n*-propyl alcohol?

Let us assume for the moment that these reactions proceed by the S_N1 mechanism. According to this mechanism the rate-determining step is heterolysis to give a carbocation; and, as we have seen, it is the nature of this carbocation that

$$R-X \longrightarrow R^+X^- \xrightarrow{:Z} R-Z + X^-$$

Substrate   Carbocation   Product

largely controls the course of reaction. In these cases, since the substrates are allyl substrates, the intermediate cation will be the allyl cation. From what we have just

$$CH_2=CH-CH_2-X \longrightarrow CH_2=CH-CH_2{}^+X^- \xrightarrow{:Z} CH_2=CH-CH_2-Z + X^-$$

Allyl substrate   Allyl cation   Allyl product

learned about the allyl cation, what can we predict about these S_N1 reactions?

In S_N1 it is the rate of formation of the carbocation that determines the overall rate of reaction. So far, we have found that the rate of formation of carbocations—whether in nucleophilic substitution, elimination, or addition—parallels their stability. The allyl cation, we concluded in the preceding section, is about as stable as a secondary cation. We expect, therefore, that *allyl substrates will react about as fast by S_N1 as secondary substrates.*

Once more our prediction is correct. Solvolysis of allyl substrates (allyl chloride, for example, or allyl tosylate), which appears to follow an S_N1 mechanism, takes place roughly as fast as the corresponding reaction of secondary substrates. (Often, it is somewhat faster.) Secondary substrates, we have seen, react *much* faster by S_N1 than primary substrates. (*Remember*: It is very likely that all or nearly all of the rate measured for solvolysis of primary substrates (Sec. 5.22) is for reaction by S_N2.) In nucleophilic substitution, then, the allyl cation—like the secondary cation—is formed perhaps a million times as fast as its saturated analog, the *n*-propyl cation. We can now expand our sequence of Sec. 5.22.

Rate of formation
of carbocations
$$3° > \frac{allyl}{2°} > 1° > CH_3{}^+$$

The reactivity we have just been discussing is that of substrates containing the simple allyl group itself, $CH_2{=}CH{-}CH_2{-}$. The presence of alkyl substituents—*at either end* of the allylic system—increases the reactivity still further.

Problem 10.10 For solvolysis of several allylic chlorides in formic acid (HCOOH) containing a small amount of water, the following relative rates have been measured:

$CH_2{=}CHCH_2Cl$	$CH_2{=}CHCHCl$ $\overset{\displaystyle CH_3}{\vert}$	$CH_2{=}CCH_2Cl$ $\overset{\displaystyle CH_3}{\vert}$	$CH{=}CHCH_2Cl$ $\overset{\displaystyle CH_3}{\vert}$
I	II	III	IV
1.0	5670	0.5	3550

Drawing all pertinent structures, account for the fact that, while the methyl group in III actually deactivates slightly, the methyl group in IV (farther from the center of reaction) activates powerfully, and nearly as much as the methyl group in II.

Let us make one more prediction: we expect that *S_N1 reactions of allylic substrates can show allylic rearrangement.* In the second step of S_N1, the combining of the carbocation with the nucleophile should take place at either terminal carbon of the allylic system and thus, if the structure permits, give two different products.

What are the facts? Consider the conversion of the isomeric allylic chlorides IV and V into the ethyl ethers, VI and VII. In a concentrated solution of sodium ethoxide in ethanol, each chloride reacts by second-order kinetics; IV gives exclusively VI, and V gives exclusively VII. With a high concentration of a strong nucleophile, reaction is straightforward S_N2; the incoming nucleophile attaches itself to the same carbon that is losing the chloride ion.

$$CH_3CH{=}CHCH_2 \quad \xrightarrow[C_2H_5OH]{C_2H_5O^-} \quad CH_3CH{=}CHCH_2$$
$$\quad\quad\quad\;\; | \quad\quad\quad\quad\quad\quad\quad\quad\quad\quad\quad\quad\quad | $$
$$\quad\quad\quad Cl \quad\quad\quad\quad\quad\quad\quad\quad\quad\quad\quad\quad OC_2H_5$$

IV

1-Chloro-2-butene

VI

1-Ethoxy-2-butene

Only product

$$CH_3CHCH{=}CH_2 \quad \xrightarrow[C_2H_5OH]{C_2H_5O^-} \quad CH_3CHCH{=}CH_2$$
$$\quad\quad | \quad\quad\quad\quad\quad\quad\quad\quad\quad\quad\quad\quad\quad | $$
$$\quad\quad Cl \quad\quad\quad\quad\quad\quad\quad\quad\quad\quad\quad\quad OC_2H_5$$

V

3-Chloro-1-butene

VII

3-Ethoxy-1-butene

Only product

If, now, the same chlorides are heated in ethanol in the absence of added base, the course of reaction changes dramatically. Whichever allylic substrate one starts with, *both* ethers are found in the product. Under solvolytic conditions, reaction

$$\textit{Either} \quad CH_3CH{=}CHCH_2 \quad \textit{or} \quad CH_3CHCH{=}CH_2 \quad \xrightarrow{C_2H_5OH}$$
$$\quad\quad\quad\quad\quad\quad\quad | \quad\quad\quad\quad\quad\quad\quad\quad | $$
$$\quad\quad\quad\quad\quad\quad\quad Cl \quad\quad\quad\quad\quad\quad\quad Cl$$

$$CH_3CH{=}CH{-}CH_2 \quad \textit{and} \quad CH_3CHCH{=}CH_2$$
$$\quad\quad\quad\quad\quad\quad\quad\quad | \quad\quad\quad\quad\quad\quad\quad\quad\quad | $$
$$\quad\quad\quad\quad\quad\quad\quad\quad OC_2H_5 \quad\quad\quad\quad\quad\quad OC_2H_5$$

shifts to the S_N1 mechanism; and, as we predicted, allylic rearrangement takes place.

In the example above, we have shown the isomeric chlorides as forming the same hybrid allylic cation. If that were exactly the case, we would expect identical mixtures of products from the two substrates. That is *not true*: not here or, indeed, in most cases of allylic rearrangement. Although both chlorides give mixtures of the same isomeric products, the exact composition of the mixture—the *proportions* of the two products—differs somewhat, and depends upon which chloride it is obtained from.

If we think about this, we find that it is not, after all, surprising. The cations— *as formed by heterolysis*—are *not* identical, and for two reasons. First, there is evidence that the intermediates are ion pairs. (Indeed, some of the earliest and strongest evidence for the intermediacy of ion pairs in solvolysis came from the study of just such allylic systems.) These ion pairs will, of course, not be identical: the exact structure of the ion pair—the location of the chloride ion—will depend upon which carbon the chloride ion has just left, C–1 or C–3.

Second, it is likely that, like secondary cations, allylic cations are formed with nucleophilic assistance (Sec. 6.9). This, too, should differ depending upon the substrate, and the structure of the two nucleophilically solvated cations will differ. (It may even be that a fraction of the reaction occurs by a full-fledged S_N2 mechanism.)

Problem 10.11 On treatment with HCl, 1,3-butadiene yields a mixture of 3-chloro-1-butene and 1-chloro-2-butene. How do you account for the formation of these two products?

10.14   Stabilization of carbocations: the resonance effect

In our introduction to carbocations (Sec. 5.19) we spoke of their relative stabilities as being the property of greatest importance to our understanding of their chemistry. Since then we have seen the remarkable parallel between the stability of carbocations and the stability of transition states leading to their formation—transition states of reactions of many different kinds.

The stabilization of a carbocation, we said (Sec. 5.21), depends upon *dispersal of charge*. This, in turn, depends upon how well the electron-deficient carbon can get electrons from elsewhere in the molecule. One way in which this can happen, we saw, was through the *inductive effect* of a substituent: electron release acting through the molecular framework or through space, and steadily weakening with increasing distance between the substituent and the reaction center. (Electron withdrawal, of course, has the opposite effect: it intensifies the charge and destabilizes the carbocation.)

Now we have encountered a second way in which charge can be dispersed: a **resonance effect** (or *conjugative effect*), due to overlap between certain orbitals. Unlike the inductive effect, the resonance effect does not vary in strength in a gradual way depending upon distance. It is an all-or-nothing effect, which depends upon a specific relationship between the interacting atoms: the relationship we have called *conjugation*.

We said earlier (Sec. 5.18) that two features of a carbocation's structure are intimately involved in determining its stability: the *p* orbital, even though formally empty; and the flatness about the electron-deficient carbon. We see now how this comes about. In a conjugated carbocation, the empty *p* orbital is available for the overlap that provides electrons to the electron-deficient carbon; and the flatness makes this overlap geometrically possible.

The electron-deficient carbon, we shall find, can be conjugated with atoms or groups other than a simple carbon–carbon double bond: most notably with an *aryl* group in what are called *benzylic* cations (Sec. 15.17). The empty *p* orbital can overlap orbitals other than π orbitals: unshared electrons on a properly located atom (Sec. 10.15); even, perhaps, σ orbitals of carbon–hydrogen bonds (Sec. 10.16). And, in all these cases, the resulting delocalization of electrons and dispersal of charge results in stabilization of the carbocation.

But in some reactions it is not a positive but a negative charge that develops. The most stable, most easily formed, and most important of these anionic compounds, we shall find, are conjugated; and they owe their stability and, ultimately, their importance to dispersal of their charge through resonance.

Delocalization of electrons through resonance is the most powerful of the polar factors affecting the stability of charged molecules, positive or negative, and as such plays a leading part in determining orientation and reactivity in a wide variety of organic reactions, and even the course of reaction itself.

10.15   Stabilization of carbocations: the role of unshared pairs

Let us look at an example of conjugation of an electron-deficient carbon with an atom bearing unshared pairs of electrons.

The chloroether, methoxymethyl chloride, undergoes solvolysis (evidently by the S_N1 mechanism) more than 10^{14} times as fast as methyl chloride—faster, even,

$$CH_3-O-CH_2-Cl \longrightarrow CH_3-O-CH_2^{\oplus} + Cl^-$$

Methoxymethyl chloride

$$\textvisiblespace\longrightarrow \text{products}$$

than simple alkyl chlorides of any class. How are we to account for the enormous rate enhancement brought about by the CH_3O- group?

As always with S_N1 reactions, let us look at the structure of the intermediate carbocation. To the electron-deficient carbon is attached an oxygen atom. Oxygen is electronegative and, like halogen, should exert an electron-withdrawing inductive effect—an effect that, as we have seen (Sec. 5.25), tends to destabilize a carbocation. Yet here, if the parallel between rate of formation and stability holds, we have evidence of powerful *stabilization*.

The parallel *does* hold: measurements have shown the methoxymethyl cation to be 76 kcal/mol more stable than the methyl cation—more stable, even, than *tert*-butyl. Can oxygen, then, release electrons? The answer is *yes, by its resonance effect*.

Although electronegative, the oxygen of the CH_3O- group is basic; it has unshared pairs of electrons that it tends to share, thus acquiring a positive charge. Just as water accepts a proton to form the hydronium (oxonium) ion,

$$H_2\overset{..}{O} + H^+ \longrightarrow H_3O^+$$

so alcohols and ethers, as we have found, accept protons to form substituted oxonium ions.

$$R\overset{..}{O}H + H^+ \longrightarrow ROH_2^+ \qquad R\overset{..}{O}R + H^+ \longrightarrow R\overset{\overset{\displaystyle H}{|}}{O}R^+$$

The effects of properly placed oxygen on the stability of carbocations—here, and in other kinds of reactions—can be accounted for by assuming that oxygen can share more than one pair of electrons with electron-deficient carbon and can accommodate a positive charge. Fundamentally, it is the *basicity* of oxygen that is involved.

With that background, let us return to the structure of the methoxymethyl cation. We have written its structure as I, but we could just as well have written it

$$CH_3-\overset{..}{\underset{..}{O}}-\overset{\overset{\displaystyle H}{|}}{\underset{\oplus}{C}}-H \qquad\qquad CH_3-\overset{..}{O}=\overset{\overset{\displaystyle H}{|}}{\underset{\oplus}{C}}-H$$

I   II

Especially stable:
every atom has octet

as II. Again the conditions for resonance have been met: two structures that differ only in the arrangement of electrons.

Of the two structures, we would expect II to be by far the more stable, since in it *every atom* (except hydrogen, of course) *has a complete octet of electrons*. By sharing *two* pairs of electrons with carbon—and thus acquiring the positive charge itself—oxygen has provided carbon with the electrons needed to complete its octet. Structure II is so much more stable than I that, by itself, it must pretty well represent the structure of the cation. The cation is thus hardly a *carbo*cation at all, but an *oxonium ion*. This oxonium ion formed from methoxymethyl chloride is enormously more stable than the carbocation that would be formed from a simple alkyl chloride, and it is formed at a vastly faster rate.

(Compare, for example, the structures of H_3O^+ and CH_3^+. Here it is not a matter of which atom, oxygen or carbon, can better accommodate a positive charge; it is a matter of complete *vs.* incomplete octets.)

This is a conjugated system: the double bond is formed by overlap of the empty *p* orbital of carbon with a filled *p* orbital of oxygen. The electron-deficient carbon of a carbocation can be conjugated with an unshared pair on atoms other than oxygen: nitrogen, sulfur, even halogen. There, too, we shall find, the resulting

$$\overset{\oplus}{-\overset{|}{C}}-\overset{..}{A} \qquad \qquad -\overset{|}{C}=\overset{\oplus}{A}$$

Conjugation with an unshared pair

stabilization of the carbocation can have spectacular effects on the rate, not just of heterolysis, but of reactions of many other types.

In the case of the oxonium ion, then, electron release through the resonance effect is clearly much more powerful than electron withdrawal through the inductive effect, and controls reactivity. Now let us look at a situation where the two factors are more closely balanced. To do this, let us turn to the formation of a carbocation in a reaction of a different type, electrophilic addition, and examine the effect of a different element, chlorine. Let us take as our example the addition of hydrogen iodide to vinyl chloride, and consider both reactivity and orientation in this reaction. The facts are these: the addition takes place more slowly than to ethylene itself, and yields 1-chloro-1-iodoethane.

$$CH_2{=}CH{-}Cl + HI \longrightarrow CH_2{-}CH{-}Cl$$

Vinyl chloride

Less reactive than ethylene

    H   I

1-Chloro-1-iodoethane

Markovnikov addition

Let us examine first the matter of *reactivity*. Electrophilic addition, we have seen (Sec. 8.9), is a two-step process. The first, slow step is formation of a carbocation; the stability of this carbocation determines how fast it is formed and hence how fast addition occurs (Sec. 8.12). Addition to ethylene gives the ethyl cation, III; addition to vinyl chloride gives the 1-chloroethyl cation, IV. The

$$CH_2{=}CH_2 + HI \longrightarrow CH_2{-}CH_2^{\oplus} + I^-$$

    H

III

$$CH_2{=}CH{-}Cl + HI \longrightarrow CH_2{-}\overset{\oplus}{CH}{-}Cl + I^-$$

    H

IV

electron-withdrawing inductive effect of chlorine intensifies the positive charge in IV, makes the cation less stable, and causes a slower reaction.

So far, so good. But now let us turn to the *orientation* of this reaction. Orientation in electrophilic addition, we have seen (Sec. 8.12), is determined by which of two possible carbocations is formed in the first step; and, again, the more stable carbocation is formed faster. Addition to vinyl chloride could involve either of two cations, IV or V. The product actually obtained, 1-chloro-1-iodoethane,

shows that IV is formed preferentially and hence, presumably, is the more stable. Yet in IV the positive charge is located on C–1, the position closest to the chlorine and where we would expect the inductive effect to be strongest and most destabilizing. How are we to account for this puzzling orientation?

The answer is found in conjugation: conjugation between the electron-deficient carbon and an unshared pair of electrons on chlorine. The cation actually formed is not adequately represented by structure IV; it is a hybrid of IV with structure VI,

in which carbon and chlorine are joined by a double bond, and chlorine carries the positive charge. Like the oxonium ion we discussed earlier, structure VI is comparatively stable because in it every atom (except hydrogen) has a *complete octet*. In the alternative cation V, with the positive charge on C–2, there is no comparable conjugation, and no possible contribution from a structure like VI. To the extent, then, that structure VI contributes to the hybrid, it makes the cation with the charge on C–1 the more stable one.

Through its inductive effect chlorine tends to withdraw electrons and thus to destabilize the intermediate carbocation. This effect is felt at both carbons, but more strongly at C–1. Through its resonance effect chlorine tends to release electrons and thus to stabilize the intermediate carbocation. This electron release is effective *only* at C–1.

The inductive effect is stronger than the resonance effect and causes net electron withdrawal and hence deactivation of the molecule relative to unsubstituted ethylene. The resonance effect tends to oppose the inductive effect for formation of the cation with charge on C–1, and hence makes this less destabilized than the alternative cation.

Reactivity is thus controlled by the stronger inductive effect, and orientation is controlled by the resonance effect, which, although weaker, is more selective.

(In Sec. 14.19 we shall find exactly the same interplay of inductive and resonance effects determining reactivity and orientation in a reaction—electrophilic aromatic substitution—which on the surface seems to be quite different from this one, but which is basically quite similar.)

To account for the stereochemistry of addition of halogens to alkenes, we saw (Secs. 8.14 and 9.6), an intermediate halonium ion has been proposed: halogen is pictured as sharing two pairs of electrons and acquiring a positive charge. Despite the high electronegativity of halogen, such an intermediate is—on the basis of the stereochemical evidence—more stable than the open cation. In the resonance effect we have just described, as in the formation of cyclic halonium ions, relative stability is a matter, not of the electronegativity of the atom carrying the positive charge, but of completeness vs. incompleteness of octets.

10.16   Resonance stabilization of alkyl cations: hyperconjugation

Let us look at one more place where the electron-deficient carbon of a carbocation may get electrons: carbon–hydrogen bonds. It has been proposed that, like the p orbital of a free radical, the empty p orbital of a carbocation can overlap σ orbitals of alkyl groups to which it is attached (Fig. 10.4).

(a)   (b)

Figure 10.4 Hyperconjugation in a carbocation. (a) Separate σ and p orbitals. (b) Overlapping orbitals.

As we discussed for free radicals, this kind of overlap permits individual electrons to help bind together three nuclei, two carbons and one hydrogen. This kind of delocalization we recognize as hyperconjugation.

In resonance language, the ethyl cation, for example, would be described as a hybrid of not only structure I, but also the three structures II, III, and IV, in which a double bond joins the two carbons, and the positive charge is carried by a

$$
\begin{array}{cccc}
\begin{array}{c} H\ \ H \\ |\ \ \ | \\ H-C-C^+ \\ |\ \ \ | \\ H\ \ H \end{array}
&
\begin{array}{c} H\ \ H \\ |\ \ \ | \\ H^+\ C{=}C \\ |\ \ \ | \\ H\ \ H \end{array}
&
\begin{array}{c} H^+ H \\ |\ \ \ | \\ H-C{=}C \\ |\ \ \ | \\ H\ \ H \end{array}
&
\begin{array}{c} H\ \ H \\ |\ \ \ | \\ H-C{=}C \\ |\ \ \ | \\ H^+ H \end{array}
\\
I & II & III & IV
\end{array}
$$

hydrogen. To the extent that the carbon–carbon bond has acquired double-bond character, the electron-deficient carbon has gained electrons, and the positive charge is dispersed over the three hydrogens.

With increased branching of the carbocation, dispersal of charge and the resulting stabilization is greater: dispersal over six hydrogens for the isopropyl cation and nine for the *tert*-butyl cation.

Earlier (Sec. 5.21), we described electron release by alkyl groups as an inductive effect; here, we see it as a resonance effect. Despite a great deal of work on the problem, the relative importance of these two factors is not clear. One frequently finds the two lumped together as the "inductive–hyperconjugative" effect of alkyl groups. In this book we may often refer to the inductive effect of alkyl groups, but it should be understood that this may well include a contribution from hyperconjugation.

10.17　Nucleophilic substitution in allylic substrates: S_N2

Let us return to the behavior of allyl substrates in nucleophilic substitution. In reactions by S_N1, we saw, they are about as reactive as saturated secondary substrates. We attributed this to dispersal, through resonance, of the positive charge developing in the transition state of the rate-determining step.

Now, in nucleophilic substitution by S_N2, it has been found, allyl substrates are roughly as reactive as saturated primary substrates. (Often the allyl substrates are somewhat—up to 40 times or so—more reactive.) This, too, is understandable: the chief factor governing S_N2 reactivity is steric hindrance, and the allyl group is about as bulky as an unbranched primary group.

In a sense, the allyl group has the best of both worlds: a capacity for charge dispersal comparable to that of a secondary group, but without the bulkiness that would hinder direct nucleophilic attack.

Problem 10.12　In solvolysis, we saw (Problem 10.10, p. 383), 3-chloro-1-butene is some 5600 times as reactive as allyl chloride. In the second-order reaction with sodium ethoxide in ethanol, by contrast, it is only about one-twentieth as reactive as allyl chloride. How do you account for this dramatic switch in relative reactivities?

10.18　Nucleophilic substitution in vinylic substrates

Let us continue our examination of the effect of the double bond on nucleophilic substitution, and look at substrates in which the leaving group is attached to one of the doubly bonded carbons, that is, *vinylic substrates*.

$$-\overset{|}{C}=\overset{|}{C}-W \ + \ :Z \ \longrightarrow \ -\overset{|}{C}=\overset{|}{C}-Z \ + \ :W$$

A vinylic substrate

We have seen (Sec. 5.26) that an alkyl halide is conveniently detected by the precipitation of insoluble silver halide when it is warmed with alcoholic silver nitrate. This reaction is an example of nucleophilic substitution—solvolysis—with the silver ion lending assistance by pulling away halide ion. The reaction occurs instantaneously with tertiary, allylic, and benzylic (Sec. 15.18) bromides, and within five minutes or so with primary and secondary bromides.

Vinylic halides (or aryl halides, Sec. 29.5), however, do *not* yield silver halide under these conditions. Vinyl bromide can be heated with alcoholic $AgNO_3$ for days without AgBr being detected. Toward nucleophilic substitution in general vinylic halides are *very much less reactive* than their saturated counterparts. They are not ordinarily used in the array of syntheses (Sec. 5.8) based upon nucleophilic substitution reactions of alkyl halides.

How are we to account for this low reactivity of vinylic halides? Fundamental to the understanding of these compounds is the fact that they contain *an unusually strong carbon–halogen bond*. Table 1.3 (p. 22) shows that the heterolytic bond dissociation energy for vinyl chloride is 207 kcal, as compared with 191 kcal for ethyl chloride and 227 kcal for methyl chloride. Values for the fluorides, bromides, and iodides show similar differences. It takes 16 to 18 kcal more energy to break the carbon–halogen bond in a vinyl halide than in the corresponding ethyl halide. Except for the bond in methyl halides, this is the strongest carbon–halogen bond we have so far encountered.

Two factors have been proposed as being responsible for the unusual strength of the vinyl–halogen bond: (a) it results from overlap with an sp^2 orbital of carbon rather than the sp^3 orbital of saturated carbon (Secs. 7.4 and 10.3); (b) it has, through resonance, some double-bond character. We shall discuss this matter in more detail later (Sec. 29.6). For now, we can say this much: it seems quite likely that both factors are at work.

It also seems quite likely that these same two factors strengthen the bond between doubly bonded carbon and the oxygen of such leaving groups as tosylates. In any case, vinylic substrates containing other commonly used leaving groups share in the low reactivity of vinylic halides. Thus, (*E*)-2-buten-2-yl tosylate reacts with aqueous methanol only *one-millionth as fast* as *sec*-butyl tosylate.

$$\begin{array}{c} \mathrm{CH_3} \qquad\quad \mathrm{CH_3} \\ \diagdown\quad\;\; \diagup \\ \mathrm{C}\!=\!\mathrm{C} \\ \diagup\quad\;\; \diagdown \\ \mathrm{H}\qquad\quad \mathrm{OTs} \end{array}$$

(*E*)-2-Buten-2-yl tosylate

The fact is, then, the vinyl–halogen bond is a very strong one. Now, whether nucleophilic substitution takes place by S_N2 or S_N1, the rate-determining step involves breaking of the carbon–halogen bond. The stronger bond in vinyl halides is harder to break, and reaction is slower. With this as a starting point, let us examine these reactions more closely.

10.19   Nucleophilic substitution in vinylic substrates: vinylic cations

As we shall shortly find, vinylic substrates of certain special kinds do undergo nucleophilic substitution by S_N1—but not, evidently, by S_N2. Let us therefore consider first substitution of the S_N1 kind. Such reactions of vinylic substrates would, of course, involve *vinylic cations*.

$$\begin{array}{ccc} \diagdown\quad\;\; \diagup & & \diagdown\quad\;\; \overset{\oplus}{} \\ \mathrm{C}\!=\!\mathrm{C} & \longrightarrow & \mathrm{C}\!=\!\mathrm{C}\!- \\ \diagup\quad\;\; \diagdown & & \diagup \\ & \mathrm{W} & \end{array}$$

A vinylic substrate   A vinylic cation

The extremely low reactivity of vinylic halides by S_N1 indicates that, from these substrates, vinylic cations are formed extremely slowly. In view of the parallel we have so far observed between the stability of carbocations and the rate of their formation, we ask: are vinylic cations relatively unstable?

We have just seen evidence that this is so. If it takes 16–18 kcal more energy to make vinyl cations from vinyl halides than to make ethyl cations from ethyl halides, then, by definition, vinyl cations are less stable than ethyl cations—*relative to the halide from which each is formed*. In the same way, vinyl cations are more stable than methyl cations by 16–20 kcal. We can therefore expand our sequence of Sec. 10.12.

Stability of
carbocations
$$3° > \frac{allyl}{2°} > 1° > vinyl > CH_3{}^+$$

(It must be emphasized that, until we find differently, the position of the vinyl cation in this sequence applies only to its stability *relative to substrates for heterolysis*. We shall see the importance of this limitation in Sec. 11.9.)

How do we account for this low stability of the vinyl cation? Let us remind ourselves of how we approach this kind of problem. The measure of stability of a carbocation is simply the amount of energy it takes to generate it from whatever substrate we have selected. This amount of energy varies from cation to cation. So much is fact.

Now the intellectual process starts. We try to explain these variations—to find the broader pattern into which they fit. We look for factors that will account for these variations and at the same time be consistent with the rest of the structural theory. So far in our mental analysis of carbocation stability, we have focused our attention on differences between the cations themselves. For example, the energy climb from *tert*-butyl chloride to the *tert*-butyl cation is less than from isopropyl chloride to the isopropyl cation because, we say, the energy level of the *tert*-butyl cation is *lowered* by dispersal of charge. There is no particular factor that we can see to make *tert*-butyl chloride very different from isopropyl chloride; but there *is* a factor to make the cations different from each other: charge dispersal. And this approach has worked very well.

But in the generation of vinyl cations, the situation is different. First, as we shall see in Sec. 11.9, vinyl cations *can* be generated without undue difficulty by a reaction other than heterolysis. Second, it is clear that vinyl halides differ considerably from saturated alkyl halides of all classes. Despite the differences in stability among 1°, 2°, and 3° cations, the carbon–halogen bonds in the parent halides are of almost identical length. In vinyl halides, by contrast, the carbon–halogen bonds are significantly shorter: consistent with the picture of a tighter, stronger bond due to overlap of an sp^2 orbital of carbon and to partial double-bond character (Sec. 29.6).

Our interpretation, then, is this. When starting from the organic halides, the energy climb to the vinyl cation is greater than the climb to the ethyl cation, not because factors are at work raising the energy level of the vinyl cation, but because factors are at work *lowering* the energy level of the vinyl halide.

This is an illustration of a fundamental chemical principle. In interpreting differences in reactivity we tend to look at factors that stabilize one transition state more than another—generally reflecting the product character it possesses—and thus decrease E_{act}; and, more often than not, this approach works well. But we must always have an eye open for factors that might stabilize one reactant more than another and thus increase E_{act}.

What we have said so far has had to do with the difficulty of generating vinylic cations by heterolysis. Not surprisingly, this difficulty has been taken as a challenge to the organic chemist, and, in work done mostly since about 1970, vinylic cations have emerged as accessible intermediates with fascinating properties. Many people from many countries have been involved in this research, among them being Michael Hanack (University of Tübingen), Zvi Rappoport (Hebrew University of Jerusalem), Giorgio Modena (University of Padua), and Peter Stang (University of Utah).

Vinylic cations can readily be made through solvolysis of the S_N1 kind if two conditions are met: (a) the leaving group is an *extremely* good one; and (b) the vinylic group contains electron-releasing substituents.

Most commonly used for this purpose is the "super" leaving group, trifluoro-methanesulfonate, $-OSO_2CF_3$, known as *triflate*.

A vinylic triflate   A vinylic cation   Triflate anion

Trifluoromethanesulfonyl

The powerfully electron-withdrawing fluorine atoms (through dispersal of the *negative* charge) help to stabilize the triflate anion, $CF_3SO_2O^-$, and make the parent acid CF_3SO_2OH one of the strongest Lowry–Brønsted acids known—much stronger than the familiar H_2SO_4 or $HClO_4$. The triflate anion is, correspondingly, an extremely weak base, and one of the best leaving groups in organic chemistry. Towards solvolysis, saturated alkyl triflates have been found to be 10 000 to 100 000 times as reactive as the corresponding tosylates, and as much as a *billion times as reactive* as the chlorides or bromides!

The electron-releasing substituents in the vinylic moiety are very commonly aryl groups (Sec. 15.18), but alkyl groups are sufficient to allow reaction by S_N1. For example:

3-Methyl-2-buten-2-yl triflate   3-Methyl-2-buten-2-yl trifluoroethyl ether

We cannot take more time here to discuss the properties of vinylic cations, except to say that, like saturated alkyl cations, they have a rich and varied chemistry. They can be made from different kinds of substrates in different kinds of reactions; they can lead to elimination as well as substitution; they can rearrange. We shall encounter them again in Sec. 11.9.

Perhaps the most important lesson we can learn from all this is not the chemistry of vinyl cations as such—interesting as it is—but that a problem was solved in a logical way by recognizing straightforward principles that we have learned: the importance to the heterolytic process of (a) a good—weakly basic—leaving group and (b) electron release in the cation being formed.

Problem 10.13 On treatment with CF_3CH_2OH, the triflate I gives not only II, but also III, IV, and V. (a) How do you account for the formation of III and IV? (b) The formation of V?

$$(CH_3)_2CHC{=}CH_2$$
$$|$$
$$OTf$$

I

$$(CH_3)_2CHC{=}CH_2$$
$$|$$
$$OCH_2CF_3$$

II

$$(CH_3)_2CHC{\equiv}CH$$

III

$$\begin{array}{c}CH_3\\|\\CH_3C{=}C{=}CH_2\end{array}$$

IV

$$\begin{array}{c}CH_3\\|\\CH_3CCH{=}CH_2\\|\\OCH_2CF_3\end{array}$$

V

What we have just discussed is unimolecular nucleophilic substitution, S_N1. Reaction by S_N2, we indicated earlier, does not appear to happen with vinylic substrates. This lower reactivity has been attributed to steric hindrance to backside attack by the nucleophile: hindrance due to the π cloud.

If strong nucleophiles are used, and if the vinylic substrate contains electron-withdrawing groups, a bimolecular substitution reaction does occur. But it is not S_N2. With reaction by S_N2 difficult, reaction takes an easier course: it follows a mechanism of a different type which, significantly, *does not* involve breaking of the bond to the leaving group in the rate-determining step. Also significantly, this mechanism closely resembles the kind of nucleophilic substitution characteristic of aryl halides (Chap. 29), another class of substrates with unusually strong carbon–halogen bonds. (See Problem 15, p. 412.)

10.20 Dienes: structure and properties

So far in this chapter, we have discussed the effect of the double bond, acting as a substituent, on certain reactions taking place elsewhere in the molecule: free-radical substitution and nucleophilic substitution. Now let us look at its effect on alkene chemistry. That is, let us look at the effect of a double bond on the chemistry of *another* double bond in the same molecule: on its formation and on the reactions it undergoes.

To do this we shall study chiefly **dienes**, alkenes that contain two carbon–carbon double bonds. What we shall say applies equally well to compounds with more than two double bonds. The double bond in a diene has essentially the same properties as a double bond in the alkenes we have already studied. But in certain of the dienes, these properties are *modified* by the presence of the second double bond; we shall focus our attention on these modifications.

Dienes are divided into three classes according to the arrangement of the double bonds. Double bonds that alternate with single bonds are **conjugated**.

$$-\overset{|}{C}=\overset{|}{C}-\overset{|}{C}=\overset{|}{C}- \qquad -\overset{|}{C}=\overset{|}{C}-\overset{|}{\underset{|}{C}}-\overset{|}{C}=\overset{|}{C}- \qquad -\overset{|}{C}=C=\overset{|}{C}-$$

| Conjugated | Isolated | Cumulated |
| double bonds | double bonds | double bonds |

Double bonds that are separated by more than one single bond are **isolated**. Double bonds that share a carbon are *cumulated*, and the compounds are called **allenes**. For example:

$$CH_2=CH-CH=CH_2 \qquad CH_2=CH-CH_2-CH=CH_2 \qquad CH_3CH=C=CH_2$$

| 1,3-Butadiene | 1,4-Pentadiene | 1,2-Butadiene |
| *A conjugated diene* | | *An allene* |

The chemical properties of a diene depend upon this arrangement of its double bonds. Isolated double bonds exert little effect on each other, and hence each reacts as though it were the only double bond in the molecule. Except for the consumption of larger amounts of reagents, then, the chemical properties of the non-conjugated dienes are identical with those of the simple alkenes. Allenes are of increasing interest to organic chemists, but we shall have time to do very little with them.

We shall concentrate our attention on conjugated dienes. They differ from simple alkenes in four ways: (a) they are *more stable*; (b) they are the *preferred products of elimination*; (c) they undergo *1,4-addition*, both electrophilic and free-radical; and (d) toward free-radical addition, they are *more reactive*.

10.21 Stability of conjugated dienes

If we look closely at Table 8.1 (p. 304), we find that the heats of hydrogenation of alkenes having similar structures are remarkably constant. For monosubstituted alkenes ($RCH=CH_2$) the values are very close to 30 kcal/mol; for disubstituted alkenes ($R_2C=CH_2$ or $RCH=CHR$), 28 kcal/mol; and for trisubstituted alkenes ($R_2C=CHR$), 27 kcal/mol. For a compound containing more than one double bond we might expect a heat of hydrogenation that is the sum of the heats of hydrogenation of the individual double bonds.

For non-conjugated dienes this additive relationship is found to hold. As shown in Table 10.1, 1,4-pentadiene and 1,5-hexadiene, for example, have heats of hydrogenation very close to 2×30 kcal, or 60 kcal/mol.

Table 10.1   HEATS OF HYDROGENATION OF DIENES

Diene	ΔH of hydrogenation, kcal/mol
1,4-Pentadiene	60.8
1,5-Hexadiene	60.5
1,3-Butadiene	57.1
1,3-Pentadiene	54.1
2-Methyl-1,3-butadiene (Isoprene)	53.4
2,3-Dimethyl-1,3-butadiene	53.9
1,2-Propadiene (Allene)	71.3

For conjugated dienes, however, the measured values are slightly lower than expected. For 1,3-butadiene we might expect 2×30, or 60 kcal: the actual value, 57 kcal, is 3 kcal lower. In the same way the values for 1,3-pentadiene and 2,3-dimethyl-1,3-butadiene are also below the expected values by 2–4 kcal.

Heats of hydrogenation

$$CH_2\!=\!CH\!-\!CH\!=\!CH_2 \qquad\qquad CH_3\!-\!CH\!=\!CH\!-\!CH\!=\!CH_2$$

Expected: $30 + 30 = 60$ kcal   *Expected:* $28 + 30 = 58$ kcal

Observed: 57   *Observed:* 54

$$\overset{\displaystyle CH_3\ \ CH_3}{\underset{\displaystyle |\ \ \ \ \ |}{CH_2\!=\!C\!-\!\!-\!C\!=\!CH_2}}$$

Expected: $28 + 28 = 56$ kcal

Observed: 54

What do these heats of hydrogenation tell us about the conjugated dienes? Using the approach of Sec. 8.4, let us compare, for example, 1,3-pentadiene (heat of hydrogenation, 54 kcal) and 1,4-pentadiene (heat of hydrogenation, 61 kcal). They both consume two moles of hydrogen and yield the same product, *n*-pentane. If 1,3-pentadiene *evolves* less energy than 1,4-pentadiene, it can only mean that it *contains* less energy; that is to say, the conjugated 1,3-pentadiene is more stable than the non-conjugated 1,4-pentadiene.

In the next three sections we shall see how two different factors have been invoked to account for the relative stabilities of conjugated dienes, and of simple alkenes as well: (a) delocalization of π electrons, and (b) strengthening of σ bonds through changes in hybridization of carbon.

Problem 10.14 (a) Predict the heat of hydrogenation of *allene*, $CH_2\!=\!C\!=\!CH_2$. (b) The actual value is 71 kcal. What can you say about the stability of a *cumulated* diene?

10.22   Resonance in conjugated dienes

Let us focus our attention on the four key carbon atoms of any conjugated diene system. We ordinarily write the C(1)–C(2) and C(3)–C(4) bonds as double, and the C(2)–C(3) bond as single:

$$\overset{\displaystyle 1\ \ \ 2\ \ \ 3\ \ \ 4}{-\!\underset{|}{C}\!=\!\underset{|}{C}\!-\!\underset{|}{C}\!=\!\underset{|}{C}\!-}$$

This would correspond to an orbital picture of the molecule (see Fig. 10.5*a*) in which π bonds are formed by overlap of the *p* orbitals of C–1 and C–2, and overlap of the *p* orbitals of C–3 and C–4.

In the allyl radical and cation we saw that resonance resulted from the overlap of the *p* orbital of a carbon atom with *p* orbitals on *both* sides. We might expect that, in the same way, there could be a certain amount of overlap between the *p* orbitals of C–2 and C–3, as shown in Fig. 10.5*b*. The resulting delocalization of the π electrons makes the molecule more stable: each pair of electrons attracts—and is attracted by—not just two carbon nuclei, but *four*.

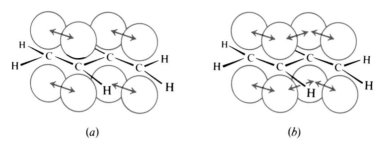

(a)   (b)

Figure 10.5   Conjugated diene. (*a*) Overlap of *p* orbitals to form two double bonds. (*b*) Overlap of *p* orbitals to form conjugated system: delocalization of π electrons.

Using the language of conventional valence-bond structures, we say that a conjugated diene is a resonance hybrid of I and II. The dotted line in II represents

$$ \overset{1}{-}\overset{2}{C}=\overset{3}{C}-\overset{4}{C}=C- \qquad\qquad -\dot{C}-C=C-\dot{C}- $$

I   II

a *formal bond*, and simply means that an electron on C–1 and an electron on C–4 have opposite spins, that is to say, are *paired*.

To the extent that II contributes to the structure, it gives a certain double-bond character to the C(2)–C(3) bond and a certain single-bond character to the C(1)–C(2) and C(3)–C(4) bonds; most important, it makes the molecule more stable than we would expect I (the most stable contributing structure) to be.

Formation of a bond releases energy and stabilizes a system; all other things being equal, the more bonds, the more stable a structure. Consideration of *number of bonds* is one of the criteria (Sec. 10.10) that can be used to estimate relative stability and hence relative importance of a contributing structure. On this basis we would expect II with 10 bonds (the formal bond does not count) to be less stable than I with 11 bonds. The resonance energy for such a hybrid of non-equivalent structures should be less than for a hybrid made up of equivalent structures. The structure of a conjugated diene should resemble I more than II, since the more stable structure I makes the larger contribution to the hybrid.

Consistent with partial double-bond character, the C(2)–C(3) bond in 1,3-butadiene is 1.48 Å long, as compared with 1.53 Å for a pure single bond. The resonance energy of a conjugated diene is only 2–4 kcal/mol, compared with 10 kcal/mol for the allyl radical. (However, for an alternative interpretation, see Sec. 10.24.)

10.23   Resonance in alkenes. Hyperconjugation

Heats of hydrogenation showed us (Sec. 8.4) that alkenes are stabilized not only by conjugation but also by the presence of alkyl groups: *the greater the number of alkyl groups attached to the doubly bonded carbon atoms, the more stable the alkene.*

To take the simplest example, the heat of hydrogenation of propylene is 2.7 kcal lower than that of ethylene, indicating that (relative to the corresponding alkane) propylene is 2.7 kcal more stable than ethylene.

Stabilization by alkyl groups has been attributed to the same fundamental factor as stabilization by a second double bond: *delocalization of electrons*, this time through overlap between a π orbital and a σ orbital of the alkyl group (Fig. 10.6).

Figure 10.6 Hyperconjugation in an alkene: overlap between σ and π orbitals.

Through this overlap, individual electrons can, to an extent, help bind together four nuclei. Delocalization of this kind, involving σ bond orbitals, we recognize as *hyperconjugation* (Sec. 10.11).

Translated into resonance terminology, such hyperconjugation is represented by contribution from structures like II. (As before, the dotted line in II represents

$$
\begin{array}{ccc}
\text{H} & \text{H} & \text{H} \\
| & | & | \\
\text{H}-\text{C}-\text{C}=\text{C}-\text{H} \\
| \\
\text{H} \\
\text{3} \quad \text{2} \quad \text{1}
\end{array}
\qquad\qquad
\begin{array}{ccc}
\text{H} & \text{H} & \text{H} \\
| & | & | \\
\text{H}-\text{C}=\text{C}-\text{C}-\text{H} \\
\text{H}\cdots\cdots\vdots \\
\text{3} \quad \text{2} \quad \text{1}
\end{array}
$$

I

II

and two more equivalent structures

a formal bond, indicating that electrons on the two atoms are paired.) Considered by itself, a structure like II is indeed strange, since there is no real bond joining the hydrogen to carbon. This is, however, simply a rough way of indicating that the carbon–hydrogen bond is something *less* than a single bond, that the C(2)–C(3) bond has some double-bond character, and that the C(1)–C(2) bond has some single-bond character.

Consistent with partial double-bond character, the carbon–carbon "single" bond in propylene is 1.50 Å long, as compared to 1.53 Å for a pure single bond.

The greater the number of alkyl groups attached to the doubly bonded carbons, the greater the number of contributing structures like II, the greater the delocalization of electrons, and the more stable the alkene.

Hyperconjugation of the kind described above is called *sacrificial hyperconjugation*, since there is one less real bond in structures like II than in I. In contrast, the kind of hyperconjugation we encountered in connection with free radicals and carbocations involves no "sacrifice" of a bond and is called *isovalent hyperconjugation*.

10.24 Stability of dienes and alkenes: an alternative interpretation

We have seen that the carbon–hydrogen bond length decreases as we proceed along the series ethane, ethylene, acetylene, and we attributed this to changes in hybridization of carbon (see Table 10.2). As the p character of the bonding orbital decreases, the orbital size decreases, and the bond becomes shorter (Sec. 7.4).

Table 10.2 CARBON–HYDROGEN SINGLE-BOND
LENGTHS AND HYBRIDIZATION

Compound	Length, Å	Hybridization
$CH_3{-}CH_3$	1.112	$sp^3{-}s$
$CH_2{=}CH_2$	1.103	$sp^2{-}s$
$HC{\equiv}CH$	1.079	$sp{-}s$

The carbon–carbon single-bond length also decreases along an analogous series, ethane, propylene, propyne (Table 10.3). We notice that these differences are bigger than for carbon–hydrogen bonds. Here, the bond shortening has been attributed to hyperconjugation, as discussed in Sec. 10.23.

Table 10.3 CARBON–CARBON SINGLE-BOND
LENGTHS AND HYBRIDIZATION

Compound	Length, Å	Hybridization
$CH_3{-}CH_3$	1.53	$sp^3{-}sp^3$
$CH_2{=}CH{-}CH_3$	1.50	$sp^2{-}sp^3$
$HC{\equiv}C{-}CH_3$	1.46	$sp{-}sp^3$

It has been argued, most notably by M. J. S. Dewar of the University of Texas, that there is no need to invoke hyperconjugation in molecules like these, and that the changes in C—C bond length—like the changes in C—H bond length—are due simply to changes in hybridization of carbon.

Furthermore, Dewar has proposed that such shortening of bonds is accompanied by a proportional increase in bond energies (E); that is, shortening a bond makes the molecule more stable. Change in hybridization affects bond lengths more—and hence affects molecular stability more—when carbon–carbon bonds are involved than when carbon–hydrogen bonds are involved. An alkyl substituent stabilizes an alkene, relative to the corresponding alkane, because sp^2 hybridization strengthens a carbon–carbon bond more than a carbon–hydrogen bond.

In a similar way, the unusual stability of conjugated dienes is attributed, not to delocalization of the π electrons, but to the fact that $sp^2{-}sp^2$ hybridization makes the C(2)–C(3) bond short (1.48 Å) and strong.

There is little doubt that both factors, delocalization of π electrons and change in σ bonds, are at work. The question is: what is the relative importance of each? The answer may well turn out to be: *both* are important.

In the case of molecules like the allyl radical, where clearly no single structure is acceptable, Dewar has not questioned the importance of π-electron delocalization, although he considers σ-bond stability to play a larger part than has been recognized. He also accepts a more important role for isovalent hyperconjugation—in free radicals and carbocations—than for the sacrificial hyperconjugation we have so far discussed.

10.25 Ease of formation of conjugated dienes: orientation of elimination

The greater stability of conjugated dienes is reflected in their greater ease of formation. Where possible, they are the preferred diene products of elimination reactions. For example:

$$\underset{\text{3-Bromocyclohexene}}{\underset{\text{CHBr}}{\underset{\text{H}_2\text{C}}{\overset{\text{CH}}{\overset{\text{H}_2\text{C}}{\bigcirc}}}}} \xrightarrow{\text{KOH (alc)}} \underset{\text{1,3-Cyclohexadiene}}{\underset{\text{CH}}{\underset{\text{H}_2\text{C}}{\overset{\text{CH}}{\overset{\text{H}_2\text{C}}{\bigcirc}}}}}$$

The most important diene, **1,3-butadiene** (used to make rubber substitutes, Sec. 10.30), is obtained industrially in very large amounts by the cracking of hydrocarbons:

$$CH_3CH_2CH=CH_2 \quad \text{1-Butene}$$

$$CH_3CH_2CH_2CH_3 \xrightarrow[\text{catalyst}]{\text{heat}}$$

$$n\text{-Butane}$$

$$\xrightarrow[\text{catalyst}]{\text{heat}} CH_2=CH-CH=CH_2$$

$$\text{1,3-Butadiene}$$

$$CH_3CH=CHCH_3 \xrightarrow[\text{catalyst}]{\text{heat}}$$

$$\text{2-Butene}$$

Problem 10.15 Predict the major product of dehydrohalogenation of 4-bromo-1-hexene.

10.26 Electrophilic addition to conjugated dienes. 1,4-Addition

When 1,4-pentadiene is treated with bromine under conditions (what are they?) that favor formation of the *dihalide*, there is obtained the expected product, 4,5-dibromo-1-pentene. Addition of more bromine yields the 1,2,4,5-tetrabromo-

$$CH_2=CH-CH_2-CH=CH_2 \xrightarrow{Br_2}$$

$$\underset{Br \quad\; Br}{CH_2-CH-CH_2-CH=CH_2} \xrightarrow{Br_2} \underset{Br\quad\; Br \qquad\quad Br\quad\; Br}{CH_2-CH-CH_2-CH-CH_2}$$

pentane. This is typical of the behavior of dienes containing isolated double bonds: the double bonds react independently, as though they were in different molecules.

When 1,3-butadiene is treated with bromine under similar conditions, there is obtained not only the expected 3,4-dibromo-1-butene, but also 1,4-dibromo-2-butene. Treatment with HCl yields not only 3-chloro-1-butene, but also 1-chloro-2-butene. Hydrogenation yields not only 1-butene but also 2-butene.

$$\xrightarrow{\text{Br}_2} \quad \underset{\underset{\text{Br} \quad \text{Br}}{|\quad\;\;|}}{CH_2-CH-CH=CH_2} \quad + \quad \underset{\underset{\text{Br} \qquad\qquad \text{Br}}{|\quad\qquad\qquad|}}{CH_2-CH=CH-CH_2}$$

1,2-Addition   1,4-Addition

$$CH_2=CH-CH=CH_2 \xrightarrow{\text{HCl}} \quad \underset{\underset{\text{H} \quad \text{Cl}}{|\quad\;|}}{CH_2-CH-CH=CH_2} \quad + \quad \underset{\underset{\text{H} \qquad\qquad \text{Cl}}{|\qquad\qquad\quad|}}{CH_2-CH=CH-CH_2}$$

1,3-Butadiene   1,2-Addition   1,4-Addition

$$\xrightarrow[\text{cat.}]{\text{H}_2} \quad \underset{\underset{\text{H} \quad \text{H}}{|\quad\;|}}{CH_2-CH-CH=CH_2} \quad + \quad \underset{\underset{\text{H} \qquad\qquad \text{H}}{|\qquad\qquad\;\;|}}{CH_2-CH=CH-CH_2}$$

1,2-Addition   1,4-Addition

Study of many conjugated dienes and many reagents shows that such behavior is typical: *in* **additions to conjugated dienes**, *a reagent may attach itself not only to a pair of adjacent carbons* **(1,2-addition)**, *but also to the carbons at the two ends of the conjugated system* **(1,4-addition)**. Very often the 1,4-addition product is the major one.

$$\underset{1 \quad\;\; 2 \quad\;\; 3 \quad\;\; 4}{-C=C-C=C-} \xrightarrow{\text{YZ}} \underset{\underset{\text{Y} \quad \text{Z}}{|\quad\;|}}{-C-C-C=C-} \quad + \quad \underset{\underset{\text{Y} \qquad\qquad \text{Z}}{|\qquad\qquad\;|}}{-C-C=C-C-}$$

1,2-Addition   1,4-Addition

How can we account for the products that are obtained and, particularly, for the occurrence of 1,4-addition? We have seen (Secs. 8.9 and 8.12) that electrophilic addition is a two-step process, and that the first step takes place in the way that yields the more stable carbocation. Let us apply this principle to the addition, for example, of HCl to 2,4-hexadiene, which yields 4-chloro-2-hexene and 2-chloro-3-hexene:

$$CH_3-CH=CH-CH=CH-CH_3 \xrightarrow{\text{HCl}} \underset{\underset{\text{H} \quad \text{Cl}}{|\quad\;|}}{CH_3-CH-CH-CH=CH-CH_3}$$

2,4-Hexadiene   4-Chloro-2-hexene

$$+ \; \underset{\underset{\text{H} \qquad\qquad\qquad \text{Cl}}{|\qquad\qquad\qquad|}}{CH_3-CH-CH=CH-CH-CH_3}$$

2-Chloro-3-hexene

These products show, first of all, that hydrogen adds to C–2 to yield carbocation I, rather than to C–3 to yield carbocation II:

$$CH_3-CH=CH-CH=CH-CH_3 + H^+ \longrightarrow$$

$$\longrightarrow \; \underset{\underset{\text{H}}{|}}{CH_3-CH-\overset{\oplus}{C}H-CH=CH-CH_3}$$

I

$$\xcancel{\longrightarrow} \; \underset{\underset{\text{H}}{|}}{CH_3-\overset{\oplus}{C}H-CH-CH=CH-CH_3}$$

II

Since both I and II are secondary cations, how can we account for this preference? The answer is, of course, found in the structure of I: it is not simply a secondary cation, but is an *allylic* cation as well, since the carbon bearing the positive charge is attached to a doubly bonded carbon. It is, then, a resonance hybrid:

$$\left[\begin{array}{cc} CH_3\!-\!CH\!-\!CH\!-\!CH\!=\!CH\!-\!CH_3 & CH_3\!-\!CH\!-\!CH\!=\!CH\!-\!CH\!-\!CH_3 \\ \quad\;\; | \quad\; \oplus & \qquad\; | \qquad\qquad\quad \oplus \\ \quad\;\; H & \qquad\; H \\ \qquad\quad I & \qquad\qquad\quad III \end{array} \right]$$

equivalent to

$$CH_3\!-\!CH\!-\!CH\text{---}CH\text{---}CH\!-\!CH_3$$
$$| \qquad\qquad\quad$$
$$H \qquad \underbrace{\qquad\qquad}_{\oplus}$$
$$IV$$

As a cation that is both secondary and allylic, I is more stable than II, and is the preferred cationic intermediate.

The products obtained from addition to conjugated dienes are always consistent with the formation of the most stable intermediate carbocation: an allylic cation. This requires the first step to be *addition to one of the ends* of the conjugated

$$Y\!:\!Z + -\!\overset{|}{C}\!=\!\overset{|}{C}\!-\!\overset{|}{C}\!=\!\overset{|}{C}\!- \;\rightleftharpoons\; -\!\overset{|}{C}\!-\!\overset{|}{C}\text{---}\overset{|}{C}\text{---}\overset{|}{C}\!- \; + \;:\!Z$$

Adds to end of conjugated system $\qquad\qquad\qquad \underset{Y \qquad\; \oplus}{}$

An allylic cation

system. Now, in the second step cation IV combines with chloride ion to form the product. The chloride ion can attach itself to either end of the allylic system and thus yield the 1,2- or 1,4-product.

$$CH_3\!-\!CH\!-\!CH\text{---}CH\text{---}CH\!-\!CH_3 \;\longrightarrow\; CH_3\!-\!CH\!-\!CH\!-\!CH\!=\!CH\!-\!CH_3$$
$$\qquad | \qquad\qquad\qquad\qquad\qquad\qquad\qquad | \quad\; |$$
$$\qquad H \qquad\;\; \oplus \qquad\qquad\qquad\qquad\qquad H \quad Cl$$

IV $\qquad\qquad\qquad\qquad\qquad\qquad\qquad\qquad\qquad$ 1,2-Addition product

$$+ \;\; CH_3\!-\!CH\!-\!CH\!=\!CH\!-\!CH\!-\!CH_3$$
$$\qquad\qquad | \qquad\qquad\qquad |$$
$$\qquad\qquad H \qquad\qquad\qquad Cl$$

1,4-Addition product

Like allylic rearrangement (Sec. 10.13), we see, the occurrence of 1,4-addition is a natural consequence of the hybrid nature of the intermediate allylic cation.

$$Y\!:\!Z$$
$$+$$
$$-\!\overset{|}{\underset{1}{C}}\!=\!\overset{|}{\underset{2}{C}}\!-\!\overset{|}{\underset{3}{C}}\!=\!\overset{|}{\underset{4}{C}}\!- \;\longrightarrow\; -\!\overset{|}{C}\!-\!\overset{|}{C}\text{---}\overset{|}{C}\text{---}\overset{|}{C}\!- \;\longrightarrow\; -\!\overset{|}{C}\!-\!\overset{|}{C}\!-\!\overset{|}{C}\!=\!\overset{|}{C}\!- \; + \; -\!\overset{|}{C}\!-\!\overset{|}{C}\!=\!\overset{|}{C}\!-\!\overset{|}{C}\!-$$

Addition to end of $\qquad\quad$ Allylic cation $\qquad\quad$ 1,2-Addition $\qquad\quad$ 1,4-Addition
conjugated system

Thus the hybrid nature of the allylic cation governs both steps of electrophilic addition to conjugated dienes: the first, through stabilization of the cation; the second, by permitting attachment to either of two carbon atoms.

Problem 10.16   Account for the fact that 2-methyl-1,3-butadiene reacts (a) with HCl to yield only 3-chloro-3-methyl-1-butene and 1-chloro-3-methyl-2-butene; (b) with bromine to yield only 3,4-dibromo-3-methyl-1-butene and 1,4-dibromo-2-methyl-2-butene.

10.27   1,2- *vs.* 1,4-Addition. Rate *vs.* equilibrium

A very important principle emerges when we look at the relative amounts of 1,2- and 1,4-addition products obtained.

Addition of HBr to 1,3-butadiene yields both the 1,2- and the 1,4-products; the *proportions* in which they are obtained are markedly affected by the temperature

$$
\begin{array}{c}
\text{HBr} \\
+ \\
\text{CH}_2\!\!=\!\!\text{CH}\!\!-\!\!\text{CH}\!\!=\!\!\text{CH}_2
\end{array}
$$

$-80\,°C$ (left branch) $40\,°C$ (right branch)

Left (−80 °C):
$$
\left\{
\begin{array}{l}
80\%\ \underset{\underset{\text{H}}{|}}{\text{CH}_2}\!\!-\!\!\underset{\underset{\text{Br}}{|}}{\text{CH}}\!\!-\!\!\text{CH}\!\!=\!\!\text{CH}_2 \\[2mm]
20\%\ \underset{\underset{\text{H}}{|}}{\text{CH}_2}\!\!-\!\!\text{CH}\!\!=\!\!\text{CH}\!\!-\!\!\underset{\underset{\text{Br}}{|}}{\text{CH}_2}
\end{array}
\right.
$$

$\xrightarrow{\ 40\,°C\ }$

Right (40 °C):
$$
\left\{
\begin{array}{l}
20\%\ \underset{\underset{\text{H}}{|}}{\text{CH}_2}\!\!-\!\!\underset{\underset{\text{Br}}{|}}{\text{CH}}\!\!-\!\!\text{CH}\!\!=\!\!\text{CH}_2 \\[2mm]
80\%\ \underset{\underset{\text{H}}{|}}{\text{CH}_2}\!\!-\!\!\text{CH}\!\!=\!\!\text{CH}\!\!-\!\!\underset{\underset{\text{Br}}{|}}{\text{CH}_2}
\end{array}
\right.
$$

at which the reaction is carried out. Reaction at a low temperature ($-80\,°C$) yields a mixture containing 20% of the 1,4-product and 80% of the 1,2-product. Reaction at a higher temperature ($40\,°C$) yields a mixture of quite different composition, 80% 1,4- and 20% 1,2-product. At intermediate temperatures, mixtures of intermediate compositions are obtained. Although each isomer is quite stable at low temperatures, prolonged heating of either the 1,4- or the 1,2-compound yields the same mixture. How are these observations to be interpreted?

The fact that either compound is converted into the same mixture by heating indicates that this mixture is the result of equilibrium between the two compounds. The fact that the 1,4-compound predominates in the equilibrium mixture indicates that it is the more stable of the two.

The fact that more 1,2- than 1,4-product is obtained at $-80\,°C$ indicates that the 1,2-product is formed *faster* than the 1,4-product; since each compound remains unchanged at $-80\,°C$, the proportions in which they are isolated show the proportions in which they were initially formed. As the reaction temperature is raised, the proportions in which the products are initially formed may remain the same, but there is faster conversion of the initially formed products into the equilibrium mixture.

The proportions of products actually isolated from the low-temperature addition are determined by the **rates** of addition, whereas for the high-temperature addition they are determined by the **equilibrium** between the two isomers.

Let us examine the matter of 1,2- and 1,4-addition more closely by drawing a potential energy curve for the reactions involved (Fig. 10.7). The carbocation initially formed reacts to yield the 1,2-product faster than the 1,4-product; consequently, the energy of activation leading to the 1,2-product must be less than that

Figure 10.7 Potential energy changes during progress of reaction: 1,2- *vs.* 1,4-addition.

leading to the 1,4-product. We represent this by the lower hill leading from the cation to the 1,2-product. More collisions have enough energy to climb the low hill than the high hill, so that the 1,2-compound is formed faster than the 1,4-compound. The 1,4-product, however, is more stable than the 1,2-product, and hence we must place its valley at a lower level than that of the 1,2-product.

As we know (Sec. 10.13), allylic halides readily undergo heterolysis, that is, ionization. Now ionization of either bromo compound yields the same carbocation; the most likely—and simplest—way in which the 1,2- and 1,4-products reach equilibrium is through this cation.

$$CH_2\text{---}CH\text{---}CH\text{==}CH_2 \rightleftarrows CH_2\text{---}CH\text{--}\overset{\oplus}{\overset{\frown}{CH}}\text{--}CH_2 \rightleftarrows CH_2\text{---}CH\text{==}CH\text{---}CH_2$$

H   Br   H   H   Br

1,2-   + Br⁻   1,4-

$$CH_2\text{==}CH\text{---}CH\text{==}CH_2$$

+ HBr

Ionization of the bromides involves climbing the potential hills back toward this carbocation. But there is a higher hill separating the cation from the 1,4-product than from the 1,2-product; consequently, the 1,4-product will ionize more slowly than the 1,2-product. Equilibrium is reached when the rates of the opposing reactions are equal. The 1,2-product is formed rapidly, but ionizes rapidly. The 1,4-product is formed slowly, but ionizes even more slowly; once formed, the 1,4-product tends to persist. At temperatures high enough for equilibrium to be reached—that is, high enough for significantly fast ionization—the more stable 1,4-product predominates.

We have not tried to account for the fact that the 1,2-product is formed faster than the 1,4-product, or for the fact that the 1,4-product is more stable than the 1,2-product (although we notice that this is consistent with our generalization that disubstituted alkenes are more stable than monosubstituted alkenes). We have accepted these facts and have simply tried to show what they mean in terms of energy considerations. Similar relationships have been observed for other dienes and reagents.

These facts illustrate two important points. First, we must be cautious when we interpret product composition in terms of rates of reaction; we must be sure that one product is not converted into the other *after* its formation. Second, the more stable product is by no means *always* formed faster. *On the basis of much evidence*, we have concluded that *generally* the more stable a carbocation or free radical, the faster it is formed; a consideration of the transition states for the various reactions has shown that this is reasonable. *We must not, however, extend this principle to other reactions unless the evidence warrants it.*

Problem 10.17   Addition of one mole of bromine to 1,3,5-hexatriene yields only 5,6-dibromo-1,3-hexadiene and 1,6-dibromo-2,4-hexadiene. (a) Are these products consistent with the formation of the most stable intermediate carbocation? (b) What other product or products would also be consistent? (c) Actually, which factor appears to be in control, rate or position of equilibrium?

10.28   Free-radical addition to conjugated dienes: orientation

Like other alkenes, conjugated dienes undergo addition not only by electrophilic reagents but also by free radicals. In free-radical addition, conjugated dienes show two special features: they undergo **1,4-addition** as well as 1,2-addition, and they are **much more reactive** than ordinary alkenes. We can account for both features—*orientation* and *reactivity*—by examining the structure of the intermediate free radical.

Let us take, as an example, addition of $BrCCl_3$ to 1,3-butadiene in the presence of a peroxide. As we have seen (Sec. 8.20), the peroxide decomposes (step 1) to yield a free radical, which abstracts bromine from $BrCCl_3$ (step 2) to generate a $\cdot CCl_3$ radical.

(1)   Peroxide $\longrightarrow$ Rad$\cdot$

(2)   Rad$\cdot$ + $BrCCl_3$ $\longrightarrow$ Rad—Br + $\cdot CCl_3$

The $\cdot CCl_3$ radical thus formed adds to the butadiene (step 3). The products obtained show that this addition is to one of the *ends* of the conjugated system. Why is this? In free-radical addition to conjugated dienes, it seems clear that orientation (as well as reactivity) is controlled by the stability of the radical being formed, and not by polar factors (Sec. 8.19). Thus $\cdot CCl_3$ adds where it does because in this way a resonance-stabilized allylic free radical is formed.

(3)
$$\overset{\overset{\textstyle\cdot CCl_3}{\curvearrowleft}}{\underset{\underset{1 \quad\quad 2 \quad\quad 3 \quad\quad 4}{}}{CH_2{=}CH{-}CH{=}CH_2}} \longrightarrow \left[\begin{array}{l} Cl_3C{-}CH_2{-}\overset{\cdot}{C}H{-}CH{=}CH_2 \\ Cl_3C{-}CH_2{-}CH{=}CH{-}\overset{\cdot}{C}H_2 \end{array} \right]$$

Addition to end
of conjugated system

equivalent to

$$Cl_3C{-}CH_2{-}CH{\cdots}CH{\cdots}CH_2$$

Allylic free radical

The allylic free radical then abstracts bromine from a molecule of $BrCCl_3$ (step 4) to complete the addition, and in doing so forms a new $\cdot CCl_3$ radical which can carry on the chain. In step (4) bromine can become attached to either C–2 or C–4 to yield either the 1,2- or 1,4-product.

(4) $Cl_3C{-}CH_2{-}CH{\cdots}CH{\cdots}CH_2 \xrightarrow{\text{BrCCl}_3} Cl_3C{-}CH_2{-}\underset{\underset{Br}{|}}{C}H{-}CH{=}CH_2$

Allylic free radical

1,2-Addition product

$+ \; Cl_3C{-}CH_2{-}CH{=}CH{-}CH_2{-}Br$
1,4-Addition product

10.29   Free-radical addition to conjugated dienes: reactivity

If $BrCCl_3$ is allowed to react with a 50:50 mixture of 1,3-butadiene and a simple alkene like 1-octene, addition occurs almost exclusively to the 1,3-butadiene. Evidently the $\cdot CCl_3$ radical adds much more rapidly to the conjugated diene than to the simple alkene. Similar results have been observed in a great many free-radical additions.

How can we account for the unusual reactivity of conjugated dienes? In our discussion of halogenation of the simple alkanes (Sec. 3.27), we found that not only orientation but also relative reactivity was related to the stability of the free radical formed in the first step. On this basis alone, we might expect addition to a conjugated diene, which yields a stable allylic free radical, to occur faster than addition to a simple alkene. (There is evidence showing that polar factors (Sec. 8.19) are of only minor importance here.)

On the other hand, we have just seen (Sec. 10.21) that conjugated dienes are more stable than simple alkenes. On this basis alone, we might expect addition to conjugated dienes to occur more slowly than to simple alkenes.

The relative rates of the two reactions depend chiefly upon the E_{act} values. Stabilization of the incipient allylic free radical lowers the energy level of the transition state; stabilization of the diene lowers the energy of the reactants.

Whether the net E_{act} is larger or smaller than for addition to a simple alkene depends upon *which* is stabilized *more* (see Fig. 10.8).

Figure 10.8   Molecular structure and rate of reaction. The transition state from the diene is stabilized more than the diene itself: E_{act} is lowered. (The plots are aligned with each other for easy comparison.)

The fact is that conjugated dienes are more reactive than simple alkenes. In the present case, then—and in most cases involving alkenes and free radicals, or alkenes and carbocations—the factors stabilizing the transition state are more important than the factors stabilizing the reactant. However, this is *not always* true. (It does not seem to be true, for example, in electrophilic addition to conjugated dienes.)

10.30   Free-radical polymerization of dienes. Rubber and rubber substitutes

Like substituted ethylenes, conjugated dienes, too, undergo free-radical polymerization. From 1,3-butadiene, for example, there is obtained a polymer

$$CH_2{=}CH{-}CH{=}CH_2 \qquad\qquad [{-}CH_2{-}CH{=}CH{-}CH_2{-}]_n$$

1,3-Butadiene   Polybutadiene

whose structure indicates that 1,4-addition occurs predominantly:

Rad· $CH_2{=}CH{-}CH{=}CH_2$ $CH_2{=}CH{-}CH{=}CH_2$ $CH_2{=}CH{-}CH{=}CH_2$

1,3-Butadiene

$$\downarrow$$

Rad$-CH_2{-}CH{=}CH{-}CH_2{-}CH_2{-}CH{=}CH{-}CH_2{-}CH_2{-}CH{=}CH{-}CH_2\sim$

Such a polymer differs from the polymers of simple alkenes in one very important way: each unit still contains one double bond.

Natural rubber has a structure that strongly resembles these synthetic poly-dienes. We could consider it to be a polymer of the conjugated diene 2-methyl-1,3-butadiene, **isoprene**.

$$CH_2=\overset{\displaystyle \overset{CH_3}{|}}{C}-CH=CH_2 \qquad \left[-CH_2-\overset{\displaystyle \overset{CH_3}{|}}{C}=CH-CH_2-\right]_n$$

Isoprene   *cis*-Polyisoprene

Natural rubber

The double bonds in the rubber molecule are highly important, since—apparently by providing reactive allylic hydrogens—they permit *vulcanization*, the formation of sulfur bridges between different chains. These *cross-links* make the rubber harder and stronger, and do away with the tackiness of the untreated rubber.

Natural rubber   Vulcanized rubber

Polymerization of dienes to form substitutes for rubber was the forerunner of the enormous present-day plastics industry. *Polychloroprene* (Neoprene, Duprene) was the first commercially successful rubber substitute in the United States.

$$CH_2=\overset{\displaystyle \overset{Cl}{|}}{C}-CH=CH_2 \qquad \left[-CH_2-\overset{\displaystyle \overset{Cl}{|}}{C}=CH-CH_2-\right]_n$$

Chloroprene   Polychloroprene

The properties of rubber substitutes—like those of other polymers—are determined, in part, by the nature of the substituent groups. Polychloroprene, for example, is inferior to natural rubber in some properties, but superior in its resistance to oil, gasoline, and other organic solvents.

Polymers of isoprene, too, can be made artificially: they contain the same unsaturated chain and the same substituent (the $-CH_3$ group) as natural rubber. But polyisoprene made by the free-radical process we have been talking about was—in the properties that really matter—a far cry from natural rubber. It differed in *stereochemistry*: natural rubber has the *cis* configuration at (nearly) every double bond; the artificial material was a mixture of *cis* and *trans*. Not until 1955 could a true synthetic *rubber* be made; what was needed was an entirely new kind of catalyst and an entirely new mechanism of polymerization (Sec. 36.6). With these, it became possible to carry out a stereoselective polymerization of isoprene to a material virtually identical with natural rubber: *cis*-1,4-polyisoprene.

Natural rubber
All-cis configuration

10.31   Isoprene and the isoprene rule

The isoprene unit is one of nature's favorite building blocks. It occurs not only in rubber, but in a wide variety of compounds isolated from plant and animal sources. For example, nearly all the *terpenes* (found in the essential oils of many plants) have carbon skeletons made up of isoprene units joined in a regular, head-to-tail way. Recognition of this fact—the so-called **isoprene rule**—has been of great help in working out structures of terpenes.

Vitamin A₁

Citronellol: *a terpene*
(found in oil of geranium)

γ-Terpinene: *a terpene*
(found in coriander oil)

A fascinating area of research linking organic chemistry and biology is the study of the *biogenesis* of natural products: the detailed sequence of reactions by which a compound is formed in living systems, plant or animal. All the isoprene units in nature, it appears, originate from the same compound, "isopentenyl" pyrophosphate.

Isopentenyl pyrophosphate

Work done since about 1950 has shown how compounds as seemingly different from rubber as *cholesterol* (p. 650) are built up, step by step, from isoprene units.

Isopentenyl
pyrophosphate

Squalene

Lanosterol

Cholesterol

Problem 10.18 (a) Mark off the isoprene units making up the squalene molecule. (b) There is one deviation from the head-to-tail sequence. Where is it? Does its particular location suggest anything to you—in general terms—about the biogenesis of this molecule? (c) What skeletal changes, if any, accompany the conversion of squalene into lanosterol? Of lanosterol into cholesterol?

10.32 Analysis of dienes

Dienes respond to characterization tests in the same way as alkenes: they decolorize bromine in carbon tetrachloride without evolution of hydrogen bromide, and they decolorize cold, neutral, dilute permanganate; they are not oxidized by chromic anhydride. They are, however, more unsaturated than alkenes. This property can be detected by determination of their molecular formulas (C_nH_{2n-2}) and by a quantitative hydrogenation (two moles of hydrogen are taken up per mole of hydrocarbon).

Proof of structure is best accomplished by the same degradative methods that are used in studying alkenes. Ozonolysis of dienes yields aldehydes and ketones, including double-ended ones containing two C=O groups per molecule. For example:

(Spectroscopic analysis of dienes is discussed in Chapter 16.)

Problem 10.19   Contrast the ozonolysis products of the following isomers:
(a) 1,3-pentadiene, (b) 1,4-pentadiene, (c) isoprene (2-methyl-1,3-butadiene).

Problem 10.20   Predict the ozonolysis products from polybutadiene, $(C_4H_6)_n$:
(a) if 1,2-addition is involved in the polymerization; (b) if 1,4-addition is involved.

Problem 10.21   Ozonolysis of natural rubber yields chiefly (90%) the compound

$$\begin{array}{ccccccc} H & & & & & CH_3 \\ | & & & & & | \\ O=C-CH_2-CH_2-C=O \end{array}$$

What does this tell us about the structure of rubber?

PROBLEMS

1. Draw the structure of 6-methyl-2-heptene. Label each set of hydrogen atoms to show their relative reactivities toward chlorine atoms, using (1) for the most reactive, (2) for the next, etc.

2. (a) Draw structures of all isomeric dienes of formula C_6H_{10}, omitting cumulated dienes. (b) Name each one. (c) Indicate which ones are conjugated. (d) Indicate which ones can show geometric isomerism, and draw the isomeric structures. (e) Draw structures of the ozonolysis products expected from each. (f) Which isomers (other than *cis–trans* pairs) could not be distinguished on the basis of (e)?

3. Give structures and names of the organic products expected from the reaction (if any) of 1,3-butadiene with:

(a) 1 mol H_2, Ni   (c) 1 mol Br_2 (e) 1 mol HCl   (g) O_3, then H_2O
(b) 2 mol H_2, Ni   (d) 2 mol Br_2 (f) 2 mol HCl   (h) hot $KMnO_4/NaIO_4$

4. Answer Problem 3 for 1,4-pentadiene instead of 1,3-butadiene.

5. Give structures and names of the products from dehydrohalogenation by strong base of each of the following halides. Where more than one product is expected, indicate which will be the major product.

(a) 1-chlorobutane; 2-chlorobutane
(b) 1-chlorobutane; 4-chloro-1-butene
(c) 2-bromo-2-methylbutane; 2-bromo-3-methylbutane
(d) 1-bromo-2-methylbutane; 1-bromo-3-methylbutane
(e) 1-chloro-2,3-dimethylbutane; 2-chloro-2,3-dimethylbutane
(f) 4-chloro-1-butene; 5-chloro-1-pentene

6. Which alkyl halide of each set in Problem 5 would you expect to undergo dehydrohalogenation faster?

7. Give structures of the chief product or products expected from addition of one mole of HCl to each of the following compounds:

(a) 1,3-butadiene; 1-butene   (c) 1,3-butadiene; 2-methyl-1,3-butadiene
(b) 1,3-butadiene; 1,4-pentadiene   (d) 1,3-butadiene; 1,3-pentadiene

8. Answer Problem 7 for the addition of $BrCCl_3$ in the presence of peroxides instead of addition of HCl.

9. Which compound of each pair in Problem 7 would you expect to be more reactive toward addition of $BrCCl_3$?

10. Arrange the compounds of each set in order of reactivity toward S_N1 substitution. (If you expect two of them to be of about the same reactivity, say so.)

(a) 1-chloropropene, 3-chloropropene, *n*-propyl chloride
(b) 2-bromobutane, 3-bromo-1-butene, 2-bromo-1-butene
(c) 4-bromo-2-pentene, 4-bromo-2-methyl-2-pentene, 4-bromo-3-methyl-2-pentene
(d) 2-buten-1-yl tosylate, 2-penten-4-yl tosylate, 4-methyl-2-penten-4-yl tosylate
(e) *sec*-butyl tosylate, *sec*-butyl triflate, *sec*-butyl chloride, *sec*-butyl bromide

11. Answer Problem 10 for reactivity toward S_N2 substitution.

12. How do you account for the following facts: formic acid, HCOOH, contains one carbon–oxygen bond of 1.36 Å and another of 1.23 Å, yet sodium formate, HCOO⁻ Na⁺, contains two equal carbon–oxygen bonds, each of 1.27 Å. (Check your answer in Sec. 23.13.)

$$H-C\overset{\displaystyle O}{\underset{\displaystyle OH}{\big<}}$$

Formic acid

13. (a) Make a model of *allene*, $CH_2{=}C{=}CH_2$, a cumulated diene. What is the spatial relationship between the pair of hydrogens at one end of the molecule and the pair of hydrogens at the other end? (b) Substituted allenes of the type $RCH{=}C{=}CHR$ have been obtained in optically active form. Is this consistent with the shape of the molecule in (a)? Where are the chiral centers in the substituted allene? (c) Work out the electronic configuration of allene. (*Hint:* How many atoms are attached to the middle carbon? To each of the end carbons?) Does this lead to the same shape of molecule that you worked out in (a) and (b)?

14. When allowed to react with aqueous HBr, 3-buten-2-ol ($CH_3CHOHCH{=}CH_2$) yields not only 3-bromo-1-butene ($CH_3CHBrCH{=}CH_2$) but also 1-bromo-2-butene ($CH_3CH{=}CHCH_2Br$). (a) How do you account for these results? (b) Predict the product of the reaction between HBr and 2-buten-1-ol ($CH_3CH{=}CHCH_2OH$).

15. Treatment of $CF_3(C_6H_5)C{=}CF_2$ with EtONa/EtOH yields chiefly $CF_3(C_6H_5)C{=}CF(OEt)$. Similar treatment of $CF_2Cl(C_6H_5)C{=}CF_2$ yields $EtOCF_2(C_6H_5)C{=}CF_2$. The rates of the two reactions are almost identical. It has been suggested that both reactions proceed by the same mechanism.
   Show all steps in a mechanism that is consistent with the nature of these reactants, and that accounts for the similarity in rate despite the difference in final product.

16. Treatment with phosphoric acid converts 2,7-dimethyl-2,6-octadiene into I.

$$\begin{array}{c}
H_3C \quad CH_3 \\
\diagdown C \diagup \\
\quad \quad \quad CH_3 \\
H_2C \quad \quad CH-C \diagdown \\
\quad \quad \quad \quad CH_2 \\
H_2C-CH_2
\end{array}$$

I

1,1-Dimethyl-2-isopropenylcyclopentane

Using reaction steps already familiar to you, suggest a mechanism for this reaction.

17. Describe simple chemical tests that would distinguish between:

(a) 1,3-pentadiene and *n*-pentane
(b) allyl bromide and 2,3-dimethyl-1,3-butadiene
(c) 1-chloro-2-butene and 2-chloro-2-butene

   Tell exactly what you would *do* and *see*.

18. A hydrocarbon of formula C_6H_{10} absorbs only *one* mole of H_2, upon catalytic hydrogenation. Upon ozonolysis the hydrocarbon yields

$$\underset{\text{H}}{\overset{\text{H}}{\underset{|}{\text{O=C}}}}-\text{CH}_2-\text{CH}_2-\text{CH}_2-\text{CH}_2-\overset{\text{H}}{\underset{|}{\text{C}}}=\text{O}$$

What is the structure of the hydrocarbon? (Check your answer in Sec. 12.18.)

19. A hydrocarbon was found to have a molecule weight of 80–85. A 10.02-mg sample took up 8.40 mL of H_2 gas measured at 0 °C and 760 mm pressure. Ozonolysis yielded only

$$\underset{\text{O}}{\overset{}{\text{H}-\text{C}-\text{H}}} \quad \text{and} \quad \underset{\text{O}\ \ \text{O}}{\overset{}{\text{H}-\text{C}-\text{C}-\text{H}}}$$

What was the hydrocarbon?

20. *Myrcene*, $C_{10}H_{16}$, a terpene isolated from oil of bay, absorbs three moles of hydrogen to form $C_{10}H_{22}$. Upon ozonolysis myrcene yields:

$$\underset{\text{O}}{\overset{}{\text{CH}_3-\text{C}-\text{CH}_3}} \quad \underset{\text{O}}{\overset{}{\text{H}-\text{C}-\text{H}}} \quad \underset{\text{O}\ \ \ \ \ \ \ \ \ \ \ \ \ \text{O}\ \ \text{O}}{\overset{}{\text{H}-\text{C}-\text{CH}_2-\text{CH}_2-\text{C}-\text{C}-\text{H}}}$$

(a) What structures are consistent with these facts?
(b) On the basis of the isoprene rule (Sec. 10.31), what is the most likely structure for myrcene?

21. *Dihydromyrcene*, $C_{10}H_{18}$, formed from myrcene (Problem 20), absorbs two moles of hydrogen to form $C_{10}H_{22}$. Upon cleavage by $KMnO_4$, dihydromyrcene yields:

$$\underset{\text{O}}{\overset{}{\text{CH}_3-\text{C}-\text{CH}_3}} \quad \underset{\text{O}}{\overset{}{\text{CH}_3-\text{C}-\text{OH}}} \quad \underset{\text{O}\ \text{O}}{\overset{}{\text{CH}_3-\text{C}-\text{CH}_2-\text{CH}_2-\text{C}-\text{OH}}}$$

(a) Keeping in mind the isoprene rule, what is the most likely structure for dihydromyrcene?
(b) Is it surprising that a compound of this structure is formed by reduction of myrcene?

22. At the beginning of the biogenesis of squalene (Sec. 10.31) isopentenyl pyrophosphate, $CH_2=C(CH_3)CH_2CH_2OPP$, is enzymatically isomerized to dimethylallyl pyrophosphate, $(CH_3)_2C=CHCH_2OPP$. These two compounds then react together to yield *geranyl pyrophosphate*, $(CH_3)_2C=CHCH_2CH_2C(CH_3)=CHCH_2OPP$. (a) Assuming that the weakly basic pyrophosphate anion is, like the protonated hydroxyl group, a good leaving group,

$$\text{R}-\text{OPP} \longrightarrow \text{R}\oplus + \text{OPP}^-$$

can you suggest a series of familiar steps by which geranyl pyrophosphate might be formed?
(b) Geranyl pyrophosphate then reacts with another molecule of isopentenyl pyrophosphate to form *farnesyl pyrophosphate*. What is the structure of farnesyl pyrophosphate?
(c) What is the relationship between farnesyl pyrophosphate and squalene? (d) An enzyme system from the rubber plant catalyzes the conversion of isopentenyl pyrophosphate into rubber; dimethylallyl pyrophosphate appears to act as an initiator for the process. Can you suggest a "mechanism" for the formation of natural rubber?

11

Alkynes

11.1 Introduction

So far we have discussed two kinds of carbon–carbon bonds: the single bond and the double bond. The carbon–carbon single bond is of low reactivity; its main function is to act as the principal cement holding most organic compounds together. The carbon–carbon double bond is unsaturated and hence highly reactive toward a wide variety of reagents; as a substituent it can exert remarkable effects on the rest of the molecule.

Now we come to the *carbon–carbon triple bond*, the functional group of the family called *alkynes*. Like the double bond it is unsaturated and highly reactive: toward the reagents that double bonds react with, and toward some others besides. It also can exert remarkable effects on the rest of the molecule, and in its own particular way. Through a combination of its characteristic properties, the carbon–carbon triple bond plays a special role—one of increasing importance—in organic synthesis.

11.2 Structure of acetylene. The carbon–carbon triple bond

The simplest member of the alkyne family is **acetylene**, C_2H_2. Using the methods we applied to the structure of ethylene (Sec. 7.2), we arrive at a structure in which the carbon atoms share *three* pairs of electrons, that is, are joined by a *triple bond*. *The* **carbon–carbon triple bond** *is the distinguishing feature of the alkyne structure.*

$$H:C:::C:H \qquad\qquad H-C\equiv C-H$$
Acetylene

Again, quantum mechanics tells us a good deal more about acetylene, and about the carbon–carbon triple bond. To form bonds with two other atoms, carbon makes use of two equivalent hybrid orbitals: *sp* orbitals, formed by the mixing of *one s* and *one p* orbital (Sec. 1.9). These *sp* orbitals lie along a straight line that passes through the carbon nucleus; the angle between the two orbitals is thus 180°. This **linear** arrangement (Fig. 1.5) permits the hybrid orbitals to be as far apart as possible. Just as mutual repulsion among orbitals gives four tetrahedral bonds or three trigonal bonds, so it gives two linear bonds.

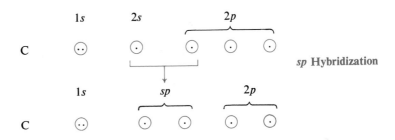

sp **Hybridization**

If we arrange the two carbons and the two hydrogens of acetylene to permit maximum overlap of orbitals, we obtain the structure shown in Fig. 11.1.

Figure 11.1 Acetylene molecule: only σ bonds shown.

Acetylene is a *linear molecule*, all four atoms lying along a single straight line. Both carbon–hydrogen and carbon–carbon bonds are cylindrically symmetrical about a line joining the nuclei, and are therefore σ bonds.

The molecule is not yet complete, however. In forming the *sp* orbitals already described, each carbon atom has used only one of its three *p* orbitals; it has two remaining *p* orbitals. Each of these consists of two equal lobes, whose axis lies at right angles both to the axis of the other *p* orbital and to the line of the *sp* orbitals; each *p* orbital is occupied by a single electron. But the sum of two perpendicular *p* orbitals is not four spherical lobes, but a single doughnut-shaped cloud (Fig. 11.2).

Figure 11.2 Acetylene molecule: two *p* orbitals on one carbon (dough-nut-shaped cloud) can overlap two *p* orbitals on the other carbon.

Overlap of the *p* orbitals on one carbon with the *p* orbitals on the other carbon permits pairing of electrons. Two π bonds are formed, which together make a single cylindrical sheath about the line joining the nuclei (Fig. 11.3).

Figure 11.3 Acetylene molecule: carbon–carbon triple bond. The π cloud forms a cylindrical sheath.

The carbon–carbon "triple bond" is thus made up of one strong σ bond and two weaker π bonds; it has a total strength of 198 kcal. It is stronger than the carbon–carbon double bond of ethylene (163 kcal) or the carbon–carbon single bond of ethane (88 kcal), and therefore is shorter than either.

Figure 11.4 Acetylene molecule: shape and size.

$$H - C \equiv C - H$$

1.21 Å   1.08 Å   180°

Again, the quantum mechanical structure is verified by direct evidence. Electron diffraction, x-ray diffraction, and spectroscopy show acetylene (Fig. 11.4) to be a linear molecule. The C—C distance is 1.21 Å, as compared with 1.34 Å in ethylene and 1.53 Å in ethane. Figure 11.5 shows a model of the acetylene molecule.

Figure 11.5 Electronic configuration and molecular shape. Model of the acetylene molecule: two views.

As in the case of the double bond, the structure of the triple bond is verified—although this time in a negative way—by the evidence of isomer number. As we can readily see from models, the linearity of the bonding should not permit geometric isomerism; no such isomers have ever been found.

The C—H distance in acetylene is 1.08 Å, even shorter than in ethylene (1.103 Å); because of their greater s character, sp orbitals are smaller than sp^2 orbitals, and sp-hybridized carbon forms shorter bonds than sp^2-hybridized carbon.

The C—H bond dissociation energy in acetylene is not known, but we would expect it to be even greater than in ethylene. Oddly enough, the same *sp* hybridization that almost certainly makes cleavage of the C—H bond to form free radicals (*homolysis*) more difficult, makes cleavage to form ions (*heterolysis*) easier, as we shall see (Sec. 11.11).

$$HC \equiv C:H \longrightarrow HC \equiv C \cdot + H \cdot$$

Homolysis:
*one electron to
each fragment*

$$HC \equiv C:H \longrightarrow HC \equiv C:^- + H^+$$

Heterolysis:
*both electrons to
one fragment*

Problem 11.1 Compare the electronic configurations of CO_2, which is a linear molecule (check your answer to Problem 1.6, p. 26), and H_2O, which has a bond angle of 105°.

11.3 Higher alkynes. Nomenclature

Like the alkanes and alkenes, the alkynes form a homologous series, the increment again being CH_2.

The alkynes are named according to two systems. In one, they are considered to be derived from acetylene by replacement of one or both hydrogen atoms by alkyl groups.

$H-C \equiv C-C_2H_5$	$CH_3-C \equiv C-CH_3$	$CH_3-C \equiv C-CH(CH_3)_2$
Ethylacetylene	Dimethylacetylene	Isopropylmethylacetylene
1-Butyne	2-Butyne	4-Methyl-2-pentyne

For more complicated alkynes the **IUPAC** names are used. The rules are exactly the same as for the naming of alkenes, except that the ending **-yne** replaces *-ene*. The parent structure is the longest continuous chain that contains the triple bond, and the positions both of substituents and of the triple bond are indicated by numbers. The triple bond is given the number of the *first* triply bonded carbon encountered, starting from the end of the chain nearest the triple bond.

11.4 Physical properties of alkynes

Being compounds of low polarity, the alkynes have physical properties that are essentially the same as those of the alkanes and alkenes. They are insoluble in water but quite soluble in the usual organic solvents of low polarity: ligroin, ether, benzene, carbon tetrachloride. They are less dense than water. Their boiling points

(Table 11.1) show the usual increase with increasing carbon number, and the usual effects of chain-branching; they are very nearly the same as the boiling points of alkanes or alkenes with the same carbon skeletons.

Table 11.1   ALKYNES

Name	Formula	M.p., °C	B.p., °C	Relative density (at 20 °C)
Acetylene	$HC{\equiv}CH$	− 82	−75	
Propyne	$HC{\equiv}CCH_3$	−101.5	−23	
1-Butyne	$HC{\equiv}CCH_2CH_3$	−122	9	
1-Pentyne	$HC{\equiv}C(CH_2)_2CH_3$	− 98	40	0.695
1-Hexyne	$HC{\equiv}C(CH_2)_3CH_3$	−124	72	.719
1-Heptyne	$HC{\equiv}C(CH_2)_4CH_3$	− 80	100	.733
1-Octyne	$HC{\equiv}C(CH_2)_5CH_3$	− 70	126	.747
1-Nonyne	$HC{\equiv}C(CH_2)_6CH_3$	− 65	151	.763
1-Decyne	$HC{\equiv}C(CH_2)_7CH_3$	− 36	182	.770
2-Butyne	$CH_3C{\equiv}CCH_3$	− 24	27	.694
2-Pentyne	$CH_3C{\equiv}CCH_2CH_3$	−101	55	.714
3-Methyl-1-butyne	$HC{\equiv}CCH(CH_3)_2$		29	.665
2-Hexyne	$CH_3C{\equiv}C(CH_2)_2CH_3$	− 92	84	.730
3-Hexyne	$CH_3CH_2C{\equiv}CCH_2CH_3$	− 51	81	.725
3,3-Dimethyl-1-butyne	$HC{\equiv}CC(CH_3)_3$	− 81	38	.669
4-Octyne	$CH_3(CH_2)_2C{\equiv}C(CH_2)_2CH_3$		131	.748
5-Decyne	$CH_3(CH_2)_3C{\equiv}C(CH_2)_3CH_3$		175	.769

11.5   Industrial source of acetylene

The alkyne of chief industrial importance is the simplest member of the family, **acetylene**. Traditionally, it has been prepared by the action of water on calcium carbide, CaC_2, which itself is prepared by the reaction between calcium oxide and coke at the very high temperatures of the electric furnace. The calcium oxide and coke are in turn obtained from limestone and coal, respectively.

$$\text{Coal} \longrightarrow \text{coke}$$
$$\text{Limestone} \longrightarrow \text{CaO}$$
$$\xrightarrow{2000\ °C} CaC_2 \xrightarrow{H_2O} H{-}C{\equiv}C{-}H$$

An alternative synthesis, based on petroleum, has displaced the carbide process. This involves the controlled, high-temperature partial oxidation of methane.

$$6CH_4 + O_2 \xrightarrow{1500\ °C} 2HC{\equiv}CH + 2CO + 10H_2$$

Because of the high cost of acetylene, its formerly huge market has dwindled, and most of the chemicals once made from it are now made from ethylene. It is still the source, however, of a number of compounds used to make polymers.

11.6   Preparation of alkynes

Making alkynes can involve either of two processes: generating a carbon–carbon triple bond, or increasing the size of a molecule that already contains a triple bond.

PREPARATION OF ALKYNES

1. **Dehydrohalogenation of alkyl dihalides.** Discussed in Sec. 11.6.

$$\left[\begin{array}{c} \text{H H} \\ -\overset{|}{\text{C}}=\overset{|}{\text{C}}- \end{array}\right] \xrightarrow{X_2} \begin{array}{c}\text{H H}\\ -\overset{|}{\underset{X}{\text{C}}}-\overset{|}{\underset{X}{\text{C}}}-\end{array} \xrightarrow{\text{KOH (alc)}} \begin{array}{c}\text{H}\\-\overset{|}{\text{C}}=\overset{|}{\underset{X}{\text{C}}}-\end{array} \xrightarrow{\text{NaNH}_2} -\text{C}\equiv\text{C}-$$

Example:

$$\text{CH}_3\text{CH}=\text{CH}_2 \xrightarrow{\text{Br}_2} \underset{\substack{\text{Br}\quad\text{Br}\\ \text{1,2-Dibromopropane}\\(\text{Propylene bromide})}}{\text{CH}_3\text{CH}-\text{CH}_2} \xrightarrow{\text{KOH (alc)}} \underset{\text{1-Bromo-1-propene}}{\text{CH}_3\text{CH}=\text{CHBr}} \xrightarrow{\text{NaNH}_2} \underset{\text{Propyne}}{\text{CH}_3\text{C}\equiv\text{CH}}$$

2. **Reaction of metal acetylides with primary alkyl halides.** Discussed in Sec. 11.12.

$$\left[-\text{C}\equiv\text{C}-\text{H} \xrightarrow{\text{LiNH}_2}\right] -\text{C}\equiv\text{C:}^-\text{Li}^+ + \text{RX} \longrightarrow -\text{C}\equiv\text{C}-\text{R} + \text{LiX}$$
$$\text{R } must\ be\ 1°$$

Examples:

$$\text{HC}\equiv\text{C:}^-\text{Li}^+ + \text{CH}_3\text{CH}_2\text{CH}_2\text{CH}_2\text{Br} \longrightarrow \text{HC}\equiv\text{C}-\text{CH}_2\text{CH}_2\text{CH}_2\text{CH}_3$$
$$\underset{\text{Lithium acetylide}}{} \quad \underset{n\text{-Butyl bromide}}{} \qquad \underset{\substack{\text{1-Hexyne}\\(n\text{-Butylacetylene})}}{}$$

$$\text{CH}_3(\text{CH}_2)_4\text{C}\equiv\text{C:}^-\text{Li}^+ + \text{CH}_3(\text{CH}_2)_3\text{CH}_2\text{Cl} \longrightarrow \text{CH}_3(\text{CH}_2)_4\text{C}\equiv\text{C}-\text{CH}_2(\text{CH}_2)_3\text{CH}_3$$
$$\underset{\text{Lithium }n\text{-pentylacetylide}}{}\quad \underset{n\text{-Pentyl chloride}}{} \qquad \underset{\text{6-Dodecyne}}{}$$

A carbon–carbon triple bond is formed in the same way as a double bond: elimination of atoms or groups from two adjacent carbons. The groups eliminated and the reagents used are essentially the same as in the preparation of alkenes.

$$\begin{array}{c}\text{W X}\\ -\overset{|}{\underset{Y}{\text{C}}}-\overset{|}{\underset{Z}{\text{C}}}-\end{array} \longrightarrow \begin{array}{c}\text{W X}\\-\overset{|}{\text{C}}=\overset{|}{\text{C}}-\end{array} \longrightarrow -\text{C}\equiv\text{C}-$$

Dehydrohalogenation of vicinal dihalides is particularly useful since the dihalides themselves are readily obtained from the corresponding alkenes by addition of halogen. This amounts to conversion—by several steps—of a double bond into a triple bond.

Dehydrohalogenation can generally be carried out in two stages as shown.

$$\underset{\substack{| \ |\\X\ X}}{-\overset{\substack{H\ H\\| \ |}}{C}-\overset{}{C}-} \quad \xrightarrow{\text{KOH (alc)}} \quad \underset{\substack{|\\X}}{-\overset{\substack{H\\|}}{C}=\overset{}{C}-} \quad \xrightarrow{\text{NaNH}_2} \quad -C\equiv C-$$

A vinylic halide
Very unreactive

Carried through only the first stage, it is a valuable method for preparing unsaturated halides. The halides thus obtained, with halogen attached directly to double-bonded carbon, we recognize as vinylic halides (Sec. 10.18); as we know, they are very unreactive. Under mild conditions, therefore, dehydrohalogenation stops at the vinylic halide stage; more vigorous conditions—use of a stronger base—are required for alkyne formation.

Conversion of smaller alkynes into larger ones is done by use of **metal acetylides**. These are particularly easy to generate because of a special property of certain alkynes and, once made, are highly versatile reagents.

11.7   Reactions of alkynes

Just as alkene chemistry is the chemistry of the carbon–carbon double bond, so alkyne chemistry is the chemistry of the carbon–carbon triple bond. Like alkenes, alkynes undergo electrophilic addition, and for the same reason: availability of the loosely held π electrons.

Addition of hydrogen, halogens, and hydrogen halides to alkynes is very much like addition to alkenes, except that here *two* molecules of reagent can be consumed for each triple bond. As shown, it is generally possible, by proper selection of conditions, to limit reaction to the first stage of addition, formation of alkenes. In some cases at least, this is made simpler because of the way that the substituents introduced in the first stage affect the second stage.

Problem 11.2   (a) Write the equation for the two-stage addition of bromine to 2-butyne. (b) How will the first two bromine atoms affect the reactivity of the double bond? (c) How will this influence the competition for halogen between 2-butyne and 2,3-dibromo-2-butene? (d) In what proportions would you mix the reagents to help limit reaction to the first stage? (e) Would you bubble 2-butyne into a solution of Br_2 in CCl_4, or drip the bromine solution into a solution of 2-butyne?

Besides addition, alkynes undergo certain reactions that are due to the acidity of a hydrogen atom held by triply bonded carbon.

REACTIONS OF ALKYNES _____

Addition Reactions

$$-C\equiv C- + YZ \quad \longrightarrow \quad \underset{\substack{| \ |\\Y\ Z}}{-\overset{}{C}=\overset{}{C}-} \quad \xrightarrow{YZ} \quad \underset{\substack{| \ |\\Y\ Z}}{-\overset{\substack{Y\ Z\\| \ |}}{C}-\overset{}{C}-}$$

CONTINUED

1. **Addition of hydrogen.** Discussed in Sec. 11.8.

$$-C\equiv C- \xrightarrow{2H_2} \underset{\underset{\displaystyle H\ \ H}{|\ \ \ |}}{\overset{\overset{\displaystyle H\ \ H}{|\ \ \ |}}{-C-C-}}$$

Alkyne   Alkane

Anti

Syn

Examples:

$$CH_3-C\equiv C-CH_3 \xrightarrow{2H_2,\ Ni} CH_3CH_2CH_2CH_3$$

2-Butyne   *n*-Butane

$C_2H_5C\equiv CC_2H_5$
3-Hexyne

trans-3-Hexene
Chief product

cis-3-Hexene
98–99% pure

2. **Addition of halogens.** Discussed in Sec. 11.9.

$$-C\equiv C- \xrightarrow{X_2} \underset{\underset{\displaystyle X\ \ X}{|\ \ \ |}}{-C=C-} \xrightarrow{X_2} \underset{\underset{\displaystyle X\ \ X}{|\ \ \ |}}{\overset{\overset{\displaystyle X\ \ X}{|\ \ \ |}}{-C-C-}} \qquad X_2 = Cl_2,\ Br_2$$

Example:

$$CH_3C\equiv CH \xrightarrow{Br_2} \underset{\underset{\displaystyle Br\ \ Br}{|\ \ \ \ |}}{CH_3-C=CH} \xrightarrow{Br_2} \underset{\underset{\displaystyle Br\ \ Br}{|\ \ \ \ |}}{\overset{\overset{\displaystyle Br\ \ Br}{|\ \ \ \ |}}{CH_3-C-CH}}$$

CONTINUED

3. Addition of hydrogen halides. Discussed in Sec. 11.9.

$$-C\equiv C- \xrightarrow{\text{HX}} \underset{\underset{\text{H}\ \ \text{X}}{|}}{-C}=\underset{}{C}- \xrightarrow{\text{HX}} \underset{\underset{\text{H}\ \ \text{X}}{|\ \ |}}{\overset{\overset{\text{H}\ \ \text{X}}{|\ \ |}}{-C-C-}} \qquad \text{HX} = \text{HCl, H Br, HI}$$

Example:

$$\text{CH}_3\text{C}\equiv\text{CH} \xrightarrow{\text{HCl}} \underset{\underset{\text{Cl}}{|}}{\text{CH}_3\text{C}}=\text{CH}_2 \xrightarrow{\text{HI}} \underset{\underset{\text{Cl}}{|}}{\overset{\overset{\text{I}}{|}}{\text{CH}_3-\text{C}-\text{CH}_3}}$$

4. Addition of water. Hydration. Discussed in Sec. 11.10.

$$-C\equiv C- + \text{H}_2\text{O} \xrightarrow{\text{H}_2\text{SO}_4,\ \text{HgSO}_4} \left[\underset{\underset{\text{H}\ \ \text{OH}}{|\ \ \ |}}{-C}=\underset{}{C}-\right] \rightleftharpoons \underset{\underset{\text{H}\ \ \text{O}}{|\ \ \ ||}}{\overset{\overset{\text{H}}{|}}{-C-C-}}$$

Examples:

$$\text{H}-\text{C}\equiv\text{C}-\text{H} + \text{H}_2\text{O} \xrightarrow{\text{H}_2\text{SO}_4,\ \text{HgSO}_4} \underset{\underset{\text{H}\ \ \text{O}}{|\ \ \ ||}}{\overset{\overset{\text{H}}{|}}{\text{H}-\text{C}-\text{C}-\text{H}}}$$

Acetaldehyde

$$\text{CH}_3-\text{C}\equiv\text{C}-\text{H} + \text{H}_2\text{O} \xrightarrow{\text{H}_2\text{SO}_4,\ \text{HgSO}_4} \underset{\underset{\text{H}\ \ \text{O}\ \ \text{H}}{|\ \ \ ||\ \ |}}{\overset{\overset{\text{H}\ \ \ \ \ \text{H}}{|\ \ \ \ \ \ |}}{\text{H}-\text{C}-\text{C}-\text{C}-\text{H}}}$$

Acetone

Reactions as Acids

$$-C\equiv C-H + \text{base} \longrightarrow -C\equiv C:^-$$

5. Formation of metal acetylides. Discussed in Sec. 11.11.

Examples:

$$\underset{\underset{\text{CH}_3}{|}}{\text{CH}_3-\text{CH}-\text{C}\equiv\text{C}-\text{H}} + \text{LiNH}_2 \xrightarrow{\text{ether}} \underset{\underset{\text{CH}_3}{|}}{\text{CH}_3-\text{CH}-\text{C}\equiv\text{C}:^-\text{Li}^+} + \text{NH}_3$$

Lithium isopropylacetylide

$$\text{CH}_3-\text{C}\equiv\text{C}-\text{H} + n\text{-BuLi} \longrightarrow \text{CH}_3\text{C}\equiv\text{C}^-\text{Li}^+ + n\text{-BuH}$$

Lithium methylacetylide

$$\text{HC}\equiv\text{CH} + \text{C}_2\text{H}_5\text{MgBr} \longrightarrow \text{HC}\equiv\text{CMgBr} + \text{C}_2\text{H}_6$$

Ethynylmagnesium
bromide

11.8 Reduction to alkenes

Reduction of an alkyne to the double-bond stage can—unless the triple bond is at the end of a chain—yield either a *cis* alkene or a *trans* alkene. Just which isomer predominates depends upon the choice of reducing agent.

Predominantly *trans* alkene is obtained by reduction of alkynes with sodium or lithium in liquid ammonia. Almost entirely *cis* alkene (as high as 98%) is obtained by hydrogenation of alkynes with several different catalysts: a specially prepared palladium called the *Lindlar catalyst*, for example.

$$
R-C\equiv C-R
\begin{cases}
\xrightarrow{\text{Na, NH}_3\text{(liq)}} & \underset{R}{\overset{H}{>}}C=C\underset{H}{\overset{R}{<}} \quad \textit{Anti} \\[2em]
\xrightarrow[\text{Lindlar catalyst}]{\text{H}_2} & \underset{H}{\overset{R}{>}}C=C\underset{H}{\overset{R}{<}} \quad \textit{Syn}
\end{cases}
$$

Each of these reactions is, then, highly stereoselective. The stereoselectivity in the *syn*-reduction of alkynes is attributed, in a general way, to the attachment of two hydrogens to the same side of an alkyne sitting on the catalyst surface; presumably this same stereochemistry holds for the hydrogenation of terminal alkynes, $RC\equiv CH$, which cannot yield *cis* and *trans* alkenes.

The mechanism that gives rise to *anti*-reduction will not be taken up here.

We have discussed many times the stereospecificity of biological systems. Where alkenes are concerned, this takes the form of diastereospecificity: organisms discriminating between geometric isomers. We saw a number of examples of this in the response of insects toward pheromones (Sec. 9.3). For synthetic materials to be effective in a living organism, we said, the stereospecificity of biological action demands an equal stereoselectivity in the synthesis of these materials (Sec. 9.4). Many new methods have been developed to generate double bonds stereoselectively; but the simplest and the one most often turned to is the hydrogenation of alkynes.

The matter goes much further than this. These alkenes may be the final products desired, as with some pheromones. But more often they are simply an intermediate stage. Alkenes undergo a variety of reactions, many of them diastereoselective and even (as we shall see in Sec. 20.7) enantioselective; if the stereoselectivity of these reactions is to be utilized fully, one must start with a stereochemically pure alkene.

Problem 11.3 Most methods of making alkenes (Secs. 7.12 and 7.25) yield predominantly the more stable isomer, usually the *trans*. Outline all steps in the conversion of a mixture of 75% *trans*-2-pentene and 25% *cis*-2-pentene into essentially pure *cis*-2-pentene.

11.9   Electrophilic addition to alkynes

Addition of acids like the hydrogen halides is electrophilic addition, and it appears to follow the same mechanism with alkynes as with alkenes (Sec. 8.9): via an intermediate carbocation. The difference is that here the intermediate is a *vinylic cation*.

$$-C\equiv C- \;+\; H:Z \;\longrightarrow\; \underset{\oplus}{-\overset{\overset{\displaystyle H}{|}}{C}=C-} \;+\; :Z \;\longrightarrow\; -\overset{\overset{\displaystyle H}{|}}{C}=\overset{\overset{\displaystyle Z}{|}}{C}-$$

A vinylic cation

In Sec. 10.19 we learned that—relative to the substrates for heterolysis—vinyl cations are even less stable than primary alkyl cations; and we saw that, by heterolysis, they are formed comparatively slowly and can be generated only by the departure of "super" leaving groups.

Now, in electrophilic addition to alkenes, we saw (Sec. 8.12), reactivity depends upon the stability of the intermediate carbocation: the more stable the carbocation, the faster it is formed. Does this mean, then, that addition to alkynes will be a great deal slower than to alkenes?

The fact is, it is *not* very much slower: addition of protic acids to alkynes takes place at very much the same rate as to alkenes. The explanation is found in our definition of stability of a carbocation: *relative to the substrate from which it is generated.* Relative to substrates for heterolysis, vinylic cations *are* unstable, and we have attributed this to the unusually strong bond holding the leaving group in vinylic substrates—not to any inherent instability in the cations themselves. And by heterolysis vinylic cations are slow to form. But in addition reactions the substrates are alkenes and alkynes, and these compounds must be the standards for comparison of carbocation stabilities: an alkene for a saturated carbocation, and an alkyne for a vinylic cation. Relative to the substrate from which each is generated *in an addition reaction*, the two are of about the same stability. The energy climb from alkyne to a vinylic cation is about the same as the climb from an alkene to a saturated cation.

Toward the addition of halogens, alkynes are considerably less reactive than alkenes. For alkenes, as we have seen (Sec. 8.14), this reaction involves the initial formation of a cyclic halonium ion. The lower reactivity of alkynes has been attributed to the greater difficulty of forming such cyclic intermediates.

Problem 11.4   The addition of HCl to 3,3-dimethyl-1-butyne gives the following products: 3,3-dichloro-2,2-dimethylbutane (44%), 2,3-dichloro-2,3-dimethylbutane (18%), 1,3-dichloro-2,3-dimethylbutane (34%). Account in detail for the formation of each of these products. (*Hint*: If you have trouble, see Sec.10.19 *and* Sec.10.13.)

11.10   Hydration of alkynes. Tautomerism

Like alkenes, alkynes can be hydrated. In the presence of acid—and, for simple alkynes, HgSO$_4$ as well—a molecule of water adds to the triple bond. Like hydration of alkenes, this involves electrophilic addition, and proceeds by way of carbocations. But at first glance this does not appear to be the case.

Let us consider the simplest example, hydration of acetylene itself. The product obtained is acetaldehyde, CH_3CHO, which seems a strange product from the attachment of groups to the two triply bonded carbons. Actually, however, the product can be accounted for in a rather simple way.

$$H-C\equiv C-H \xrightarrow{\ H_2O,\ H_2SO_4,\ HgSO_4\ } \underset{\text{Vinyl alcohol}}{H-\overset{\displaystyle}{C}=\overset{\displaystyle}{C}-H} \rightleftharpoons \underset{\text{Acetaldehyde}}{H-\overset{\displaystyle H}{\underset{\displaystyle H}{C}}-\overset{\displaystyle H}{\underset{\displaystyle O}{C}}}$$

Acetylene

If hydration of acetylene followed the same pattern as hydration of alkenes, we would expect addition of H— and —OH to the triple bond to yield the structure that we would call *vinyl alcohol*. But all attempts to prepare vinyl alcohol result—like hydration of acetylene—in the formation of acetaldehyde.

A structure with —OH attached to doubly bonded carbon is called an **enol** (*-ene* for the carbon–carbon double bond, *-ol* for *alcohol*). It is almost always true that when we try to make a compound with the enol structure, we obtain instead a compound with the **keto** structure (one that contains a $C=O$ group). There is an

$$\underset{\text{Enol structure}}{-\overset{\displaystyle|}{C}=\overset{\displaystyle|}{C}-O-H} \rightleftharpoons \underset{\text{Keto structure}}{-\overset{\displaystyle|}{C}-\underset{\displaystyle H}{\overset{\displaystyle|}{C}}=O} \qquad \text{Keto–enol tautomerism}$$

equilibrium between the two structures, but it generally lies very much in favor of the keto form. Thus, vinyl alcohol is formed initially by hydration of acetylene, but it is rapidly converted into an equilibrium mixture that is almost all acetaldehyde.

Rearrangements of this enol–keto kind take place particularly easily because of the polarity of the —O—H bond. A hydrogen ion separates readily from oxygen to form a hybrid anion; but when a hydrogen ion (most likely a *different* one) returns, it may attach itself either to oxygen or to carbon of the anion. When it returns to oxygen, it may readily come off again; but when it attaches itself to

$$\underset{\text{Stronger acid}}{-\overset{\displaystyle|}{C}=\overset{\displaystyle|}{C}-O-H} \rightleftharpoons \left[-\overset{\displaystyle|}{C}\cdots\overset{\displaystyle|}{C}\cdots O\right]^{\ominus} + H^+ \rightleftharpoons \underset{\text{Weaker acid}}{-\overset{\displaystyle|}{C}-\underset{\displaystyle H}{\overset{\displaystyle|}{C}}=O} \qquad \begin{array}{l}\text{Keto–enol}\\ \text{tautomerism}\end{array}$$

carbon, it tends to stay there. This is actually an example of the tendency of a stronger acid to displace a weaker acid from its salts (Sec. 11.11).

Compounds whose structures differ markedly in arrangement of atoms, but which exist in easy and rapid equilibrium, are called **tautomers**. The most common kind of **tautomerism** involves structures that differ in the point of attachment of *hydrogen*. In these cases, as in **keto–enol tautomerism**, the tautomeric equilibrium generally favors the structure in which hydrogen is bonded to carbon rather than to a more electronegative atom; that is, equilibrium favors the weaker acid.

Problem 11.5 Hydration of propyne yields the ketone *acetone*, CH_3COCH_3, rather than the aldehyde CH_3CH_2CHO. What does this suggest about the orientation of the initial addition?

11.11 Acidity of alkynes. Very weak acids

In our earlier consideration of acids (in the Lowry–Brønsted sense, Sec. 1.22), we took *acidity* to be a measure of the tendency of a compound to lose a hydrogen ion. Appreciable acidity is generally shown by compounds in which hydrogen is attached to a rather electronegative atom (e.g., N, O, S, X). The bond holding the hydrogen is polar, and the relatively positive hydrogen can separate as the positive ion; considered from another viewpoint, an electronegative element can better accommodate the pair of electrons left behind. In view of the electronegativity series, F > O > N > C, it is not surprising to find that HF is a fairly strong acid, H_2O a comparatively weak one, NH_3 still weaker, and CH_4 so weak that we would not ordinarily consider it an acid at all.

In organic chemistry we are frequently concerned with the acidities of compounds that do not turn litmus red or neutralize aqueous bases, yet have a tendency—even though small—to lose a hydrogen ion.

A triply bonded carbon acts as though it were an entirely different element—a more electronegative one—from a carbon having only single or double bonds. As a result, hydrogen attached to triply bonded carbon, as in acetylene or any alkyne with the triple bond at the end of the chain ($RC\equiv C-H$), shows appreciable acidity. For example, sodium reacts with acetylene to liberate hydrogen gas and form the compound *sodium acetylide*.

$$HC\equiv C-H + Na \longrightarrow HC\equiv C:^- Na^+ + \tfrac{1}{2}H_2$$
$$\text{Sodium acetylide}$$

Just how strong an acid is acetylene? Let us compare it with two familiar compounds, ammonia and water.

Lithium metal reacts with ammonia to form lithium amide, $LiNH_2$, which is the salt of the weak acid, $H-NH_2$.

$$NH_3 + Li \longrightarrow Li^+NH_2^- + \tfrac{1}{2}H_2$$
$$\text{Lithium}$$
$$\text{amide}$$

Addition of acetylene to lithium amide dissolved in ether yields ammonia and lithium acetylide.

$$HC\equiv C-H \ + \ Li^+\,NH_2^- \ \rightleftarrows \ H-NH_2 \ + \ HC\equiv C^-\,Li^+$$

| Stronger acid | Stronger base | Weaker acid | Weaker base |

The weaker acid, $H-NH_2$, is displaced from its salt by the stronger acid, $HC\equiv C-H$. In other language, the strong base, NH_2^-, pulls the proton away from the weaker base, $HC\equiv C^-$; if NH_2^- holds the proton more tightly than $HC\equiv C^-$, then $H-NH_2$ must necessarily be a weaker acid than $HC\equiv C-H$.

Addition of water to lithium acetylide forms lithium hydroxide and regenerates the acetylene.

$$\text{H---OH} \ + \ \text{HC}\equiv\text{C}^-\text{Li}^+ \ \rightleftharpoons \ \text{HC}\equiv\text{C---H} \ + \ \text{Li}^+\text{OH}^-$$

| Stronger acid | Stronger base | Weaker acid | Weaker base |

The weaker acid, $HC\equiv C$—H, is displaced from its salt by the stronger acid, H—OH.

Thus we see that acetylene is a stronger acid than ammonia, but a weaker acid than water.

Relative acidities $H_2O > HC\equiv CH > NH_3 > RH$

Relative basicities $OH^- \ < \ HC\equiv C^- \ < \ NH_2^- \ < \ R^-$

Other alkynes that contain a hydrogen attached to triply bonded carbon—that is, *terminal alkynes*—show comparable acidity.

The method we have just described for comparing the acidities of alkanes, ammonia, and water is a general one, and has been used to determine the relative acidities of a number of extremely weak acids. *One compound is shown to be a stronger acid than another by its ability to displace the second compound from salts.*

$$\text{A---H} + \text{B}^-\text{M}^+ \ \longrightarrow \ \text{B---H} + \text{A}^-\text{M}^+$$

| Stronger acid | Weaker acid |

According to our sequence, acetylene should be a stronger acid than an alkane, RH. This is quite true, and the difference in acidity is of considerable use in synthesis. If a terminal acetylene is treated with an alkylmagnesium halide or an alkyllithium, the alkane is displaced from its "salt", and the metal acetylide is obtained. For example:

$$CH_3C\equiv C\text{---H} \ + \ C_2H_5MgBr \ \longrightarrow \ C_2H_6 \ + \ CH_3C\equiv C\text{---MgBr}$$

Propyne   Propyn-1-ylmagnesium bromide

$$CH_3C\equiv C\text{---H} \ + \ CH_3CH_2CH_2CH_2Li \ \longrightarrow \ CH_3CH_2CH_2CH_3 \ + \ CH_3C\equiv C\text{---Li}$$

Propyne   Propyn-1-yllithium

Such reactions provide the best route to these important organometallic compounds.

How can we account for the fact that hydrogen attached to triply bonded carbon is especially acidic? How can we account for the fact that acetylene is a stronger acid than, say, ethane? A possible explanation can be found in the electronic configurations of the anions.

If acetylene is a stronger acid than ethane, then the acetylide ion must be a weaker base than the ethide ion, $C_2H_5^-$. In the acetylide anion the unshared pair

$$HC{\equiv}C{:}H \ \xrightleftharpoons{} \ H^+ \ + \ HC{\equiv}C{:}^-$$

Acetylene   Acetylide ion

Stronger   *Weaker*
acid   *base*

$$CH_3CH_2{:}H \ \longleftarrow \ H^+ \ + \ CH_3CH_2{:}^-$$

Ethane   Ethide ion

Weaker   *Stronger*
acid   *base*

of electrons occupies an *sp* orbital; in the ethide anion the unshared pair of electrons occupies an *sp³* orbital. The availability of this pair for sharing with acids determines the basicity of the anion. Now, compared with an *sp³* orbital, an *sp* orbital has less *p* character and more *s* character (Sec. 7.4). An electron in a *p* orbital is at some distance from the nucleus and is held relatively loosely; an electron in an *s* orbital, on the other hand, is close to the nucleus and is held more tightly. The acetylide ion is the weaker base since its pair of electrons is held more tightly, in an *sp* orbital.

Problem 11.6   What do you suppose the structure of calcium carbide is? Can you suggest another name for it? What is the nature of its reaction with water?

Problem 11.7   Addition of water to a sodium alkoxide, RONa, yields sodium hydroxide and the parent alcohol, ROH. Addition of an alcohol to sodamide, NaNH₂, yields ammonia and the sodium alkoxide. Expand the sequences of relative acidity and basicity (p. 428) to include alcohols.

11.12   Reactions of metal acetylides. Synthesis of alkynes

We have described metal acetylides as important organometallic compounds. Now, why is this? Because *they enable us to convert little alkynes into big ones.*

Like lithium dialkylcoppers (Sec. 3.17), lithium or sodium acetylides can react with primary alkyl halides.

$$RC{\equiv}C{-}Li \ + \ R'X \ \longrightarrow \ RC{\equiv}C{-}R' \ + \ LiX$$

A 1° alkyl halide

The alkyl group becomes attached to the triply bonded carbon, and a new, larger alkyne has been generated. For example:

$$HC{\equiv}C{-}Li \ + \ CH_3CH_2Br \ \longrightarrow \ HC{\equiv}C{-}CH_2CH_3 \ + \ LiBr$$

Ethynyllithium   1-Butyne

$$CH_3CH_2C{\equiv}C{-}Li \ + \ CH_3Br \ \longrightarrow \ CH_3CH_2C{\equiv}C{-}CH_3 \ + \ LiBr$$

1-Butyn-1-yllithium   2-Pentyne

This reaction gives acceptable yields only with primary alkyl halides, and for a familiar reason (Sec. 7.24): acetylide ions are strong bases, and with secondary or tertiary alkyl halides, elimination is the predominant reaction.

As we shall find in Chapters 17 and 18, complicated alcohols can be readily synthesized by use of Grignard reagents and organolithium compounds. These syntheses can involve not only alkyl organometallics but their alkynyl counterparts as well, and thus yield molecules that contain two highly reactive groups, the carbon–carbon triple bond and —OH.

11.13 Formation of carbon–carbon bonds. Role played by organometallic compounds

There is something very special about the kind of synthesis we have just described. We have taken two organic molecules and converted them into a bigger molecule. We have done something that lies at the heart of organic synthesis: *we have formed a carbon–carbon bond*. Let us look more closely at this process, and at the special role played by organometallic compounds.

Carbon–carbon bonds are most commonly formed heterolytically. This means that one of the carbons furnishes a pair of electrons, and the other carbon accepts them: that is, reaction occurs between a nucleophilic carbon and an electrophilic carbon.

Except for hydrogen or another carbon, the elements that we generally find attached to carbon are more electronegative than carbon, and pull electrons away from it: halogen in alkyl halides, for example, or, as we shall see, oxygen in aldehydes and ketones. The carbon in such compounds is electron-deficient and hence *electrophilic*; it tends to react with nucleophiles. And so, we find alkyl halides typically undergoing nucleophilic substitution, and aldehydes and ketones typically undergoing nucleophilic addition.

Now, if such reactions are to result in the formation of a carbon–carbon bond, we must use reagents in which the nucleophilic element is carbon. Where are we to find such reagents? The answer is: *in organometallic compounds*. Just as electronegative elements make carbon electrophilic, so electropositive elements— metals—make carbon nucleophilic. It is reaction between the nucleophilic carbon of an organometallic reagent and the electrophilic carbon of a substrate that gives rise to a new carbon–carbon bond.

$$-\overset{|}{\underset{|}{C}}\overset{\delta_- \;\; \delta_+}{-M}$$

Organometallic compounds are most commonly synthesized from organic halides. In this synthesis, the nature of carbon is changed, from electrophilic to nucleophilic. This reaction is perhaps the oldest and simplest example of what is called *umpolung*, that is, the reversal of polarity of carbon. The concept of *umpolung* is applied today in a variety of ways in an effort to create nucleophilic carbon. In the formation of an organometallic compound, the electrons that make carbon electron-rich come ultimately from the free metal, as it does what metals, by their very nature, do: give up electrons.

We saw earlier one use of organometallic reagents in making carbon–carbon bonds: the Corey–House synthesis of hydrocarbons (Sec. 3.17). Compared with that reaction the use of metal acetylides has a special advantage: not only is a new carbon–carbon bond formed, but the product contains a highly reactive functional group.

It is little wonder that the carbon–carbon triple bond has become an important building block for organic synthesis. The acidity of a terminal alkyne permits its easy conversion into a metal acetylide. Through this acetylide, a triply bonded structural unit can be introduced into molecules of many kinds. Through addition, this triple bond can then be converted into many other compounds: in particular, into a double bond, and with a high degree of stereoselectivity. We now have a double bond of known stereochemistry at a specific location in the molecule, and the door is open to the many reactions that take place at this functional group.

11.14   Analysis of alkynes

In their response to characterization tests, alkynes resemble alkenes: they de-colorize bromine in carbon tetrachloride without evolution of hydrogen bromide, and they decolorize cold, neutral, dilute permanganate; they are not oxidized by chromic anhydride. Like dienes, however, they are more unsaturated than alkenes. This property can be detected by determination of their molecular formulas (C_nH_{2n-2}) and by a quantitative hydrogenation (two moles of hydrogen are taken up per mole of hydrocarbon).

Proof of structure is best accomplished by the same degradative methods that are used in studying alkenes. Upon ozonolysis alkynes yield carboxylic acids, whereas alkenes yield aldehydes and ketones. For example:

$$CH_3CH_2C\equiv CCH_3 \xrightarrow{\ O_3\ } \xrightarrow{\ H_2O\ } CH_3CH_2COOH + HOOCCH_3$$

2-Pentyne   Carboxylic acids

Acidic alkynes react with certain heavy metal ions, chiefly Ag^+ and Cu^+, to form insoluble acetylides. Formation of a precipitate upon addition of an alkyne to a solution of $AgNO_3$ in alcohol, for example, is an indication of hydrogen attached to triply bonded carbon. This reaction can be used to differentiate *terminal* alkynes (those with the triple bond at the *end* of the chain) from *non-terminal* alkynes.

$$CH_3CH_2C\equiv C-H \xrightarrow{\ Ag^+\ } CH_3CH_2C\equiv C-Ag \left[\xrightarrow{\ HNO_3\ } CH_3CH_2C\equiv C-H + Ag^+\right]$$

1-Butyne   Precipitate   1-Butyne

A terminal alkyne

$$CH_3-C\equiv C-CH_3 \xrightarrow{\ Ag^+\ } \text{no reaction}$$

2-Butyne

A non-terminal alkyne

If allowed to dry, these heavy metal acetylides are likely to explode. They should be destroyed while still wet by warming with nitric acid; the strong mineral acid regenerates the weak acid, acetylene.

(Spectroscopic analysis of alkynes is discussed in Chapter 16.)

Problem 11.8 Contrast the ozonolysis products of the following isomers: (a) 1-pentyne, (b) 2-pentyne, (c) 3-methyl-1-butyne, (d) 1,3-pentadiene, (e) 1,4-penta-diene.

PROBLEMS

1. (a) Draw structures of the seven isomeric alkynes of formula C_6H_{10}. (b) Give the IUPAC and derived name of each. (c) Indicate which ones will react with Ag^+ or $Cu(NH_3)_2{}^+$. (d) Draw structures of the ozonolysis products expected from each.

2. Outline all steps in the synthesis of propyne from each of the following compounds, using any needed organic or inorganic reagents. Follow the other directions given on page 221.

(a) 1,2-dibromopropane
(b) propylene
(c) isopropyl bromide

(d) *n*-propyl alcohol
(e) 1,1-dichloropropane
(f) acetylene

3. Outline all steps in the synthesis from acetylene of each of the following compounds, using any needed organic or inorganic reagents.

(a) ethylene
(b) ethane
(c) 1,1-dibromoethane
(d) vinyl chloride
(e) 1,2-dichloroethane
(f) acetaldehyde
(g) propyne

(h) 1-butyne
(i) 2-butyne
(j) *cis*-2-butene
(k) *trans*-2-butene
(l) 1-pentyne
(m) 2-pentyne
(n) 3-hexyne

4. Give structures and names of the organic products expected from the reaction (if any) of 1-butyne with:

(a) 1 mol H_2, Ni
(b) 2 mol H_2, Ni
(c) 1 mol Br_2
(d) 2 mol Br_2
(e) 1 mol HCl
(f) 2 mol HCl
(g) H_2O, H^+, Hg^{2+}
(h) Ag^+

(i) product (h) + HNO_3
(j) $LiNH_2$
(k) product (j) + C_2H_5Br
(l) product (j) + *tert*-butyl chloride
(m) C_2H_5MgBr
(n) product (m) + H_2O
(o) O_3, then H_2O
(p) hot $KMnO_4$

5. Outline all steps in the synthesis from 2-butyne of each of the following compounds, using any needed organic or inorganic reagents.

(a) *cis*-2-butene
(b) *trans*-2-butene
(c) *meso*-2,3-dibromobutane
(d) racemic *threo*-3-chloro-2-butanol

(e) *meso*-2,3-butanediol
(f) racemic 2,3-butanediol
(g) 2-butanone, $CH_3CH_2COCH_3$

6. Outline all steps in a possible laboratory synthesis of each of the following, using acetylene and alcohols of four carbons or fewer as your only organic source, and any necessary inorganic reagents. (*Remember*: Work backwards.)

(a) *meso*-3,4-dibromohexane   (b) racemic (2*R*,3*R*;2*S*,3*S*)-2,3-heptanediol

7. Describe simple chemical tests that would distinguish between:

(a) 2-pentyne and *n*-pentane
(b) 1-pentyne and 1-pentene
(c) 1-pentyne and 2-pentyne
(d) 1,3-pentadiene and *n*-pentane

(e) 1,3-pentadiene and 1-pentyne
(f) 2-hexyne and isopropyl alcohol
(g) allyl bromide and 2,3-dimethyl-1,3-butadiene

Tell exactly what you would *do* and *see*.

8. Describe chemical methods (not necessarily simple tests) that would distinguish between:

(a) 2-pentyne and 2-pentene
(b) 1,4-pentadiene and 2-pentene

(c) 1,4-pentadiene and 2-pentyne
(d) 1,4-pentadiene and 1,3-pentadiene

9. On the basis of physical properties, an unknown compound is believed to be one of the following:

n-pentane (b.p. 36 °C)
2-pentene (b.p. 36 °C)
1-chloropropene (b.p. 37 °C)
trimethylethylene (b.p. 39 °C)

1-pentyne (b.p. 40 °C)
methylene chloride (b.p. 40 °C)
3,3-dimethyl-1-butene (b.p. 41 °C)
1,3-pentadiene (b.p. 42 °C)

Describe how you would go about finding out which of the possibilities the unknown actually is. Where possible, use simple chemical tests; where necessary, use more elaborate chemical methods like quantitative hydrogenation and cleavage. Tell exactly what you would *do* and *see*.

10. *Muscalure* is the sex pheromone (Sec. 9.3) of the common house fly. On the basis of the following synthesis, give the structure of muscalure (and, of course, of the intermediates A and B).

$$n\text{-}C_{13}H_{27}C{\equiv}CH + n\text{-BuLi} \longrightarrow A\ (C_{13}H_{27}Li)$$
$$A + n\text{-}C_8H_{17}Br \longrightarrow B\ (C_{23}H_{44})$$
$$B + H_2,\ \text{Lindlar catalyst} \longrightarrow muscalure\ (C_{23}H_{46})$$

12

Alicyclic Hydrocarbons

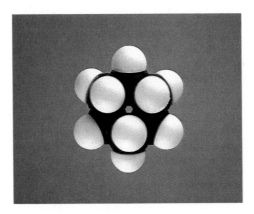

12.1 Open-chain and cyclic compounds

In the compounds that we have studied so far, the carbon atoms are attached to one another to form *chains*; these are called **open-chain** compounds. In many compounds, however, the carbon atoms are arranged to form *rings*; these are called **cyclic** compounds.

In this chapter we shall take up *alicyclic* compounds (*aliphatic cyclic* compounds). Much of their chemistry we already know, since it is essentially the chemistry of their open-chain counterparts: a cycloalkane, for example, *is* an alkane, and in general acts like one. But the cyclic nature of some of these compounds confers very special properties on them. It was here that the study of conformational analysis had its real beginnings, and it is here that we, in turn, can begin to appreciate the practical importance of this branch of stereochemistry. Being a ring—or several rings—places restrictions on the shape a molecule can assume, or on just where solvent molecules can cluster; steric hindrance can be increased—or decreased; the attack by a reagent can be limited to just one direction. It is because of these special properties that, over the years, alicyclic chemistry has become what Professor Lloyd Ferguson, of the California State College at Los Angeles, has called "the playground for organic chemists". It is on some of these special properties that we shall focus our attention.

435

12.2   Nomenclature

Cyclic aliphatic hydrocarbons are named by prefixing **cyclo-** to the name of the corresponding open-chain hydrocarbon having the same number of carbons as the ring. For example:

$$
\begin{array}{ccc}
\text{H}_2\text{C} & \text{H}_2\text{C—CH}_2 & \text{CH} \\
\quad \diagdown\text{CH}_2 & | \quad | & \diagup \quad \diagdown \\
\text{H}_2\text{C}\diagup & \text{H}_2\text{C—CH}_2 & \text{H}_2\text{C} \quad \text{CH} \\
& & \text{H}_2\text{C—CH}_2 \\
\text{Cyclopropane} & \text{Cyclobutane} & \text{Cyclopentene}
\end{array}
$$

Table 12.1   CYCLIC ALIPHATIC HYDROCARBONS

Name	M.p., °C	B.p., °C	Relative density (at 20 °C)
Cyclopropane	− 127	− 33	
Cyclobutane	− 80	13	
Cyclopentane	− 94	49	0.746
Cyclohexane	6.5	81	.778
Cycloheptane	− 12	118	.810
Cyclooctane	14	149	.830
Methylcyclopentane	− 142	72	.749
cis-1,2-Dimethylcyclopentane	− 62	99	.772
trans-1,2-Dimethylcyclopentane	− 120	92	.750
Methylcyclohexane	− 126	100	.769
Cyclopentene	− 93	46	.774
1,3-Cyclopentadiene	− 85	42	.798
Cyclohexene	− 104	83	.810
1,3-Cyclohexadiene	− 98	80.5	.840
1,4-Cyclohexadiene	− 49	87	.847

Substituents on the ring—alkyl groups, halogens—are named, and their positions are indicated by numbers. We assign position 1 to a particular carbon and then number either clockwise or counterclockwise around the ring; we do all this in such a way as to give the lowest combination of numbers. For example:

$$
\begin{array}{ccc}
\text{H}_2\text{C}\diagdown \quad \diagup\text{H} & \text{CH}_3\diagdown \quad \diagup\text{CH}_3 & \text{CH}_3 \\
\quad \text{C} & \text{C} & | \\
\text{H}_2\text{C}\diagup \quad \diagdown\text{Cl} & \text{H}_2\text{C} \quad \text{CH}_2 & \text{CH} \\
& \text{H}_2\text{C—CH}_2 & \\
\text{Chlorocyclopropane} & \text{1,1-Dimethylcyclopentane} & \text{1,3-Dimethylcyclohexane}
\end{array}
$$

In simple cycloalkenes and cycloalkynes the doubly and triply bonded carbons are considered to occupy positions 1 and 2. For example:

3-Ethylcyclopentene　　　　　　　　　　　　　1,3-Cyclohexadiene

For convenience, aliphatic rings are often represented by simple geometric figures: a triangle for cyclopropane, a square for cyclobutane, a pentagon for cyclopentane, a hexagon for cyclohexane, and so on. It is understood that two hydrogens are located at each corner of the figure unless some other group is indicated. For example:

Cyclopentane　　　　　　3-Ethylcyclopentene　　　　　　1,3-Cyclopentadiene

Cyclohexane　　　1,3-Dimethylcyclohexane　　　2-Chloro-1-methylcyclohexane

Polycyclic compounds contain two or more rings that share two or more carbon atoms. We can illustrate the naming system with *norbornane*, whose systematic name is bicyclo[2.2.1]heptane: (a) *heptane*, since it contains a total of *seven* carbon atoms; (b) *bicyclo*, since it contains *two* rings, that is, breaking two carbon–carbon

Bicyclo[2.2.1]heptane　　　Bicyclo[2.2.2]octa-2-ene　　Tricyclo[2.2.1.0^{2,6}]heptane
Norbornane　　　　　　　　　　　　　　　　　　　　Nortriclene

bonds converts it into an open-chain compound; (c) [2.2.1], since the number of carbons between bridgeheads (shared carbons) is *two* (C–2 and C–3), *two* (C–5 and C–6), and *one* (C–7).

Polycyclic compounds in a variety of strange and wonderful shapes have been made, and their properties have revealed unexpected facets of organic chemistry. Underlying much of this research there has always been the challenge: *can such a compound be made?*

Cubane　　　　　　　　　　Basketane　　　　　　　　　Adamantane

The ultimate polycyclic aliphatic system is *diamond* which is, of course, not a hydro-carbon at all, but one of the allotropic forms of elemental carbon. In diamond each carbon

Diamond

atom is attached to four others by tetrahedral bonds of the usual single bond length, 1.54 Å. (Note the cyclohexane chairs, Sec. 12.11.)

12.3   Industrial source

We have already mentioned (Sec. 3.13) that petroleum from certain areas (in particular California) is rich in cycloalkanes, known to the petroleum industry as *naphthenes*. Among these are cyclohexane, methylcyclohexane, methylcyclo-pentane, and 1,2-dimethylcyclopentane.

These cycloalkanes are converted by *catalytic reforming* into aromatic hydro-carbons, and thus provide one of the major sources of these important compounds (Sec. 15.5). In this reaction there is *elimination* of hydrogen from the molecules. For example:

Methylcyclohexane
Aliphatic

$$\xrightarrow[\text{300 lb/in.}^2]{\text{Mo}_2\text{O}_3 \cdot \text{Al}_2\text{O}_3,\ 560\ °\text{C}}$$

$$C_6H_5CH_3 + 3H_2 \qquad \text{Dehydrogenation}$$

Toluene *Elimination*
Aromatic

Just as elimination of hydrogen from cyclic aliphatic compounds yields aromatic compounds, so *addition* of hydrogen to aromatic compounds yields cyclic aliphatic compounds, specifically cyclohexane derivatives. An important example of this is the hydrogenation of benzene to yield pure cyclohexane.

$$C_6H_6 + 3H_2 \xrightarrow[\text{25 atm.}]{\text{Ni, 150–250 °C}}$$

Benzene
Aromatic

Cyclohexane
Aliphatic

Hydrogenation
Addition

As we might expect, hydrogenation of substituted benzenes yields substituted cyclohexanes. For example:

$$C_6H_5OH + 3H_2 \xrightarrow[\text{150 atm.}]{\text{Ni, 150-200 °C}}$$

Phenol
Aromatic

CH₂
H₂C   CHOH
H₂C   CH₂
CH₂

Cyclohexanol
Aliphatic

From cyclohexanol many other cyclic compounds containing a six-membered ring can be made.

12.4 Preparation

Preparation of alicyclic compounds from other aliphatic compounds generally involves two stages: (a) conversion of some open-chain compound or compounds into a compound that contains a ring, a process called *cyclization*; (b) conversion of the cyclic compound thus obtained into the kind of compound that we want: for example, conversion of a cyclic alcohol into a cyclic alkyl halide, or of a cyclic alkene into a cyclic alkane.

Very often, cyclic compounds are made by the *adapting* of a standard method of preparation to the job of closing a ring. For example, we have seen (Sec. 3.17) that the alkyl groups of two alkyl halides can be coupled together through conversion of one halide into an organometallic compound (a lithium dialkylcopper):

$$CH_3CH_2-Cl \longrightarrow CH_3CH_2-M$$
$$CH_3CH_2-Cl$$

Ethyl chloride
2 moles

$$\begin{array}{c} CH_3CH_2 \\ | \\ CH_3CH_2 \end{array}$$

n-Butane

The same method applied to a *di*halide can bring about coupling between two alkyl groups *that are part of the same molecule*:

$$H_2C \begin{array}{c} CH_2-Cl \\ \\ CH_2-Cl \end{array} \xrightarrow{\text{Zn, NaI}} H_2C \begin{array}{c} CH_2-ZnX \\ \\ CH_2-X \end{array} \longrightarrow H_2C \begin{array}{c} CH_2 \\ | \\ CH_2 \end{array}$$

1,3-Dichloropropane   Cyclopropane

In this case zinc happens to do a good job. Although this particular method works well only for the preparation of cyclopropane, it illustrates an important technique: the carrying out of what is normally an *intermolecular* (between-molecules) reaction under such circumstances that it becomes an *intramolecular* (within-a-molecule) reaction. As we can see, it involves tying together the ends of a difunctional molecule.

Alicyclic hydrocarbons are prepared from other cyclic compounds (e.g., halides or alcohols) by exactly the same methods that are used for preparing open-chain hydrocarbons from other open-chain compounds. Because we are dealing with cyclic molecules, however, there can sometimes be special stereochemical features to these reactions; we shall look at one of these in Sec. 12.15, after we have learned something about the stereoisomerism of alicyclic compounds.

Problem 12.1 Starting with cyclohexanol (Sec. 12.3) how would you prepare: (a) cyclohexene, (b) 3-bromocyclohexene, (c) 1,3-cyclohexadiene?

Problem 12.2 Bromocyclobutane can be obtained from open-chain compounds. How would you prepare cyclobutane from it?

The most important route to rings of many different sizes is through the important class of reactions called **cycloadditions**: *reactions in which molecules are added together to form rings.* We shall see many examples of cycloaddition (Secs. 12.16–12.17, 31.8, and 33.9).

12.5 Reactions

With certain very important and interesting exceptions, alicyclic hydrocarbons undergo the same reactions as their open-chain analogs.

Cycloalkanes undergo chiefly free-radical substitution (compare Sec. 3.19). For example:

$$\underset{\text{Cyclopropane}}{\overset{H_2C}{\underset{H_2C}{\diagup}}\!\!\!>\!CH_2} + Cl_2 \xrightarrow{\text{light}} \underset{\text{Chlorocyclopropane}}{\overset{H_2C}{\underset{H_2C}{\diagup}}\!\!\!>\!CH\!-\!Cl} + HCl$$

$$\underset{\text{Cyclopentane}}{\overset{CH_2}{\underset{H_2C-CH_2}{\overset{H_2C}{\diagup}\diagdown CH_2}}} + Br_2 \xrightarrow{300\,°C} \underset{\text{Bromocyclopentane}}{\overset{CH_2}{\underset{H_2C-CH_2}{\overset{H_2C}{\diagup}\diagdown CH\!-\!Br}}} + HBr$$

Cycloalkenes undergo chiefly addition reactions, both electrophilic and free-radical (compare Sec. 8.2); like other alkenes, they can also undergo cleavage and allylic substitution. For example:

$$\underset{\text{Cyclohexene}}{\overset{CH}{\underset{CH_2}{\overset{H_2C}{\diagup}\diagdown{\overset{\parallel}{CH}}}}} + Br_2 \longrightarrow \underset{\text{1,2-Dibromocyclohexane}}{\overset{CHBr}{\underset{CH_2}{\overset{H_2C}{\diagup}\diagdown CHBr}}}$$

$$\underset{\text{1-Methylcyclopentene}}{\overset{CH_3}{\underset{H_2C-CH_2}{\overset{H_2C}{\diagup}\diagdown CH}}} + HI \longrightarrow \underset{\text{1-Iodo-1-methylcyclopentane}}{\overset{I\diagdown\;\diagup CH_3}{\underset{H_2C-CH_2}{\overset{H_2C}{\diagup}\diagdown CH_2}}}$$

3,5-Dimethylcyclopentene   A di-aldehyde

The two smallest cycloalkanes, cyclopropane and cyclobutane, show certain chemical properties that are entirely different from those of the other members of their family. Some of these exceptional properties fit into a pattern and, as we shall see, can be understood in a general way.

The chemistry of bicyclic compounds is even more remarkable, and is one of the most intensively studied areas of organic chemistry (Sec. 32.10).

12.6 Reactions of small-ring compounds. Cyclopropane and cyclobutane

Besides the free-radical substitution reactions that are characteristic of cyclo-alkanes and of alkanes in general, cyclopropane and cyclobutane undergo certain addition reactions. These addition reactions destroy the cyclopropane and cyclo-butane ring systems, and yield open-chain products. For example:

In each of these reactions a carbon–carbon bond is broken, and the two atoms of the reagent appear at the ends of the propane chain:

In general, cyclopropane undergoes addition less readily than propylene: chlorination, for example, requires a Lewis acid catalyst to polarize the chlorine molecule (compare Sec. 14.11). Yet the reaction with sulfuric acid and other aqueous protic acids takes place considerably faster for cyclopropane than for an alkene. (Odder still, treatment with bromine and $FeBr_3$ yields a grand mixture of bromopropanes.)

Cyclobutane does not undergo most of the ring-opening reactions of cyclo-propane; it is hydrogenated, but only under more vigorous conditions than those required for cyclopropane. Thus cyclobutane undergoes addition less readily than cyclopropane and, with some exceptions, cyclopropane less readily than an alkene. The remarkable thing is that these cycloalkanes undergo addition at all.

12.7   Baeyer strain theory

In 1885 Adolf von Baeyer (of the University of Munich) proposed a theory to account for certain aspects of the chemistry of cyclic compounds. The part of his theory dealing with the ring-opening tendencies of cyclopropane and cyclobutane is generally accepted today, although it is dressed in more modern language. Other parts of his theory have been shown to be based on false assumptions, and have been discarded.

Baeyer's argument was essentially the following. In general, when carbon is bonded to four other atoms, the angle between any pair of bonds is the tetrahedral angle 109.5°. But the ring of cyclopropane is a triangle with three angles of 60°, and the ring of cyclobutane is a square with four angles of 90°. In cyclopropane or cyclobutane, therefore, one pair of bonds to each carbon cannot assume the tetrahedral angle, but must be compressed to 60° or 90° to fit the geometry of the ring.

These deviations of bond angles from the "normal" tetrahedral value cause the molecules to be *strained*, and hence to be unstable compared with molecules in which the bond angles are tetrahedral. Cyclopropane and cyclobutane undergo ring-opening reactions since these relieve the strain and yield the more stable open-chain compounds. Because the deviation of the bond angles in cyclopropane (109.5° − 60° = 49.5°) is greater than in cyclobutane (109.5° − 90° = 19.5°), cyclopropane is more highly strained, more unstable, and more prone to undergo ring-opening reactions than is cyclobutane.

The angles of a regular pentagon (108°) are very close to the tetrahedral angle (109.5°), and hence cyclopentane should be virtually free of angle strain. The angles of a regular hexagon (120°) are somewhat larger than the tetrahedral angle, and hence, Baeyer proposed (incorrectly), there should be a certain amount of strain in cyclohexane. Further, he suggested (incorrectly) that as one proceeded to cyclohéptane, cyclooctane, etc., the deviation of the bond angles from 109.5° would become progressively larger, and the molecules would become progressively more strained.

Thus Baeyer considered that rings smaller or larger than cyclopentane or cyclohexane were unstable; it was because of this instability that the three- and four-membered rings underwent ring-opening reactions; it was because of this instability that great difficulty had been encountered in the synthesis of the larger rings. How does Baeyer's strain theory agree with the facts?

12.8   Heats of combustion and relative stabilities of the cycloalkanes

We recall (Sec. 2.6) that the heat of combustion is the quantity of heat evolved when one mole of a compound is burned to carbon dioxide and water. Like heats of hydrogenation (Secs. 8.4 and 10.21), heats of combustion can often furnish valuable information about the relative stabilities of organic compounds. Let us see if the heats of combustion of the various cycloalkanes support Baeyer's proposal that rings smaller or larger than cyclopentane and cyclohexane are unstable.

Examination of the data for a great many compounds has shown that the heat of combustion of an aliphatic hydrocarbon agrees rather closely with that calculated by assuming a certain characteristic contribution from each structural unit. For open-chain alkanes each methylene group, $-CH_2-$, contributes very close to 157.4 kcal/mol to the heat of combustion. Table 12.2 lists the heats of combustion that have been measured for some of the cycloalkanes.

Table 12.2   HEATS OF COMBUSTION OF CYCLOALKANES

Ring size	Heat of combustion per CH_2, kcal/mol	Ring size	Heat of combustion per CH_2, kcal/mol
3	166.6	10	158.6
4	164.0	11	158.4
5	158.7	12	157.6
6	157.4	13	157.8
7	158.3	14	157.4
8	158.6	15	157.5
9	158.8	17	157.2
	Open-chain 157.4		

We notice that for cyclopropane the heat of combustion per $-CH_2-$ group is 9 kcal higher than the open-chain value of 157.4; for cyclobutane it is 7 kcal higher than the open-chain value. Whatever the compound in which it occurs, a $-CH_2-$ group yields the same products on combustion: carbon dioxide and water.

$$-CH_2- + \tfrac{3}{2}O_2 \longrightarrow CO_2 + H_2O + heat$$

If cyclopropane and cyclobutane evolve more energy per $-CH_2-$ group than an open-chain compound, it can mean only that they *contain* more energy per $-CH_2-$ group. In agreement with the Baeyer angle-strain theory, then, cyclopropane and cyclobutane are less stable than open-chain compounds; it is reasonable to suppose that their tendency to undergo ring-opening reactions is related to this instability.

According to Baeyer, rings larger than cyclopentane and cyclohexane also should be unstable, and hence also should have high heats of combustion; furthermore, relative instability—and, with it, heat of combustion—should increase steadily with ring size. However, we see from Table 12.2 that almost exactly the opposite is true. For none of the rings larger than four carbons does the heat of combustion per $-CH_2-$ deviate much from the open-chain value of 157.4. Indeed, one of the biggest deviations is for Baeyer's "most stable" compound, cyclopentane: 1.3 kcal per $-CH_2-$, or 6.5 kcal for the molecule. Rings containing seven to eleven carbons have about the same value as cyclopentane, and when we reach rings of twelve carbons or more, heats of combustion are indistinguishable from the open-chain values. Contrary to Baeyer's theory, then, none of these rings is appreciably less stable than open-chain compounds, and the larger ones are completely free of strain. Furthermore, once they have been synthesized, these large-ring cycloalkanes show little tendency to undergo the ring-opening reactions characteristic of cyclopropane and cyclobutane.

What is wrong with Baeyer's theory that it does not apply to rings larger than four members? Simply this: the angles that Baeyer used for each ring were based on the assumption that the rings were *flat*. For example, the angles of a regular (flat) hexagon are 120°, the angles for a regular decagon are 144°. But the cyclohexane ring is not a regular hexagon, and the cyclodecane ring is not a regular decagon. These rings are not flat, but are puckered (see Fig. 12.1) so that each bond angle of carbon can be 109.5°.

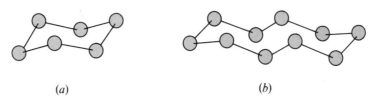

(*a*)   (*b*)

Figure 12.1   Puckered rings. (*a*) Cyclohexane. (*b*) Cyclodecane.

A three-membered ring must be planar, since three points (the three carbon nuclei) define a plane. A four-membered ring need not be planar, but puckering here would increase (angle) strain. A five-membered ring need not be planar, but in this case a planar arrangement would permit the bond angles to have nearly the tetrahedral value. All rings larger than this are puckered. (Actually, as we shall see, cyclobutane and cyclopentane are puckered, too, but this is *in spite of* increased angle strain.)

If large rings are stable, why are they difficult to synthesize? Here we encounter Baeyer's second false assumption. The fact that a compound is difficult to synthesize does not necessarily mean that it is unstable. The closing of a ring requires that two ends of a chain be brought close enough to each other for a bond to form. The larger the ring one wishes to synthesize, the longer must be the chain from which it is made, and the less is the likelihood of the two ends of the chain approaching each other. Under these conditions the end of one chain is more likely to encounter the end of a *different* chain, and thus yield an entirely different product (see Fig. 12.2).

Figure 12.2   Ring closure (upper) *vs.* chain lengthening (lower).

The methods that are used successfully to make large rings take this fact into consideration. Reactions are carried out in highly dilute solutions where collisions between two different chains are unlikely; under these conditions the ring-closing reaction, although slow, is the principal one. Five- and six-membered rings are the kind most commonly encountered in organic chemistry because they are large enough to be free of angle strain, and small enough that ring closure is likely.

12.9   Orbital picture of angle strain

What is the meaning of Baeyer's angle strain in terms of the modern picture of the covalent bond?

We have seen (Sec. 1.8) that, for a bond to form, two atoms must be located so that an orbital of one overlaps an orbital of the other. For a given pair of atoms, the greater the overlap of atomic orbitals, the stronger the bond. When carbon is bonded to four other atoms, its bonding orbitals (sp^3 orbitals) are directed to the corners of a tetrahedron; the angle between any pair of orbitals is thus 109.5°. Formation of a bond with another carbon atom involves overlap of one of these sp^3 orbitals with a similar sp^3 orbital of the other carbon atom. This overlap is most effective, and hence the bond is strongest, when the two atoms are located so that an sp^3 orbital of each atom points toward the other atom. This means that when carbon is bonded to two other carbon atoms the C—C—C bond angle should be 109.5°.

In cyclopropane, however, the C—C—C bond angle cannot be 109.5°, but instead must be 60°. As a result, the carbon atoms cannot be located to permit their sp^3 orbitals to point toward each other (see Fig. 12.3). There is less overlap and the bond is weaker than the usual carbon–carbon bond.

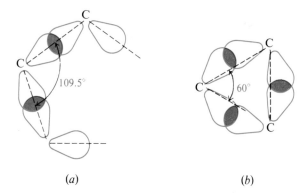

(a)   (b)

Figure 12.3   Angle strain. (a) Maximum overlap permitted for open-chain or large-ring compounds. (b) Poor overlap for the cyclopropane ring. Bent bonds have much p character.

The decrease in stability of a cyclic compound attributed to *angle strain* is due to poor overlap of atomic orbitals in the formation of the carbon–carbon bonds.

On the basis of quantum mechanical calculations, C. A. Coulson and W. A. Moffitt (of Oxford University) proposed *bent bonds* between carbon atoms of cyclopropane rings; this idea is supported by electron density maps based on x-ray studies. Carbon uses sp^2 orbitals for carbon–hydrogen bonds (which are short and strong), and orbitals with much p character (sp^4 to sp^5) for the carbon–carbon bonds. The high p character of these carbon–carbon bonds, and their location— largely outside the ring—seems to underlie much of the unusual chemistry of these rings. The carbon–carbon bond orbitals can overlap orbitals on adjacent atoms; the resulting delocalization is responsible for the effects of cyclopropyl as a substituent. The carbon–carbon bond orbitals provide a site for the attack by acids that is the first step of ring-opening. (Indeed, "edge-protonated" cyclopropanes seem to be key intermediates in many reactions that do not, on the surface, seem to involve cyclopropane rings.)

Ring-opening is *due to* the weakness of the carbon–carbon bonds, but the *way in which it happens* reflects the unusual nature of the bonds; all this stems ultimately from the geometry of the rings and angle strain.

12.10   Factors affecting stability of conformations

To go more deeply into the chemistry of cyclic compounds, we must use conformational analysis (Sec. 4.20). As preparation for that, let us review the factors that determine the stability of a conformation.

Any atom tends to have bond angles that match those of its bonding orbitals: tetrahedral (109.5°) for sp^3-hybridized carbon, for example. Any deviations from the "normal" bond angles are accompanied by **angle strain** (Secs. 12.8–12.9).

Any pair of tetrahedral carbons attached to each other tend to have their bonds staggered. That is to say, any ethane-like portion of a molecule tends, like ethane, to take up a staggered conformation. Any deviations from the staggered arrangement are accompanied by **torsional strain** (Sec. 3.3).

Any two atoms (or groups) that are not bonded to each other can interact in several ways, depending on their size and polarity, and how closely they are brought together. These non-bonded interactions can be either repulsive or attractive, and the result can be either destabilization or stabilization of the conformation.

Non-bonded atoms (or groups) that just touch each other—that is, that are about as far apart as the sum of their van der Waals radii—attract each other. If brought any closer together, they repel each other: such crowding together is accompanied by **van der Waals strain (steric strain)** (Secs. 1.19 and 3.5).

Non-bonded atoms (or groups) tend to take positions that result in the most favorable **dipole–dipole interactions**: that is, positions that minimize dipole–dipole repulsions or maximize dipole–dipole attractions. (A particularly powerful attraction results from the special kind of dipole–dipole interaction called the **hydrogen bond** (Secs. 1.19 and 6.2).)

All these factors, working together or opposing each other, determine the net stability of a conformation. To figure out what the most stable conformation of a particular molecule should be, one ideally should consider all possible combinations of bond angles, angles of rotation, and even bond lengths, and see which combination results in the lowest energy content. Such calculations have become quite feasible through the use of computers.

Both calculations and experimental measurements show that the final result is a compromise, and that few molecules have the idealized conformations that we assign them and, for convenience, usually work with. For example, probably no tetravalent carbon compound—except one with four identical substituents—has *exactly* tetrahedral bond angles: a molecule accepts a certain amount of angle strain to relieve van der Waals strain or dipole–dipole interaction. In the *gauche* conformer of *n*-butane (Sec. 3.5), the dihedral angle between the methyl groups is not 60°, but almost certainly larger: the molecule accepts some torsional strain to ease van der Waals strain between the methyl groups.

12.11   Conformations of cycloalkanes

Let us look more closely at the matter of puckered rings, starting with cyclohexane, the most important of the cycloalkanes. Let us make a model of the molecule, and examine the conformations that are free of angle strain.

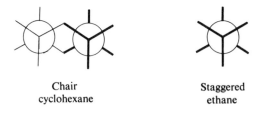

Chair conformation	Boat conformation	Twist-boat conformation
	An energy maximum	

Figure 12.4   Conformations of cyclohexane that are free of angle strain.

First, there is the **chair form** (Fig. 12.4). If we sight along each of the carbon–carbon bonds in turn, we see in every case perfectly staggered bonds:

Chair
cyclohexane

Staggered
ethane

The conformation is thus not only free of angle strain but free of torsional strain as well. It lies at an energy minimum, and is therefore a conformational isomer. *The chair form is the most stable conformation of cyclohexane, and, indeed, of nearly every derivative of cyclohexane.* (See Fig. 12.5.)

(a)   (b)

Figure 12.5 Electronic configuration and molecular shape. A model of cyclohexane in the chair conformation: two views.

Only a scale model can show us what a marvelous creation chair cyclohexane is: symmetrical, compact, and completely free of strain—angle, torsional, van der Waals. Every angle is the tetrahedral angle. About every carbon–carbon bond there is precise staggering. There is no crowding of hydrogen atoms. Indeed, the hydrogens closest together may well feel mild van der Waals attraction for each other: hydrogens on adjacent carbons, and certain hydrogens on alternate carbons—the three facing us in (b) and the three corresponding ones on the opposite face of the molecule.

This architectural perfection results naturally—inevitably—from a happy coincidence: bond angles, bond lengths, and atomic sizes happen to match exactly the geometrical demands of this six-membered ring. *Everything just fits.* Small wonder that we see cyclohexane chairs in the structure of diamond (p. 438), the most stable form of carbon and the hardest substance known. Small wonder that—with an oxygen atom replacing one methylene group—similar chairs make up the most abundant building block of the organic world, D-glucose.

Now, let us take chair cyclohexane and flip the "left" end of the molecule up (Fig. 12.4) to make the *boat conformation*. (Like all the transformations we shall carry out in this section, this involves only rotations about single bonds; what we are making are indeed conformations.) This is not a very happy arrangement. Sighting along either of two carbon–carbon bonds, we see sets of exactly eclipsed bonds,

Flagpole bonds

Boat
cyclohexane

Eclipsed
ethane

and hence we expect considerable torsional strain: as much as in *two* ethane molecules. In addition, there is van der Waals strain due to crowding between the "flagpole" hydrogens, which lie only 1.83 Å apart, considerably closer than the sum of their van der Waals radii (2.5 Å). The boat conformation is a good deal less stable (7.1 kcal/mol, it has been calculated) than the chair conformation. It is believed to lie, not at an energy minimum, but at an energy maximum; it is thus not a conformer, but a transition state between two conformers. (See Fig. 12.6a.)

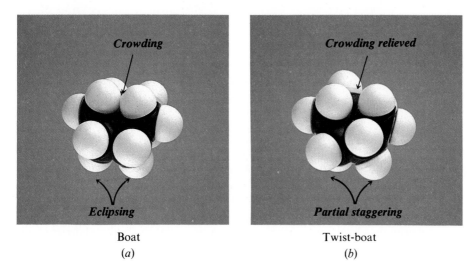

Boat	Twist-boat
(a)	(b)

Figure 12.6   Cyclohexane in the boat and twist-boat conformations. (a) In the boat conformation there is not only eclipsing of bonds, but crowding of flagpole hydrogens. (b) In the twist-boat conformation bonds are partially staggered, and the flagpole hydrogens have moved apart.

Now, what are these two conformers that lie—energetically speaking—on either side of the boat conformation? To see what they are, let us hold a model of the boat conformation with the flagpole hydrogens (H_a and H_b) pointing up, and look down through the ring. We grasp C–2 and C–3 in the right hand and C–5 and

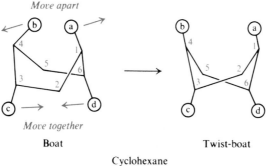

Boat	Twist-boat

Cyclohexane

C–6 in the left hand, and *twist* the molecule so that, say, C–3 and C–6 go *down*, and C–2 and C–5 come *up*. As we do this, H_a and H_b move diagonally apart, and we see (below the ring) a pair of hydrogens, H_c and H_d (on C–3 and C–6, respectively), begin to approach each other. (If this motion is continued, we make a new boat conformation with H_c and H_d becoming the flagpole hydrogens.) When the H_a–H_b distance is equal to the H_c–H_d distance, we stop and examine the molecule. We have minimized the flagpole–flagpole interactions, and at the same time have partly relieved the torsional strain at the C(2)–C(3) and C(5)–C(6) bonds. (See Fig. 12.6*b*.)

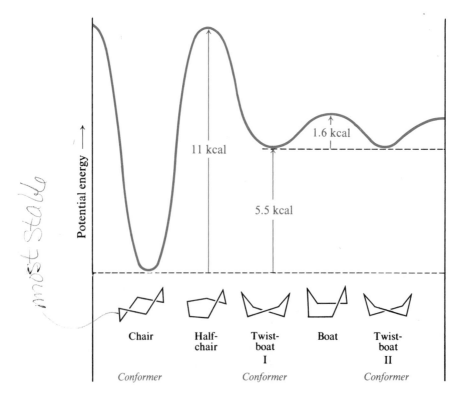

This new configuration is the **twist-boat form**. It is a conformer, lying at an energy minimum 5.5 kcal above the chair conformation. The twist-boat conformer is separated from another, enantiomeric twist-boat conformer by an energy barrier 1.6 kcal high, at the top of which is the boat conformation.

Between the chair form and the twist-boat form lies the highest barrier of all: a transition state conformation (the *half-chair*) which, with angle strain and torsional strain, lies about 11 kcal above the chair form.

The overall relationships are summarized in Fig. 12.7. Equilibrium exists between the chair and twist-boat forms, with the more stable chair form being favored—10 000 to 1 at room temperature.

Figure 12.7 Potential energy relationships among conformations of cyclohexane.

If chair cyclohexane is, conformationally speaking, the perfect specimen of a cycloalkane, planar cyclopentane (Fig. 12.8) must certainly be one of the poorest: there is exact bond eclipsing between every pair of carbons. To (partially) relieve this torsional strain, cyclopentane takes on a slightly puckered conformation, even at the cost of a little angle strain. (See also Problem 10, p. 473.)

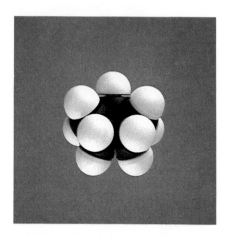

Figure 12.8 Planar cyclopentane: much torsional strain. The molecule is actually puckered.

Evidence of many kinds strongly indicates that cyclobutane is not planar, but rapidly changes between equivalent, slightly folded conformations (Fig. 12.9). Here, too, torsional strain is partially relieved at the cost of a little angle strain.

Figure 12.9 Cyclobutane: rapid transformation between equivalent non-planar "folded" conformations.

Rings containing seven to twelve carbon atoms, too, are less stable than cyclohexane. They are subject to torsional strain and, as scale models reveal, to serious crowding of hydrogens inside the rings (see Fig. 12.10). Only quite large ring systems are as stable as cyclohexane.

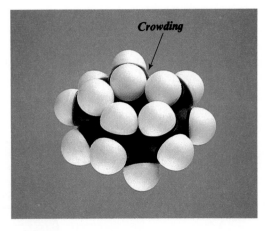

Figure 12.10 Model of cyclodecane. Staggering about carbon–carbon bonds
is achieved only at the expense of crowding of hydrogens inside the ring.

12.12   Equatorial and axial bonds in cyclohexane

Let us return to the model of the chair conformation of cyclohexane (see Fig.
12.11). Although the cyclohexane ring is not flat, we can consider that the carbon
atoms lie roughly in a plane. If we look at the molecule in this way, we see that the
hydrogen atoms occupy two kinds of position: six hydrogens lie in the plane, while

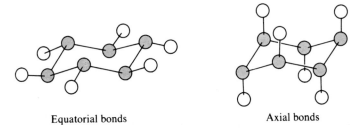

Equatorial bonds   Axial bonds

Figure 12.11   Chair cyclohexane: equatorial and axial bonds.

six hydrogens lie above or below the plane. The bonds holding the hydrogens that
are in the plane of the ring lie in a belt about the "equator" of the ring, and are
called **equatorial bonds**. The bonds holding the hydrogen atoms that are above and
below the plane are pointed along an axis perpendicular to the plane and are called
axial bonds. In the chair conformation each carbon atom has one equatorial bond
and one axial bond.

Cyclohexane itself, in which only hydrogens are attached to the carbon atoms,
is not only free of angle strain and torsional strain, but free of van der Waals strain
as well (see Fig. 12.12). Hydrogens on adjacent carbons are the same distance apart
(2.3 Å) as in (staggered) ethane and, if anything, feel mild van der Waals attraction
for each other. We notice that the three axial hydrogens on the same side of the
molecule are thrown rather closely together, despite the fact that they are attached
to alternate carbon atoms; as it happens, however, they are the same favorable
distance apart (2.3 Å) as the other hydrogens are.

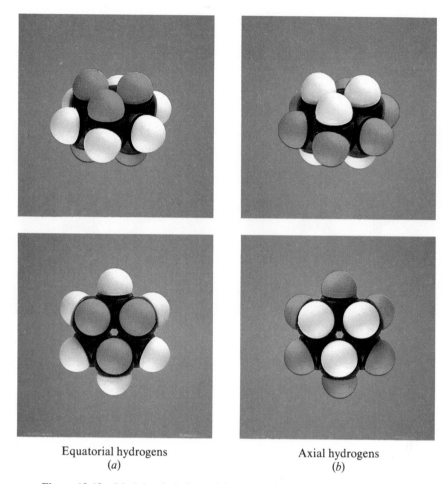

Equatorial hydrogens
(a)

Axial hydrogens
(b)

Figure 12.12   Models of chair cyclohexane. (*a*) Equatorial hydrogens are shown in white. (*b*) Axial hydrogens are shown in white.

If, now, a hydrogen is replaced by a larger atom or group, crowding occurs. The most severe crowding is among atoms held by the three axial bonds on the same side of the molecule; the resulting interaction is called **1,3-diaxial interaction**. Except for hydrogen, *a given atom or group has more room in an equatorial position than in an axial position.*

As a simple example of the importance of 1,3-diaxial interactions, let us consider methylcyclohexane. In estimating relative stabilities of various conformations of this compound, we must focus our attention on methyl, since it is the largest substituent on the ring and hence the one most subject to crowding. There are two possible chair conformations (see Fig. 12.13), one with —CH_3 in an equa-

Equatorial —CH_3

Axial —CH_3

Figure 12.13   Chair conformations of methylcyclohexane.

torial position, the other with —CH_3 in an axial position. As shown in Fig. 12.14, the two axial hydrogens (on C–3 and C–5) approach the axial —CH_3 (on C–1) more closely than any hydrogens approach the equatorial —CH_3. We would expect the equatorial conformation to be the more stable, and it is, by about 1.8 kcal. Most

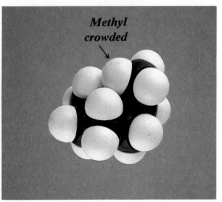

Equatorial —CH_3 Axial —CH_3

Figure 12.14 1,3-Diaxial interaction in methylcyclohexane. An axial —CH_3 is more crowded than an equatorial —CH_3.

molecules (about 95% at room temperature) exist in the conformation with methyl in the uncrowded equatorial position.

In an equatorial position, we see, —CH_3 points *away from* its nearest neighbors: the two hydrogens—one axial, and one equatorial—on the adjacent carbons. This is not true of —CH_3 in an axial position, since it is held by a bond that is *parallel to* the bonds holding its nearest neighbors, the two axial hydrogens.

Conformational analysis can account not only for the fact that one conformation is more stable than another, but often—with a fair degree of accuracy—for just *how much* more stable it is. We have attributed the 1.8-kcal energy difference between the two conformations of methylcyclohexane to 1,3-diaxial interactions between a methyl group and *two* hydrogens. If, on that basis, we assign a value of 0.9 kcal/mol to each 1,3-diaxial methyl–hydrogen interaction, we shall find that we can account amazingly well for the energy differences between conformations of a variety of cyclohexanes containing more than one methyl group.

We notice that 0.9 kcal is nearly the same value that we earlier (Sec. 3.5) assigned to a *gauche* interaction in *n*-butane; examination of models shows that this is not just accidental.

Let us make a model of the conformation of methylcyclohexane with axial methyl. If we hold it so that we can sight along the C(1)–C(2) bond, we see something like this, represented by a Newman projection:

Axial —CH₃

Gauche
n-Butane

The methyl group and C–3 of the ring have the same relative locations as the two methyl groups in the *gauche* conformation of *n*-butane (Sec. 3.5). If we now sight along the C(1)–C(6) bond, we see a similar arrangement but with C–5 taking the place of C–3.

Next, let us make a model of the conformation with equatorial methyl. This time, if we sight along the C(1)–C(2) bond, we see this:

Equatorial —CH₃

Anti
n-Butane

Here, methyl and C–3 of the ring have the same relative locations as the two methyl groups in the *anti* conformation of *n*-butane. And if we sight along the C(1)–C(6) bond, we see methyl and C–5 in the *anti* relationship.

Thus, for each 1,3-diaxial methyl–hydrogen interaction there is a "butane-*gauche*" interaction between the methyl group and a carbon atom of the ring. Of the two approaches, however, looking for 1,3-diaxial interactions is much the easier and has the advantage, when we study substituents other than methyl, of focusing our attention on the sizes of the groups being crowded together.

In general, then, it has been found that (a) chair conformations are more stable than twist conformations, and (b) the most stable chair conformations are those in which the largest groups are in equatorial positions. There are exceptions to both these generalizations (which we shall encounter later in problems), but the exceptions are understandable ones.

Problem 12.3   For other alkylcyclohexanes the difference in energy between equatorial and axial conformations has been found to be: ethyl, 1.9 kcal/mol; isopropyl, 2.1 kcal/mol; and *tert*-butyl, more than 5 kcal/mol. Using models, can you account for the big increase at *tert*-butyl? (*Hint*: Don't forget freedom of rotation about *all* the single bonds.)

12.13   Stereoisomerism of cyclic compounds: *cis* and *trans* isomers

Let us turn for the moment from conformational analysis, and look at configurational isomerism in cyclic compounds.

We shall begin with the compound 1,2-cyclopentanediol. Using models, we find that we can arrange the atoms of this molecule as in I, in which both hydroxyls lie below (or above) the plane of the ring, and as in II, in which one hydroxyl lies above and the other lies below the plane of the ring.

I	II
cis-1,2-Cyclopentanediol	trans-1,2-Cyclopentanediol

I and II cannot be superimposed, and hence are isomers. They differ only in the way their atoms are oriented in space, and hence are stereoisomers. No amount of rotation about bonds can interconvert I and II, and hence they are not conformational isomers. They are configurational isomers; they are interconverted only by breaking of bonds, and hence are isolable. They are not mirror images, and hence are diastereomers; they should, therefore, have different physical properties, as the two diols actually have. Configuration I is designated the *cis* configuration, and II is designated the *trans* configuration. (Compare *cis* and *trans* alkenes, Sec. 7.6.)

Problem 12.4   You have two bottles labeled "1,2-Cyclopentanediol", one containing a compound of m.p. 30 °C, the other a compound of m.p. 55 °C; both compounds are optically inactive. How could you decide, beyond any doubt, which bottle should be labeled "*cis*" and which "*trans*"?

Problem 12.5   (a) Starting from cyclopentanol, outline a synthesis of stereochemically pure *cis*-1,2-cyclopentanediol. (b) Of stereochemically pure *trans*-1,2-cyclopentanediol.

Stereoisomerism of this same sort should be possible for compounds other than diols, and for rings other than cyclopentane. Some examples of isomers that have been isolated are:

cis-1,2-Dibromocyclopentane	trans-1,2-Dibromocyclopentane

cis-1,3-Cyclopentanedicarboxylic acid	trans-1,3-Cyclopentanedicarboxylic acid

cis-1,3-Cyclobutanedicarboxylic acid

trans-1,3-Cyclobutanedicarboxylic acid

cis-1,2-Dimethylcyclopropane

trans-1,2-Dimethylcyclopropane

If we examine models of *cis*- and *trans*-1,2-cyclopentanediol more closely, we find that each compound contains two chiral centers. We know (Sec. 4.18) that compounds containing more than one chiral center are often—but not always—chiral. Are these diols chiral? As always, to test for possible chirality, we construct a model of the molecule and a model of its mirror image, and see if the two are superimposable. When we do this for the *trans* diol, we find that the models are not superimposable. The *trans* diol is chiral, and the two models we have constructed therefore correspond to enantiomers. Next, we find that the models are not interconvertible by rotation about single bonds. They therefore represent, not

mirror

Not superimposable
Enantiomers: resolvable
trans-1,2-Cyclopentanediol

conformational isomers, but configurational isomers; they should be capable of isolation—*resolution*—and, when isolated, each should be optically active.

Next let us look at *cis*-1,2-cyclopentanediol. This, too, contains two chiral centers; is it also chiral? This time we find that a model of the molecule and a

mirror

Superimposable
A meso compound
cis-1,2-Cyclopentanediol

model of its mirror image *are* superimposable. In spite of its chiral centers, *cis*-1,2-cyclopentanediol is not chiral; it cannot exist in two enantiomeric forms, and cannot be optically active. It is a *meso* compound.

We might have recognized *cis*-1,2-cyclopentanediol as a *meso* structure on sight from the fact that one half of the molecule is the mirror image of the other half (Sec. 4.18):

A meso compound
cis-1,2-Cyclopentanediol

Thus, of the two 1,2-cyclopentanediols obtainable by ordinary synthesis, only one is separable into enantiomers, that is, is *resolvable*; this must necessarily be the *trans* diol. The other diol is a single, inactive, non-resolvable compound, and it must have the *cis* configuration.

What is the relationship between the *meso cis* diol and either of the enantiomeric *trans* diols? They are *diastereomers*, since they are stereoisomers that are not enantiomers.

Problem 12.6   Five of the eight structures shown at the bottom of page 456 and the top of page 457 are achiral. Which are these?

12.14   Stereoisomerism of cyclic compounds. Conformational analysis

So far, we have described the relative positions of groups in *cis* and *trans* isomers in terms of flat rings: both groups are below (or above) the plane of the ring, or one group is above and the other is below the plane of the ring. In view of what we have said about puckering, however, we realize that this is a highly simplified picture even for four- and five-membered rings, and for six-membered rings is quite inaccurate.

Let us apply the methods of conformational analysis to the stereochemistry of cyclohexane derivatives; and, since we are already somewhat familiar with interactions of the methyl group, let us use the dimethylcyclohexanes as our examples.

If we consider only the more stable, chair conformations, we find that a particular molecule of *trans*-1,2-dimethylcyclohexane, to take our first example, can exist in two conformations (see Fig. 12.15). In one, both —CH_3 groups are in equatorial positions, and in the other, both —CH_3 groups are in axial positions. Thus, we see, the two —CH_3 groups of the *trans* isomer are not necessarily on opposite sides of the ring; in fact, because of lesser crowding between —CH_3 groups and axial hydrogens of the ring (less 1,3-diaxial interaction), the more stable conformation is the diequatorial one.

Figure 12.15   Chair conformations of *trans*-1,2-dimethylcyclohexane.

A molecule of *cis*-1,2-dimethylcyclohexane can also exist in two conformations (see Fig. 12.16). In this case, the two are of equal stability (they are mirror images) since in each there is one equatorial and one axial $-CH_3$ group.

Figure 12.16   Chair conformations of *cis*-1,2-dimethylcyclohexane.

In the most stable conformation of *trans*-1,2-dimethylcyclohexane, both $-CH_3$ groups occupy uncrowded equatorial positions. In either conformation of the *cis*-1,2-dimethylcyclohexane, only one $-CH_3$ group can occupy an equatorial position. It is not surprising to find that *trans*-1,2-dimethylcyclohexane is more stable than *cis*-1,2-dimethylcyclohexane.

It is interesting to note that in the most stable conformation (diequatorial) of the *trans* isomer, the $-CH_3$ groups are exactly the same distance apart as they are in either conformation of the *cis* isomer. Clearly, it is not repulsion between the $-CH_3$ groups—as one might incorrectly infer from planar representations—that causes the difference in stability between the *trans* and *cis* isomers: the cause is 1,3-diaxial interactions (Sec. 12.12).

Now, just *how much* more stable is the *trans* isomer? In the *cis*-1,2-dimethylcyclohexane there is one axial methyl group, which means *two* 1,3-diaxial methyl–hydrogen interactions: one with each of two hydrogen atoms. (Or, what is equivalent (Sec. 12.12), there are two butane-*gauche* interactions between the methyl groups and carbon atoms of the ring.) In addition, there is one butane-*gauche* interaction between the two methyl groups. On the basis of 0.9 kcal for each 1,3-diaxial methyl–hydrogen interaction or butane-*gauche* interaction, we calculate a total of 2.7 kcal of van der Waals strain for the *cis*-1,2-dimethylcyclohexane. In the (diequatorial) *trans* isomer there are no 1,3-diaxial methyl–hydrogen interactions, but there is one butane-*gauche* interaction between the methyl groups; this confers 0.9 kcal of van der Waals strain on the molecule. We subtract 0.9 kcal from 2.7 kcal and conclude that the *trans* isomer should be more stable than the *cis* isomer by 1.8 kcal/mol, in excellent agreement with the measured value of 1.87 kcal.

Problem 12.7   Compare stabilities of the possible chair conformations of:

(a) *cis*-1,2-dimethylcyclohexane   (d) *trans*-1,3-dimethylcyclohexane
(b) *trans*-1,2-dimethylcyclohexane   (e) *cis*-1,4-dimethylcyclohexane
(c) *cis*-1,3-dimethylcyclohexane   (f) *trans*-1,4-dimethylcyclohexane

(g) On the basis of 0.9 kcal/mol per 1,3-diaxial methyl–hydrogen interaction, predict (where you can) the potential energy difference between the members of each pair of conformations.

Problem 12.8   On theoretical grounds, K. S. Pitzer (then at the University of California) calculated that the energy difference between the conformations of *cis*-1,3-dimethylcyclohexane should be about 5.4 kcal, much larger than that between the chair conformations of *trans*-1,2-dimethylcyclohexane or of *trans*-1,4-dimethylcyclohexane. (a) What special factor must Pitzer have recognized in the *cis*-1,3 isomer? (b) Using the 0.9-kcal value where it applies, what value must you assign to the factor you invoked in (a), if you are to arrive at the energy difference of 5.4 kcal for the *cis*-1,3 conformations? (c) The potential energy difference between *cis*- and *trans*-1,1,3,5-tetramethylcyclohexane was then measured by Norman L. Allinger (University of Georgia) as 3.7 kcal/mol. This measurement was carried out because of its direct bearing on the matter of *cis*-1,3-dimethylcyclohexane. What is the connection between this measurement and parts (a) and (b)? Does Allinger's measurement support Pitzer's calculation?

Problem 12.9   Predict the relative stabilities of the *cis* and *trans* isomers of: (a) 1,3-dimethylcyclohexane; (b) 1,4-dimethylcyclohexane. (c) On the basis of 0.9 kcal/mol per 1,3-diaxial methyl–hydrogen interaction or butane-*gauche* interaction, and assuming that each stereoisomer exists exclusively in its more stable conformation, predict the potential energy difference between members of each pair of stereoisomers.

Conformational analysis of cyclohexane derivatives containing several *different* substituents follows along the same lines as that of the dimethylcyclohexanes. We need to keep in mind that, of two groups, the larger one will tend to call the tune. Because of its very large 1,3-diaxial interactions (Problem 12.3, p. 455), the bulky *tert*-butyl group is particularly prone to occupy an equatorial position. If— as is usually the case—other substituents are considerably smaller than *tert*-butyl, the molecule is virtually locked in a single conformation: the one with an equatorial *tert*-butyl group (see Fig. 12.17). Following a suggestion by Professor Saul Winstein (at the University of California, Los Angeles), *tert*-butyl has been widely used as a holding group, to permit the study of physical and chemical properties associated with a purely axial or purely equatorial substituent.

Problem 12.10   Use the energy differences given in Problem 12.3 (p. 455) to calculate values for the various alkyl–hydrogen 1,3-diaxial interactions, and from these calculate the difference in energy between the two conformations of:

(a) *cis*-4-*tert*-butylmethylcyclohexane
(b) *trans*-4-*tert*-butylmethylcyclohexane
(c) *trans*-3-*cis*-4-dimethyl-*tert*-butylcyclohexane

Now, what can we say about the possible chirality of the 1,2-dimethylcyclohexanes? Let us make a model of *trans*-1,2-dimethylcyclohexane—in the more stable diequatorial conformation, say—and a model of its mirror image. We find they are not superimposable, and therefore are enantiomers. We find that they are not interconvertible, and hence are configurational isomers. (When we flip one of

cis Isomer $trans$ Isomer

Figure 12.17   Diastereomeric cyclohexanes containing a 4-$tert$-butyl group cis or $trans$ to another substituent, —G. In each diastereomer the very large $tert$-butyl group occupies an equatorial position. (Compare the size of $tert$-butyl with that of the brown bromine—itself a fairly large atom—in the CPK models.) The $tert$-butyl group holds —G exclusively in the axial or in the equatorial position, yet, because of its distance, exerts little electronic effect on —G.

these into the opposite chair conformation, it is converted, not into its mirror image, but into a diaxial conformation.) Thus, $trans$-1,2-dimethylcyclohexane should, in principle, be resolvable into (configurational) enantiomers, each of which should be optically active.

mirror

Not superimposable; not interconvertible
$trans$-1,2-Dimethylcyclohexane
A resolvable racemic modification

Next, let us make a model of cis-1,2-dimethylcyclohexane and a model of its mirror image. We find they are not superimposable, and hence are enantiomers. In

mirror

Not superimposable, but interconvertible
cis-1,2-Dimethylcyclohexane
A non-resolvable racemic modification

contrast to what we have said for the *trans* compound, however, we find that these models *are* interconvertible by flipping one chair conformation into the other. These are conformational enantiomers and hence, except possibly at low temperatures, should interconvert too rapidly for resolution and measurement of optical activity.

Thus, just as with the *cis*- and *trans*-1,2-cyclopentanediols (Sec. 12.13), we could assign configurations to the *cis*- and *trans*-1,2-dimethylcyclohexanes by finding out which of the two is resolvable. The *cis*-1,2-dimethylcyclohexane is not literally a *meso* compound, but it is a non-resolvable racemic modification, which for most practical purposes amounts to the same thing.

To summarize, then, 1,2-dimethylcyclohexane exists as a pair of (configurational) diastereomers: the *cis* and *trans* isomers. The *cis* isomer exists as a pair of conformational enantiomers. The *trans* isomer exists as a pair of configurational enantiomers, each of which in turn exists as two conformational diastereomers (axial–axial and equatorial–equatorial).

Because of the ready interconvertibility of chair conformations, it is possible to use planar drawings to predict the configurational stereoisomerism of cyclohexane derivatives. To understand the true geometry of such molecules, however,

mirror

Superimposable
cis-1,2-Cyclohexanediol

mirror

Not superimposable
trans-1,2-Cyclohexanediol

and with it the matter of stability, one must use models and formulas like those in Figs. 12.15 and 12.16.

Problem 12.11 Which of the following compounds are resolvable, and which are non-resolvable? Which are truly *meso* compounds? Use models as well as drawings.

(a) *cis*-1,2-cyclohexanediol   (d) *trans*-1,3-cyclohexanediol
(b) *trans*-1,2-cyclohexanediol   (e) *cis*-1,4-cyclohexanediol
(c) *cis*-1,3-cyclohexanediol   (f) *trans*-1,4-cyclohexanediol

Problem 12.12 Tell which, if any, of the compounds of Problem 12.11 exist as:

(a) a single conformation;
(b) a pair of conformational enantiomers;
(c) a pair of conformational diastereomers;
(d) a pair of (configurational) enantiomers, each of which exists as a single conformation;
(e) a pair of (configurational) enantiomers, each of which exists as a pair of conformational diastereomers;
(f) none of the above answers. (Give the correct answer.)

Problem 12.13 Draw structural formulas for all stereoisomers of the following. Label any *meso* compounds and indicate pairs of enantiomers. Do any (like *cis*-1,2-dimethyl-cyclohexane) exist as a non-resolvable racemic modification?

(a) *cis*-2-chlorocyclohexanol   (d) *trans*-3-chlorocyclopentanol
(b) *trans*-2-chlorocyclohexanol   (e) *cis*-4-chlorocyclohexanol
(c) *cis*-3-chlorocyclopentanol   (f) *trans*-4-chlorocyclohexanol

12.15 Stereochemistry of elimination from alicyclic compounds

So far, we have been concerned with the stereoisomerism of these cyclic compounds, and the relative stabilities of various isomers and conformations. Now let us apply what we have learned, and see how the cyclic nature of these molecules can sometimes determine *the way they react*. That is to say, let us see how the factors we have discussed can affect the relative stabilities of transition states.

As our example, let us return to the stereochemistry of E2 elimination (Sec. 9.7). This reaction, we saw, is stereoselective, and typically involves *anti*-elimination: in the transition state the hydrogen and the leaving group are located in the *anti* relationship, as contrasted to the *gauche* or *eclipsed* (Fig. 12.18).

Figure 12.18 The E2 reaction of alkyl halides: *anti*-elimination. Hydrogen and the leaving group, —X, are as far apart as possible, in the *anti* relationship.

Just how strong the preference for *anti*-elimination from halides can be is best shown by study of cyclic compounds. In cyclohexane rings, 1,2-substituents can take up the *anti* conformation only by occupying axial positions; this, in turn, is possible only if they are *trans* to each other (see Fig. 12.19).

Figure 12.19   Only *trans*-1,2-substituents can assume the *anti* relationship.

To take a specific example: E2 elimination converts *neomenthyl chloride* into a mixture of 75% 3-menthene and 25% 2-menthene. This is about what we might expect, the more stable—because more highly substituted—3-menthene being the preferred product. But, in marked contrast, E2 elimination converts the diastereomeric *menthyl chloride* exclusively into the less stable 2-menthene.

| Neomenthyl chloride | 2-Menthene 25% | 3-Menthene 75% |

| Menthyl chloride | 2-Menthene *Only product* |

How are we to account for these differences in behavior? In neomenthyl chloride there is a hydrogen on either side of the chlorine which is *trans* to the chlorine, and which can take up a conformation *anti* to it. Either hydrogen *can* be eliminated, and the ratio of products is determined in the usual way, by the relative stabilities of the alkenes being formed. In menthyl chloride, on the other hand, only one hydrogen is *trans* to the chlorine, and it is the only one that is eliminated, despite the fact that this yields the less stable alkene.

It is clear that E2 reactions can also proceed by *syn*-elimination: in the transition state the hydrogen and leaving group are in the *eclipsed* (or *gauche*) relationship. Although uncommon for alkyl halides, *syn*-elimination is often observed for quaternary ammonium salts (Sec. 27.6) and sometimes for alkyl sulfonates. On electronic grounds, the most stable transition states seem to be those in which the hydrogen and leaving group are *periplanar* (in the same plane) to permit overlap of incipient *p* orbitals in the partially formed double bond. Of the two periplanar eliminations, the *anti* is probably easier than the *syn*—other things being equal. But various factors may throw the stereochemistry one way or the other. Conformational effects enter in, and the degree of carbanion character; the stereochemistry is affected by the strength of the base and by its bulk and the bulk of the leaving group. Ring systems present special situations: it is difficult for *cis*-1,2-substituents to become *syn*-periplanar in cyclohexanes, but easy in cyclopentanes.

Problem 12.14 Of the various isomeric 1,2,3,4,5,6-hexachlorocyclohexanes, one isomer undergoes dehydrohalogenation by base much more slowly than the others. Which isomer is probably the unreactive one, and why is it unreactive?

Problem 12.15 The behavior of menthyl chloride described above is that observed in reaction with sodium ethoxide in ethanol. By contrast, when menthyl chloride is heated in ethanol in the absence of added base, it yields both 3-menthene (68%) and 2-menthene (32%). How do you account for this difference in behavior?

Problem 12.16 Using models, suggest explanations for the following.
   (a) Attached to a doubly bonded carbon, phenyl greatly stabilizes an alkene (Sec. 15.19), and hence exerts a powerful effect on the orientation of elimination. On E2 elimination with *t*-BuOK/*t*-BuOH, both *cis*- and *trans*-2-phenylcyclopentyl tosylates give 1-phenylcyclopentene as the only alkene; the *cis* isomer reacts nine times as fast as the *trans*.
   (b) On E2 elimination with n-$C_5H_{11}ONa$/n-$C_5H_{11}OH$ to give 2-chloronorbornene, II reacts about 100 times as fast as its diastereomer, I.

I   II

endo-cis-2,3-Dichloronorbornane   *trans*-2,3-Dichloronorbornane

12.16   Carbenes. Methylene. Cycloaddition

The most important route to cyclic compounds, we said in Sec. 12.4, is via the class of reactions called *cycloaddition*. Let us look at one kind of cycloaddition and, at the same time, become acquainted with a highly unusual class of reagents.

The difference between successive members of a homologous series, we have seen, is the CH_2 unit, or *methylene*. But methylene is more than just a building block for the mental construction of compounds; it is an actual molecule, and its chemistry and the chemistry of its derivatives, the **carbenes**, has become one of the most exciting and productive fields of organic research.

Methylene is formed by the photolysis of either *diazomethane*, CH_2N_2, or *ketene*, CH_2=C=O. (Notice that the two starting materials and the two other

$$CH_2=\overset{+}{N}=\overset{-}{N} \xrightarrow{\text{ultraviolet light}} CH_2 + N_2$$

Diazomethane   Methylene

$$CH_2=C=O \xrightarrow{\text{ultraviolet light}} CH_2 + CO$$

Ketene   Methylene

products, nitrogen and carbon monoxide, are pairs of *isoelectronic* molecules, that is, molecules containing the same number of valence electrons.)

Methylene as a highly reactive molecule was first proposed in the 1930s to account for the fact that something formed by the above reactions was capable of removing certain metal mirrors (compare Problem 15, p. 73). Its existence was definitely established in 1959 by spectroscopic studies.

Figure 12.20 Evidence of early (1944) research on methylene, CH_2, by D. Duck. (As unearthed by Professors P. P. Gaspar and G. S. Hammond of the California Institute of Technology.)

These studies revealed that methylene not only exists but exists in two different forms (different spin states), generally referred to by their spectroscopic designations: *singlet* methylene, in which the unshared electrons are paired:

$$CH_2: \qquad H:\overset{\overset{H}{\cdot\cdot}}{C}:$$

Singlet methylene
Unshared electrons paired

and *triplet* methylene, in which the unshared electrons are *not* paired.

$$\cdot CH_2\cdot \qquad H:\overset{\cdot}{C}:H$$

Triplet methylene
*Unshared electrons not paired:
a diradical*

Triplet methylene is thus a free radical; in fact, it is a *di*radical. As a result of the difference in electronic configuration, the two kinds of molecules differ in shape and in chemical properties. Singlet methylene is the less stable form, and is often the form first generated, in the initial photolysis.

The exact chemical properties observed depend upon which form of methylene is reacting, and this in turn depends upon the experimental conditions. In the liquid phase, the first-formed singlet methylene reacts rapidly with the abundant solvent molecules before it loses energy. In the gas phase—especially in the presence of an inert gas like nitrogen or argon—singlet methylene loses energy through collisions and is converted into triplet methylene, which then reacts.

When methylene is generated in the presence of alkenes, there are obtained cyclopropanes. For example:

$$CH_3CH{=}CHCH_3 + CH_2N_2 \xrightarrow{\text{light}} CH_3CH{-}CHCH_3 + N_2$$
$$\underset{CH_2}{\diagdown \diagup}$$

2-Butene   Diazomethane

1,2-Dimethylcyclopropane

This is an example of the most important reaction of methylene and other carbenes: *addition to the carbon–carbon double bond. This particular kind of addition, in which a ring is generated*, is called **cycloaddition**. In its various forms (Secs. 31.8 and 33.9) cycloaddition provides the most important route to rings of various sizes.

$$\diagup\diagdown C{=}C\diagup\diagdown + CH_2 \longrightarrow {-}C{-}C{-} \qquad \textbf{Cycloaddition}$$
$$\underset{CH_2}{\diagdown\diagup}$$

The most striking feature of the addition of methylene is that it can occur with two different kinds of stereochemistry. For example, photolysis of diazomethane in liquid *cis*-2-butene gives only *cis*-1,2-dimethylcyclopropane, and in liquid *trans*-2-butene gives only *trans*-1,2-dimethylcyclopropane. Addition here is stereoselective and stereospecific, and *syn*. Photolysis of diazomethane in gaseous 2-butene—either *cis* or *trans*—gives *both cis*- and *trans*-1,2-dimethylcyclopropanes. Addition here is non-stereoselective and non-stereospecific.

There seems to be little doubt that the following interpretation (due to P. S. Skell of Pennsylvania State University) is, in broad outline, the correct one.

It is **singlet** methylene that undergoes the *stereoselective addition*. Although neutral, singlet methylene is electron-deficient and hence *electrophilic*; like other

$$CH_2{:} + \diagup\diagdown C{=}C\diagup\diagdown \longrightarrow \left[{-}C{\text{===}}C{-} \right] \longrightarrow {-}C{-}C{-}$$

Singlet methylene
*Stereoselective
electrophilic
addition*

electrophiles, it can find electrons at the carbon–carbon double bond. The stereochemistry strongly indicates simultaneous attachment to both doubly bonded carbon atoms. Reaction involves overlap of the π cloud of the alkene with the empty *p* orbital of the carbene. Electron density flows into this empty orbital, and the alkene carbons become relatively positive in the transition state. Electron-releasing substituents in the alkene disperse this developing charge, stabilize the transition state, and speed up reaction. The reactivity pattern of alkenes is quite similar to that observed for addition of halogens, another reaction we have pictured as involving simultaneous attachment of the electrophile to both alkene carbons.

It is **triplet** methylene that undergoes the *non-stereoselective addition*. Triplet methylene is a diradical, and it adds by a *free-radical* two-step mechanism: actually,

$$\cdot CH_2\cdot \ + \ \underset{/}{\overset{\backslash}{C}}{=}\underset{\backslash}{\overset{/}{C}} \ \longrightarrow \ \underset{\underset{CH_2\cdot}{|}}{-\overset{|}{C}{-}\overset{\cdot}{C}{-}} \ \longrightarrow \ \underset{CH_2}{-C\underset{\diagup}{\diagdown}C-}$$

Triplet methylene
Non-stereoselective
free-radical
addition

I

addition followed by combination. The intermediate diradical I lasts long enough for rotation to occur about the central carbon–carbon bond, and both *cis* and *trans* products are formed. (*Problem*: Using the approach of Sec. 9.6, assure yourself that this is so.)

Besides addition, methylene undergoes another reaction which, quite literally, belongs in a class of itself: *insertion*.

$$-\overset{|}{\underset{|}{C}}{-}H + CH_2 \ \longrightarrow \ -\overset{|}{\underset{|}{C}}{-}CH_2{-}H$$ Insertion

Methylene can *insert itself* into every carbon–hydrogen bond of most kinds of molecules. We cannot take time to say more here about this remarkable reaction, except that when addition is the desired reaction, insertion becomes an annoying side reaction.

Problem 12.17 In the gas phase, with low alkene concentration and in the presence of an inert gas, addition of methylene to the 2-butenes is, we have seen, non-stereoselective. If, however, there is present in this system a little oxygen, addition becomes almost completely stereoselective (*syn*). Account in detail for the effect of oxygen. (*Hint*: See Sec. 2.14.)

12.17 Addition of substituted carbenes. 1,1-Elimination

The addition of carbenes to alkenes is used principally to make cyclopropanes. For this purpose one seldom uses methylene itself, but rather various substituted carbenes. These are often generated in ways quite different from the photochemical reactions described in the preceding section.

A common method for making cyclopropanes is illustrated by the reaction of 2-butene with chloroform in the presence of potassium *tert*-butoxide:

$$CH_3CH{=}CHCH_3 + CHCl_3 \ \xrightarrow{\textit{t}\text{-BuO}^-\text{K}^+} \ CH_3CH\underset{\underset{Cl\quad Cl}{\diagup}}{\overset{\diagdown}{C}}CHCH_3 + \textit{t}\text{-BuOH} + KCl$$

2-Butene　　　Chloroform

3,3-Dichloro-1,2-dimethylcyclopropane

The dichlorocyclopropanes obtained can be reduced to hydrocarbons (Sec. 5.8) or hydrolyzed to *ketones*, the starting point for many syntheses (Chap. 21).

Here, too, reaction involves a divalent carbon compound, a derivative of methylene: *dichlorocarbene*, $:CCl_2$. It is generated in two steps, initiated by attack on chloroform by the very strong base, *tert*-butoxide ion, and then adds to the alkene.

(1) $t\text{-BuO:}^- + H\text{:}CCl_3 \;\rightleftharpoons\; \text{:}CCl_3^- + t\text{-BuO:}H$

(2) $\text{:}CCl_3^- \longrightarrow \text{:}CCl_2 + Cl^-$

  Dichlorocarbene

(3) $CH_3CH{=}CHCH_3 + \text{:}CCl_2 \longrightarrow CH_3CH{-}CHCH_3$

  $\underset{Cl \quad Cl}{\overset{\diagdown C \diagup}{}}$

It is believed that, because of the presence of the halogen atoms, the singlet form, with the electrons paired, is the more stable form of dichlorocarbene, and is the one adding to the double bond. (Stabilization by the halogen atoms is presumably one reason why dihalocarbenes do not generally undergo the insertion reaction that is so characteristic of unsubstituted singlet methylene.)

The addition of dihalocarbenes, like that of singlet methylene, is *stereoselective* and *stereospecific*, and *syn*.

Problem 12.18 (a) Addition of $\text{:}CCl_2$ to cyclopentene yields a single compound. What is it? (b) Addition of $\text{:}CBrCl$ to cyclopentene yields a mixture of stereoisomers. In light of (a), how do you account for this? What are the isomers likely to be? (*Hint:* Use models.)

In dehydrohalogenation of alkyl halides (Sec. 7.12), we encountered a reaction in which hydrogen ion and halide ion are eliminated from a molecule by the action of base; there —H and —X are lost from adjacent carbons, and so the process is called *1,2-elimination* (or *β-elimination*). In the generation of the carbene shown here, both —H and —X are eliminated from the same carbon, and the process is called *1,1-elimination* (or *α-elimination*). (Later on, in Sec. 28.12, we shall see some of the evidence for the mechanism of 1,1-elimination shown above.)

$$-\overset{\overset{\displaystyle X}{|}}{\underset{\underset{\displaystyle H}{|}}{C}}-\overset{|}{\underset{|}{C}}- \;\xrightarrow{\text{base}}\; \overset{\diagdown}{\diagup}C{=}C\overset{\diagup}{\diagdown} \qquad \text{1,2-elimination}$$

$$-\overset{|}{\underset{\underset{\displaystyle X}{|}}{C}}-H \;\xrightarrow{\text{base}}\; -\overset{|}{C}\text{:} \qquad \text{1,1-elimination}$$

There are many ways of generating what appear to be carbenes. But in some cases at least, it seems clear that no *free* carbene is actually an intermediate; instead, a *carbenoid* (carbene-like) reagent transfers a carbene unit directly to a double bond. For example, in the extremely useful Simmons–Smith reaction (H. E. Simmons and R. D. Smith of the du Pont Company) the carbenoid is an organozinc

$$CH_2I_2 + Zn(Cu) \longrightarrow ICH_2ZnI$$

$$\overset{\diagdown}{\diagup}C{=}C\overset{\diagup}{\diagdown} + ICH_2ZnI \longrightarrow \left[\underset{I{-}Zn\text{----}I}{-\overset{|}{\underset{CH_2}{C}}{-}\overset{|}{C}-} \right] \longrightarrow -\overset{|}{\underset{CH_2}{C}}{-}\overset{|}{C}- + ZnI_2$$

compound which delivers methylene stereoselectively (and without competing insertion) to the double bond.

Problem 12.19 (a) Why does $CHCl_3$ not undergo β-elimination through the action of base? (b) What factor would you expect to make α-elimination from $CHCl_3$ easier than from, say, CH_3Cl?

12.18 Analysis of alicyclic hydrocarbons

A cyclopropane readily dissolves in concentrated sulfuric acid, and in this resembles an alkene or alkyne. It can be differentiated from these unsaturated hydrocarbons, however, by the fact that it is not oxidized by cold, dilute, neutral permanganate.

Other alicyclic hydrocarbons have the same kind of properties as their open-chain counterparts, and they are characterized in the same way: cycloalkanes by their general inertness, and cycloalkenes and cycloalkynes by their response to tests for unsaturation (bromine in carbon tetrachloride, and aqueous permanganate). That one is dealing with cyclic hydrocarbons is shown by molecular formulas and by degradation products.

The properties of cyclohexane, for example, show clearly that it is an alkane. However, combustion analysis and molecular weight determination show its molecular formula to be C_6H_{12}. Only a cyclic structure (although not necessarily a six-membered ring) is consistent with both sets of data.

Similarly, the absorption of only one mole of hydrogen shows that cyclohexene contains only one carbon–carbon double bond; yet its molecular formula is C_6H_{10}, which in an open-chain compound would correspond to two carbon–carbon double bonds or one triple bond. Again, only a cyclic structure fits the facts.

Problem 12.20 Compare the molecular formulas of: (a) n-hexane and cyclohexane; (b) n-pentane and cyclopentane; (c) 1-hexene and cyclohexene; (d) dodecane, n-hexylcyclohexane, and cyclohexylcyclohexane. (e) In general, how can you deduce the number of rings in a compound from its molecular formula and degree of unsaturation?

Problem 12.21 What is the molecular formula of: (a) cyclohexane; (b) methylcyclopentane; (c) 1,2-dimethylcyclobutane? (d) Does the molecular formula give any information about the *size* of ring in a compound?

Problem 12.22 The yellow plant pigments α-, β-, and γ-*carotene*, and the red pigment of tomatoes, *lycopene*, are converted into vitamin A_1 in the liver. All four have the molecular formula $C_{40}H_{56}$. Upon catalytic hydrogenation, α- and β-carotene yield $C_{40}H_{78}$, γ-carotene yields $C_{40}H_{80}$, and lycopene yields $C_{40}H_{82}$. How many rings, if any, are there in each compound?

Cleavage products of cycloalkenes and cycloalkynes also reveal the cyclic structure. Ozonolysis of cyclohexene, for example, does not break the molecule into two aldehydes of lower carbon number, but simply into a single six-carbon compound containing *two* aldehyde groups.

A di-aldehyde

Cyclohexene

A diacid

Problem 12.23　Predict the ozonolysis products of: (a) cyclopentane; (b) 1-methyl-cyclopentene; (c) 3-methylcyclopentene; (d) 1,3-cyclohexadiene; (e) 1,4-cyclohexadiene.

Problem 12.24　Both cyclohexene and 1,7-octadiene yield the di-aldehyde $OHC(CH_2)_4CHO$ upon ozonolysis. What other facts would enable you to distinguish between the two compounds?

(Analysis of cyclic aliphatic hydrocarbons by spectroscopy will be discussed in Secs. 16.18–16.19.)

PROBLEMS

1. Draw structural formulas of:

(a) methylcyclopentane
(b) 1-methylcyclohexene
(c) 3-methylcyclopentene
(d) trans-1,3-dichlorocyclobutane
(e) cis-1-bromo-2-methylcyclopentane

(f) cyclohexylcyclohexane
(g) cyclopentylacetylene
(h) 4-chloro-1,1-dimethylcycloheptane
(i) bicyclo[2.2.1]hepta-2,5-diene
(j) 1-chlorobicyclo[2.2.2]octane

2. Give structures and names of the principal organic products expected from each of the following reactions:

(a) cyclopropane + Cl_2, $FeCl_3$
(b) cyclopropane + Cl_2 (300 °C)
(c) cyclopropane + conc. H_2SO_4
(d) cyclopentane + Cl_2, $FeCl_3$
(e) cyclopentane + Cl_2 (300 °C)
(f) cyclopentane + conc. H_2SO_4
(g) cyclopentene + Br_2/CCl_4
(h) cyclopentene + Br_2 (300 °C)
(i) 1-methylcyclohexene + HCl
(j) 1-methylcyclohexene + Br_2(aq)
(k) 1-methylcyclohexene + HBr
　　　(peroxides)

(l) 1,3-cyclohexadiene + HCl
(m) cyclopentanol + H_2SO_4 (heat)
(n) bromocyclohexane + KOH(alc)
(o) cyclopentene + cold $KMnO_4$
(p) cyclopentene + HCO_2OH
(q) cyclopentene + hot $KMnO_4$
(r) cyclopentene + NBS
(s) 3-bromocyclopentene + KOH (hot)
(t) 1,4-cyclohexanediol + H_2SO_4 (hot)
(u) cyclohexene + H_2SO_4 ⟶ $C_{12}H_{20}$
(v) cyclopentene + $CHCl_3$ + t-BuOK
(w) cyclopentene + CH_2I_2 + Zn(Cu)
(x) chlorocyclopentane + $(C_2H_5)_2CuLi$
(y) 1-methylcyclopentene + cold conc. H_2SO_4
(z) 3-methylcyclopentene + O_3, then H_2O/Zn

3. Outline all steps in the laboratory synthesis of each of the following from cyclohexanol.

(a) cyclohexene
(b) cyclohexane
(c) *trans*-1,2-dibromocyclohexane
(d) *cis*-1,2-cyclohexanediol
(e) *trans*-1,2-cyclohexanediol
(f) OHC(CH$_2$)$_4$CHO

(g) adipic acid, HOOC(CH$_2$)$_4$COOH
(h) bromocyclohexane
(i) 2-chlorocyclohexanol
(j) 3-bromocyclohexene
(k) 1,3-cyclohexadiene
(l) cyclohexylcyclohexane

(m) *norcarane*, bicyclo[4.1.0]heptane

4. Give structures for all isomers of the following. For cyclohexane derivatives, planar formulas (p. 462) will be sufficient here. Label pairs of enantiomers, and *meso* compounds.

(a) dichlorocyclopropanes
(b) dichlorocyclobutanes
(c) dichlorocyclopentanes

(d) dichlorocyclohexanes
(e) chloro-1,1-dimethylcyclohexanes
(f) 1,3,5-trichlorocyclohexanes

(g) There are a number of stereoisomeric 1,2,3,4,5,6-hexachlorocyclohexanes. Without attempting to draw all of them, give the structure of the most stable isomer, and show its preferred conformation.

5. (a) 2,5-Dimethyl-1,1-cyclopentanedicarboxylic acid (I) can be prepared as two optically inactive substances (A and B) of different m.p. Draw their structures. (b) Upon heating, A yields two 2,5-dimethylcyclopentanecarboxylic acids (II), and B yields only one. Assign structures to A and B.

6. (a) The following compounds can be resolved into optically active enantiomers.

3,3'-Diaminospiro[3.3]heptane   4-Methylcyclohexylideneacetic acid

Using models and then drawing three-dimensional formulas, account for this. Label the chiral center in each compound.

(b) Addition of bromine to optically active 4-methylcyclohexylideneacetic acid yields two optically active dibromides. Assuming a particular configuration for the starting material, draw stereochemical formulas for the products.

7. (a) *trans*-1,2-Dimethylcyclohexane exists about 99% in the diequatorial conformation. *trans*-1,2-Dibromocyclohexane (or *trans*-1,2-dichlorocyclohexane), on the other hand, exists about equally in the diequatorial and diaxial conformations; furthermore, the fraction of the diaxial conformation decreases with increasing polarity of the solvent. How do you account for the contrast between the dimethyl and dibromo (or dichloro) compounds? (*Hint*: See Problem 11, p. 161.)

(b) If *trans*-3-*cis*-4-dibromo-*tert*-butylcyclohexane is subjected to prolonged heating, it is converted into an equilibrium mixture (about 50:50) of itself and a diastereomer. What is the diastereomer likely to be? How do you account for the approximately equal stability of these two diastereomers? (Here, and in (c), consider the more stable conformation of each diastereomer to be the one with an equatorial *tert*-butyl group.)

(c) There are two more diastereomeric 3,4-dibromo-*tert*-butylcyclohexanes. What are they? How do you account for the fact that neither is present to an appreciable extent in the equilibrium mixture?

8. The compound *decalin*, $C_{10}H_{18}$, consists of two fused cyclohexane rings:

Decalin

(a) Using models, show how there can be two isomeric decalins, *cis* and *trans*. (b) How many different conformations free of angle strain are possible for *cis*-decalin? For *trans*-decalin? (c) Which is the most stable conformation of *cis*-decalin? Of *trans*-decalin? (d) Account for the fact that *trans*-decalin is more stable than *cis*-decalin. (*Hint*: Consider each ring in turn. What are the largest substituents on each ring?) (e) The difference in stability between *cis*- and *trans*-decalin is about 2 kcal/mol; conversion of one into the other takes place only under very vigorous conditions. The chair and twist-boat forms of cyclohexane, on the other hand, differ in stability by about 6 kcal/mol, yet are readily interconverted at room temperature. How do you account for the contrast? Draw energy curves to illustrate your answer.

9. Allinger (p. 460) found the energy difference between *cis*- and *trans*-1,3-di-*tert*-butylcyclohexane to be 5.9 kcal/mol, and considers that this value represents the energy difference between the chair and twist-boat forms of cyclohexane. Defend Allinger's position.

10. It has been suggested that in certain substituted cyclopentanes the ring exists preferentially in the "envelope" form:

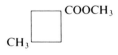

Using models, suggest a possible explanation for each of the following facts:
   (a) The attachment of a methyl group to the badly strained cyclopentane ring raises the heat of combustion very little more than attachment of a methyl group to the unstrained cyclohexane ring. (*Hint*: Where is the methyl group located in the "envelope" form?)
   (b) Of the 1,2-dimethylcyclopentanes, the *trans* isomer is more stable than the *cis*. Of the 1,3-dimethylcyclopentanes, on the other hand, the *cis* isomer is more stable than the *trans*.
   (c) The *cis* isomer of methyl 3-methylcyclobutanecarboxylate

$$CH_3 \quad \square \quad COOCH_3$$

is more stable than the *trans* isomer.

11. Arrange the compounds of each set in order of reactivity toward the indicated reaction:
   (a) bromocyclohexane, 1-bromo-1-methylcyclohexane, (bromomethyl)cyclohexane toward S_N2 displacement
   (b) the compounds of part (a) toward S_N1 displacement
   (c) 5-bromo-1,3-cyclohexadiene, bromocyclohexane, 3-bromocyclohexene, 1-bromocyclohexene toward dehydrohalogenation by strong base
   (d) *cis*- and *trans*-2-bromo-1-methylcyclohexane toward dehydrohalogenation by strong base

12. Each of the following reactions is carried out, and the products are separated by careful distillation, recrystallization, or chromatography. For each reaction tell how many fractions will be collected. Draw a stereochemical formula of the compound or compounds making up each fraction. Tell whether each fraction, as collected, will be optically active or optically inactive.

(a) (*R*)-3-hydroxycyclohexene + $KMnO_4$ $\longrightarrow$ $C_6H_{12}O_3$
(b) (*R*)-3-hydroxycyclohexene + HCO_2OH $\longrightarrow$ $C_6H_{12}O_3$
(c) (*S,S*)-1,2-dichlorocyclopropane + Cl_2 (300 °C) $\longrightarrow$ $C_3H_3Cl_3$
(d) *racemic* 4-methylcyclohexene + Br_2/CCl_4
(e) *trans*-4-methyl-1-bromocyclohexane + OH^- (second-order) $\longrightarrow$ $C_7H_{14}O + C_7H_{12}$

13. When *trans*-2-methylcyclopentanol is heated with acid, it gives chiefly 1-methyl-cyclopentene. When the same alcohol is treated with tosyl chloride,

$$R-O-H \ + \ Ts-Cl \ \longrightarrow \ R-O-Ts \ + \ HCl$$

and the product is treated with potassium *tert*-butoxide, the only alkene obtained is 3-methylcyclohexene. Account in detail for the contrast between these two synthetic routes.

14. Outline the sequence of steps that best accounts for the following facts.

(a) 2,2-dimethylcyclohexanol $\xrightarrow{H^+}$ 1,2-dimethylcyclohexene + 1-isopropylcyclopentene. (*Hint*: Use models.)

(b) 3-cyclobutyl-3-pentanol $\xrightarrow{H^+}$ 1,2-diethylcyclopentene

(c)

$\xrightarrow{H^+}$ 1,2-dimethylcyclohexene

(d)

Cyclohexen-1-yl   Cyclohexanone
triflate

15. Treatment of the triflate III with CF_3CH_2OH gives not only IV but also V and VI. (a) How do you account for the formation of V? (b) The formation of VI?

III   IV   V   VI

16. When neomenthyl chloride undergoes E2 elimination, 2-menthene makes up one-fourth of the reaction product (Sec. 12.15). Since menthyl chloride can yield *only* 2-menthene, we might expect it to react at one-fourth of the rate of neomenthyl chloride. Actually, however, it reacts only 1/200 as fast as neomenthyl chloride: that is, only 1/50 as fast as we would have expected. How do you account for this unusually slow elimination from menthyl chloride? (*Hint*: Use models.)

17. *cis*-4-*tert*-Butylcyclohexyl tosylate reacts rapidly with NaOEt in EtOH to yield 4-*tert*-butylcyclohexene; the rate of reaction is proportional to the concentration of both tosylate and ethoxide ion. Under the same conditions, *trans*-4-*tert*-butylcyclohexyl tosylate reacts slowly to yield the alkene (plus 4-*tert*-butylcyclohexyl ethyl ether); the rate of reaction depends only on the concentration of the tosylate.

How do you account for these observations?

18. When 1-bromocyclohexene (ordinary bromine) is allowed to react with radioactive Br_2, and the resulting tribromide is treated with iodide ion, there is obtained 1-bromocyclo-hexene that contains less than 0.3% of radioactive bromine. Explain in detail. (*Hint*: See Problem 11, p. 364.)

19. Outline all steps in a possible laboratory synthesis of each of the following from acetylene, chloroform, diiodomethane, and alcohols of four carbons or fewer, using any needed inorganic reagents.

(a) *cis*-1,2-dimethylcyclopropane
(b) *trans*-1,2-dimethylcyclopropane
(c) *cis*-1,2-di(*n*-propyl)cyclopropane
(d) racemic *trans*-1,1-dichloro-2-ethyl-3-methylcyclopropane

20. Describe simple chemical tests that would distinguish between:

(a) cyclopropane and propane
(b) cyclopropane and propylene
(c) 1,2-dimethylcyclopropane and cyclopentane
(d) cyclobutane and 1-butene
(e) cyclopentane and 1-pentene
(f) cyclopentane and cyclopentene
(g) cyclohexanol and *n*-butylcyclohexane
(h) 1,2-dimethylcyclopentene and cyclopentanol
(i) cyclohexane, cyclohexene, cyclohexanol, and bromocyclohexane

Tell exactly what you would *do* and *see*.

21. How many rings does each of the following contain?
(a) *Camphane*, $C_{10}H_{18}$, a terpene related to camphor, takes up no hydrogen. (b) *Cholestane*, $C_{27}H_{48}$, a steroid of the same ring structure as cholesterol, cortisone, and the sex hormones, takes up no hydrogen. (c) *β-Phellandrene*, $C_{10}H_{16}$, a terpene, reacts with bromine to form $C_{10}H_{16}Br_4$. (d) *Ergocalciferol* (so-called "Vitamin D_2"), $C_{28}H_{44}O$, an alcohol, gives $C_{28}H_{52}O$ upon catalytic hydrogenation. (e) How many double bonds does ergocalciferol contain?

22. On the basis of the results of catalytic hydrogenation, how many rings does each of the following aromatic hydrocarbons contain?

(a) *benzene* (C_6H_6) ⟶ C_6H_{12}
(b) *naphthalene* $(C_{10}H_8)$ ⟶ $C_{10}H_{18}$
(c) *toluene* (C_7H_8) ⟶ C_7H_{14}
(d) *anthracene* $(C_{14}H_{10})$ ⟶ $C_{14}H_{24}$

(e) *phenanthrene* $(C_{14}H_{10})$ ⟶ $C_{14}H_{24}$
(f) *3,4-benzpyrene* $(C_{20}H_{12})$ ⟶ $C_{20}H_{32}$
(g) *chrysene* $(C_{18}H_{12})$ ⟶ $C_{18}H_{30}$

(Check your answers by use of the index.)

23. (a) A hydrocarbon of formula $C_{10}H_{16}$ absorbs only one mole of H_2 upon hydro-genation. How many rings does it contain? (b) Upon ozonolysis it yields 1,6-cyclodecane-dione (VII). What is the hydrocarbon?

$$O=C \overset{\displaystyle -(CH_2)_4-}{\underset{\displaystyle -(CH_2)_4-}{}} C=O$$

VII

$$CH_3-C-CH_2-CH_2-CH\overset{\displaystyle C-CH_3}{\underset{\displaystyle CH_2COOH}{}}$$
(with $\overset{O}{\|}$ on middle C and $\overset{O}{\|}$ on the C–CH_3)

VIII

24. *Limonene*, $C_{10}H_{16}$, a terpene found in orange, lemon, and grapefruit peel, absorbs only two moles of hydrogen, forming *p-menthane*, $C_{10}H_{20}$. Oxidation by permanganate converts limonene into VIII. (a) How many rings, if any, are there in limonene? (b) What structures are consistent with the oxidation product? (c) On the basis of the isoprene rule, which structure is most likely for limonene? For *p*-menthane? (d) Addition of one mole of H_2O converts limonene into an alcohol. What are the most likely structures for this alcohol? (e) Addition of two moles of H_2O to limonene yields *terpin hydrate*. What is the most likely structure for terpin hydrate?

25. α-*Terpinene*, $C_{10}H_{16}$, a terpene found in coriander oil, absorbs only two moles of hydrogen, forming *p-menthane*, $C_{10}H_{20}$. Ozonolysis of α-terpinene yields IX; permanganate cleavage yields X.

$$CH_3-\overset{\displaystyle \underset{\displaystyle O}{\|}}{C}-CH_2-CH_2-\overset{\displaystyle \underset{\displaystyle O}{\|}}{C}-CH(CH_3)_2$$

IX

$$HOOC-\overset{\displaystyle \overset{\textstyle CH_3}{|}}{\underset{\displaystyle \underset{\textstyle OH}{|}}{C}}-CH_2-CH_2-\overset{\displaystyle \overset{\textstyle CH(CH_3)_2}{|}}{\underset{\displaystyle \underset{\textstyle OH}{|}}{C}}-COOH$$

X

(a) How many rings, if any, are there in α-terpinene? (b) On the basis of the cleavage products, IX and X, and the isoprene rule, what is the most likely structure for α-terpinene? (c) How do you account for the presence of the —OH groups in X?

26. Using only chemistry that you have already encountered, can you suggest a mechanism for the conversion of *nerol* ($C_{10}H_{18}O$) into α-terpineol ($C_{10}H_{18}O$) in the presence of dilute H_2SO_4?

$$CH_3-\overset{\displaystyle \overset{\textstyle CH_3}{|}}{C}=CH-CH_2-CH_2-\overset{\displaystyle \overset{\textstyle CH_3}{|}}{C}=CH-CH_2OH \xrightarrow{\quad H_2O,\ H^+ \quad}$$

Nerol (found in bergamot)

α-Terpineol

13

Aromaticity

Benzene

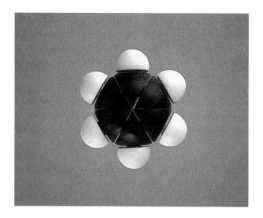

13.1 Aliphatic and aromatic compounds

Chemists have found it useful to divide all organic compounds into two broad classes: **aliphatic** compounds and **aromatic** compounds. The original meanings of the words "aliphatic" (*fatty*) and "aromatic" (*fragrant*) no longer have any significance.

Aliphatic compounds are open-chain compounds and those cyclic compounds that resemble open-chain compounds. Except for the occasional appearance of a phenyl (C_6H_5) group, the hydrocarbon portions of the compounds that we have studied so far have been aliphatic.

Aromatic compounds *are benzene and compounds that resemble benzene in chemical behavior*. Aromatic properties are those properties of benzene that distinguish it from aliphatic hydrocarbons. The benzene molecule is a *ring*: a ring of a very special kind. There are certain compounds—other ring compounds—which seem to differ from benzene in structure, yet which behave very much like benzene. These other compounds, it turns out, actually do resemble benzene in structure—in basic electronic configuration—and they are aromatic, too.

Aliphatic hydrocarbons—alkanes, alkenes, alkynes, and their cyclic analogs—undergo chiefly addition and free-radical substitution: addition at multiple bonds, and free-radical substitution at other points along the aliphatic chain. These same reactions, as we have seen, take place in the hydrocarbon portions of other aliphatic

477

compounds. The reactivity of these hydrocarbon portions is affected by the presence of other functional groups, and the reactivity of these other functional groups is affected by the presence of the hydrocarbon portions.

In contrast to aliphatic hydrocarbons, we shall find, *aromatic hydrocarbons are characterized by a tendency to undergo heterolytic substitution.* Furthermore, these same substitution reactions are characteristic of aromatic rings wherever they appear, regardless of other functional groups the molecule may contain. These other functional groups affect the reactivity of the aromatic rings, and the aromatic rings affect the reactivity of these other functional groups.

In this chapter we shall examine the fundamental quality of *aromaticity*: just how aromatic compounds differ in behavior from aliphatic compounds, and what there is in their structure that makes them different. In Chapter 14 we shall see how these characteristic aromatic reactions take place, and how they are affected by substituents on the aromatic ring. In Chapter 15 we shall take the opposite viewpoint, and look at the remarkable effects that aromatic rings, acting themselves as substituents, exert on reactions taking place in other parts of the molecule.

In the remainder of the book we shall do as organic chemists do, and deal with both aliphatic molecules and aromatic molecules as they happen to appear— or, as is commonly the case, with molecules that are *both* aliphatic *and* aromatic. It is important not to attach undue weight to the division between aliphatic and aromatic compounds. Although extremely useful, it is often less important than some other classification. The similarities between aliphatic and aromatic acids, for example, or between aliphatic and aromatic amines, are more important than the differences.

13.2   Structure of benzene

It is obvious from our definition of aromatic compounds that any study of their chemistry must begin with a study of benzene. Benzene has been known since 1825; its chemical and physical properties are perhaps better known than those of any other single organic compound. In spite of this, no satisfactory structure for benzene had been advanced until about 1931, and it was ten to fifteen years before this structure was generally used by organic chemists.

The difficulty was not the complexity of the benzene molecule, but rather the limitations of the structural theory as it had so far developed. Since an understanding of the structure of benzene is important both in our study of aromatic compounds and in extending our knowledge of the structural theory, we shall examine in some detail the facts upon which this structure of benzene is built.

13.3   Molecular formula. Isomer number. Kekulé structure

(a) *Benzene has the molecular formula* C_6H_6. From its elemental composition and molecular weight, benzene was known to contain six carbon atoms and six hydrogen atoms. The question was: how are these atoms arranged?

In 1858, August Kekulé had proposed that carbon atoms can join to one another to form *chains*. Then, in 1865, he offered an answer to the question of benzene: these carbon chains can sometimes be closed, to form *rings*. As he described it later:

"I was sitting writing at my textbook, but the work did not progress; my thoughts were elsewhere. I turned my chair to the fire, and dozed. Again the atoms were gamboling before my eyes. This time the smaller groups kept modestly in the background. My mental eye, rendered more acute by repeated visions of this kind, could now distinguish larger structures of manifold conformations; long rows, sometimes more closely fitted together; all twisting and turning in snake-like motion. But look! What was that? One of the snakes had seized hold of its own tail, and the form whirled mockingly before my eyes. As if by a flash of lightning I woke; ... I spent the rest of the night working out the consequences of the hypothesis. Let us learn to dream, gentlemen, and then perhaps we shall learn the truth."—August Kekulé, 1890.

Kekulé's structure of benzene was one that we would represent today as I.

I
Kekulé formula

II
"Dewar" formula

III

$$CH_3-C{\equiv}C-C{\equiv}C-CH_3$$
IV

$$CH_2{=}CH-C{\equiv}C-CH{=}CH_2$$
V

Other structures are, of course, consistent with the formula C_6H_6: for example, II–V. Of all these, Kekulé's structure was accepted as the most nearly satisfactory; the evidence was of a kind with which we are already familiar: **isomer number** (Sec. 4.2).

(b) *Benzene yields only one monosubstitution product,* C_6H_5Y. Only one bromobenzene, C_6H_5Br, is obtained when one hydrogen atom is replaced by bromine; similarly, only one chlorobenzene, C_6H_5Cl, or one nitrobenzene, $C_6H_5NO_2$, etc., has ever been made. This fact places a severe limitation on the structure of benzene: each hydrogen must be exactly equivalent to every other hydrogen, since the replacement of any one of them yields the same product.

Structure V, for example, must now be rejected, since it would yield two isomeric monobromo derivatives, the 1-bromo and the 2-bromo compounds; all hydrogens are not equivalent in V. Similar reasoning shows us that II and III are likewise unsatisfactory. (How many monosubstitution products would each of these yield?)   I and IV, among others, are still possibilities, however.

(c) *Benzene yields three isomeric disubstitution products,* $C_6H_4Y_2$ or C_6H_4YZ. Three and only three isomeric dibromobenzenes, $C_6H_4Br_2$, three chloronitrobenzenes, $C_6H_4ClNO_2$, etc., have ever been made. This fact further limits our choice of a structure; for example, IV must now be rejected. (How many disubstitution products would IV yield?)

At first glance, structure I seems to be consistent with this new fact; that is, we can expect three isomeric dibromo derivatives, the 1,2-, the 1,3-, and the 1,4-dibromo compounds shown:

1,2-Dibromobenzene   1,3-Dibromobenzene   1,4-Dibromobenzene

Closer examination of structure I shows, however, that *two* 1,2-dibromo isomers (VI and VII), differing in the positions of bromine relative to the double bonds, should be possible:

VI   VII

But Kekulé visualized the benzene molecule as a dynamic thing: ". . . the form whirled mockingly before my eyes . . ." He described it in terms of two structures, VIII and IX, between which the benzene molecule alternates. As a consequence, the two 1,2-dibromobenzenes (VI and VII) would be in rapid equilibrium and hence could not be separated.

VIII   IX

VI   VII

Later, when the idea of tautomerism (Sec. 11.10) became defined, it was assumed that Kekulé's "alternation" essentially amounted to tautomerism.

On the other hand, it is believed by some that Kekulé had intuitively anticipated by some 75 years our present concept of delocalized electrons, and drew two pictures (VIII and IX)—as we shall do, too—as a crude representation of something that neither picture alone satisfactorily represents. Rightly or wrongly, the term "Kekulé structure" has come to mean a (hypothetical) molecule with alternating single and double bonds—just as the term "Dewar benzene" has come to mean a structure (II) that James Dewar devised in 1867 as an example of what benzene was *not*.

13.4 Stability of the benzene ring. Reactions of benzene

Kekulé's structure, then, accounts satisfactorily for facts (a), (b), and (c) in Sec. 13.3. But there are a number of facts that are still not accounted for by this structure; most of these unexplained facts seem related to unusual stability of the benzene ring. The most striking evidence of this stability is found in the chemical reactions of benzene.

(d) *Benzene undergoes substitution rather than addition.* Kekulé's structure of benzene is one that we would call "cyclohexatriene". We would expect this cyclohexatriene, like the very similar compounds, cyclohexadiene and cyclohexene, to undergo readily the addition reactions characteristic of the alkene structure. As the examples in Table 13.1 show, this is not the case; under conditions that cause an alkene to undergo rapid addition, benzene reacts either not at all or very slowly.

Table 13.1 CYCLOHEXENE *vs.* BENZENE

Reagent	Cyclohexene gives	Benzene gives
$KMnO_4$ (cold, dilute, aqueous)	Rapid oxidation	No reaction
Br_2/CCl_4 (in the dark)	Rapid addition	No reaction
HI	Rapid addition	No reaction
$H_2 + Ni$	Rapid hydrogenation at 25 °C, 20 lb/in.^{2}	Slow hydrogenation at 100–200 °C, 1500 lb/in.^{2}

In place of addition reactions, benzene readily undergoes a new set of reactions, all involving **substitution**. The most important are shown below.

REACTIONS OF BENZENE ⎯⎯⎯⎯⎯⎯⎯⎯⎯⎯⎯⎯⎯⎯⎯⎯⎯⎯

1. **Nitration.** Discussed in Sec. 14.8.

$$C_6H_6 + HONO_2 \xrightarrow{H_2SO_4} C_6H_5{-}NO_2 + H_2O$$
Nitrobenzene

CONTINUED

CONTINUED

2. **Sulfonation.** Discussed in Sec. 14.9.

$$C_6H_6 + HOSO_3H \xrightarrow{SO_3} C_6H_5{-}SO_3H + H_2O$$
Benzenesulfonic acid

3. **Halogenation.** Discussed in Sec. 14.11.

$$C_6H_6 + Cl_2 \xrightarrow{Fe} C_6H_5{-}Cl + HCl$$
Chlorobenzene

$$C_6H_6 + Br_2 \xrightarrow{Fe} C_6H_5{-}Br + HBr$$
Bromobenzene

4. **Friedel–Crafts alkylation.** Discussed in Secs. 14.10 and 15.7.

$$C_6H_6 + RCl \xrightarrow{AlCl_3} C_6H_5{-}R + HCl$$
An alkylbenzene

5. **Friedel–Crafts acylation.** Discussed in Sec. 21.5.

$$C_6H_6 + RCOCl \xrightarrow{AlCl_3} C_6H_5{-}COR + HCl$$
An acyl chloride   A ketone

In each of these reactions an atom or group has been substituted for one of the hydrogen atoms of benzene. The product can itself undergo further substitution of the same kind; the fact that it has retained the characteristic properties of benzene indicates that it has retained the characteristic structure of benzene.

It would appear that benzene resists addition, in which the benzene ring system would be destroyed, whereas it readily undergoes substitution, in which the ring system is preserved.

13.5   Stability of the benzene ring. Heats of hydrogenation and combustion

Besides the above qualitative indications that the benzene ring is more stable than we would expect cyclohexatriene to be, there exist quantitative data which show *how much* more stable.

(e) *Heats of hydrogenation and combustion of benzene are lower than expected.* We recall (Sec. 8.3) that heat of hydrogenation is the quantity of heat evolved when one mole of an unsaturated compound is hydrogenated. In most cases the value is about 28–30 kcal for each double bond the compound contains. It is not surprising, then, that cyclohexene has a heat of hydrogenation of 28.6 kcal and cyclohexadiene has one about twice that (55.4 kcal).

We might reasonably expect cyclohexatriene to have a heat of hydrogenation about three times as large as cyclohexene, that is, about 85.8 kcal. Actually, the value for benzene (49.8 kcal) is *36 kcal less* than this expected amount.

This can be more easily visualized, perhaps, by means of an energy diagram (Fig. 13.1), in which the height of a horizontal line represents the potential energy content of a molecule. The broken lines represent the expected values, based upon three equal steps of 28.6 kcal. The final product, cyclohexane, is the same in all three cases.

Figure 13.1   Heats of hydrogenation and stability: benzene, cyclohexadiene, and cyclohexene.

The fact that benzene *evolves* 36 kcal less energy than predicted can only mean that benzene *contains* 36 kcal less energy than predicted; in other words, benzene is more stable by 36 kcal than we would have expected cyclohexatriene to be. The heat of combustion of benzene is also lower than that expected, and by about the same amount.

Problem 13.1   From Fig. 13.1 determine the ΔH of the following reactions:

(a) benzene + H_2 ⟶ 1,3-cyclohexadiene
(b) 1,3-cyclohexadiene + H_2 ⟶ cyclohexene

Problem 13.2   For a large number of organic compounds, the heat of combustion actually measured agrees rather closely with that calculated by assuming a certain characteristic contribution from each kind of bond, e.g., 54.0 kcal for each C—H bond, 49.3 kcal for each C—C bond, and 117.4 kcal for each C=C bond (*cis*-1,2-disubstituted). (a) On this basis, what is the calculated heat of combustion for cyclohexatriene? (b) How does this compare with the measured value of 789.1 kcal for benzene?

13.6   Carbon–carbon bond lengths in benzene

(f) *All carbon–carbon bonds in benzene are equal and are intermediate in length between single and double bonds.* Carbon–carbon double bonds in a wide variety of compounds are found to be about 1.34 Å long. Carbon–carbon single bonds, in which the nuclei are held together by only one pair of electrons, are considerably longer: 1.53 Å in ethane, for example, 1.50 Å in propylene, 1.48 Å in 1,3-butadiene.

If benzene actually possessed three single and three double bonds, as in a Kekulé structure, we would expect to find three short bonds (1.34 Å) and three long bonds (1.48 Å, probably, as in 1,3-butadiene). Actually, x-ray diffraction studies show that the six carbon–carbon bonds in benzene are equal and have a length of 1.39 Å, and are thus intermediate between single and double bonds.

13.7 Resonance structure of benzene

The Kekulé structure of benzene, while admittedly unsatisfactory, was generally used by chemists as late as 1945. The currently accepted structure did not arise from the discovery of new facts about benzene, but is the result of an extension or modification of the structural theory; this extension is the concept of *resonance* (Sec. 10.6).

The Kekulé structures I and II, we now immediately recognize, meet the conditions for resonance: *structures that differ only in the arrangement of electrons.*

Benzene is a *hybrid* of I and II. Since I and II are exactly equivalent, and hence of exactly the same stability, they make equal contributions to the hybrid. And, also since I and II are exactly equivalent, stabilization due to resonance should be large.

The puzzling aspects of benzene's properties now fall into place. The six bond lengths are identical because the six bonds are identical: they are one-and-a-half bonds and their length, 1.39 Å, is intermediate between the lengths of single and double bonds.

When it is realized that all carbon–carbon bonds in benzene are equivalent, there is no longer any difficulty in accounting for the number of isomeric disubstitution products. It is clear that there should be just three, in agreement with experiment:

1,2-Dibromobenzene   1,3-Dibromobenzene   1,4-Dibromobenzene

Finally, the "unusual" stability of benzene is not unusual at all: it is what one would expect of a hybrid of equivalent structures. The 36 kcal of energy that benzene does not contain—compared with cyclohexatriene—is resonance energy. It is the 36 kcal of resonance stabilization that is responsible for the new set of properties we call *aromatic properties.*

Addition reactions convert an alkene into a more stable saturated compound. Hydrogenation of cyclohexene, for example, is accompanied by the evolution of 28.6 kcal; the product lies 28.6 kcal lower than the reactants on the energy scale (Fig. 13.1).

But addition would convert benzene into a *less* stable product by destroying the resonance-stabilized benzene ring system; for example, according to Fig. 13.1 the first stage of hydrogenation of benzene requires 5.6 kcal to convert benzene into the less stable cyclohexadiene. As a consequence, it is easier for reactions of benzene to take an entirely different course, one in which the ring system is retained: *substitution*.

(This is not quite all of the story in so far as stability goes. As we shall see in Sec. 13.10, an additional factor besides resonance is necessary to make benzene what it is.)

13.8 Orbital picture of benzene

A more detailed picture of the benzene molecule is obtained from a consideration of the bond orbitals in this molecule.

Since each carbon is bonded to three other atoms, it uses sp^2 orbitals (as in ethylene, Sec. 7.2). These lie in the same plane, that of the carbon nucleus, and are directed toward the corners of an equilateral triangle. If we arrange the six carbons and six hydrogens of benzene to permit maximum overlap of these orbitals, we obtain the structure shown in Fig. 13.2a.

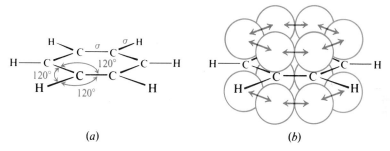

(a)   (b)

Figure 13.2 Benzene molecule. (a) Only the σ bonds are shown. (b) The p orbitals overlap to form π bonds.

Benzene is a *flat molecule*, with every carbon and every hydrogen lying in the same plane. It is a very *symmetrical molecule*, too, with each carbon atom lying at the angle of a regular hexagon; every bond angle is 120°. Each bond orbital is cylindrically symmetrical about the line joining the atomic nuclei, and hence, as before, these bonds are designated as σ bonds.

The molecule is not yet complete, however. There are still six electrons to be accounted for. In addition to the three orbitals already used, each carbon atom has a fourth orbital, a p orbital. As we know, this p orbital consists of two equal lobes, one lying above and the other lying below the plane of the other three orbitals, that is, above and below the plane of the ring; it is occupied by a single electron.

As in the case of ethylene, the p orbital of one carbon can overlap the p orbital of an adjacent carbon atom, permitting the electrons to pair and an additional π bond to be formed (see Fig. 13.2b). But the overlap here is not limited to a pair of p orbitals as it was in ethylene; the p orbital of any one carbon atom overlaps equally well the p orbitals of *both* carbon atoms to which it is bonded. The result (see Fig. 13.3) is two continuous doughnut-shaped electron clouds, one lying above and·the other below the plane of the atoms.

As with the allyl radical, it is the overlap of the p orbitals in both directions, and the resulting participation of each electron in several bonds, that corresponds to our description of the molecule as a resonance hybrid of two structures. Again

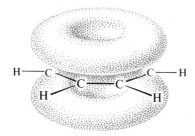

Figure 13.3 Benzene molecule: π clouds above and below the plane of the ring.

it is the *delocalization* of the π electrons—their participation in several bonds— that makes the molecule more stable.

To accommodate six π electrons, there must be *three* orbitals (Sec. 33.5). Their sum is, however, the symmetrical π clouds we have described.

The orbital approach reveals the importance of the planarity of the benzene ring. The ring is flat because the trigonal (sp^2) bond angles of carbon just fit the $120°$ angles of a regular hexagon; it is this flatness that permits the overlap of the p orbitals in both directions, with the resulting delocalization and stabilization.

Figure 13.4 Benzene molecule: shape and size.

The facts are consistent with the orbital picture of the benzene molecule. X-ray and electron diffraction show benzene (Fig. 13.4) to be a completely flat, symmetrical molecule with all carbon–carbon bonds equal, and all bond angles $120°$. (See Fig. 13.5.)

As we shall see, the chemical properties of benzene are just what we would expect of this structure. Despite delocalization, the π electrons are nevertheless more loosely held than the σ electrons. The π electrons are thus particularly available to a reagent that is seeking electrons: *the typical reactions of the benzene ring are those in which it serves as a source of electrons for electrophilic (acidic) reagents*. Because of the resonance stabilization of the benzene ring, *these reactions lead to substitution*, in which the aromatic character of the benzene ring is preserved.

Problem 13.3 The carbon–hydrogen homolytic bond dissociation energy for benzene (110 kcal) is considerably larger than for cyclohexane. On the basis of the orbital picture of benzene, what is one factor that may be responsible for this? What piece of physical evidence tends to support your answer? (*Hint*: Look at Fig. 13.4 and see Sec. 7.4.)

Problem 13.4 The molecules of *pyridine*, C_5H_5N, are flat, with all bond angles about $120°$. All carbon–carbon bonds are 1.39 Å long and the two carbon–nitrogen bonds are 1.36 Å long. The measured heat of combustion is 23 kcal lower than that calculated by the method of Problem 13.2 on page 483. Pyridine undergoes such substitution reactions as nitration and sulfonation (Sec. 13.4). (a) Is pyridine adequately represented by formula I? (b) Account for the properties of pyridine by both valence-bond and orbital structures. (Check your answer in Sec. 35.6.)

I

Problem 13.5 The compound *borazole*, $B_3N_3H_6$, is shown by electron diffraction to have a flat cyclic structure with alternating boron and nitrogen atoms, and all boron–nitrogen bond lengths the same. (a) How would you represent borazole by valence-bond structures? (b) In terms of orbitals? (c) How many π electrons are there, and which atoms have they "come from"?

Figure 13.5 Electronic configuration and molecular shape. Model of benzene: two views. Like cyclohexane (Fig. 12.5, p. 448), benzene is a marvelous creation: flat, compact, symmetrical, and of a stability that makes it the model for an entire class of organic molecules. Here, too, the architectural perfection is the result of a happy coincidence. Six carbons make a hexagon, whose angles happen to match exactly the trigonal angle. Six carbons provide six π electrons; and *six*, as we shall see, is a "magic" number of π electrons.

13.9   Representation of the benzene ring

For convenience we shall represent the benzene ring by a regular hexagon containing a circle (I); it is understood that a hydrogen atom is attached to each angle of the hexagon unless another atom or group is indicated.

I   II   III

Formula I represents a resonance hybrid of the Kekulé structures II and III. The straight lines stand for the σ bonds joining carbon atoms. The circle stands for the cloud of six delocalized π electrons. (From another viewpoint, the straight lines stand for single bonds, and the circle stands for the extra *half-bonds*.)

Formula I is a particularly useful representation of the benzene ring, since it emphasizes the equivalence of the various carbon–carbon bonds. The presence of the circle distinguishes the benzene ring from the cyclohexane ring, which is generally represented today by a plain hexagon.

There is no complete agreement among chemists about how to represent the benzene ring. The student should expect to encounter it often as one of the Kekulé formulas. The representation adopted in this book has certain advantages, and its use is gaining ground. It is interesting that very much the same representation was advanced as long ago as 1899 by Johannes Thiele (of the University of Munich), who used a broken circle to stand for partial bonds ("partial valences").

13.10   Aromatic character. The Hückel $4n + 2$ rule

We have defined aromatic compounds as those that resemble benzene. But just which properties of benzene must a compound possess before we speak of it as being aromatic? Besides the compounds that contain benzene rings, there are many other substances that are called aromatic; yet some of these superficially bear little resemblance to benzene.

What properties do all aromatic compounds have in common?

From the experimental standpoint, aromatic compounds are compounds whose molecular formulas would lead us to expect a high degree of unsaturation, and yet which are resistant to the addition reactions generally characteristic of unsaturated compounds. Instead of addition reactions, we often find that these aromatic compounds undergo electrophilic substitution reactions like those of benzene. Along with this resistance toward addition—and presumably the cause of it—we find evidence of unusual stability: low heats of hydrogenation and low heats of combustion. Aromatic compounds are cyclic—generally containing five-, six-, or seven-membered rings—and when examined by physical methods, they are found to have flat (or nearly flat) molecules. Their protons show the same sort of *chemical shift* in NMR spectra (Sec. 16.8) as the protons of benzene and its derivatives.

From a theoretical standpoint, to be aromatic a compound must have a molecule that contains *cyclic clouds of delocalized π electrons above and below the plane of the molecule*; furthermore, *the π clouds must contain a total of* **$(4n + 2)$ π electrons**. That is to say, for the particular degree of stability that characterizes an aromatic compound, delocalization alone is not enough. There must be a particular number of π electrons: 2, or 6, or 10, etc. This requirement, called the $4n + 2$ *rule* or *Hückel rule* (after Erich Hückel, of the Institut für theoretische Physik, Stuttgart), is based on quantum mechanics, and has to do with the filling up of the various orbitals that make up the π cloud (Sec. 33.6). The Hückel rule is strongly supported by the facts.

Let us look at some of the evidence supporting the Hückel rule. Benzene has six π electrons, the *aromatic sextet*; six is, of course, a Hückel number, corresponding to $n = 1$. Besides benzene and its relatives (naphthalene, anthracene, phenanthrene, Chap. 34), we shall encounter a number of heterocyclic compounds (Chap. 35) that are clearly aromatic; these aromatic heterocycles, we shall see, are just the ones that can provide an aromatic sextet.

Or consider these six compounds, for each of which just one contributing structure is shown:

Cyclopentadienyl
cation

Four π electrons

Cyclopentadienyl
radical

Five π electrons

Cyclopentadienyl
anion

Six π electrons
Aromatic

Cycloheptatrienyl
cation
(Tropylium ion)

Six π electrons
Aromatic

Cycloheptatrienyl
radical

Seven π electrons

Cycloheptatrienyl
anion

Eight π electrons

Each molecule is a hybrid of either five or seven equivalent structures, with the charge or odd electron on each carbon. Yet, of the six compounds, only *two* give evidence of *unusually* high stability: the cyclopentadienyl anion and the cyclo-heptatrienyl cation (*tropylium ion*).

For a hydrocarbon, cyclopentadiene is an unusually strong acid ($K_a = 10^{-15}$), indicating that loss of a hydrogen ion gives a particularly stable anion. (It is, for example, a much stronger acid than cycloheptatriene, $K_a = 10^{-45}$, despite the fact that the latter gives an anion that is stabilized by seven contributing structures.) Dicyclopentadienyliron (*ferrocene*), $[(C_5H_5)^-]_2Fe^{2+}$, is a stable molecule that has been shown to be a "sandwich" of an iron atom between two flat five-membered rings. All carbon–carbon bonds are 1.4 Å long. The rings of ferrocene undergo two typically aromatic substitution reactions: sulfonation and the Friedel–Crafts reaction.

Ferrocene

Of the cycloheptatrienyl derivatives, on the other hand, it is the cation that is unusual. Tropylium bromide, C_7H_7Br, melts above 200 °C, is soluble in water but insoluble in non-polar solvents, and gives an immediate precipitate of AgBr when treated with silver nitrate. This is strange behavior for an organic bromide, and strongly suggests that, even in the solid, we are dealing with an ionic compound, R^+Br^-, the cation of which is actually a *stable* carbocation.

Consider the electronic configuration of the cyclopentadienyl anion (Fig. 13.6).
Each carbon, trigonally hybridized, is held by a σ bond to two other carbons and

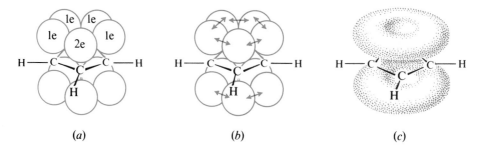

(a)   (b)   (c)

Figure 13.6   Cyclopentadienyl anion. (a) Two electrons in the p orbital of
one carbon; one electron in the p orbital of each of the other carbons.
(b) Overlap of the p orbitals to form π bonds. (c) The π clouds above and
below the plane of the ring; a total of six π electrons, the aromatic sextet.

one hydrogen. The ring is a regular pentagon, whose angles (108°) are not a bad fit
for the 120° trigonal angle; any instability due to imperfect overlap (angle strain)
is more than made up for by the delocalization that is to follow. Four carbons have
one electron each in p orbitals; the fifth carbon (the "one" that lost the proton,
but actually, of course, indistinguishable from the others) has two electrons. Overlap
of the p orbitals gives rise to π clouds containing a total of six electrons, the aromatic
sextet. (See Fig. 13.7.)

Figure 13.7   Electronic configuration and molecular shape. Model of the
cyclopentadienyl anion: two views.

In ferrocene, each cyclopentadienyl anion is held to iron through overlap of the π cloud
with an empty orbital of the metal. This kind of bonding has been shown to exist between
compounds with π electrons—alkenes, aromatics—and acidic molecules of many kinds:
silver ions, for example, or halogens. Such **π-complexes** have been detected spectroscopically
and sometimes, as with ferrocene, isolated. Reversible formation of π-complexes has been
postulated as a step preliminary to the reaction of many electrophiles with alkenes and
aromatic compounds.

The isolation of ferrocene in 1951 aroused great interest, not just because it was an
example of a non-benzenoid aromatic compound consistent with the Hückel rule. More

important, the strength of this π bonding sparked a revolution in the field of organic complexes of transition metals (Sec. 20.5)—a revolution that has given organic chemists catalysts of unprecedented selectivity and power.

In a similar way, we arrive at the configuration of the tropylium ion. It is a regular heptagon (angles 128.5°). Six carbons contribute one p electron each, and the seventh contributes only an empty p orbital. Result: the aromatic sextet.

The ions are conveniently represented as:

Cyclopentadienyl
anion

Cycloheptatrienyl
cation
(Tropylium ion)

Six is the Hückel number most often encountered, and for good reason. To provide p orbitals, the atoms of the aromatic ring must be trigonally (sp^2) hybridized, which means, ideally, bond angles of 120°. To permit the overlap of the p orbitals that gives rise to the π cloud, the aromatic compound must be flat, or nearly so. The number of trigonally hybridized atoms that will fit a flat ring without undue angle strain (i.e., with reasonably good overlap for π bond formation) is five, six, or seven. Six is the Hückel number of π electrons that can be provided—as we have just seen—by these numbers of atoms. (Benzene, as we have seen, is the "perfect" specimen: six carbons to provide six π electrons and to make a hexagon whose angles exactly match the trigonal angle.)

Now, what evidence is there that other Hückel numbers—2, 10, 14, etc.—are also "magic" numbers? We cannot expect aromatic character necessarily to appear here in the form of highly stable compounds comparable to benzene and its derivatives. The rings will be too small or too large to accommodate trigonally hybridized atoms very well, so that any stabilization due to aromaticity may be largely offset by angle strain or poor overlap of p orbitals, or both.

We must look for stability on a *comparative* basis—as was done above with the cyclopentadienyl and cycloheptatrienyl derivatives—and may find evidence of aromaticity only in the fact that one molecular species is *less unstable* than its relatives. The net effect of a great deal of elegant work is strongly to support the $4n + 2$ rule. The question now seems rather to be: over how unfavorable a combination of angle strain and multiple charge can aromaticity manifest itself?

Problem 13.6 Ronald Breslow (of Columbia University) found that treatment of 3-chlorocyclopropene with $SbCl_5$ yields a stable crystalline solid, I, of formula

H   Cl
3-Chlorocyclopropene

$C_3H_3SbCl_6$, insoluble in non-polar solvents but soluble in polar solvents like nitromethane, acetonitrile, or sulfur dioxide. The NMR spectrum of I shows three exactly equivalent protons. 3-Chlorocyclopropene reacts with $AgBF_4$ to give AgCl and a

solution with an NMR spectrum identical to that of I. Treatment of I with chloride
ion regenerates 3-chlorocyclopropene.

Conversion of I into $C_3H_3^+$ requires 153 kcal/mol, as compared with 173 kcal/
mol for conversion of allyl chloride into $C_3H_5^+$.

(a) Give in detail the most likely structure of I, and show how this structure
accounts for the various observations. (b) Of what theoretical significance are these
findings?

Problem 13.7 *1,3,5,7-Cyclooctatetraene*, C_8H_8, has a heat of combustion (compare
Problem 13.2, p. 483) of 1095 kcal; it rapidly decolorizes cold aqueous $KMnO_4$ and
reacts with Br_2/CCl_4 to yield $C_8H_8Br_8$. (a) How should its structure be represented?
(b) Upon what theoretical grounds might one have predicted its structure and
properties? (c) Treatment of cyclooctatetraene with potassium metal has been found
to yield a stable compound $2K^+ C_8H_8^{2-}$. Of what significance is the formation of this
salt? (d) Using models, suggest a possible shape (or shapes) for cyclooctatetraene.
What shape would you predict for the $C_8H_8^{2-}$ anion?

13.11 Nomenclature of benzene derivatives

In later chapters we shall consider in detail the chemistry of many of the
derivatives of benzene. Nevertheless, for our present discussion of the reactions of
the benzene ring it will be helpful for us to learn to name some of the more
important of these derivatives.

For many of these derivatives we simply prefix the name of the substituent
group to the word *-benzene*, as, for example, in *chlorobenzene*, *bromobenzene*,
iodobenzene, or *nitrobenzene*. Other derivatives have special names which may show

| Chlorobenzene | Bromobenzene | Iodobenzene | Nitrobenzene |

no resemblance to the name of the attached substituent group. For example,
methylbenzene is always known as *toluene*, aminobenzene as *aniline*, hydroxy-
benzene as *phenol*, and so on. The most important of these special compounds are:

| Toluene | Aniline | Phenol | Benzoic acid | Benzenesulfonic acid |

If several groups are attached to the benzene ring, we must not only tell what
they are, but also indicate their relative positions. The three possible isomers of a
disubstituted benzene are differentiated by the use of the names **ortho**, **meta**, and
para, abbreviated as *o-*, *m-*, and *p-*. For example:

o-Dibromobenzene   m-Dibromobenzene   p-Dibromobenzene
      ortho   meta   para

If the two groups are different, and neither is a group that gives a special name to the molecule, we simply name the two groups successively and end the word with -*benzene*, as, for example, *chloronitrobenzene*, *bromoiodobenzene*, etc. If one of the two groups is the kind that gives a special name to the molecule, then the compound is named as a derivative of that special compound, as, for example, *nitrotoluene*, *bromophenol*, etc.

p-Bromoiodobenzene   m-Chloronitrobenzene   p-Chlorobenzenesulfonic
                                                                   acid

o-Nitrotoluene   p-Bromophenol   m-Nitrobenzoic   o-Iodoaniline
                                                acid

If more than two groups are attached to the benzene ring, numbers are used to indicate their relative positions. For example:

1,2,4-Tribromobenzene   2-Chloro-4-nitrophenol   2,6-Dinitrotoluene

3-Bromo-5-chloronitrobenzene   2,4,6-Tribromoaniline

If all the groups are the same, each is given a number, the sequence being the one that gives the lowest combination of numbers; if the groups are different, then the last-named group is understood to be in position 1 and the other numbers conform to that, as, for example, in *3-bromo-5-chloronitrobenzene*. If one of the groups that gives a special name is present, then the compound is named as having the special group in position 1; thus in *2,6-dinitrotoluene* the methyl group is considered to be at the 1-position.

Problem 13.8   You have three bottles containing the three isomeric dibromo-benzenes; they have the melting points +87 °C, +6 °C, and −7 °C. By a great deal of work, you prepare six dibromonitrobenzenes ($C_6H_3Br_2NO_2$) and find that, of the six, *one* is related to (derived from or convertible into) the dibromobenzene of m.p. +87 °C, *two* to the isomer of m.p. +6 °C, and *three* to the isomer of m.p. −7 °C.
    Label each bottle with the correct name of *ortho*, *meta*, or *para*.
    (This work was actually carried out by Wilhelm Körner, of the University of Milan, and was the first example of the **Körner method of absolute orientation**.)

13.12   Quantitative elemental analysis: nitrogen and sulfur

This chapter has dealt with the structure of benzene and with some of its reactions. It is well to remind ourselves again that all this discussion has meaning only because it is based upon solid facts. As we saw earlier (Sec. 2.25), we can discuss the structure and reactions of a compound only when we know its molecular formula and the molecular formulas of its products.

To know a molecular formula we must know what elements are present in the compound, and in what proportions. In Sec. 2.26 we saw how various elements can be detected in an organic compound, and in Sec. 2.27 how the percentage of carbon, hydrogen, and halogen can be measured.

Quantitative analysis for nitrogen is carried out either (a) by the *Dumas method* or (b) by the *Kjeldahl method*. The Kjeldahl method is somewhat more convenient, particularly if many analyses must be carried out; however, it cannot be used for all kinds of nitrogen compounds.

In the Dumas method, the organic compound is passed through a tube containing, first, hot copper oxide and, next, hot copper metal gauze. The copper oxide oxidizes the compound (as in the carbon–hydrogen combustion, Sec. 2.27), converting combined nitrogen into molecular nitrogen. The copper gauze reduces any nitrogen oxides that may be formed, also to molecular nitrogen. The nitrogen gas is collected and its volume is measured. For example, an 8.32-mg sample of *aniline* yields 1.11 mL of nitrogen at 21 °C and 743 mm pressure (corrected for the vapor pressure of water). We calculate the volume at standard temperature and pressure,

$$\text{vol. } N_2 \text{ at S.T.P.} = 1.11 \times \frac{273}{273 + 21} \times \frac{743}{760} = 1.01 \text{ mL}$$

and, from it, the weight of nitrogen,

$$\text{wt. } N = \frac{1.01}{22400} \times (2 \times 14.01) = 0.00126 \text{ g } or \text{ 1.26 mg}$$

and, finally, the percentage of nitrogen in the sample

$$\%N = \frac{1.26}{8.32} \times 100 = 15.2\%$$

Problem 13.9   Why is the nitrogen in the Dumas analysis collected over 50% aqueous KOH rather than, say, pure water, aqueous NaCl, or mercury?

In the Kjeldahl method, the organic compound is digested with concentrated sulfuric acid, which converts combined nitrogen into ammonium sulfate. The solution is then made alkaline. The ammonia thus liberated is distilled, and its amount is determined by titration with standard acid. For example, the ammonia formed from a 3.51-mg sample of aniline neutralizes 3.69 mL of 0.0103 N acid. For every milliequivalent of acid there is a milliequivalent of ammonia, and a

$$\text{milligram-atoms N} = \text{milliequivalents NH}_3 = \text{milliequivalents acid}$$

$$= 3.69 \times 0.0103 = 0.0380$$

milligram-atom of nitrogen. From this, the weight and, finally, the percentage of nitrogen in the compound can be calculated.

$$\text{wt. N} = \text{milligram-atoms N} \times 14.01 = 0.0380 \times 14.01 = 0.53 \text{ mg}$$

$$\%\text{N} = \frac{0.53}{3.51} \times 100 = 15.1\%$$

Sulfur in an organic compound is converted into sulfate ion by the methods used in halogen analysis (Sec. 2.27): treatment with sodium peroxide or with nitric acid (*Carius method*). This is then converted into barium sulfate, which is weighed.

Problem 13.10   A Dumas nitrogen analysis of a 5.72-mg sample of *p-phenylenediamine* gave 1.31 mL of nitrogen at 20 °C and 746 mm. The gas was collected over saturated aqueous KOH solution (the vapor pressure of water, 6 mm). Calculate the percentage of nitrogen in the compound.

Problem 13.11   A Kjeldahl nitrogen analysis of a 3.88-mg sample of *ethanolamine* required 5.73 mL of 0.0110 N hydrochloric acid for titration of the ammonia produced. Calculate the percentage of nitrogen in the compound.

Problem 13.12   A Carius sulfur analysis of a 4.81-mg sample of *p-toluenesulfonic acid* gave 6.48 mg of BaSO$_4$. Calculate the percentage of sulfur in the compound.

Problem 13.13   How does each of the above answers compare with the theoretical value calculated from the formula of the compound? (Each compound is listed in the index.)

PROBLEMS

1. Draw structures of:

(a) *p*-dinitrobenzene
(b) *m*-bromonitrobenzene
(c) *o*-chlorobenzoic acid
(d) *m*-nitrotoluene
(e) *p*-bromoaniline
(f) *m*-iodophenol

(g) mesitylene (1,3,5-trimethylbenzene)
(h) 3,5-dinitrobenzenesulfonic acid
(i) 4-chloro-2,3-dinitrotoluene
(j) 2-amino-5-bromo-3-nitrobenzoic acid
(k) *p*-hydroxybenzoic acid
(l) 2,4,6-trinitrophenol (picric acid)

2. Give structures and names of all the possible isomeric:

(a) xylenes (dimethylbenzenes)
(b) aminobenzoic acids ($H_2NC_6H_4COOH$)
(c) trimethylbenzenes

(d) dibromonitrobenzenes
(e) bromochlorotoluenes
(f) trinitrotoluenes

3. (a) How many isomeric monosubstitution products are theoretically possible from each of the following structures of formula C_6H_6? (b) How many disubstitution products? (c) Which structures, if any, would be acceptable for benzene on the basis of isomer number?

$$HC\equiv C-CH_2-CH_2-C\equiv CH \qquad HC\equiv C-CH_2-C\equiv C-CH_3$$
$$\text{I} \qquad\qquad\qquad\qquad\qquad \text{II}$$

$$HC\equiv C-C\equiv C-CH_2-CH_3$$
$$\text{III}$$

IV   V

4. Give structures and names of all theoretically possible products of the ring mono-nitration of:

(a) *o*-dichlorobenzene
(b) *m*-dichlorobenzene
(c) *p*-dichlorobenzene
(d) *o*-bromochlorobenzene
(e) *m*-bromochlorobenzene
(f) *p*-bromochlorobenzene

(g) *o*-chloronitrobenzene
(h) *m*-chloronitrobenzene
(i) *p*-chloronitrobenzene
(j) 1,3,5-trimethylbenzene
(k) 4-bromo-1,2-dimethylbenzene
(l) *p*-ethyltoluene

5. Give structures and names of all benzene derivatives that *theoretically* can have the indicated number of isomeric ring-substituted derivatives.

(a) C_8H_{10}: one monobromo derivative
(b) C_8H_{10}: two monobromo derivatives
(c) C_8H_{10}: three monobromo derivatives
(d) C_9H_{12}: one mononitro derivative

(e) C_9H_{12}: two mononitro derivatives
(f) C_9H_{12}: three mononitro derivatives
(g) C_9H_{12}: four mononitro derivatives

6. There are three known tribromobenzenes, of m.p. 44 °C, 87 °C, and 120 °C. Could these isomers be assigned structures by use of the Körner method (Problem 13.8, p. 494)? Justify your answer.

7. For a time the prism formula VI, proposed in 1869 by Albert Ladenburg of Germany, was considered as a possible structure for benzene, on the grounds that it would yield one monosubstitution product and three isomeric disubstitution products.

$$\begin{array}{c} CH\!-\!\!-\!\!-\!CH \\ CH \\ CH\!-\!|\!-\!CH \\ CH \end{array}$$

VI

(a) Draw Ladenburg structures of three possible isomeric dibromobenzenes.
(b) On the basis of the Körner method of absolute orientation, label each Ladenburg structure in (a) as *ortho*, *meta*, or *para*.

(c) In light of Chapter 4, can the Ladenburg formula actually pass the test of isomer number?
(Derivatives of Ladenburg "benzene", called *prismanes*, have actually been made.)

8. In 1874 Griess (p. 1287) reported that he had decarboxylated the six known diaminobenzoic acids, $C_6H_3(NH_2)_2COOH$, to the diaminobenzenes. Three acids gave a diamine of m.p. 63 °C, two acids gave a diamine of m.p. 104 °C, and one acid gave a diamine of m.p. 142 °C. Draw the structural formulas for the three isomeric diamino-benzenes and label each with its melting point.

9. For which of the following might you expect aromaticity (geometry permitting)?

(a) The annulenes containing up to 20 carbons. (*Annulenes* are monocyclic compounds of the general formula $[-CH=CH-]_n$.)
(b) The monocyclic polyenes C_9H_{10}, $C_9H_9{}^+$, $C_9H_9{}^-$.

10. The properties of *pyrrole*, commonly represented by VII,

VII

show that it is aromatic. Account for its aromaticity on the basis of orbital theory. (*Hint*: See Sec. 13.10. Check your answer in Sec. 35.2.)

11. When benzene is treated with chlorine under the influence of ultraviolet light, a solid material of mol. wt. 291 is formed. Quantitative analysis gives an empirical formula of CHCl. (a) What is the molecular formula of the product? (b) What is a possible structural formula? (c) What kind of reaction has taken place? (d) Is the product aromatic? (e) Actually, the product can be separated into six isomeric compounds, one of which has been used as an insecticide (Gammexane or Lindane). How do these isomers differ from each other? (f) Are more than six isomers possible?

12. Can you account for the following order of acidity. (*Hint*: See Sec. 11.11.)

acetylene > benzene > *n*-pentane

13. *Naphthalene*, $C_{10}H_8$, is a polynuclear hydrocarbon (Chap. 34) whose properties clearly show that it is aromatic. One of the structures contributing to the hybrid molecule is VIII.

VIII

(a) Account for the aromaticity of the compound.
(b) In contrast to the six equivalent bonds in benzene, the carbon–carbon bonds in naphthalene come in two lengths: C(1)–C(2), for example, is 1.365 Å long, while C(2)–C(3) is 1.404 Å long. How do you account for this?

14

Electrophilic Aromatic
Substitution

14.1 Introduction

We have already seen that the characteristic reactions of benzene involve substitution, in which the resonance-stabilized ring system is preserved. What kind of reagents bring about this substitution? What is the mechanism by which these reactions take place?

Above and below the plane of the benzene ring there is a cloud of π electrons (Fig. 14.1). Through resonance, these π electrons are more involved in holding together carbon nuclei than are the π electrons of a carbon–carbon double bond. Still, in comparison with σ electrons, these π electrons are loosely held and are available to a reagent that is seeking electrons.

Figure 14.1 Benzene ring: the π cloud is a source of electrons.

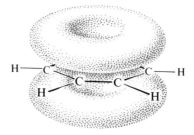

It is not surprising that *in its typical reactions the benzene ring serves as a* **source of electrons**, that is, as a **base**. The compounds with which it reacts are deficient in electrons, that is, are electrophilic reagents or acids. Just as the typical reactions of the alkenes are electrophilic addition reactions, so *the typical reactions of the benzene ring are* **electrophilic substitution reactions**.

These reactions are characteristic not only of benzene itself, but of the benzene ring wherever it is found—and, indeed, of many aromatic rings, benzenoid and non-benzenoid.

Electrophilic aromatic substitution includes a wide variety of reactions: nitration, halogenation, sulfonation, and Friedel–Crafts reactions, undergone by nearly all aromatic rings; reactions like nitrosation and diazo coupling, undergone only by rings of high reactivity; and reactions like desulfonation, isotopic exchange, and many ring closures which, although apparently unrelated, are found on closer examination to be properly and profitably viewed as reactions of this kind. In synthetic importance electrophilic aromatic substitution is probably unequaled by any other class of organic reactions. It is the initial route of access to nearly all aromatic compounds: it permits the direct introduction of certain substituent groups which can then be converted, by replacement or by transformation, into other substituents, including even additional aromatic rings.

_____ ELECTROPHILIC AROMATIC SUBSTITUTION _____

Ar = *aryl*, any aromatic group with attachment directly to ring carbon

1. Nitration. Discussed in Sec. 14.8.

$$ArH + HONO_2 \xrightarrow{H_2SO_4} Ar-NO_2 + H_2O$$
A nitro compound

2. Sulfonation. Discussed in Sec. 14.9.

$$ArH + HOSO_3H \xrightarrow{SO_3} Ar-SO_3H + H_2O$$
A sulfonic acid

3. Halogenation. Discussed in Sec. 14.11.

$$ArH + Cl_2 \xrightarrow{Fe} Ar-Cl + HCl$$
An aryl chloride

$$ArH + Br_2 \xrightarrow{Fe} Ar-Br + HBr$$
An aryl bromide

4. Friedel–Crafts alkylation. Discussed in Sec. 14.10.

$$ArH + RCl \xrightarrow{AlCl_3} Ar-R + HCl$$
An alkylbenzene

CONTINUED ——

—— CONTINUED ——

5. **Friedel–Crafts acylation.** Discussed in Sec. 21.5.

$$\text{ArH} + \text{RCOCl} \xrightarrow{\text{AlCl}_3} \text{Ar—COR} + \text{HCl}$$
$$\qquad\qquad \text{An acyl chloride} \qquad\quad \text{A ketone}$$

6. **Protonation.** Discussed in Sec. 14.12.

$$\text{ArSO}_3\text{H} + \text{H}^+ \xrightarrow{\text{H}_2\text{O}} \text{Ar—H} + \text{H}_2\text{SO}_4 \qquad\qquad \textit{Desulfonation}$$

$$\text{ArH} + \text{D}^+ \xrightarrow{\text{D}_2\text{O}} \text{Ar—D} + \text{H}^+ \qquad\qquad \textit{Hydrogen exchange}$$

7. **Nitrosation.** Discussed in Secs. 27.11 and 28.10.

$$\text{ArH} + \text{HONO} \longrightarrow \text{Ar—N}{=}\text{O} + \text{H}_2\text{O} \qquad \textit{Only for highly}$$
$$\qquad\qquad\qquad \text{A nitroso compound} \qquad\qquad \textit{reactive ArH}$$

8. **Diazo coupling.** Discussed in Sec. 27.18.

$$\text{ArH} + \text{Ar}'\text{N}_2{}^+\text{X}^- \longrightarrow \text{Ar—N}{=}\text{NAr}' + \text{HX} \qquad \textit{Only for highly}$$
$$\quad\text{A diazonium salt} \qquad\quad \text{An azo compound} \qquad\qquad \textit{reactive ArH}$$

9. **Kolbe reaction.** Discussed in Sec. 28.11.   *Only for phenols*

10. **Reimer–Tiemann reaction.** Discussed in Sec. 28.12.   *Only for phenols* ∎

14.2 Effect of substituent groups

Like benzene, toluene undergoes electrophilic aromatic substitution: sulfonation, for example. Although there are three possible monosulfonation products, this reaction actually yields appreciable amounts of only two of them: the *ortho* and *para* isomers.

Toluene	*p*-Toluene-sulfonic acid	*o*-Toluene-sulfonic acid
	62%	32%

Benzene and toluene are insoluble in sulfuric acid, whereas the sulfonic acids are readily soluble; completion of reaction is indicated simply by disappearance of the hydrocarbon layer. When shaken with fuming sulfuric acid at room temperature, benzene reacts completely within 20 to 30 minutes, whereas toluene is found to react within only a minute or two.

Studies of nitration, halogenation, and Friedel–Crafts alkylation of toluene give analogous results. In some way the methyl group makes the ring more reactive than unsubstituted benzene, and *directs* the attacking reagent to the *ortho* and *para* positions of the ring.

On the other hand, nitrobenzene, to take a different example, has been found to undergo substitution more slowly than benzene, and to yield chiefly the *meta* isomer.

Like methyl or nitro, any group attached to a benzene ring affects the **reactivity** of the ring and determines the **orientation** of substitution. When an electrophilic reagent attacks an aromatic ring, it is the group already attached to the ring that determines *how readily* the attack occurs and *where* it occurs.

A group that makes the ring more reactive than benzene is called an **activating group**. A group that makes the ring less reactive than benzene is called a **deactivating group**.

A group that causes attack to occur chiefly at positions *ortho* and *para* to it is called an *ortho,para* **director**. A group that causes attack to occur chiefly at positions *meta* to it is called a *meta* **director**.

In this chapter we shall examine the methods that are used to measure these effects on reactivity and orientation, the results of these measurements, and a theory that accounts for these results. The theory is, of course, based on the most likely mechanism for electrophilic aromatic substitution; we shall see what this mechanism is, and some of the evidence supporting it. First let us look at the facts.

14.3   Determination of orientation

To determine the effect of a group on orientation is, in principle, quite simple: the compound containing this group attached to benzene is allowed to undergo substitution and the product is analyzed for the proportions of the three isomers. Identification of each isomer as *ortho*, *meta*, or *para* generally involves comparison with an authentic sample of that isomer prepared by some other method from a compound whose structure is known. In the last analysis, of course, all these identifications go back to absolute determinations of the Körner type (Problem 13.8, p. 494).

In this way it has been found that every group can be put into one of two classes: *ortho,para* directors or *meta* directors. Table 14.1 summarizes the orienta-

Table 14.1   ORIENTATION OF NITRATION OF C_6H_5Y

Y	Ortho	Para	Ortho plus para	Meta
—OH	50–55	45–50	100	trace
—NHCOCH$_3$	19	79	98	2
—CH$_3$	58	38	96	4
—F	12	88	100	trace
—Cl	30	70	100	trace
—Br	37	62	99	1
—I	38	60	98	2
—NO$_2$	6.4	0.3	6.7	93.3
—N(CH$_3$)$_3$$^+$	0	11	11	89
—CN	—	—	19	81
—COOH	19	1	20	80
—SO$_3$H	21	7	28	72
—CHO	—	—	28	72

tion of nitration in a number of substituted benzenes. Of the five positions open to attack, three (60%) are *ortho* and *para* to the substituent group, and two (40%) are *meta* to the group; if there were no selectivity in the substitution reaction, we would expect the *ortho* and *para* isomers to make up 60% of the product, and the *meta* isomer to make up 40%. We see that seven of the groups direct 96–100% of nitration to the *ortho* and *para* positions; the other six direct 72–94% to the *meta* positions.

A given group causes the same general kind of orientation—predominantly *ortho*,*para* or predominantly *meta*—whatever the electrophilic reagent involved. The actual distribution of isomers may vary, however, from reaction to reaction. In Table 14.2, for example, compare the distribution of isomers obtained from toluene by sulfonation or bromination with that obtained by nitration.

Table 14.2   ORIENTATION OF SUBSTITUTION IN TOLUENE

	Ortho	Meta	Para
Nitration	58	4	38
Sulfonation	32	6	62
Bromination	33	—	67

14.4   Determination of relative reactivity

A group is classified as *activating* if the ring it is attached to is more reactive than benzene, and is classified as *deactivating* if the ring it is attached to is less reactive than benzene. The reactivities of benzene and a substituted benzene are compared in one of the following ways.

The **time required** for reactions to occur under identical conditions can be measured. Thus, as we just saw, toluene is found to react with fuming sulfuric acid in about one-tenth to one-twentieth the time required by benzene. Toluene is more reactive than benzene, and $-CH_3$ is therefore an activating group.

The **severity of conditions** required for comparable reaction to occur within the same period of time can be observed. For example, benzene is nitrated in less than an hour at 60 °C by a mixture of concentrated sulfuric acid and concentrated nitric acid; comparable nitration of nitrobenzene requires treatment at 90 °C with fuming nitric acid and concentrated sulfuric acid. Nitrobenzene is evidently less reactive than benzene, and the nitro group, $-NO_2$, is a deactivating group.

For an exact, quantitative comparison under identical reaction conditions, **competitive reactions** can be carried out, in which the compounds to be compared are allowed to compete for a limited amount of a reagent (Sec. 3.22). For example, if equimolar amounts of benzene and toluene are treated with a small amount of nitric acid (in a solvent like nitromethane or acetic acid, which will dissolve both organic and inorganic reactants), about 25 times as much nitrotoluene as nitro-benzene is obtained, showing that toluene is 25 times as reactive as benzene. On the other hand, a mixture of benzene and chlorobenzene yields a product in which nitrobenzene exceeds the nitrochlorobenzenes by 30:1, showing that chloro-benzene is only one-thirtieth as reactive as benzene. The chloro group is therefore classified as deactivating, the methyl group as activating. The activation or deacti-vation caused by some groups is extremely powerful: aniline, $C_6H_5NH_2$, is roughly one million times as reactive as benzene, and nitrobenzene, $C_6H_5NO_2$, is roughly one-millionth as reactive as benzene.

14.5   Classification of substituent groups

The methods described in the last two sections have been used to determine the effects of a great number of groups on electrophilic substitution. As shown in Table 14.3, nearly all groups fall into one of two classes: activating and *ortho,para*-directing, or deactivating and *meta*-directing. The halogens are in a class by themselves, being deactivating but *ortho,para*-directing.

Table 14.3   EFFECT OF GROUPS ON ELECTROPHILIC AROMATIC SUBSTITUTION

Activating: *Ortho, para* directors	Deactivating: *Meta* directors
Strongly activating	—NO_2
—NH_2 (—NHR, —NR$_2$)	—$N(CH_3)_3{}^+$
—OH	—CN
	—COOH (—COOR)
	—SO_3H
Moderately activating	—CHO, —COR
—OCH_3 (—OC_2H_5, etc.)	
—$NHCOCH_3$	
	Deactivating: *Ortho, para* directors
Weakly activating	—F, —Cl, —Br, —I
—C_6H_5	
—CH_3 (—C_2H_5, etc.)	

Just by knowing the effects summarized in these short lists, we can now predict fairly accurately the course of hundreds of aromatic substitution reactions. We now know, for example, that bromination of nitrobenzene will yield chiefly the *meta* isomer and that the reaction will go more slowly than the bromination of benzene itself; indeed, it will probably require severe conditions to go at all. We now know that nitration of $C_6H_5NHCOCH_3$ (*acetanilide*) will yield chiefly the *ortho* and *para* isomers and will take place more rapidly than nitration of benzene.

Although, as we shall see, it is possible to account for these effects in a reasonable way, it is necessary for you to memorize the classifications in Table 14.3 so that you may deal rapidly with synthetic problems involving aromatic compounds.

14.6   Orientation in disubstituted benzenes

The presence of two substituents on a ring makes the problem of orientation more complicated, but even here we can frequently make very definite predictions.

First of all, the two substituents may be located so that the directive influence of one *reinforces* that of the other; for example, in I, II, and III the orientation clearly must be that indicated by the arrows.

    I   II   III

On the other hand, when the directive effect of one group *opposes* that of the other, it may be difficult to predict the major product; in such cases complicated mixtures of several products are often obtained.

Even where there are opposing effects, however, it is still possible in certain cases to make predictions in accordance with the following generalizations.

(a) *Strongly activating groups generally win out over deactivating or weakly activating groups.* The differences in directive power in the sequence

$$-NH_2, -OH > -OCH_3, -NHCOCH_3 > -C_6H_5, -CH_3 > meta\ directors$$

are great enough to be used in planning feasible syntheses. For example:

Sole product

Chief product

Chief product

There must be, however, a fairly large difference in the effects of the two groups for clear-cut results; otherwise one gets results like these:

                                     58%   *42%*

(b) *There is often little substitution between two groups that are meta to each other.* In many cases it seems as though there just is not enough room between two groups located *meta* to each other for appreciable substitution to occur there, as illustrated by IV and V:

Nitration

IV

Nitration

V

14.7 Orientation and synthesis

As we discussed earlier (Sec. 3.14), a laboratory synthesis is generally aimed at obtaining a single, pure compound. Whenever possible we should avoid use of a reaction that produces a mixture, since this lowers the yield of the compound we want and causes difficult problems of purification. With this in mind, let us see some of the ways in which we can apply our knowledge of orientation to the synthesis of pure aromatic compounds.

First of all, *we must consider the order in which we introduce these various substituents into the ring.* In the preparation of the bromonitrobenzenes, for example, it is obvious that if we nitrate first and then brominate, we will obtain the *meta* isomer; whereas if we brominate first and then nitrate, we will obtain a mixture of the *ortho* and *para* isomers. The order in which we decide to carry out the two steps, then, depends upon which isomer we want.

m-Bromonitrobenzene

Bromonitrobenzene

ortho-
38%

para-
62%

Next, if our synthesis involves conversion of one group into another, *we must consider the proper time for this conversion.* For example, oxidation of a methyl group yields a carboxyl group (Sec. 15.11). In the preparation of nitrobenzoic acids from toluene, the particular product obtained depends upon whether oxidation or nitration is carried out first.

Substitution controlled by an activating group yields a mixture of *ortho* and *para* isomers; nevertheless, we must often make use of such reactions, as in the examples just shown. It is usually possible to obtain the pure *para* isomer from the mixture by fractional crystallization. As the more symmetrical isomer, it is the less soluble (Sec. 15.4), and crystallizes while the solvent still retains the soluble *ortho* isomer. Some *para* isomer, of course, remains in solution to contaminate the *ortho* isomer, which is therefore difficult to purify. As we shall see, special approaches are often used to prepare *ortho* isomers.

m-Nitrobenzoic acid

Toluene

o-Nitrobenzoic acid

p-Nitrobenzoic acid

In the special case of nitro compounds, the difference in boiling points is often large enough that both *ortho* and *para* isomers can be obtained pure by fractional distillation. As a result, many aromatic compounds are best prepared not by direct substitution but by conversion of one group into another, in the last analysis starting from an original nitro compound; we shall take up these methods of conversion later.

14.8 Mechanism of nitration

Now that we have seen the effects that substituent groups exert on orientation and reactivity in electrophilic aromatic substitution, let us see how we can account for these effects. The first step in doing this is to examine the mechanism for the reaction. Let us begin with nitration, using benzene as the aromatic substrate.

The commonly accepted mechanism for nitration with a mixture of nitric and sulfuric acids (the widely used "mixed acid" of the organic chemist) involves the following sequence of reactions:

(1) $$HONO_2 + 2H_2SO_4 \rightleftarrows H_3O^+ + 2HSO_4^- + \oplus NO_2$$

Nitronium ion

(2) $$\oplus NO_2 + C_6H_6 \longrightarrow C_6H_5 \overset{\oplus}{\underset{NO_2}{\overset{H}{\diagdown}}} \qquad\qquad Slow$$

(3) $$C_6H_5 \overset{\overset{H}{\underset{\oplus}{\diagup}}}{\underset{NO_2}{\diagdown}} + HSO_4^- \longrightarrow C_6H_5-NO_2 + H_2SO_4 \qquad\qquad Fast$$

Step (1) generates the **nitronium ion**, $\oplus NO_2$, which is the electrophilic particle that actually attacks the benzene ring. This reaction is simply an acid–base equilibrium in which sulfuric acid serves as the acid and the much weaker nitric acid serves as a base. We may consider that the very strong acid, sulfuric acid, causes nitric acid to ionize in the sense, $HO^- \cdots ^+NO_2$, rather than in the usual way, $H^+ \cdots ^-ONO_2$. The nitronium ion is well known, existing in salts such as nitronium perchlorate, $NO_2^+ClO_4^-$, and nitronium fluoroborate, $NO_2^+BF_4^-$. Indeed, solutions of these stable nitronium salts in solvents like nitromethane or acetic acid have been found by George Olah (p. 194) to nitrate aromatic compounds smoothly and in high yield at room temperature.

Needing electrons, the nitronium ion finds them particularly available in the π cloud of the benzene ring, and so in step (2) attaches itself to one of the carbon atoms by a covalent bond. This forms the carbocation,

$$C_6H_5 \overset{\oplus}{\underset{NO_2}{<}} \overset{H}{}$$

often called a *benzenonium ion*.

Just what is the structure of this carbocation? We find that we can represent it by three structures (I, II, and III) that differ from each other only in position of double bonds and positive charge. The actual ion must then be a resonance hybrid of these three structures.

This means, of course, that the positive charge is not localized on one carbon atom, but is distributed over the molecule, being particularly strong on the carbon atoms *ortho* and *para* to the carbon bearing the —NO_2 group. (As we shall see later, this *ortho,para* distribution is significant.) The dispersal of the positive charge over the molecule by resonance makes this ion more stable than an ion with a localized positive charge. It is probably because of this stabilization that the carbocation forms at all, in view of the stability of the original benzene itself. Sometimes the hybrid carbocation is represented as IV, where the broken line stands for the fractional bonds due to the delocalized π electrons.

Thus far the reaction is like addition to alkenes: an electrophilic particle, attracted by the π electrons, attaches itself to the molecule to form a carbocation. But the fate of this carbocation is different from the fate of the ion formed from an alkene. Attachment of a basic group to the benzenonium ion to yield the addition product would destroy the aromatic character of the ring. Instead, the basic ion, HSO_4^-, abstracts a proton (step 3) to yield the substitution product, which retains the resonance-stabilized ring. Loss of a proton, as we have seen, is one of the reactions typical of a carbocation (Sec. 7.15); it is the *preferred* reaction in this case.

As with other carbocation reactions we have studied, it is the *formation* of the carbocation (step 2) that is the more difficult step; once formed, the carbocation rapidly loses a proton (step 3) to form the products. (We shall see proof of this in Sec. 14.14.)

Electrophilic substitution, then, like electrophilic addition, is a stepwise process involving an intermediate carbocation. The two reactions differ, however, in the fate of the carbocation. While the mechanism of nitration is, perhaps, better established than the mechanisms for other aromatic substitution reactions, it seems clear that all these reactions follow the same course.

Problem 14.1   Nitration by nitric acid alone is believed to proceed by essentially the same mechanism as nitration in the presence of sulfuric acid. Write an equation for the generation of NO_2^+ from nitric acid alone.

14.9   Mechanism of sulfonation

Sulfonation of many aromatic compounds involves the following steps:

(1) $$2H_2SO_4 \rightleftharpoons H_3O^+ + HSO_4^- + SO_3$$

(2) $$SO_3 + C_6H_6 \rightleftharpoons C_6H_5\!\!\begin{array}{c}\overset{\oplus}{\diagup} H \\ \diagdown SO_3^-\end{array} \qquad \textit{Slow}$$

(3) $$C_6H_5\!\!\begin{array}{c}\overset{\oplus}{\diagup} H \\ \diagdown SO_3^-\end{array} + HSO_4^- \rightleftharpoons C_6H_5\!\!-\!\!SO_3^- + H_2SO_4 \qquad \textit{Fast}$$

(4) $$C_6H_5\!\!-\!\!SO_3^- + H_3O^+ \rightleftharpoons C_6H_5\!\!-\!\!SO_3H + H_2O \qquad \textit{Equilibrium far to the left}$$

Again the first step, which generates the electrophilic sulfur trioxide, is simply an acid–base equilibrium, this time between molecules of sulfuric acid. For sulfonation we commonly use sulfuric acid containing an excess of SO_3; even if this is not done, it appears that SO_3 formed in step (1) can be the electrophile.

$$\begin{array}{c} :\!\ddot{O}\!: \\ \ddot{S}\!:\!\ddot{O}\!: \\ :\!\ddot{O}\!: \end{array}$$

In step (2) the electrophilic reagent, SO_3, attaches itself to the benzene ring to form the intermediate carbocation. Although sulfur trioxide is not positively charged, it is electron-deficient, and hence an acid, nevertheless.

Step (3) is the loss of a proton to form the resonance-stabilized substitution product, this time the anion of benzenesulfonic acid which, being a strong acid, is highly dissociated (step 4).

With some aromatic substrates and at certain acidities, the electrophile may be HSO_3^+ or molecules that can readily transfer SO_3 or HSO_3^+ to the aromatic ring.

Problem 14.2   Write an equation for the formation from H_2SO_4 of each of the following sulfonating electrophiles: (a) $H_3SO_4^+$; (b) HSO_3^+; (c) $H_2S_2O_7$.

14.10   Mechanism of Friedel–Crafts alkylation

In Friedel–Crafts alkylation, the electrophile is typically a carbocation. It, too, is formed in an acid–base equilibrium, this time in the Lewis sense:

$$(1) \qquad RCl + AlCl_3 \; \rightleftarrows \; AlCl_4^- + R^{\oplus}$$

$$(2) \qquad R^{\oplus} + C_6H_6 \; \rightleftarrows \; C_6H_5 \overset{\oplus}{\underset{R}{\overset{H}{\diagup}}} \qquad \qquad \textit{Slow}$$

$$(3) \qquad C_6H_5 \overset{\oplus}{\underset{R}{\overset{H}{\diagup}}} + AlCl_4^- \; \rightleftarrows \; C_6H_5\text{—}R + HCl + AlCl_3 \qquad \textit{Fast}$$

In certain cases, there is no free carbocation involved. Instead, the alkyl group is transferred—without a pair of electrons—directly to the aromatic ring from the polar complex, I, between $AlCl_3$ and the alkyl halide:

$$Cl^{\ominus}\overset{\overset{\displaystyle Cl}{|}}{\underset{\underset{\displaystyle Cl}{|}}{Al}}\text{—}\overset{\oplus}{Cl}\text{—}R + C_6H_6 \; \longrightarrow \; C_6H_5 \overset{\oplus}{\underset{R}{\overset{H}{\diagup}}} + AlCl_4^- \qquad \textit{Slow}$$

$$\text{I}$$

The electrophile is thus either (a) R^+ or (b) a molecule like I that can readily *transfer* R^+ to the aromatic ring. *This duality of mechanism is common in electrophilic aromatic substitution.* In either case, the Lewis acid R^+ is displaced from RCl by the other Lewis acid, $AlCl_3$.

We speak of the Friedel–Crafts reaction as electrophilic substitution and, from the viewpoint of the aromatic ring, it is. But, just as an acid reacts with a base, so an electrophile reacts with a nucleophile, a molecule that provides the electrons that the electrophile seeks. From the opposite point of view, then, this reaction involves nucleophilic attack by the aromatic ring on the alkyl group of complex I. The $AlCl_4^-$ ion is a better leaving group than Cl^- would be; the Lewis acid, $AlCl_3$, serves the same purpose here that a Lowry–Brønsted acid does in protonation of an alcohol (Sec. 5.25).

As we shall find out when we take up the Friedel–Crafts reaction as a synthetic tool (Sec. 15.7), the Friedel–Crafts reaction in its widest sense involves reactants other than alkyl halides and Lewis acids other than aluminum chloride: BF_3, $SnCl_4$, HF, and even H^+.

Problem 14.3   How do you account for the fact that benzene in the presence of $AlCl_3$ reacts: (a) with *n*-propyl bromide to give isopropylbenzene; (b) with isobutyl bromide to yield *tert*-butylbenzene; (c) with neopentyl bromide to yield *tert*-pentylbenzene? (d) By which of the alternative mechanisms for the Friedel–Crafts reaction are these products probably formed?

Problem 14.4   Write all steps in the most likely mechanism for the reaction of benzene: (a) with *tert*-butyl alcohol in the presence of H_2SO_4 to yield *tert*-butylbenzene; (b) with propylene in the presence of H_3PO_4 to form isopropylbenzene.

14.11   Mechanism of halogenation

Aromatic halogenation, illustrated for chlorination, involves the following steps.

(1) $$Cl_2 + FeCl_3 \rightleftharpoons Cl_3\overset{\ominus}{Fe}\!-\!\overset{\oplus}{Cl}\!-\!Cl$$
$$II$$

(2) $$Cl_3\overset{\ominus}{Fe}\!-\!\overset{\oplus}{Cl}\!-\!Cl + C_6H_6 \longrightarrow C_6\overset{\oplus}{H}_5\!\!\overset{H}{\underset{Cl}{\diagup}} + FeCl_4^- \qquad Slow$$
$$II$$

(3) $$C_6\overset{\oplus}{H}_5\!\!\overset{H}{\underset{Cl}{\diagup}} + FeCl_4^- \longrightarrow C_6H_5\!-\!Cl + HCl + FeCl_3 \qquad Fast$$

The key step (2) is the attachment of positive chlorine to the aromatic ring. It seems unlikely, though, that an actually free Cl^+ ion is involved. Instead, ferric chloride combines with Cl_2 to form complex II, from which chlorine is transferred, without its electrons, directly to the ring.

Addition of halogens to alkenes, we have seen (Sec. 8.14), similarly involves attack by positive halogen to form an intermediate cation. The loosely held π electrons of an alkene make it more reactive, however, and positive halogen is transferred from the halogen molecule itself, X_2, with loss of Cl^-. The less reactive benzene molecule needs the assistance of a Lewis acid; reaction occurs with the loss of the better leaving group, $FeCl_4^-$. Indeed, more highly reactive aromatic compounds, that is, those whose π electrons are more available, do react with halogens in the absence of any added Lewis acid.

Problem 14.5   Certain activated benzene rings can be chlorinated by hypochlorous acid, HOCl, and this reaction is catalyzed by H^+. In light of the above discussion, can you suggest a possible function of H^+?

Problem 14.6   Aromatic bromination catalyzed by the Lewis acid thallium acetate, $Tl(OOCCH_3)_3$, gives only the *para* isomer. Suggest an explanation for this regioselectivity. (*Hint*: See Sec. 14.7.)

14.12   Desulfonation. Mechanism of protonation

When an aromatic sulfonic acid is heated to 100–175 °C with aqueous acid, it is converted into sulfuric acid and an aromatic hydrocarbon. This *desulfonation* is the exact reverse of the sulfonation process by which the sulfonic acid was originally made.

$$C_6H_6 + H_2SO_4 \overset{H^+}{\rightleftharpoons} C_6H_5SO_3H + H_2O$$

Hydrocarbon	Sulfonic acid
Volatile	*Non-volatile*

By applying the usual equilibrium principles, we can select conditions that will drive the reaction in the direction we want it to go. To sulfonate we use a large excess of concentrated or fuming sulfuric acid; high concentration of sulfonating agent and low concentration of water (or its removal by reaction with SO_3) shift the equilibrium toward sulfonic acid. To desulfonate we use dilute acid and often pass superheated steam through the reaction mixture; high concentration of water and removal of the relatively volatile hydrocarbon by steam distillation shift the equilibrium toward hydrocarbon.

(1) $$C_6H_5{-}SO_3{}^- + H^+ \rightleftharpoons \overset{\oplus}{C_6H_5}\diagup^H \diagdown SO_3{}^-$$

(2) $$\overset{\oplus}{C_6H_5}\diagup^H \diagdown SO_3{}^- \rightleftharpoons C_6H_6 + SO_3$$

According to the principle of microscopic reversibility (Sec. 7.25), the mechanism of desulfonation must be the exact reverse of the mechanism of sulfonation. The reaction is simply another example of electrophilic aromatic substitution. The electrophile is the proton, H^+, and the reaction is *protonation* or, more specifically, *protodesulfonation*.

Sulfonation is unusual among electrophilic aromatic substitution reactions in its reversibility. It is also unusual in another way: in sulfonation, ordinary hydrogen (protium) is displaced from an aromatic ring about twice as fast as deuterium. These two facts are related to each other and, as we shall see in Sec. 14.14, give us a more detailed picture of sulfonation and of electrophilic aromatic substitution in general.

Problem 14.7   Predict the product or products of: (a) monobromination of toluene; (b) monobromination of *p*-toluenesulfonic acid followed by treatment with acid and superheated steam. (c) Using the principle of (b), and following the guidelines of Sec. 14.7, outline a synthesis from benzene of *o*-dibromobenzene; of *o*-bromochlorobenzene.

14.13   Mechanism of electrophilic aromatic substitution: a summary

Electrophilic aromatic substitution reactions seem, then, to proceed by a single mechanism, whatever the particular reagent involved. This can be summarized for the reagent YZ as follows:

(1) $$C_6H_6 + Y^+ \longrightarrow \overset{\oplus}{C_6H_5}\diagup^H \diagdown Y \qquad \textit{Slow}$$

(2) $$\overset{\oplus}{C_6H_5}\diagup^H \diagdown Y + :Z^- \longrightarrow C_6H_5{-}Y + H:Z \qquad \textit{Fast}$$

Two essential steps are involved: (1) attack by an electrophilic reagent upon the ring to form a carbocation, $\overset{\oplus}{C_6H_5}\diagup^H \diagdown Y$

and (2) abstraction of a proton from this carbocation by some base. In each case there is a preliminary acid–base reaction which generates the attacking particle; the actual substitution, however, is contained in these two steps.

Most of the support for this mechanism comes from evidence about the nature of the attacking particle in each of these reactions: evidence, that is, that substitution is *electrophilic*. This evidence, in turn, comes largely from kinetics, augmented by various other observations: the nitrating power of preformed nitronium salts (Sec. 14.8), for example, or carbocation-like rearrangements in some Friedel–Crafts alkylations (Problem 14.3, p. 510). The electrophilic nature of these reactions is supported in a very broad way by the fact that other reactions which show the same reactivity and orientation features also fit into the same mechanistic pattern.

Now let us turn to evidence of another kind: the evidence of isotope effects.

Problem 14.8 In each of the following reactions, groups on the ring under attack exert the kinds of effects summarized in Sec. 14.5. Suggest a likely electrophile in each case, and write a likely mechanism.

(a) $ArH + R-\underset{\underset{O}{\|}}{C}-Cl \xrightarrow{AlCl_3} Ar-\underset{\underset{O}{\|}}{C}-R$

(b) $ArH + Ar'N_2{}^+Cl^- \longrightarrow Ar-N{=}N-Ar'$

(c) $ArH + HONO \xrightarrow{H^+} Ar-NO + H_2O$

Problem 14.9 When phenol is treated with D_2SO_4 in D_2O (deuterium sulfate in heavy water), there is formed phenol containing deuterium instead of hydrogen at positions *ortho* and *para* to the —OH group. Benzene undergoes similar exchange but at a much lower rate; under the same conditions benzenesulfonic acid does not undergo exchange at all. (a) Outline the most probable mechanism for hydrogen–deuterium exchange in aromatic compounds. (b) What is the attacking reagent in each case, and to what general class does this reaction belong?

14.14 Mechanism of electrophilic aromatic substitution: the two steps

So far we have spoken only of evidence indicating that these reactions are electrophilic, and revealing what the actual electrophiles are likely to be. But this is only part of the mechanism. Granting that substitution is electrophilic, how do we know that electrophilic aromatic substitution involves *two* steps,

(1) $\qquad ArH + Y^+ \longrightarrow Ar\overset{\oplus}{\underset{Y}{\diagup}}{}^H \qquad\qquad$ Slow: *rate-determining*

(2) $\qquad Ar\overset{\oplus}{\underset{Y}{\diagup}}{}^H + :Z \longrightarrow Ar-Y + H:Z \qquad$ Fast

instead of just *one,*

(1a) $\qquad$ $ArH + Y^+ \longrightarrow \left[Ar \underset{Y}{\overset{H}{\diagdown}} \right]^+ \longrightarrow Ar{-}Y + H^+$

and how do we know that, of these two steps, the first is much slower than the second?

The answer is found in a series of studies begun by Lars Melander (University of Gothenberg) and extended by many other workers. A variety of aromatic compounds labeled with deuterium or tritium were subjected to nitration, bromination, and Friedel–Crafts alkylation. It was found that in these reactions deuterium or tritium is replaced at the *same* rate as protium; *there is no significant isotope effect.*

We have seen (Sec. 7.17) that a carbon–deuterium bond is broken more slowly than a carbon–protium bond, and a carbon–tritium bond more slowly yet. Such primary isotope effects are sizable: k^H/k^D may be 5 to 8, and k^H/k^T about twice that large. How, then, are we to interpret the fact that there is no isotope effect here? If the rates of replacement of the various hydrogen isotopes are the same, it can only mean that the reactions *whose rates we are comparing* do not involve the breaking of a carbon–hydrogen bond.

This interpretation is consistent with our mechanism. The rate of the overall substitution is determined by the slow attachment of the electrophilic reagent to the aromatic ring to form the carbocation. Once formed, the carbocation rapidly loses a hydrogen ion—a proton or deuteron—to form the products. Step (1) is thus the *rate-determining step.* Since it does not involve the breaking of a carbon–hydrogen bond, its rate—and hence the rate of the overall reaction—is independent of the particular hydrogen isotope that is present.

If substitution involved a *single* step, as in (1a), this step would necessarily be the rate-determining step and, since it involves breaking of the carbon–hydrogen bond, an isotope effect would be observed. Or, if step (2) of the two-step sequence were slow enough relative to step (1) to affect the overall rate, again we would expect an isotope effect. (Indeed, sulfonation *does* show a small isotope effect and, as we shall see, for just this reason. Even in sulfonation, however, the overall rate is controlled chiefly by step (1).)

Thus the absence of isotope effects establishes not only the two-step nature of electrophilic aromatic substitution, but also the relative speeds of the steps. Attachment of the electrophile to a carbon atom of the ring is the difficult step (see Fig. 14.2); but it is equally difficult whether the carbon carries protium or deuterium. The next step, loss of hydrogen ion, is easy. Although it occurs more slowly for deuterium than for protium, this really makes no difference; slightly faster or slightly slower, its speed has no effect on the overall rate.

Let us look at this matter more closely (Fig. 14.2, inset). Every carbocation formed, whether I(H) or I(D), goes on to product, since the energy barrier to the

$$Ar \underset{NO_2}{\overset{\oplus}{\diagup}} H \qquad\qquad Ar \underset{NO_2}{\overset{\oplus}{\diagup}} D$$

$$\text{I(H)} \qquad\qquad\qquad \text{I(D)}$$

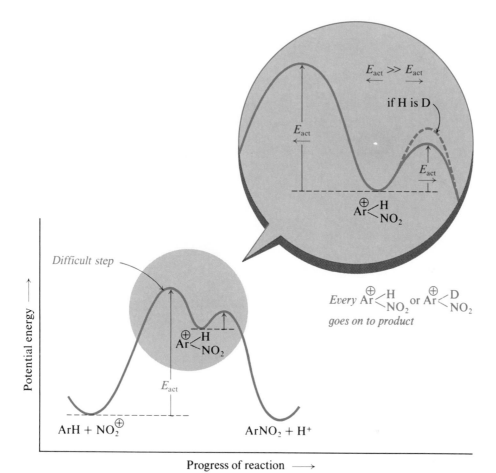

Figure 14.2   Nitration. Formation of the carbocation is the rate-controlling step; it occurs equally rapidly whether protium (H) or deuterium (D) is at the point of attack. All the carbocations go on to product. There is no isotope effect, and nitration is irreversible.

right (ahead of the carbocation)—whether slightly higher for deuterium or slightly lower for protium—is still considerably lower than the barrier to the left (behind the carbocation). But the barrier behind the carbocation is the E_{act} for the *reverse* of step (1). It is this reverse reaction that must be much slower than step (2) if step (1) is to be truly rate-determining (Sec. 5.16). Summarized in terms of the *rate constants*, k, for the various steps, we have:

$$\text{ArH} + {}^{+}\text{NO}_2 \underset{k_{-1}}{\overset{k_1}{\rightleftarrows}} \text{Ar}\underset{\text{NO}_2}{\overset{\oplus\ \ \text{H}}{\diagup\diagdown}} \overset{k_2}{\longrightarrow} \text{Ar}-\text{NO}_2 + \text{H}^{+} \qquad k_2 \gg k_{-1}$$

$$(1) \qquad\qquad\qquad\qquad\quad \text{I} \qquad (2)$$

We can see why nitration and reactions like it are not reversible. In the reverse of nitration, nitrobenzene is protonated (the reverse of reaction 2) to form carbocation I; but this is, of course, no different from the ion I formed in the nitration process, and it does the same thing: (re)forms nitrobenzene.

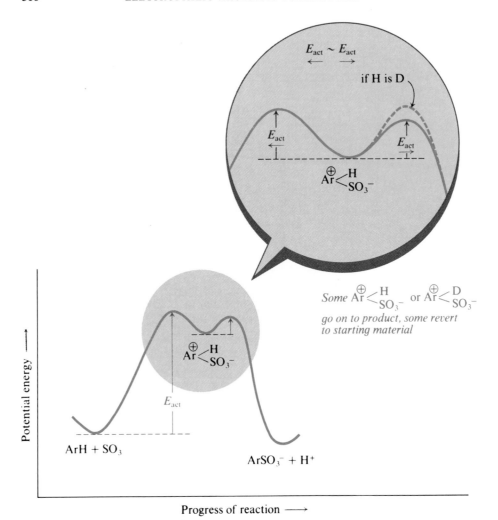

Figure 14.3 Sulfonation. Some carbocations go on to product, some revert to starting material. There is an isotope effect, and sulfonation is reversible.

Unlike most other electrophilic substitution reactions, sulfonation shows a moderate isotope effect: ordinary hydrogen (protium) is displaced from an aromatic ring about twice as fast as deuterium. Does this mean that sulfonation takes place by a different mechanism than nitration, one involving a single step? Almost certainly not.

$$\text{ArH} + \text{SO}_3 \underset{k_{-1}}{\overset{k_1}{\rightleftharpoons}} \overset{\underset{\displaystyle \text{SO}_3^-}{\displaystyle \big|}}{\underset{\text{II}}{\overset{\oplus}{\text{Ar}}\diagup\text{H}}} \overset{k_2}{\longrightarrow} \text{Ar}-\text{SO}_3^- + \text{H}^+ \qquad k_2 \sim k_{-1}$$

(1)   (2)

Unlike most other electrophilic substitution reactions, sulfonation is reversible, and this fact gives us our clue. Reversibility means that carbocation II can lose SO₃ to form the hydrocarbon. Evidently here reaction (2) is *not* much faster than the reverse of reaction (1). In sulfonation, the energy barriers on either side of the carbocation II must be roughly the same height; some ions go one way, some go the other (Fig. 14.3). Now, whether the carbocation is II(D) or II(H), the barrier to the left (behind it) is the same height. But to climb the barrier to the right (ahead), a carbon–hydrogen bond must be broken, so this barrier is higher for carbocation II(D) than for carbocation II(H). More deuterated ions than ordinary ions revert to starting material, and so overall sulfonation is slower for the deuterated benzene. Thus, the particular shape of potential energy curve that makes sulfonation reversible also permits an isotope effect to be observed.

By use of especially selected aromatic substrates—highly hindered ones—isotope effects can be detected in other kinds of electrophilic aromatic substitution, even in nitration. In certain reactions the *size* of the isotope can be deliberately varied by changes in experimental conditions—and in a way that shows dependence on the relative rates of (2) and the reverse of (1). There can be little doubt that all these reactions follow the same two-step mechanism, but with differences in the shape of potential energy curves. In isotope effects the chemist has an exceedingly delicate probe for the examination of organic reaction mechanisms.

Problem 14.10   From the reaction of mesitylene (1,3,5-trimethylbenzene) with HF and BF_3, Olah (see p. 194) isolated at low temperatures a bright-yellow solid whose elemental composition corresponds to mesitylene:HF:BF_3 in the ratio 1:1:1. The compound was poorly soluble in organic solvents and, when molten, conducted an electric current; chemical analysis showed the presence of the BF_4^- ion. When heated, the compound evolved BF_3 and regenerated mesitylene.

What is a likely structure for the yellow compound? The isolation of this and related compounds is considered to be strong support for the mechanism of electrophilic aromatic substitution. Why should this be so?

14.15   Reactivity and orientation

We have seen that certain groups activate the benzene ring and direct substitution to *ortho* and *para* positions, and that other groups deactivate the ring and (except halogens) direct substitution to *meta* positions. Let us see if we can account for these effects on the basis of principles we have already learned.

First of all, we must remember that reactivity and orientation are both matters of relative rates of reaction. Methyl is said to activate the ring because it makes the ring react *faster* than benzene; it causes *ortho,para* orientation because it makes the *ortho* and *para* positions react *faster* than the *meta* positions.

Now, we know that, whatever the specific reagent involved, the rate of electrophilic aromatic substitution is determined by the same slow step—attack of the electrophile on the ring to form a carbocation:

$$C_6H_6 + Y^+ \longrightarrow C_6H_5{\overset{\displaystyle \oplus}{\underset{\displaystyle Y}{}}}H \qquad \text{Slow: } \textit{rate-determining}$$

Any differences in rate of substitution must therefore be due to differences in the rate of this step.

For closely related reactions, a difference in rate of formation of carbocations is largely determined by a difference in E_{act}, that is, by a difference in stability of transition states. As with other carbocation reactions we have studied, factors that stabilize the ion by dispersing the positive charge should for the same reason stabilize the incipient carbocation of the transition state. Here again we expect the more stable carbocation to be formed more rapidly. We shall therefore concentrate on the relative stabilities of the carbocations.

In electrophilic aromatic substitution the intermediate carbocation is a hybrid of structures I, II, and III, in which the positive charge is distributed about the ring, being strongest at the positions *ortho* and *para* to the carbon atom being attacked.

A group already attached to the benzene ring should affect the stability of the carbocation by dispersing or intensifying the positive charge, depending upon its electron-releasing or electron-withdrawing nature. It is evident from the structure of the ion (I–III) that this stabilizing or destabilizing effect should be especially important when the group is attached *ortho* or *para* to the carbon being attacked.

14.16   Theory of reactivity

To compare rates of substitution in benzene, toluene, and nitrobenzene, we compare the structures of the carbocations formed from the three compounds:

By releasing electrons, the methyl group (II) tends to neutralize the positive charge of the ring and so become more positive itself; this dispersal of the charge stabilizes the carbocation. In the same way the inductive effect stabilizes the developing positive charge in the transition state and thus leads to a faster reaction.

Transition state:
developing positive charge

Carbocation:
full positive charge

The $-NO_2$ group, on the other hand, has an electron-withdrawing inductive effect (III); this tends to intensify the positive charge, destabilizes the carbocation, and thus causes a slower reaction.

Reactivity in electrophilic aromatic substitution depends, then, upon the tendency of a substituent group to release or withdraw electrons. **A group that releases electrons activates the ring; a group that withdraws electrons deactivates the ring.**

Electrophilic Aromatic Substitution

G *releases electrons:*
stabilizes carbocation,
activates

$$G = -NH_2$$
$$-OH$$
$$-OCH_3$$
$$-NHCOCH_3$$
$$-C_6H_5$$
$$-CH_3$$

G *withdraws electrons:*
destabilizes carbocation,
deactivates

$$G = -N(CH_3)_3{}^+$$
$$-NO_2$$
$$-CN$$
$$-SO_3H$$
$$-COOH$$
$$-CHO$$
$$-COR$$
$$-X$$

Like $-CH_3$, other alkyl groups release electrons, and like $-CH_3$ they activate the ring. For example, *tert*-butylbenzene is 16 times as reactive as benzene toward nitration. Electron release by $-NH_2$ and $-OH$, and by their derivatives $-OCH_3$ and $-NHCOCH_3$, is due not to their inductive effect but to resonance, and is discussed later (Sec. 14.18).

We are already familiar with the electron-withdrawing effect of the halogens (Sec. 5.25). The full-fledged positive charge of the $-N(CH_3)_3{}^+$ group has, of course, a powerful attraction for electrons. In the other deactivating groups (e.g., $-NO_2$, $-CN$, $-COOH$), the atom next to the ring is attached by a multiple bond to oxygen or nitrogen. These electronegative atoms attract the mobile π electrons, making the atom next to the ring electron-deficient; to make up this deficiency, the atom next to the ring withdraws electrons from the ring.

We might expect replacement of hydrogen in $-CH_3$ by halogen to decrease the electron-releasing tendency of the group, and perhaps to convert it into an electron-withdrawing group. This is found to be the case. Toward nitration, toluene

| Activating | Weakly deactivating | Moderately deactivating | Strongly deactivating |

is 25 times as reactive as benzene; benzyl chloride is only one-third as reactive as benzene. The $-CH_2Cl$ group is thus weakly deactivating. Further replacement of hydrogen by halogen to yield the $-CHCl_2$ and the $-CCl_3$ groups results in stronger deactivation.

14.17   Theory of orientation

Before we try to account for orientation in electrophilic substitution, let us look more closely at the facts.

An activating group activates all positions of the benzene ring; even the positions *meta* to it are more reactive than any single position in benzene itself. It directs *ortho* and *para* simply because it activates the *ortho* and *para* positions much *more* than it does the *meta*.

A deactivating group deactivates all positions in the ring, even the positions *meta* to it. It directs *meta* simply because it deactivates the *ortho* and *para* positions even *more* than it does the *meta*.

Thus both *ortho,para* orientation and *meta* orientation arise in the same way: **the effect of any group—whether activating or deactivating—is strongest at the *ortho* and *para* positions.**

To see if this is what we would expect, let us compare, for example, the carbocations formed by attack at the *para* and *meta* positions of toluene, a compound that contains an activating group. Each of these is a hybrid of three structures, I–III for *para*, IV–VI for *meta*. In one of these six structures, II, the positive charge is located on the carbon atom to which —CH$_3$ is attached. Although —CH$_3$ releases electrons to all positions of the ring, it does so most strongly to the

Para attack

Especially stable:
charge on carbon
carrying substituent

carbon atom nearest it; consequently, structure II is a particularly stable one. Because of contribution from structure II, the hybrid carbocation resulting from

Meta attack

attack at the *para* position is more stable than the carbocation resulting from attack at a *meta* position. *Para* substitution, therefore, occurs faster than *meta* substitution.

In the same way, it can be seen that attack at an *ortho* position (VII–IX) also

Ortho attack

Especially stable:
charge on carbon
carrying substituent

yields a more stable carbocation, through contribution from IX, than attack at a *meta* position.

In toluene, *ortho,para* substitution is thus faster than *meta* substitution because electron release by —CH$_3$ is more effective during attack at the positions *ortho* and *para* to it.

Next, let us compare the carbocations formed by attack at the *para* and *meta* positions of nitrobenzene, a compound that contains a deactivating group. Each of these is a hybrid of three structures, X–XII for *para* attack, XIII–XV for *meta* attack. In one of the six structures, XI, the positive charge is located on the carbon

Para attack

X   XI   XII

Especially unstable:
charge on carbon
carrying substituent

atom to which —NO$_2$ is attached. Although —NO$_2$ withdraws electrons from all positions, it does so most from the carbon atom nearest it, and hence this carbon atom, already positive, has little tendency to accommodate the positive charge of the carbocation. Structure XI is thus a particularly unstable one and does little to help stabilize the ion resulting from attack at the *para* position. The ion for *para* attack is virtually a hybrid of only two structures, X and XII; the positive charge is mainly restricted to only *two* carbon atoms. It is less stable than the ion resulting from attack at a *meta* position, which is a hybrid of three structures, and in which the positive charge is accommodated by *three* carbon atoms. *Para* substitution, therefore, occurs more slowly than *meta* substitution.

Meta attack

XIII   XIV   XV

In the same way it can be seen that attack at an *ortho* position (XVI–XVIII) yields a less stable carbocation, because of the instability of XVIII, than attack at a *meta* position.

Ortho attack

XVI   XVII   XVIII

Especially unstable:
charge on carbon
carrying substituent

In nitrobenzene, *ortho,para* substitution is thus slower than *meta* substitution because electron withdrawal by $-NO_2$ is more effective during attack at the positions *ortho* and *para* to it.

Thus we see that both *ortho,para* orientation by activating groups and *meta* orientation by deactivating groups follow logically from the structure of the intermediate carbocation. The charge of the carbocation is strongest at the positions *ortho* and *para* to the point of attack, and hence a group attached to one of these positions can exert the strongest effect, whether activating or deactivating.

The unusual behavior of the halogens, which direct *ortho* and *para* although deactivating, results from a combination of two opposing factors, and will be taken up in Sec. 14.19.

14.18   Electron release via resonance

We have seen that a substituent group affects both reactivity and orientation in electrophilic aromatic substitution by its tendency to release or withdraw electrons. So far, we have considered electron release and electron withdrawal only as inductive effects, that is, as effects due to the electronegativity of the group concerned.

But certain groups ($-NH_2$ and $-OH$, and their derivatives) act as powerful activators toward electrophilic aromatic substitution, even though they contain electronegative atoms and can be shown in other ways to have electron-withdrawing inductive effects. If our approach to the problem is correct, these groups must release electrons in some other way than through their inductive effects; they are believed to do this by a resonance effect. But before we discuss this, let us review a little of what we know about nitrogen and oxygen.

Although electronegative, the nitrogen of the $-NH_2$ group is basic and tends to share its last pair of electrons and acquire a positive charge. Just as ammonia accepts a proton to form the ammonium ($NH_4{}^+$) ion, so organic compounds related to ammonia accept protons to form substituted ammonium ions.

$$\ddot{N}H_3 + H^+ \longrightarrow NH_4{}^+ \qquad R\ddot{N}H_2 + H^+ \longrightarrow RNH_3{}^+$$

$$R_2\ddot{N}H + H^+ \longrightarrow R_2NH_2{}^+ \qquad R_3\ddot{N} + H^+ \longrightarrow R_3NH^+$$

The $-OH$ group shows similar but weaker basicity; we are already familiar with oxonium ions, $ROH_2{}^+$.

$$H_2\ddot{O} + H^+ \longrightarrow H_3O^+ \qquad R\ddot{O}H + H^+ \longrightarrow ROH_2{}^+$$

The effects of $-NH_2$ and $-OH$ on electrophilic aromatic substitution can be accounted for by assuming that nitrogen and oxygen can share more than a pair of electrons with the ring and can accommodate a positive charge.

The carbocation formed by attack *para* to the $-NH_2$ group of aniline, for example, is considered to be a hybrid not only of structures I, II, and III, with positive charges located on carbons of the ring, but also of structure IV in which the positive charge is carried by nitrogen. Structure IV is especially stable, since in it *every atom* (except hydrogen, of course) *has a complete octet of electrons*. This carbocation is much more stable than the one obtained by attack on benzene itself, or the one obtained (V–VII) from attack *meta* to the $-NH_2$ group of aniline; in neither of these cases is a structure like IV possible. (Compare, for example, the

Para attack

I II III IV

Especially stable:
every atom has octet

Meta attack

V VI VII

stabilities of the ions NH_4^+ and CH_3^+. Here it is not a matter of which atom, nitrogen or carbon, can better accommodate a positive charge; it is a matter of which atom has a complete octet of electrons.)

Examination of the corresponding structures (VIII–XI) shows that *ortho* attack is much like *para* attack:

Ortho attack

VIII IX X XI

Especially stable:
every atom has octet

Thus substitution in aniline occurs faster than substitution in benzene, and occurs predominantly at the positions *ortho* and *para* to $—NH_2$.

In the same way activation and *ortho,para* orientation by the $—OH$ group is accounted for by contribution of structures like XII and XIII, in which every atom has a complete octet of electrons:

XII

Para attack

XIII

Ortho attack

The similar effects of the derivatives of $—NH_2$ and $—OH$ are accounted for by similar structures (shown only for *para* attack):

$—NHCH_3$ $—N(CH_3)_2$ $—NHCOCH_3$ $—OCH_3$

The tendency of oxygen and nitrogen in groups like these to share more than a pair of electrons with an aromatic ring is shown in a number of other ways, which will be discussed later (Secs. 27.3 and 28.7).

Much of what has just been said should sound familiar. In Sec. 10.15 we saw how the filled p orbital of an oxygen atom can overlap the empty p orbital of an adjacent electron-deficient carbon, and thus provide it with the electrons it needs. Basically the same thing is involved here, except that through overlap with the conjugated π system of the benzenonium ring an element like oxygen can provide electrons to more remote electron-deficient carbons.

14.19 Effect of halogen on electrophilic aromatic substitution

Halogens are unusual in their effect on electrophilic aromatic substitution: they are deactivating yet *ortho,para*-directing. Deactivation is characteristic of electron withdrawal, whereas *ortho,para* orientation is characteristic of electron release. Can halogen both withdraw and release electrons?

The answer is *yes*. Halogen withdraws electrons through its inductive effect, and releases electrons through its resonance effect. So, presumably, can the —NH$_2$ and —OH groups, but there the much stronger resonance effect greatly outweighs the other. For halogen, the two effects are more evenly balanced, and we observe the operation of both.

Let us first consider **reactivity**. Electrophilic attack on benzene yields carbo-

—Cl *withdraws electrons:*
destabilizes carbocation,
deactivates ring

cation I, attack on chlorobenzene yields carbocation II. The electron-withdrawing inductive effect of chlorine intensifies the positive charge in carbocation II, makes the ion less stable, and causes a slower reaction.

Next, to understand **orientation**, let us compare the structures of the carbocations formed by attack at the *para* and *meta* positions of chlorobenzene. Each of

Para attack

Especially unstable:
charge on carbon
bearing substituent

Meta attack

these is a hybrid of three structures, III–V for *para*, VI–VIII for *meta*. In one of these six structures, IV, the positive charge is located on the carbon atom to which chlorine is attached. Through its inductive effect chlorine withdraws electrons most from the carbon to which it is joined, and thus makes structure IV especially unstable. As before, we expect IV to make little contribution to the hybrid, which should therefore be less stable than the hybrid ion resulting from attack at the *meta* positions. If only the inductive effect were involved, then, we would expect not only deactivation but also *meta* orientation.

But the existence of halonium ions (Sec. 9.6) has shown us that halogen can share more than a pair of electrons and can accommodate a positive charge. If we apply that idea to the present problem, what do we find? The ion resulting from *para* attack is a hybrid not only of structures III–V, but also of structure IX, in which chlorine bears a positive charge and is joined to the ring by a double bond.

Para attack

IX

Comparatively stable:
every atom has octet

This structure should be comparatively stable, since in it every atom (except hydrogen, of course) has a *complete octet of electrons.* (Structure IX is exactly analogous to those proposed to account for activation and *ortho,para* direction by —NH_2 and —OH.) No such structure is possible for the ion resulting from *meta* attack. To the extent that structure IX contributes to the hybrid, it makes the ion resulting from *para* attack more stable than the ion resulting from *meta* attack. Although we could not have predicted the relative importance of the two factors— the instability of IV and the stabilization by IX—the result indicates that the contribution from IX is the more important.

In the same way it can be seen that attack at an *ortho* position also yields an ion (X–XIII) that can be stabilized by accommodation of the positive charge by chlorine.

Ortho attack

X   XI   XII   XIII

Especially unstable: *Comparatively stable:*
charge on carbon   *every atom has octet*
bearing substituent

Through its inductive effect halogen tends to withdraw electrons and thus to destabilize the intermediate carbocation. This effect is felt for attack at all positions, but particularly for attack at the positions *ortho* and *para* to the halogen.

Through its resonance effect halogen tends to release electrons and thus to stabilize the intermediate carbocation. This electron release is effective only for attack at the positions *ortho* and *para* to the halogen.

The inductive effect is stronger than the resonance effect and causes net electron withdrawal—and hence deactivation—for attack at all positions. The resonance effect tends to oppose the inductive effect for attack at the *ortho* and *para* positions, and hence makes the deactivation less for *ortho,para* attack than for *meta*.

Reactivity is thus controlled by the stronger inductive effect, and orientation is controlled by the resonance effect, which, although weaker, is more selective.

In electrophilic addition to vinyl halides (Sec. 10.15) we saw halogen playing the same dual role as a substituent, again in the formation of a carbocation. There, too, reactivity is controlled by the inductive effect, and orientation by the resonance effect.

Thus we find that a single structural concept—partial double-bond formation between halogen and carbon—helps to account for unusual chemical properties of such seemingly different compounds as aryl halides and vinyl halides. The structures involving doubly bonded halogen, which probably make important contributions not only to benzenonium ions but to the parent aryl halides as well (Sec. 29.6), certainly do not seem to meet our usual standard of reasonableness (Sec. 10.10). The sheer weight of evidence forces us to accept the idea that certain carbon–halogen bonds possess double-bond character. If this idea at first appears strange to us, it simply shows how much, after all, we still have to learn about molecular structure.

14.20 Relation to other carbocation reactions

In summary, we can say that both reactivity and orientation in electrophilic aromatic substitution are determined by the rates of formation of the intermediate carbocations concerned. These rates parallel the stabilities of the carbocations, which are determined by the electron-releasing or electron-withdrawing tendencies of the substituent groups.

A group may release or withdraw electrons by an inductive effect, a resonance effect, or both. These effects oppose each other only for the —NH_2 and —OH groups (and their derivatives) and for the halogens, —X. For —NH_2 and —OH the resonance effect is much the more important; for —X the effects are more evenly matched. It is because of this that the halogens occupy the unusual position of being deactivating groups but *ortho,para* directors.

We have accounted for the facts of electrophilic aromatic substitution in exactly the way that we accounted for reactivity in substitution by S_N1 and elimination by E1, for the relative ease of dehydration of alcohols, and for reactivity and orientation in electrophilic addition to alkenes: the more stable the carbocation, the faster it is formed; the faster the carbocation is formed, the faster the reaction goes.

In all this we have estimated the stability of carbocations on the same basis: **the dispersal or concentration of the charge** due to electron release or electron withdrawal by the substituent groups. As we shall see, this approach, which has worked so well for these reactions in which a positive charge develops, works equally well for *nucleophilic aromatic substitution* (Sec. 29.9), in which a *negative* charge develops. Finally, we shall find that this approach will help us to understand *acidity* or *basicity* of such compounds as carboxylic acids, sulfonic acids, amines, and phenols.

PROBLEMS

1. Give structures and names of the principal products expected from the ring monobromination of each of the following compounds. In each case, tell whether bromination will occur faster or slower than with benzene itself.

(a) acetanilide ($C_6H_5NHCOCH_3$)
(b) iodobenzene
(c) sec-butylbenzene
(d) N-methylaniline ($C_6H_5NHCH_3$)
(e) ethyl benzoate ($C_6H_5COOC_2H_5$)
(f) acetophenone ($C_6H_5COCH_3$)

(g) ethyl phenyl ether ($C_6H_5OC_2H_5$)
(h) diphenylmethane ($C_6H_5CH_2C_6H_5$)
(i) benzonitrile (C_6H_5CN)
(j) benzotrifluoride ($C_6H_5CF_3$)
(k) biphenyl (C_6H_5—C_6H_5)

2. Give structures and names of the principal organic products expected from mono-nitration of:

(a) o-nitrotoluene
(b) m-dibromobenzene
(c) p-nitroacetanilide
    (p-$O_2NC_6H_4NHCOCH_3$)
(d) m-dinitrobenzene
(e) m-cresol (m-$CH_3C_6H_4OH$)
(f) o-cresol

(g) p-cresol
(h) m-nitrotoluene
(i) p-xylene (p-$C_6H_4(CH_3)_2$)
(j) terephthalic acid (p-$C_6H_4(COOH)_2$)
(k) anilinium hydrogen sulfate
    ($C_6H_5NH_3{}^+HSO_4{}^-$)

3. Give structures and names of the principal organic products expected from the monosulfonation of:

(a) cyclohexylbenzene
(b) nitrobenzene
(c) anisole ($C_6H_5OCH_3$)
(d) benzenesulfonic acid
(e) salicylaldehyde (o-HOC_6H_4CHO)
(f) m-nitrophenol

(g) o-fluoroanisole
(h) o-nitroacetanilide
    (o-$O_2NC_6H_4NHCOCH_3$)
(i) o-xylene
(j) m-xylene
(k) p-xylene

4. Arrange the following in order of reactivity toward ring nitration, listing by structure the most reactive at the top, the least reactive at the bottom.

(a) benzene, mesitylene (1,3,5-$C_6H_3(CH_3)_3$), toluene, m-xylene, p-xylene
(b) benzene, bromobenzene, nitrobenzene, toluene
(c) acetanilide ($C_6H_5NHCOCH_3$), acetophenone ($C_6H_5COCH_3$), aniline, benzene
(d) terephthalic acid, toluene, p-toluic acid (p-$CH_3C_6H_4COOH$), p-xylene
(e) chlorobenzene, p-chloronitrobenzene, 2,4-dinitrochlorobenzene
(f) 2,4-dinitrochlorobenzene, 2,4-dinitrophenol
(g) m-dinitrobenzene, 2,4-dinitrotoluene

5. For each of the following compounds, indicate which ring you would expect to be attacked in nitration, and give structures of the principal products.

(a) O_2N⟨◯⟩—⟨◯⟩

p-Nitrobiphenyl

(b) ⟨◯⟩—CH_2—⟨◯⟩
   O_2N

m-Nitrodiphenylmethane

(c) ⟨◯⟩—C(=O)—O—⟨◯⟩

Phenyl benzoate

6. Arrange the compounds of each set in order of reactivity toward electrophilic substitution. Indicate in each set which would yield the highest percentage of *meta* isomer, and which would yield the lowest.

(a) $C_6H_5N(CH_3)_3{}^+$, $C_6H_5CH_2N(CH_3)_3{}^+$, $C_6H_5CH_2CH_2N(CH_3)_3{}^+$,
    $C_6H_5CH_2CH_2CH_2N(CH_3)_3{}^+$
(b) $C_6H_5NO_2$, $C_6H_5CH_2NO_2$, $C_6H_5CH_2CH_2NO_2$
(c) $C_6H_5CH_3$, $C_6H_5CH_2COOC_2H_5$, $C_6H_5CH(COOC_2H_5)_2$, $C_6H_5C(COOC_2H_5)_3$

7. There is evidence that the phenyl group, $C_6H_5—$, has an electron-withdrawing inductive effect. Yet each ring of biphenyl, $C_6H_5—C_6H_5$, is more reactive than benzene toward electrophilic substitution, and the chief products are *ortho* and *para* isomers. Show how reactivity and orientation can be accounted for on the basis of resonance.

8. There is evidence that the reaction between HNO_3 and H_2SO_4 to generate $^+NO_2$ (which we have summarized in one equation, Sec. 14.8) actually involves three steps, the second of which is the slowest one and the one that actually produces $^+NO_2$. Can you suggest a reasonable sequence of reactions? (*Hint*: See Sec. 5.25.)

9. Treatment of *sulfanilic acid* ($p\text{-}H_2NC_6H_4SO_3H$) with three moles of bromine yields 2,4,6-tribromoaniline. Treatment of 4-hydroxy-1,3-benzenedisulfonic acid with nitric acid yields picric acid, 2,4,5-trinitrophenol. (a) Outline the most probable mechanism for the replacement of $—SO_3H$ by $—Br$ and by $—NO_2$. (b) To what general class of organic reactions do those reactions belong?

10. Using only individual steps with which you are already familiar, outline a likely mechanism for the following reaction.

$$C_6H_5C(CH_3)_3 + Br_2(AlBr_3) \longrightarrow C_6H_5Br + HBr + (CH_3)_2C=CH_2$$

11. In light of what you have learned in this chapter, predict the major products of each of the following reactions.

(a) $(CH_3)_3\overset{+}{N}CH=CH_2 + HI$
(b) $CH_2=CHCF_3 + HBr(AlBr_3)$
(c) What is the function of $AlBr_3$ in (b)? Why is it needed here?

12. You are trying to find out whether or not there is an isotope effect in a particular kind of substitution in which the electrophile Y replaces a hydrogen of an aromatic ring. In each of the following cases, tell what you would *do*, and what you would *expect to observe* if there were an isotope effect. (You can quantitatively analyze mixtures of isomers. Your mass spectrometer will tell you what percentage of the hydrogen in a compound is deuterium, but not the location of deuterium in a molecule.)

(a) C_6H_6 and C_6D_6 are allowed to react separately but under identical conditions.
(b) A 50:50 mixture of C_6H_6 and C_6D_6 is allowed to react with a limited amount of the reagent.
(c) Anisole and [4-D]anisole are allowed to react separately. (Both your watch and your mass spectrometer are under repair when this particular experiment is carried out.)
(d) [1,3,5-D_3]benzene (1,3,5-trideuteriobenzene) is allowed to react.

13. In Problem 10 (p. 497) you accounted for the aromaticity of the heterocyclic compound *pyrrole*.

Pyrrole

Among its aromatic properties is the tendency to undergo electrophilic aromatic substitution, which it does extremely readily (like the most reactive of benzene derivatives) and predominantly at the 2-position. Drawing all pertinent resonance structures, account in detail for (a) its high reactivity and (b) the orientation of substitution. (*Hint*: See Sec. 14.18.)

14. In Problem 13 (p. 497) you accounted for the aromaticity of the polynuclear hydrocarbon *naphthalene*. This compound readily undergoes electrophilic aromatic substi-

Naphthalene

tution and, in the case of bromination and nitration, predominantly at the 1-position. How do you account for the observed orientation? (*Hint*: Draw all pertinent resonance structures, and examine them closely.)

15. Outline all steps in the laboratory synthesis of the following compounds from benzene and/or toluene, using any needed aliphatic or inorganic reagents. (Review the general instructions on page 221. Assume that a pure *para* isomer can be separated from an *ortho,para* mixture.)

(a) *p*-nitrotoluene
(b) *p*-bromonitrobenzene
(c) *p*-dichlorobenzene
(d) *m*-bromobenzenesulfonic acid
(e) *p*-bromobenzenesulfonic acid
(f) *p*-bromobenzoic acid
(g) *m*-bromobenzoic acid

(h) 1,3,5-trinitrobenzene
(i) 2-bromo-4-nitrotoluene
(j) 2-bromo-4-nitrobenzoic acid
(k) 4-bromo-3-nitrobenzoic acid
(l) 3,5-dinitrobenzoic acid
(m) 4-nitro-1,2-dibromobenzene
(n) 2-nitro-1,4-dichlorobenzene

16. Outline all steps in the following laboratory syntheses, using any needed aliphatic or inorganic reagents. (Follow the other instructions in Problem 15.)

(a) 4-nitro-2,6-dibromoanisole from anisole ($C_6H_5OCH_3$)
(b) 4-bromo-2-nitrobenzoic acid from *o*-nitrotoluene
(c) 2,4,6-tribromoaniline from aniline
(d) 2,4-dinitroacetanilide from acetanilide ($C_6H_5NHCOCH_3$)
(e) 5-nitroisophthalic acid from *m*-xylene
(f) 4-nitroisophthalic acid from *m*-xylene
(g) 2-nitroterephthalic acid from *p*-xylene (two ways)
(h) Which way in (g) is preferable? Why?

15

Aromatic–Aliphatic Compounds

Arenes and Their Derivatives

15.1 The aromatic ring as a substituent

In the two preceding chapters we looked at the aromatic ring—in benzene, mostly, and its simple derivatives—as the site of reaction: typically, electrophilic substitution. We saw how this reaction takes place, and how reactivity and orientation are affected by substituents attached to the ring.

Now, as we did for the carbon–carbon double bond, let us change our point of view; let us look at the aromatic ring, not as a functional group, but as a *substituent*. Like the double bond, we shall find, the aromatic ring exerts powerful effects. These effects resemble in many ways those of the double bond, and for a very good reason: they, too, are the result of *conjugation*. To the aromatic ring there can be attached any one—or two, or three—of dozens of different substituents; these *modify* the effects of the ring, and make substituted phenyl groups the most widely used of probes into the electronic demands of organic reactions.

Let us return to each of the families of compounds and types of reactions that we have already discussed, and look at the effects exerted by the aromatic ring. Let us begin with hydrocarbons.

15.2 Aromatic–aliphatic hydrocarbons: arenes

From our study so far, we know what kind of chemical properties to expect of an aliphatic hydrocarbon, that is, of an alkane, alkene, or alkyne. We know what kind of chemical behavior to expect of the parent aromatic hydrocarbon, benzene. Many important hydrocarbons are not just aliphatic or just aromatic, however, but contain both aliphatic and aromatic units; hydrocarbons of this kind are known collectively as **arenes**. *Ethylbenzene*, for example, contains a benzene ring and an aliphatic side chain.

$$\text{Ethylbenzene}: \quad C_6H_5\text{—}CH_2CH_3$$

Ethylbenzene

What kind of chemical properties might we expect of one of these mixed aromatic–aliphatic hydrocarbons? First, we might expect it to show *two* sets of chemical properties. The ring of ethylbenzene should undergo the electrophilic

o-Nitroethylbenzene *p*-Nitroethylbenzene *More readily than for benzene*

Chief products

1-Bromo-1-phenylethane *More readily than for ethane*
(α-Phenylethyl bromide)

Only product

substitution characteristic of benzene, and the side chain should undergo the free-radical substitution characteristic of ethane. Second, the properties of each portion of the molecule should be modified by the presence of the other portion. The ethyl group should modify the aromatic properties of the ring, and the ring should modify the aliphatic properties of the side chain.

These predictions are correct. Treatment of ethylbenzene with nitric acid and sulfuric acid, for instance, introduces a nitro group into the ring; treatment with bromine in the presence of light introduces a bromine atom into the side chain. But because of the ethyl group, nitration takes place more readily than with benzene itself, and occurs chiefly at the positions *ortho* and *para* to the ethyl group; and because of the ring, bromination takes place more readily than with ethane, and occurs exclusively on the carbon nearer the ring. Thus *each portion of the molecule affects the **reactivity** of the other portion and determines the **orientation** of attack*.

Of the arenes, we shall examine first those which, like ethylbenzene, are made up of aromatic and alkane units, the *alkylbenzenes*. We have already (Chap. 14) discussed the effects of alkyl groups on reactions that, like nitration, take place in the aromatic ring. Now we shall concentrate on the other aspect of this structural interplay, and discuss the effects of the aromatic ring on reactions that take place in the side chain.

The reactions of alkylbenzenes will lead us to derivatives of alkylbenzenes with halogen in the side chain, and we shall then discuss their chemistry. This, in turn, will lead us to aromatic–alkene compounds (*alkenylbenzenes*) and aromatic–alkyne compounds (*alkynylbenzenes*). In all this, we shall be dealing with what is basically familiar chemistry—free-radical substitution, nucleophilic substitution, elimination, electrophilic and free-radical addition—and our main concern will be to see how this is affected by the presence of the aromatic ring.

15.3 Structure and nomenclature of arenes and their derivatives

The simplest of the alkylbenzenes, methylbenzene, is given the special name of **toluene**. Compounds containing longer side chains are named by prefixing the name of the alkyl group to the word *-benzene*, as, for example, in *ethylbenzene*, *n-propylbenzene*, and *isobutylbenzene*.

Toluene   Ethylbenzene   *n*-Propylbenzene   Isobutylbenzene

The simplest of the dialkylbenzenes, the dimethylbenzenes, are given the special name of **xylenes**; we have, then, *o-xylene, m-xylene*, and *p-xylene*. Dialkylbenzenes containing one methyl group are named as derivatives of toluene, while others are named by prefixing the names of both alkyl groups to the word *-benzene*. A compound containing a very complicated side chain might be named as a

o-Xylene   *m*-Xylene   *p*-Xylene

p-Ethyltoluene   *m*-Ethylisopropylbenzene   2-Methyl-3-phenylpentane

phenylalkane (C_6H_5 = **phenyl**). Compounds containing more than one benzene ring are nearly always named as derivatives of alkanes.

Diphenylmethane   1,2-Diphenylethane

The simplest alkenylbenzene has the special name **styrene**. Others are generally named as substituted alkenes, occasionally as substituted benzenes. Alkynylbenzenes are named as substituted alkynes.

Styrene	Allylbenzene	2-Phenyl-2-butene	Phenylacetylene
(Vinylbenzene)	(3-Phenylpropene)		
(Phenylethylene)			

Organic halides derived from arenes are of two kinds. If the halogen is attached directly to the aromatic ring, the compound is an *aryl halide*; its general formula is ArX, where Ar = phenyl or substituted phenyl. Aryl halides differ so much from the alkyl halides in their preparation and properties that they will be taken up in a separate chapter (Chap. 29).

If the halogen is not attached directly to the ring, the compound is an *aralkyl halide*, an alkyl halide that happens to contain an aromatic group; its chemistry is essentially the chemistry of other alkyl halides, but modified by the presence of the aryl substituent. It is this kind that we shall take up in this chapter.

An aryl or aralkyl halide is named according to the pattern we have seen before for halogen compounds (Secs. 3.10, 5.4, and 13.11): we name the parent compound, and then prefix this with *chloro-*, *bromo-*, etc. Certain aralkyl halides are commonly named as substituted alkyl halides. The $C_6H_5CH_2-$ group is given the special name of **benzyl**.

Bromobenzene	Benzyl bromide	*p*-Nitrobenzyl bromide
An aryl halide		

α-Phenylethyl chloride	β-Phenylethyl chloride
1-Phenyl-1-chloroethane	2-Phenyl-1-chloroethane

In a similar way, hydroxy compounds derived from arenes are of two kinds. If —OH is attached directly to the aromatic ring, the compound is a *phenol*, with special properties all its own (Chap. 28). If —OH is not attached directly to the ring, the compound is an aryl-substituted *alcohol*, and has the properties we have so far come to expect of an alcohol.

Benzyl alcohol	α-Phenylethyl alcohol	β-Phenylethyl alcohol
	1-Phenylethanol	2-Phenylethanol

m-Bromobenzyl alcohol	Diphenylmethanol
	(Diphenylcarbinol)

15.4   Physical properties

As compounds of low polarity, the alkylbenzenes possess physical properties that are essentially the same as those of the hydrocarbons we have already studied. They are insoluble in water, but quite soluble in non-polar solvents like ether, carbon tetrachloride, or ligroin. They are almost always less dense than water. As we can see from Table 15.1, boiling points rise with increasing molecular weight, the boiling point increment being the usual 20–30 degrees for each carbon atom.

Table 15.1   AROMATIC–ALIPHATIC HYDROCARBONS

Name	Formula	M.p., °C	B.p., °C	Relative density (at 20 °C)
Benzene	C_6H_6	5.5	80	0.879
Toluene	$C_6H_5CH_3$	−95	111	.866
o-Xylene	$1,2\text{-}C_6H_4(CH_3)_2$	−25	144	.880
m-Xylene	$1,3\text{-}C_6H_4(CH_3)_2$	−48 '	139	.864
p-Xylene	$1,4\text{-}C_6H_4(CH_3)_2$	13	138	.861
Hemimellitene	$1,2,3\text{-}C_6H_3(CH_3)_3$	−25	176	.895
Pseudocumene	$1,2,4\text{-}C_6H_3(CH_3)_3$	−44	169	.876
Mesitylene	$1,3,5\text{-}C_6H_3(CH_3)_3$	−45	165	.864
Prehnitene	$1,2,3,4\text{-}C_6H_2(CH_3)_4$	−6.5	205	.902
Isodurene	$1,2,3,5\text{-}C_6H_2(CH_3)_4$	−24	197	
Durene	$1,2,4,5\text{-}C_6H_2(CH_3)_4$	80	195	
Pentamethylbenzene	$C_6H(CH_3)_5$	53	231	
Hexamethylbenzene	$C_6(CH_3)_6$	165	264	
Ethylbenzene	$C_6H_5C_2H_5$	−95	136	.867
n-Propylbenzene	$C_6H_5CH_2CH_2CH_3$	−99	159	.862
Cumene	$C_6H_5CH(CH_3)_2$	−96	152	.862
n-Butylbenzene	$C_6H_5(CH_2)_3CH_3$	−81	183	.860
Isobutylbenzene	$C_6H_5CH_2CH(CH_3)_2$		171	.867
sec-Butylbenzene	$C_6H_5CH(CH_3)C_2H_5$	−83	173.5	.864
tert-Butylbenzene	$C_6H_5C(CH_3)_3$	−58	169	.867
p-Cymene	$1,4\text{-}CH_3C_6H_4CH(CH_3)_2$	−70	177	.857
Biphenyl	$C_6H_5C_6H_5$	70	255	
Diphenylmethane	$C_6H_5CH_2C_6H_5$	26	263	
Triphenylmethane	$(C_6H_5)_3CH$	93	360	
1,2-Diphenylethane	$C_6H_5CH_2CH_2C_6H_5$	52	284	
Styrene	$C_6H_5CH{=}CH_2$	−31	145	.907
trans-Stilbene	$trans\text{-}C_6H_5CH{=}CHC_6H_5$	124	307	
cis-Stilbene	$cis\text{-}C_6H_5CH{=}CHC_6H_5$	6		
1,1-Diphenylethene	$(C_6H_5)_2C{=}CH_2$	9	277	1.02
Triphenylethene	$(C_6H_5)_2C{=}CHC_6H_5$	73		
Tetraphenylethene	$(C_6H_5)_2C{=}C(C_6H_5)_2$	227	425	
Phenylacetylene	$C_6H_5C{\equiv}CH$	−45	142	0.930
Diphenylacetylene	$C_6H_5C{\equiv}CC_6H_5$	62.5	300	

Since melting points depend not only on molecular weight but also on molecular shape, their relationship to structure is a very complicated one. One important general relationship does exist, however, between melting point and structure of aromatic compounds: *among isomeric disubstituted benzenes, the para isomer generally melts considerably higher than the other two.* The xylenes, for example, boil within six degrees of one another; yet they differ widely in melting point, the *ortho* and *meta* isomers melting at $-25\ °C$ and $-48\ °C$, and the *para* isomer melting at $+13\ °C$. Since dissolution, like melting, involves overcoming the intermolecular forces of the crystal, it is not surprising to find that *generally the para isomer is also the least soluble in a given solvent.*

The higher melting point and lower solubility of a *para* isomer are only special examples of the general effect of molecular symmetry on intracrystalline forces. The more symmetrical a compound, the better it fits into a crystal lattice and hence the higher the melting point and the lower the solubility. *Para* isomers are simply the most symmetrical of disubstituted benzenes (see Fig. 15.1). We can see (Table 15.1) that 1,2,4,5-tetramethylbenzene melts 85–100 degrees higher than the less

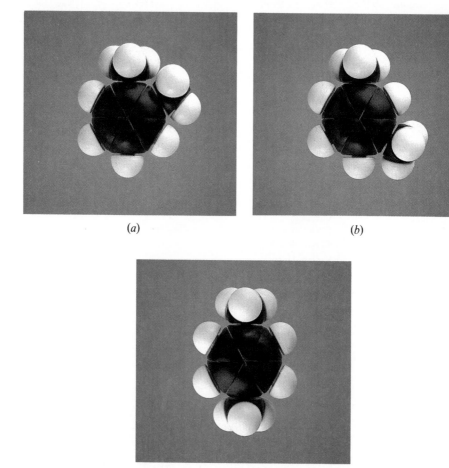

(a) (b)

(c)

Figure 15.1 Molecular symmetry and physical properties: effect of symmetry. The xylenes: (a) *ortho*, m.p. $-25\ °C$; (b) *meta*, m.p. $-48\ °C$; (c) *para*, m.p. $+13\ °C$. The *para* isomer is the most symmetrical, fits into a crystal lattice best, and has the highest melting point and lowest solubility.

symmetrical 1,2,3,5 and 1,2,3,4 isomers. A particularly striking example of the effect of symmetry on melting point is that of benzene and toluene. The introduction

of a single methyl group into the extremely symmetrical benzene molecule lowers
the melting point from 5 °C to −95 °C (see Fig. 15.2).

 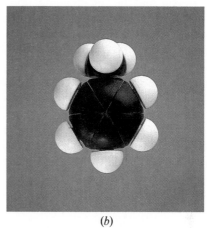

(a)   (b)

Figure 15.2 Molecular symmetry and physical properties: effect of symmetry. Benzene and toluene: (a) benzene, m.p. 5 °C; (b) toluene, m.p.
−95 °C. The highly symmetrical benzene melts 100 degrees higher than
toluene.

The effects on physical properties of attaching a halogen to an arene molecule
is just what we would expect from our discussion of alkyl halides (Sec. 5.6).

15.5 Industrial source of alkylbenzenes

It would be hard to exaggerate the importance to the chemical industry and
to our entire economy of the large-scale production of benzene and the alkylbenzenes. Just as the alkanes obtained from petroleum are ultimately the source of
nearly all our aliphatic compounds, so benzene and the alkylbenzenes are ultimately
the source of nearly all our aromatic compounds. When chemists wish to make a
complicated aromatic compound, whether in the laboratory or in industry, they do
not make a benzene ring; they take a simpler compound already containing a
benzene ring and then add to it, piece by piece, until they have built the structure
they want.

Just where do the enormous quantities of simple aromatic compounds come
from? There are two large reservoirs of organic material, **coal** and **petroleum**, and
aromatic compounds are obtained from both. Aromatic compounds are separated
as such from coal tar, and are synthesized from the alkanes of petroleum.

By far the larger portion of coal that is mined today is converted into coke,
which is needed for the smelting of iron ore to steel. When coal is heated in the
absence of air, it is partly broken down into simpler, volatile compounds which
are driven out; the residue is *coke*. The volatile materials consist of *coal gas* and a
liquid known as **coal tar**.

From coal tar by distillation there are obtained a number of aromatic
compounds. Upon coking, a ton of soft coal may yield about 120 pounds of coal
tar. From this 120 pounds the following aromatic compounds can be separated:

benzene, 2 pounds; toluene, 0.5 pound; xylenes, 0.1 pound; phenol, 0.5 pound; cresols, 2 pounds; naphthalene, 5 pounds. Two pounds of benzene from a ton of coal does not represent a very high percentage yield, yet so much coal is coked every year that the annual production of benzene from coal tar is very large.

But still larger quantities of aromatic hydrocarbons are needed, and these are synthesized from alkanes through the process of **catalytic reforming** (Sec. 12.3). This can bring about not only *dehydrogenation*, as in the formation of toluene from methylcyclohexane, but also *cyclization* and *isomerization*, as in the formation of toluene from *n*-heptane or 1,2-dimethylcyclopentane. In an analogous way, benzene is obtained from cyclohexane and methylcyclopentane, as well as from the *hydrodealkylation* of toluene.

Today, petroleum is the *chief* source of the enormous quantities of benzene, toluene, and the xylenes required for chemicals and fuels. Half of the toluene and xylenes are utilized in high-test gasoline where, in a sense, they replace the aliphatic compounds—inferior as fuels—from which they were made. (A considerable fraction even of naphthalene, the major component of coal tar distillate, is now being produced from petroleum hydrocarbons.)

15.6   Preparation of alkylbenzenes

Although a number of the simpler alkylbenzenes are available from industrial sources, the more complicated compounds must be synthesized in one of the ways outlined below.

PREPARATION OF ALKYLBENZENES

1. **Attachment of alkyl groups: Friedel–Crafts alkylation.** Discussed in Secs. 15.7–15.9.

$$\text{Benzene} + \text{R—X} \xrightarrow{\text{Lewis acid}} \text{C}_6\text{H}_5\text{R} + \text{HX} \qquad R \textit{ may rearrange}$$

Lewis acid: $AlCl_3$, BF_3, HF, etc.

Ar—X *cannot be used in place of* R—X

2. **Conversion of side chain.** Discussed in Sec. 21.9.

$$\text{Ar—C(=O)—R} \xrightarrow[\text{or } N_2H_4,\ \text{base, heat}]{\text{Zn(Hg), HCl, heat}} \text{Ar—CH}_2\text{R} \qquad \textbf{Clemmensen or Wolff–Kishner} \text{ reduction}$$

A ketone

$$\text{Ar—CH=CHR} \xrightarrow{H_2,\ Ni} \text{Ar—CH}_2\text{CH}_2\text{R} \qquad \blacksquare$$

Friedel–Crafts alkylation is extremely useful since it permits the direct attachment of an alkyl group to the aromatic ring. There are, however, a number of limitations to its use (Sec. 15.9), including the fact that the alkyl group that becomes attached to the ring is not always the same as the alkyl group of the parent halide; this **rearrangement** of the alkyl group is discussed in Sec. 15.8.

There are frequently available aromatic compounds containing aliphatic side chains that are not simple alkyl groups. An alkylbenzene can be prepared from one of these compounds by converting the side chain into an alkyl group. Although there is an aromatic ring in the molecule, this conversion is essentially the preparation of an alkane from some other aliphatic compound. The methods used are those that we have already learned for the preparation of alkanes: hydrogenation of a carbon–carbon double bond in a side chain, for example. Many problems of the alkylbenzenes are solved by a consideration of simple alkane chemistry.

The most important side-chain conversion involves **reduction of ketones** either by amalgamated zinc and HCl (*Clemmensen reduction*) or by hydrazine and strong base (*Wolff–Kishner reduction*). This method is important because the necessary ketones are readily available through a modification of the Friedel–Crafts reaction that involves acid chlorides (see Sec. 21.5). Unlike alkylation by the Friedel–Crafts reaction, this method does not involve rearrangement.

Much of the importance of alkylbenzenes lies in a fact that will become apparent to us as we go through this chapter: unlike alkanes, alkylbenzenes are extremely useful precursors of the compounds that are formally their derivatives— halides, alcohols, and related compounds.

Problem 15.1   How might you prepare ethylbenzene from: (a) benzene and ethyl alcohol; (b) acetophenone, $C_6H_5COCH_3$; (c) styrene, $C_6H_5CH=CH_2$; (d) α-phenylethyl alcohol, $C_6H_5CHOHCH_3$; and (e) β-phenylethyl chloride, $C_6H_5CH_2CH_2Cl$?

15.7   Friedel–Crafts alkylation

If a small amount of anhydrous aluminum chloride is added to a mixture of benzene and methyl chloride, a vigorous reaction occurs, hydrogen chloride gas is

$$\text{C}_6\text{H}_6 + CH_3Cl \xrightarrow{\ AlCl_3\ } \text{C}_6\text{H}_5CH_3 + HCl$$

Toluene

evolved, and toluene can be isolated from the reaction mixture. This is the simplest example of the reaction discovered in 1877 at the University of Paris by the French–American team of chemists, Charles Friedel and James Crafts. *Considered in its various modifications, the Friedel–Crafts reaction is by far the most important method for attaching alkyl side chains to aromatic rings.*

Each of the components of the simple example just given can be varied. The alkyl halide may contain an alkyl group more complicated than methyl, and a halogen atom other than chlorine; in some cases alcohols are used or—especially in industry—alkenes. Substituted alkyl halides, like benzyl chloride, $C_6H_5CH_2Cl$, also can be used. Because of the low reactivity of halogen attached to an aromatic ring (Sec. 29.5), aryl halides *cannot* be used in place of alkyl halides.

The aromatic ring to which the side chain becomes attached may be that of benzene itself, certain substituted benzenes (chiefly alkylbenzenes and halobenzenes), or more complicated aromatic ring systems like naphthalene and anthracene (Chap. 34).

In place of aluminum chloride, other Lewis acids can be used, in particular BF_3, HF, and phosphoric acid.

The reaction is carried out by simply mixing together the three components; usually the only problems are those of moderating the reaction by cooling and of trapping the hydrogen halide gas. Since the attachment of an alkyl side chain makes the ring more susceptible to further attack (Sec. 14.5), steps must be taken to limit substitution to *mono*alkylation. As in halogenation of alkanes (Sec. 2.8), this is accomplished by using an *excess* of the hydrocarbon. In this way an alkyl carbocation seeking an aromatic ring is more likely to encounter an unsubstituted ring than a substituted one. Frequently the aromatic compound does double duty, serving as solvent as well as reactant.

From polyhalogenated alkanes it is possible to prepare compounds containing more than one aromatic ring:

$$2C_6H_6 + CH_2Cl_2 \xrightarrow{\text{AlCl}_3} C_6H_5CH_2C_6H_5 + 2HCl$$
$$\text{Diphenylmethane}$$

$$2C_6H_6 + ClCH_2CH_2Cl \xrightarrow{\text{AlCl}_3} C_6H_5CH_2CH_2C_6H_5 + 2HCl$$
$$\text{1,2-Diphenylethane}$$

$$3C_6H_6 + CHCl_3 \xrightarrow{\text{AlCl}_3} C_6H_5-\underset{\underset{H}{|}}{\overset{\overset{C_6H_5}{|}}{C}}-C_6H_5 + 3HCl$$
$$\text{Triphenylmethane}$$

$$3C_6H_6 + CCl_4 \xrightarrow{\text{AlCl}_3} C_6H_5-\underset{\underset{Cl}{|}}{\overset{\overset{C_6H_5}{|}}{C}}-C_6H_5 + 3HCl$$
$$\text{Triphenylchloromethane}$$

15.8　Mechanism of Friedel–Crafts alkylation

In Sec. 14.10 we said that two mechanisms are possible for Friedel–Crafts alkylation. Both involve electrophilic aromatic substitution, but they differ as to the nature of the electrophile.

One mechanism for Friedel–Crafts alkylation involves the following steps,

(1)　　　　　　　　$R-Cl + AlCl_3 \rightleftarrows AlCl_4^- + R\oplus$

(2)　　　　　　　　$R\oplus + C_6H_6 \rightleftarrows C_6H_5\overset{\oplus}{\underset{\diagdown H}{\diagup R}}$

(3)　　　　　　$C_6H_5\overset{\oplus}{\underset{\diagdown H}{\diagup R}} + AlCl_4^- \rightleftarrows C_6H_5-R + HCl + AlCl_3$

in which the electrophile is an alkyl cation. The function of the aluminum chloride is to generate this carbocation by abstracting the halogen from the alkyl halide. It is not surprising that other Lewis acids can function in the same way and thus take the place of aluminum chloride:

$$\text{R:X:} + \text{Al:Cl:} \rightleftharpoons \text{R}\oplus + \text{:X:Al:Cl:} \ominus$$

$$\text{R:X:} + \text{B:F:} \rightleftharpoons \text{R}\oplus + \text{:X:B:F:} \ominus \qquad \textit{Carbocations from alkyl halides}$$

$$\text{R:X:} + \text{Fe:Cl:} \rightleftharpoons \text{R}\oplus + \text{:X:Fe:Cl:} \ominus$$

$$\text{R:X:} + \text{H:F:} \rightleftharpoons \text{R}\oplus + \text{:X:---H:F:} \ominus$$

Judging from the mechanism just described, we might expect the benzene ring to be attacked by carbocations generated in other ways: by the action of acid on alcohols (Sec. 5.25) and on alkenes (Sec. 8.9).

$$\text{R—OH} + \text{H}^+ \rightleftharpoons \text{R—OH}_2 \oplus \rightleftharpoons \text{R}\oplus + \text{H}_2\text{O}$$

Carbocations from alcohols and from alkenes

$$-\text{C}=\text{C}- + \text{H}^+ \rightleftharpoons -\underset{\underset{\text{H}}{|}}{\text{C}}-\overset{|}{\text{C}}\oplus$$

This expectation is correct: alcohols and alkenes, in the presence of acids, alkylate aromatic rings in what we may consider to be a modification of the Friedel–Crafts reaction.

$$\text{C}_6\text{H}_6 + (\text{CH}_3)_3\text{COH} \xrightarrow{\text{H}_2\text{SO}_4} \text{C}_6\text{H}_5-\text{C}(\text{CH}_3)_3$$

tert-Butyl alcohol   *tert*-Butylbenzene

$$\text{C}_6\text{H}_6 + (\text{CH}_3)_2\text{C}=\text{CH}_2 \xrightarrow{\text{H}_2\text{SO}_4} \text{C}_6\text{H}_5-\text{C}(\text{CH}_3)_3$$

Isobutylene   *tert*-Butylbenzene

Also judging from the mechanism, we might expect Friedel–Crafts alkylation to be accompanied by the kind of rearrangement that is characteristic of carbocation reactions (Sec. 5.23). This expectation, too, is correct. As the following examples show, alkylbenzenes containing rearranged alkyl groups not only are formed but are sometimes the sole products. In each case, we see that the particular

$$\text{C}_6\text{H}_6 + \text{CH}_3\text{CH}_2\text{CH}_2\text{Cl} \xrightarrow[-18\,°\text{C to 80}\,°\text{C}]{\text{AlCl}_3} \text{C}_6\text{H}_5\text{CH}_2\text{CH}_2\text{CH}_3 + \text{C}_6\text{H}_5\overset{\overset{\text{CH}_3}{|}}{\text{C}}\text{HCH}_3$$

n-Propyl chloride   *n*-Propylbenzene   Isopropylbenzene

35–31%   65–69%

$$C_6H_6 + CH_3CH_2CH_2CH_2Cl \xrightarrow[0\,°C]{AlCl_3} C_6H_5CH_2CH_2CH_2CH_3 \quad + \quad \underset{\substack{CH_3 \\ |}}{C_6H_5CHCH_2CH_3}$$

n-Butyl chloride   n-Butylbenzene   sec-Butylbenzene
                              34%   66%

$$C_6H_6 + \underset{\substack{CH_3 \\ |}}{CH_3CHCH_2Cl} \xrightarrow[-18\,°C\ to\ 80\,°C]{AlCl_3} \underset{\substack{CH_3 \\ | \\ CH_3}}{C_6H_5CCH_3}$$

Isobutyl chloride   tert-Butylbenzene
                                        Only product

$$C_6H_6 + \underset{\substack{CH_3 \\ | \\ CH_3}}{CH_3CCH_2OH} \xrightarrow[60\,°C]{BF_3} \underset{\substack{CH_3 \\ | \\ CH_3}}{C_6H_5CCH_2CH_3}$$

Neopentyl alcohol   tert-Pentylbenzene
                                 Only product

kind of rearrangement corresponds to what we would expect if a less stable (1°) carbocation were to rearrange by a 1,2-shift to a more stable (2° or 3°) carbocation.

We can now make another addition to our list of carbocation reactions (Sec. 8.17). **A carbocation may:**

    (a) combine with a negative ion or other basic molecule;
    (b) rearrange to a more stable carbocation;
    (c) eliminate a hydrogen ion to form an alkene;
    (d) add to an alkene to form a larger carbocation;
    (e) abstract a hydride ion from an alkane;
    (f) alkylate an aromatic ring.

A carbocation formed by (b) or (d) can subsequently undergo any of the reactions.

In alkylation, as in its other reactions, the carbocation gains a pair of electrons to complete the octet of the electron-deficient carbon—this time from the π cloud of an aromatic ring.

Problem 15.2 *tert*-Pentylbenzene is the major product of the reaction of benzene in the presence of BF₃ with each of the following alcohols: (a) 2-methyl-1-butanol, (b) 3-methyl-2-butanol, (c) 3-methyl-1-butanol, and (d) neopentyl alcohol. Account for its formation in each case.

In some of the examples given above, we see that *part* of the product is made up of *unrearranged* alkylbenzenes. Must we conclude that part of the reaction does not go by way of carbocations? Not *necessarily*. Attack on an aromatic ring is probably one of the most difficult jobs a carbocation is called on to do; that is to say, toward carbocations an aromatic ring is a reagent of low reactivity and hence high selectivity. Although there may be present a higher concentration of the more stable, rearranged carbocations, the aromatic ring may tend to seek out the scarce unrearranged ions because of their higher reactivity. In some cases, it is quite possible that some of the carbocations react with the aromatic ring before they have time to rearrange; the same low stability that makes primary carbocations, for example, prone to rearrangement also makes them highly reactive.

On the other hand, there is additional evidence (of a kind we cannot go into here) that makes it very likely that there *is* a second mechanism for Friedel–Crafts alkylation. In this mechanism, the electrophile is not an alkyl cation but an acid–base complex of alkyl halide and Lewis acid, from which the alkyl group is transferred *in one step* from halogen to the aromatic ring.

$$\underset{\text{Cl}}{\overset{\text{Cl}}{\text{Cl}\overset{\ominus}{-}\!\overset{\oplus}{\text{Al}}\!-\!\text{Cl}\!-\!\text{R}}} + C_6H_6 \longrightarrow \left[\underset{\text{R----ClAlCl}_3}{\overset{\delta+}{C_6H_5}\overset{H}{\diagdown}}^{\delta-} \right] \longrightarrow \overset{\oplus}{C_6H_5}\underset{R}{\overset{H}{\diagup}} + AlCl_4^-$$

This duality of mechanism does not reflect exceptional behavior, but is usual for electrophilic aromatic substitution. It also fits into a familiar pattern for nucleophilic aliphatic substitution (Secs. 5.12 and 5.24), which—from the standpoint of the alkyl halide—is the kind of reaction taking place. Furthermore, the particular halides (1° and methyl) which appear to react by this second, bimolecular mechanism are just the ones that would have been *expected* to do so.

15.9　Limitations of Friedel–Crafts alkylation

We have encountered three limitations to the use of Friedel–Crafts alkylation: (a) the danger of polysubstitution; (b) the possibility that the alkyl group will rearrange; and (c) the fact that aryl halides cannot take the place of alkyl halides. Besides these, there are several other limitations.

(d) An aromatic ring less reactive than that of the halobenzenes does not undergo the Friedel–Crafts reaction; evidently the carbocation, R^+, is a less powerful electrophile than NO_2^+ and the other electron-deficient reagents that bring about electrophilic aromatic substitution.

Next, (e) aromatic rings containing the $-NH_2$, $-NHR$, or $-NR_2$ group do not undergo Friedel–Crafts alkylation, partly because the strongly basic nitrogen ties up the Lewis acid needed for ionization of the alkyl halide:

$$C_6H_5\overset{..}{N}H_2 + AlCl_3 \longrightarrow C_6H_5\underset{\oplus}{\overset{\overset{\ominus AlCl_3}{..}}{N}H_2}$$

I

> **Problem 15.3**　Tying up of the acidic catalyst by the basic nitrogen is not the only factor that prevents alkylation, since even when excess catalyst is used, reaction does not occur. Looking at the structure of the complex (I) shown for aniline, can you suggest another factor? (*Hint*: See Sec. 14.16.)

Despite these numerous limitations, the Friedel–Crafts reaction, in its various modifications (for example, acylation, Sec. 21.5), is an extremely useful synthetic tool.

15.10　Reactions of alkylbenzenes

The most important reactions of the alkylbenzenes are outlined below, with toluene and ethylbenzene as specific examples; essentially the same behavior is shown by compounds bearing other side chains. Except for hydrogenation and oxidation, these reactions involve either **electrophilic substitution in the aromatic ring** or **free-radical substitution in the aliphatic side chain**.

In following sections we shall be mostly concerned with (a) how experimental conditions determine which portion of the molecule—aromatic or aliphatic—is attacked, and (b) how each portion of the molecule modifies the reactions of the other portion.

_____ REACTIONS OF ALKYLBENZENES _____

1. Hydrogenation

Example:

2. Oxidation. Discussed in Sec. 15.11.

Example:

3. **Substitution in the ring. Electrophilic aromatic substitution.** Discussed in Sec. 15.12.

Examples:

———— CONTINUED ————

4. Substitution in the side chain. Free-radical halogenation. Discussed in Secs. 15.13–15.15.

Examples:

| Toluene | Benzyl chloride | Benzal chloride | Benzotrichloride |

Ethylbenzene   α-Phenylethyl   β-Phenylethyl
                                            chloride   chloride
                                     Chief product

Note: Competition between ring and side chain. Discussed in Sec. 15.13.

Free-radical
substitution

Electrophilic
substitution   ■

15.11 Oxidation of alkylbenzenes

Although benzene and alkanes are quite unreactive toward the usual oxidizing agents ($KMnO_4$, $K_2Cr_2O_7$, etc.), the benzene ring renders an aliphatic side chain quite susceptible to oxidation. The side chain is oxidized down to the ring, only a carboxyl group (—COOH) remaining to indicate the position of the original side chain. Potassium permanganate is generally used for this purpose, although potassium dichromate or dilute nitric acid also can be used. (Oxidation of a side chain is more difficult, however, than oxidation of an alkene, and requires prolonged treatment with hot $KMnO_4$.)

n-Butylbenzene   Benzoic acid

This reaction is used for two purposes: (a) synthesis of carboxylic acids, and (b) identification of alkylbenzenes.

(a) Synthesis of carboxylic acids. One of the most useful methods of preparing an aromatic carboxylic acid involves oxidation of the proper alkylbenzene. For example:

p-Xylene

Terephthalic acid
(1,4-Benzenedicarboxylic acid)

p-Nitrotoluene

p-Nitrobenzoic acid

(b) Identification of alkylbenzenes. The number and relative positions of side chains can frequently be determined by oxidation to the corresponding acids. Suppose, for example, that we are trying to identify an unknown liquid of formula C_8H_{10} and boiling point 137–139 °C that we have shown in other ways to be an alkylbenzene (Sec. 15.23). Looking in Table 15.1 (p. 535), we find that it could be any one of four compounds: o-, m-, or p-xylene, or ethylbenzene. As shown below, oxidation of each of these possible hydrocarbons yields a different acid, and these acids can readily be distinguished from each other by their melting points or the melting points of derivatives.

o-Xylene
(b.p. 144 °C)

Phthalic acid, m.p. 231 °C
(p-nitrobenzyl ester, m.p. 155 °C)

m-Xylene
(b.p. 139 °C)

Isophthalic acid, m.p. 348 °C
(p-nitrobenzyl ester, m.p. 215 °C)

p-Xylene
(b.p. 138 °C)

Terephthalic acid, m.p. 300 °C subl.
(p-nitrobenzyl ester, m.p. 263 °C)

Ethylbenzene
(b.p. 136 °C)

Benzoic acid, m.p. 122 °C
(p-nitrobenzyl ester, m.p. 89 °C)

15.12 Electrophilic aromatic substitution in alkylbenzenes

Because of its electron-releasing effect, an alkyl group activates a benzene ring to which it is attached, and directs *ortho* and *para* (Secs. 14.16 and 14.17).

Problem 15.4 Treatment with methyl chloride and $AlCl_3$ at 0 °C converts toluene chiefly into *o-* and *p-*xylenes; at 80 °C, however, the chief product is *m-*xylene. Furthermore, either *o-* or *p-*xylene is readily converted into *m-*xylene by treatment with $AlCl_3$ and HCl at 80 °C.

How do you account for this effect of temperature on orientation? Suggest a role for the HCl.

Problem 15.5 Why is polysubstitution a complicating factor in Friedel–Crafts alkylation but not in aromatic nitration, sulfonation, or halogenation?

15.13 Halogenation of alkylbenzenes: ring *vs.* side chain

Alkylbenzenes clearly offer two sites where halogen can attack: the ring and the side chain. If we think about the reactions involved, we find that we should be able to direct the attack to either one of these sites by our choice of reaction conditions.

The side chain is alkane-like, and should undergo halogenation as alkanes do: via free-radical substitution. This reaction requires conditions under which halogen atoms are formed, that is, high temperatures or light.

$$CH_4 + Cl_2 \xrightarrow{\text{heat or light}} CH_3Cl + HCl$$

The ring is benzene-like, and should undergo substitution as benzene does: via electrophilic substitution. This reaction involves transfer of positive halogen, which is promoted by acid catalysts like ferric chloride.

$$C_6H_6 + Cl_2 \xrightarrow{\text{FeCl}_3,\text{ cold}} C_6H_5Cl + HCl$$

We must expect, then, that the position of attack in, say, toluene would be governed by which attacking particle is involved, and therefore by the conditions employed. This is so. If chlorine is bubbled into boiling toluene that is exposed to

Atom: *attacks side chain*

Ion: *attacks ring*

ultraviolet light, substitution occurs almost exclusively in the side chain. In the absence of light and in the presence of ferric chloride, substitution occurs mostly in the ring. We saw a similar competition between homolytic and heterolytic reactions in the halogenation of alkenes. There, free radicals brought about substitution, as they do here; electrophilic attack led to addition, the characteristic reaction of alkenes, just as it leads here to ring substitution, the characteristic reaction of aromatic compounds.

Like nitration and sulfonation, ring halogenation yields chiefly the *ortho*

Toluene   *o*-Chlorotoluene *p*-Chlorotoluene
                                      58%   42%

and *para* isomers. Similar results are obtained with other alkylbenzenes, and with bromine as well as chlorine.

Side-chain halogenation, like halogenation of alkanes, may yield polyhalogenated products; even when reaction is limited to monohalogenation, it may yield a mixture of isomers.

Side-chain chlorination of toluene can yield successively the mono-, di-, and trichloro compounds. These are known as *benzyl chloride, benzal chloride,* and

Toluene   Benzyl chloride   Benzal chloride   Benzotrichloride

benzotrichloride; such compounds are important intermediates in the synthesis of alcohols, aldehydes, and acids.

15.14   Side-chain halogenation of alkylbenzenes

Chlorination and bromination of side chains differ from one another in orientation and reactivity in one very significant way. Let us look first at bromination, and then at chlorination.

An alkylbenzene with a side chain more complex than methyl may offer more than one position for attack, and so we must consider the likelihood of obtaining a mixture of isomers. Bromination of ethylbenzene, for example, could theoretically yield two products: 1-bromo-1-phenylethane and 2-bromo-1-phenylethane. Despite

1-Bromo-1-phenylethane
Only product

Ethylbenzene

2-Bromo-1-phenylethane

a probability factor that favors 2-bromo-1-phenylethane by 3:2, the *only* product found is 1-bromo-1-phenylethane. Evidently abstraction of the hydrogens attached to the carbon next to the aromatic ring is greatly preferred.

Hydrogen atoms attached to carbon joined directly to an aromatic ring are called **benzylic hydrogens**.

Benzylic hydrogen:
easy to abstract

The relative ease with which benzylic hydrogens are abstracted is shown not only by orientation of bromination but also—and in a more exact way—by comparison of reactivities of different compounds. Competition experiments (Sec. 3.22) show, for example, that at 40 °C a benzylic hydrogen of toluene is 3.3 times as reactive toward bromine atoms as the tertiary hydrogen of an alkane—and nearly 100 million times as reactive as a hydrogen of methane!

Examination of reactions that involve attack not only by halogen atoms but by other free radicals as well has shown that this is a general rule: benzylic hydrogens are extremely easy to abstract and thus resemble allylic hydrogens. We can now expand the reactivity sequence of Sec. 10.3:

Ease of abstraction of hydrogen atoms $\begin{array}{c}\text{allylic}\\\text{benzylic}\end{array} > 3° > 2° > 1° > CH_4 > \text{vinylic}$

Side-chain halogenation of alkylbenzenes proceeds by the same mechanism as halogenation of alkanes. Bromination of toluene, for example, includes the following steps:

Toluene   Benzyl radical   Benzyl bromide

The fact that benzylic hydrogens are unusually easy to abstract means that benzyl radicals are unusually easy to form.

Ease of formation of free radicals $\begin{array}{c}\text{allyl}\\\text{benzyl}\end{array} > 3° > 2° > 1° > CH_3· > \text{vinyl}$

Again we ask the question: are these findings in accord with our rule that *the more stable the radical, the more rapidly it is formed*? Is the rapidly formed benzyl radical relatively stable?

The bond dissociation energies in Table 1.2 (p. 21) show that only 85 kcal is needed for formation of benzyl radicals from a mole of toluene, as compared with 92 kcal for formation of *tert*-butyl radicals and 88 kcal for formation of allyl radicals. Relative to the hydrocarbon from which each is formed, then, a benzyl radical contains less energy and is more stable than a *tert*-butyl radical.

We can now expand the sequence of radical stabilities (Sec. 10.3). Relative to the hydrocarbon from which each is formed, the relative stability of free radicals is:

Stability of free radicals $\begin{array}{c}\text{allyl}\\\text{benzyl}\end{array} > 3° > 2° > 1° > CH_3· > \text{vinyl}$

Orientation of chlorination shows that chlorine atoms, like bromine atoms, preferentially attack benzylic hydrogen; but, as we see, the preference is less marked:

Ethylbenzene   1-Chloro-1-phenylethane   2-Chloro-1-phenylethane
                      Major product, 91%   9%

Furthermore, competition experiments show that, under conditions where 3°, 2°, and 1° hydrogens show relative reactivities of 5.0:3.8:1.0, the relative rate per benzylic hydrogen of toluene is only 1.3. As in its attack on alkanes (Sec. 3.28), the more reactive chlorine atom is less selective than the bromine atom: less selective between hydrogens in a single molecule, and less selective between hydrogens in different molecules.

In the attack by the comparatively unreactive bromine atom, we have said (Sec. 2.24), the transition state is reached late in the reaction process: the carbon–hydrogen bond is largely broken, and the organic group has acquired a great deal of free-radical character. The factors that stabilize the benzyl free radical stabilize the incipient benzyl free radical in the transition state.

In contrast, in the attack by the highly reactive chlorine atom, the transition state is reached early in the reaction process: the carbon–hydrogen bond is only slightly broken, and the organic group has acquired little free-radical character. The factors that stabilize the benzyl radical have little effect on this transition state.

Just why benzylic hydrogens are *less* reactive toward chlorine atoms than even secondary hydrogens is not understood. It has been attributed to *polar factors* (Sec. 8.19), but this hypothesis has been questioned.

15.15 Resonance stabilization of the benzyl radical

How are we to account for the stability of the benzyl radical? Bond dissociation energies indicate that 19 kcal/mol less energy $(104 - 85)$ is needed to form the benzyl radical from toluene than to form the methyl radical from methane.

$$C_6H_5CH_3 \longrightarrow C_6H_5CH_2\cdot + H\cdot \qquad \Delta H = +85 \text{ kcal}$$

Toluene   Benzyl radical

As we did for the allyl radical (Sec. 10.7), let us examine the structures involved. Toluene contains the benzene ring and is therefore a hybrid of the two Kekulé structures, I and II:

I   II

Similarly, the benzyl radical is a hybrid of the two Kekulé structures, III and IV:

This resonance causes stabilization, that is, lowers the energy content. However, resonance involving Kekulé structures presumably stabilizes both molecule and radical to the same extent, and hence does not affect the *difference* in their energy contents. If there were no other factors involved, then we might reasonably expect the bond dissociation energy for a benzylic hydrogen to be about the same as that of a methane hydrogen (see Fig. 15.3).

$CH_3\cdot + H\cdot$

Extra stabilization of
benzyl radical: 19 kcal

$C_6H_5CH_2\cdot + H\cdot$

Potential energy ⟶

CH_4

$C_6H_5CH_3$

Progress of reaction ⟶

Figure 15.3 Molecular structure and reactivity. The resonance-stabilized benzyl radical is formed faster than the methyl radical. (The plots are aligned with each other for easy comparison.)

Considering further, however, we find that we can draw three additional structures for the radical: V, VI, and VII. In these structures there is a double bond between the side chain and the ring, and the odd electron is located on the carbon atoms *ortho* and *para* to the side chain. Drawing these pictures is, of course, our

$$\begin{bmatrix} CH_2 & CH_2 & CH_2 \\ \end{bmatrix} \quad \text{equivalent to} \quad CH_2$$

V   VI   VII

way of indicating that the odd electron is not localized on the side chain but is *delocalized*, being distributed about the ring. We cannot draw comparable structures for the toluene molecule.

Contribution from the three structures V–VII stabilizes the radical in a way that is not possible for the molecule. Resonance thus lowers the energy content of the benzyl radical more than it lowers the energy content of toluene. This extra stabilization of the radical evidently amounts to 19 kcal/mol (Fig. 15.3).

We say, then, that the benzyl radical is *stabilized by resonance*. When we use this expression, we must always bear in mind that we actually mean that the benzyl radical is stabilized by resonance *to a greater extent than* the hydrocarbon from which it is formed.

In terms of orbitals, delocalization results from overlap of the *p* orbital occupied by the odd electron with the π cloud of the ring (Fig. 15.4).

Figure 15.4   Benzyl radical. The *p* orbital occupied by the odd electron overlaps the π cloud of the ring.

Like the allyl radical, we see, the benzyl radical is a *conjugated* molecule (Sec. 10.9). Here the *p* orbital on the carbon bearing the odd electron is conjugated, not just with one double bond, but with the entire π system of the benzene ring. The conjugation of the aromatic ring has been *extended* to include the side-chain carbon; and with this extension comes greater stabilization.

Problem 15.6   It is believed that the side-chain hydrogens of the benzyl radical lie in the same plane as the ring. Why should they?

Problem 15.7   The strength of the bond holding side-chain hydrogen in *m*-xylene is the same as in toluene; in *o*- and *p*-xylene it is 3–4 kcal lower. How do you account for these differences?

15.16   Triphenylmethyl: a stable free radical

We have said that benzyl and allyl free radicals are stabilized by resonance; but we must realize, of course, that they are stable only in comparison with simple alkyl radicals like methyl or ethyl. Benzyl and allyl free radicals are extremely reactive, unstable particles, whose fleeting existence (a few thousandths of a second) has been proposed simply because it is the best way to account for certain experimental observations. We do not find bottles on the laboratory shelf labeled "benzyl radicals" or "allyl radicals". Is there, then, any direct evidence for the existence of free radicals?

In 1900 a remarkable paper appeared in the *Journal of the American Chemical Society* and in the *Berichte der deutschen chemischen Gesellschaft*; its author was the young Russian-born chemist Moses Gomberg, who was at that time an instructor at the University of Michigan. Gomberg was interested in completely phenylated alkanes. He had prepared tetraphenylmethane (a synthesis a number of eminent chemists had previously attempted, but unsuccessfully), and he had now set himself the task of synthesizing hexaphenylethane. Having available triphenylchloro-methane (Sec. 15.7), he went about the job in just the way we might today: he tried to couple together two triphenylmethyl groups by use of a metal (Sec. 12.4). Since sodium did not work very well, he used instead finely divided silver, mercury, or, best of all, zinc dust. He allowed a benzene solution of triphenylchloromethane to stand over one of these metals, and then filtered the solution free of the metal

$$\text{—C—Cl} + \text{Zn} + \text{Cl—C—} \longrightarrow \text{Hexaphenylethane} + ZnCl_2$$
Expected product

Triphenylchloromethane
2 moles

halide. When the benzene was evaporated, there was left behind a white crystal-line solid which after recrystallization melted at 185 °C; this he thought was hexaphenylethane.

As a chemist always does with a new compound, Gomberg analyzed his product for its carbon and hydrogen content. To his surprise, the analysis showed 88% carbon and 6% hydrogen, a total of only 94%. Thinking that combustion had not been complete, he carried out the analysis again, this time more carefully and under more vigorous conditions; he obtained the same results as before. Repeated analysis of samples prepared from both triphenylchloromethane and triphenyl-bromomethane, and purified by recrystallization from a variety of solvents, finally convinced him that he had prepared not a hydrocarbon—not hexaphenylethane—but a compound containing 6% of some other element, probably oxygen.

Oxygen could have come from impure metals; but extremely pure samples of metals, carefully freed of oxygen, gave the same results.

Oxygen could have come from the air, although he could not see how molecular oxygen could react at room temperature with a hydrocarbon. He carried out the reaction again, this time under an atmosphere of carbon dioxide. When he filtered the solution (also under carbon dioxide) and evaporated the solvent, there was left behind not his compound of m.p. 185 °C but an entirely different substance, much more soluble in benzene than his first product, and having a much lower melting point. This new substance was eventually purified, and on analysis it gave the correct composition for hexaphenylethane: 93.8% carbon, 6.2% hydrogen.

Dissolved in benzene, the new substance gave a yellow solution. When a small amount of air was admitted to the container, the yellow color disappeared, and then after a few minutes reappeared. When more oxygen was admitted, the same thing happened: disappearance of the color and slow reappearance. Finally the color disappeared for good; evaporation of the solvent yielded the original com-pound of m.p. 185 °C.

Not only oxygen but also halogens were rapidly absorbed by ice-cold solutions of this substance; even solutions of normally unreactive iodine were instantly decolorized.

The compound of m.p. 185 °C was the peroxide,

$$(C_6H_5)_3C—O—O—C(C_6H_5)_3$$

as Gomberg showed by preparing it in an entirely different way. The products of the halogen reactions were the triphenylhalomethanes, $(C_6H_5)_3C—X$.

If this new substance he had made was indeed hexaphenylethane, it was behaving very strangely. Cleavage of a carbon–carbon bond by such mild reagents as oxygen and iodine was unknown to organic chemists.

Triphenylchloromethane   Triphenylmethyl

Triphenylmethyl
Yellow
⇌ Dimer
Colorless

"The experimental evidence presented above forces me to the conclusion that we have to deal here with a free radical, triphenylmethyl, $(C_6H_5)_3C$. On this assumption alone do the results described above become intelligible and receive an adequate explanation." Gomberg was proposing that he had prepared a *stable* free radical.

Figure 15.5 The triphenylmethyl free radical. Crowding between *ortho* hydrogens prevents coplanarity of the rings and, as a result, the rings are turned, like the blades of a propeller.

It was nearly ten years before Gomberg's proposal was generally accepted. It now seems clear that what happens is the following: the metal abstracts a chlorine atom from triphenylchloromethane to form the free radical triphenylmethyl; two of these radicals then combine to form a dimeric hydrocarbon. But the carbon–carbon bond in the dimer is a very weak one, and even at room temperature can break to regenerate the radicals. Thus an equilibrium exists between the free radicals and the hydrocarbon. Although this equilibrium tends to favor the hydrocarbon, any solution of the dimer contains an appreciable concentration of free triphenylmethyl radicals. The fraction of material existing as free radicals is about 2% in a 1 M solution, 10% in a 0.01 M solution, and nearly 100% in very dilute solutions. We could quite correctly label a bottle containing a dilute solution of this substance as "triphenylmethyl radicals".

Triphenylmethyl is yellow; both the dimer and the peroxide are colorless. A solution of the dimer is yellow because of the triphenylmethyl present in the equilibrium mixture. When oxygen is admitted, the triphenylmethyl rapidly reacts to form the peroxide, and the yellow color disappears. More dimer dissociates to restore equilibrium and the yellow color reappears. Only when all the dimer–triphenylmethyl mixture is converted into the peroxide does the yellow color fail to appear. In a similar way it is triphenylmethyl that reacts with iodine.

$$\text{Dimer} \rightleftharpoons 2(C_6H_5)_3C\cdot \quad
\begin{cases}
\xrightarrow{O_2,\,0\,°C} & (C_6H_5)_3C-O-O-C(C_6H_5)_3 \\
\\
\xrightarrow{I_2,\,0\,°C} & 2(C_6H_5)_3C-I
\end{cases}$$

Triphenylmethyl
radical

Thus the dimer undergoes its surprising reactions by first dissociating into triphenylmethyl, which, although unusually stable for a free radical, is nevertheless an exceedingly reactive particle.

Now, what *is* this dimer? For nearly 70 years it was believed to be hexaphenylethane. It—and dozens of analogs—were studied exhaustively, and the equilibria between them and triarylmethyl radicals were interpreted on the basis of the hexaarylethane structure. Then, in 1968, the dimer was shown to have

$$(C_6H_5)_3C\diagdown\underset{H}{\diagup}\!\!\!\!\!\!\bigcirc\!\!=\!\!C(C_6H_5)_2$$

I

the structure I. Gomberg's original task is still unaccomplished: hexaphenylethane, it seems, has never been made.

The basic significance of Gomberg's work remains unchanged. Many dimers have been prepared, and the existence of free triarylmethyl radicals has been substantiated in a number of ways; indeed, certain of these compounds seem to exist entirely as the free radical even in the solid state. The most convincing evidence for the free-radical nature of these substances lies in properties that arise directly from the odd electron that characterizes a free radical. Two electrons that occupy the same orbital and thus make up a pair have opposite spins (Sec. 1.6); the magnetic moments corresponding to their spins exactly cancel each other. But, by definition (Sec. 2.12), the odd electron of a free radical is not paired, and hence the effect of its spin is not canceled. This spin gives to the free radical a net magnetic moment. This magnetic moment reveals itself in two ways: (a) the compound is *paramagnetic*; that is, unlike most matter, it is attracted by a magnetic field; and (b) the compound gives a characteristic *paramagnetic resonance absorption* spectrum (or *electron spin resonance* spectrum, Sec. 16.17), which depends upon the orientation of the spin of an unpaired electron in a changing external magnetic field. This latter property permits the detection not only of stable free radicals but of low concentrations of short-lived intermediates in chemical reactions, and can even give information about their structure. (See, for example, Sec. 8.18.)

The remarkable dissociation to form free radicals is the result of two factors. First, triphenylmethyl radicals are unusually stable because of resonance of the sort we have proposed for the benzyl radical. Here, of course, there are an even larger number of structures (36 of them) that stabilize the radical but not the hydrocarbon; the odd electron is highly delocalized, being distributed over three aromatic rings.

Second, crowding among the large aromatic rings tends to stretch and weaken the carbon–carbon bond joining the triphenylmethyl groups in the dimer. Once the radicals are formed, the bulky groups make it difficult for the carbon atoms to approach each other closely enough for bond formation: *so* difficult, in fact, that hexaphenylethane is not formed at all, but instead dimer I—even with the sacrifice of aromaticity of one ring. Even so, there is crowding in the dimer, and the total effect is to lower the dissociation energy to only 11 kcal/mol, as compared with a dissociation energy of 80–90 kcal for most carbon–carbon single bonds.

It would be hard to overestimate the importance of Gomberg's contribution to the field of free radicals and to organic chemistry as a whole. Although triphenylmethyl was isolable only because it was *not a typical* free radical, its chemical properties showed what kind of behavior to expect of free radicals *in general*; most important of all, it proved that such things as free radicals could exist.

Problem 15.8 The ΔH for dissociation of the dimer I has been measured as 11 kcal/mol, the E_{act} as 19 kcal/mol. (a) Draw the potential energy curve for the reaction. (b) What is the energy of activation for the reverse reaction, combination of triphenylmethyl radicals? (c) How do you account for this unusual fact? (Compare Sec. 2.17.)

Problem 15.9 When 1.5 g of "diphenyltetra(*o*-tolyl)ethane" is dissolved in 50 g of benzene, the freezing point of the solvent is lowered 0.5 °C (the cryoscopic constant for benzene is 5 °C). Interpret these results.

15.17 Stability of the benzyl cation

Now let us turn to heterolytic chemistry, and that key intermediate, the carbocation. The conjugation that stabilizes the allyl free radical, we saw (Sec. 10.12), also stabilizes the allyl cation. Does the same thing hold for the benzyl particles? Is the benzyl cation, like the free radical, unusually stable?

Table 1.3 (p. 22) shows that the heterolytic bond dissociation energy for benzyl chloride is 166 kcal/mol, somewhat less than for allyl chloride (173 kcal) or isopropyl

$$C_6H_5CH_2Cl \longrightarrow C_6H_5CH_2{}^+ + Cl^- \qquad \Delta H = +166 \text{ kcal}$$

Benzyl chloride   Benzyl cation

chloride (170 kcal), and 61 kcal less than for methyl chloride (277 kcal). Comparison of the alkyl bromides or iodides or the alcohols reveals exactly the same pattern. Relative to the substrate from which each cation is generated, the benzyl cation is about as stable as the allyl or isopropyl cation. We can now expand our sequence of Sec. 10.12 to include the benzyl cation.

Stability of carbocations

$$3° > \begin{array}{c} \text{benzyl} \\ \text{allyl} \\ 2° \end{array} > 1° > CH_3{}^+$$

The presence of a phenyl group in place of a hydrogen of methyl chloride thus stabilizes the cation by 61 kcal/mol. As we did for the benzyl free radical, we attribute the stabilization to conjugation with the benzene ring, and account for it on the basis of resonance. Both the benzyl cation and the substrate from which it is made are hybrids of Kekulé structures. In addition, the carbocation can be represented by three other structures, I, II, and III, in which the positive charge is

located on the *ortho* and *para* carbon atoms. Whether considered as resonance stabilization or simply as dispersal of charge, contribution from these structures stabilizes the carbocation.

The orbital picture of the benzyl cation is similar to that of the benzyl free radical (Sec. 15.15) except that the p orbital that overlaps the π cloud is an *empty* one. The p orbital contributes no electrons, but permits further delocalization of the π electrons to include the carbon nucleus of the side chain.

Problem 15.10   How do you account for the following facts? (a) Triphenylchloromethane is completely ionized in certain solvents (e.g., liquid SO_2); (b) triphenylcarbinol, $(C_6H_5)_3COH$, dissolves in concentrated H_2SO_4 to give a solution that has the same intense yellow color as triphenylchloromethane solutions. (*Note*: This yellow color is different from that of solutions of triphenylmethyl.)

Problem 15.11   In light of Problem 15.10, can you suggest a possible reason, besides steric hindrance, why the reaction of CCl_4 with benzene stops at triphenylchloromethane? (See Secs. 15.7–15.8.)

Problem 15.12   Suggest an explanation for the following order of acidity:

triphenylmethane > diphenylmethane > toluene > *n*-pentane

15.18   Nucleophilic substitution in benzylic substrates

How do benzylic substrates behave in nucleophilic aliphatic substitution? Let us begin with substitution of the S_N1 type, in which the rate of reaction depends upon the rate of formation of a carbocation. Although formally primary, a benzyl cation is about as stable as a secondary cation. If our parallel between stability of carbocations and the rate of their formation holds here, we would expect benzyl cations to be formed at about the same rate as secondary cations. And they are. Benzyl substrates undergo S_N1 reactions about as fast as secondary substrates.

The rate of an S_N2 reaction, we have seen (Sec. 5.15), depends chiefly upon steric factors. Here benzyl substrates enjoy, to an extent, the same advantage as an allyl substrate: they are primary, and offer relatively little steric hindrance to nucleophilic attack. And so they undergo S_N2 about as fast as primary substrates.

Substituents on the α-carbon of benzylic substrates have the kind of effects that we would expect. Additional phenyl groups raise the stability of the cation still further, and speed up its formation by S_N1. At the same time, they increase steric hindrance to nucleophilic attack and slow down S_N2. The result is a familiar

$$\text{RX} = \overset{\overleftarrow{\qquad S_N2 \text{ increases} \qquad}}{\underset{\underrightarrow{\qquad S_N1 \text{ increases} \qquad}}{C_6H_5CH_2X \quad (C_6H_5)_2CHX \quad (C_6H_5)_3CX}} \qquad \begin{matrix} S_N2 \\ vs. \\ S_N1 \end{matrix}$$

one (Sec. 5.24): the tendency to undergo a shift in mechanism from bimolecular to unimolecular as branching increases.

Problem 15.13   Predict the order of reactivity for the set of substrates, $C_6H_5CH_2Cl$, $C_6H_5CHClCH_3$, $C_6H_5CCl(CH_3)_2$: (a) by S_N1; (b) by S_N2.

In solvolysis, we saw (Sec. 6.9), the formation of secondary carbocations has need of considerable nucleophilic assistance by the solvent. In addition to the ionizing effect of a solvent cluster, one individual molecule plays a special role, and helps to push out the leaving group. There is formed a carbocation with the solvent molecule clinging to its back side, and it is this *nucleophilically solvated* carbocation that then undergoes further reaction to yield the product.

Now, if a benzyl cation is about as stable as a secondary cation, its formation should have a similar need for nucleophilic assistance by the solvent. But, as we have just seen, a benzyl substrate is less branched than a secondary, and hence should be more open to this nucleophilic attack by the solvent. In agreement with these expectations, the rate of solvolysis of benzylic substrates is found to be particularly sensitive not only to the ionizing power of a solvent, but to its nucleophilic power as well. As a result, it is the study of benzylic substrates that has revealed most about the role of the solvent in nucleophilic substitution.

This brings us to the most important aspect of the chemistry of benzylic compounds: by introducing various substituents into the aromatic ring, we can prepare scores of different benzylic substrates. Substituents at the *meta* or *para* position have no effect on steric hindrance at the benzylic carbon, but can change the polar effect of the aryl group in either direction and to varying degrees. From the *para* position, for example, —OCH_3 exerts powerful electron release, and —NO_2 powerful electron withdrawal; —CH_3 exerts weak electron release, and —X weak electron withdrawal. As we would expect, electron release increases the stability of a benzylic cation, and electron withdrawal decreases its stability. With these changes in cation stability there occur corresponding changes in the rate at which substrates undergo S_N1; and there occur changes in the need for nucleophilic assistance by the solvent.

G *releases electrons*:
stabilizes carbocation,
activates substrate

G *withdraws electrons*:
destabilizes carbocation,
deactivates substrate

The effects of these substituents here parallel their effects on electrophilic aromatic substitution (Sec. 14.16), and for a very good reason: in both kinds of reaction, a positive charge is developing in the aromatic ring; a substituent can either disperse or intensify the charge, and thus either stabilize or destabilize the incipient carbocation.

Problem 15.14 Benzyl bromide reacts with H_2O in formic acid solution to yield benzyl alcohol; the rate is independent of $[H_2O]$. Under the same conditions *p*-methylbenzyl bromide reacts 58 times as fast.

Benzyl bromide reacts with NaOEt in EtOH to yield benzyl ethyl ether; the rate depends upon both [RBr] and [OEt$^-$]. Under the same conditions *p*-methylbenzyl bromide reacts 1.5 times as fast.

Account in detail for these observations.

Problem 15.15 (a) Benzylic tosylates have been found to undergo solvolysis in aqueous acetone at the following relative rates: benzyl, 1.00; *m*-methylbenzyl, 1.8; *p*-methylbenzyl, 30. How do you account for the greater activation by —CH_3 from the *para* position, *farther* from the center of reaction? (*Hint*: See Sec. 14.17.)

(b) Under the same conditions, the following relative rates were measured for bromobenzyl tosylates: benzyl, 1.00; *m*-bromobenzyl, 0.082; *p*-bromobenzyl, 0.41. How do you account for the fact that, unlike methyl, the —Br exerts a greater effect from the *meta* position? (*Hint*: See Sec. 14.19.)

(c) Under the same conditions, the following relative rates were measured for methoxybenzyl tosylates: benzyl, 1.00; *m*-methoxybenzyl, 0.61; *p*-methoxybenzyl, 25 000. How do you account for the fact that —OCH_3 activates powerfully from the *para* position, yet actually deactivates from the *meta* position? (*Hint*: See Sec. 14.18.)

Problem 15.16 Arrange the alcohols of each set in order of reactivity toward aqueous HBr.

(a) 1-phenyl-1-propanol, 3-phenyl-1-propanol, 1-phenyl-2-propanol
(b) benzyl alcohol, *p*-cyanobenzyl alcohol, *p*-hydroxybenzyl alcohol
(c) benzyl alcohol, diphenylmethanol, triphenylmethanol

15.19 Preparation of alkenylbenzenes. Conjugation with the ring

An aromatic hydrocarbon with a side chain containing a double bond can be prepared by essentially the same methods as simple alkenes, that is, by

1,2-elimination (Sec. 7.11). The presence of the aromatic ring in the molecule may affect the orientation of elimination and the ease with which it takes place.

On an industrial scale, the elimination generally involves *dehydrogenation*. For example, **styrene**, the most important of these compounds—and perhaps the most important synthetic aromatic compound—can be prepared by simply heating ethylbenzene to about 600 °C in the presence of a catalyst. The ethylbenzene,

$$\text{C}_6\text{H}_6 + \text{CH}_2{=}\text{CH}_2 \xrightarrow{\text{H}_3\text{PO}_4} \underset{\text{Ethylbenzene}}{\text{C}_6\text{H}_5\text{CH}_2\text{CH}_3} \xrightarrow{\text{Cr}_2\text{O}_3 \cdot \text{Al}_2\text{O}_3,\, 600\,^\circ\text{C}} \underset{\text{Styrene}}{\text{C}_6\text{H}_5\text{CH}{=}\text{CH}_2} + \text{H}_2$$

in turn, is prepared by a Friedel–Crafts reaction between two simple hydrocarbons, benzene and ethylene.

In the laboratory, however, we are most likely to use dehydrohalogenation or dehydration.

$$\underset{\text{1-Phenyl-1-chloroethane}}{\text{C}_6\text{H}_5{-}\underset{\underset{\text{Cl}}{|}}{\text{CH}}{-}\text{CH}_3} \xrightarrow{\text{KOH (alc), heat}} \underset{\text{Styrene}}{\text{C}_6\text{H}_5\text{CH}{=}\text{CH}_2}$$

$$\underset{\text{1-Phenylethanol}}{\text{C}_6\text{H}_5{-}\underset{\underset{\text{OH}}{|}}{\text{CH}}{-}\text{CH}_3} \xrightarrow{\text{acid, heat}} \underset{\text{Styrene}}{\text{C}_6\text{H}_5\text{CH}{=}\text{CH}_2}$$

Dehydrohalogenation of 1-phenyl-2-chloropropane, or dehydration of 1-phenyl-2-propanol, could yield two products: 1-phenylpropene or 3-phenylpropene. Actually, only the first of these products is obtained. We saw earlier (Secs. 7.20, 7.25, and 10.25) that where isomeric alkenes can be formed by such

$$\underset{\text{1-Phenyl-2-chloropropane}}{\text{C}_6\text{H}_5\text{CH}_2\underset{\underset{\text{Cl}}{|}}{\text{CH}}\text{CH}_3} \overset{\text{KOH}}{\underset{\text{heat}}{\xrightarrow{\text{(alc)}}}} \begin{array}{c} \to \underset{\substack{\text{1-Phenylpropene} \\ \textit{Only product}}}{\text{C}_6\text{H}_5\text{CH}{=}\text{CH}{-}\text{CH}_3} \leftarrow \\[8pt] \not\to \underset{\text{3-Phenylpropene}}{\text{C}_6\text{H}_5\text{CH}_2\text{CH}{=}\text{CH}_2} \not\leftarrow \end{array} \overset{\text{acid}}{\underset{\text{heat}}{\xleftarrow{\hspace{1em}}}} \underset{\text{1-Phenyl-2-propanol}}{\text{C}_6\text{H}_5\text{CH}_2\underset{\underset{\text{OH}}{|}}{\text{CH}}\text{CH}_3}$$

elimination, the preferred product is generally the more stable alkene. This is the case here, too. That 1-phenylpropene is much more stable than its isomer is shown by the fact that 3-phenylpropene is rapidly converted into 1-phenylpropene by treatment with hot alkali.

$$\underset{\substack{\text{3-Phenylpropene} \\ \text{(Allylbenzene)}}}{\text{C}_6\text{H}_5\text{CH}_2{-}\text{CH}{=}\text{CH}_2} \xrightarrow{\text{KOH, heat}} \underset{\text{1-Phenylpropene}}{\text{C}_6\text{H}_5\text{CH}{=}\text{CH}{-}\text{CH}_3}$$

A double bond that is separated from a benzene ring by one single bond is said to be *conjugated with the ring*. Such conjugation confers unusual stability on a

Double bond conjugated with ring: *unusually stable system*

molecule. This stability is reflected in a faster rate of formation, which affects not only orientation of elimination, but also the ease with which elimination takes place.

Problem 15.17 Account for the stability of alkenes like styrene on the basis of: (a) delocalization of π electrons, showing both resonance structures and orbital overlap; and (b) change in hybridization.

Problem 15.18 Considering the nature of the reagent, can you suggest a possible mechanism for the conversion of 3-phenylpropene into 1-phenylpropene described above?

15.20 Reactions of alkenylbenzenes

As we might expect, alkenylbenzenes undergo two sets of reactions: **substitution in the ring**, and **addition to the double bond in the side chain**. Since both ring and double bond are good sources of electrons, there may be competition between the two sites for certain electrophilic reagents; it is not surprising that, in general, the double bond shows higher reactivity than the resonance-stabilized benzene ring. Our main interest in these reactions will be the way in which the aromatic ring affects the reactions of the double bond.

Although both the benzene ring and the carbon–carbon double bond can be hydrogenated catalytically, the conditions required for the double bond are much

milder; by proper selection of conditions it is quite easy to hydrogenate the side chain without touching the aromatic ring.

Mild oxidation of the double bond yields a 1,2-diol; more vigorous oxidation cleaves the carbon–carbon double bond and generally gives a carboxylic acid in which the —COOH group is attached directly to the ring.

Both double bond and ring react with halogens by heterolytic mechanisms that have essentially the same first step: attack on the π cloud by positively charged halogen. Halogen is consumed by the double bond first, and only after the side chain is completely saturated does substitution on the ring occur. Ring-halogenated alkenylbenzenes must be prepared, therefore, by generation of the double bond after halogen is already present on the ring. For example:

$$\underset{\text{(ring with } C_2H_5)}{C_6H_5C_2H_5} \xrightarrow{\text{Cl}_2,\ \text{FeCl}_3} \underset{\text{(ring with } C_2H_5,\ Cl)}{} \xrightarrow{\text{Cl}_2,\ \text{heat}} \underset{\text{(ring with CHClCH}_3,\ Cl)}{} \xrightarrow{\text{KOH}} \underset{\text{(ring with CH=CH}_2,\ Cl)}{}$$

p-Chlorostyrene

In a similar way, alkenylbenzenes undergo the other addition reactions characteristic of the carbon–carbon double bond. Let us look further at the reactions of *conjugated* alkenylbenzenes, and the way in which the ring affects *orientation* and *reactivity*.

15.21 Addition to conjugated alkenylbenzenes

Addition of an unsymmetrical reagent to a carbon–carbon double bond may in general yield two different products, that is to say, may take place with either of two different orientations. In conjugated alkenylbenzenes, the aromatic ring is attached to one of the doubly bonded carbons, and determines what this orientation will be.

This effect can be well illustrated by a single example, addition of HBr to 1-phenylpropene. In the absence of peroxides, bromine becomes attached to the carbon adjacent to the ring; in the presence of peroxides, bromine becomes attached to the carbon once removed from the ring. According to the mechanisms proposed for these two reactions, these products are formed as follows:

$$C_6H_5CH{=}CHCH_3 \xrightarrow{\text{HBr}} \underset{\oplus}{C_6H_5CHCH_2CH_3} \xrightarrow{\text{Br}^-} \underset{\overset{|}{Br}}{C_6H_5CHCH_2CH_3} \qquad \begin{array}{l}\text{No}\\ \text{peroxides}\end{array}$$

A benzyl cation

$$C_6H_5CH{=}CHCH_3 \xrightarrow{\text{Br}\cdot} \underset{\overset{|}{Br}}{C_6H_5CHCHCH_3} \xrightarrow{\text{HBr}} \underset{\overset{|}{Br}}{C_6H_5CH_2CHCH_3} \qquad \begin{array}{l}\text{Peroxides}\\ \text{present}\end{array}$$

A benzyl free radical

The first step of each of these reactions takes place in the way that yields the *benzyl* cation or the *benzyl* free radical rather than the alternative secondary cation or secondary free radical. Once more, we see, *the first step of addition takes place in the way that yields the more stable particle, carbocation or free radical.* The same fundamental factor, conjugation with the aromatic ring, which determines orientation in the formation of alkenylbenzenes, also determines orientation in their reactions.

Now, on the basis of the greater stability of the benzylic particle being formed, we might expect addition to a conjugated alkenylbenzene to occur faster than addition to a simple alkene.

On the other hand, we have seen (Sec. 15.19) that conjugated alkenylbenzenes are more stable than simple alkenes. On this basis alone, we might expect addition to conjugated alkenylbenzenes to occur more slowly than to simple alkenes.

The situation is exactly analogous to the one discussed for addition to conjugated dienes (Sec. 10.29). Both *reactant* and *transition state* are stabilized by resonance; whether reaction is faster or slower than for simple alkenes depends upon *which* is stabilized *more* (see Fig. 10.8, p. 407).

The fact is that conjugated alkenylbenzenes are much more reactive than simple alkenes toward both ionic and free-radical addition. Here again—as in *most* cases of this sort—resonance stabilization of the transition state leading to a carbocation or free radical is more important than resonance stabilization of the reactant. We must realize, however, that this is *not always* true.

Problem 15.19 Draw a potential energy diagram similar to Fig. 10.8 (p. 407) to summarize what has been said in this section.

Problem 15.20 Suggest one reason why tetraphenylethylene does not react with bromine in carbon tetrachloride.

15.22 Alkynylbenzenes

The preparations and properties of the alkynylbenzenes are just what we might expect from our knowledge of benzene and the alkynes.

Problem 15.21 Outline all steps in the conversion of: (a) ethylbenzene into phenylacetylene; (b) *trans*-1-phenylpropene into *cis*-1-phenylpropene.

15.23 Analysis of arenes

Aromatic hydrocarbons with saturated side chains are distinguished from alkenes by their failure to decolorize bromine in carbon tetrachloride (without evolution of hydrogen bromide) and by their failure to decolorize cold, dilute, neutral permanganate solutions. (Oxidation of the side chains requires more vigorous conditions; see Sec. 15.11.)

They are distinguished from alkanes by the readiness with which they are sulfonated by—and thus dissolve in—cold fuming sulfuric acid (Sec. 14.4).

They are distinguished from alcohols and other oxygen-containing compounds by their failure to dissolve immediately in cold concentrated sulfuric acid, and from primary and secondary alcohols by their failure to give a positive chromic anhydride test (Sec. 8.24).

Upon treatment with chloroform and aluminum chloride, alkylbenzenes give orange to red colors. These colors are due to triarylmethyl cations, Ar_3C^+, which are probably produced by a Friedel–Crafts reaction followed by a transfer of hydride ion (Sec. 5.23):

$$ArH \xrightarrow{CHCl_3, \ AlCl_3} ArCHCl_2 \xrightarrow{ArH, \ AlCl_3} Ar_2CHCl \xrightarrow{ArH, \ AlCl_3} Ar_3CH$$

$$Ar_2CHCl \xrightarrow{AlCl_3} Ar_2CH^+AlCl_4^-$$

$$\left. \begin{array}{c} Ar_2CH^+AlCl_4^- \\ Ar_3CH \end{array} \right\} \longrightarrow Ar_2CH_2 + Ar_3C^+AlCl_4^-$$

Orange to red color

This test is given by any aromatic compound that can undergo the Friedel–Crafts reaction, with the particular color produced being characteristic of the aromatic system involved: orange to red from halobenzenes, blue from *naphthalene*, purple from *phenanthrene*, green from *anthracene* (Chap. 34).

Problem 15.22 Describe simple chemical tests (if any) that would distinguish between: (a) *n*-propylbenzene and *o*-chlorotoluene; (b) benzene and toluene; (c) *m*-chlorotoluene and *m*-dichlorobenzene; (d) bromobenzene and bromocyclohexane; (e) bromobenzene and 3-bromo-1-hexene; (f) ethylbenzene and benzyl alcohol ($C_6H_5CH_2OH$). Tell exactly what you would *do* and *see*.

The number and orientation of side chains in an alkylbenzene is shown by the carboxylic acid produced upon vigorous oxidation (Sec. 15.11).

Problem 15.23 On the basis of characterization tests and physical properties, an unknown compound of b.p. 182 °C is believed to be either *m*-diethylbenzene or *n*-butylbenzene. How could you distinguish between the two possibilities?

Aromatic hydrocarbons with unsaturated side chains undergo the reactions characteristic of aromatic rings and of the carbon–carbon double or triple bond.

Problem 15.24 Predict the response of allylbenzene to the following test reagents: (a) cold concentrated sulfuric acid; (b) Br_2 in CCl_4; (c) cold, dilute, neutral permanganate; (d) $CHCl_3$ and $AlCl_3$; (e) CrO_3 and H_2SO_4.

Problem 15.25 Describe simple chemical tests (if any) that would distinguish between: (a) styrene and ethylbenzene; (b) styrene and phenylacetylene; (c) allylbenzene and 1-nonene; (d) allylbenzene and allyl alcohol (CH_2=CH—CH_2OH). Tell exactly what you would *do* and *see*.

(Analysis of arenes by spectroscopic methods will be discussed in Chapter 16, especially Secs. 16.18–16.19.)

PROBLEMS

1. Draw the structure of:

(a) *m*-xylene
(b) mesitylene
(c) *o*-ethyltoluene
(d) *p*-di-*tert*-butylbenzene
(e) cyclohexylbenzene
(f) 3-phenylpentane

(g) isopropylbenzene (cumene)
(h) (*Z*)-1,2-diphenylethene
(i) 1,4-diphenyl-1,3-butadiene
(j) *p*-dibenzylbenzene
(k) *m*-bromostyrene
(l) diphenylacetylene

2. Outline all steps in the synthesis of ethylbenzene from each of the following compounds, using any needed aliphatic or inorganic reagents.

(a) benzene
(b) styrene
(c) phenylacetylene
(d) α-phenylethyl alcohol
(e) β-phenylethyl alcohol

(f) 1-chloro-1-phenylethane
(g) 2-chloro-1-phenylethane
(h) *p*-bromoethylbenzene
(i) acetophenone ($C_6H_5\overset{\text{O}}{\underset{||}{C}}CH_3$)

3. Give structures and names of the principal organic products expected from reaction (if any) of *n*-propylbenzene with each of the following. Where more than one product is to be expected, indicate which will predominate.

(a) H_2, Ni, room temperature, low pressure
(b) H_2, Ni, 200 °C, 100 atm
(c) cold dilute $KMnO_4$
(d) hot $KMnO_4$
(e) $K_2Cr_2O_7$, H_2SO_4, heat
(f) boiling NaOH(aq)
(g) boiling HCl(aq)
(h) HNO_3, H_2SO_4
(i) H_2SO_4; SO_3
(j) Cl_2, Fe

(k) Br_2, Fe
(l) I_2, Fe
(m) Br_2, heat, light
(n) CH_3Cl, $AlCl_3$, 0 °C **F.C.**
(o) $C_6H_5CH_2Cl$, $AlCl_3$, 0 °C **F.C**
(p) C_6H_5Cl, $AlCl_3$, 80 °C
(q) isobutylene, HF
(r) *tert*-butyl alcohol, H_2SO_4
(s) cyclohexene, HF

4. Give structures and names of the principal organic products expected from reaction (if any) of *trans*-1-phenyl-1-propene with:

(a) H_2, Ni, room temperature, low pressure
(b) H_2, Ni, 200 °C, 100 atm
(c) Br_2 in CCl_4
(d) excess Br_2, Fe
(e) HCl
(f) HBr
(g) HBr (peroxides)
(h) cold conc. H_2SO_4

(i) Br_2, H_2O
(j) cold dilute $KMnO_4$
(k) hot $KMnO_4$
(l) HCO_2OH
(m) O_3, then H_2O/Zn
(n) Br_2, 300 °C
(o) $CHBr_3$, *t*-BuOK
(p) product (c), KOH(alc)

5. Give structures and names of the principal organic products expected from each of the following reactions:

(a) benzene + cyclohexene + HF
(b) phenylacetylene + alcoholic $AgNO_3$
(c) *m*-nitrobenzyl chloride + $K_2Cr_2O_7$ + H_2SO_4 + heat
(d) allylbenzene + HCl
(e) *p*-chlorotoluene + hot $KMnO_4$
(f) *eugenol* ($C_{10}H_{12}O_2$, 2-methoxy-4-allylphenol) + hot KOH $\longrightarrow$ *isoeugenol* ($C_{10}H_{12}O_2$)
(g) benzyl chloride + Mg + dry ether
(h) product of (g) + D_2O
(i) *p*-xylene + Br_2 + Fe
(j) 1-phenyl-1,3-butadiene + 1 mol H_2 + Ni, 2 atm, 30 °C
(k) *cis*-1,2-diphenylethene + O_3, then H_2O/Zn
(l) 1,3-diphenylpropyne + H_2, Pd $\longrightarrow C_{15}H_{14}$
(m) 1,3-diphenylpropyne + Li, NH_3(liq) $\longrightarrow C_{15}H_{14}$
(n) *p*-$CH_3OC_6H_4CH{=}CHC_6H_5$ + HBr

6. Label each set of hydrogens in each of the following compounds in order of expected ease of abstraction by bromine atoms. Use (1) for the most reactive, (2) for the next, etc.

(a) 1-phenyl-2-hexene

(b) CH_3⟨◯⟩CH_2⟨◯⟩$CH_2CH_2CH_3$

(c) 1,2,4-trimethylbenzene (*Hint*: See Problem 15.7, p. 552.)
(d) What final monobromination product or products would abstraction of each kind of hydrogen in (a) lead to?

7. Give structures and names of the products expected from dehydrohalogenation of each of the following. Where more than one product can be formed, predict the major product.

(a) 1-chloro-1-phenylbutane
(b) 1-chloro-2-phenylbutane
(c) 2-chloro-2-phenylbutane

(d) 2-chloro-1-phenylbutane
(e) 3-chloro-2-phenylbutane

8. Answer Problem 7 for dehydration of the alcohol corresponding to each of the halides given. (*Hint*: Do not forget Sec. 5.23.)

9. Arrange in order of ease of dehydration:

(a) the alcohols of Problem 8
(b) $C_6H_5CH_2CH_2OH$, $C_6H_5CHOHCH_3$, $(C_6H_5)_2C(OH)CH_3$
(c) α-phenylethyl alcohol, α-(*p*-bromophenyl)ethyl alcohol, α-(*p*-tolyl)ethyl alcohol

10. Arrange the compounds of each set in order of reactivity toward the indicated reaction.

(a) *addition of HCl:* styrene, *p*-chlorostyrene, *p*-methylstyrene
(b) *dehydration:* α-phenylethyl alcohol, α-(*p*-nitrophenyl)ethyl alcohol, α-(*p*-amino-phenyl)ethyl alcohol
(c) S_N1 *solvolysis:* benzyl chloride, *p*-chlorobenzyl chloride, *p*-methoxybenzyl chloride, *p*-methylbenzyl chloride, *p*-nitrobenzyl chloride
(d) S_N1 *solvolysis:* benzyl bromide, α-phenylethyl bromide, β-phenylethyl bromide
(e) *elimination by KOH(alc):* 1-phenyl-2-bromopropane, 1-phenyl-3-bromopropane

11. (a) Draw structures of all possible products of addition of one mole of Br_2 to 1-phenyl-1,3-butadiene. (b) Which of these possible products are consistent with the intermediate formation of the most stable carbocation? (c) Actually, only 1-phenyl-3,4-dibromo-1-butene is obtained. What is the most likely explanation of this fact?

12. (a) The heats of hydrogenation of the stereoisomeric stilbenes (1,2-diphenylethenes) are: *cis*-, 26.3 kcal; *trans*-, 20.6 kcal. Which isomer is the more stable? (b) *cis*-Stilbene is converted into *trans*-stilbene (but not vice versa) either (i) by action of a very small amount of Br_2 in the presence of light, or (ii) by action of a very small amount of HBr (but not HCl) in the presence of peroxides. What is the agent that probably brings about the conversion? Can you suggest a way in which the conversion might take place? (c) Why is *trans*-stilbene not converted into *cis*-stilbene?

13. One mole of triphenylmethanol lowers the freezing point of 1000 g of 100% sulfuric acid twice as much as one mole of methanol. How do you account for this?

14. When a mixture of toluene and $CBrCl_3$ was irradiated with ultraviolet light, there were obtained, in almost exactly equimolar amounts, benzyl bromide and $CHCl_3$. (a) Show in detail all steps in the most likely mechanism for this reaction. (b) There were also obtained, in small amounts, HBr and C_2Cl_6; the ratio of $CHCl_3$ to HBr was 20:1. How do you account for the formation of HBr? Of C_2Cl_6? What, specifically, does the 20:1 ratio tell you about the reaction?

15. When the product of the HF-catalyzed reaction of benzene with 1-dodecene, previously reported to be pure 2-phenyldodecane, was analyzed by gas chromatography, five evenly spaced peaks of about the same size were observed, indicating the presence of five components, probably closely related in structure. What five compounds most likely make up this mixture, and how could you have anticipated their formation?

16. Upon addition of bromine, *cis*-1-phenyl-1-propene gives a mixture of 17% *erythro* dibromide and 83% *threo*; *trans*-1-phenyl-1-propene gives 88% *erythro*, 12% *threo*; and *trans*-1-(*p*-methoxyphenyl)propene gives 63% *erythro*, 37% *threo*.

How do these results compare with those obtained with the 2-butenes (Sec. 9.5)? Suggest a possible explanation for the difference. What is the effect of the *p*-methoxy group, and how might you account for this?

17. The three xylenes are obtained as a mixture from the distillation of coal tar;

further separation by distillation is difficult because of the closeness of their boiling points (see Table 15.1, p. 535), and so a variety of chemical methods have been used. In each case below tell which isomer you would expect to react preferentially, and why.

(a) An old method: treatment of the mixture at room temperature with 80% sulfuric acid.

(b) Another old method: sulfonation of all three xylenes, and then treatment of the sulfonic acids with dilute aqueous acid.

(c) A current method: extraction of one isomer into a BF_3/HF layer.

(d) A proposed method:

$$C_6H_5C(CH_3)_2{}^- Na^+ + xylenes \xrightleftharpoons{} C_6H_5CH(CH_3)_2 + X + two\ xylenes$$

(*Hint* to part (d): See Secs. 11.11, 5.21, and 14.16.)

18. Outline all steps in a possible laboratory synthesis of each of the following compounds from benzene and/or toluene, using any necessary aliphatic or inorganic reagents. Follow instructions on p. 221. Assume a pure *para* isomer can be separated from an *ortho,para* mixture.

(a) ethylbenzene
(b) styrene
(c) phenylacetylene
(d) isopropylbenzene
(e) 2-phenylpropene
(f) 3-phenylpropene (allylbenzene)
(g) 1-phenylpropyne (two ways)
(h) (*E*)-1-phenylpropene
(i) (*Z*)-1-phenylpropene

(j) *p-tert*-butyltoluene
(k) *p*-nitrostyrene
(l) *p*-bromobenzyl bromide
(m) *p*-nitrobenzal bromide
(n) *p*-bromobenzoic acid
(o) *m*-bromobenzoic acid
(p) 1,2-diphenylethane
(q) *p*-nitrodiphenylmethane
        (p-$O_2NC_6H_4CH_2C_6H_5$)

19. Describe simple chemical tests that would distinguish between:

(a) benzene and cyclohexane
(b) benzene and 1-hexene
(c) toluene and *n*-heptane
(d) cyclohexylbenzene and 1-phenylcyclohexene **Bromine**
(e) benzyl alcohol ($C_6H_5CH_2OH$) and *n*-pentylbenzene
(f) cinnamyl alcohol ($C_6H_5CH=CHCH_2OH$) and 3-phenyl-1-propanol
       ($C_6H_5CH_2CH_2CH_2OH$)
(g) chlorobenzene and ethylbenzene **Zinc nitrate — sodium fusion**
(h) nitrobenzene and *m*-dibromobenzene

Tell exactly what you would *do* and *see*.

20. Describe chemical methods (not necessarily simple tests) that would enable you to distinguish between the compounds of each of the following sets. (For example, make use of Table 23.1, page 819.)

(a) 1-phenylpropene, 2-phenylpropene, 3-phenylpropene (allylbenzene)
(b) all alkylbenzenes of formula C_9H_{12}
(c) *m*-chlorotoluene and benzyl chloride
(d) *p*-divinylbenzene (p-$C_6H_4(CH=CH_2)_2$) and 1-phenyl-1,3-butadiene
(e) $C_6H_5CHClCH_3$, p-$CH_3C_6H_4CH_2Cl$, and p-$ClC_6H_4C_2H_5$

21. An unknown compound is believed to be one of the following. Describe how you would go about finding out which of the possibilities the unknown actually is. Where possible, use simple chemical tests; where necessary, use more elaborate chemical methods like quantitative hydrogenation, cleavage, etc. (Where necessary, make use of Table 23.1, page 819.)

	b.p., °C		b.p., °C
bromobenzene	156	*p*-chlorotoluene	162
3-phenylpropene	157	*o*-ethyltoluene	162
m-ethyltoluene	158	*p*-ethyltoluene	163
n-propylbenzene	159	mesitylene	165
o-chlorotoluene	159	2-phenylpropene	165
m-chlorotoluene	162		

22. The compound *indene*, C_9H_8, found in coal tar, rapidly decolorizes Br_2/CCl_4 and dilute $KMnO_4$. Only one mole of hydrogen is absorbed readily to form *indane*, C_9H_{10}. More vigorous hydrogenation yields a compound of formula C_9H_{16}. Vigorous oxidation of indene yields phthalic acid. What is the structure of indene? Of indane? (*Hint*: See Problem 12.20, p. 470.)

23. A solution of 0.01 mole *tert*-butyl peroxide (p. 122) in excess ethylbenzene was irradiated with ultraviolet light for several hours. Gas chromatographic analysis of the product showed the presence of nearly 0.02 mole of *tert*-butyl alcohol. Evaporation of the alcohol and unreacted ethylbenzene left a solid residue which was separated by chromatography into just two products: A (1 g) and B (1 g). A and B each had the empirical formula C_8H_9 and mol. wt. 210; each was inert toward cold dilute $KMnO_4$ and toward Br_2/CCl_4.

When isopropylbenzene was substituted for ethylbenzene in the above reaction, exactly similar results were obtained, except that the single compound C (2.2 g) was obtained instead of A and B. C had the empirical formula C_9H_{11}, mol. wt. 238, and was inert toward cold, dilute $KMnO_4$ and toward Br_2/CCl_4.

What are the most likely structures for A, B, and C, and what is the most likely mechanism by which they are formed?

24. The bond dissociation energy for the central C—C bond of hexacyclopropylethane is only 45 kcal/mol. Besides steric interaction, what is a second factor that may contribute to the weakness of this bond? (*Hint*: See Sec. 12.9.)

16

Spectroscopy and Structure

16.1 Determination of structure: spectroscopic methods

Near the beginning of our study (Sec. 3.32), we outlined the general steps an organic chemist takes when he is confronted with an unknown compound and sets out to find the answer to the question: *what is it?* We have seen, in more detail, some of the ways in which he carries out the various steps: determination of molecular weight and molecular formula; detection of the presence—or absence—of certain functional groups; degradation to simpler compounds; conversion into derivatives; synthesis by an unambiguous route.

At every stage of structure determination—from the isolation and purification of the unknown substance to its final comparison with an authentic sample—the use of instruments has, since World War II, revolutionized organic chemical practice. Instruments not only help an organic chemist to do what he does *faster* but, more important, let him do what could not be done *at all* before: to analyze complicated mixtures of closely related compounds; to describe the structure of molecules in detail never imagined before; to detect, identify, and measure the concentration of short-lived intermediates whose very existence was, not so long ago, only speculation.

By now, we are familiar with some of the features of the organic chemical landscape; so long as we do not wander too far from home, we can find our way about without becoming lost. We are ready to learn a little about how to interpret the kind of information these modern instruments give, so that they can help us to see more clearly the new things we shall meet, and to recognize them more readily when we encounter them again. The instruments most directly concerned with our primary interest, molecular structure, are the *spectrometers*—measurers of spectra. Of the various spectra, we shall actually work with only two kinds: *infrared* (*IR*) and *nuclear magnetic resonance* (*proton NMR* and *CMR*), since they are the workhorses of the organic chemical laboratory today; of these, we shall spend most of our time with nuclear magnetic resonance. We shall look very briefly at three other kinds of spectra: *mass, ultraviolet* (*UV*), and *electron spin resonance* (*ESR*).

In all this, we must constantly keep in mind that what we learn at this stage must be greatly simplified. There are many exceptions to the generalizations we shall learn; there are many pitfalls into which we can stumble. Our ability to apply spectroscopic methods to the determination of organic structure is limited by our understanding of organic chemistry as a whole—and in this we are, of course, only beginners. But so long as we are aware of the dangers of a little learning, and are willing to make mistakes and profit from them, it is worthwhile for us to become beginners in this area of organic chemistry, too.

Let us look first at the mass spectrum, and then at the others, which, as we shall see, are all parts—different ranges of wavelengths—of a single spectrum: that of electromagnetic radiation.

16.2   The mass spectrum

In the mass spectrometer, molecules are bombarded with a beam of energetic electrons. The molecules are ionized and broken up into many fragments, some of which are positive ions. Each kind of ion has a particular ratio of mass to charge, or *m/e value*. For most ions, the charge is 1, so that *m/e* is simply the mass of the ion. Thus, for neopentane:

$$CH_3-\overset{\overset{\displaystyle CH_3}{|}}{\underset{\underset{\displaystyle CH_3}{|}}{C}}-CH_3$$

$$\downarrow e^-$$

$$2e^- + (C_5H_{12})^{\ddagger} \ m/e = 72$$

Molecular ion

	$(C_4H_9)^+$	$(C_3H_5)^+$	$(C_2H_5)^+$	$(C_2H_3)^+$ and others
m/e:	57	41	29	27
Relative intensity:	100	41.5	38.5	15.7
	Base peak			

The set of ions is analyzed in such a way that a signal is obtained for each value of *m/e* that is represented; the intensity of each signal reflects the relative abundance of the ion producing the signal. The largest peak is called the *base peak*; its intensity is taken as 100, and the intensities of the other peaks are expressed relative to it. A plot—or even a list—showing the relative intensities of signals at the various *m/e* values is called a *mass spectrum*, and is highly characteristic of a particular compound. Compare, for example, the spectra of two isomers shown in Fig. 16.1.

Figure 16.1 Mass spectra of two isomeric alkanes. (*a*) *n*-Octane; (*b*) 2,2,4-trimethylpentane.

Mass spectra can be used in two general ways: (a) to prove the identity of two compounds, and (b) to help establish the structure of a new compound.

Two compounds are shown to be identical by the fact that they have identical physical properties: melting point, boiling point, density, refractive index, etc. The greater the number of physical properties measured, the stronger the evidence. Now, a single mass spectrum amounts to dozens of physical properties, since it shows the relative abundances of dozens of different fragments. If we measure the mass spectrum of an unknown compound and find it to be identical with the spectrum of a previously reported compound of known structure, then we can conclude that—almost beyond the shadow of doubt—the two compounds are identical.

The mass spectrum helps to establish the structure of a *new* compound in several different ways: it can give an exact molecular weight; it can give a molecular formula—or at least narrow the possibilities to a very few; and it can indicate the presence in a molecule of certain structural units.

If one electron is removed from the parent molecule, there is produced M^+, the *molecular ion* (or *parent ion*), whose m/e value is, of course, the molecular weight of the compound. Sometimes the M^+ peak is the base peak, and is easily recognized; often, though, it is not the base peak—it may even be very small—and considerable work is required to locate it. Once identified, it gives the most accurate molecular weight obtainable.

$$M + e^- \longrightarrow M^+ + 2e^-$$

Molecular ion

(Parent ion)

m/e = mol. wt.

We might at first think that the M^+ peak would be the peak of highest m/e value. This is not so, however. Most elements occur naturally as several isotopes; generally the lightest one greatly predominates, and the heavier ones occur to lesser extent. Table 16.1 lists the relative abundances of several heavy isotopes.

Table 16.1   ABUNDANCE OF SOME HEAVY ISOTOPES

Heavy isotope	Abundance relative to isotope of lowest atomic weight
^2H	0.015%
^{13}C	1.11
^{15}N	0.37
^{18}O	0.20
^{33}S	0.78
^{34}S	4.4
^{37}Cl	32.5
^{81}Br	98.0

The molecular weight that one usually measures and works with is the sum of the average atomic weights of the elements, and reflects the presence of these heavy isotopes. But this is not true of the molecular weight obtained from the mass spectrum; here, the M^+ peak is due to molecules containing only the commonest isotope of each element.

Consider benzene, for example. The M^+ peak, m/e 78, is due only to ions of formula $C_6H_6{}^+$. There is a peak at m/e 79, the $M + 1$ peak, which is due to $C_5{}^{13}CH_6{}^+$ and $C_6H_5D^+$. There is an $M + 2$ peak at m/e 80, due to $C_4{}^{13}C_2H_6{}^+$, $C_5{}^{13}CH_5D^+$, and $C_6H_4D_2{}^+$. Now, because of the low natural abundance of most heavy isotopes, these isotopic peaks are generally much less intense than the M^+ peak; just how much less intense depends upon which elements they are due to. In the case of benzene, the $M + 1$ and $M + 2$ peaks are, respectively, 6.75% and 0.18% as intense as the M^+ peak. (Table 16.1 shows us, however, that a monochloro compound would have an $M + 2$ peak about one-third as intense as the M^+ peak, and a monobromo compound would have M and $M + 2$ peaks of about equal intensity.)

It is these isotopic peaks that make it possible for us to determine the molecular formula of the compound. Knowing the relative natural abundances of isotopes, one can calculate for any molecular formula the relative intensity to be expected for each isotopic peak: M + 1, M + 2, etc. The results of such calculations are available in tables. Consider, for example, a compound for which M^+ is 44. The compound might be (among other less likely possibilities) N_2O, CO_2, C_2H_4O, or C_3H_8. By use of Table 16.2, we clearly could pick out the most likely formula from the mass spectral data.

Table 16.2   CALCULATED INTENSITIES OF ISOTOPIC PEAKS

	M	M + 1	M + 2
N_2O	100	0.80	0.20
CO_2	100	1.16	0.40
C_2H_4O	100	1.91	0.01
C_3H_8	100	3.37	0.04

Finally, study of compounds of known structure is beginning to reveal the factors that determine which fragments a particular structure is likely to break into. In this we can find much that is familiar to us: the preferential formation of carbocations that we recognize as being relatively stable ones; elimination of small, stable molecules like water, ammonia, and carbon monoxide. Under the energetic conditions, extensive rearrangement can occur, complicating the interpretation; but here, too, patterns are emerging. The *direction* of rearrangement is, as we would expect, toward more stable ions. As this knowledge accumulates, the process is reversed: from the kind of fragmentation an unknown compound gives, its structure is deduced.

Problem 16.1   (a) Referring to the neopentane fragmentation (p. 570), what is a likely structure for $C_4H_9{}^+$; $C_3H_5{}^+$; $C_2H_5{}^+$; $C_2H_3{}^+$? (b) Write a balanced equation for the formation of $C_4H_9{}^+$ from the molecular ion $C_5H_{12}{}^+$.

16.3   The electromagnetic spectrum

We are already familiar with various kinds of electromagnetic radiation: light (visible, ultraviolet, infrared), x-rays, radio and radar waves. These are simply different parts of a broad spectrum that stretches from gamma rays, whose wavelengths are measured in fractions of an angstrom, to radio waves, whose wavelengths are measured in meters or even kilometers. All these waves have the same velocity, 3×10^{10} centimeters per second. Their frequency is related to the wavelength by the expression

$$v = c/\lambda$$

where

v = frequency, in Hz (*hertz*, cycles/sec)
λ = wavelength, in cm
c = velocity, 3×10^{10} cm/sec

The shorter the wavelength, the higher the frequency.

When a beam of electromagnetic radiation is passed through a substance, the radiation can be either absorbed or transmitted, depending upon its frequency and the structure of the molecules it encounters. Electromagnetic radiation is energy, and hence when a molecule absorbs radiation, it gains energy. Just how much energy it gains depends upon the frequency of the radiation: the higher the frequency (the shorter the wavelength), the greater the gain in energy.

$$\Delta E = h\nu$$

where
ΔE = gain in energy, in ergs
h = Planck's constant, 6.5×10^{-27} erg sec
ν = frequency, in Hz

The energy gained by the molecule in this way may bring about increased vibration or rotation of the atoms, or may raise electrons to higher energy levels. The particular frequency of radiation that a given molecule can absorb depends upon the changes in vibrations or rotations or electronic states that are permitted to a molecule of that structure. The spectrum of a compound is a plot that shows how much electromagnetic radiation is absorbed (or transmitted) at each frequency. It can be highly characteristic of the compound's structure.

16.4 The infrared spectrum

Of all the properties of an organic compound, the one that, by itself, gives the most information about the compound's structure is its infrared spectrum.

A molecule is constantly vibrating: its bonds *stretch* (and contract), and *bend* with respect to each other. Changes in vibrations of a molecule are caused by absorption of infrared light: light lying beyond (lower frequency, longer wavelength, less energy) the red of the visible spectrum.

A particular part of the infrared spectrum is referred to either by its wavelength or—and this is considered preferable—by its frequency. Wavelength is expressed in microns, μ ($1\ \mu = 10^{-4}$ cm or 10^4 Å). Frequency is expressed, not in hertz, but in *wavenumbers*, cm^{-1}, often called *reciprocal centimeters*; the wavenumber is simply the number of waves per centimeter, and is equal to the reciprocal of the wavelength in centimeters.

Like the mass spectrum, an infrared spectrum is a highly characteristic property of an organic compound—see, for example, the spectra in Fig. 16.2—and can be used both to establish the identity of two compounds and to reveal the structure of a new compound.

Two substances that have identical infrared spectra are, in effect, identical in thousands of different physical properties—the absorption of light at thousands of different frequencies—and must almost certainly be the same compound. (One region of the infrared spectrum is called, appropriately, the *fingerprint* region.)

The infrared spectrum helps to reveal the structure of a new compound by telling us what groups are present in—or absent from—the molecule. A particular group of atoms gives rise to *characteristic absorption bands*; that is to say, a particular group absorbs light of certain frequencies that are much the same from compound to compound. For example, the —OH group of alcohols absorbs strongly at 3200–3600 cm^{-1}; the C=O group of ketones at 1710 cm^{-1}; the —C≡N group at 2250 cm^{-1}; the —CH$_3$ group at 1450 and 1375 cm^{-1}.

Figure 16.2   Infrared spectra. (*a*) 1-Octene; (*b*) isopropyl bromide; (*c*) *n*-butylbenzene.

Interpretation of an infrared spectrum is not a simple matter. Bands may be obscured by the overlapping of other bands. Overtones (harmonics) may appear at just twice the frequency of the fundamental band. The absorption band of a particular group may be *shifted* by various structural features—conjugation, electron withdrawal by a neighboring substituent, angle strain or van der Waals strain, hydrogen bonding—and be mistaken for a band of an entirely different group. (On the other hand, recognized for what they are, such shifts *reveal* the structural features that cause them.)

In our work we shall have modest aims: to learn to recognize a few of the more striking absorption bands, and to gain a little practice in correlating infrared data with other kinds of information. We must realize that we shall be taking from an infrared spectrum only a tiny fraction of the information that is there, and which can be gotten from it by an experienced person with a broad understanding of organic structure.

Table 16.3 lists infrared absorption frequencies characteristic of various groups. We shall look more closely at the infrared spectra of hydrocarbons in Sec. 16.18 and, in following chapters, at the infrared spectra of other families of compounds.

Table 16.3   CHARACTERISTIC INFRARED ABSORPTION FREQUENCIES[a]

Bond	Compound type	Frequency range, cm^{-1}	Reference
C—H	Alkanes	2850–2960 1350–1470	Sec. 16.18
C—H	Alkenes	3020–3080 (*m*) 675–1000	Sec. 16.18
C—H	Aromatic rings	3000–3100 (*m*) 675–870	Sec. 16.18
C—H	Alkynes	3300	Sec. 16.18
C=C	Alkenes	1640–1680 (*v*)	Sec. 16.18
C≡C	Alkynes	2100–2260 (*v*)	Sec. 16.18
C⋯C	Aromatic rings	1500, 1600 (*v*)	Sec. 16.18
C—O	Alcohols, ethers, carboxylic acids, esters	1080–1300	Sec. 18.11 Sec. 19.18 Sec. 23.22 Sec. 24.25
C=O	Aldehydes, ketones, carboxylic acids, esters	1690–1760	Sec. 21.16 Sec. 23.22 Sec. 24.25
O—H	Monomeric alcohols, phenols	3610–3640 (*v*)	Sec. 18.11 Sec. 28.14
	Hydrogen-bonded alcohols, phenols	3200–3600 (*broad*)	Sec. 18.11 Sec. 28.14
	Carboxylic acids	2500–3000 (*broad*)	Sec. 23.22
N—H	Amines	3300–3500 (*m*)	Sec. 27.21
C—N	Amines	1180–1360	Sec. 27.21
C≡N	Nitriles	2210–2260 (*v*)	
—NO$_2$	Nitro compounds	1515–1560 1345–1385	

[a] All bands strong unless marked: *m*, moderate; *v*, variable.

16.5   The ultraviolet spectrum

Light of wavelength between about 400 nm and 750 nm is visible. (A *nanometer*, nm, is 10^{-7} cm, and equals one mμ.) Just beyond the red end of the visible spectrum (λ greater than 750 nm) lies the infrared region. Just beyond the violet end of the visible spectrum (λ less than 400 nm) lies the ultraviolet region.

The ultraviolet spectrometers commonly used measure absorption of light in the visible and "near" ultraviolet region, that is, in the 200–750 nm range. This light is of higher frequency (and greater energy) than infrared light and, when it is absorbed by a molecule, the changes it produces are, naturally, ones that require greater energy: changes in electronic states.

In a transition to a higher electronic level, a molecule can go *from* any of a number of sub-levels—corresponding to various vibrational and rotational states— *to* any of a number of sub-levels; as a result, ultraviolet absorption bands are broad. Where an infrared spectrum shows many sharp peaks, a typical ultraviolet spectrum shows only a few broad humps. One can conveniently describe such a spectrum in terms of the *position* of the top of the hump (λ_{max}) and the *intensity* of that absorption (ϵ_{max}, the extinction coefficient).

When we speak of a molecule as being raised to a higher electronic level, we mean that an electron has been changed from one orbital to another orbital of higher energy. This electron can be of any of the kinds we have encountered: a σ electron, a π electron, or an n electron (a non-bonding electron—that is, one of an unshared pair). A σ electron is held tightly, and a good deal of energy is required to excite it: energy corresponding to ultraviolet light of short wavelength, in a region—"far" ultraviolet—outside the range of the usual spectrometer. It is chiefly excitations of the comparatively loosely held n and π electrons that appear in the (near) ultraviolet spectrum, and, of these, only jumps to the lower—more stable— excited states.

The electronic transitions of most concern to the organic chemist are: (a) $n \rightarrow \pi^*$, in which the electron of an unshared pair goes to an unstable (*anti-bonding*) π orbital, as, for example,

$$\text{C=Ö:} \longrightarrow \text{C=Ö:} \qquad\qquad n \longrightarrow \pi^*$$

and (b) $\pi \rightarrow \pi^*$, in which an electron goes from a stable (*bonding*) π orbital to an unstable π orbital, as, for example,

$$\text{C=Ö:} \longrightarrow \text{C÷Ö:} \qquad\qquad \pi \longrightarrow \pi^*$$

A $\pi \rightarrow \pi^*$ transition can occur for even a simple alkene, like ethylene, but absorption occurs in the far ultraviolet. Conjugation of double bonds, however, lowers the energy required for the transition, and absorption moves to longer wavelengths, where it can be more conveniently measured. If there are enough double bonds in conjugation, absorption will move into the visible region, and the compound will be colored. β-Carotene, for example, is a yellow pigment found in carrots and green leaves, and is a precursor of vitamin A; it contains eleven carbon–carbon double bonds in conjugation, and owes its color to absorption at the violet end of the visible spectrum (λ_{max} 451 nm).

How does conjugation bring about this effect? We have seen (Sec. 10.22) that 1,3-butadiene, for example, is stabilized by contribution from structures involving formal bonds. Stabilization is not very great, however, since such structures—and additional, ionic structures—are not very stable and make only small contribution to the hybrid. Similar structures contribute to an excited state of butadiene, too, but here, because of the instability of the molecule, they make much larger contribution. Resonance stabilizes the excited state *more* than it stabilizes the ground state, and thus reduces the difference between them.

In contrast to the infrared spectrum, the ultraviolet spectrum is not used primarily to show the presence of individual functional groups, but rather to show relationships between functional groups, chiefly conjugation: conjugation between two or more carbon–carbon double (or triple) bonds; between carbon–carbon and carbon–oxygen double bonds; between double bonds and an aromatic ring; and even the presence of an aromatic ring itself. It can, in addition, reveal the number and location of substituents attached to the carbons of the conjugated system.

Problem 16.2 In Problem 12.22, page 470, you calculated the number of rings in β-carotene. Taking into account also the molecular formula, the number of double bonds, conjugation, its natural occurrence, and its conversion into vitamin A_1 (p. 409), what possible structure for β-carotene occurs to you?

Problem 16.3 Compounds A, B, and C have the formula C_5H_8, and on hydrogenation all yield *n*-pentane. Their ultraviolet spectra show the following values of λ_{max}: A, 176 nm; B, 211 nm; C, 215 nm. (1-Pentene has λ_{max} 178 nm.) (a) What is a likely structure for A? For B and C? (b) What kind of information might enable you to assign specific structures to B and C?

16.6   The nuclear magnetic resonance (NMR) spectrum

Like electrons, the nuclei of certain atoms are considered to *spin*. The spinning of these charged particles—the circulation of charge—generates a *magnetic moment* along the axis of spin, so that these nuclei act like tiny bar magnets. One such nucleus—and the one we shall be mostly concerned with—is the *proton*, the nucleus of ordinary hydrogen, ^1H.

Now, if a proton is placed in an external magnetic field, its magnetic moment, according to quantum mechanics, can be aligned in either of two ways: *with* or *against* the external field. Alignment with the field is the more stable, and energy must be absorbed to "flip" the tiny proton magnet over to the less stable alignment, against the field.

Just how much energy is needed to flip the proton over depends, as we might expect, on the strength of the external field: the stronger the field, the greater the tendency to remain lined up with it, and the higher the frequency (*Remember*: $\Delta E = h\nu$) of the radiation needed to do the job.

$$\nu = \frac{\gamma H_0}{2\pi}$$

where ν = frequency, in Hz

H_0 = strength of the magnetic field, in gauss

γ = a nuclear constant, the *magnetogyric ratio*, 26 750 for the proton

In a field of 14 092 gauss, for example, the energy required corresponds to electromagnetic radiation of frequency 60 MHz (60 megahertz or 60 million cycles per second): radiation in the radiofrequency range, and of much lower energy (lower frequency, longer wavelength) than even infrared light.

In principle, we could place a substance in a magnetic field of constant strength, and then obtain a spectrum in the same way we obtain an infrared or an ultraviolet spectrum: pass radiation of steadily changing frequency through the substance, and observe the frequency at which radiation is absorbed. In practice, however, it has been found more convenient to keep the radiation frequency constant, and to vary the strength of the magnetic field; at some value of the field strength the energy required to flip the proton matches the energy of the radiation, absorption occurs, and a signal is observed. Such a spectrum is called a *nuclear magnetic resonance* (NMR) *spectrum* (Fig. 16.3).

Since the nucleus involved is the proton, the spectrum is sometimes called a *PMR* (proton magnetic resonance) spectrum, to differentiate it from spectra involving such nuclei as ^{13}C (called *CMR* or ^{13}C *NMR* spectra) or ^{19}F.

Figure 16.3   The NMR spectrum.

Now, if the situation were as simple as we have so far described it, all the protons in an organic molecule would absorb at exactly the same field strength, and the spectrum would consist of a single signal that would tell us little about the structure of the molecule. But the frequency at which a proton absorbs depends on the magnetic field which that proton *feels*, and this *effective* field strength is not exactly the same as the *applied* field strength. The effective field strength at each proton depends on the environment of that proton—on, among other things, the electron density at the proton, and the presence of other, nearby protons. Each proton—or, more precisely, each set of equivalent protons—will have a slightly different environment from every other set of protons, and hence will require a slightly *different applied* field strength to produce the *same effective* field strength: the particular field strength at which absorption takes place.

At a given radiofrequency, then, *all protons absorb at the same effective field strength, but they absorb at different applied field strengths.* It is this applied field strength that is measured, and against which the absorption is plotted.

The result is a spectrum showing many absorption peaks, whose relative positions, reflecting as they do differences in environment of protons, can give almost unbelievably detailed information about molecular structure.

In the following sections, we shall look at various aspects of the NMR spectrum:

(a) the *number of signals*, which tells us how many different "kinds" of protons there are in a molecule;

(b) the *positions of the signals*, which tell us something about the electronic environment of each kind of proton;

(c) the *intensities of the signals*, which tell us how many protons of each kind there are; and

(d) the *splitting of a signal* into several peaks, which tells us about the environment of a proton with respect to other, nearby protons.

16.7   NMR. Number of signals. Equivalent and non-equivalent protons

In a given molecule, protons with the same environment absorb at the same (applied) field strength; protons with different environments absorb at different (applied) field strengths. A set of protons with the same environment are said to be *equivalent*; the number of signals in the NMR spectrum tells us, therefore, how many sets of equivalent protons—how many "kinds" of protons—a molecule contains.

For our purposes here, equivalent protons are simply chemically equivalent protons, and we have already had considerable practice in judging what these are. Looking at each of the following structural formulas, for example, we readily pick out as equivalent the protons designated with the same letter:

$$CH_3-CH_2-Cl \qquad CH_3-CHCl-CH_3 \qquad CH_3-CH_2-CH_2-Cl$$

a	*b*		*a*	*b*	*a*	*a*	*b*	*c*

2 *NMR signals*   2 *NMR signals*   3 *NMR signals*

Ethyl chloride   Isopropyl chloride   *n*-Propyl chloride

Realizing that, to be chemically equivalent, protons must also be *stereochemically* equivalent, we find we can readily analyze the following formulas, too:

2 *NMR signals*   3 *NMR signals*   3 *NMR signals*   4 *NMR signals*

Isobutylene   2-Bromopropene   Vinyl chloride   Methylcyclopropane

1,2-Dichloropropane (optically active or optically inactive) gives four NMR signals, and it takes only a little work with models or stereochemical formulas to see that this should indeed be so.

$$CH_3\text{---}CHCl\text{---}\underset{\underset{H}{\overset{c}{\overset{H}{|}}}}{C}\text{---}Cl$$

$$
\begin{array}{c}
Cl \\
| \\
H\text{---}C\text{---}H \\
| \\
H\text{---}C\text{---}Cl \\
| \\
CH_3
\end{array}
$$

a   b   d

4 NMR signals
1,2-Dichloropropane

The environments of the two protons on C–1 are *not* the same (and no amount of rotation about single bonds will make them so); the protons are not equivalent, and will absorb at different field strengths.

We can tell from a formula which protons are in different environments and hence should give different signals. We cannot always tell—particularly with stereochemically different protons—just *how* different these environments are; they may not be different enough for the signals to be noticeably separated, and we may see *fewer* signals than we predict.

Now, just how did we arrive at the conclusions of the last few paragraphs? Most of us—perhaps without realizing it—judge the equivalence of protons by following the approach of isomer number (Sec. 4.2). This is certainly the easiest way to do it. We imagine each proton in turn to be replaced by some other atom Z. If replacement of either of two protons by Z would yield the same product—or enantiomeric products—then the two protons are chemically equivalent in an achiral medium. We ignore the existence of conformational isomers and, as we shall see in Sec. 16.13, this is just what we should do.

Take, for example, ethyl chloride. Replacement of a methyl proton would give $CH_2Z\text{---}CH_2Cl$; replacement of a methylene proton would give $CH_3\text{---}CHZCl$. These are, of course, different products, and we easily recognize the methyl protons as being non-equivalent to the methylene protons.

The product $CH_2Z\text{---}CH_2Cl$ is the same regardless of *which one* of the three methyl protons is replaced. The (average) environment of the three protons is identical, and hence we expect one NMR signal for all three.

Replacement of either of the two methylene protons would give one of a pair of enantiomers:

$$
\begin{array}{c}
CH_3 \\
| \\
H\text{---}C\text{---}H \\
| \\
Cl
\end{array}
\qquad
\left[
\begin{array}{cc}
\begin{array}{c}
CH_3 \\
| \\
H\text{---}C\text{---}Z \\
| \\
Cl
\end{array}
&
\begin{array}{c}
CH_3 \\
| \\
Z\text{---}C\text{---}H \\
| \\
Cl
\end{array}
\end{array}
\right]
$$

Enantiotopic
protons
Ethyl chloride

Such pairs of protons are called **enantiotopic protons**. The environments of these two protons are mirror images of each other; in an achiral medium, these protons behave *as if* they were equivalent, and we see one NMR signal for the pair.

Turning to 2-bromopropene, we see that replacement of either of the vinylic protons gives one of a pair of diastereomers (geometric isomers, in this case):

$$
\begin{array}{ccc}
\underset{\underset{Br}{}}{CH_3}\!\!\diagdown C\!=\!C\diagup\underset{H}{}H
&
\left[\;\underset{\underset{Br}{}}{CH_3}\!\!\diagdown C\!=\!C\diagup\underset{H}{}Z\right.
&
\left.\underset{\underset{Br}{}}{CH_3}\!\!\diagdown C\!=\!C\diagup\underset{Z}{}H\;\right]
\end{array}
$$

Diastereotopic
protons

2-Bromopropene

Such pairs of protons are called **diastereotopic protons**. The environments of these two protons are neither identical nor mirror images of each other; these protons are non-equivalent, and we expect an NMR signal from each one.

Similarly, in 1,2-dichloropropane the two protons on C–1 are diastereotopic, non-equivalent, and give separate NMR signals.

$$
\begin{array}{ccc}
\begin{array}{c}Cl\\H\!-\!C\!-\!H\\H\!-\!C\!-\!Cl\\CH_3\end{array}
&
\left[\begin{array}{c}Cl\\H\!-\!C\!-\!Z\\H\!-\!C\!-\!Cl\\CH_3\end{array}\right.
&
\left.\begin{array}{c}Cl\\Z\!-\!C\!-\!H\\H\!-\!C\!-\!Cl\\CH_3\end{array}\right]
\end{array}
$$

Diastereotopic
protons

1,2-Dichloropropane

We shall return to these concepts of enantiotopic and diastereotopic ligands in Chapter 22, and see their fundamental importance to our understanding of stereochemistry.

In Sec. 16.13, we shall take a closer look at equivalence. The guidelines we have laid down here, however—based on rapid rotation about single bonds—hold for most spectra taken under ordinary conditions: specifically, at room temperature.

Problem 16.4 Draw the structural formula of each of the following compounds (disregarding enantiomerism), and label all sets of equivalent protons. How many NMR signals would you expect to see from each?

(a) the two isomers of formula $C_2H_4Cl_2$
(b) the four isomers of $C_3H_6Br_2$
(c) ethylbenzene and *p*-xylene
(d) mesitylene, *p*-ethyltoluene, isopropylbenzene
(e) CH_3CH_2OH and CH_3OCH_3
(f) $CH_3CH_2OCH_2CH_3$, $CH_3OCH_2CH_2CH_3$, $CH_3OCH(CH_3)_2$, $CH_3CH_2CH_2CH_2OH$
(g) $CH_2\!-\!CH_2$, $CH_3\!-\!CH\!-\!CH_2$ (*Hint*: Make models.)
  $\underset{}{CH_2\!-\!O}$ $\underset{}{\diagup O\diagdown}$
(h) $CH_3CH_2\underset{\underset{O}{\|}}{C}\!-\!H$, $CH_3\underset{\underset{O}{\|}}{C}CH_3$, and $CH_2\!=\!CHCH_2OH$

Problem 16.5   Three isomeric dimethylcyclopropanes give, respectively, two, three, and four NMR signals. Draw a stereoisomeric formula for the isomer giving rise to each number of signals.

Problem 16.6   How many NMR signals would you expect from cyclohexane? Why?

16.8   NMR. Positions of signals. Chemical shift

Just as the number of signals in an NMR spectrum tells us how many kinds of protons a molecule contains, so the *positions of the signals* help to tell us *what kinds* of protons they are: aromatic, aliphatic, primary, secondary, tertiary; benzylic, vinylic, acetylenic; adjacent to halogen or to other atoms or groups. These different kinds of protons have different electronic environments, and it is the electronic environment that determines just where in the spectrum a proton absorbs.

When a molecule is placed in a magnetic field—as it is when one determines an NMR spectrum—its electrons are caused to circulate and, in circulating, they generate secondary magnetic fields: *induced* magnetic fields.

Circulation of electrons *about the proton itself* generates a field aligned in such a way that—at the proton—it opposes the applied field. The field felt by the proton is thus diminished, and the proton is said to be **shielded**.

Circulation of electrons—specifically, π electrons—*about nearby nuclei* generates a field that can either oppose or reinforce the applied field at the proton, depending on the proton's location (Fig. 16.4). If the induced field opposes the applied field, the proton is shielded, as before. If the induced field reinforces the applied field, then the field felt by the proton is augmented, and the proton is said to be **deshielded**.

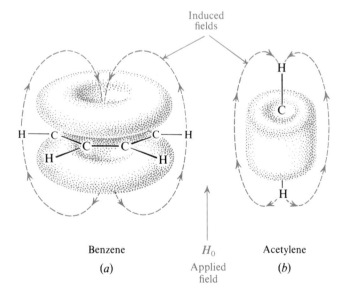

Benzene H_0 Acetylene

(*a*)   Applied   (*b*)
                         field

Figure 16.4   Induced field (*a*) reinforces the applied field at the aromatic protons, and (*b*) opposes the applied field at the acetylenic protons. Aromatic protons are deshielded; acetylenic protons are shielded.

Compared with a naked proton, a shielded proton requires a higher applied field strength—and a deshielded proton requires a lower applied field strength—to provide the particular effective field strength at which absorption occurs. Shielding thus shifts the absorption upfield, and deshielding shifts the absorption downfield. Such shifts in the position of NMR absorptions, arising from shielding and deshielding by electrons, are called **chemical shifts**.

How are the direction and magnitude—the *value*—of a particular chemical shift to be measured and expressed?

The unit in which a chemical shift is most conveniently expressed is parts per million (ppm) of the total applied magnetic field. Since shielding and deshielding arise from *induced* secondary fields, the magnitude of a chemical shift is proportional to the strength of the applied field—or, what is equivalent, proportional to the radiofrequency the field must match. If, however, it is expressed as a *fraction* of the applied field—that is, if the observed shift is divided by the particular radiofrequency used—then a chemical shift has a constant value that is independent of the radiofrequency and the magnetic field that the NMR spectrometer employs.

The **reference point** from which chemical shifts are measured is, for practical reasons, not the signal from a naked proton, but the signal from an actual compound: usually tetramethylsilane, $(CH_3)_4Si$. Because of the low electronegativity of silicon, the shielding of the protons in the silane is greater than in most other organic molecules; as a result, most NMR signals appear in the same direction from the tetramethylsilane signal: *downfield*.

The most commonly used scale is the δ (*delta*) scale. The position of the tetramethylsilane signal is taken as 0.0 ppm. Most chemical shifts have δ values between 0 and 10 (minus 10, actually). A *small* δ value represents a *small* downfield shift, and a *large* δ value represents a *large* downfield shift.

One sometimes encounters another scale: the τ (*tau*) scale, on which the tetramethylsilane signal is taken as 10.0 ppm. Most τ values lie between 0 and 10. The two scales are related by the expression $\tau = 10 - \delta$.

An NMR signal from a particular proton appears at a different field strength than the signal from tetramethylsilane. This difference—the chemical shift—is measured not in gauss, as we might expect, but in the equivalent frequency units (*Remember:* $v = \gamma H_0/2\pi$), and it is divided by the frequency of the spectrometer used. Thus, for a spectrometer operating at 60 MHz, that is, at 60×10^6 Hz:

$$\delta = \frac{\text{observed shift (Hz)} \times 10^6}{60 \times 10^6 \text{ (Hz)}}$$

The chemical shift for a proton is determined, then, by the electronic environment of the proton. In a given molecule, protons with different environments—non-equivalent protons—have different chemical shifts. Protons with the same environment—equivalent protons—have the same chemical shift. (So, too, do protons with mirror-image environments—enantiotopic protons.) We have already seen what the equivalence of protons means in terms of molecular structure.

Furthermore, it has been found that a proton with a particular environment shows much the same chemical shift, whatever the molecule it happens to be part of (see Table 16.4). Take, for example, our familiar classes of hydrogens: primary, secondary, and tertiary. In the absence of other nearby substituents, absorption occurs at about these values:

$$\begin{array}{ccc} \overset{\displaystyle H}{\underset{\displaystyle H}{R-\overset{|}{\underset{|}{C}}-H}} & \overset{\displaystyle H}{\underset{\displaystyle R}{R-\overset{|}{\underset{|}{C}}-H}} & \overset{\displaystyle R}{\underset{\displaystyle R}{R-\overset{|}{\underset{|}{C}}-H}} \\ \delta\ 0.9 & \delta\ 1.3 & \delta\ 1.5 \end{array}$$

All these protons, in turn, differ widely from aromatic protons which, because of the powerful deshielding due to the circulation of the π electrons (see Fig. 16.4, p. 583), absorb far downfield:

$$Ar-H \qquad \delta\ 6\text{--}8.5$$

Table 16.4 CHARACTERISTIC PROTON CHEMICAL SHIFTS

Type of proton		Chemical shift, ppm
		δ
Cyclopropane		0.2
Primary	$R\overset{H}{\underset{H}{\overset{|}{\underset{|}{C}}}}-H$	0.9
Secondary	R_2C-H	1.3
Tertiary	R_3C-H	1.5
Vinylic	$C=C-H$	4.6–5.9
Acetylenic	$C\equiv C-H$	2–3
Aromatic	$Ar-H$	6–8.5
Benzylic	$Ar-C-H$	2.2–3
Allylic	$C=C-\overset{H}{\underset{H}{\overset{|}{\underset{|}{C}}}}-H$	1.7
Fluorides	$H-C-F$	4–4.5
Chlorides	$H-C-Cl$	3–4
Bromides	$H-C-Br$	2.5–4
Iodides	$H-C-I$	2–4
Alcohols	$H-C-OH$	3.4–4
Ethers	$H-C-OR$	3.3–4
Esters	$RCOO-C-H$	3.7–4.1
Esters	$H-C-COOR$	2–2.2
Acids	$H-C-COOH$	2–2.6
Carbonyl compounds	$H-C-C=O$	2–2.7
Aldehydic	$R\overset{H}{\overset{|}{C}}=O$	9–10
Hydroxylic	$RO-H$	1–5.5
Phenolic	$ArO-H$	4–12
Enolic	$C=C-O-H$	15–17
Carboxylic	$RCOO-H$	10.5–12
Amino	$R\overset{H}{\overset{|}{N}}-H$	1–5

Figure 16.5 NMR spectra: chemical shift. (*a*) Toluene; (*b*) *p*-xylene; (*c*) mesitylene.

Attachment of chlorine to the carbon bearing the proton causes a downfield shift. If the chlorine is attached to the carbon once removed from the carbon bearing the proton, there is again a downfield shift, but this time much weaker.

$$
\begin{array}{ccc}
\overset{\displaystyle H}{\underset{\displaystyle H}{H-C-Cl}} & \overset{\displaystyle H}{\underset{\displaystyle H}{R-C-Cl}} & \overset{\displaystyle H}{\underset{\displaystyle R}{R-C-Cl}} \\
\delta\,3.0 & \delta\,3.4 & \delta\,4.0
\end{array}
$$

$$
\begin{array}{ccc}
\overset{\displaystyle H}{\underset{\displaystyle H}{H-C-C-Cl}} & \overset{\displaystyle H}{\underset{\displaystyle H}{R-C-C-Cl}} & \overset{\displaystyle H}{\underset{\displaystyle R}{R-C-C-Cl}} \\
\delta\,1.5 & \delta\,1.7 & \delta\,1.6
\end{array}
$$

Two chlorines cause a greater downfield shift. Other halogens show similar effects.

The downfield shift caused by chlorine is what we might have expected from its inductive effect: electron withdrawal lowers the electron density in the vicinity of the proton and thus causes deshielding. The effect of a substituent on the chemical shift is unquestionably the net result of many factors; yet we shall often observe chemical shifts which strongly suggest that an inductive effect is at least one of the factors at work.

The NMR spectra (Fig. 16.5) of the alkylbenzenes *toluene, p-xylene,* and *mesitylene* illustrate the points we have just made. In each spectrum there are two signals: one for the side-chain protons, and one for the ring protons. (Here, as in some—though not most—aromatic compounds, the *ortho, meta,* and *para* protons have nearly the same chemical shifts.)

In each spectrum, the ring protons show the low-field absorption we have said is characteristic of aromatic protons. Absorption is not only at low field, but at nearly the *same* field strength for the three compounds: at δ 7.17, 7.05, and 6.78. (These values are not *exactly* the same, however, since the environments of the aromatic protons are not exactly the same in the three compounds.)

In each compound, side-chain protons—benzylic protons—are close enough to the ring to feel a little of the deshielding effect of the π electrons (Fig. 16.4, p. 583), and hence absorb somewhat downfield from ordinary alkyl protons: at δ 2.32, 2.30, and 2.25. In all three compounds, the environment of the side-chain protons is almost identical, and so are the chemical shifts.

The similarity in structure among these three alkylbenzenes is thus reflected in the similarity of their NMR spectra. There is, however, a major difference in their structures—a difference in *numbers* of aromatic and side-chain protons—and, as we shall see in the next section, this is reflected in a major difference in their NMR spectra.

The chemical shift is fundamental to the NMR spectrum since, by separating the absorption peaks due to the various protons of a molecule, it reveals all the other features of the spectrum. The *numerical values* of chemical shifts, although significant, do not have the overriding importance that absorption frequencies have in the infrared spectrum. In our work with NMR, we shall escape much of the uncertainty that accompanies the beginner's attempts to identify precisely infrared absorption bands; at the same time, we have a greater *variety* of concepts to learn about—but these, at our present level, we may find more satisfying and intellectually more stimulating.

Problem 16.7 What is a possible explanation for the following differences in chemical shift for aromatic protons? Benzene δ 7.37; toluene δ 7.17; p-xylene δ 7.05; mesitylene δ 6.78.

16.9 NMR. Peak area and proton counting

Let us look again at the NMR spectra (Fig. 16.5, p. 586) of toluene, p-xylene, and mesitylene, and this time focus our attention, not on the positions of the signals, but on their relative *intensities*, as indicated by the sizes of the absorption peaks.

Judging roughly from the peak heights, we see that the (high-field) peak for side-chain protons is smaller than the (low-field) peak for aromatic protons in the case of toluene, somewhat larger in the case of p-xylene, and considerably larger in the case of mesitylene. More exact comparison, based on the *areas under the peaks*, shows that the peaks for side-chain and aromatic protons have sizes in the ratio 3:5 for toluene; 3:2 (or 6:4) for p-xylene; and 3:1 (or 9:3) for mesitylene.

This illustrates a general quality of all NMR spectra. *The area under an NMR signal is directly proportional to the number of protons giving rise to the signal.*

It is not surprising that this is so. The absorption of every quantum of energy is due to exactly the same thing: the flipping over of a proton in the same effective magnetic field. The more protons flipping, the more the energy absorbed, and the greater is the area under the absorption peak.

Areas under NMR signals are measured by an electronic integrator, and are often given on the spectrum chart in the form of a stepped curve; heights of steps are proportional to peak areas. NMR chart paper is cross-hatched, and we can conveniently estimate step heights by simply counting squares. We arrive at a set of numbers that are in the same ratio as the numbers of different kinds of protons. We convert this set of numbers into a set of smallest whole numbers just as we did in calculating empirical formulas (Sec. 2.28). The number of protons giving rise to each signal is equal to the whole number for that signal—or to some multiple of it. See, for example, Fig. 16.6.

We take any shortcuts we can. If we know the molecular formula and hence the total number of protons, we can calculate from the combined step heights the number of squares per proton. If we suspect a particular structural feature that gives a characteristic signal— an aldehydic (—CHO) or carboxylic (—COOH) proton, say, which gives a far-downfield peak—we can use this step height as a starting point.

Working the following problems will give us some idea of the tremendous help "proton counting" by NMR can be in assigning a structure to a compound.

Problem 16.8 Go back to Problem 16.4 (p. 582), where you predicted the number of NMR signals from several compounds. Tell, where you can, the relative positions of the signals (that is, their sequence as one moves downfield) and, roughly, the δ value expected for each. For each signal tell the number of protons giving rise to it.

Problem 16.9 Give a structure or structures consistent with each of the NMR spectra shown in Fig. 16.7 (p. 590).

Figure 16.6 NMR spectrum of *p-tert*-butyltoluene. Proton counting. The ratio of step heights $a:b:c$ is

$$8.8:2.9:3.8 = 3.0:1.0:1.3 = 9.0:3.0:3.9$$

Alternatively, since the molecular formula $C_{11}H_{16}$ is known,

$$\frac{16\ H}{15.5\ \text{units}} = 1.03\ H\ \text{per unit}$$

$$a = 1.03 \times 8.8 = 9.1 \qquad b = 1.03 \times 2.9 = 3.0 \qquad c = 1.03 \times 3.8 = 3.9$$

Either way, we find: a, 9H; b, 3H; c, 4H.

The 4H of c (δ 7.1) are in the aromatic range, suggesting a disubstituted benzene —C_6H_4—. The 3H of b (δ 2.28) have a shift expected for benzylic protons, giving CH_3—C_6H_4—. There is left C_4H_9 which, in view of the 9H of a (δ 1.28) must be —$C(CH_3)_3$; since these are once removed from the ring, their shift is nearly normal for an alkyl group. The compound is *tert*-butyltoluene (actually, as shown by the absorption pattern of the aromatic protons, the *para* isomer).

16.10   NMR. Splitting of signals. Spin–spin coupling

An NMR spectrum, we have said, shows a signal for each kind of proton in a molecule; the few spectra we have examined so far bear this out. If we look much further, however, we soon find that most spectra are—or *appear* to be—much more complicated than this. Figure 16.8 (p. 591), for example, shows the NMR spectra for three compounds,

$$CH_2Br—CHBr_2 \qquad\qquad CH_3—CHBr_2 \qquad\qquad CH_3—CH_2Br$$

1,1,2-Tribromoethane   1,1-Dibromoethane   Ethyl bromide

each of which contains only two kinds of protons; yet, instead of two peaks, these spectra show *five*, *six*, and *seven* peaks, respectively.

What does this multiplicity of peaks mean? How does it arise, and what can it tell us about molecular structure?

Figure 16.7   NMR spectra for Problem 16.9, p. 588.

(a) 1,1,2-Tribromoethane

CHBr₂—CH₂Br
  b   a

(b) 1,1,-Dibromoethane

CH₃—CHBr₂
 a   b

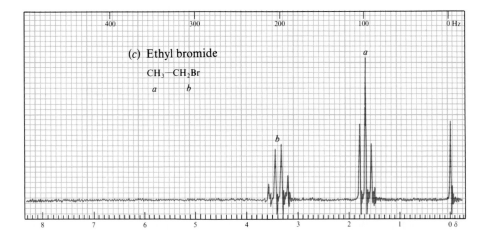

(c) Ethyl bromide

CH₃—CH₂Br
 a   b

Figure 16.8 NMR spectra: splitting of signals. (a) 1,1,2-Tribromoethane; (b) 1,1-dibromoethane; (c) ethyl bromide.

The answer is that we are observing the *splitting* of NMR signals caused by spin–spin coupling. The signal we expect from each set of equivalent protons is appearing, not as a single peak, but as a *group* of peaks. Splitting reflects the environment of the absorbing protons: not with respect to electrons, but with respect to other, nearby protons. It is as though we were permitted to sit on a proton and look about in all directions: we can *see* and *count* the protons attached to the carbon atoms next to our own carbon atom and, sometimes, even see protons still farther away.

Let us take the case of adjacent carbon atoms carrying, respectively, a pair of secondary protons and a tertiary proton, and consider first the absorption by one of the secondary protons:

$$-\overset{\displaystyle}{C}H-\overset{\displaystyle H}{\underset{\displaystyle H}{C}}-$$

The magnetic field that a secondary proton feels at a particular instant is slightly increased or slightly decreased by the spin of the neighboring tertiary proton: *increased* if the tertiary proton happens at that instant to be aligned *with* the applied field; or *decreased* if the tertiary proton happens to be aligned *against* the applied field.

For half the molecules, then, absorption by a secondary proton is shifted slightly downfield, and for the other half of the molecules the absorption is shifted slightly upfield. The signal is split into *two* peaks: a *doublet*, with equal peak intensities (Fig. 16.9).

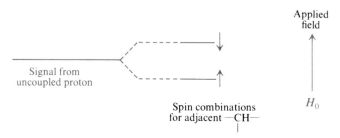

Figure 16.9 Spin–spin coupling. Coupling with one proton gives a 1:1 doublet.

Next, what can we say about the absorption by the tertiary proton?

$$-\overset{\displaystyle}{\underset{\displaystyle H}{C}}-CH_2-$$

It is, in its turn, affected by the spin of the neighboring secondary protons. But now there are *two* protons whose alignments in the applied field we must consider. There are four equally probable combinations of spin alignments for these two protons, of which two are equivalent. At any instant, therefore, the tertiary proton feels any one of three fields, and its signal is split into three equally spaced peaks: a *triplet*, with relative peak intensities 1:2:1, reflecting the combined (double) probability of the two equivalent combinations (Fig. 16.10).

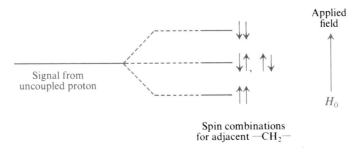

Spin combinations
for adjacent —CH$_2$—

Figure 16.10   Spin–spin coupling. Coupling with two protons gives a 1:2:1 triplet.

Figure 16.11 shows an idealized NMR spectrum due to the grouping —CH—CH$_2$—. We see a 1:1 doublet (from the —CH$_2$—) and a 1:2:1 triplet (from the —CH—). The total area (both peaks) under the doublet is *twice* as big as the total area (all three peaks) of the triplet, since the doublet is due to absorption by twice as many protons as the triplet.

A little measuring shows us that the separation of peaks (the *coupling constant*, J, Sec. 16.11) in the doublet is exactly the same as the separation of peaks in the triplet. (Spin–spin coupling is a *reciprocal* affair, and the effect of the secondary

CH$_2$Br—CHBr$_2$
 a   *b*

Figure 16.11   Spin–spin splitting. Signal *a* is split into a doublet by coupling with one proton; signal *b* is split into a triplet by two protons. Spacings in both sets are the same (J_{ab}).

protons on the tertiary proton must be identical with the effect of the tertiary proton on the secondary protons.) Even if they were to appear in a complicated spectrum of many absorption peaks, the identical peak separations would tell us that this doublet and triplet were related: that the (two) protons giving the doublet and the (one) proton giving the triplet are coupled, and hence are attached to adjacent carbon atoms.

We have seen that an NMR signal is split into a doublet by one nearby proton, and into a triplet by two (equivalent) nearby protons. What splitting can we expect more than two protons to produce? In Fig. 16.12, we see that three equivalent protons split a signal into four peaks—a quartet—with the intensity pattern 1:3:3:1.

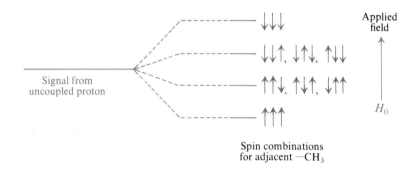

Figure 16.12 Spin–spin coupling. Coupling with three protons gives a 1:3:3:1 quartet.

It can be shown that, in general, *a set of n equivalent protons will split an NMR signal into n + 1 peaks.*

If we turn once more to Fig. 16.8 (p. 591), we no longer find these spectra so confusing. We now see not just five or six or seven peaks, but instead a doublet and a triplet, or a doublet and a quartet, or a triplet and a quartet. We recognize each of these multiplets from the even spacings within it, and from its symmetrical intensity pattern (1:1, or 1:2:1, or 1:3:3:1). Each spectrum does show absorption by just two kinds of protons; but clearly it shows a great deal more than that.

If we keep in mind that the peak area reflects the number of *absorbing* protons, and the multiplicity of splittings reflects the number of *neighboring* protons, we find in each spectrum just what we would expect.

In the spectrum of $CHBr_2$—CH_2Br we see

Downfield triplet **and** *Upfield doublet*
Area: 1   *Area: 2*

In the spectrum of CH_3—$CHBr_2$ we see

Downfield quartet   **and**   *Upfield doublet*
Area: 1   *Area: 3*

$$-\overset{\displaystyle |}{\underset{\displaystyle H}{C}}-CH_3 \qquad\qquad -CH-\overset{\displaystyle H}{\underset{\displaystyle H}{C}}-H$$

and in the spectrum of CH_3—CH_2Br we see

Downfield quartet   **and**   *Upfield triplet*
Area: 2   *Area: 3*

$$-\overset{\displaystyle H}{\underset{\displaystyle H}{C}}-CH_3 \qquad\qquad -CH_2-\overset{\displaystyle H}{\underset{\displaystyle H}{C}}-H$$

We see chemical shifts that are consistent with the deshielding effect of halogens: in each spectrum, the protons on the carbon carrying the greater number of halogens absorb farther downfield (larger δ).

In each spectrum, we see that the spacing of the peaks within one multiplet is the same as within the other, so that even in a spectrum with many other peaks, we could pick out these two multiplets as being coupled.

Finally, we see a feature that we have not yet discussed: the various multiplets do not show quite the symmetry we have attributed to them. In spectrum (*a*), we see

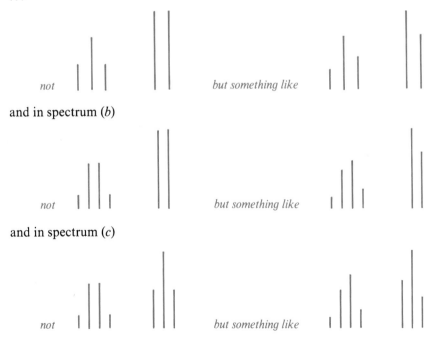

and in spectrum (*b*)

and in spectrum (*c*)

In each case, the inner peaks—the peaks nearer the other, coupled multiplets—are larger than the outer peaks.

Perfectly symmetrical multiplets are to be expected only when the separation between multiplets is very large relative to the separation within multiplets—that is, when the chemical shift is much larger than the coupling constant (Sec. 16.11). The patterns we see here are very commonly observed, and are helpful in matching up multiplets: we know in which direction—upfield or downfield—to look for the second multiplet.

We have not yet answered a very basic question: just which protons in a molecule can be coupled? *We may expect to observe spin–spin splitting only between non-equivalent neighboring protons.* By "non-equivalent" protons we mean protons with different chemical shifts, as we have already discussed (Sec. 16.8). By "neighboring" protons we mean most commonly protons on *adjacent* carbons, as in the examples we have just looked at (Fig. 16.8, p. 591); sometimes protons farther removed from each other may also be coupled, particularly if π bonds intervene. (If protons on the *same* carbon are non-equivalent—as they sometimes are—they may show coupling.)

We do *not* observe splitting due to coupling between the protons making up the same —CH_3 group, since they are equivalent. We do *not* observe splitting due to coupling between the protons on C–1 and C–2 of 1,2-dichloroethane

$$\begin{array}{cc} CH_2 & CH_2 \\ | & | \\ Cl & Cl \end{array}$$ *No splitting*

1,2-Dichloroethane

since, although on different carbons, they, too, are equivalent.

In the spectrum of 1,2-dibromo-2-methylpropane,

$$\begin{array}{c} CH_3 \\ | \\ CH_3{-}C{-}CH_2Br \\ | \\ Br \end{array}$$ *No splitting*

1,2-Dibromo-2-methylpropane

we do *not* observe splitting between the six methyl protons, on the one hand, and the two —CH_2— protons on the other hand. They are non-equivalent, and give rise to different NMR signals, but they are not on adjacent carbons, and their spins do not (noticeably) affect each other. The NMR spectrum contains two singlets, with a peak area ratio of 3:1 (or 6:2). For the same reason, we do *not* observe splitting due to coupling between ring and side-chain protons in alkylbenzenes (Fig. 16.5, p. 586).

We do *not* observe splitting between the two vinyl protons of isobutylene since

$$\begin{array}{cc} CH_3 & H \\ & \diagdown\ \ \diagup \\ & C{=}C \\ & \diagup\ \ \diagdown \\ CH_3 & H \end{array}$$ *No splitting*

Isobutylene

they are equivalent. On the other hand, we may observe splitting between the two vinyl protons on the same carbon if, as in 2-bromopropene, they are non-equivalent.

$$\begin{array}{cc} CH_3 & H_a \\ & \diagdown\ \ \diagup \\ & C{=}C \\ & \diagup\ \ \diagdown \\ Br & H_b \end{array}$$

2-Bromopropene

Figures 16.13 and 16.14 and Figure 16.15 (p. 598) illustrate some of the kinds of splitting we are likely to encounter in NMR spectra.

Figure 16.13   NMR spectrum of isopropyl bromide. Absorption by the six methyl protons H_a appears upfield, split into a doublet by the single adjacent proton H_b. Absorption by the lone proton H_b appears downfield (the inductive effect of bromine) split into a septet by the six adjacent protons—with the small outside peaks typically hard to see.

Figure 16.14   NMR spectrum of *n*-propylbenzene. Moving downfield, we see the expected sequence of signals: *a*, primary (3H); *b*, secondary (2H); *c*, benzylic (2H); and *d*, aromatic (5H). Signals *a* and *c* are each split into a triplet by the two secondary protons H_b. The five protons adjacent to the secondary protons—three on one side and two on the other—are, of course, not equivalent; but the coupling constants, J_{ab} and J_{bc}, are nearly the same, and signal *b* appears as a sextet (5 + 1 peaks). The coupling constants are not *exactly* the same, however, as shown by the broadening of the six peaks.

Figure 16.15 NMR spectrum of 1,2-dibromo-1-phenylethane. The diastereotopic protons H_a and H_b give different signals, each split into a doublet by H_c; the downfield peaks of the doublets happen to coincide. (The above spectrum shows no splitting due to coupling between H_a and H_b. Run at higher gain, however, the spectrum shows this coupling: each doublet is split into a quartet.)

The four-line pattern of c is due to successive splittings by H_a and H_b. (If J_{ac} and J_{bc} were equal—as they would have to be if, for example, H_a and H_b were equivalent—the middle peaks of c would merge to give the familiar $1:2:1$ triplet.)

The fluorine (^{19}F) nucleus has magnetic properties of the same kind as the proton. It gives rise to NMR spectra, although at a quite different frequency–field strength combination than the proton. Fluorine nuclei can be coupled not only with each other, but also with protons. *Absorption by* fluorine does not appear in the proton NMR spectrum—it is far off the scale—but the *splitting by* fluorine of proton signals can be seen. The signal for the two protons of 1,2-dichloro-1,1-difluoroethane, for example,

$$\begin{array}{c} H \quad F \\ | \quad\quad | \\ Cl-C-C-Cl \\ | \quad\quad | \\ H \quad F \end{array}$$

appears as a $1:2:1$ triplet with peak spacings of 11 Hz. (What would you expect to see in the fluorine NMR spectrum?)

16.11   NMR. Coupling constants

The distance between peaks in a multiplet is a measure of the effectiveness of spin–spin coupling, and is called the **coupling constant**, J. Coupling (unlike chemical shift) is not a matter of induced magnetic fields. The value of the coupling constant—as measured, in Hz—remains the same, whatever the applied magnetic field (that is, whatever the radiofrequency used). In this respect, of course, spin–spin splitting differs from chemical shift, and, when necessary, the two can be distinguished on this basis: the spectrum is run at a second, different radiofrequency; when measured in Hz, peak separations due to splitting remain constant, whereas peak separations due to chemical shifts change. (When divided by the radiofrequency and thus converted into ppm, the numerical value of the chemical shift would, of course, remain constant.)

As we can see from the following summary, the size of a coupling constant depends markedly on the structural relationship between the coupled protons. For

example, in any substituted ethylene—or in any pair of geometric isomers—J is always larger between *trans* protons than between *cis* protons; furthermore, the size of J varies in a regular way with the electronegativity of substituents, so that one can often assign configuration without having both isomers in hand.

Although we shall not work very much with the *values* of coupling constants, we should realize that, to an experienced person, they can often be the most important feature of an NMR spectrum: the feature that gives exactly the kind of information about molecular structure that is being looked for.

Problem 16.10   Go back to Problem 16.8 (p. 588), and tell, where you can, the kind of splitting expected for each signal.

Problem 16.11   In Problem 16.9 (p. 588) you analyzed some NMR spectra. Does the absence of splitting in these spectra now lead you to change any of your answers?

Problem 16.12   Give a structure or structures consistent with each of the NMR spectra shown in Fig. 16.16 (p. 600).

(a) C$_8$H$_{10}$

[handwritten] 5

[handwritten structure: benzene ring with D inside] —CH$_2$—CH$_3$

[handwritten] 3

[handwritten] 2

[handwritten] The CH$_3$ splits the CH$_2$ into 4 peaks

(b) C$_3$H$_6$Br$_2$

[handwritten] CH$_2$ CH$_2$ CH$_2$
[handwritten] Br   Br

[handwritten] 4

[handwritten] 2

(c) C$_3$H$_7$Br

[handwritten] 2

[handwritten] 3

[handwritten] 2

Figure 16.16   NMR spectra for Problem 16.12, p. 599.

[handwritten] CH$_3$CHCH$_3$
[handwritten] NO! Br

[handwritten] Yes CH$_3$CH$_2$CH$_2$Br
[handwritten] 2 on top of each other

16.12   NMR. Complicated spectra. Deuterium labeling

Most NMR spectra that the organic chemist is likely to encounter are considerably more complicated than the ones given in this book. How are these analyzed?

First of all, many spectra showing a large number of peaks can be completely analyzed by the same general methods we shall use here. It just takes practice.

Then again, in many cases complete analysis is not necessary for the job at hand. Evidence of other kinds may already have limited the number of possible structures, and all that is required of the NMR spectrum is that it let us choose among these. Sometimes all that we need to know is how many kinds of protons there are—or, perhaps, how many kinds and how many of each kind. Sometimes only one structural feature is still in doubt—for example, does the molecule contain two methyl groups or one ethyl group?—and the answer is given in a set of peaks standing clear from the general confusion. (See, for example, Fig. 16.17, below.)

Instrumental techniques are available, and others are being developed, to help

4.93   10.0

τ

Courtesy of *The Journal of the American Chemical Society*

Figure 16.17   NMR spectrum of 2-α-acetoxycholestan-3-one, taken by K. L. Williamson and W. S. Johnson at the University of Wisconsin and Stanford University. The four downfield peaks are due to the proton on C–2, whose signal is split successively by the axial proton and the equatorial proton on C–1.

Courtesy of *The Journal of the American Chemical Society*

Figure 16.18 NMR spectra of (*top*) cyclohexanol and (*bottom*) 3,3,4,4,5,5-hexadeuteriocyclohexanol, taken by F. A. L. Anet of the University of California, Los Angeles. With absorption and splitting by six protons eliminated, the pattern due to the five remaining protons can be analyzed.

The diastereotopic sets of protons, H_a and H_b, give different signals. Signal *a* is split successively into doublets by H_b (only *one* H_b splits each H_a) and by H_c. Signal *b* is split similarly by H_a and H_c. Downfield signal *c* is split successively into triplets by H_a (both protons) and H_b (both protons).

in the analysis of complicated spectra, and to simplify the spectra actually measured. By the method of *double resonance* (or *double irradiation*), for example, the spins of two sets of protons can be *decoupled*, and a simpler spectrum obtained.

The molecule is irradiated with two radiofrequency beams: the usual one, whose absorption is being measured; and a second, much stronger beam, whose frequency differs from that of the first in such a way that the following happens.

When the field strength is reached at which the proton we are interested in absorbs and gives a signal, the splitting protons are absorbing the other, very strong radiation. These splitting protons are "stirred up" and flip over very rapidly—so rapidly that the signalling proton sees them, not in the various combinations of spin alignments (Sec. 16.10), but in a single *average* alignment. The spins are decoupled, and the signal appears as a single, unsplit peak.

A particularly elegant way to simplify an NMR spectrum—and one that is easily understood by an organic chemist—is the use of *deuterium labeling*.

Because a deuteron has a much smaller magnetic moment than a proton, it absorbs at a much higher field and so gives no signal in the proton NMR spectrum. Furthermore, its coupling with a proton is weak and it ordinarily broadens, but does not split, a proton's signal; even this effect can be eliminated by double irradiation.

As a result, then, the replacement of a proton by a deuteron removes from an NMR spectrum both the signal from that proton and the splitting by it of signals of other protons; it is as though there were no hydrogen at all at that position in the molecule. For example:

CH_3-CH_2-	CH_2D-CH_2-	CH_3-CHD-
Triplet Quartet	Triplet Triplet	Doublet Quartet
3H 2H	2H 2H	3H 1H

One can use deuterium labeling to find out which signal is produced by which proton or protons: one observes the disappearance of a particular signal when a proton in a known location is replaced by deuterium. One can use deuterium labeling to simplify a complicated spectrum so that a certain set of signals can be seen more clearly: see, for example, Fig. 16.18, p. 602. (This figure also illustrates a point made at the beginning of this section: the formidable looking nine-peak multiplet is analyzed without too much difficulty.)

16.13 Equivalence of protons: a closer look

We have seen that equivalence—or non-equivalence—of protons is fundamental to the NMR spectrum, since it affects both the number of signals and their splitting. Let us look more closely at equivalence, and see how it is affected by the rate at which certain molecular changes occur:

(a) *rotations about single bonds*, as in the interconversion between conformations of substituted ethanes or cyclohexanes;

(b) *inversion of molecules*, that is, the turning inside out of pyramidal molecules like amines (Sec. 26.6);

$$H \mathbin{-} N \rightleftharpoons N \mathbin{-} H$$

(c) *proton exchange*, as, for example, of alcohols (Sec. 18.11).

$$R^*O\mathbin{-}H^* + RO\mathbin{-}H \rightleftharpoons R^*O\mathbin{-}H + RO\mathbin{-}H^*$$

Each of these molecular changes can change the environment—both electronic and protonic—of a given proton, and hence can affect both its chemical shift and its coupling with other protons. The basic question that arises is whether or not the NMR spectrometer sees the proton in *each* environment or in an *average of all* of them. The answer is, in short, that it can often see the proton in either way, depending upon the temperature, and in this ability lies much of the usefulness of NMR spectroscopy.

In comparing it with other spectrometers, Professor John D. Roberts of the California Institute of Technology has likened the NMR spectrometer to a camera with a relatively slow shutter speed. Such a camera photographs the spokes of a wheel in different ways depending upon the speed with which the wheel spins: as sharp, individual spokes if spinning is slow; as blurred spokes if spinning is faster; and as a single circular smear if spinning is faster yet. In the same way, if the molecular change is relatively fast, the NMR spectrometer sees a proton in its average environment—a smeared-out picture; if the molecular process is slow, the spectrometer sees the proton in each of its environments.

In this section we shall examine the effects on the NMR spectrum of rotations about single bonds, and in later sections the effects of other molecular changes.

Let us return to ethyl chloride (Sec. 16.7), and focus our attention on the methyl protons. If, at any instant, we could look at an individual molecule, we would almost certainly see it in conformation I. One of the methyl protons is *anti* to the chlorine and two protons are *gauche*; quite clearly, the *anti* proton is in a

I

different environment from the others, and—for the moment—is not equivalent to them. Yet, we have seen, the three methyl protons of ethyl chloride give a single NMR signal (a triplet, because of the adjacent methylene group), and hence must be magnetically equivalent. How can this be? The answer is, of course, that rotation about the single bond is—compared with the NMR "shutter speed"—a fast process; the NMR "camera" takes a smeared-out picture of the three protons. Each proton is seen in an *average* environment, which is exactly the same as the average environment of each of the other two: one-third *anti*, and two-thirds *gauche*.

There are three conformations of ethyl chloride, II, III, and IV, identical except that a different individual proton occupies the *anti* position. Being of equal stability, the three conformations are exactly equally populated: one-third of the molecules in each. In one of these conformations a given proton is *anti* to chlorine, and in two it is *gauche*.

| II | III | IV |

1,1,2-Trichloroethane, to take another example, presents a somewhat different conformational picture, but the net result is the same: identical average environments and hence equivalence for the two methylene protons.

| V | VI | VII |

The environments of the two protons are the same in V. The environments are different for the two in VI and VII, but average out the same because of the equal populations of these enantiomeric conformations. (Here, however, we cannot say just *what* the average environment is, unless we know the ratio of V to the racemic modification (VI plus VII).)

With diastereotopic protons, on the other hand, the situation is different: diastereotopic protons are non-equivalent and no rotation will change this. We decided (Sec. 16.7) that the two C–1 protons of 1,2-dichloropropane, $CH_3CHClCH_2Cl$, are diastereotopic, since replacement of either one by an atom Z would yield diastereomers:

1,2-Dichloropropane

Rotation cannot interconvert the diastereomers, nor can it make the protons, H_a and H_b, equivalent. In none of the conformations (VIII, IX, or X)

is the environment of the two protons the same; nor is there a pair of mirror-image conformations to balance out their environments. (This holds true whether the compound is optically active or inactive; the presence or absence of an enantiomeric molecule has no effect on the environment of a proton in any individual molecule.) These diastereotopic protons give different signals, couple with the proton on C–2 (with different coupling constants), and couple with each other.

Cyclohexane presents an exactly analogous situation, since the transformation of one chair form into another involves rotations about single bonds. In any chair conformation there are two kinds of protons: six equatorial protons and six axial protons. Yet there is a single NMR signal for all twelve, since their *average* environments are identical: half equatorial, half axial.

If, however, we replace a proton by, say, bromine, the picture changes. Now, the axial and equatorial protons on each carbon are diastereotopic protons: replacement of one would give a *cis* diastereomer, replacement of the other a *trans* diastereomer. Protons H_a and H_b—or any other geminal pair on the ring—have different environments. When H_a is equatorial, so is —Br, and when H_a is axial, so is —Br; H_b always occupies a position opposite to that of —Br. Furthermore, the stabilities and hence populations of the two conformations will, in general, be different, and H_a and H_b will spend different fractions of their time in axial and equatorial positions; however, even if by coincidence the conformations are of equal stability, H_a and H_b are still not equivalent.

So far, we have discussed situations in which the speed of rotation about single bonds is so fast that the NMR spectrometer sees protons in their average environment. This is the *usual* situation. It is this situation in which our earlier test for equivalence would work: if replacement of either of two protons by Z would give the same (or enantiomeric) products, the protons are equivalent. We ignore conformers in judging the identity of two products.

Now, if—by lowering the temperature—we could sufficiently slow down rotations about single bonds, we would expect an NMR spectrum that reflects the "instantaneous" environments of protons in each conformation. *This is exactly what happens.* As cyclohexane, for example, is cooled down, the single sharp peak observed at room temperature is seen to broaden and then, at about $-70\,°C$, to split into two peaks, which at $-100\,°C$ are clearly separated: one peak is due to axial protons, and the other peak is due to equatorial protons.

This does *not* mean that the molecule is frozen into a single conformation; it still flips back and forth between two (equivalent) chair conformations; a given proton is axial one moment and equatorial the next. It is just that now the time between interconversions is long enough that we "photograph" the molecule, not as a blur, but sharply as one conformation or the other.

By study of the broadening of the peak, or of the coalescence of the two peaks, it is possible to estimate the E_{act} for rotation. Indeed, it was by this method that the barrier of 11 kcal/mol (Sec. 12.11) was calculated.

Problem 16.13 The fluorine NMR spectrum (Sec. 16.10) of 1,2-difluorotetrachloro-ethane, $CFCl_2CFCl_2$, shows a single peak at room temperature, but at $-120\,°C$ shows two peaks (singlets) of unequal area. Interpret each spectrum, and account for the difference. What is the significance of the unequal areas of the peaks in the low-temperature spectrum? Why is there no splitting in either spectrum?

Problem 16.14 At room temperature, the fluorine NMR spectrum of CF_2BrCBr_2CN (3,3-difluoro-2,2,3-tribromopropanenitrile) shows a single sharp peak. As the temperature is lowered this peak broadens and, at $-98\,°C$, is split into two doublets (equal spacing) and a singlet. The combined area of the doublets is considerably larger than—more than twice as large as—the area of the singlet. Interpret each spectrum, and account for the relative peak areas in the low-temperature spectrum.

16.14   Carbon-13 NMR (CMR) spectroscopy

Among the atoms that, like the proton, give rise to NMR spectra is one of the isotopes of carbon, ^{13}C. The ^{13}C NMR (CMR) spectrum is generated in the same fundamental way as the proton NMR spectrum (PMR), and the same basic principles that we learned before apply here, too. Practically, however, obtaining a useable spectrum is more difficult for CMR than for proton NMR, and requires more sophisticated instrumentation. Such instrumental methods have been developed in the years since about 1970, and today CMR spectroscopy is used routinely to complement proton NMR spectroscopy.

To distinguish these two kinds of nuclear magnetic resonance, we shall generally use the terms "CMR" and "proton NMR". When used alone, "NMR" should be taken to mean "proton NMR".

The isotope ^{13}C makes up only 1.1% of naturally occurring carbon, but the sensitivity of modern spectrometers makes this level adequate for the measurement of CMR spectra. Indeed, as we shall see in the following section, this low natural abundance is actually an advantage.

The CMR spectrum gives much the same kinds of information as proton NMR, but now the information is directly *about the carbon skeleton*—not just the protons attached to it.

(a) The *number of signals* tells us how many different carbons—or different sets of equivalent carbons—there are in a molecule.

(b) The *splitting of a signal* tells us how many hydrogens are attached to each carbon.

(c) The *chemical shift* tells us the hybridization (sp^3, sp^2, sp) of each carbon.

(d) The *chemical shift* tells us about the electronic environment of each carbon with respect to other, nearby carbons or functional groups.

(For reasons we shall not go into, we cannot in general use signal intensities in CMR to tell us how many nuclei are giving rise to each signal. In most cases, this is not a serious handicap, and if necessary can, with some trouble, be overcome.)

In the following sections, we shall look at various aspects of the CMR spectrum. Let us begin with the splitting of signals.

16.15   CMR. Splitting

One of the major practical problems in CMR spectroscopy is the splitting of signals: too much splitting, potentially, and hence spectra that would be too complicated to interpret easily.

Part of the problem is already solved for us by the low natural abundance of ^{13}C. Only occasionally is a ^{13}C near enough to another ^{13}C for ^{13}C–^{13}C spin–spin coupling to occur. As a result *CMR spectra do not ordinarily show carbon–carbon splitting*, and are thus enormously simplified. (An equal blessing: proton spectra do not show splitting by ^{13}C!)

But there remains splitting of ^{13}C signals by protons. In a CMR spectrum we cannot see the *absorption* by protons because these signals are far off the scale. But we can see *splitting* of a carbon signal by protons: protons on the carbon itself, and on more distant carbons as well. Unless something is done about this, the spectrum will consist of many overlapping multiplets very difficult to interpret.

Unwanted splitting is removed by decoupling (Sec. 16.12) the ^{13}C spin from that of the proton. This decoupling can be done in either of two principal ways, depending upon the frequency of the radiation used in the double resonance.

One method of decoupling gives a *completely* **proton-decoupled** spectrum. This spectrum shows *no splitting at all*; it consists of a set of single peaks, one for each carbon—or each set of equivalent carbons—in the molecule. Even for very complicated molecules, such a spectrum is amazingly simple. This is the kind of spectrum most commonly run for structural analysis, and (except for Fig. 16.21) is the kind shown throughout this book.

Look, for example, at Fig. 16.19, which shows the proton-decoupled spectrum of *sec*-butyl bromide. There are four carbons in this molecule, all different—that is, non-equivalent.

$$\underset{a}{CH_3}-\underset{c}{CH_2}-\underset{d}{\underset{|}{\underset{Br}{CH}}}-\underset{b}{CH_3}$$

4 CMR signals
sec-Butyl bromide

In the spectrum we see four peaks, one for each of these four carbons.

a 12.1
b 26.1
c 34.3
d 52.8

a   c   d   b
CH₃—CH₂—CH—CH₃
             |
            Br

Sadtler 613C   © Sadtler Research Laboratories, Division of Bio-Rad, Inc., (1976).

Figure 16.19   Proton-decoupled CMR spectrum of *sec*-butyl bromide.

We judge the equivalence or non-equivalence of carbons in the same way that we dealt with protons (Sec. 16.7). We must not forget that, to be chemically equivalent, carbons, like protons, must be *stereo*chemically equivalent: in an achiral medium diastereotopic carbons will give different signals, and enantiotopic carbons will not.

Problem 16.15   Draw the structural formula for each of the following compounds (disregarding enantiomerism), and label all sets of equivalent carbons.

(i) How many signals would you expect in the CMR spectrum of each?
(ii) Which compounds, if any, could you distinguish from all the others of the same set simply on the basis of the numbers of signals?
(a) the two isomers of C₃H₇Cl   (c) the five isomers of C₆H₁₄
(b) the three isomers of C₅H₁₂   (d) *o*-, *m*-, and *p*-xylenes
          (e) *n*-octane, 2-methylheptane, 3-methylheptane, 4-methylheptane
          (f) mesitylene, *p*-ethyltoluene, isopropylbenzene
          (g) 1-hexene, *cis*-3-hexene, *trans*-3-hexene, 2-methyl-2-pentene

Problem 16.16   Give a structure or structures consistent with each of the CMR spectra shown in Fig. 16.20 (p. 610).

A second method of decoupling (called *off-resonance*) gives a spectrum which shows splitting of the carbon signal only *by protons attached to that carbon itself.* That is, we see only ¹³C—H coupling and not ¹³C—C—H or ¹³C—C—C—H coupling. We shall refer to this kind of spectrum as a **proton-coupled** spectrum.

For each carbon, then, the multiplicity of the signal depends upon how many protons are attached to it:

—C—	—C—	—C—	—C—H
	H	H, H	H, H, H
No protons	One proton	Two protons	Three protons
Singlet	*Doublet*	*Triplet*	*Quartet*

a 9.5 q
b 32.0 q
c 39.0 t
d 70.9 s

(*a*) C$_5$H$_{11}$Cl

Sadtler 1014C   © Sadtler Research Laboratories, Division of Bio-Rad, Inc., (1976).

a 22.1 q
b 25.8 d
c 41.8 t
d 43.2 t

(*b*) C$_5$H$_{11}$Cl

Sadtler 6450C   © Sadtler Research Laboratories, Division of Bio-Rad, Inc., (1979).

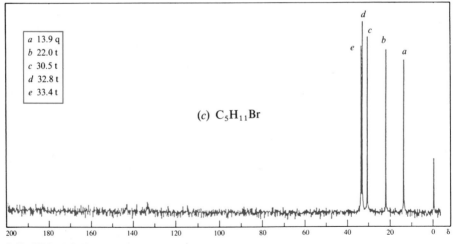

a 13.9 q
b 22.0 t
c 30.5 t
d 32.8 t
e 33.4 t

(*c*) C$_5$H$_{11}$Br

Sadtler 252C   © Sadtler Research Laboratories, Division of Bio-Rad, Inc., (1976).

Figure 16.20   CMR spectra for Problem 16.16 (p. 609).

Figure 16.21   Proton-coupled CMR spectrum of *sec*-butyl bromide.

In Fig. 16.21, for example, we see again a CMR spectrum of *sec*-butyl bromide, but this time proton-coupled. Now each peak is a multiplet: we see a doublet, a triplet, and two quartets.

$$\underset{\text{Br}}{\overset{\begin{array}{cccc} q & t & d & q \end{array}}{CH_3-CH_2-\underset{|}{CH}-CH_3}}$$

sec-Butyl bromide

So now, we have two kinds of CMR spectra which give us two kinds of information about the structure of a molecule. The proton-decoupled spectrum tells us how many different carbons there are (and much besides), and the proton-coupled spectrum tells us how many protons are attached to each of these carbons. Together, these spectra give us a remarkably detailed picture of the molecule.

Most of the CMR spectra in this book will *display* the proton-decoupled spectra, and will *list* the multiplicity of the peaks in the upper left-hand corner. (See, for example, Fig. 16.20). We shall thus have both kinds of information given within a single frame.

Problem 16.17   (a) Go back to Problem 16.15 (p. 609) and tell the kind of splitting expected for each signal. (b) Would this information enable you to settle any ambiguities in your answer to part (ii)?

Problem 16.18   In Problem 16.16 (p. 609) you analyzed some CMR spectra on the basis of the number of signals alone. (a) Does the splitting summarized in the corner of the spectrum lead you to change any of your answers? (b) Assign as many peaks as you can to specific carbons in the compound.

16.16   CMR. Chemical shift

Chemical shifts in the CMR spectrum arise in basically the same way as in the proton NMR spectrum (Sec. 16.6). Each carbon nucleus has its own electronic environment, different from the environment of other, non-equivalent nuclei; it feels a different magnetic field, and absorbs at a different applied field strength. But the shifts in CMR differ in several ways from those in proton NMR.

To begin with, chemical shifts are much larger in CMR than in proton NMR. Figure 16.22 summarizes shifts for carbons of various "kinds". As we see, the scale extends from δ 0 to beyond δ 200, more than 30 times as wide as in NMR.

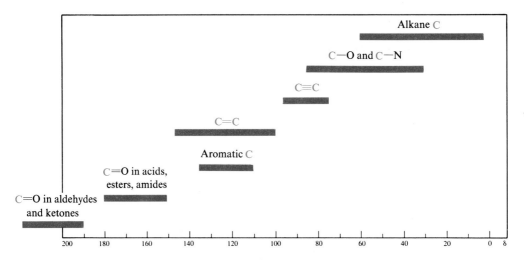

Figure 16.22   Chemical shifts for ^{13}C in various kinds of compounds.

Of these shifts, the biggest and most important are determined by the **hybridization** of the carbon: something that, of course, is not a factor for the proton. Look, for example, at the spectrum of 1-octene (Fig. 16.23). We see the peaks for the sp^3-hybridized carbons upfield, between δ 14.1 and δ 34.0, and for the sp^2-hybridized carbons *over 100 ppm downfield* from them, at δ 113 and δ 140!

$$\overset{14.1}{CH_3}-\overset{22.9}{CH_2}-\overset{32.1}{CH_2}-\overset{29.3}{CH_2}-\overset{29.1}{CH_2}-\overset{34.1}{CH_2}-\overset{139}{CH}=\overset{114}{CH_2}$$

1-Octene

Aromatic carbons are also sp^2-hybridized, and also absorb downfield, in much the same region as alkene carbons do. In the spectrum of ethylbenzene (Fig. 16.23) we again see two widely separated sets of peaks: upfield, a set from the side-chain (sp^3-hybridized) carbons, and downfield, after a 100-ppm gap, a set from the ring (sp^2-hybridized) carbons.

$$\overset{29.1}{CH_2}-\overset{15.6}{CH_3}$$

144.2
127.9
128.4
125.7

Ethylbenzene

a 14.1 q	e 32.1 t
b 22.9 t	f 34.1 t
c 29.1 t	g 114.2 t
d 29.3 t	h 139.1 d

(a) 1-Octene

CH₃—CH₂—CH₂—CH₂—CH₂—CH₂—CH=CH₂

Sadtler 1621C   © Sadtler Research Laboratories, Division of Bio-Rad, Inc., (1977).

a 15.6 q	d 127.9 d
b 29.1 t	e 128.4 d
c 125.7 d	f 144.2 s

(b) Ethylbenzene

—CH₂—CH₃

Sadtler 600C   © Sadtler Research Laboratories, Division of Bio-Rad, Inc., (1976).

a 13.7 q
b 18.3 t
c 22.1 t
d 30.9 t
e 68.4 d
f 84.5 s

(c) 1-Hexyne

CH₃—CH₂—CH₂—CH₂—C≡CH

Sadtler 2785C   © Sadtler Research Laboratories, Division of Bio-Rad, Inc., (1977).

Figure 16.23   CMR spectra: chemical shift. (a) 1-Octene; (b) ethylbenzene; (c) 1-hexyne.

Absorption by triply bonded (*sp*-hybridized) carbon falls between the regions for *sp*^{3}-hybridized and *sp*^{2}-hybridized carbon, as shown for 1-hexyne (Fig. 16.23).

$$\underset{\text{1-Hexyne}}{\overset{\displaystyle 13.7 \quad 22.1 \quad 30.9 \quad 18.3 \quad 84.5 \quad 68.4}{CH_3-CH_2-CH_2-CH_2-C\equiv\!\equiv CH}}$$

In relating structure to chemical shift, then, *we begin with the* **hybridization of carbon**.

Next, we consider the *effects of substituents*, which are superimposed on the hybridization effects. As in proton NMR, most substituents in most positions deshield the nucleus, and shift the signal downfield. But with carbon these effects are bigger, are felt from farther away, and fall into different patterns. To get some idea of what these patterns are like, let us examine the effects of several substituents on absorption by *sp*^{3}-*hybridized carbons*.

Let us look first at the effects of chlorine on the absorption by various carbons of a saturated chain. The spectra of *n*-pentane and 1-chloropentane, for example, give data that we can summarize like this:

$$\underset{\text{\textit{n}-Pentane}}{\overset{\displaystyle 13.7 \quad 22.6 \quad 34.5 \quad 22.6 \quad 13.7}{CH_3-CH_2-CH_2-CH_2-\underset{\displaystyle H}{\overset{\displaystyle |}{CH_2}}}} \qquad \underset{\text{1-Chloropentane}}{\overset{\displaystyle 13.6 \quad 22.1 \quad 29.2 \quad 32.7 \quad 44.3}{CH_3-CH_2-CH_2-CH_2-\underset{\displaystyle Cl}{\overset{\displaystyle |}{CH_2}}}}$$

Let us compare, carbon by carbon, the δ values for the two compounds.

In the signal for C–1, chlorine causes a very large downfield shift, from δ 13.7 to δ 44.3, a difference of +30.6 ppm. Such a shift, *for the carbon bearing the substituent*, is called an **α-effect**.

An α-effect $\qquad\qquad\qquad\qquad\qquad C-C-\overset{\displaystyle +30.6}{\underset{\displaystyle \overset{\displaystyle |}{Cl}}{C}}$

In the signal for C–2, chlorine again exerts a downfield shift, from δ 22.6 to δ 32.7, a difference of +10.1 ppm. Such a shift, *for the carbon once removed from the carbon bearing the substituent*, is called a **β-effect**.

A β-effect $\qquad\qquad\qquad\qquad\qquad C-\overset{\displaystyle +10.1}{C}-\underset{\displaystyle \overset{\displaystyle |}{Cl}}{C}$

At C–3 we see a reversal of the substituent effect. Absorption here is shifted *upfield*, from δ 34.5 to δ 29.2, a difference of −5.3 ppm. Such a shift, *for the carbon twice removed from the carbon bearing the substituent*, is called a **γ-effect**.

A γ-effect $\qquad\qquad\qquad\qquad\qquad \overset{\displaystyle -5.3}{C}-C-\underset{\displaystyle \overset{\displaystyle |}{Cl}}{C}$

Beyond the γ-carbon, we see, the effect of chlorine, like that of other substituents, is very small.

Each substituent chlorine exerts effects of about these same sizes on absorption by saturated carbon in a wide variety of compounds.

Nearly all substituent effects on absorption by sp^3-hybridized carbon follow the same pattern as those for chlorine: α- and β-effects *downfield*, with α greater than β; and γ-effects still smaller, and *upfield*. And for most of these substituents, the sizes of the effects are, like those of chlorine, quite large. Consider, for example, the α-effects exerted by these substituents attached to C–1 of pentane: F, $+70.1$ ppm; Br, $+19.3$ ppm; NH_2, $+29.7$ ppm; OH, $+48.3$ ppm; NO_2, $+64.5$ ppm.

Alkyl groups exert smaller effects than other substituents, and follow a somewhat different pattern. Let us look, for example, at the effects of a methyl group, using data from the spectra of *n*-pentane and *n*-hexane. If we display the δ values as before, considering *n*-hexane to be *n*-pentane with a methyl substituent on C–1,

13.7 22.6 34.5 22.6 13.7   13.9 22.9 32.0 32.0 22.9
CH_3—CH_2—CH_2—CH_2—CH_2 CH_3—CH_2—CH_2—CH_2—CH_2
                         | |
                         H CH_3

n-Pentane   *n*-Hexane

we can calculate the following substituent effects of methyl:

γ β α
-2.5 $+9.5$ $+9.1$
C——C——C
          |
          CH_3

These effects are typical of alkyl groups acting on the absorption by saturated carbon: α- and β-effects *downfield*, and of about the same size; and γ-effects smaller and *upfield*.

What we have discussed so far are substituent effects on chemical shifts for sp^3-hybridized carbon. For absorption by sp^2-hybridized carbons—in alkenes and aromatic rings—the pattern of effects is somewhat different. Many substituents exert alternating effects, which we shall not go into, but which will be evident in some of the spectra we shall encounter.

The presence of a carbon–carbon double bond in a molecule can introduce a new factor, *geometric isomerism*, which has important effects on the absorption by sp^3-hybridized carbons. To see how this stereochemical factor works, let us compare the absorption data for propylene with the data for *cis*- and *trans*-2-butene.

18.7   11.4   11.4   16.8
CH_3 H CH_3 CH_3 H CH_3
   \136 115/   \124 124/   \125 125/
    C══C   C══C   C══C
   /   \   /   \   /   \
  H   H   H   H CH_3 H
                                                   16.8

Propylene   *cis*-2-Butene   *trans*-2-Butene

Let us focus our attention on the methyl carbon of propylene, and see how its absorption is affected by the substitution of a methyl group on one or the other of the vinylic hydrogens. We are, of course, looking at γ-effects.

Propylene   *cis*-2-Butene   *trans*-2-Butene

As we see, these γ-effects are both upfield: -7.3 ppm for the *cis* isomer and -1.9 ppm for the *trans*. But *the γ-effect for the cis isomer is stronger, by 5.4 ppm.*

Size of γ-effects

cis > *trans*

The stereochemical influence on γ-effects in alkenes is a general one, and is extremely useful in assigning configuration to geometric isomers.

In our work here with CMR, we shall make use of chemical shifts in only a general way: to tell us the hybridization of various carbons in a molecule, and something about their environments. We shall be concerned with the *regions* of the spectrum in which absorption occurs, and not—except where geometric isomers are involved—with precise δ values. We shall return to this point in Sec. 16.19 where we discuss the analysis of hydrocarbons.

Problem 16.19 Using the δ values for *n*-pentane given on p. 615, and the α-, β-, and γ-effects calculated there for a methyl group, predict the values for each carbon in 2-methylpentane. How do these values compare with the actual δ values shown below?

2-Methylpentane

Problem 16.20 (a) Using the δ values for propylene and *cis*- and *trans*-2-butene shown above, calculate the α- and β-effects of the added methyl group on the doubly bonded carbons. (b) Compare the δ values for propylene with those shown above for 1-butene. Calculate the β- and γ-effects of this added methyl group on the doubly bonded carbons. How does the β-effect here compare with that in part (a)? (c) From the standpoint of structure, should the β-effects in (a) and (b) be strictly comparable?

Problem 16.21 In Problem 16.18 (p. 611) you analyzed some CMR spectra on the basis of numbers of signals and of splitting. Re-examine these spectra in light of what you know about chemical shifts. (a) Can you draw any further conclusions about the structures of the unknown compounds? (b) Can you assign more peaks to specific carbons?

Problem 16.22   The CMR spectra of the two stereoisomeric 3-hexenes show the following δ values.

| 13.9 | 25.8 | 131 | 131 | 25.8 | 13.9 | | 14.3 | 20.6 | 131 | 131 | 20.6 | 14.3 |
| CH_3 | CH_2 | CH | CH | CH_2 | CH_3 | | CH_3 | CH_2 | CH | CH | CH_2 | CH_3 |

$$CH_3-CH_2-CH=CH-CH_2-CH_3 \qquad CH_3-CH_2-CH=CH-CH_2-CH_3$$

Isomer A   Isomer B

Which is the *cis* isomer, and which is the *trans*?

16.17   The electron spin resonance (ESR) spectrum

Let us consider a free radical placed in a magnetic field and subjected to electromagnetic radiation; and let us focus our attention, not on the nuclei, but on the odd, unpaired electron. This electron spins and thus generates a magnetic moment, which can be lined up with or against the external magnetic field. Energy is required to change the spin state of the electron, from alignment with the field to the less stable alignment, against the field. This energy is provided by absorption of radiation of the proper frequency. An absorption spectrum is produced, which is called an *electron spin resonance (ESR) spectrum* or an *electron paramagnetic resonance (EPR) spectrum*.

The ESR spectrum is thus analogous to the NMR spectrum. An electron, however, has a much larger magnetic moment than the nucleus of a proton, and more energy is required to reverse the spin. In a field of 3200 gauss, for example, where NMR absorption would occur at about 14 MHz, ESR absorption occurs at a much higher frequency: 9000 MHz, in the *microwave* region.

Like NMR signals, ESR signals show splitting, and from exactly the same cause, coupling with the spins of certain nearby nuclei: for example, protons near carbon atoms that carry—or help to carry—the odd electron. For this reason, ESR spectroscopy can be used not only to detect the presence of free radicals and to measure their concentration, but also to give evidence about their structure: what free radicals they are, and how the odd electron is spread over the molecule.

Problem 16.23   Although all electrons spin, only molecules containing unpaired electrons—only free radicals—give ESR spectra. Why is this? (*Hint*: Consider the possibility (a) that one electron of a pair has its spin reversed, or (b) that both electrons of a pair have their spins reversed.)

Problem 16.24   In each of the following cases, tell what free radical is responsible for the ESR spectrum, and show how the observed splitting arises. (a) X-irradiation of methyl iodide at low temperatures: a four-line signal. (b) γ-irradiation at 77 K of propane and of *n*-butane: symmetrical signals of, respectively, 8 lines and 7 lines. (c) Triphenylmethyl chloride + zinc: a very complex signal.

16.18   Spectroscopic analysis of hydrocarbons. Infrared spectra

In this first encounter with infrared spectra, we shall see absorption bands due to vibrations of carbon–hydrogen and carbon–carbon bonds: bands that will constantly reappear in all the spectra we meet, since along with their various functional groups, compounds of all kinds contain carbon and hydrogen. We must expect to find these spectra complicated and, at first, confusing. Our aim is to learn to pick out of the confusion those bands that are most characteristic of certain structural features.

Let us look first at the various kinds of vibration, and see how the positions of the bands associated with them vary with structure.

Bands due to *carbon–carbon stretching* may appear at about 1500 and 1600 cm^{-1} for aromatic bonds, at 1650 cm^{-1} for double bonds (shifted to about 1600 cm^{-1} by conjugation), and at 2100 cm^{-1} for triple bonds. These bands, however, are often unreliable. (They may disappear entirely for fairly symmetrically substituted alkynes and alkenes, because the vibrations do not cause the change in dipole moment that is essential for infrared absorption.) More generally useful bands are due to the various carbon–hydrogen vibrations.

Absorption due to *carbon–hydrogen stretching*, which occurs at the high-frequency end of the spectrum, is characteristic of the hybridization of the carbon holding the hydrogen: at 2800–3000 cm^{-1} for tetrahedral carbon; at 3000–3100 cm^{-1} for trigonal carbon (alkenes and aromatic rings); and at 3300 cm^{-1} for digonal carbon (alkynes).

Absorption due to various kinds of *carbon–hydrogen bending*, which occurs at lower frequencies, can also be characteristic of structure. Methyl and methylene groups absorb at about 1430–1470 cm^{-1}; for methyl, there is another band, quite characteristic, at 1375 cm^{-1}. The isopropyl "split" is characteristic: a doublet, with equal intensity of the two peaks, at 1370 and 1385 cm^{-1} (confirmed by a band at 1170 cm^{-1}). *tert*-Butyl gives an unsymmetrical doublet: 1370 cm^{-1} (*strong*) and 1395 cm^{-1} (*moderate*).

Carbon–hydrogen bending in alkenes and aromatic rings is both in-plane and out-of-plane, and of these the latter kind is more useful. For **alkenes**, out-of-plane bending gives strong bands in the 800–1000 cm^{-1} region, the exact location depending upon the nature and number of substituents, and the stereochemistry:

RCH=CH$_2$	910–920 cm^{-1} 990–1000	*cis*-RCH=CHR	675–730 cm^{-1} (*variable*)
R$_2$C=CH$_2$	880–900	*trans*-RCH=CHR	965–975

For **aromatic rings**, out-of-plane C—H bending gives strong absorption in the 675–870 cm^{-1} region, the exact frequency depending upon the number and location of substituents; for many compounds absorption occurs at:

monosubstituted	690–710 cm^{-1} 730–770	*m*-disubstituted	690–710 cm^{-1} 750–810
o-disubstituted	735–770	*p*-disubstituted	810–840

Now, what do we look for in the infrared spectrum of a hydrocarbon? To begin with, we can readily tell whether the compound is aromatic or purely aliphatic. The spectra in Fig. 16.2 (p. 575) show the contrast that is typical: aliphatic absorption is strongest at higher frequency and is essentially missing below 900 cm^{-1}; aromatic absorption is strong at lower frequencies (C—H out-of-plane bending) between 650 and 900 cm^{-1}. In addition, an aromatic ring will show C—H stretching at 3000–3100 cm^{-1}; often, there is carbon–carbon stretching at 1500 and 1600 cm^{-1} and C—H in-plane bending in the 1000–1100 cm^{-1} region.

An alkene shows C—H stretching at 3000–3100 cm^{-1} and, most characteristically, strong out-of-plane C—H bending between 800 and 1000 cm^{-1}, as discussed above.

A terminal alkyne, RC≡CH, is characterized by its C—H stretching band, a strong and sharp band at 3300 cm^{-1}, and by carbon–carbon stretching at 2100 cm^{-1}. A disubstituted alkyne, on the other hand, does not show the 3300 cm^{-1} band and, if the two groups are fairly similar, the 2100 cm^{-1} band may be missing, too.

Some of these characteristic bands are labeled in the spectra of Fig. 16.2 (p. 575).

Problem 16.25   What is a likely structure for a hydrocarbon of formula C_6H_{12} that shows strong absorption at 2920 and 2840 cm^{-1}, and at 1450 cm^{-1}; none above 2920 cm^{-1}; and below 1450 cm^{-1} none until about 1250 cm^{-1}?

16.19   Spectroscopic analysis of hydrocarbons. NMR and CMR

For hydrocarbons as for other kinds of compounds, we shall find that, where the infrared spectrum helps to tell us what *kind* of compound we are dealing with, nuclear magnetic resonance helps to tell us *which* compound.

What do we look for in the NMR or CMR spectrum of a hydrocarbon? To begin with, there is the matter of *where* the peaks appear, which depends upon **chemical shifts**. We have not attempted to memorize lists of δ values, but we know in a general way where certain "kinds" of carbon and certain "kinds" of protons absorb. And so we look first at the disposition of peaks.

Alkanes and alkane-like, saturated groups will give *upfield* peaks in both CMR and NMR: absorption by sp^3-hybridized carbons and the protons attached to them.

An *aromatic ring* will show *downfield* absorption in both CMR and NMR. An **alkene** will show similar absorption in CMR, but *not* in NMR; vinylic protons absorb well upfield from aromatic protons, and this will often let us distinguish between the two kinds of structure. (This distinction is, of course, clearly shown by the infrared spectrum.)

The triply bonded, sp-hybridized carbons of an **alkyne** will give peaks between those from sp^3- and sp^2-hybridized carbons.

Attachment of electronegative atoms—halogen, oxygen, nitrogen—will shift peaks downfield in both CMR and NMR, but not usually outside the region where we expect to see them.

Like infrared spectra, then, chemical shifts in nuclear magnetic resonance help to tell us what *kind* of compound we are dealing with.

To find out *which* compound we have, we turn to the **number of signals** and their **splitting**. As we have already seen in this chapter, it is these that give us our closest look at the exact structure of the molecule. We see how many "kinds" of carbons there are, and how many "kinds" of protons. For each carbon we see how many protons are attached to it; for each proton we see how many other protons are nearby.

At this point we can often arrive at a single structure for an unknown compound. We may have to return to chemical shifts to clear up remaining doubts: the point of attachment of a substituent, for example. To establish the configuration about a carbon–carbon double bond we may use γ-effects in CMR or coupling constants in proton NMR (Sec. 16.11). Sometimes we can do all this using only the molecular formula and one or the other of the nuclear magnetic resonance spectra. Together, NMR and CMR make an even more powerful team—especially if we have in addition the infrared spectrum.

16.20   Spectroscopic analysis of alkyl halides

Throughout this chapter, we have used many alkyl halides as examples, and know pretty well how the presence of a halogen changes the spectrum of a molecule: in general it causes a downfield shift, both in CMR and in NMR.

About Analyzing Spectra

In problems you will be given the molecular formula of a compound and asked to deduce its structure from its spectroscopic properties: sometimes from its infrared or NMR or CMR spectrum alone, sometimes from two or three of these. The compound will generally be a simple one, and you may need to look only at a few features of the spectra to find the answer. To confirm your answer, however, and to gain experience, see how much information you can get from the spectra: try to identify as many infrared bands as you can, to assign all NMR and CMR signals to specific protons and carbons, and to analyze the various spin–spin splittings. Above all, look at as many spectra as you can find: in the laboratory, in other books, in catalogs of spectra in the library.

PROBLEMS

1. Give a structure or structures consistent with each of the following sets of NMR data.

(a) $C_3H_3Cl_5$
   a triplet, δ 4.52, 1H
   b doublet, δ 6.07, 2H

(b) $C_3H_5Cl_3$
   a singlet, δ 2.20, 3H
   b singlet, δ 4.02, 2H

(c) C_4H_9Br
   a doublet, δ 1.04, 6H
   b multiplet, δ 1.95, 1H
   c doublet, δ 3.33, 2H

(d) $C_{10}H_{14}$
   a singlet, δ 1.30, 9H
   b singlet, δ 7.28, 5H

(e) $C_{10}H_{14}$
   a doublet δ 0.88, 6H
   b multiplet, δ 1.86, 1H
   c doublet, δ 2.45, 2H
   d singlet, δ 7.12, 5H

(f) C_9H_{10}
   a quintet, δ 2.04, 2H
   b triplet, δ 2.91, 4H
   c singlet, δ 7.17, 4H

(g) $C_{10}H_{13}Cl$
   a singlet, δ 1.57, 6H
   b singlet, δ 3.07, 2H
   c singlet, δ 7.27, 5H

(h) $C_{10}H_{12}$
   a multiplet, δ 0.65, 2H
   b multiplet, δ 0.81, 2H
   c singlet, δ 1.37, 3H
   d singlet, δ 7.17, 5H

(i) $C_9H_{11}Br$
   a quintet, δ 2.15, 2H
   b triplet, δ 2.75, 2H
   c triplet, δ 3.38, 2H
   d singlet, δ 7.22, 5H

(j) $C_3H_5ClF_2$
   a triplet, δ 1.75, 3H
   b triplet, δ 3.63, 2H

2. Give a structure or structures consistent with each of the following sets of CMR data.

(a) $C_3H_5Cl_3$
   a triplet δ 45.3
   b doublet δ 59.0

(b) C_4H_9Br
   a quartet δ 20.9
   b doublet δ 30.7
   c triplet δ 42.2

(c) $C_3H_6Cl_2$
   a quartet δ 22.4
   b triplet δ 49.5
   c doublet δ 55.8

(d) C_3H_5Br
   a triplet δ 32.6
   b triplet δ 118.8
   c doublet δ 134.2

(e) C_6H_{10}
   a triplet δ 22.9
   b triplet δ 25.3
   c doublet δ 127.2

(f) $C_4H_8Br_2$
   a quartet δ 10.9
   b triplet δ 29.0
   c triplet δ 35.5
   d doublet δ 54.3

3. Identify the stereoisomeric 1,3-dibromo-1,3-dimethylcyclobutanes on the basis of their NMR spectra.

Isomer X: singlet, δ 2.13, 6H
              singlet, δ 3.21, 4H

Isomer Y: singlet, δ 1.88, 6H
              doublet, δ 2.84, 2H
              doublet, δ 3.54, 2H
              doublets have equal spacing

4. When mesitylene (NMR spectrum, Fig. 16.5, p. 586) is treated with HF and SbF_5 in liquid SO_2 solution, the following peaks, all singlets, are observed in the NMR spectrum: δ 2.8, 6H; δ 2.9, 3H; δ 4.6, 2H; and δ 7.7, 2H. To what compound is the spectrum due? Assign all peaks in the spectrum.
   Of what general significance to chemical theory is such an observation?

5. (a) On catalytic hydrogenation, compound A, C_5H_8, gave *cis*-1,2-dimethylcyclo-propane. On this basis, three isomeric structures were considered possible for A. What were they? (b) Absence of infrared absorption at 890 cm^{-1} made one of the structures unlikely. Which one was it? (c) The NMR spectrum of A showed signals at δ 2.22 and δ 1.04 with intensity ratio 3:1. Which of the three structures in (a) is consistent with this? (d) The base peak in the mass spectrum was found at m/e 67. What ion was this peak probably due to, and how do you account for its abundance? (e) Compound A was synthesized in one step from open-chain compounds. How do you think this was done?

6. X-ray analysis shows that the [18]annulene (Problem 9, p. 497, $n = 9$) is planar. The NMR spectrum shows two broad bands: τ 1.1 and τ 11.8, peak area ratio 2:1. (a) Are these properties consistent with aromaticity? Explain. (b) Would you have predicted aromaticity for this compound? Explain. (*Hint*: Carefully draw a structural formula for the compound, keeping in mind bond angles and showing all hydrogen atoms.)

7. At room temperature, the proton NMR spectrum of 2,2,3,3-tetrachlorobutane shows a single sharp peak. Lowering the temperature causes a broadening of this peak until finally, at about $-45\ °\text{C}$, it separates into two peaks (singlets) of unequal intensities. (a) Account for the effect of temperature. (b) What is the significance of the unequal intensities in the low-temperature spectrum? (c) The larger peak in the low-temperature spectrum is downfield from the smaller. Given that halogens exert a deshielding effect from the *gauche* position, can you be more specific in your answer to (b)?

8. Hydrocarbon B, C_6H_6, gave an NMR spectrum with two signals: δ 6.55 and δ 3.84, peak area ratio 2:1. When warmed in pyridine for three hours, B was quantitatively converted into benzene.

Mild hydrogenation of B yielded C, whose spectra showed the following: mass spectrum, mol. wt. 82; infrared spectrum, no double bonds; NMR spectrum, one broad peak at δ 2.34.

(a) How many rings are there in C? (See Problem 12.20, p. 470.) (b) How many rings are there (probably) in B? How many double bonds in B? (c) Can you suggest a structure for B? For C?

(d) In the NMR spectrum of B, the upfield signal was a quintet, and the downfield signal was a triplet. How must you account for these splittings?

9. The five known 1,2,3,4,5,6-hexachlorocyclohexanes can be described in terms of the equatorial (e) or axial (a) disposition of successive chlorines: eeeeee, eeeeea, eeeeaa, eeaeea, eeeaaa. Their NMR spectra have been measured.

Which of these would give: (a) only one peak (two isomers); (b) two peaks, 5H:1H (one isomer); (c) two peaks, 4H:2H (two isomers)?

(d) Which one of the isomers in (a) would you expect to show no change in NMR spectrum at low temperature? Which one would show a split into two peaks? Predict the relative peak areas for the latter case.

10. (a) Although the NMR spectrum of *trans*-4-*tert*-butyl-1-bromocyclohexane is complicated, the signal from one proton stands clear (δ 3.83), downfield from the rest. Which proton is this, and why? (b) The *cis* isomer shows a corresponding peak, but at δ 4.63. Assuming that the *tert*-butyl group exerts no direct magnetic effect, to what do you attribute the difference in chemical shifts between the two spectra? These data are typical, and are the basis of a generalization relating conformation and chemical shift. What is that generalization?

11. The NMR spectrum of bromocyclohexane shows a downfield peak (1H) at δ 4.16. This signal is a single peak at room temperature, but at $-75\ °\text{C}$ separates into two peaks of *unequal* area (but totalling *one* proton): δ 3.97 and δ 4.64 in the ratio 4.6:1.0. How do you account for the separation of peaks? On the basis of your generalization of the previous problem, which conformation of the molecule predominates, and (at $-75\ °\text{C}$) what percentage of molecules does it account for?

12. At $90\ °\text{C}$ the CMR spectrum of *cis*-decalin (Problem 8, p. 473) shows three peaks of relative intensities 2:2:1. At $-50\ °\text{C}$ the spectrum shows five peaks of equal intensity.

(a) Account for the effect of temperature.

(b) The smaller peak in the high-temperature spectrum is not shifted by the drop in temperature; the other peaks are. To which carbons is this smaller peak due?

(c) Assign, as definitely as you can, the peaks in both spectra to carbons in *cis*-decalin.

13. (a) In the liquid form, *tert*-butyl fluoride and isopropyl fluoride gave the following NMR spectra.

> *tert*-butyl fluoride: doublet, δ 1.30, $J = 20$ Hz
> isopropyl fluoride: two doublets, δ 1.23, 6H, $J = 23$ Hz and 4 Hz
> two multiplets, δ 4.64, 1H, $J = 48$ Hz and 4 Hz

How do you account for each of these spectra? (*Hint*: See Sec. 16.10.)

(b) When the alkyl fluorides were dissolved in liquid SbF_5, the following NMR spectra were obtained.

> from *tert*-butyl fluoride: singlet, δ 4.35
> from isopropyl fluoride: doublet, δ 5.06, 6H, $J = 4$ Hz
> multiplet, δ 13.5, 1H, $J = 4$ Hz

To what molecule is each of these spectra due? (*Hint*: What does the disappearance of just half the peaks observed in part (a) suggest?) Is the very large downfield shift what you might have expected for molecules like these? Of what fundamental significance to organic theory are these observations?

14. Treatment of neopentyl chloride with the strong base sodamide ($NaNH_2$) yields a hydrocarbon of formula C_5H_{10}, which readily dissolves in concentrated sulfuric acid, but is not oxidized by cold, dilute, neutral permanganate. Its NMR spectrum shows absorption at δ 0.20 and δ 1.05 with peak area ratio 2:3. When the same reaction is carried out using the labeled alkyl halide, $(CH_3)_3CCD_2Cl$, the product obtained has its M^+ peak at m/e 71. What is a likely structure for the hydrocarbon, and how is it probably formed? Is the result of the labeling experiment consistent with your mechanism? (*Hint*: See Sec. 12.17.)

15. When methallyl chloride, $CH_2=C(CH_3)CH_2Cl$, was treated with sodamide in tetrahydrofuran solution, there was obtained a hydrocarbon, C_4H_6, which gave the following NMR spectrum:

> *a* doublet, δ 0.83, 2H, $J = 2$ Hz
> *b* doublet, δ 2.13, 3H, $J = 1$ Hz
> *c* multiplet, δ 6.40, 1H

(a) What is a likely structure for this hydrocarbon, and by what mechanism was it probably formed? (b) What product would you expect to obtain by the same reaction from allyl chloride?

16. Hydrocarbon D has been prepared in two different ways:

(i)

$$Cl\!-\!\square\!-\!Br + Na \longrightarrow D$$

1-Bromo-3-chlorocyclobutane

(ii)

$$CH_2=CH-CH_2-CHN_2 \xrightarrow{\text{light}} D$$

Allyldiazomethane

Mass spectrometry shows a molecular weight of 54 for D. (What is its molecular formula?) On gas chromatography, D was found to have a different retention time from cyclobutene, butadiene, or methylenecyclopropane. D was stable at 180 °C (unlike cyclobutene), but was converted into butadiene at 225 °C. The NMR spectrum of D showed: *a*, singlet, δ 0.45, 2H; *b*, multiplet, δ 1.34, 2H; *c*, multiplet, δ 1.44, 2H.

(a) What single structure for D is consistent with all these facts? (*Hint*: In analyzing the NMR spectrum, take stereochemistry into consideration.) (b) By what familiar reaction is D formed in (i)? In (ii)?

17. Give a structure or structures consistent with each of the infrared spectra in Fig. 16.24, p. 625.

18. Give a structure or structures consistent with each of the proton NMR spectra in Fig. 16.25, p. 626.

19. Give a structure or structures consistent with each of the CMR spectra in Fig. 16.26, p. 627.

20. Give a structure or structures consistent with each of the proton NMR spectra in Fig. 16.27, p. 628.

21. Give a structure or structures consistent with each of the CMR spectra in Fig. 16.28, p. 629.

22. Give a structure or structures consistent with each of the CMR spectra in Fig. 16.29, p. 630.

23. Give a structure or structures consistent with the CMR spectrum in Fig. 16.30, p. 631.

24. Give a structure or structures for the compound E, whose infrared and proton NMR spectra are shown in Fig. 16.31, p. 631.

Sadtler 8023 K

Sadtler 8514 K

Sadtler 8339 K

Figure 16.24 Infrared spectra for Problem 17, p. 623.

Figure 16.25   Proton NMR spectra for Problem 18, p. 624.

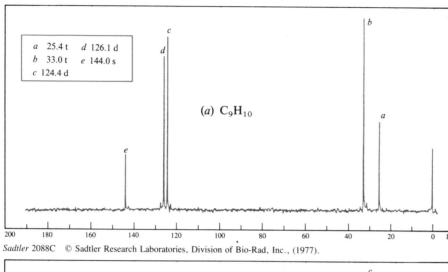

a 25.4 t   d 126.1 d
b 33.0 t   e 144.0 s
c 124.4 d

(a) C$_9$H$_{10}$

Sadtler 2088C   © Sadtler Research Laboratories, Division of Bio-Rad, Inc., (1977).

a 24.8 d
b 25.5 q
c 30.1 q
d 31.0 s
e 53.3 t

(b) C$_8$H$_{18}$

Sadtler 1613C   © Sadtler Research Laboratories, Division of Bio-Rad, Inc., (1977).

a 20.8 q   e 128.8 d
b 31.4 q   f 134.4 s
c 34.2 s   g 147.9 s
d 125.0 d

(c) C$_{11}$H$_{16}$

Sadtler 1477C   © Sadtler Research Laboratories, Division of Bio-Rad, Inc., (1977).

Figure 16.26   CMR spectra for Problem 19, p. 624.

Figure 16.27   Proton NMR spectra for Problem 20, p. 624.

a	19.1 q	f	130.5 d
b	19.6 q	g	133.1 s
c	20.9 q	h	135.1 s
d	126.6 d	i	136.1 s
e	129.6 d		

(a) C_9H_{12}

200   180   160   140   120   100   80   60   40   20   0   δ

Sadtler 170C   © Sadtler Research Laboratories, Division of Bio-Rad, Inc., (1976).

a	21.2 q
b	127.1 d
c	137.6 s

(b) C_9H_{12}

200   180   160   140   120   100   80   60   40   20   0   δ

Sadtler 319C   © Sadtler Research Laboratories, Division of Bio-Rad, Inc., (1976).

a	24.1 q	d	126.4 d
b	34.2 d	e	128.4 d
c	125.8 d	f	148.8 s

(c) C_9H_{12}

200   180   160   140   120   100   80   60   40   20   0   δ

Sadtler 58C   © Sadtler Research Laboratories, Division of Bio-Rad, Inc., (1976).

Figure 16.28   CMR spectra for Problem 21, p. 624.

Figure 16.29   CMR spectra for Problem 22, p. 624.

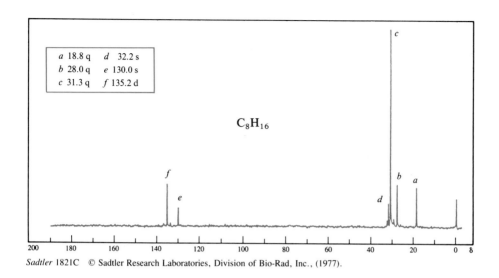

a 18.8 q   d 32.2 s
b 28.0 q   e 130.0 s
c 31.3 q   f 135.2 d

C_8H_{16}

Sadtler 1821C   © Sadtler Research Laboratories, Division of Bio-Rad, Inc., (1977).

Figure 16.30   CMR spectrum for Problem 23, p. 624.

E C_9H_{10}

Sadtler 8020 K

E C_9H_{10}

Figure 16.31   Infrared and proton NMR spectra for Problem 24, p. 624.

17

Alcohols I. Preparation and Physical Properties

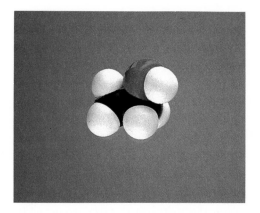

17.1 Introduction

If, as an organic chemist, you were allowed to choose the ten aliphatic compounds with which to be stranded on a desert island, you would almost certainly pick alcohols. From them you could make nearly every other kind of aliphatic compound: alkyl halides, alkenes, ethers, aldehydes, ketones, acids, esters, and a host of others. From the alkyl halides, you could make Grignard reagents, and from the reaction between these and the aldehydes and ketones obtain more complicated alcohols and so on. On your desert island you would use your alcohols not only as raw materials, but frequently as the solvents in which reactions are carried out and from which products are recrystallized. Finally, hot and tired after a long day in the laboratory, you could refresh yourself with an (isopropyl) alcohol rub and perhaps relax over a cool (ethyl) alcoholic drink.

We have already encountered alcohols playing a variety of roles: as *substrates* in nucleophilic substitution (Sec. 5.25) and in elimination (Sec. 7.25); as *nucleophiles* (Sec. 5.9) and as *bases* (Sec. 7.12); and, nearly everywhere, as solvents. We know that they are basic, and can be protonated (Sec. 5.5). We know that they are acidic and can be converted into alkoxides, and that these alkoxides can serve as nucleophilic and basic reagents (Secs. 5.9 and 7.12).

In this chapter and the following one, we shall study alcohols in a systematic way: review and consolidate what we have already learned about them, and look at new aspects of their rich and varied chemistry.

17.2 Structure

Alcohols are compounds of the general formula R—OH, where R is any alkyl or substituted alkyl group. The group may be primary, secondary, or tertiary; it may be open-chain or cyclic; it may contain a double bond, a halogen atom, an aromatic ring, or additional hydroxyl groups. For example:

$$CH_3{-}\underset{\underset{\textstyle OH}{|}}{\overset{\overset{\textstyle CH_3}{|}}{C}}{-}CH_3$$

2-Methyl-2-propanol
tert-Butyl alcohol

$$CH_2{=}CH{-}CH_2OH$$

2-Propen-1-ol
Allyl alcohol

Cyclohexanol

Benzyl alcohol

$$\underset{\underset{\textstyle Cl}{|}}{CH_2}{-}\underset{\underset{\textstyle OH}{|}}{CH_2}$$

2-Chloroethanol
Ethylene chlorohydrin

$$\underset{\underset{\textstyle OH}{|}}{CH_2}{-}\underset{\underset{\textstyle OH}{|}}{CH}{-}\underset{\underset{\textstyle OH}{|}}{CH_2}$$

1,2,3-Propanetriol
Glycerol

All alcohols contain the hydroxyl (—OH) group, which, as the functional group, determines the properties characteristic of this family. Variations in structure of the R group may affect the rate at which the alcohol undergoes certain reactions, and even, in a few cases, may affect the kind of reaction.

Compounds in which the hydroxyl group is attached directly to an aromatic ring are not alcohols; they are *phenols,* and differ so markedly from the alcohols that we shall consider them in a separate chapter.

17.3 Classification

We classify a carbon atom as *primary, secondary,* or *tertiary* according to the number of other carbon atoms attached to it (Sec. 3.11). An alcohol is classified according to the kind of carbon that bears the —OH group:

$$R{-}\underset{\underset{\textstyle H}{|}}{\overset{\overset{\textstyle H}{|}}{C}}{-}OH$$

Primary
(1°)

$$R{-}\underset{\underset{\textstyle H}{|}}{\overset{\overset{\textstyle R}{|}}{C}}{-}OH$$

Secondary
(2°)

$$R{-}\underset{\underset{\textstyle R}{|}}{\overset{\overset{\textstyle R}{|}}{C}}{-}OH$$

Tertiary
(3°)

One reaction, oxidation, which directly involves the hydrogen atoms attached to the carbon bearing the —OH group, takes an entirely different course for each class of alcohol. Usually, however, alcohols of different classes differ only in *rate* or *mechanism* of reaction, and in a way consistent with their structures. Certain substituents may affect reactivity in such a way as to make an alcohol of one class resemble the members of a different class. The presence of —Cl, we have already seen (Sec. 5.25), makes the secondary alcohol 1-chloro-2-propanol act like a primary alcohol. This effect of chlorine we attributed to its powerful electron-withdrawing tendency. Other variations in properties of alcohols, we shall find, are consistent with the structures involved.

17.4   Nomenclature

Alcohols are named by two principal systems. For the simpler alcohols the **common names**, which we have already encountered (Sec. 5.5), are most often used. A common name consists simply of the name of the alkyl group followed by the word *alcohol*. For example:

CH_3OH
Methyl alcohol

CH_3CHCH_3
$\overset{|}{OH}$
Isopropyl alcohol

$\overset{CH_3}{\overset{|}{CH_3CHCH_2OH}}$
Isobutyl alcohol

$CH_3CH_2-\overset{\overset{\displaystyle CH_3}{|}}{\underset{\underset{\displaystyle OH}{|}}{C}}-CH_3$
tert-Pentyl alcohol

$O_2N\langle\bigcirc\rangle CH_2OH$
p-Nitrobenzyl alcohol

$\bigcirc\overset{\alpha\ \ \beta}{\underset{\underset{\displaystyle OH}{|}}{CHCH_3}}$
α-Phenylethyl alcohol

The most versatile system is, of course, the **IUPAC**. The rules are:

1. Select as the parent structure the longest continuous carbon chain *that contains the —OH group*; then consider the compound to have been derived from this structure by replacement of hydrogen by various groups. The parent structure is known as *ethanol, propanol, butanol*, etc., depending upon the number of carbon atoms; each name is derived by replacing the terminal -*e* of the corresponding alkane name by **-ol**.

2. Indicate by a number the position of the —OH group in the parent chain, generally using the lowest possible number for this purpose.

3. Indicate by numbers the positions of other groups attached to the parent chain.

$CH_3CH_2CH_2OH$
1-Propanol

CH_3CHCH_2Br
$\overset{|}{OH}$
1-Bromo-2-propanol

$\overset{CH_3}{\underset{\text{2-Methylcyclopentanol}}{\bigcirc\!\!-OH}}$

$C_6H_5CH_2CH_2OH$
2-Phenylethanol

$CH_3-\overset{\overset{\displaystyle CH_3}{|}}{\underset{\underset{\displaystyle CH_3}{|}}{C}}-CH_2OH$
2,2-Dimethyl-1-propanol

$CH_3CHCHCH_3$
$\overset{CH_3}{}\qquad\overset{|}{OH}$
3-Methyl-2-butanol

cis-1,2-Cyclopentanediol

An alcohol containing a double bond is named as an **alkenol**, with numbers to indicate the positions of the double bond and the hydroxyl group.

$$\overset{3}{C}H_2=\overset{2}{C}H-\overset{1}{C}H_2OH$$

2-Propen-1-ol

$$\overset{4}{C}H_2=\overset{3}{C}H-\overset{2}{C}H-\overset{1}{C}H_3$$
$$\hspace{3.5cm}|$$
$$\hspace{3.5cm}OH$$

3-Buten-2-ol

4-Cyclohexenol

Note that *-ol* takes priority over *-ene*; *-ol* appears last in the name, and, where possible, is given the lower number. (See also the names of the pheromones shown on p. 347.)

17.5   Physical properties

The physical properties of an alcohol are best understood, we have seen (Sec. 6.3), if we recognize this simple fact: structurally, an alcohol is a composite of an alkane and water. It contains a lipophilic, alkane-like group and a hydrophilic,

R—H   H—OH   R—OH

An alkane   Water   An alcohol

water-like hydroxyl group. Of these two structural units, it is the —OH group that gives the alcohol its characteristic physical properties and the alkyl group that, depending upon its size and shape, modifies these properties.

The —OH group is highly polar and, most important, is capable of hydrogen bonding: hydrogen bonding to its fellow alcohol molecules (Sec. 1.20), to other

$$R-O\cdots H-\underset{\underset{H}{|}}{\overset{\overset{R}{|}}{O}} \qquad R-O\cdots H-\underset{\underset{H}{|}}{\overset{\overset{H}{|}}{O}} \qquad A^-\cdots H-\overset{\overset{R}{|}}{O}$$

neutral molecules (Sec. 6.3), and to anions (Sec. 6.4). The physical properties (Table 17.1) show some of the effects of this hydrogen bonding.

Let us look first at **boiling points.** Among hydrocarbons the factors that determine boiling point seem to be chiefly molecular weight and shape; this is to be expected of molecules that are held together chiefly by van der Waals forces. Alcohols, too, show increase in boiling point with increasing carbon number, and

Table 17.1 ALCOHOLS

Name	Formula	M.p., °C	B.p., °C	Relative density (at 20 °C)	Solubility, g/100 g H_2O
Methyl	CH_3OH	− 97	64.5	0.793	∞
Ethyl	CH_3CH_2OH	−115	78.3	.789	∞
n-Propyl	$CH_3CH_2CH_2OH$	−126	97	.804	∞
n-Butyl	$CH_3(CH_2)_2CH_2OH$	− 90	118	.810	7.9
n-Pentyl	$CH_3(CH_2)_3CH_2OH$	− 78.5	138	.817	2.3
n-Hexyl	$CH_3(CH_2)_4CH_2OH$	− 52	156.5	.819	0.6
n-Heptyl	$CH_3(CH_2)_5CH_2OH$	− 34	176	.822	0.2
n-Octyl	$CH_3(CH_2)_6CH_2OH$	− 15	195	.825	0.05
n-Decyl	$CH_3(CH_2)_8CH_2OH$	6	228	.829	
n-Dodecyl	$CH_3(CH_2)_{10}CH_2OH$	24			
n-Tetradecyl	$CH_3(CH_2)_{12}CH_2OH$	38			
n-Hexadecyl	$CH_3(CH_2)_{14}CH_2OH$	49			
n-Octadecyl	$CH_3(CH_2)_{16}CH_2OH$	58.5			
Isopropyl	$CH_3CHOHCH_3$	− 86	82.5	.789	∞
Isobutyl	$(CH_3)_2CHCH_2OH$	−108	108	.802	10.0
sec-Butyl	$CH_3CH_2CHOHCH_3$	−114	99.5	.806	12.5
tert-Butyl	$(CH_3)_3COH$	25.5	83	.789	∞
Isopentyl	$(CH_3)_2CHCH_2CH_2OH$	−117	132	.813	2
active-Amyl	$(-)\text{-}CH_3CH_2CH(CH_3)CH_2OH$		128	.816	3.6
tert-Pentyl	$CH_3CH_2C(OH)(CH_3)_2$	− 12	102	.809	12.5
Cyclopentanol	$cyclo\text{-}C_5H_9OH$		140	.949	
Cyclohexanol	$cyclo\text{-}C_6H_{11}OH$	24	161.5	.962	
Allyl	$CH_2{=}CHCH_2OH$	−129	97	.855	∞
Crotyl	$CH_3CH{=}CHCH_2OH$		118	.853	16.6
Methylvinyl-methanol	$CH_2{=}CHCHOHCH_3$		97		
Benzyl	$C_6H_5CH_2OH$	− 15	205	1.046	4
α-Phenylethyl	$C_6H_5CHOHCH_3$		205	1.013	
β-Phenylethyl	$C_6H_5CH_2CH_2OH$	− 27	221	1.02	1.6
Diphenylmethanol (Benzhydrol)	$(C_6H_5)_2CHOH$	69	298		0.05
Triphenylmethanol	$(C_6H_5)_3COH$	162.5			
Cinnamyl	$C_6H_5CH{=}CHCH_2OH$	33	257.5		
1,2-Ethanediol	CH_2OHCH_2OH	− 16	197	1.113	
1,2-Propanediol	$CH_3CHOHCH_2OH$		187	1.040	
1,3-Propanediol	$HOCH_2CH_2CH_2OH$		215	1.060	
Glycerol	$HOCH_2CHOHCH_2OH$	18	290	1.261	
Pentaerythritol	$C(CH_2OH)_4$	260			6

decrease in boiling point with branching. But the unusual thing about alcohols is that they boil so *high*: as Table 17.2 (on the next page) shows, much higher than hydrocarbons of the same molecular weight, and higher, even, than many other compounds of considerable polarity. How are we to account for this?

The answer is, of course, that alcohols, like water, are *associated liquids* (Sec. 1.20): their abnormally high boiling points are due to the greater energy needed to

Table 17.2   Structure and Boiling Point

Name	Structure	Mol. wt.	Dipole moment, D	B.p., °C
n-Pentane	$CH_3CH_2CH_2CH_2CH_3$	72	0	36
Diethyl ether	$CH_3CH_2-O-CH_2CH_3$	74	1.18	35
n-Propyl chloride	$CH_3CH_2CH_2Cl$	79	2.10	47
n-Butyraldehyde	$CH_3CH_2CH_2CHO$	72	2.72	76
n-Butyl alcohol	$CH_3CH_2CH_2CH_2OH$	74	1.63	118

break the hydrogen bonds that hold the molecules together. Although ethers and aldehydes contain oxygen, they contain hydrogen that is bonded only to carbon; these hydrogens are not positive enough to bond appreciably with oxygen.

Infrared spectroscopy (Sec. 16.4) has played a key role in the study of hydrogen bonding. In dilute solution in a non-polar solvent like carbon tetrachloride (or in the gas phase), where association between molecules is minimal, ethanol, for example, shows an O—H stretching band at 3640 cm^{-1}. As the concentration of ethanol is increased, this band is gradually replaced by a broader band at 3350 cm^{-1}. The bonding of hydrogen to the second oxygen weakens the O—H bond, and lowers the energy and hence the frequency of vibration.

Problem 17.1   The infrared spectrum of *cis*-1,2-cyclopentanediol has an O—H stretching band at a lower frequency than for a free —OH group, and this band does not disappear even at high dilution. *trans*-1,2-Cyclopentanediol shows no such band. Can you suggest a possible explanation?

The behavior of alcohols as **solutes** also reflects their ability to form hydrogen bonds. In sharp contrast to hydrocarbons, the lower alcohols are miscible with water. Since alcohol molecules are held together by the same sort of intermolecular forces as water molecules, there can be mixing of the two kinds of molecules: the energy required to break a hydrogen bond between two water molecules or two alcohol molecules is provided by formation of a hydrogen bond between a water molecule and an alcohol molecule.

But, as we saw in Sec. 6.3, this is true only for the lower alcohols, where the hydrophilic —OH group constitutes a large part of the molecule. As the lipophilic alkyl group becomes larger, water solubility decreases. For practical purposes we consider that the borderline between solubility and insolubility in water occurs at about four to five carbon atoms for normal primary alcohols.

Polyhydroxy alcohols provide more than one site per molecule for hydrogen bonding, and their properties reflect this. The simplest diol, 1,2-ethanediol (ethylene glycol), boils at 197 °C. The lower diols are miscible with water, and those containing as many as seven carbon atoms show appreciable solubility in water. (Ethylene glycol owes its use as an antifreeze—e.g., Prestone—to its high boiling point, low freezing point, and high solubility in water.)

Problem 17.2   The disaccharide *sucrose*, $C_{12}H_{22}O_{11}$, is a big molecule and yet (it is ordinary table sugar) is extremely soluble in water. What might you guess about its structure? (Check your answer on p. 1330.)

Problem 17.3 How do you account for the fact that, although diethyl ether has a much lower boiling point than *n*-butyl alcohol, it has the same solubility (8 g per 100 g) in water?

We have already (Secs. 6.3 and 6.4) discussed the behavior of alcohols as *solvents*. Through their lipophilic alkyl groups, they can dissolve non-ionic compounds: organic substrates, for example. Through their —OH groups, they can dissolve ionic compounds: inorganic reagents, for example. As *protic* solvents, they solvate anions especially strongly through hydrogen bonding. They solvate cations through unshared pairs of electrons on oxygen.

As solvents, we have seen, alcohols are far from being innocent by-standers. Their oxygen is basic and nucleophilic. In elimination of the E1 kind, alcohols can serve as the base as well as the solvent. In nucleophilic substitution, alcohols can act as the nucleophile in S_N2 reactions, and render nucleophilic assistance in the formation of cationic intermediates.

17.6   Industrial source

For alcohols to be such important starting materials in aliphatic chemistry, they must be not only versatile in their reactions but also available in large amounts and at low prices. There are three principal ways to get the simple alcohols that are the backbone of aliphatic organic synthesis, ways that can utilize all our sources of organic raw material—petroleum, natural gas, coal, and the biomass. These methods are: (a) by *hydration of alkenes* obtained from the cracking of petroleum; (b) by the *oxo process* from alkenes, carbon monoxide, and hydrogen; and (c) by *fermentation of carbohydrates*. In addition to these three chief methods, there are others that have more limited application (see Fig. 17.1). Methanol, for example,

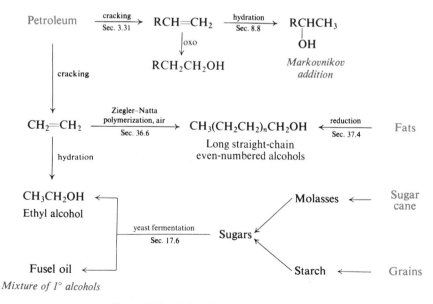

Figure 17.1   Industrial sources of alcohols.

is made by the catalytic hydrogenation of carbon monoxide; the necessary mixture of hydrogen and carbon monoxide is obtained from the high-temperature reaction of water with methane, higher alkanes, or coal.

(a) Hydration of Alkenes   We have already seen (Sec. 3.31) that alkenes containing up to four or five carbon atoms can be separated from the mixture obtained from the cracking of petroleum. We have also seen (Secs. 8.7 and 8.8) that alkenes are readily converted into alcohols either by direct addition of water, or by addition of sulfuric acid followed by hydrolysis. By this process there can be obtained only those alcohols whose formation is consistent with the application of Markovnikov's rule: for example, isopropyl but not *n*-propyl, *sec*-butyl but not *n*-butyl, *tert*-butyl but not isobutyl. Thus the *only* primary alcohol obtainable in this way is ethyl alcohol.

$$CH_3-CH=CH_2 + H_2SO_4 \longrightarrow CH_3-\underset{\underset{OSO_3H}{|}}{CH}-CH_3 \xrightarrow{H_2O} CH_3-\underset{\underset{OH}{|}}{CH}-CH_3$$

Isopropyl alcohol
(2°)

$$CH_3-\underset{\overset{|}{CH_3}}{C}=CH_2 + H_2O \xrightarrow{H^+} CH_3-\underset{\underset{^+OH_2}{|}}{\overset{\overset{CH_3}{|}}{C}}-CH_3 \xrightarrow{-H^+} CH_3-\underset{\underset{OH}{|}}{\overset{\overset{CH_3}{|}}{C}}-CH_3$$

tert-Butyl alcohol
(3°)

(b) Oxo Process   Primary alcohols *can* be obtained from alkenes, however, by the oxo process.

In the presence of the proper catalyst, alkenes react with carbon monoxide and hydrogen to yield *aldehydes*, compounds containing the —CHO group.

$$\underset{/}{\overset{\backslash}{C}}=\underset{\backslash}{\overset{/}{C}} + CO + H_2 \xrightarrow{\text{catalyst}} -\underset{\underset{H}{|}}{C}-\underset{\underset{CHO}{|}}{C}-$$

An aldehyde

Aldehydes can be readily reduced by catalytic hydrogenation to primary alcohols; indeed, the oxo process is often carried out in such a way that this reduction process takes place as the aldehydes are formed, to give the alcohols directly.

$$R-\underset{\overset{|}{\text{ }}}{\overset{\overset{H}{|}}{C}}=O + H_2 \xrightarrow{\text{catalyst}} R-\underset{\underset{H}{|}}{\overset{\overset{H}{|}}{C}}-OH$$

An aldehyde   A 1° alcohol

The classical oxo catalyst is octacarbonyldicobalt, $Co_2(CO)_8$, formed by reaction of metallic cobalt with carbon monoxide. The oxo process was discovered in Germany during World War II, and was the first industrial application of catalysis by a transition metal complex. We have already encountered such catalysis in homogeneous hydrogenation (Sec. 8.3). In Chapter 20, we shall find out that in the oxo process the catalyst exerts its effect in basically the same manner as in hydrogenation and other examples of such catalysis.

The oxo process amounts to addition of —H and —CHO across a carbon–carbon double bond. The —CHO group is the *formyl* group, and so the process is called *hydroformylation*. Like other such addition reactions, it can take place with either of two orientations and, if the structure permits, yield either of two products.

$$\underset{\diagdown}{\overset{\diagup}{C}}{=}\underset{\diagup}{\overset{\diagdown}{C}} \xrightarrow[\text{catalyst}]{\text{CO, H}_2} \quad -\underset{|}{\overset{|}{C}}-\underset{|}{\overset{|}{C}}- \quad or \quad -\underset{|}{\overset{|}{C}}-\underset{|}{\overset{|}{C}}-$$
$$ \text{H} \quad \text{CHO} \qquad \text{OHC} \quad \text{H}$$

Propylene, for example, can yield either a straight-chain or a branched-chain aldehyde:

$$CH_3CH{=}CH_2 \xrightarrow[\text{catalyst}]{\text{CO, H}_2} \quad CH_3CH_2CH_2CHO \quad or \quad CH_3\underset{|}{CH}CH_3$$
$$ \textit{n}\text{-Butyraldehyde} \text{CHO}$$
$$ \text{Isobutyraldehyde}$$

Straight-chain aldehydes are usually the more desired products. The process tends to form these preferentially, and this tendency can be increased by modifications in reaction conditions.

There are a number of attractive features to the oxo process. First, it can give primary alcohols, not available through hydration of alkenes. Next, the aldehydes need not be reduced to alcohols, but can be converted into other products: oxidized to carboxylic acids (Chap. 23), for example, or allowed to undergo the aldol condensation (Sec. 25.5). Finally, part of the carbon in oxo products comes from carbon monoxide, which can be made from coal rather than scarce petroleum or even scarcer natural gas.

Problem 17.4 "Nonanol", a mixture of 3,5,5-trimethyl-1-hexanol and some of its isomers, is used as a plasticizer for polymers. It is manufactured via the oxo process. Can you suggest how it might be made starting from hydrocarbons of four carbons or fewer?

(c) Fermentation of Carbohydrates   Fermentation of sugars by yeast, the oldest synthetic chemical process used by man, is still of enormous importance for the preparation of **ethyl alcohol** and certain other alcohols. The sugars come from a variety of sources, mostly molasses from sugar cane, or starch obtained from various grains; the name "grain alcohol" has been given to ethyl alcohol for this reason.

When starch is the starting material, there is obtained, in addition to ethyl alcohol, a smaller amount of *fusel oil* (German: *Fusel*, inferior liquor), a mixture of primary alcohols: mostly isopentyl alcohol with smaller amounts of *n*-propyl alcohol, isobutyl alcohol, and 2-methyl-1-butanol, known as *active amyl alcohol* (*amyl = pentyl*).

In the future there will undoubtedly be a shift toward carbohydrates as our source of carbon: carbon for organic chemicals and carbon in the form of fuels. With this shift, fermentation processes will take on greater and greater importance. But there is no free lunch. A great deal of energy is required for the distillation that separates fermentation products from the dilute solutions in which they are formed. And all the carbohydrates being grown today to feed mankind could supply only a small fraction of the energy now provided by petroleum.

Problem 17.5 The isopentyl and active amyl alcohols are formed by enzymatic transformation of the amino acids *leucine* and *isoleucine*, which come from hydrolysis of protein material in the starch.

$$(CH_3)_2CHCH_2CH(NH_3{}^+)COO^- \qquad\qquad CH_3CH_2CH(CH_3)CH(NH_3{}^+)COO^-$$

Leucine   Isoleucine

(a) Which amino acid gives which alcohol? (b) Although both amino acids are optically active, and the transformation processes are analogous, only one gives an alcohol that is optically active. Why is this?

17.7 Ethyl alcohol

Ethyl alcohol is not only the oldest synthetic organic chemical used by man, but it is also one of the most important.

In industry ethyl alcohol is widely used as a solvent for lacquers, varnishes, perfumes, and flavorings; as a medium for chemical reactions; and in recrystallizations. In addition, it is an important raw material for synthesis; after we have learned more about the reactions of alcohols (Chap. 18), we can better appreciate the role played by the leading member of the family. For these industrial purposes ethyl alcohol is prepared both by hydration of ethylene and by fermentation of sugar from molasses (or sometimes starch); thus its ultimate sources are petroleum, sugar cane, and various grains.

Ethyl alcohol is the alcohol of "alcoholic" beverages. For this purpose it is prepared by fermentation of sugar from a truly amazing variety of vegetable sources. The particular beverage obtained depends upon what is fermented (rye or corn, grapes or elderberries, cactus pulp or dandelions), how it is fermented (whether carbon dioxide is allowed to escape or is bottled up, for example), and what is done after fermentation (whether or not it is distilled). The special flavor of a beverage is not due to the ethyl alcohol but to other substances, either characteristic of the particular source, or deliberately added.

Medically, ethyl alcohol is classified as a *hypnotic* (sleep producer); it is less toxic than other alcohols. (Methanol, for example, is quite *poisonous*: drinking it, breathing it for prolonged periods, or allowing it to remain long on the skin can lead to blindness or death.) Because of its unique position as both a highly taxed beverage and an important industrial chemical, ethyl alcohol poses a special problem: it must be made available to the chemical industry in a form that is unfit to drink. This problem is solved by addition of a *denaturant*, a substance that makes it unpalatable or even poisonous. Two of the eighty-odd legal denaturants, for example, are methanol and high-test gasoline. When necessary, pure undenatured ethyl alcohol is available for chemical purposes, but its use is strictly controlled by the Federal Government.

Except for alcoholic beverages, nearly all the ethyl alcohol used is a mixture of 95% alcohol and 5% water, known simply as *95% alcohol*. What is so special about the concentration of 95%? Whatever the method of preparation, ethyl alcohol is obtained first mixed with water; this mixture is then concentrated by fractional distillation. But it happens that the component of lowest boiling point is not ethyl alcohol (b.p. 78.3 °C) but a *binary azeotrope* containing 95% alcohol and 5% water (b.p. 78.15 °C). As an azeotrope, it of course gives a vapor of the same composition, and hence cannot be further concentrated by distillation no matter how efficient the fractionating column used.

Pure ethyl alcohol is known as *absolute alcohol*. Although more expensive than 95% alcohol, it is available for use when specifically required. It is obtained by taking advantage of the existence of another azeotrope, this time a *ternary* one of b.p. 64.9 °C; this contains 7.5% water, 18.5% ethyl alcohol, and 74% benzene.

Problem 17.6   Describe exactly what will happen if one distills a mixture of 150 g of 95% alcohol and 74 g of benzene.

For certain special purposes (Secs. 30.2 and 30.3) even the slight trace of water found in commercial absolute alcohol must be removed. This can be accomplished by treatment of the alcohol with metallic magnesium; water is converted into insoluble $Mg(OH)_2$, from which the alcohol is then distilled.

17.8   Preparation of alcohols

Most of the simple alcohols and a few of the complicated ones are available from the industrial sources described in Sec. 17.6. Other alcohols must be prepared by one of the methods outlined below.

PREPARATION OF ALCOHOLS ─────

1. **Oxymercuration–demercuration.** Discussed in Sec. 17.9.

Markovnikov addition

Examples:

3,3-Dimethyl-1-butene

3,3-Dimethyl-2-butanol
No rearrangement

Norbornene

exo-Norborneol

CONTINUED

CONTINUED

2. Hydroboration–oxidation. Discussed in Secs. 17.10–17.12.

$$\text{C=C} + (BH_3)_2 \longrightarrow \underset{H \quad B}{-\overset{|}{C}-\overset{|}{C}-} \xrightarrow{H_2O_2,\ OH^-} \underset{H \quad OH}{-\overset{|}{C}-\overset{|}{C}-} + B(OH)_3$$

Diborane   Alkylborane   *Anti-Markovnikov orientation*

Examples:

1-Methylcyclopentene   *trans*-2-Methyl-1-cyclopentanol   *syn*-Addition

$$\underset{\underset{CH_3}{|}}{\overset{\overset{CH_3}{|}}{CH_3-C-CH=CH_2}} \xrightarrow{(BH_3)_2} \xrightarrow{H_2O_2,\ OH^-} \underset{\underset{CH_3}{|}}{\overset{\overset{CH_3}{|}}{CH_3-C-CH_2-CH_2OH}} \quad No\ rearrangement$$

3,3-Dimethyl-1-butene   3,3-Dimethyl-1-butanol

3. Grignard synthesis. Discussed in Secs. 17.14–17.17.

$$\text{C=O} + R-MgX \longrightarrow \underset{R}{-\overset{|}{C}-\bar{O}\overset{+}{M}gX} \xrightarrow{H_2O} \underset{R}{-\overset{|}{C}-OH} + Mg^{2+} + X^-$$

$$\underset{H}{\overset{H}{\diagdown}}C=O + R-MgX \longrightarrow \underset{R}{H-\overset{|}{C}-\bar{O}\overset{+}{M}gX} \xrightarrow{H_2O} \underset{R}{H-\overset{|}{C}-OH} \quad 1° \text{ alcohol}$$

Formaldehyde

$$\underset{R'}{\overset{H}{\diagdown}}C=O + R-MgX \longrightarrow \underset{R}{R'-\overset{|}{C}-\bar{O}\overset{+}{M}gX} \xrightarrow{H_2O} \underset{R}{R'-\overset{|}{C}-OH} \quad 2° \text{ alcohol}$$

Higher aldehydes

$$\underset{R'}{\overset{R''}{\diagdown}}C=O + R-MgX \longrightarrow \underset{R}{R'-\overset{R''}{\overset{|}{C}}-\bar{O}\overset{+}{M}gX} \xrightarrow{H_2O} \underset{R}{R'-\overset{R''}{\overset{|}{C}}-OH} \quad 3° \text{ alcohol}$$

Ketones

$$\underset{\diagdown O \diagup}{H_2C-CH_2} + R-MgX \longrightarrow R-CH_2CH_2\bar{O}\overset{+}{M}gX \xrightarrow{H_2O} R-CH_2CH_2OH \quad 1° \text{ alcohol}$$

Ethylene oxide   *two carbons added*

CONTINUED

—— CONTINUED ——

$$R'COOC_2H_5 + 2R\text{—}MgX \longrightarrow R'\text{—}\overset{\displaystyle R}{\underset{\displaystyle R}{\overset{|}{\underset{|}{C}}}}\text{—}\overset{-}{O}\overset{+}{M}gX \xrightarrow{H_2O} R'\text{—}\overset{\displaystyle R}{\underset{\displaystyle R}{\overset{|}{\underset{|}{C}}}}\text{—}OH \qquad 3° \text{ alcohol}$$

An ester

Discussed in Sec. 24.21.

4. Hydrolysis of alkyl halides. Discussed in Sec. 17.8.

$$R\text{—}X + OH^- \text{ (or } H_2O) \longrightarrow R\text{—}OH + X^- \text{ (or } HX)$$

Examples:

$$\text{C}_6\text{H}_5\text{—CH}_2\text{Cl} \xrightarrow{\text{aqueous NaOH}} \text{C}_6\text{H}_5\text{—CH}_2\text{OH}$$

Benzyl chloride   Benzyl alcohol

$$CH_2\text{=}CH_2 \xrightarrow{Cl_2, H_2O} \underset{\underset{\displaystyle Cl \quad OH}{|\quad\;\; |}}{CH_2\text{—}CH_2} \xrightarrow{Na_2CO_3, H_2O} \underset{\underset{\displaystyle OH \quad OH}{|\quad\;\; |}}{CH_2\text{—}CH_2}$$

Ethylene   Ethylene chlorohydrin   Ethylene glycol

5. Aldol condensation. Discussed in Sec. 25.7.

6. Reduction of carbonyl compounds. Discussed in Sec. 21.9.

7. Reduction of acids and esters. Discussed in Secs. 23.18 and 24.22.

8. Hydroxylation of alkenes. Discussed in Secs. 8.22 and 19.13.

$$\begin{array}{l} \xrightarrow{KMnO_4} \underset{\underset{\displaystyle HO \quad OH}{|\quad\;\; |}}{\text{—}C\text{—}C\text{—}} \qquad\qquad \textit{syn-}\text{Hydroxylation} \\[3em] \text{C=C} \\[3em] \xrightarrow{RCO_2OH} \underset{\underset{\displaystyle O}{\diagdown\,\diagup}}{\text{—}C\text{—}C\text{—}} \xrightarrow{H_2O, H^+} \underset{\underset{\displaystyle HO}{|}}{\text{—}C\text{—}\overset{\overset{\displaystyle OH}{|}}{C}\text{—}} \qquad \textit{anti-}\text{Hydroxylation} \quad \blacksquare \end{array}$$

We can follow either of two approaches to the synthesis of alcohols—or, for that matter, of most other kinds of compounds. (a) We can retain the original carbon skeleton, and simply convert one functional group into another until we arrive at an alcohol; or (b) we can generate a new, bigger carbon skeleton and at the same time produce an alcohol.

By far the most important method of preparing alcohols is the **Grignard synthesis.** This is an example of the second approach, since it leads to the formation of carbon–carbon bonds. In the laboratory a chemist is chiefly concerned with preparing the more complicated alcohols that cannot be bought; these are prepared by the Grignard synthesis from rather simple starting materials. The alkyl halides from which the Grignard reagents are made, as well as the aldehydes and ketones themselves, are most conveniently prepared from alcohols; thus the method ultimately involves the synthesis of alcohols from less complicated alcohols.

Alcohols can be conveniently made from compounds containing carbon–carbon double bonds in two ways; by **oxymercuration–demercuration** and by **hydroboration–oxidation.** Both amount to addition of water to the double bond, but with *opposite orientation*—Markovnikov and anti-Markovnikov—and hence the two methods neatly complement each other.

Hydrolysis of alkyl halides is severely limited as a method of synthesizing alcohols, since alcohols are usually more available than the corresponding halides; indeed, the best general preparation of halides is from alcohols. The synthesis of benzyl alcohol from toluene, however, is an example of a useful application of this method (Sec. 15.13).

$$
\underset{\text{Toluene}}{\boxed{\text{CH}_3}} \xrightarrow{\text{Cl}_2,\ \text{heat, light}} \underset{\text{Benzyl chloride}}{\boxed{\text{CH}_2-\text{Cl}}} \xrightarrow{\text{aqueous NaOH}} \underset{\text{Benzyl alcohol}}{\boxed{\text{CH}_2-\text{OH}}}
$$

For those halides that can undergo elimination, the formation of alkene must always be considered a possible side reaction.

17.9   Oxymercuration–demercuration

Alkenes react with mercuric acetate in the presence of water to give hydroxy-mercurial compounds which on reduction yield alcohols.

$$
\underset{\substack{\text{Alkene} \quad \text{Water} \quad \text{Mercuric acetate}}}{\overset{\text{Oxymercuration}}{\underset{}{\smash{\Big\langle}}\text{C}=\text{C}\smash{\Big\rangle} + \text{H}_2\text{O} + \text{Hg(OAc)}_2} \longrightarrow \underset{\text{OH \quad HgOAc}}{-\overset{|}{\text{C}}-\overset{|}{\text{C}}-} \overset{\text{Demercuration}}{\xrightarrow{\text{NaBH}_4}} \underset{\substack{\text{OH \quad H} \\ \text{Alcohol}}}{-\overset{|}{\text{C}}-\overset{|}{\text{C}}-}}
$$

$$-\text{OAc} = \text{CH}_3\text{COO}-$$

The first stage, *oxymercuration*, involves addition to the carbon–carbon double bond of —OH and —HgOAc. Then, in *demercuration*, the —HgOAc is replaced by —H. The reaction sequence amounts to hydration of the alkene, but is much more widely applicable than direct hydration.

The two-stage process of oxymercuration–demercuration is fast and convenient, takes place under mild conditions, and gives excellent yields—often over 90%. The alkene is added at room temperature to an aqueous solution of mercuric acetate diluted with the solvent tetrahydrofuran. Reaction is generally complete within minutes. The organomercurial compound is not isolated but is simply reduced *in situ* by sodium borohydride, NaBH$_4$. (The mercury is recovered as a ball of elemental mercury.)

Oxymercuration–demercuration is highly regioselective, and gives alcohols corresponding to *Markovnikov* addition of water to the carbon–carbon double bond. For example:

$$CH_3(CH_2)_3CH{=}CH_2 \xrightarrow{\text{Hg(OAc)}_2,\ H_2O} \xrightarrow{\text{NaBH}_4} CH_3(CH_2)_3\underset{\underset{\displaystyle OH}{|}}{CH}CH_3$$

1-Hexene

2-Hexanol

$$CH_3CH_2\underset{\underset{\displaystyle CH_3}{|}}{\overset{\overset{\displaystyle CH_3}{|}}{C}}{=}CH_2 \xrightarrow{\text{Hg(OAc)}_2,\ H_2O} \xrightarrow{\text{NaBH}_4} CH_3CH_2\underset{\underset{\displaystyle OH}{|}}{\overset{\overset{\displaystyle CH_3}{|}}{C}}CH_3$$

2-Methyl-1-butene

tert-Pentyl alcohol

1-Methylcyclopentene $\xrightarrow{\text{Hg(OAc)}_2,\ H_2O}$ $\xrightarrow{\text{NaBH}_4}$ 1-Methylcyclopentanol

$$CH_3{-}\underset{\underset{\displaystyle CH_3}{|}}{\overset{\overset{\displaystyle CH_3}{|}}{C}}{-}CH{=}CH_2 \xrightarrow{\text{Hg(OAc)}_2,\ H_2O} \xrightarrow{\text{NaBH}_4} CH_3{-}\underset{\underset{\displaystyle CH_3}{|}}{\overset{\overset{\displaystyle CH_3}{|}}{C}}{-}\underset{\underset{\displaystyle OH}{|}}{CH}{-}CH_3$$

3,3-Dimethyl-1-butene

3,3-Dimethyl-2-butanol

Oxymercuration involves electrophilic addition to the carbon–carbon double bond, with the mercuric ion acting as electrophile. The absence of rearrangement and the high degree of stereoselectivity (typically *anti*)—*in the oxymercuration step*—argues against an open carbocation as intermediate. Instead, it has been proposed, there is formed a cyclic *mercurinium ion*, analogous to the bromonium

$$\left[-\overset{|}{C}\underset{\underset{\displaystyle Hg}{\diagdown\,\diagup}}{-}\overset{|}{C}- \right]^{2+}$$

and chloronium ions involved in the addition of halogens. In 1971, Olah (p. 194) reported spectroscopic evidence for the preparation of stable solutions of such mercurinium ions, and they have since been observed in the gas phase.

The mercurinium ion is attacked by the nucleophilic solvent—water, in the present case—to yield the addition product. This attack is back-side (unless prevented by some structural feature) and the net result is *anti*-addition, as in the addition of halogens (Secs. 9.5–9.6). Attack is thus of the S_N2 type; yet the orientation of addition shows that the nucleophile becomes attached to the more highly substituted carbon—as though there were a carbocation intermediate. Here, as with a halonium ion, we have an unstable three-membered ring; when it reacts, the transition state evidently has much S_N1 character (Sec. 8.15), and orientation is controlled by polar factors, not steric hindrance.

Although the demercuration reaction is not really understood, free radicals have been proposed as intermediates. Whatever the mechanism, demercuration is generally not stereoselective and can, in certain special cases, be accompanied by rearrangement.

Despite the stereoselectivity of the first stage, then, the overall process is not, in general, stereoselective. Rearrangements *can* occur, but are not common. The reaction of 3,3-dimethyl-1-butene shown above illustrates the absence of the rearrangements that are typical of intermediate carbocations.

Mercuration can be carried out in different solvents to yield products other than alcohols. This use of *solvomercuration* as a general synthetic tool is due largely to H. C. Brown (p. 650).

Problem 17.7   Predict the product of the reaction of propylene with mercuric acetate in methanol solution, followed by reduction with $NaBH_4$.

17.10   Hydroboration–oxidation

With the reagent *diborane*, $(BH_3)_2$, alkenes undergo *hydroboration* to yield alkylboranes, R_3B, which on oxidation give alcohols. For example:

$$(BH_3)_2 \xrightarrow{\ H_2C=CH_2\ } CH_3CH_2BH_2 \xrightarrow{\ H_2C=CH_2\ }$$

Diborane

$$(CH_3CH_2)_2BH \xrightarrow{\ H_2C=CH_2\ } (CH_3CH_2)_3B$$

Triethylboron

$$(CH_3CH_2)_3B + 3H_2O_2 \xrightarrow{\ OH^-\ } 3CH_3CH_2OH + B(OH)_3$$

Triethylboron   Hydrogen   Ethyl alcohol   Boric acid
peroxide

The reaction procedure is simple and convenient, the yields are exceedingly high, and, as we shall see, the products are ones difficult to obtain from alkenes in any other way.

Diborane is the dimer of the hypothetical BH_3 (*borane*) and, in the reactions that concern us, acts much as though it were BH_3. Indeed, in tetrahydrofuran, one of the solvents used for hydroboration, the reagent exists as the monomer, in the form of an acid–base complex with the solvent.

Borane   Diborane   Borane–tetrahydrofuran
complex

Hydroboration involves addition of BH_3 (or, in following stages, BH_2R and BHR_2) to the double bond, with hydrogen becoming attached to one doubly bonded carbon, and boron to the other. The alkylborane can then undergo oxidation, in which the boron is replaced by —OH (by a mechanism we shall encounter in Sec. 32.6).

Hydroboration   Oxidation

Alkene

Alcohol

$$H-B \Big< \ = \ H-BH_2, \ H-BHR, \ H-BR_2$$

Thus, the two-stage reaction process of hydroboration–oxidation permits, in effect, the addition to the carbon–carbon double bond of the elements of H—OH.

Reaction is carried out in an ether, commonly tetrahydrofuran or "diglyme" (*di*ethylene *gly*col *me*thyl ether, $CH_3OCH_2CH_2OCH_2CH_2OCH_3$). Diborane is commercially available in tetrahydrofuran solution. The alkylboranes are not isolated, but are simply treated *in situ* with alkaline hydrogen peroxide.

17.11   Orientation and stereochemistry of hydroboration

Hydroboration–oxidation, then, converts alkenes into alcohols. Addition is highly regioselective; the preferred product here, however, is exactly *opposite* to the one formed by oxymercuration–demercuration or by direct acid-catalyzed hydration. For example:

$$CH_3CH{=}CH_2 \xrightarrow{(BH_3)_2} \xrightarrow{H_2O_2,\ OH^-} CH_3CH_2CH_2OH$$

Propylene   *n*-Propyl alcohol
              (1°)

$$CH_3CH_2CH{=}CH_2 \xrightarrow{(BH_3)_2} \xrightarrow{H_2O_2,\ OH^-} CH_3CH_2CH_2CH_2OH$$

1-Butene   *n*-Butyl alcohol
              (1°)

$$\underset{\text{Isobutylene}}{CH_3{-}\overset{\overset{\textstyle CH_3}{|}}{C}{=}CH_2} \xrightarrow{(BH_3)_2} \xrightarrow{H_2O_2,\ OH^-} \underset{\substack{\text{Isobutyl alcohol}\\(1°)}}{CH_3{-}\overset{\overset{\textstyle CH_3}{|}}{C}H{-}CH_2OH}$$

$$\underset{\text{2-Methyl-2-butene}}{CH_3{-}CH{=}\overset{\overset{\textstyle CH_3}{|}}{C}{-}CH_3} \xrightarrow{(BH_3)_2} \xrightarrow{H_2O_2,\ OH^-} \underset{\substack{\text{3-Methyl-2-butanol}\\(2°)}}{CH_3{-}\overset{\overset{\textstyle CH_3}{|}}{C}H{-}\underset{\underset{\textstyle OH}{|}}{C}H{-}CH_3}$$

$$\underset{\text{3,3-Dimethyl-1-butene}}{CH_3{-}\overset{\overset{\textstyle CH_3}{|}}{\underset{\underset{\textstyle CH_3}{|}}{C}}{-}CH{=}CH_2} \xrightarrow{(BH_3)_2} \xrightarrow{H_2O_2,\ OH^-} \underset{\substack{\text{3,3-Dimethyl-1-butanol}\\(1°)}}{CH_3{-}\overset{\overset{\textstyle CH_3}{|}}{\underset{\underset{\textstyle CH_3}{|}}{C}}{-}CH_2{-}CH_2OH}$$

The hydroboration–oxidation process gives products corresponding to **anti-Markovnikov** *addition of water to the carbon–carbon double bond.*

The reaction of 3,3-dimethyl-1-butene illustrates a particular advantage of the method. *Rearrangement does not occur in hydroboration*—evidently because carbocations are not intermediates—and hence the method can be used without the complications that often accompany other addition reactions.

The reaction of 1,2-dimethylcyclopentene illustrates the stereochemistry of the synthesis: *hydroboration–oxidation involves overall* **syn-addition.**

1,2-Dimethylcyclopentene   *cis*-1,2-Dimethylcyclopentanol

Through a combination of features of which we take up only three—orientation, stereochemistry, and freedom from rearrangements—hydroboration–oxidation gains its great synthetic utility: it gives a set of alcohols not obtainable from alkenes by other methods and, through these alcohols (Sec. 18.8), provides a convenient route to corresponding members of many chemical families.

We catch here a brief glimpse of just one of the many applications of hydroboration to organic synthesis that have been discovered by H. C. Brown (of Purdue University). Although generally recognized as an outstanding organic chemist, Professor Brown was originally trained as an inorganic chemist, in the laboratory of H. I. Schlesinger at the University of Chicago. It was in this laboratory—in the course of a search for volatile uranium compounds, during World War II—that lithium aluminum hydride and sodium borohydride (Sec. 21.9) were first made and their reducing properties first observed; and it was here that Brown's interest in borohydrides originated—an interest that culminated in his receiving the Nobel Prize in 1979.

The examples we have used to show the fundamentals of hydroboration–oxidation have been, necessarily, simple ones. In practice, synthesis generally involves more complicated molecules, but the principles remain the same. For example:

Cholesterol   Cholestane-3β,6α-diol

Problem 17.8 Predict the products of hydroboration–oxidation of:
(a) (E)-2-phenyl-2-butene; (b) (Z)-2-phenyl-2-butene; (c) 1-methylcyclohexene.

Problem 17.9 The stereochemistry of hydroboration–oxidation is the *net* result of the stereochemistry of the two steps. The hydroboration step has been shown to involve *syn*-addition. What, then, must the stereochemistry of the oxidation step be?

17.12   Mechanism of hydroboration

Much of the usefulness of hydroboration–oxidation lies in the "unusual" orientation of the hydration. The —OH takes the position occupied by boron in the intermediate alkylborane, and hence the final product reflects the orientation of the hydroboration step. But is this orientation really "unusual"?

The orientation appears to be unusual because hydrogen adds to the opposite end of the double bond from where it adds in ordinary electrophilic addition. But the fundamental idea in electrophilic addition is that the *electrophilic* part of the reagent—the *acidic* part—becomes attached, using the π electrons, in such a way that the carbon being deprived of the π electrons is the one best able to stand the deprivation. In the addition of HZ to propylene, for example, the proton attaches itself to C–1; in that way the positive charge develops on C–2, where it can be dispersed by the methyl group. A secondary carbocation is formed instead of a primary.

$$CH_3-CH{=}CH_2 \xrightarrow{\text{HZ}} \left[CH_3-\overset{\delta+}{CH}{\cdots}\underset{\underset{\overset{|}{\overset{|}{Z}}}{\overset{|}{\underset{|}{H}}}}{CH_2} \right] \longrightarrow CH_3-\underset{\oplus}{CH}-\underset{\overset{|}{H}}{CH_2} + :Z$$

Now, what is the center of acidity in BH_3? Clearly, *boron*, with only six electrons. It is not at all surprising that boron should seek out the π electrons of the double bond and begin to attach itself to carbon. In doing this, it attaches itself in such a way that the positive charge can develop on the carbon best able to accommodate it. Thus:

$$CH_3-\overset{\delta+}{CH}{\cdots}CH_2$$
$$\underset{\underset{\overset{|}{H}}{H-B-H}}{\vdots\,{\scriptstyle\delta-}}$$

Unlike ordinary electrophilic addition, however, the reaction does not proceed to give a carbocation. As the transition state is approached, the carbon that is losing the π electrons becomes itself increasingly acidic: electron-deficient boron is acidic but so, too, is electron-deficient carbon. Not too far away is a hydrogen atom held to boron by a pair of electrons. Carbon begins to take that hydrogen, with its electron pair; boron, as it gains the π electrons, is increasingly willing to release that hydrogen. Boron and hydrogen both add to the doubly bonded carbons in the same transition state:

$$CH_3-\overset{\delta+}{CH}{\cdots}CH_2$$
$$\underset{H{-}{-}{-}{-}\underset{\overset{|}{}}{B}{-}}{\vdots\qquad\vdots\,{\scriptstyle\delta-}}$$

Transition state for hydroboration

In view of the basic nature of alkenes and the acidic nature of BH_3, the principal driving force of the reaction is almost certainly *attachment of boron to carbon*. In the transition state attachment of boron to C–1 has proceeded to a greater extent than attachment of hydrogen to C–2. Thus loss of (π) electrons by C–2 to the

C(1)—B bond exceeds its gain of electrons from hydrogen, and so C–2, the carbon that can best accommodate the charge, has become positive.

On theoretical grounds (Chap. 33) it has been postulated that the step we have described must follow a preliminary step in which boron attaches itself to both carbon atoms, or perhaps to the π electrons.

Thus orientation of addition in hydroboration is controlled in fundamentally the same way as in two-step electrophilic addition. Hydrogen becomes attached to opposite ends of the double bond in the two reactions because it adds without electrons in one case (as a *proton*, an acid), and with electrons in the other case (as a *hydride ion*, a base).

Because of the Lowry–Brønsted treatment of acids and bases, we tend to think of hydrogen chiefly in its proton character. Actually, its hydride character has considerably more *reality*. Solid lithium hydride, for example, has an ionic crystalline lattice made up of Li^+ and H^-; by contrast, a naked unsolvated proton is not encountered by the organic chemist.

We are already familiar with the facile transfer of hydride from carbon to carbon: within a single molecule (hydride shift in rearrangements), and between molecules (abstraction by carbocation, Sec. 8.17). Later on we shall encounter a set of remarkably versatile reducing agents (hydrides like *lithium aluminum hydride*, $LiAlH_4$, and *sodium borohydride*, $NaBH_4$) that function by transfer of hydride ion to organic molecules.

The orientation of hydroboration is affected, not just by the polar factor we have just described, but also by a steric factor: attachment of the boron moiety of the reagent (not just —BH_2, remember, but the larger —BHR and —BR_2) takes place more readily to the less crowded carbon of the double bond. Since this in general would lead to the same orientation as would the polar factor alone, it is not easy to tell which factor is in control. We can, however, expect this much: the bulkier the substituents on the alkene, the more important the steric factor; the more strongly electron-releasing or electron-withdrawing the substituents, the more important the polar factor.

Problem 17.10   Identify the acids and bases (Lewis or Lowry–Brønsted) in each of the following reactions:

(a) $Li^+H^- + H_2O \longrightarrow H_2 + Li^+OH^-$

(b) $(C_2H_5)_3B + NH_3 \longrightarrow (C_2H_5)_3\overset{-}{B}{:}\overset{+}{N}H_3$

(c) $(BH_3)_2 + 2(CH_3)_3N \longrightarrow 2H_3\overset{-}{B}{:}\overset{+}{N}(CH_3)_3$

(d) $2Li^+H^- + (BH_3)_2 \longrightarrow 2Li^+BH_4^-$

17.13   Aldehydes and ketones: an introduction

To discuss the chemistry of alkyl halides and alkenes, we found it necessary in Chapter 5 to learn something about alcohols. In the same way, to discuss the chemistry of alcohols, we need to know something about another class of compounds, the *aldehydes* and *ketones*.

Aldehydes and ketones have the general formulas:

$$\begin{array}{c} H \\ | \\ R\!-\!\overset{|}{C}\!=\!O \end{array} \qquad\qquad \begin{array}{c} R' \\ | \\ R\!-\!\overset{|}{C}\!=\!O \end{array}$$

An aldehyde   A ketone

The functional group of both is the **carbonyl group**, $-\overset{|}{C}\!=\!O$, and, as we shall see later (Chap. 21), aldehydes and ketones resemble each other closely in most of their reactions. Like the carbon–carbon double bond, the carbonyl group is unsaturated, and like the carbon–carbon double bond, it undergoes addition. But this addition differs in several important ways from addition to the carbon–carbon double bond.

Since the electrons of the carbonyl double bond hold together atoms of quite different electronegativity, they are not shared equally; in particular, the mobile π cloud is pulled strongly toward the more electronegative atom, oxygen. As a result, carbonyl carbon is electron-deficient and carbonyl oxygen is electron-rich.

$$\begin{array}{c} R' \\ \diagdown_{\delta+}\;\;{}_{\delta-} \\ C\!=\!O \\ \diagup \\ R \end{array}$$

Carbonyl group:
strongly polarized

This strong polarization of the carbonyl group has two important consequences. First, we are in no doubt as to the orientation of addition to a carbonyl group: whatever the mechanism involved, addition of an unsymmetrical reagent is oriented so that the nucleophilic (basic) portion attaches itself to carbon, and the electrophilic (acidic) portion attaches itself to oxygen. Second, the electron-deficient carbonyl carbon is especially susceptible to attack by nucleophiles. Where the typical reaction of alkenes is electrophilic addition, *the typical reaction of aldehydes and ketones is nucleophilic addition.*

At this point we shall be concerned with just one example of such addition.

17.14   Grignard synthesis of alcohols

The Grignard reagent, we recall, has the formula RMgX, and is prepared by the reaction of metallic magnesium with the appropriate organic halide (Sec. 3.16). This halide can be alkyl (1°, 2°, 3°), allylic, aralkyl (e.g., benzyl), or aryl (phenyl or

$$RX + Mg \xrightarrow{\text{anhydrous ether}} RMgX$$

A Grignard
reagent

substituted phenyl). The halogen may be —Cl, —Br or —I. (Arylmagnesium *chlorides* must be made in the cyclic ether tetrahydrofuran instead of diethyl ether.)

One of the most important uses of the Grignard reagent lies in its reaction with aldehydes and ketones. The carbon–magnesium bond of the Grignard reagent is a highly polar bond, carbon being negative relative to electropositive magnesium. It is not surprising, then, that in the addition to carbonyl compounds, the organic group becomes attached to carbon and magnesium to oxygen. The product is the

$$\overset{\delta+}{\underset{}{C}}\!\!=\!\!\overset{\delta-}{O} + \overset{}{R}\!\!-\!\!MgX \longrightarrow -\overset{|}{\underset{R}{C}}-\bar{O}\overset{+}{M}gX \xrightarrow{H_2O} -\overset{|}{\underset{R}{C}}-OH + Mg(OH)X$$

An alcohol

$$\downarrow H^+$$

$$Mg^{2+} + X^- + H_2O$$

magnesium salt of the weakly acidic alcohol and is easily converted into the alcohol itself by the addition of the stronger acid, water. Since the Mg(OH)X thus formed is a gelatinous material difficult to handle, dilute mineral acid (HCl, H_2SO_4) is commonly used instead of water, so that water-soluble magnesium salts are formed.

Grignard reagents are the classical reagents for such syntheses. Increasingly, however, *organolithium* compounds are being used instead, chiefly because they are less prone to unwanted side reactions. Organolithiums can be prepared in the same way as Grignard reagents, by reaction of the metal with organic

$$RX + 2Li \xrightarrow{\text{anhydrous ether}} RLi + LiX$$

An organolithium

halides. Because lithium is more electropositive than magnesium, carbon–lithium bonds are more polar than carbon–magnesium bonds; carbon is more negative—more carbanion-like—and organolithiums are in general somewhat more reactive than Grignard reagents.

Organolithiums react with aldehydes and ketones in the same manner that we have shown for Grignard reagents, and yield the same kinds of products. We shall consider this reaction to be an extension of Grignard's original synthesis. We shall refer to the general method as the *Grignard synthesis of alcohols*, and often

$$\overset{\delta+}{\underset{}{C}}\!\!=\!\!\overset{\delta-}{O} \quad \overset{}{R}\!\!-\!\!Li \longrightarrow -\overset{|}{\underset{R}{C}}-O^-Li^+ \xrightarrow{H_2O} -\overset{|}{\underset{R}{C}}-OH$$

An aldehyde
or ketone

An alcohol

discuss it in terms of organomagnesium reagents; it should be understood, however, that most of what we say applies to the analogous synthesis involving organolithiums.

> **Problem 17.11** Write equations for the reaction of *n*-butyllithium with: (a) H_2O; (b) D_2O; (c) C_2H_5OH; (d) CH_3NH_2; (e) $C_2H_5C\!\!\equiv\!\!CH$; (f) CH_3CCH_3.
> $$\underset{O}{\overset{||}{}}$$

Now, why is the Grignard synthesis so important? Because it enables us to take two organic molecules and convert them into a bigger one. To do this, *we form a carbon–carbon bond.* Once again (Sec. 11.13) we join together electrophilic carbon and nucleophilic carbon. This time, electrophilic carbon is furnished by the carbonyl group. For nucleophilic carbon we turn again to the carbanion-like organic group of an organometallic compound: a Grignard reagent or an organolithium. The Grignard reaction is thus an example of the typical reaction of aldehydes and ketones: nucleophilic addition.

But this is only half the story. Not only does the Grignard synthesis involve formation of a carbon–carbon bond, but the product contains the highly versatile group, —OH. And now, as we shall soon see, the way is open to further synthesis, and the building of still bigger and more complicated structures.

17.15   Products of the Grignard synthesis

The class of alcohol that is obtained from a Grignard synthesis depends upon the type of carbonyl compound used: *formaldehyde*, HCHO, *yields primary alcohols*; *other aldehydes*, RCHO, *yield secondary alcohols*; and *ketones*, R_2CO, *yield tertiary alcohols*.

This relationship arises directly from our definitions of aldehydes and ketones, and our definitions of primary, secondary, and tertiary alcohols. The number of hydrogens attached to the carbonyl carbon defines the carbonyl compound as formaldehyde, higher aldehyde, or ketone. The carbonyl carbon is the one that finally bears the —OH group in the product; here the number of hydrogens defines the alcohol as primary, secondary, or tertiary. For example:

$$
\underset{\substack{\text{Isobutyl magnesium}\\\text{bromide}}}{CH_3\overset{\overset{\displaystyle CH_3}{|}}{C}HCH_2-MgBr} + \underset{\text{Acetaldehyde}}{CH_3-\overset{\overset{\displaystyle H}{|}}{C}=O} \longrightarrow CH_3\overset{\overset{\displaystyle CH_3}{|}}{C}HCH_2-\underset{\underset{\displaystyle OMgBr}{|}}{C}HCH_3
$$

$$\Big\downarrow H_2O$$

$$
CH_3\overset{\overset{\displaystyle CH_3}{|}}{C}HCH_2-\underset{\underset{\displaystyle OH}{|}}{C}HCH_3
$$

A 2° alcohol
4-Methyl-2-pentanol

$$
\underset{\substack{n\text{-Butylmagnesium}\\\text{bromide}}}{n\text{-}C_4H_9-MgBr} + \underset{\text{Acetone}}{CH_3-\overset{\overset{\displaystyle CH_3}{|}}{C}=O} \longrightarrow n\text{-}C_4H_9-\overset{\overset{\displaystyle CH_3}{|}}{\underset{\underset{\displaystyle CH_3}{|}}{C}}-OMgBr \xrightarrow{H_2O} n\text{-}C_4H_9-\overset{\overset{\displaystyle CH_3}{|}}{\underset{\underset{\displaystyle CH_3}{|}}{C}}-OH
$$

A 3° alcohol
2-Methyl-2-hexanol

A related synthesis utilizes *ethylene oxide* (Sec. 19.15) to make *primary alcohols containing two more carbons* than the Grignard reagent. Here, too, the organic group

$$
\underset{\text{Ethylene oxide}}{H_2C-CH_2 \atop \diagdown\mkern-6mu O\mkern-6mu\diagup} + R-MgX \longrightarrow R-CH_2CH_2OMgX \xrightarrow{H_2O} R-CH_2CH_2OH
$$

A 1° alcohol:
two carbons added

becomes attached to carbon and magnesium to oxygen, this time with the breaking of a carbon–oxygen σ bond in a highly strained three-membered ring (Sec. 8.15). For example:

$$
\underset{\substack{\text{Phenylmagnesium}\\\text{bromide}}}{\text{⬡}MgBr} + \underset{\substack{\text{Ethylene}\\\text{oxide}}}{H_2C-CH_2 \atop \diagdown\mkern-6mu O\mkern-6mu\diagup} \longrightarrow \text{⬡}CH_2CH_2OMgBr
$$

$$\Big\downarrow H_2O$$

$$\text{⬡}CH_2CH_2OH$$

2-Phenylethanol
(β-Phenylethyl alcohol)

Like their alkyl counterparts, both lithium acetylides and alkynyl Grignard reagents can add to aldehydes and ketones to generate alcohols. There are thus

$$
RC\equiv C-Li + \underset{}{\overset{}{>}}C=O \longrightarrow RC\equiv C-\overset{|}{\underset{|}{C}}-O^-Li^+ \xrightarrow{H^+} RC\equiv C-\overset{|}{\underset{|}{C}}-OH
$$

$$
RC\equiv C-MgX + \underset{}{\overset{}{>}}C=O \longrightarrow RC\equiv C-\overset{|}{\underset{|}{C}}-O^-MgX^+ \xrightarrow{H^+} RC\equiv C-\overset{|}{\underset{|}{C}}-OH
$$

formed compounds that contain not only —OH but a second highly reactive group, the carbon–carbon triple bond. For example:

$$
\begin{array}{cccccc}
 & & & CH_3 & & CH_3 \\
 & & & | & & | \\
HC\equiv C-Li & + & CH_3-C=O & \longrightarrow & HC\equiv C-C-CH_3 & \xrightarrow{H^{\cdot}} & HC\equiv C-C-CH_3 \\
 & & & & | & & | \\
\text{Ethynyllithium} & & \text{Acetone} & & OLi & & OH
\end{array}
$$

2-Methyl-3-butyn-2-ol

$$
\begin{array}{cccc}
 & & H & \\
 & & | & \\
CH_3C\equiv C-MgBr & + & CH_3C=O & \longrightarrow & CH_3C\equiv C-CHCH_3 & \xrightarrow{H^{\cdot}} & CH_3C\equiv C-CHCH_3 \\
 & & & & | & & | \\
\text{Propyn-1-ylmagnesium} & & \text{Acetaldehyde} & & OMgBr & & OH \\
\text{bromide} & & & & &
\end{array}
$$

3-Pentyn-2-ol

17.16   Planning a Grignard synthesis

How do we decide which Grignard reagent and which carbonyl compound to use in preparing a particular alcohol? We have only to look at the structure of the alcohol we want. Of groups attached to the carbon bearing the —OH group, one must come from the Grignard reagent, the other two (including any hydrogens) must come from the carbonyl compound.

Most alcohols can be obtained from more than one combination of reagents; we usually choose the combination that is most readily available. Consider, for example, the synthesis of 2-methyl-2-hexanol:

$$
\begin{array}{ccc}
 & CH_3 & \\
 & | & \\
CH_3CH_2CH_2CH_2\!-\!\!\!\!-\!C\!-\!CH_3 & \longleftarrow & CH_3CH_2CH_2CH_2-MgBr + \begin{array}{c} CH_3 \\ | \\ C-CH_3 \\ \| \\ O \end{array} \\
 & | & \\
 & OH &
\end{array}
$$

| 2-Methyl-2-hexanol | *n*-Butylmagnesium bromide | Acetone |

$$
\begin{array}{ccc}
 & CH_3 & \\
 & | & \\
CH_3CH_2CH_2CH_2-C\!\!\prec\!\! CH_3 & \longleftarrow & CH_3CH_2CH_2CH_2-\begin{array}{c} CH_3 \\ | \\ C \\ \| \\ O \end{array} + BrMg-CH_3 \\
 & | & \\
 & OH &
\end{array}
$$

| 2-Methyl-2-hexanol | Methyl *n*-butyl ketone | Methylmagnesium bromide |

As shown, we could make this either from the four-carbon Grignard reagent and acetone, or from the methyl Grignard reagent and the six-carbon aliphatic ketone. As we shall know when we have studied aldehydes and ketones (Chap. 21), the first route uses the more readily available carbonyl compound and is the one actually used to make this alcohol.

17.17   Limitations of the Grignard synthesis

The very reactivity that makes a Grignard reagent so useful strictly limits how we may use it. We must keep this reactivity in mind when we plan the experimental conditions of the synthesis, when we select the halide that is to become the Grignard reagent, and when we select the compound with which it is to react.

In our first encounter with the Grignard reagent (Sec. 3.16), we allowed it to react with water to form an alkane; the stronger acid, water, displaced the extremely weak acid, the alkane, from its salt. In the same way, *any* compound containing hydrogen attached to an electronegative element—oxygen, nitrogen, sulfur, or even triply bonded carbon—is acidic enough to decompose a Grignard reagent. A Grignard reagent reacts rapidly with oxygen and carbon dioxide, and with nearly every organic compound containing a carbon–oxygen or carbon–nitrogen multiple bond.

How does all this affect our reaction between a Grignard reagent and, say, an aldehyde? First of all, alkyl halide, aldehyde, and the ether used as solvent must be scrupulously dried and freed of the alcohol from which each was very probably made; a Grignard reagent will not even form in the presence of water. Our apparatus must be completely dry before we start. We must protect the reaction system from the water vapor, oxygen, and carbon dioxide of the air: water vapor can be kept out by use of calcium chloride tubes, and oxygen and carbon dioxide can be swept out of the system with dry nitrogen. Having done all this we may hope to obtain a good yield of product—providing we have properly chosen the halide and the aldehyde.

We cannot prepare a Grignard reagent from a compound (e.g., $HOCH_2CH_2Br$) that contains, in addition to halogen, some group (e.g., —OH) that will react with a Grignard reagent; if this were tried, as fast as a molecule of Grignard reagent formed it would react with the active group (—OH) in another molecule to yield an undesired product ($HOCH_2CH_2$—H).

We must be particularly watchful in the preparation of an arylmagnesium halide, in view of the wide variety of substituents that might be present on the benzene ring. Carboxyl (—COOH), hydroxyl (—OH), amino (—NH$_2$), and —SO$_3$H all contain hydrogen attached to oxygen or nitrogen, and therefore are so acidic that they will decompose a Grignard reagent. We have just learned that a Grignard reagent adds to the carbonyl group (C=O), and we shall learn that it adds similarly to —COOR and —C≡N groups. The nitro (—NO$_2$) group oxidizes a Grignard reagent. It turns out that only a comparatively few groups may be present in the halide molecule from which we prepare a Grignard reagent; among these are —R, —Ar, —OR, and —Cl (of an aryl chloride).

By the same token, the aldehyde (or other compound) with which a Grignard reagent is to react may not contain other groups that are reactive toward a Grignard reagent. For example, a Grignard reagent would be decomposed before it could add to the carbonyl group of:

m-Nitrobenzaldehyde   *p*-Aminoacetophenone

p-Benzoylbenzoic acid

These may seem like severe limitations, and they are. Nevertheless, the number of acceptable combinations is so great that the Grignard reagent is one of our most valuable synthetic tools. The kind of precautions described here must be taken in any kind of organic synthesis: we must not restrict our attention to the group we happen to be interested in, but must look for possible interference by other functional groups.

Even when we do see the possibility of interference, there is often something positive that we can do. We may be able to introduce—temporarily—a *protecting group* to prevent an unwanted reaction. (See, for example, Sec. 19.9.)

17.18   Steroids

Cholesterol (p. 650), notorious as the substance deposited on the walls of arteries and as the chief constituent of gallstones, is the kind of alcohol called a *sterol*. Sterols belong, in turn, to the class of compounds called **steroids**: compounds of the general formula

A steroid

The rings are (generally) aliphatic. Lines like the vertical ones attached to the 10- and 13-positions represent *angular methyl* groups. Commonly, in cholesterol, for example,

Stereochemistry is indicated by solid lines (β-bonds, coming *out* of the plane of the paper) and broken lines (α-bonds, going *behind* the plane of the paper).

A 3β,6α-diol

I

Thus in I the —H and —OH at the 5- and 6-positions are *cis* to each other, but *trans* to the 3-OH and to the angular methyl at the 10-position. Fusion of the rings to each other can be *cis* or *trans*, thus increasing the complications of the stereochemistry.

trans Fusion

cis Fusion

Finally, in any rigid cyclic system like this, conformational effects are marked, and often completely control the course of reaction.

Steroids include sex hormones and adrenal cortical hormones (*cortisone* is one), cardiac glycosides, and bile acids. Because of their biological importance—and, undoubtedly, because of the fascinating complexity of the chemistry—the study of steroids has been, and is now, one of the most active areas of organic chemical research.

Estrone

An *estrogen*, or
female sex hormone

Testosterone

An *androgen*, or
male sex hormone

Cortisone

An adrenocortical
hormone

Ergosterol

A precursor of
vitamin D

PROBLEMS

---

1. (a) Ignoring enantiomerism, draw the structures of the eight isomeric pentyl alcohols, $C_5H_{11}OH$. (b) Name each by the IUPAC system. (c) Label each as primary, secondary, or tertiary. (d) Which one is isopentyl alcohol? *n*-Pentyl alcohol? *tert*-Pentyl alcohol? (e) Give the structure of a primary, a secondary, and a tertiary alcohol of the formula $C_6H_{13}OH$. (f) Give the structure of a primary, a secondary, and a tertiary *cyclic* alcohol of the formula C_5H_9OH.

2. Without referring to tables, arrange the following compounds in order of decreasing boiling point:

(a) 3-hexanol   (d) *n*-octyl alcohol
(b) *n*-hexane   (e) *n*-hexyl alcohol
(c) 2-methyl-2-pentanol

3. Looking at the beginning of each chapter for the structure involved, tell which families of compounds discussed in this book can: (a) form hydrogen bonds with other molecules of the same kind; (b) form hydrogen bonds with water.

4. Which compound would you expect to have the higher boiling point? (Check your answers in the proper tables.)

(a) *p*-cresol (*p*-CH$_3$C$_6$H$_4$OH) or anisole (C$_6$H$_5$OCH$_3$)

(b) methyl acetate, $CH_3C\overset{O}{\underset{OCH_3}{\diagup\diagdown}}$, or propionic acid, $CH_3CH_2C\overset{O}{\underset{OH}{\diagup\diagdown}}$

(c) propionic acid or *n*-pentyl alcohol.

5. Consider the possible synthesis of the eight isomeric pentyl alcohols of Problem 1(a) by oxymercuration–demercuration and hydroboration–oxidation. For each alcohol show the alkene or alkenes (if any) from which it could be made in pure form, and the synthetic method that would be used in each case.

6. Give structures of the Grignard reagent and the substrate (aldehyde, ketone, or ethylene oxide) that would react to yield each of the following alcohols. If more than one combination of reactants is possible, show each of the combinations.

(a)–(h) each of the isomeric pentyl alcohols of Problem 1(a)
(i) 1-phenyl-1-propanol   (o) 1-cyclohexylethanol
(j) 2-phenyl-2-propanol   (p) 2,4-dimethyl-3-pentanol
(k) 1-phenyl-2-propanol   (q) 1-(*p*-tolyl)ethanol
(l) 3-phenyl-1-propanol   (r) 1-pentyn-3-ol
(m) 1-methylcyclohexanol   (s) 3-pentyn-2-ol
(n) cyclohexylmethanol

7. For many 2-substituted ethanols, GCH_2CH_2OH, the *gauche* conformation is more stable than the *anti*:

   G = —OH, —NH$_2$, —F, —Cl, —Br, —OCH$_3$, —NHCH$_3$, —N(CH$_3$)$_2$, and —NO$_2$.

How might this be accounted for?

8. (a) As shown in Sec. 17.11, cholesterol is converted into cholestane-3β,6α-diol through *syn*-hydration by hydroboration–oxidation. What stereoisomeric product could also have been formed by *syn*-hydration? Actually, the reaction gives a 78% yield of cholestane-3β,6α-diol, and only a small amount of its stereoisomer. What factor do you think is responsible for this particular stereoselectivity? (*Hint*: See Sec. 17.18.)

(b) Hydroboration of androst-9(11)-ene (on the next page) gives 90% of a single stereoisomer. Which would you expect this to be?

Androst-9(11)-ene

9. (a) Using models and then drawing formulas, show the possible chair conformations for *cis*-1,3-cyclohexanediol. (b) On the basis solely of 1,3-interaction, which would you expect to be the more stable conformation? (c) Infrared evidence indicates intramolecular hydrogen bonding in *cis*-1,3-cyclohexanediol. Just how would the infrared spectrum show this? Which conformation in (a) is indicated by this evidence, and what is the source of its stability?

10. The infrared spectrum of the stereoisomer of 2,5-di-*tert*-butyl-1,4-cyclohexanediol in which all four substituents are *cis* to each other shows the presence of an intramolecular hydrogen bond. In what conformation does the molecule exist? (*Hint*: Use models.)

11. The carbon–metal bond in *n*-propyllithium is more polar than that in *n*-propyl-magnesium bromide, but is still essentially covalent. Spectroscopic studies of allyllithium show that it contains four equivalent hydrogens. What does this suggest about the structure of the molecule? How do you account for this?

12. In each of the following examples of carbon–carbon bond formation, show (where you can) the step in which this bond is formed. Identify the nucleophilic carbon and the electrophilic carbon.

(a) *n*-butyl bromide + KCN $\longrightarrow$ pentanenitrile ($CH_3CH_2CH_2CH_2CN$)

(b) *n*-propyllithium + HCHO $\longrightarrow$ $CH_3CH_2CH_2CH_2OLi$

(c) isobutylene + isobutane $\xrightarrow{H^+}$ "isooctane"

(d) isopropylmagnesium bromide + CO_2 $\longrightarrow$ $(CH_3)_2CHCOOMgBr$

(e) $HC{\equiv}CH$ $\xrightarrow{NaNH_2}$ $HC{\equiv}CNa$ $\xrightarrow{EtBr}$ $HC{\equiv}CC_2H_5$

(f) propylene + $CHCl_3$ $\xrightarrow{t\text{-BuOK}}$ 1,1-dichloro-2-methylcyclopropane

(g) benzene + isopropyl chloride $\xrightarrow{AlCl_3}$ isopropylbenzene

13. (a) What are the two diastereomeric products that could be formed by *anti*-addition of bromine to cholesterol? To 2-cholestene? (b) Actually, one product greatly predominates in each case, as shown:

Cholesterol

5α,6β-Dibromo-3β-hydroxycholestane
85% yield

2-Cholestene

$2\beta,3\alpha$-Dibromocholestane

70% yield

How do you account for the observed stereochemistry? (It is *not* a matter of relative stability of the diastereomers.) (*Hint*: Consider carefully the stereochemical possibilities at each step of the mechanism.)

14. When (*E*)-3-methyl-2-pentene reacts with CO and H_2 in the presence of $RhH(CO)(PPh_3)_3$, there is obtained (to the extent of 95%) racemic *threo*-2,3-dimethyl-1-pentanol. What is the stereochemistry of this reaction?

15. On treatment with a variety of reagents (water, for example), borate esters of the kind shown are converted into alkenes:

$$(RO)_2BCH_2CH_2Br \longrightarrow CH_2=CH_2$$

The *cis* and *trans* esters (I) were prepared, and their configurations were assigned by NMR. Each ester was treated with bromine, and the resulting dibromide was treated with water; *cis*-I gave only *trans*-II as the final product, and *trans*-I gave only *cis*-II.

$$CH_3CH=C(CH_3)B(OR)_2 \qquad CH_3CH=CBrCH_3$$

I (*cis* or *trans*) II (*cis* or *trans*)

Making use of what you know about the addition of bromine to alkenes, what do you conclude about the stereochemistry of this elimination reaction? Show the most likely mechanism for the elimination, including the part played by water.

18

Alcohols II. Reactions

18.1 Chemistry of the —OH group

The chemical properties of an alcohol, ROH, are determined by its functional group, —OH, the hydroxyl group. When we have learned the chemistry of the alcohols, we shall have learned much of the chemistry of the hydroxyl group in whatever compound it may occur; we shall know, in part at least, what to expect of hydroxy halides, hydroxy acids, hydroxy aldehydes, etc.

Reactions of an alcohol can involve the breaking of either of two bonds: the C—OH bond, with removal of the —OH group; or the O—H bond, with removal of —H. Either kind of reaction can involve substitution, in which a group replaces the —OH or —H, or elimination, in which a double bond is formed.

We are already acquainted with some of the chemical properties of alcohols: their acidity and basicity, their nucleophilic power, their conversion into alkyl halides, alkyl sulfonates, and alkenes. We have seen how the —OH group—either directly or, more often, via an alkyl halide or sulfonate—can be replaced by a host of other groups, or be eliminated to form a carbon–carbon double bond. In this chapter we shall review that familiar chemistry, going more deeply into some of it.

But we shall take up a good deal of new chemistry, too. We shall look at an entirely different facet of the chemistry of alcohols: their conversion into oxygen compounds of higher oxidation states: aldehydes and ketones, and carboxylic acids. We shall see how, by combining the chemistry of this chapter with that of the preceding one, we can broaden our approach to organic synthesis; we shall begin to appreciate the choice of desert island companions made by our chemical Crusoe of Sec. 17.1.

18.2 Reactions

Some of the more important reactions of alcohols are listed below, and are discussed in following sections.

REACTIONS OF ALCOHOLS

C---OH Bond Cleavage

$$R \!\!\mid\!\! OH$$

1. Reaction with hydrogen halides. Discussed in Secs. 5.25 and 18.3.

$$R-OH + HX \longrightarrow R-X + H_2O \qquad R \text{ may rearrange}$$

Reactivity of HX: $HI > HBr > HCl$

Reactivity of ROH: allyl, benzyl $> 3° > 2° > 1°$

Examples:

$$\underset{\underset{OH}{|}}{CH_3CHCH_3} \xrightarrow[\substack{\text{or NaBr, H}_2\text{SO}_4 \\ \text{reflux}}]{\text{conc. HBr}} \underset{\underset{Br}{|}}{CH_3CHCH_3}$$

Isopropyl alcohol   Isopropyl bromide

$$CH_3CH_2CH_2CH_2CH_2OH \xrightarrow[\text{heat}]{\text{HCl, ZnCl}_2} CH_3CH_2CH_2CH_2CH_2Cl$$

n-Pentyl alcohol   *n*-Pentyl chloride

$$\underset{\underset{OH}{|}}{\overset{\overset{CH_3}{|}}{CH_3-C-CH_3}} \xrightarrow[\text{room temp.}]{\text{conc. HCl}} \underset{\underset{Cl}{|}}{\overset{\overset{CH_3}{|}}{CH_3-C-CH_3}}$$

tert-Butyl alcohol   *tert*-Butyl chloride

2. Reaction with phosphorus trihalides. Discussed in Sec. 18.8.

$$3R-OH + PX_3 \longrightarrow 3R-X + H_3PO_3$$
$$(PX_3 = PBr_3, PI_3)$$

Examples:

$$\underset{}{\overset{\overset{CH_3}{|}}{CH_3CH_2CHCH_2OH}} \xrightarrow{PBr_3} \underset{}{\overset{\overset{CH_3}{|}}{CH_3CH_2CHCH_2Br}}$$

2-Methyl-1-butanol   2-Methyl-1-bromobutane

_____ CONTINUED _____

——— CONTINUED ———

$$C_6H_5\underset{\underset{OH}{|}}{C}HCH_3 \xrightarrow{PBr_3} C_6H_5\underset{\underset{Br}{|}}{C}HCH_3$$

1-Phenylethanol   1-Bromo-1-phenylethane

$$CH_3CH_2OH \xrightarrow{P+I_2} CH_3CH_2I$$

Ethyl alcohol   Ethyl iodide

3. Dehydration. Discussed in Secs. 7.25 and 18.3.

$$-\underset{\underset{H}{|}}{C}-\underset{\underset{OH}{|}}{C}- \xrightarrow{acid} -\underset{}{C}=\underset{}{C}- + H_2O \qquad \textit{Rearrangement may occur}$$

Reactivity of ROH: 3° > 2° > 1°

Examples:

$$CH_3CH_2CH_2CH_2OH \xrightarrow{H_2SO_4,\ heat} CH_3CH{=}CHCH_3 + CH_3CH_2CH{=}CH_2$$

n-Butyl alcohol   2-Butene   1-Butene
                                Major product

Cyclohexanol   Cyclohexene

$$C_6H_5-\underset{\underset{OH}{|}}{\overset{\overset{CH_3}{|}}{C}}-CH_3 \xrightarrow{H_2SO_4,\ heat} C_6H_5-\overset{\overset{CH_3}{|}}{C}{=}CH_2$$

2-Phenyl-2-propanol   2-Phenylpropene

O---H Bond Cleavage

$$RO{+}H$$

4. Reaction as acids: reaction with active metals. Discussed in Sec. 18.4.

$$RO{-}H + M \longrightarrow RO^-M^+ + \tfrac{1}{2}H_2 \qquad M = Na, K, Mg, Al, etc.$$

Reactivity of ROH: CH₃OH > 1° > 2° > 3°

Examples:

$$CH_3CH_2OH \xrightarrow{Na} CH_3CH_2O^-Na^+ + \tfrac{1}{2}H_2$$

Ethyl alcohol   Sodium ethoxide

——— CONTINUED ———

CONTINUED

$$CH_3-\underset{\underset{H}{|}}{\overset{\overset{CH_3}{|}}{C}}-OH \xrightarrow{Al} CH_3-\underset{\underset{H}{|}}{\overset{\overset{CH_3}{|}}{C}}-O)_3Al$$

Isopropyl alcohol   Aluminum isopropoxide

$$CH_3-\underset{\underset{CH_3}{|}}{\overset{\overset{CH_3}{|}}{C}}-OH \xrightarrow{K} CH_3-\underset{\underset{CH_3}{|}}{\overset{\overset{CH_3}{|}}{C}}-O^-K^+$$

tert-Butyl alcohol   Potassium *tert*-butoxide

5. **Ester formation.** Discussed in Secs. 18.5 and 23.16.

Examples:

$$CH_3CH_2O-H + CH_3\langle\bigcirc\rangle SO_2Cl \xrightarrow{base} CH_3\langle\bigcirc\rangle SO_2-OCH_2CH_3$$

Tosyl chloride   Ethyl tosylate
(*p*-Toluenesulfonyl chloride)

$$CH_3CH_2O-H + CH_3C\overset{O}{\underset{OH}{\diagdown}} \xrightarrow{H^+} CH_3C\overset{O}{\underset{OC_2H_5}{\diagdown}} + H_2O$$

Acetic acid   Ethyl acetate

6. **Oxidation.** Discussed in Sec. 18.6.

Primary $R-CH_2OH$
$\xrightarrow{C_5H_5NH^+CrO_3Cl^-} R-\overset{\overset{H}{|}}{C}=O$
An aldehyde

$\xrightarrow{KMnO_4} R-COOH$
A carboxylic acid

Secondary $R-\overset{\overset{R}{|}}{C}HOH \xrightarrow{K_2Cr_2O_7 \text{ or } CrO_3} R-\overset{\overset{R}{|}}{C}=O$
A ketone

Tertiary $R-\underset{\underset{R}{|}}{\overset{\overset{R}{|}}{C}}-OH \xrightarrow{neut. \ KMnO_4}$ no reaction

Examples:

$$CH_3CH_2CH_2OH \xrightarrow{C_5H_5NH^+CrO_3Cl^-} CH_3CH_2\overset{\overset{H}{|}}{C}=O$$

n-Propyl alcohol   Propionaldehyde
(1°)

CONTINUED

$$\underset{\substack{\text{2-Methyl-1-butanol}\\(1°)}}{\text{CH}_3\text{CH}_2\overset{\displaystyle\text{CH}_3}{\underset{\displaystyle|}{\text{CH}}}\text{CH}_2\text{OH}} \xrightarrow{\text{KMnO}_4} \underset{\substack{\text{2-Methylbutanoic acid}}}{\text{CH}_3\text{CH}_2\overset{\displaystyle\text{CH}_3}{\underset{\displaystyle|}{\text{CH}}}\text{COOH}}$$

Cyclohexanol
(2°)

Cyclohexanone

3-Cholestanol
(2°)

3-Cholestanone

We can see that alcohols undergo many kinds of reactions, to yield many kinds of products. Because of the availability of alcohols, each of these reactions is one of the best ways to make the particular kind of product. After we have learned a little more about the reactions themselves, we shall look at some of the ways in which they can be applied to synthetic problems.

18.3   Cleavage of the C---OH bond

We have already studied two important reactions of alcohols: their reaction with hydrogen halides to form alkyl halides,

$$\underset{\text{Alcohol}}{\text{R—OH}} \overset{\text{H}^+}{\rightleftharpoons} \underset{\substack{\text{Protonated}\\\text{alcohol}}}{\text{R—OH}_2{}^+} \xrightarrow{\text{X}} \underset{\substack{\text{Alkyl}\\\text{halide}}}{\text{R—X}} + \text{H}_2\text{O}$$

and their dehydration to form alkenes.

$$\underset{\substack{\text{Alcohol}}}{\underset{\displaystyle\text{H}\ \ \text{OH}}{-\overset{|}{\underset{|}{\text{C}}}-\overset{|}{\underset{|}{\text{C}}}-}} \overset{\text{H}^+}{\rightleftharpoons} \underset{\substack{\text{Protonated alcohol}}}{\underset{\displaystyle\text{H}\ \ \text{OH}_2{}^+}{-\overset{|}{\underset{|}{\text{C}}}-\overset{|}{\underset{|}{\text{C}}}-}} \overset{-\text{H}_2\text{O}}{\rightleftharpoons} \underset{\substack{\text{Carbocation}}}{\underset{\displaystyle\text{H}}{-\overset{|}{\underset{|}{\text{C}}}-\overset{\oplus}{\underset{|}{\text{C}}}-}} \overset{-\text{H}^+}{\rightleftharpoons} \underset{\substack{\text{Alkene}}}{-\overset{|}{\text{C}}\!=\!\overset{|}{\text{C}}-}$$

These are, of course, examples of two of the principal types of reactions that we have studied so far, nucleophilic substitution and elimination.

Each of these reactions requires the presence of acid to convert the alcohol into the actual substrate, the *protonated* alcohol. Whether the reaction is substitution or elimination, and whether it follows a bimolecular or unimolecular mechanism, the carbon–oxygen bond must undergo heterolytic cleavage—the substrate must lose a leaving group. The protonated alcohol readily loses the weakly basic water molecule. The unprotonated alcohol would have to lose the strongly basic hydroxide ion, a process so difficult that it seldom if ever happens.

Alcohols are the precursors of a wide variety of compounds in which the —OH group has been lost: replaced by some other group, or eliminated with formation of a double bond. But this loss of —OH is not brought about directly, in a single step, from the alcohol itself: it must be brought about indirectly, by first converting the alcohol into something else; the very poor leaving group must be converted into a good leaving group. The simplest way to accomplish this is through protonation. But the acidic medium required for protonation severely limits us in our choice of nucleophiles or bases; any appreciably basic reagent will rapidly neutralize the acid. With only a weak nucleophile or a weak base present, reaction tends to follow a unimolecular mechanism: via carbocations, and hence with the likelihood of rearrangement. As we shall discuss in Sec. 18.5, there is another way to change the —OH into a good leaving group: more elaborate than protonation, but with certain important advantages.

18.4 Alcohols as acids and bases

When we first met alcohols (Sec. 5.5), we learned that they are bases, and of about the same strength as water. Like water they contain oxygen, and it is this oxygen, with its unshared pairs, that makes them basic. Since that initial meeting, we have seen many times how their basicity plays a central role in determining their chemical behavior, both as substrates and as reagents. They accept protons from acids to form protonated alcohols, and this protonation permits them to serve as substrates in nucleophilic substitution and elimination—something they could not do in their unprotonated form. They accept protons from carbocations and thus act as basic reagents in bringing about elimination. Their basicity makes them nucleophilic as well, and able to bring about a substitution: in full-fledged S_N2 reactions, or by rendering nucleophilic assistance to the formation of carbo-cations, or by combining with carbocations once they are formed.

We learned earlier that alcohols are acids, too, and again of roughly the same strength as water. Hydrogen is bonded to the very electronegative element oxygen. The polarity of the O—H bond facilitates the separation of a relatively positive proton; viewed differently, electronegative oxygen readily accommodates the negative charge of electrons left behind.

The acidity of alcohols is shown by their reaction with active metals to liberate hydrogen gas and form alkoxides.

$$ROH + Na \longrightarrow RO^-Na^+ + \tfrac{1}{2}H_2$$

A sodium alkoxide

Just *how* acidic are alcohols? With the possible exception of methanol, they are somewhat weaker acids than water,

$$RO^-Na^+ + H—OH \longrightarrow Na^+OH^- + RO—H$$

| Stronger base | Stronger acid | Weaker base | Weaker acid |

but stronger acids than acetylene

$$HC\equiv C^-\ Na^+\ +\ RO{-}H\ \longrightarrow\ RO^-\ Na^+\ +\ HC\equiv C{-}H$$

| Stronger | Stronger | Weaker | Weaker |
| base | acid | base | acid |

or ammonia.

$$H_2N^-\ Na^+\ +\ RO{-}H\ \longrightarrow\ RO^-\ Na^+\ +\ H_2N{-}H$$

| Stronger | Stronger | Weaker | Weaker |
| base | acid | base | acid |

Alcohols—like water and ammonia—are enormously stronger acids than alkanes, and readily displace them from their "salts": from Grignard reagents, for example.

$$ROH\ +\ R'MgX\ \longrightarrow\ R'H\ +\ Mg(OR)X$$

| Stronger | Weaker |
| acid | acid |

As before, these relative acidities are determined by the method of displacement (Sec. 11.11). We may expand our series of acidities and basicities, then, to the following:

Relative acidities $H_2O > ROH > HC\equiv CH > NH_3 > RH$

Relative basicities $OH^- < OR^- < HC\equiv C^- < NH_2^- < R^-$

Let us look more closely at the relative acidities of alcohols and water. The difference between an alcohol and water is, of course, the alkyl group. Not only does the alkyl group make an alcohol less acidic than water, but the *bigger* the alkyl group, the less acidic the alcohol; methanol is the strongest and tertiary alcohols are the weakest. This acid-weakening effect of alcohols is not a polar effect, as was once believed, with electron release destabilizing the anion and making it a stronger base. In the gas phase, the relative acidities of the various alcohols and of alcohols and water are reversed; evidently, the easily polarized alkyl groups are helping to accommodate the negative charge, just as they help to accommodate the positive charge in carbocations (Sec. 5.21). Alcohols *are* weaker acids than water *in solution*—which is where we are normally concerned with acidity—and this is a solvation effect; a bulky group interferes with ion–dipole interactions that stabilize the anion.

Since an alcohol is a weaker acid than water, an alkoxide is not prepared by the reaction of the alcohol with sodium hydroxide, but rather by reaction of the alcohol with the active metal itself.

Alkoxides are extremely useful reagents. They are powerful bases—stronger than hydroxide—and, by varying the alkyl group, we can vary their degree of basicity, their steric requirements, and their solubility properties. As nucleophiles, they can be used to introduce the alkoxy group into molecules. We have already made use of alkoxides both as bases and as nucleophiles, and will continue to encounter them throughout our study of organic chemistry.

Problem 18.1 Which would you expect to be the stronger acid:

(a) β-Chloroethyl alcohol or ethyl alcohol?
(b) Isopropyl alcohol or hexafluoroisopropyl alcohol?
(c) *n*-Propyl alcohol or glycerol, $HOCH_2CHOHCH_2OH$?
(d) Benzyl alcohol or *p*-nitrobenzyl alcohol?

(e) Which alcohol of each pair would you expect to be the stronger nucleophile?

Problem 18.2 Sodium metal was added to *tert*-butyl alcohol and allowed to react. When the metal was consumed, ethyl bromide was added to the resulting mixture. Work-up of the reaction mixture yielded a compound of formula $C_6H_{14}O$.

In a similar experiment, sodium metal was allowed to react with ethanol. When *tert*-butyl bromide was added, a gas was evolved, and work-up of the remaining mixture gave ethanol as the only organic material.

(a) Write equations for all reactions. (b) What familiar reaction type is involved in each case? (c) Why did the reactions take different courses?

18.5 Formation of alkyl sulfonates

Sulfonyl chlorides (the acid chlorides of sulfonic acids) are prepared by the action of phosphorus pentachloride or thionyl chloride on sulfonic acids or their salts:

$$ArSO_2OH + PCl_5 \xrightarrow{\text{heat}} ArSO_2Cl + POCl_3 + HCl$$

$$\text{(or } ArSO_3Na) \qquad\qquad \underset{\text{chloride}}{\underset{\text{A sulfonyl}}{}} \qquad \text{(or } NaCl)$$

Alcohols react with these sulfonyl chlorides to form esters, *alkyl sulfonates*:

$$ArSO_2Cl + ROH \xrightarrow[\text{or pyridine}]{\text{aqueous OH}^-} ArSO_2OR + Cl^- + H_2O$$

$$\text{An alkyl sulfonate}$$

We have already seen (Secs. 5.9 and 7.12) that the weak basicity of sulfonate anions, $ArSO_3{}^-$, makes them good leaving groups, and that as a result alkyl sulfonates undergo nucleophilic substitution and elimination in much the same manner as alkyl halides.

Alkyl sulfonates offer a very real advantage over alkyl halides in reactions where stereochemistry is important; this advantage lies not so much in the reactions of alkyl sulfonates as in their *preparation*. Whether we use an alkyl halide or sulfonate, and whether we let it undergo substitution or elimination, our starting point for the study is almost certainly the alcohol. The sulfonate *must* be prepared from the alcohol; the halide nearly always *will* be. It is at the alcohol stage that any resolution will be carried out, or any diastereomers separated; the alcohol is then converted into the halide or sulfonate, the reaction we are studying is carried out, and the products are examined.

Now, any preparation of a halide from an alcohol must involve breaking of the carbon–oxygen bond, and hence is accompanied by the likelihood of stereo-

$$R\!-\!\!\!\mid\!\!O\!-\!H \xrightarrow{\text{HX or PX}_3} R\!-\!X$$

chemical inversion and the possibility of racemization. Preparation of a sulfonate, on the other hand, does not involve the breaking of the carbon–oxygen bond, and hence proceeds with complete retention; when we carry out a reaction with this sulfonate, we know exactly what we are starting with.

$$
\text{R—O}\!\mid\!\text{H} + \text{Cl—}\overset{\displaystyle O}{\underset{\displaystyle O}{\text{S}}}\text{—Ar} \longrightarrow \text{R—O—}\overset{\displaystyle O}{\underset{\displaystyle O}{\text{S}}}\text{—Ar}
$$

As a way of changing the —OH group of an alcohol into a good leaving group, conversion into sulfonates is just about ideal. We do not disturb the stereochemistry of the alkyl group. We can vary the structure of the sulfonate group and thus vary its leaving ability over a tremendous range. (See, for example, Sec. 10.19.) Although protonation of alcohols also generates a good leaving group, it limits our choice of reagents to those compatible with an acidic medium; but we can allow these alkyl sulfonates to react with just about any nucleophile or base we care to use.

Problem 18.3 Outline all steps in the synthesis of *sec*-butyl tosylate, starting with benzene, toluene, and any necessary aliphatic and inorganic reagents.

Problem 18.4 You prepare *sec*-butyl tosylate from alcohol of [α] +6.9°. On hydrolysis with aqueous base, this ester gives *sec*-butyl alcohol of [α] −6.9°. Without knowing the configuration or optical purity of the starting alcohol, what (if anything) can you say about the stereochemistry of the hydrolysis step?

18.6   Oxidation of alcohols

The oxidation of an alcohol involves the loss of one or more hydrogens (*α-hydrogens*) from the carbon bearing the —OH group. The kind of product that is formed depends upon how many of these α-hydrogens the alcohol contains, that is, upon whether the alcohol is primary, secondary, or tertiary.

A **primary alcohol** contains two α-hydrogens, and can either lose one of them to form an *aldehyde*,

$$
\text{R—}\overset{\displaystyle H}{\underset{\displaystyle H}{\text{C}}}\text{—OH} \longrightarrow \text{R—}\overset{\displaystyle H}{\text{C}}\text{=O}
$$

$$
\text{A } 1° \text{ alcohol} \qquad\qquad \text{An aldehyde}
$$

or both of them to form a *carboxylic acid*.

$$
\text{R—}\overset{\displaystyle H}{\underset{\displaystyle H}{\text{C}}}\text{—OH} \longrightarrow \text{R—}\overset{\displaystyle OH}{\text{C}}\text{=O}
$$

$$
\text{A } 1° \text{ alcohol} \qquad\qquad \text{A carboxylic acid}
$$

(Under the proper conditions, as we shall find, an aldehyde can itself be oxidized to a carboxylic acid.)

A **secondary alcohol** can lose its only α-hydrogen to form a *ketone*.

$$
\begin{array}{ccc}
\underset{\underset{\text{H}}{|}}{\overset{\overset{\text{R}}{|}}{\text{R}-\text{C}-\text{OH}}} & \longrightarrow & \overset{\overset{\text{R}}{|}}{\text{R}-\text{C}=\text{O}} \\
& & \text{A ketone}
\end{array}
$$

A 2° alcohol

A **tertiary alcohol** contains no α-hydrogens and is *not oxidized*. (An acidic oxidizing agent can, however, dehydrate the alcohol to an alkene and then oxidize *this*.)

$$
\underset{\underset{\text{R}}{|}}{\overset{\overset{\text{R}}{|}}{\text{R}-\text{C}-\text{OH}}} \longrightarrow \quad \text{no oxidation}
$$

A 3° alcohol

We have already encountered these oxidation products—aldehydes, ketones, and carboxylic acids—and should recognize them from their structures even though we have not yet discussed much of their chemistry. They are important compounds, and their preparation by the oxidation of alcohols is of great value in organic synthesis (Secs. 18.7–18.8).

The number of oxidizing agents available to the organic chemist is growing at a tremendous rate. As with all synthetic methods, emphasis is on the development of highly *selective* reagents, which will operate on only one functional group in a complex molecule, and leave the other functional groups untouched. Of the many reagents that can be used to oxidize alcohols, we can consider only the most common ones, those containing Mn(VII) or Cr(VI). We have already (Sec. 8.22) encountered heptavalent manganese in the form of potassium permanganate, $KMnO_4$. Also widely used is hexavalent chromium, chromic acid essentially, in a form selected for the job at hand: acidic aqueous $K_2Cr_2O_7$, CrO_3 in glacial acetic acid, CrO_3 in pyridine, etc.

Oxidation of primary alcohols to carboxylic acids is usually accomplished by use of potassium permanganate. Best yields are obtained if the permanganate and the alcohol are brought together in a non-polar solvent by use of phase-transfer catalysis (see Sec. 6.7). When reaction is complete, an aqueous solution of the

$$
\begin{array}{ccccc}
RCH_2OH & + & KMnO_4 & \longrightarrow & RCOO^-K^+ + MnO_2 + KOH \\
\text{1° alcohol} & & \textit{Purple} & & \textit{Sol. in } H_2O \qquad \textit{Brown}
\end{array}
$$

$$\downarrow H^+$$

RCOOH

A carboxylic acid

Insol. in H_2O

soluble potassium salt of the carboxylic acid is filtered from MnO_2, and the acid is liberated by the addition of a stronger mineral acid.

Oxidation of alcohols to the aldehyde or ketone stage is usually accomplished by the use of Cr(VI) in one of the forms described above. Oxidation of secondary alcohols to ketones is generally straightforward.

$$\underset{\text{A 2° alcohol}}{R-\overset{\overset{\displaystyle R'}{|}}{C}HOH} \quad \xrightarrow{\text{K}_2\text{Cr}_2\text{O}_7 \text{ or CrO}_3} \quad \underset{\text{A ketone}}{R-\overset{\overset{\displaystyle R'}{|}}{C}=O}$$

Because aldehydes are susceptible to further oxidation, the conversion of primary alcohols to aldehydes can be troublesome. One of the best and most convenient reagents for this purpose is pyridinium chlorochromate $(C_5H_5NH^+CrO_3Cl^-)$ formed by the reaction between chromic acid and pyridinium chloride (Sec. 35.11).

$$RCH_2OH \quad \xrightarrow[\text{in CH}_2\text{Cl}_2]{\text{C}_5\text{H}_5\text{NHCrO}_3\text{Cl}} \quad RCHO + Cr^{3+}$$

In connection with analysis, we shall encounter two reagents used to oxidize alcohols of special kinds: (a) *hypohalite* (Sec. 18.9), and (b) *periodic acid* (Sec. 18.10).

18.7   Synthesis of alcohols

Let us try to get a broader picture of the synthesis of complicated alcohols. We learned (Sec. 17.14) that they are most often prepared by the reaction of Grignard reagents with aldehydes or ketones. In this chapter we have learned that aldehydes and ketones, as well as the alkyl halides from which the Grignard reagents are made, are themselves most often prepared from alcohols. Finally, we know that the simple alcohols are among our most readily available compounds. We have available to us, then, a synthetic route leading from simple alcohols to more complicated ones.

alcohol $\longrightarrow$ alkyl halide $\longrightarrow$ Grignard reagent

alcohol $\longrightarrow$ aldehyde or ketone

more complicated alcohol

As a simple example, consider conversion of the two-carbon ethyl alcohol into the four-carbon *sec*-butyl alcohol:

CH_3CH_2OH
Ethyl alcohol

$\xrightarrow{\text{HBr}} CH_3CH_2-Br \xrightarrow{\text{Mg}} CH_3CH_2-MgBr$

$\xrightarrow{\text{pyridinium chlorochromate}} \underset{\text{Acetaldehyde}}{CH_3-\overset{\overset{\displaystyle H}{|}}{C}=O}$

$CH_3CH_2-\underset{\overset{\displaystyle |}{OMgBr}}{CHCH_3}$

$\downarrow \text{H}_2\text{O, H}^+$

$\underset{\text{sec-Butyl alcohol}}{CH_3CH_2-\underset{\overset{\displaystyle |}{OH}}{CHCH_3}}$

Using the *sec*-butyl alcohol thus obtained, we could prepare even larger alcohols:

By combining our knowledge of alcohols with what we know about alkylbenzenes and aromatic substitution, we can extend our syntheses to include aromatic alcohols. Starting from benzene we can make, for example, 1-phenylethanol,

and from toluene, 2-methyl-1-phenyl-2-propanol.

Granting that we know the chemistry of the individual steps, how do we go about planning a route to these more complicated alcohols? In almost every organic synthesis it is best to **work backwards** from the compound we want. There are relatively few ways to make a complicated alcohol; there are relatively few ways to make the Grignard reagent or the aldehyde or ketone; and so on back to our primary starting materials. On the other hand, alcohols can undergo so many different reactions that, if we go at the problem the other way around, we find a bewildering number of paths, few of which take us where we want to go.

Let us suppose (and this is quite reasonable) that we have available all alcohols of four carbons or fewer, and that we want to make, say, 2-methyl-2-hexanol. Let us set down the structure of this *target molecule*, and see what we need to make it.

$$\underset{\text{2-Methyl-2-hexanol}}{CH_3CH_2CH_2CH_2-\overset{\overset{\displaystyle CH_3}{|}}{\underset{\underset{\displaystyle OH}{|}}{C}}-CH_3}$$

Since it is a tertiary alcohol, we must use a Grignard reagent and a ketone. But which Grignard reagent? And which ketone? Using the same approach as before (Sec. 17.16), we see that there are two possibilities:

$$\underset{\text{2-Methyl-2-hexanol}}{CH_3CH_2CH_2CH_2\!-\!\overset{\overset{\displaystyle CH_3}{|}}{\underset{\underset{\displaystyle OH}{|}}{C}}\!-\!CH_3} \longleftarrow \underset{\substack{\text{n-Butylmagnesium}\\\text{bromide}}}{CH_3CH_2CH_2CH_2-MgBr} + \underset{\text{Acetone}}{\overset{\overset{\displaystyle CH_3}{|}}{\underset{\underset{\displaystyle O}{\|}}{C}}-CH_3}$$

$$\underset{\text{2-Methyl-2-hexanol}}{CH_3CH_2CH_2CH_2-\overset{\overset{\displaystyle CH_3}{|}}{\underset{\underset{\displaystyle OH}{|}}{C}}\!\!\!\!\diagup\!\!CH_3} \longleftarrow \underset{\substack{\text{Methyl n-butyl}\\\text{ketone}}}{CH_3CH_2CH_2CH_2-\overset{\overset{\displaystyle CH_3}{|}}{\underset{\underset{\displaystyle O}{\|}}{C}}} + \underset{\substack{\text{Methylmagnesium}\\\text{bromide}}}{BrMg-CH_3}$$

Of these two possibilities we would select the one involving the four-carbon Grignard reagent and the three-carbon ketone; now how are we to make *them*? The Grignard reagent can be made only from the corresponding alkyl halide, *n*-butyl bromide, and that in turn most likely from an alcohol, *n*-butyl alcohol. Acetone requires, of course, isopropyl alcohol. Putting together the entire synthesis, we have the following sequence:

$$CH_3CH_2CH_2CH_2-MgBr \xleftarrow{\text{Mg}} CH_3CH_2CH_2CH_2-Br$$

$$\Big\uparrow \text{HBr}$$

$$\underset{\text{n-Butyl alcohol}}{CH_3CH_2CH_2CH_2-OH}$$

$$\underset{\text{2-Methyl-2-hexanol}}{CH_3CH_2CH_2CH_2-\overset{\overset{\displaystyle CH_3}{|}}{\underset{\underset{\displaystyle OH}{|}}{C}}-CH_3} \longleftarrow$$

$$\underset{}{\overset{\overset{\displaystyle CH_3}{|}}{\underset{\underset{\displaystyle O}{\|}}{C}}-CH_3} \xleftarrow[\text{heat}]{K_2Cr_2O_7} \underset{\text{Isopropyl alcohol}}{H-\overset{\overset{\displaystyle CH_3}{|}}{\underset{\underset{\displaystyle OH}{|}}{C}}-CH_3}$$

Let us consider that, in addition to our alcohols of four carbons or fewer, we have available benzene and toluene, another reasonable assumption, and that we wish to make, say, 3-methyl-1-phenyl-2-butanol. Again we set down the structure of the alcohol we want and work backwards to the starting materials. To make a

1-Phenyl-3-methyl-2-butanol

secondary alcohol, we use a Grignard reagent and an aldehyde, and, as usual, there are two choices: we may consider the molecule to be put together either (a) between C–1 and C–2 or (b) between C–2 and C–3. Of the two possibilities we select the

first, since this requires a compound with only one carbon attached to the benzene ring, which we have available in toluene. We need, then, a four-carbon aldehyde and benzylmagnesium chloride. The aldehyde can be made from isobutyl alcohol. The benzylmagnesium chloride is, of course, made from benzyl chloride, which in turn is made from toluene by free-radical chlorination. Our synthesis is complete:

Our basic synthesis can be varied still further. Like their alkyl and aryl counterparts, both lithium acetylides and alkynyl Grignard reagents can add to aldehydes and ketones to give alcohols. For example:

$$CH_3C{\equiv}C{-}MgBr \;+\; CH_3\overset{\overset{\displaystyle H}{|}}{C}{=}O \;\longrightarrow\; CH_3C{\equiv}C{-}\underset{\underset{\displaystyle OMgBr}{|}}{C}HCH_3 \;\xrightarrow{\;H^+\;}\; CH_3C{\equiv}C{-}\underset{\underset{\displaystyle OH}{|}}{C}HCH_3$$

Propyn-1-ylmagnesium   Acetaldehyde
bromide

3-Pentyn-2-ol

We have now formed compounds that contain not only —OH, but a *second* highly reactive group, the carbon–carbon triple bond. As we have discussed (Sec. 11.13), the triple bond can be converted with a high degree of stereoselectivity into a double bond; and at this double bond there can occur various addition reactions, also highly stereoselective, to yield a wide variety of products—each of which also contains the —OH group, too!

Now that we know how to make complicated alcohols from simple ones, what can we use them for?

18.8 Syntheses using alcohols

The alcohols that we have learned to make can be converted into other kinds of compounds having the same carbon skeleton; from complicated alcohols we can make complicated aldehydes, ketones, acids, halides, alkenes, alkynes, alkanes, etc.

Alkyl halides are prepared from alcohols by use of hydrogen halides or phosphorus halides. Phosphorus halides are often preferred because they tend less to bring about rearrangement (Sec. 5.25).

Alkenes are prepared from alcohols either by direct dehydration or by dehydrohalogenation of intermediate alkyl halides; to avoid rearrangement we often select dehydrohalogenation of halides even though this route involves an extra step. Or, sometimes better, we use elimination from alkyl sulfonates.

Alkanes, we learned (Sec. 3.15), are best prepared from the corresponding alkenes by hydrogenation, so that now we have a route from complicated alcohols to complicated alkanes.

Complicated aldehydes and ketones are made by oxidizing complicated alcohols. By reaction with Grignard reagents these aldehydes and ketones can be converted into even more complicated alcohols, and so on.

Given the time, necessary inorganic reagents, and the single alcohol ethanol, our chemical Crusoe of Sec. 17.1 could synthesize all the aliphatic compounds that have ever been made—and for that matter the aromatic ones, too.

In planning the synthesis of these other kinds of compounds, we again follow our system of working backwards. We try to limit the synthesis to as few steps as possible, but nevertheless do not sacrifice purity for time. For example, where rearrangement is likely to occur we prepare an alkene in two steps via the halide or sulfonate rather than by the single step of dehydration.

Assuming again that we have available benzene, toluene and alcohols of four

$$CH_3$$
$$|$$
$$CH_3\!-\!CH\!-\!CH\!=\!CH_2$$
3-Methyl-1-butene

carbons or fewer, let us take as our target 3-methyl-1-butene. It could be prepared by dehydrohalogenation of an alkyl halide of the same carbon skeleton, or by dehydration of an alcohol. If the halogen or hydroxyl group were attached to C–2, we would obtain some of the desired product, but much more of its isomer, 2-methyl-2-butene:

We would select, then, a compound with the functional group attached to C–1. Even so, if we were to use the alcohol, there would be extensive rearrangement to yield, again, the more stable 2-methyl-2-butene:

Only dehydrohalogenation of 1-bromo-3-methylbutane would yield the desired product in pure form:

How do we prepare the necessary alkyl halide? Certainly not by bromination of an alkane, since even if we could make the proper alkane in some way, bromination would occur almost entirely at the tertiary position to give the wrong product. (Chlorination would give the proper chloride—but as a minor component of a grand mixture.) As usual, then, we would prepare the halide from the corresponding alcohol, in this case 3-methyl-1-butanol. Since this is a primary alcohol (without branching near the —OH group), and hence does not form the halide via the carbocation, rearrangement is not likely; we might use, then, either hydrogen bromide or PBr_3.

$$CH_3\text{—}\overset{\underset{\displaystyle |}{CH_3}}{CH}\text{—}CH_2\text{—}CH_2\text{—}Br \xleftarrow{PBr_3} CH_3\text{—}\overset{\underset{\displaystyle |}{CH_3}}{CH}\text{—}CH_2\text{—}CH_2\text{—}OH$$
3-Methyl-1-butanol

Now, how do we make 3-methyl-1-butanol? It is a primary alcohol and contains one carbon more than our largest available alcohol; therefore we would use the reaction of a Grignard reagent with formaldehyde. The necessary Grignard reagent is isobutylmagnesium bromide, which we could have prepared from

$$CH_3\text{—}\overset{\underset{\displaystyle |}{CH_3}}{CH}\text{—}CH_2\text{—}CH_2OH \longleftarrow \left\{ \begin{array}{l} H\text{—}\overset{\underset{\displaystyle \|}{H}}{C}\text{=}O \\ \text{Formaldehyde} \\[1em] CH_3\text{—}\overset{\underset{\displaystyle |}{CH_3}}{CH}\text{—}CH_2\text{—}MgBr \\ \text{Isobutylmagnesium bromide} \end{array} \right.$$

3-Methyl-1-butanol

isobutyl bromide, and that in turn from isobutyl alcohol. The formaldehyde is made by oxidation of methanol. The entire sequence, from which we could expect to obtain quite pure 3-methyl-1-butene, is the following:

$$CH_3\text{—}\overset{\underset{\displaystyle |}{CH_3}}{CH}\text{—}CH\text{=}CH_2 \xleftarrow{KOH} CH_3\text{—}\overset{\underset{\displaystyle |}{CH_3}}{CH}\text{—}CH_2\text{—}CH_2\text{—}Br \xleftarrow{PBr_3} CH_3\text{—}\overset{\underset{\displaystyle |}{CH_3}}{CH}\text{—}CH_2\text{—}CH_2\text{—}OH$$
3-Methyl-1-butene   3-Methyl-1-butanol

$$CH_3\text{—}\overset{\underset{\displaystyle |}{CH_3}}{CH}\text{—}CH_2\text{—}CH_2OH \longleftarrow \left\{ \begin{array}{l} H\text{—}\overset{\underset{\displaystyle \|}{H}}{C}\text{=}O \xleftarrow{K_2Cr_2O_7} CH_3OH \\ \qquad\qquad\qquad\quad \text{Methanol} \\[1em] CH_3\text{—}\overset{\underset{\displaystyle |}{CH_3}}{CH}\text{—}CH_2\text{—}MgBr \xleftarrow{Mg} CH_3\text{—}\overset{\underset{\displaystyle |}{CH_3}}{CH}\text{—}CH_2\text{—}Br \end{array} \right.$$

3-Methyl-1-butanol

$$\big\uparrow PBr_3$$

$$CH_3\text{—}\overset{\underset{\displaystyle |}{CH_3}}{CH}\text{—}CH_2\text{—}OH$$
Isobutyl alcohol

18.9   Analysis of alcohols. Characterization. Iodoform test

Alcohols dissolve in cold concentrated sulfuric acid. This property they share with alkenes, amines, practically all compounds containing oxygen, and easily sulfonated compounds. (Alcohols, like other oxygen-containing compounds, form oxonium salts, which dissolve in the highly polar sulfuric acid.)

Alcohols are not oxidized by cold, dilute, neutral permanganate (although primary and secondary alcohols are, of course, oxidized by permanganate under more vigorous conditions). However, as we have seen (Sec. 8.24), alcohols often contain impurities that *are* oxidized under these conditions, and so the permanganate test must be interpreted with caution.

Alcohols do not decolorize bromine in carbon tetrachloride. This property serves to distinguish them from alkenes and alkynes.

Alcohols are further distinguished from alkenes and alkynes—and, indeed, from nearly every other kind of compound—by their oxidation by chromic anhydride, CrO_3, in aqueous sulfuric acid: within *two seconds*, the clear orange solution turns blue-green and becomes opaque.

$$ROH + HCrO_4{}^- \longrightarrow \textit{Opaque, blue-green}$$
$$\text{1° or 2°} \quad \textit{Clear}$$
$$\textit{orange}$$

Tertiary alcohols do not give this test. Aldehydes do, but are easily differentiated in other ways (Sec. 21.15).

Reaction of alcohols with sodium metal, with the evolution of hydrogen gas, is of some use in characterization; a *wet* compound of any kind, of course, will do the same thing, until the water is used up.

The presence of the —OH group in a molecule is often indicated by the formation of an ester upon treatment with an acid chloride or anhydride (Sec. 23.16). Some esters are sweet-smelling; others are solids with sharp melting points, and can be derivatives in identifications. (If the molecular formulas of starting material and product are determined, it is possible to calculate *how many* —OH groups are present.)

Problem 18.5 Make a table to show the response of each kind of compound we have studied so far toward the following reagents:

(a) cold concentrated H_2SO_4
(b) cold, dilute, neutral $KMnO_4$
(c) Br_2 in CCl_4
(d) CrO_3 in H_2SO_4

(e) cold fuming sulfuric acid
(f) $CHCl_3$ and $AlCl_3$
(g) sodium metal

Whether an alcohol is primary, secondary, or tertiary is shown by the **Lucas test**, which is based upon the difference in reactivity of the three classes toward hydrogen halides (Sec. 5.25). Alcohols (of not more than six carbons) are soluble in the *Lucas reagent*, a mixture of concentrated hydrochloric acid and zinc chloride. (Why are they more soluble in this than in water?) The corresponding alkyl chlorides are insoluble. Formation of a chloride from an alcohol is indicated by the cloudiness that appears when the chloride separates from the solution; hence, the time required for cloudiness to appear is a measure of the reactivity of the alcohol.

A tertiary alcohol reacts immediately with the Lucas reagent, and a secondary alcohol reacts within five minutes; a primary alcohol does not react appreciably at room temperature. Allyl alcohol reacts as rapidly as tertiary alcohols with the Lucas reagent; allyl chloride, however, is soluble in the reagent. (Why?)

Whether or not an alcohol contains one particular structural unit is shown by the **iodoform test**. The alcohol is treated with iodine and sodium hydroxide (sodium hypoiodite, NaOI); an alcohol of the structure

$$R-\overset{\overset{\displaystyle H}{|}}{\underset{\underset{\displaystyle OH}{|}}{C}}-CH_3 \quad \textit{where R is H or an alkyl or aryl group}$$

yields a yellow precipitate of iodoform (CHI$_3$, m.p. 119 °C). For example:

Gives positive iodoform test	Gives negative iodoform test
$CH_3-\overset{\overset{\displaystyle H}{|}}{\underset{\underset{\displaystyle OH}{|}}{C}}-H$	Any other primary alcohol
$CH_3-\overset{\overset{\displaystyle H}{|}}{\underset{\underset{\displaystyle OH}{|}}{C}}-CH_3$	$CH_3-\overset{\overset{\displaystyle CH_3}{|}}{\underset{\underset{\displaystyle OH}{|}}{C}}-CH_3$
$CH_3-\overset{\overset{\displaystyle H}{|}}{\underset{\underset{\displaystyle OH}{|}}{C}}-CH_2CH_2CH_3$	$CH_3CH_2-\overset{\overset{\displaystyle H}{|}}{\underset{\underset{\displaystyle OH}{|}}{C}}-CH_2CH_3$
$C_6H_5-\overset{\overset{\displaystyle H}{|}}{\underset{\underset{\displaystyle OH}{|}}{C}}-CH_3$	$C_6H_5-CH_2-CH_2OH$

The reaction involves oxidation, halogenation, and cleavage.

$$R-\overset{\overset{\displaystyle H}{|}}{\underset{\underset{\displaystyle OH}{|}}{C}}-CH_3 + NaOI \longrightarrow R-\overset{\overset{\displaystyle O}{\|}}{C}-CH_3 + NaI + H_2O$$

$$R-\overset{\overset{\displaystyle O}{\|}}{C}-CH_3 + 3NaOI \longrightarrow R-\overset{\overset{\displaystyle O}{\|}}{C}-CI_3 + 3NaOH$$

$$R-\overset{\overset{\displaystyle O}{\|}}{C}\ CI_3 + NaOH \longrightarrow RCOO^-Na^+ + CHI_3$$
$$\qquad\qquad\qquad\qquad\qquad\qquad\quad \textit{Yellow precipitate}$$

As would be expected from the equations, a compound of structure

$$R-\overset{\overset{\displaystyle }{}}{\underset{\underset{\displaystyle O}{\|}}{C}}-CH_3 \quad \textit{where R is H or an alkyl or aryl group}$$

also gives a positive test (Sec. 21.15).

   In certain special cases this reaction is used not as a test, but to synthesize the carboxylic acid, RCOOH. Here, hypobromite or the cheaper hypochlorite would probably be used.

18.10   Analysis of 1,2-diols. Periodic acid oxidation

Upon treatment with periodic acid, HIO_4, compounds containing two or more $-OH$ or $=O$ groups attached to *adjacent* carbon atoms undergo oxidation with cleavage of carbon–carbon bonds. For example:

$$R-\underset{\underset{OH}{|}}{CH}-\underset{\underset{OH}{|}}{CH}-R' + HIO_4 \longrightarrow RCHO + R'CHO \quad (+ HIO_3)$$

$$R-\underset{\overset{||}{O}}{C}-\underset{\overset{||}{O}}{C}-R' + HIO_4 \longrightarrow RCOOH + R'COOH$$

$$R-\underset{\underset{OH}{|}}{CH}-\underset{\overset{||}{O}}{C}-R' + HIO_4 \longrightarrow RCHO + R'COOH$$

$$R-\underset{\underset{OH}{|}}{CH}-\underset{\underset{OH}{|}}{CH}-\underset{\underset{OH}{|}}{CH}-R' + 2HIO_4 \longrightarrow RCHO + HCOOH + R'CHO$$

$$R-\underset{\underset{OH}{|}}{\overset{\overset{R}{|}}{C}}-\underset{\underset{OH}{|}}{CH}-R' + HIO_4 \longrightarrow R_2CO + R'CHO$$

$$R-\underset{\underset{OH}{|}}{CH}-CH_2-\underset{\underset{OH}{|}}{CH}-R' + HIO_4 \longrightarrow \text{no reaction}$$

The oxidation is particularly useful in determination of structure. Qualitatively, oxidation by HIO_4 is indicated by formation of a white precipitate ($AgIO_3$) upon addition of silver nitrate. Since the reaction is usually quantitative, valuable information is given by the nature and amounts of the products, and by the quantity of periodic acid consumed.

Problem 18.6   When one mole of each of the following compounds is treated with HIO_4, what will the products be, and how many moles of HIO_4 will be consumed?

(a) $CH_3CHOHCH_2OH$
(b) $CH_3CHOHCHO$
(c) $CH_2OHCHOHCH_2OCH_3$
(d) $CH_2OHCH(OCH_3)CH_2OH$

(e) *cis*-1,2-cyclopentanediol
(f) $CH_2OH(CHOH)_3CHO$
(g) $CH_2OH(CHOH)_3CH_2OH$

Problem 18.7   Assign a structure to each of the following compounds:

$$
\begin{array}{ll}
A + 1 \text{ mol } HIO_4 & \longrightarrow \quad CH_3COCH_3 + HCHO \\
B + 1 \text{ mol } HIO_4 & \longrightarrow \quad OHC(CH_2)_4CHO \\
C + 1 \text{ mol } HIO_4 & \longrightarrow \quad HOOC(CH_2)_4CHO \\
D + 1 \text{ mol } HIO_4 & \longrightarrow \quad 2HOOC-CHO \\
E + 3HIO_4 & \longrightarrow \quad 2HCOOH + 2HCHO \\
F + 3HIO_4 & \longrightarrow \quad 2HCOOH + HCHO + CO_2 \\
G + 5HIO_4 & \longrightarrow \quad 5HCOOH + HCHO
\end{array}
$$

18.11   Spectroscopic analysis of alcohols

Infrared   In the infrared spectrum of a hydrogen-bonded alcohol—and this is the kind that we commonly see—the most conspicuous feature is a strong, broad band in the 3200–3600 cm^{-1} region due to O—H stretching (see Fig. 18.1).

O—H stretching, *strong, broad*

Alcohols, ROH (or phenols, ArOH) 3200–3600 cm^{-1}

(A monomeric alcohol gives a sharp, variable band at 3610–3640 cm^{-1}.)

Another strong, broad band, due to C—O stretching, appears in the 1000–1200 cm^{-1} region, the exact frequency depending on the nature of the alcohol:

C—O stretching, *strong, broad*

1° ROH   about 1050 cm^{-1} 3° ROH   about 1150 cm^{-1}

2° ROH   about 1100 cm^{-1} ArOH   about 1230 cm^{-1}

(Compare the locations of this band in the spectra of Fig. 18.1.)

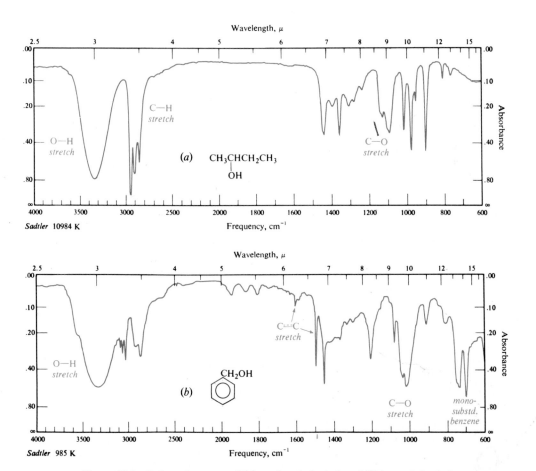

Figure 18.1   Infrared spectra of (*a*) *sec*-butyl alcohol and (*b*) benzyl alcohol.

Phenols (ArOH) also show both these bands, but the C—O stretching appears at somewhat higher frequencies. Ethers show C—O stretching, but the O—H band is absent. Carboxylic acids and esters show C—O stretching, but give absorption characteristic of the carbonyl group, C=O, as well. (For a comparison of certain oxygen compounds, see Table 24.3, p. 890.)

NMR   NMR absorption by a hydroxylic proton (O—H) is shifted downfield by hydrogen bonding. The chemical shift that is observed depends, therefore, on the degree of hydrogen bonding, which in turn depends on temperature, concentration, and the nature of the solvent (Sec. 17.5). As a result, the signal can appear anywhere in the range δ 1–5. It may be hidden among the peaks due to alkyl protons, although its presence there is often revealed through proton counting.

A hydroxyl proton ordinarily gives rise to a singlet in the NMR spectrum: its signal is not split by nearby protons, not does it split their signals. Proton exchange between two (identical) molecules of alcohol

$$R^*—O—H^* + R—O—H \rightleftharpoons R^*—O—H + R—O—H^*$$

is so fast that the proton—now in one molecule and in the next instant in another—cannot see nearby protons in their various combinations of spin alignments, but in a single *average* alignment.

Presumably through its inductive effect, the oxygen of an alcohol causes a downfield shift for nearby protons: a shift of about the same size as other electronegative atoms (Table 16.4, p. 585).

Problem 18.8   Can you suggest a procedure that might move a hidden O—H peak into the open? (*Hint*: See Sec. 17.5)

Problem 18.9   (a) Very dry, pure samples of alcohols show spin–spin splitting of the O—H signals. What splitting would you expect for a primary alcohol? A secondary alcohol? A tertiary alcohol? (b) This splitting disappears on the addition of a trace of acid or base. Write equations to show just how proton exchange would be speeded up by an acid (H:B); by a base (:B).

CMR   In the CMR spectrum the hydroxyl group exerts strong effects following the common pattern for electronegative substituents: α-effects, in the range $+40$ to $+50$ ppm; β-effects, $+7$ to $+10$ ppm; and γ-effects, -2 to -6 ppm.

PROBLEMS

1. Refer to the isomeric pentyl alcohols of Problem 1(a), p. 661. (a) Indicate which (if any) will give a positive iodoform test. (b) Describe how each will respond to the Lucas reagent. (c) Describe how each will respond to chromic anhydride. (d) Outline all steps in a possible synthesis of each, starting from alcohols of four carbons or fewer, and using any necessary inorganic reagents.

2. Give structures and names of the chief products expected from the reaction (if any) of cyclohexanol with:

(a) cold conc. H_2SO_4
(b) H_2SO_4, heat
(c) cold dilute $KMnO_4$
(d) CrO_3, H_2SO_4
(e) Br_2/CCl_4
(f) conc. aqueous HBr
(g) $P + I_2$
(h) Na
(i) CH_3COOH, H^+
(j) H_2, Ni
(k) CH_3MgBr
(l) NaOH(aq)

(m) product (f) + Mg
(n) product (m) + product (d)
(o) product (b) + cold alk. $KMnO_4$
(p) product (b) + Br_2/CCl_4
(q) product (b) + C_6H_6, HF
(r) product (b) + H_2, Ni
(s) product (q) + HNO_3/H_2SO_4
(t) product (b) + N-bromosuccinimide
(u) product (b) + $CHCl_3$ + t-BuOK
(v) product (d) + C_6H_5MgBr
(w) tosyl chloride, OH^-
(x) product (w) + t-BuOK

3. Outline all steps in a possible laboratory synthesis of each of the following compounds from n-butyl alcohol, using any necessary inorganic reagents. Follow the general instructions on p. 221.

(a) n-butyl bromide
(b) 1-butene
(c) n-butyl hydrogen sulfate
(d) potassium n-butoxide
(e) n-butyraldehyde,
     $CH_3CH_2CH_2CHO$
(f) n-butyric acid,
     $CH_3CH_2CH_2COOH$
(g) n-butane
(h) 1,2-dibromobutane
(i) 1-chloro-2-butanol
(j) 1-butyne
(k) ethylcyclopropane
(l) 1,2-butanediol

(m) n-octane
(n) 3-octyne
(o) cis-3-octene
(p) trans-3-octene
(q) 4-octanol
(r) 4-octanone,
     $CH_3CH_2CH_2CH_2CCH_2CH_2CH_3$
     with C=O shown as $\overset{\|}{O}$
(s) 5-(n-propyl)-5-nonanol
(t) n-butyl n-butyrate,
     $CH_3CH_2CH_2C-OCH_2CH_2CH_2CH_3$
     with C=O shown as $\overset{\|}{O}$

4. Give structures and (where possible) names of the principal organic products of the following:

(a) benzyl alcohol + Mg
(b) isobutyl alcohol + C_6H_5COOH + H^+
(c) ethylene bromide + excess NaOH(aq)
(d) n-butyl alcohol + H_2, Pt
(e) crotyl alcohol ($CH_3CH=CHCH_2OH$) + Br_2/H_2O
(f) CH_3OH + C_2H_5MgBr
(g) p-bromobenzyl bromide + NaOH(aq)
(h) tert-butyl alcohol + C_6H_6 + H_2SO_4

5. In Great Britain during the past years, thousands of motorists have been (politely) stopped by the police and asked to blow into a "breathalyser": a glass tube containing silica gel impregnated with certain chemicals, and leading into a plastic bag. If, for more than half the length of the tube, the original yellow color turns green, the motorist looks very unhappy and often turns red. What chemicals are impregnated on the silica gel, why does the tube turn green, and why does the motorist turn red?

6. Arrange the alcohols of each set in order of reactivity toward aqueous HBr:

(a) the isomeric pentyl alcohols of Problem 1(a), p. 661. (*Note*: It may be necessary to list these in groups of about the same reactivity.)
(b) 1-phenyl-1-propanol, 3-phenyl-1-propanol, 1-phenyl-2-propanol
(c) benzyl alcohol, p-cyanobenzyl alcohol, p-hydroxybenzyl alcohol
(d) 2-buten-1-ol, 3-buten-1-ol
(e) cyclopentylmethanol, 1-methylcyclopentanol, trans-2-methylcyclopentanol
(f) benzyl alcohol, diphenylmethanol, methanol, and triphenylmethanol

7. Outline all steps in a possible laboratory synthesis of each of the following compounds from cyclohexanol and any necessary aliphatic, aromatic, or inorganic reagents.

(a) cyclohexanone ($C_6H_{10}O$)
(b) bromocyclohexane
(c) 1-methylcyclohexanol
(d) 1-methylcyclohexene
(e) *trans*-2-methylcyclohexanol
(f) 1-cyclohexylethanol

(g) *trans*-1,2-dibromocyclohexane
(h) cyclohexylmethanol
(i) 1-bromo-1-phenylcyclohexane
(j) cyclohexanecarboxylic acid
(k) adipic acid, $HOOC(CH_2)_4COOH$
(l) norcarane (see p. 176)

8. Outline all steps in a possible laboratory synthesis of each of the following compounds from benzene, toluene, and alcohols of four carbons or fewer.

(a) 2,3-dimethyl-2-butanol
(b) 2-phenyl-2-propanol
(c) 2-phenylpropene
(d) 2-methyl-1-butene
(e) isopentane
(f) 1,2-dibromo-2-methylbutane
(g) 3-hexanol
(h) 3-hexanone (I)
(i) 4-ethyl-4-heptanol
(j) 2-bromo-2-methylhexane
(k) methylacetylene

(l) *trans*-1,2-dimethylcyclopropane
(m) 1-chloro-1-phenylethane
 (α-phenylethyl chloride)
(n) *sec*-butylbenzene
(o) isopropyl methyl ketone (II)
(p) 2-methylhexane
(q) benzyl methyl ketone (III)
(r) 2,2-dimethylhexane
(s) 2-bromo-1-phenylpropane
(t) 3-heptyne
(u) ethyl propionate (IV)

$$CH_3CH_2CH_2\underset{\underset{O}{\|}}{C}CH_2CH_3 \qquad CH_3\underset{\underset{O}{\|}}{C}CH(CH_3)_2$$

I   II

$$C_6H_5CH_2\underset{\underset{O}{\|}}{C}CH_3 \qquad CH_3CH_2\underset{\underset{O}{\|}}{C}\!-\!OCH_2CH_3$$

III   IV

9. Compounds "labeled" at various positions by isotopic atoms are useful in determining reaction mechanisms and in following the fate of compounds in biological systems. Outline a possible synthesis of each of the following labeled compounds using $^{14}CH_3OH$ as the source of ^{14}C, and D_2O as the source of deuterium.

(a) 2-methyl-1-propanol-1-^{14}C, $(CH_3)_2CH^{14}CH_2OH$
(b) 2-methyl-1-propanol-2-^{14}C, $(CH_3)_2{}^{14}CHCH_2OH$
(c) 2-methyl-1-propanol-3-^{14}C, $^{14}CH_3CH(CH_3)CH_2OH$
(d) propene-1-^{14}C, $CH_3CH{=}^{14}CH_2$
(e) propene-2-^{14}C, $CH_3{}^{14}CH{=}CH_2$
(f) propene-3-^{14}C, $^{14}CH_3CH{=}CH_2$
(g) C_6H_5D
(h) $CH_3CH_2CHD^{14}CH_3$

10. When *trans*-2-methylcyclopentanol is treated with tosyl chloride and the product with potassium *tert*-butoxide, the only alkene obtained is 3-methylcyclopentene. (a) What is the stereochemistry of this reaction? (b) This is the final step of a general synthetic route to 3-alkylcyclopentenes, starting from cyclopentanone. Outline all steps in this route,

carefully choosing your reagents in each step. (c) What advantage does this sequence have over an analogous one involving an intermediate halide instead of a tosylate?

11. Making use of any necessary organic or inorganic reagents, outline all steps in the conversion of:

(a) androst-9(11)-ene (p. 662) into the saturated 11-keto derivative.

(b)

Conessine
(3β-dimethylaminocon-5-enine)
An alkaloid

into

3β-Dimethylaminoconanin-6-one

(c)

5α-Pregnane-3α-ol-20-one
(acetate ester)

into

3-Cholestanone

where R = —CH(CH$_3$)CH$_2$CH$_2$CH$_2$CH(CH$_3$)$_2$

(*Hint*: CH$_3$COOCH$_3$ + H$_2$O $\xrightarrow{\text{OH}^-,\ \text{heat}}$ CH$_3$COO$^-$ + CH$_3$OH.)

12. Assign structures to the compounds A through HH.

(a) ethylene + Cl$_2$(aq) $\longrightarrow$ A (C$_2$H$_5$OCl)
    A + NaHCO$_3$(aq) $\longrightarrow$ B (C$_2$H$_6$O$_2$)
(b) ethylene + Cl$_2$(aq) $\longrightarrow$ A (C$_2$H$_5$OCl)
    A + HNO$_3$ $\longrightarrow$ C (C$_2$H$_3$O$_2$Cl)
    C + H$_2$O $\longrightarrow$ D (C$_2$H$_4$O$_3$)
(c) E + 6HIO$_4$ $\longrightarrow$ 6HCOOH
(d) F (C$_{18}$H$_{34}$O$_2$) + HCO$_2$OH $\longrightarrow$ G (C$_{18}$H$_{36}$O$_4$)
    G + HIO$_4$ $\longrightarrow$ CH$_3$(CH$_2$)$_7$CHO + OHC(CH$_2$)$_7$COOH
(e) allyl alcohol + Br$_2$/CCl$_4$ $\longrightarrow$ H (C$_3$H$_6$OBr$_2$)
    H + HNO$_3$ $\longrightarrow$ I (C$_3$H$_4$O$_2$Br$_2$)
    I + Zn $\longrightarrow$ J (C$_3$H$_4$O$_2$)
(f) 1,2,3-tribromopropane + KOH(alc) $\longrightarrow$ K (C$_3$H$_4$Br$_2$)
    K + NaOH(aq) $\longrightarrow$ L (C$_3$H$_5$OBr)
    L + KOH(alc) $\longrightarrow$ M (C$_3$H$_4$O)
(g) 2,2-dichloropropane + NaOH(aq) $\longrightarrow$ [N (C$_3$H$_8$O$_2$)] $\longrightarrow$ O (C$_3$H$_6$O)
(h) propyne + Cl$_2$(aq) $\longrightarrow$ [P (C$_3$H$_6$O$_2$Cl$_2$)] $\longrightarrow$ Q (C$_3$H$_4$OCl$_2$)
    Q + Cl$_2$(aq) $\longrightarrow$ R (C$_3$H$_3$OCl$_3$)
    R + NaOH(aq) $\longrightarrow$ CHCl$_3$ + S (C$_2$H$_3$O$_2$Na)
(i) cyclohexene + KMnO$_4$ $\longrightarrow$ T (C$_6$H$_{12}$O$_2$)
    T + CH$_3$COOH, H$^+$ $\longrightarrow$ U (C$_{10}$H$_{16}$O$_4$)
(j) V (C$_3$H$_8$O$_3$) + CH$_3$COOH, H$^+$ $\longrightarrow$ W (C$_9$H$_{14}$O$_6$)
(k) cyclohexanol + K$_2$Cr$_2$O$_7$, H$^+$ $\longrightarrow$ X (C$_6$H$_{10}$O)
    X + C$_6$H$_5$MgBr, followed by H$_2$O $\longrightarrow$ Y (C$_{12}$H$_{16}$O)
    Y + heat $\longrightarrow$ Z (C$_{12}$H$_{24}$)
    Z + KMnO$_4$/NaIO$_4$ $\longrightarrow$ AA (C$_{12}$H$_{14}$O$_3$)

(l) (R)-(+)-1-bromo-2,4-dimethylpentane + Mg $\longrightarrow$ BB
BB + $(CH_3)_2CHCH_2CHO$, then H_2O $\longrightarrow$ CC $(C_{12}H_{26}O)$, *a mixture*
CC + CrO_3 $\longrightarrow$ DD $(C_{12}H_{24}O)$
DD + CH_3MgBr, then H_2O $\longrightarrow$ EE $(C_{13}H_{28}O)$, *a mixture*
EE + I_2, heat $\longrightarrow$ FF $(C_{13}H_{26})$, *a mixture*
FF + H_2, Ni $\longrightarrow$ GG $(C_{13}H_{28})$ + HH $(C_{13}H_{28})$
                              Optically   Optically
                               active   inactive

13. When 1,4-hexadien-3-ol is dissolved in H_2SO_4, it is converted completely into 3,5-hexadien-2-ol. How do you account for this?

14. When 1,3,5,5-tetramethyl-1,3-cyclohexadiene is dissolved in cold concentrated H_2SO_4, the solution shows a freezing-point lowering that corresponds to two particles for each molecule of diene dissolved. On addition of water to the solution, the diene is completely regenerated. How do you account for these observations? Just what is happening and why?

15. The sex attractant of the Douglas-fir tussock moth has been synthesized in the following way. Give the structure of the sex attractant and all intermediate compounds.

1-heptyne + $LiNH_2$ $\longrightarrow$ II $(C_7H_{11}Li)$
II + 1-chloro-3-bromopropane $\longrightarrow$ JJ $(C_{10}H_{17}Cl)$
JJ + Mg; then n-$C_{10}H_{21}CHO$; then H^+ $\longrightarrow$ KK $(C_{21}H_{40}O)$
KK + H_2, Lindlar catalyst $\longrightarrow$ LL $(C_{21}H_{42}O)$
LL + CrO_3 $\longrightarrow$ sex attractant $(C_{21}H_{40}O)$

16. Tricyclopropylcarbinol (R_3COH, R = cyclopropyl) gives a complex NMR spectrum in the region δ 0.2–1.1, and is transparent in the near ultraviolet. A solution of the alcohol in concentrated H_2SO_4 has the following properties:

(i) A freezing-point lowering corresponding to four particles for each molecule dissolved;
(ii) intense ultraviolet absorption (λ_{max} 270 nm, ϵ_{max} 22 000);
(iii) an NMR spectrum with one peak, a singlet, δ 2.26.

When the solution is diluted and neutralized, the original alcohol is recovered.

(a) What substance is formed in sulfuric acid solution? Show how its formation accounts for each of the facts (i)–(iii). How do you account for the evident stability of this substance? (*Hint*: See Secs. 12.9 and 15.17.)
(b) A solution of 2-cyclopropyl-2-propanol in strong acid gives the following NMR spectrum:

a singlet, δ 2.60, 3H
b singlet, δ 3.14, 3H
c multiplet, δ 3.5–4, 5H

A similar solution of 2-cyclopropyl-1,1,1-trideuterio-2-propanol gives a similar spectrum except that a and b are each reduced to one-half their former area.
What general conclusion about the relative locations of the two methyl groups must you make? Can you suggest a specific geometry for the molecule that is consistent not only with this spectrum but also with your answer to part (a)? (*Hint*: Use models.)

17. Describe simple chemical tests that would serve to distinguish between:

(a) n-butyl alcohol and n-octane
(b) n-butyl alcohol and 1-octene
(c) n-butyl alcohol and n-pentyl bromide
(d) n-butyl alcohol and 3-buten-1-ol
(e) 3-buten-1-ol and 2-buten-1-ol

(f) 3-pentanol and 1-pentanol
(g) 3-pentanol and 2-pentanol
(h) 2-bromoethanol and n-butyl alcohol
(i) 1,2-propanediol and 1,3-propanediol
(j) n-butyl alcohol and *tert*-pentyl alcohol

Tell exactly what you would *do* and *see*.

18. Identify each of the following isomers of formula $C_{20}H_{18}O$:

Isomer MM (m.p. 88 °C)   *a* singlet, δ 2.23, 1H
                                b doublet, δ 3.92, 1H, $J = 7$ Hz
                                c doublet, δ 4.98, 1H, $J = 7$ Hz
                                d singlet, δ 6.81, 10H
                                e singlet, δ 6.99, 5H

Isomer NN (m.p. 88 °C)   *a* singlet, δ 2.14, 1H
                                b singlet, δ 3.55, 2H
                                c broad peak, δ 7.25, 15H

What single simple chemical test would distinguish between these two isomers?

19. Give a structure or structures consistent with each of the proton NMR spectra in Fig. 18.2, p. 693.

20. Give a structure or structures consistent with each of the proton NMR spectra in Fig. 18.3, p. 694.

21. Give a structure or structures consistent with each of the CMR spectra in Fig. 18.4, p. 695.

22. Upon hydrogenation, compound OO (C_4H_8O) is converted into PP ($C_4H_{10}O$). On the basis of their infrared spectra (Fig. 18.5, p. 696), give the structural formulas of OO and PP.

23. Give a structure or structures for the compound QQ, whose infrared and proton NMR spectra are shown in Fig. 18.6, p. 696, and Fig. 18.7, p. 697.

24. Give a structure or structures for the compound RR, whose CMR and proton NMR spectra are shown in Fig. 18.8, p. 697.

25. *Geraniol*, $C_{10}H_{18}O$, a terpene found in rose oil, gives the infrared, CMR, and proton NMR spectra shown in Fig. 18.9 (p. 698). In the next problem, chemical evidence is given from which its structure can be deduced; before working that problem, however, let us see how much information we can get from the spectra alone.

(a) Examine the infrared spectrum. Is geraniol aliphatic or aromatic? What functional group is clearly present? From the molecular formula, what other groupings must also be present in the molecule? Is their presence confirmed by the infrared spectrum?

(b) Examine the CMR spectrum. How many signals are there? How does this compare with the number of carbons given in the molecular formula? What does this tell you about the geraniol molecule?

(c) Now look at the multiplicities listed for the CMR signals. How many hydrogens are attached to each carbon? How many (if any) methyl groups are indicated? How many methylene groups? How many carbons carry only one hydrogen? How many carry none?

(d) At this point, how many carbons and hydrogens have you been able to account for through the CMR spectrum? What atoms, if any, are still missing? What functional group is most likely present? Does this agree with what you saw in the infrared spectrum? Judging from the δ values, which carbon in the CMR spectrum carries this functional group? How many hydrogens are on that carbon?

(e) Clearly, signals *g*, *h*, *i*, and *j* in the CMR spectrum are far removed from the other six. Considering the molecular formula of geraniol, the δ values, and the conclusions you have so far drawn from the CMR spectrum, is geraniol aliphatic or aromatic? If aromatic, how many rings? If aliphatic, how many (if any) double bonds? Are there any carbon–oxygen double bonds?

(f) In the proton NMR spectrum, assign the number of protons to each signal on the basis of the integration curve. From the chemical shift values, and keeping in mind the infrared information, what kind of proton probably gives rise to each signal?

(g) When geraniol is shaken with D_2O, the proton NMR peak at δ 3.32 disappears. Why?

(h) Write down likely groupings in the molecule as indicated by the proton NMR spectrum. How many (if any) methyl groups are there? Methylene groups? Vinylic or allylic protons? What relationships among these groupings are suggested by chemical shift values, splittings, etc.?

(i) Now, using all the information you have, try to put the pieces together and draw possible structures for geraniol. Taking into account the source of geraniol, are any of these more likely than others?

26. *Geraniol*, $C_{10}H_{18}O$, a terpene found in rose oil, adds two moles of bromine to form a tetrabromide, $C_{10}H_{18}OBr_4$. It can be oxidized to a ten-carbon aldehyde or to a ten-carbon carboxylic acid. Upon vigorous oxidation, geraniol yields:

$$CH_3\!-\!\overset{\displaystyle}{\underset{\displaystyle O}{C}}\!-\!CH_3 \qquad CH_3\!-\!\overset{\displaystyle}{\underset{\displaystyle O}{C}}\!-\!CH_2\!-\!CH_2\!-\!\overset{\displaystyle}{\underset{\displaystyle O}{C}}\!-\!OH \qquad HO\!-\!\overset{\displaystyle}{\underset{\displaystyle O}{C}}\!-\!\overset{\displaystyle}{\underset{\displaystyle O}{C}}\!-\!OH$$

(a) Keeping in mind the isoprene rule (Sec. 10.31), what is the most likely structure for geraniol? (b) Nerol (Problem 26, p. 476) can be converted into the same saturated alcohol as geraniol, and yields the same oxidation products as geraniol, yet has different physical properties. What is the most probable structural relationship between geraniol and nerol? (c) Like nerol, geraniol is converted by sulfuric acid into α-terpineol (Problem 26, p. 476), but much more slowly than nerol. On this basis, what structures might you assign to nerol and geraniol? (*Hint*: Use models.)

27. Upon treatment with HBr, both geraniol (preceding problem) and *linalool* (from oil of lavender, bergamot, coriander) yield the same bromide, of formula $C_{10}H_{17}Br$. How do you account for this fact?

$$CH_3\!-\!\overset{\overset{\displaystyle CH_3}{|}}{C}\!=\!CH\!-\!CH_2\!-\!CH_2\!-\!\overset{\overset{\displaystyle CH_3}{|}}{\underset{\underset{\displaystyle OH}{|}}{C}}\!-\!CH\!=\!CH_2$$

Linalool

(a) C$_4$H$_{10}$O

(b) C$_4$H$_{10}$O

(c) C$_4$H$_{10}$O

Figure 18.2 Proton NMR spectra for Problem 19, p. 691.

Figure 18.3   Proton NMR spectra for Problem 20, p. 691.

a 22.7 q
b 24.9 d
c 41.8 t
d 60.7 t

(*a*) $C_5H_{12}O$

Sadtler 135C   © Sadtler Research Laboratories, Division of Bio-Rad, Inc., (1976).

a 11.2 q
b 23.1 t
c 43.8 d
d 64.7 t

(*b*) $C_6H_{14}O$

Sadtler 30C   © Sadtler Research Laboratories, Division of Bio-Rad, Inc., (1976).

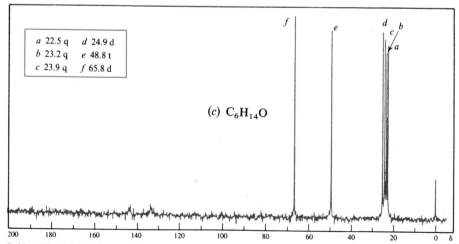

a 22.5 q   *d* 24.9 d
b 23.2 q   *e* 48.8 t
c 23.9 q   *f* 65.8 d

(*c*) $C_6H_{14}O$

Sadtler 15C   © Sadtler Research Laboratories, Division of Bio-Rad, Inc., (1976).

Figure 18.4   CMR spectra for Problem 21, p. 691.

Sadtler 12158

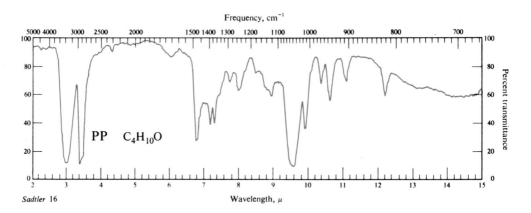

Sadtler 16

Figure 18.5   Infrared spectra for Problem 22, p. 691.

Sadtler 3436

Figure 18.6   Infrared spectrum for Problem 23, p. 691.

Figure 18.7   Proton NMR spectrum for Problem 23, p. 691.

Sadtler 5738C   © Sadtler Research Laboratories, Division of Bio-Rad, Inc., (1978).

Sadtler 22726M   © Sadtler Research Laboratories, Division of Bio-Rad, Inc., (1976).

Figure 18.8   CMR and proton NMR spectra for Problem 24, p. 691.

IRDC 1773

a 16.2 q	f 59.0 t
b 17.6 q	g 124.2 d
c 25.6 q	h 124.3 d
d 26.7 t	i 131.4 s
e 39.7 t	j 138.4 s

Geraniol
$C_{10}H_{18}O$

Sadtler 1249C   © Sadtler Research Laboratories, Division of Bio-Rad, Inc., (1976).

Geraniol
$C_{10}H_{18}O$

Figure 18.9   Infrared, CMR, and proton NMR spectra for Problem 25, p. 691.

19

Ethers and Epoxides

ETHERS

19.1 Structure and nomenclature of ethers

Ethers are compounds of the general formula R—O—R, Ar—O—R, or Ar—O—Ar.

To name ethers we usually name the two groups that are attached to oxygen, and follow these names by the word *ether*:

$C_2H_5OC_2H_5$

Diethyl ether

$$CH_3\overset{\displaystyle CH_3}{\underset{}{CH}}-O-\overset{\displaystyle CH_3}{\underset{}{CH}}CH_3$$

Diisopropyl ether

$$CH_3-O-\overset{\displaystyle CH_3}{\underset{\displaystyle CH_3}{C}}-CH_3$$

tert-Butyl methyl ether

$$CH_3-\overset{\displaystyle CH_3}{\underset{\displaystyle H}{C}}-O\bigcirc$$

Isopropyl phenyl ether

If one group has no simple name, the compound may be named as an *alkoxy* derivative:

$$CH_3CH_2CH_2CHCH_2CH_3$$
$$|$$
$$OCH_3$$

3-Methoxyhexane

$$C_2H_5O\langle\bigcirc\rangle COOH$$

p-Ethoxybenzoic acid

$$CH_2CH_2$$
$$|\quad\quad|$$
$$HO\quad OC_2H_5$$

2-Ethoxyethanol

The simplest alkyl aryl ether, methyl phenyl ether, has the special name of *anisole*.

$$\langle\bigcirc\rangle OCH_3$$

Anisole

If the two groups are identical, the ether is said to be *symmetrical* (e.g., *diethyl ether*, *diisopropyl ether*); if different, *unsymmetrical* (e.g., *tert-butyl methyl ether*).

19.2 Physical properties of ethers

Since the C—O—C bond angle is not 180°, the dipole moments of the two C—O bonds do not cancel each other; consequently, ethers possess a small net dipole moment (e.g., 1.18 D for diethyl ether).

110°   net dipole moment

This weak polarity does not appreciably affect the boiling points of ethers, which are about the same as those of alkanes having comparable molecular weights, and much lower than those of isomeric alcohols. Compare, for example, the boiling points of *n*-heptane (98 °C), methyl *n*-pentyl ether (100 °C), and *n*-hexyl alcohol (157 °C). The hydrogen bonding that holds alcohol molecules strongly together is not possible for ethers, since they contain hydrogen bonded only to carbon (Sec. 17.5).

On the other hand, ethers show a solubility in water comparable to that of the alcohols, both diethyl ether and *n*-butyl alcohol, for example, being soluble to the extent of about 8 g per 100 g water. We attributed the water solubility of the

Table 19.1 ETHERS

Name	M.p., °C	B.p., °C	Name	M.p., °C	B.p., °C
Dimethyl ether	−140	− 24	Anisole	− 37	154
Diethyl ether	−116	34.6	(Methyl phenyl ether)		
Di-*n*-propyl ether	−122	91	Phenetole	− 33	172
Diisopropyl ether	− 60	69	(Ethyl phenyl ether)		
Di-*n*-butyl ether	− 95	142	Diphenyl ether	27	259
Divinyl ether		35	1,4-Dioxane	11	101
Diallyl ether		94	Tetrahydrofuran	−108	66

lower alcohols to hydrogen bonding between water molecules and alcohol mole-cules; the water solubility of ether arises in the same way.

$$
\begin{array}{c}
\text{H} \\
|\\
\text{R}-\text{O}\text{-}\text{-}\text{H}-\text{O} \\
|\\
\text{R}
\end{array}
$$

(We shall look at the properties of ethers as solvents in Sec. 19.10.)

19.3   Industrial sources of ethers. Dehydration of alcohols

A number of symmetrical ethers containing the lower alkyl groups are prepared on a large scale, chiefly for use as solvents. The most important of these is **diethyl ether**, the familiar solvent we use in extractions and in the preparation of Grignard reagents; others include diisopropyl ether and di-*n*-butyl ether.

These ethers are prepared by reactions of the corresponding alcohols with sulfuric acid. Since a molecule of water is lost for every pair of alcohol molecules, the reaction is a kind of *dehydration*. Dehydration to ethers rather than to alkenes

$$
2\text{R}-\text{O}-\text{H} \xrightarrow{\text{H}_2\text{SO}_4,\ \text{heat}} \text{R}-\text{O}-\text{R} + \text{H}_2\text{O}
$$

is controlled by the choice of reaction conditions. For example, ethylene is prepared by heating ethyl alcohol with concentrated sulfuric acid to 180 °C; diethyl ether is prepared by heating a mixture of ethyl alcohol and concentrated sulfuric acid to 140 °C, alcohol being continuously added to keep it in excess.

Dehydration is generally limited to the preparation of symmetrical ethers, because, as we might expect, a combination of two alcohols usually yields a mixture of three ethers.

Ether formation by dehydration is an example of nucleophilic substitution, with the protonated alcohol as substrate and a second molecule of alcohol as nucleophile.

Problem 19.1   (a) Give all steps of a likely mechanism for the dehydration of an alcohol to an ether. (b) Is this the only possibility? Give all steps of an alternative mechanism. (c) Dehydration of *n*-butyl alcohol gives di-*n*-butyl ether. Which of your alternatives appears to be operating here?

Problem 19.2   In ether formation by dehydration, as in other cases of substitution, there is a competing elimination reaction. What is this reaction, and what products does it yield? For what alcohols would elimination be most important?

Problem 19.3   (a) Upon treatment with sulfuric acid, a mixture of ethyl and *n*-propyl alcohols yields a mixture of three ethers. What are they? (b) On the other hand, a mixture of *tert*-butyl alcohol and ethyl alcohol gives a good yield of a single ether. What ether is this likely to be? How do you account for the good yield?

On standing in contact with air, most aliphatic ethers are converted slowly into unstable peroxides. Although present in only low concentrations, these peroxides are very dangerous, since they can cause violent explosions during the distillations that normally follow extractions with ether.

The presence of peroxides is indicated by formation of a red color when the ether is shaken with an aqueous solution of ferrous ammonium sulfate and potassium thiocyanate; the peroxide oxidizes ferrous ion to ferric ion, which reacts with thiocyanate ion to give the characteristic blood-red color of the complex.

$$\text{peroxide} + Fe^{2+} \longrightarrow Fe^{3+} \xrightarrow{\;SCN^-\;} Fe(SCN)_n^{(3-n)-} \qquad (n = 1 \text{ to } 6)$$
$$\text{Red}$$

Peroxides can be removed from ethers in a number of ways, including washing with solutions of ferrous ion (which reduces peroxides), or distillation from concentrated H_2SO_4 (which oxidizes peroxides).

For use in the preparation of Grignard reagents, the ether (usually diethyl) must be free of traces of water and alcohol. This so-called **absolute ether** can be prepared by distillation of ordinary ether from concentrated H_2SO_4 (which removes not only water and alcohol but also peroxides), and subsequent storing over metallic sodium. There is available today commercial anhydrous ether of such high quality that only the treatment with sodium is needed to make it ready for the Grignard reaction.

It is hard to overemphasize the hazards met in using diethyl ether, even when it is free of peroxides: it is highly volatile, and the flammability of its vapors makes explosions and fires ever-present dangers unless proper precautions are observed.

19.4 Preparation of ethers

The following methods are generally used for the laboratory preparation of ethers. (The Williamson synthesis is used for the preparation of alkyl aryl ethers industrially, as well.)

PREPARATION OF ETHERS

1. **Williamson synthesis.** Discussed in Sec. 19.5.

$$RX + \begin{matrix} R'O^-Na^+ \\ \text{or} \\ ArO^-Na^+ \end{matrix} \longrightarrow \begin{matrix} R-OR' \\ \text{or} \\ R-OAr \end{matrix} \qquad \begin{matrix} \textit{Yield from } RX: \\ CH_3 > 1° > 2° (>3°) \end{matrix}$$

Examples:

$$\left[(CH_3)_2CHO-H \xrightarrow{\;Na\;} \right] (CH_3)_2CHO^-Na^+ + CH_3CH_2CH_2Br$$
$$\begin{matrix} \text{Sodium} \\ \text{isopropoxide} \end{matrix} \qquad\qquad \begin{matrix} n\text{-Propyl} \\ \text{bromide} \end{matrix}$$

$$\longrightarrow CH_3CH_2CH_2O-CH(CH_3)_2$$
$$\text{Isopropyl } n\text{-propyl ether}$$

$$\langle \bigcirc \rangle O-H + CH_3CH_2Br \xrightarrow{\;aq.\ NaOH\;} \langle \bigcirc \rangle O-CH_2CH_3$$

Phenol   Ethyl bromide   Ethyl phenyl ether

CONTINUED

——— CONTINUED ———

2. **Alkoxymercuration–demercuration.** Discussed in Sec. 19.6.

$$\text{C=C} + ROH + Hg(OOCCF_3)_2 \longrightarrow -\underset{RO}{C}-\underset{HgOOCCF_3}{C}- \xrightarrow{NaBH_4} -\underset{RO}{C}-\underset{H}{C}-$$

Mercuric
trifluoroacetate

*Markovnikov
orientation*

Example:

$$CH_3-\underset{CH_3}{\overset{CH_3}{C}}-CH=CH_2 + CH_3CH_2OH \xrightarrow{Hg(OOCCF_3)_2} \xrightarrow{NaBH_4} CH_3-\underset{CH_3}{\overset{CH_3}{C}}-\underset{OC_2H_5}{CH}-CH_3$$

3,3,-Dimethyl-1-butene

3-Ethoxy-2,2-dimethylbutane
No rearrangement ■

19.5 Preparation of ethers. Williamson synthesis

In the laboratory, the Williamson synthesis of ethers is important because of its versatility: it can be used to make unsymmetrical ethers as well as symmetrical ethers.

In the Williamson synthesis an alkyl halide (or substituted alkyl halide) is allowed to react with a sodium alkoxide or a sodium phenoxide:

$$R\!-\!X + Na^+ \ ^-O\!-\!R' \longrightarrow R\!-\!O\!-\!R' + Na^+X^-$$

$$R\!-\!X + Na^+ \ ^-O\!-\!Ar \longrightarrow R\!-\!O\!-\!Ar + Na^+X^-$$

For the preparation of aryl methyl ethers, *methyl sulfate*, $(CH_3)_2SO_4$, is frequently used instead of the more expensive methyl halides.

$$CH_3Br + Na^+ \ ^-O\!-\!\underset{CH_3}{\overset{CH_3}{C}}\!-\!CH_3 \longrightarrow CH_3\!-\!O\!-\!\underset{CH_3}{\overset{CH_3}{C}}\!-\!CH_3$$

Sodium
tert-butoxide

tert-Butyl methyl ether

$$\text{⬡}O\!-\!H + CH_3OSO_2OCH_3 \xrightarrow{aq.\ NaOH} \text{⬡}O\!-\!CH_3 + CH_3OSO_3^-Na^+$$

Phenol   Methyl sulfate

Anisole

$$\text{⬡}CH_2Br + H\!-\!O\text{⬡} \xrightarrow{aq.\ NaOH} \text{⬡}CH_2\!-\!O\text{⬡}$$

Benzyl bromide   Phenol

Benzyl phenyl ether

The Williamson synthesis involves nucleophilic substitution of alkoxide ion or phenoxide ion for halide ion; it is strictly analogous to the preparation of alcohols by treatment of alkyl halides with aqueous hydroxide (Sec. 17.8). *Aryl halides cannot in general be used, because of their low reactivity toward nucleophilic sub-stitution.*

Problem 19.4 (a) On what basis could you have predicted that methyl sulfate would be a good methylating agent in reactions like those presented above? (*Hint*: What is the *leaving group*? See Sec. 5.9.) (b) Can you suggest another class of compounds that might serve in place of alkyl halides in the Williamson synthesis?

Sodium alkoxides are made by direct action of sodium metal on dry alcohols:

$$\text{ROH} + \text{Na} \longrightarrow \text{RO}^-\text{Na}^+ + \tfrac{1}{2}\text{H}_2$$

An alkoxide

Sodium phenoxides, on the other hand, because of the appreciable acidity of phenols (Sec. 28.7), are made by the action of aqueous sodium hydroxide on phenols:

$$\text{ArOH} + \text{Na}^+\,{}^-\text{OH} \longrightarrow \text{ArO}^-\text{Na}^+ + \text{H}_2\text{O}$$

A phenol	A phenoxide	Water
Stronger acid	*Weaker acid*	

If we wish to make an unsymmetrical dialkyl ether, we have a choice of two combinations of reagents; one of these is nearly always better than the other. In the preparation of *tert*-butyl ethyl ether, for example, the following combinations are conceivable:

$$\text{CH}_3\text{CH}_2\text{Br} + \text{Na}^+\,{}^-\text{O}-\underset{\underset{\text{CH}_3}{|}}{\overset{\overset{\text{CH}_3}{|}}{\text{C}}}-\text{CH}_3 \qquad \textit{Feasible}$$

$$\text{CH}_3\text{CH}_2-\text{O}-\underset{\underset{\text{CH}_3}{|}}{\overset{\overset{\text{CH}_3}{|}}{\text{C}}}-\text{CH}_3$$

tert-Butyl ethyl ether

$$\text{CH}_3-\underset{\underset{\text{CH}_3}{|}}{\overset{\overset{\text{CH}_3}{|}}{\text{C}}}-\text{Cl} + \text{Na}^+\,{}^-\text{OCH}_2\text{CH}_3 \qquad \textit{Not feasible}$$

Which do we choose? As always, we must consider the danger of elimination competing with the desired substitution; elimination should be particularly serious here because of the strong basicity of the alkoxide reagent. We therefore reject the use of the tertiary halide, which we expect to yield mostly—or all—elimination product; we must use the other combination. The disadvantage of the slow reaction

$$\text{CH}_3\text{CH}_2\text{Br} + {}^-\text{O}-\underset{\underset{\text{CH}_3}{|}}{\overset{\overset{\text{CH}_3}{|}}{\text{C}}}-\text{CH}_3 \longrightarrow \text{CH}_3\text{CH}_2-\text{O}-\underset{\underset{\text{CH}_3}{|}}{\overset{\overset{\text{CH}_3}{|}}{\text{C}}}-\text{CH}_3 + \text{Br}^- \qquad \text{Substitution}$$

tert-Butyl ethyl ether

$$\text{CH}_3-\underset{\underset{\text{CH}_3}{|}}{\overset{\overset{\text{CH}_3}{|}}{\text{C}}}-\text{Cl} + {}^-\text{OC}_2\text{H}_5 \longrightarrow \text{CH}_3-\underset{\underset{\text{CH}_3}{|}}{\overset{}{\text{C}}}{=}\text{CH}_2 + \text{C}_2\text{H}_5\text{O}-\text{H} + \text{Cl}^- \qquad \text{Elimination}$$

between sodium and *tert*-butyl alcohol (Sec. 18.4) in the preparation of the alkoxide is more than offset by the tendency of the primary halide to undergo substitution rather than elimination. In planning a Williamson synthesis of a dialkyl ether, we must always keep in mind that the tendency for alkyl halides to undergo dehydro-halogenation is $3° > 2° > 1°$.

For the preparation of an alkyl aryl ether there are again two combinations to be considered; here, one combination can usually be rejected out of hand. Phenyl *n*-propyl ether, for example, can be prepared only from the alkyl halide and the phenoxide, since the aryl halide is quite unreactive toward alkoxides.

$$CH_3CH_2CH_2Br \; + \; Na^+ \; ^-O\!\!\left\langle\bigcirc\right\rangle \; \longrightarrow \; CH_3CH_2CH_2\!-\!O\!\!\left\langle\bigcirc\right\rangle \; + \; Na^+ \, Br^-$$

n-Propyl bromide	Sodium phenoxide	Phenyl *n*-propyl ether

$$\left\langle\bigcirc\right\rangle\!Br \; + \; Na^+ \; ^-OCH_2CH_2CH_3 \; \longrightarrow \; \text{no reaction}$$

Bromobenzene   Sodium *n*-propoxide

Since alkoxides and phenoxides are prepared from the corresponding alcohols and phenols, and since alkyl halides are commonly prepared from the alcohols, the Williamson method ultimately involves the synthesis of an ether from two hydroxy compounds.

Problem 19.5   Outline the synthesis, from alcohols and/or phenols, of:

(a) *tert*-butyl ethyl ether   (c) *sec*-butyl isobutyl ether
(b) phenyl *n*-propyl ether   (d) cyclohexyl methyl ether

Problem 19.6   Outline the synthesis of *p*-nitrobenzyl phenyl ether from any of these starting materials: toluene, bromobenzene, phenol. (*Caution*: Double-check the nitration stage.)

Problem 19.7   When optically active 2-octanol of specific rotation $-8.24°$ is converted into its sodium salt, and the salt is then treated with ethyl bromide, there is obtained the optically active ether, 2-ethoxyoctane, with specific rotation $-15.6°$. Making use of the configuration and maximum rotation of 2-octanol given on page 184, what, if anything, can you say about: (a) the configuration of $(-)$-2-ethoxy octane? (b) the maximum rotation of 2-ethoxyoctane?

Problem 19.8   (*Work this after Problem 19.7.*) When $(-)$-2-bromooctane of specific rotation $-30.3°$ is treated with ethoxide ion in ethyl alcohol, there is obtained 2-ethoxyoctane of specific rotation $+15.3°$. Using the configuration and maximum rotation of the bromide given on page 184, answer the following questions. (a) Does this reaction involve complete retention of configuration, complete inversion, or inversion plus racemization? (b) By what mechanism does this reaction appear to proceed? (c) In view of the reagent and solvent, is this the mechanism you would have expected to operate? (d) What mechanism do you suppose is involved in the alternative synthesis (Problem 19.7) of 2-ethoxyoctane from the salt of 2-octanol and ethyl bromide? (e) Why, then, do the products of the two syntheses have *opposite* rotations?

Problem 19.9   A mixture of *n*-butyl chloride and *n*-octyl alcohol gives a 95% yield of *n*-butyl *n*-octyl ether when brought into contact with a concentrated aqueous solution of NaOH containing a little quat salt.

Account in detail for the formation of the ether. What advantage does this method offer over what we have described in this section?

19.6   Preparation of ethers. Alkoxymercuration–demercuration

Alkenes react with mercuric trifluoroacetate in the presence of an alcohol to give alkoxymercurial compounds which on reduction yield ethers.

We recognize this two-stage process as the exact analog of the oxymercuration–demercuration synthesis of alcohols (Sec. 17.9). In the place of water we use an alcohol which, not surprisingly, can play exactly the same role. Instead of introducing the hydroxy group to make an alcohol, we introduce an *alkoxy* group to make an ether. This example of *solvomercuration–demercuration* amounts to Markovnikov addition of an alcohol to a carbon–carbon double bond.

Problem 19.10   Write all steps of a likely mechanism for alkoxymercuration.

Alkoxymercuration–demercuration has all the advantages we saw for its counterpart: speed, convenience, high yield, and the virtual absence of rearrangement. Compared with the Williamson synthesis, it has one tremendous advantage: there is no competing elimination reaction. As a result, it can be used for the synthesis of nearly every kind of alkyl ether except—evidently for steric reasons—di-*tert*-alkyl ethers. For example:

3,3-Dimethyl-1-butene   3-Isopropoxy-2,2-dimethylbutane

We notice that, instead of the mercuric acetate which was used in the preparation of alcohols, here mercuric *trifluoro*acetate is used. With a bulky alcohol—secondary or tertiary—as solvent, the trifluoroacetate is required for a good yield of ether.

Problem 19.11   In the presence of a secondary or tertiary alcohol, mercuric acetate adds to alkenes to give much—or even chiefly—organic acetate instead of ether as the

$$\begin{array}{c} \underset{\underset{AcO}{|}\ \underset{HgOAc}{|}}{-C-C-} \qquad \textit{instead of} \qquad \underset{\underset{RO}{|}\ \underset{HgOAc}{|}}{-C-C-} \end{array}$$

product. How do you account for the advantage of using mercuric trifluoroacetate? (*Hint*: Trifluoroacetic acid is a much stronger acid than acetic.)

Problem 19.12 Starting with any alcohols, outline all steps in the synthesis of each of the following ethers, using the Williamson synthesis or alkoxymercuration–demercuration, whichever you think is best suited for the particular job.

(a) *n*-hexyl isopropyl ether
(b) 2-hexyl isopropyl ether
(c) *tert*-butyl cyclohexyl ether
(d) dicyclohexyl ether

19.7 Reactions of ethers. Cleavage by acids

Ethers are comparatively unreactive compounds. The ether linkage is quite stable toward bases, oxidizing agents, and reducing agents. In so far as the ether linkage itself is concerned, ethers undergo just one kind of reaction, **cleavage by acids**:

$$R-O-R' + HX \longrightarrow R-X + R'-OH \xrightarrow{HX} R'-X$$

$$Ar-O-R + HX \longrightarrow R-X + Ar-OH$$

$$\text{Reactivity of HX:}\quad HI > HBr > HCl$$

Cleavage takes place only under quite vigorous conditions: concentrated acids (usually HI or HBr) and high temperatures.

A dialkyl ether yields initially an alkyl halide and an alcohol; the alcohol may react further to form a second mole of alkyl halide. Because of the low reactivity at the bond between oxygen and an aromatic ring, an alkyl aryl ether undergoes cleavage of the alkyl–oxygen bond and yields a phenol and an alkyl halide. For example:

$$\underset{\text{Diisopropyl ether}}{\underset{\underset{CH_3}{|}}{CH_3-CH}-O-\underset{\underset{CH_3}{|}}{CH}-CH_3} \xrightarrow[130-140\ ^\circ C]{48\%\ HBr} \underset{\text{Isopropyl bromide}}{2CH_3-\underset{\underset{CH_3}{|}}{CH}-Br}$$

$$\underset{\text{Anisole}}{\text{⬡}O-CH_3} \xrightarrow[120-130\ ^\circ C]{57\%\ HI} \underset{\text{Phenol}}{\text{⬡}O-H} + \underset{\text{Methyl iodide}}{CH_3I}$$

Cleavage involves nucleophilic attack by halide ion on the protonated ether, with displacement of the weakly basic alcohol molecule:

$$R-\overset{..}{\underset{..}{O}}-R' + HX \rightleftharpoons R-\overset{\overset{\displaystyle H}{|}}{\underset{..}{O}}-R' + X^- \xrightarrow[\substack{or \\ S_N2}]{S_N1} R-X + R'OH$$

Weak base:
good leaving group

Such a reaction occurs much more readily than displacement of the strongly basic alkoxide ion from the neutral ether.

$$R\text{---}O\text{---}R' + X^- \xrightarrow{\quad\quad} R\text{---}X + R'O^-$$

Strong base:

poor leaving group

Reaction of a protonated ether with halide ion, like the corresponding reaction of a protonated alcohol, can proceed either by an S_N1 mechanism,

(1)
$$R\overset{\overset{\displaystyle H}{|}}{\underset{\oplus}{O}}\text{---}R' \xrightarrow{\text{slow}} R^+ + HOR'$$

S_N1

(2)
$$R^+ + X^- \xrightarrow{\text{fast}} R\text{---}X$$

or by an S_N2 mechanism,

$$R\overset{\overset{\displaystyle H}{|}}{\underset{\oplus}{O}}\text{---}R' + X^- \longrightarrow \left[\overset{\delta-}{X}\text{----}R\text{----}\overset{\overset{\displaystyle H}{|}}{\underset{\delta+}{O}R'}\right] \longrightarrow R\text{---}X + HOR' \quad S_N2$$

depending upon conditions and the structure of the ether. As we might expect, a primary alkyl group tends to undergo S_N2 displacement, whereas a tertiary alkyl group tends to undergo S_N1 displacement.

Problem 19.13   Cleavage of optically active *sec*-butyl methyl ether by anhydrous HBr yields chiefly methyl bromide and *sec*-butyl alcohol; the *sec*-butyl alcohol has the same configuration and optical purity as the starting material. How do you interpret these results?

19.8   Electrophilic substitution in aromatic ethers

The alkoxy group, —OR, was listed (Sec. 14.5) as *ortho,para*-directing toward electrophilic aromatic substitution, and moderately activating. It is a much stronger activator than —R, but much weaker than —OH.

The carbocations resulting from *ortho* and *para* attack were considered (Sec. 14.20) to be stabilized by contribution from structures I and II. These structures

⊕OR   ⊕OR

I   II

are especially stable ones, since in them every atom (except hydrogen, of course) has a complete octet of electrons.

The ability of the oxygen to share more than a pair of electrons with the ring and to accommodate a positive charge is consistent with the basic character of ethers.

Problem 19.14   Predict the principal products of: (a) bromination of *p*-methylanisole; (b) nitration of *m*-nitroanisole; (c) nitration of benzyl phenyl ether.

19.9   Cyclic ethers

In their preparation and properties, most cyclic ethers are just like the ethers we have already studied: the chemistry of the ether linkage is essentially the same whether it forms part of an open chain or part of an aliphatic ring.

Problem 19.15   *1,4-Dioxane* is prepared industrially (for use as a water-soluble solvent) by dehydration of an alcohol. What alcohol is used?

$$\underset{\text{1,4-Dioxane}}{\underset{\displaystyle H_2C \diagdown_{O} \diagup CH_2}{H_2C \diagup^{O} \diagdown CH_2}} \qquad \underset{\text{Furan}}{\underset{\displaystyle HC \diagdown_{O} \diagup CH}{HC \!=\! CH}} \qquad \underset{\text{Tetrahydrofuran}}{\underset{\displaystyle H_2C \diagdown_{O} \diagup CH_2}{H_2C \!-\! CH_2}}$$

Problem 19.16   The unsaturated cyclic ether *furan* can readily be made from substances isolated from oat hulls and corncobs; one of its important uses involves its conversion into (a) *tetrahydrofuran*, and (b) 1,4-dichlorobutane. Using your knowledge of alkene chemistry and ether chemistry, show how these conversions can be carried out.

The unsaturated cyclic ether 2,3-dihydro-4*H*-pyran (DHP) reacts readily with alcohols (ROH) in the presence of acid to give alkyl *tetrahydropyranyl ethers* (RO–THP).

$$\text{DHP} \quad + \quad \text{RO} \!-\! \text{H} \quad \xrightarrow{\;H^+\;} \quad \text{A THP ether}$$

2,3-Dihydro-4*H*-pyran	A tetrahydropyranyl ether
DHP	A THP ether

Like other ethers, a THP ether is resistant to base and many other reagents, and is cleaved by acid. However, because of its special structure—there are two ether oxygens attached to the same carbon, making it an *acetal* (Sec. 21.13)—a THP ether is *very readily* cleaved by dilute aqueous acid.

$$\underset{\text{A THP ether}}{\text{RO}\!-\!\text{THP}} \quad + \quad H_2O \quad \xrightarrow{\;H^+\;} \quad \text{RO}\!-\!\text{H}$$

The THP group thus has the qualities necessary for a *protecting group*: it is easily attached and easily removed, and under conditions that will not harm other functional groups in the molecule; and while it is present it is resistant to certain reagents that would otherwise attack the group it protects. The —OH group is, for example, acidic, and rapidly destroys organometallic compounds like the Grignard reagent or organolithiums (Sec. 17.17). We cannot, therefore, prepare a Grignard reagent from an organic halide that contains —OH, or allow a Grignard reagent to react with an aldehyde or ketone that contains an —OH. But if the —OH is first converted into —OTHP, we *can* carry out such reactions; and then, when they are over, simply remove the THP group.

Problem 19.17 (a) To what class of reactions does the formation of a THP ether belong? (Simply look at the structures involved.)

(b) Show all steps in a likely mechanism for this reaction.

(c) Why does it take place so readily? Why does it yield the product it does and not an isomer of that product? (*Hint*: See Sec. 10.15.)

(d) Starting from ethanol and making use of DHP, outline all steps in a possible synthesis of 1,3-butanediol.

Cyclic ethers of two particular kinds deserve special attention because of their unusual properties: the *crown ethers* (Sec. 19.10) and the *epoxides* (Secs. 19.11–19.16).

19.10   Crown ethers. Host–guest relationship

As we have seen, ethers cannot furnish an acidic proton for hydrogen bonding. They are thus aprotic solvents, but—the simple ones, at least—not very polar, and are essentially insoluble in water. Diethyl ether is very commonly used to extract organic materials from an aqueous solution, leaving ionic compounds behind in the water layer.

But the oxygen of ethers carries unshared electrons, and through these unshared pairs ethers can solvate cations (Sec. 6.4). Diethyl ether and tetrahydrofuran are, for example, the solvents in which Grignard reagents (Sec. 3.16) are usually prepared and used. They are able to dissolve these important reagents because they strongly solvate the magnesium of the RMg^+ cation.

$$
\begin{array}{c}
Et \quad Et \\
\diagdown \diagup \\
\overset{..}{O} \\
\vdots \\
R - \overset{\cdot\cdot}{Mg}{}^+ \quad X^- \\
\vdots \\
\overset{..}{O} \\
\diagup \diagdown \\
Et \quad Et
\end{array}
$$

Now, crown ethers are cyclic ethers containing several—four, five, six, or more—oxygen atoms. Let us take as our example the crown ether I, which is one of the most effective and widely used of these catalysts. It is called 18-crown-6, to show that there are 18 atoms in the ring, of which 6 are oxygen. The ring contains more than one kind of atom, and hence is a *heterocyclic* ring (Greek: *hetero*, different). Since divalent oxygen has bond angles not very different from those of carbon (Sec. 1.12), the rings of crown ethers can exist in much the same conformations as the alicyclic rings we have already discussed in Chapter 12. The rings of crown ethers are therefore puckered. The name of "crown" was given to the

$$H_2C-CH_2$$

$$H_2C-O \qquad O-CH_2$$

$$H_2C \qquad\qquad CH_2$$

$$O \qquad\qquad O$$

$$H_2C \qquad\qquad CH_2$$

$$H_2C-O \qquad O-CH_2$$

$$H_2C-CH_2$$

I

first of these because, as its discoverer Charles J. Pedersen (E. I. Du Pont De Nemours) has said, "its molecular model looks like one and, with it, cations could be crowned and uncrowned without physical damage to either . . . ".

This brings us to the function of these crown ethers. They are phase-transfer catalysts, and very powerful ones. They are used to transfer ionic compounds into an organic phase either from a water phase or, more commonly, from the solid crystal. Unlike the quat ions we studied earlier (Sec. 6.7), crown ethers are neutral molecules; yet they do the same job. Now, how do they work?

Let us examine the structure of 18-crown-6 (Fig. 19.1, on the next page). Unfolded, the molecule is shaped like a doughnut, and has a hole in the middle. Facing into the hole are the oxygen atoms; facing outward are the twelve CH_2 groups. There is thus a hydrophilic interior and a lipophilic exterior. The hole has a diameter of 2.7 Å.

Now, to this crown ether let us add a potassium ion, K^+. It has a diameter of 2.66 Å and *just fits* into the hole in the crown, where it is held by unshared pairs of electrons on the six oxygen atoms. Because of the close fit and because there are six oxygens, K^+ is bound very tightly. The crown ether is not a solvent, but it holds K^+ by the same forces that a solvent uses; the forces are simply much stronger here.

Together, K^+ and the crown ether make up a new cation. This new cation is much like a quat ion, except that it is held together by ion–dipole bonds instead of covalent bonds. Like a quat ion, it is lipophilic on the outside, and has the positive charge buried within the molecule. The lipophilicity makes it soluble in organic solvents of low polarity. When it enters such solvents, it takes an anion with it. This anion is shielded from the positive charge on K^+ by the bulky crown, thus forming only loose ion pairs, and is highly reactive.

Crown ethers have been made in a wide variety of shapes and sizes, and their ability to hold cations has been extensively studied. The hole in the ether can be larger than the cation and still bind it: Na^+, for example, is smaller than K^+, but is still bound by 18-crown-6, although less strongly than K^+. (The best size of hole for sodium is provided by 15-crown-5.) The hole can be smaller than the cation; in this case, the cation is simply seated in the cavity on one face or the other of the crown.

What we are seeing here is an example of the **host–guest relationship**. The crown ether is the *host*; the cation is the *guest*. This kind of relationship is of intense interest to the organic chemist, and is the subject of much research: for the practical purpose of designing new and better reagents; and for theoretical reasons, to understand better a wide range of interactions that extends all the way to that ultimate host–guest relationship, the one between enzyme and substrate.

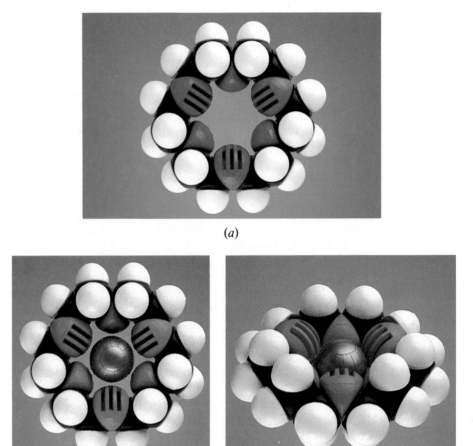

Figure 19.1 Host–guest relationship: crown ether–cation. (*a*) 18-Crown-6, unfolded. The hole is lined with oxygens, and has a diameter of 2.7 Å. (*b*) and (*c*) 18-Crown-6 holding a potassium ion through ion–dipole bonds to the oxygens. Diameter of K^+, 2.66 Å. The outside of the crown ether is carbon and hydrogen, and lipophilic.

Let us look at one example of a host–guest relationship involving hosts that are made, not by organic chemists, but by microorganisms. For various enzyme systems to function properly, cells must maintain certain concentrations of cations like K^+ and Na^+. Such maintenance is made feasible by the normally slow passage of these hydrated inorganic ions through the fatty (lipophilic) core of the cell membranes (Sec. 37.8). A large number of antibiotics (*gramicidin*, *valinomycin*, *nonactin*, for example) upset this ionic balance: in their presence cations escape rapidly through the membrane, and the enzyme system must expend its energy forcing them back. It seems clear that these antibiotics exert their effect by *transporting* the cations through the membrane. Like crown ethers they wrap around the cation, holding it through ion–dipole bonds; then, with their lipophilic parts turned outward and the cation hidden within, they pass easily through the membrane. See, for example, nonactin in Fig. 19.2.

Figure 19.2 Host–guest relationship: the antibiotic nonactin holding a K^+ ion. The cation is held by ion–dipole bonds to inward-turning oxygens. The lipophilic parts of nonactin are turned outward.

EPOXIDES

19.11 Preparation of epoxides

Epoxides are compounds containing the three-membered ring:

Epoxide ring
(Oxirane ring)

They are ethers, but the three-membered ring gives them unusual properties.

By far the most important epoxide is the simplest one, ethylene oxide. It is prepared on an industrial scale by catalytic oxidation of ethylene by air.

$$CH_2\!=\!CH_2 \xrightarrow{\ O_2,\ Ag,\ 250\,°C\ } CH_2\!-\!CH_2$$

Ethylene   Ethylene oxide

Other epoxides are prepared by the following methods.

PREPARATION OF EPOXIDES _____

1. **From halohydrins.** Discussed in Sec. 19.11.

$$-\!\overset{|}{\underset{|}{C}}\!=\!\overset{|}{\underset{|}{C}}\!- \xrightarrow{\ X_2,\ H_2O\ } -\!\overset{|}{\underset{X}{C}}\!-\!\overset{|}{\underset{OH}{C}}\!- \xrightarrow{\ OH^-\ } -\!\overset{|}{C}\!-\!\overset{|}{C}\!- + H_2O + X^-$$

CONTINUED

Example:

$$CH_3-CH=CH_2 \xrightarrow{Cl_2, H_2O} CH_3-\underset{\underset{OH}{|}}{C}H-\underset{\underset{Cl}{|}}{C}H_2 \xrightarrow{\text{conc. aq. OH}^-} CH_3-CH-CH_2$$

Propylene
chlorohydrin

Propylene oxide

2. **Peroxidation of carbon–carbon double bonds.** Discussed in Sec. 19.11.

$$-\overset{|}{C}=\overset{|}{C}- + C_6H_5CO_2OH \longrightarrow -\overset{|}{C}-\overset{|}{C}- + C_6H_5COOH$$

Peroxybenzoic
acid

Examples:

$$\text{Styrene} -CH=CH_2 \xrightarrow{\text{peroxybenzoic acid}} -CH-CH_2$$

Styrene

Styrene oxide

$$\text{Cyclohexene} \xrightarrow{\text{peroxybenzoic acid}} \text{Cyclohexene oxide}$$

Cyclohexene

Cyclohexene oxide

The conversion of halohydrins into epoxides by the action of base is simply an adaptation of the Williamson synthesis (Sec. 19.5); a cyclic compound is obtained because both alcohol and halide happen to be part of the same molecule. In the presence of hydroxide ion a small proportion of the alcohol exists as alkoxide; this alkoxide displaces halide ion from another portion of the same molecule to yield the cyclic ether.

$$(1) \quad \underset{\underset{OH}{|}}{\overset{\overset{Br}{|}}{C}}H_2-CH_2 + OH^- \;\rightleftharpoons\; H_2O + \overset{\overset{Br}{|}}{C}H_2-CH_2\underset{O^{\ominus}}{}$$

$$(2) \quad \overset{Br}{C}H_2-CH_2 \longrightarrow \left[\overset{\delta_-}{\underset{H}{\overset{Br}{H-C}}}\cdots\underset{\underset{\delta_-}{O}}{\overset{H}{C}}-H \right] \longrightarrow CH_2-CH_2 + Br^-$$

Since halohydrins are nearly always prepared from alkenes by addition of halogen and water to the carbon–carbon double bond (Sec. 8.15), this method amounts to the conversion of an alkene into an epoxide.

Alternatively, the carbon–carbon double bond may be oxidized directly to the epoxide group by peroxy compounds, such as peroxybenzoic acid:

Peroxybenzoic acid

When allowed to stand in ether or chloroform solution, the peroxy acid and the unsaturated compound—which need not be a simple alkene—react to yield benzoic acid and the epoxide. For example:

| Cyclopentene | Peroxybenzoic acid | | Cyclopentene oxide | Benzoic acid |

$$C_6H_5-\overset{\overset{\displaystyle H}{|}}{C}=\overset{\overset{\displaystyle H}{|}}{C}-CH_2OH + C_6H_5C\overset{O}{\underset{O-OH}{}} \longrightarrow C_6H_5-\overset{\overset{\displaystyle H}{|}}{C}-\overset{\overset{\displaystyle H}{|}}{C}-CH_2OH + C_6H_5C\overset{O}{\underset{OH}{}}$$

3-Phenyl-2-propen-1-ol
Cinnamyl alcohol

19.12 Reactions of epoxides

Epoxides owe their importance to their high reactivity, which is due to the ease of opening of the highly strained three-membered ring. The bond angles of the ring, which average 60°, are considerably less than the normal tetrahedral carbon angle of 109.5°, or the divalent oxygen angle of 110° for open-chain ethers (Sec. 19.2). Since the atoms cannot be located to permit maximum overlap of orbitals (Sec. 12.9), the bonds are weaker than in an ordinary ether, and the molecule is less stable.

Epoxides undergo acid-catalyzed reactions with extreme ease, and—unlike ordinary ethers—can even be cleaved by bases. Some of the important reactions are outlined below.

REACTIONS OF EPOXIDES _____

1. **Acid-catalyzed cleavage.** Discussed in Sec. 19.13.

Examples:

$$H_2O + \underset{O}{CH_2-CH_2} \xrightarrow{H^+} \underset{OH\quad OH}{CH_2-CH_2}$$

1,2-Ethanediol

CONTINUED

CONTINUED

$$C_2H_5OH + CH_2\!-\!CH_2 \xrightarrow{\ H^+\ } \begin{matrix} CH_2\!-\!CH_2 \\ | \qquad | \\ C_2H_5O \quad OH \end{matrix}$$

2-Ethoxyethanol

Phenol $\;+\; CH_2\!-\!CH_2 \xrightarrow{\ H^+\ }$ ⬡O$-CH_2CH_2OH$

Phenol   2-Phenoxyethanol

$$HBr + CH_2\!-\!CH_2 \longrightarrow \begin{matrix} CH_2\!-\!CH_2 \\ | \qquad | \\ Br \quad OH \end{matrix}$$

Ethylene bromohydrin
(2-Bromoethanol)

2. Base-catalyzed cleavage. Discussed in Sec. 19.14.

$$Z\!:\!\overset{\frown}{+} =\!\!\overset{|}{C}\!-\!\overset{|}{C}\!- \longrightarrow \overset{\overset{\displaystyle Z}{|}}{-\!\overset{|}{C}\!-\!\overset{|}{C}\!-} \xrightarrow{\ HZ\ } \overset{\overset{\displaystyle Z}{|}}{-\!\overset{|}{C}\!-\!\overset{|}{C}\!-} + \; :\!Z$$
$$\qquad\qquad\quad O \qquad\qquad\quad O^- \qquad\qquad OH$$

Examples:

$$C_2H_5O^- Na^+ + CH_2\!-\!CH_2 \longrightarrow C_2H_5O\!-\!CH_2CH_2OH$$
Sodium ethoxide   O   2-Ethoxyethanol

⬡O$^-$ Na$^+$ + $CH_2\!-\!CH_2 \longrightarrow$ ⬡O$-CH_2CH_2OH$

Sodium phenoxide   2-Phenoxyethanol

$$NH_3 + CH_2\!-\!CH_2 \longrightarrow H_2N\!-\!CH_2CH_2OH$$
$$\qquad\qquad O$$

2-Aminoethanol
(Ethanolamine)

3. Reaction with Grignard reagents. Discussed in Sec. 19.15.

$$R\!\!\overset{\frown}{-}\!MgX + CH_2\!-\!CH_2 \longrightarrow R\!-\!CH_2CH_2O^- Mg^+ \xrightarrow{\ H^+\ } R\!-\!CH_2CH_2OH$$
$$\qquad\qquad\qquad\qquad O$$

Primary alcohol:
*chain has been lengthened
by two carbons*

Examples:

$$CH_3CH_2CH_2CH_2\!-\!MgBr + CH_2\!-\!CH_2 \longrightarrow CH_3CH_2CH_2CH_2\!-\!CH_2CH_2OH$$
$$\qquad\qquad\qquad\qquad\qquad O$$

1-Hexanol

⬡MgBr + $CH_2\!-\!CH_2 \longrightarrow$ ⬡CH_2CH_2OH

2-Phenylethanol
(β-Phenylethyl alcohol)

19.13   Acid-catalyzed cleavage of epoxides. *anti*-Hydroxylation

Like other ethers, an epoxide is protonated by acid; the protonated epoxide can then undergo attack by any of a number of nucleophilic reagents.

An important feature of the reactions of epoxides is the formation of compounds that contain *two* functional groups. Thus, reaction with water yields a 1,2-diol; reaction with an alcohol yields a compound that is both ether and alcohol.

A 1,2-diol

An alkoxy alcohol
(A hydroxy ether)

Problem 19.18   The following compounds are commercially available for use as water-soluble solvents. How could each be made?

(a) CH_3CH_2—O—CH_2CH_2—O—CH_2CH_2—OH   Carbitol
(b) C_6H_5—O—CH_2CH_2—O—CH_2CH_2—OH   Phenyl carbitol
(c) HO—CH_2CH_2—O—CH_2CH_2—OH   Diethylene glycol
(d) HO—CH_2CH_2—O—CH_2CH_2—O—CH_2CH_2—OH   Triethylene glycol

Problem 19.19   Show in detail (including structures and transition states) the steps in the acid-catalyzed hydrolysis of ethylene oxide by an S_N1 mechanism; by an S_N2 mechanism.

The two-stage process of epoxidation followed by hydrolysis is stereoselective, and gives 1,2-diols corresponding to *anti*-addition to the carbon–carbon double bond. Exactly the same stereochemistry was observed (Problem 9.1, p. 352) for hydroxylation of alkenes by peroxyformic acid—and for good reason: an epoxide is formed there, too, but is rapidly cleaved in the acidic medium, formic acid. The interpretation is exactly the same as that given to account for *anti*-addition of halogens (Sec. 9.6); indeed, epoxides and their hydrolysis served as a model on which the halonium ion mechanism was patterned.

Hydroxylation with permanganate gives *syn*-addition (Problem 9.1, p. 352). To account for this stereochemistry it has been suggested that an intermediate like I is involved:

$$
\underset{\text{I}}{
\begin{array}{c}
-\overset{|}{\text{C}}-\overset{|}{\text{C}}- \\
\text{O} \qquad \text{O} \\
\diagdown \; \diagup \\
\text{Mn} \\
\diagup \quad \diagdown \\
\text{O} \qquad \text{O} \\
\ominus
\end{array}
}
\quad \xrightarrow{\text{H}_2\text{O}} \quad
\begin{array}{c}
-\overset{|}{\text{C}}-\overset{|}{\text{C}}- \\
\text{HO} \quad \text{OH}
\end{array}
$$

Hydrolysis of such an intermediate would yield the *cis* diol. This mechanism is supported by the fact that osmium tetroxide, OsO$_4$, which also yields the *cis* diol, actually forms stable intermediates of structure II.

$$
\underset{\text{II}}{
\begin{array}{c}
-\overset{|}{\text{C}}-\overset{|}{\text{C}}- \\
\text{O} \qquad \text{O} \\
\diagdown \; \diagup \\
\text{Os} \\
\diagup \quad \diagdown \\
\text{O} \qquad \text{O}
\end{array}
}
$$

Thus, the two methods of hydroxylation—by peroxy acids and by permanganate—differ in stereochemistry because they differ in mechanism.

> **Problem 19.20** Using both models and drawings of the kind in Sec. 9.6, show all steps in the formation and hydrolysis of the epoxide of: (a) cyclopentene; (b) *cis*-2-butene; (c) *trans*-2-butene; (d) *cis*-2-pentene; (e) *trans*-2-pentene. (f) Which (if any) of the above products, as obtained, would be optically active?

19.14 Base-catalyzed cleavage of epoxides

Unlike ordinary ethers, epoxides can be cleaved under alkaline conditions. Here it is the epoxide itself, not the protonated epoxide, that undergoes nucleophilic attack:

$$
\text{Z:} \; + \;
\begin{array}{c}
-\overset{|}{\text{C}}-\overset{|}{\text{C}}- \\
\diagdown \; \diagup \\
\text{O}
\end{array}
\; \longrightarrow \;
\begin{array}{c}
\overset{\text{Z}}{\overset{|}{}} \\
-\overset{|}{\text{C}}-\overset{|}{\text{C}}- \\
\text{O}_-
\end{array}
\; \xrightarrow{\text{HZ}} \;
\begin{array}{c}
\overset{\text{Z}}{\overset{|}{}} \\
-\overset{|}{\text{C}}-\overset{|}{\text{C}}- \\
\text{OH}
\end{array}
\; + \; \text{:Z}
$$

The lower reactivity of the non-protonated epoxide is compensated for by the more basic, more strongly nucleophilic reagents that are compatible with the alkaline solution: alkoxides, phenoxides, ammonia, etc.

Let us look, for example, at the reaction of ethylene oxide with phenol. Acid catalyzes reaction by converting the epoxide into the highly reactive protonated epoxide. Base catalyzes reaction by converting the phenol into the more strongly nucleophilic phenoxide ion.

$$\text{C}_6\text{H}_5\text{OH} + \overset{\oplus}{\underset{\underset{H}{O}}{CH_2\text{—}CH_2}} \longrightarrow \text{C}_6\text{H}_5\overset{\underset{\oplus}{H}}{O}\text{—}CH_2CH_2OH \xrightarrow{-H^+}$$

Weakly
nucleophilic
reagent

Protonated
epoxide

Highly reactive

$$\longrightarrow \text{C}_6\text{H}_5\text{O—}CH_2CH_2OH$$

$$\text{C}_6\text{H}_5\text{O}^- + \underset{\underset{O}{}}{CH_2\text{—}CH_2} \longrightarrow \text{C}_6\text{H}_5\text{O—}CH_2CH_2O^- \xrightarrow{H^+}$$

Strongly
nucleophilic
reagent

Non-protonated
epoxide

Like alkyl halides and sulfonates, and like carbonyl compounds, epoxides are an important source of *electrophilic* carbon: of carbon that is highly susceptible to attack by a wide variety of nucleophiles. (As we shall see in Sec. 34.20, epoxides generated from carcinogenic hydrocarbons are even attacked by the nucleophilic portion of the genetic material DNA and thereby induce mutation and tumors.)

Problem 19.21   Write equations for the reaction of ethylene oxide with: (a) methanol in the presence of a little H_2SO_4; (b) methanol in the presence of a little $CH_3O^-\,Na^+$; (c) aniline.

Problem 19.22   Using the reaction between phenol and ethylene oxide as an example, show why it is not feasible to bring about reaction between the protonated epoxide and the highly nucleophilic reagent phenoxide ion. (*Hint*: Consider what would happen if one started with a solution of sodium phenoxide and ethylene oxide and added acid to it.)

Problem 19.23   Poly(oxypropylene)glycols,

$$\underset{HO\text{—}CH\text{—}CH_2\text{—}O}{\overset{CH_3}{|}}\left[\underset{CH_2\text{CH}\text{—}O}{\overset{CH_3}{|}}\right]_n\underset{\text{—}CH_2CHOH}{\overset{CH_3}{|}}$$

which are used in the manufacture of polyurethane foam rubber, are formed by the action of base (e.g. hydroxide ion) on propylene oxide in the presence of 1,2-propanediol as an initiator. Write all steps in a likely mechanism for their formation.

19.15   Reaction of ethylene oxide with Grignard reagents

Reaction of Grignard reagents with ethylene oxide is an important method of preparing primary alcohols since the product contains two carbons more than the alkyl or aryl group of the Grignard reagent. As in reaction with the carbonyl group (Sec. 17.14), we see the nucleophilic alkyl or aryl group of the Grignard reagent attach itself to the electrophilic carbon of the epoxide, with the formation of a carbon–carbon bond. Use of higher epoxides is complicated by rearrangements and formation of mixtures.

$$R\text{—}MgX + \underset{\underset{O}{}}{CH_2\text{—}CH_2} \longrightarrow R\text{—}CH_2CH_2O^-MgX^+ \xrightarrow{H^+} R\text{—}CH_2CH_2OH$$

19.16 Orientation of cleavage of epoxides

There are two carbon atoms in an epoxide ring and, in principle, either one can suffer nucleophilic attack. In a symmetrical epoxide like ethylene oxide, the two carbons are equivalent, and attack occurs randomly at either. But in an unsymmetrical epoxide, the carbons are *not* equivalent, and the product we obtain depends upon which one is preferentially attacked. Just what is the orientation of cleavage of epoxides, and how does one account for it?

The preferred point of attack, it turns out, depends chiefly on whether the reaction is acid-catalyzed or base-catalyzed. Consider, for example, two reactions of isobutylene oxide:

$$\underset{\overset{\displaystyle |}{\underset{\displaystyle \text{O}}{\text{CH}_3-\text{C}-\text{CH}_2}}}{\overset{\displaystyle \text{CH}_3}{}} + \text{H}_2{}^{18}\text{O} \xrightarrow{\text{H}^+} \underset{\overset{\displaystyle |}{{}^{18}\text{OH}}}{\overset{\displaystyle \text{CH}_3}{\text{CH}_3-\text{C}-\text{CH}_2\text{OH}}}$$

$$\underset{\overset{\displaystyle |}{\underset{\displaystyle \text{O}}{\text{CH}_3-\text{C}-\text{CH}_2}}}{\overset{\displaystyle \text{CH}_3}{}} + \text{CH}_3\text{OH} \xrightarrow{\text{CH}_3\text{ONa}} \underset{\overset{\displaystyle |}{\text{OH}}}{\overset{\displaystyle \text{CH}_3}{\text{CH}_3-\text{C}-\text{CH}_2-\text{OCH}_3}}$$

Here, as in general, the nucleophile attacks the more substituted carbon in acid-catalyzed cleavage, and the less substituted carbon in base-catalyzed cleavage.

Our first thought is that two different mechanisms are involved here, S_N1 and S_N2. But the evidence indicates pretty clearly that both are of the S_N2 type: cleavage of the carbon–oxygen bond and attack by the nucleophile occur in a single step. (There is not only stereochemical evidence—complete inversion—but also evidence of several kinds that we cannot go into here.) How, then, are we to account for the difference in orientation—in particular, for S_N2 attack at the *more hindered* position in acid-catalyzed cleavage?

We encountered this same kind of orientation in the formation of halohydrins (Sec. 8.15), and the explanation we gave there applies in the present case, too. In the transition state of most S_N2 reactions, bond-breaking and bond-making have proceeded to about the same extent, and carbon has not become appreciably positive or negative; as a result steric factors, not electronic factors, chiefly determine reactivity. But in acid-catalyzed cleavage of an epoxide, the carbon–oxygen bond, already weak because of the angle strain of the three-membered ring, is further weakened by protonation: the leaving group is a very good one, the weakly basic alcohol hydroxyl. The nucleophile, on the other hand, is a poor one (water, alcohol). In the transition state bond-breaking has proceeded further than bond-making, and carbon has acquired a considerable positive charge.

Acid-catalyzed S_N2 cleavage

$$\text{Z}: + \underset{\overset{\displaystyle |}{\underset{\displaystyle \text{H}}{\underset{\displaystyle \text{O} \oplus}{}}}}{-\text{C}-\text{C}-} \longrightarrow \left[\underset{\overset{\displaystyle |}{\underset{\displaystyle \text{H}}{\text{O}}}}{\overset{\displaystyle \text{Z}}{\overset{\delta+}{-}\text{C}-\text{C}-}}\ {}^{\delta+} \right] \longrightarrow \underset{\overset{\displaystyle |}{\text{OH}}}{\overset{\displaystyle \text{Z}}{-\text{C}-\text{C}-}}$$

Bond-breaking exceeds
bond-making:
positive charge on carbon

Since both leaving group and nucleophile are far away, crowding is relatively unimportant. As in halohydrin formation, the stability of the transition state is determined chiefly by electronic factors and not steric factors, and the reaction has considerable S_N1 character. *Attack occurs at the carbon that can best accommodate the positive charge.*

In base-catalyzed cleavage, the leaving group is a poorer one—a strongly basic alkoxide oxygen—and the nucleophile is a good one (hydroxide, alkoxide). Bond-breaking and bond-making are more nearly balanced, and reactivity is controlled in the more usual way, by steric factors. *Attack occurs at the less hindered carbon.*

Base-catalyzed S_N2 cleavage

Bond-making balances bond-breaking:

no particular charge on carbon

Problem 19.24   Predict the chief product of each of the following reactions:

(a) styrene oxide + dry HCl
(b) styrene oxide + CH_3OH + a little CH_3ONa
(c) propylene oxide + ammonia
(d) trimethylethylene oxide + HCl

One further point. We have encountered the two-step addition of unsymmetrical reagents in which the first step is attack by positive halogen: formation of halohydrins (Sec. 8.15) and heterolytic addition of IN_3 and BrN_3 (Problem 5, p. 362). There, we saw, the orientation is like that for acid-catalyzed (*not* base-catalyzed) cleavage of epoxides: the orientation is what would be expected if a carbocation were the intermediate. Propylene chlorohydrin, for example, is $CH_3CHOHCH_2Cl$; the reagent IN_3 adds to terminal alkenes to yield $RCH(N_3)CH_2I$. Yet the exclusively *anti* stereochemistry (Problem 3, p. 362, and Problem 5, p. 362) indicates that the intermediate in these reactions is not an open cation but a *halonium ion*; cleavage of this ring must involve attack by the nucleophile (H_2O or N_3^-) at the more hindered carbon. This is not really surprising, in view of what we have just said about epoxides. The halonium ring is even less stable than that of a protonated epoxide, and bond-breaking should be even easier; cleavage has much S_N1 character, and takes place at the carbon that can best accommodate the positive charge. (Consider, too, the orientation of solvomercuration, in which the intermediate is a cyclic *mercurinium ion*.)

19.17   Analysis of ethers

Because of the low reactivity of the functional group, the chemical behavior of ethers—both aliphatic and aromatic—resembles that of the hydrocarbons to which they are related. They are distinguished from hydrocarbons, however, by their solubility in cold concentrated sulfuric acid through formation of oxonium salts.

Problem 19.25 Because of their highly reactive benzene rings, aryl ethers may decolorize bromine in carbon tetrachloride. How could this behavior be distinguished from the usual unsaturation test? (*Hint*: See Sec. 8.24.)

Problem 19.26 Expand the table you made in Problem 18.5, p. 682, to include ethers.

Problem 19.27 Describe simple chemical tests (if any) that would distinguish between an aliphatic ether and (a) an alkane; (b) an alkene; (c) an alkyne; (d) an alkyl halide; (e) a primary or secondary alcohol; (f) a tertiary alcohol; (g) an alkyl aryl ether.

Identification as a previously reported ether is accomplished through the usual comparison of physical properties. This can be confirmed by cleavage with hot concentrated hydriodic acid (Sec. 19.7) and identification of one or both products. Aromatic ethers can be converted into solid bromination or nitration products whose melting points can then be compared with those of previously reported derivatives. Proof of structure of a new ether would involve cleavage by hydriodic acid and identification of the products formed.

Figure 19.3   Infrared spectra of (a) di-*n*-propyl ether and (b) ethyl phenyl ether.

19.18   Spectroscopic analysis of ethers

Infrared   The infrared spectrum of an ether does not, of course, show the O—H band characteristic of alcohols; but the strong band due to C—O stretching is still present, in the 1060–1300 cm^{-1} range, and is the striking feature of the spectrum. (See Fig. 19.3.)

C—O stretching, *strong, broad*

Alkyl ethers   1060–1150 cm^{-1}

Aryl and vinyl ethers   1200–1275 cm^{-1} (and, weaker, at 1020–1075 cm^{-1})

Carboxylic acids and esters show C—O stretching, but show carbonyl absorption as well. (For a comparison of certain oxygen compounds, see Table 24.3, p. 890.)

NMR and CMR   The oxygen of ethers, like that of alcohols, deshields both protons and carbons, and shifts absorption downfield.

PROBLEMS

1. Write structural formulas for:

(a)   dimethyl ether
(b)   diisopropyl ether
(c)   *n*-butyl methyl ether
(d)   *tert*-butyl isobutyl ether
(e)   3-methoxyhexane
(f)   divinyl ether
(g)   diallyl ether

(h)   di-β-chloroethyl ether
(i)   anisole
(j)   phenetole
(k)   diphenyl ether
(l)   cyclohexene oxide
(m)   *p*-nitrobenzyl *n*-propyl ether
(n)   1,2-epoxypentane

2. Name the following structures:

(a) $(CH_3)_2CHCH_2$—O—$CH_2CH(CH_3)_2$
(b) CH_3—O—$CH(CH_3)_2$
(c) $(CH_3)_3C$—O—CH_2CH_3
(d) $CH_3CH_2CH_2CH(OCH_3)CH_2CH_2CH_3$

(e) p-$BrC_6H_4OC_2H_5$
(f) o-$O_2NC_6H_4CH_2OC_6H_5$
(g)   2,4-$Br_2C_6H_3OCH_3$

3. Outline a possible laboratory synthesis of each of the following compounds from alcohols and phenols:

(a)   *tert*-butyl methyl ether
(b)   phenetole ($C_6H_5OC_2H_5$)
(c)   *n*-butyl cyclohexyl ether

(d)   benzyl *p*-tolyl ether
(e)   isobutyl isopropyl ether
(f)   *tert*-butyl isopropyl ether

(g)   resorcinol dimethyl ether (1,3-dimethoxybenzene)

4. Arrange the compounds of each set in order of reactivity toward bromine:

(a)   anisole, benzene, chlorobenzene, nitrobenzene, phenol
(b)   anisole, *m*-hydroxyanisole, *o*-methylanisole, *m*-methylanisole
(c) p-$C_6H_4(OH)_2$, p-$CH_3OC_6H_4OH$, p-$C_6H_4(OCH_3)_2$

5. Write a balanced equation for each of the following. (If no reaction occurs, indicate "no reaction".)

(a)   potassium *tert*-butoxide + ethyl iodide
(b)   *tert*-butyl iodide + potassium ethoxide
(c)   ethyl alcohol + H_2SO_4 (140 °C)
(d)   di-*n*-butyl ether + boiling aqueous NaOH

(e) ethyl methyl ether + excess HI (hot)
(f) dimethyl ether + Na
(g) diethyl ether + cold conc. H_2SO_4
(h) diethyl ether + hot conc. H_2SO_4
(i) $C_6H_5OC_2H_5$ + hot conc. HBr
(j) $C_6H_5OC_2H_5$ + HNO_3, H_2SO_4
(k) $p\text{-}CH_3C_6H_4OCH_3$ + $KMnO_4$ + KOH + heat
(l) $C_6H_5OCH_2C_6H_5$ + Br_2, Fe

6. Like other oxygen-containing compounds, *n*-butyl *tert*-butyl ether dissolves in cold concentrated H_2SO_4. On standing, however, an acid-insoluble layer, made up of high-boiling hydrocarbon material, slowly separates from the solution. What is this material likely to be, and how is it formed?

7. The propylene oxide used to make the polymers described in Problem 19.23 (p. 719) can be manufactured from propylene by the action of *tert*-butyl hydroperoxide, *t*-BuO—O—OH.

(a) As is often the case with industrial processes, this one is economical only if there is a marketable co-product. What co-product will be formed here?

(b) The *tert*-butyl hydroperoxide is made by a high-temperature chain reaction of isobutane with O_2. Suggest a likely series of steps for this reaction.

(c) The air oxidation of alkene to epoxide that is used to manufacture ethylene oxide (Sec. 19.11) cannot be used to make propylene oxide. Tell why this is probably so. (*Hint*: Look at your answer to part (b).)

8. Describe simple chemical tests that would distinguish between:

(a) di-*n*-butyl ether and *n*-pentyl alcohol
(b) diethyl ether and methyl iodide
(c) methyl *n*-propyl ether and 1-pentene
(d) diisopropyl ether and diallyl ether
(e) anisole and toluene
(f) divinyl ether and diethyl ether
(g) *n*-butyl *tert*-butyl ether and *n*-octane

Tell exactly what you would *do* and *see*.

9. An unknown compound is believed to be one of the following. Describe how you would go about finding out which of the possibilities the unknown actually is. Where possible, use simple chemical tests; where necessary, use more elaborate chemical methods like quantitative hydrogenation, cleavage, etc. Make use of any needed tables of physical constants.

(a) di-*n*-propyl ether (b.p. 91 °C) and 2-methylhexane (b.p. 91 °C)
(b) benzyl ethyl ether (b.p. 188 °C) and allyl phenyl ether (b.p. 192 °C)
(c) methyl *p*-tolyl ether (b.p. 176 °C) and methyl *m*-tolyl ether (b.p. 177 °C)
(d) ethyl *n*-propyl ether (b.p. 64 °C), 1-hexene (b.p. 64 °C), and methanol (b.p. 65 °C)
(e) anisole (b.p. 154 °C), bromobenzene (b.p. 156 °C), *o*-chlorotoluene (b.p. 159 °C), *n*-propylbenzene (b.p. 159 °C), and cyclohexanol (b.p. 162 °C)
(f) diethyl ether (b.p. 35 °C), *n*-pentane (b.p. 36 °C), and isoprene (b.p. 34 °C)
(g) methyl *o*-tolyl ether (b.p. 171 °C), phenetole (b.p. 172 °C), and diisopentyl ether (b.p. 173 °C)

10. Give the structures and names of the products you would expect from the reaction of ethylene oxide with:

(a) H_2O, H^+
(b) H_2O, OH^-
(c) C_2H_5OH, H^+
(d) product of (c), H^+
(e) $HOCH_2CH_2OH$, H^+
(f) product of (e), H^+
(g) anhydrous HBr
(h) HCN

(i) HCOOH
(j) C_6H_5MgBr
(k) NH_3
(l) diethylamine ($C_2H_5NHC_2H_5$)
(m) phenol, H^+
(n) phenol, OH^-
(o) $HC{\equiv}C^-Na^+$

11. Propylene oxide can be converted into 1,2-propanediol by the action of either dilute acid or dilute base. When optically active propylene oxide is used, the 1,2-diol obtained from acidic hydrolysis has a rotation opposite to that obtained from alkaline hydrolysis. What is the most likely interpretation of these facts?

12. (a) Draw formulas for all the stereoisomers of I.

I

(b) Indicate which isomers, when separated from all others, will be optically active, and which will be optically inactive. (c) One of these stereoisomers is very readily converted into an ether, $C_{10}H_{18}O$. Which isomer is this, and what is the structure of the ether?

13. Give the structures (including configurations where pertinent) of compounds A–M:

(a) $CH_2=CH_2 + Cl_2/H_2O \longrightarrow$ A (C_2H_5OCl)
A $+ H_2SO_4 +$ heat $\longrightarrow$ B $(C_4H_8OCl_2)$
B $+$ alc. KOH $\longrightarrow$ C (C_4H_6O)
(b) $ClCH_2CH-CH_2 + CH_3OH + H_2SO_4 \longrightarrow$ D $(C_4H_9O_2Cl)$
$\underset{O}{\diagdown\diagup}$

D $+$ NaOCl $\longrightarrow$ $CHCl_3 +$ E $(C_3H_6O_3)$
D $+$ NaOH(aq) $\longrightarrow$ F $(C_4H_8O_2)$
(c) $ClCH_2CH_2CH_2OH + KOH \longrightarrow$ G (C_3H_6O)
(d) $CH_2=CHCH_2CH_2CH_2OH + Hg(OAc)_2 + H_2O$, then $NaBH_4 \longrightarrow$ H $(C_5H_{10}O)$
(e) cyclohexane oxide $+$ anhydrous HCl $\longrightarrow$ I $(C_6H_{11}OCl)$
(f) 1-methylcyclohexene $+ HCO_2OH \longrightarrow$ J $(C_7H_{14}O_2)$
(g) racemic 3,4-epoxy-1-butene $+$ cold alkaline $KMnO_4$, then
dilute acid $\longrightarrow$ K $(C_4H_{10}O_4)$
(h) *cis*-2-butene $+ Cl_2/H_2O$, then OH^-, then dilute acid $\longrightarrow$ L $(C_4H_{10}O_2)$
(i) *trans*-2-butene treated as in (h) $\longrightarrow$ M $(C_4H_{10}O_2)$

14. An insect pheromone that we have already encountered has been made in the following way. (*Useful information*: An alcohol, ROH, is often converted into its acetate, CH_3COOR, by treatment with acetyl chloride, CH_3COCl.)

1,8-octanediol $+$ HBr $\longrightarrow$ N $(C_8H_{17}OBr)$
N $+$ DHP $+$ H$^+$ $\longrightarrow$ O $(C_{13}H_{25}O_2Br)$
O $+$ HC≡CLi $\longrightarrow$ P $(C_{15}H_{26}O_2)$
P $+$ LiNH$_2$; then C_2H_5Br $\longrightarrow$ Q $(C_{17}H_{30}O_2)$
Q $+$ H$_2$O, H$^+$ $\longrightarrow$ R $(C_{12}H_{22}O)$
R $+$ CH$_3$COCl $\longrightarrow$ S $(C_{14}H_{24}O_2)$
S $+$ H$_2$, Lindlar catalyst $\longrightarrow$ the pheromone $(C_{14}H_{26}O_2)$

(a) Give the structure of the pheromone and all intermediate compounds.
(b) For maximum biological activity there should be present 4% of its geometric isomer. How could you modify the above synthesis to obtain this isomer?

15. The sex attractant of the pink bollworm moth is an approximately 50:50 mixture of two geometric isomers, and is known as *gossyplure*. One component has been synthesized in the following way. (*Useful information*: An alcohol, ROH, is often converted into its acetate, CH_3COOR, by treatment with acetyl chloride, CH_3COCl.)

1-hexyne + *n*-BuLi; then ethylene oxide; then H$^+$ $\longrightarrow$ T (C$_8$H$_{14}$O)
T + H$_2$, Lindlar catalyst $\longrightarrow$ U (C$_8$H$_{16}$O)
U + PBr$_3$ $\longrightarrow$ V (C$_8$H$_{15}$Br)
6-bromo-1-hexanol + DHP + H$^+$ $\longrightarrow$ W (C$_{11}$H$_{21}$O$_2$Br)
W + HC$\equiv$CLi $\longrightarrow$ X (C$_{13}$H$_{22}$O$_2$)
X + *n*-BuLi; then V $\longrightarrow$ Y (C$_{21}$H$_{36}$O$_2$)
Y + H$_2$, Lindlar catalyst $\longrightarrow$ Z (C$_{21}$H$_{38}$O$_2$)
Z + H$_2$O, H$^+$ $\longrightarrow$ AA (C$_{16}$H$_{30}$O)
AA + CH$_3$COCl $\longrightarrow$ the pheromone (C$_{18}$H$_{32}$O$_2$)

(a) What is the structure of the pheromone just formed?
(b) This synthesis has been modified to obtain each of the geometric isomers of the compound in (a), one of which is the other component of the natural pheromone. Show how this could be done.

16. Give a structure or structures for the compound whose infrared spectrum is shown in Fig. 19.4, p. 727. If you find more than one structure consistent with the spectrum, could you decide among the possibilities on the basis of the proton NMR spectrum? Of the CMR spectrum? Tell what you would expect to see in each case.

17. Give a structure or structures for compound BB, whose infrared and proton NMR spectra are shown in Fig. 19.5, p. 727.

18. Give a structure or structures consistent with each of the proton NMR spectra shown in Fig. 19.6, p. 728.

19. Give a structure or structures consistent with each of the CMR spectra shown in Fig. 19.7, p. 729.

20. Give the structures of compounds CC, DD, and EE on the basis of their infrared spectra (Fig. 19.8, p. 730) and their proton NMR spectra (Fig. 19.9, p. 731).

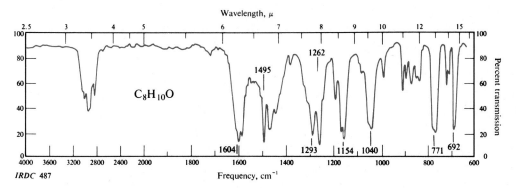

Figure 19.4   Infrared spectrum for Problem 16, p. 726.

Figure 19.5   Infrared and proton NMR spectra for Problem 17, p. 726.

(a) $C_6H_{14}O$

(b) $C_6H_{14}O$

(c) $C_6H_{14}O$

Figure 19.6 Proton NMR spectra for Problem 18, p. 726.

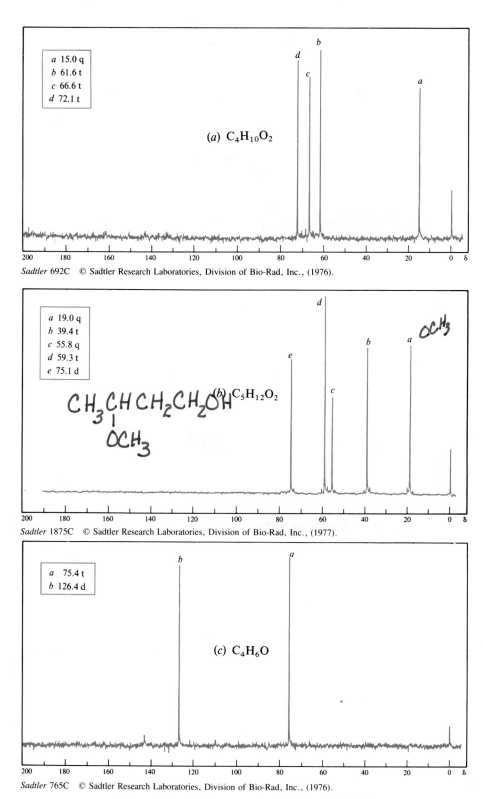

a 15.0 q
b 61.6 t
c 66.6 t
d 72.1 t

(a) $C_4H_{10}O_2$

200 180 160 140 120 100 80 60 40 20 0 δ

Sadtler 692C   © Sadtler Research Laboratories, Division of Bio-Rad, Inc., (1976).

a 19.0 q
b 39.4 t
c 55.8 q
d 59.3 t
e 75.1 d

CH₃CHCH₂CH₂OH   (b) $C_5H_{12}O_2$
        |
       OCH₃

OCH₃

200 180 160 140 120 100 80 60 40 20 0 δ

Sadtler 1875C   © Sadtler Research Laboratories, Division of Bio-Rad, Inc., (1977).

a 75.4 t
b 126.4 d

(c) C_4H_6O

200 180 160 140 120 100 80 60 40 20 0 δ

Sadtler 765C   © Sadtler Research Laboratories, Division of Bio-Rad, Inc., (1976).

Figure 19.7   CMR spectra for Problem 19, p. 726.

CC $C_9H_{12}O$

Sadtler 7188

$H_3C \bigcirc OCH_2CH_3$ *p-ethoxytoluene*

DD $C_9H_{12}O$

1456 C→H   1100   736 696

IRDC 4167

$\bigcirc -CH_2-OCH_2CH_3$ *mono substituted*

EE $C_9H_{12}O$

OH   1498   1455   1057 1028   743 697

IRDC 4325

Figure 19.8   Infrared spectra for Problem 20, p. 726.

$\bigcirc -CH_2CH_2CH_2OH$

Figure 19.9   Proton NMR spectra for Problem 20, p. 726.

20

Symphoria

Neighboring Group Effects. Catalysis by Transition Metal Complexes

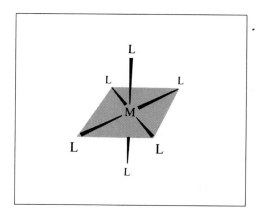

20.1 Symphoria

We have, so far, seen something of the effects on reactivity of polar factors, steric factors, and the solvent. But there is another structural feature to be considered: the spatial relationship among reacting atoms and molecules. *Being in the right place*, we shall find, can be the most powerful factor of all in determining how fast a reaction goes—and what product it yields.

In this chapter we shall take up certain reactions from quite different areas: nucleophilic substitution in (seemingly) ordinary substrates; catalysis by transition metal complexes; the action of enzymes in living cells. Although, on the surface, these reactions appear quite dissimilar, they all share one quality. Judged by what we have studied so far, they all seem highly *unusual*: they take place at unexpectedly high rates and with unexpected stereochemistry.

Now, how are we to account for this unusual behavior? The underlying feature in all these reactions, it turns out, is this: prior to reaction, the reactants are *brought together* and *held* in exactly the right positions for reaction to occur. This bringing together can take place in various ways. The substrate and the reagent may be held by secondary bonding to an enzyme molecule; they may be held in the coordination sphere of a transition metal. They may even be two functional groups in a single molecule; in that case they are brought into the proper spatial relationship by simple rotation about a carbon–carbon bond.

Now, once they have been brought together, the substrate and the reagent are—if only temporarily—*parts of the same molecule*. When they react they have a tremendous advantage over ordinary, separated reactants. They do not have to wait until their paths happen to cross. They do not have to give up precious freedom of motion (translational entropy) when they are locked into a transition state. Between the reactants there are no tightly clinging solvent molecules to be stripped away as reaction occurs. The result is reaction with an enormously enhanced rate, reaction with a special stereochemistry.

The factor that makes all this possible we shall call **symphoria**: *the bringing together of reactants into the proper spatial relationship.*

We have taken the word symphoria from the Greek *symphoro*, to bring together usefully (*sym*, together + *phero*, to bring).

In the following sections we shall look closely at several reactions in which symphoria is at work. In each case, we shall find, reaction occurs the way it does because the reacting atoms are *near* each other and *in exactly the right positions*. Clearly, symphoria is three-dimensional chemistry, but of a kind that goes far beyond what is generally thought of as stereochemistry. Let us begin by examining a set of reactions in which we can most readily *see* and *measure* symphoric effects: that is, reactions where we can see evidence that such a thing as symphoria actually exists.

We shall look first at an example of a familiar reaction—acid-catalyzed substitution in an alcohol—a very special example, which changed the course of organic chemistry.

20.2   Neighboring group effects: the discovery. Stereochemistry

When treated with concentrated hydrobromic acid, the bromohydrin-3-bromo-2-butanol is converted into 2,3-dibromobutane. This, we say, involves nothing out of the ordinary; it is simply nucleophilic attack (S_N1 or S_N2) by bromide ion on the protonated alcohol (Sec. 5.25). But in 1939 Saul Winstein (p. 244) and Howard J. Lucas (California Institute of Technology) described the stereochemistry of this reaction and, in doing this, opened the door to an entirely new concept in organic chemistry: the *neighboring group effect*.

$$CH_3-\underset{\underset{\textstyle Br}{|}}{CH}-\underset{\underset{\textstyle OH}{|}}{CH}-CH_3 \xrightarrow{\text{HBr}} CH_3-\underset{\underset{\textstyle Br}{|}}{CH}-\underset{\underset{\textstyle Br}{|}}{CH}-CH_3$$

3-Bromo-2-butanol   2,3-Dibromobutane

First, Winstein and Lucas found (Fig. 20.1) that (racemic) *erythro* bromohydrin yields only the *meso* dibromide, and (racemic) *threo* bromohydrin yields only the *racemic* dibromide. Apparently, then, reaction proceeds with complete retention of configuration—unusual for nucleophilic substitution. But something even more unusual was still to come.

(Figure 20.1 structures: Erythro Racemic → HBr → Meso; Threo Racemic → HBr → Racemic)

Figure 20.1 Conversion of racemic 3-bromo-2-butanols into 2,3-dibromobutanes.

They carried out the same reaction again but this time used *optically active* starting materials (Fig. 20.2). From optically active *erythro* bromohydrin they obtained, of course, optically inactive product: the *meso* dibromide. But *optically active threo bromohydrin also yielded optically inactive product: the racemic dibromide.*

(Figure 20.2 structures: Erythro Optically active → HBr → Meso; Threo Optically active → HBr → I, II Racemic)

Figure 20.2 Conversion of optically active 3-bromo-2-butanols into 2,3-dibromobutanes.

In one of the products (I) from the *threo* bromohydrin, there is retention of configuration. But in the other product (II), there is inversion, not only at the carbon that held the hydroxyl group, but also at the carbon that held bromine—a carbon that, on the surface, is not even involved in the reaction. How is one to account for the fact that exactly half the molecules react with complete retention, and the other half with this strange double inversion?

Winstein and Lucas gave the following interpretation of these facts. In step (1) the protonated bromohydrin loses water to yield, not the open carbocation, but a bridged bromonium ion. In step (2) bromide ion attacks this bromonium ion to

(1)

$$-\overset{|}{\underset{|}{C}}-\overset{Br}{\underset{OH_2{}^+}{\underset{|}{C}}}- \longrightarrow -\overset{Br^\oplus}{\underset{|}{\overset{|}{C}}}-\overset{}{\underset{|}{C}}- \;+\; H_2O$$

A bromonium ion

(2)

$$-\overset{Br^\oplus}{\underset{|}{\overset{|}{C}}}-\overset{}{\underset{|}{C}}- \;+\; Br^- \longrightarrow -\overset{Br}{\underset{Br}{\overset{|}{C}}}-\overset{}{\underset{|}{C}}- \;+\; -\overset{Br}{\underset{Br}{\overset{|}{C}}}-\overset{}{\underset{|}{C}}-$$

give the dibromide. But it can attack the bromonium ion *at either of two carbon atoms*: attack at one gives the product with retention at both chiral centers; attack at the other gives the product with inversion about both centers. Figure 20.3 depicts the reaction of the optically active *threo* bromohydrin.

Protonated
threo-3-bromo-2-butanol
Optically active

Bromonium ion

I

II

I *and* II *are enantiomers*

Racemic 2,3-dibromobutane

Figure 20.3 Conversion of optically active *threo*-3-bromo-2-butanol into racemic 2,3-dibromobutane via a cyclic bromonium ion. Opposite-side attacks *a* and *b* are equally likely, and give enantiomers in equal amounts.

The bromonium ion has the same structure as that proposed two years earlier by Roberts and Kimball (Sec. 9.6) as an intermediate in the addition of bromine to alkenes. Here it is formed in a different way, but its reaction is the same, and so is the final product.

Reaction consists of two successive nucleophilic substitutions. In the first one the nucleophile is the neighboring bromine; in the second, it is bromide ion from outside the molecule. Both substitutions are pictured as being S_N2-like; that is, single-step processes with attachment of the nucleophile and loss of the leaving group taking place in the same transition state. This is consistent with the complete stereoselectivity: an open carbocation in either (1) or (2) might be expected to result in the formation of a mixture of diastereomers.

(As we shall see, a neighboring bromine can affect more than the stereochemistry of such a reaction.)

Problem 20.1   Drawing structures like those in Fig. 20.3, show the stereochemical course of reaction of optically active *erythro*-3-bromo-2-butanol with hydrogen bromide.

Problem 20.2   Actually, the door opened by Winstein and Lucas was already ajar. In 1937, E. D. Hughes, Ingold (p. 183), and their co-workers reported that, in contrast to the neutral acid or its ester, sodium α-bromopropionate undergoes hydrolysis with *retention* of configuration.

$$CH_3CHBrCOO^- Na^+ \xrightarrow[\text{H}_2\text{O}]{\text{OH}^-} CH_3CHOHCOO^- Na^+$$

Sodium α-bromopropionate   Sodium lactate

Give a likely interpretation of these findings.

20.3   Neighboring group effects: intramolecular nucleophilic attack

Let us see just what is involved in neighboring group effects. The basic process, it turns out, is closely related to a process we have already spent some time with: rearrangement of carbocations. And so, let us begin by taking another look at rearrangement.

Carbocations, we know, can rearrange through migration of an organic group or a hydrogen atom, with its pair of electrons, to the electron-deficient carbon.

Indeed, when carbocations were first postulated as reactive intermediates (Sec. 5.17), it was to account for rearrangements of a particular kind. Such rearrangements still provide the best single clue that we are dealing with a carbocation reaction.

The driving force behind all carbocation reactions is the need to provide electrons to the electron-deficient carbon. When an electron-deficient carbon is generated, a nearby group may help to relieve this deficiency. It may, of course, remain in place and release electrons through space or through the molecular framework, inductively or by resonance. Or—and this is what we are concerned with here—it may actually *carry the electrons* to where they are needed.

An electron-deficient carbon is most commonly generated by the departure of a leaving group which takes the bonding electrons with it. The migrating group is, of course, a nucleophile, and so a rearrangement of this sort amounts to *intramolecular nucleophilic substitution*. Now, as we have seen, nucleophilic substitution can be of two kinds, S_N2 and S_N1. Exactly the same possibilities exist for a rearrangement. As we have described rearrangement so far, it is S_N1-like, with the migrating group waiting for the departure of the leaving group before it

S_N1-like migration

S_N2-like migration

G = migrating group
S = migration source
T = migration terminus

moves. But it *could* be S_N2-like, with the neighboring group helping to push out the leaving group in a single-step reaction. This matter of *timing* of bond-breaking and bond-making is—as it is with all reactions—of major concern in the study of rearrangements.

When the migrating group helps to expel the leaving group, it is said to give *anchimeric assistance* (Greek: *anchi + meros*, adjacent parts).

Now, in a rearrangement, a nearby group carries electrons to an electron-deficient atom, and then *stays there*. But sometimes, it happens, a group brings electrons and then *goes back to where it came from*. This gives rise to what are called **neighboring group effects**: intramolecular effects exerted on a reaction through direct participation—that is, through movement to within bonding distance—by a group near the reaction center.

Neighboring group effects involve the same basic process as rearrangement. Indeed, in many cases there *is* rearrangement, but it is *hidden*. What we see on the surface may be this:

But what is actually happening may be this:

The neighboring group, acting as an internal nucleophile, attacks carbon at the reaction center; the leaving group is lost, and there is formed a *bridged intermediate* (I), usually a cation. This undergoes attack by an external nucleophile to yield the product. The overall stereochemistry is determined by the way in which the bridged ion is formed and the way in which it reacts, and typically differs from the stereochemistry observed for simple attack by an external nucleophile.

In the rearrangement of 3-bromo-2-butanol we saw a typical example of this "abnormal" stereochemistry. The basic process there was the same as in a rearrangement: intramolecular (1,2) nucleophilic attack. Indeed, rearrangement *did* occur there: in half the molecules formed, the bromine migrated from one carbon to the next.

But something besides stereochemistry can be involved here. If a neighboring group helps to push out the leaving group—that is, gives anchimeric assistance—it may accelerate the reaction, sometimes tremendously. Thus, neighboring group participation is most often revealed by a *special kind of stereochemistry* or by an *unusually fast rate of reaction*—and often by both.

If a neighboring group is to form a bridged cation, it must have electrons to form the extra bond. These may be *unshared pairs* on atoms like sulfur, nitrogen, oxygen, or bromine; π *electrons* of a double bond or aromatic ring; or even, in some cases, σ *electrons*.

In making its nucleophilic attack, a neighboring group competes with outside molecules that are often intrinsically much stronger nucleophiles. Yet the evidence clearly shows that the neighboring group enjoys—for its nucleophilic power—a very great advantage over these outside nucleophiles. Why is this? The answer is quite simple: *because it is there.*

The neighboring group is there, in the same molecule, poised in the proper position for attack. And with this comes those advantages we spoke of earlier: high "effective concentration", favorable entropy of activation, lack of interference by solvent molecules. The electronic reorganization—changes in overlap—that accompany reaction undoubtedly happen more easily in this cyclic system. Clearly, a neighboring group effect is a symphoric effect, and of the simplest kind. The substrate and the reagent are already part of the same molecule; all that is needed to bring them into the right spatial relationship is rotation about a carbon–carbon bond.

Problem 20.3 Draw the structure of the bridged intermediate (I, p. 738) expected if each of the following were to act as a neighboring group. To what class of compounds does each intermediate belong?

(a) —N(CH$_3$)$_2$

(b) —SCH$_3$

(c) —OH

(d) —O$^-$

(e) —Br

(f) —C$_6$H$_5$

(g) —C$_6$H$_4$OCH$_3$-*p*

(h) —C$_6$H$_4$O$^-$-*p*

(i) —CH=CHR

Now let us return to nucleophilic substitution and look at evidence that anchimeric assistance does indeed exist. To do this we shall turn, not to alcohols, but to other substrates.

20.4 Neighboring group effects: rate of reaction. Anchimeric assistance

Like other alkyl halides, mustard gas (2,2′-dichlorodiethyl sulfide) undergoes hydrolysis. But this hydrolysis is unusual in several ways: (a) the kinetics is first-order, with the rate independent of added base; and (b) although the substrate is primary, it is *enormously* faster than hydrolysis of ordinary primary alkyl chlorides.

$$\text{ClCH}_2\text{CH}_2\text{—S—CH}_2\text{CH}_2\text{Cl} \xrightarrow{\text{H}_2\text{O}} \text{ClCH}_2\text{CH}_2\text{—S—CH}_2\text{CH}_2\text{OH}$$

2,2′-Dichlorodiethyl sulfide   2-Hydroxy-2′-chlorodiethyl sulfide

We have encountered this kind of kinetics before in S_N1 reactions and know, in a general way, what it must mean: in the rate-determining step, the substrate is reacting unimolecularly to form an intermediate, which then reacts rapidly with solvent or other nucleophile. But what is this intermediate? It can hardly be the carbocation. A primary cation is highly unstable and hard to form, so that primary alkyl chlorides ordinarily react by S_N2 reactions instead; and here we have electron-withdrawing sulfur further to destabilize a carbocation.

This is another example of a neighboring group effect, one that shows itself not in stereochemistry but in *rate of reaction*. Sulfur helps to push out chloride ion, forming a cyclic *sulfonium ion* in the process. As fast as it is formed, this intermediate reacts with water to yield the product.

A sulfonium ion
$$k_2 \gg k_1$$

Reaction thus involves formation of a cation, but not a highly unstable carbocation with its electron-deficient carbon; instead, it is a cation in which *every atom has an octet of electrons*. Open-chain sulfonium ions, R_3S^+, are well-known, stable molecules; here, because of angle strain, the sulfonium ion is less stable and highly reactive—but still enormously more stable and easier to form than a carbocation.

The first, rate-determining step is unimolecular, but it is S_N2-like. As with other primary halides, a nucleophile is needed to help push out the leaving group. Here the nucleophile happens to be part of the same molecule. Sulfur has unshared electrons it is willing to share, and hence is highly nucleophilic. Most important, *it is there*: poised in just the right position for attack. The result is an enormous increase in rate.

There is much additional evidence to support the postulate that the effect of neighboring sulfur is due to anchimeric assistance. Cyclohexyl chloride undergoes solvolysis in ethanol–water to yield a mixture of alcohol and ether. As usual for secondary alkyl substrates, reaction is S_N1 with nucleophilic assistance from the solvent (see Sec. 6.9). A C_6H_5S- group on the adjacent carbon can speed up

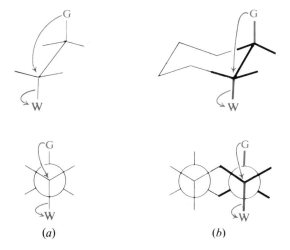

Relative rates of reaction

$$G: \textit{trans-}C_6H_5S \gg H > \textit{cis-}C_6H_5S$$
$$70\ 000 \qquad\quad 1.00 \qquad 0.16$$

reaction powerfully—*but only if it is trans to chlorine.* The *cis* substituted chloride actually reacts more slowly than the unsubstituted compound.

The *trans* sulfide group evidently gives strong anchimeric assistance. Why cannot the *cis* sulfide? The answer is found in the examination of molecular models. Like other nucleophiles, a neighboring group attacks carbon at the side away from the leaving group. In an open-chain compound like mustard gas—or like either diastereomer of 3-bromo-2-butanol—rotation about a carbon–carbon bond can bring the neighboring group into the proper position for back-side attack: *anti* to the leaving group (Fig. 20.4*a*). But in cyclohexane derivatives, 1,2-substituents are *anti* to each other only when they both occupy axial positions—possible only for *trans* substituents (Fig. 20.4*b*). Hence, only the *trans* chloride shows the neighboring group effect, anchimeric assistance from sulfur. The *cis* isomer reacts without anchimeric assistance; through its electron-withdrawing inductive effect, sulfur slows down formation of the carbocation, and thus the rate of reaction.

Figure 20.4 Anchimeric assistance. (*a*) An *anti* relationship between the neighboring group and the leaving group is required for back-side attack. (*b*) In cyclohexane derivatives, only *trans*-1,2-substituents can assume an *anti* relationship.

Let us look at another example of solvolysis. A very commonly studied system is one in which the solvent is acetic acid, CH_3COOH (often represented as HOAc), and the substrate is one already familiar to us, an alkyl ester of a sulfonic acid: a tosylate, ROTs; a brosylate, ROBs; etc. Loss of the weakly basic sulfonate anion,

$$R\text{—OTs} \xrightarrow[\text{NaOAc}]{\text{HOAc}} R\text{—OAc} + \text{OTs}^-$$

Alkyl tosylate   Alkyl acetate

with more or less nucleophilic assistance from the solvent, generates a cation—as part of an ion pair—which combines with the solvent to yield the product. The product is an alkyl ester of acetic acid, an alkyl *acetate*. Such solvolysis is called *acetolysis*, that is, cleavage by acetic acid.

Now, let us consider the special case of a substrate that is not only a tosylate but also an acetate. It is the tosylate that is the leaving group in the reaction. The

An acetoxy tosylate   A diacetate

strongly basic acetate is a very poor leaving group and remains in the molecule—doing nothing, apparently. And so, the product of acetolysis is a *di*acetate.

When 2-acetoxycyclohexyl tosylate is heated in acetic acid there is obtained, as expected, the diacetate of 1,2-cyclohexanediol. The reactant exists as diastereomers, and just what happens—and how fast it happens—depends upon which

2-Acetoxycyclohexyl   1,2-Cyclohexanediol
tosylate   diacetate

diastereomer we start with. The *cis* tosylate yields chiefly the *trans* diacetate. Reaction takes the usual course for nucleophilic substitution, predominant inversion. But the *trans* tosylate also yields *trans* diacetate. Here, apparently, reaction takes place with *retention*, unusual for nucleophilic substitution, and in contrast to what is observed for the *cis* isomer. Two pieces of evidence show us clearly what

(+)-*trans*-2-Acetoxycyclohexyl   *trans*-1,2-Cyclohexanediol
tosylate   diacetate
Optically active   *Racemic*

is happening here: (a) optically active *trans* tosylate yields *optically inactive trans* diacetate; and (b) the *trans* tosylate reacts *800 times as fast as the cis isomer*. Here we see a special kind of stereochemistry *and* an unusually fast rate of reaction: both of the manifestations of a neighboring group effect.

The neighboring group is acetoxy, containing oxygen with unshared electrons. Through back-side nucleophilic attack, acetoxy helps to push out the tosylate anion (step 1) and, in doing this, inverts the configuration at the carbon under attack.

(1)

An acetoxonium ion

(2)

There is formed an *acetoxonium ion*. This symmetrical intermediate undergoes nucleophilic attack (step 2) by the solvent at either of two carbons—again with inversion—and yields the product. The result: in half the molecules, retention at both carbons; in the other half, inversion at both carbons.

The *cis* tosylate cannot assume the diaxial conformation needed for back-side attack by acetoxy, and there is no neighboring group effect. Stereochemistry is normal, and reaction is much slower than for the *trans* tosylate.

Compared with unsubstituted cyclohexyl tosylate, the 2-acetoxycyclohexyl tosylates show the following relative reactivities toward acetolysis:

Cyclohexyl tosylate		*trans*-2-Acetoxycyclohexyl tosylate		*cis*-2-Acetoxycyclohexyl tosylate
1.00	>	0.30	>	0.000 45

Reaction of the *cis* tosylate is much slower than that of cyclohexyl tosylate, and this we can readily understand: powerful electron withdrawal by acetoxy slows down formation of the carbocation in the S_N1 process. Reaction of the *trans* tosylate, although much faster than that of its diastereomer, is still somewhat slower than that of cyclohexyl tosylate. But should not the anchimerically assisted reaction be much *faster* than the unassisted reaction of the unsubstituted tosylate? The answer is, *not necessarily*. We must not forget the polar effect of the acetoxy substituent. Although S_N2-like, attack by acetoxy has considerable S_N1 character (see Sec. 8.15); deactivation by electron withdrawal tends to offset activation by anchimeric assistance. The *cis* tosylate is electronically similar to the *trans*, and is a much better standard by which to measure anchimeric assistance. (This point will be discussed further in the next section.)

In Sec. 8.15 we said that the orientation of opening of strained rings like halonium ions and protonated epoxides indicates considerable S_N1 character in the transition state. But if ring-*opening* has S_N1 character so, according to the principle of microscopic reversibility, must ring-*closing*—as in the intramolecular attack by the acetoxy group. And it is the ring-closing step, remember, that determines the overall rate of reaction.

Problem 20.4 How do you account for the following relative rates of acetolysis of 2-substituted cyclohexyl brosylates? In which cases is there evidence of a neighboring group effect?

Relative rates

G	cis	trans
Cl	1.6	5.9
Br	1.5	1250
I		2.2×10^8
H	1.2×10^4	

We should note once again the basic similarity between a neighboring group effect and the stabilization of an incipient carbocation by resonance (Secs. 10.14–10.16). In both cases a nearby atom or group provides electrons to a carbon that is becoming electron-deficient through the departure of a leaving group. In both cases the electrons can be an unshared pair on the neighboring atom: an atom like oxygen or sulfur or halogen. And in both cases this atom, even though electronegative, can accommodate the developing positive charge better than carbon can because of the preservation of an octet of electrons (Sec. 10.15). The difference between the two effects lies in the *way* the electrons are delivered to where they are needed: by sideways overlap of orbitals in a resonance effect; by being carried to the reaction site in a neighboring group effect.

The similarity goes even further. A resonance effect, we have seen, can involve not only unshared pairs on an atom like oxygen, but also electrons supplied by carbon and hydrogen: π electrons and even σ electrons. And, as we shall see in Chapter 32, carbon and hydrogen can furnish electrons in neighboring group effects, too.

Now let us turn to entirely different reactions involving different kinds of compounds, and see how, if we look beneath the surface, we can find a pattern of behavior similar to what we have just discussed.

20.5 Homogeneous hydrogenation. Transition metal complexes

In Section 8.3 we described very briefly homogeneous hydrogenation: the hydrogenation of carbon–carbon double bonds catalyzed by organic complexes of transition metals. Let us look more closely at this reaction: to see how catalysts of this kind work, and to see another example of symphoria.

In 1951 the organoiron compound *ferrocene* (p. 489) was first prepared, and by 1952 its structure had been worked out. The unexpected stability of this compound, and the (then) unusual kind of bonding holding iron to carbon, caught the imagination of chemists and set off a revolution in the field of organic complexes of the transition metals. A broad theory of the mechanisms of reaction of these "inorganic" compounds has grown up; in its form and rapid growth, this theory has been likened to the theory of organic reactions, whose origins go back more than 20 years earlier. Technically, of course, these compounds *are* organic, since they contain carbon. The distinction lies in the element about which reaction centers: a transition metal, or carbon.

Inorganic compounds or not, these transition metal complexes are of increasing importance to organic chemists today as catalysts of unprecedented power and selectivity. We shall be concerned with them because of their usefulness and because, as we shall see, their mode of action fits into a basic pattern of chemical reactivity that extends all the way to the action of enzymes in living organisms: that is, from "inorganic" chemistry to the most "organic"—in the old sense—of all chemistry.

Now, how do these metal complexes work? As always, we begin by examining their *structure*.

By definition, transition metals have outer shells (*d* and sometimes *f*) that are only partly filled, and it is these vacant bonding sites—this "unsaturation"—that enable the metals to act as catalysts.

A metal complex is made up of the metal and certain ions and molecules, called *ligands* (from the Latin, *ligare*, to bind), that are held by it. Each ligand (L) is bonded to the metal by overlap of an empty orbital on the metal with a filled

M = transition metal   L = ligand

orbital on the ligand. (Sometimes, besides this σ bonding there is π bonding as well, involving overlap of a filled orbital on the metal with an empty orbital on the ligand: so-called *back-bonding*.) The bonding is thus covalent, with varying degrees of ionic character depending upon the extent to which positive and negative charges on the metal and ligand help to hold them together.

In some ligand molecules, more than one atom has an electron pair to share with the metal. The ligand has more than one binding site, and is said to be *bidentate*, *tridentate*, etc. (that is to say, "two-toothed", "three-toothed", etc.). Such a ligand can, by forming a ring, hold the metal by two (or more) of its binding sites—between its "teeth":

Chelation:

makes complex ion especially stable

Binding of this kind is called **chelation** (Greek: *chele*, claw). In general chelation gives a much more stable complex than one formed by binding of analogous separate ligands. (For examples of the chelation of metals, see *chlorophyll* (p. 1207) and *heme* (p. 1367).)

The spatial arrangement of a metal complex depends upon the orbitals used to hold the ligands, which in turn depend upon the particular metal involved and the number of ligands it holds. Some of the ways in which ligands (L) are commonly arranged, and the ways these configurations are often represented, are shown in Fig. 20.5.

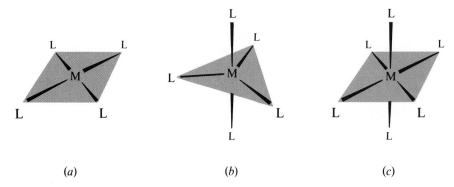

(a)   (b)   (c)

Figure 20.5 Some configurations commonly observed for transition metal complexes. (*a*) Square planar: four ligands (L). (*b*) Trigonal bipyramid: five ligands. (*c*) Octahedral (two square pyramids base-to-base): six ligands.

These ligands take no direct part in the reaction that is being catalyzed, but their presence on the metal is absolutely necessary. Like substituents in an organic molecule, ligands can—through their electronic or steric effect, their lipophilicity, or their chirality—help to determine the course of reaction. They stabilize the complex, modify its reactivity, make it soluble in organic solvents, and can even bring about stereoselectivity in the product formed.

Finally, and outside this *coordination sphere*, there are whatever counter-ions are needed to balance any net charge, positive or negative, that may reside on the metal complex.

The metal exerts its catalytic effect by, first, bringing the substrate and reagent near to each other. It does this by forming bonds to them. The reactants thus become ligands in a new metal complex, often by taking the place of some of the old ligands. In forming these bonds the metal may change one or both of the reactants profoundly: it may, for example, break the bond of molecular hydrogen, H_2, and bind two hydride ions as separate ligands.

Now, while holding the reactants in just the right spatial relationship to each other, the metal allows them to interact, often in several steps, to yield the product. The product then departs, and the metal complex is available to begin the catalytic cycle all over again.

As an illustration of what has just been said, let us look fairly closely at one catalyst and see how it is believed to exert its effect. It is *Wilkinson's catalyst*, the most widely used of all catalysts for homogeneous hydrogenation. From this single example we can learn a good deal about how all these catalysts work: to promote, not just hydrogenation, but many other reactions as well.

It is especially fitting that we begin with this particular catalyst, discovered by the inorganic chemist Sir Geoffrey Wilkinson (Imperial College London). Wilkinson received the Nobel Prize for his work on the elucidation of the structure of ferrocene: work which, as we said earlier, opened the door to the "new" chemistry of transition metal complexes. Under Wilkinson's name in *Who's Who* is found the entry, "Leisure interest: organic chemistry".

Wilkinson's catalyst is a complex of the transition metal rhodium called tris(triphenylphosphine)chlororhodium(I). Its formula is $RhCl(PPh_3)_3$, where Ph stands for phenyl, C_6H_5. The ligand Ph_3P is triphenylphosphine. Phosphorus

Tris(triphenylphosphine)chlororhodium(I)
Wilkinson's catalyst

belongs to the same family of the Periodic Table as nitrogen, and phosphines, R_3P, are structurally similar to amines, R_3N, which in turn are derived from ammonia. Like the nitrogen of ammonia and amines, the phosphorus of phosphines has an unshared electron pair, which confers basicity—although weaker than that of the nitrogen analogs—on these molecules. It is through this electron pair that triphenylphosphine is bonded to rhodium.

In solution the complex $RhCl(PPh_3)_3$ is believed to exchange reversibly one Ph_3P for a loosely held solvent molecule, to give the complex $RhCl(PPh_3)_2$(solvent).

Now the catalyst is brought into contact with the reactants, the alkene and molecular hydrogen, H_2. It reacts (step 1) with the hydrogen to form the *dihydrido* complex, $RhH_2Cl(PPh_3)_2$. The H—H bond is broken, and each hydrogen becomes

(1)

Dihydrido complex

bonded separately to rhodium. (To do this the metal uses one of its electron pairs, and is thereby oxidized to the rhodium(III) state.)

Next, the alkene reacts (step 2) with the complex and, perhaps by replacing a solvent molecule, attaches itself to rhodium. The alkene–metal bond involves

(2)

π-Complex with alkene

overlap of an empty orbital of the metal with the π cloud of the alkene; rhodium is bonded, not just to one of the alkene carbons, but to both.

This kind of bonding has been shown to exist between compounds with π electrons—alkenes, aromatics—and acidic molecules of many kinds—silver ion, for example, or halogens. Such **π-complexes** have been detected spectroscopically and, in some cases, isolated. The ferrocene referred to above is a π-complex, and the interest it aroused lay primarily in just that fact: the strong binding between iron and the π cloud of the organic moiety (p. 490). Reversible formation of π-complexes has been postulated as a step preliminary to the reaction of many electrophiles with alkenes and aromatic compounds.

At this point both reactants are bonded to rhodium, and the stage is set for hydrogenation to occur. The two hydrogens are transferred to the two doubly bonded carbons—not simultaneously, but one at a time, in two separate reactions.

In step (3) a hydrogen migrates from the metal to one of the doubly bonded

(3)

carbons. The other doubly bonded carbon becomes attached to the metal by a straightforward σ bond, and a metal alkyl has been formed.

One can view this step in several ways: as a 1,2-shift of hydrogen from the metal to carbon, for example; or addition of hydrogen and the metal to the carbon–carbon double bond. It is often considered as *insertion* of the alkene into the metal–hydrogen bond. Such insertion of an alkene into a metal–ligand bond is a key step in important catalyzed processes other than hydrogenation (Secs. 20.8 and 36.6).

Now, in step (4), the second hydrogen migrates from the metal to carbon: this time to the carbon that is still bonded to the metal—the second of the two original

(4)

alkene carbons. Addition of hydrogen is complete, and the saturated product leaves the coordination sphere of the regenerated catalyst. (The metal has regained the electrons it used to cleave H_2, and is reduced to its original rhodium(I) state.)

The mechanism we have given is strongly supported by evidence of many kinds, including kinetics studies. The postulated intermediates have been detected in solution and even in some cases isolated, and their structures have been established by spectroscopic methods (chiefly NMR, Chap. 16) and x-ray analysis. One of the enormous advantages of homogeneous catalysis over heterogeneous catalysis is that mechanisms of reaction *can* readily be studied and, by use of the power this knowledge gives, catalysis can be modified to accomplish things never before possible.

The reaction we have just discussed, addition, is different from the nucleophilic substitution of the preceding sections, and the central element here is a transition metal instead of carbon. But the factor at work is *exactly the same*: *symphoria*. The reacting atoms are being held—whether by carbon or by a transition metal—in the proper positions for reaction to occur.

So far we have discussed the reactivity aspect of homogeneous hydrogenation: why reaction occurs at all. But, as in neighboring group effects, there is a stereochemical aspect as well. Let us look at that.

20.6   Stereochemistry of homogeneous hydrogenation: diastereoselectivity

Consider the homogeneous hydrogenation with Wilkinson's catalyst of the unsaturated carboxylic acid *butenedioic acid*. The hydrogen used is not ordinary hydrogen but deuterium, D_2. There is formed the saturated acid *butanedioic acid* containing two deuterium atoms.

$$HOOC-CH=CH-COOH + D_2 \xrightarrow[\text{catalyst}]{\text{Wilkinson's}} HOOC-\overset{*}{C}H-\overset{*}{C}H-COOH$$

$$\underset{\text{Butenedioic acid}}{} \qquad\qquad \underset{D\quad D}{}$$

$$(2,3\text{-}D_2)\text{Butanedioic acid}$$

Two chiral centers are generated in the reaction, and the product, we can easily show, can exist as a *meso* compound and a pair of enantiomers.

The reactant, too, exists as stereoisomers: a pair of geometric isomers. These are nearly always given their common names of *maleic acid* (the *cis* isomer) and *fumaric acid* (the *trans* isomer).

If we start with maleic acid, we obtain *only* the *meso* butanedioic acid; none of the racemic compound is obtained.

Maleic acid
cis-Butenedioic acid

$$\xrightarrow[\substack{\text{Wilkinson's} \\ \text{catalyst}}]{D_2}$$

I
meso-(2,3-D$_2$)Butanedioic acid

The reaction is thus completely *stereoselective* (Sec. 9.2). Since the selectivity is between diastereomeric products, it is of the kind called *diastereoselectivity*.

If, on the other hand, we start with fumaric acid, we obtain *only* the racemic butanedioic acid.

Fumaric acid
trans-Butenedioic acid

II
(2R,3R)

III
(2S,3S)

rac-(2,3-D$_2$)Butanedioic acid

Stereochemically different starting materials thus react differently: they yield stereochemically different products. Reaction is thus not only stereoselective, but *stereospecific*, too.

We have seen (Sec. 9.5) that stereoselective additions are of two kinds, *syn* and *anti*, depending upon whether the added groups become attached to the same face or opposite faces of the double bond. In Chapter 9 we saw examples of both *syn*- and *anti*-addition. Which kind is homogeneous hydrogenation?

Examination of the structures involved shows us that hydrogenation with Wilkinson's catalyst involves *syn*-addition. Let us see that this is so. If we start (Fig. 20.6) with maleic acid, we can attach the two hydrogens to the same face of the double bond in two different ways: either from the top as in (*a*) or from the bottom as in (*b*). Whichever way we choose, we obtain I, which we recognize as the *meso* product.

Maleic acid

I

meso-(2,3-D$_2$)Butanedioic acid

Figure 20.6   *syn*-Addition to maleic acid. Attachment as in (*a*) or (*b*) gives the *meso* product.

Starting with fumaric acid (Fig. 20.7), we can again attach the hydrogens to the same face of the double bond in two different ways. Attachment from the top, path (c), gives enantiomer II; attachment from the bottom, path (d), gives enantiomer III. Since, whatever the mechanism, (c) and (d) are equally likely, we obtain the racemic modification.

II and III are enantiomers
rac-(2,3-D$_2$)Butanedioic acid

Figure 20.7 *syn*-Addition to fumaric acid. Attachment as in (c) or (d) is equally likely, and gives the racemic modification.

This observed overall *syn*-addition is interpreted the following way. In step (3) the metal and a hydrogen attach themselves simultaneously to the doubly bonded carbons. Because of the juxtaposition of metal and hydrogen—they are bonded to each other in the reactant—this addition of the two atoms necessarily, on geometric grounds, takes place to the same face of the double bond. This step thus involves *syn*-addition.

In step (4) hydrogen migrates from the metal to carbon, and in doing this attaches itself to the same face of carbon that was attached to the metal. That is, there is *front-side* attack leading to *retention* of configuration about the carbon. The alternative, back-side attack is impossible, again on geometric grounds; the hydrogen is held near the front side of carbon by the metal, which in the transition state is bonded to both hydrogen and carbon.

(4)

The net result of *syn*-addition in step (3) and retention of configuration in step (4) is overall *syn*-addition in the hydrogenation. One face or the other of the alkene is held toward the metal in the initial metal–alkene complex, and it is to that face that both hydrogens become attached.

20.7   Stereochemistry of homogeneous hydrogenation: enantioselectivity

Now let us turn to another aspect of the stereochemistry of homogeneous hydrogenation. Let us use as our example a reaction of great practical importance: the synthesis of amino acids. Amino acids are the building blocks from which those vital giant molecules, the proteins (Chap. 40), are made. They have the general formula:

$$^{+}H_{3}N{-}\underset{\underset{\text{H}}{|}}{\overset{\overset{\text{R}}{|}}{C}}{-}COO^{-}$$

An amino acid

Except for the simplest case, where R is H, an amino acid contains a chiral center— the α-carbon (*alpha* carbon)—and hence can exist as a pair of enantiomers. As ordinarily prepared, from an optically inactive substrate and optically inactive reagents, an amino acid is of course obtained as equal amounts of the two enantiomers, that is, as the racemic modification.

But naturally occurring amino acids—the ones that help to make up proteins— are not racemic, but optically active. They occur in just one enantiomeric form; with rare exceptions, this has the absolute configuration,

$$H_{3}\overset{+}{N}{-}\underset{\underset{\text{R}}{\vdots}}{\overset{\overset{\text{COO}^{-}}{\vdots}}{C}}{\rightarrow} H$$

L-Amino acid

designated, for reasons that we shall see later, as L.

And so, to prepare a synthetic protein, one needs to start with amino acids that are optically active and of the correct configuration. If the synthetic amino acids have been prepared by ordinary methods, they must be resolved (Sec. 4.27) before they can be used: an additional, often lengthy, step which involves loss of half the material.

Clearly, what is needed is a synthesis of amino acids that yields directly only one enantiomer—an *enantioselective* synthesis. In work carried out since about 1970, just such a synthesis has been developed, based upon hydrogenation with homogeneous catalysis.

Consider the synthesis of the simple amino acid *alanine* from the unsaturated (and achiral) starting material IV.

1-Acetamidopropenoic
acid

IV

Acetylalanine

Alanine

(The acetyl group, —$COCH_3$, is often used to *protect* the amino group, and is readily removed in the final step by hydrolysis.) In the hydrogenation step a chiral center is generated; just which enantiomer is obtained depends upon which face of IV the hydrogen adds to. (Compound IV contains enantiotopic faces, Sec. 22.6.)

As shown in Fig. 20.8, attachment from the top, by path (*e*), gives enantiomer V; attachment from the bottom, by path (*f*), gives enantiomer VI. Examination shows that enantiomer V has the L configuration and is the naturally occurring isomer.

Now, if we use the ordinary, optically inactive Wilkinson's catalyst that we have so far described, we can expect that attachment to the two faces will be equally likely, and that we will obtain the racemic modification. And this, of course, is exactly what happens.

V and VI are enantiomers
Acetylalanine

Figure 20.8   Generation of a chiral center by addition to enantiotopic faces of a carbon–carbon double bond.

Suppose, however, that we were to modify the catalyst to make it optically active. Reagents attached to the metal would now be reacting *in a chiral medium*: the chiral coordination sphere of the metal. Under these conditions, we might expect—to a degree, in any case—preferential formation of one of the two enantiomers, that is, enantioselectivity.

But how could one make a complex optically active? The answer, of course, is to prepare a catalyst containing an *optically active ligand*. A number of such ligands have been developed, many of them by Brice Bosnich (University of Toronto). For example:

$$CH_2-CH-CH_3 \qquad CH_3-CH-CH-CH_3 \qquad CH_3-CH-CH_2-CH-CH_3$$
$$Ph_2P \quad\;\; PPh_2 \qquad\quad Ph_2P \quad\;\; PPh_2 \qquad\quad Ph_2P \qquad\qquad PPh_2$$

1,2-Bis(diphenylphosphino)-propane	2,3-Bis(diphenylphosphino)-butane	2,4-Bis(diphenylphosphino)-pentane
Prophos	Chiraphos	Skewphos

These are bidentate ligands, and chelate with rhodium to give optically active catalysts. The results obtained by use of these catalysts have been spectacular. Amino acids and other kinds of compounds have been prepared from achiral unsaturated compounds with a degree of enantioselectivity that rivals that of enzymes: optical purities in some cases of nearly 100%!

One encounters several terms used in connection with enantioselective syntheses: *optical purity*, *optical yield*, and *enantiomeric excess*. These all mean the same thing: percentage of the total that the predominant enantiomer makes up, with the other component considered to be the racemic modification. To calculate these values, we simply divide the observed optical rotation by the rotation of optically pure material, and multiply by 100.

We can visualize two different ways in which the enantioselectivity could arise. First, there could be preferential binding of the metal to one of the faces of the alkene rather than to the other: there is a *better fit* of the alkene in the coordination sphere, and one diastereomeric π-complex is formed in preference to the other. Alternatively, both π-complexes are formed, but one is more reactive than the other: transfer of hydrogen occurs more readily within it, perhaps because of a better spatial relationship between hydrogen and the bound alkene. Recent evidence suggests that, in one case at least, the latter explanation is the correct one.

Just which enantiomer is obtained can be controlled by the choice of configuration of the chiral ligand: (*R*)-prophos, for example, gives amino acids of the natural, L configuration; (*S*,*S*)-chiraphos gives amino acids of the opposite configuration. This method is even being used on an industrial scale, to make, for example, L-dopa, an amino acid used in the treatment of Parkinson's disease.

$$CH_3COO-\langle\bigcirc\rangle-CH_2-CH-COO^-$$
$$CH_3O \qquad\qquad\qquad NH_3{}^+$$

3-Methoxy-4-acetoxyphenylalanine
L-Dopa
(*S* configuration)

Metal complexes have been developed to catalyze many reactions other than hydrogenation with, in many cases, the special advantage of stereoselectivity. And this is only a beginning; the field is growing by leaps and bounds.

20.8 The oxo process

In Sec. 17.6, we saw that an important industrial method of making alcohols is the *oxo process*. In the presence of the proper catalyst, alkenes react with carbon monoxide and hydrogen to yield aldehydes.

$$\text{C=C} + \text{CO} + \text{H}_2 \xrightarrow{\text{catalyst}} \underset{\text{H}\quad\text{CHO}}{-\text{C}-\text{C}-}$$

An aldehyde

Upon reduction, the aldehydes give primary alcohols.

The classical catalyst for this reaction is octacarbonyldicobalt, $Co_2(CO)_8$, formed by reaction of metallic cobalt with carbon monoxide. Let us see how this catalyst brings about reaction.

$$2\text{Co} + 8\text{CO} \longrightarrow$$

Octacarbonyldicobalt

The following steps are believed to take place during the oxo process. The octacarbonyldicobalt reacts with hydrogen (step 1) to form the hydrido complex $CoH(CO)_4$, the active catalyst. (This is soluble in hydrocarbons, so that once again

(1)

Octacarbonyldicobalt   Hydridotetracarbonylcobalt

we are dealing with a case of homogeneous catalysis.) Next, the alkene replaces (step 2) one molecule of carbon monoxide to form our familiar π-complex.

(2)

I

At this point, cobalt holds as ligands the three units that must react with each other: the alkene, carbon monoxide, and a hydrogen. Now, as in homogeneous hydrogenation, the hydrogen migrates (step 3) from cobalt to one of the doubly bonded carbons; simultaneously the other doubly bonded carbon attaches itself to cobalt, and a metal alkyl has been formed. This acquires an additional molecule of carbon monoxide.

(3)

Next, the newly formed alkyl group migrates (step 4) to the carbon of a carbon monoxide ligand. This is the key step, since in it a carbon–carbon bond is formed.

(4)

Now hydrogen is absorbed (step 5) to form a dihydrido complex. One of these hydrogens migrates to carbon of the C=O group to form an aldehyde molecule, which leaves the coordination sphere of the regenerated catalyst.

(5)

An aldehyde

I

Wilkinson has found that the complex $RhH(CO)(Ph_3P)_3$, which is very like his hydrogenation catalyst (Sec. 20.5), is even more efficient than the cobalt complex at promoting the oxo process. Using his catalyst, he has found evidence for a series of steps analogous to those we have just outlined.

In the oxo process, we see, the catalyst exerts its effect in basically the same manner as in homogeneous hydrogenation. We shall find another, analogous process in Ziegler–Natta polymerization (Sec. 36.6).

20.9 Enzyme action

Enzymes are catalysts for the reactions involved in life processes. They speed up these reactions enormously, and with a high degree of selectivity. An enzyme accomplishes this, first, by binding the substrate. The substrate molecule just fits in a pocket in the giant convoluted enzyme molecule, where it is held by a combination of forces: van der Waals; dipole–dipole interaction, particularly hydrogen bonding; ionic bonding. Now, held in just the right position, the substrate is attacked by the reagent: a functional group that is either a permanent part of the enzyme molecule, or a molecule temporarily bonded to the enzyme. (See, for example, Sec. 41.2.)

The similarity to the action of metal complexes is striking. A metal complex has a much simpler structure than an enzyme, and it binds the substrate by different forces. But fundamentally the mode of action is the same: to bring together into just the right spatial relationship the substrate and the reagent. Underlying both enzyme action and catalysis by metal complexes is *symphoria*.

PROBLEMS

1. On treatment with concentrated aqueous sodium hydroxide, 2-chloroethanol is converted into the epoxide, ethylene oxide.

Ethylene oxide   Cyclohexene oxide

(a) Showing all steps, suggest a likely mechanism for this reaction.
(b) Using models, show the steric course it probably follows.
(c) Suggest a reason why sodium hydroxide readily converts *trans*-2-chlorocyclohexanol into cyclohexene oxide, but converts the *cis* isomer into entirely different products.

2. Account in detail for each of the following observations:

(a) $CH_3CHClCH_2NEt_2$ $\xrightarrow{OH^-}$ $CH_3CH(NEt_2)CH_2OH$

(b) *Either* $CH_3CHOHCH_2SEt$ *or* $CH_3CH(SEt)CH_2OH$ $\xrightarrow{HCl}$ $CH_3CHClCH_2SEt$

(c) Treatment of *either* epoxide I *or* epoxide II with aqueous OH^- gives the same product III.

        I   II   III

3. Upon reaction with bromine water, as we have seen (Problem 10, p. 341), allyl bromide gives chiefly the primary alcohol, $CH_2BrCHBrCH_2OH$, evidently because of electron withdrawal by the —Br already in the molecule. When allyl bromide labeled with [82]Br undergoes this reaction, the following products are obtained:

$$CH_2\text{---}CH\text{=}CH_2 \xrightarrow{\ Br_2,\ H_2O\ } CH_2\text{---}CH\text{---}CH_2 \ + \ CH_2\text{---}CH\text{---}CH_2 \ + \ CH_2\text{---}CH\text{---}CH_2$$

| ^{82}Br | | ^{82}Br | Br | OH | | Br | ^{82}Br | OH | | ^{82}Br | OH | Br |

(55%)　　　　　　　　(24%)　　　　　　　　(21%)

(a) How do you account for the formation of the 24% of product with ^{82}Br attached to C-2?

(b) When similarly labeled allyl chloride is used there is obtained only 4% of the product with the label attached to C-2. How do you account for this difference between the chloride and bromide?

4. On treatment with aqueous HBr both *cis-* and *trans-*2-bromocyclohexanol are converted into the same product. What would you expect this product to be, and how do you account for its formation from both substrates?

5. Account for the fact that addition of chlorine and water to oleic acid (*cis-*9-octadecenoic acid) followed by treatment with base gives the same epoxide (same stereoisomer) as does treatment of oleic acid with a peroxy acid.

6. Go back to Problem 14 (p. 663) and account for the stereochemistry observed there for the oxo process.

7. (a) *Disparlure*, the sex pheromone of the gypsy moth, has been synthesized in the following way.

$$n\text{-}C_{10}H_{21}Br \ + \ HC\text{≡}CNa \ \longrightarrow \ A\ (C_{12}H_{22})$$
$$A \ + \ n\text{-}BuLi \ \longrightarrow \ B\ (C_{12}H_{21}Li)$$
$$5\text{-methyl-1-hexene} \ + \ HBr,\ peroxides \ \longrightarrow \ C\ (C_7H_{15}Br)$$
$$C \ + \ B \ \longrightarrow \ D\ (C_{19}H_{36})$$
$$D \ + \ H_2,\ Lindlar\ catalyst \ \longrightarrow \ E\ (C_{19}H_{38})$$
$$E \ + \ a\ peroxybenzoic\ acid \ \longrightarrow \ disparlure\ (C_{19}H_{38}O)$$

Optically inactive

What is the structure of disparlure?

(b) Unlike the product obtained above, the natural pheromone is optically active. Examining the structure of the molecule, tell just what the optical activity is due to. Account for the formation of optically inactive material in (a).

(c) An alternative route to disparlure involves the following intermediate step, carried out in the presence of titanium tetraisopropoxide and diethyl tartrate.

$$(Z)\text{-}2\text{-tridecen-1-ol} \ + \ t\text{-}BuO_2H \ \longrightarrow \ F\ (C_{13}H_{26}O_2)$$

tert-Butyl hydroperoxide

What is the structure of F?

(d) When (−)-diethyl tartrate is used in (c), the F obtained is optically active, and it yields, ultimately, the natural (+)-disparlure. What is the function of (−)-diethyl tartrate in this synthesis?

(e) The (+)-disparlure is a more powerful attractant than the optically inactive material obtained by synthesis (a). In general terms, how do you account for this?

21

Aldehydes and Ketones

Nucleophilic Addition

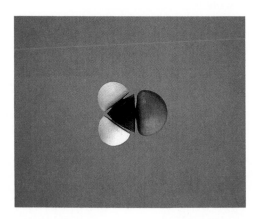

21.1 Structure

Aldehydes are compounds of the general formula RCHO; ketones are compounds of the general formula RR′CO. The groups R and R′ may be aliphatic or aromatic. (In one aldehyde, HCHO, R is H.)

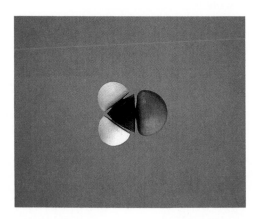

Aldehydes   A ketone

Both aldehydes and ketones contain the carbonyl group, C=O, and are often referred to collectively as **carbonyl compounds**. *It is the carbonyl group that largely determines the chemistry of aldehydes and ketones.*

It is not surprising to find that aldehydes and ketones resemble each other closely in most of their properties. However, there is a hydrogen atom attached to the carbonyl group of aldehydes, and there are two organic groups attached to the carbonyl group of ketones. This difference in structure affects their properties in two ways: (a) aldehydes are quite easily oxidized, whereas ketones are oxidized only with difficulty; (b) aldehydes are usually more reactive than ketones toward nucleophilic addition, the characteristic reaction of carbonyl compounds.

Let us examine the structure of the carbonyl group. Carbonyl carbon is joined to three other atoms by σ bonds; since these bonds utilize sp^2 orbitals (Sec. 1.10), they lie in a plane, and are 120° apart. The remaining p orbital of the carbon overlaps a p orbital of oxygen to form a π bond; carbon and oxygen are thus joined

$$\overset{R'}{\underset{R}{\diagdown}}\overset{\delta+}{C}\!=\!\!=\!\overset{\delta-}{O}$$

by a double bond. The part of the molecule immediately surrounding carbonyl carbon is *flat*; oxygen, carbonyl carbon, and the two atoms directly attached to carbonyl carbon lie in a plane.

The electrons of the carbonyl double bond hold together atoms of quite different electronegativity, and hence the electrons are not equally shared; in particular, the mobile π cloud is pulled strongly toward the more electronegative atom, oxygen.

The facts are consistent with the orbital picture of the carbonyl group. Electron diffraction and spectroscopic studies of aldehydes and ketones show that carbon, oxygen, and the two other atoms attached to carbonyl carbon lie in a plane; the three bond angles of carbon are very close to 120° (see Fig. 21.1). The large dipole

Figure 21.1 Electronic configuration and molecular shape: the carbonyl group. Model of formaldehyde, HCHO: two views.

moments (2.3–2.8 D) of aldehydes and ketones indicate that the electrons of the carbonyl group are quite unequally shared. We shall see how the physical and chemical properties of aldehydes and ketones are determined by the structure of the carbonyl group.

21.2 Nomenclature

The common names of **aldehydes** are derived from the names of the corresponding carboxylic acids by replacing *-ic acid* by *-aldehyde*. (For the common names of carboxylic acids, see Sec. 23.2.) Branched-chain aldehydes are named as derivatives of straight-chain aldehydes. To indicate the point of attachment, the

Greek letters, α-, β-, γ-, δ-, etc., are used; the α-carbon is the one bearing the —CHO group.

$$\overset{\delta}{C}-\overset{\gamma}{C}-\overset{\beta}{C}-\overset{\alpha}{C}-CHO \qquad \textit{Used in common names}$$

The IUPAC names of aldehydes follow the usual pattern. The longest chain carrying the —CHO group is considered the parent structure and is named by replacing the -*e* of the corresponding alkane by -*al*. The position of a substituent is indicated by a number, the carbonyl carbon always being considered as C–1. We

$$\overset{5}{C}-\overset{4}{C}-\overset{3}{C}-\overset{2}{C}-\overset{1}{C}HO \qquad \textit{Used in IUPAC names}$$

notice that C–2 of the IUPAC name corresponds to *alpha* of the common name.

H−C=O	CH₃C=O	CH₃CH₂C=O	CH₃CH₂CH₂C=O
Formaldehyde	Acetaldehyde	Propionaldehyde	*n*-Butyraldehyde
Methanal	Ethanal	Propanal	Butanal

Benzaldehyde

p-Nitrobenzaldehyde

p-Tolualdehyde

Salicylaldehyde
o-Hydroxybenzaldehyde

Phenylacetaldehyde
Phenylethanal

$$CH_3CH_2CH_2\overset{\alpha}{C}H\overset{H}{\underset{\underset{CH_3}{|}}{C}}=O$$

α-Methylvaleraldehyde
2-Methylpentanal

$$CH_3CH_2\overset{\beta}{C}H\overset{}{\underset{\underset{CH_3}{|}}{C}}H_2\overset{H}{C}=O$$

β-Methylvaleraldehyde
3-Methylpentanal

$$CH_3\overset{\gamma}{C}H\overset{}{\underset{\underset{CH_3}{|}}{C}}H_2CH_2\overset{H}{C}=O$$

Isocaproaldehyde
γ-Methylvaleraldehyde
4-Methylpentanal

The simplest aliphatic ketone has the common name of *acetone*. For most other aliphatic **ketones** we name the two groups that are attached to carbonyl carbon, and follow these names by the word *ketone*. A ketone in which the carbonyl group is attached to a benzene ring is named as a -*phenone*, as illustrated below.

According to the IUPAC system, the longest chain carrying the carbonyl group is considered the parent structure, and is named by replacing the -*e* of the corresponding alkane with -*one*. The positions of various groups are indicated by numbers, the carbonyl carbon being given the lowest possible number.

In certain polyfunctional compounds, the presence of a carbonyl group can be indicated by the prefix *oxo-*, with a number to show its position in the molecule.

$$CH_3-\overset{\underset{\displaystyle O}{\|}}{C}-CH_3$$

Acetone

Propanone

$$CH_3CH_2-\overset{\underset{\displaystyle O}{\|}}{C}-CH_3 \quad (MEK)$$

Ethyl methyl ketone

Butanone

$$CH_3CH_2CH_2-\overset{\underset{\displaystyle O}{\|}}{C}-CH_3$$

Methyl *n*-propyl ketone

2-Pentanone

$$CH_3CH_2-\overset{\underset{\displaystyle O}{\|}}{C}-CH_2CH_3$$

Diethyl ketone

3-Pentanone

$$CH_3\overset{\displaystyle CH_3}{\underset{}{CH}}-\overset{\underset{\displaystyle O}{\|}}{C}-CH_3$$

Isopropyl methyl ketone

3-Methyl-2-butanone

$$\text{(ring)}CH_2-\overset{\underset{\displaystyle O}{\|}}{C}-CH_3$$

Benzyl methyl ketone

1-Phenyl-2-propanone

$$\text{(ring)}-\overset{\underset{\displaystyle O}{\|}}{C}-CH_3$$

Acetophenone

$$\text{(ring)}-\overset{\underset{\displaystyle O}{\|}}{C}-CH_2CH_2CH_3$$

n-Butyrophenone

$$\text{(ring)}-\overset{\underset{\displaystyle O}{\|}}{C}-\text{(ring)}$$

Benzophenone

$$CH_3\text{(ring)}-\overset{\underset{\displaystyle O}{\|}}{C}-\text{(ring)}NO_2$$

3-Nitro-4'-methylbenzophenone

21.3 Physical properties

The polar carbonyl group makes aldehydes and ketones polar compounds, and hence they have higher boiling points than non-polar compounds of comparable molecular weight. By themselves, they are not capable of intermolecular hydrogen bonding since they contain hydrogen bonded only to carbon; as a result they have lower boiling points than comparable alcohols or carboxylic acids. For example, compare *n*-butyraldehyde (b.p. 76 °C) and ethyl methyl ketone (b.p. 80 °C) with *n*-pentane (b.p. 36 °C) and diethyl ether (b.p. 35 °C) on the one hand, and with *n*-butyl alcohol (b.p. 118 °C) and propionic acid (b.p. 141 °C) on the other.

The lower aldehydes and ketones are appreciably soluble in water, presumably because of hydrogen bonding between solute and solvent molecules; borderline solubility is reached at about five carbons. Aldehydes and ketones are soluble in the usual organic solvents.

Formaldehyde is a gas (b.p. -21 °C), and is handled either as an aqueous solution (*Formalin*), or as one of its solid polymers: *paraformaldehyde*, $(CH_2O)_n$, or *trioxane*, $(CH_2O)_3$. When dry formaldehyde is desired, as, for example, for reaction with a Grignard reagent, it is obtained by heating paraformaldehyde or trioxane.

Table 21.1 ALDEHYDES AND KETONES

	M.p., °C	B.p., °C	Solubility, g/100 g H_2O
Formaldehyde	− 92	− 21	v.sol.
Acetaldehyde	−121	20	∞
Propionaldehyde	− 81	49	16
n-Butyraldehyde	− 99	76	7
n-Valeraldehyde	− 91	103	sl.s.
Caproaldehyde		131	sl.s.
Heptaldehyde	− 42	155	0.1
Phenylacetaldehyde		194	sl.s.
Benzaldehyde	− 26	178	0.3
o-Tolualdehyde		196	
m-Tolualdehyde		199	
p-Tolualdehyde		205	
Salicylaldehyde	2	197	1.7
(*o*-Hydroxybenzaldehyde)			
p-Hydroxybenzaldehyde	116		1.4
Anisaldehyde	3	248	0.2
Vanillin	82	285	1
Piperonal	37	263	0.2
Acetone	− 94	56	∞
Ethyl methyl ketone	− 86	80	26
2-Pentanone	− 78	102	6.3
3-Pentanone	− 41	101	5
2-Hexanone	− 35	150	2.0
3-Hexanone		124	sl.s
Isobutyl methyl ketone	− 85	119	1.9
Acetophenone	21	202	
Propiophenone	21	218	
n-Butyrophenone	11	232	
Benzophenone	48	306	

~CH$_2$OCH$_2$OCH$_2$O~
Paraformaldehyde

Trioxane

Acetaldehyde (b.p. 20 °C) is often generated from its higher-boiling trimer by heating the trimer with acid:

3CH$_3$CH=O

Acetaldehyde
b.p. 20 °C

Paraldehyde
b.p. 125 °C

21.4   Preparation

A few of the many laboratory methods of preparing aldehydes and ketones are outlined below; most of these are already familiar to us. Some of the methods involve oxidation or reduction in which an alcohol, hydrocarbon, or acid chloride is converted into an aldehyde or ketone of the same carbon number. Other methods involve the formation of new carbon–carbon bonds, and yield aldehydes or ketones of higher carbon number than the starting materials.

Industrial preparations often involve special methods, or the modification of laboratory methods by use of cheaper reagents: formaldehyde and acetone are made by oxidation of methanol and isopropyl alcohol, respectively, but by air in the presence of a catalyst. Some aldehydes are obtained by the oxo process, in which they are the initial products (Sec. 17.6).

___ PREPARATION OF ALDEHYDES _____

1. **Oxidation of primary alcohols.** Discussed in Secs. 18.6 and 21.4.

$$R-CH_2OH \xrightarrow{\text{pyridinium chlorochromate}} R-\overset{H}{\underset{|}{C}}=O$$

$$1° \text{ alcohol} \qquad\qquad\qquad\qquad\qquad \text{Aldehyde}$$

Example:

$$CH_3CH_2CH_2CH_2OH \xrightarrow{\text{pyridinium chlorochromate}} CH_3CH_2CH_2CHO$$

$$\text{\textit{n}-Butyl alcohol} \qquad\qquad\qquad\qquad \text{\textit{n}-Butyraldehyde}$$
$$\text{1-Butanol} \qquad\qquad\qquad\qquad\qquad \text{Butanal}$$

2. **Oxidation of methylbenzenes.** Discussed in Sec. 21.4.

$$Ar-CH_3 \quad\begin{cases} \xrightarrow{Cl_2,\ heat} & Ar-CHCl_2 \xrightarrow{H_2O} \\ \xrightarrow{CrO_3,\ acetic\ anhydride} & Ar-CH(OOCCH_3)_2 \xrightarrow{H_2O} \end{cases} \quad Ar-CHO$$

Examples:

$$Br\langle\bigcirc\rangle CH_3 \xrightarrow{Cl_2,\ heat,\ light} Br\langle\bigcirc\rangle CHCl_2 \xrightarrow{CaCO_3,\ H_2O} Br\langle\bigcirc\rangle CHO$$

$$\text{\textit{p}-Bromotoluene} \qquad\qquad\qquad\qquad\qquad\qquad \text{\textit{p}-Bromobenzaldehyde}$$

$$O_2N\langle\bigcirc\rangle CH_3 \xrightarrow{CrO_3,\ Ac_2O} O_2N\langle\bigcirc\rangle CH(OAc)_2 \xrightarrow{H_2O,\ H_2SO_4} O_2N\langle\bigcirc\rangle CHO$$

$$\text{\textit{p}-Nitrotoluene} \qquad\qquad\qquad\qquad\qquad\qquad \text{\textit{p}-Nitrobenzaldehyde}$$

___ CONTINUED ___

— CONTINUED —

3. **Reduction of acid chlorides.** Discussed in Sec. 21.4.

$$\text{R—COCl} \quad \text{or} \quad \text{Ar—COCl} \xrightarrow{\text{LiAlH(OBu-}t)_3} \text{R—CHO} \quad \text{or} \quad \text{Ar—CHO}$$

Acid chloride   Aldehyde

Example:

$$O_2N\text{—}\langle O \rangle\text{—COCl} \xrightarrow{\text{LiAlH(OBu-}t)_3} O_2N\text{—}\langle O \rangle\text{—CHO}$$

p-Nitrobenzoyl chloride   p-Nitrobenzaldehyde

4. **Reimer–Tiemann reaction. Phenolic aldehydes.** Discussed in Sec. 28.12.

PREPARATION OF KETONES

1. **Oxidation of secondary alcohols.** Discussed in Sec. 18.6.

$$\underset{\underset{\text{OH}}{|}}{\text{R—CH—R}'} \xrightarrow{\text{CrO}_3 \text{ or } \text{K}_2\text{Cr}_2\text{O}_7} \underset{\underset{\text{O}}{\|}}{\text{R—C—R}'}$$

2° alcohol   Ketone

Example:

(−)-Menthol   (−)-Menthone

2. **Friedel–Crafts acylation.** Discussed in Sec. 21.5.

$$\underset{\text{Cl}}{\overset{\overset{\text{O}}{\|}}{\text{R—C}}} + \text{Ar—H} \xrightarrow[\substack{\text{or other} \\ \text{Lewis acid}}]{\text{AlCl}_3} \underset{\underset{\text{O}}{\|}}{\text{R—C—Ar}} + \text{HCl}$$

Acid chloride   Ketone

Examples:

$$n\text{-C}_5\text{H}_{11}\text{COCl} + \langle O \rangle \xrightarrow{\text{AlCl}_3} n\text{-C}_5\text{H}_{11}\text{—}\underset{\underset{\text{O}}{\|}}{\text{C}}\text{—}\langle O \rangle + \text{HCl}$$

Caproyl chloride

n-Pentyl phenyl ketone

No rearrangement of n-pentyl group

— CONTINUED —

CONTINUED

Benzoyl chloride

Benzophenone
(Diphenyl ketone)

Acetic anhydride

Acetophenone
(Methyl phenyl ketone)

3. **Reaction of acid chlorides with organocopper compounds.** Discussed in Sec. 21.6.

Examples:

Lithium di-*n*-butylcuprate   Isobutyryl chloride

$$2CH_3CH_2CH_2CH_2 \!-\! \underset{\underset{O}{\|}}{\overset{\overset{CH_3}{|}}{C}} CHCH_3$$

n-Butyl isopropyl ketone
2-Methyl-3-heptanone

m-Bromotoluene

Butyryl chloride

n-Propyl *m*-tolyl ketone

4. **Acetoacetic ester synthesis.** Discussed in Sec. 30.3.

Depending upon the availability of starting materials, **aliphatic aldehydes** can be prepared from alcohols or acid chlorides of the same carbon skeleton, and **aromatic aldehydes** can be prepared from methylbenzenes or aromatic acid

$$R{-}CH_2OH$$
$$R{-}COOH \longrightarrow R{-}COCl$$
$$\longrightarrow R{-}CHO$$
Preparation of aliphatic aldehydes

$$Ar{-}CH_3$$
$$Ar{-}COOH \longrightarrow Ar{-}COCl$$
$$\longrightarrow Ar{-}CHO$$
Preparation of aromatic aldehydes

chlorides. There are, in addition, a number of methods by which the aldehyde group is introduced into an aromatic ring: for example, the Reimer–Tiemann synthesis of phenolic aldehydes (Sec. 28.12).

Aliphatic ketones are readily prepared from the corresponding secondary alcohols, if these are available. More complicated aliphatic ketones can be prepared by the reaction of acid chlorides with organocopper compounds. A particularly

$$R{-}\underset{\underset{OH}{|}}{C}H{-}R'$$

$$R{-}COOH \longrightarrow R{-}COCl \xrightarrow{R'_2CuLi}$$

$$\longrightarrow R{-}\underset{\underset{O}{\|}}{C}{-}R'$$
Preparation of aliphatic ketones

useful method for making complicated aliphatic ketones, the acetoacetic ester synthesis, will be discussed later (Sec. 30.3). **Aromatic ketones** containing a carbonyl group attached directly to an aromatic ring are conveniently prepared by Friedel–Crafts acylation (Sec. 21.5).

$$Ar{-}H \xrightarrow{RCOCl\ (Ar'COCl),\ AlCl_3}$$

$$Ar{-}Br \longrightarrow Ar{-}Li \longrightarrow Ar_2CuLi \xrightarrow{RCOCl\ (Ar'COCl)}$$

$$Ar{-}COOH \longrightarrow Ar{-}COCl \xrightarrow{R_2CuLi\ (Ar'_2CuLi)}$$

$$\longrightarrow Ar{-}\underset{\underset{O}{\|}}{C}{-}R \quad (Ar{-}\underset{\underset{O}{\|}}{C}{-}Ar')$$
Preparation of aromatic ketones

As we see, important precursors of both aldehydes and ketones are *acid chlorides*. These are conveniently made from the corresponding carboxylic acids by treatment with thionyl chloride ($SOCl_2$), phosphorus trichloride (PCl_3), or phosphorus pentachloride (PCl_5). Since we already know several of the most

$$R{-}\overset{\overset{O}{\|}}{C}{-}OH \ + \ \left\{ \begin{array}{l} SOCl_2 \\ PCl_3 \\ PCl_5 \end{array} \right\} \ \longrightarrow \ R{-}\overset{\overset{O}{\|}}{C}{-}Cl$$

Carboxylic acid   Acid chloride

important ways of making carboxylic acids—oxidation of primary alcohols (Sec. 18.6) and oxidation of toluenes (Sec. 15.11)—we can begin to fit these syntheses of carbonyl compounds into the overall framework of organic chemistry.

21.5 Preparation of ketones by Friedel–Crafts acylation

One of the most important modifications of the Friedel–Crafts reaction involves the use of acid chlorides rather than alkyl halides. An acyl group, RCO—, becomes attached to the aromatic ring, thus forming a ketone; the process is called **acylation**. As usual for the Friedel–Crafts reaction (Sec. 15.9), the aromatic ring undergoing substitution must be at least as reactive as that of a halobenzene; catalysis by aluminum chloride or another Lewis acid is required.

$$Ar{-}H + R{-}C\overset{O}{\underset{Cl}{\diagdown}} \quad \xrightarrow{AlCl_3} \quad Ar{-}\underset{\underset{O}{\|}}{C}{-}R + HCl$$

An acid chloride　　　　　A ketone

The most likely mechanism for Friedel–Crafts acylation is analogous to the carbocation mechanism for Friedel–Crafts alkylation (Sec. 14.10), and involves the following steps:

$$(1) \qquad RCOCl + AlCl_3 \longrightarrow RC{\equiv}\overset{\oplus}{O} + AlCl_4{}^-$$

$$(2) \qquad ArH + RC{\equiv}\overset{\oplus}{O} \longrightarrow \overset{\oplus}{Ar}\overset{H}{\underset{COR}{\diagup}}$$

$$(3) \qquad \overset{\oplus}{Ar}\overset{H}{\underset{COR}{\diagup}} + AlCl_4{}^- \longrightarrow Ar{-}\underset{\underset{O}{\|}}{C}{-}R + HCl + AlCl_3$$

This fits the pattern of electrophilic aromatic substitution, the attacking reagent this time being the **acylium ion**, $R{-}C{\equiv}\overset{\oplus}{O}$. The acylium ion is considerably more stable than ordinary carbocations since in it every atom has an octet of electrons.

Alternatively, it may be that the electrophile is a complex between acid chloride and Lewis acid:

$$R{-}C\overset{\overset{+}{O}{-}\bar{A}lCl_3}{\underset{Cl}{\diagdown}}$$

In this case, from the standpoint of the acid chloride, reaction is acid-catalyzed nucleophilic acyl substitution, of the kind discussed in Sec. 24.4, with the aromatic ring acting as the nucleophile.

In planning the synthesis of diaryl ketones, ArCOAr′, it is particularly important to select the right combination of ArCOCl and Ar′H. As shown below, in the preparation of *m*-nitrobenzophenone, for example, the nitro group can be present in the acid chloride but not in the ring undergoing substitution, since as a strongly deactivating group it prevents the Friedel–Crafts reaction (Sec. 15.9).

Friedel–Crafts acylation is one of the most important methods of preparing ketones in which the carbonyl group is attached to an aromatic ring. Once formed, these ketones may be converted into secondary alcohols by reduction, into tertiary alcohols by reaction with Grignard reagents, and into many other important classes of compounds, as we shall see.

Of particular importance is the conversion of the acyl group into an alkyl group. This can be accomplished by the **Clemmensen reduction** (amalgamated zinc and concentrated hydrochloric acid), or the **Wolff–Kishner reduction** (hydrazine and base). For example:

A straight-chain alkyl group longer than ethyl generally cannot be attached in good yield to an aromatic ring by Friedel–Crafts alkylation because of rearrangement (Sec. 15.8). Such a group is readily introduced, however, in two steps: (1) formation of a ketone by Friedel–Crafts acylation (or by the reaction of an organocopper compound with an acyl chloride, described in the following section); (2) Clemmensen or Wolff–Kishner reduction of the ketone.

21.6   Preparation of ketones by use of organocopper compounds

Treatment of alkyl or aryl halides with lithium metal gives organolithium compounds (Sec. 17.14) which, on treatment with a cuprous halide, form lithium organocuprates, R_2CuLi or Ar_2CuLi. Since the late 1960s such organocopper

$$RX \xrightarrow{Li} RLi \xrightarrow{CuX} R_2CuLi$$

A lithium dialkylcuprate

$$ArX \xrightarrow{Li} ArLi \xrightarrow{CuX} Ar_2CuLi$$

A lithium diarylcuprate

compounds have found rapidly increasing application to organic synthesis because of their remarkable ability to form carbon–carbon bonds. We have already (Sec. 3.17) encountered their reaction with alkyl halides to form alkanes.

Lithium organocuprates react readily with acid chlorides to yield ketones. Here, as in its other reactions (Sec. 24.4), the acid chloride is undergoing nucleo-

$$R'{-}CuLi \ + \ R{-}\overset{\displaystyle O}{\underset{\displaystyle Cl}{C}} \longrightarrow R{-}\overset{\displaystyle \ }{\underset{\displaystyle O}{C}}{-}R' \qquad \begin{array}{l}R \ and \ R' \\ may \ be \\ alkyl \ or \ aryl\end{array}$$

An acid chloride   A ketone

philic substitution, the nucleophile being the basic alkyl or aryl group of the organometallic compound.

Grignard reagents (or organolithiums) react readily with acid chlorides, too, but the products are usually tertiary alcohols, formed by reaction of initially formed ketones with additional Grignard reagents. (If tertiary alcohols are desired, they are better prepared from esters than from acid chlorides (Sec. 24.21).) Organocopper reagents are less reactive than Grignard reagents toward the carbonyl group of ketones, and reaction stops at the ketone stage.

It is interesting that organocopper compounds are *more* reactive than Grignard reagents toward many kinds of compounds—alkyl halides, for example, which in general are not attacked by Grignard reagents. Organocopper compounds are highly *selective* toward different functional groups, and this selectivity is a major factor in determining their usefulness.

This lower reactivity of organocopper compounds not only makes the synthesis of ketones possible, but in addition widens the applicability of the method. Organocopper reagents do not react with many of the functional groups with which Grignard reagents and organolithiums do react: $-NO_2$, $-CN$, $-CO-$, $-COOR$, for example. Consequently, the presence of one of these groups in the acid chloride does not interfere with the synthesis of a ketone (compare with Sec. 17.17). For example:

$$2O_2N{-}\langle\bigcirc\rangle{-}COCl \ + \ (CH_3)_2CuLi \longrightarrow 2O_2N{-}\langle\bigcirc\rangle{-}\overset{\displaystyle \ }{\underset{\displaystyle O}{C}}{-}CH_3$$

p-Nitrobenzoyl   Lithium
chloride   dimethylcuprate   p-Nitroacetophenone
(Methyl p-nitrophenyl ketone)

CH$_3$OCCH$_2$CH$_2$CCl + [(CH$_3$)$_2$CHCH$_2$CH$_2$]$_2$CuLi $\longrightarrow$

|| ||

O   O

Lithium
diisopentylcuprate

CH$_3$OCCH$_2$CH$_2$CCH$_2$CH$_2$CH(CH$_3$)$_2$

|| ||

O   O

Methyl 4-oxo-7-methyloctanoate
(A γ-keto ester)

Problem 21.1   Would it be feasible to make *p*-nitroacetophenone via a reaction between lithium di(*p*-nitrophenyl)cuprate, (*p*-O$_2$NC$_6$H$_4$)$_2$CuLi, and acetyl chloride?

21.7   Reactions. Nucleophilic addition

The carbonyl group, C=O, governs the chemistry of aldehydes and ketones. It does this in two ways: (a) by providing a site for nucleophilic addition, and (b) by increasing the acidity of the hydrogen atoms attached to the *alpha* carbon. Both these effects are quite consistent with the structure of the carbonyl group and, in fact, are due to the same thing: the ability of oxygen to accommodate a negative charge.

In this section, we shall examine the carbonyl group as a site for nucleophilic addition; in Sec. 25.1, we shall see how the acid-strengthening effect arises.

The carbonyl group contains a carbon–oxygen double bond; since the mobile π electrons are pulled strongly toward oxygen, carbonyl carbon is electron-deficient and carbonyl oxygen is electron-rich. Because it is flat, this part of the molecule is open to relatively unhindered attack from above or below, in a direction perpendicular to the plane of the group. It is not surprising that this accessible, polarized group is highly reactive.

What kind of reagents will attack such a group? Since the important step in these reactions is the formation of a bond to the electron-deficient (electrophilic) carbonyl carbon, the carbonyl group is most susceptible to attack by electron-rich, nucleophilic reagents, that is, by bases. **The typical reaction of aldehydes and ketones is nucleophilic addition.**

Nucleophilic addition

Reactant	Transition state	Product	
Trigonal	*Becoming tetrahedral*	*Tetrahedral*	
	Partial negative charge on oxygen	*Negative charge on oxygen*	

As might be expected, we can get a much truer picture of the reactivity of the carbonyl group by looking at the transition state for attack by a nucleophile. In the reactant, carbon is trigonal. In the transition state, carbon has begun to acquire the tetrahedral configuration it will have in the product; the attached groups are thus being brought closer together. We might expect moderate steric hindrance in this reaction; that is, larger groups (R and R′) will tend to resist crowding more than smaller groups. But the transition state is a relatively roomy one compared, say, with the transition state for an S_N2 reaction, with its pentavalent carbon; it is this comparative uncrowdedness that we are really referring to when we say that the carbonyl group is "accessible" to attack (see Fig. 21.2).

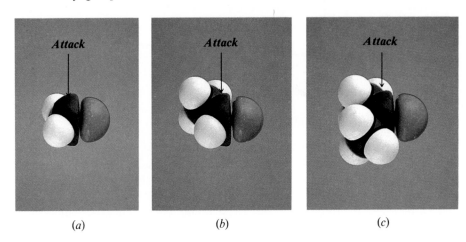

Figure 21.2 Molecular structure and reactivity: nucleophilic attack on the carbonyl group. Models of: (a) formaldehyde, HCHO; (b) acetaldehyde, CH_3CHO; (c) acetone, CH_3COCH_3. The flat carbonyl group is open to attack from above (or below). As hydrogen is replaced by the larger methyl, moderate crowding lowers reactivity.

In the transition state, oxygen has started to acquire the electrons—and the negative charge—that it will have in the product. *It is the tendency of oxygen to acquire electrons—its ability to carry a negative charge—that is the real cause of the reactivity of the carbonyl group toward nucleophiles.* (The polarity of the carbonyl group is not the *cause* of the reactivity; it is simply another *manifestation* of the electronegativity of oxygen.)

Aldehydes generally undergo nucleophilic addition more readily than ketones. This difference in reactivity is consistent with the transition states involved, and seems to be due to a combination of electronic and steric factors. A ketone contains a second alkyl or aryl group where an aldehyde contains a hydrogen atom. A second alkyl or aryl group of a ketone is larger than the hydrogen of an aldehyde, and resists more strongly the crowding together in the transition state (Fig. 21.2). An alkyl group releases electrons, and thus destabilizes the transition state by intensifying the negative charge developing on oxygen.

An aryl group has an electron-withdrawing inductive effect (Sec. 23.14), and we might have expected it to stabilize the transition state and thus speed up reaction; however, it seems to stabilize the *reactant* even more, by resonance (contribution by I), and thus causes net deactivation.

$$\left\langle \underset{}{+} \right\rangle = \overset{\overset{\text{R}}{|}}{\text{C}} - \ddot{\text{O}} \text{:}^{-}$$

I

If acid is present, hydrogen ion becomes attached to carbonyl oxygen. This prior protonation lowers the E_{act} for nucleophilic attack, since it permits oxygen to

Acid-catalyzed nucleophilic addition

$$\underset{R}{\overset{R'}{\diagdown}}C=O \;\underset{}{\overset{H^+}{\rightleftarrows}}\; \underset{R}{\overset{R'}{\diagdown}}\overset{:Z}{\underset{}{C=\overset{\oplus}{O}H}} \longrightarrow \left[\underset{R}{\overset{Z}{\underset{|}{R'-C\cdots\overset{\delta+}{OH}}}} \right] \longrightarrow \underset{R}{\overset{Z}{\underset{|}{R'-C}}}\diagdown OH$$

*Undergoes nucleophilic
attack more readily*

acquire the π electrons without having to accept a negative charge. Thus nucleophilic addition to aldehydes and ketones can be catalyzed by acids (sometimes, by *Lewis* acids).

REACTIONS OF ALDEHYDES AND KETONES

 1. **Oxidation.** Discussed in Sec. 21.8.

  (a) **Aldehydes**

$$\text{R—CHO or Ar—CHO} \quad \boxed{\begin{array}{c} Ag(NH_3)_2{}^+ \\ \hline KMnO_4 \\ \hline K_2Cr_2O_7 \end{array}} \longrightarrow \text{R—COOH or Ar—COOH}$$

Used chiefly for detection of aldehydes

Example:

$$CH_3CHO + 2Ag(NH_3)_2{}^+ + 3OH^- \longrightarrow 2Ag + CH_3COO^- + 4NH_3 + 2H_2O$$

$\qquad\qquad\quad$ *Colorless* $\qquad\qquad\qquad\qquad$ *Silver*
$\qquad\qquad\quad$ *solution* $\qquad\qquad\qquad\qquad$ *mirror*

Tollens' test

  (b) **Methyl ketones**

$$\underset{O}{\overset{}{R-\overset{\|}{C}-CH_3}} \text{ or } \underset{O}{\overset{}{Ar-\overset{\|}{C}-CH_3}} \xrightarrow{OX^-} R-COO^- \text{ or } Ar-COO^- + CHX_3$$

Haloform reaction

Examples:

$$\underset{O}{\overset{}{C_2H_5-\overset{\|}{C}-CH_3}} + 3OI^- \longrightarrow C_2H_5COO^- + CHI_3 + 2OH^-$$

$\qquad\qquad\qquad\qquad\qquad\qquad\qquad$ Iodoform
$\qquad\qquad\qquad\qquad\qquad\qquad\qquad$ *Yellow; m.p. 119 °C*

$$\underset{O}{\overset{CH_3}{\underset{}{CH_3\overset{|}{C}=CHC\overset{\|}{C}H_3}}} \xrightarrow[60\ C]{KOCl} CHCl_3 + \overset{CH_3}{\underset{}{CH_3\overset{|}{C}=CHCOOK}} \xrightarrow{H_2SO_4} \overset{CH_3}{\underset{}{CH_3\overset{|}{C}=CHCOOH}}$$

Mesityl oxide $\qquad\qquad\qquad\qquad\qquad\qquad\qquad\qquad\qquad\qquad\qquad$ 3-Methyl-2-butenoic acid
(4-Methyl-3-penten-2-one)

CONTINUED

CONTINUED

2. Reduction

(a) Reduction to alcohols. Discussed in Sec. 21.9.

Examples:

Cyclopentanone   Cyclopentanol

Acetophenone   α-Phenylethyl alcohol

(b) Reduction to hydrocarbons. Discussed in Sec. 21.9.

Clemmensen reduction
for compounds sensitive to base

Wolff–Kishner reduction
for compounds sensitive to acid

Examples:

n-Butyrophenone
(Phenyl *n*-propyl ketone)

n-Butylbenzene

Cyclopentanone   Cyclopentane

(c) Reductive amination. Discussed in Sec. 26.11.

CONTINUED

—— CONTINUED ——

3. **Addition of Grignard reagents.** Discussed in Secs. 17.14–17.17 and 21.10.

$$\underset{O}{\overset{\diagup}{\underset{\parallel}{C}}} + R{-}MgX \longrightarrow \underset{OMgX}{\overset{\mid}{\underset{\mid}{-C-R}}} \xrightarrow{H_2O} \underset{OH}{\overset{\mid}{\underset{\mid}{-C-R}}}$$

4. **Addition of cyanide. Cyanohydrin formation.** Discussed in Sec. 21.11.

$$\underset{O}{\overset{\diagup}{\underset{\parallel}{C}}} + CN^{-} \xrightarrow{H^{+}} \underset{OH}{\overset{\mid}{\underset{\mid}{-C-CN}}}$$

$$\text{Cyanohydrin}$$

Examples:

Benzaldehyde $\xrightarrow[\text{NaHSO}_3]{\text{NaCN}}$ Mandelonitrile $\xrightarrow{\text{H}_2\text{O, HCl}}$ Mandelic acid

$$CH_3{-}\underset{O}{\overset{\parallel}{C}}{-}CH_3 + NaCN \xrightarrow{H_2SO_4} CH_3{-}\underset{OH}{\overset{CH_3}{\underset{\mid}{C}}}{-}CN \xrightarrow{H_2O,\,H_2SO_4} \left[CH_3{-}\underset{OH}{\overset{CH_3}{\underset{\mid}{C}}}{-}COOH \right]$$

Acetone   Acetone cyanohydrin

$$\downarrow$$

$$CH_2{=}\underset{}{\overset{CH_3}{\underset{\mid}{C}}}{-}COOH$$

Methacrylic acid
(2-Methylpropenoic acid)

5. **Addition of derivatives of ammonia.** Discussed in Sec. 21.12.

$$\underset{O}{\overset{\diagup}{\underset{\parallel}{C}}} + H_2N{-}G \longrightarrow \left[\underset{OH}{\overset{\mid}{\underset{\mid}{-C-NH-G}}} \right] \longrightarrow \overset{\diagup}{\underset{\diagdown}{C}}{=}N{-}G + H_2O \qquad \textit{Used for identification}$$

$H_2N{-}G$		Product	
$H_2N{-}OH$	Hydroxylamine	$\diagdown$C$=$N${-}$OH	Oxime
$H_2N{-}NH_2$	Hydrazine	$\diagdown$C$=$N${-}$NH$_2$	Hydrazone
$H_2N{-}NHC_6H_5$	Phenylhydrazine	$\diagdown$C$=$N${-}$NHC$_6$H$_5$	Phenylhydrazone
$H_2N{-}NHCONH_2$	Semicarbazide	$\diagdown$C$=$N${-}$NHCONH$_2$	Semicarbazone

—— CONTINUED ——

_____ CONTINUED _____

Examples:

$$\underset{\text{Acetaldehyde}}{\overset{\displaystyle\text{H}}{\underset{\displaystyle}{\text{CH}_3\text{C}}}=\text{O}} + \underset{\text{Hydroxylamine}}{\text{H}_2\text{N}-\text{OH}} \xrightarrow{\text{H}^+} \underset{\text{Acetaldoxime}}{\overset{\displaystyle\text{H}}{\text{CH}_3\text{C}}=\text{NOH}} + \text{H}_2\text{O}$$

$$\underset{\text{Benzaldehyde}}{\overset{\displaystyle\text{H}}{\bigcirc\!\!-\text{C}}=\text{O}} + \underset{\text{Phenylhydrazine}}{\text{H}_2\text{N}-\text{NHC}_6\text{H}_5} \xrightarrow{\text{H}^+} \underset{\text{Benzaldehyde phenylhydrazone}}{\overset{\displaystyle\text{H}}{\bigcirc\!\!-\text{C}}=\text{NNHC}_6\text{H}_5} + \text{H}_2\text{O}$$

6. **Addition of alcohols. Acetal formation.** Discussed in Sec. 21.13.

$$\underset{\text{O}}{\overset{\displaystyle}{\text{C}}} + 2\text{ROH} \underset{\displaystyle}{\overset{\text{H}^+}{\rightleftharpoons}} \underset{\underset{\text{An acetal}}{\displaystyle\text{OR}}}{-\text{C}-\text{OR}} + \text{H}_2\text{O}$$

Example:

$$\underset{\text{Acetaldehyde}}{\text{CH}_3-\overset{\displaystyle\text{H}}{\text{C}}=\text{O}} + 2\text{C}_2\text{H}_5\text{OH} \xrightarrow{\text{HCl}} \underset{\underset{\substack{\text{Acetal}\\ \text{(Acetaldehyde}\\ \text{diethyl acetal)}}}{\displaystyle\text{OC}_2\text{H}_5}}{\text{CH}_3-\overset{\displaystyle\text{H}}{\underset{\displaystyle}{\text{C}}}-\text{OC}_2\text{H}_5} + \text{H}_2\text{O}$$

7. **Cannizzaro reaction.** Discussed in Sec. 21.14.

$$2\underset{\substack{\textit{An aldehyde with}\\ \textit{no } \alpha\textit{-hydrogens}}}{-\overset{\displaystyle\text{H}}{\text{C}}=\text{O}} \xrightarrow{\text{strong base}} \underset{\underset{\text{salt}}{\text{Acid}}}{-\text{COO}^+} + \underset{\text{Alcohol}}{-\text{CH}_2\text{OH}}$$

Examples:

$$2\text{HCHO} \xrightarrow{50\% \text{ NaOH, room temperature}} \text{HCOO}^- + \text{CH}_3\text{OH}$$
$$\underset{\text{Formaldehyde}}{} \qquad\qquad\qquad\qquad \underset{\text{Formate ion}}{} \ \ \underset{\text{Methanol}}{}$$

$$2\ \underset{\text{m-Chlorobenzaldehyde}}{\overset{\displaystyle\text{CHO}}{\bigcirc\!\!\!-\text{Cl}}} \xrightarrow{50\% \text{ KOH}} \underset{\substack{\text{m-Chlorobenzoate}\\ \text{ion}}}{\overset{\displaystyle\text{COO}^-}{\bigcirc\!\!\!-\text{Cl}}} + \underset{\substack{\text{m-Chlorobenzyl}\\ \text{alcohol}}}{\overset{\displaystyle\text{CH}_2\text{OH}}{\bigcirc\!\!\!-\text{Cl}}}$$

CONTINUED

CHO ⟨benzene ring⟩ OCH₃ OCH₃ + HCHO →(50% NaOH, 65 °C)→ CH₂OH ⟨benzene ring⟩ OCH₃ OCH₃ + HCOO⁻   Crossed Cannizzaro reaction

Veratraldehyde   3,4-Dimethoxybenzyl alcohol
3,4-Dimethoxybenzaldehyde

8. **Halogenation of ketones.** Discussed in Secs. 25.3–25.4.

$$-\overset{O}{\overset{\|}{C}}-\overset{H}{\underset{\|}{C}}- + X_2 \xrightarrow{\text{acid or base}} -\overset{O}{\overset{\|}{C}}-\overset{X}{\underset{\|}{C}}- + HX \qquad \alpha\text{-}Halogenation$$

$$X_2 = Cl_2, Br_2, I_2$$

9. **Addition of carbanions.**

   (a) **Aldol condensation.** Discussed in Secs. 25.5–25.8.

   (b) **Reactions related to aldol condensation.** Discussed in Sec. 25.9.

   (c) **Wittig reaction.** Discussed in Sec. 25.10.   ■

21.8 Oxidation

Aldehydes are easily oxidized to carboxylic acids; ketones are not. Oxidation is the reaction in which aldehydes differ most from ketones, and this difference stems directly from their difference in structure: by definition, an aldehyde has a hydrogen atom attached to the carbonyl carbon, and a ketone has not. Regardless of exact mechanism, this hydrogen is abstracted in oxidation, either as a proton or as an atom, and the analogous reaction for a ketone—abstraction of an alkyl or aryl group—does not take place.

Aldehydes are oxidized not only by the same reagents that oxidize primary and secondary alcohols—permanganate and dichromate—but also by the very mild oxidizing agent silver ion. Oxidation by silver ion requires an alkaline medium; to prevent precipitation of the insoluble silver oxide, a complexing agent is added: ammonia.

Tollens' reagent contains the silver ammonia ion, $Ag(NH_3)_2^+$. Oxidation of the aldehyde is accompanied by reduction of silver ion to free silver (in the form of a *mirror* under the proper conditions).

$$RCHO + Ag(NH_3)_2^+ \longrightarrow RCOO^- + Ag$$

Colorless   *Silver*
solution   *mirror*

(Oxidation by complexed cupric ion is a characteristic of certain substituted carbonyl compounds, and will be taken up with *carbohydrates* in Sec. 38.6.)

Oxidation by Tollens' reagent is useful chiefly for detecting aldehydes, and in particular for differentiating them from ketones (see Sec. 21.15). The reaction is of value in synthesis in those cases where aldehydes are more readily available than the corresponding acids: in particular, for the synthesis of unsaturated acids from the unsaturated aldehydes obtained from the aldol condensation (Sec. 25.6), where advantage is taken of the fact that Tollens' reagent does not attack carbon–carbon double bonds.

$$\underset{\alpha,\beta\text{-Unsaturated aldehyde}}{\overset{\overset{\text{H}}{|}}{\underset{\beta}{\text{R}}\text{CH}\overset{}{=}\underset{\alpha}{\text{CH}}\text{—C}\text{=O}}} \xrightarrow{\text{Tollens' reagent}} \underset{\alpha,\beta\text{-Unsaturated acid}}{\underset{\beta}{\text{R}}\text{CH}\overset{}{=}\underset{\alpha}{\text{CH}}\text{—COOH}}$$

Oxidation of ketones requires breaking of carbon–carbon bonds, and (except for the haloform reaction) takes place only under vigorous conditions. Cleavage involves the double bond of the *enol* form (Sec. 11.10) and, where the structure

$$\underset{\text{Enol}}{\overset{\text{OH}}{\underset{|}{\overset{|}{\text{—C—C}}}}\overset{}{=}\overset{|}{\underset{\text{H}}{\text{C—}}}} \rightleftharpoons \underset{\text{Ketone}}{\overset{\text{O}}{\overset{||}{\underset{\text{H}}{\text{—C—C—C—}}}}\underset{\text{H}}{}} \rightleftharpoons \underset{\text{Enol}}{\overset{\text{OH}}{\overset{|}{\underset{\text{H}}{\text{—C}=\text{C—C—}}}}}$$

permits, occurs on either side of the carbonyl group; in general, then, mixtures of carboxylic acids are obtained (see Sec. 8.23).

Problem 21.2 Predict the product(s) of vigorous oxidation of: (a) 3-hexanone; (b) cyclohexanone.

Methyl ketones are oxidized smoothly by means of hypohalite in the haloform reaction (Sec. 18.19). Besides being commonly used to detect these ketones (Sec. 21.15), this reaction is often useful in synthesis, hypohalite having the special advantage of not attacking carbon–carbon double bonds. For example:

Available by aldol condensation
(Sec. 25.8)

α-Methylcinnamic acid

21.9 Reduction

Aldehydes can be reduced to primary alcohols, and ketones to secondary alcohols, either by catalytic hydrogenation or by use of chemical reducing agents like lithium aluminum hydride, $LiAlH_4$. Such reduction is useful for the preparation of certain alcohols that are less available than the corresponding carbonyl compounds, in particular carbonyl compounds that can be obtained by the aldol condensation (Sec. 25.7). For example:

Cyclopentanone   Cyclopentanol

$$CH_3CH=CHCHO \xrightarrow{H_2, \, Ni} CH_3CH_2CH_2CH_2OH$$

Crotonaldehyde   *n*-Butyl alcohol
From aldol condensation
of acetaldehyde

$CH=CHCHO \xrightarrow{9\text{-BBN}} \xrightarrow{HOCH_2CH_2NH_2}$ $CH=CHCH_2OH$

Cinnamaldehyde   Cinnamyl alcohol
From aldol condensation
of benzaldehyde and acetaldehyde
(Sec. 25.8)

To reduce a carbonyl group that is conjugated with a carbon–carbon double bond without reducing the carbon–carbon double bond, too, requires a *regioselective* reducing agent. One of these is shown above, and will be discussed in Sec. 25.7.

Aldehydes and ketones can be reduced to hydrocarbons by the action (a) of amalgamated zinc and concentrated hydrochloric acid, the **Clemmensen reduction**; or (b) of hydrazine, NH_2NH_2, and a strong base like KOH or potassium *tert*-butoxide, the **Wolff–Kishner reduction**. These are particularly important when applied to the alkyl aryl ketones obtained from Friedel–Crafts acylation, since this reaction sequence permits, indirectly, the attachment of straight alkyl chains to the benzene ring. For example:

Resorcinol   4-*n*-Hexylresorcinol
Used as an antiseptic

A special sort of oxidation and reduction, the *Cannizzaro reaction*, will be discussed in Sec. 21.14.

Let us look a little more closely at reduction by metal hydrides. Alcohols are formed from carbonyl compounds, smoothly and in high yield, by the action of such compounds as lithium aluminum hydride, $LiAlH_4$. Here again, we see

$$4R_2C=O + LiAlH_4 \longrightarrow (R_2CHO)_4AlLi \xrightarrow{H_2O} 4R_2CHOH + LiOH + Al(OH)_3$$

nucleophilic addition: this time the nucleophile is hydrogen transferred with a pair of electrons—as a hydride ion, $H:^-$—from the metal to carbonyl carbon:

21.10 Addition of Grignard reagents

The addition of Grignard reagents to aldehydes and ketones has already been discussed as one of the most important methods of preparing complicated alcohols (Secs. 17.14–17.17).

The organic group, transferred *with a pair of electrons* from magnesium to carbonyl carbon, is a powerful nucleophile.

$$\text{C=O} + \text{R—MgX} \longrightarrow \overset{\text{R}}{\underset{}{-\text{C}-\bar{\text{O}}\overset{+}{\text{MgX}}}}$$

21.11 Addition of cyanide

The elements of HCN add to the carbonyl group of aldehydes and ketones to yield compounds known as **cyanohydrins**:

$$\text{C} + \text{CN}^- \xrightarrow{\text{H}^+} -\overset{}{\underset{\text{OH}}{\text{C}}}-\text{CN}$$

A cyanohydrin

The reaction is often carried out by adding mineral acid to a mixture of the carbonyl compound and aqueous sodium cyanide.

Addition appears to involve nucleophilic attack on carbonyl carbon by the strongly basic cyanide ion; subsequently (or possibly simultaneously) oxygen accepts a hydrogen ion to form the cyanohydrin product:

$$\text{C} \longrightarrow -\overset{}{\underset{\text{O}^-}{\text{C}}}-\text{CN} \xrightarrow{\text{H}^+} -\overset{}{\underset{\text{OH}}{\text{C}}}-\text{CN}$$

:CN⁻   Cyanohydrin

Nucleophilic reagent

Although it is the elements of HCN that become attached to the carbonyl group, a highly acidic medium—in which the concentration of un-ionized HCN is highest—actually retards reaction. This is to be expected, since the very weak acid HCN is a poor source of cyanide ion.

Cyanohydrins are *nitriles* (see Sec. 23.8), and their principal use is based on the fact that, like other nitriles, they undergo hydrolysis; in this case the products are α-hydroxy acids or unsaturated acids. For example:

$$\underset{\text{O}_2\text{N}}{\text{⬡}}-\overset{\text{H}}{\underset{}{\text{C}}}=\text{O} \xrightarrow{\text{CN}^-,\ \text{H}^+} \underset{\text{O}_2\text{N}}{\text{⬡}}-\overset{\text{H}}{\underset{\text{OH}}{\text{C}}}-\text{CN} \xrightarrow{\text{HCl, heat}} \underset{\text{O}_2\text{N}}{\text{⬡}}-\overset{\text{H}}{\underset{\text{OH}}{\text{C}}}-\text{COOH}$$

m-Nitrobenzaldehyde   *m*-Nitromandelic acid

$$\underset{\substack{\text{Ethyl methyl ketone}\\\text{2-Butanone}}}{CH_3CH_2-\overset{\overset{\displaystyle CH_3}{|}}{C}=O} \xrightarrow{CN^-,\ H^+} \underset{\substack{\\OH}}{CH_3CH_2-\overset{\overset{\displaystyle CH_3}{|}}{\underset{|}{C}}-CN} \xrightarrow{H_2SO_4,\ heat} \left[CH_3CH_2-\overset{\overset{\displaystyle CH_3}{|}}{\underset{\underset{\displaystyle OH}{|}}{C}}-COOH \right]$$

$$\downarrow$$

$$\underset{\text{2-Methyl-2-butenoic acid}}{CH_3CH=\overset{\overset{\displaystyle CH_3}{|}}{C}-COOH}$$

Problem 21.3 Each of the following is converted into the cyanohydrin, and the products are separated by careful fractional distillation, crystallization, or chromatography. For each reaction tell how many fractions will be collected, and whether each fraction, as collected, will be optically active or inactive, resolvable or non-resolvable.
  (a) Acetaldehyde; (b) benzaldehyde; (c) acetone.
  (d) R-(+)-glyceraldehyde, $CH_2OHCHOHCHO$; (e) ($\pm$)-glyceraldehyde.
  (f) How would your answer to each of the above be changed if each mixture were subjected to hydrolysis to hydroxy acids before fractionation?

21.12   Addition of derivatives of ammonia

Certain compounds related to ammonia add to the carbonyl group to form derivatives that are important chiefly for the characterization and identification of aldehydes and ketones (Sec. 21.15). The products contain a carbon–nitrogen double bond resulting from elimination of a molecule of water from the initial addition products. Some of these reagents and their products are:

$$\underset{O}{\overset{\diagdown}{\underset{\diagup}{C}}} + :NH_2OH \xrightarrow{H^+} \left[-\overset{|}{\underset{\underset{\displaystyle OH}{|}}{C}}-NHOH \right] \longrightarrow \underset{\text{Oxime}}{\overset{\diagdown}{\underset{\diagup}{C}}=NOH} + H_2O$$

Hydroxylamine

$$\underset{O}{\overset{\diagdown}{\underset{\diagup}{C}}} + :NH_2NHC_6H_5 \xrightarrow{H^+} \left[-\overset{|}{\underset{\underset{\displaystyle OH}{|}}{C}}-NHNHC_6H_5 \right] \longrightarrow \underset{\text{Phenylhydrazone}}{\overset{\diagdown}{\underset{\diagup}{C}}=NNHC_6H_5} + H_2O$$

Phenylhydrazine

$$\underset{O}{\overset{\diagdown}{\underset{\diagup}{C}}} + :NH_2NHCONH_2 \xrightarrow{H^+} \left[-\overset{|}{\underset{\underset{\displaystyle OH}{|}}{C}}-NHNHCONH_2 \right] \longrightarrow$$

Semicarbazide

$$\underset{\text{Semicarbazone}}{\overset{\diagdown}{\underset{\diagup}{C}}=NNHCONH_2} + H_2O$$

Like ammonia, these derivatives of ammonia are basic, and therefore react with acids to form salts: hydroxylamine hydrochloride, $HONH_3^+Cl^-$; phenylhydrazine hydrochloride, $C_6H_5NHNH_3^+Cl^-$; and semicarbazide hydrochloride, $NH_2CONHNH_3^+Cl^-$. The salts are less easily oxidized by air than the free bases, and it is in this form that the reagents are best preserved and handled. When needed, the basic reagents are liberated from their salts in the presence of the carbonyl compound by addition of a base, usually sodium acetate.

$$C_6H_5NHNH_3{}^+Cl^- + CH_3COO^-Na^+ \rightleftarrows C_6H_5NHNH_2 + CH_3COOH + Na^+Cl^-$$

Phenylhydrazine hydrochloride	Sodium acetate	Phenylhydrazine	Acetic acid
Stronger acid	*Stronger base*	*Weaker base*	*Weaker acid*

It is often necessary to adjust the reaction medium to just the right acidity. Addition involves nucleophilic attack by the basic nitrogen compound on carbonyl carbon. Protonation of carbonyl oxygen makes carbonyl carbon more susceptible to nucleophilic attack; in so far as the carbonyl compound is concerned, then, addition will be favored by high acidity. But the ammonia derivative, H_2N—G, can also undergo protonation to form the ion, ^+H_3N—G, which lacks unshared electrons and is no longer nucleophilic; in so far as the nitrogen compound is concerned, then, addition is favored by low acidity. The conditions under which

$$\underset{O}{\overset{\diagdown}{C}}{\diagup} \quad \underset{H^+}{\rightleftarrows} \quad \underset{\oplus OH}{\overset{\diagdown}{C}}{\diagup} \longrightarrow \left[\begin{array}{c} H \\ | \overset{\oplus}{} \\ -C-N-G \\ | \quad | \\ HO \quad H \end{array} \right] \longrightarrow \underset{}{\overset{\diagdown}{C}}=N-G + H_2O + H^+$$

$$H_2\overset{..}{N}-G \quad \underset{H^+}{\rightleftarrows} \quad {}^+H_3N-G$$

Free base: Salt:
nucleophilic   *not nucleophilic*

addition proceeds most rapidly are thus the result of a compromise: the solution must be acidic enough for an appreciable fraction of the carbonyl compound to be protonated, but not so acidic that the concentration of the free nitrogen compound is too low. The exact conditions used depend upon the basicity of the reagent, and upon the reactivity of the carbonyl compound.

Problem 21.4 Semicarbazide (1 mol) is added to a mixture of cyclohexanone (1 mol) and benzaldehyde (1 mol). If the product is isolated immediately, it consists almost entirely of the semicarbazone of cyclohexanone; if the product is isolated after several hours, it consists almost entirely of the semicarbazone of benzaldehyde. How do you account for these observations? (*Hint*: See Sec. 10.27.)

21.13 Addition of alcohols. Acetal formation

Alcohols add to the carbonyl group of aldehydes in the presence of anhydrous acids to yield **acetals**:

$$\underset{\text{Aldehyde}}{\overset{H}{\underset{|}{R'-C}}=O} + \underset{\text{Alcohol}}{2ROH} \quad \underset{\text{dry HCl}}{\rightleftarrows} \quad \underset{\underset{\text{Acetal}}{\overset{|}{OR}}}{\overset{H}{\underset{|}{R'-C-OR}}} + H_2O$$

The reaction is carried out by allowing the aldehyde to stand with an excess of the anhydrous alcohol and a little anhydrous acid, usually hydrogen chloride. In the preparation of ethyl acetals the water is often removed as it is formed by means of the azeotrope of water, benzene, and ethyl alcohol (b.p. 64.9 °C, Sec. 17.7). (Simple *ketals* are usually difficult to prepare by reaction of ketones with alcohols, and are made in other ways.)

$$\text{Benzaldehyde} \qquad \text{Ethyl alcohol}$$

$$\underset{\text{Benzaldehyde}}{\overset{H}{\underset{|}{\text{C}}}{=}O} + 2C_2H_5OH \xrightarrow[]{\text{dry HCl}} \underset{\text{OC}_2H_5}{\overset{H}{\underset{|}{\text{C}}}{-}OC_2H_5} + H_2O$$

Diethyl acetal of benzaldehyde

There is good evidence that in alcoholic solution an aldehyde exists in equilibrium with a compound called a **hemiacetal**:

$$R'{-}\overset{H}{\underset{|}{\text{C}}}{=}O + ROH \underset{}{\overset{H^+}{\rightleftarrows}} R'{-}\underset{OH}{\overset{H}{\underset{|}{\text{C}}}{-}OR}$$

A hemiacetal

A hemiacetal is formed by the addition of the nucleophilic alcohol molecule to the carbonyl group; it is both an ether and an alcohol. With a few exceptions, hemiacetals are too unstable to be isolated.

In the presence of acid the hemiacetal, acting as an alcohol, reacts with more of the solvent alcohol to form the acetal, an ether:

$$R'{-}\underset{OH}{\overset{H}{\underset{|}{\text{C}}}{-}OR} + ROH \overset{H^+}{\rightleftarrows} R'{-}\underset{OR}{\overset{H}{\underset{|}{\text{C}}}{-}OR} + H_2O$$

Hemiacetal   Acetal
(An alcohol)   (An ether)

The reaction involves the formation (step 1) of the ion I, which then combines (step 2) with a molecule of alcohol to yield the protonated acetal. As we can see,

(1) $R'{-}\underset{OH}{\overset{H}{\underset{|}{\text{C}}}{-}OR} + H^+ \rightleftarrows R'{-}\underset{\oplus OH_2}{\overset{H}{\underset{|}{\text{C}}}{-}OR} \rightleftarrows R'{-}\underset{I}{\overset{H}{\underset{|}{\text{C}}}{=}\overset{\oplus}{O}R} + H_2O$

Hemiacetal

(2) $R'{-}\underset{I}{\overset{H}{\underset{|}{\text{C}}}{=}\overset{\oplus}{O}R} + ROH \rightleftarrows R'{-}\underset{\underset{H}{\overset{|}{\oplus OR}}}{\overset{H}{\underset{|}{\text{C}}}{-}OR} \rightleftarrows R'{-}\underset{OR}{\overset{H}{\underset{|}{\text{C}}}{-}OR} + H^+$

Acetal

this mechanism is strictly analogous to the S_N1 route we have previously encountered (Sec. 19.3) for the formation of ethers.

Acetal formation thus involves (a) nucleophilic addition to a carbonyl group, and (b) ether formation via a carbocation.

Acetals have the structure of ethers and, like ethers, are cleaved by acids and are stable toward bases. Acetals differ from ethers, however, in the extreme *ease* with which they undergo acidic cleavage; they are rapidly converted even at room

$$R'-\underset{\underset{OR}{|}}{\overset{\overset{H}{|}}{C}}-OR + H_2O \xrightarrow[\text{fast}]{H^+} R'-\overset{\overset{H}{|}}{C}=O + 2ROH$$

Acetal   Aldehyde   Alcohol

temperature into the aldehyde and alcohol by dilute mineral acids. The mechanism of hydrolysis is exactly the reverse of the one by which acetals are formed.

Problem 21.5 Account for the fact that anhydrous acids bring about formation of acetals whereas aqueous acids bring about hydrolysis of acetals.

The heart of the chemistry of acetals is the "carbocation",

$$\left[R-\underset{\oplus}{\overset{\overset{H}{|}}{C}}-OR \qquad R-\underset{\oplus}{\overset{\overset{H}{|}}{C}}=OR \right]$$

Ia   Ib

*Especially stable:
every atom has octet*

which is a hybrid of structures Ia and Ib. Contribution from Ib, in which every atom has an octet of electrons, makes this ion considerably more stable than ordinary carbocations. (Indeed, Ib *alone* may pretty well represent the ion, in which case it is not a carbocation at all but an *oxonium* ion.)

Now, generation of this cation is the rate-determining step both in formation of acetals (reading to the right in equation 1) and in their hydrolysis (reading to the left in equation 2). The same factor—the providing of electrons by oxygen—that stabilizes the ion also stabilizes the transition state leading to its formation. Generation of the ion is speeded up, and along with it the entire process: formation or hydrolysis of the acetal.

(Oddly enough, oxygen causes activation toward *nucleophilic* substitution here in precisely the same way it activates aromatic ethers toward *electrophilic* substitution (Sec. 14.18); the common feature is, of course, development of a positive charge in the transition state of the rate-determining step.)

We shall find the chemistry of hemiacetals and acetals to be fundamental to the study of carbohydrates (Chaps. 38 and 39).

Problem 21.6 (a) The following reaction is an example of what familiar synthesis?

(b) What structural factor favors this course of reaction? (c) To what family of compounds does II belong? (d) What will II yield upon treatment with acid? With base?

Problem 21.7   *Glyceraldehyde*, $CH_2OHCHOHCHO$, is commonly made from the acetal of acrolein, $CH_2=CH—CHO$. Show how this could be done. Why is acrolein itself not used?

Problem 21.8   How do you account for the following differences in ease of hydrolysis?

(a) $RC(OR')_3 \gg RCH(OR')_2 \gg RCH_2OR'$

An ortho   An acetal   An ether
ester

(b) $R_2C(OR')_2 > RCH(OR')_2 > H_2C(OR')_2$

A ketal   An acetal   A formal

Problem 21.9   The simplest way to prepare an aldehyde, $RCH^{18}O$, labeled at the carbonyl oxygen, is to allow an ordinary aldehyde to stand in $H_2^{18}O$ in the presence of a little acid. Suggest a detailed mechanism for this oxygen exchange.

21.14   Cannizzaro reaction

In the presence of concentrated alkali, aldehydes containing no α-hydrogens undergo self-oxidation-and-reduction to yield a mixture of an alcohol and a salt of a carboxylic acid. This reaction, known as the **Cannizzaro reaction**, is generally brought about by allowing the aldehyde to stand at room temperature with concentrated aqueous or alcoholic hydroxide. (Under these conditions an aldehyde containing α-hydrogens would undergo aldol condensation faster, Sec. 25.5.)

$$2HCHO \xrightarrow{50\% \ NaOH} CH_3OH + HCOO^- Na^+$$

Formaldehyde   Methanol   Sodium formate

$$O_2N\langle\bigcirc\rangle CHO \xrightarrow{35\% \ NaOH} O_2N\langle\bigcirc\rangle CH_2OH + O_2N\langle\bigcirc\rangle COO^- Na^+$$

p-Nitrobenzaldehyde   *p*-Nitrobenzyl alcohol   Sodium *p*-nitrobenzoate

In general, a mixture of two aldehydes undergoes a Cannizzaro reaction to yield all possible products. If one of the aldehydes is formaldehyde, however, reaction yields almost exclusively sodium formate and the alcohol corresponding to the other aldehyde:

$$ArCHO + HCHO \xrightarrow{conc. \ NaOH} ArCH_2OH + HCOO^- Na^+$$

Such a reaction is called a **crossed Cannizzaro reaction**. For example:

$$\underset{\substack{\text{CHO} \\ \\ \text{OCH}_3}}{\bigcirc} + HCHO \xrightarrow{conc. \ NaOH} \underset{\substack{\text{CH}_2\text{OH} \\ \\ \text{OCH}_3}}{\bigcirc} + HCOO^- Na^+$$

Anisaldehyde   *p*-Methoxybenzyl alcohol
(*p*-Methoxybenzaldehyde)

Evidence, chiefly from kinetics and experiments with isotopically labeled compounds, indicates that even this seemingly different reaction follows the familiar pattern for carbonyl compounds: nucleophilic addition. Two successive additions are involved: addition of hydroxide ion (step 1) to give intermediate I;

$$(1) \qquad \underset{\substack{\text{H} \\ |}}{\text{Ar—C=O}} + \text{OH}^- \; \rightleftharpoons \; \text{Ar—}\underset{\substack{| \\ \text{OH} \\ \\ \text{I}}}{\overset{\substack{\text{H} \\ |}}{\text{C}}}\text{—O}^-$$

$$(2) \qquad \underset{\substack{\text{H} \\ |}}{\text{Ar—C=O}} + \text{Ar—}\underset{\substack{| \\ \text{OH} \\ \\ \text{I}}}{\overset{\substack{\text{H} \\ |}}{\text{C}}}\text{—O}^- \; \longrightarrow \; \text{Ar—}\underset{\substack{| \\ \text{H}}}{\overset{\substack{\text{H} \\ |}}{\text{C}}}\text{—O}^- + \text{Ar—}\underset{\substack{| \\ \text{OH}}}{\overset{\substack{\text{H} \\ |}}{\text{C}}}\text{=O}$$

$$\qquad\qquad\qquad\qquad\qquad\qquad\qquad\qquad\qquad \downarrow {\scriptstyle +\text{H}^+} \qquad\qquad \downarrow {\scriptstyle -\text{H}^+}$$

$$\qquad\qquad\qquad\qquad\qquad\qquad\qquad\qquad\qquad \text{ArCH}_2\text{OH} \qquad\quad \text{ArCOO}^-$$

and addition of a hydride ion from I (step 2) to a second molecule of aldehyde. The presence of the negative charge on I aids in the loss of hydride ion.

Problem 21.10   In the case of some aldehydes there is evidence that intermediate II is the hydride donor in the Cannizzaro reactions. (a) How would II be formed from I?

$$\underset{\substack{| \\ \text{O}_- \\ \\ \text{II}}}{\overset{\substack{\text{H} \\ |}}{\text{R—C—O}^-}}$$

(b) Why would you expect II to be a better hydride donor than I? (*Hint:* What is one product of the hydride transfer from II?)

Problem 21.11   Suggest an experiment to prove that a hydride transfer of the kind shown in step (2) is actually involved, that is, that hydrogen is transferred from I and not from the solvent.

Problem 21.12   From examination of the mechanism, can you suggest one factor that would tend to make a crossed Cannizzaro reaction involving formaldehyde take place in the particular way it does?

Problem 21.13   Phenylglyoxal, C_6H_5COCHO, is converted by aqueous sodium hydroxide into sodium mandelate, $C_6H_5CHOHCOONa$. Suggest a likely mechanism for this conversion.

Problem 21.14   In the **benzilic acid rearrangement**, the diketone *benzil* is converted by sodium hydroxide into the salt of *benzilic acid*.

$$\underset{\text{Benzil}}{C_6H_5COCOC_6H_5} \xrightarrow{\text{OH}^-} (C_6H_5)_2C(OH)COO^- \xrightarrow{\text{H}^+} \underset{\text{Benzilic acid}}{(C_6H_5)_2C(OH)COOH}$$

If sodium methoxide is used instead of sodium hydroxide, the ester $(C_6H_5)_2C(OH)COOCH_3$ is obtained. Suggest a possible mechanism for this rearrangement.

21.15   Analysis of aldehydes and ketones

Aldehydes and ketones are characterized through the addition to the carbonyl group of nucleophilic reagents, especially derivatives of ammonia (Sec. 21.12). An aldehyde or ketone will, for example, react with 2,4-dinitrophenylhydrazine to form an insoluble yellow or red solid.

Aldehydes are characterized, and in particular are differentiated from ketones, through their ease of oxidation: aldehydes give a positive test with Tollens' reagent (Sec. 21.8); ketones do not. A positive Tollens' test is also given by a few other kinds of easily oxidized compounds, e.g., certain phenols and amines; these compounds do not, however, give positive tests with 2,4-dinitrophenylhydrazine.

Aldehydes are also, of course, oxidized by many other oxidizing agents: by cold, dilute, neutral $KMnO_4$ and by CrO_3 in H_2SO_4 (Sec. 8.24).

A highly sensitive test for aldehydes is the *Schiff test*. An aldehyde reacts with the fuchsin–aldehyde reagent to form a characteristic magenta color.

Aliphatic aldehydes and ketones having an α-hydrogen react with Br_2 in CCl_4. This reaction is generally too slow to be confused with a test for unsaturation, and moreover it liberates HBr.

Aldehydes and ketones are generally identified through the melting points of derivatives like 2,4-dinitrophenylhydrazones, oximes, and semicarbazones (Sec. 21.12).

Methyl ketones are characterized through the iodoform test (see Sec. 18. 9).

Problem 21.15 Make a table to summarize the behavior of each class of compound we have studied toward each of the oxidizing agents we have studied.

Problem 21.16 A convenient test for aldehydes and most ketones depends upon the fact that a carbonyl compound generally causes a change in color when it is added to a solution of hydroxylamine hydrochloride and an acid–base indicator. What is the basis of this test?

Problem 21.17 Expand the table you made in Problem 19.26, p. 722, to include aldehydes and ketones, and, in particular, to emphasize oxidizing agents.

21.16   Spectroscopic analysis of aldehydes and ketones

Infrared   Infrared spectroscopy is by far the best way to detect the presence of a carbonyl group in a molecule. The strong band due to C=O stretching appears at about 1700 cm^{-1}, where it is seldom obscured by other strong absorptions; it is one of the most useful bands in the infrared spectrum, and is often the first one looked for (see Fig. 21.3, on the next page).

C=O stretching, *strong*

| RCHO 1725 cm^{-1} | R_2CO 1710 cm^{-1} | Cyclobutanones 1780 cm^{-1} |
| ArCHO 1700 cm^{-1} | ArCOR 1690 cm^{-1} | Cyclopentanones 1740 cm^{-1} |

—C=C—CHO 1685 cm^{-1} —C=C—C=O 1675 cm^{-1} —C=C—C— 1540–1640 cm^{-1}
OH-----O
(enols)

IRDC 6137

IRDC 82

Figure 21.3   Infrared spectra of (a) n-butyraldehyde and (b) acetophenone.

The carbonyl band is given not only by aldehydes and ketones, but also by carboxylic acids and their derivatives. Once identified as arising from an aldehyde or ketone (see below), its exact frequency can give a great deal of information about the structure of the molecule.

The —CHO group of an aldehyde has a characteristic C—H stretching band near 2720 cm^{-1}; this, in conjunction with the carbonyl band, is fairly certain evidence for an aldehyde (see Fig. 21.3).

Carboxylic acids (Sec. 23.22) and esters (Sec. 24.25) also show carbonyl absorption, and in the same general region as aldehydes and ketones. Acids, however, also show the broad O—H band. Esters usually show the carbonyl band at somewhat higher frequencies than ketones of the same general structure; furthermore, esters show characteristic C—O stretching bands. (For a comparison of certain oxygen compounds, see Table 24.3, p. 890.)

NMR   The proton of an aldehyde group, —CHO, absorbs far downfield, at δ 9–10. Coupling of this proton with adjacent protons has a small constant (J 1–3 Hz), and the fine splitting is often seen superimposed on other splittings.

CMR   Carbonyl carbon is both sp^2-hybridized and attached to electronegative oxygen, and as a result is powerfully deshielded. It absorbs *farther downfield than any other kind of carbon*: in the range δ 190–220 for aldehydes and ketones.

Carboxylic acids and their derivatives also contain carbonyl carbon, and here, too, absorption is far downfield, although not so far as for aldehydes and ketones: in the range δ 150–185.

As an electronegative substituent, the carbonyl group strongly deshields adjacent carbons.

Ultraviolet   The ultraviolet spectrum can tell a good deal about the structure of carbonyl compounds: particularly, as we might expect from our earlier discussion (Sec. 16.5), about conjugation of the carbonyl group with a carbon–carbon double bond.

Saturated aldehydes and ketones absorb weakly in the near ultraviolet. Conjugation moves this weak band (the R band) to longer wavelengths (*Why?*) and, more important, moves a very intense band (the K band) from the far ultraviolet to the near ultraviolet.

$$-\overset{|}{\underset{}{C}}=O$$

λ_{max} 270–300 nm

ϵ_{max} 10–20

$$-\overset{|}{\underset{}{C}}=\overset{|}{\underset{}{C}}-\overset{|}{\underset{}{C}}=O$$

λ_{max} 300–350 nm λ_{max} 215–250 nm

ϵ_{max} 10–20 ϵ_{max} 10 000–20 000

The exact position of this K band gives information about the number and location of substituents in the conjugated system.

PROBLEMS

1. Neglecting enantiomerism, give structural formulas, common names, and IUPAC names for:

(a) the seven carbonyl compounds of formula $C_5H_{10}O$
(b) the five carbonyl compounds of formula C_8H_8O that contain a benzene ring

2. Give the structural formula of:

(a) acetone
(b) benzaldehyde
(c) isobutyl methyl ketone
(d) trimethylacetaldehyde
(e) acetophenone
(f) cinnamaldehyde
(g) 4-methylpentanal
(h) phenylacetaldehyde
(i) benzophenone
(j) α,γ-dimethylcaproaldehyde

(k) 3-methyl-2-pentanone
(l) 2-butenal
(m) 4-methyl-3-penten-2-one (mesityl oxide)
(n) 1,3-diphenyl-2-propen-1-one (benzal-acetophenone)
(o) 3-hydroxypentanal
(p) benzyl phenyl ketone
(q) salicylaldehyde
(r) *p,p'*-dihydroxybenzophenone
(s) *m*-tolualdehyde

3. Write balanced equations, naming all organic products, for the reaction (if any) of phenylacetaldehyde with:

(a) Tollens' reagent
(b) CrO_3/H_2SO_4
(c) cold dilute $KMnO_4$
(d) $KMnO_4$, H^+, heat
(e) H_2, Ni, 20 lb/in², 30 °C
(f) LiAlH₄
(g) NaBH₄
(h) C_6H_5MgBr, then H_2O

(i) isopropylmagnesium chloride, then H_2O
(j) HC≡CLi, then H_2O
(k) CN^-, H^+
(l) hydroxylamine
(m) phenylhydrazine
(n) 2,4-dinitrophenylhydrazine
(o) semicarbazide
(p) ethyl alcohol, dry HCl(g)

4. Answer Problem 3 for cyclohexanone.

5. Write balanced equations, naming all organic products, for the reaction (if any) of benzaldehyde with:

(a) conc. NaOH

(b) formaldehyde, conc. NaOH

(c) CN^-, H^+

(d) product (c) + H_2O, H^+, heat

(e) CH_3MgI, then H_2O

(f) product (e) + H^+, heat

(g) $(CH_3)_2{}^{14}CHMgBr$, then H_2O

(h) $H_2{}^{18}O$, H^+

6. Write equations for all steps in the synthesis of the following from propionaldehyde, using any other needed reagents:

(a) *n*-propyl alcohol

(b) propionic acid

(c) α-hydroxybutyric acid

(d) *sec*-butyl alcohol

(e) 1-phenyl-1-propanol

(f) ethyl methyl ketone

(g) *n*-propyl propionate

(h) 2-methyl-3-pentanol

7. Write equations for all steps in the synthesis of the following from acetophenone, using any other needed reagents:

(a) ethylbenzene

(b) benzoic acid

(c) α-phenylethyl alcohol

(d) 2-phenyl-2-butanol

(e) 1,1-diphenylethanol

(f) α-hydroxy-α-phenylpropionic acid

8. Outline all steps in a possible laboratory synthesis of each of the following from benzene, toluene, and alcohols of four carbons or fewer, using any needed inorganic reagents:

(a) isobutyraldehyde

(b) phenylacetaldehyde

(c) *p*-bromobenzaldehyde

(d) ethyl methyl ketone

(e) 2,4-dinitrobenzaldehyde

(f) *p*-nitrobenzophenone

(g) 2-methyl-3-pentanone

(h) benzyl methyl ketone

(i) *m*-nitrobenzophenone

(j) *n*-propyl *p*-tolyl ketone

(k) α-methylbutyraldehyde

(l) *n*-butyl isobutyl ketone

(m) *p*-nitroacetophenone

(n) 3-nitro-4'-methylbenzophenone

(o) *p*-nitropropiophenone

9. Outline all steps in a possible laboratory synthesis of each of the following from benzene, toluene, and alcohols of four carbons or fewer, using any needed inorganic reagents:

(a) *n*-butylbenzene

(b) α-hydroxy-*n*-valeric acid

(c) 2-methylheptane

(d) 2,3,5-trimethyl-3-hexanol

(e) *p*-nitro-α-hydroxyphenylacetic acid

(f) 1,2-diphenyl-2-propanol

(g) 1-*p*-bromophenyl-1-phenyl-1-propanol

(h) 3-methyl-2-butenoic acid

10. (a) What are A, B, and C?

$$C_6H_5C(CH_3)_2CH_2COOH + PCl_3 \longrightarrow A\ (C_{11}H_{13}OCl)$$
$$A + AlCl_3/CS_2 \longrightarrow B\ (C_{11}H_{12}O)$$
$$B + N_2H_4, OH^-, heat, high\text{-}boiling\ solvent \longrightarrow C\ (C_{11}H_{14})$$

C gave the following proton NMR spectrum:

a singlet, δ 1.22, 6H

b triplet, δ 1.85, 2H, $J = 7$ Hz

c triplet, δ 2.83, 2H, $J = 7$ Hz

d singlet, δ 7.02, 4H

(b) C was also formed by treatment of the alcohol D ($C_{11}H_{16}O$) with concentrated sulfuric acid. What is the structure of D?

11. Give stereochemical formulas for compounds E–J.

R-(+)-*glyceraldehyde* ($CH_2OHCHOHCHO$) + CN^-, H^+ $\longrightarrow$ E + F

(both E and F have the formula $C_4H_7O_3N$)

E + F + OH^-, H_2O, heat; then H^+ $\longrightarrow$ G + H (both $C_4H_8O_5$)

G + HNO_3 $\longrightarrow$ I ($C_4H_6O_4$), *optically active*

H + HNO_3 $\longrightarrow$ J ($C_4H_6O_4$), *optically inactive*

12. (a) *cis*-1,2-Cyclopentanediol reacts with acetone in the presence of dry HCl to yield compound K, $C_8H_{14}O_2$, which is resistant to boiling alkali, but which is readily converted into the starting materials by aqueous acids. What is the most likely structure of K? To what class of compounds does it belong?

(b) *trans*-1,2-Cyclopentanediol does not form an analogous compound. How do you account for this fact?

13. The oxygen exchange described in Problem 21.9 (p. 785) can be carried out by use of hydroxide ion instead of hydrogen ion as catalyst. Suggest a detailed mechanism for exchange under these conditions. (*Hint*: See Sec. 21.14.)

14. Vinyl alkyl ethers, $RCH=CHOR'$, are very rapidly hydrolyzed by dilute aqueous acid to form the alcohol $R'OH$ and the aldehyde RCH_2CHO. Hydrolysis in $H_2^{18}O$ gives alcohol $R'OH$ containing only ordinary oxygen. Outline all steps in the most likely mechanism for the hydrolysis. Show how this mechanism accounts not only for the results of the tracer experiment, but also for the extreme ease with which hydrolysis takes place.

15. On treatment with bromine, certain diarylmethanols (I) are converted into a 50:50 mixture of aryl bromide (II) and aldehyde (III).

$$CH_3O-\bigcirc-CHOH-\bigcirc-G + Br_2 \longrightarrow CH_3O-\bigcirc-Br + G-\bigcirc-CHO$$

I   II   III

Whether G is $-NO_2$, $-H$, $-Br$, or $-CH_3$, bromine appears *only* in the ring containing the $-OCH_3$ group. The rate of reaction is affected moderately by the nature of G, decreasing along the series: $G = -CH_3 > -H > -Br > -NO_2$. The rate of reaction is slowed down by the presence of added bromide ion.

Outline all steps in the most likely mechanism for this reaction. Show how your mechanism accounts for each of the above facts.

16. A naïve graduate student needed a quantity of benzhydrol, $(C_6H_5)_2CHOH$, and decided to prepare it by the reaction between phenylmagnesium bromide and benzaldehyde. He prepared a mole of the Grignard reagent. To insure a good yield, he then added, not one, but *two* moles of the aldehyde. On working up the reaction mixture, he was at first gratified to find he had obtained a good yield of a crystalline product, but his hopes were dashed when closer examination revealed that he had made, not benzhydrol, but the ketone benzophenone. Bewildered, the student made the first of many trips to his research director's office.

He returned shortly, red-faced, to the laboratory, carried out the reaction again using equimolar amounts of the reactants, and obtained a good yield of the compound he wanted.

What had gone wrong in his first attempt? How had his generosity with benzaldehyde betrayed him? (*Hint*: See Sec. 21.14. Examine the structure of the initial addition product.) (In Problem 20, p. 928, we shall follow his further adventures.)

17. (a) How do you account for the extreme ease with which *tetrahydropyranyl ethers* (Sec. 19.9) undergo hydrolysis in dilute aqueous acid? (b) Predict the products of such hydrolysis of EtO–THP.

18. (a) Give structural formulas of compounds L and M, and of *isoeugenol* and *vanillin*.

eugenol (below) + KOH, 225 °C $\longrightarrow$ isoeugenol $(C_{10}H_{12}O_2)$

isoeugenol + $(CH_3CO)_2O$ $\longrightarrow$ L $(C_{12}H_{14}O_3)$ (See Sec. 24.10)

L + $K_2Cr_2O_7$, H_2SO_4, 75 °C $\longrightarrow$ M $(C_{10}H_{10}O_4)$

M + HSO_3^-, H_2O, boil $\longrightarrow$ vanillin $(C_8H_8O_3)$

(b) Account for the conversion of eugenol into isoeugenol.

OH
⬡ OCH₃
CH₂CH=CH₂   CH₂CH=CH₂   CHO
Eugenol   Safrole   Piperonal

(c) Suggest a way to convert *safrole* into *piperonal* (above).

19. Suggest a mechanism for the following reaction.

$$(CH_3)_2C=CHCH_2CH_2C(CH_3)=CHCHO + H_3O^+ \longrightarrow$$

3,8-Carvomenthenediol

The ring-closing step can be considered as either nucleophilic addition or electrophilic addition depending on one's point of view. Show how this is so, identifying both the electrophile and the nucleophile.

20. The trimer of trichloroacetaldehyde (compare *paraldehyde*, p. 763) exists in two forms, N and O, which give the following proton NMR data.

N: singlet, δ 4.28
O: two singlets, δ 4.63 and δ 5.50, peak area ratio 2:1

Show in as much detail as you can the structure of each of these.

21. How do you account for the difference in behavior between diastereomers IV and V? (*Hint*: Draw Newman projections. Which are the bulkiest groups?)

IV

V

22. The acetal (VI) of glycerol and benzaldehyde has been found to exist in two configurations. (a) Draw them. (b) One of these exists preferentially in a conformation in

VI

which the phenyl group occupies an axial position. Which configuration is this, and what counterbalances the unfavorable steric factor?

23. Describe a simple chemical test that would serve to distinguish between:

(a) *n*-valeraldehyde and diethyl ketone
(b) phenylacetaldehyde and benzyl alcohol
(c) cyclohexanone and cyclohexyl methyl ether
(d) 2-pentanone and 3-pentanone
(e) propionaldehyde and diethyl ether

(f) diethyl acetal and *n*-valeraldehyde
(g) diethyl acetal and di-*n*-propyl ether
(h) methyl *m*-tolyl ketone and propiophenone
(i) 2-pentanone and 2-pentanol
(j) paraldehyde and diisobutyl ether
(k) dioxane and trioxane

   Tell exactly what you would *do* and *see*.

24. An unknown compound is believed to be one of the following, all of which boil within a few degrees of each other. Describe how you would go about finding out which of the possibilities the unknown actually is. Where possible use simple chemical tests; where necessary use more elaborate chemical methods such as quantitative hydrogenation, cleavage, neutralization equivalent, saponification equivalent, etc. Make use of any needed tables of physical constants.

(a) phenylacetaldehyde
   m-tolualdehyde
   o-tolualdehyde
   acetophenone
   p-tolualdehyde
(b) methyl *β*-phenylethyl ketone
   cyclohexylbenzene
   benzyl *n*-butyrate
   γ-phenylpropyl alcohol

(c) isophorone (3,5,5-trimethyl-2-cyclo-
   hexen-1-one)
   n-dodecane
   benzyl *n*-butyl ether
   n-nonyl alcohol
(d) *p*-chloroacetophenone
   methyl *o*-chlorobenzoate
   p-chlorobenzyl chloride
   m-chloronitrobenzene

25. *Citral*, $C_{10}H_{16}O$, is a terpene that is the major constituent of lemongrass oil. It reacts with hydroxylamine to yield a compound of formula $C_{10}H_{17}ON$, and with Tollens' reagent to give a silver mirror and a compound of formula $C_{10}H_{16}O_2$. Upon vigorous oxidation citral yields acetone, oxalic acid (HOOC—COOH), and levulinic acid ($CH_3COCH_2CH_2COOH$).

   (a) Propose a structure for citral that is consistent with these facts and with the isoprene rule (Sec. 10.31).

   (b) Actually citral seems to consist of two isomers, citral *a* (*geranial*) and citral *b* (*neral*), which yield the same oxidation products. What is the most likely structural difference between these two isomers?

   (c) Citral *a* is obtained by mild oxidation of geraniol (Problem 26, p. 692); citral *b* is obtained in a similar way from nerol. On this basis assign structures to citral *a* and citral *b*.

26. (+)-*Carvotanacetone*, $C_{10}H_{16}O$, is a terpene found in thuja oil. It reacts with hydroxylamine and semicarbazide to form crystalline derivatives. It gives negative tests with Tollens' reagent, but rapidly decolorizes cold dilute $KMnO_4$.

   Carvotanacetone can be reduced successively to *carvomenthone*, $C_{10}H_{18}O$, and *carvomenthol*, $C_{10}H_{20}O$. Carvomenthone reacts with hydroxylamine but not with cold dilute $KMnO_4$. Carvomenthol does not react with hydroxylamine or cold dilute $KMnO_4$, but gives a positive test with CrO_3/H_2SO_4.

   One set of investigators found that oxidation of carvotanacetone gave isopropylsuccinic acid and pyruvic acid, $CH_3COCOOH$; another set of investigators isolated acetic acid and *β*-isopropylglutaric acid.

HOOCCHCH$_2$COOH   HOOCCH$_2$CHCH$_2$COOH
    | |
  CH(CH$_3$)$_2$ CH(CH$_3$)$_2$
Isopropylsuccinic acid   *β*-Isopropylglutaric acid

What single structure for carvotanacetone is consistent with all these facts?

27. Which (if any) of the following compounds could give rise to each of the infrared spectra shown in Fig. 21.4 (p. 795)?

isobutyraldehyde
2-butanone
tetrahydrofuran

ethyl vinyl ether
cyclopropylmethanol
3-buten-2-ol

28. Give a structure or structures consistent with each of the CMR spectra in Fig. 21.5, p. 796.

29. Give a structure or structures consistent with each of the proton NMR spectra in Fig. 21.6, p. 797.

30. Give the structures of compounds P, Q, and R on the basis of their infrared spectra (Fig. 21.7, p. 798) and their proton NMR spectra (Fig. 21.8, p. 799).

31. Give the structure of compound S on the basis of its infrared, CMR, and proton NMR spectra (Fig. 21.9, p. 800).

Wavelength, μ

IRDC 7302

CH₃ CH₂ C CH₃
　　　　 ‖
　　　　 O　27

Wavelength, μ

IRDC 3577

CH₃ CH — CHO
　 |
　CH₃

Wavelength, μ

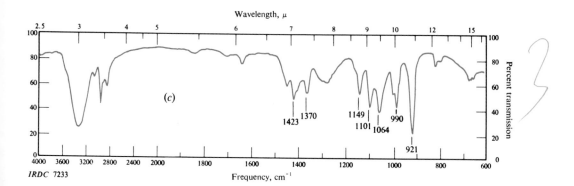

IRDC 7233

Figure 21.4　Infrared spectra for Problem 27, p. 793.

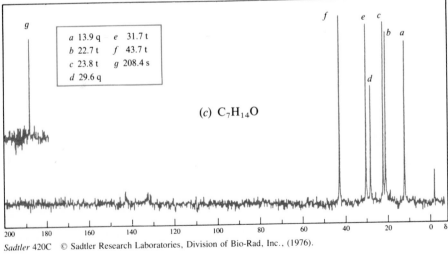

Figure 21.5   CMR spectra for Problem 28, p. 794.

(a) $C_5H_{10}O$

290 C CH_3 CH_2 CH_2 $C^{\overset{O}{\underset{\parallel}{}}}$ CH_3

(b) $C_5H_{10}O$

CH_3 CH_3 $CH - C^{\overset{O}{\underset{\parallel}{}}} - CH_3$

Figure 21.6   Proton NMR spectra for Problem 29, p. 794.

(c) C_4H_8O

CH_3 CH_2 $C^{\overset{O}{\underset{\parallel}{}}}$ $O - CH_3$

Figure 21.7   Infrared spectra for Problem 30, p. 794.

P $C_8H_8O_2$

CH₃ –O– ⬡ CHO
    4   7   8

Q $C_9H_{10}O_2$

R $C_{10}H_{12}O$

Figure 21.8   Proton NMR spectra for Problem 30, p. 794.

⬡ C(=O)–CH–CH₃ (with CH₂ notation)

Sadtler 15514K   © Sadtler Research Laboratories, Division of Bio-Rad, Inc., (1969).

a 17.7 q	f 37.1 t
b 19.9 q	g 51.1 t
c 25.4 t	h 124.4 d
d 25.6 q	i 131.5 s
e 28.0 d	j 202.1 d

Sadtler 2918C   © Sadtler Research Laboratories, Division of Bio-Rad, Inc., (1977).

Sadtler 10936M   © Sadtler Research Laboratories, Division of Bio-Rad, Inc., (1971).

Figure 21.9 Infrared, CMR, and proton NMR spectra for Problem 31, p. 794.

22

Stereochemistry III. Enantiotopic and Diastereotopic Ligands and Faces

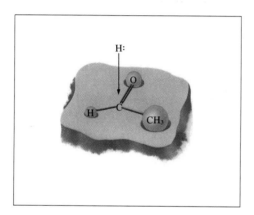

22.1 Introduction

So far in this book we have discussed the stereochemical relationships between molecules; some, we have found, are enantiomers, some are diastereomers, and some are not different at all, but are identical. We have discussed stereospecific reactions: reactions in which stereochemically different molecules react differently.

Yet, as we shall see in this chapter, this is not enough—not if we are to understand the stereochemistry of the reactions that take place in living organisms or an increasing number of reactions brought about in the laboratory. We must dig deeper, and examine the stereochemical relationships *between different parts of the same molecule*. We must expand our definition of stereospecific reactions to accommodate this broader view of stereoisomerism.

For our introduction to these new concepts let us turn to a familiar reaction system, the interconversion of alcohols and aldehydes through oxidation and reduction—but oxidation and reduction in a rather special environment.

22.2 Biological oxidation and reduction. Ethanol and acetaldehyde

Alcohols can be oxidized, not only in the test tube, but in living organisms. Let us examine just one example of such an oxidation, the conversion of ethanol into acetaldehyde. This is a simple example, as biological reactions go, but from it we can learn the basic ideas of a concept that is fundamental to an understanding of stereospecificity: the concept of *enantiotopic* and *diastereotopic* ligands and faces.

$$CH_3CH_2OH \underset{reduction}{\overset{oxidation}{\rightleftarrows}} CH_3CHO$$

Ethanol   Acetaldehyde

Like almost all biological reactions, this one requires catalysis by an enzyme: in this case, *alcohol dehydrogenase*. The oxidizing agent is a very common one in biological systems, *nicotinamide adenine dinucleotide*, or NAD. It is a coenzyme, an organic molecule that works with an enzyme to cause a particular chemical change. Here, the enzyme brings together the ethanol and the coenzyme, and the coenzyme does the actual oxidizing.

NAD is a much smaller molecule than the enzyme. Its structure is known, and so is the change in structure that takes place when it acts as an oxidizing agent (Sec. 40.15). The mechanism of the oxidation process has been the subject of much study. For our present purpose we need only know that NAD oxidizes by abstracting a hydrogen and a pair of electrons—a hydride ion, in effect—from the substrate. We shall represent the oxidized form of NAD as NAD^+, and the reduced form as NADH.

The oxidation of ethanol thus becomes:

$$CH_3CH_2OH + NAD^+ \underset{}{\overset{alcohol\ dehydrogenase}{\rightleftarrows}} CH_3CHO + NADH + H^+$$

Oxidized NAD   Reduced NAD

Ethanol loses one of its α-hydrogens with a pair of electrons, and then—or probably simultaneously—loses a proton from oxygen to give the aldehyde.

$$CH_3-\overset{\overset{H}{|}}{\underset{\underset{H}{|}}{C}}-O-H \rightleftarrows CH_3-\overset{\overset{H}{|}}{C}=O + NAD-H + H^+$$

NAD^+

Like all catalysts, enzymes speed up reaction in both directions: under the proper conditions, alcohol dehydrogenase catalyzes the reduction of acetaldehyde to ethanol by NADH.

$$CH_3CHO + NADH + H^+ \underset{}{\overset{alcohol\ dehydrogenase}{\rightleftarrows}} CH_3CH_2OH + NAD^+$$

Reduced NAD   Oxidized NAD

The reduction, of course, follows exactly the same path as the oxidation, but in the opposite direction. Acetaldehyde gains a hydride ion from NADH, and a proton from the solvent.

$$
\begin{array}{c}
\text{H} \\
| \\
\text{CH}_3\text{—C}{=}\text{O} \quad \text{H}^+ \\
| \\
\text{NAD—H}
\end{array}
\rightleftarrows
\begin{array}{c}
\text{H} \\
| \\
\text{CH}_3\text{—C—OH} + \text{NAD}^+ \\
| \\
\text{H}
\end{array}
$$

We recognize this as an example of the typical reaction of aldehydes and ketones, nucleophilic addition, with hydride as the nucleophile. We shall be concerned here with this reduction reaction, too.

22.3 Biological oxidation and reduction. Deuterium labeling experiments

Now let us look at certain classic experiments carried out on this reaction, with alcohol dehydrogenase from yeast as the catalyst, and NAD as the oxidizing and reducing agent.

When ordinary ethanol, CH_3CH_2OH, is oxidized in D_2O solution, there is obtained ordinary, unlabeled NADH.

$$CH_3CH_2OH + NAD^+ \xrightleftharpoons{D_2O} CH_3CHO + NAD{-}H + H^+$$
$$\textit{Unlabeled}$$

When the dideuterated ethanol CH_3CD_2OH is oxidized in ordinary water, there is obtained NADD, that is, reduced NAD containing one deuterium per molecule.

$$CH_3CD_2OH + NAD^+ \xrightleftharpoons{H_2O} CH_3CDO + NAD{-}D + H^+$$
$$\textit{Labeled}$$

Evidently transfer of hydrogen occurs *directly* from ethanol to NAD^+, and not via the solvent.

When the NADD obtained in this way is allowed to reduce ordinary acetaldehyde, CH_3CHO, there is obtained the monodeuterated ethanol I, CH_3CHDOH.

$$
CH_3CHO + NAD{-}D + H^+ \rightleftarrows
\begin{array}{c}
\text{H} \\
| \\
\text{CH}_3\text{—C—OH} + \text{NAD}^+ \\
| \\
\text{D}
\end{array}
$$
$$\text{I}$$

Thus, the NADD transfers back the kind of hydrogen it received earlier, deuterium.

Now, something surprising: when ethanol I is oxidized, *all* of its deuterium is found transferred to the NAD$^+$ to give NADD. Only ordinary acetaldehyde is formed. There are two α-hydrogens in ethanol I, one protium and one deuterium.

$$CH_3\!-\!\overset{\displaystyle H}{\underset{\displaystyle D}{\overset{|}{\underset{|}{C}}}}\!-\!OH \;+\; NAD^+ \;\rightleftarrows\; CH_3CHO \;+\; \underset{\textit{Labeled}}{NAD\!-\!D} \;+\; H^+$$

$$I$$

Yet, of these, only *one* is transferred to NAD$^+$: the deuterium, the *same hydrogen* that the NADD previously transferred to the acetaldehyde. How can the NAD molecule "remember" which hydrogen it transferred to acetaldehyde? Clearly, the ethanol must keep this particular hydrogen in a different "drawer" from the other one. But how can this be, if both are α-hydrogens?

Let us look at another experiment. Labeled acetaldehyde, CH_3CDO, is prepared. On reduction by ordinary NADH, there is obtained monodeuterated ethanol II.

$$CH_3CDO \;+\; NAD\!-\!H \;+\; H^+ \;\rightleftarrows\; CH_3\!-\!\overset{\displaystyle H}{\underset{\displaystyle D}{\overset{|}{\underset{|}{C}}}}\!-\!OH \;+\; NAD^+$$

$$II$$

When *this* ethanol is oxidized, *none* of its deuterium is transferred to the NAD$^+$; only ordinary NADH is obtained. The deuterium remains in the acetaldehyde.

$$CH_3\!-\!\overset{\displaystyle H}{\underset{\displaystyle D}{\overset{|}{\underset{|}{C}}}}\!-\!OH \;+\; NAD^+ \;\rightleftarrows\; CH_3CDO \;+\; \underset{\textit{Unlabeled}}{NAD\!-\!H} \;+\; H^+$$

$$II$$

Once more the NAD molecule takes back only the hydrogen that had previously been transferred to the aldehyde. This time the ethanol has protium, not deuterium, in its special "drawer".

22.4 Biological oxidation and reduction. Stereochemistry

So now we have two kinds of monodeuterated ethanol, I and II. One transfers only deuterium to NAD$^+$, and the other transfers only protium. How can I and II differ? To find the answer, all we need do is examine the structure of the molecule. With one of the hydrogens deuterium, the molecule is chiral, and can exist as a pair of enantiomers. Ethanol I is one of these enantiomers and ethanol II is the other.

$$
\begin{array}{cc}
CH_3 & CH_3 \\
| & | \\
D\!-\!C\!\rightarrow\!H & H\!-\!C\!\rightarrow\!D \\
| & | \\
OH & OH \\
I & II
\end{array}
$$

Most of this elegant work was done by Frank H. Westheimer and Birget Vennesland at the University of Chicago, and was reported in 1953. The formation of the enantiomeric alcohols I and II was demonstrated unequivocally without measurement of optical rotation—there was not enough material available! It was not until four years later that the optical activity of the CH_3CHDOH was measured directly. The enantiomers have the configurations shown: I is the (R)-$(+)$-isomer, and II is the (S)-$(-)$-isomer.

Clearly, then, the special "drawer" in which ethanol keeps its transferrable hydrogen is a particular stereochemical location in the molecule. I can lose only D,

$$CH_3$$
$$D-\overset{\displaystyle CH_3}{\underset{\displaystyle OH}{C}}\rightarrow H \qquad\qquad H-\overset{\displaystyle CH_3}{\underset{\displaystyle OH}{C}}\rightarrow D$$

I	II
Loses only D	*Loses only H*

and II can lose only H; as stereochemical formulas show, these atoms occupy the same relative positions in the two molecules—on the "left", as we have drawn them here.

Now, the most important point. There can be no doubt that unlabeled ethanol—the kind that the organism deals with regularly—behaves in exactly the same way as the labeled molecules: on oxidation by NAD^+ it gives up only H^a, the

$$H^a-\overset{\displaystyle CH_3}{\underset{\displaystyle OH}{C}}\rightarrow H^b$$

Ordinary ethanol
Loses only H^a

hydrogen on the "left", and retains H^b. The use of deuterium and the formation of enantiomers is only a technique used to show the stereochemical course of the reaction: that the biological oxidizing agent discriminates completely between two seemingly equivalent hydrogens in the ethanol molecule.

But this is only half the story. The evidence we have described shows that there is stereochemical discrimination in the *formation* of ethanol, too—in the reduction as well as in the oxidation. Let us return to our labeled alcohols. Only I is formed by reduction of unlabeled acetaldehyde by NADD, and only II is formed by reduction of labeled acetaldehyde by NADH.

Let us examine this reduction, starting with the formation of I. The carbonyl carbon of acetaldehyde is bonded to three other atoms, and this portion of the molecule is flat. Reduction involves transfer of D from NADD to the carbonyl carbon, that is, to one face or the other of this flat molecule. As Fig. 22.1 (on the next page) shows, just which product is formed depends upon *which* face D becomes attached to. Attack by path (*a*) would give ethanol I, and attack by path (*b*) would give ethanol II. The fact that only I is actually obtained shows that attack occurs only by path (*a*): NADD transfers D to only one of the faces of the aldehyde, and completely shuns the other.

Figure 22.1 Enzymatic reduction of CH_3CHO by NADD. Attachment of D could take place by either path (*a*) or path (*b*), to give either I or II. Path (*a*) is the one actually followed.

If we examine the reduction of labeled acetaldehyde by NADH in the same way (Fig. 22.2), we find exactly the same sort of thing: NADH transfers H to only one of the faces of the flat molecule. Furthermore, the face it selects is the *same* face that is selected by NADD in the other reaction. In both cases reduced NAD attacks only along path (*a*)—from the "left", as we have oriented the molecules.

Figure 22.2 Enzymatic reduction of CH_3CDO by NADH. Attachment of H could take place by either path (*a*) or path (*b*), to give either II or I. Path (*a*) is the one actually followed.

Here, too, as in the oxidation, there can be no doubt that the reaction of unlabeled molecules follows exactly the same stereochemical course as the reaction of the labeled ones, and that NADH discriminates absolutely between the two faces of the acetaldehyde molecule. Again the use of deuterium in the formation of enantiomers is simply a device used to reveal the stereochemistry of the reaction.

$$\underset{NAD}{\overset{CH_3}{\underset{H}{\overset{H}{\underset{O}{C}}}}} \quad \xrightarrow{H^{\cdot}} \quad \overset{CH_3}{\underset{OH}{H-C-H}}$$

We call this "stereochemistry"—but it is a strange kind of stereochemistry. Both acetaldehyde and unlabeled ethanol are achiral; so far as they are concerned no chiral center is generated or destroyed in the reactions; so far as they are concerned reaction does not involve the formation or the reaction of stereoisomers. Yet we *are* dealing with stereochemistry: the enzyme–coenzyme system discriminates on a *three-dimensional* basis between two seemingly equivalent positions in ethanol and between two seemingly equivalent faces of acetaldehyde. We must expand our definition of stereospecificity. A **stereospecific reaction** is one in which stereochemically different molecules—or *stereochemically different parts of molecules*—react differently.

In biological systems such discrimination—such stereospecificity—is the rule, not the exception; and it is being achieved in man-made systems, too. To see just what is involved—which substrates are susceptible, and what special conditions are required—we must turn to a stereochemical concept that we have so far only touched on.

22.5   Enantiotopic and diastereotopic ligands

In Sec. 16.7 we found that, to account for the number of signals in NMR spectra, we had to consider the stereochemical relationship between different parts of the same molecule. We had to decide whether certain ligands—atoms or groups— that were equivalent in chemical composition and location on a chain or ring were or were not stereochemically equivalent.

Now let us go more deeply into this matter, using as before our old tool, the concept of isomer number (Sec. 4.2). If two atoms in a molecule are truly equivalent, replacement of either will give the same product. Let us take a simple molecule like ethyl chloride, and focus our attention on C—1 and the pair of hydrogens attached to it. Let us imagine that one of these hydrogens is replaced by some other

$$\underset{\text{Enantiotopic hydrogens}}{\overset{CH_3}{\underset{Cl}{H-C-H}}} \qquad \left[\underset{III}{\overset{CH_3}{\underset{Cl}{H-C-Z}}} \qquad \underset{IV}{\overset{CH_3}{\underset{Cl}{Z-C-H}}} \right]$$

atom or group, Z. Depending upon which hydrogen we replace, we obtain either III or IV. These, we can easily see, are enantiomeric. We have mentally generated a new chiral center.

Since the products are not identical, but are stereoisomeric, the two hydrogens are *not stereochemically equivalent*. Such pairs of ligands are said to be **enantiotopic**: *replacement of one or the other of them gives one or the other of a pair of enantiomers.*

In ethyl chloride the pair of enantiotopic ligands are attached to the same carbon, but this does not have to be the case. In *meso*-2,3-dichlorobutane, for example, the hydrogens shown in blue are on different carbons; yet our technique of imaginary replacement shows us beyond question that they are enantiotopic. (To ensure yourself that V and VI are indeed enantiomers, simply rotate VI end-for-end.)

Enantiotopic hydrogens
On different carbons   V   VI

"Replacement"—either imaginary or actual—does not necessarily mean removing an entire ligand and putting a new one in its place. For example, the ligand $-CH_2OH$ is, in effect, *replaced by* $-CH_2Z$ if Z is substituted for OH.

A carbon to which a pair of enantiotopic ligands are attached is called a *prochiral center*, since replacement of one of its ligands would convert the carbon into a chiral center. Just as the carbon of CWXYZ is a chiral center, so the carbon of CWWXY is a prochiral center. This concept can sometimes be useful in detecting enantiotopic ligands, but not all enantiotopic pairs fit into this formulation: the hydrogens of *meso*-2,3-dichlorobutane, for example, as we have just seen. *The safest and easiest way to detect enantiotopic ligands is through imaginary replacement.*

How can we refer to a particular ligand of an enantiotopic pair without having to draw a stereochemical formula and label the ligand? We use the Cahn–Ingold–Prelog procedure (Secs. 4.15–4.16) in a special way. We assign the particular ligand concerned a priority higher than that of its partner: we imagine, for example, that an atom has been replaced by a heavier isotope—deuterium for protium, say. We then specify this imaginary molecule in the usual way as *R* or *S*. If replacement of the ligand concerned gives the *R* configuration, the ligand is specified as pro-*R*; if the *S* configuration, then as pro-*S*. Using this procedure with ethyl chloride, for example, we specify the enantiotopic hydrogens in the following way, where H_R is pro-*R* and H_S is pro-*S*.

Enantiotopic hydrogens:
pro-*R* and pro-*S*

We must not confuse these specifications with the specification as *R* and *S* of any real enantiomers formed by actual replacement of a ligand. This latter specification would depend upon the priority of the new ligand. Actual replacement of the pro-*R* hydrogen of ethyl chloride by D would give the (*R*)-enantiomer; but replacement by OH would give the (*S*)-enantiomer.

Problem 22.1   Draw a stereochemical formula for each of the following compounds. Identify all pairs of enantiotopic ligands, and specify each as pro-*R* or pro-*S*.

(a) propane
(b) *n*-butane
(c) isobutane
(d) *n*-propyl chloride
(e) isopropyl chloride
(f) isobutyl chloride

(g) 1,1-dichloroethane
(h) ethanol
(i) *n*-propyl alcohol
(j) isopropyl alcohol
(k) *meso*-2,3-butanediol
(l) (*R*,*R*)-2,3-butanediol

Now, why are we concerned with this concept? The word *enantiotopic* means "in mirror-image places", and it is here that we find the importance of the relationship: enantiotopic ligands exist in environments that are mirror-images of each other. An ordinary, optically inactive reagent will not feel this difference in environment, and will not distinguish between the enantiotopic ligands. An attacking bromine atom, for example, will find the enantiotopic hydrogens of ethyl chloride identical, and will not discriminate between them in its attack: the two will be abstracted at identical rates.

But not all reagents are optically inactive. As we have seen (Secs. 4.11 and 9.3), the enzymes of biological systems *are* optically active, and so are a rapidly growing number of man-made catalysts used in organic synthesis (Sec. 20.7). Optically active reagents (or reagents in the presence of optically active catalysts) do feel the difference between mirror-image environments, and do distinguish between enantiotopic ligands. We cannot hope to understand the important chemistry of such reagents unless we can recognize the enantiotopic relationship.

We must have one point absolutely clear. We use the imaginary replacement of ligands simply as the most convenient way of finding out whether or not they are enantiotopic; it is a purely intellectual process. In an actual reaction, discrimination between enantiotopic ligands by a chiral reagent is not limited to formation of one or the other of a pair of enantiomers. A new chiral center need not be generated at all.

Oxidation of ethanol to acetaldehyde, for example, involves loss of one of the α-hydrogens of the alcohol. By applying our replacement test, we can easily see that these hydrogens are enantiotopic. (Indeed, we have *already* applied this test; in the

$$H_R - \underset{\underset{\displaystyle OH}{|}}{\overset{\overset{\displaystyle CH_3}{|}}{C}} - H_S$$

Enantiotopic hydrogens:
pro-*R* and pro-*S*
Loses only H_R to NAD^+

preceding section we saw that replacement by deuterium gives rise to enantiomers.) When the oxidation is enzymatic, we know, the coenzyme NAD^+ abstracts only a particular one of these hydrogens, the hydrogen which we would now specify as pro-*R*. Both the substrate, ethanol, and the product, acetaldehyde, are achiral; yet the optically active oxidizing agent discriminates totally between the mirror-image environments of these enantiotopic hydrogens.

Now let us take another molecule, 1,2-dichloropropane—the *R* isomer, say—

and focus our attention on the hydrogens attached to C–1. Again let us imagine that one of these hydrogens is replaced by Z. Depending upon which of the hydrogens is replaced, we obtain either VII or VIII. These, we see, are diastereomeric.

Diastereotopic hydrogens   VII   VIII

Again the two hydrogens being replaced are stereochemically non-equivalent, but in a way that is different from what we saw before. Such pairs of ligands are said to be **diastereotopic**: *replacement of one or the other of them gives one or the other of a pair of diastereomers.*

By applying our replacement test, we find that a set of diastereotopic ligands are sometimes attached to different carbons within a molecule, and are sometimes attached to the same doubly bonded carbon—geometric isomers, after all, are one kind of diastereomer. For example:

Diastereotopic ligands

Problem 22.2   Draw a stereochemical formula for each of the following compounds, and identify all sets of diastereotopic ligands.

(a) propylene
(b) isobutylene

(c) 2-methyl-2-butene
(d) vinyl chloride

(e) (S)-*sec*-butyl chloride
(f) (S)-2-chloropentane

Diastereotopic ligands exist in environments that are neither identical nor mirror-images of each other. Whether an attacking reagent is optically active or optically inactive, it will feel the difference between these environments, and will distinguish between the ligands. Even an attacking bromine atom, for example, will find the diastereotopic hydrogens of 1,2-dichloropropane different, and will discriminate between them in its attack; the two will be abstracted at different rates.

As before, however, the imaginary replacement of ligands is simply our way of detecting the stereochemical relationship. No actual replacement need take place, and no stereoisomers need be formed. A reagent will distinguish between diastereotopic ligands regardless of the kind of reaction that takes place.

Together, enantiotopic and diastereotopic ligands are known as **heterotopic** ligands, that is, ligands "in different places". Sets of ligands in identical stereochemical environments—ligands whose imaginary replacement leads to identical products—are called **homotopic** ligands, that is, ligands "in the same place".

22.6   Enantiotopic and diastereotopic faces

We started out to examine the stereochemical relationship between different parts of the same molecule, and so far we have discussed different ligands—different atoms or groups attached to the molecule. Now, when we are dealing with a flat molecule—or, rather, a molecule containing a flat portion—we speak of it as having *faces*. We did this, for example, in discussing the addition reactions of alkenes (Sec. 9.5) and of carbonyl compounds (Sec. 22.4). So now let us turn to the stereochemical relationship between different faces of the same molecule.

As before, let us use the concept of isomer number. If two faces of a molecule are truly equivalent, attachment of an atom or group to either will give the same product. Let us consider the simple molecule, acetaldehyde. As an aldehyde, it typically undergoes nucleophilic addition. Let us imagine, then, that some nucleophile, :Z, becomes attached to the carbonyl carbon and, to complete the reaction, a proton becomes attached to the carbonyl oxygen. Acetaldehyde, we

Enantiotopic faces

IX   X

saw, is a flat molecule. We can attach :Z to one or the other of its faces and, depending upon which face, obtain either IX or X. These we recognize as enantiomers. We have mentally generated a new chiral center.

Since the products are not identical, but are stereoisomeric, the two faces are not stereochemically equivalent. They are examples of **enantiotopic faces**: *attachment of a ligand to one or the other of them gives rise to one or the other of a pair of enantiomers.*

The central atom of enantiotopic faces is another kind of *prochiral centre*. Just as the carbon of CWWXY is a tetrahedral prochiral centre, so the carbon of CWXY is a trigonal prochiral center.

To specify a particular face of an enantiotopic pair, we again adapt the Cahn–Ingold–Prelog procedure. We assign priority on the usual basis (Sec. 4.16) to the three ligands attached to the (trigonal) prochiral center, and visualize it as lying flat on the paper before us. If, in proceeding in the usual way in order of decreasing priority of ligands, our eye travels in a clockwise direction, the face turned upward toward us is specified as the **Re face** (*rectus*); if counterclockwise, it is specified as the **Si face** (*sinister*). For example:

Re face   *Si* face

Now, approaching one face or the other of an enantiotopic pair, an optically inactive reagent will feel no difference and will attack them indiscriminately. But an optically active reagent will feel the difference and will discriminate; to a greater or lesser extent, it will attach itself preferentially to one or the other.

Once more, a reminder: the imaginary attachment of a ligand is simply an intellectual test—the easiest way to identify enantiotopic faces. A chiral reagent will discriminate between these faces even if enantiomers are not being formed; in the actual chemical reaction, a new chiral center need not be generated.

In the enzymatic reduction of acetaldehyde to ethanol, we have seen, NADH transfers H to only one of the faces of the aldehyde—a face that we now recognize as the *Re* face. Since the ligand being attached, H, is the same as one already

Attacks only Re face

present in the molecule, enantiomers are not formed. Yet these are enantiotopic faces, and there is total discrimination between them.

Depending upon the structure of the rest of the molecule, there can also be **diastereotopic faces**: *attachment of a ligand to one or the other of them gives rise to one or the other of a pair of diastereomers.* Like diastereotopic ligands, these diastereotopic faces are discriminated between by any reagent, optically active or inactive.

Together, enantiotopic and diastereotopic faces are called **heterotopic** faces. Faces that are not heterotopic—that are stereochemically equivalent—are called **homotopic**.

Problem 22.3   Draw a structural formula for each of the following compounds. Identify all pairs of heterotopic faces and tell whether the members of each pair are enantiotopic or diastereotopic. Specify each face as *Re* or *Si*.

(a) propionaldehyde
(b) acetone
(c) (*R*)-3-methyl-2-pentanone
(d) 2-bromopropene

(e) isobutylene
(f) 3-bromo-2-methylpropene
(g) (*Z*)-1-chloropropene
(h) (*S*)-3-phenyl-1-butene

22.7   Origin of enantiospecificity

So far we have spoken only in general terms: an optically active reagent "feels the difference between mirror-image environments", we have said, and thus "distinguishes between" enantiotopic ligands or faces. Now let us be more specific. Let us take as an example the enzymatic reduction of acetaldehyde, and see the kind of thing that must be involved.

An enzyme, as we shall see, is an enormous molecule, wholly or mostly protein. It is a long chain, looped, coiled, or folded in a complicated, irregular way: at first glance a random arrangement, but actually highly characteristic of every molecule of that enzyme. (See, for example, the three-dimensional structure of α-chymotrypsin on p. 1381.) This characteristic shape enables the enzyme to do its special job.

Like any substrate, acetaldehyde is bound by the enzyme. At a particular location in the giant molecule there is a site of a size, shape, and chemical nature just right to hold the acetaldehyde. (As Emil Fischer (p. 1374) put it, enzyme and substrate "must fit together like a lock and key".) Let us represent the site *in a purely schematic way* (Fig. 22.3) as three holes on the surface of the enzyme: a big

(a)

(b)   (c)

Figure 22.3 Schematic representation of enzymatic reduction of acetaldehyde. (a) A binding site on the surface of the enzyme, with holes for CH_3, O, and H. (b) Acetaldehyde bound to the site in the only way it can fit: with the *Re* face upward. Only this face can receive H from NADH.

hole for CH_3, a middle-sized hole for O, and a little hole for H. As shown, we can place the acetaldehyde on the site so that each ligand fits into its own hole—but *only if* a particular face of the molecule (the *Re* face) is turned upward. If we flip the molecule over, it can no longer be made to fit: we can, say, place O in its middle-sized hole, but then CH_3 lies over the little hole, and H lies in the big hole. (To fit the lock, the key must be right side up.)

The reducing agent, the coenzyme NADH, is also bound to the enzyme, and also in a very specific way: in just the right position for it to transfer H to acetaldehyde. According to our schematic representation, this transfer can take place only to the upper, exposed face of the acetaldehyde. Since the bound aldehyde must have its *Re* face turned upward, it is to this face only that transfer can occur. (The key can be turned in the lock only when grasped by its projecting end.)

And thus, through its own chirality, the reagent—the enzyme–coenzyme—recognizes the mirror-image environments of the two faces, and discriminates between them.

All this is possible, we realize, because of the particular way the enzyme has brought together the substrate and the reagent. Underlying this enantiospecificity is symphoria.

PROBLEMS

1. Draw a stereochemical formula for each of the following compounds. Identify all pairs of heterotopic ligands and faces, and tell whether the members of each pair are enantiotopic or diastereotopic. Specify each ligand as pro-R or pro-S, and each face as Re or Si. (*Caution*: In most of these compounds there may be considerably more than first meets the eye.)

(a) 3-chloropentane

(b) 1,3-propanediol

(c) 2-butanone

(d) (R,S)-2,4-pentanediol

(e) (R,R)-2,4-pentanediol

(f) (S)-$CH_3CHOHCHO$

(g) methylcyclopropane

(h) 3-methylcyclopropene

(i) *cis*-2-butene

(j) 2-methyl-2-butene

2. In a further experiment Westheimer and Vennesland (Sec. 22.4) treated labeled ethanol II in the following way,

$$CH_3CHDOH \xrightarrow[\text{pyridine}]{\text{TsCl}} CH_3CHDOTs \xrightarrow{\text{OH}} CH_3CHDOH$$

II

and oxidized the product, CH_3CHDOH, with (unlabeled) NAD^+. (a) What is the product, CH_3CHDOH? (b) Which would you expect to get, NADH or NADD?

3. Let us look at the work on the enzymatic oxidation–reduction of ethanol–acetaldehyde from the standpoint of the NAD. The following observations were made. (The reaction in (v) is ordinary chemical reduction. All the others are enzyme-catalyzed. The oxidation by glucose in (iii) involves an enzyme different from alcohol dehydrogenase.)

(i) $NAD^+ \xrightarrow{CH_3CD_2OH} NADD(i)$

One D per molecule

(ii) $NADD(i) \xrightarrow{CH_3CHO} NAD^+(ii)$

No D

(iii) $NADD(i) \xrightarrow{glucose} NAD^+(iii)$

One D per molecule

(iv) $NAD^+(iii) \xrightarrow{CH_3CH_2OH} NADD(iv)$

One D per molecule

(v) $NAD^+ \xrightarrow[D_2O]{Na_2S_2O_3} NADD(v)$

One D per molecule

(vi) $NADD(v) \xrightarrow{CH_3CHO} NAD^+(vi)$

0.44 D per molecule

There are thus three samples of NADD, each containing one deuterium atom per molecule: NADD(i), NADD(iv), NADD(v). When oxidized by CH_3CHO, each NADD gives NAD^+ containing a different number of deuterium atoms per molecule: none, one, and 0.44.

Regardless of the exact structural formula of NADH, what can you conclude about its structure from these observations? Using letters to stand for unknown groups, draw a structural formula for NADH. What is the relationship between NADD(i) and NADD(iv)? What is NADD(v)? Account in detail for each of the observations.

4. Figure 20.8 (p. 753) shows the hydrogenation with Wilkinson's catalyst of 1-acetamidopropenoic acid to give acetylalanine, which on hydrolysis yields the amino acid alanine.

(a) When (*R*)-prophos is part of the catalyst, this synthesis gives, ultimately, the natural amino acid, (*S*)-alanine. Considering the nature of the hydrolysis step (look at the structures involved), which of the acetylalanines in Fig. 20.8 would you expect to give this amino acid?

(b) Identify the enantiotopic faces in 1-acetamidopropenoic acid. Label each as *Re* or *Si*.

(c) To which of these faces does hydrogen preferentially add?

5. Using the same approach as in Fig. 22.3 (p. 813), show in a schematic way how NAD^+ discriminates between the enantiotopic hydrogens of ethanol.

6. The system of reactions called *respiration* (Sec. 41.5) is the final and aerobic stage in the biological utilization of food—carbohydrates, proteins, and fats—as fuel. Central to respiration is the interconversion of certain carboxylic acids, a process that occurs in nearly every aerobic organism. In 1937 H. A. Krebs proposed a particular sequence of chemical changes—the *Krebs tricarboxylic acid cycle*—for this interconversion. Part of this cycle involves the following transformation via the key intermediate, *citric acid*.

$$
\begin{array}{ccccc}
\text{C*OOH} & & \text{CH}_2\text{C*OOH} & & \text{C*OOH} \\
| & & | & & | \\
\text{CH}_2 & & \text{HO-C-COOH} & & \text{C=O} \\
| & \longrightarrow & | & \longrightarrow & | \\
\text{C=O} & & \text{CH}_2\text{COOH} & & \text{CH}_2 \\
| & & \text{Citric acid} & & | \\
\text{COOH} & & & & \text{CH}_2 \\
\text{Oxaloacetic acid} & & & & | \\
& & & & \text{COOH} \\
& & & & \alpha\text{-Ketoglutaric acid}
\end{array}
$$

Krebs made the following prediction: if citric acid were indeed the intermediate, then oxaloacetic acid isotopically labeled on one of the carboxyl groups (C^*OOH) would yield α-ketoglutaric acid with the label evenly divided between its carboxyl groups. In 1941 the results of such labeling experiments were reported: contrary to Krebs' prediction, the α-ketoglutaric acid formed was found to contain the label in *only one* of its carboxyl groups, as shown above.

(a) What, do you think, was the line of reasoning that led Krebs to make his prediction?

(b) Would this line of reasoning be valid? Explain your answer.

(c) In light of the labeling experiments, can citric acid be an intermediate in the process shown, or can it not?

7. Following are some enzyme-catalyzed transformations. For each one, tell whether or not there *could* be discrimination on a stereochemical basis (that is, discrimination *other than* any indicated below). Explain your answer in each case by use of stereochemical formulas.

(a)   (*R*)-CH$_3$CHOHCHO $\longrightarrow$ CH$_3$CHOHCH$_2$OH

     (*R*)-Lactaldehyde   1,2-Propanediol

(b)   CH$_2$OHCHOHCH$_2$OH $\longrightarrow$ CH$_2$OHCHOHCH$_2$OPO$_3$H$_2$

       Glycerol   Glycerol 1-phosphate

(c)   (*S,S*)-HOCH$_2$CHOHCHOHCH$_2$OH $\longrightarrow$ HOCH$_2$CHOHCHOHCH$_2$OPO$_3$H$_2$

          L-Threitol   L-Threitol 1-phosphate

(d)   (*R,S*)-HOCH$_2$CHOHCHOHCH$_2$OH $\longrightarrow$ HOCH$_2$CHOHCHOHCH$_2$OPO$_3$H$_2$

          L-Erythritol   L-Erythritol 1-phosphate

(e) *trans*-HOOCCH=CHCOOH $\longrightarrow$ HOOCCH$_2$CHOHCOOH
         Fumaric acid   Malic acid

(f) HOOCCH$_2$CH$_2$COOH $\longrightarrow$ *trans*-HOOCCH=CHCOOH
         Succinic acid   Fumaric acid

23

Carboxylic Acids

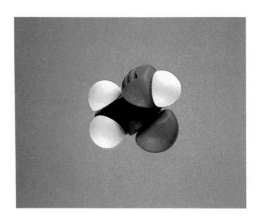

23.1 Structure

Of the organic compounds that show appreciable acidity, by far the most important are the carboxylic acids. These compounds contain the **carboxyl group**

$$H-C\overset{O}{\underset{OH}{\Big\langle}} \qquad R-C\overset{O}{\underset{OH}{\Big\langle}} \qquad Ar-C\overset{O}{\underset{OH}{\Big\langle}}$$

attached to hydrogen (HCOOH), an alkyl group (RCOOH), or an aryl group (ArCOOH). (See Fig. 23.1, p. 818.) For example:

HCOOH	CH_3COOH	$CH_3(CH_2)_{10}COOH$	$CH_3(CH_2)_7CH{=}CH(CH_2)_7COOH$
Formic acid	Acetic acid	Lauric acid	Oleic acid
Methanoic acid	Ethanoic acid	Dodecanoic acid	*cis*-9-Octadecenoic acid

Benzoic acid $\langle\!\!\langle\bigcirc\rangle\!\!\rangle$COOH

$O_2N\langle\!\!\langle\bigcirc\rangle\!\!\rangle$COOH *p*-Nitrobenzoic acid

$\langle\!\!\langle\bigcirc\rangle\!\!\rangle CH_2COOH$ Phenylacetic acid

CH₃—CH—COOH

| |

     Br

$CH_3\text{—}CH\text{—}COOH$

Br

α-Bromopropionic acid

2-Bromopropanoic acid

⬡COOH

Cyclohexanecarboxylic acid

$CH_2\text{=}CHCOOH$

Acrylic acid

Propenoic acid

Whether the group is aliphatic or aromatic, saturated or unsaturated, substituted or unsubstituted, the properties of the carboxyl group are essentially the same.

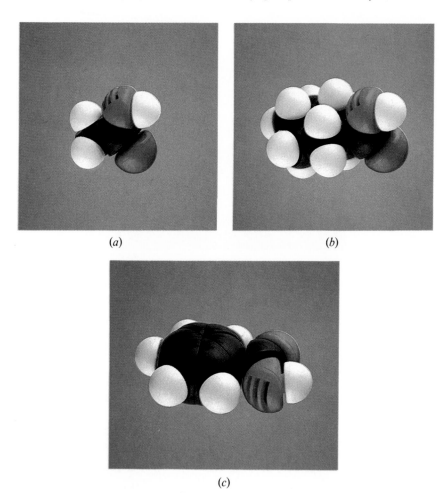

(a)   (b)

(c)

Figure 23.1   Models of some carboxylic acids: (*a*) acetic acid, CH_3COOH; (*b*) cyclohexanecarboxylic acid, *cyclo*-$C_6H_{11}COOH$; (*c*) benzoic acid, C_6H_5COOH.

23.2   Nomenclature

The aliphatic carboxylic acids have been known for a long time, and as a result have common names that refer to their sources rather than to their chemical structures. The **common names** of the more important acids are shown in Table 23.1. *Formic acid*, for example, adds the sting to the bite of an ant (Latin: *formica*,

Table 23.1 CARBOXYLIC ACIDS

Name	Formula	M.p., °C	B.p., °C	Solubility g/100 g H_2O
Formic	HCOOH	8	100.5	∞
Acetic	CH_3COOH	16.6	118	∞
Propionic	CH_3CH_2COOH	−22	141	∞
Butyric	$CH_3(CH_2)_2COOH$	−6	164	∞
Valeric	$CH_3(CH_2)_3COOH$	−34	187	3.7
Caproic	$CH_3(CH_2)_4COOH$	−3	205	1.0
Caprylic	$CH_3(CH_2)_6COOH$	16	239	0.7
Capric	$CH_3(CH_2)_8COOH$	31	269	0.2
Lauric	$CH_3(CH_2)_{10}COOH$	44	225^{100}	i.
Myristic	$CH_3(CH_2)_{12}COOH$	54	251^{100}	i.
Palmitic	$CH_3(CH_2)_{14}COOH$	63	269^{100}	i.
Stearic	$CH_3(CH_2)_{16}COOH$	70	287^{100}	i.
Oleic	cis-9-Octadecenoic	16	223^{10}	i.
Linoleic	cis,cis-9,12-Octadecadienoic	−5	230^{16}	i.
Linolenic	cis,cis,cis-9,12,15-Octadecatrienoic	−11	232^{17}	i.
Cyclohexanecarboxylic	cyclo-$C_6H_{11}COOH$	31	233	0.20
Phenylacetic	$C_6H_5CH_2COOH$	77	266	1.66
Benzoic	C_6H_5COOH	122	250	0.34
o-Toluic	o-$CH_3C_6H_4COOH$	106	259	0.12
m-Toluic	m-$CH_3C_6H_4COOH$	112	263	0.10
p-Toluic	p-$CH_3C_6H_4COOH$	180	275	0.03
o-Chlorobenzoic	o-ClC_6H_4COOH	141		0.22
m-Chlorobenzoic	m-ClC_6H_4COOH	154		0.04
p-Chlorobenzoic	p-ClC_6H_4COOH	242		0.009
o-Bromobenzoic	o-BrC_6H_4COOH	148		0.18
m-Bromobenzoic	m-BrC_6H_4COOH	156		0.04
p-Bromobenzoic	p-BrC_6H_4COOH	254		0.006
o-Nitrobenzoic	o-$O_2NC_6H_4COOH$	147		0.75
m-Nitrobenzoic	m-$O_2NC_6H_4COOH$	141		0.34
p-Nitrobenzoic	p-$O_2NC_6H_4COOH$	242		0.03
Phthalic	o-$C_6H_4(COOH)_2$	231		0.70
Isophthalic	m-$C_6H_4(COOH)_2$	348		0.01
Terephthalic	p-$C_6H_4(COOH)_2$	300 subl.		0.002
Salicylic	o-HOC_6H_4COOH	159		0.22
p-Hydroxybenzoic	p-HOC_6H_4COOH	213		0.65
Anthranilic	o-$H_2NC_6H_4COOH$	146		0.52
m-Aminobenzoic	m-$H_2NC_6H_4COOH$	179		0.77
p-Aminobenzoic	p-$H_2NC_6H_4COOH$	187		0.3
o-Methoxybenzoic	o-$CH_3OC_6H_4COOH$	101		0.5
m-Methoxybenzoic	m-$CH_3OC_6H_4COOH$	110		s. hot
p-Methoxybenzoic (Anisic)	p-$CH_3OC_6H_4COOH$	184		0.04

ant); *butyric acid* gives rancid butter its typical smell (Latin: *butyrum*, butter); and *caproic, caprylic,* and *capric acids* are all found in goat fat (Latin: *caper*, goat).

Branched-chain acids and substituted acids are named as derivatives of the straight-chain acids. To indicate the position of attachment, the Greek letters, α-, β-, γ-, δ-, etc., are used; the α-carbon is the one bearing the carboxyl group.

$$\overset{\delta}{C}-\overset{\gamma}{C}-\overset{\beta}{C}-\overset{\alpha}{C}-COOH$$ *Used in common names*

For example:

$$CH_3CH_2CHCOOH \quad\quad CH_3CH_2CH-CHCOOH$$
$$| \quad\quad\quad\quad\quad\quad\quad\quad | \quad\quad |$$
$$CH_3 \quad\quad\quad\quad\quad\quad\quad CH_3 \quad CH_3$$

$$CH_2CH_2CH_2COOH$$

α-Methylbutyric acid   α,β-Dimethylvaleric acid   γ-Phenylbutyric acid

$$CH_2CH_2CHCOOH \quad\quad\quad CH_3CHCOOH$$
$$| \quad\quad | \quad\quad\quad\quad\quad\quad\quad\quad |$$
$$Cl \quad\quad CH_3 \quad\quad\quad\quad\quad\quad\quad OH$$

γ-Chloro-α-methylbutyric acid   α-Hydroxypropionic acid
Lactic acid

Generally the parent acid is taken as the one of longest carbon chain, although some compounds are named as derivatives of acetic acid.

Aromatic acids, ArCOOH, are usually named as derivatives of the parent acid, **benzoic acid**, C_6H_5COOH. The methylbenzoic acids are given the special name of *toluic acids*.

COOH   COOH   COOH
              NO_2 CH_3

Br NO_2

p-Bromobenzoic acid   2,4-Dinitrobenzoic acid   *m*-Toluic acid

The **IUPAC names** follow the usual pattern. The longest chain carrying the carboxyl group is considered the parent structure, and is named by replacing the *-e* of the corresponding alkane with **-oic acid**. For example:

$$CH_3CH_2CH_2CH_2COOH \quad\quad\quad CH_3CH_2CHCOOH$$
$$| $$
$$CH_3$$

$$CH_2CH_2COOH$$

Pentanoic acid   2-Methylbutanoic acid   3-Phenylpropanoic acid

$$CH_3$$
$$|$$
$$Cl-\bigcirc-CHCH_2COOH \quad\quad\quad CH_3CH=CHCOOH$$

3-(*p*-Chlorophenyl)butanoic acid   2-Butenoic acid

The position of a substituent is indicated as usual by a number. We should notice

$$\overset{5}{C}-\overset{4}{C}-\overset{3}{C}-\overset{2}{C}-\overset{1}{COOH} \quad\quad\quad\quad \textit{Used in IUPAC names}$$

that the carboxyl carbon is always considered as C–1, and hence C–2 corresponds to α of the common names, C–3 to β, and so on. (*Caution*: Do not mix Greek letters with IUPAC names, or Arabic numerals with common names.)

The name of a **salt** of a carboxylic acid consists of the name of the cation (*sodium, potassium, ammonium,* etc.) followed by the name of the acid with the ending -*ic acid* changed to -**ate**. For example:

⬡⟩COONa $(CH_3COO)_2Ca$ $HCOONH_4$

Sodium benzoate   Calcium acetate   Ammonium formate

$$CH_2-CH-COOK$$
$$\ \ |\ \ \ \ \ |$$
$$Br\ \ \ Br$$

Potassium α,β-dibromopropionate
Potassium 2,3-dibromopropanoate

23.3   Physical properties

As we would expect from their structure, carboxylic acid molecules are polar, and like alcohol molecules can form hydrogen bonds with each other and with other kinds of molecules. The aliphatic acids therefore show very much the same solubility behavior as the alcohols: the first four are miscible with water, the five-carbon acid is partly soluble, and the higher acids are virtually insoluble. Water solubility undoubtedly arises from hydrogen bonding between the carboxylic acid and water. The simplest aromatic acid, benzoic acid, contains too many carbon atoms to show appreciable solubility in water.

Carboxylic acids are soluble in less polar solvents like ether, alcohol, benzene, etc.

We can see from Table 23.1 that as a class the carboxylic acids are even higher boiling than alcohols. For example, propionic acid (b.p. 141 °C) boils more than 20 °C higher than the alcohol of comparable molecular weight, *n*-butyl alcohol (b.p. 118 °C). These very high boiling points are due to the fact that a pair of carboxylic acid molecules are held together not by one but by two hydrogen bonds:

$$\begin{array}{ccc} & O\text{---}H-O & \\ & \diagup\diagup \qquad \diagdown & \\ R-C & & C-R \\ & \diagdown \qquad \diagup\diagup & \\ & O-H\text{---}O & \end{array}$$

Problem 23.1   At 110 °C and 454 mm pressure, 0.11 g acetic acid vapor occupies 63.7 mL; at 156 °C and 458 mm, 0.081 g occupies 66.4 mL. Calculate the molecular weight of acetic acid in the vapor phase at each temperature. How do you interpret these results?

The odors of the lower aliphatic acids progress from the sharp, irritating odors of formic and acetic acids to the distinctly unpleasant odors of butyric, valeric, and caproic acids; the higher acids have little odor because of their low volatility.

23.4 Salts of carboxylic acids

Although much weaker than the strong mineral acids (sulfuric, hydrochloric, nitric), the carboxylic acids are tremendously more acidic than the very weak organic acids (alcohols, acetylene) we have so far studied; they are much stronger acids than water. Aqueous hydroxides therefore readily convert carboxylic acids into their salts; aqueous mineral acids readily convert the salts back into the carboxylic acids. Since we can do little with carboxylic acids without encountering

$$\underset{\text{Acid}}{\text{RCOOH}} \quad \overset{\text{OH}^-}{\underset{\text{H}^+}{\rightleftarrows}} \quad \underset{\text{Salt}}{\text{RCOO}^-}$$

this conversion into and from their salts, it is worthwhile for us to examine the properties of these salts.

Salts of carboxylic acids—like all salts—are crystalline non-volatile solids made up of positive and negative ions; their properties are what we would expect of such structures. The strong electrostatic forces holding the ions in the crystal lattice can be overcome only by heating to a high temperature, or by a very polar solvent. The temperature required for melting is so high that before it can be reached carbon–carbon bonds break and the molecule decomposes, generally in the neighborhood of 300–400 °C. A decomposition point is seldom useful for the identification of a compound, since it usually reflects the rate of heating rather than the identity of the compound.

The alkali metal salts of carboxylic acids (sodium, potassium, ammonium) are soluble in water but insoluble in non-polar solvents; most of the heavy metal salts (iron, silver, copper, etc.) are insoluble in water.

Thus we see that, except for the acids of four carbons or fewer, which are soluble both in water and in organic solvents, *carboxylic acids and their alkali metal salts show exactly opposite solubility behavior*. Because of the ready interconversion of acids and their salts, this difference in solubility behavior may be used in two important ways; for *identification* and for *separation*.

A water-insoluble organic compound that dissolves in cold dilute aqueous sodium hydroxide must be either a carboxylic acid or one of the few other kinds of organic compounds more acidic than water; that it is indeed a carboxylic acid can then be shown in other ways.

$$\underset{\substack{\text{Stronger acid}\\ \textit{Insoluble in } H_2O}}{\text{RCOOH}} + \text{NaOH} \longrightarrow \underset{\substack{\textit{Soluble in}\\ H_2O}}{\text{RCOONa}} + \underset{\substack{\text{Weaker}\\ \text{acid}}}{\text{H}_2\text{O}}$$

Instead of sodium hydroxide, we can use aqueous sodium bicarbonate; even if the unknown is water-soluble, its acidity is shown by the evolution of bubbles of CO_2.

$$\underset{\textit{Insoluble in } H_2O}{\text{RCOOH}} + \text{NaHCO}_3 \longrightarrow \underset{\textit{Soluble in } H_2O}{\text{RCOONa}} + \text{H}_2\text{O} + \text{CO}_2\uparrow$$

We can separate a carboxylic acid from non-acidic compounds by taking advantage of its solubility and their insolubility in aqueous base; once the separation has been accomplished, we can regenerate the acid by acidification of the aqueous solution. If we are dealing with solids, we simply stir the mixture with aqueous base and then filter the solution from insoluble, non-acidic materials; addition of mineral acid to the filtrate precipitates the carboxylic acid, which can be collected on a filter. If we are dealing with liquids, we shake the mixture with aqueous base in a separatory funnel and separate the aqueous layer from the insoluble organic layer; addition of acid to the aqueous layer again liberates the carboxylic acid, which can then be separated from the water. For completeness of separation and ease of handling, we often add a water-insoluble solvent like ether to the acidified mixture. The carboxylic acid is extracted from the water by the ether, in which it is more soluble; the volatile ether is readily removed by distillation from the comparatively high-boiling acid.

For example, an aldehyde prepared by the oxidation of a primary alcohol (Sec. 18.6) may very well be contaminated with the carboxylic acid; this acid can be simply washed out with dilute aqueous base. The carboxylic acid prepared by oxidation of an alkylbenzene (Sec. 15.11) may very well be contaminated with unreacted starting material; the carboxylic acid can be taken into solution by aqueous base, separated from the insoluble hydrocarbon, and regenerated by addition of mineral acid.

Since separations of this kind are more clear-cut and less wasteful of material, they are preferred wherever possible over recrystallization or distillation.

23.5   Industrial source

Acetic acid, by far the most important of all carboxylic acids, has been prepared chiefly by catalytic air oxidation of various hydrocarbons or of acetaldehyde. A newer method involves reaction between methanol and carbon monoxide in the

presence of an iodine–rhodium catalyst—still another example of catalysis by a transition metal complex (see Secs. 8.3, 17.6, and 20.5–20.8).

Large amounts of acetic acid are also produced as the dilute aqueous solution known as *vinegar*. Here, too, the acetic acid is prepared by air oxidation; the compound that is oxidized is ethyl alcohol, and the catalysts are bacterial (*Acetobacter*) enzymes.

The most important sources of aliphatic carboxylic acids are the animal and vegetable **fats** (Secs. 37.2–37.4). From fats there can be obtained, in purity of over 90%, straight-chain carboxylic acids of even carbon number ranging from six to eighteen carbon atoms. These acids can be converted into the corresponding alcohols (Sec. 23.18), which can then be used, in the ways we have already studied (Sec. 18.8), to make a great number of other compounds containing long, straight-chain units.

The most important of the aromatic carboxylic acids, **benzoic acid** and the **phthalic acids**, are prepared on an industrial scale by a reaction we have already encountered: oxidation of alkylbenzenes (Sec. 15.11). The toluene and xylenes required are readily obtained from petroleum by catalytic reforming of aliphatic hydrocarbons (Sec. 15.5); much smaller amounts of these arenes are isolated directly from coal tar. Another precursor of phthalic acid (the *ortho* isomer) is the aromatic hydrocarbon *naphthalene*, also found in coal tar. Cheap oxidizing agents like chlorine or even air (in the presence of catalysts) are used.

Problem 23.2 In the presence of peroxides, carboxylic acids (or esters) react with 1-alkenes to yield more complicated acids. For example:

$$n\text{-}C_4H_9CH{=}CH_2 + CH_3CH_2CH_2COOH \xrightarrow{\text{peroxides}} n\text{-}C_4H_9CH_2CH_2CHCOOH$$

1-Hexene   *n*-Butyric acid C_2H_5

2-Ethyloctanoic acid
(70% yield)

(a) Outline all steps in a likely mechanism for this reaction. (*Hint*: See Sec. 8.20.) Predict the products of similar reactions between: (b) 1-octene and propionic acid; (c) 1-decene and isobutyric acid; (d) 1-octene and ethyl malonate, $CH_2(COOC_2H_5)_2$.

Problem 23.3 (a) Carbon monoxide converts a sulfuric acid solution of each of the following into 2,2-dimethylbutanoic acid: 2-methyl-2-butene, *tert*-pentyl alcohol, neopentyl alcohol. Suggest a likely mechanism for this method of synthesizing carboxylic acids. (b) *n*-Butyl alcohol and *sec*-butyl alcohol give the same product. What would you expect it to be?

23.6 Preparation

The straight-chain aliphatic acids up to C_6 and those of even carbon number up to C_{18} are commercially available, as are the simple aromatic acids. Other carboxylic acids can be prepared by the methods outlined below.

PREPARATION OF CARBOXYLIC ACIDS

1. **Oxidation of primary alcohols.** Discussed in Sec. 18.6.

$$R-CH_2OH \xrightarrow{KMnO_4} R-COOH$$

Examples:

$$CH_3CH_2\overset{\overset{\displaystyle CH_3}{|}}{C}HCH_2OH \xrightarrow{KMnO_4} CH_3CH_2\overset{\overset{\displaystyle CH_3}{|}}{C}HCOOH$$

2-Methyl-1-butanol   2-Methylbutanoic acid

$$CH_3\overset{\overset{\displaystyle CH_3}{|}}{C}HCH_2OH \xrightarrow{KMnO_4} CH_3\overset{\overset{\displaystyle CH_3}{|}}{C}HCOOH$$

Isobutyl alcohol   Isobutyric acid

2. **Oxidation of alkylbenzenes.** Discussed in Sec. 15.11.

$$Ar-R \xrightarrow{KMnO_4 \text{ or } K_2Cr_2O_7} Ar-COOH$$

Examples:

$$O_2N\langle\bigcirc\rangle CH_3 \xrightarrow{K_2Cr_2O_7, H_2SO_4, \text{ heat}} O_2N\langle\bigcirc\rangle COOH$$

p-Nitrotoluene   *p*-Nitrobenzoic acid

$$\underset{Br}{\overset{CH_3}{\bigcirc}} \xrightarrow[\text{heat}]{KMnO_4, OH^-} \underset{Br}{\overset{COOH}{\bigcirc}}$$

o-Bromotoluene   *o*-Bromobenzoic acid

3. **Carbonation of Grignard reagents.** Discussed in Sec. 23.7.

$$\underset{\text{(or Ar-X)}}{R-X} \xrightarrow{Mg} R-MgX \xrightarrow{CO_2} R-COOMgX \xrightarrow{H^+} \underset{\text{(or Ar-COOH)}}{R-COOH}$$

Examples:

$$\underset{\underset{C_2H_5}{\overset{|}{\underset{}{CH_3-CH}}}}{\overset{Br}{\bigcirc}} \xrightarrow{Mg} \underset{\underset{C_2H_5}{\overset{|}{\underset{}{CH_3-CH}}}}{\overset{MgBr}{\bigcirc}} \xrightarrow{CO_2} \underset{\underset{C_2H_5}{\overset{|}{\underset{}{CH_3-CH}}}}{\overset{COOMgBr}{\bigcirc}} \xrightarrow{H^+} \underset{\underset{C_2H_5}{\overset{|}{\underset{}{CH_3-CH}}}}{\overset{COOH}{\bigcirc}}$$

p-Bromo-*sec*-butylbenzene   *p*-*sec*-Butylbenzoic acid

CONTINUED

CONTINUED

$$
\underset{\substack{\text{\textit{tert}-Pentyl}\\\text{chloride}}}{\text{C}_2\text{H}_5-\overset{\displaystyle\text{CH}_3}{\underset{\displaystyle\text{CH}_3}{\text{C}}}-\text{Cl}}
\xrightarrow{\text{Mg}}
\text{C}_2\text{H}_5-\overset{\displaystyle\text{CH}_3}{\underset{\displaystyle\text{CH}_3}{\text{C}}}-\text{MgCl}
\xrightarrow{\text{CO}_2}
\xrightarrow{\text{H}^+}
\underset{\substack{\text{Ethyldimethylacetic}\\\text{acid}\\\text{(2,2-Dimethylbutanoic}\\\text{acid)}}}{\text{C}_2\text{H}_5-\overset{\displaystyle\text{CH}_3}{\underset{\displaystyle\text{CH}_3}{\text{C}}}-\text{COOH}}
$$

4. Hydrolysis of nitriles. Discussed in Sec. 23.8.

$$
\begin{array}{l}\text{R}-\text{C}\equiv\text{N}\\\text{or}\\\text{Ar}-\text{C}\equiv\text{N}\end{array}
+\ \text{H}_2\text{O}\ \xrightarrow{\text{acid or base}}\
\begin{array}{l}\text{R}-\text{COOH}\\\text{or}\\\text{Ar}-\text{COOH}\end{array}
+\ \text{NH}_3
$$

Examples:

Benzyl chloride $\xrightarrow{\text{NaCN}}$ Phenylacetonitrile $\xrightarrow{70\% \text{ H}_2\text{SO}_4,\ \text{reflux}}$ Phenylacetic acid $+\ \text{NH}_4^+$

$$
\underset{\substack{\text{\textit{n}-Butyl bromide}}}{\textit{n}\text{-C}_4\text{H}_9\text{Br}}
\xrightarrow{\text{NaCN}}
\underset{\substack{\textit{n}\text{-Valeronitrile}\\\text{(Pentanenitrile)}}}{\textit{n}\text{-C}_4\text{H}_9\text{CN}}
\xrightarrow{\text{aq. alc. NaOH, reflux}}
\textit{n}\text{-C}_4\text{H}_9\text{COO}^-\ +\ \text{NH}_3
$$

$$
\downarrow \text{H}^+
$$

$$
\underset{\substack{\textit{n}\text{-Valeric acid}\\\text{(Pentanoic acid)}}}{\textit{n}\text{-C}_4\text{H}_9\text{COOH}\ +\ \text{NH}_4^+}
$$

Diazonium salt $\longrightarrow$ *o*-Tolunitrile $\xrightarrow{75\% \text{ H}_2\text{SO}_4,\ 150-160\ ^\circ\text{C}}$ *o*-Toluic acid $+\ \text{NH}_4^+$

5. Malonic ester synthesis. Discussed in Sec. 30.2.

6. Special methods for phenolic acids. Discussed in Sec. 28.11.   ■

All the methods listed are important; our choice is governed by the availability of starting materials.

Oxidation is the most direct and is generally used when possible, some lower aliphatic acids being made from the available alcohols, and substituted aromatic acids from substituted toluenes.

The **Grignard synthesis** and the **nitrile synthesis** have the special advantage of increasing the length of a carbon chain, and thus extending the range of available materials. In the aliphatic series both Grignard reagents and nitriles are prepared from halides, which in turn are usually prepared from alcohols. The syntheses thus amount to the preparation of acids from alcohols containing one less carbon atom.

$$\text{KMnO}_4 \longrightarrow \text{R—COOH} \quad \textit{Same carbon number}$$

R—CH$_2$OH

$$\textit{Higher carbon number}$$

$$\xrightarrow{\text{Mg}} \text{RCH}_2\text{MgBr} \xrightarrow{\text{CO}_2} \xrightarrow{\text{H}^+} \text{R—CH}_2\text{COOH}$$

$$\xrightarrow{\text{PBr}_3} \text{RCH}_2\text{Br}$$

$$\xrightarrow{\text{CN}^-} \text{RCH}_2\text{CN} \xrightarrow{\text{H}_2\text{O}} \text{R—CH}_2\text{COOH}$$

Problem 23.4 Which carboxylic acid can be prepared from *p*-bromotoluene: (a) by direct oxidation? (b) by free-radical chlorination followed by the nitrile synthesis?

Aromatic nitriles generally cannot be prepared from the unreactive aryl halides (Sec. 29.5). Instead they are made from diazonium salts by a reaction we shall discuss later (Sec. 27.14). Diazonium salts are prepared from aromatic amines, which in turn are prepared from nitro compounds. Thus the carboxyl group eventually occupies the position on the ring where a nitro group was originally introduced by direct nitration (Sec. 14.8).

$$\text{Ar—H} \longrightarrow \underset{\substack{\text{Nitro} \\ \text{compound}}}{\text{Ar—NO}_2} \longrightarrow \underset{\text{Amine}}{\text{Ar—NH}_2} \longrightarrow \underset{\substack{\text{Diazonium} \\ \text{ion}}}{\text{Ar—N}_2{}^+} \longrightarrow \underset{\text{Nitrile}}{\text{Ar—C}\equiv\text{N}} \longrightarrow \underset{\text{Acid}}{\text{Ar—COOH}}$$

For the preparation of quite complicated acids, the most versatile method of all is used, the *malonic ester synthesis* (Sec. 30.2).

23.7 Grignard synthesis

The Grignard synthesis of a carboxylic acid is carried out by bubbling gaseous CO_2 into the ether solution of the Grignard reagent, or by pouring the Grignard reagent on crushed Dry Ice (solid CO_2); in the latter method Dry Ice serves not only as reagent but also as cooling agent.

The Grignard reagent adds to the carbon–oxygen double bond just as in the reaction with aldehydes and ketones (Sec. 17.14). The product is the magnesium salt of the carboxylic acid, from which the free acid is liberated by treatment with mineral acid.

$$\text{R—MgX} + \overset{\text{O}}{\underset{\text{O}}{\text{C}}} \longrightarrow \text{R—COO}^-\text{MgX}^+ \xrightarrow{\text{H}^+} \text{R—COOH} + \text{Mg}^{2+} + \text{X}^-$$

The Grignard reagent can be prepared from primary, secondary, tertiary, or aromatic halides; the method is limited only by the presence of other reactive groups in the molecule (Sec. 17.17). The following syntheses illustrate the application of this method:

$$CH_3\text{-}\underset{\underset{CH_3}{|}}{\overset{\overset{CH_3}{|}}{C}}\text{-}OH \xrightarrow{HCl} CH_3\text{-}\underset{\underset{CH_3}{|}}{\overset{\overset{CH_3}{|}}{C}}\text{-}Cl \xrightarrow{Mg} CH_3\text{-}\underset{\underset{CH_3}{|}}{\overset{\overset{CH_3}{|}}{C}}\text{-}MgCl \xrightarrow[]{CO_2} \xrightarrow{H^+} CH_3\text{-}\underset{\underset{CH_3}{|}}{\overset{\overset{CH_3}{|}}{C}}\text{-}COOH$$

tert-Butyl alcohol	*tert*-Butyl chloride		Trimethylacetic acid

Mesitylene $\xrightarrow{Br_2}$ Bromomesitylene $\xrightarrow{Mg}$ (MgBr) $\xrightarrow[]{CO_2}$ $\xrightarrow{H^+}$ Mesitoic acid (2,4,6-Trimethylbenzoic acid)

23.8 Nitrile synthesis

Aliphatic nitriles are prepared by treatment of alkyl halides with sodium cyanide in a solvent that will dissolve both reactants; in dimethyl sulfoxide, reaction occurs rapidly and exothermically at room temperature. The resulting nitrile is then hydrolyzed to the acid by boiling aqueous alkali or acid.

$$R\text{-}X + CN^- \longrightarrow R\text{-}C\equiv N + X^-$$

$$R\text{-}C\equiv N + H_2O \left\{ \begin{array}{l} \xrightarrow{H^+} R\text{-}COOH + NH_4^+ \\ \\ \xrightarrow{OH^-} R\text{-}COO^- + NH_3 \end{array} \right.$$

The reaction of an alkyl halide with cyanide ion involves nucleophilic substitution (Sec. 5.8). The fact that HCN is a very weak acid tells us that cyanide ion is a strong base; as we might expect, this strongly basic ion can abstract hydrogen ion and thus cause elimination as well as substitution. Indeed, with tertiary halides

$CH_3CH_2CH_2CH_2Br + CN^- \longrightarrow CH_3CH_2CH_2CH_2CN$		1° halide:
n-Butyl bromide Valeronitrile		*substitution*

$CH_3\text{-}\underset{\underset{CH_3}{\|}}{\overset{\overset{CH_3}{\|}}{C}}\text{-}Br + CN^- \longrightarrow CH_3\text{-}\underset{}{\overset{\overset{CH_3}{\|}}{C}}=CH_2 + HCN$		3° halide:
tert-Butyl bromide Isobutylene		*elimination*

elimination is the principal reaction; even with secondary halides the yield of substitution product is poor. Here again we find a nucleophilic substitution reaction that is of synthetic importance *only when primary halides are used*.

As already mentioned, aromatic nitriles are made, not from the unreactive aryl halides, but from diazonium salts (Sec. 27.14).

Although nitriles are sometimes named as *cyanides* or as *cyano* compounds, they generally take their names from the acids they yield upon hydrolysis. They are named by dropping *-ic acid* from the common name of the acid and adding **-nitrile**; usually for euphony an "o" is inserted between the root and the ending (e.g., *acetonitrile*). In the IUPAC system they are named by adding *-nitrile* to the name of the parent hydrocarbon (e.g., *ethanenitrile*). For example:

$$CH_3C{\equiv}N \qquad CH_3(CH_2)_3C{\equiv}N$$

Acetonitrile   *n*-Valeronitrile   Benzonitrile   *p*-Tolunitrile
(Ethanenitrile)   (Pentanenitrile)

23.9 Reactions

The characteristic chemical behavior of carboxylic acids is, of course, determined by their functional group, **carboxyl**, —COOH. This group is made up of a carbonyl group ($C{=}O$) and a hydroxyl group (—OH). As we shall see, it is the —OH that actually undergoes nearly every reaction—loss of H^+, or replacement by another group—but *it does so in a way that is possible only because of the effect of the* $C{=}O$.

The rest of the molecule undergoes reactions characteristic of its structure; it may be aliphatic or aromatic, saturated or unsaturated, and may contain a variety of other functional groups.

REACTIONS OF CARBOXYLIC ACIDS

1. Acidity. Salt formation. Discussed in Secs. 23.4, 23.10–23.14.

$$RCOOH \rightleftharpoons RCOO^- + H^+$$

Examples:

$$2CH_3COOH + Zn \longrightarrow (CH_3COO^-)_2Zn^{2+} + H_2$$

Acetic acid   Zinc acetate

$$CH_3(CH_2)_{10}COOH + NaOH \longrightarrow CH_3(CH_2)_{10}COO^-Na^+ + H_2O$$

Lauric acid   Sodium laurate

COOH + NaHCO₃ ⟶ COO⁻Na⁺ + CO₂ + H₂O

Benzoic acid   Sodium benzoate

2. Conversion into functional derivatives

$$R-C \underset{OH}{\overset{O}{\Vert}} \longrightarrow R-C \underset{Z}{\overset{O}{\Vert}} \qquad (Z = -Cl, \ -OR', \ -NH_2)$$

CONTINUED

CONTINUED

(a) Conversion into acid chlorides. Discussed in Sec. 23.15.

$$R-\overset{\displaystyle O}{\overset{\|}{C}}-OH \quad + \begin{Bmatrix} SOCl_2 \\ PCl_3 \\ PCl_5 \end{Bmatrix} \quad \longrightarrow \quad R-\overset{\displaystyle O}{\overset{\|}{C}}-Cl$$

Acid chloride

Examples:

$$\text{\Large⬡}COOH + PCl_5 \xrightarrow{\;100\ ^\circ C\;} \text{\Large⬡}COCl + POCl_3 + HCl$$

Benzoic acid   Benzoyl chloride

$$n\text{-}C_{17}H_{35}COOH + SOCl_2 \xrightarrow{\;\text{reflux}\;} n\text{-}C_{17}H_{35}COCl + SO_2 + HCl$$

Stearic acid   Thionyl   Stearoyl chloride
                chloride

$$3CH_3COOH + PCl_3 \xrightarrow{\;50\ ^\circ C\;} 3CH_3COCl + H_3PO_3$$

Acetic acid   Acetyl chloride

(b) Conversion into esters. Discussed in Secs. 23.16 and 24.15.

$$R-\overset{\displaystyle O}{\overset{\|}{C}}-OH + R'OH \underset{\longleftarrow}{\overset{H^+}{\longrightarrow}} R-\overset{\displaystyle O}{\overset{\|}{C}}-OR' + H_2O \quad \text{Reactivity of } R'OH: 1^\circ > 2^\circ (>3^\circ)$$

An ester

$$R-\overset{\displaystyle O}{\overset{\|}{C}}-OH \xrightarrow{\;SOCl_2\;} R-\overset{\displaystyle O}{\overset{\|}{C}}-Cl \xrightarrow{\;R'OH\;} R-\overset{\displaystyle O}{\overset{\|}{C}}-OR'$$

An acid chloride   An ester

Examples:

$$\text{\Large⬡}COOH + CH_3OH \underset{\longleftarrow}{\overset{H^+}{\longrightarrow}} \text{\Large⬡}COOCH_3 + H_2O$$

Benzoic acid   Methanol   Methyl benzoate

$$CH_3COOH + \text{\Large⬡}CH_2OH \underset{\longleftarrow}{\overset{H^+}{\longrightarrow}} CH_3COOCH_2\text{\Large⬡} + H_2O$$

Acetic acid   Benzyl alcohol   Benzyl acetate

$$(CH_3)_3CCOOH \xrightarrow{\;SOCl_2\;} (CH_3)_3CCOCl \xrightarrow{\;C_2H_5OH\;} (CH_3)_3CCOOC_2H_5$$

Trimethylacetic acid   Ethyl trimethylacetate

CONTINUED

_____ CONTINUED _____

(c) **Conversion into amides.** Discussed in Sec. 23.17.

$$R-\overset{\displaystyle O}{\underset{\displaystyle OH}{C}} \xrightarrow{\text{SOCl}_2} R-\overset{\displaystyle O}{\underset{\displaystyle Cl}{C}} \xrightarrow{\text{NH}_3} R-\overset{\displaystyle O}{\underset{\displaystyle NH_2}{C}}$$

$$\text{An acid chloride} \qquad\qquad \text{An amide}$$

Example:

$$\underset{\text{Phenylacetic acid}}{C_6H_5CH_2COOH} \xrightarrow{\text{SOCl}_2} \underset{\text{Phenylacetyl chloride}}{C_6H_5CH_2COCl} \xrightarrow{\text{NH}_3} \underset{\text{Phenylacetamide}}{C_6H_5CH_2CONH_2}$$

3. **Reduction.** Discussed in Sec. 23.18.

$$R-COOH \xrightarrow{\text{LiAlH}_4} \underset{1° \text{ alcohol}}{R-CH_2OH} \qquad \textit{Also reduced via esters} \textbf{ (Sec. 24.22)}$$

Examples:

$$\underset{\substack{\text{Trimethylacetic} \\ \text{acid}}}{4(CH_3)_3CCOOH} + 3LiAlH_4 \xrightarrow{\text{ether}} \underset{+\ 2LiAlO_2 + 4H_2}{[(CH_3)_3CCH_2O]_4AlLi} \xrightarrow{\text{H}^+} \underset{\substack{\text{Neopentyl alcohol} \\ \text{(2,2-Dimethyl-} \\ \text{1-propanol)}}}{(CH_3)_3CCH_2OH}$$

$$\underset{m\text{-Toluic acid}}{\text{COOH, CH}_3 \text{ ring}} \xrightarrow{\text{LiAlH}_4} \underset{m\text{-Methylbenzyl alcohol}}{\text{CH}_2\text{OH, CH}_3 \text{ ring}}$$

4. **Substitution in alkyl or aryl group**

(a) **Alpha-halogenation of aliphatic acids. Hell–Volhard–Zelinsky reaction.** Discussed in Sec. 23.19.

$$RCH_2COOH + X_2 \xrightarrow{P} \underset{\underset{X}{|}}{RCHCOOH} + HX \qquad X_2 = Cl_2, Br_2$$

$$\text{An } \alpha\text{-halo acid}$$

Examples:

$$\underset{\substack{\text{Acetic} \\ \text{acid}}}{CH_3COOH} \xrightarrow{\text{Cl}_2,\,\text{P}} \underset{\substack{\text{Chloroacetic} \\ \text{acid}}}{ClCH_2COOH} \xrightarrow{\text{Cl}_2,\,\text{P}} \underset{\substack{\text{Dichloroacetic} \\ \text{acid}}}{Cl_2CHCOOH} \xrightarrow{\text{Cl}_2,\,\text{P}} \underset{\substack{\text{Trichloroacetic} \\ \text{acid}}}{Cl_3CCOOH}$$

$$\underset{\text{Isovaleric acid}}{\overset{\displaystyle CH_3}{\underset{\displaystyle |}{CH_3CHCH_2COOH}}} \xrightarrow{\text{Br}_2,\,\text{P}} \underset{\substack{| \\ Br \\ \alpha\text{-Bromoisovaleric acid}}}{\overset{\displaystyle CH_3}{\underset{\displaystyle |}{CH_3CHCHCOOH}}}$$

_____ CONTINUED _____

—————— CONTINUED ——————

(b) **Ring substitution in aromatic acids.** Discussed in Secs. 14.5 and 14.15.

—COOH: deactivates, and directs *meta* in electrophilic substitution.

Example:

COOH   COOH

$\xrightarrow{\text{HNO}_3,\ \text{H}_2\text{SO}_4,\ \text{heat}}$ NO$_2$

Benzoic acid   *m*-Nitrobenzoic acid

The most characteristic property of the carboxylic acids is the one that gives them their name: **acidity.** Their tendency to give up a hydrogen ion is such that in aqueous solution a measurable equilibrium exists between acid and ions; they are thus much more acidic than any other class of organic compounds we have studied so far.

$$\text{RCOOH} + \text{H}_2\text{O} \rightleftharpoons \text{RCOO}^- + \text{H}_3\text{O}^+$$

The OH of an acid can be replaced by a Cl, OR′, or NH$_2$ group to yield an *acid chloride,* an *ester,* or an *amide.* These compounds are called **functional derivatives** of acids; they all contain the **acyl group**:

$$\text{R}-\text{C}\underset{\diagdown}{\overset{\diagup O}{}}$$

The functional derivatives are all readily reconverted into the acid by simple hydrolysis, and are often converted one into another.

One of the few reducing agents capable of reducing an acid directly to an alcohol is *lithium aluminum hydride,* LiAlH$_4$.

The hydrocarbon portion of an aliphatic acid can undergo the free-radical halogenation characteristic of alkanes, but because of the random nature of the substitution it is seldom used. The presence of a small amount of phosphorus, however, causes halogenation (by a heterolytic mechanism) to take place *exclusively at the alpha position.* This reaction is known as the **Hell–Volhard–Zelinsky reaction,** and it is of great value in synthesis.

An aromatic ring bearing a carboxyl group undergoes the aromatic electrophilic substitution reactions expected of a ring carrying a deactivating, *meta*-directing group. Deactivation is so strong that the Friedel–Crafts reaction does not take place. We have already accounted for this effect of the —COOH group on the basis of its strong electron-withdrawing tendencies (Sec. 14.16).

COOH

—COOH withdraws electrons:
deactivates, directs meta in
electrophilic substitution

Decarboxylation, that is, elimination of the —COOH group as CO_2, is of limited importance for aromatic acids, and highly important for certain substituted aliphatic acids: malonic acids (Sec. 30.2) and β-keto acids (Sec. 30.3). It is worthless for most simple aliphatic acids, yielding a complicated mixture of hydrocarbons.

23.10   Ionization of carboxylic acids. Acidity constant

In aqueous solution a carboxylic acid exists in equilibrium with the carboxylate anion and the hydrogen ion (actually, of course, the hydronium ion, H_3O^+).

$$RCOOH + H_2O \rightleftharpoons RCOO^- + H_3O^+$$

As for any equilibrium, the concentrations of the components are related by the expression

$$K_{eq} = \frac{[RCOO^-][H_3O^+]}{[H_2O][RCOOH]}$$

Since the concentration of water, the solvent, remains essentially constant, we can combine it with K_{eq} to obtain the expression

$$K_a = \frac{[RCOO^-][H_3O^+]}{[RCOOH]}$$

in which K_a equals $K_{eq}[H_2O]$. This new constant, K_a, is called the **acidity constant**.

Every carboxylic acid has its characteristic K_a, which indicates how strong an acid it is. Since the acidity constant is the ratio of ionized to un-ionized material, the larger the K_a the greater the extent of the ionization (under a given set of conditions) and the stronger the acid. We use the K_a values, then, to compare in an exact way the strengths of different acids.

We see in Table 23.2 (p. 839) that unsubstituted aliphatic and aromatic acids have K_a values of about 10^{-4} to 10^{-5} (0.0001 to 0.00001). This means that they are weakly acidic, with only a slight tendency to release protons.

By the same token, carboxylate anions are moderately basic, with an appreciable tendency to combine with protons. They react with water to increase the concentration of hydroxide ions, a reaction often referred to as *hydrolysis*. As a

$$RCOO^- + H_2O \rightleftharpoons RCOOH + OH^-$$

result aqueous solutions of carboxylate salts are slightly alkaline. (The basicity of an aqueous solution of a carboxylate salt is due chiefly, of course, to the carboxylate anions, not to the comparatively few hydroxide ions they happen to generate.)

We may now expand the series of relative acidities and basicities:

Acidity $RCOOH > HOH > ROH > HC{\equiv}CH > NH_3 > RH$

Basicity $RCOO^- < HO^- < RO^- < HC{\equiv}C^- < NH_2^- < R^-$

Certain substituted acids are much stronger or weaker than a typical acid like CH_3COOH. We shall see that the acid-strengthening or acid-weakening effect of a substituent can be accounted for in a reasonable way; however, we must first learn a little more about equilibrium in general.

23.11 Equilibrium

So far we have dealt very little with the problem of equilibrium. Under the conditions employed, most of our reactions have been essentially irreversible; that is, they have been one-way reactions. With a few exceptions—1,4-addition, for example (Sec. 10.27)—the products obtained, and their relative yields, have been determined by how fast reactions go and not by how nearly to completion they proceed before equilibrium is reached. Consequently, we have been concerned with the relationship between structure and rate; now we shall turn to the relationship between structure and equilibrium.

Let us consider the reversible reaction between A and B to form C and D. The

$$A + B \; \rightleftarrows \; C + D$$

yield of C and D does not depend upon how fast A and B react, but rather upon how completely they have reacted when equilibrium is reached.

The concentrations of the various components are related by the familiar expression

$$K_{eq} = \frac{[C][D]}{[A][B]}$$

in which K_{eq} is the equilibrium constant. The more nearly a reaction has proceeded to completion when it reaches equilibrium, the larger is [C][D] compared with [A][B], and hence the larger the K_{eq}. The value of K_{eq} is therefore a measure of the tendency of the reaction to go to completion.

The value of K_{eq} is determined by the change in *free energy*, G, on proceeding from reactants to products (Fig. 23.2). The exact relationship is given by the expression

$$\Delta G^\circ = -2.303RT \log K_{eq}$$

where ΔG° is the *standard free energy change*.

Free energy change is related to our familiar quantity ΔH (precisely ΔH°, which is only slightly different) by the expression,

$$\Delta G^\circ = \Delta H - T\Delta S^\circ$$

where ΔS° is the *standard entropy change*. Entropy corresponds, roughly, to the *randomness* of the system. To the extent that $T\Delta S^\circ$ contributes to ΔG°, equilibrium tends to shift toward the side in which fewer restrictions are placed on the positions of atoms and molecules. ("Die Energie der Welt ist constant. Die Entropie der Welt strebt einem Maximum zu."—*Clausius, 1865.*)

Under the same experimental conditions two reversible reactions have K_{eq} values of different sizes because of a difference in ΔG°. In attempting to understand the effect of structure on position of equilibrium, we shall estimate differences in relative stabilities of reactants and products. Now, what we estimate in this way are not differences in free energy change but differences in potential energy change. It turns out that very often these differences are *proportional to* differences in ΔG°. So long as we compare closely related compounds, the predictions we make by this approach are generally good ones.

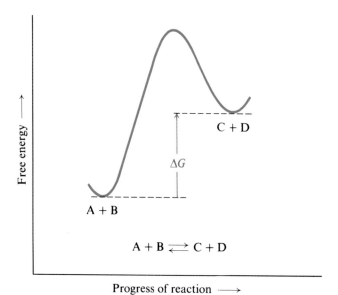

Figure 23.2   Free energy curve for a reversible reaction.

These predictions are good ones despite the fact that the free energy changes on which they depend are made up to varying degrees of ΔH and $\Delta S°$. For example, p-nitrobenzoic acid is a stronger acid than benzoic acid. We attribute this (Sec. 23.14) to stabilization of the p-nitrobenzoate anion (relative to the benzoate anion) through dispersal of charge by the electron-withdrawing nitro group. Yet, in this case, the greater acidity is due about as much to a more favorable $\Delta S°$ as to a more favorable ΔH. How can our simple "stabilization by dispersal of charge" account for an effect that involves the randomness of a system?

Stabilization *is* involved, but it appears partly in the $\Delta S°$ for this reason. Ionization of an acid is possible only because of solvation of the ions produced: the many ion–dipole bonds provide the energy needed for dissociation. But solvation requires that molecules of solvent leave their relatively unordered arrangement to cluster in some ordered fashion about the ions. This is good for the ΔH but bad for the $\Delta S°$. Now, because of its greater intrinsic stability, the p-nitrobenzoate anion does not *need* as many solvent molecules to help stabilize it as the benzoate anion does. The $\Delta S°$ is thus more favorable. We can visualize the p-nitrobenzoate ion accepting only as many solvent molecules as it has to, and stopping when the gain in stability (decrease in enthalpy) is no longer worth the cost in entropy.

(In the same way, it has been found that very often a more polar solvent speeds up a reaction—as, for example, an S_N1 reaction of alkyl halides (Sec. 6.5)—not so much by lowering E_{act} as by bringing about a more favorable entropy of activation. A more polar solvent is already rather ordered, and its clustering about the ionizing molecule amounts to very little loss of randomness—indeed, it may even amount to an *increase* in randomness.)

In dealing with rates, we compare the stability of the reactants with the stability of the transition state. In dealing with equilibria, we shall compare the stability of the reactants with the stability of the products. For closely related reactions, we are justified in assuming that the more stable the products relative to the reactants, the further reaction proceeds toward completion.

By the organic chemist's approach we can make *very* good predictions indeed. We can not only account for, say, the relative acidities of a set of acids, but we can correlate these acidities *quantitatively* with the relative acidities of another set of acids, or even with the relative rates of a set of reactions. These relationships are summarized in the Hammett equation (named for Louis P. Hammett of Columbia University),

$$\log \frac{K}{K_0} = \rho\sigma \quad \text{or} \quad \log \frac{k}{k_0} = \rho\sigma$$

where K or k refers to the reaction of a *m*- or *p*-substituted phenyl compound (say, ionization of a substituted benzoic acid) and K_0 or k_0 refers to the same reaction of the unsubstituted compound (say, ionization of benzoic acid).

The *substituent constant* (σ, *sigma*) is a number (+ or −) indicating the relative electron-withdrawing or electron-releasing effect of a particular substituent. The *reaction constant* (ρ, *rho*) is a number (+ or −) indicating the relative *need* of a particular reaction for electron withdrawal or electron release.

A vast amount of research has shown that the Hammett relationship holds for *hundreds of sets of reactions*. (Ionization of 40-odd *p*-substituted benzoic acids, for example, is *one* set.) By use of just two tables—one of σ constants and one of ρ constants—we can calculate the relative K_{eq} values or relative rates for thousands of individual reactions. For example, from the σ value for *m*-NO$_2$ (+ 0.710) and the ρ value for ionization of benzoic acids in water at 25 °C (+ 1.000), we can calculate that K_a for *m*-nitrobenzoic acid is 5.13 times as big as the K_a for benzoic acid. Using the same σ value, and the ρ value for acid-catalyzed hydrolysis of benzamides in 60% ethanol at 80 °C (− 0.298), we can calculate that *m*-nitrobenzamide will be hydrolyzed only 0.615 times as fast as benzamide.

The Hammett relationship is called a *linear free energy relationship* since it is based on—and reveals—the fact that a linear relationship exists between free energy change and the effect exerted by a substituent. Other linear free energy relationships are known, which take into account steric as well as electronic effects, and which apply to *ortho* substituted phenyl compounds as well as *meta* and *para*, and to aliphatic as well as aromatic compounds. Together they make up what is perhaps the greatest accomplishment of physical organic chemistry.

23.12 Acidity of carboxylic acids

Let us see how the acidity of carboxylic acids is related to structure. In doing this we shall assume that acidity is determined chiefly by the difference in stability between the acid and its anion.

First, and most important, there is the fact that carboxylic acids are acids at all. How can we account for the fact that the —OH of a carboxylic acid tends to release a hydrogen ion so much more readily than the —OH of, say, an alcohol? Let us examine the structures of the reactants and products in these two cases.

We see that the alcohol and alkoxide ion are each represented satisfactorily by a single structure. However, we can draw two reasonable structures (I and II) for the carboxylic acid and two reasonable structures (III and IV) for the carboxylate anion. Both acid and anion are resonance hybrids. But is resonance equally

$$R{-}O{-}H \rightleftharpoons H^+ + R{-}O^-$$

Non-equivalent: resonance less important

Equivalent: resonance more important

important in the two cases? By the principles of Sec. 10.10 we know that resonance is much more important between the exactly equivalent structures III and IV than between the non-equivalent structures I and II. As a result, although both acid and anion are stabilized by resonance, stabilization is far greater for the anion than for the acid (see Fig. 23.3). Equilibrium is shifted in the direction of increased ionization, and K_a is increased.

Strictly speaking, resonance is less important for the acid because the contributing structures are of *different stability*, whereas the equivalent structures for the ion must necessarily be of *equal stability*. In structure II two atoms of similar electronegativity carry opposite charges; since energy must be supplied to separate opposite charges, II should contain more energy and hence be less stable than I. Consideration of *separation of charge* is one of the rules of thumb (Sec. 10.10) that can be used to estimate relative stability and hence relative importance of a contributing structure.

The acidity of a carboxylic acid is thus due to the powerful resonance stabilization of its anion. *This stabilization and the resulting acidity are possible only because of the presence of the carbonyl group.*

Figure 23.3   Molecular structure and position of equilibrium. A carboxylic acid yields a resonance-stabilized anion; it is a stronger acid than an alcohol. (The plots are aligned with each other for easy comparison.)

23.13   Structure of carboxylate ions

According to the resonance theory, then, a carboxylate ion is a hybrid of two structures which, being of equal stability, contribute equally. Carbon is joined to each oxygen by a "one-and-a-half" bond. The negative charge is evenly distributed over both oxygen atoms.

That the anion is indeed a resonance hybrid is supported by the evidence of bond length. Formic acid, for example, contains a carbon–oxygen double bond and a carbon–oxygen single bond; we would expect these bonds to have different lengths. Sodium formate, on the other hand, if it is a resonance hybrid, ought to contain two equivalent carbon–oxygen bonds; we would expect these to have the same length, intermediate between double and single bonds. X-ray and electron diffraction show that these expectations are correct. Formic acid contains one carbon–oxygen bond of 1.36 Å (single bond) and another of 1.23 Å (double bond); sodium formate contains two equal carbon–oxygen bonds, each 1.27 Å long.

1.23 Å ⟋O
H—C
1.36 Å ⟍OH

Formic acid

1.27 Å ⟋O⟍
H—C   } – Na⁺
1.27 Å ⟍O⟋

Sodium formate

Problem 23.5 How do you account for the fact that the three carbon–oxygen bonds in $CaCO_3$ have the same length, and that this length (1.31 Å) is greater than that found in sodium formate?

What does this resonance mean in terms of orbitals? Carboxyl carbon is joined to the three other atoms by σ bonds (Fig. 23.4); since these bonds utilize sp^2 orbitals (Sec. 7.2), they lie in a plane and are 120° apart. The remaining p orbital of the carbon overlaps equally well p orbitals from *both* of the oxygens, to form hybrid bonds (compare benzene, Sec. 13.8). In this way the electrons are bound not just

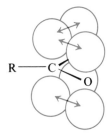

Figure 23.4 Carboxylate ion. Overlap of p orbitals in both directions: delocalization of π electrons, and dispersal of charge.

to one or two nuclei but to *three* nuclei (one carbon and two oxygens); they are therefore held more tightly, the bonds are stronger, and the anion is more stable. This participation of electrons in more than one bond, this smearing-out or delocalization of the electron cloud, is what is meant by representing the anion as a resonance hybrid of two structures.

Problem 23.6 How do you account for the fact that the α-hydrogens of an aldehyde (say, *n*-butyraldehyde) are much more acidic than any other hydrogens in the molecule? (Check your answer in Sec. 25.1.)

$\overset{\gamma}{CH_3}\overset{\beta}{CH_2}\overset{\alpha}{CH_2}\overset{H}{C}{=}O$

n-Butyraldehyde

23.14 Effect of substituents on acidity

Next, let us see how changes in the structure of the group bearing the —COOH affect the acidity. Any factor that stabilizes the anion more than it stabilizes the acid should increase the acidity; any factor that makes the anion less stable should decrease acidity. From what we have learned about carbocations, we know what we might reasonably expect. Electron-withdrawing substituents should disperse the negative charge, stabilize the anion, and thus increase acidity. Electron-releasing substituents should intensify the negative charge, destabilize the anion, and thus decrease acidity.

Acid strength

G withdraws electrons: *stabilizes anion, strengthens acid*

G releases electrons: *destabilizes anion, weakens acid*

The K_a values listed in Table 23.2 are in agreement with this prediction.

Table 23.2 ACIDITY CONSTANTS OF CARBOXYLIC ACIDS

	K_a			K_a	
HCOOH	17.7×10^{-5}		$CH_3CHClCH_2COOH$	8.9×10^{-5}	
CH_3COOH	1.75	,,	$ClCH_2CH_2CH_2COOH$	2.96	,,
$ClCH_2COOH$	136	,,	FCH_2COOH	260	,,
$Cl_2CHCOOH$	5530	,,	$BrCH_2COOH$	125	,,
Cl_3CCOOH	23200	,,	ICH_2COOH	67	,,
$CH_3CH_2CH_2COOH$	1.52	,,	$C_6H_5CH_2COOH$	4.9	,,
$CH_3CH_2CHClCOOH$	139	,,	$p\text{-}O_2NC_6H_4CH_2COOH$	14.1	,,

ACIDITY CONSTANTS OF SUBSTITUTED BENZOIC ACIDS

K_a of benzoic acid = 6.3×10^{-5}

	K_a			K_a			K_a	
$p\text{-}NO_2$	36×10^{-5}		$m\text{-}NO_2$	32×10^{-5}		$o\text{-}NO_2$	670×10^{-5}	
$p\text{-}Cl$	10.3	,,	$m\text{-}Cl$	15.1	,,	$o\text{-}Cl$	120	,,
$p\text{-}CH_3$	4.2	,,	$m\text{-}CH_3$	5.4	,,	$o\text{-}CH_3$	12.4	,,
$p\text{-}OCH_3$	3.3	,,	$m\text{-}OCH_3$	8.2	,,	$o\text{-}OCH_3$	8.2	,,
$p\text{-}OH$	2.6	,,	$m\text{-}OH$	8.3	,,	$o\text{-}OH$	105	,,
$p\text{-}NH_2$	1.4	,,	$m\text{-}NH_2$	1.9	,,	$o\text{-}NH_2$	1.6	,,

Looking first at the aliphatic acids, we see that the electron-withdrawing halogens strengthen acids: chloroacetic acid is 100 times as strong as acetic acid, dichloroacetic acid is still stronger, and trichloroacetic acid is more than 10 000 times as strong as the unsubstituted acid. The other halogens exert similar effects.

Problem 23.7 (a) What do the K_a values of the monohaloacetic acids tell us about the relative strengths of the inductive effects of the different halogens? (b) On the basis of Table 23.2, what kind of inductive effect does the phenyl group, $-C_6H_5$, appear to have?

α-Chlorobutyric acid is about as strong as chloroacetic acid. As the chlorine is moved away from the —COOH, however, its effect rapidly dwindles: β-chloro-butyric acid is only six times as strong as butyric acid, and γ-chlorobutyric acid is only twice as strong. It is typical of inductive effects that they decrease rapidly with distance, and are seldom important when acting through more than four atoms.

$$Cl \longleftarrow CH_2 \longleftarrow CH_2 \longleftarrow CH_2 \longleftarrow C \overset{\displaystyle O}{\underset{\displaystyle O}{\Big\}} \ominus$$

Inductive effect: *decreases with distance*

The aromatic acids are similarly affected by substituents: $-CH_3$, $-OH$, and $-NH_2$ make benzoic acid weaker, and $-Cl$ and $-NO_2$ make benzoic acid stronger. We recognize the acid-weakening groups as the ones that activate the ring toward electrophilic substitution (and deactivate toward nucleophilic substitution). The acid-strengthening groups are the ones that deactivate toward electrophilic substitution (and activate toward nucleophilic substitution). Furthermore, the groups that have the largest effects on reactivity—whether activating or deactivating—have the largest effects on acidity.

The $-OH$ and $-OCH_3$ groups display both kinds of effect we have attributed to them (Sec. 14.18): from the *meta* position, an electron-withdrawing acid-strengthening inductive effect; and from the *para* position, an electron-releasing acid-weakening resonance effect (which at this position outweighs the inductive effect). Compare the two effects exerted by halogen on electrophilic aromatic substitution (Sec. 14.19).

ortho-Substituted acids do not fit into the pattern set by their *meta* and *para* isomers, and by aliphatic acids. Nearly all *ortho* substituents exert an effect of the same kind—acid-strengthening—whether they are electron-withdrawing or electron-releasing, and the effect is unusually large. (Compare, for example, the effects of o-NO_2 and o-CH_3, of o-NO_2 and m- or p-NO_2.) This *ortho* effect undoubtedly has to do with the *nearness* of the groups involved, but is more than just steric hindrance arising from their bulk.

Thus we see that the same concepts—inductive effect and resonance—that we found so useful in dealing with rates of reaction are also useful in dealing with equilibria. By using these concepts to estimate the stabilities of anions, we are able to predict the relative strengths of acids; in this way we can account not only for the effect of substituents on the acid strength of carboxylic acids but also for the very fact that the compounds are acids.

Problem 23.8 There is evidence that certain groups like p-methoxy weaken the acidity of benzoic acids not so much by destabilizing the anion as by stabilizing the acid. Draw structures to show the kind of resonance that might be involved. Why would you expect such resonance to be more important for the acid than for the anion?

23.15   Conversion into acid chlorides

A carboxylic acid is perhaps more often converted into the acid chloride than into any other of its functional derivatives. From the highly reactive acid chloride there can then be obtained many other kinds of compounds, including esters and amides (Sec. 24.8).

An acid chloride is prepared by substitution of —Cl for the —OH of a carboxylic acid. Three reagents are commonly used for this purpose: *thionyl chloride*, $SOCl_2$; *phosphorus trichloride*, PCl_3; and *phosphorus pentachloride*, PCl_5. (Of what inorganic acids might we consider these reagents to be the acid chlorides?) For example:

Benzoic acid   Benzoyl chloride

3,5-Dinitrobenzoic   3,5-Dinitrobenzoyl
acid   chloride

Thionyl chloride is particularly convenient, since the products formed besides the acid chloride are gases and thus easily separated from the acid chloride; any excess of the low-boiling thionyl chloride (79 °C) is easily removed by distillation.

23.16   Conversion into esters

Acids are frequently converted into their esters via the acid chlorides:

Acid   Acid chloride   Ester

A carboxylic acid is converted directly into an ester when heated with an alcohol in the presence of a little mineral acid, usually concentrated sulfuric acid or dry hydrogen chloride. This reaction is reversible, and generally reaches equilibrium when there are appreciable quantities of both reactants and products present.

Acid   Alcohol   Ester

For example, when we allow one mole of acetic acid and one mole of ethyl alcohol to react in the presence of a little sulfuric acid until equilibrium is reached (after several hours), we obtain a mixture of about two-thirds mole each of ester and water, and one-third mole each of acid and alcohol. We obtain this same equilibrium mixture, of course, if we start with one mole of ester and one mole of water, again in the presence of sulfuric acid. *The same catalyst, hydrogen ion, that catalyzes the forward reaction, esterification, necessarily catalyzes the reverse reaction, hydrolysis.*

This reversibility is a disadvantage in the preparation of an ester directly from an acid; the preference for the acid chloride route is due to the fact that both steps—preparation of acid chloride from acid, and preparation of ester from acid chloride—are essentially irreversible and go to completion.

Direct esterification, however, has the advantage of being a single-step synthesis; it can often be made useful by application of our knowledge of equilibria. If either the acid or the alcohol is cheap and readily available, it can be used in large excess to shift the equilibrium toward the products and thus to increase the yield of ester. For example, it is worthwhile to use eight moles of cheap ethyl alcohol to convert one mole of valuable γ-phenylbutyric acid more completely into the ester:

$$\underset{\substack{\text{γ-Phenylbutyric acid} \\ \textit{1 mole}}}{\bigcirc\!\!-CH_2CH_2CH_2C\!\!\begin{matrix}O\\ \\OH\end{matrix}} + \underset{\substack{\text{Ethyl alcohol} \\ \textit{8 moles}}}{C_2H_5OH} \xrightarrow[\;]{H_2SO_4,\ reflux} \underset{\substack{\text{Ethyl γ-phenylbutyrate} \\ \textit{85–88\% yield}}}{\bigcirc\!\!-CH_2CH_2CH_2C\!\!\begin{matrix}O\\ \\OC_2H_5\end{matrix}}$$

$$+ H_2O$$

Sometimes the equilibrium is shifted by removing one of the products. An elegant way of doing this is illustrated by the preparation of ethyl adipate. The dicarboxylic acid adipic acid, an excess of ethyl alcohol, and toluene are heated with a little sulfuric acid under a distillation column. The lowest boiling component (b.p. 75 °C) of the reaction mixture is an azeotrope of water, ethyl alcohol, and toluene (compare Sec. 17.7); consequently, as fast as water is formed it is removed as the azeotrope by distillation. In this way a 95–97% yield of ester is obtained:

$$\underset{\substack{\text{Adipic acid} \\ \textit{Non-volatile}}}{HOOC(CH_2)_4COOH} + \underset{\substack{\text{Ethyl alcohol} \\ \textit{b.p. 78 °C}}}{2C_2H_5OH} \xrightarrow[\substack{H_2SO_4}]{\substack{\text{toluene (b.p. 111 °C),}}} \underset{\substack{\text{Ethyl adipate} \\ \textit{b.p. 245 °C}}}{C_2H_5OOC(CH_2)_4COOC_2H_5}$$

$$+ 2H_2O$$
$$\textit{Removed as}$$
$$\textit{azeotrope, b.p. 75 °C}$$

The equilibrium is particularly unfavorable when phenols (ArOH) are used instead of alcohols; yet, if water is removed during the reaction, phenolic esters (RCOOAr) are obtained in high yield.

The presence of bulky groups near the site of reaction, whether in the alcohol or in the acid, slows down esterification (as well as its reverse, hydrolysis). This

Reactivity $CH_3OH > 1° > 2° (>3°)$
in esterifi-
cation $HCOOH > CH_3COOH > RCH_2COOH > R_2CHCOOH > R_3CCOOH$

steric hindrance can be so marked that special methods are required to prepare esters of tertiary alcohols or esters of acids like 2,4,6-trimethylbenzoic acid (mesitoic acid). (See Fig. 23.5.)

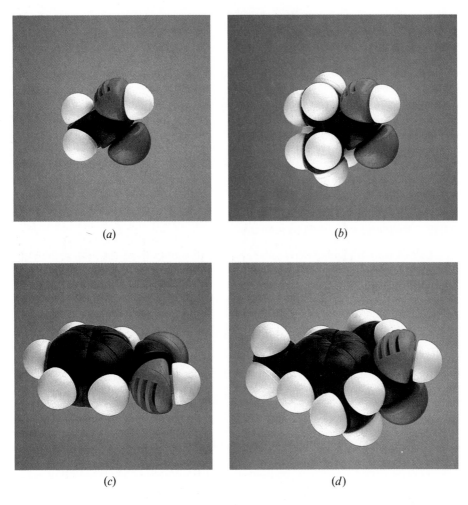

(a)   (b)

(c)   (d)

Figure 23.5   Molecular structure and reactivity: the steric factor in esterification. Crowding about the carboxyl group. Compare (a) acetic acid with (b) trimethylacetic acid, and (c) benzoic acid with (d) 2,4,6-trimethylbenzoic acid.

The mechanism of esterification is necessarily the exact reverse of the mechanism of hydrolysis of esters. We shall discuss both mechanisms when we take up the chemistry of esters (Sec. 24.18), after we have learned a little more about the carbonyl group.

Problem 23.9 (a) In the formation of an acid chloride, which bond of a carboxylic acid is broken, C—OH or CO—H? (b) When labeled methanol, $CH_3^{18}OH$, was allowed to react with ordinary benzoic acid, the methyl benzoate produced was found to be enriched in ^{18}O, whereas the water formed contained only ordinary oxygen. In this esterification, which bond of the carboxylic acid is broken, C—OH or CO—H? Which bond of the alcohol?

23.17 Conversion into amides

Amides are compounds in which the —OH of the carboxylic acid has been

$$
\underset{\text{Acid}}{R-C\!\!\overset{\displaystyle O}{\underset{\displaystyle OH}{\big\langle}}} \longrightarrow \underset{\text{Acid chloride}}{R-C\!\!\overset{\displaystyle O}{\underset{\displaystyle Cl}{\big\langle}}} \overset{NH_3}{\longrightarrow} \underset{\text{Amide}}{R-C\!\!\overset{\displaystyle O}{\underset{\displaystyle NH_2}{\big\langle}}}
$$

replaced by —NH_2. These are generally prepared by reaction of ammonia with acid chlorides.

23.18 Reduction of acids to alcohols

Conversion of alcohols into acids (Sec. 23.6) is important because, in general, alcohols are more available than acids. This is not always true, however; long straight-chain acids from fats are more available than are the corresponding alcohols, and here the reverse process becomes important: reduction of acids to alcohols.

Lithium aluminum hydride, $LiAlH_4$, is one of the few reagents that can reduce an acid to an alcohol; the initial product is an alkoxide from which the alcohol is liberated by hydrolysis:

$$4R-COOH + 3LiAlH_4 \longrightarrow 4H_2 + 2LiAlO_2 + (RCH_2O)_4AlLi \xrightarrow{H_2O} \underset{1^\circ \text{ alcohol}}{4R-CH_2OH}$$

Because of the excellent yields it gives, $LiAlH_4$ is widely used in the laboratory for the reduction of not only acids but many other classes of compounds. Since it is somewhat expensive, it can be used in industry only for the reduction of small amounts of valuable raw materials, as in the synthesis of certain drugs and hormones.

As an alternative to direct reduction, acids are often converted into alcohols by a two-step process: esterification, and reduction of the ester. Esters can be reduced in a number of ways (Sec. 24.22) that are adaptable to both laboratory and industry.

We have seen (Sec. 23.5) that in the carboxylic acids obtained from fats we have available long straight-chain units for use in organic synthesis. Reduction of these acids to alcohols (either directly or as esters) is a fundamental step in the utilization of these raw materials, since from the alcohols, as we know, a host of other compounds can be prepared (Sec. 18.8). Although only acids of even carbon number are available, it is possible, of course, to increase the chain length and thus prepare compounds of odd carbon number. (For an alternative source of long, straight-chain, primary alcohols, see Sec. 36.6.)

Problem 23.10   Outline the synthesis from lauric acid (n-$C_{11}H_{23}COOH$, dodecanoic acid) of the following compounds:

    (a) 1-bromododecane   (c) 1-tetradecanol
    (b) tridecanoic acid (C_{13} acid)   (d) 1-dodecene

(e) dodecane   (i) undecanoic acid
(f) 1-dodecyne   (j) 2-tetradecanol
(g) *n*-decyl methyl ketone   (k) 2-methyl-2-tetradecanol
(h) 2-dodecanol

23.19   Halogenation of aliphatic acids. Substituted acids

In the presence of a small amount of phosphorus, aliphatic carboxylic acids react smoothly with chlorine or bromine to yield a compound in which α-hydrogen has been replaced by halogen. This is the **Hell–Volhard–Zelinsky reaction**. Because of its specificity—*only alpha halogenation*—and the readiness with which it takes place, it is of considerable importance in synthesis.

$$CH_3COOH \xrightarrow{Cl_2,\ P} ClCH_2COOH \xrightarrow{Cl_2,\ P} Cl_2CHCOOH \xrightarrow{Cl_2,\ P} Cl_3CCOOH$$

$$CH_3CH_2COOH \xrightarrow{Br_2,\ P} CH_3CHBrCOOH \xrightarrow{Br_2,\ P} CH_3CBr_2COOH$$
$$\downarrow Br_2,\ P$$
$$\text{no further substitution}$$

The function of the phosphorus is ultimately to convert a little of the acid into acid halide. In this form (for reasons we cannot go into here) each molecule of acid sooner or later undergoes α-halogenation.

$$P + X_2 \longrightarrow PX_3$$

$$RCH_2COOH + PX_3 \longrightarrow RCH_2COX$$

$$RCH_2COX + X_2 \longrightarrow \underset{X}{RCHCOX} + HX$$

$$\underset{X}{RCHCOX} + RCH_2COOH \rightleftarrows \underset{X}{RCHCOOH} + RCH_2COX$$
$$\text{α-Halo acid}$$

The halogen of these halogenated acids undergoes *nucleophilic displacement* and *elimination* much as it does in the simpler alkyl halides (Secs. 5.8 and 7.12). Halogenation is therefore the first step in the conversion of a carboxylic acid into many important substituted carboxylic acids:

$$\underset{Br}{RCHCOOH} + \text{large excess of } NH_3 \longrightarrow \underset{NH_2}{RCHCOOH}$$

An α-halogenated acid   An α-amino acid

$$\underset{Br}{RCHCOOH} + NaOH \longrightarrow \underset{OH}{RCHCOONa} \xrightarrow{H^+} \underset{OH}{RCHCOOH}$$

An α-hydroxy acid

$$\text{RCH}_2\underset{\underset{\text{Br}}{|}}{\text{CHCOOH}} + \text{KOH (alc)} \longrightarrow \text{RCH=CHCOO}^- \xrightarrow{\text{H}^+} \text{RCH=CHCOOH}$$

$$\text{An } \alpha,\beta\text{-unsaturated acid}$$

These new substituents can, in turn, undergo *their* characteristic reactions.

Problem 23.11 Predict the product of each of the following reactions:
(a) $CH_2{=}CHCOOH + H_2/Ni$
(b) *trans*-$CH_3CH{=}CHCOOH + Br_2/CCl_4$
(c) $C_6H_5CH(OH)CH_2COOH + H^+$, heat $\longrightarrow C_9H_8O_2$
(d) *o*-$HOOCC_6H_4CH_2OH + H^+$, heat $\longrightarrow C_8H_6O_2$

23.20   Dicarboxylic acids

If the substituent is a second carboxyl group, we have a *dicarboxylic acid*. For example:

$HOOCCH_2COOH$	$HOOCCH_2CH_2COOH$	$HOOCCH_2CH_2CH_2CH_2COOH$
Malonic acid	Succinic acid	Adipic acid
Propanedioic acid	Butanedioic acid	Hexanedioic acid

	$\overset{\overset{\displaystyle CH_3}{\|}}{}$	
$HOOCCH_2CH_2\underset{\underset{\text{Br}}{\|}}{\text{CHCOOH}}$	$HOOCCH_2\underset{\underset{\text{CH}_3}{\|}}{\text{CCH}_2}COOH$	$HOOC\underset{\underset{\text{Cl}}{\|}}{\text{CHCH}_2}\underset{\underset{\text{Cl}}{\|}}{\text{CHCOOH}}$
α-Bromoglutaric acid	β,β-Dimethylglutaric acid	α,α-Dichloroglutaric acid
2-Bromopentanedioic acid	3,3-Dimethylpentanedioic acid	2,4-Dichloro-pentanedioic acid

We have already encountered the benzenedicarboxylic acids, the *phthalic acids* (Sec. 15.11).

Table 23.3   DICARBOXYLIC ACIDS

Name	Formula	M.p., °C	Solubility g/100 g H_2O at 20 °C	K_1	K_2
Oxalic	$HOOC{-}COOH$	189	9	5400×10^{-5}	5.2×10^{-5}
Malonic	$HOOCCH_2COOH$	136	74	140	0.20
Succinic	$HOOC(CH_2)_2COOH$	185	6	6.4	0.23
Glutaric	$HOOC(CH_2)_3COOH$	98	64	4.5	0.38
Adipic	$HOOC(CH_2)_4COOH$	151	2	3.7	0.39
Maleic	*cis*-$HOOCCH{=}CHCOOH$	130.5	79	1000	0.055
Fumaric	*trans*-$HOOCCH{=}CHCOOH$	302	0.7	96	4.1
Phthalic	$1,2\text{-}C_6H_4(COOH)_2$	231	0.7	110	0.4
Isophthalic	$1,3\text{-}C_6H_4(COOH)_2$	348.5	0.01	24	2.5
Terephthalic	$1,4\text{-}C_6H_4(COOH)_2$	300 *subl*	0.002	29	3.5

Most dicarboxylic acids are prepared by adaptation of methods used to prepare monocarboxylic acids. Where hydrolysis of a nitrile yields a monocarboxylic acid, hydrolysis of a dinitrile or a cyanocarboxylic acid yields a dicarboxylic acid; where oxidation of a methylbenzene yields a benzoic acid, oxidation of a dimethylbenzene yields a phthalic acid. For example:

$$Cl{-}CH_2COO^-Na^+ \xrightarrow{CN^-} \begin{matrix} COO^-Na^+ \\ | \\ CH_2 \\ | \\ CN \end{matrix}$$

Sodium
chloroacetate

Sodium
cyanoacetate

$$\xrightarrow{H_2O,\ H^+} \begin{matrix} COOH \\ | \\ CH_2 \\ | \\ COOH \end{matrix} + NH_4^+$$

Malonic acid

$$\xrightarrow{C_2H_5OH,\ H^+} \begin{matrix} COOC_2H_5 \\ | \\ CH_2 \\ | \\ COOC_2H_5 \end{matrix} + NH_4^+$$

Ethyl malonate

Problem 23.12 Why is chloroacetic acid converted into its salt before treatment with cyanide in the above preparation?

Problem 23.13 Outline a synthesis of: (a) pentanedioic acid from 1,3-propanediol (available from a fermentation of glycerol); (b) nonanedioic acid from *cis*-9-octadecenoic acid (oleic acid, obtained from fats); (c) succinic acid from 1,4-butynediol (available from acetylene and formaldehyde).

In general, dicarboxylic acids show the same chemical behavior as monocarboxylic acids. It is possible to prepare compounds in which only one of the carboxyl groups has been converted into a derivative; it is possible to prepare compounds in which the two carboxyl groups have been converted into different derivatives.

Problem 23.14 Predict the products of the following reactions:
(a) adipic acid (146 g) + 95% ethanol (146 g) + benzene + conc. H_2SO_4, 100 °C
(b) adipic acid (146 g) + 95% ethanol (50 g) + benzene + conc. H_2SO_4, 100 °C
(c) succinic acid + $LiAlH_4$
(d) pentanedioic acid + 1 mol Br_2, P
(e) terephthalic acid + excess $SOCl_2$
(f) maleic acid (*cis*-butenedioic acid) + Br_2/CCl_4

As with other acids containing more than one ionizable hydrogen (H_2SO_4, H_2CO_3, H_3PO_4, etc.), ionization of the second carboxyl group occurs less readily than ionization of the first (compare K_1 values with K_2 values in Table 23.3). More energy is required to separate a positive hydrogen ion from the doubly charged anion than from the singly charged anion.

$$\begin{matrix} COOH \\ | \\ | \\ COOH \end{matrix} \xrightleftharpoons{K_1} H^+ + \begin{matrix} COO^- \\ | \\ | \\ COOH \end{matrix} \xrightleftharpoons{K_2} H^+ + \begin{matrix} COO^- \\ | \\ | \\ COO^- \end{matrix} \qquad K_1 > K_2$$

Problem 23.15 Compare the acidity (first ionization) of oxalic acid with that of formic acid; of malonic acid with that of acetic acid. How do you account for these differences?

Problem 23.16 Arrange oxalic, malonic, succinic, and glutaric acids in order of acidity (first ionization). How do you account for this order?

Certain reactions of dicarboxylic acids, while fundamentally the same as those undergone by any carboxylic acid, lead to unusual results simply because there *are* two carboxyl groups in each molecule (Sec. 36.7). In addition, some dicarboxylic acids undergo certain special reactions that are possible only because the two carboxyl groups are located in a particular way with respect to each other (Sec. 30.4).

Problem 23.17 Give a likely structure for the product of each of the following reactions:
(a) oxalic acid + ethylene glycol $\longrightarrow$ $C_4H_4O_4$
(b) succinic acid + heat $\longrightarrow$ $C_4H_4O_3$
(c) terephthalic acid + ethylene glycol $\longrightarrow$ $(C_{10}H_8O_4)_n$, the polymer Dacron

23.21 Analysis of carboxylic acids. Neutralization equivalent

Carboxylic acids are recognized through their acidity. They dissolve in aqueous sodium hydroxide and in aqueous sodium bicarbonate. The reaction with bicarbonate releases bubbles of carbon dioxide (see Sec. 23.4).

(Phenols, Sec. 28.7, are more acidic than water, but—with certain exceptions—are considerably weaker than carboxylic acids; they dissolve in aqueous sodium hydroxide, but *not* in aqueous sodium bicarbonate. Sulfonic acids are even more acidic than carboxylic acids, but they contain sulfur, which can be detected by elemental analysis.)

Once characterized as a carboxylic acid, an unknown is identified as a particular acid on the usual basis of its physical properties and the physical properties of derivatives. The derivatives commonly used are *amides* (Secs. 24.11 and 27.7) and *esters* (Sec. 24.15).

Problem 23.18 Expand the table you made in Problem 21.17, p. 787, to include. carboxylic acids.

Particularly useful both in identification of previously studied acids and in proof of structure of new ones is the **neutralization equivalent**: *the equivalent weight of the acid as determined by titration with standard base.* A weighed sample of the acid is dissolved in water or aqueous alcohol, and the volume of standard base needed to neutralize the solution is measured. For example, a 0.224-g sample of an unknown acid (m.p. 139–140 °C) required 13.6 mL of 0.104 N sodium hydroxide solution for neutralization (to a phenolphthalein end point). Since each 1000 mL of the base contains 0.104 equivalents, and since the number of equivalents of base required equals the number of equivalents of acid present,

$$\frac{13.6}{1000} \times 0.104 \text{ equivalent of acid} = 0.224 \text{ g}$$

and

$$1 \text{ equivalent of acid} = 0.224 \times \frac{1000}{13.6} \times \frac{1}{0.104} = 158 \text{ g}$$

Problem 23.19   Which of the following compounds might the above acid be: (a) *o*-chlorobenzoic acid (m.p. 141 °C) or (b) 2,6-dichlorobenzoic acid (m.p. 139 °C)?

Problem 23.20   A 0.187-g sample of an acid (b.p. 203–205 °C) required 18.7 mL of 0.0972 N NaOH for neutralization. (a) What is the neutralization equivalent? (b) Which of the following acids might it be: *n*-caproic acid (b.p. 205 °C), methoxy-acetic acid (b.p. 203 °C), or ethoxyacetic acid (b.p. 206 °C)?

Problem 23.21   (a) How many equivalents of base would be neutralized by one mole of phthalic acid? What is the neutralization equivalent of phthalic acid? (b) What is the relation between neutralization equivalent and the number of acidic hydrogens per molecule of acid? (c) What is the neutralization equivalent of 1,3,5-benzene-tricarboxylic acid? Of mellitic acid, $C_6(COOH)_6$?

A metal salt of a carboxylic acid is recognized through these facts: (a) it leaves a residue when strongly heated (*ignition test*); (b) it decomposes at a fairly high temperature, instead of melting; and (c) it is converted into a carboxylic acid upon treatment with dilute mineral acid.

Problem 23.22   The residue left upon ignition of a sodium salt of a carboxylic acid was white, soluble in water, turned moist litmus blue, and reacted with dilute hydrochloric acid with the formation of bubbles. What was its probable chemical composition?

23.22   Spectroscopic analysis of carboxylic acids

Infrared   The carboxyl group is made up of a carbonyl group (C=O) and a hydroxyl group (OH), and the infrared spectrum of carboxylic acids reflects both these structural units. For hydrogen-bonded (dimeric) acids, O—H stretching gives a strong, broad band in the 2500–3000 cm^{-1} range (see Fig. 23.6, on the next page).

O—H stretching, *strong, broad*

—COOH and enols	2500–3000 cm^{-1}
ROH and ArOH	3200–3600 cm^{-1}

With acids we encounter again absorption due to stretching of the carbonyl group. As we saw for aldehydes and ketones (Sec. 21.16), this strong band appears in a region that is usually free of other strong absorption, and by its exact frequency

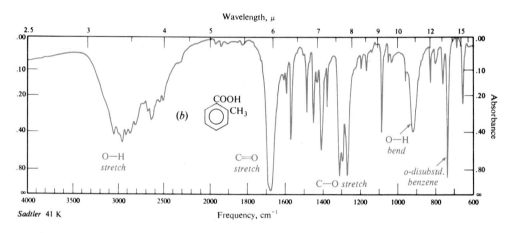

Figure 23.6   Infrared spectra of (*a*) propionic acid and (*b*) *o*-toluic acid.

gives much information about structure. For (hydrogen-bonded) acids, the C=O band is at about 1700 cm^{-1}.

C=O **stretching**, *strong*

R—C—OH   1700–1725 cm^{-1} —C=C—C—OH   1680–1700 cm^{-1}
‖
O   O

Ar—C—OH   1680–1700 cm^{-1} —C=CH—C—   1540–1640 cm^{-1}
‖
O   OH-------O

(*enols*)

Acids also show a C—O stretching band at about 1250 cm^{-1} (compare alcohols, Sec. 18.11, and ethers, Sec. 19.18), and bands for O—H bending near 1400 cm^{-1} and 920 cm^{-1} (*broad*).

Enols, too, show both O—H and C=O absorption; these can be distinguished by the particular frequency of the C=O band. Aldehydes, ketones, and esters show carbonyl absorption, but the O—H band is missing. (For a comparison of certain oxygen compounds, see Table 24.3, p. 890.)

NMR   The outstanding feature of the NMR spectrum of a carboxylic acid is the absorption far downfield (δ 10.5–12) by the proton of —COOH. (Compare the absorption by the acid proton of phenols, ArOH, in Sec. 28.14.)

CMR   In the CMR spectrum of a carboxylic acid we see the far downfield absorption by carbonyl carbon, δ 165–185. This is in the same general region as for functional derivatives of carboxylic acids, but somewhat upfield from the absorption by aldehydes and ketones.

PROBLEMS

1. Give the common names and IUPAC names for the straight-chain saturated carboxylic acids containing the following numbers of carbon atoms: 1, 2, 3, 4, 5, 6, 8, 10, 12, 16, 18.

2. Give the structural formula and, where possible, a second name (by a different system) for each of the following:

(a) isovaleric acid
(b) trimethylacetic acid
(c) α,β-dimethylcaproic acid
(d) 2-methyl-4-ethyloctanoic acid
(e) phenylacetic acid
(f) γ-phenylbutyric acid
(g) adipic acid
(h) p-toluic acid
(i) phthalic acid

(j) isophthalic acid
(k) terephthalic acid
(l) p-hydroxybenzoic acid
(m) potassium α-methylbutyrate
(n) magnesium 2-chloropropanoate
(o) maleic acid
(p) α,α'-dibromosuccinic acid
(q) isobutyronitrile
(r) 2,4-dinitrobenzonitrile

3. Write equations to show how each of the following compounds could be converted into benzoic acid:

(a) toluene
(b) bromobenzene
(c) benzonitrile

(d) benzyl alcohol
(e) benzotrichloride
(f) acetophenone (*Hint*: See Sec. 18.9.)

4. Write equations to show how each of the following compounds could be converted into *n*-butyric acid:

(a) *n*-butyl alcohol
(b) *n*-propyl alcohol

(c) *n*-propyl alcohol (a second way)
(d) methyl *n*-propyl ketone

Which of the above methods could be used to prepare trimethylacetic acid?

5. Write equations to show how tetrahydrofuran could be converted into:

(a) succinic acid   (b) glutaric acid   (c) adipic acid

6. Write equations to show the reaction (if any) of benzoic acid with:

(a) KOH
(b) Al
(c) CaO
(d) Na_2CO_3
(e) NH_3(aq)
(f) H_2, Ni, 20 °C, 1 atm.

(g) $LiAlH_4$
(h) hot $KMnO_4$
(i) PCl_5
(j) PCl_3
(k) $SOCl_2$
(l) Br_2/Fe

(m) $Br_2 + P$
(n) HNO_3/H_2SO_4
(o) fuming sulfuric acid
(p) CH_3Cl, $AlCl_3$
(q) *n*-propyl alcohol, H^+

7. Answer Problem 6 for *n*-valeric acid.

8. Write equations to show how isobutyric acid could be converted into each of the following, using any needed reagents:

(a) ethyl isobutyrate
(b) isobutyryl chloride
(c) isobutyramide

(d) magnesium isobutyrate
(e) isobutyl alcohol

9. Write equations to show all steps in the conversion of benzoic acid into:

(a) sodium benzoate
(b) benzoyl chloride
(c) benzamide
(d) benzene

(e) n-propyl benzoate
(f) p-tolyl benzoate
(g) m-bromophenyl benzoate
(h) benzyl alcohol

10. Write equations to show how phenylacetic acid could be converted into each of the following, using any needed reagents.

(a) sodium phenylacetate
(b) ethyl phenylacetate
(c) phenylacetyl chloride
(d) phenylacetamide
(e) p-bromophenylacetic acid
(f) p-nitrophenylacetic acid

(g) β-phenylethyl alcohol
(h) α-bromophenylacetic acid
(i) α-aminophenylacetic acid
(j) α-hydroxyphenylacetic acid
(k) phenylmalonic acid,
$C_6H_5CH(COOH)_2$

11. Complete the following, giving the structures and names of the principal organic products.

(a) $C_6H_5CH=CHCOOH + KMnO_4 + OH^- + heat$
(b) $p-CH_3C_6H_4COOH + HNO_3 + H_2SO_4$
(c) succinic acid + $LiAlH_4$, followed by H^+
(d) $C_6H_5COOH + C_6H_5CH_2OH + H^+$
(e) product (d) + $HNO_3 + H_2SO_4$
(f) n-butyric acid + Br_2,P
(g) cyclo-$C_6H_{11}MgBr + CO_2$, followed by H_2SO_4
(h) product (g) + $C_2H_5OH + H^+$
(i) product (g) + $SOCl_2$ + heat
(j) $m-CH_3C_6H_4OCH_3 + KMnO_4 + OH^-$
(k) mesitylene + $K_2Cr_2O_7 + H_2SO_4$
(l) isobutyric acid + isobutyl alcohol + H^+
(m) salicylic acid ($o-HOC_6H_4COOH$) + Br_2, Fe
(n) sodium acetate + p-nitrobenzyl bromide
(o) linolenic acid + excess H_2, Ni
(p) oleic acid + $KMnO_4$, heat
(q) linoleic acid + O_3, then H_2O, Zn
(r) benzoic acid ($C_7H_6O_2$) + H_2, Ni, heat, pressure $\longrightarrow C_7H_{12}O_2$
(s) benzoic acid + ethylene glycol + $H^+ \longrightarrow C_{16}H_{14}O_4$
(t) phthalic acid + ethyl alcohol + $H^+ \longrightarrow C_{12}H_{14}O_4$
(u) oleic acid + Br_2/CCl_4
(v) product (u) + KOH (alcoholic)
(w) oleic acid + HCO_2OH

12. Outline a possible laboratory synthesis of the following labeled compounds, using $Ba^{14}CO_3$ or $^{14}CH_3OH$ as the source of ^{14}C.

(a) $CH_3CH_2CH_2^{14}COOH$
(b) $CH_3CH_2^{14}CH_2COOH$

(c) $CH_3^{14}CH_2CH_2COOH$
(d) $^{14}CH_3CH_2CH_2COOH$

13. Outline all steps in a possible laboratory synthesis of each of the following compounds from toluene and any needed aliphatic and inorganic reagents.

(a) benzoic acid
(b) phenylacetic acid
(c) p-toluic acid
(d) m-chlorobenzoic acid

(e) p-chlorobenzoic acid
(f) p-bromophenylacetic acid
(g) α-chlorophenylacetic acid

14. Outline a possible laboratory synthesis of each of the following compounds from benzene, toluene, and alcohols of four carbons or fewer, using any needed inorganic reagents.

(a) ethyl α-methylbutyrate
(b) 3,5-dinitrobenzoyl chloride
(c) α-amino-p-bromophenylacetic acid
(d) α-hydroxypropionic acid
(e) p-HO$_3$SC$_6$H$_4$COOH
(f) 2-pentenoic acid

(g) p-toluamide
(h) n-hexyl benzoate
(i) 3-bromo-4-methylbenzoic acid
(j) α-methylphenylacetic acid
(k) 2-bromo-4-nitrobenzoic acid
(l) 1,2,4-benzenetricarboxylic acid

15. Without referring to tables, arrange the compounds of each set in order of acidity:

(a) butanoic acid, 2-bromobutanoic acid, 3-bromobutanoic acid, 4-bromobutanoic acid
(b) benzoic acid, p-chlorobenzoic acid, 2,4-dichlorobenzoic acid, 2,4,6-trichlorobenzoic acid
(c) benzoic acid, p-nitrobenzoic acid, p-toluic acid   *aminobenzoic*
(d) α-chlorophenylacetic acid, p-chlorophenylacetic acid, phenylacetic acid, α-phenylpropionic acid
(e) p-nitrobenzoic acid, p-nitrophenylacetic acid, β-(p-nitrophenyl)propionic acid
(f) acetic acid, acetylene, ammonia, ethane, ethanol, sulfuric acid, water
(g) acetic acid, malonic acid, succinic acid

16. Arrange the monosodium salts of the acids in Problem 15(f) in order of basicity.

17. The two water-insoluble solids, benzoic acid and o-chlorobenzoic acid, can be separated by treatment with an aqueous solution of sodium formate. What reaction takes place? (*Hint*: Look at the K_a values in Table 23.2.)

18. Arrange the compounds of each set in order of reactivity in the indicated reaction:

(a) esterification by benzoic acid: sec-butyl alcohol, methanol, tert-pentyl alcohol, n-propyl alcohol
(b) esterification by ethyl alcohol: benzoic acid, 2,6-dimethylbenzoic acid, o-toluic acid
(c) esterification by methanol: acetic acid, formic acid, isobutyric acid, propionic acid, trimethylacetic acid

19. Give stereochemical formulas of compounds A–F:

(a) racemic β-bromobutyric acid + one mole Br$_2$, P $\longrightarrow$ A + B
(b) fumaric acid + HCO$_2$OH $\longrightarrow$ C (C$_4$H$_6$O$_6$)
(c) 1,4-cyclohexadiene + CHBr$_3$/t-BuOK $\longrightarrow$ D (C$_7$H$_8$Br$_2$)
    D + KMnO$_4$ $\longrightarrow$ E (C$_7$H$_8$Br$_2$O$_4$)
    E + H$_2$, Ni(base) $\longrightarrow$ F (C$_7$H$_{10}$O$_4$)

20. Give structures of compounds G–J:

$$\text{acetylene} + CH_3MgBr \longrightarrow G + CH_4$$

$$G + CO_2 \longrightarrow H \xrightarrow{H^+} I \ (C_3H_2O_2)$$

$$I \xrightarrow{H_2O, \ H_2SO_4, \ HgSO_4} J \ (C_3H_4O_3)$$

$$J + KMnO_4 \longrightarrow CH_2(COOH)_2$$

21. Describe simple chemical tests (other than color change of an indicator) that would serve to distinguish between:

(a) propionic acid and n-pentyl alcohol
(b) isovaleric acid and n-octane
(c) ethyl n-butyrate and isobutyric acid
(d) propionyl chloride and propionic acid
(e) p-aminobenzoic acid and benzamide
(f) C$_6$H$_5$CH=CHCOOH and C$_6$H$_5$CH=CHCH$_3$

Tell exactly what you would *do* and *see*.

22. Compare benzoic acid and sodium benzoate with respect to:

(a) volatility
(b) melting point
(c) solubility in water and (d) in ether

(e) degree of ionization of solid
(f) degree of ionization in water
(g) acidity and basicity

Does this comparison hold generally for acids and their salts?

23. Tell how you would separate by chemical means the following mixtures, recovering each component in reasonably pure form:

(a) caproic acid and ethyl caproate
(b) di-*n*-butyl ether and *n*-butyric acid

(c) isobutyric acid and 1-hexanol
(d) sodium benzoate and triphenylmethanol

Tell exactly what you would *do* and *see*.

24. An unknown compound is believed to be one of the following. Describe how you would go about finding out which of the possibilities the unknown actually is. Where possible, use simple chemical tests; where necessary, use more elaborate chemical methods like quantitative hydrogenation, cleavage, neutralization equivalent, etc. Make use of any needed tables of physical constants.

(a) acrylic acid (CH$_2$=CHCOOH, b.p. 142 °C) and propionic acid (b.p. 141 °C)
(b) mandelic acid (C$_6$H$_5$CHOHCOOH, m.p. 120 °C) and benzoic acid (m.p. 122 °C)
(c) *o*-chlorobenzoic acid (m.p. 141 °C), mesotartaric acid (m.p. 140 °C), *m*-nitrobenzoic acid (m.p. 141 °C), and suberic acid (HOOC(CH$_2$)$_6$COOH, m.p. 144 °C)
(d) chloroacetic acid (b.p. 189 °C), α-chloropropionic acid (b.p. 186 °C), dichloroacetic acid (b.p. 194 °C), and *n*-valeric acid (b.p. 187 °C)
(e) 3-nitrophthalic acid (m.p. 220 °C) and 2,4,6-trinitrobenzoic acid (m.p. 220 °C)
(f) *p*-chlorobenzoic acid (m.p. 242 °C), *p*-nitrobenzoic acid (m.p. 242 °C), *o*-nitrocinnamic acid (*o*-O$_2$NC$_6$H$_4$CH=CHCOOH, m.p. 240 °C)
(g) The following compounds, all of which boil within a few degrees of each other:

o-chloroanisole	isodurene
β-chlorostyrene	linalool (see Problem 27, p. 692)
p-cresyl ethyl ether	4-methylpentanoic acid
cis-decalin (see Problem 8, p. 473)	α-phenylethyl chloride
2,4-dichlorotoluene	*o*-toluidine (*o*-CH$_3$C$_6$H$_4$NH$_2$)

25. By use of Table 23.4 (below) tell which acid or acids each of the following is likely to be. Tell what further steps you would take to identify it or to confirm your identification.

K: m.p. 155–7 °C; positive halogen test; *p*-nitrobenzyl ester, m.p. 104–6 °C; neutralization equivalent, 158 ± 2

Table 23.4   DERIVATIVES OF SOME CARBOXYLIC ACIDS

	Acid M.p., °C	Amide M.p., °C	Anilide M.p., °C	*p*-Nitrobenzyl ester M.p., °C
trans-Crotonic (CH$_3$CH=CHCOOH)	72	161	118	67
Phenylacetic	77	156	118	65
Arachidic (*n*-C$_{19}$H$_{39}$COOH)	77	108	92	—
α-Hydroxyisobutyric	79	98	136	80
Glycolic (HOCH$_2$COOH)	80	120	97	107
β-Iodopropionic	82	101	—	—
Iodoacetic	83	95	143	—
Adipic (HOOC(CH$_2$)$_4$COOH)	151	220	241	106
p-Nitrophenylacetic	153	198	198	—
2,5-Dichlorobenzoic	153	155	—	—
m-Chlorobenzoic	154	134	122	107
2,4,6-Trimethylbenzoic	155	—	—	188
m-Bromobenzoic	156	155	136	105
p-Chlorophenoxyacetic	158	133	125	—
Salicylic (*o*-HOC$_6$H$_4$COOH)	159	142	136	98

L: m.p. 152–4 °C; negative tests for halogen and nitrogen
M: m.p. 153–5 °C; positive chlorine test; neutralization equivalent, 188 ± 4
N: m.p. 72–3 °C; anilide, m.p. 117–8 °C; amide, m.p. 155–7 °C
O: m.p. 79–80 °C; amide, m.p. 97–9 °C
P: m.p. 78–80 °C; negative tests for halogen and nitrogen; positive test with CrO_3/H_2SO_4

26. An unknown acid was believed to be either *o*-nitrobenzoic acid (m.p. 147 °C) or anthranilic acid (m.p. 146 °C). A 0.201-g sample neutralized 12.4 mL of 0.098 N NaOH. Which acid was it?

27. Carboxylic acid Q contained only carbon, hydrogen, and oxygen, and had a neutralization equivalent of 149 ± 3. Vigorous oxidation by $KMnO_4$ converted Q into R, m.p. 345–50 °C, neutralization equivalent 84 ± 2.
   When Q was heated strongly with soda lime a liquid S of b.p. 135–7 °C distilled. Vigorous oxidation by $KMnO_4$ converted S into T, m.p. 121–2 °C, neutralization equivalent 123 ± 2.
   U, an isomer of Q, gave upon oxidation V, m.p. 375–80 °C, neutralization equivalent 70 ± 2.
   What were compounds Q through V? (Make use of any needed tables of physical constants.)

28. Tropic acid (obtained from the alkaloid atropine, found in deadly nightshade, *Atropa belladona*), $C_9H_{10}O_3$, gives a positive CrO_3/H_2SO_4 test and is oxidized by hot $KMnO_4$ to benzoic acid. Tropic acid is converted by the following sequence of reactions into hydratropic acid:

tropic acid $\xrightarrow{\text{HBr}}$ $C_9H_9O_2Br$ $\xrightarrow{\text{OH}^-}$ $C_9H_8O_2$ (atropic acid)

atropic acid $\xrightarrow{\text{H}_2,\ \text{Ni}}$ hydratropic acid $(C_9H_{10}O_2)$

   (a) What structure or structures are possible at this point for hydratropic acid? For tropic acid?
   (b) When α-phenylethyl chloride is treated with magnesium in ether, the resulting solution poured over dry ice, and the mixture then acidified, there is obtained an acid whose amide has the same melting point as the amide of hydratropic acid. A mixed melting point determination shows no depression. Now what is the structure of hydratropic acid? Of tropic acid?

29. Give a structure or structures consistent with each of the following sets of proton NMR data:

(a) $C_3H_5ClO_2$
   a doublet, δ 1.73, 3H
   b quartet, δ 4.47, 1H
   c singlet, δ 11.22, 1H

(b) $C_3H_5ClO_2$
   a singlet, δ 3.81, 3H
   b singlet, δ 4.08, 2H

(c) $C_4H_7BrO_2$
   a triplet, δ 1.30, 3H
   b singlet, δ 3.77, 2H
   c quartet, δ 4.23, 2H

(d) $C_4H_7BrO_2$
   a triplet, δ 1.08, 3H
   b quintet, δ 2.07, 2H
   c triplet, δ 4.23, 1H
   d singlet, δ 10.97, 1H

(e) $C_4H_8O_3$
   a triplet, δ 1.27, 3H
   b quartet, δ 3.66, 2H
   c singlet, δ 4.13, 2H
   d singlet, δ 10.95, 1H

30. Which (if any) of the following compounds could give rise to each of the infrared spectra shown in Fig. 23.7 (p. 856)?

n-butyric acid
crotonic acid $(CH_3CH{=}CHCOOH)$
malic acid $(HOOCCHOHCH_2COOH)$
benzoic acid

p-nitrobenzoic acid
mandelic acid $(C_6H_5CHOHCOOH)$
p-nitrobenzyl alcohol

Figure 23.7　Infrared spectra for Problem 30, p. 855.

24

Functional Derivatives of Carboxylic Acids

Nucleophilic Acyl Substitution

Closely related to the carboxylic acids and to each other are a number of chemical families known as **functional derivatives of carboxylic acids**: *acid chlorides*, *anhydrides*, *amides*, and *esters*. These derivatives are compounds in which the —OH of a carboxyl group has been replaced by —Cl, —OOCR, —NH$_2$, or —OR'.

$$R-C\overset{O}{\underset{Cl}{\diagup}} \qquad R-C\overset{O}{\diagdown}_{O} \atop R-C\diagdown_{O} \qquad R-C\overset{O}{\underset{NH_2}{\diagup}} \qquad R-C\overset{O}{\underset{OR'}{\diagup}} \qquad \begin{array}{l}R \textit{ may be}\\ \textit{alkyl or}\\ \textit{aryl}\end{array}$$

Acid chloride   Anhydride   Amide   Ester

They all contain the **acyl group**, $R-C\overset{O}{\diagdown}$

Like the acid to which it is related, an acid derivative may be aliphatic or aromatic, substituted or unsubstituted; whatever the structure of the rest of the molecule, the properties of the functional group remain essentially the same.

24.2 Nomenclature

The names of acid derivatives are taken in simple ways from either the common name or the IUPAC name of the corresponding carboxylic acid. For example:

$$CH_3-C\underset{OH}{\overset{O}{\big\langle}}$$

Acetic acid
Ethanoic acid

Benzoic acid

$$CH_3-C\underset{Cl}{\overset{O}{\big\langle}}$$

Acetyl chloride
Ethanoyl chloride

Benzoyl chloride

Change:
-ic acid to *-yl chloride*

Acetic anhydride
Ethanoic anhydride

Benzoic anhydride

acid to *anhydride*

$$CH_3-C\underset{NH_2}{\overset{O}{\big\langle}}$$

Acetamide
Ethanamide

Benzamide

-ic acid of common name
(or *-oic acid* of IUPAC name)
to *-amide*

$$CH_3-C\underset{OC_2H_5}{\overset{O}{\big\langle}}$$

Ethyl acetate
Ethyl ethanoate

Ethyl benzoate

-ic acid to *-ate*,
preceded by name of
alcohol or phenol group

24.3 Physical properties

The presence of the $C{=}O$ group makes the acid derivatives polar compounds. Acid chlorides and anhydrides (Table 24.1) and esters (Table 24.2, p. 873) have boiling points that are about the same as those of aldehydes or ketones of comparable molecular weight (see Sec. 17.5). Amides (Table 24.1) have quite high boiling points because they are capable of strong intermolecular hydrogen bonding.

The border line for solubility in water ranges from three to five carbons for the esters to five or six carbons for the amides. The acid derivatives are soluble in the usual organic solvents.

Volatile esters have pleasant, rather characteristic odors; they are often used in the preparation of perfumes and artificial flavorings. Acid chlorides have sharp, irritating odors, at least partly due to their ready hydrolysis to HCl and carboxylic acids.

Table 24.1   ACID CHLORIDES, ANHYDRIDES, AND AMIDES

Name	M.p., °C	B.p., °C	Name	M.p., °C	B.p., °C
Acetyl chloride	−112	51	Succinic anhydride	120	
Propionyl chloride	− 94	80	Maleic anhydride	60	
n-Butyryl chloride	− 89	102			
n-Valeryl chloride	−110	128	Formamide	3	200d
Stearoyl chloride	23	215^{15}	Acetamide	82	221
Benzoyl chloride	− 1	197	Propionamide	79	213
p-Nitrobenzoyl chloride	72	154^{15}	n-Butyramide	116	216
			n-Valeramide	106	232
3,5-Dinitrobenzoyl chloride	74	196^{12}	Stearamide	109	251^{12}
			Benzamide	130	290
Acetic anhydride	− 73	140	Succinimide	126	
Phthalic anhydride	131	284	Phthalimide	238	

24.4 Nucleophilic acyl substitution. Role of the carbonyl group

Before we take up each kind of acid derivative separately, it will be helpful to outline certain general patterns into which we can then fit the rather numerous individual facts.

Each derivative is nearly always prepared—directly or indirectly—from the corresponding carboxylic acid, and can be readily converted into the carboxylic acid by simple hydrolysis. Much of the chemistry of acid derivatives involves their conversion one into another, and into the parent acid. In addition, each derivative has certain characteristic reactions of its own.

The derivatives of carboxylic acids, like the acids themselves, contain the carbonyl group, $C=O$. This group is retained in the products of most reactions undergone by these compounds, and does not suffer any permanent changes itself. But by its presence in the molecule it determines the characteristic reactivity of these compounds, and is the key to the understanding of their chemistry.

Here, too, as in aldehydes and ketones, the carbonyl group performs two functions: (a) it provides a site for nucleophilic attack, and (b) it increases the acidity of hydrogens attached to the *alpha* carbon.

(We shall discuss reactions resulting from the acidity of α-hydrogens in Secs. 25.11–25.12, and 30.1–30.3.)

Acyl compounds—carboxylic acids and their derivatives—typically undergo **nucleophilic substitution** in which —OH, —Cl, —OOCR, —NH₂, or —OR′ is replaced by some other basic group. Substitution takes place much more readily than at a saturated carbon atom; indeed, many of these substitutions do not usually take place at all in the absence of the carbonyl group, as, for example, replacement of —NH₂ by —OH.

$$R-C\overset{\displaystyle O}{\underset{\displaystyle W}{\diagup}} \; + \; :Z \;\longrightarrow\; R-\overset{\displaystyle O^-}{\underset{\displaystyle W}{C}}-Z \;\longrightarrow\; R-C\overset{\displaystyle O}{\underset{\displaystyle Z}{\diagup}} \; + \; :W$$

$$-W = -OH, -Cl, -OOCR, -NH_2, -OR'$$

To account for the properties of acyl compounds, let us turn to the carbonyl group. We have encountered this group in our study of aldehydes and ketones (Secs. 21.1 and 21.7), and we know what it is like and what in general to expect of it.

Carbonyl carbon is joined to three other atoms by σ bonds; since these bonds utilize sp^2 orbitals (Sec. 1.10), they lie in a plane and are 120° apart. The remaining p orbital of the carbon overlaps a p orbital of oxygen to form a π bond; carbon and oxygen are thus joined by a double bond. The part of the molecule immediately surrounding carbonyl carbon is *flat*; oxygen, carbonyl carbon, and the two atoms directly attached to carbonyl carbon lie in a plane:

$$\underset{R}{\overset{W}{>}}C \overset{\delta_+}{=\!=\!=}\overset{\delta_-}{O} \quad 120°$$

We saw before that both electronic and steric factors make the carbonyl group particularly susceptible to nucleophilic attack at the carbonyl carbon: (a) the tendency of oxygen to acquire electrons even at the expense of gaining a negative charge; and (b) the relatively unhindered transition state leading from the trigonal reactant to the tetrahedral intermediate. These factors make acyl compounds, too, susceptible to nucleophilic attack (Fig. 24.1).

It is in the second step of the reaction that acyl compounds differ from aldehydes and ketones. The tetrahedral intermediate from an aldehyde or ketone gains a proton, and the result is *addition*. The tetrahedral intermediate from an acyl compound ejects the :W group, returning to a trigonal compound, and thus the result is *substitution*.

$$R-C\overset{\displaystyle O}{\underset{\displaystyle R'}{\diagup}} \; + \; :Z \;\longrightarrow\; R-\overset{\displaystyle O^-}{\underset{\displaystyle R'}{C}}-Z \;\overset{H^+}{\longrightarrow}\; R-\overset{\displaystyle OH}{\underset{\displaystyle R'}{C}}-Z \qquad \text{Aldehyde or ketone} \\ \textit{Addition}$$

$$R-C\overset{\displaystyle O}{\underset{\displaystyle W}{\diagup}} \; + \; :Z \;\longrightarrow\; R-\overset{\displaystyle O^-}{\underset{\displaystyle W}{C}}-Z \;\longrightarrow\; R-C\overset{\displaystyle O}{\underset{\displaystyle Z}{\diagup}} \; + \; :W \qquad \text{Acyl compound} \\ \textit{Substitution}$$

We can see why the two classes of compounds differ as they do. The ease with which :W is lost depends upon its basicity: the weaker the base, the better the

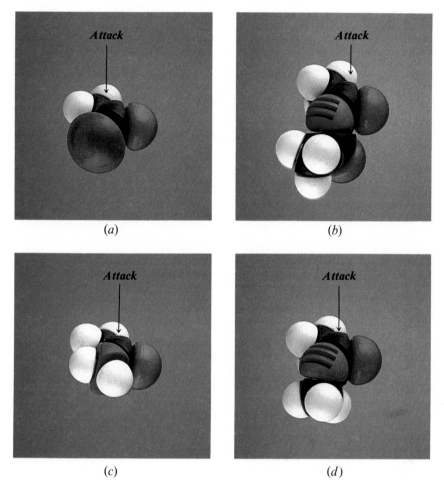

Figure 24.1 Molecular structure and reactivity: nucleophilic attack on the acyl group. Models of: (*a*) acetyl chloride, CH_3COCl; (*b*) acetic anhydride, $(CH_3CO)_2O$; (*c*) acetamide, CH_3CONH_2; (*d*) methyl acetate, CH_3COOCH_3. The flat carbonyl group is open to attack from above (or below).

leaving group. For acid chlorides, acid anhydrides, esters, and amides, :W is, respectively: the very weak base Cl^-; the moderately weak base $RCOO^-$; and the strong bases $R'O^-$ and NH_2^-. But for an aldehyde or ketone to undergo substitution, the leaving group would have to be hydride ion ($:H^-$) or alkide ion ($:R^-$) which, as we know, are the strongest bases of all. (Witness the low acidity of H_2 and RH.) And so with aldehydes and ketones addition almost always takes place instead.

Problem 24.1 Suggest a likely mechanism for each of the following reactions, and account for the behavior shown:

(a) The last step in the haloform reaction (Sec. 18.9),

$$OH^- + R-\overset{\displaystyle O}{\underset{\displaystyle \|}{C}}-CX_3 \xrightarrow{\text{H}_2\text{O}} RCOO^- + CHX_3$$

(b) The reaction of *o*-fluorobenzophenone with amide ion,

Thus, nucleophilic acyl substitution proceeds by two steps, with the intermediate formation of a tetrahedral compound. Generally, the overall rate is affected by the rate of both steps, but the *first* step is the more important. The first step, formation of the tetrahedral intermediate, is affected by the same factors as in

Nucleophilic acyl substitution

Reactant	Transition state	Intermediate	Product	Leaving group
Trigonal	*Becoming tetrahedral*	*Tetrahedral*	*Trigonal*	*Weaker base leaves more readily*
	Partial negative charge on oxygen	*Negative charge on oxygen*		

addition to aldehydes and ketones (Sec. 21.7): it is favored by electron withdrawal, which stabilizes the developing negative charge; and it is hindered by the presence of bulky groups, which become crowded together in the transition state. The second step depends, as we have seen, on the basicity of the leaving group, :W.

If acid is present, H^+ becomes attached to carbonyl oxygen, thus making the

Acid-catalyzed nucleophilic acyl substitution

Undergoes nucleophilic attack more readily

carbonyl group even more susceptible to the nucleophilic attack; oxygen can now acquire the π electrons without having to accept a negative charge.

It is understandable that acid derivatives are hydrolyzed more readily in either alkaline or acidic solution than in neutral solution: alkaline solutions provide hydroxide ion, which acts as a strongly nucleophilic reagent; acid solutions provide hydrogen ion, which attaches itself to carbonyl oxygen and thus renders the molecule vulnerable to attack by the weakly nucleophilic reagent, water.

Alkaline hydrolysis

Acidic hydrolysis

24.5 Nucleophilic substitution: alkyl *vs.* acyl

As we have said, nucleophilic substitution takes place much more readily at an acyl carbon than at saturated carbon. Thus, toward nucleophilic attack acid chlorides are more reactive than alkyl chlorides, amides are more reactive than amines (RNH_2), and esters are more reactive than ethers.

Acid chloride   *more reactive than*   R—Cl

Alkyl chloride

Amide   *more reactive than* $R—NH_2$

Amine

Ester   *more reactive than*   R—OR′

Ether

Reactivity in nucleophilic displacement

It is, of course, the carbonyl group that makes acyl compounds more reactive than alkyl compounds. Nucleophilic attack (S_N2) on a tetrahedral alkyl carbon involves a badly crowded transition state containing pentavalent carbon; a bond must be partly broken to permit the attachment of the nucleophile:

Alkyl nucleophilic substitution

$$Z: \quad -\overset{|}{\underset{|}{C}}-W \quad \xrightarrow{S_N2} \quad Z\cdots\overset{|}{\underset{|}{C}}\cdots W \quad \longrightarrow \quad Z-\overset{|}{\underset{|}{C}}- \quad + \quad :W$$

     Tetrahedral C   Pentavalent C
     Attack hindered   *Unstable*

Nucleophilic attack on a flat acyl compound involves a relatively unhindered transition state leading to a tetrahedral intermediate that is actually a compound; since the carbonyl group is unsaturated, attachment of the nucleophile requires

Acyl nucleophilic substitution

$$Z: \quad \overset{R \quad W}{\underset{\underset{O}{\|}}{C}} \quad \longrightarrow \quad Z-\overset{R}{\underset{\underset{O^-}{|}}{C}}-W \quad \longrightarrow \quad \overset{R \quad Z}{\underset{\underset{O}{\|}}{C}} \quad + \quad :W$$

     Trigonal C   Tetrahedral C
     Attack relatively   *Stable*
     unhindered

breaking only of the weak π bond, and places a negative charge on an atom quite willing to accept it, oxygen.

ACID CHLORIDES

24.6   Preparation of acid chlorides

Acid chlorides are prepared from the corresponding acids by reaction with thionyl chloride, phosphorus trichloride, or phosphorus pentachloride, as discussed in Sec. 23.15.

24.7   Reactions of acid chlorides

Like other acid derivatives, acid chlorides typically undergo nucleophilic substitution. Chlorine is expelled as chloride ion or hydrogen chloride, and its place is taken by some other basic group. Because of the carbonyl group these reactions take place much more rapidly than the corresponding nucleophilic substitution reactions of the alkyl halides. Acid chlorides are the most reactive of the derivatives of carboxylic acids.

―――― REACTIONS OF ACID CHLORIDES ――――――――――――――――――

1. Conversion into acids and derivatives. Discussed in Sec. 24.8.

$$R-\overset{\displaystyle O}{\underset{\displaystyle Cl}{C}} \quad + \quad HZ \quad \longrightarrow \quad R-\overset{\displaystyle O}{\underset{\displaystyle Z}{C}} \quad + \quad HCl$$

―――――――――――――――― CONTINUED ――

CONTINUED

(a) **Conversion into acids. Hydrolysis**

$$R-\overset{\displaystyle O}{\underset{\displaystyle Cl}{C}} + H_2O \longrightarrow R-\overset{\displaystyle O}{\underset{\displaystyle OH}{C}} + HCl$$

An acid

Example:

$$\text{⬡}COCl + H_2O \longrightarrow \text{⬡}COOH + HCl$$

Benzoyl chloride   Benzoic acid

(b) **Conversion into amides. Ammonolysis**

$$R-\overset{\displaystyle O}{\underset{\displaystyle Cl}{C}} + 2NH_3 \longrightarrow R-\overset{\displaystyle O}{\underset{\displaystyle NH_2}{C}} + NH_4Cl$$

An amide

Example:

$$\text{⬡}COCl + 2NH_3 \longrightarrow \text{⬡}CONH_2 + NH_4Cl$$

Benzoyl chloride   Benzamide

(c) **Conversion into esters. Alcoholysis**

$$R-\overset{\displaystyle O}{\underset{\displaystyle Cl}{C}} + R'OH \longrightarrow R-\overset{\displaystyle O}{\underset{\displaystyle OR'}{C}} + HCl$$

An ester

Example:

$$\text{⬡}COCl + C_2H_5OH \longrightarrow \text{⬡}COOC_2H_5 + HCl$$

Benzoyl chloride   Ethyl   Ethyl benzoate
                     alcohol

2. **Formation of ketones. Friedel–Crafts acylation.** Discussed in Sec. 21.5.

$$R-\overset{\displaystyle O}{\underset{\displaystyle Cl}{C}} + ArH \xrightarrow[\substack{\text{or other} \\ \text{Lewis acid}}]{AlCl_3} R-\overset{\displaystyle}{\underset{\displaystyle O}{C}}-Ar + HCl$$

A ketone

CONTINUED

CONTINUED

3. **Formation of ketones. Reaction with organocopper compounds.** Discussed in Sec. 21.6.

$$R'\!-\!Li \xrightarrow{\text{CuX}} R'\!-\!CuLi$$

$$R\!-\!\overset{O}{\underset{Cl}{C}} \quad or \quad Ar\!-\!\overset{O}{\underset{Cl}{C}}$$

$$\longrightarrow R\!-\!\overset{R'}{\underset{O}{C}} \quad or \quad Ar\!-\!\overset{R'}{\underset{O}{C}}$$

Ketone

4. **Formation of aldehydes by reduction.** Discussed in Sec. 21.4.

$$R\!-\!COCl \quad or \quad Ar\!-\!COCl \xrightarrow{\text{LiAlH(OBu-}t)_3} R\!-\!CHO \quad or \quad Ar\!-\!CHO$$

Aldehyde ∎

24.8 Conversion of acid chlorides into acid derivatives

In the laboratory, amides and esters are usually prepared from the acid chloride rather than from the acid itself. Both the preparation of the acid chloride and its reactions with ammonia or an alcohol are rapid, essentially irreversible reactions. It is more convenient to carry out these two steps than the single slow, reversible reaction with the acid. For example:

$$n\text{-}C_{17}H_{35}COOH \xrightarrow[\text{heat}]{\text{SOCl}_2} n\text{-}C_{17}H_{35}COCl \xrightarrow[\text{cold}]{\text{NH}_3} n\text{-}C_{17}H_{35}CONH_2$$

Stearic acid   Stearoyl chloride   Stearamide

3,5-Dinitrobenzoic acid $\xrightarrow[\text{heat}]{\text{PCl}_5}$ 3,5-Dinitrobenzoyl chloride $\xrightarrow[\text{pyridine}]{n\text{-}C_3H_7OH}$ n-Propyl 3,5-dinitrobenzoate

Benzoyl chloride + HO—(Phenol) $\xrightarrow[\text{NaOH}]{\text{aqueous}}$ Phenyl benzoate

Aromatic acid chlorides (ArCOCl) are considerably less reactive than the aliphatic acid chlorides. With cold water, for example, acetyl chloride reacts almost explosively, whereas benzoyl chloride reacts only very slowly. The reaction of aromatic acid chlorides with an alcohol or a phenol is often carried out using the **Schotten–Baumann** technique: the acid chloride is added in portions (followed by vigorous shaking) to a mixture of the hydroxy compound and a base, usually aqueous sodium hydroxide or pyridine (an organic base, Sec. 35.11). Base serves not only to neutralize the hydrogen chloride that would otherwise be liberated, but also to catalyze the reaction. Pyridine, in particular, seems to convert the acid chloride into an even more powerful acylating agent.

ACID ANHYDRIDES

24.9 Preparation of acid anhydrides

Only one monocarboxylic acid anhydride is encountered very often; however, this one, **acetic anhydride**, is immensely important. It is prepared by the reaction of acetic acid with **ketene**, $CH_2=C=O$, which itself is prepared by high-temperature dehydration of acetic acid.

$$CH_3COOH \xrightarrow[700\ °C]{AlPO_4} H_2O + \underset{\text{Ketene}}{CH_2=C=O} \xrightarrow{CH_3COOH} \underset{\text{Acetic anhydride}}{(CH_3CO)_2O}$$

Ketene is an extremely reactive, interesting compound, which we have already encountered as a source of *methylene* (Sec. 12.16). It is made in the laboratory by

$$CH_3COCH_3 \xrightarrow{700–750\ °C} CH_4 + \underset{\text{Ketene}}{CH_2=C=O}$$

pyrolysis of acetone, and ordinarily used as soon as it is made.

In contrast to monocarboxylic acids, certain *di*carboxylic acids yield anhydrides on simple heating: in those cases where a five- or six-membered ring is produced. For example:

Succinic
anhydride

Phthalic
anhydride

Ring size is crucial: with adipic acid, for example, anhydride formation would produce a seven-membered ring, and does not take place. Instead, carbon dioxide is lost and cyclopentanone, a ketone with a five-membered ring, is formed:

Cyclopentanone

Problem 24.2 Cyclic anhydrides can be formed from only the *cis*-1,2-cyclopentanedicarboxylic acid, but from both the *cis*- and *trans*-1,2-cyclohexanedicarboxylic acids. How do you account for this?

24.10   Reactions of acid anhydrides

Acid anhydrides undergo the same reactions as acid chlorides, but a little more slowly; where acid chlorides yield a molecule of HCl, anhydrides yield a molecule of carboxylic acid.

Compounds containing the acetyl group are often prepared from acetic anhydride; it is cheap, readily available, less volatile and more easily handled than acetyl chloride, and it does not form corrosive hydrogen chloride. It is widely used industrially for the esterification of the polyhydroxy compounds known as *carbohydrates*, especially cellulose (Chap. 39).

___ REACTIONS OF ACID ANHYDRIDES _____

1. **Conversion into acids and acid derivatives.** Discussed in Sec. 24.10.

$$(RCO)_2O + HZ \longrightarrow RCOZ + RCOOH$$

(a) **Conversion into acids. Hydrolysis**

Example:

$$(CH_3CO)_2O \quad + H_2O \longrightarrow \quad 2CH_3COOH$$
$$\text{Acetic anhydride} \qquad\qquad\qquad \text{Acetic acid}$$

(b) **Conversion into amides. Ammonolysis**

Examples:

$$(CH_3CO)_2O \quad + 2NH_3 \longrightarrow \quad CH_3CONH_2 + CH_3COO^-NH_4^+$$
$$\text{Acetic anhydride} \qquad\qquad\qquad \text{Acetamide} \qquad \text{Ammonium acetate}$$

$$\begin{array}{c} & O \\ & \| \\ H_2C\!-\!C \\ \quad | \qquad\ \ \diagdown O + 2NH_3 \\ H_2C\!-\!C\diagup \\ & \| \\ & O \end{array} \longrightarrow \begin{array}{c} CH_2CONH_2 \\ | \\ CH_2COONH_4 \end{array} \xrightarrow{\ H^+\ } \begin{array}{c} CH_2CONH_2 \\ | \\ CH_2COOH \end{array}$$

$$\text{Succinic anhydride} \qquad\qquad \text{Ammonium succinamate} \qquad \text{Succinamic acid}$$

(c) **Conversion into esters. Alcoholysis**

Examples:

$$(CH_3CO)_2O + CH_3OH \longrightarrow CH_3COOCH_3 + CH_3COOH$$
$$\text{Acetic anhydride} \qquad\qquad\qquad \text{Methyl acetate} \qquad \text{Acetic acid}$$
$$\text{(An ester)}$$

CONTINUED ____

Phthalic anhydride   sec-Butyl alcohol   sec-Butyl hydrogen phthalate

2. **Formation of ketones. Friedel–Crafts acylation.** Discussed in Sec. 21.5.

$$(RCO_2)O + ArH \xrightarrow[\substack{\text{or other} \\ \text{Lewis acid}}]{AlCl_3} R-\underset{\underset{O}{\|}}{C}-Ar + RCOOH$$

A ketone

Acetic anhydride   Mesitylene   Methyl mesityl ketone   Acetic acid

Phthalic anhydride   o-Benzoylbenzoic acid

Only "half" of the anhydride appears in the acyl product; the other "half" forms a carboxylic acid. A cyclic anhydride, we see, undergoes exactly the same reactions as any other anhydride. However, since both "halves" of the anhydride are attached to each other by carbon–carbon bonds, the acyl compound and the carboxylic acid formed will have to be part of the same molecule. Cyclic anhydrides can thus be used to make compounds containing both the acyl group and the carboxyl group: compounds that are, for example, both acids and amides, both acids and esters, etc. These difunctional compounds are of great value in further synthesis.

Problem 24.3 (a) The two 1,3-cyclobutanedicarboxylic acids (p. 457) have been assigned configurations on the basis of the fact that one can be converted into an anhydride and the other cannot. Which configuration would you assign to the one that can form the anhydride, and why? (b) The method of (a) cannot be used to assign configurations to the 1,2-cyclohexanedicarboxylic acids, since *both* give anhydrides. Why is this? (c) Could the method of (a) be used to assign configurations to the 1,3-cyclohexanedicarboxylic acids?

Problem 24.4 Give structural formulas for compounds A through G.

$$\text{Benzene} + \text{succinic anhydride} \xrightarrow{\text{AlCl}_3} \text{A} (C_{10}H_{10}O_3)$$

$$\text{A} + \text{Zn(Hg)} \xrightarrow{\text{HCl}} \text{B} (C_{10}H_{12}O_2)$$

$$\text{B} + \text{SOCl}_2 \longrightarrow \text{C} (C_{10}H_{11}OCl)$$

$$\text{C} \xrightarrow{\text{AlCl}_3} \text{D} (C_{10}H_{10}O)$$

$$\text{D} + \text{H}_2 \xrightarrow{\text{Pt}} \text{E} (C_{10}H_{12}O)$$

$$\text{E} + \text{H}_2SO_4 \xrightarrow{\text{heat}} \text{F} (C_{10}H_{10})$$

$$\text{F} \xrightarrow{\text{Pt, heat}} \text{G} (C_{10}H_8) + H_2$$

Problem 24.5 (a) What product will be obtained if D of the preceding problem is treated with C_6H_5MgBr and then water? (b) What will you finally get if the product from (a) replaces E in the preceding problem?

Problem 24.6 When heated with acid (e.g., concentrated H_2SO_4), *o*-benzoylbenzoic acid yields a product of formula $C_{14}H_8O_2$. What is the structure of this product? What general type of reaction has taken place?

Problem 24.7 Predict the products of the following reactions:
(a) toluene + phthalic anhydride + $AlCl_3$
(b) the product from (a) + conc. H_2SO_4 + heat

Problem 24.8 Alcohols are the class of compounds most commonly resolved (Sec. 4.27), despite the fact that they are not acidic enough or basic enough to form (stable) salts. Outline all steps in a procedure for the resolution of *sec*-butyl alcohol, using as resolving agent the base $(-)$-B.

AMIDES

24.11 Preparation of amides

In the laboratory amides are prepared by the reaction of ammonia with acid chlorides or, when available, acid anhydrides (Secs. 24.8 and 24.10). In industry they are often made by heating the ammonium salts of carboxylic acids.

24.12 Reactions of amides

An amide is hydrolyzed when heated with aqueous acids or aqueous bases. The products are ammonia and the carboxylic acid, although one product or the other is obtained in the form of a salt, depending upon the acidity or basicity of the medium.

Another reaction of importance, the Hofmann degradation of amides, will be discussed later (Secs. 26.12 and 32.2–32.5).

REACTIONS OF AMIDES

1. **Hydrolysis.** Discussed in Sec. 24.13.

$$R-C\overset{O}{\underset{NH_2}{}} + H_2O \longrightarrow \begin{cases} \xrightarrow{H^+} R-C\overset{O}{\underset{OH}{}} + NH_4^+ \\ \xrightarrow{OH^-} R-COO^- + NH_3 \end{cases}$$

CONTINUED

——— CONTINUED ———

Examples:

$$\langle\bigcirc\rangle\!-\!CONH_2 + H_2SO_4 + H_2O \longrightarrow \langle\bigcirc\rangle\!-\!COOH + NH_4{}^+HSO_4$$

       Benzamide   Benzoic acid

$$CH_3CH_2CH_2CONH_2 + NaOH + H_2O \longrightarrow CH_3CH_2CH_2COO^-Na^+ + NH_3$$

    Butyramide   Sodium butyrate

   2. **Conversion into imides.** Discussed in Sec. 24.14.

   3. **Hofmann degradation of amides.** Discussed in Secs. 26.12 and 32.2–32.5.

$$R\!-\!CONH_2 \text{ or } Ar\!-\!CONH_2 \xrightarrow{\text{OBr}^-} R\!-\!NH_2 \text{ or } Ar\!-\!NH_2 + CO_3{}^{2-}$$

               Amide   1° amine ∎

24.13 Hydrolysis of amides

Hydrolysis of amides is typical of the reactions of carboxylic acid derivatives. It involves nucleophilic substitution, in which the —NH$_2$ group is replaced by —OH. Under acidic conditions hydrolysis involves attack by water on the protonated amide:

$$R\!-\!\underset{NH_2}{\overset{O}{C}} \xrightarrow{H^+} \left. R\!-\!\underset{NH_2}{\overset{OH}{C}} \right\}\oplus \xrightarrow{H_2O} R\!-\!\underset{NH_2}{\overset{OH}{\underset{|}{C}}}\!-\!OH_2{}^+ \longrightarrow$$

$$NH_3 + R\!-\!\underset{OH}{\overset{O}{C}} \longrightarrow RCOO^-NH_4{}^+$$

Under alkaline conditions hydrolysis involves attack by the strongly nucleophilic hydroxide ion on the amide itself:

$$R\!-\!\underset{NH_2}{\overset{O}{C}} \xrightarrow{OH^-} R\!-\!\underset{NH_2}{\overset{O^-}{\underset{|}{C}}}\!-\!OH \longrightarrow RCOO^- + NH_3$$

24.14 Imides

Like other anhydrides, cyclic anhydrides react with ammonia to yield amides; in this case the product contains both —CONH$_2$ and —COOH groups. If this acid–amide is heated, a molecule of water is lost, a ring forms, and a product is obtained in which two acyl groups have become attached to nitrogen; compounds of this sort are called **imides**. Phthalic anhydride gives *phthalamic acid* and *phthalimide*:

Phthalic anhydride   Ammonium phthalamate   Phthalamic acid

Phthalimide

Problem 24.9   Outline all steps in the synthesis of *succinimide* from succinic acid.

Problem 24.10   Account for the following sequence of acidities. (*Hint*: See Sec. 23.12.)

	K_a
Ammonia	10^{-33}
Benzamide	10^{-14} to 10^{-15}
Phthalimide	5×10^{-9}

ESTERS

24.15   Preparation of esters

Esters are usually prepared by the reaction of alcohols or phenols with acids or acid derivatives. The most common methods are outlined below.

PREPARATION OF ESTERS

1. **From acids.** Discussed in Secs. 23.16 and 24.15.

Reactivity of R'OH:
1° > 2° (> 3°)

Examples:

$$CH_3COOH + HOCH_2 \bigcirc \overset{H^+}{\rightleftharpoons} CH_3COOCH_2 \bigcirc$$

Acetic acid   Benzyl alcohol   Benzyl acetate

CONTINUED

——— CONTINUED ———

$$\text{Benzoic acid} \quad \text{COOH} + \text{HOCH}_2\overset{\overset{\displaystyle CH_3}{|}}{C}HCH_3 \underset{\text{Isobutyl alcohol}}{\overset{H^+}{\rightleftharpoons}} \quad \text{COOCH}_2\overset{\overset{\displaystyle CH_3}{|}}{C}HCH_3 \quad \text{Isobutyl benzoate}$$

2. From acid chlorides or anhydrides. Discussed in Secs. 24.8 and 24.10.

$$R-\overset{\overset{\displaystyle O}{\|}}{C}-Cl + R'O-H \text{ (or ArO-H)} \longrightarrow R-\overset{\overset{\displaystyle O}{\|}}{C}-OR' \text{ (or RCO-OAr)} + HCl$$

$$(RCO)_2O + R'OH \text{ (or ArOH)} \longrightarrow RCOOR' \text{ (or RCOOAr)} + RCOOH$$

Examples:

$$\underset{\substack{\text{o-Bromobenzoyl} \\ \text{chloride}}}{\overset{Br}{\text{COCl}}} + C_2H_5OH \overset{\text{pyridine}}{\longrightarrow} \underset{\text{Ethyl o-bromobenzoate}}{\overset{Br}{\text{COOC}_2H_5}} + HCl$$

$$\underset{\substack{\text{Acetic} \\ \text{anhydride}}}{(CH_3CO)_2O} + \underset{\text{p-Nitrophenol}}{HO-\bigcirc-NO_2} \overset{\text{NaOH}}{\longrightarrow} \underset{\text{p-Nitrophenyl acetate}}{CH_3COO-\bigcirc-NO_2} + CH_3COOH$$

3. From esters. Transesterification. Discussed in Sec. 24.20. ∎

The direct reaction of alcohols or phenols with acids involves an equilibrium and—especially in the case of phenols—requires effort to drive to completion (see Sec. 23.16). In the laboratory, reaction with an acid chloride or anhydride is more commonly used.

The effect of the structure of the alcohol and of the acid on ease of esterification has already been discussed (Sec. 23.16).

Table 24.2   ESTERS OF CARBOXYLIC ACIDS

Name	M.p., °C	B.p., °C	Name	M.p., °C	B.p., °C
Methyl acetate	−98	57.5	Ethyl formate	−80	54
Ethyl acetate	−84	77	Ethyl acetate	−84	77
n-Propyl acetate	−92	102	Ethyl propionate	−74	99
n-Butyl acetate	−77	126	Ethyl n-butyrate	−93	121
n-Pentyl acetate		148	Ethyl n-valerate	−91	146
Isopentyl acetate	−78	142	Ethyl stearate	34	215[15]
Benzyl acetate	−51	214	Ethyl phenylacetate		226
Phenyl acetate		196	Ethyl benzoate	−35	213

As was mentioned earlier, esterification using aromatic acid chlorides, ArCOCl, is often carried out in the presence of base (the Schotten–Baumann technique, Sec. 24.8).

Problem 24.11 When benzoic acid is esterified by methanol in the presence of a little sulfuric acid, the final reaction mixture contains five substances: benzoic acid, methanol, water, methyl benzoate, sulfuric acid. Outline a procedure for the separation of the pure ester.

A hydroxy acid is both alcohol and acid. In those cases where a five- or six-membered ring can be formed, *intramolecular* esterification occurs. Thus, a γ- or δ-hydroxy acid loses water spontaneously to yield a cyclic ester known as a **lactone**. Treatment with base (actually hydrolysis of an ester) rapidly opens the lactone ring

$$RCHCH_2CH_2COO^-Na^+ \underset{OH^-}{\overset{H^+}{\rightleftharpoons}}$$

OH

Salt of a
γ-hydroxy acid

A γ-lactone

A cyclic ester: five-membered ring

$$RCHCH_2CH_2CH_2COO^-Na^+ \underset{OH^-}{\overset{H^+}{\rightleftharpoons}}$$

OH

Salt of a
δ-hydroxy acid

A δ-lactone

A cyclic ester: six-membered ring

to give the open-chain salt. We shall encounter lactones again in our study of carbohydrates (Sec. 38.8).

Problem 24.12 Suggest a likely structure for the product formed by heating each of these acids. (a) *Lactic acid*, $CH_3CHOHCOOH$, gives *lactide*, $C_6H_8O_4$. (b) 10-Hydroxy-decanoic acid gives a material of high molecular weight (1000–9000).

24.16 Reactions of esters

Esters undergo the nucleophilic substitution that is typical of carboxylic acid derivatives. Attack occurs at the electron-deficient carbonyl carbon, and results in the replacement of the —OR′ group by —OH, —OR″, or —NH₂:

$$R-C\overset{O}{\underset{OR'}{\big\langle}} + :Z \longrightarrow R-\overset{O^-}{\underset{OR'}{C}}-Z \longrightarrow R-C\overset{O}{\underset{Z}{\big\langle}} + :OR'^-$$

$$:Z = :OH^-, \quad :OR''^-, \quad :NH_3$$

These reactions are sometimes carried out in the presence of acid. In these acid-catalyzed reactions, H^+ attaches itself to the oxygen of the carbonyl group, and thus renders carbonyl carbon even more susceptible to nucleophilic attack.

$$R-C\overset{O}{\underset{OR'}{}} + H^+ \rightleftharpoons R-C\overset{OH}{\underset{OR'}{}} \Big\}\oplus \quad :Z$$

Acid catalysis:

makes carbon more susceptible to nucleophilic attack

REACTIONS OF ESTERS

1. Conversion into acids and acid derivatives

    (a) Conversion into acids. Hydrolysis. Discussed in Secs. 24.17 and 24.18.

$$R-C\overset{O}{\underset{OR'}{}} + H_2O \longrightarrow \begin{cases} \xrightarrow{H^+} R-C\overset{O}{\underset{OH}{}} + R'O-H \\ \xrightarrow{OH^-} R-COO^- + R'O-H \end{cases}$$

Example:

$$\bigcirc\!\!-COOC_2H_5 + H_2O \longrightarrow \begin{cases} \xrightarrow{H_2SO_4} \bigcirc\!\!-COOH + C_2H_5OH \\ \xrightarrow{NaOH} \bigcirc\!\!-COO^-Na^+ + C_2H_5OH \end{cases}$$

Ethyl benzoate

Benzoic acid   Ethyl alcohol

Sodium benzoate   Ethyl alcohol

    (b) Conversion into amides. Ammonolysis. Discussed in Sec. 24.19.

$$R-C\overset{O}{\underset{OR'}{}} + NH_3 \longrightarrow R-C\overset{O}{\underset{NH_2}{}} + R'O-H$$

Example:

$$CH_3COOC_2H_5 + NH_3 \longrightarrow CH_3CONH_2 + C_2H_5OH$$

    Ethyl acetate   Acetamide   Ethyl alcohol

    (c) Conversion into esters. Transesterification. Alcoholysis. Discussed in Sec. 24.20.

$$R-C\overset{O}{\underset{OR'}{}} + R''O-H \xrightleftharpoons{\text{acid or base}} R-C\overset{O}{\underset{OR''}{}} + R'O-H$$

CONTINUED

CONTINUED

Example:

$$
\begin{array}{c}
CH_2-O-C-R \\
\quad\quad \| \\
\quad\quad O \\
| \\
CH-O-C-R' + CH_3OH \\
\quad\quad \| \\
\quad\quad O \\
| \\
CH_2-O-C-R'' \\
\quad\quad \| \\
\quad\quad O
\end{array}
\xrightarrow{\text{acid or base}}
\begin{array}{c}
RCOOCH_3 \quad CH_2OH \\
+ \quad\quad\quad | \\
R'COOCH_3 + CHOH \\
+ \quad\quad\quad | \\
R''COOCH_3 \quad CH_2OH \\
\text{Mixture of} \quad \text{Glycerol} \\
\text{methyl esters}
\end{array}
$$

A glyceride
(A fat)

2. Reaction with Grignard reagents. Discussed in Sec. 24.21.

$$
R-C\overset{\displaystyle O}{\underset{\displaystyle OR'}{<}} \; + \; 2R''-MgX \; \longrightarrow \; R-\overset{\displaystyle R''}{\underset{\displaystyle OH}{C}}-R''
$$

Tertiary alcohol

Example:

$$
\overset{\displaystyle CH_3}{\underset{}{|}}
CH_3CHCOOC_2H_5 + 2CH_3MgI \; \longrightarrow \; CH_3CH-\overset{\displaystyle CH_3\;CH_3}{\underset{\displaystyle OH}{C}}-CH_3
$$

Ethyl   Methylmagnesium
isobutyrate   iodide
2 moles

2,3-Dimethyl-2-butanol

3. Reduction to alcohols. Discussed in Sec. 24.22.

   (a) Catalytic hydrogenation. Hydrogenolysis

$$
R-COOR' \; + \; 2H_2 \; \xrightarrow[\substack{250\ ^\circ C \\ 3000-6000\ lb/in.^2}]{CuO \cdot CuCr_2O_4} \; R-CH_2OH \; + \; R'OH
$$

1° alcohol

Example:

$$
\overset{\displaystyle CH_3}{\underset{\displaystyle CH_3}{CH_3-C-COOC_2H_5}} + 2H_2 \; \xrightarrow[250\ ^\circ C,\ 3300\ lb/in.^2]{CuO \cdot CuCr_2O_4} \; \overset{\displaystyle CH_3}{\underset{\displaystyle CH_3}{CH_3-C-CH_2OH}} + C_2H_5OH
$$

Ethyl trimethylacetate   Neopentyl alcohol   Ethyl alcohol
(Ethyl 2,2-dimethylpropanoate)   (2,2-Dimethylpropanol)

CONTINUED

—— CONTINUED ——

(b) **Chemical reduction**

$$4R\text{—COOR}' + 2LiAlH_4 \xrightarrow[\text{ether}]{\text{anhyd.}} \left\{ \begin{array}{c} LiAl(OCH_2R)_4 \\ + \\ LiAl(OR')_4 \end{array} \right\} \xrightarrow{H^+} \left\{ \begin{array}{c} R\text{—CH}_2OH \\ + \\ R'OH \end{array} \right\}$$

Example:

$$CH_3(CH_2)_7CH{=}CH(CH_2)_7COOCH_3 \xrightarrow{LiAlH_4} \xrightarrow{H^+} CH_3(CH_2)_7CH{=}CH(CH_2)_7CH_2OH$$

Methyl oleate	Oleyl alcohol
(Methyl *cis*-9-octadecenoate)	(*cis*-9-Octadecen-1-ol)

4. **Reaction with carbanions. Claisen condensation.** Discussed in Secs. 25.11 and 25.12.

A β-keto ester

24.17 Alkaline hydrolysis of esters

A carboxylic ester is hydrolyzed to a carboxylic acid and an alcohol or phenol when heated with aqueous acid or aqueous base. Under alkaline conditions, of course, the carboxylic acid is obtained as its salt, from which it can be liberated by addition of mineral acid.

Base promotes hydrolysis of esters by providing the strongly nucleophilic

Ester	Hydroxide			Salt	Alcohol

reagent OH^-. This reaction is essentially irreversible, since a resonance-stabilized carboxylate anion (Sec. 23.13) shows little tendency to react with an alcohol.

Let us look at the various aspects of the mechanism we have written, and see what evidence there is for each of them.

First, reaction involves attack on the ester by hydroxide ion. This is consistent with the **kinetics**, which is second-order, with the rate depending on both ester concentration and hydroxide concentration.

Next hydroxide attacks at the carbonyl carbon and displaces alkoxide ion. That is to say, reaction involves cleavage of the bond between oxygen and the acyl group, $RCO{\dagger}OR'$. For this there are two lines of evidence, the first being the **stereochemistry**.

Let us consider, for example, the formation and subsequent hydrolysis of an ester of optically active *sec*-butyl alcohol. Reaction of (+)-*sec*-butyl alcohol with benzoyl chloride must involve cleavage of the hydrogen–oxygen bond and hence cannot change the configuration about the chiral center (see Sec. 4.23). If hydrolysis of this ester involves cleavage of the bond between oxygen and the *sec*-butyl group, we would expect almost certainly inversion (or inversion plus racemization if the reaction goes by an S_N1 type of mechanism):

$$C_6H_5COO^- +$$

(+)-*sec*-Butyl alcohol   Cleavage between oxygen and alkyl group: *inversion*   (−)-*sec*-Butyl alcohol

If, on the other hand, the bond between oxygen and the *sec*-butyl group remains intact during hydrolysis, then we would expect to obtain *sec*-butyl alcohol of the same configuration as the starting material:

$$C_6H_5COO^- +$$

(+)-*sec*-Butyl alcohol   Cleavage between oxygen and acyl group: *retention*   (+)-*sec*-Butyl alcohol

When *sec*-butyl alcohol of rotation $+13.8°$ was actually converted into the benzoate and the benzoate was hydrolyzed in alkali, there was obtained *sec*-butyl alcohol of rotation $+13.8°$. This complete retention of configuration strongly indicates that bond cleavage occurs between oxygen and the acyl group.

Tracer studies have confirmed the kind of bond cleavage indicated by the stereochemical evidence. When ethyl propionate labeled with ^{18}O was hydrolyzed by base in ordinary water, the ethanol produced was found to be enriched in ^{18}O; the propionic acid contained only the ordinary amount of ^{18}O:

The alcohol group retained the oxygen that it held in the ester; cleavage occurred between oxygen and the acyl group.

The study of a number of other hydrolyses by both tracer and stereochemical methods has shown that cleavage between oxygen and the acyl group is the usual one in ester hydrolysis. This behavior indicates that the preferred point of nucleophilic attack is the carbonyl carbon rather than the alkyl carbon; this is, of course, what we might have expected in view of the generally greater reactivity of carbonyl carbon (Sec. 24.5).

Finally, according to the mechanism, attack by hydroxide ion on carbonyl carbon does not displace alkoxide ion in one step,

but rather in *two steps* with the intermediate formation of a tetrahedral compound. These alternative mechanisms were considered more or less equally likely until 1950 when elegant work on **isotopic exchange** was reported by Myron Bender (now at Northwestern University).

Bender carried out the alkaline hydrolysis of carbonyl-labeled ethyl benzoate, $C_6H_5C^{18}OOC_2H_5$, in ordinary water, and focused his attention, not on the product, but on the *reactant*. He interrupted the reaction after various periods of time, and isolated the unconsumed ester and analyzed it for ^{18}O content. He found that in the alkaline solution the ester was undergoing not only hydrolysis but also *exchange of its ^{18}O for ordinary oxygen from the solvent.*

Oxygen exchange is not consistent with the one-step mechanism, which provides no way for it to happen. Oxygen exchange is consistent with a two-step mechanism in which intermediate I is not only formed, but partly reverts into starting material and partly is converted (probably via the neutral species II) into III—an intermediate that is equivalent to I except for the position of the label. If all this is so, the "reversion" of intermediate III into "starting material" yields ester that has lost its ^{18}O.

Bender's work does not *prove* the mechanism we have outlined. Conceivably, oxygen exchange—and hence the tetrahedral intermediate—simply represent a blind-alley down which ester molecules venture but which does not lead to hydrolysis. Such coincidence is unlikely, however, particularly in light of certain kinetic relationships between oxygen exchange and hydrolysis.

Similar experiments have indicated the reversible formation of tetrahedral intermediates in hydrolysis of other esters, amides, anhydrides, and acid chlorides, and are the basis of the general mechanism we have shown for nucleophilic acyl substitution.

Exchange experiments are also the basis of our estimate of the relative importance of the two steps: differences in rate of hydrolysis of acyl derivatives depend chiefly on how fast intermediates are formed, and also on what fraction of the intermediate goes on to product. As we have said, the rate of formation of the intermediate is affected by both electronic and steric factors: in the transition state, a negative charge is developing and carbon is changing from trigonal toward tetrahedral.

Even in those cases where oxygen exchange cannot be detected, we cannot rule out the possibility of an intermediate; it may simply be that it goes on to hydrolysis products much faster than it does anything else.

Problem 24.13 The relative rates of alkaline hydrolysis of ethyl p-substituted benzoates, p-$GC_6H_4COOC_2H_5$, are:

$$G = NO_2 > Cl > H > CH_3 > OCH_3$$
$$110 \quad 4 \quad 1 \quad 0.5 \quad 0.2$$

(a) How do you account for this order of reactivity? (b) What kind of effect, activating or deactivating, would you expect from p-Br? from p-NH_2? from p-$C(CH_3)_3$? (c) Predict the order of reactivity toward alkaline hydrolysis of: p-aminophenyl acetate, p-methylphenyl acetate, p-nitrophenyl acetate, phenyl acetate.

Problem 24.14 The relative rates of alkaline hydrolysis of alkyl acetates, CH_3COOR, are:

$$R = CH_3 > C_2H_5 > (CH_3)_2CH > (CH_3)_3C$$
$$1 \quad 0.6 \quad 0.15 \quad 0.008$$

(a) What two factors might be at work here? (b) Predict the order of reactivity toward alkaline hydrolysis of: methyl acetate, methyl formate, methyl isobutyrate, methyl propionate, and methyl trimethylacetate.

Problem 24.15 Exchange experiments show that the fraction of the tetrahedral intermediate that goes on to products follows the sequence:

$$\text{acid chloride} > \text{acid anhydride} > \text{ester} > \text{amide}$$

What is one factor that is probably at work here?

24.18 Acidic hydrolysis of esters

Hydrolysis of esters is promoted not only by base but also by acid. Acidic hydrolysis, as we have seen (Sec. 23.16), is reversible,

$$R-C\overset{O}{\underset{OR'}{\big\backslash}} + H_2O \underset{H^+}{\overset{H^-}{\rightleftarrows}} R-C\overset{O}{\underset{OH}{\big\backslash}} + R'O-H$$

and hence the mechanism for hydrolysis is also—taken in the opposite direction—the mechanism for esterification. Any evidence about one reaction must apply to both.

The mechanism for acid-catalyzed hydrolysis and esterification is contained in the following equilibria:

$$
\underset{\text{Ester}}{\underset{\overset{\displaystyle +}{\overset{\displaystyle H^+}{}}}{R-\overset{\displaystyle O}{\overset{\|}{C}}-OR'}} \rightleftarrows \underset{}{\underset{\overset{\displaystyle +}{\overset{\displaystyle \text{Water}}{\overset{\displaystyle H_2O}{}}}}{R-\overset{\displaystyle OH}{\overset{\cdots}{C}{=}OR'}}} \rightleftarrows \underset{\underset{\displaystyle \oplus OH_2}{}}{\underset{\displaystyle I}{R-\overset{\displaystyle OH}{\underset{|}{\overset{|}{C}}}-OR'}} \rightleftarrows
$$

$$
\underset{\text{II}}{R-\underset{\displaystyle OH}{\overset{\displaystyle OH}{\underset{|}{\overset{|}{C}}}}\underset{\displaystyle H}{\overset{\oplus}{\underset{|}{OR'}}} \rightleftarrows \underset{\begin{array}{c}+\\ R'O-H\\ \text{Alcohol}\end{array}}{\left.\begin{array}{c}OH\\ R-C\\ OH\end{array}\right\}}\oplus \rightleftarrows \underset{\begin{array}{c}\text{Acid}\\+\\ H^+\end{array}}{R-\overset{\displaystyle O}{\overset{\|}{C}}-OH}
$$

Mineral acid speeds up both processes by protonating carbonyl oxygen and thus rendering carbonyl carbon more susceptible to nucleophilic attack (Sec. 24.4). In hydrolysis, the nucleophile is a water molecule and the leaving group is an alcohol; in esterification, the roles are exactly reversed.

As in alkaline hydrolysis, there is almost certainly a tetrahedral intermediate—or, rather, several of them. The existence of more than one intermediate is required by, among other things, the reversible nature of the reaction. Looking only at hydrolysis, intermediate II is *likely*, since it permits separation of the weakly basic alcohol molecule instead of the strongly basic alkoxide ion; but consideration of esterification shows that II almost certainly *must* be involved, since it is the product of attack by alcohol on the protonated acid.

The evidence for the mechanism is much the same as in alkaline hydrolysis.

The position of cleavage, RCO$+$OR' and RCO$+$OH, has been shown by ^{18}O studies of both hydrolysis and esterification. The existence of the tetrahedral intermediates was demonstrated, as in the alkaline reaction, by ^{18}O exchange between the carbonyl oxygen of the ester and the solvent.

Problem 24.16 Write the steps to account for exchange between RC^{18}OOR' and H$_2$O in acidic solution. There is reason to believe that a key intermediate here is identical with one in alkaline hydrolysis. What might this intermediate be?

Problem 24.17 Account for the fact (Sec. 23.16) that the presence of bulky substituents in either the alcohol group or the acid group slows down both esterification and hydrolysis.

Problem 24.18   Acidic hydrolysis of *tert*-butyl acetate in water enriched in ^{18}O has been found to yield *tert*-butyl alcohol enriched in ^{18}O and acetic acid containing ordinary oxygen. Acidic hydrolysis of the acetate of optically active 3,7-dimethyl-3-octanol has been found to yield alcohol of much lower optical purity than the starting alcohol, and having the opposite sign of rotation. (a) How do you interpret these two sets of results? (b) Is it surprising that these particular esters should show this kind of behavior?

24.19   Ammonolysis of esters

Treatment of an ester with ammonia, generally in ethyl alcohol solution, yields the amide. This reaction involves nucleophilic attack by a base, ammonia, on the electron-deficient carbon; the alkoxy group, —OR′, is replaced by —NH$_2$. For example:

$$CH_3-C\overset{O}{\underset{OC_2H_5}{\big\backslash}} \quad + NH_3 \longrightarrow \quad CH_3-C\overset{O}{\underset{NH_2}{\big\backslash}} \quad + C_2H_5OH$$

Ethyl acetate   Acetamide

24.20   Transesterification

In the esterification of an acid, an alcohol acts as a nucleophilic reagent; in hydrolysis of an ester, an alcohol is displaced by a nucleophilic reagent. Knowing this, we are not surprised to find that one alcohol is capable of displacing another alcohol from an ester. This *alcoholysis* (cleavage by an alcohol) of an ester is called **transesterification**.

$$R-C\overset{O}{\underset{OR'}{\big\backslash}} \quad + R''O-H \underset{}{\overset{H^+ \text{ or } OR''^-}{\rightleftarrows}} R-C\overset{O}{\underset{OR''}{\big\backslash}} \quad + R'O-H$$

Transesterification is catalyzed by acid (H$_2$SO$_4$ or dry HCl) or base (usually alkoxide ion). The mechanisms of these two reactions are exactly analogous to those we have already studied. For acid-catalyzed transesterification:

R″OH
Alcohol B

H$^+$ +
+
O $\overset{+}{OH}$ OH
‖   ‖ |
R—C—OR′ $\rightleftarrows$ R—C≡OR′ $\rightleftarrows$ R—C—OR′ $\rightleftarrows$
                                    |
Ester A $\overset{\oplus}{OR''}$
                                    H

OH   OH   O
| $\oplus$ ‖   ‖
R—C—OR′ $\rightleftarrows$ R—C $\Big\}$ $\oplus$ $\rightleftarrows$ R—C—OR′
| H   ‖   Ester B
OR″   OR″
                          +   +
                         R′OH   H$^+$
                       Alcohol A

For base-catalyzed transesterification:

| Ester A | Alkoxide B | | | Ester B | Alkoxide A |

Transesterification is an equilibrium reaction. To shift the equilibrium to the right, it is necessary to use a large excess of the alcohol whose ester we wish to make, or else to remove one of the products from the reaction mixture. The second approach is the better one when feasible, since in this way the reaction can be driven to completion.

24.21   Reaction of esters with Grignard reagents

The reaction of carboxylic esters with Grignard reagents is an excellent method for preparing tertiary alcohols. As in the reaction with aldehydes and ketones (Sec. 21.10), the nucleophilic (basic) alkyl or aryl group of the Grignard reagent attaches itself to the electron-deficient carbonyl carbon. Expulsion of the alkoxide group would yield a ketone, and in certain special cases ketones are indeed isolated from this reaction. However, as we know, ketones themselves readily react with Grignard reagents to yield tertiary alcohols (Sec. 17.15); in the present case the products obtained correspond to the addition of the Grignard reagent to such a ketone:

Two of the three groups attached to the carbon bearing the hydroxyl group in the alcohol come from the Grignard reagent and hence must be identical; this, of course, places limits upon the alcohols that can be prepared by this method. But, where applicable, reaction of a Grignard reagent with an ester is preferred to reaction with a ketone because esters are generally more accessible.

Problem 24.19   Starting from valeric acid, and using any needed reagents, outline the synthesis of 3-ethyl-3-heptanol via the reaction of a Grignard reagent with: (a) a ketone; (b) an ester.

Problem 24.20   (a) Esters of which acid would yield *secondary* alcohols on reaction with Grignard reagents? (b) Starting from alcohols of four carbons or fewer, outline all steps in the synthesis of 4-heptanol.

24.22 Reduction of esters

Like many organic compounds, esters can be reduced in two ways: (a) by catalytic hydrogenation using molecular hydrogen, or (b) by chemical reduction. In either case, the ester is cleaved to yield (in addition to the alcohol or phenol from which it was derived) a primary alcohol corresponding to the acid portion of the ester.

$$R-COOR' \xrightarrow{\text{reduction}} R-CH_2OH + R'OH$$

$$\text{Ester} \qquad\qquad \text{1}° \text{ alcohol}$$

Hydrogenolysis (cleavage by hydrogen) of an ester requires more severe conditions than simple hydrogenation of (addition of hydrogen to) a carbon–carbon double bond. High pressures and elevated temperatures are required: the catalyst used most often is a mixture of oxides known as *copper chromite*, of approximately the composition $CuO.CuCr_2O_4$. For example:

$$CH_3(CH_2)_{10}COOCH_3 \xrightarrow[\text{150 °C, 5000 lb/in.}^2]{\text{H}_2, \text{ CuO}.\text{CuCr}_2\text{O}_4} CH_3(CH_2)_{10}CH_2OH + CH_3OH$$

Methyl laurate   Lauryl alcohol
(Methyl dodecanoate)   (1-Dodecanol)

Chemical reduction is carried out by use of sodium metal and alcohol, or more usually by use of lithium aluminum hydride. For example:

$$CH_3(CH_2)_{14}COOC_2H_5 \xrightarrow{\text{LiAlH}_4} \xrightarrow{\text{H}^+} CH_3(CH_2)_{14}CH_2OH$$

Ethyl palmitate   1-Hexadecanol
(Ethyl hexadecanoate)

Problem 24.21 Predict the products of the hydrogenolysis of *n*-butyl oleate over copper chromite.

24.23 Functional derivatives of carbonic acid

Much of the chemistry of the functional derivatives of carbonic acid is already quite familiar to us through our study of carboxylic acids. The first step in dealing with one of these compounds is to recognize just how it is related to the parent acid. Since carbonic acid is bifunctional, each of its derivatives, too, contains two functional groups; these groups can be the same or different. For example:

$$\left[\begin{array}{c} HO-\underset{\underset{O}{\|}}{C}-OH \end{array} \right] \qquad Cl-\underset{\underset{O}{\|}}{C}-Cl \qquad H_2N-\underset{\underset{O}{\|}}{C}-NH_2 \qquad C_2H_5O-\underset{\underset{O}{\|}}{C}-OC_2H_5$$

Carbonic acid   Phosgene   Urea   Ethyl carbonate
Acid   (Carbonyl chloride)   (Carbamide)   *Ester*
            Acid chloride   *Amide*

$$H_2N-\underset{\underset{O}{\|}}{C}-OC_2H_5$$

Urethane

$$C_2H_5O{-}\underset{\underset{O}{\|}}{C}{-}Cl \qquad H_2N{-}C{\equiv}N \qquad H_2N{-}\underset{\underset{O}{\|}}{C}{-}OC_2H_5$$

Ethyl chlorocarbonate   Cyanamide   Urethane
Acid chloride–ester   *Amide–nitrile*   (Ethyl carbamate)
Ester–amide

We use these functional relationships to carbonic acid simply for convenience. Many of these compounds could just as well be considered as derivatives of other acids, and, indeed, are often so named. For example:

$$\left[H_2N{-}\underset{\underset{O}{\|}}{C}{-}OH \right] \qquad H_2N{-}\underset{\underset{O}{\|}}{C}{-}NH_2 \qquad H_2N{-}\underset{\underset{O}{\|}}{C}{-}OC_2H_5$$

Carbamic acid   Carbamide   Ethyl carbamate
Acid   *Amide*   *Ester*

$$[HO{-}C{\equiv}N] \qquad H_2N{-}C{\equiv}N$$

Cyanic acid   Cyanamide
Acid   *Amide*

In general, a derivative of carbonic acid containing an —OH group is unstable, and decomposes to carbon dioxide. For example:

$$\left[HO{-}\underset{\underset{O}{\|}}{C}{-}OH \right] \longrightarrow CO_2 + H_2O$$

Carbonic acid

$$\left[RO{-}\underset{\underset{O}{\|}}{C}{-}OH \right] \longrightarrow CO_2 + ROH$$

Alkyl hydrogen
carbonate

$$\left[H_2N{-}\underset{\underset{O}{\|}}{C}{-}OH \right] \longrightarrow CO_2 + NH_3$$

Carbamic acid

$$\left[Cl{-}\underset{\underset{O}{\|}}{C}{-}OH \right] \longrightarrow CO_2 + HCl$$

Chlorocarbonic
acid

Most derivatives of carbonic acid are made from one of three industrially available compounds: phosgene, urea, or cyanamide.

Phosgene, $COCl_2$, a highly poisonous gas, is manufactured by the reaction between carbon monoxide and chlorine.

$$CO + Cl_2 \xrightarrow{\text{activated charcoal, 200 °C}} Cl{-}\underset{\underset{O}{\|}}{C}{-}Cl$$

Phosgene

It undergoes the usual reactions of an acid chloride.

$$\text{Cl—C—Cl} \xrightarrow{}$$

(Phosgene)

$$\xrightarrow{H_2O} \quad \text{Cl—C—OH} \xrightarrow{} CO_2 + HCl$$

$$\xrightarrow{NH_3} \quad H_2N\text{—C—}NH_2$$

Urea

$$\xrightarrow{ROH} \quad \text{Cl—C—OR} \xrightarrow{ROH} \text{RO—C—OR}$$

Alkyl
chlorocarbonate   Alkyl carbonate

$$\xrightarrow{NH_3} \quad H_2N\text{—C—}OR$$

Alkyl carbamate
(A urethane)

Problem 24.22   Suggest a possible synthesis of
(a) 2-pentylurethane, $H_2NCOOCH(CH_3)(n\text{-}C_3H_7)$, used as a hypnotic;
(b) benzyl chlorocarbonate (*carbobenzoxy chloride*), $C_6H_5CH_2OCOCl$, used in the synthesis of peptides (Sec. 40.10).

Urea, H_2NCONH_2, is excreted in the urine as the chief nitrogen-containing end product of protein metabolism. It is synthesized on a large scale for use as a fertilizer and as a raw material in the manufacture of urea–formaldehyde plastics and of drugs.

$$CO_2 + 2NH_3 \; \rightleftharpoons \; H_2NCOONH_4 \; \underset{}{\overset{\text{heat, pressure}}{\rightleftharpoons}} \; H_2N\text{—C—}NH_2$$

Ammonium carbamate   Urea

Urea is weakly basic, forming salts with strong acids. The fact that it is a stronger base than ordinary amides is attributed to resonance stabilization of the cation:

$$H_2N\text{—C—}NH_2 + H^+ \; \rightleftharpoons \; \left[H_2N\overset{+}{\underset{\overset{\|}{\oplus OH}}{C}}NH_2 \;\; H_2\overset{+}{N}\text{=C—}NH_2 \;\; H_2N\text{—C=}\overset{+}{N}H_2 \right]$$

$$\qquad\qquad\qquad\qquad\qquad\qquad OH \qquad\qquad OH$$

$$equivalent\ to \qquad \left. H_2N\text{⋯C⋯}NH_2 \right\}\oplus$$

$$\qquad\qquad\qquad\qquad\qquad\qquad OH$$

Problem 24.23   Account for the fact that *guanidine*, $(H_2N)_2C=NH$, is *strongly* basic.

Urea undergoes hydrolysis in the presence of acids, bases, or the enzyme *urease* (isolable from jack beans; generated by many bacteria, such as *Micrococcus ureae*).

$$H_2N-\underset{\underset{O}{\|}}{C}-NH_2 \xrightarrow{H_2O} \begin{cases} \xrightarrow{H^+} NH_4^+ + CO_2 \\ \xrightarrow{OH^-} NH_3 + CO_3^{2-} \\ \xrightarrow{urease} NH_3 + CO_2 \end{cases}$$

Urea

Urea reacts with nitrous acid to yield carbon dioxide and nitrogen; this is a useful way to destroy excess nitrous acid in diazotizations (Sec. 27.12).

$$H_2N-\underset{\underset{O}{\|}}{C}-NH_2 \xrightarrow{HONO} CO_2 + N_2$$

Urea is converted by hypohalites into nitrogen and carbonate.

$$H_2N-\underset{\underset{O}{\|}}{C}-NH_2 \xrightarrow{Br_2,\ OH^-} N_2 + CO_3^{2-} + Br^-$$

Treatment of urea with acid chlorides or anhydrides yields **ureides**. Of special

$$H_2N-\underset{\underset{O}{\|}}{C}-NH_2 + CH_3COCl \longrightarrow CH_3CONH-\underset{\underset{O}{\|}}{C}-NH_2$$

Acetylurea
A ureide

importance are the cyclic ureides formed by reaction with malonic esters; these are known as **barbiturates** and are important hypnotics (sleep-producers). For example:

Urea   Ethyl malonate   Barbituric acid
                                                   (Malonylurea)

Cyanamide, $H_2N-C\equiv N$, is obtained in the form of its calcium salt by the high-temperature reaction between calcium carbide and nitrogen. This reaction is

$$CaC_2 + N_2 \xrightarrow{1000\ °C} CaNCN + C$$

Calcium   Calcium
carbide   cyanamide

important as a method of nitrogen fixation; calcium cyanamide has been used as a fertilizer, releasing ammonia by the action of water.

Problem 24.24 Give the electronic structure of the cyanamide anion, $(NCN)^{2-}$. Discuss its molecular shape, bond lengths, and location of charge.

Problem 24.25 Give equations for the individual steps probably involved in the conversion of calcium cyanamide into ammonia in the presence of water. What other product or products will be formed in this process? Label each step with the name of the fundamental reaction type to which it belongs.

Problem 24.26 Cyanamide reacts with water in the presence of acid or base to yield urea; with methanol in the presence of acid to yield methylisourea, $H_2NC(\!=\!NH)OCH_3$; with hydrogen sulfide to yield *thiourea*, $H_2NC(\!=\!S)NH_2$; and with ammonia to yield *guanidine*, $H_2NC(\!=\!NH)NH_2$. (a) What functional group of cyanamide is involved in each of these reactions? (b) To what general class of reaction do these belong? (c) Show the most probable mechanisms for these reactions, pointing out the function of acid or base wherever involved.

24.24 Analysis of carboxylic acid derivatives. Saponification equivalent

Functional derivatives of carboxylic acids are recognized by their hydrolysis—under more or less vigorous conditions—to carboxylic acids. Just *which kind* of derivative it is is indicated by the other products of the hydrolysis.

Problem 24.27 Which kind (or kinds) of acid derivative: (a) rapidly forms a white precipitate (insoluble in HNO_3) upon treatment with alcoholic silver nitrate? (b) reacts with boiling aqueous NaOH to liberate a gas that turns moist litmus paper blue? (c) reacts immediately with cold NaOH to liberate a gas that turns moist litmus blue? (d) yields *only* a carboxylic acid upon hydrolysis? (e) yields an alcohol when heated with acid or base?

Identification or proof of structure of an acid derivative involves the identification or proof of structure of the carboxylic acid formed upon hydrolysis (Sec. 23.21). In the case of an ester, the alcohol that is obtained is also identified (Sec. 18.9). (In the case of a substituted amide, Sec. 27.7, the amine obtained is identified, Sec. 27.20.)

If an ester is hydrolyzed in a known amount of base (taken in excess), the amount of base used up can be measured and used to give the **saponification equivalent**: the equivalent weight of the ester, which is similar to the neutralization equivalent of an acid (see Sec. 23.21).

$$RCOOR' + OH^- \longrightarrow RCOO^- + R'OH$$

$$\underset{\substack{one \\ equivalent}}{} \quad \underset{\substack{one \\ equivalent}}{}$$

Problem 24.28 (a) What is the saponification equivalent of *n*-propyl acetate? (b) There are eight other simple aliphatic esters that have the same saponification equivalent. What are they? (c) In contrast, how many simple aliphatic acids have this equivalent weight? (d) Is saponification equivalent as helpful in identification as neutralization equivalent?

Problem 24.29 (a) How many equivalents of base would be used up by one mole of methyl phthalate, *o*-$C_6H_4(COOCH_3)_2$? What is the saponification equivalent of methyl phthalate? (b) What is the relation between saponification equivalent and the number of ester groups per molecule? (c) What is the saponification equivalent of glyceryl stearate (tristearoylglycerol)?

24.25 Spectroscopic analysis of carboxylic acid derivatives

Infrared   The infrared spectrum of an acyl compound shows the strong band in the neighborhood of 1700 cm^{-1} that we have come to expect of C=O stretching (see Fig. 24.2).

Figure 24.2   Infrared spectra of (*a*) methyl acetate and (*b*) benzamide.

The exact frequency depends on the family the compound belongs to (see Table 24.3) and, for a member of a particular family, on its exact structure. For esters, for example:

C=O stretching, *strong*

RCOOR 1740 cm^{-1} ArCOOR 1715–1730 cm^{-1} RCOOAr 1770 cm^{-1}
                             or   or

$$-\underset{|}{C}=\underset{|}{C}-COOR$$ $$RCOO\underset{|}{C}=\underset{|}{C}-$$

Table 24.3   INFRARED ABSORPTION BY SOME OXYGEN COMPOUNDS

Compound	O—H	C—O	C=O
Alcohols	3200–3600 cm^{-1}	1000–1200 cm^{-1}	—
Phenols	3200–3600	1140–1230	—
Ethers, aliphatic	—	1060–1150	—
Ethers, aromatic	—	1200–1275	—
		1020–1075	
Aldehydes, ketones	—	—	1675–1725 cm^{-1}
Carboxylic acids	2500–3000	1250	1680–1725
Esters	—	1050–1300	1715–1740
		(*two bands*)	
Acid chlorides	—	—	1750–1810
Amides (RCONH$_2$)	(N—H 3050–3550)	—	1650–1690

Esters are distinguished from acids by the absence of the O—H band. They are distinguished from ketones by two strong C—O stretching bands in the 1050–1300 cm^{-1} region; the exact position of these bands, too, depends on the ester's structure.

Besides the carbonyl band, amides (RCONH$_2$) show absorption due to N—H stretching in the 3050–3550 cm^{-1} region (the number of bands and their location depending on the degree of hydrogen bonding), and absorption due to N—H bending in the 1600–1640 cm^{-1} region.

NMR   As we can see in Table 16.4 (p. 585), the protons in the alkyl portion of an ester (RCOOCH$_2$R′) absorb farther downfield than the protons in the acyl portion (RCH$_2$COOR′).

Absorption by the —CO—NH protons of an amide appears in the range δ 5–8, typically as a broad, low hump.

CMR   The carbonyl carbon in these functional derivatives absorbs in the range δ 150–180, roughly the same region as for carboxylic acid.

PROBLEMS

1. Draw structures and give names of:

(a) nine isomeric esters of formula $C_5H_{10}O_2$
(b) six isomeric esters of formula $C_8H_8O_2$
(c) three isomeric methyl esters of formula $C_7H_{12}O_4$

2. Write balanced equations, naming all organic products, for the reaction (if any) of *n*-butyryl chloride with:

(a) H_2O
(b) isopropyl alcohol
(c) *p*-nitrophenol
(d) ammonia
(e) toluene, $AlCl_3$

(f) nitrobenzene, $AlCl_3$
(g) $NaHCO_3$(aq)
(h) alcoholic $AgNO_3$
(i) CH_3NH_2
(j) $(CH_3)_2NH$

(k) $(CH_3)_3N$
(l) $C_6H_5NH_2$
(m) $(C_6H_5)_2CuLi$
(n) C_6H_5MgBr

(Check your answers to (i) through (l) in Sec. 27.7.)

3. Answer Problem 2, parts (a) through (l), for acetic anhydride.

4. Write equations to show the reaction (if any) of succinic anhydride with:

(a) hot aqueous NaOH
(b) aqueous ammonia
(c) aqueous ammonia, then cold dilute HCl

(d) aqueous ammonia, then strong heat
(e) benzyl alcohol
(f) toluene, $AlCl_3$, heat

5. Write balanced equations, naming all organic products, for the reaction (if any) of phenylacetamide with: (a) hot HCl(aq), (b) hot NaOH(aq).

6. Answer Problem 5 for phenylacetonitrile.

7. Write balanced equations, naming all organic products, for the reaction (if any) of methyl *n*-butyrate with:

(a) hot H_2SO_4(aq)
(b) hot KOH(aq)
(c) isopropyl alcohol + H_2SO_4
(d) benzyl alcohol + $C_6H_5CH_2ONa$

(e) ammonia
(f) phenylmagnesium bromide
(g) isobutylmagnesium bromide
(h) $LiAlH_4$, then acid

8. Outline the synthesis of each of the following labeled compounds, using $H_2{}^{18}O$ as the source of ^{18}O.

$$\text{(a) } C_6H_5\overset{\displaystyle O}{\overset{\displaystyle \|}{-C}}-{}^{18}OCH_3 \qquad \text{(b) } C_6H_5\overset{\displaystyle {}^{18}O}{\overset{\displaystyle \|}{-C}}-OCH_3 \qquad \text{(c) } C_6H_5\overset{\displaystyle {}^{18}O}{\overset{\displaystyle \|}{-C}}-{}^{18}OCH_3$$

Predict the products obtained from each upon alkaline hydrolysis in ordinary H_2O.

9. Outline the synthesis of each of the following labeled compounds, using $^{14}CO_2$ or $^{14}CH_3OH$ and $H_2{}^{18}O$ as the source of the "tagged" atoms.

(a) $CH_3CH_2{}^{14}COCH_3$
(b) $CH_3CH_2CO^{14}CH_3$
(c) $CH_3{}^{14}CH_2COCH_3$
(d) $^{14}CH_3CH_2COCH_3$

(e) $C_6H_5{}^{14}CH_2CH_3$
(f) $C_6H_5CH_2{}^{14}CH_3$
(g) $CH_3CH_2C^{18}OCH_3$

10. Predict the product of the reaction of γ-butyrolactone with (a) ammonia, (b) $LiAlH_4$, (c) $C_2H_5OH + H_2SO_4$.

11. When *sec*-butyl alcohol of rotation $+13.8°$ was treated with tosyl chloride, and the resulting tosylate was allowed to react with sodium benzoate, there was obtained *sec*-butyl benzoate. Alkaline hydrolysis of this ester gave *sec*-butyl alcohol of rotation $-13.4°$. In which step must inversion have taken place? How do you account for this?

12. Account for the following observations.

$$C_6H_5-\underset{\underset{O}{\overset{\|}{OCR}}}{CH}-CH=CH-CH_3$$

optically active

$\xrightarrow{\text{5N NaOH}}$

$$C_6H_5-\underset{\overset{|}{OH}}{CH}-CH=CH-CH_3$$

complete retention

$\xrightarrow{\text{dil. NaOH}}$

$$C_6H_5-CH=CH-\underset{\overset{|}{OH}}{CH}-CH_3$$

inactive

$$C_6H_5-CH=CH-\underset{\underset{O}{\overset{\|}{OCR}}}{CH}-CH_3$$

optically active

$\xrightarrow{\text{dil. NaOH}}$

$$C_6H_5-CH=CH-\underset{\overset{|}{OH}}{CH}-CH_3$$

inactive

$\xrightarrow{\text{5N NaOH}}$

$$C_6H_5-CH=CH-\underset{\overset{|}{OH}}{CH}-CH_3$$

complete retention

13. An unknown compound is believed to be one of the following, all of which boil within a few degrees of each other. Describe how you would go about finding out which of the possibilities the unknown actually is. Where possible use simple chemical tests; where necessary use more elaborate chemical methods like quantitative hydrogenation, cleavage, neutralization equivalent, saponification equivalent, etc. Make use of any needed tables of physical constants.

benzyl acetate	isopropyl benzoate	methyl *o*-toluate	methyl *p*-toluate
ethyl benzoate	methyl phenylacetate	methyl *m*-toluate	

14. Describe simple chemical tests that would serve to distinguish between:

(a) propionic acid and methyl acetate
(b) *n*-butyryl chloride and *n*-butyl chloride
(c) *p*-nitrobenzamide and ethyl *p*-nitrobenzoate
(d) glyceryl tristearate and glyceryl trioleate
(e) benzonitrile and nitrobenzene
(f) acetic anhydride and *n*-butyl alcohol
(g) glyceryl monopalmitate and glyceryl tripalmitate
(h) ammonium benzoate and benzamide
(i) *p*-bromobenzoic acid and benzoyl bromide

Tell exactly what you would *do* and *see*.

15. Tell how you would separate by chemical means the following mixtures, recovering each component in reasonably pure form: (a) benzoic acid and ethyl benzoate; (b) *n*-valeronitrile and *n*-valeric acid; (c) ammonium benzoate and benzamide. Tell exactly what you would *do* and *see*.

16. Carboxyl groups are often masked by reaction with dihydropyran (Sec. 19.9), which yields esters that are stable toward base but easily hydrolyzed by dilute aqueous acids. Account in detail both for the formation of these esters and for their ease of hydrolysis.

$$\text{(structure)} + \text{RCOOH} \xrightarrow[\text{dioxane}]{\text{TsOH}} \text{RCOO}\text{(structure)}$$

Dihydropyran   Tetrahydropyranyl ester
(DHP)   (RCOOTHP)

17. Treatment of 2,4-pentanedione with KCN and acetic acid, followed by hydrolysis, gives two products, A and B. Both A and B are dicarboxylic acids of formula $C_7H_{12}O_6$. A melts at 98 °C. When heated, B gives first a lactonic acid ($C_7H_{10}O_5$, m.p. 90 °C) and finally a dilactone ($C_7H_8O_4$, m.p. 105 °C). (a) What structure must B have that permits ready formation of both a monolactone and a dilactone? (b) What is the structure of A? (*Hint*: Use models.)

18. For many years esters have sometimes been synthesized by the reaction of sodium carboxylates with alkyl halides, but the method has not been a particularly good one. Recently, however, a simple modification in the experimental procedure has been found to increase yields dramatically. Can you suggest what this change is likely to have been?

19. Give the structures (including configurations where pertinent) of components C through O.

(a) urea (H_2NCONH_2) + hot dilute NaOH $\longrightarrow$ C + NH_3
(b) phosgene ($COCl_2$) + 1 mol C_2H_5OH, then + NH_3 $\longrightarrow$ D ($C_3H_7O_2N$)
(c) bromobenzene + Mg, ether $\longrightarrow$ E (C_6H_5MgBr)
    E + ethylene oxide, followed by H^+ $\longrightarrow$ F ($C_8H_{10}O$)
    F + PBr_3 $\longrightarrow$ G (C_8H_9Br)
    G + NaCN $\longrightarrow$ H (C_9H_9N)
    H + H_2SO_4, H_2O, heat $\longrightarrow$ I ($C_9H_{10}O_2$)
    I + $SOCl_2$ $\longrightarrow$ J (C_9H_9OCl)
    J + anhydrous HF $\longrightarrow$ K (C_9H_8O)
    K + H_2, catalyst $\longrightarrow$ L ($C_9H_{10}O$)
    L + H_2SO_4, warm $\longrightarrow$ M (C_9H_8)
(d) *trans*-2-methylcyclohexanol + acetyl chloride $\longrightarrow$ N
    N + NaOH(aq) + heat $\longrightarrow$ O + sodium acetate

20. *Progesterone* is a hormone, secreted by the corpus luteum, that is involved in the control of pregnancy. Its structure was established, in part, by the following synthesis from the steroid *stigmasterol*, obtained from soybean oil.

Stigmasterol

stigmasterol ($C_{29}H_{48}O$) + ($CH_3CO)_2O$ $\longrightarrow$ P ($C_{31}H_{50}O_2$)
P + Br_2 $\longrightarrow$ Q ($C_{31}H_{50}O_2Br_2$)
Q + O_3, then Ag_2O $\longrightarrow$ R ($C_{24}H_{36}O_4Br_2$)
R + Zn/CH_3COOH $\longrightarrow$ S ($C_{24}H_{36}O_4$)
S + C_2H_5OH, H^+ $\longrightarrow$ T ($C_{26}H_{40}O_4$)
T + C_6H_5MgBr, then H_2O $\longrightarrow$ U ($C_{36}H_{46}O_3$)
U + acid, warm $\longrightarrow$ V ($C_{36}H_{44}O_2$)
V + Br_2; then CrO_3, H^+ $\longrightarrow$ W ($C_{23}H_{34}O_3Br_2$)
W + Zn/CH_3COOH $\longrightarrow$ X ($C_{23}H_{34}O_3$)
X + H_2O, H^+, heat $\longrightarrow$ Y ($C_{21}H_{32}O_2$), *pregnenolone*
Y + Br_2; then CrO_3, H^+ $\longrightarrow$ Z ($C_{21}H_{30}O_2Br_2$)
Z + Zn/CH_3COOH $\longrightarrow$ *progesterone* ($C_{21}H_{30}O_2$)

(a) Give structures for progesterone and the intermediates P–Z.
(b) Progesterone shows strong absorption in the near ultraviolet: λ_{max} 240 nm, ϵ_{max} 17 600. On this basis, what is the structure for progesterone?

21. On the basis of the following evidence assign structures to: (a) compounds AA to DD, isomers of formula $C_3H_8O_2$; (b) compounds EE to MM, isomers of formula $C_3H_6O_2$. (*Note*: α-Hydroxy ketones, —CHOH—CO—, give positive tests with Tollens' reagent and with Fehling's and Benedict's solutions (p. 1286), but negative Schiff's tests.)

	$NaHCO_3$	Acetic anhydride	Tollens'	Schiff's	HIO_4
(a) AA	–	$C_7H_{12}O_4$	–	–	–
BB	–	$C_7H_{12}O_4$	–	–	+
CC	–	$C_5H_{10}O_3$	–	–	–
DD	–		$-^1$	$-^1$	–
(b) EE	–	$C_5H_8O_3$	+	+	+
FF	–	$C_5H_8O_3$	+	–	+
GG	–	$C_5H_8O_3$	+	+	–
HH	CO_2		–	–	–
II	$-^2$		+	–	–
JJ	–		–	–	–
KK	–	$C_7H_{10}O_4$	–	–	+
LL	–		$-^1$	$-^1$	$-^1$
MM	–	$C_5H_8O_3$	–	–	$-^1$

[1] After treatment with dilute acid, solution gives positive test.
[2] After treatment with NaOH, solution gives positive iodoform test.

22. 2,5-Dimethyl-1,1-cyclopentanedicarboxylic acid can be prepared as a mixture of two optically inactive substances of different physical properties, NN and OO. When each is heated and the reaction mixture worked up by fractional crystallization, NN yields a single product, PP, of formula $C_8H_{14}O_2$, and OO yields two products, QQ and RR, both of formula $C_8H_{14}O_2$.

(a) Give stereochemical formulas for NN, OO, PP, QQ, and RR. (b) Describe another method by which you could assign configurations to NN and OO.

23. (a) (−)-*Erythrose*, $C_4H_8O_4$, gives tests with Tollens' reagent and Benedict's solution (p. 1286), and is oxidized by bromine water to an optically active acid, $C_4H_8O_5$. Treatment with acetic anhydride yields $C_{10}H_{14}O_7$. Erythrose consumes three moles of HIO_4 and yields three moles of formic acid and one mole of formaldehyde. Oxidation of erythrose by nitric acid yields an *optically inactive* compound of formula $C_4H_6O_6$.

(−)-*Threose*, an isomer of erythrose, shows similar chemical behavior except that nitric acid oxidation yields an *optically active* compound of formula $C_4H_6O_6$.

On the basis of this evidence what structure or structures are possible for (−)-erythrose? For (−)-threose?

(b) When *R*-glyceraldehyde, $CH_2OHCHOHCHO$, is treated with cyanide and the resulting product is hydrolyzed, two monocarboxylic acids are formed (see Problem 11, p. 790). These acids are identical with the acids obtained by oxidation with bromine water of (−)-threose and (−)-erythrose.

Assign a single structure to (−)-erythrose and to (−)-threose.

24. At room temperature, *N,N*-dimethylacetamide gives three sharp singlets of equal area in the proton NMR spectrum. As the temperature is raised, two of the peaks (but not the third) broaden and finally, at 110 °C, form one sharp singlet. (a) How do you account for this? What does it indicate about the structure of the amide? (b) What would you expect to see in the CMR spectrum of this compound at room temperature? At 110 °C?

25. Which (if any) of the following compounds could give rise to each of the infrared spectra shown in Fig. 24.3 (p. 896)?

ethyl acetate
ethyl acrylate (CH_2=$CHCOOC_2H_5$)
isobutyric acid

methacrylic acid [CH_2=$C(CH_3)COOH$]
methacrylamide [CH_2=$C(CH_3)CONH_2$]
phenylacetamide

26. Give a structure or structures consistent with each of the proton NMR spectra shown in Fig. 24.4 (p. 897).

27. Give a structure or structures consistent with the proton NMR spectrum shown in Fig. 24.5, p. 898.

28. Give a structure or structures consistent with each of the CMR spectra shown in Fig. 24.6, p. 898.

29. Give a structure or structures consistent with each of the CMR spectra shown in Fig. 24.7, p. 899.

30. Give the structure of compounds SS, TT, and UU on the basis of their infrared spectra (Fig. 24.8, p. 900) and their proton NMR spectra (Fig. 24.9, p. 901).

31. Give the structure of compound VV on the basis of its infrared, CMR, and proton NMR spectra shown in Fig. 24.10, p. 902.

32. Give a structure or structures consistent with each of the NMR spectra shown in Fig. 24.11, p. 903.

IRDC 353

IRDC 114

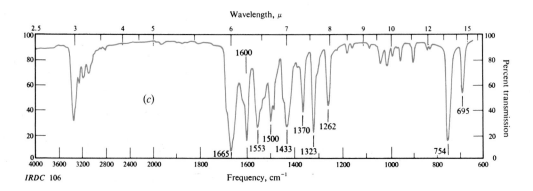

IRDC 106

Figure 24.3   Infrared spectra for Problem 25, p. 894.

Figure 24.4 Proton NMR spectra for Problem 26, p. 895.

$C_{10}H_{12}O_3$

Figure 24.5   Proton NMR spectrum for Problem 27, p. 895.

a	21.2 q
b	21.9 q
c	67.5 d
d	170.0 s

(a) $C_5H_{10}O_2$

Sadtler 2830C   © Sadtler Research Laboratories, Division of Bio-Rad, Inc., (1977).

a	22.3 t
b	27.9 t
c	68.9 t
d	178.2 s

(b) $C_4H_6O_2$

Sadtler 706C   © Sadtler Research Laboratories, Division of Bio-Rad, Inc., (1976).

Figure 24.6   CMR spectra for Problem 28, p. 895.

Sadtler 2062C   © Sadtler Research Laboratories, Division of Bio-Rad, Inc., (1977).

a 13.8 q	e 64.5 t
b 18.3 q	f 124.7 t
c 19.5 t	g 137.1 s
d 31.1 t	h 167.2 s

(a) $C_8H_{14}O_2$

a 21.1 q	d 31.9 t
b 23.9 t	e 72.5 s
c 25.7 t	f 170.1 s

(b) $C_8H_{14}O_2$

Sadtler 1149C   © Sadtler Research Laboratories, Division of Bio-Rad, Inc., (1976).

a	14.1 q
b	61.3 t
c	133.7 d
d	164.9 s

(c) $C_8H_{12}O_4$

Sadtler 154C   © Sadtler Research Laboratories, Division of Bio-Rad, Inc., (1976).

Figure 24.7   CMR spectra for Problem 29, p. 895.

Figure 24.8   Infrared spectra for Problem 30, p. 895.

Figure 24.9   Proton NMR spectra for Problem 30, p. 895.

Figure 24.10   Infrared, CMR, and proton NMR spectra for Problem 31, p. 895.

Figure 24.11 Proton NMR spectra for Problem 32, p. 895.

25

Carbanions I

Aldol and Claisen Condensations

$$-\overset{|}{\underset{H}{C}}-\overset{|}{\underset{O}{C}}- + :B \;\rightleftharpoons\; -\overset{|}{C}\!=\!\!\overset{|}{\underset{\underset{\ominus}{\overset{|}{O}}}{C}}- + B:H$$

I

25.1 Acidity of α-hydrogens

In our introduction to aldehydes and ketones, we learned that it is the carbonyl group that largely determines the chemistry of aldehydes and ketones. At that time, we saw in part how the carbonyl group does this: by providing a site at which nucleophilic addition can take place. Now we are ready to learn another part of the story: how the carbonyl group strengthens the acidity of the hydrogen atoms attached to the α-carbon and, by doing this, gives rise to a whole set of chemical reactions.

Ionization of an α-hydrogen,

$$-\overset{|}{\underset{H}{C}}-\overset{|}{\underset{O}{C}}- + :B \;\rightleftharpoons\; -\overset{|}{C}\!=\!\!\overset{|}{\underset{\underset{\ominus}{\overset{|}{O}}}{C}}- + B:H$$

I

yields a carbanion I that is a resonance hybrid of two structures, II and III,

$$\left[-\overset{|}{\underset{\underset{\ddot{O}:}{\cdot\cdot}}{C}}-\overset{|}{C}- \quad -\overset{|}{C}\!=\!\overset{|}{\underset{:\ddot{O}:^-}{C}}- \right] \quad \textit{equivalent to} \quad -\overset{|}{C}\!=\!\!\overset{|}{\underset{\underset{\ominus}{\overset{|}{O}}}{C}}-$$

II   III   I

resonance that is possible only through participation by the carbonyl group. Resonance of this kind is *not* possible for carbanions formed by ionization of β-hydrogens, γ-hydrogens, etc., from saturated carbonyl compounds.

Problem 25.1   Which structure, II or III, would you expect to make the larger contribution to the carbanion I? Why?

Problem 25.2   Account for the fact that the diketone acetylacetone (2,4-pentane-dione) is about as acidic as phenol, and much more acidic than, say, acetone. Which hydrogens are the most acidic?

Problem 25.3   How do you account for the following order of acidity?

$$(C_6H_5)_3CH > (C_6H_5)_2CH_2 > C_6H_5CH_3 > CH_4$$

The carbonyl group thus affects the acidity of α-hydrogens in just the way it affects the acidity of carboxylic acids: by helping to accommodate the negative charge of the anion.

I

Resonance in I involves structures (II and III) of quite different stabilities, and hence is much less important than the resonance involving equivalent structures in a carboxylate ion. Compared with the hydrogen of a —COOH group, the α-hydrogen atoms of an aldehyde or ketone are very weakly acidic; the important thing is that they are considerably more acidic than hydrogen atoms anywhere else in the molecule, and that they are acidic enough for *significant*—even though very low—concentrations of carbanions to be generated.

We call I a *carbanion* since it is the conjugate base of a *carbon acid*, that is, an acid which loses its proton from carbon. The stability that gives these ions their importance is due, however, to the very fact that most of the charge is carried *not* by carbon but by oxygen.

A carbanion like this, stabilized by an adjacent carbonyl group, is often called an *enolate anion*, since the anion is, formally, the conjugate base not only of the *keto* form of the carbonyl compound but of the *enol* form as well (Sec. 11.10). For example:

$$\text{CH}_3\text{CCH}_3 + :B \rightleftharpoons H:B + \text{CH}_3\text{C}\cdots\text{CH}_2 \rightleftharpoons \text{CH}_3\text{C}=\text{CH}_2 + :B$$

O	O⁻	OH
Keto	Enolate anion	Enol
Acid	*Conjugate base*	*Acid*

We saw before (Sec. 21.7) that the susceptibility of the carbonyl group to nucleophilic attack is due to the ability of oxygen to accommodate the negative charge that develops as a result of the attack,

$$Z: + \diagdown C=O \quad \longrightarrow \quad \left[-\overset{|}{\underset{Z}{C}}\overset{\delta}{=\!\!:\!\!O} \right] \quad \longrightarrow \quad -\overset{|}{\underset{Z}{C}}-O^-$$

precisely the same property of oxygen that underlies the acidity of α-hydrogens. We have started with two apparently unrelated chemical properties of carbonyl compounds and have traced them to a common origin—an indication of the simplicity underlying the seeming confusion of organic chemistry.

Problem 25.4 In the reaction of aqueous NaCN with an α,β-unsaturated ketone like

$$\underset{5}{CH_3}-\underset{4}{\overset{|}{\underset{CH_3}{C}}}=\underset{3}{CH}-\underset{2}{\overset{O}{\underset{\|}{C}}}-\underset{1}{CH_3}$$

CN⁻ adds, not to C-2, but to C-4. (a) How do you account for this behavior? (b) What product would you expect to isolate from the reaction mixture? (*Hint*: See Secs. 21.11 and 10.26.) (Check your answers in Sec. 31.5.)

25.2 Reactions involving carbanions

The carbonyl group occurs in compounds other than aldehydes and ketones—in esters, for example—and, wherever it is, it makes any α-hydrogens acidic and thus aids in formation of carbanions. Since these α-hydrogens are only weakly acidic, however, the carbanions are highly basic, exceedingly reactive particles. In their reactions they behave as we would expect: as *nucleophiles*. As nucleophiles, carbanions can attack carbon and, in doing so, form carbon–carbon bonds. *From the standpoint of synthesis, acid-strengthening by carbonyl groups is probably the most important structural effect in organic chemistry.*

We shall take up first the behavior of ketones toward the halogens, and see evidence that carbanions do indeed exist; at the same time, we shall see an elegant example of the application of kinetics, stereochemistry, and isotopic tracers to the understanding of reaction mechanisms. And while we are at it, we shall see something of the role that keto–enol tautomerism plays in the chemistry of carbonyl compounds.

Next, we shall turn to reactions in which the carbonyl group plays *both* its roles: the *aldol condensation*, in which a carbanion generated from one molecule of aldehyde or ketone adds, as a nucleophile, to the carbonyl group of a second molecule; and the *Claisen condensation*, in which a carbanion generated from one molecule of ester attacks the carbonyl group of a second molecule, with acyl substitution as the final result.

But, as we know, nucleophiles can attack not only carbonyl carbon but also the carbon of alkyl halides and related compounds, to bring about nucleophilic aliphatic substitution (Chap. 5). Carbanions can do this, too, as we shall see in the *malonic ester* and *acetoacetic ester syntheses* (Chap. 30). Then, in the *Michael addition* (Sec. 31.7), we shall find carbanions undergoing—as other nucleophiles do—nucleophilic conjugate addition to α,β-unsaturated carbonyl compounds (Chap. 31).

___ REACTIONS INVOLVING CARBANIONS ___

1. **Halogenation of ketones.** Discussed in Secs. 25.3–25.4.

$$-\overset{|}{\underset{H}{C}}-\overset{|}{\underset{\parallel}{C}}- + X_2 \xrightarrow{H^+ \text{ or } OH^-} -\overset{|}{\underset{X}{C}}-\overset{|}{\underset{\parallel}{C}}- + HX \quad X_2 = Cl_2, Br_2, I_2$$

Ketone   α-Halo ketone

Examples:

Cyclohexanone   2-Bromocyclohexanone

$$CH_3-\overset{CH_3}{\underset{CH_3}{C}}-\overset{O}{\underset{\parallel}{C}}-CH_3 + I_2 + OH^- \longrightarrow \left[CH_3-\overset{CH_3}{\underset{CH_3}{C}}-\overset{O}{\underset{\parallel}{C}}-CI_3 \right] \xrightarrow{OH^-}$$

tert-Butyl methyl ketone
(3,3-Dimethyl-2-butanone)

$$CH_3-\overset{CH_3}{\underset{CH_3}{C}}-COO^- + CHI_3$$

Iodoform

Trimethylacetate ion

2. **Nucleophilic addition to carbonyl compounds.**

   (a) **Aldol condensation.** Discussed in Secs. 25.5–25.8.

An aldol
(A β-hydroxy carbonyl compound)

Examples:

Acetaldehyde   Acetaldol   Crotonaldehyde
2 moles   (3-Hydroxybutanal)   (2-Butenal)

Acetone   Diacetone alcohol   Mesityl oxide
2 moles   (4-Methyl-3-penten-2-one)

_____ CONTINUED _____

CONTINUED

Benzaldehyde   Acetaldehyde

Cinnamaldehyde
(3-Phenyl-2-propenal)

Benzaldehyde   Acetone

Benzalacetone
(*Benzal* is $C_6H_5CH=$)
(4-Phenyl-3-buten-2-one)

Benzaldehyde   Acetophenone

Benzalacetophenone
(1,3-Diphenyl-2-propen-1-one)

(b) **Reactions related to aldol condensation.** Discussed in Sec. 25.9.

(c) Addition of Grignard reagents. Discussed in Sec. 21.10.

(d) **Wittig reaction.** Discussed in Sec. 25.10.

An ylide

A betaine

Examples:

$C_6H_5CH=CHCHO$ + $Ph_3P=CH_2$ $\longrightarrow$ $C_6H_5CH=CH-\overset{H}{\underset{O^{\ominus}}{C}}-\overset{H}{\underset{\oplus PPh_3}{CH}}$ $\longrightarrow$

Cinnamaldehyde   Methylenetriphenylphosphorane

$C_6H_5CH=CH-CH=CH_2$
1-Phenyl-1,3-butadiene
(*69%*)

Cyclohexanone

Methylenecyclohexane
(*48%*)

CONTINUED

CONTINUED

3. **Nucleophilic acyl substitution.**

(a) **Claisen condensation.** Discussed in Secs. 25.11–25.12.

A β-keto ester

Examples:

Ethyl acetate
2 moles

Ethyl acetoacetate

Ethyl benzoate   Ethyl acetate

Ethyl benzoylacetate

(b) Acylation of organocopper compounds. Discussed in Sec. 21.6.

4. **Nucleophilic aliphatic substitution.**

(a) **Coupling of alkyl halides with organometallic compounds.** Discussed in Sec. 3.17.
(b) **Synthesis of acetylides.** Discussed in Sec. 11.13.
(c) **Alkylation of malonic ester and acetoacetic ester.** Discussed in Secs. 30.2–30.3.

5. **Addition to α,β-unsaturated carbonyl compounds. Michael addition.** Discussed in Sec. 31.7. ∎

25.3 Base-promoted halogenation of ketones

Acetone reacts with bromine to form bromoacetone; the reaction is accelerated by bases (e.g., hydroxide ion, acetate ion, etc.). Study of the kinetics shows that

$$CH_3COCH_3 + Br_2 + :B \longrightarrow CH_3COCH_2Br + Br^- + H:B$$

Acetone   Bromoacetone

the rate of reaction depends upon the concentration of acetone, [acetone], and of base, [:B], but is *independent of bromine concentration*:

$$\text{rate} = k\,[\text{acetone}][:B]$$

We have encountered this kind of situation before (Sec. 5.16) and know, in a general way, what it must mean: if the rate of reaction does not depend upon [Br_2], it can only mean that the reaction *whose rate we are measuring* does not involve Br_2.

The kinetics is quite consistent with the following mechanism. The base slowly abstracts a proton (step 1) from acetone to form carbanion I, which then reacts rapidly with bromine (step 2) to yield bromoacetone. Step (1), generation of the carbanion, is the rate-determining step, since its rate determines the overall rate of the reaction sequence. As fast as carbanions are generated, they are snapped up by bromine molecules.

(1) $CH_3CCH_3 + :B$ $\rightleftharpoons$ $H:B + CH_3C\text{---}CH_2$ **Slow**: *rate-determining*
           $\overset{\|}{O}$ $\overset{\|}{O}$

                                                                              I

(2) $CH_3C\text{---}CH_2 + Br_2 \longrightarrow CH_3CCH_2Br + Br^-$ **Fast**
      $\overset{\|}{O}$ $\overset{\|}{O}$

      I

Strong support for this interpretation comes from the kinetics of iodination. Here, too, the rate of reaction depends upon [acetone] and [:B] but is independent of [I_2]. Furthermore, and most significant, at a given [acetone] and [:B], bromination and iodination *proceed at identical rates*. That is, in the rate expression

$$\text{rate} = k\,[\text{acetone}][:\text{B}]$$

the value of k is the same regardless of which halogen is involved. It *should* be, of course, according to the proposed mechanism, since in both cases it is the rate constant for the same reaction, abstraction of a proton from the ketone.

Study of the bromination of acetone, done by A. Lapworth (of the University of Manchester) in 1904, showed for the first time how kinetics could be used to reveal the mechanism of an organic reaction. The carbanion mechanism has since been confirmed not only by the iodination work, but also by studies of stereochemistry and isotopic exchange.

Problem 25.5 Show in detail exactly how each of the following facts provides evidence for the carbanion mechanism of base-promoted halogenation of ketones.

(a) In basic solution, (+)-*sec*-butyl phenyl ketone undergoes racemization; the rate constant for loss of optical activity is identical with the rate constant for bromination of this ketone.

(b) Ketone II undergoes racemization in basic solution, but ketone III does not.

                  CH_3 CH_3
                   | |
      $C_6H_5-C-CH-C_6H_5$ $C_6H_5-C-C-C_6H_5$
                  $\overset{\|}{O}$ $\overset{\|}{O}$ $\overset{|}{C_4H_9}$-n

              II   III

(c) When (+)-*sec*-butyl phenyl ketone is allowed to stand in D_2O containing OD^-, it not only undergoes racemization, but also becomes labeled with deuterium at the α-position; the rate constants for racemization and hydrogen exchange are identical.

Problem 25.6 (a) Suggest a mechanism for the base-catalyzed racemization of the optically active ester, ethyl mandelate, $C_6H_5CHOHCOOC_2H_5$.

(b) How do you account for the fact that optically active mandelic acid undergoes racemization in base *much more slowly* than the ester? (*Hint*: See Sec. 23.20.)

(c) What would you predict about the rate of base-catalyzed racemization of α-methylmandelic acid, $C_6H_5C(CH_3)(OH)COOH$?

Problem 25.7 Suppose, as an alternative to the carbanion mechanism, that hydrogen exchange and racemization were both to arise by some kind of direct displacement of one hydrogen (H) by another (D) with inversion of configuration. What relationship would you then expect between the rates of racemization and exchange? (*Hint*: Take one molecule at a time, and see what happens when H is replaced by D with inversion.)

25.4 Acid-catalyzed halogenation of ketones. Enolization

Acids, like bases, speed up the halogenation of ketones. Acids are not, how-ever, consumed, and hence we may properly speak of acid-*catalyzed* halogenation (as contrasted to base-*promoted* halogenation). Although the reaction is not, strictly

$$CH_3COCH_3 + Br_2 \xrightarrow{\ \text{acid}\ } CH_3COCH_2Br + HBr$$

Acetone   Bromoacetone

speaking, a part of carbanion chemistry, this is perhaps the best place to take it up, since it shows a striking parallel in every aspect to the base-promoted reaction we have just left.

Here, too, the kinetics shows the rate of halogenation to be independent of halogen concentration, but dependent upon ketone concentration and, this time, acid concentration. Here, too, we find the remarkable identity of rate constants for apparently different reactions: for bromination and iodination of acetone, and exchange of its hydrogens for deuterium; for iodination and racemization of *sec*-butyl phenyl ketone.

The interpretation, too, is essentially the same as the one we saw before: *preceding* the step that involves halogen, there is a rate-determining reaction that can lead not only to halogenation but also to racemization and to hydrogen exchange.

The rate-determining reaction here is the formation of the *enol*, which involves two steps: rapid, reversible protonation (step 1) of the carbonyl oxygen, followed by the slow loss of an α-hydrogen (step 2).

(1) $CH_3{-}\underset{\underset{\displaystyle O}{\|}}{C}{-}CH_3 + H{:}B \ \rightleftarrows \ CH_3{-}\underset{\underset{\displaystyle _+OH}{\|}}{C}{-}CH_3 + {:}B$ Fast

(2) $CH_3{-}\underset{\underset{\displaystyle _+OH}{\|}}{C}{-}CH_3 + {:}B \ \longrightarrow \ CH_3{-}\underset{\underset{\displaystyle OH}{|}}{C}{=}CH_2 + H{:}B$ Slow

                                                                Enol

(3) $CH_3{-}\underset{\underset{\displaystyle OH}{|}}{C}{=}CH_2 + X_2 \ \longrightarrow \ CH_3{-}\underset{\underset{\displaystyle _+OH}{\|}}{C}{-}CH_2X + X^-$ Fast

                                                I

(4)　　　CH₃—C—CH₂X + :B ⇌ CH₃—C—CH₂X + H:B　　Fast

Once formed, the enol reacts rapidly with halogen (step 3). We might have expected the unsaturated enol to undergo addition and, indeed, the reaction starts out exactly as though this were going to happen: positive halogen attaches itself to form a cation. As usual (Sec. 8.12), attachment occurs in the way that yields the more stable cation.

The ion formed in this case, I, is an exceedingly stable one, owing its stability to the fact that it is hardly a "carbocation" at all, since oxygen can carry the charge and still have an octet of electrons. The ion is, actually, a protonated ketone; loss of the proton yields the product, bromoacetone.

We may find it odd, considering that we call this reaction "acid-catalyzed", that the rate-determining step (2) is really the same as in the base-promoted reaction: abstraction of an α-hydrogen by a base—here, by the conjugate base of the catalyzing acid. Actually, what we see here must always hold true: a reaction that is truly *catalyzed* by acid or base is catalyzed by *both acid and base*. In our case, transfer of the proton from the acid H:B to carbonyl oxygen (step 1) makes the ketone more reactive and hence speeds up enolization. But, if this is truly catalysis, the acid must not be *consumed*. Regeneration of the acid H:B requires that the conjugate base :B get a proton from somewhere; it takes it from the α-carbon (step 2), and thus completes the enolization. Both acid and base speed up the rate-determining step (2): base directly, as one of the reactants, and acid indirectly, by increasing the concentration of the other reactant, the protonated ketone. Using a strong mineral acid in aqueous solution, we would not be aware of the role played by the base; the acid is H_3O^+ and the conjugate base, H_2O, is the solvent.

Problem 25.8 Show in detail how the enolization mechanism accounts for the following facts: (a) the rate constants for acid-catalyzed hydrogen–deuterium exchange and bromination of acetone are identical; (b) the rate constants for acid-catalyzed racemization and iodination of *sec*-butyl phenyl ketone are identical.

Problem 25.9 (a) In the acid-catalyzed dehydration of alcohols (Sec. 7.25), what is the base involved? (b) In the base-catalyzed racemization and hydrogen exchange of *sec*-butyl phenyl ketone (Problem 25.5, p. 911), what is the acid involved?

25.5 Aldol condensation

Under the influence of dilute base or dilute acid, two molecules of an aldehyde or a ketone may combine to form a β-hydroxy aldehyde or β-hydroxy ketone. This reaction is called the **aldol condensation**. In every case the product results from addition of one molecule of aldehyde (or ketone) to a second molecule in such a way that the α-carbon of the first becomes attached to the carbonyl carbon of the second. For example:

$$CH_3—C=O + H—C—C=C \xrightarrow{\ OH^-\ } CH_3—C—C—C=O$$

Acetaldehyde　　　　　　　　　Aldol

2 moles　　　　　　　　β-Hydroxybutyraldehyde

(3-Hydroxybutanal)

$$\underset{\substack{\text{Propionaldehyde} \\ \text{2 moles}}}{\text{CH}_3\text{CH}_2\overset{\displaystyle H}{\underset{\displaystyle H}{\text{C}}}=\text{O} + \text{CH}_3\overset{\displaystyle H}{\underset{\displaystyle H}{\text{C}}}\overset{\displaystyle H}{\text{C}}=\text{O}} \xrightarrow{\text{OH}^-} \underset{\substack{\beta\text{-Hydroxy-}\alpha\text{-methylvaleraldehyde} \\ \text{(3-Hydroxy-2-methylpentanal)}}}{\text{CH}_3\text{CH}_2\overset{\displaystyle H}{\underset{\displaystyle \text{OH}}{\text{C}}}\overset{\displaystyle \text{CH}_3}{\underset{\displaystyle H}{\text{C}}}\overset{\displaystyle H}{\text{C}}=\text{O}}$$

$$\underset{\substack{\text{Acetone} \\ \text{2 moles}}}{\text{CH}_3\overset{\displaystyle \text{CH}_3}{\text{C}}=\text{O} + \text{H}\overset{\displaystyle H}{\underset{\displaystyle H}{\text{C}}}\overset{\displaystyle}{\underset{\displaystyle O}{\text{C}}}\overset{\displaystyle H}{\underset{\displaystyle H}{\text{C}}}\text{H}} \xrightarrow{\text{OH}^-} \underset{\substack{\text{4-Hydroxy-4-methyl-2-pentanone} \\ \text{(Diacetone alcohol)}}}{\text{CH}_3\overset{\displaystyle \text{CH}_3}{\underset{\displaystyle \text{OH}}{\text{C}}}\overset{\displaystyle H}{\underset{\displaystyle H}{\text{C}}}\overset{\displaystyle}{\underset{\displaystyle O}{\text{C}}}\overset{\displaystyle H}{\underset{\displaystyle H}{\text{C}}}\text{H}}$$

If the aldehyde or ketone does not contain an α-hydrogen, a simple aldol condensation cannot take place. For example:

$$\left.\begin{array}{l} \text{ArCHO} \\ \text{HCHO} \\ (\text{CH}_3)_3\text{CCHO} \\ \text{ArCOAr} \\ \text{ArCOCR}_3 \end{array}\right\} \xrightarrow{\text{dilute OH}^-} \text{no reaction}$$

No
α-hydrogen
atoms

(In concentrated base, however, such aldehydes may undergo the Cannizzaro reaction, Sec. 21.14.)

The generally accepted mechanism for the base-catalyzed condensation involves the following steps, acetaldehyde being used as an example. Hydroxide ion

(1) $$\text{CH}_3\text{CHO} + \text{OH}^- \underset{\text{catalyst}}{\overset{\displaystyle\longrightarrow}{\underset{\text{Basic}}{\rightleftharpoons}}} \text{H}_2\text{O} + \underset{\text{I}}{[\text{CH}_2\text{CHO}]^-}$$

(2) $$\underset{\substack{\text{I} \\ \textit{Nucleophilic} \\ \textit{reagent}}}{\text{CH}_3\overset{\displaystyle H}{\text{C}}=\text{O} + [\text{CH}_2\text{CHO}]^-} \rightleftharpoons \underset{\text{II}}{\text{CH}_3\overset{\displaystyle H}{\underset{\displaystyle \text{O}_-}{\text{C}}}\text{CH}_2\text{CHO}}$$

(3) $$\underset{\text{II}}{\text{CH}_3\overset{\displaystyle H}{\underset{\displaystyle \text{O}_-}{\text{C}}}\text{CH}_2\text{CHO} + \text{H}_2\text{O}} \rightleftharpoons \underset{\text{III}}{\text{CH}_3\overset{\displaystyle H}{\underset{\displaystyle \text{OH}}{\text{C}}}\text{CH}_2\text{CHO} + \text{OH}^-}$$

abstracts (step 1) a hydrogen ion from the α-carbon of the aldehyde to form carbanion I, which attacks (step 2) carbonyl carbon to form ion II. Ion II (an alkoxide) abstracts (step 3) a hydrogen ion from water to form the β-hydroxy-aldehyde III, regenerating hydroxide ion. The purpose of hydroxide ion is thus to produce the carbanion I, which is the actual nucleophilic reagent.

Problem 25.10   Illustrate these steps for:

    (a) propionaldehyde   (d) cyclohexanone
    (b) acetone   (e) phenylacetaldehyde
    (c) acetophenone

Problem 25.11   The aldol condensation of unsymmetrical ketones (ethyl methyl ketone, for example) is usually of little value in synthesis. Why do you think this is so?

The carbonyl group plays two roles in the aldol condensation. It not only provides the unsaturated linkage at which addition (step 2) occurs, but also makes the α-hydrogens acidic enough for carbanion formation (step 1) to take place.

Problem 25.12   In *acid-catalyzed aldol condensations*, acid is believed to perform two functions: to catalyze conversion of carbonyl compound into the enol form, and to provide protonated carbonyl compound with which the enol can react. The reaction that then takes place can, depending upon one's point of view, be regarded either as acid-catalyzed nucleophilic addition to a carbonyl group, or as electrophilic addition to an alkene. On this basis, write all steps in the mechanism of acid-catalyzed aldol condensation of acetaldehyde. In the actual *condensation* step, identify the nucleophile and the electrophile.

Problem 25.13   (a) When acetaldehyde at fairly high concentration was allowed to undergo base-catalyzed aldol condensation in heavy water (D_2O), the product was found to contain almost no deuterium bound to carbon. This finding has been taken as one piece of evidence that the slow step in this aldol condensation is formation of the carbanion. How would you justify this conclusion? (b) The kinetics also supports this conclusion. What kinetics would you expect if this were the case? (*Remember:* Two molecules of acetaldehyde are involved in aldol condensation.) (c) When the experiment in part (a) was carried out at low acetaldehyde concentration, the product was found to contain considerable deuterium bound to carbon. How do you account for this? (*Hint*: See Sec. 7.18.) (d) In contrast to acetaldehyde, acetone was found to undergo base-catalyzed hydrogen–deuterium exchange much faster than aldol condensation. What is one important factor contributing to this difference in behavior?

Problem 25.14   In alkaline solution, 4-hydroxy-4-methyl-2-pentanone is partly converted into acetone. What does this reaction amount to? Show all steps in the most likely mechanism. (*Hint*: See Sec. 7.25.)

25.6   Dehydration of aldol products

The β-hydroxy aldehydes and β-hydroxy ketones obtained from aldol condensations are very easily dehydrated; the major products have the carbon–carbon double bond between the α- and β-carbon atoms. For example:

                Aldol   Crotonaldehyde
                                          (2-Butenal)

$$CH_3-\underset{\underset{\displaystyle HO}{|}}{C}-\underset{\underset{\displaystyle H}{|}}{CH}-\underset{\underset{\displaystyle O}{\|}}{C}-CH_3 \xrightarrow{\text{I}_2 \text{ (a Lewis acid), distill}} CH_3-\underset{CH_3}{\overset{|}{C}}=CH-\underset{\underset{\displaystyle O}{\|}}{C}-CH_3 + H_2O$$

Diacetone alcohol
(4-Hydroxy-4-methyl-2-pentanone)

Mesityl oxide
(4-Methyl-3-penten-2-one)

Both the ease and the orientation of elimination are related to the fact that the alkene obtained is a particularly stable one, since the carbon–carbon double bond is conjugated with the carbon–oxygen double bond of the carbonyl group (compare Sec. 10.21).

> **Problem 25.15** Draw resonance structures to account for the unusual stability of an α,β-unsaturated aldehyde or ketone. What is the significance of these structures in terms of orbitals? (See Sec. 10.22.)

As we know, an alkene in which the carbon–carbon double bond is conjugated with an aromatic ring is particularly stable (Sec. 15.19); in those cases where elimination of water from the aldol product can form such a conjugated alkene, the unsaturated aldehyde or ketone is the product actually isolated from the reaction. For example:

Acetophenone
2 moles

1,3-Diphenyl-2-buten-1-one

25.7 Use of aldol condensation in synthesis

Catalytic hydrogenation of α,β-unsaturated aldehydes and ketones yields saturated alcohols, addition of hydrogen occurring both at carbon–carbon and at carbon–oxygen double bonds. It is for the purpose of ultimately preparing saturated alcohols that the aldol condensation is often carried out. For example, *n*-butyl alcohol and 2-ethyl-1-hexanol are both prepared on an industrial scale in this way:

$$2CH_3CHO \xrightarrow{OH^-} CH_3CHOH—CH_2CHO \xrightarrow{-H_2O} CH_3CH=CHCHO$$

Acetaldehyde   Aldol   Crotonaldehyde
                                                                      (2-Butenal)

$$\downarrow H_2, Ni$$

$$CH_3CH_2—CH_2CH_2OH$$
n-Butyl alcohol

$$CH_3CH_2CH_2CHO \xrightarrow{OH^-} CH_3CH_2CH_2CHOH—\underset{\underset{C_2H_5}{|}}{C}HCHO \longrightarrow \downarrow -H_2O$$

n-Butyraldehyde

$$\uparrow Cu,\ 250\ ^\circ C$$

$$CH_3CH_2CH_2CH_2OH$$
n-Butyl alcohol

$$CH_3CH_2CH_2CH=\underset{\underset{C_2H_5}{|}}{C}CHO$$

$$\downarrow H_2, Ni$$

$$CH_3CH_2CH_2CH_2—\underset{\underset{C_2H_5}{|}}{C}HCH_2OH$$

2-Ethyl-1-hexanol

To prepare an unsaturated alcohol from an α,β-unsaturated aldehyde or ketone, we need a reagent that reduces only the carbonyl group and leaves the carbon–carbon double bond intact. A reagent that, in performing its particular job, selectively attacks one of several different functional groups is called a *chemoselective* reagent. It is a major aim of synthetic chemistry today to find highly selective reagents—regioselective, stereoselective, chemoselective—and nowhere is that aim more evident than in the development of oxidizing and reducing agents. The particular job facing us here can be done by the hydroborane known

$$RCH=CH—\underset{\underset{O}{\|}}{C}—R' + H—B \longrightarrow RCH=CH—\underset{\underset{O—B}{|}}{\overset{\overset{H}{|}}{C}}—R'$$

Unsaturated   9-BBN
carbonyl compound   (Sec. 30.7)

$$\downarrow HOCH_2CH_2NH_2$$

$$RCH=CH—\underset{\underset{OH}{|}}{C}H—R' + NH_2CH_2CH_2O—B$$

Unsaturated
alcohol

as 9-BBN. Just as boranes add to carbon–carbon double bonds (Sec. 17.10), so this one adds to the carbonyl double bond; and, evidently because of the bulky organic group attached to boron, it does this very much faster than it adds to the carbon–carbon double bond. The result is an extraordinarily high degree of selectivity. (We shall encounter 9-BBN again, in Sec. 30.7, involved in a synthesis of a quite different kind: another example of the extreme versatility of the organoboranes.)

Problem 25.16 Outline the synthesis of the following alcohols starting from alcohols of smaller carbon number:

(a) 2-methyl-1-pentanol (d) 2,4-diphenyl-1-butanol
(b) 4-methyl-2-pentanol (e) 4-methyl-3-penten-2-ol
(c) 2-cyclohexylcyclohexanol

Problem 25.17 The insect repellent "6-12" (2-ethyl-1,3-hexanediol) is produced by the same chemical company that produces *n*-butyl alcohol and 2-ethyl-1-hexanol; suggest a method for its synthesis. How could you synthesize 2-methyl-2,4-pentanediol?

Problem 25.18 The reagent 9-BBN has the structure shown below. It is made by the reaction of diborane with a diene. Can you suggest a possible structure for this diene?

B—H

9-Borabicyclo[3.3.1]nonane
9-BBN

25.8 Crossed aldol condensation

An aldol condensation between two different carbonyl compounds—a so-called **crossed aldol condensation**—is not always feasible in the laboratory, since a mixture of the four possible products may be obtained. On a commercial scale, however, such a synthesis may be worthwhile if the mixture can be separated and the components marketed.

Under certain conditions, a good yield of a single product can be obtained from a crossed aldol condensation: (a) one reactant contains no α-hydrogens and therefore is incapable of condensing with itself (e.g., aromatic aldehydes or formaldehyde); (b) this reactant is mixed with the catalyst; and then (c) a carbonyl

$$\underset{\text{Benzaldehyde}}{\underset{\text{H}}{\overset{}{\bigcirc}-\overset{}{\underset{}{C}}=O}} \xrightarrow{\quad OH^- \quad}$$

CH₃CHO, 20 °C →

H H H
⬡—C=C—C=O
Cinnamaldehyde

CH₃COCH₃, 100 °C →

H H
⬡—C=C—C—CH₃
      ‖
      O
Benzalacetone

CH₃COC₆H₅, 20 °C →

H H
⬡—C=C—C—⬡
      ‖
      O
Benzalacetophenone

Crossed aldol condensations

compound that contains α-hydrogens is added slowly to this mixture. There is thus present at any time only a very low concentration of the ionizable carbonyl compound, and the carbanion it forms reacts almost exclusively with the other carbonyl compound, which is present in large excess.

Problem 25.19 Outline the synthesis of each of the following from benzene or toluene and any readily available alcohols:

(a) 4-phenyl-2-butanol
(b) 1,3-diphenyl-1-propanol
(c) 1,3-diphenylpropane

(d) 2,3-diphenyl-1-propanol
(e) 1,5-diphenyl-1,4-pentadien-3-one
   (dibenzalacetone)

Problem 25.20 (a) What prediction can you make about the acidity of the γ-hydrogens of α,β-unsaturated carbonyl compounds,

$$\overset{\gamma}{-\underset{|}{\overset{|}{C}}}-\overset{\beta}{\underset{|}{C}}=\overset{\alpha}{\underset{|}{C}}-C=O$$
$$\qquad\quad H$$

as, for example, in crotonaldehyde? (b) In view of your answer to (a), suggest a way to synthesize 5-phenyl-2,4-pentadienal, $C_6H_5CH=CH-CH=CH-CHO$.

25.9 Reactions related to the aldol condensation

There are a large number of condensations that are closely related to the aldol condensation. Each of these reactions has its own name—*Perkin, Knoevenagel, Doebner, Claisen, Dieckmann*, for example—and at first glance each may seem quite different from the others. Closer examination shows, however, that like the aldol condensation each of these involves attack by a carbanion on a carbonyl group. In each case the carbanion is generated in very much the same way: the abstraction by base of a hydrogen ion *alpha* to a carbonyl group. Different bases may be used—sodium hydroxide, sodium ethoxide, sodium acetate, amines—and the carbonyl group to which the hydrogen is *alpha* may vary—aldehyde, ketone, anhydride, ester—but the chemistry is essentially the same as that of the aldol condensation. We shall take up a few of these condensations in the following problems and in following sections; in doing this, we must not lose sight of the fundamental resemblance of each of them to the aldol condensation.

Problem 25.21 Esters can be condensed with aromatic aldehydes in the presence of alkoxides; thus benzaldehyde and ethyl acetate, in the presence of sodium ethoxide, give ethyl cinnamate, $C_6H_5CH=CHCOOC_2H_5$. Show all steps in the most likely mechanism for this condensation.

Problem 25.22 Account for the following reactions:

(a) $C_6H_5CHO + CH_3NO_2 \xrightarrow{\text{KOH}} C_6H_5CH=CHNO_2 + H_2O$

(b) $C_6H_5CHO + C_6H_5CH_2CN \xrightarrow{\text{NaOC}_2H_5} C_6H_5CH=\underset{\underset{C_6H_5}{|}}{C}-CN + H_2O$

(c) $C_6H_5CHO + CH_3$⟨NO$_2$⟩NO$_2$ $\xrightarrow{2° \text{ amine}}$ $C_6H_5CH=CH$⟨NO$_2$⟩NO$_2$ + H$_2$O

(d) $CH_3CHO + NaC\equiv CH$ $\xrightarrow{NH_3(l)}$ $CH_3\underset{\underset{ONa}{|}}{C}HC\equiv CH$ $\xrightarrow{NH_4Cl}$ $CH_3\underset{\underset{OH}{|}}{C}HC\equiv CH$

(e) **A Perkin condensation:**

$C_6H_5CHO + (CH_3CO)_2O$ $\xrightarrow{CH_3COONa}$ $C_6H_5CH=CHCOOH$

                Acetic anhydride   Cinnamic acid

(f) **A Knoevenagel reaction:**

$C_6H_5CHO + CH_2(COOC_2H_5)_2$ $\xrightarrow{2° \text{ amine}}$ $C_6H_5CH=C(COOC_2H_5)_2$

(g) **A Cope reaction:**

⟨⟩=O + N$\equiv$CCH$_2$COOC$_2$H$_5$ $\xrightarrow{CH_3COONH_4}$ ⟨⟩=$\underset{\underset{CN}{|}}{C}$—COOC$_2H_5$

25.10 The Wittig reaction

In 1954, Georg Wittig (then at the University of Tübingen) reported a method of synthesizing alkenes from carbonyl compounds, which amounts to the replacement of carbonyl oxygen, =O, by the group =CRR'. The heart of the synthesis is

$$\underset{}{\overset{}{\diagdown}}C=O + Ph_3P=\underset{\underset{R}{|}}{\overset{\overset{R'}{|}}{C}}—R \longrightarrow -\underset{\underset{O}{|}}{C}—\underset{\underset{PPh_3}{|}}{\overset{\overset{R'}{|}}{C}}—R \longrightarrow -C=\overset{\overset{R'}{|}}{C}—R + Ph_3PO$$

          An ylide   Triphenylphosphine oxide

                           A betaine

the nucleophilic attack on carbonyl carbon by an *ylide* to form a *betaine* which—often spontaneously—undergoes elimination to yield the product. For example:

$(C_6H_5)_2C=O + Ph_3P=CH_2$ $\longrightarrow$ $(C_6H_5)_2\underset{\underset{O}{|}}{C}—\underset{\underset{PPh_3}{|}}{C}H_2$ $\longrightarrow$ $(C_6H_5)_2C=CH_2$

Benzophenone   Methylenetriphenyl-phosphorane   1,1-Diphenylethene

$C_6H_5CHO + C_6H_5CH=CH—CH=PPh_3$ $\longrightarrow$ $C_6H_5\underset{\underset{O}{|}}{C}H—\underset{\underset{PPh_3}{|}}{C}H—CH=CHC_6H_5$ $\longrightarrow$

Benzaldehyde

$$C_6H_5CH=CH—CH=CHC_6H_5$$

1,4-Diphenyl-1,3-butadiene

The reaction is carried out under mild conditions, and the position of the carbon–carbon double bond is not in doubt. Carbonyl compounds may contain a wide variety of substituents, and so may the ylide. (Indeed, in its broadest form, the Wittig reaction involves reactants other than carbonyl compounds, and may lead to products other than substituted alkenes.)

The phosphorus ylides have hybrid structures, and it is the negative charge on

$$\left[Ph_3P{=}\overset{\displaystyle R'}{\underset{\displaystyle |}{C}}{-}R \qquad \overset{+}{Ph_3P}{-}\overset{\displaystyle R'}{\underset{\displaystyle |}{\underset{\displaystyle -}{C}}}{-}R \right]$$

carbon—the carbanion character of ylides—that is responsible for their character-istic reactions: in this case, nucleophilic attack on carbonyl carbon.

The preparation of ylides is a two-stage process, each stage of which belongs to a familiar reaction type: nucleophilic attack on an alkyl halide, and abstraction of a proton by a base.

$$R{-}\overset{\displaystyle R'}{\underset{\displaystyle |}{C}}HX + Ph_3P \longrightarrow \overset{+}{Ph_3P}{-}\overset{\displaystyle R'}{\underset{\displaystyle |}{C}}H{-}R \quad X^- \xrightarrow{\text{base}} Ph_3P{=}\overset{\displaystyle R'}{\underset{\displaystyle |}{C}}{-}R + \text{base:H}$$

Triphenyl-phosphine   A phosphonium salt   An ylide

Many different bases have been used—chiefly alkoxides and organometallics—and in a variety of solvents. For example:

$$CH_3Br + Ph_3P \longrightarrow \overset{+}{Ph_3P}{-}CH_3 \quad Br^- \xrightarrow[\text{THF}]{C_6H_5Li} Ph_3P{=}CH_2 + C_6H_6 + LiBr$$

$$CH_2{=}CHCH_2Cl + Ph_3P \longrightarrow \overset{+}{Ph_3P}{-}CH_2CH{=}CH_2 \quad Cl^- \xrightarrow[\text{DMF}]{\text{NaOEt}}$$

$$Ph_3P{=}CHCH{=}CH_2$$

In 1979, the Nobel Prize was awarded to Georg Wittig and to H.C. Brown (p. 648), in recognition of their remarkable contributions to synthetic organic chemistry: Brown's centering about the element boron and Wittig's about phos-phorus.

Problem 25.23 What side reactions would you expect to encounter in the prepar-ation of an ylide like $Ph_3P{=}C(CH_3)CH_2CH_3$?

Problem 25.24 Give the structure of an ylide and a carbonyl compound from which each of the following could be made.
(a) $CH_3CH_2CH_2CH{=}C(CH_3)CH_2CH_3$
(b) $C_6H_5C(CH_3){=}CHCH_2C_6H_5$
(c) $C_6H_5CH{=}CHC_6H_5$

(d) ⬠=CHCH_3

(e) 1,4-diphenyl-1,3-butadiene (an alternative to the set of reagents used on p. 920)
(f) $CH_2{=}CHCH{=}C(CH_3)COOCH_3$

Problem 25.25 Outline all steps in a possible laboratory synthesis of each ylide and each carbonyl compound in the preceding problem, starting from benzene, toluene, alcohols of four carbons or fewer, acetic anhydride, triphenylphosphine, and cyclo-pentanol, and using any needed inorganic reagents.

Problem 25.26   Give the structures of compounds A–C.

$C_6H_5OCH_2Cl + Ph_3P$, then t-BuOK $\longrightarrow$ A $(C_{25}H_{21}OP)$
A + ethyl methyl ketone $\longrightarrow$ Ph_3PO + B $(C_{11}H_{14}O)$
B + dilute aqueous acid $\longrightarrow$ C $(C_5H_{10}O)$

   The above sequence offers a general route to what class of compounds?

Problem 25.27   Give the structures of compounds D–F.

(a) $C_6H_5COCH_2CH_2CH_2CH_2Br + Ph_3P$, then NaOEt $\longrightarrow$ D $(C_{11}H_{12})$
(b) $BrCH_2CH_2CH_2Br + Ph_3P$, then base $\longrightarrow$ E $(C_{39}H_{34}P_2)$
   E + o-$C_6H_4(CHO)_2$ $\longrightarrow$ F $(C_{11}H_{10})$

Problem 25.28   Give the structures of compounds G and H, and account for the stereochemistry of each step.

$trans$-2-octene + $C_6H_5CO_2OH$ $\longrightarrow$ G $(C_8H_{16}O)$
G + Ph_2PLi, then CH_3I $\longrightarrow$ H $(C_{21}H_{29}OP)$
H $\longrightarrow$ cis-2-octene

25.11   Claisen condensation. Formation of β-keto esters

An α-hydrogen in an ester, like an α-hydrogen in an aldehyde or ketone, is weakly acidic, and for the same reason: through resonance, the carbonyl group helps accommodate the negative charge of the carbanion. Let us look at an exceedingly important reaction of esters that depends upon the acidity of α-hydrogens. It is—for esters—the exact counterpart of the aldol condensation; reaction takes a different turn at the end, but a turn that is typical of the chemistry of acyl compounds.

When ethyl acetate is treated with sodium ethoxide, and the resulting mixture is acidified, there is obtained ethyl β-ketobutyrate (ethyl 3-oxobutanoate), generally known as **ethyl acetoacetate** or **acetoacetic ester**:

$$2CH_3COOC_2H_5 + Na^{+\ -}OC_2H_5 \xrightarrow{C_2H_5OH} \overset{\displaystyle O}{\overset{\displaystyle \|}{CH_3C}}{-}CHCOOC_2H_5\ ^-Na^+ + 2C_2H_5OH$$

Ethyl acetate   Sodium   Sodioacetoacetic ester
2 moles   ethoxide

$$\downarrow H^+$$

$$\overset{\displaystyle O}{\overset{\displaystyle \|}{CH_3C}}{-}\overset{\displaystyle \beta}{CH_2}COOC_2H_5$$

Ethyl acetoacetate
Acetoacetic ester
A β-keto ester

Ethyl acetoacetate is the ester of a *β-keto acid*; its preparation illustrates the reaction known as the **Claisen condensation**.

The generally accepted mechanism for the Claisen condensation (shown here for ethyl acetate) is:

$$(1) \qquad CH_3COOC_2H_5 + \ ^-OC_2H_5 \ \rightleftharpoons \ C_2H_5OH + \ ^-CH_2COOC_2H_5$$

$$\qquad\qquad\qquad\qquad\qquad\qquad\qquad\qquad\qquad\qquad\qquad\qquad\qquad I$$

(2) CH_3—$\overset{\overset{\displaystyle O}{\|}}{C}$—$OC_2H_5$ + $^-CH_2COOC_2H_5$ $\underset{}{\overset{\longrightarrow}{\longleftarrow}}$ CH_3—$\overset{\overset{\displaystyle O^-}{|}}{\underset{\underset{\displaystyle OC_2H_5}{|}}{C}}$—$CH_2COOC_2H_5$

I

$\updownarrow$

$CH_3\overset{\overset{\displaystyle O}{\|}}{C}$—$CH_2COOC_2H_5$ + $^-OC_2H_5$

(3) $CH_3\overset{\overset{\displaystyle O}{\|}}{C}$—$CH_2COOC_2H_5$ + $^-OC_2H_5$ $\overset{\longrightarrow}{\longleftarrow}$ $CH_3\overset{\overset{\displaystyle O}{\|}}{C}$—$CHCOOC_2H_5^-$ + C_2H_5OH

Stronger acid   Weaker acid

Ethoxide ion abstracts (step 1) a hydrogen ion from the α-carbon of the ester to form carbanion I. The powerfully nucleophilic carbanion I attacks (step 2) the carbonyl carbon of a second molecule of ester to displace ethoxide ion and yield the keto ester.

Like the aldol condensation and related reactions, the Claisen condensation involves nucleophilic attack by a carbanion on an electron-deficient carbonyl carbon. *In the aldol condensation, nucleophilic attack leads to addition, the typical reaction of aldehydes and ketones; in the Claisen condensation, nucleophilic attack leads to substitution, the typical reaction of acyl compounds* (Sec. 24.4).

When reaction is complete there is present, not acetoacetic ester, but its sodium salt, *sodioacetoacetic ester.* The α-hydrogens of acetoacetic ester are located *alpha* to *two* carbonyl groups, and hence ionization yields a particularly stable carbanion in which two carbonyl groups help accommodate the charge. As a result acetoacetic ester is a much stronger acid than ordinary esters or other carbon acids containing a single carbonyl group. It is considerably stronger than ethyl alcohol, and hence it reacts (step 3) with ethoxide ion to form ethyl alcohol and the anion

Acetoacetic ester   *equivalent to*

of sodioacetoacetic ester. Formation of the salt of acetoacetic ester is essential to the success of the reaction; of the various equilibria involved in the reaction, only (3) is favorable to the product we want.

Problem 25.29   Better yields are obtained if the Claisen condensation is carried out in ether with alcohol-free sodium ethoxide as catalyst instead of in ethyl alcohol solution. How do you account for this?

As we might expect, the Claisen condensation of more complicated esters yields the products resulting from ionization of an α-hydrogen of the ester; as a result, it is always the α-carbon of one molecule that becomes attached to the carbonyl carbon of another. For example:

$$2CH_3CH_2COOC_2H_5 + {}^-OC_2H_5 \longrightarrow CH_3CH_2\underset{\underset{CH_3}{||}}{C}-COOC_2H_5{}^- + 2C_2H_5OH$$

Ethyl propionate

$$\downarrow H^+$$

$$CH_3CH_2\overset{\beta}{\underset{\underset{O}{||}}{C}}-\overset{\alpha}{\underset{CH_3}{C}}HCOOC_2H_5$$

Ethyl 3-oxo-2-methylpentanoate
Ethyl α-methyl-β-ketovalerate
A β-keto ester

Problem 25.30   Sodium ethoxide converts ethyl adipate into 2-carbethoxycyclopentanone (II). This is an example of the **Dieckmann condensation**.

II

(a) How do you account for formation of II? (b) What product would you expect from the action of sodium ethoxide on ethyl pimelate (ethyl heptanedioate)? (c) Would you expect similar behavior from ethyl glutarate or ethyl succinate? Actually, ethyl succinate reacts with sodium ethoxide to yield a compound of formula $C_{12}H_{16}O_6$ containing a six-membered ring. What is the likely structure for this last product?

25.12   Crossed Claisen condensation

Like a crossed aldol condensation (Sec. 25.8), a **crossed Claisen condensation** is generally feasible only when one of the reactants has no α-hydrogens and thus is incapable of undergoing self-condensation. For example:

Ethyl benzoate   Ethyl acetate

Ethyl benzoylacetate

$$HCOOC_2H_5 + CH_3COOC_2H_5 \xrightarrow{^-OC_2H_5} \underset{\underset{O}{\|}}{H-C}-CH_2COOC_2H_5 + C_2H_5OH$$

Ethyl formate   Ethyl acetate

Ethyl formylacetate
(known only as the Na salt)

$$\underset{\underset{COOC_2H_5}{|}}{COOC_2H_5} + CH_3COOC_2H_5 \xrightarrow{^-OC_2H_5} C_2H_5OOC-\underset{\underset{O}{\|}}{C}-CH_2COOC_2H_5 + C_2H_5OH$$

Ethyl oxalate   Ethyl acetate

Ethyl oxaloacetate

$$\underset{\underset{O}{\|}}{C_2H_5O-C}-OC_2H_5 + C_6H_5CH_2COOC_2H_5 \xrightarrow{^-OC_2H_5}$$

Ethyl carbonate   Ethyl phenylacetate

$$\underset{\underset{O}{\|}}{C_2H_5O-C}-\overset{\overset{C_6H_5}{|}}{C}HCOOC_2H_5 + C_2H_5OH$$

Ethyl phenylmalonate
Phenylmalonic ester

Problem 25.31 In what order should the reactants be mixed in each of the above crossed Claisen condensations? (*Hint*: See Sec. 25.8.)

Problem 25.32 Ketones (but not aldehydes) undergo a crossed Claisen condensation with esters. For example:

$$CH_3COOC_2H_5 + CH_3COCH_3 \xrightarrow{NaOC_2H_5} CH_3COCH_2COCH_3 + C_2H_5OH$$

Ethyl acetate   Acetone   Acetylacetone

(a) Outline all steps in the most likely mechanism for this reaction. (b) Predict the principal products expected from the reaction in the presence of sodium ethoxide of ethyl propionate and acetone; (c) of ethyl benzoate and acetophenone; (d) of ethyl oxalate and cyclohexanone.

Problem 25.33 Outline the synthesis from simple esters of:
(a) ethyl α-phenylbenzoylacetate, $C_6H_5COCH(C_6H_5)COOC_2H_5$
(b) ethyl 2,3-dioxo-1,4-cyclopentanedicarboxylate (I). (*Hint*: Use ethyl oxalate as one ester.)
(c) ethyl 1,3-dioxo-2-indanecarboxylate (II)

PROBLEMS

    1. Write balanced equations, naming all organic products, for the reaction (if any) of phenylacetaldehyde with:

(a) dilute NaOH
(b) dilute HCl
(c) aqueous Na_2CO_3

(d) Br_2/CCl_4
(e) $Ph_3P{=}CH_2$

2. Answer Problem 1 for cyclohexanone.

3. Write balanced equations, naming all organic products, for the reaction (if any) of benzaldehyde with:

(a) dilute NaOH
(b) conc. NaOH
(c) acetaldehyde, dilute NaOH
(d) propionaldehyde, dilute NaOH
(e) acetone, dilute NaOH
(f) product (e), dilute NaOH
(g) acetophenone, NaOH
(h) acetic anhydride, sodium acetate, heat

(i) ethyl acetate, sodium ethoxide
(j) ethyl phenylacetate, sodium ethoxide
(k) formaldehyde, conc. NaOH
(l) crotonaldehyde, NaOH
(m) $Ph_3P=CHCH=CH_2$
(n) $Ph_3P=CH(OC_6H_5)$
(o) product (n), dilute acid

4. Write equations for all steps in the synthesis of the following from propionaldehyde, using any other needed reagents:

(a) β-hydroxy-α-methylvaleraldehyde
(b) 2-methyl-1-pentanol
(c) 2-methyl-2-pentenal
(d) 2-methyl-2-penten-1-ol
(e) 2-methyl-1,3-pentanediol

(f) α-methylvaleric acid
(g) 2-methyl-3-phenylpropenal
(h) CH_3CD_2CHO
(i) $CH_3CH_2CH^{18}O$
(j) 2-methyl-3-hexene

5. Write equations for all steps in the synthesis of the following from acetophenone, using any other needed reagents:

(a) benzoic acid
(b) 1,3-diphenyl-2-buten-1-one
(c) 1,3-diphenyl-1-butanol
(d) 1,3-diphenyl-2-buten-1-ol

(e) 1,3-diphenyl-2-propen-1-one
(f) α-phenylpropionaldehyde (*Hint*: See Problem 25.26.)

6. Give the structures of the principal products expected from the reaction in the presence of sodium ethoxide of:

(a) ethyl *n*-butyrate
(b) ethyl phenylacetate
(c) ethyl isovalerate
(d) ethyl formate and ethyl propionate
(e) ethyl oxalate and ethyl succinate

(f) ethyl benzoate and ethyl phenylacetate
(g) ethyl propionate and cyclohexanone
(h) ethyl phenylacetate and acetophenone
(i) ethyl carbonate and acetophenone

7. Sodium ethoxide is added to a mixture of ethyl acetate and ethyl propionate. (a) Give the structures of the products expected. (b) Would this reaction be a good method of synthesizing any one of these?

8. Outline all steps in a possible synthesis of each of the following via the Claisen condensation, using any needed reagents:

(a) $C_6H_5COCH(CH_3)COOC_2H_5$
(b) $C_6H_5CH_2COCH(C_6H_5)COOC_2H_5$
(c) $C_2H_5OOCCOCH(CH_3)COOC_2H_5$
(d) $C_6H_5CH(CHO)COOC_2H_5$

(e) $(CH_3)_2CHCOCH_2COCH_3$
(f) $C_6H_5COCH_2COCH_3$
(g) 2-benzoylcyclohexanone
(h) $C_2H_5OOCCH(CHO)CH_2COOC_2H_5$

9. The cinnamic acid obtained by the Perkin condensation is the more stable *trans* isomer. Suggest a method of preparing *cis*-cinnamic acid.

10. Outline all steps in a possible laboratory synthesis of each of the following from benzene, toluene, acetic anhydride, triphenylphosphine, and alcohols of four carbons or fewer, using any needed inorganic reagents:

(a) 4-hydroxy-4-methyl-2-pentanone
(b) 4-methyl-2-pentanol
(c) crotonaldehyde, $CH_3CH=CHCHO$
(d) cinnamyl alcohol, $C_6H_5CH=CHCH_2OH$
(e) *p*-nitrocinnamaldehyde
(f) 1,3-butanediol
(g) 3-methyl-2-butenoic acid (via aldol condensation)

(h) 3-methyl-1-pentyn-3-ol (*Oblivon*, a hypnotic)
(i) 1-phenyl-1,3,5-hexatriene
(j) 1,6-diphenyl-1,3,5-hexatriene
(k) 2,3-dimethyl-2-pentenoic acid
(l) α,α-dimethylcaproic acid
(m) indanone (I)
(n) racemic *erythro*-2,3-dihydroxy-3-phenylpropanoic acid (II and its enantiomer)

11. How do you account for the formation of γ-methylparaconic acid from the reaction of acetaldehyde with succinic acid?

12. Considerable quantities of acetone are consumed in the manufacture of methyl isobutyl ketone (MIBK). How do you think the synthesis of MIBK is accomplished?

13. Methyl ethyl ketone can be made to undergo the Claisen condensation with a given ester to yield either of two products, depending upon experimental conditions. (a) What are these two products? (b) How could you tell quickly and simply which product you had obtained? (*Note*: Use ethyl benzoate as the ester.)

14. The acetylenic ester CH_3—$C\equiv C$—$COOC_2H_5$ can be converted into ethyl acetoacetate. (a) How? (b) Outline a synthesis of the acetylenic ester from acetylene and any needed reagents.

15. The compound *pentaerythritol*, $C(CH_2OH)_4$, used in making explosives, is obtained from the reaction of acetaldehyde and formaldehyde in the presence of calcium hydroxide. Outline the probable steps in this synthesis.

16. The labeled alkene, 1,3,3-trideuteriocyclohexene, needed for a particular stereochemical study, was prepared from cyclohexanone. Outline all steps in such a synthesis.

17. (a) The haloform test (Sec. 18.9) depends upon the fact that three hydrogens on the same carbon atom are successively replaced by halogen. Using acetone as an example, show why the carbon that suffers the initial substitution should be the preferred site of further substitution. (*Hint*: See Sec. 23.14.)

(b) The haloform test also depends upon the ease with which the trihalomethyl ketone produced in (a) is cleaved by base. What is the most likely mechanism for this cleavage? What factor makes such a reaction possible in this particular case?

18. Upon treatment with dilute NaOH, β-methylcrotonaldehyde, $(CH_3)_2C\!=\!CHCHO$, yields a product of formula $C_{10}H_{14}O$, called *dehydrocitral*. What is a likely structure for this product, and how is it formed? (*Hint*: See *citral*, Problem 25, p. 793.)

19. As part of the total synthesis of vitamin D_3, compound III was converted into IV by a number of stages, two of which involved use of the Wittig reaction. Show how this conversion might have been carried out.

20. Meanwhile, back at the laboratory, our naïve graduate student (Problem 16, p. 79) had need of the hydroxy ester $(CH_3)_2C(OH)CH_2COOC_2H_5$. Turning once again to the Grignard reaction, he prepared methylmagnesium iodide and to it he added acetoacetic ester. Everything went well; indeed, even without the application of heat, the reaction mixture bubbled merrily. Working carefully and with great skill, he isolated an excellent yield of the starting material, acetoacetic ester. He poured this down the sink and fled, sobbing, to his research director's office, where he begged for a new research problem.

What reaction had taken place? What was the bubbling due to? (In Problem 15, p. 1097, we shall see how he made out with his new research problem.)

21. (a) The sex attractant of the Egyptian cotton leafworm has been prepared in the following way. On the basis of this synthesis what structure or structures can you assign to this pheromone (and to all intermediates)? (b) At one point in the synthesis, it is necessary to separate a pair of isomers. At which point is this, and what are the isomers?

$$9\text{-bromo-1-nonanol} \;+\; DHP, H^+ \;\longrightarrow\; A\;(C_{14}H_{27}O_2Br)$$

$$A \;+\; Ph_3P;\ \text{then base} \;\longrightarrow\; B\;(C_{32}H_{41}O_2P)$$

$$B \;+\; (E)\text{-2-penten-1-al} \;\longrightarrow\; C\;(C_{19}H_{34}O_2)$$

$$C \;+\; H_2O, HCl \;\longrightarrow\; D\;(C_{14}H_{26}O)$$

$$D \;+\; Ac_2O, \text{pyridine} \;\longrightarrow\; E,\ \text{the pheromone}\;(C_{16}H_{28}O_2)$$

22. *Bombykol*, the sex pheromone of the silkworm moth, has been prepared in the following way. What is the structure of bombykol? What uncertainties, if any, are there in your answer?

$$1\text{-pentyne} \;+\; n\text{-}C_4H_9MgBr \;\longrightarrow\; F\;(C_5H_7MgBr)$$

$$F \;+\; HCHO;\ \text{then } H^+ \;\longrightarrow\; G\;(C_6H_{10}O)$$

$$G \;+\; PBr_3 \;\longrightarrow\; H\;(C_6H_9Br)$$

$$H \;+\; Ph_3P, \text{base} \;\longrightarrow\; I\;(C_{24}H_{23}P)$$

$$I \;+\; \text{ethyl 10-oxodecanoate} \;\longrightarrow\; J\;(C_{18}H_{30}O_2)$$

$$J \;+\; H_2, \text{Lindlar catalyst} \;\longrightarrow\; K\;(C_{18}H_{32}O_2)$$

$$K \;+\; LiAlH_4 \;\longrightarrow\; \text{bombykol}\;(C_{16}H_{30}O)$$

23. In contrast to simple carbonyl compounds, 1,3-dicarbonyl compounds—like acetoacetic ester or 2,4-pentanedione (acetylacetone)—exist to an appreciable extent in the enol form.

(a) Pure samples of keto and enol forms of acetoacetic ester have been isolated. Each retained its identity for weeks if acids and bases were carefully excluded. Write equations to show exactly how keto–enol interconversion is speeded up by a base. By an acid.

(b) Draw the structure of the enol form of, say, 2,4-pentanedione. Can you suggest one factor that would tend to stabilize the enol form of such a compound?

(c) Although the enol form of acetoacetic ester is an alcohol, it does *not* have a higher boiling point than the keto form. (Actually, it boils somewhat *lower*.) Can you suggest a second factor that would tend to stabilize the enol form of a 1,3-dicarbonyl compound?

24. Draw the structures (stereochemical where pertinent) of products L and M.

(a)

$+ \text{NaOEt, then } H_2O \;\longrightarrow\; L\;(C_{10}H_{14}O_2),\ \textit{highly enolic}$

(b) ethyl methyl ketone + ethyl oxalate + NaOEt $\;\longrightarrow\;$ M $(C_6H_6O_3)$

25. (a) Figure 25.1(a) shows the proton NMR spectrum of a solution of acetylacetone, $CH_3COCH_2COCH_3$, in chloroform. Besides the peaks shown, there is a small hump, e, near δ 15 of about the same area as the peak d at δ 5.5. How do you interpret this spectrum? What *quantitative* conclusion can you draw?

(b) Figure 25.1(b) shows the proton NMR spectrum of benzoylacetone, $C_6H_5COCH_2COCH_3$. There is an additional peak, d, near δ 16 of about the same area as the peak b at δ 6.1. How do you interpret this spectrum? How do you account for the difference between it and the spectrum in (a)?

Figure 25.1   Proton NMR spectra of (a) acetylacetone and (b) benzoylacetone.

26

Amines I. Preparation and Physical Properties

26.1 Structure

Nearly all the organic compounds that we have studied so far are bases, although very weak ones. Much of the chemistry of alcohols, ethers, esters, and even of alkenes and aromatic hydrocarbons is understandable in terms of the basicity of these compounds.

Of the organic compounds that show appreciable basicity (for example, those strong enough to turn litmus blue), by far the most important are the **amines**. An amine has the general formula RNH_2, R_2NH, or R_3N, where R is any alkyl or aryl group. For example:

CH_3NH_2	$(CH_3)_2NH$	$(CH_3)_3N$	$H_2NCH_2CH_2NH_2$
Methylamine	Dimethylamine	Trimethylamine	Ethylenediamine
(1°)	(2°)	(3°)	(1°)

NH₂	NHCH₃	N(CH₃)₂
⬡	⬡	⬡
Aniline	N-Methylaniline	N,N-Dimethylaniline
(1°)	(2°)	(3°)

931

26.2 Classification

Amines are classified as **primary**, **secondary**, or **tertiary**, according to the number of groups attached to the nitrogen atom.

$$
\begin{array}{c}
\text{H} \\
\mid \\
\text{R—N—H}
\end{array}
\qquad
\begin{array}{c}
\text{R}' \\
\mid \\
\text{R—N—H}
\end{array}
\qquad
\begin{array}{c}
\text{R}' \\
\mid \\
\text{R—N—R}''
\end{array}
$$

Primary   Secondary   Tertiary

$1°$ $2°$ $3°$

In their fundamental properties—*basicity* and the accompanying *nucleophilicity*—amines of different classes are very much the same. In many of their reactions, however, the final products depend upon the number of hydrogen atoms attached to the nitrogen atom, and hence are different for amines of different classes.

26.3 Nomenclature

Aliphatic amines are named by naming the alkyl group or groups attached to nitrogen, and following these by the word -*amine*. More complicated ones are often named by prefixing *amino*- (or *N-methylamino*-, *N,N-diethylamino*-, etc.) to the name of the parent chain. For example:

$$
\begin{array}{c}
\text{CH}_3 \\
\mid \\
\text{CH}_3\text{—C—CH}_3 \\
\mid \\
\text{NH}_2
\end{array}
\qquad
\begin{array}{c}
\text{H} \\
\mid \\
\text{CH}_3\text{CH}_2\text{—N—CH}_3
\end{array}
\qquad
\begin{array}{c}
\text{CH}_3 \\
\mid \\
\text{CH}_3\text{—N—CHCH}_2\text{CH}_3 \\
\mid \\
\text{CH}_3
\end{array}
$$

tert-Butylamine   Ethylmethylamine   *sec*-Butyldimethylamine

$(1°)$ $(2°)$ $(3°)$

$$
\text{H}_2\text{NCH}_2\text{CH}_2\text{CH}_2\text{COOH}
\qquad
\text{H}_2\text{NCH}_2\text{CH}_2\text{OH}
\qquad
\begin{array}{c}
\text{H} \\
\mid \\
\text{CH}_3\text{—N—CH(CH}_2)_4\text{CH}_3 \\
\mid \\
\text{CH}_3
\end{array}
$$

γ-Aminobutyric acid   2-Aminoethanol   (Ethanolamine)   2-(*N*-Methylamino)heptane

$(1°)$ $(1°)$ $(2°)$

Aromatic amines—those in which nitrogen is attached directly to an aromatic ring—are generally named as derivatives of the simplest aromatic amine, **aniline**. An aminotoluene is given the special name of *toluidine*. For example:

2,4,6-Tribromoaniline
$(1°)$

N-Ethyl-*N*-methylaniline
$(3°)$

N(CH$_3$)$_2$

NO

p-Nitroso-N,N-dimethylaniline
(3°)

CH$_3$

NH$_2$

p-Toluidine
(1°)

H
N

Diphenylamine
(2°)

O$_2$N⟨⟩—N—⟨⟩NO$_2$
          H

4,4'-Dinitrodiphenylamine
(2°)

Salts of amines are generally named by replacing *-amine* by *-ammonium* (or *-aniline* by *-anilinium*), and adding the name of the anion (*chloride, nitrate, sulfate,* etc.). For example:

(C$_2$H$_5$NH$_3$$^+$)$_2SO_4$$^{2-}$
Ethylammonium
sulfate

(CH$_3$)$_3$NH$^+$NO$_3$$^-$
Trimethylammonium
nitrate

C$_6$H$_5$NH$_3$$^+Cl^-$
Anilinium
chloride

26.4 Physical properties of amines

Like ammonia, amines are polar compounds and, except for tertiary amines, can form intermolecular hydrogen bonds. Amines have higher boiling points than

$$CH_3-N-H\cdots N-H\cdots N-CH_3$$

H   CH$_3$ H

H   H

non-polar compounds of the same molecular weight, but lower boiling points than alcohols or carboxylic acids.

Amines of all three classes are capable of forming hydrogen bonds with water. As a result, smaller amines are quite soluble in water, with borderline solubility

Table 26.1 AMINES

Name	M.p., °C	B.p., °C	Solubility, g/100 g H$_2$O	K_b
Methylamine	− 92	− 7.5	v.sol.	4.5×10^{-4}
Dimethylamine	− 96	7.5	v.sol.	5.4
Trimethylamine	−117	3	91	0.6
Ethylamine	− 80	17	∞	5.1
Diethylamine	− 39	55	v.sol.	10.0
Triethylamine	−115	89	14	5.6
n-Propylamine	− 83	49	∞	4.1
Di-n-propylamine	− 63	110	s.sol.	10
Tri-n-propylamine	− 93	157	s.sol.	4.5
Isopropylamine	−101	34	∞	4
n-Butylamine	− 50	78	v.sol.	4.8

CONTINUED

Table 26.1 AMINES (*continued*)

Name	M.p., °C	B.p., °C	Solubility, g/100 g H_2O	K_b
Isobutylamine	− 85	68	∞	3×10^{-4}
sec-Butylamine	−104	63	∞	4
tert-Butylamine	− 67	46	∞	5
Cyclohexylamine	− 18	134	s.sol.	5
Benzylamine	10	185	∞	0.2
α-Phenylethylamine	33	187	4.2	1.2
β-Phenylethylamine		195	s.	1.5
Ethylenediamine	8	117	s.	0.85
Tetramethylenediamine $[H_2N(CH_2)_4NH_2]$	27	158	v.sol.	
Hexamethylenediamine	39	196	v.sol.	5
Tetramethylammonium hydroxide	63	135*d*	220	strong base
Aniline	− 6	184	3.7	4.2×10^{-10}
Methylaniline	− 57	196	v.sl.sol.	7.1
Dimethylaniline	3	194	1.4	11.7
Diphenylamine	53	302	i.	0.0006
Triphenylamine	127	365	i.	
o-Toluidine	− 28	200	1.7	2.6
m-Toluidine	− 30	203	s.sol.	5
p-Toluidine	44	200	0.7	12
o-Anisidine (o-$CH_3OC_6H_4NH_2$)	5	225	s.sol.	3
m-Anisidine		251	s.sol.	2
p-Anisidine	57	244	v.sl.sol.	20
o-Chloroaniline	− 2	209	i.	0.05
m-Chloroaniline	− 10	236		0.3
p-Chloroaniline	70	232		1
o-Bromoaniline	32	229	s.sol.	0.03
m-Bromoaniline	19	251	v.sl.sol.	0.4
p-Bromoaniline	66	*d*	i.	0.7
o-Nitroaniline	71	284	0.1	0.00006
m-Nitroaniline	114	307*d*	0.1	0.029
p-Nitroaniline	148	332	0.05	0.001
2,4-Dinitroaniline	187		s.sol.	
2,4,6-Trinitroaniline (picramide)	188		0.1	
o-Phenylenediamine [o-$C_6H_4(NH_2)_2$]	104	252	3	3
m-Phenylenediamine	63	287	25	10
p-Phenylenediamine	142	267	3.8	140
Benzidine	127	401	0.05	9
p-Aminobenzoic acid	187		0.3	0.023
Sulfanilic acid	288*d*		1	17
Sulfanilamide	163		0.4	

Name	Formula	M.p., °C
Acetanilide	$C_6H_5NHCOCH_3$	114
Benzanilide	$C_6H_5NHCOC_6H_5$	163
Aceto-*o*-toluidide	o-$CH_3C_6H_4NHCOCH_3$	110
Aceto-*m*-toluidide	m-$CH_3C_6H_4NHCOCH_3$	66
Aceto-*p*-toluidide	p-$CH_3C_6H_4NHCOCH_3$	147
o-Nitroacetanilide	o-$O_2NC_6H_4NHCOCH_3$	93
m-Nitroacetanilide	m-$O_2NC_6H_4NHCOCH_3$	154
p-Nitroacetanilide	p-$O_2NC_6H_4NHCOCH_3$	216

being reached at about six carbon atoms. Amines are soluble in less polar solvents like ether, alcohol, benzene, etc. The methylamines and ethylamines smell very much like ammonia; the higher alkylamines have decidedly "fishy" odors.

Aromatic amines are generally very toxic; they are readily absorbed through the skin, often with fatal results.

Aromatic amines are very easily oxidized by air, and although most are colorless when pure, they are often encountered discolored by oxidation products.

26.5   Salts of amines

Aliphatic amines are about as basic as ammonia; aromatic amines are considerably less basic. Although amines are much weaker bases than hydroxide ion or ethoxide ion, they are much stronger bases than alcohols, ethers, esters, etc.; they are much stronger bases than water. Aqueous mineral acids or carboxylic acids readily convert amines into their salts; aqueous hydroxide ion readily converts the salts back into the free amines. As with the carboxylic acids, we can do little

$$
\begin{array}{ll}
\left.\begin{array}{l} RNH_2 \\ 1° \text{ amine} \end{array}\right\} & \left\{\begin{array}{l} RNH_3{}^+ \\ \text{salt} \end{array}\right. \\[2ex]
\left.\begin{array}{l} R_2NH \\ 2° \text{ amine} \end{array}\right\} \quad \underset{OH^-}{\overset{H^+}{\rightleftarrows}} \quad & \left\{\begin{array}{l} R_2NH_2{}^+ \\ \text{salt} \end{array}\right. \\[2ex]
\left.\begin{array}{l} R_3N \\ 3° \text{ amine} \end{array}\right\} & \left\{\begin{array}{l} R_3NH^+ \\ \text{salt} \end{array}\right. \\[1ex]
\textit{Insoluble} & \textit{Soluble} \\
\textit{in water} & \textit{in water}
\end{array}
$$

with amines without encountering this conversion into and from their salts; it is therefore worthwhile to look at the properties of these salts.

In Sec. 23.4 we contrasted physical properties of carboxylic acids with those of their salts; amines and their salts show the same contrast. Amine salts are typical ionic compounds. They are non-volatile solids, and when heated generally decompose before the high temperature required for melting is reached. The halides, nitrates, and sulfates are soluble in water but are insoluble in non-polar solvents.

The difference in solubility behavior between amines and their salts can be used both to detect amines and to separate them from non-basic compounds. A water-insoluble organic compound that dissolves in cold, dilute aqueous hydrochloric acid must be appreciably basic, which means almost certainly that it is an amine. An amine can be separated from non-basic compounds by its solubility in acid; once separated, the amine can be regenerated by making the aqueous solution alkaline. (See Sec. 23.4 for a comparable situation for carboxylic acids.)

Problem 26.1   Describe exactly how you would go about separating a mixture of the three water-insoluble liquids, aniline (b.p. 184 °C), *n*-butylbenzene (b.p. 183 °C), and *n*-valeric acid (b.p. 187 °C), recovering each compound pure and in essentially quantitative yield. Do the same for a mixture of the three water-insoluble solids, *p*-toluidine, *o*-bromobenzoic acid, and *p*-nitroanisole.

26.6   Stereochemistry of nitrogen

So far in our study of organic chemistry, we have devoted considerable time to the spatial arrangement of atoms and groups attached to carbon atoms, that is, to the stereochemistry of carbon. Now let us look briefly at the stereochemistry of nitrogen.

Amines are simply ammonia in which one or more hydrogen atoms have been replaced by organic groups. Nitrogen uses sp^3 orbitals, which are directed to the corners of a tetrahedron. Three of these orbitals overlap s orbitals of hydrogen or carbon; the fourth contains an unshared pair of electrons (see Fig. 1.12, p. 18). Amines, then, are like ammonia, pyramidal, and with very nearly the same bond angles: 108° in trimethylamine, for example. (See Fig. 26.1.)

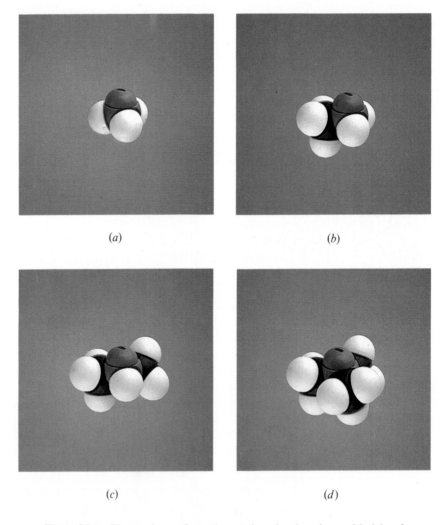

(a)   (b)

(c)   (d)

Figure 26.1   Electronic configuration and molecular shape. Models of: (a) ammonia, NH_3; (b) methylamine, CH_3NH_3; (c) dimethylamine, $(CH_3)_2NH$; (d) trimethylamine, $(CH_3)_3N$. Like ammonia, amines are pyramidal, with the unshared pair of electrons occupying the fourth sp^3 orbital of nitrogen.

From an examination of models, we can see that a molecule in which nitrogen carries three different groups is not superimposable on its mirror image; it is chiral and should exist in two enantiomeric forms (I and II) each of which—separated

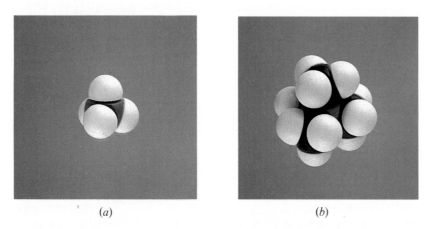

from the other—might be expected to show optical activity.

But such enantiomers have not yet been isolated—for simple amines—and spectroscopic studies have shown why: the energy barrier between the two pyramidal arrangements about nitrogen is ordinarily so low that they are rapidly interconverted. Just as rapid rotation about carbon–carbon single bonds prevents isolation of conformational enantiomers (Sec. 4.20), so rapid *inversion* about nitrogen prevents isolation of enantiomers like I and II. Evidently, an unshared pair of electrons of nitrogen cannot ordinarily serve as a fourth group to maintain configuration.

Next, let us consider the quaternary ammonium salts, compounds in which four alkyl groups are attached to nitrogen. Here all four sp^3 orbitals are used to form bonds, and quaternary nitrogen is tetrahedral. (See, for example, Fig. 26.2.)

<table>
<tr><td>(a)</td><td>(b)</td></tr>
</table>

Figure 26.2   Electronic configuration and molecular shape. Models of: (a) ammonium ion, NH_4^+; (b) tetramethylammonium ion, $(CH_3)_4N^+$. Like the ammonium ion, quaternary ammonium ions are tetrahedral, with nitrogen using four sp^3 orbitals.

Quaternary ammonium salts in which nitrogen holds four different groups have been found to exist as *configurational* enantiomers, capable of showing optical activity: methylallylphenylbenzylammonium iodide, for example.

Problem 26.2   Racemization in certain free-radical and carbocation reactions has been attributed (Secs. 4.28 and 5.19) to loss of configuration in a flat intermediate. Account for the fact that the formation of alkyl carbanions, $R:^-$—which are believed to be *pyramidal*—can also lead to racemization.

Problem 26.3 At room temperature, the NMR spectrum of 1-ethylaziridine (III) shows the triplet–quartet of an ethyl group, and two other signals of equal peak area. When the temperature is raised to 120 °C, the latter two signals merge into a single signal. How do you interpret these observations?

$$
\begin{array}{cc}
\underset{\text{III}}{
\begin{array}{l}
H_2C \\
\quad\ \diagdown \\
\quad\quad N\!-\!C_2H_5 \\
\quad\ \diagup \\
H_2C
\end{array}
} &
\underset{\text{IV}}{
\begin{array}{l}
H_2C \\
\quad\ \diagdown \\
\quad\quad N\!-\!Cl \\
\quad\ \diagup \\
HC \\
\quad | \\
\quad CH_3
\end{array}
}
\end{array}
$$

Problem 26.4 Account for the following, drawing all pertinent stereochemical formulas. (a) 1-Chloro-2-methylaziridine (IV, above) was prepared in two isomeric forms separable at 25 °C by ordinary gas chromatography. (b) The reaction of the *imine* $(C_6H_5)_2C=NCH_3$ with (R)-$(+)$-2-phenylperoxypropionic acid gave a product, $C_{14}H_{13}ON$, with $[\alpha]+12.5°$, which showed no loss of optical activity up to (at least) 90 °C.

26.7 Industrial source

Some of the simplest and most important amines are prepared on an industrial scale by processes that are not practicable as laboratory methods.

The most important of all amines, **aniline**, is prepared in several ways: (a) reduction of nitrobenzene by the cheap reagents, iron and dilute hydrochloric acid (or by catalytic hydrogenation, Sec. 26.9); (b) treatment of chlorobenzene with

Nitrobenzene   Anilinium chloride   Aniline

Chlorobenzene   Aniline

ammonia at high temperatures and high pressures in the presence of a catalyst. Process (b), we shall see (Chap. 29), involves nucleophilic aromatic substitution.

Methylamine, dimethylamine, and trimethylamine are synthesized on an industrial scale from methanol and ammonia:

$$
NH_3 \xrightarrow[\text{Al}_2\text{O}_3,\,450\,°C]{CH_3OH} CH_3NH_2 \xrightarrow[\text{Al}_2\text{O}_3,\,450\,°C]{CH_3OH} (CH_3)_2NH \xrightarrow[\text{Al}_2\text{O}_3,\,450\,°C]{CH_3OH} (CH_3)_3N
$$

Ammonia   Methylamine   Dimethylamine   Trimethylamine

Alkyl halides are used to make some higher alkylamines, just as in the laboratory (Sec. 26.10). The acids obtained from fats (Sec. 37.4) can be converted into long-chain 1-aminoalkanes of even carbon number via reduction of nitriles (Sec. 26.8).

$$
R\!-\!COOH \xrightarrow{NH_3,\,\text{heat}} R\!-\!CONH_2 \xrightarrow{\text{heat}} R\!-\!C\!\equiv\!N \xrightarrow{H_2,\,\text{cat.}} R\!-\!CH_2NH_2
$$

Acid   Amide   Nitrile   Amine

26.8 Preparation

Some of the many methods that are used to prepare amines in the laboratory are outlined on the following pages.

PREPARATION OF AMINES _____

1. **Reduction of nitro compounds.** Discussed in Sec. 26.9.

$$\begin{array}{c} Ar-NO_2 \\ \text{or} \\ R-NO_2 \end{array} \xrightarrow{\text{metal, } H^+; \text{ or } H_2, \text{ catalyst}} \begin{array}{c} Ar-NH_2 \\ \text{or} \\ R-NH_2 \end{array}$$

Nitro compound   1° amine

Chiefly for aromatic amines

Examples:

Ethyl *p*-nitrobenzoate ($COOC_2H_5$, NO_2) $\xrightarrow{H_2, Pt}$ Ethyl *p*-aminobenzoate ($COOC_2H_5$, NH_2)

p-Nitroaniline (NH_2, NO_2) $\xrightarrow[\text{heat}]{\text{Sn, HCl}}$ *p*-Phenylenediamine (NH_2, NH_2)

$$CH_3CH_2CH_2NO_2 \xrightarrow{\text{Fe, HCl}} CH_3CH_2CH_2NH_2$$

1-Nitropropane   *n*-Propylamine

2. **Reaction of halides with ammonia or amines.** Discussed in Secs. 26.10 and 26.13.

$$NH_3 \xrightarrow{RX} R-NH_2 \xrightarrow{RX} R-\underset{\underset{R}{|}}{N}H \xrightarrow{RX} R-\underset{\underset{R}{|}}{\overset{\overset{R}{|}}{N}} \xrightarrow{RX} R-\underset{\underset{R}{|}}{\overset{\overset{R}{|}}{N}}-R^{\oplus}X^-$$

      1° amine   2° amine   3° amine   Quaternary ammonium salt (4°)

RX must be alkyl, or aryl with electron-withdrawing substituents

Examples:

$$CH_3COOH \xrightarrow[P]{Cl_2} \underset{\underset{Cl}{|}}{CH_2COOH} \xrightarrow{NH_3} \underset{\underset{NH_2}{|}}{CH_2COO^- NH_4^+} \xrightarrow{H^+} \underset{\underset{NH_2}{|}}{CH_2COOH} \text{ (or } \underset{\underset{+NH_3}{|}}{CH_2COO^-})$$

Acetic acid   Chloroacetic acid   Aminoacetic acid (Glycine: an amino acid) (1°)

_____ CONTINUED _____

—— CONTINUED ——

$$C_2H_5Cl \xrightarrow{NH_3} C_2H_5NH_2 \xrightarrow{CH_3Cl} C_2H_5\overset{H}{\underset{|}{N}}CH_3$$

Ethyl chloride   Ethylamine   Ethylmethylamine
                          (1°)   (2°)

$$\text{⬡}CH_2Cl \xrightarrow{NH_3} \text{⬡}CH_2NH_2 \xrightarrow{2CH_3Cl} \text{⬡}CH_2\overset{CH_3}{\underset{|}{N}}CH_3$$

Benzyl chloride   Benzylamine   Benzyldimethylamine
                          (1°)   (3°)

$$\text{⬡}N(CH_3)_2 \xrightarrow{CH_3I} \text{⬡}N(CH_3)_3{}^+I^-$$

N,N-Dimethylaniline   Phenyltrimethylammonium iodide
      (3°)   (4°)

2,4-Dinitrochlorobenzene   N-Methyl-2,4-dinitroaniline
                                    (2°)

3. **Reductive amination.** Discussed in Sec. 26.11.

$$\overset{}{\underset{}{}}C{=}O + NH_3 \xrightarrow[\text{or NaBH}_3\text{CN}]{H_2,\ Ni} \overset{}{\underset{}{}}CH{-}NH_2 \qquad 1° \text{ amine}$$

$$+ RNH_2 \xrightarrow[\text{or NaBH}_3\text{CN}]{H_2,\ Ni} \overset{}{\underset{}{}}CH{-}NHR \qquad 2° \text{ amine}$$

$$+ R_2NH \xrightarrow[\text{or NaBH}_3\text{CN}]{H_2,\ Ni} \overset{}{\underset{}{}}CH{-}NR_2 \qquad 3° \text{ amine}$$

Examples:

$$CH_3{-}\overset{}{\underset{O}{C}}{-}CH_3 + NH_3 + H_2 \xrightarrow{Ni} CH_3{-}\overset{}{\underset{NH_2}{CH}}{-}CH_3$$

      Acetone   Isopropylamine
                                            (1°)

$$(CH_3)_2CH\overset{H}{\underset{}{C}}{=}O + \text{⬡}NH_2 \xrightarrow{NaBH_3CN} \text{⬡}\overset{H}{\underset{}{N}}CH_2CH(CH_3)_2$$

Isobutyraldehyde   Aniline   N-Isobutylaniline
                        (1°)   (2°)

$$CH_3\overset{H}{\underset{}{C}}{=}O + (CH_3)_2NH + H_2 \xrightarrow{Ni} CH_3CH_2{-}\overset{CH_3}{\underset{}{N}}{-}CH_3$$

Acetaldehyde   Dimethylamine   Dimethylethylamine
                    (2°)   (3°)

—— CONTINUED ——

_____ CONTINUED _____

4. **Reduction of nitriles.** Discussed in Sec. 26.8.

$$R-C\equiv N \xrightarrow{\text{2H}_2, \text{ catalyst}} R-CH_2NH_2$$

Nitrile $\qquad\qquad\qquad$ 1° amine

Examples:

(⬡)CH$_2$Cl $\xrightarrow{\text{NaCN}}$ (⬡)CH$_2$CN $\xrightarrow{\text{H}_2, \text{ Ni, 140 °C}}$ (⬡)CH$_2$CH$_2$NH$_2$

Benzyl chloride $\qquad$ Phenylacetonitrile $\qquad$ β-Phenylethylamine

$\qquad\qquad\qquad$ (Benzyl cyanide) $\qquad\qquad\qquad$ (1°)

$ClCH_2CH_2CH_2CH_2Cl \xrightarrow{\text{NaCN}} NC(CH_2)_4CN \xrightarrow{\text{H}_2, \text{ Ni}} H_2NCH_2(CH_2)_4CH_2NH_2$

1,4-Dichlorobutane $\qquad$ Adiponitrile $\qquad$ Hexamethylenediamine

$\qquad\qquad\qquad\qquad\qquad\qquad\qquad\qquad$ (1,6-Diaminohexane)

$\qquad\qquad\qquad\qquad\qquad\qquad\qquad\qquad\qquad$ (1°)

5. **Hofmann degradation of amides.** Discussed in Secs. 26.12 and 32.2–32.5.

$$R-CONH_2 \text{ or } Ar-CONH_2 \xrightarrow{\text{OBr}^-} R-NH_2 \text{ or } Ar-NH_2 + CO_3{}^{2-}$$

Amide $\qquad\qquad\qquad\qquad\qquad\qquad$ 1° amine

Examples:

$$CH_3(CH_2)_4CONH_2 \xrightarrow{\text{KOBr}} CH_3(CH_2)_4NH_2$$

Caproamide $\qquad\qquad\qquad$ *n*-Pentylamine

(Hexanamide)

(⬡)CONH$_2$ $\xrightarrow{\text{KOBr}}$ (⬡)NH$_2$

Br $\qquad\qquad\qquad\qquad$ Br

m-Bromobenzamide $\qquad$ *m*-Bromoaniline $\qquad\qquad\qquad$ ∎

Reduction of aromatic nitro compounds is by far the most useful method of preparing amines, since it uses readily available starting materials, and yields the most important kind of amines, *primary aromatic amines*. These amines can be converted into aromatic diazonium salts, which are among the most versatile class of organic compounds known (see Secs. 27.12–27.18). The sequence

nitro compound $\longrightarrow$ amine $\longrightarrow$ diazonium salt

provides the best possible route to dozens of kinds of aromatic compounds.

Reduction of aliphatic nitro compounds is limited by the availability of the starting materials.

Ammonolysis of halides is usually limited to the aliphatic series, because of the generally low reactivity of aryl halides toward nucleophilic substitution. (However, see Chap. 29.) Ammonolysis has the disadvantage of yielding a mixture of different classes of amines. It is important to us as one of the most general methods of introducing the amino (—NH$_2$) group into molecules of all kinds; it can be used, for example, to convert bromo acids into amino acids. The exactly analogous reaction of halides with amines permits the preparation of every class of amine (as well as quaternary ammonium salts, $R_4N^+X^-$).

Reductive amination, the catalytic chemical reduction of aldehydes (RCHO) and ketones (R_2CO) in the presence of ammonia or an amine, accomplishes much the same purpose as the reaction of halides. It too can be used to prepare any class of amine, and has certain advantages over the halide reaction. The formation of mixtures is more readily controlled in reductive amination than in ammonolysis of halides. Reductive amination of ketones yields amines containing a *sec*-alkyl group; these amines are difficult to prepare by ammonolysis because of the tendency of *sec*-alkyl halides to undergo elimination rather than substitution.

Synthesis via **reduction of nitriles** has the special feature of *increasing the length of a carbon chain*, producing a primary amine that has one more carbon atom than the alkyl halide from which the nitrile was made. The **Hofmann degradation of amides** has the feature of *decreasing the length of a carbon chain* by one carbon atom; it is also of interest as an example of an important class of reactions involving rearrangement.

Problem 26.5   Show how *n*-pentylamine can be synthesized from available materials by the four routes just outlined.

26.9   Reduction of nitro compounds

Like many organic compounds, nitro compounds can be reduced in two general ways: (a) by catalytic hydrogenation using molecular hydrogen, or (b) by chemical reduction, usually by a metal and acid.

Hydrogenation of a nitro compound to an amine takes place smoothly when a solution of the nitro compound in alcohol is shaken with finely divided nickel or platinum under hydrogen gas. For example:

o-Nitroacetanilide   *o*-Aminoacetanilide

This method cannot be used when the molecule also contains some other easily hydrogenated group, such as a carbon–carbon double bond.

Chemical reduction in the laboratory is most often carried out by adding hydrochloric acid to a mixture of the nitro compound and a metal, usually granulated tin. In the acidic solution, the amine is obtained as its salt; the free amine is liberated by the addition of base, and is steam-distilled from the reaction mixture.

The crude amine is generally contaminated with some unreduced nitro compound, from which it can be separated by taking advantage of the basic properties of the amine; the amine is soluble in aqueous mineral acid, and the nitro compound is not.

Reduction of nitro compounds to amines is an essential step in what is probably the most important synthetic route in aromatic chemistry. Nitro compounds are readily prepared by direct nitration; when a mixture of *ortho* and *para* isomers is obtained, it can generally be separated to yield the pure isomers. The primary aromatic amines obtained by the reduction of these nitro compounds are readily converted into diazonium salts; the diazonium group, in turn, can be replaced by a large number of other groups (Sec. 27.12). In most cases this sequence is the best method of introducing these other groups into the aromatic ring. In addition, diazonium salts can be used to prepare the extremely important class of compounds, the *azo dyes*.

26.10 Ammonolysis of halides

Many organic halogen compounds are converted into amines by treatment with aqueous or alcoholic solutions of ammonia. The reaction is generally carried out either by allowing the reactants to stand together at room temperature or by heating them under pressure. Displacement of halogen by NH_3 yields the amine salt, from which the free amine can be liberated by treatment with hydroxide ion.

$$RX + NH_3 \longrightarrow RNH_3^+X^-$$

$$RNH_3^+X^- + OH^- \longrightarrow RNH_2 + H_2O + X^-$$

Ammonolysis of halides belongs to the class of reactions that we have called nucleophilic substitution. The organic halide is attacked by the nucleophilic ammonia molecule in the same way that it is attacked by hydroxide ion, alkoxide ion, cyanide ion, acetylide ion, and water:

$$H_3N: + R-X \longrightarrow \left[H_3\overset{\delta+}{N}\text{---}R\text{---}\overset{\delta-}{X} \right] \longrightarrow H_3\overset{+}{N}-R + X^-$$

Like these other nucleophilic substitution reactions, ammonolysis is limited chiefly to alkyl halides or substituted alkyl halides. As with other reactions of this kind, elimination tends to compete (Sec. 7.24) with substitution: ammonia can attack hydrogen to form alkene as well as attack carbon to form amine. Ammonolysis thus gives the highest yields with primary halides (where substitution predominates) and is virtually worthless with tertiary halides (where elimination predominates).

$$CH_3CH_2CH_2CH_2Br \xrightarrow{NH_3} CH_3CH_2CH_2CH_2NH_3{}^+Br^- \qquad Substitution$$

$$CH_3-\underset{\underset{Br}{|}}{\overset{\overset{CH_3}{|}}{C}}-CH_3 \xrightarrow{NH_3} CH_3-\underset{}{\overset{\overset{CH_3}{|}}{C}}=CH_2 + NH_4Br \qquad Elimination$$

Because of their generally low reactivity, aryl halides are converted into amines only (a) if the ring carries —NO$_2$ groups, or other strongly electron-withdrawing groups, at positions *ortho* and *para* to the halogen, or (b) if a high temperature or a strongly basic reagent is used (Chap. 29).

Some examples of the application of ammonolysis to synthesis are:

$$\bigcirc CH_3 \xrightarrow[heat]{Cl_2} \bigcirc CH_2Cl \xrightarrow{NH_3} \bigcirc CH_2NH_2$$

| Toluene | Benzyl chloride | Benzylamine |

$$CH_3CH_2COOH \xrightarrow{Br_2}{}_{P} CH_3CHCOOH \xrightarrow{NH_3} CH_3CHCOOH$$
$$\qquad\qquad\qquad\qquad\qquad\quad \underset{Br}{|} \qquad\qquad\qquad \underset{NH_2}{|}$$

| Propionic acid | α-Bromopropionic acid | Alanine |
| | | (α-Aminopropionic acid) |

$$CH_2=CH_2 \xrightarrow{Cl_2} ClCH_2CH_2Cl \xrightarrow{2NH_3} H_2NCH_2CH_2NH_2$$

| Ethylene | Ethylene chloride | Ethylenediamine |

A serious disadvantage to the synthesis of amines by ammonolysis is the formation of more than one class of amine. The primary amine salt, formed by the

$$RX + NH_3 \longrightarrow RNH_3{}^+X^-$$
$$1° \text{ amine salt}$$

initial substitution, reacts with the reagent ammonia to yield the ammonium salt and the free primary amine; the following equilibrium thus exists:

$$RNH_3{}^+ + NH_3 \rightleftarrows RNH_2 + NH_4{}^+$$
$$\qquad\qquad\qquad 1° \text{ amine}$$

The free primary amine, like the ammonia from which it was made, is a nucleophilic reagent; it too can attack the alkyl halide, to yield the salt of a secondary amine:

$$RNH_2 + RX \longrightarrow R_2NH_2{}^+X^- \xrightarrow{NH_3} R_2NH$$
$$1° \text{ amine} \qquad\qquad\qquad\qquad\qquad 2° \text{ amine}$$

The secondary amine, which is in equilibrium with its salt, can in turn attack the alkyl halide to form the salt of a tertiary amine:

$$R_2NH + RX \longrightarrow R_3NH^+X^- \underset{}{\overset{NH_3}{\rightleftharpoons}} R_3N$$

2° amine   3° amine

Finally, the tertiary amine can attack the alkyl halide to form a compound of the formula $R_4N^+X^-$, called a *quaternary ammonium salt* (discussed in Sec. 27.5):

$$R_3N + RX \longrightarrow R_4N^+X^-$$

3° amine   Quaternary ammonium salt
                          (4°)

The presence of a large excess of ammonia lessens the importance of these last reactions and increases the yield of primary amine; under these conditions, a molecule of alkyl halide is more likely to encounter, and be attacked by, one of the numerous ammonia molecules rather than one of the relatively few amine molecules. At best, the yield of primary amine is always cut down by the formation of the higher classes of amines. Except in the special case of methylamine, the primary amine can be separated from these by-products by distillation.

26.11   Reductive amination

Many aldehydes (RCHO) and ketones (R_2CO) are converted into amines by **reductive amination**: reduction in the presence of ammonia. Reduction can be accomplished catalytically or by use of sodium cyanohydridoborate, $NaBH_3CN$. Reaction involves reduction of an intermediate compound (an *imine*, RCH=NH or R_2C=NH) that contains a carbon–nitrogen double bond.

$$\underset{\text{An aldehyde}}{R-\overset{\overset{\text{H}}{|}}{C}=O} + NH_3 \longrightarrow \left[\underset{\text{An imine}}{R-\overset{\overset{\text{H}}{|}}{C}=NH} \right] \xrightarrow[\text{or NaBH}_3\text{CN}]{H_2, Ni} \underset{\underset{\text{A 1° amine}}{}}{R-\overset{\overset{\text{H}}{|}}{\underset{\underset{\text{H}}{|}}{C}}-NH_2}$$

$$\underset{\text{A ketone}}{R-\overset{\overset{\text{R}'}{|}}{C}=O} + NH_3 \longrightarrow \left[\underset{\text{An imine}}{R-\overset{\overset{\text{R}'}{|}}{C}=NH} \right] \xrightarrow[\text{or NaBH}_3\text{CN}]{H_2, Ni} \underset{\underset{\text{A 1° amine}}{}}{R-\overset{\overset{\text{R}'}{|}}{\underset{\underset{\text{H}}{|}}{C}}-NH_2}$$

Reductive amination has been used successfully with a wide variety of aldehydes and ketones, both aliphatic and aromatic. For example:

$$CH_3(CH_2)_5CHO \xrightarrow{NH_3, H_2, Ni} CH_3(CH_2)_5CH_2NH_2$$

Heptaldehyde   *n*-Heptylamine
(Heptanal)   (1-Aminoheptane)

Benzaldehyde → Benzylamine

2-Pentanone
(Methyl *n*-propyl ketone) → 2-Aminopentane

Acetophenone
(Methyl phenyl ketone) → α-Phenylethylamine

Reductive amination of ketones yields amines containing a *sec*-alkyl group; such amines are difficult to obtain by ammonolysis because of the tendency for *sec*-alkyl halides to undergo elimination. For example, cyclohexanone is converted into cyclohexylamine in good yield, whereas ammonolysis of bromocyclohexane yields only cyclohexene.

Cyclohexanol
Cyclohexanone → Cyclohexylamine
Bromocyclohexane → Cyclohexene

During reductive amination the aldehyde or ketone can react not only with ammonia but also with the primary amine that has already been formed, and thus yield a certain amount of secondary amine. The tendency for the reaction to go

Aldehyde   1° amine   Imine   2° amine

beyond the desired stage can be fairly well limited by the proportions of reactants employed and is seldom a serious handicap.

26.12 Hofmann degradation of amides

As a method of synthesis of amines, the Hofmann degradation of amides has the special feature of yielding a product containing one less carbon than the starting material. As we can see, reaction involves migration of a group from carbonyl

$$R-C\overset{O}{\underset{NH_2}{\diagdown}} \xrightarrow{OBr^-} R-NH_2 + CO_3^{2-}$$

An amide   A 1° amine

carbon to the adjacent nitrogen atom, and thus is an example of a *molecular rearrangement*. (We shall return to the Hofmann degradation in Secs. 32.2–32.5 and discuss its mechanism in detail.)

Problem 26.6 Using a different method in each case, show how the following amines could be prepared from *toluene* and any aliphatic reagents:

(a) $\bigcirc$ CH_2NH_2

(c) $\bigcirc$ $CH_2CH_2NH_2$

(b) CH_3 $\bigcirc$ $CHCH_3$ / NH_2

(d) CH_3 $\bigcirc$ NH_2

(e) $\bigcirc$ NH_2

26.13 Synthesis of secondary and tertiary amines

So far we have been chiefly concerned with the synthesis of primary amines. Secondary and tertiary amines are prepared by adaptations of one of the processes already described: ammonolysis of halides or reductive amination. For example:

$$CH_3CH_2CH_2CH_2NH_2 + CH_3CH_2Br \longrightarrow CH_3CH_2CH_2CH_2\overset{H}{\underset{}{-N-}}CH_2CH_3$$

n-Butylamine   Ethyl bromide   *n*-Butylethylamine
(1°)   (2°)

$$CH_3CH_2CCH_3 + CH_3NH_2 \xrightarrow{H_2, Ni} CH_3CH_2CHCH_3$$
$$\underset{O}{\|} \qquad \text{Methylamine} \qquad \underset{NHCH_3}{|}$$

Butanone   (1°)   *sec*-Butylmethylamine
(Ethyl methyl ketone)   (2°)

$\bigcirc$ NH_2 $\xrightarrow{CH_3Cl}$ $\bigcirc$ $NHCH_3$ $\xrightarrow{CH_3Cl}$ $\bigcirc$ $N(CH_3)_2$

Aniline   *N*-Methylaniline   *N,N*-Dimethylaniline
(1°)   (2°)   (3°)

$$\underset{\substack{\text{n-Butylethylamine} \\ (2°)}}{CH_3CH_2CH_2CH_2\overset{\overset{\displaystyle H}{|}}{N}CH_2CH_3} + \underset{\substack{\text{Methyl} \\ \text{bromide.}}}{CH_3Br} \longrightarrow \underset{\substack{\text{n-Butylethylmethylamine} \\ (3°)}}{CH_3CH_2CH_2CH_2\overset{\overset{\displaystyle CH_3}{|}}{N}CH_2CH_3}$$

Where ammonia has been used to produce a primary amine, a primary amine can be used to produce a secondary amine, or a secondary amine can be used to produce a tertiary amine. In each of these syntheses there is a tendency for reaction to proceed beyond the first stage and to yield an amine of a higher class than the one that is wanted.

26.14   Heterocyclic amines

A particularly important kind of amino compound is one in which the nitrogen makes up part of a ring. Since such a ring contains more than one kind of atom— nitrogen plus the usual carbon—the compound of which it is a part is said to be *heterocyclic*. (Compare, for example, the heterocyclic oxygen compounds in Secs. 19.9–19.10.) We shall discuss heterocyclic compounds in detail in Chapter 35. But it is hard to go very far in organic chemistry without encountering heterocyclic nitrogen compounds—indeed, we have already encountered them as reagents— and so we shall look briefly at some of them here.

These heterocyclic amines can be saturated or unsaturated, aliphatic or aromatic; a nitrogen may share the ring with another nitrogen or with a hetero atom of a different kind—oxygen, say, or sulfur. For example:

| Pyrrole | Pyridine | Pyrimidine |
| Pyrrolidine | Piperidine | Purine |

The important thing for us to realize at this point is that, part of a ring or not, nitrogen is still nitrogen. It retains its most important property, *basicity*; and this basicity, as we shall see in the next chapter, is the property that determines the chemical behavior of amines.

We have all heard of the *bases* whose sequence along the DNA molecule constitutes the genetic code. These bases are heterocyclic bases, and their basicity comes from nitrogen.

PROBLEMS

1. Draw structures, give names, and classify as primary, secondary, or tertiary:

(a) the eight isomeric amines of formula $C_4H_{11}N$
(b) the five isomeric amines of formula C_7H_9N that contain a benzene ring

2. Give the structural formulas of the following compounds:

(a) *sec*-butylamine
(b) *o*-toluidine
(c) anilinium chloride
(d) diethylamine
(e) *p*-aminobenzoic acid
(f) benzylamine
(g) isopropylammonium benzoate
(h) *o*-phenylenediamine

(i) *N,N*-dimethylaniline
(j) ethanolamine (2-aminoethanol)
(k) β-phenylethylamine
(l) *N,N*-dimethylaminocyclohexane
(m) diphenylamine
(n) 2,4-dimethylaniline
(o) tetra-*n*-butylammonium iodide
(p) *p*-anisidine

3. Show how *n*-propylamine could be prepared from each of the following:

(a) *n*-propyl bromide
(b) *n*-propyl alcohol + HBr
(c) propionaldehyde
(d) 1-nitropropane

(e) propionitrile
(f) *n*-butyramide
(g) *n*-butyl alcohol
(h) ethyl alcohol

Which of these methods can be applied to the preparation of aniline? Of benzylamine?

4. Outline all steps in a possible laboratory synthesis of each of the following compounds from benzene, toluene, and alcohols of four carbons or fewer using any needed inorganic reagents.

(a) isopropylamine
(b) *n*-pentylamine
(c) *p*-toluidine
(d) ethylisopropylamine
(e) α-phenylethylamine
(f) β-phenylethylamine
(g) *m*-chloroaniline

(h) *p*-aminobenzoic acid
(i) 3-aminoheptane
(j) *N*-ethylaniline
(k) 2,4-dinitroaniline
(l) the drug *benzedrine* (2-amino-1-phenylpropane)
(m) *p*-nitrobenzylamine
(n) 2-amino-1-phenylethanol

5. Outline all steps in a possible laboratory synthesis from palmitic acid, $n\text{-}C_{15}H_{31}COOH$, of:

(a) $n\text{-}C_{16}H_{33}NH_2$
(b) $n\text{-}C_{17}H_{35}NH_2$

(c) $n\text{-}C_{15}H_{31}NH_2$
(d) $n\text{-}C_{15}H_{31}CH(NH_2)\text{-}n\text{-}C_{16}H_{33}$

6. On the basis of the following synthesis give the structures of *putrescine* and *cadaverine*, found in rotting flesh:

(a) ethylene bromide $\xrightarrow{\text{KCN}}$ $C_4H_4N_2$ $\xrightarrow{\text{Na, } C_2H_5OH}$ putrescine $(C_4H_{12}N_2)$

(b) $Br(CH_2)_5Br$ $\xrightarrow{\text{NH}_3}$ cadaverine $(C_5H_{14}N_2)$

7. One of the raw materials for the manufacture of nylon-6,6 is *hexamethylenediamine*, $NH_2(CH_2)_6NH_2$. Much of this amine is made by a process that begins with the 1,4-addition of chlorine to 1,3-butadiene. What do you think might be the subsequent steps in this process?

8. Outline all steps in a possible synthesis of β-alanine (β-aminopropionic acid) from succinic anhydride.

9. Using models and then drawing formulas, show the stereoisomeric forms in which each of the following compounds can exist. Tell which stereoisomers when separated from all others would be optically active and which would be optically inactive.

(a) α-phenylethylamine
(b) *N*-ethyl-*N*-methylaniline

(c) ethylmethylphenyl-*n*-propylammonium bromide

(d)

$$\text{Br}^- \quad \underset{\underset{C_2H_5}{|}}{\overset{\overset{CH_3}{\diagup}}{N^{\oplus}}} \diagdown CH_2{-}CH_2 \diagdown \underset{\underset{C_2H_5}{|}}{\overset{\overset{CH_3}{\diagup}}{N^{\oplus}}} \diagdown \text{Br}^-$$

(e)

$$\underset{C_6H_5}{\overset{H}{\diagdown}}C\underset{CH_2{-}CH_2}{\diagup} \diagup N^{\oplus} \diagdown CH_2{-}CH_2 \diagdown \underset{COOC_2H_5}{\overset{H}{\diagup}}C \diagup \quad \text{Br}^-$$

(f) ethylmethylphenylamine oxide, $(CH_3)(C_2H_5)(C_6H_5)N{-}O$

10. Two geometric isomers of benzaldoxime, $C_6H_5CH{=}NOH$, are known. (a) Draw their structures, showing the geometry of the molecules. (b) Show how this geometry results from their electronic configurations. (c) Would you predict geometric isomerism for benzophenoneoxime, $(C_6H_5)_2C{=}NOH$? For acetophenoneoxime, $C_6H_5C(CH_3){=}NOH$? For azobenzene, $C_6H_5N{=}NC_6H_5$?

11. (a) Give structural formulas of compounds A through D.

$$\text{phthalimide (Sec. 24.14)} + \text{KOH(alc.)} \longrightarrow \text{A} (C_8H_4O_2NK)$$

$$\text{A} + CH_3CH_2CH_2Br, \text{ heat} \longrightarrow \text{B} (C_{11}H_{13}O_2N)$$

$$\text{B} + H_2O, OH^-, \text{ heat} \longrightarrow \text{C} (C_3H_9N) + \text{D}$$

(b) This sequence illustrates the **Gabriel synthesis**. What class of compounds does it produce? What particular advantage does it have over alternative methods for the production of these compounds? On what special property of phthalimide does the synthesis depend?

27

Amines II. Reactions

27.1 Reactions

Like ammonia, the three classes of amines contain nitrogen that bears an unshared pair of electrons; as a result, amines closely resemble ammonia in chemical properties (Fig. 27.1, on the next page). *The tendency of nitrogen to share this pair of electrons underlies the entire chemical behavior of amines*: their basicity, their action as nucleophiles—in both aliphatic and acyl substitution—and the unusually high reactivity of aromatic rings bearing amino or substituted amino groups.

REACTIONS OF AMINES _____

    1. Basicity. Salt formation. Discussed in Secs. 26.5 and 27.2–27.4.

$$RNH_2 + H^+ \rightleftharpoons RNH_3^+$$

$$R_2NH + H^+ \rightleftharpoons R_2NH_2^+$$

$$R_3N + H^+ \rightleftharpoons R_3NH^+$$

CONTINUED

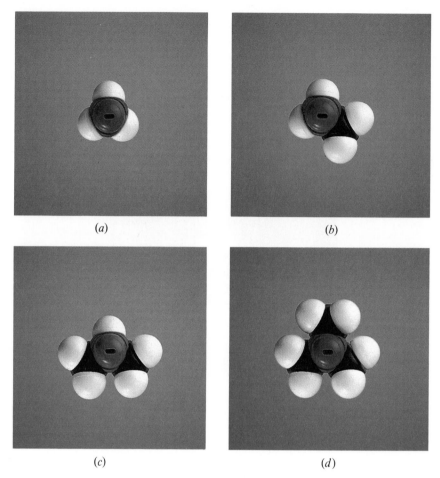

(a)   (b)

(c)   (d)

Figure 27.1 Molecular structure and chemical reactivity. Models of:
(a) ammonia, NH_3; (b) methylamine, CH_3NH_2; (c) dimethylamine,
$(CH_3)_2NH$; (d) trimethylamine, $(CH_3)_3N$. The chemical behavior of amines
depends upon the tendency of nitrogen to share its unshared pair of electrons,
shown facing us in each model.

—— CONTINUED ——

Examples:

NH$_2$ + HCl ⇌ NH$_3^+$Cl$^-$

Aniline   Anilinium chloride
                                 (Aniline hydrochloride)

$(CH_3)_2NH + HNO_3$ ⇌ $(CH_3)_2NH_2^+NO_3^-$

Dimethylamine   Dimethylammonium nitrate

N(CH$_3$)$_2$ + CH$_3$COOH ⇌ $\overset{H}{N}$(CH$_3$)$_2^+$ $^-$OOCCH$_3$

N,N-Dimethylaniline   N,N-Dimethylanilinium acetate

—— CONTINUED ——

CONTINUED

2. Alkylation. Discussed in Secs. 26.13 and 27.5.

$$RNH_2 \xrightarrow{RX} R_2NH \xrightarrow{RX} R_3N \xrightarrow{RX} R_4N^+X^-$$

$$ArNH_2 \xrightarrow{RX} ArNHR \xrightarrow{RX} ArNR_2 \xrightarrow{RX} ArNR_3^+X^-$$

Examples:

$$(n\text{-}C_4H_9)_2NH + \bigcirc CH_2Cl \longrightarrow (n\text{-}C_4H_9)_2NCH_2\bigcirc$$

Di-*n*-butylamine   Benzyl chloride   Benzyldi(*n*-butyl)amine
(2°)   (3°)

$$n\text{-}C_3H_7NH_2 \xrightarrow{CH_3I} n\text{-}C_3H_7\overset{\overset{\displaystyle H}{|}}{N}CH_3 \xrightarrow{CH_3I} n\text{-}C_3H_7\overset{\overset{\displaystyle CH_3}{|}}{N}CH_3 \xrightarrow{CH_3I} n\text{-}C_3H_7\overset{\overset{\displaystyle CH_3}{|}}{\underset{\underset{\displaystyle CH_3}{|}}{N}}CH_3{}^+I^-$$

n-Propylamine   Methyl-*n*-propylamine   Dimethyl-*n*-propylamine   Trimethyl-*n*-propylammonium
(1°)   (2°)   (3°)   iodide
                                                                                        (4°)

3. Conversion into amides. Discussed in Sec. 27.7.

Primary   RNH$_2$ — $\xrightarrow{R'COCl}$ R'CO—NHR   An *N*-substituted amide

— $\xrightarrow{ArSO_2Cl}$ ArSO$_2$—NHR   An *N*-substituted sulfonamide

Secondary   R$_2$NH — $\xrightarrow{R'COCl}$ R'CO—NR$_2$ An *N,N*-disubstituted amide

— $\xrightarrow{ArSO_2Cl}$ ArSO$_2$—NR$_2$ An *N,N*-disubstituted sulfonamide

Tertiary   R$_3$N — $\xrightarrow{R'COCl}$ No reaction

— $\xrightarrow{ArSO_2Cl}$ No reaction under conditions of
Hinsberg test (*but see* Sec. 27.19)

CONTINUED

CONTINUED

Examples:

4. **Ring substitution in aromatic amines.** Discussed in Secs. 27.8, 27.11 and 27.18.

$$\left.\begin{array}{c} -NH_2 \\ -NHR \\ -NR_2 \end{array}\right\}$$ Activate powerfully, and direct *ortho,para* in electrophilic aromatic substitution

$-NHCOR:$ Less powerful activator than $-NH_2$

Examples:

CONTINUED

_____ CONTINUED _____

$$N,N\text{-Dimethyl-} \atop \text{aniline} \quad \xrightarrow{\text{NaNO}_2,\ \text{HCl}} \quad p\text{-Nitroso-}N,N\text{-dimethylaniline}$$

$$(CH_3)_2N\text{—} \quad + \quad \text{—}N_2{}^+Cl^- \quad \xrightarrow{\text{acid}} \quad (CH_3)_2N\text{—}\text{—}N{=}N\text{—}\text{—} \quad + \quad HCl$$

| N,N-Dimethyl-
aniline | Benzenediazonium
chloride | An azo compound |

5. **Hofmann elimination from quaternary ammonium salts.** Discussed in Secs. 27.5–27.6.

$$\begin{matrix} H \\ | \\ \text{—C—C—} \\ | \\ \oplus NR_3 \end{matrix} \quad \xrightarrow{\text{OH}^-,\ \text{heat}} \quad \underset{\text{Alkene}}{\diagup C{=}C\diagdown} \quad + \quad \underset{3^\circ\ \text{amine}}{R_3N} \quad + \quad H_2O$$

Quaternary
ammonium ion

6. **Reactions with nitrous acid.** Discussed in Secs. 27.11–27.12.

Primary aromatic $ArNH_2 \xrightarrow{\text{HONO}} Ar{-}N{\equiv}N^+$ Diazonium salt

Primary aliphatic $RNH_2 \xrightarrow{\text{HONO}} [R{-}N{\equiv}N^+] \xrightarrow{\text{H}_2\text{O}} N_2 +$ mixture of alcohols and alkenes

Secondary aromatic or aliphatic $\begin{matrix}ArNHR\\ \text{or}\\ R_2NH\end{matrix} \xrightarrow{\text{HONO}} \begin{matrix}R\\|\\Ar{-}N{-}N{=}O\\ \text{or}\\ R_2N{-}N{=}O\end{matrix}$ N-Nitrosoamine

Tertiary aromatic $\text{—}NR_2 \xrightarrow{\text{HONO}} O{=}N\text{—}\text{—}NR_2$ p-Nitroso compound ■

27.2 Basicity of amines. Basicity constant

Like ammonia, amines are converted into their salts by aqueous mineral acids and are liberated from their salts by aqueous hydroxides. Like ammonia, therefore, amines are more basic than water and less basic than hydroxide ion:

$$RNH_2 + H_3O^+ \longrightarrow RNH_3{}^+ + H_2O$$

Stronger base ··· Weaker base

$$RNH_3{}^+ + OH^- \longrightarrow RNH_2 + H_2O$$

Stronger base ··· Weaker base

We found it convenient to compare acidities of carboxylic acids by measuring the extent to which they give up hydrogen ion to water; the equilibrium constant for this reaction we combined with $[H_2O]$ to obtain the acidity constant, K_a. In the same way, it is convenient to compare basicities of amines by measuring the extent to which they accept hydrogen ion from water; the equilibrium constant for this reaction we combine with $[H_2O]$ to obtain the **basicity constant, K_b**.

$$RNH_2 + H_2O \;\underset{\longleftarrow}{\overset{\longrightarrow}{}}\; RNH_3^+ + OH^-$$

$$K_b = K_{eq}[H_2O] = \frac{[RNH_3^+][OH^-]}{[RNH_2]}$$

Each amine has its characteristic K_b; the larger the K_b, the stronger the base.

We must not lose sight of the fact that the principal base in an aqueous solution of an amine (or of ammonia, for that matter) is the *amine* itself, not hydroxide ion. Measurement of $[OH^-]$ is simply a convenient way to compare basicities.

We see in Table 26.1 (p. 933) that aliphatic amines of all three classes have K_b values of about 10^{-3} to 10^{-4} (0.001 to 0.0001); they are thus somewhat stronger bases than ammonia ($K_b = 1.8 \times 10^{-5}$). Aromatic amines, on the other hand, are considerably weaker bases than ammonia, having K_b values of 10^{-9} or less. Substituents on the ring have a marked effect on the basicity of aromatic amines, *p*-nitroaniline, for example, being only 1/4000 as basic as aniline (Table 27.1).

Table 27.1   BASICITY CONSTANTS OF SUBSTITUTED ANILINES

			K_b of aniline $= 4.2 \times 10^{-10}$		
	K_b		K_b		K_b
p-NH$_2$	140×10^{-10}	*m*-NH$_2$	10×10^{-10}	*o*-NH$_2$	3×10^{-10}
p-OCH$_3$	20	*m*-OCH$_3$	2	*o*-OCH$_3$	3
p-CH$_3$	12	*m*-CH$_3$	5	*o*-CH$_3$	2.6
p-Cl	1	*m*-Cl	0.3	*o*-Cl	0.05
p-NO$_2$	0.001	*m*-NO$_2$	0.029	*o*-NO$_2$	0.00006

27.3   Structure and basicity

Let us see how basicity of amines is related to structure. We shall handle basicity just as we handled acidity: we shall compare the stabilities of amines with the stabilities of their ions; the more stable the ion relative to the amine from which it is formed, the more basic the amine.

First of all, amines are more basic than alcohols, ethers, esters, etc., for the same reason that ammonia is more basic than water: nitrogen is less electronegative than oxygen, and can better accommodate the positive charge of the ion.

An aliphatic amine is more basic than ammonia because the electron-releasing alkyl groups tend to disperse the positive charge of the substituted ammonium ion, and therefore stabilize it in a way that is not possible for the unsubstituted ammonium ion. Thus an ammonium ion is stabilized by electron release in the same way as a carbocation (Sec. 5.21). From another point of view, we can consider that an alkyl group pushes electrons toward nitrogen, and thus makes the fourth pair more available for sharing with an acid. (The differences in basicity among primary, secondary, and tertiary aliphatic amines are due to a combination of solvation and polar factors.)

$$
\begin{array}{ccc}
\text{H} & & \text{H} \\
| & & | \\
\text{R} \rightarrow \text{N:} + \text{H}^+ & \rightleftharpoons & \text{R} \rightarrow \text{N} - \text{H}^+ \\
| & & | \\
\text{H} & & \text{H}
\end{array}
$$

R *releases electrons:*   R *releases electrons:*
makes unshared pair   *stabilizes ion,*
more available   *increases basicity*

How can we account for the fact that aromatic amines are weaker bases than ammonia? Let us compare the structures of aniline and the anilinium ion with the structures of ammonia and the ammonium ion. We see that ammonia and the ammonium ion are each represented satisfactorily by a single structure:

$$
\begin{array}{cc}
\text{H} & \text{H} \\
\text{H:N:H} & \text{H:N:H}^+ \\
& \text{H}
\end{array}
$$

Ammonia   Ammonium ion

Aniline and anilinium ion contain the benzene ring and therefore are hybrids of the Kekulé structures I and II, and III and IV. This resonance presumably stabilizes

I   II   III   IV
Aniline   Anilinium ion

both amine and ion to the same extent. It lowers the energy content of each by the same number of kilocalories per mole, and hence does not affect the *difference* in their energy contents, that is, does not affect ΔG of ionization. If there were no other factors involved, then, we might expect the basicity of aniline to be about the same as the basicity of ammonia.

However, there are additional structures to be considered. To account for the powerful activating effect of the —NH$_2$ group on electrophilic aromatic substitution (Sec. 14.18), we considered that the intermediate carbocation is stabilized by structures in which there is a double bond between nitrogen and the ring; contribution from these structures is simply a way of indicating the tendency for nitrogen to share its fourth pair of electrons and to accept a positive charge. The —NH$_2$ group tends to share electrons with the ring, not only in the carbocation that is the intermediate in electrophilic aromatic substitution, but also in the aniline molecule itself.

Thus aniline is a hybrid not only of structures I and II but also of structures V, VI, and VII. We cannot draw comparable structures of the anilinium ion. Contribution from the three structures V, VI, and VII stabilizes the amine in a

way that is not possible for the ammonium ion; resonance thus lowers the energy content of aniline more than it lowers the energy content of the anilinium ion. The net effect is to shift the equilibrium in the direction of less ionization, that is, to make K_b smaller (Fig. 27.2). (See, however, the discussion in Sec. 23.11.)

Figure 27.2 Molecular structure and position of equilibrium. A resonance-stabilized aromatic amine is a weaker base than ammonia. (The plots are aligned with each other for easy comparison.)

The low basicity of aromatic amines is thus due to the fact that the amine is stabilized by resonance to a greater extent than is the ion.

From another point of view, we can say that aniline is a weaker base than ammonia because the fourth pair of electrons is partly shared with the ring and is thus less available for sharing with a hydrogen ion. The tendency (through resonance) for the —NH_2 group to release electrons to the aromatic ring makes the ring more reactive toward electrophilic attack; at the same time this tendency necessarily makes the amine less basic. Similar considerations apply to other aromatic amines.

27.4 Effect of substituents on basicity of aromatic amines

How is the basicity of an aromatic amine affected by substituents on the ring?

In Table 27.1 (p. 956) we see that an electron-releasing substituent like $-CH_3$ increases the basicity of aniline, and an electron-withdrawing substituent like $-X$ or $-NO_2$ decreases the basicity. These effects are understandable. Electron release tends to disperse the positive charge of the anilinium ion, and thus stabilizes the ion relative to the amine. Electron withdrawal tends to intensify the positive charge of the anilinium ion, and thus destabilizes the ion relative to the amine.

Basicity of aromatic amines

G releases electrons:
stabilizes cation,
increases basicity

$G = -NH_2$
$\quad\;\; -OCH_3$
$\quad\;\; -CH_3$

G withdraws electrons:
destabilizes cation,
decreases basicity

$G = -NH_3^+$
$\quad\;\; -NO_2$
$\quad\;\; -SO_3^-$
$\quad\;\; -COOH$
$\quad\;\; -X$

We notice that the base-strengthening substituents are the ones that activate an aromatic ring toward electrophilic substitution; the base-weakening substituents are the ones that deactivate an aromatic ring toward electrophilic substitution (see Sec. 14.5). Basicity depends upon position of equilibrium, and hence on relative stabilities of reactants and products. Reactivity in electrophilic aromatic substitution depends upon rate, and hence on relative stabilities of reactants and transition state. The effect of a particular substituent is the same in both cases, however, since the controlling factor is accommodation of a positive charge.

A given substituent affects the basicity of an amine and the acidity of a carboxylic acid in opposite ways (compare Sec. 23.14). This is to be expected, since basicity depends upon ability to accommodate a positive charge, and acidity depends upon ability to accommodate a negative charge.

Once again we see the operation of the **ortho effect** (Sec. 23.14). Even electron-releasing substituents weaken basicity when they are *ortho* to the amino group, and electron-withdrawing substituents do so to a much greater extent from the *ortho* position than from the *meta* or *para* position.

From another point of view, we can consider that an electron-releasing group pushes electrons toward nitrogen and makes the fourth pair more available for sharing with an acid, whereas an electron-withdrawing group helps pull electrons away from nitrogen and thus makes the fourth pair less available for sharing.

Problem 27.1 (a) Besides destabilizing the anilinium ion, how else might a nitro group affect basicity? (*Hint*: See structures V–VII on p. 958.) (b) Why does the nitro group exert a larger base-weakening effect from the *para* position than from the nearer *meta* position?

Problem 27.2 Draw the structural formula of the product expected (if any) from the reaction of trimethylamine and BF_3.

27.5 Quaternary ammonium salts. Exhaustive methylation. Hofmann elimination

Like ammonia, an amine can react with an alkyl halide; the product is an amine of the next higher class. The alkyl halide undergoes nucleophilic substitution, with the basic amine serving as the nucleophilic reagent. We see that one of the

$$RNH_2 \xrightarrow{\text{RX}} R_2NH \xrightarrow{\text{RX}} R_3N \xrightarrow{\text{RX}} R_4N^+X^-$$
$$1° \qquad\qquad 2° \qquad\qquad 3° \qquad\qquad 4°$$

hydrogens attached to nitrogen has been replaced by an alkyl group; the reaction is therefore often referred to as *alkylation of amines*. The amine can be aliphatic or aromatic, primary, secondary, or tertiary; the halide is generally an alkyl halide.

We have already encountered alkylation of amines as a side reaction in the preparation of primary amines by the ammonolysis of halides (Sec. 26.10), and as a method of synthesis of secondary and tertiary amines (Sec. 26.13). Let us look at one further aspect of this reaction, the formation of quaternary ammonium salts.

Quaternary ammonium salts are the products of the final stage of alkylation of nitrogen. They have the formula $R_4N^+X^-$. Four organic groups are covalently bonded to nitrogen, and the positive charge of this ion is balanced by some negative ion. When the salt of a primary, secondary, or tertiary amine is treated with hydroxide ion, nitrogen gives up a hydrogen ion and the free amine is liberated. The quaternary ammonium ion, having no proton to give up, is not affected by hydroxide ion.

When a solution of a quaternary ammonium halide is treated with silver oxide, silver halide precipitates. When the mixture is filtered and the filtrate is evaporated to dryness, there is obtained a solid which is free of halogen. An aqueous solution of this substance is strongly alkaline, and is comparable to a solution of sodium hydroxide or potassium hydroxide. A compound of this sort is called a **quaternary ammonium hydroxide**. It has the structure $R_4N^+OH^-$. Its aqueous solution is basic for the same reason that solutions of sodium or potassium hydroxide are basic: the solution contains hydroxide ions.

$$
\begin{array}{c}
R \\
R:\overset{..}{N}:R^+X^- \\
\overset{..}{R}
\end{array}
\xrightarrow{\text{Ag}_2\text{O}}
\begin{array}{c}
R \\
R:\overset{..}{N}:R^+OH^- \\
\overset{..}{R}
\end{array}
+ \quad \text{AgX}
$$

Quaternary   Quaternary   *Insoluble*
ammonium salt   ammonium hydroxide

When a quaternary ammonium hydroxide is heated strongly (to 125 °C or higher), it decomposes to yield water, a tertiary amine, and an alkene. Trimethyl-*n*-propylammonium hydroxide, for example, yields trimethylamine and propylene:

$$
\begin{array}{c}
CH_3 \\
| \\
CH_3-N^+-CH_2CH_2CH_3 \; OH^- \\
| \\
CH_3
\end{array}
\xrightarrow{\text{heat}}
\begin{array}{c}
CH_3 \\
| \\
CH_3-N \\
| \\
CH_3
\end{array}
+ \; CH_2{=}CHCH_3 + H_2O
$$

Trimethyl-*n*-propylammonium   Trimethylamine   Propylene
hydroxide

This reaction, called the **Hofmann elimination**, is quite analogous to the dehydrohalogenation of an alkyl halide (Sec. 7.12). Most commonly, reaction is E2:

hydroxide ion abstracts a proton from carbon; a molecule of tertiary amine is expelled, and the double bond is generated. Bases other than hydroxide ion can be used.

$$-\overset{R_3N^{\oplus}}{\underset{\underset{\overset{|}{OH^-}}{\overset{|}{H}}}{\underset{|}{C}}}-\overset{|}{\underset{|}{C}}- \longrightarrow \ \ \ \ C=C\ \ \ + R_3N: + H-OH$$

E1 elimination from quaternary ammonium ions is also known. Competing with either E2 or E1 elimination there is, as usual, substitution: either S_N2 or S_N1. (*Problem*: What products would you expect from substitution?)

The formation of quaternary ammonium salts, followed by an elimination of the kind just described, is very useful in the determination of the structures of certain complicated nitrogen-containing compounds. The compound, which may be a primary, secondary, or tertiary amine, is converted into the quaternary ammonium hydroxide by treatment with excess methyl iodide and silver oxide. The number of methyl groups taken up by nitrogen depends upon the class of the amine; a primary amine will take up three methyl groups, a secondary amine will take up two, and a tertiary amine only one. This process is known as **exhaustive methylation of amines**.

When heated, a quaternary ammonium hydroxide undergoes elimination to an alkene and a tertiary amine. From the structures of these products it is often possible to deduce the structure of the original amine. As a simple example, contrast the products (I and II) obtained from the following isomeric cyclic amines:

2-Methylpyrrolidine

5-(Dimethylamino)-1-pentene

I

3-Methylpyrrolidine

4-(Dimethylamino)-3-methyl-1-butene

II

Problem 27.3 (a) What products would be expected from the hydrogenation of I and II? (b) How could you prepare an authentic sample of each of these expected hydrogenation products?

Problem 27.4 What products would be expected if I and II were subjected to exhaustive methylation and elimination?

27.6　E2 elimination: Hofmann orientation. The variable E2 transition state

Where the structure permits, E2 elimination can produce a mixture of isomers; which one predominates is determined by the orientation of the elimination. In dehydrohalogenation, we saw (Sec. 7.20), the orientation is *Saytzeff*: the preferred product is the more highly branched alkene which, as we saw, is the more stable one. Orientation, we said, is controlled by the alkene character of the transition state.

What is the orientation of the Hofmann elimination? A single example will show us the kind of thing that is observed:

$$CH_3CH_2CH_2CHCH_3 \xrightarrow[\text{C}_2\text{H}_5\text{OH}]{\text{C}_2\text{H}_5\text{ONa}} CH_3CH_2CH_2CH=CH_2 + CH_3CH_2CH=CHCH_3$$

$$\overset{|}{\oplus N(CH_3)_3}$$

2-Pentyltrimethyl-
ammonium ion
　　　　　　　　　　　　　　　　　　　1-Pentene　　　　　　2-Pentene
　　　　　　　　　　　　　　　　　　　96%　　　　　　　　*4%*

We see that the preferred product here is the *least* branched alkene, 1-pentene. Such orientation is called **Hofmann orientation**, since it was first observed by Hofmann in studying this particular kind of reaction.

Now, how can we account for Hofmann orientation? To see the factors at work here, let us return to dehydrohalogenation and take, as an example, elimination from the 2-hexyl halides brought about by the strong base sodium methoxide. The iodide, bromide, and chloride react with Saytzeff orientation, but the fluoride gives predominantly the less substituted alkene, 1-hexene, that is, reacts with Hofmann orientation. Furthermore, we can see that there is a steady increase in the fraction of 1-hexene along the series I, Br, Cl, F.

$$CH_3CH_2CH_2CH_2CHCH_3 \xrightarrow[\text{E2}]{\text{CH}_3\text{ONa}/\text{CH}_3\text{OH}} \text{2-hexene} + \text{1-hexene}$$

$$\overset{|}{X}$$

X =	2-hexene	1-hexene
I	81%	19%
Br	72	28
Cl	67	33
F	30	70

Such observations are best understood in terms of what Bunnett (p. 277) has called the *variable transition state* theory of E2 elimination. We are speaking, remember, of a one-step elimination; both the C—H and C—X bonds are being broken in the same transition state. But there is a whole spectrum of E2 transition states which differ in the relative *extent* to which the two bonds are broken.

Variable E2 transition state

Carbocation-like　　　　　　　Central　　　　　　　Carbanion-like
Much alkene character

At the center of the spectrum is the transition state we have described before for elimination from alkyl halides: both C—H and C—X bonds are broken to a considerable extent, the transition state has considerable alkene character, and orientation is Saytzeff.

But, if breaking of the C—H bond greatly exceeds breaking of the C—X bond, there is little alkene character to the transition state, but instead the development of negative charge on the carbon losing the proton. In this case, the transition state has *carbanion character*, and its stability is controlled as we might expect, by dispersal or intensification of the negative charge: electron-withdrawing groups stabilize, and electron-releasing groups destabilize. At one end of the spectrum, then, we have the carbanion-like transition state.

At the other end of the spectrum is the transition state in which C—X bond-breaking greatly exceeds C—H bond-breaking. Positive charge develops on the carbon losing the leaving group, giving carbocation character to the transition state. Alkene character is diminished, and we might expect orientation to be less strongly Saytzeff.

Consider elimination from the 2-hexyl halides. With the iodide, there is considerable breaking of both bonds in the transition state, much alkene character, and preferred formation of the more stable alkene: Saytzeff orientation. As we go along the series I, Br, Cl, F, the C—X bond becomes stronger, and the extent to which it is broken in the transition state decreases. At the same time, the electron-withdrawing effect of X increases, favoring the development of negative charge. With the fluoride, we have predominant C—H bond-breaking, with little alkene character but considerable carbanion character to the transition state. A primary hydrogen is preferentially abstracted by base, since that permits the negative charge to develop on a primary carbon, to which there is attached only one electron-releasing alkyl group. Orientation is Hofmann.

Bunnett believes that C—F bond-breaking lags behind C—H bond-breaking chiefly because of the strength of the C—F bond. Ingold (p. 183), who was the first to suggest carbanion character as the underlying cause of Hofmann orientation, believed that electron withdrawal by fluorine is the major factor.

On this basis, how do we account for Hofmann orientation in the E2 elimination from quaternary ammonium salts? Here, the transition state has considerable carbanion character, at least partly because powerful electron withdrawal by the positively charged nitrogen favors development of negative charge. There is preferential abstraction of a proton from the carbon that can best accommodate the partial negative charge: in the example given, from the primary carbon rather than the secondary.

Sulfonium ions, R_3S^+, react similarly to quaternary ammonium ions.

The stereochemistry of Hofmann elimination is commonly *anti*, but less so than was formerly believed. *syn*-Elimination is important for certain cyclic compounds, and can be made important even for open-chain compounds by the proper choice of base and solvent. Quaternary ammonium ions are more prone to *syn*-elimination than alkyl halides and sulfonates. Electronically, *anti* formation of the double bond is favored in eliminations; but when the alkene character of the transition state is slight—as here—other factors come into play: conformational factors, it has been postulated.

Problem 27.5 Predict the major products of E2 elimination from: (a) 2-methyl-3-pentyltrimethylammonium ion; (b) diethyldi-*n*-propylammonium ion; (c) dimethylethyl(2-chloroethyl)ammonium ion; (d) dimethylethyl-*n*-propylammonium ion.

Problem 27.6 When dimethyl-*tert*-pentylsulfonium ethoxide is heated in ethanol, the alkene obtained is chiefly (86%) 2-methyl-1-butene; when the corresponding sulfonium iodide is heated in ethanol, the alkene obtained is chiefly (86%) 2-methyl-2-butene.

(a) How do you account for the difference in products? (b) From the sulfonium iodide reaction there is also obtained considerable material identified as an ether. What ether would you expect it to be, and how is it formed? (c) What ether would you expect to obtain from the sulfonium ethoxide reaction?

Problem 27.7 2-Phenylethyl bromide undergoes E2 elimination about 10 times as fast as 1-phenylethyl bromide even though they both yield the same alkene. Suggest a possible explanation for this.

27.7 Conversion of amines into substituted amides

We have learned (Sec. 24.11) that ammonia reacts with acid chlorides of carboxylic acids to yield amides, compounds in which —Cl has been replaced by

$$NH_3 + R-C\overset{\displaystyle O}{\underset{\displaystyle Cl}{\big\langle}} \longrightarrow R-C\overset{\displaystyle O}{\underset{\displaystyle NH_2}{\big\langle}}$$

the —NH_2 group. Not surprisingly, acid chlorides of sulfonic acids react similarly.

$$NH_3 + Ar-\overset{\displaystyle O}{\underset{\displaystyle O}{\overset{|}{\underset{|}{S}}}}-Cl \longrightarrow Ar-\overset{\displaystyle O}{\underset{\displaystyle O}{\overset{|}{\underset{|}{S}}}}-NH_2$$

A sulfonyl chloride   A sulfonamide

In these reactions ammonia serves as a nucleophilic reagent, attacking the carbonyl carbon or sulfur and displacing chloride ion. In the process nitrogen loses a proton to a second molecule of ammonia or another base.

In a similar way primary and secondary amines can react with acid chlorides to form **substituted amides**, compounds in which —Cl has been replaced by the —NHR or —NR_2 group:

Primary RNH_2

           $\xrightarrow{R'COCl}$ $R'CO-NHR$ An *N*-substituted amide

           $\xrightarrow{ArSO_2Cl}$ $ArSO_2-NHR$ An *N*-substituted sulfonamide

Secondary R_2NH

           $\xrightarrow{R'COCl}$ $R'CO-NR_2$ An *N,N*-disubstituted amide

           $\xrightarrow{ArSO_2Cl}$ $ArSO_2-NR_2$ An *N,N*-disubstituted sulfonamide

Tertiary R_3N

           $\xrightarrow{R'COCl}$ No reaction

           $\xrightarrow{ArSO_2Cl}$ No reaction under conditions of Hinsberg test (*but see* Sec. 27.19)

Tertiary amines, although basic and hence nucleophilic, fail to yield amides, presumably because they cannot lose a proton (to stabilize the product) after attaching themselves to carbon or to sulfur. Here is a reaction which requires not only that amines be nucleophilic, but also that they possess a hydrogen atom attached to nitrogen. (However, see Sec. 27.20.)

Substituted amides are generally named as derivatives of the unsubstituted amides. For example:

$CH_3CNHC_2H_5$
‖
O

N-Ethylacetamide

$CH_3CH_2CH_2C\overset{\overset{\displaystyle CH_3}{|}}{\underset{\underset{\displaystyle O}{‖}}{N}}-C_2H_5$

N-Ethyl-*N*-methylbutyramide

N,N-Dimethylbenzamide

In many cases, and particularly where aromatic amines are involved, we are more interested in the amine from which the amide is derived than in the acyl group. In these cases the substituted amide is named as an acyl derivative of the amine. For example:

Acetanilide

Benzanilide

Aceto-*p*-toluidide

Substituted amides of aromatic carboxylic acids or of sulfonic acids are prepared by the Schotten–Baumann technique: the acid chloride is added to the amine in the presence of a base, either aqueous sodium hydroxide or pyridine. For example:

Aniline   Benzoyl chloride   Benzanilide

$(n\text{-}C_4H_9)_2NH$ + [benzene]SO_2Cl $\xrightarrow{NaOH}$ [benzene]$SO_2-N\overset{\displaystyle C_4H_9}{\underset{\displaystyle C_4H_9}{}}$

Di-*n*-butylamine   Benzenesulfonyl chloride   *N,N*-Di-*n*-butylbenzenesulfonamide

Acetylation is generally carried out using acetic anhydride rather than acetyl chloride. For example:

o-Toluidine   Acetic anhydride   Aceto-*o*-toluidide

Like simple amides, substituted amides undergo hydrolysis; the products are the acid and the amine, although one or the other is obtained as its salt, depending upon the acidity or alkalinity of the medium.

$$\text{C}_6\text{H}_5\text{—CO—N(C}_6\text{H}_5)\text{CH}_3 + \text{NaOH} \xrightarrow{\text{heat}} \text{C}_6\text{H}_5\text{—COO}^-\text{Na}^+ + \text{C}_6\text{H}_5\text{NHCH}_3$$

N-Methylbenzanilide   Sodium benzoate   N-Methylaniline

$$p\text{-Br-C}_6\text{H}_4\text{—NH—COCH}_3 + \text{H}_2\text{O} + \text{HCl} \xrightarrow{\text{heat}} \text{CH}_3\text{COOH} + p\text{-Br-C}_6\text{H}_4\text{—NH}_3^+\text{Cl}^-$$

p-Bromoacetanilide   Acetic acid   p-Bromoanilinium
                                                            chloride

Sulfonamides are hydrolyzed more slowly than amides of carboxylic acids; examination of the structures involved shows us what probably underlies this difference. Nucleophilic attack on a trigonal acyl carbon (Sec. 24.4) is relatively unhindered; it involves the temporary attachment of a fourth group, the nucleophilic reagent. Nucleophilic attack on tetrahedral sulfonyl sulfur is relatively hindered; it involves the temporary attachment of a *fifth* group. The tetrahedral

$$R\text{—C}(\!=\!\text{O})\text{W} + :Z \longrightarrow R\text{—C(O}^-\text{)(Z)W}$$

Trigonal C   Tetrahedral C   Acyl
Attack relatively   *Stable octet*   nucleophilic
unhindered   substitution

$$\text{Ar—S(=O)}_2\text{—W} + :Z \longrightarrow [Z\text{—S(Ar)(W)(O)(O)}]^-$$

Tetrahedral S   Pentavalent S   Sulfonyl
Attack hindered   *Unstable decet*   nucleophilic
                                                     substitution

carbon of the acyl intermediate makes use of the permitted octet of electrons; although sulfur may be able to use more than eight electrons in covalent bonding, this is a less stable system than the octet. Thus both steric and electronic factors tend to make sulfonyl compounds less reactive than acyl compounds.

There is a further contrast between the amides of the two kinds of acids. The substituted amide from a primary amine still has a hydrogen attached to nitrogen, and as a result is *acidic*: in the case of a sulfonamide, this acidity is appreciable, and much greater than for the amide of a carboxylic acid. A monosubstituted sulfonamide is less acidic than a carboxylic acid, but about the same as a phenol (Sec. 28.7); it reacts with aqueous hydroxides to form salts.

$$\text{Ar—S(=O)}_2\text{—NHR} + \text{OH}^- \longrightarrow \text{H}_2\text{O} + \{\text{Ar—S(=O)}_2\text{—NR}\}^\ominus$$

This difference in acidity, too, is understandable. A sulfonic acid is more acidic than a carboxylic acid because the negative charge of the anion is dispersed over three oxygens instead of just two. In the same way, a sulfonamide is more acidic than the amide of a carboxylic acid because the negative charge is dispersed over two oxygens plus nitrogen instead of over just one oxygen plus nitrogen.

Problem 27.8 (a) Although amides of carboxylic acids are very weakly acidic ($K_a = 10^{-14}$ to 10^{-15}), they are still enormously more acidic than ammonia ($K_a = 10^{-33}$) or amines, RNH_2. Account in detail for this.

(b) Diacetamide, $(CH_3CO)_2NH$, is much more acidic ($K_a = 10^{-11}$) than acetamide ($K_a = 8.3 \times 10^{-16}$), and roughly comparable to benzenesulfonamide ($K_a = 10^{-10}$). How can you account for this?

Problem 27.9 In contrast to carboxylic esters, we know, alkyl sulfonates undergo nucleophilic attack at alkyl carbon. What *two* factors are responsible for this difference

in behavior? (*Hint*: See Sec. 5.9.)

The conversion of an amine into a sulfonamide is used in determining the class of the amine; this is discussed in the section on analysis (Sec. 27.19).

27.8 Ring substitution in aromatic amines

We have already seen that the $-NH_2$, $-NHR$, and $-NR_2$ groups act as powerful activators and *ortho,para* directors in electrophilic aromatic substitution. These effects were accounted for by assuming that the intermediate carbocation is stabilized by structures like I and II in which nitrogen bears a positive charge and

is joined to the ring by a double bond. Such structures are especially stable since in them every atom (except hydrogen) has a complete octet of electrons; indeed, structure I or II *by itself* must pretty well represent the intermediate.

In such structures nitrogen shares more than one pair of electrons with the ring, and hence carries the charge of the "carbocation". Thus the basicity of nitrogen accounts for one more characteristic of aromatic amines.

The acetamido group, $-NHCOCH_3$, is also activating and *ortho,para*-directing, but less powerfully so than a free amino group. Electron withdrawal by oxygen of the carbonyl group makes the nitrogen of an amide a much poorer source of electrons than the nitrogen of an amine. Electrons are less available for sharing with a hydrogen ion, and therefore amides are much weaker bases than amines: amides of carboxylic acids do not dissolve in dilute aqueous acids. Electrons are less available for sharing with an aromatic ring, and therefore an acetamido group activates an aromatic ring less strongly than an amino group.

More precisely, electron withdrawal by carbonyl oxygen destabilizes a positive charge on nitrogen, whether this charge is acquired by *protonation* or by *electrophilic attack on the ring*.

(We have seen (Sec. 14.5) that the $-NR_3{}^+$ group is a powerful deactivator and *meta* director. In a quaternary ammonium salt, nitrogen no longer has electrons to share with the ring; on the contrary, the full-fledged positive charge on nitrogen makes the group strongly electron-attracting.)

In electrophilic substitution, the chief problem encountered with aromatic amines is that they are *too* reactive. In halogenation, substitution tends to occur at every available *ortho* or *para* position. For example:

p-Toluidine   3,5-Dibromo-4-aminotoluene

Nitric acid not only nitrates, but oxidizes the highly reactive ring as well, with loss of much material as tar. Furthermore, in the strongly acidic nitration medium, the amine is converted into the anilinium ion; substitution is thus controlled not by the $-NH_2$ group but by the $-NH_3{}^+$ group which, because of its positive charge, directs much of the substitution to the *meta* position.

There is, fortunately, a simple way out of these difficulties. We *protect* the amino group: we acetylate the amine, then carry out the substitution, and finally hydrolyze the amide to the desired substituted amine. For example:

p-Toluidine   Aceto-p-toluidide   3-Bromo-4-aminotoluene

Acetanilide   p-Nitroacetanilide   p-Nitroaniline

Problem 27.10 Nitration of un-acetylated aniline yields a mixture of about two-thirds *meta* and one-third *para* product. Since almost all the aniline is in the form of the anilinium ion, how do you account for the fact that even more *meta* product is not obtained?

27.9   Sulfonation of aromatic amines. Dipolar ions

Aniline is usually sulfonated by "baking" the salt, anilinium hydrogen sulfate, at 180–200 °C; the chief product is the *para* isomer. In this case we cannot discuss orientation on our usual basis of which isomer is formed *faster*. Sulfonation is

Aniline   Anilinium hydrogen   Sulfanilic acid
                      sulfate

known to be reversible, and the *para* isomer is known to be the most stable isomer; it may well be that the product obtained, the *para* isomer, is determined by the position of an equilibrium and not by relative rates of formation (see Sec. 10.27 and Sec. 15.12). It also seems likely that, in some cases at least, sulfonation of amines proceeds by a mechanism that is entirely different from ordinary aromatic substitution.

Whatever the mechanism by which it is formed, the chief product of this reaction is *p*-aminobenzenesulfonic acid, known as **sulfanilic acid**; it is an important and interesting compound.

First of all, its properties are not those we would expect of a compound containing an amino group and a sulfonic acid group. Both aromatic amines and aromatic sulfonic acids have low melting points; benzenesulfonic acid, for example, melts at 66 °C, and aniline at −6 °C. Yet sulfanilic acid has such a high melting point that on being heated it decomposes (at 280–300 °C) before its melting point can be reached. Sulfonic acids are generally very soluble in water; indeed, we have seen that the sulfonic acid group is often introduced into a molecule to make it water-soluble. Yet sulfanilic acid is not only insoluble in organic solvents, but also nearly insoluble in water. Amines dissolve in aqueous mineral acids because of their conversion into water-soluble salts. Sulfanilic acid is soluble in aqueous bases but insoluble in aqueous acids.

These properties of sulfanilic acid are understandable when we realize that sulfanilic acid actually has the structure I which contains the $-NH_3^+$ and $-SO_3^-$ groups. Sulfanilic acid is a salt, but of a rather special kind, called a **dipolar ion** (sometimes called a *zwitterion*, from the German, *Zwitter*, hermaphrodite). It is the product of reaction between an acidic group and a basic group that are part of the same molecule. The hydrogen ion is attached to nitrogen rather than oxygen

I   II

Insoluble in water   *Soluble in water*

simply because the $-NH_2$ group is a stronger base than the $-SO_3^-$ group. A high melting point and insolubility in organic solvents are properties we would expect of a salt. Insolubility in water is not surprising, since many salts are insoluble in water. In alkaline solution, the strongly basic hydroxide ion pulls hydrogen ion away from the weakly basic —NH$_2$ group to yield the *p*-aminobenzenesulfonate ion (II), which, like most sodium salts, is soluble in water. In aqueous acid, however, the sulfanilic acid structure is not changed, and therefore the compound remains insoluble; sulfonic acids are strong acids and their anions (very weak bases) show little tendency to accept hydrogen ion from H$_3$O$^+$.

We can expect to encounter dipolar ions whenever we have a molecule containing both an amino group and an acid group, providing the amine is more basic than the anion of the acid.

Problem 27.11   *p*-Aminobenzoic acid is not a dipolar ion, whereas glycine (aminoacetic acid) is a dipolar ion. How can you account for this?

27.10   Sulfanilamide. The sulfa drugs

The amide of sulfanilic acid (*sulfanilamide*) and certain related substituted amides are of considerable medical importance as the *sulfa drugs*. Although they have been supplanted to a wide extent by the antibiotics (such as penicillin, terramycin, chloromycetin, and aureomycin), the sulfa drugs still have their medical uses, and make up a considerable portion of the output of the pharmaceutical industry.

Sulfonamides are prepared by the reaction of a sulfonyl chloride with ammonia or an amine. The presence in a sulfonic acid molecule of an amino group, however, poses a special problem: if sulfanilic acid were converted to the acid chloride, the sulfonyl group of one molecule could attack the amino group of another to form an amide linkage. This problem is solved by protecting the amino group through acetylation prior to the preparation of the sulfonyl chloride. Sulfanilamide and related compounds are generally prepared in the following way:

The selective removal of the acetyl group in the final step is consistent with the general observation that amides of carboxylic acids are more easily hydrolyzed than amides of sulfonic acids.

$$CH_3-\overset{O}{\underset{\|}{C}}-NH\underset{\substack{Hydrolysis\\ occurs\ here}}{\left\langle\bigcirc\right\rangle}\overset{O}{\underset{\|}{S}}-NH_2 \xrightarrow[heat]{dilute\ HCl} H_2N\left\langle\bigcirc\right\rangle\overset{O}{\underset{\|}{S}}-NH_2$$

Sulfanilamide

The antibacterial activity—and toxicity—of a sulfanilamide stems from a rather simple fact: enzymes in the bacteria (and in the patients) confuse it for *p*-aminobenzoic acid, which is an essential metabolite. In what is known as *metabolite antagonism*, the sulfanilamide competes with *p*-aminobenzoic acid for

$$\underset{COOH}{\overset{NH_2}{\bigcirc}}$$

p-Aminobenzoic acid

$$\underset{SO_2NHR}{\overset{NH_2}{\bigcirc}}$$

Substituted sulfanilamide

reactive sites on the enzymes; deprived of the essential metabolite, the organism fails to reproduce, and dies.

Just how good a drug the sulfanilamide is depends upon the nature of the group R attached to amido nitrogen. This group must confer just the right degree of acidity to the amido hydrogen (Sec. 27.7), but acidity is clearly only one of the factors involved. Of the hundreds of such compounds that have been synthesized, only a half dozen or so have had the proper combination of high antibacterial activity and low toxicity to human beings that is necessary for an effective drug; in nearly all these effective compounds the group R contains a heterocyclic ring (Chap. 35).

$$H_2N\left\langle\bigcirc\right\rangle SO_2NH\overset{N}{\underset{N}{\left\langle\bigcirc\right\rangle}}CH_3 \qquad HOOCCH_2CH_2CONH\left\langle\bigcirc\right\rangle SO_2NH\overset{S}{\underset{N}{\left\langle\bigcirc\right\rangle}}$$

Sulfamerazine   Succinoylsulfathiazole

27.11   Reactions of amines with nitrous acid

Each class of amine yields a different kind of product in its reaction with nitrous acid, HONO. This unstable reagent is generated in the presence of the amine by the action of mineral acid on sodium nitrite.

Primary aromatic amines react with nitrous acid to yield *diazonium salts*; this is one of the most important reactions in organic chemistry. Following sections are devoted to the preparation and properties of aromatic diazonium salts.

$$Ar-NH_2 + NaNO_2 + 2HX \xrightarrow{cold} Ar-N_2{}^+X^- + NaX + 2H_2O$$

1° aromatic   A diazonium salt
amine

Primary aliphatic amines also react with nitrous acid to yield diazonium salts; but since aliphatic diazonium salts are quite unstable and break down to yield a complicated mixture of organic products (see Problem 27.12, below), this reaction

$$R—NH_2 + NaNO_2 + HX \longrightarrow [R—N_2^+X^-] \xrightarrow{\text{H}_2\text{O}} N_2 + \text{mixture of alcohols}$$
$$\underset{\text{1° aliphatic}}{} \underset{\text{amine}}{} \qquad\qquad \underset{\textit{Unstable}}{} \qquad\qquad \text{and alkenes}$$

is of little synthetic value. The fact that nitrogen is evolved quantitatively is of some importance in analysis, however, particularly of amino acids and proteins.

Problem 27.12 The reaction of *n*-butylamine with sodium nitrite and hydrochloric acid yields nitrogen and the following mixture: *n*-butyl alcohol, 25%; *sec*-butyl alcohol, 13%; 1-butene and 2-butene, 37%; *n*-butyl chloride, 5%; *sec*-butyl chloride, 3%. (a) What is the most likely intermediate common to all of these products, and how is it formed? (b) Outline reactions that account for the various products.

Problem 27.13 Predict the organic products of the reaction of: (a) isobutylamine with nitrous acid; (b) neopentylamine with nitrous acid.

Secondary amines, both aliphatic and aromatic, react with nitrous acid to yield *N*-nitrosoamines.

N-Methylaniline   N-Nitroso-N-methylaniline

Tertiary aromatic amines undergo ring substitution, to yield compounds in which a nitroso group, —N=O, is joined to carbon; thus *N,N*-dimethylaniline yields chiefly *p*-nitroso-*N,N*-dimethylaniline.

N,N-Dimethylaniline   p-Nitroso-N,N-dimethylaniline

Ring nitrosation is an electrophilic aromatic substitution reaction, in which the attacking reagent is either the *nitrosonium ion*, $^+$NO, or some species (like $H_2\overset{+}{O}$—NO or NOCl) that can easily transfer $^+$NO to the ring. The nitrosonium ion is very weakly electrophilic compared with the reagents involved in nitration, sulfonation, halogenation, and the Friedel–Crafts reaction; nitrosation ordinarily occurs only in rings bearing the powerfully activating dialkylamino (—NR$_2$) or hydroxy (—OH) group. (See Fig. 27.3.)

Despite the differences in the final product, the reaction of nitrous acid with all these amines involves the same initial step: *electrophilic attack by $^+$NO with displacement of* H$^+$. This attack occurs at the position of highest electron availability in primary and secondary amines: at nitrogen. Tertiary aromatic amines are attacked at the highly reactive ring.

Figure 27.3   Ring nitrosation of *N,N*-dimethylaniline.

Problem 27.14 (a) Write equations to show how the molecule $H_2\overset{+}{O}$—NO is formed in the nitrosating mixture. (b) Why can this transfer $^+$NO to the ring more easily than HONO can? (c) Write equations to show how NOCl can be formed from $NaNO_2$ and aqueous hydrochloric acid. (d) Why is NOCl a better nitrosating agent than HONO?

Problem 27.15 (a) Which, if either, of the following seems likely? (i) The ring of *N*-methylaniline is much less reactive toward electrophilic attack than the ring of *N,N*-dimethylaniline. (ii) Nitrogen of *N*-methylaniline is much more reactive toward electrophilic attack than nitrogen of *N,N*-dimethylaniline.
(b) How do you account for the fact that the two amines give different products with nitrous acid?

27.12   Diazonium salts. Preparation and reactions

When a primary aromatic amine, dissolved or suspended in cold aqueous mineral acid, is treated with sodium nitrite, there is formed a diazonium salt. Since

$$Ar—NH_2 + NaNO_2 + 2HX \xrightarrow{\text{cold}} Ar—N\equiv N:^+ X^- + NaX + 2H_2O$$

1° aromatic
amine   A diazonium salt

diazonium salts slowly decompose even at ice-bath temperatures, the solution is used immediately after preparation.

The large number of reactions undergone by diazonium salts may be divided into two classes: **replacement**, in which nitrogen is lost as N_2, and some other atom or group becomes attached to the ring in its place; and **coupling**, in which the nitrogen is retained in the product.

REACTIONS OF DIAZONIUM SALTS

1. **Replacement of nitrogen**

$$Ar-N_2^+ + :Z \longrightarrow Ar-Z + N_2$$

(a) **Replacement by —Cl, —Br, and —CN. Sandmeyer reaction.** Discussed in Secs. 27.13–27.14.

$$Ar-N_2^+ \begin{cases} \xrightarrow{\text{CuCl}} & Ar-Cl + N_2 \\ \xrightarrow{\text{CuBr}} & Ar-Br + N_2 \\ \xrightarrow{\text{CuCN}} & Ar-CN + N_2 \end{cases}$$

Examples:

o-Toluidine   o-Toluenediazonium   o-Chlorotoluene
                     chloride

o-Toluidine   o-Bromotoluene

o-Toluidine   o-Tolunitrile

(b) **Replacement by —I.** Discussed in Sec. 27.13.

$$Ar-N_2^+ + I^- \longrightarrow Ar-I + N_2$$

Example:

Aniline   Iodobenzene

(c) **Replacement by —F.** Discussed in Sec. 27.13.

$$Ar-N_2^+ BF_4^- \xrightarrow{\text{heat}} Ar-F + N_2 + BF_3$$

CONTINUED

——— CONTINUED ———

Example:

Aniline   Benzenediazonium   Benzenediazonium   Fluorobenzene
                        chloride   fluoroborate
                                      Isolated as crystalline salt

(d) **Replacement by —OH.** Discussed in Sec. 27.15.

$$Ar\text{—}N_2^+ + H_2O \xrightarrow{H^+} Ar\text{—}OH + N_2$$

A phenol

Example:

o-Toluidine   *o*-Cresol

(e) **Replacement by —H.** Discussed in Sec. 27.16.

$$Ar\text{—}N_2^+ + H_3PO_2 \xrightarrow{H_2O} Ar\text{—}H + H_3PO_3 + N_2$$

Example:

2,4-Dichloroaniline   *m*-Dichlorobenzene

2. **Coupling.** Discussed in Sec. 27.18.

An azo compound

G *must be a strongly electron-releasing group:*

OH, NR_2, NHR, NH_2

Example:

Benzenediazonium   Phenol   *p*-Hydroxyazobenzene
chloride   *p*-(Phenylazo)phenol

Replacement of the diazonium group is the best general way of introducing F, Cl, Br, I, CN, OH, and H into an aromatic ring. Diazonium salts are valuable in synthesis not only because they react to form so many classes of compounds, but also because they can be prepared from nearly all primary aromatic amines. There are few groups whose presence in the molecule interferes with diazotization; in this respect, diazonium salts are quite different from Grignard reagents (Sec. 17.17). The amines from which diazonium compounds are prepared are readily obtained from the corresponding nitro compounds, which are prepared by direct nitration. Diazonium salts are thus the most important link in the sequence:

$$ArH \longrightarrow ArNO_2 \longrightarrow ArNH_2 \longrightarrow Ar\!-\!N_2^+ \left\{ \begin{array}{l} \longrightarrow \ Ar\!-\!F \\ \longrightarrow \ Ar\!-\!Cl \\ \longrightarrow \ Ar\!-\!Br \\ \longrightarrow \ Ar\!-\!I \\ \longrightarrow \ Ar\!-\!CN \longrightarrow Ar\!-\!COOH \\ \longrightarrow \ Ar\!-\!OH \\ \longrightarrow \ Ar\!-\!H \end{array} \right.$$

In addition to the atoms and groups just listed, there are dozens of other groups that can be attached to an aromatic ring by replacement of the diazonium nitrogen, as, for example, $-Ar$, $-NO_2$, $-OR$, $-SH$, $-SR$, $-NCS$, $-NCO$, $-PO_3H_2$, $-AsO_3H_2$, $-SbO_3H_2$; the best way to introduce most of these groups is via diazotization.

The coupling of diazonium salts with aromatic phenols and amines yields *azo compounds*, which are of tremendous importance to the dye industry.

27.13   Diazonium salts. Replacement by halogen. Sandmeyer reaction

Replacement of the diazonium group by $-Cl$ or $-Br$ is carried out by mixing the solution of freshly prepared diazonium salt with cuprous chloride or cuprous bromide. At room temperature, or occasionally at elevated temperatures, nitrogen is steadily evolved, and after several hours the aryl chloride or aryl bromide can be isolated from the reaction mixture. This procedure, using cuprous halides, is generally referred to as the **Sandmeyer reaction**.

$$Ar\!-\!N_2^+X^- \xrightarrow{\ CuX\ } Ar\!-\!X + N_2$$

Sometimes the synthesis is carried out by a modification known as the *Gattermann reaction*, in which copper powder and hydrogen halide are used in place of the cuprous halide.

Replacement of the diazonium group by $-I$ does not require the use of a cuprous halide or copper; the diazonium salt and potassium iodide are simply mixed together and allowed to react.

$$Ar\!-\!N_2^+X^- + I^- \longrightarrow Ar\!-\!I + N_2 + X^-$$

Replacement of the diazonium group by $-F$ is carried out in a somewhat

of fluoroboric acid, HBF_4, to the solution of diazonium
tion of the diazonium fluoroborate, $ArN_2^+BF_4^-$, which
ter, washed, and dried. The diazonium fluoroborates are
m salts in being fairly stable compounds. On being heated,
roborate decomposes to yield the aryl fluoride, boron

$$\xrightarrow{\text{}BF_4} \quad Ar-N_2^+BF_4^- \xrightarrow{\text{heat}} Ar-F + BF_3 + N_2$$

. An analogous procedure involves the diazonium hexa-
PF_6^-.
the synthesis of aryl halides from diazonium salts will be
ec. 29.3. Aryl fluorides and iodides cannot generally be
genation. Aryl chlorides and bromides can be prepared by
, when a mixture of *ortho* and *para* isomers is obtained, it
e pure compounds because of their similarity in boiling
ultimately go back to nitro compounds, which are usually

ts. Replacement by —CN. Synthesis of carboxylic

e diazonium group by —CN is carried out by allowing the
with cuprous cyanide. To prevent loss of cyanide as HCN,
is neutralized with sodium carbonate before being mixed
de.

$$Ar-N_2^+X^- \xrightarrow{\text{CuCN}} Ar-CN + N_2$$

iles yields carboxylic acids. The synthesis of nitriles from
ovides us with an excellent route from nitro compounds to
xample:

ile *p*-Toluenediazonium   *p*-Toluidine   *p*-Nitrotoluene   Toluene
chloride

omatic carboxylic acids is more generally useful than either
nard reagent or oxidation of side chains. We have just seen
ounds, which are needed to prepare the Grignard reagent,
ften prepared via diazonium salts; furthermore, there are
rfere with the preparation and use of the Grignard reagent
group can generally be introduced into a molecule more
side chain; furthermore, conversion of a side chain into a
t be carried out on molecules that contain other groups
sensitive to oxidation.

27.15 Diazonium salts. Replacement by —OH. Synthesis of phenols

Diazonium salts react with water to yield phenols. This reaction takes place

$$Ar—N_2{}^+X^- + H_2O \longrightarrow Ar—OH + N_2 + H^+$$

slowly in the ice-cold solutions of diazonium salts, and is the reason diazonium salts are used immediately upon preparation; at elevated temperatures it can be made the chief reaction of diazonium salts.

As we shall see, phenols can couple with diazonium salts to form azo compounds (Sec. 27.18); the more acidic the solution, however, the more slowly this coupling occurs. To minimize coupling during the synthesis of a phenol, therefore—coupling, that is, between phenol that has been formed and diazonium ion that has not yet reacted—the diazonium solution is added slowly to a large volume of boiling dilute sulfuric acid.

This is the best general way to make the important class of compounds, the phenols.

27.16 Diazonium salts. Replacement by —H

Replacement of the diazonium group by —H can be brought about by a number of reducing agents; perhaps the most useful of these is *hypophosphorous acid*, H_3PO_2. The diazonium salt is simply allowed to stand in the presence of the hypophosphorous acid; nitrogen is lost, and hypophosphorous acid is oxidized to phosphorous acid:

$$Ar—N_2{}^+X^- + H_3PO_2 + H_2O \longrightarrow Ar—H + N_2 + H_3PO_3 + HX$$

An especially elegant way of carrying out this replacement is to use hypophosphorous acid as the diazotizing acid. The amine is dissolved in hypophosphorous acid, and sodium nitrite is added; the diazonium salt is reduced as fast as it is formed.

This reaction of diazonium salts provides a method of removing an —NH$_2$ or —NO$_2$ group from an aromatic ring. This process can be extremely useful in synthesis, as is shown in some of the examples in the following section.

27.17 Syntheses using diazonium salts

Let us look at a few examples of how diazonium salts can be used in organic synthesis.

To begin with, we might consider some rather simple compounds, the three isomeric bromotoluenes. The best synthesis of each employs diazotization, but not for the same purpose in the three cases. The *o*- and *p*-bromotoluenes are prepared from the corresponding *o*- and *p*-nitrotoluenes:

The advantage of these many-step syntheses over direct bromination is, as we have seen, that a pure product is obtained. Separation of the o- and p-bromotoluenes obtained by direct bromination is not feasible.

Synthesis of m-bromotoluene is a more complicated matter. The problem here is one of preparing a compound in which two *ortho,para*-directing groups are situated *meta* to each other. Bromination of toluene or methylation of bromobenzene would not yield the correct isomer. m-Bromotoluene is obtained by the following sequence of reactions:

The key to the synthesis is the introduction of a group that is a much stronger *ortho,para* director than —CH_3, and that can be easily removed after it has done its job of directing bromine to the correct position. Such a group is the —$NHCOCH_3$ group: it is introduced into the *para* position of toluene via nitration, reduction, and acetylation; it is readily removed by hydrolysis, diazotization, and reduction.

Problem 27.16 Outline the synthesis from benzene or toluene of the following compounds: m-nitrotoluene, m-iodotoluene, 3,5-dibromotoluene, 1,3,5-tribromobenzene, the three toluic acids ($CH_3C_6H_4COOH$), the three methylphenols (cresols).

In the synthesis of *m*-bromotoluene, advantage was taken of the fact that the diazonium group is prepared from a group that is strongly *ortho,para*-directing. Ultimately, however, the diazonium group is prepared from the —NO₂ group, which is a strongly *meta*-directing group. Advantage can be taken of this fact, too, as in the preparation of *m*-bromophenol:

$$OH \quad \xleftarrow[H^+]{H_2O} \quad N_2^+Cl^- \quad \xleftarrow[HCl]{NaNO_2} \quad NH_2 \quad \xleftarrow[HCl]{Sn} \quad NO_2 \quad \xleftarrow[Fe]{Br_2} \quad NO_2$$

m-Bromophenol	*m*-Bromoaniline	Nitrobenzene

m-Bromobenzenediazonium
chloride

m-Bromonitrobenzene

Here again there is the problem of preparing a compound with two *ortho,para* directors situated *meta* to each other. Bromination at the nitro stage gives the necessary *meta* orientation.

Problem 27.17   Outline the synthesis from benzene or toluene of the following compounds: *m*-dibromobenzene, *m*-bromoiodobenzene.

As a final example, let us consider the preparation of 1,2,3-tribromobenzene:

$$Br,Br,Br \quad \xleftarrow{H_3PO_2} \quad Cl^{-+}N_2, Br, Br, Br \quad \xleftarrow[HCl]{NaNO_2} \quad Br, Br, Br, H_2N$$

1,2,3-Tribromobenzene

3,4,5-Tribromobenzenediazonium
chloride

3,4,5-Tribromoaniline

↑ Sn, HCl

$$Br, NH_2, O_2N, Br \quad \xrightarrow[H_2SO_4]{NaNO_2} \quad Br, N_2^+HSO_4^-, O_2N, Br \quad \xrightarrow{CuBr} \quad Br, Br, O_2N, Br$$

2,6-Dibromo-4-nitroaniline	2,6-Dibromo-4-nitrobenzenediazonium hydrogen sulfate	3,4,5-Tribromonitrobenzene

↑ Br₂, CH₃COOH

O₂N, NH₂

p-Nitroaniline

In this synthesis advantage is taken of the fact that the —NO₂ group is a *meta* director, that the —NH₂ group is an *ortho,para* director, and that each of them can be converted into a diazonium group. One diazonium group is replaced by —Br, the other by —H.

Problem 27.18   Outline the synthesis from benzene or toluene of the following compounds: 2,6-dibromotoluene, 3,5-dibromonitrobenzene.

27.18 Coupling of diazonium salts. Synthesis of azo compounds

Under the proper conditions, diazonium salts react with certain aromatic compounds to yield products of the general formula Ar—N=N—Ar', called **azo compounds.** In this reaction, known as **coupling**, the nitrogen of the diazonium group is retained in the product, in contrast to the replacement reactions we have studied up to this point, in which nitrogen is lost.

$$ArN_2^+ + Ar'H \longrightarrow Ar—N=N—Ar' + H^+$$

An azo compound

The aromatic ring (Ar'H) undergoing attack by the diazonium ion must, in general, contain a powerfully electron-releasing group, generally —OH, —NR$_2$, —NHR, or —NH$_2$. Substitution usually occurs *para* to the activating group. Typically, coupling with phenols is carried out in mildly alkaline solution, and with amines in mildly acidic solution.

Activation by electron-releasing groups, as well as the evidence of kinetics studies, indicates that coupling is electrophilic aromatic substitution in which the diazonium ion is the attacking reagent:

It is significant that the aromatic compounds which undergo coupling are also the ones which undergo nitrosation. Like the nitrosonium ion, $^+$NO, the diazonium ion, ArN$_2^+$, is evidently very weakly electrophilic, and is capable of attacking only very reactive rings.

Problem 27.19 Benzenediazonium chloride couples with phenol, but not with the less reactive anisole. 2,4-Dinitrobenzenediazonium chloride, however, couples with anisole; 2,4,6-trinitrobenzenediazonium chloride even couples with the hydrocarbon mesitylene (1,3,5-trimethylbenzene). (a) How can you account for these differences in behavior? (b) Would you expect *p*-toluenediazonium chloride to be more or less reactive as a coupling reagent than benzenediazonium chloride?

In the laboratory we find that coupling involves more than merely mixing together a diazonium salt and a phenol or amine. Competing with any other reaction of diazonium salts is the reaction with water to yield a phenol. If coupling proceeds slowly because of unfavorable conditions, phenol formation may very well become the major reaction. Furthermore, the phenol formed from the diazonium salt can itself undergo coupling; even a relatively small amount of this undesired coupling product could contaminate the desired material—usually a dye whose color should be as pure as possible—to such an extent that the product would be worthless. Conditions under which coupling proceeds as rapidly as possible must therefore be selected.

It is most important that the coupling medium be adjusted to the right degree of acidity or alkalinity. This is accomplished by addition of the proper amount of hydroxide or salts like sodium acetate or sodium carbonate. It will be well to examine this matter in some detail, since it illustrates a problem that is frequently encountered in organic chemical practice.

The electrophilic reagent is the diazonium ion, ArN_2^+. In the presence of hydroxide ion, the diazonium ion exists in equilibrium with an un-ionized compound, $Ar-N=N-OH$, and salts ($Ar-N=N-O^-Na^+$) derived from it:

$$Ar-N\equiv N^+OH^- \underset{H^+}{\overset{NaOH}{\rightleftharpoons}} Ar-N=N-OH \underset{H^+}{\overset{NaOH}{\rightleftharpoons}} Ar-N=N-O^-Na^+$$

$\quad\quad\quad$ *Couples* $\quad\quad\quad\quad\quad\quad\quad\quad$ *Does not couple* $\quad\quad\quad\quad\quad$ *Does not couple*

For our purpose we need only know that hydroxide tends to convert diazonium ion, which couples, into compounds which do not couple. In so far as the electrophilic reagent is concerned, then, coupling will be favored by a low concentration of hydroxide ion, that is, by high acidity.

But what is the effect of high acidity on the amine or phenol with which the diazonium salt is reacting? Acid converts an amine into its ion, which, because of the positive charge, is relatively unreactive toward electrophilic aromatic substitution: much too unreactive to be attacked by the weakly electrophilic diazonium ion. The higher the acidity, the higher the proportion of amine that exists as its ion, and the lower the rate of coupling.

Couples $\quad\quad\quad$ *Does not couple*

An analogous situation exists for a phenol. A phenol is appreciably acidic; in aqueous solutions it exists in equilibrium with phenoxide ion:

Couples $\quad\quad\quad$ *Couples*
rapidly $\quad\quad\quad\quad$ *slowly*

The fully developed negative charge makes $-O^-$ much more powerfully electron-releasing than $-OH$; the phenoxide ion is therefore much more reactive than the un-ionized phenol toward electrophilic aromatic substitution. The higher the acidity of the medium, the higher the proportion of phenol that is un-ionized, and the lower the rate of coupling. In so far as the amine or phenol is concerned, then, coupling is favored by low acidity.

The conditions under which coupling proceeds most rapidly are the result of a compromise. The solution must not be so alkaline that the concentration of diazonium ion is too low; it must not be so acidic that the concentration of free amine or phenoxide or phenoxide ion is too low. It turns out that amines couple fastest in mildly acidic solutions, and phenols couple fastest in mildly alkaline solutions.

Problem 27.20   Suggest a reason for the use of *excess* mineral acid in the diazotization process.

Problem 27.21   (a) Coupling of diazonium salts with primary or secondary aromatic amines (but not with tertiary aromatic amines) is complicated by a side reaction that yields an isomer of the azo compound. Judging from the reaction of secondary aromatic amines with nitrous acid (Sec. 27.11), suggest a possible structure for this by-product.
(b) Upon treatment with mineral acid, this by-product regenerates the original reactants which recombine to form the azo compound. What do you think is the function of the acid in this regeneration? (*Hint*: See Sec. 7.25.)

Azo compounds are the first compounds we have encountered that as a class are strongly colored. They can be intensely yellow, orange, red, blue, or even green,

Para red
A red dye

Methyl orange
An acid–base indicator:
red in acid, *yellow* in base

depending upon the exact structure of the molecule. Because of their color, the azo compounds are of tremendous importance as dyes; about half of the dyes in industrial use today are azo dyes. Some of the acid–base indicators with which we are already familiar are azo compounds.

Problem 27.22   An azo compound is cleaved at the azo linkage by stannous chloride, SnCl₂, to form two amines. (a) What is the structure of the azo compound that is cleaved to 3-bromo-4-aminotoluene and 2-methyl-4-aminophenol? (b) Outline a synthesis of this azo compound, starting with benzene and toluene.

Problem 27.23   Show how *p*-amino-*N*,*N*-dimethylaniline can be made via an azo compound.

27.19   Analysis of amines. Hinsberg test

Amines are characterized chiefly through their basicity. A water-insoluble compound that dissolves in cold dilute hydrochloric acid—or a water-soluble compound (not a salt, Sec. 23.21) whose aqueous solution turns litmus blue—must almost certainly be an amine (Secs. 26.5 and 27.2). Elemental analysis shows the presence of nitrogen.

Whether an amine is primary, secondary, or tertiary is best shown by the **Hinsberg test**. The amine is shaken with benzenesulfonyl chloride in the presence of aqueous *potassium* hydroxide (Sec. 27.7). Primary and secondary amines form substituted sulfonamides; tertiary amines do not—*if* the test is carried out properly.

The monosubstituted sulfonamide from a primary amine has an acidic hydrogen attached to nitrogen. Reaction with potassium hydroxide converts this amide into a soluble salt which, *if the amine contained fewer than eight carbons*, is at least partly soluble. Acidification of this solution regenerates the insoluble amide.

The disubstituted sulfonamide from a secondary amine has no acidic hydrogen and remains insoluble in the alkaline reaction mixture.

Now, the all-important question: what do we actually observe when we treat an amine with benzenesulfonyl chloride and excess potassium hydroxide? A *primary amine* yields a clear solution, from which, upon acidification, an insoluble material separates. A *secondary amine* yields an insoluble compound, which is unaffected by acid. A *tertiary amine* yields an insoluble compound (the unreacted amine itself) which dissolves upon acidification of the mixture.

$$RNH_2 + C_6H_5SO_2Cl \xrightarrow{OH^-} [C_6H_5SO_2NHR] \xrightarrow{KOH} C_6H_5SO_2NR^-K^+ \xrightarrow{H^+}$$

1° amine   *Clear solution*

$$C_6H_5SO_2NHR$$

Insoluble

$$R_2NH + C_6H_5SO_2Cl \xrightarrow{OH^-} C_6H_5SO_2NR_2 \xrightarrow{KOH \text{ or } H^+} \text{No reaction}$$

2° amine   *Insoluble*

$$R_3N + C_6H_5SO_2Cl \xrightarrow{OH^-} R_3N \xrightarrow{HCl} R_3NH^+Cl^-$$

3° amine   *Insoluble*   *Clear solution*

Like all experiments, the Hinsberg test must be done *carefully* and interpreted *thoughtfully*. Among other things, misleading side reactions can occur if the proportions of reagents are incorrect, or if the temperature is too high or the time of reaction too long. Tertiary amines evidently *react*—after all, they are just as nucleophilic as other amines; but the initial product (I) has no acidic proton to

$$C_6H_5SO_2Cl + R_3N \longrightarrow C_6H_5SO_2NR_3^+Cl^- \xrightarrow{OH^-} C_6H_5SO_3^- + R_3N + Cl^-$$

I

lose, and ordinarily is hydrolyzed to regenerate the amine.

Problem 27.24 In non-aqueous medium, the product $C_6H_5SO_2N(CH_3)_3^+Cl^-$ can actually be isolated from the reaction of benzenesulfonyl chloride with one equivalent of trimethylamine. When *two* equivalents of the amine are used, there is formed, slowly, $C_6H_5SO_2N(CH_3)_2$ and $(CH_3)_4N^+Cl^-$. (a) Give all steps in a likely mechanism for this latter reaction. What fundamental type of reaction is probably involved?

(b) If, in carrying out the Hinsberg test, the reaction mixture is heated or allowed to stand, many tertiary amines give precipitates. What are these precipitates likely to be? What incorrect conclusion about the unknown amine are you likely to draw?

Problem 27.25 The sulfonamides of big primary amines are only partially soluble in aqueous KOH. (a) In the Hinsberg test, what incorrect conclusion might you draw about such an amine? (b) How might you modify the procedure to avoid this mistake?

Behavior toward nitrous acid (Sec. 27.11) is of some use in determining the class of an amine. In particular, the behavior of primary aromatic amines is quite characteristic: treatment with nitrous acid converts them into diazonium salts, which yield highly colored azo compounds upon treatment with β-naphthol (a phenol, see Sec. 27.18).

Among the numerous derivatives useful in identifying amines are: amides (e.g., acetamides, benzamides, or sulfonamides) for primary and secondary amines; quaternary ammonium salts (e.g., those from benzyl chloride or methyl iodide) for tertiary amines.

We have already discussed proof of structure by use of exhaustive methylation and elimination (Sec. 27.5).

27.20   Analysis of substituted amides

A substituted amide of a carboxylic acid is characterized by the presence of nitrogen, insolubility in dilute acid and dilute base, and hydrolysis to a carboxylic acid and an amine. It is generally identified through identification of its hydrolysis products (Secs. 23.21 and 27.19).

27.21   Spectroscopic analysis of amines and substituted amides

Infrared   The number and positions of absorption bands depend on the class to which the amine belongs (see Fig. 27.4).

An amide, substituted or unsubstituted, shows the C=O band in the 1640–1690 cm^{-1} region. In addition, if it contains a free N—H group, it will show N—H stretching at 3050–3550 cm^{-1}, and —NH bending at 1600–1640 cm^{-1} (RCONH$_2$) or 1530–1570 cm^{-1} (RCONHR′).

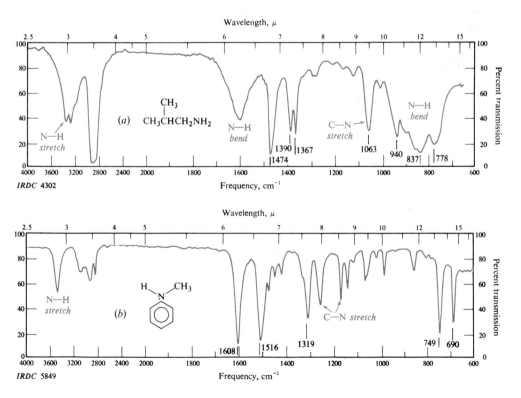

Figure 27.4   Infrared spectra of (*a*) isobutylamine and (*b*) *N*-methylaniline.

N—H stretching 3200–3500 cm^{-1}

1° Amines	2° Amines	3° Amines
Often two bands	*One band*	*No band*

N—H bending

1° Amines   Strong bands 650–900 cm^{-1} (*broad*) and 1560–1650 cm^{-1}

C—N stretching

Aliphatic   1030–1230 cm^{-1} (*weak*)   Aromatic   1180–1360 cm^{-1} (*strong*)
       (3°: *usually a doublet*)   *Two bands*

NMR   Absorption by N—H protons of amines falls in the range δ 1–5, where it is often detected only by proton counting. Absorption by —CO—NH— protons of amides (Sec. 24.25) appears as a broad, low hump farther downfield (δ 5–8).

CMR   The nitrogen of amines deshields carbon, and shifts absorption downfield.

PROBLEMS

1. Write complete equations, naming all organic products, for the reaction (if any) of *n*-butylamine with: *neutralization*

(a) dilute HCl
(b) dilute H$_2$SO$_4$
(c) acetic acid
(d) dilute NaOH
(e) acetic anhydride
(f) isobutyryl chloride
(g) *p*-nitrobenzoyl chloride + pyridine
(h) benzenesulfonyl chloride + KOH(aq)
(i) ethyl bromide

(j) benzyl bromide
(k) bromobenzene *no reaction*
(l) excess methyl iodide, then Ag$_2$O
(m) product (l) + strong heat
(n) CH$_3$COCH$_3$ + H$_2$ + Ni
(o) HONO (NaNO$_2$ + HCl)
(p) phthalic anhydride
(q) sodium chloroacetate
(r) 2,4,6-trinitrochlorobenzene

2. Without referring to tables, arrange the compounds of each set in order of basicity:

(a) ammonia, aniline, cyclohexylamine
(b) ethylamine, 2-aminoethanol, 3-amino-1-propanol
(c) aniline, *p*-methoxyaniline, *p*-nitroaniline
(d) benzylamine, *m*-chlorobenzylamine, *m*-ethylbenzylamine
(e) *p*-chloro-*N*-methylaniline, 2,4-dichloro-*N*-methylaniline,
    2,4,6-trichloro-*N*-methylaniline

3. Which is the more strongly basic, an aqueous solution of trimethylamine or an aqueous solution of tetramethylammonium hydroxide? Why? (*Hint*: What is the principal base in each solution?)

4. Compare the behavior of the three amines, aniline, *N*-methylaniline, and *N,N*-dimethylaniline, toward each of the following reagents:

(a) dilute HCl
(b) NaNO$_2$ + HCl(aq)
(c) methyl iodide
(d) benzenesulfonyl chloride + KOH(aq)

(e) acetic anhydride
(f) benzoyl chloride + pyridine
(g) bromine water

5. Answer Problem 4 for ethylamine, diethylamine, and triethylamine.

6. Give structures and names of the principal organic products expected from the action (if any) of sodium nitrite and hydrochloric acid on:

(a) *p*-toluidine
(b) *N,N*-diethylaniline
(c) *n*-propylamine
(d) sulfanilic acid

(e) *N*-methylaniline
(f) 2-amino-3-methylbutane
(g) benzidine (4,4′-diaminobiphenyl)
(h) benzylamine

7. Write equations for the reaction of *p*-nitrobenzenediazonium sulfate with:

(a) *m*-phenylenediamine   (d) *p*-cresol   (g) CuCN
(b) hot dilute H_2SO_4 (e) KI   (h) HBF_4, then heat
(c) HBr + Cu   (f) CuCl   (i) H_3PO_2

8. Give the reagents and any special conditions necessary to convert *p*-toluene-diazonium chloride into:

(a) toluene   (f) *p*-fluorotoluene
(b) *p*-cresol, *p*-$CH_3C_6H_4OH$ (g) *p*-tolunitrile, *p*-$CH_3C_6H_4CN$
(c) *p*-chlorotoluene   (h) 4-methyl-4'-(*N,N*-dimethylamino)azobenzene
(d) *p*-bromotoluene   (i) 2,4-dihydroxy-4'-methylazobenzene
(e) *p*-iodotoluene

9. Write balanced equations, naming all organic products, for the following reactions:

(a) *n*-butyryl chloride + methylamine
(b) acetic anhydride + *N*-methylaniline
(c) tetra-*n*-propylammonium hydroxide + heat
(d) isovaleryl chloride + diethylamine
(e) tetramethylammonium hydroxide + heat *no reaction*
(f) trimethylamine + acetic acid
(g) *N,N*-dimethylacetamide + boiling dilute HCl
(h) benzanilide + boiling aqueous NaOH
(i) methyl formate + aniline
(j) excess methylamine + phosgene ($COCl_2$)
(k) *m*-$O_2NC_6H_4NHCH_3$ + $NaNO_2$ + H_2SO_4
(l) aniline + Br_2 (aq) in excess
(m) *m*-toluidine + Br_2(aq) in excess
(n) *p*-toluidine + Br_2(aq) in excess
(o) *p*-toluidine + $NaNO_2$ + HCl
(p) $C_6H_5NHCOCH_3$ + HNO_3 + H_2SO_4
(q) *p*-$CH_3C_6H_4NHCOCH_3$ + HNO_3 + H_2SO_4
(r) *p*-$C_2H_5C_6H_4NH_2$ + large excess of CH_3I
(s) benzanilide + Br_2 + Fe

10. Outline all steps in a possible laboratory synthesis of each of the following compounds from benzene, toluene, and alcohols of four carbons or fewer, using any needed inorganic reagents.

(a) 4-amino-2-bromotoluene   (h) *p*-aminobenzylamine
(b) 4-amino-3-bromotoluene   (i) *N*-nitroso-*N*-isopropylaniline
(c) *p*-aminobenzenesulfonanilide   (j) *N*-ethyl-*N*-methyl-*n*-valeramide
    (*p*-$H_2NC_6H_4SO_2NHC_6H_5$)   (k) *n*-hexylamine
(d) monoacetyl *p*-phenylenediamine   (l) 1-amino-1-phenylbutane
    (*p*-aminoacetanilide)   (m) aminoacetamide
(e) *p*-nitroso-*N,N*-diethylaniline   (n) hippuric acid
(f) 4-amino-3-nitrobenzoic acid   ($C_6H_5CONHCH_2COOH$)
(g) 2,6-dibromo-4-isopropylaniline

11. Outline all steps in a possible laboratory synthesis from benzene, toluene, and any needed inorganic reagents of:

(a) the six isomeric dibromotoluenes, $CH_3C_6H_3Br_2$. (*Note*: One may be more difficult to make than any of the others.)
(b) the three isomeric chlorobenzoic acids, each one free of the others
(c) the three isomeric bromofluorobenzenes

Review the instructions on page 221. Assume that an *ortho,para* mixture of isomeric nitro compounds can be separated by distillation (see Sec. 14.7).

12. Outline all steps in a possible laboratory synthesis of each of the following compounds from benzene and toluene and any needed aliphatic and inorganic reagents.

(a) *p*-fluorotoluene
(b) *m*-fluorotoluene
(c) *p*-iodobenzoic acid
(d) *m*-bromoaniline
(e) 3-bromo-4-methylbenzoic acid
(f) 2-bromo-4-methylbenzoic acid
(g) *m*-ethylphenol

(h) 3,5-dibromoaniline
(i) 3-bromo-4-iodotoluene
(j) 2-amino-4-methylphenol
(k) 2,6-dibromoiodobenzene
(l) 4-iodo-3-nitrotoluene
(m) *p*-hydroxyphenylacetic acid
(n) 2-bromo-4-chlorotoluene

13. When adipic acid (hexanedioic acid) and hexamethylenediamine (1,6-diamino-hexane) are mixed, a salt is obtained. On heating, this salt is converted into nylon-6,6, a high-molecular-weight compound of formula $(C_{12}H_{22}O_2N_2)_n$. (a) Draw the structural formula for nylon-6,6. To what class of compounds does it belong? (b) Write an equation for the chemistry involved when a drop of hydrochloric acid makes a hole in a nylon-6,6 stocking.

14. (a) In Problem 10 (p. 497) you accounted for the aromaticity of the heterocyclic compound *pyrrole*. In light of your answer, can you suggest a reason why pyrrole is an

Pyrrole
$(K_b \sim 10^{-14})$

Pyrrolidine
$(K_b \sim 10^{-3})$

extremely weak base $(K_b \sim 2.5 \times 10^{-14})$ compared with aliphatic amines (K_b values about 10^{-3} to 10^{-4}) or even aniline (K_b 10^{-10})?

(b) Catalytic hydrogenation converts pyrrole into the corresponding saturated compound, *pyrrolidine*, which has $K_b \sim 10^{-3}$. How do you account for this enormous increase in basicity brought about by hydrogenation?

15. Account for the following reaction, making clear the role played by tosyl chloride.

16. If halide ion is present during hydrolysis of benzenediazonium ion or *p*-nitrobenzenediazonium ion, there is obtained not only the phenol, but also the aryl halide: the higher the halide ion concentration, the greater the proportion of aryl halide obtained. The presence of halide ion has no effect on the rate of decomposition of benzenediazonium ion, but speeds up decomposition of the *p*-nitrobenzenediazonium ion.

(a) Suggest a mechanism or mechanisms to account for these facts. (b) What factor is responsible for the unusually high reactivity of diazonium ions in this reaction—and, indeed, in most of their reactions? (*Hint:* See Sec. 5.9.)

17. Describe simple chemical tests (other than color reactions with indicators) that would serve to distinguish between:

(a) *N*-methylaniline and *o*-toluidine
(b) aniline and cyclohexylamine
(c) *n*-C$_4$H$_9$NH$_2$ and (*n*-C$_4$H$_9$)$_2$NH
(d) (*n*-C$_4$H$_9$)$_2$NH and (*n*-C$_4$H$_9$)$_3$N
(e) (CH$_3$)$_3$NHCl and (CH$_3$)$_4$NCl
(f) C$_6$H$_5$NH$_3$Cl and *o*-ClC$_6$H$_4$NH$_2$

(g) (C$_2$H$_5$)$_2$NCH$_2$CH$_2$OH and (C$_2$H$_5$)$_4$NOH
(h) aniline and acetanilide
(i) (C$_6$H$_5$NH$_3$)$_2$SO$_4$ and *p*-H$_3$\overset{+}{N}C_6H_4SO_3^-$
(j) ClCH$_2$CH$_2$NH$_2$ and CH$_3$CH$_2$NH$_3$Cl
(k) 2,4,6-trinitroaniline and aniline
(l) C$_6$H$_5$NHSO$_2$C$_6$H$_5$ and C$_6$H$_5$NH$_3$HSO$_4$

Tell exactly what you would *do* and *see*.

18. Describe simple chemical methods for the separation of the following mixtures, recovering each component in essentially pure form:

(a) triethylamine and *n*-heptane
(b) aniline and anisole
(c) stearamide and octadecylamine
(d) *o*-$O_2NC_6H_4NH_2$ and *p*-$H_3\overset{+}{N}C_6H_4SO_3^-$
(e) $C_6H_5NHCH_3$ and $C_6H_5N(CH_3)_2$
(f) *n*-caproic acid, tri-*n*-propylamine, and cyclohexane
(g) *o*-nitrotoluene and *o*-toluidine
(h) *p*-ethylaniline and propionanilide

Tell exactly what you would *do* and *see.*

19. The compounds in each of the following sets boil (or melt) within a few degrees of each other. Describe simple chemical tests that would serve to distinguish among the members of each set.

(a) aniline, benzylamine, and *N,N*-dimethylbenzylamine
(b) *o*-chloroacetanilide and 2,4-diaminochlorobenzene
(c) *N*-ethylbenzylamine, *N*-ethyl-*N*-methylaniline, *β*-phenylethylamine, and *o*-toluidine
(d) acetanilide and ethyl oxamate ($C_2H_5OOCCONH_2$)
(e) benzonitrile, *N,N*-dimethylaniline, and formamide
(f) *N,N*-dimethyl-*m*-toluidine, nitrobenzene, and *m*-tolunitrile
(g) *N*-(*sec*-butyl)benzenesulfonamide

p-chloroaniline	*o*-nitroaniline
N,N-dibenzylaniline	*p*-nitrobenzyl chloride
2,4-dinitroaniline	*p*-toluenesulfonyl chloride
N-ethyl-*N*-(*p*-tolyl)-*p*-toluenesulfonamide	

Tell exactly what you would *do* and *see.*

20. An unknown amine is believed to be one of those in Table 27.2. Describe how you would go about finding out which of the possibilities the unknown actually is. Where possible use simple chemical tests.

Table 27.2 DERIVATIVES OF SOME AMINES

Amine	B.p., °C	Benzene-sulfonamide M.p., °C	Acetamide M.p., °C	Benzamide M.p., °C	*p*-Toluene-sulfonamide M.p., °C
m-Toluidine	203	95	66	125	114
N-Ethylaniline	205		54	60	87
N-Methyl-*m*-toluidine	206		66		
N,N-Diethyl-*o*-toluidine	206				
N-Methyl-*o*-toluidine	207		55	66	120
N-Methyl-*p*-toluidine	207	64	83	53	60
N,N-Dimethyl-*o*-chloroaniline	207				
o-Chloroaniline	209	129	87	99	105

21. *Choline*, a constituent of *phospholipids* (fat-like phosphate esters of great physiological importance, Sec. 37.8), has the formula $C_5H_{15}O_2N$. It dissolves readily in water to form a strongly basic solution. It can be prepared by the reaction of ethylene oxide with trimethylamine in the presence of water.

(a) What is a likely structure for choline? (b) What is a likely structure for its acetyl derivative, *acetylcholine*, $C_7H_{17}O_3N$, important in nerve action?

22. *Novocaine*, a local anesthetic, is a compound of formula $C_{13}H_{20}O_2N_2$. It is insoluble in water and dilute NaOH, but soluble in dilute HCl. Upon treatment with $NaNO_2$ and HCl and then with β-naphthol, a highly colored solid is formed.

When Novocaine is boiled with aqueous NaOH, it slowly dissolves. The alkaline solution is shaken with ether and the layers are separated.

Acidification of the aqueous layer causes the precipitation of a white solid A; continued addition of acid causes A to redissolve. Upon isolation A is found to have a melting point of 185–6 °C and the formula $C_7H_7O_2N$.

Evaporation of the ether layer leaves a liquid B of formula $C_6H_{15}ON$. B dissolves in water to give a solution that turns litmus blue. Treatment of B with acetic anhydride gives C, $C_8H_{17}O_2N$, which is insoluble in water and dilute base, but soluble in dilute HCl.

B is found to be identical with the compound formed by the action of diethylamine on ethylene oxide.

(a) What is the structure of Novocaine? (b) Outline all steps in a complete synthesis of Novocaine from toluene and readily available aliphatic and inorganic reagents.

23. A solid compound D, of formula $C_{15}H_{15}ON$, was insoluble in water, dilute HCl, or dilute NaOH. After prolonged heating of D with aqueous NaOH, a liquid, E, was observed floating on the surface of the alkaline mixture. E did not solidify upon cooling to room temperature; it was steam-distilled and separated. Acidification of the alkaline mixture with hydrochloric acid caused precipitation of a white solid, F.

Compound E was soluble in dilute HCl, and reacted with benzenesulfonyl chloride and excess KOH to give a base-insoluble solid, G.

Compound F, m.p. 180 °C, was soluble in aqueous $NaHCO_3$, and contained no nitrogen.

What were compounds D, E, F, and G?

24. Give the structures of compounds H through Q:

$$\text{reduction} \longrightarrow \quad H\ (C_9H_{17}ON,\ \text{an alcohol})$$

$$H + heat \longrightarrow I\ (C_9H_{15}N)$$
$$I + CH_3I, \text{ then } Ag_2O \longrightarrow J\ (C_{10}H_{19}ON)$$
$$J + heat \longrightarrow K\ (C_{10}H_{17}N)$$
$$K + CH_3I, \text{ then } Ag_2O \longrightarrow L\ (C_{11}H_{21}ON)$$
$$L + heat \longrightarrow M\ (C_8H_{10})$$
$$M + Br_2 \longrightarrow N\ (C_8H_{10}Br_2)$$
$$N + (CH_3)_2NH \longrightarrow O\ (C_{12}H_{22}N_2)$$
$$O + CH_3I, \text{ then } Ag_2O \longrightarrow P\ (C_{14}H_{30}O_2N_2)$$
$$P + heat \longrightarrow Q\ (C_8H_8)$$

25. *Pantothenic acid*, $C_9H_{17}O_5N$, occurs in coenzyme A (p. 1389), essential to metabolism of carbohydrates and fats. It reacts with dilute NaOH to give $C_9H_{16}O_5NNa$, with ethyl alcohol to give $C_{11}H_{21}O_5N$, and with hot NaOH to give compound V (see below) and β-aminopropionic acid. Its nitrogen is non-basic. Pantothenic acid has been synthesized as follows:

$$\text{isobutyraldehyde} + \text{formaldehyde} + K_2CO_3 \longrightarrow R\ (C_5H_{10}O_2)$$
$$R + NaHSO_3, \text{ then } KCN \longrightarrow S\ (C_6H_{11}O_2N)$$
$$S + H_2O, H^+, heat \longrightarrow [T\ (C_6H_{12}O_4)] \longrightarrow U\ (C_6H_{10}O_3)$$
$$U + NaOH(aq), \text{ warm} \longrightarrow V\ (C_6H_{11}O_4Na)$$
$$U + \text{sodium } \beta\text{-aminopropionate, then } H^+ \longrightarrow \text{pantothenic acid } (C_9H_{17}O_5N)$$

What is the structure of pantothenic acid?

26. An unknown compound W contained chlorine and nitrogen. It dissolved readily in water to give a solution that turned litmus red. Titration of W with standard base gave a neutralization equivalent of 131 ± 2.

When a sample of W was treated with aqueous NaOH a liquid separated; it contained nitrogen but not chlorine. Treatment of the liquid with nitrous acid followed by β-naphthol gave a red precipitate.

What was W? Write equations for all reactions.

27. Which (if any) of the following compounds could give rise to each of the infrared spectra shown in Fig. 27.5 (p. 992)?

n-butylamine	*o*-anisidine
diethylamine	*m*-anisidine
N-methylformamide	aniline
N,*N*-dimethylformamide	*N*,*N*-dimethyl-*o*-toluidine
2-(dimethylamino)ethanol	acetanilide

28. Give a structure or structures consistent with each of the proton NMR spectra shown in Fig. 27.6, p. 993.

29. Give a structure or structures consistent with each of the CMR spectra shown in Fig. 27.7, p. 994.

30. Give the structures of compounds X, Y, and Z on the basis of their infrared spectra (Fig. 27.8, p. 995) and their proton NMR spectra (Fig. 27.9, p. 996).

IRDC 4227

IRDC 2085

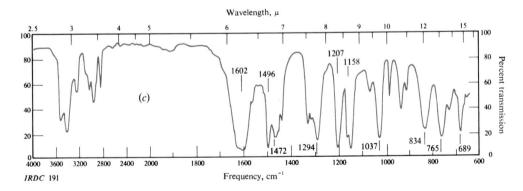

IRDC 191

Figure 27.5   Infrared spectra for Problem 27, p. 991.

Figure 27.6 Proton NMR spectra for Problem 28, p. 991.

Figure 27.7   CMR spectra for Problem 29, p. 991.

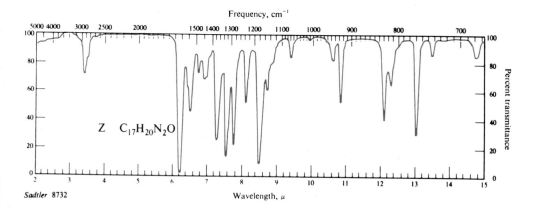

Figure 27.8 Infrared spectra for Problem 30, p. 991.

Figure 27.9 Proton NMR spectra for Problem 30, p. 991.

28

Phenols

28.1 Structure and nomenclature

Phenols are compounds of the general formula ArOH, where Ar is phenyl, substituted phenyl, or one of the other aryl groups we shall study later (e.g., naphthyl, Chap. 34). *Phenols differ from alcohols in having the* —OH *group attached directly to an aromatic ring.*

Phenols are generally named as derivatives of the simplest member of the family, **phenol**. The methylphenols are given the special name of *cresols*. Occasionally phenols are named as *hydroxy-* compounds.

Phenol   *o*-Chlorophenol   *m*-Cresol   *p*-Hydroxybenzoic acid

Catechol   Resorcinol   Hydroquinone   Salicylic acid

Both phenols and alcohols contain the —OH group, and as a result the two families resemble each other to a limited extent. We have already seen, for example, that both alcohols and phenols can be converted into ethers and esters. In most of their properties, however, and in their preparations, the two kinds of compound differ so greatly that they well deserve to be classified as different families.

28.2 Physical properties

The simplest phenols are liquids or low-melting solids; because of hydrogen bonding, they have quite high boiling points (Table 28.1). Phenol itself is somewhat soluble in water (9 g per 100 g of water), presumably because of hydrogen bonding with the water; most other phenols are essentially insoluble in water. Unless some group capable of producing color is present, phenols themselves are colorless. However, like aromatic amines, they are easily oxidized; unless carefully purified, many phenols are colored by oxidation products.

Table 28.1   PHENOLS

Name	M.p., °C	B.p., °C	Solubility, g/100 g H_2O at 25 °C	K_a
Phenol	41	182	9.3	1.1×10^{-10}
o-Cresol	31	191	2.5	0.63
m-Cresol	11	201	2.6	0.98
p-Cresol	35	202	2.3	0.67
o-Fluorophenol	16	152		15
m-Fluorophenol	14	178		5.2
p-Fluorophenol	48	185		1.1
o-Chlorophenol	9	173	2.8	77
m-Chlorophenol	33	214	2.6	16
p-Chlorophenol	43	220	2.7	6.3
o-Bromophenol	5	194		41
m-Bromophenol	33	236		14
p-Bromophenol	64	236	1.4	5.6
o-Iodophenol	43			34
m-Iodophenol	40			13
p-Iodophenol	94			6.3
o-Aminophenol	174		1.7^0	2.0
m-Aminophenol	123		2.6	69
p-Aminophenol	186		1.1^0	
o-Nitrophenol	45	217	0.2	600
m-Nitrophenol	96		1.4	50
p-Nitrophenol	114		1.7	690
2,4-Dinitrophenol	113		0.6	1 000 000
2,4,6-Trinitrophenol (picric acid)	122		1.4	very large
Catechol	104	246	45	1
Resorcinol	110	281	123	3
Hydroquinone	173	286	8	2

An important point emerges from a comparison of the physical properties of the isomeric nitrophenols (Table 28.2). We notice that o-nitrophenol has a much lower boiling point and much lower solubility in water than its isomers; it is the only one of the three that is readily steam-distillable. How can these differences be accounted for?

Table 28.2 PROPERTIES OF THE NITROPHENOLS

	B.p., °C at 70 mm	Solubility, g/100 g H_2O	
o-Nitrophenol	100	0.2	Volatile in steam
m-Nitrophenol	194	1.35	Non-volatile in steam
p-Nitrophenol	dec.	1.69	Non-volatile in steam

Let us consider first the *meta* and *para* isomers. They have very high boiling points because of intermolecular hydrogen bonding:

Intermolecular
hydrogen bonding

Their solubility in water is due to hydrogen bonding with water molecules:

Steam distillation depends upon a substance having an appreciable vapor pressure at the boiling point of water; by lowering the vapor pressure, intermolecular hydrogen bonding inhibits steam distillation of the *meta* and *para* isomers.

What is the situation for the *ortho* isomer? Examination of models (Fig. 28.1, on the next page) shows that the —NO₂ and —OH groups are located exactly right

o-Nitrophenol

Intramolecular
hydrogen bonding:
chelation

for the formation of a hydrogen bond *within a single molecule*. This **intramolecular hydrogen bonding** takes the place of *inter*molecular hydrogen bonding with other phenol molecules and with water molecules; therefore o-nitrophenol does not have the low volatility of an associated liquid, nor does it have the solubility characteristic of a compound that forms hydrogen bonds with water.

(a)

(b)   (c)

Figure 28.1 Molecular structure and physical properties: intramolecular *vs.* intermolecular hydrogen bonding. Models of the nitrophenols: (a) *ortho*, (b) *meta*, (c) *para*. The —OH and —NO₂ groups are located just right for intramolecular hydrogen bonding in the *ortho* isomer, but not in the *meta* or *para*. The *ortho* isomer has a lower boiling point and lower water solubility than its isomers.

We recognize this as an example of *chelation* (Sec. 20.5).

Intramolecular hydrogen bonding seems to occur whenever the structure of a compound permits; we shall encounter other examples of its effect on physical properties.

Problem 28.1 Interpret the following observations. The infrared O—H bands (Sec. 17.5) for the isomeric nitrophenols in solid form (KBr pellets) and in CHCl₃ solution are:

	KBr	CHCl$_3$
o-	3200 cm^{-1}	3200 cm^{-1}
m-	3330	3520
p-	3325	3530

Problem 28.2 In which of the following compounds would you expect intramolecular hydrogen bonding to occur: *o*-nitroaniline, *o*-cresol, *o*-hydroxybenzoic acid (salicylic acid), *o*-hydroxybenzaldehyde (salicylaldehyde), *o*-fluorophenol, *o*-hydroxybenzonitrile?

28.3　Salts of phenols

Phenols are fairly acidic compounds, and in this respect differ markedly from alcohols, which are even more weakly acidic than water. Aqueous hydroxides convert phenols into their salts; aqueous mineral acids convert the salts back into the free phenols. As we might expect, phenols and their salts have opposite solubility properties, the salts being soluble in water and insoluble in organic solvents.

$$ArOH \underset{H^+}{\overset{OH^-}{\rightleftarrows}} ArO^-$$

A phenol	A phenoxide ion
(acid)	(salt)
Insoluble in water	*Soluble in water*

Most phenols have K_a values in the neighborhood of 10^{-10}, and are thus considerably weaker acids than the carboxylic acids (K_a values about 10^{-5}). Most phenols are weaker than carbonic acid, and hence, unlike carboxylic acids, do not dissolve in aqueous bicarbonate solutions. Indeed, phenols are conveniently liberated from their salts by the action of carbonic acid.

$$CO_2 + H_2O \rightleftarrows H_2CO_3$$

$$H_2CO_3 + ArO^-Na^+ \longrightarrow ArOH + Na^+HCO_3^-$$

Stronger acid	Weaker acid
Soluble in water	*Insoluble in water*

The acid strength of phenols and the solubility of their salts in water are useful both in analysis and in separations. A water-insoluble substance that dissolves in aqueous hydroxide but not in aqueous bicarbonate must be more acidic than water, but less acidic than a carboxylic acid; most compounds in this range of acidity are phenols. A phenol can be separated from non-acidic compounds by means of its solubility in base; it can be separated from carboxylic acids by means of its insolubility in bicarbonate.

28.4 Industrial source

Most phenols are made industrially by the same methods that are used in the laboratory; these are described in Sec. 28.5. There are, however, special ways of obtaining certain of these compounds on a commercial scale, including the most important one, phenol. In quantity produced, phenol ranks near the top of the list of synthetic aromatic compounds. Its principal use is in the manufacture of the phenol–formaldehyde polymers (Sec. 36.2).

A certain amount of phenol, as well as the cresols, is obtained from coal tar (Sec. 15.5), but nearly all of it is synthesized. One of the synthetic processes used is the fusion of sodium benzenesulfonate with alkali (Sec. 34.12); another is the Dow process, in which chlorobenzene is allowed to react with aqueous sodium hydroxide at a temperature of about 360 °C. Like the synthesis of aniline from chlorobenzene (Sec. 26.7), this second reaction involves nucleophilic substitution under conditions that are not generally employed in the laboratory (Sec. 29.4).

Nearly all phenol is made today, however, by a newer process that starts with *cumene*, isopropylbenzene. Cumene is converted by air oxidation into cumene hydroperoxide, which is converted by aqueous acid into phenol and acetone.

Along with the large amount of phenol produced each year, a great deal of acetone is obtained, and this process is one of the principal sources of that compound, too. (The mechanism involved here is of considerable theoretical interest to us, and is discussed in detail in Secs. 32.6–32.7.)

Certain phenols and their ethers are isolated from the *essential oils* of various plants (so called because they contain the *essence*—odor or flavor—of the plants). A few of these are:

Eugenol
Oil of cloves

Isoeugenol
Oil of nutmeg

Anethole
Oil of aniseed

Vanillin
*Oil of
vanilla bean*

Thymol
*Oil of thyme
and mint*

Safrole
Oil of sassafras

28.5 Preparation

In the laboratory, phenols are generally prepared by one of the methods outlined below.

PREPARATION OF PHENOLS

1. **Hydrolysis of diazonium salts.** Discussed in Sec. 27.15.

$$Ar-N_2^+ + H_2O \longrightarrow Ar-OH + H^+ + N_2$$

Example:

m-Chlorobenzenediazonium
hydrogen sulfate

m-Chlorophenol

2. **Alkali fusion of sulfonates.** Discussed in Sec. 34.12.

Hydrolysis of diazonium salts is a highly versatile method of making phenols. It is the last step in a synthetic route that generally begins with nitration (Secs. 27.12 and 27.15).

Of limited use is the hydrolysis of aryl halides containing strongly electron-withdrawing groups *ortho* and *para* to the halogen (Sec. 29.9); 2,4-dinitrophenol and 2,4,6-trinitrophenol (*picric acid*) are produced in this way on a large scale:

Chloro-2,4-dinitrobenzene

Sodium 2,4-dinitrophenoxide

2,4-Dinitrophenol

2,4,6-Trinitrophenol
(Picric acid)

28.6   Reactions

Aside from acidity, the most striking chemical property of a phenol is the extremely high reactivity of its ring toward electrophilic substitution. Even in ring substitution, acidity plays an important part; ionization of a phenol yields the —O⁻ group, which, because of its full-fledged negative charge, is even more strongly electron-releasing than the —OH group.

Phenols undergo not only those electrophilic substitution reactions that are typical of most aromatic compounds, but also many others that are possible only because of the unusual reactivity of the ring. We shall have time to take up only a few of these reactions.

REACTIONS OF PHENOLS

1. **Acidity. Salt formation.** Discussed in Secs. 28.3 and 28.7.

$$\text{ArO—H} + \text{H}_2\text{O} \rightleftharpoons \text{ArO}^- + \text{H}_3\text{O}^+$$

Example:

Phenol   Sodium phenoxide

2. **Ether formation. Williamson synthesis.** Discussed in Secs. 19.5 and 28.8.

$$\text{ArO}^- + \text{R—X} \longrightarrow \text{ArO—R} + \text{X}^-$$

Examples:

Phenol   Ethyl iodide   Ethyl phenyl ether
                                               (Phenetole)

p-Cresol   *p*-Nitrobenzyl   *p*-Nitrobenzyl *p*-tolyl ether
                     bromide

o-Nitrophenol Methyl sulfate   *o*-Nitroanisole
                                                 (Methyl *o*-nitrophenyl ether)

CONTINUED

CONTINUED

Phenol   Chloroacetic acid   Phenoxyacetic acid

3. Ester formation. Discussed in Secs. 24.8, 24.15 and 28.9.

Examples:

Phenol   Benzoyl chloride

Phenyl benzoate

p-Nitrophenol   Acetic anhydride

p-Nitrophenyl acetate

o-Bromophenol   *p*-Toluenesulfonyl chloride

o-Bromophenyl *p*-toluenesulfonate

4. Ring substitution. Discussed in Sec. 28.10.

—OH } Activate powerfully, and direct *ortho,para*
—O⁻ } in electrophilic aromatic substitution.

—OR :   Less powerful activator than —OH.

(a) Nitration. Discussed in Sec. 28.10.

Example:

Phenol   *o*-Nitrophenol   *p*-Nitrophenol

CONTINUED

(b) **Sulfonation.** Discussed in Sec. 28.10.

Example:

o-Phenolsulfonic acid

H_2SO_4, 100 °C

p-Phenolsulfonic acid

(c) **Halogenation.** Discussed in Sec. 28.10.

Examples:

Phenol   2,4,6-Tribromophenol

Phenol   p-Bromophenol

(d) **Friedel–Crafts alkylation.** Discussed in Sec. 28.10.

Example:

Phenol   *tert*-Butyl chloride   p-*tert*-Butylphenol

(e) **Friedel–Crafts acylation. Fries rearrangement.** Discussed in Secs. 28.9 and 28.10.

Examples:

Resorcinol   Caproic acid   2,4-Dihydroxyphenyl *n*-pentyl ketone

CONTINUED

CONTINUED

OH

25 °C

2-Methyl-4-hydroxyacetophenone
Chief product

OH
CH$_3$
    (CH$_3$CO)$_2$O

m-Cresol

O
OCCH$_3$
CH$_3$
    AlCl$_3$

m-Cresyl acetate

160 °C

4-Methyl-2-hydroxyacetophenone
Chief product

(f) **Nitrosation.** Discussed in Sec. 28.10.

Example:

OH
CH$_3$ + NaNO$_2$ + H$_2$SO$_4$ ⟶

o-Cresol

OH
CH$_3$
NO

4-Nitroso-2-methylphenol

(g) **Coupling with diazonium salts.** Discussed in Secs. 27.18 and 28.10.

(h) **Carbonation. Kolbe reaction.** Discussed in Sec. 28.11.

Example:

ONa
+ CO$_2$ 125 °C, 4–7 atm ⟶

Sodium phenoxide

OH
COONa

Sodium salicylate
(Sodium *o*-hydroxybenzoate)

(i) **Aldehyde formation. Reimer–Tiemann reaction.** Discussed in Sec. 28.12.

Example:

OH
+ CHCl$_3$ aqueous NaOH ⟶

Phenol Chloroform

OH
CHO

Salicylaldehyde
(*o*-Hydroxybenzaldehyde)

(j) **Reaction with formaldehyde.** Discussed in Sec. 36.7. ■

28.7 Acidity of phenols

Phenols are converted into their salts by aqueous hydroxides, but not by aqueous bicarbonates. The salts are converted into the free phenols by aqueous mineral acids, carboxylic acids, or carbonic acid.

$$ArOH + OH^- \longrightarrow ArO^- + H_2O$$

Stronger   Weaker
acid   acid

$$ArO^- + H_2CO_3 \longrightarrow ArOH + HCO_3^-$$

Stronger   Weaker
acid   acid

Phenols must therefore be considerably stronger acids than water, but considerably weaker acids than the carboxylic acids. Table 28.1 (p. 998) shows that this is indeed so: most phenols have K_a values of about 10^{-10}, whereas carboxylic acids have K_a values of about 10^{-5}.

Although weaker than carboxylic acids, phenols are tremendously more acidic than alcohols, which have K_a values in the neighborhood of 10^{-16} to 10^{-18}. How does it happen that an —OH attached to an aromatic ring is so much more acidic than an —OH attached to an alkyl group? The answer is to be found in an examination of the structures involved. As usual we shall assume that differences in acidity are due to differences in stabilities of reactants and products (Sec. 23.12).

Let us examine the structures of reactants and products in the ionization of an alcohol and of phenol. We see that the alcohol and the alkoxide ion are each represented satisfactorily by a single structure. Phenol and the phenoxide ion contain a benzene ring and therefore must be hybrids of the Kekulé structures I and II, and III and IV. This resonance presumably stabilizes both molecule and ion to the same extent. It lowers the energy content of each by the same number of kilocalories per mole, and hence does not affect the *difference* in their energy contents. If there were no other factors involved, then, we might expect the acidity of a phenol to be about the same as the acidity of an alcohol.

$$R-\ddot{O}:H \;\rightleftharpoons\; H^+ + R-\ddot{O}:^-$$

Alcohol   Alkoxide ion

I   II   III   IV

Phenol   Phenoxide ion

However, there are additional structures to be considered. Being basic, oxygen can share more than a pair of electrons with the ring; this is indicated by contribution from structures V–VII for phenol, and VIII–X for the phenoxide ion.

V   VI   VII   VIII   IX   X

Phenol   Phenoxide ion

Now, are these two sets of structures equally important? Structures V–VII for phenol carry both positive and negative charges; structures VIII–X for phenoxide ion carry only a negative charge. Since energy must be supplied to separate opposite charges, the structures for the phenol should contain more energy and hence be less stable than the structures for phenoxide ion. (We have already encountered the effect of *separation of charge* on stability in Sec. 23.12.) The net effect of resonance is therefore to stabilize the phenoxide ion to a greater extent than the phenol, and thus to shift the equilibrium toward ionization and make K_a larger than for an alcohol (Fig. 28.2).

Figure 28.2 Molecular structure and position of equilibrium. Phenol yields a resonance-stabilized anion, and is a stronger acid than is alcohol. (The plots are aligned with each other for easy comparison.)

We have seen (Sec. 27.3) that aromatic amines are weaker bases than aliphatic amines, since resonance stabilizes the free amine to a greater extent than it does the ion. Here we have exactly the opposite situation, phenols being stronger acids than their aliphatic counterparts, the alcohols, because resonance stabilizes the ion to a greater extent than it does the free phenol. (Actually, of course, resonance with the ring exerts the *same* effect in both cases; it stabilizes—and thus weakens— the base: amine or phenoxide ion.)

In Table 28.1 (p. 998) we see that electron-attracting substituents like —X or —NO$_2$ increase the acidity of phenols, and electron-releasing substituents like —CH$_3$ decrease acidity. Thus substituents affect acidity of phenols in the same way that they affect acidity of carboxylic acids (Sec. 23.14); it is, of course, opposite to the way these groups affect basicity of amines (Sec. 27.4). Electron-attracting substituents tend to disperse the negative charge of the phenoxide ion, whereas electron-releasing substituents tend to intensify the charge.

> **Problem 28.5** How do you account for the fact that, unlike most phenols, 2,4-dinitrophenol and 2,4,6-trinitrophenol are soluble in aqueous sodium bicarbonate?

We can see that a group attached to an aromatic ring affects *position of equilibrium* in reversible reactions in the same way that it affects *rate* in irreversible reactions. An electron-releasing group favors reactions in which the ring becomes more positive, as in electrophilic substitution or in the conversion of an amine into its salt. An electron-withdrawing group favors reactions in which the ring becomes more negative, as in nucleophilic substitution (Chap. 29) or in the conversion of a phenol or an acid into its salt.

28.8 Formation of ethers. Williamson synthesis

As we have already seen (Sec. 19.5), phenols are converted into ethers by reaction in alkaline solution with alkyl halides; methyl ethers can also be prepared by reaction with methyl sulfate. In alkaline solutions a phenol exists as the phenoxide ion which, acting as a nucleophilic reagent, attacks the halide (or the sulfate) and displaces halide ion (or sulfate ion).

$$\text{ArO—H} \xrightarrow{\text{OH}^-} \text{ArO}^- \begin{cases} \xrightarrow{\text{RX}} \text{Ar—O—R} + \text{X}^- \\ \xrightarrow{\text{(CH}_3)_2\text{SO}_4} \text{Ar—O—CH}_3 + \text{CH}_3\text{OSO}_3^- \end{cases}$$

Certain ethers can be prepared by the reaction of unusually active aryl halides with sodium alkoxides. For example:

2,4-Dinitrochlorobenzene   2,4-Dinitroanisole
                                            (2,4-Dinitrophenyl methyl ether)

While alkoxy groups are activating and *ortho,para*-directing in electrophilic aromatic substitution, they are considerably less so than the —OH group. As a result, ethers do not generally undergo those reactions (Secs. 28.10–28.12) which require the especially high reactivity of phenols: coupling, Kolbe reaction, Reimer–Tiemann reaction, etc. This difference in reactivity is probably due to the fact that, unlike a phenol, an ether cannot ionize to form the extremely reactive phenoxide ion.

As a consequence of the lower reactivity of the ring, an aromatic ether is less sensitive to oxidation than a phenol. For example:

CH$_3$O—⟨◯⟩—CH$_3$ $\xrightarrow{\text{KMnO}_4,\ \text{OH}^-,\ \text{heat}}$ $\xrightarrow{\text{H}^+}$ CH$_3$O—⟨◯⟩—COOH

p-Methylanisole   Anisic acid

We have already discussed the cleavage of ethers by acids (Sec. 19.7). Cleavage of methyl aryl ethers by concentrated hydriodic acid is the basis of an important analytical procedure (MeI is formed and measured).

Problem 28.6 2,4-Dichlorophenoxyacetic acid is the important weed-killer known as 2,4-D. Outline the synthesis of this compound starting from benzene or toluene and acetic acid.

OCH$_2$COOH
◯Cl
Cl

2,4-Dichlorophenoxyacetic acid
(2,4-D)

Problem 28.7 The *n*-propyl ether of 2-amino-4-nitrophenol is one of the sweetest compounds ever prepared, being about 5000 times as sweet as the common sugar sucrose. It can be made from the dinitro compound by reduction with ammonium bisulfide. Outline the synthesis of this material starting from benzene or toluene and any aliphatic reagents.

28.9 Ester formation. Fries rearrangement

Phenols are usually converted into their esters by the action of acids, acid chlorides, or anhydrides as discussed in Secs. 23.16, 24.8, and 24.15.

Problem 28.8 Predict the products of the reaction between phenyl benzoate and one mole of bromine in the presence of iron.

When esters of phenols are heated with aluminum chloride, the acyl group migrates from the phenolic oxygen to an *ortho* or *para* position of the ring, thus yielding a ketone. This reaction, called the *Fries rearrangement*, is often used instead of direct acylation for the synthesis of phenolic ketones. For example:

OH
◯ $\xrightarrow{\text{C}_2\text{H}_5\text{COCl}}$

O
‖
O—C—C$_2$H$_5$
◯ $\xrightarrow[\text{CS}_2]{\text{AlCl}_3}$

OH O
      ‖
◯—C—C$_2$H$_5$ +

OH
◯
C—C$_2$H$_5$
‖
O

Phenol   Phenyl propionate   *o*-Hydroxyphenyl
                                              ethyl ketone
                                          (*o*-Hydroxypropiophenone)   *p*-Hydroxyphenyl
                                              Volatile in steam   ethyl ketone
                                                                          (*p*-Hydroxypropiophenone)
                                                                            Non-volatile in steam

In at least some cases, rearrangement appears to involve generation of an acylium ion, RCO^+, which then attacks the ring as in ordinary Friedel–Crafts acylation.

Problem 28.9   A mixture of *ortho* and *para* isomers obtained by the Fries rearrangement can often be separated by steam distillation, only the *ortho* isomer distilling. How do you account for this?

Problem 28.10   4-*n*-Hexylresorcinol is used in certain antiseptics. Outline its preparation starting with resorcinol and any aliphatic reagents.

OH

OH

$CH_2(CH_2)_4CH_3$

4-*n*-Hexylresorcinol

28.10   Ring substitution

Like the amino group, the phenolic group powerfully activates aromatic rings toward electrophilic substitution, and in essentially the same way. The intermediates are hardly carbocations at all, but rather oxonium ions (like I and II), in which every atom (except hydrogen) has a complete octet of electrons; they are formed tremendously faster than the carbocations derived from benzene itself. Attack on a phenoxide ion yields an even more stable—and even more rapidly formed—intermediate, an unsaturated ketone (like III and IV).

I   II   III   IV

With phenols, as with amines, special precautions must often be taken to prevent polysubstitution and oxidation.

Treatment of phenols with aqueous solutions of bromine results in replacement of every hydrogen *ortho* or *para* to the —OH group, and may even cause displacement of certain other groups. For example:

o-Cresol   4,6-Dibromo-2-methylphenol

p-Phenolsulfonic acid   2,4,6-Tribromophenol

If halogenation is carried out in a solvent of low polarity, such as chloroform, carbon tetrachloride, or carbon disulfide, reaction can be limited to monohalogenation. For example:

Phenol — Br$_2$, CS$_2$, 0 °C → *p*-Bromophenol (Chief product) + *o*-Bromophenol

Phenol is converted by concentrated nitric acid into 2,4,6-trinitrophenol (*picric acid*), but the nitration is accompanied by considerable oxidation. To obtain

Phenol — HNO$_3$ → 2,4,6-Trinitrophenol (Picric acid)

mononitrophenols, it is necessary to use dilute nitric acid at a low temperature; even then the yield is poor. (The isomeric products are readily separated by steam distillation. *Why?*)

Phenol — dilute HNO$_3$, 20 °C → *o*-Nitrophenol (40% yield) + *p*-Nitrophenol (13% yield)

Problem 28.11 Picric acid can be prepared by treatment of 2,4-phenoldisulfonic acid with nitric acid. (a) Show in detail the mechanism by which this happens. (b) What advantage does this method of synthesis have over the direct nitration of phenol?

Alkylphenols can be prepared by Friedel–Crafts alkylation of phenols, but the yields are often poor.

Although phenolic ketones can be made by direct acylation of phenols, they are more often prepared in two steps by means of the Fries rearrangement (Sec. 28.9).

Problem 28.12 The product of sulfonation of phenol depends upon the temperature of reaction: chiefly *ortho* at 15–20 °C, chiefly *para* at 100 °C. Once formed, *o*-phenolsulfonic acid is converted into the *para* isomer by sulfuric acid at 100 °C. How do you account for these facts? (*Hint:* See Sec. 10.27.)

In addition, phenols undergo a number of other reactions that also involve electrophilic substitution, and that are possible only because of the especially high reactivity of the ring.

Nitrous acid converts phenols into nitrosophenols:

Phenols are one of the few classes of compounds reactive enough to undergo attack by the weakly electrophilic nitrosonium ion, $^+$NO.

Problem 28.13 The —NO group is readily oxidized to the —NO$_2$ group by nitric acid. Suggest a better way to synthesize *p*-nitrophenol than the one given earlier in this section.

As we have seen, the ring of a phenol is reactive enough to undergo attack by diazonium salts, with the formation of azo compounds. This reaction is discussed in detail in Sec. 27.18.

28.11 Kolbe reaction. Synthesis of phenolic acids

Treatment of the salt of a phenol with carbon dioxide brings about substitution of the carboxyl group, —COOH, for hydrogen of the ring. This reaction is known as the **Kolbe reaction**; its most important application is in the conversion of phenol itself into *o*-hydroxybenzoic acid, known as *salicylic acid*. Although some *p*-hydroxybenzoic acid is formed as well, the separation of the two isomers can be

carried out readily by steam distillation, the *ortho* isomer being the more volatile. (*Why?*)

It seems likely that CO$_2$ attaches itself initially to phenoxide oxygen rather than to the ring. In any case, the final product almost certainly results from electrophilic attack by electron-deficient carbon on the highly reactive ring.

Problem 28.14 The drug *aspirin* is acetylsalicylic acid (*o*-acetoxybenzoic acid, *o*-CH$_3$COOC$_6$H$_4$COOH); *oil of wintergreen* is the ester, methyl salicylate. Outline the synthesis of these two compounds from phenol.

28.12   Reimer–Tiemann reaction. Synthesis of phenolic aldehydes. Dichlorocarbene

Treatment of a phenol with chloroform and aqueous hydroxide introduces an aldehyde group, —CHO, onto the aromatic ring, generally *ortho* to the —OH. This reaction is known as the **Reimer–Tiemann reaction.** For example:

OH ——CHCl₃, aq. NaOH / 70 °C——→ [O⁻ CHCl₂] ——→ O⁻ CHO ——HCl——→ OH CHO

Phenol   Salicylaldehyde
                                                                                          Chief product

A substituted benzal chloride is initially formed, but is hydrolyzed by the alkaline reaction medium.

The Reimer–Tiemann reaction involves electrophilic substitution on the highly reactive phenoxide ring. The electrophilic reagent is dichlorocarbene, $:CCl_2$, generated from chloroform by the action of base. Although electrically neutral, dichlorocarbene contains a carbon atom with only a sextet of electrons and hence is strongly electrophilic.

$$OH^- + CHCl_3 \rightleftarrows H_2O + {}^-:CCl_3 \longrightarrow Cl^- + :CCl_2$$

          Chloroform   Dichlorocarbene

O⁻ + $:CCl_2$ ——→ (cyclohexadienone with H and $CCl_2^{\ominus}$) ——→ O⁻ CHCl₂

          Electrophilic reagent

We encountered dichlorocarbene earlier (Sec. 12.17) as a species adding to carbon–carbon double bonds. There, as here, it is considered to be formed from chloroform by the action of a strong base.

The formation of dichlorocarbene by the sequence

(1) $CHCl_3 + OH^- \rightleftarrows CCl_3^- + H_2O$

(2) $CCl_3^- \rightleftarrows Cl^- + :CCl_2$

          $\xrightarrow{fast}$ products (addition to alkenes, Reimer–Tiemann reaction, hydrolysis, etc.)

is indicated by many lines of evidence, due mostly to elegant work by Jack Hine of The Ohio State University.

Problem 28.15   What bearing does each of the following facts have on the mechanism above? Be specific.

(a) $CHCl_3$ undergoes alkaline hydrolysis much more rapidly than CCl_4 or CH_2Cl_2.

(b) Hydrolysis of ordinary chloroform is carried out in D_2O in the presence of OD^-. When the reaction is interrupted, and unconsumed chloroform is recovered, it is found to contain deuterium. (*Hint*: See Sec. 7.18.)

(c) The presence of added Cl^- *slows down* alkaline hydrolysis of $CHCl_3$.

(d) When alkaline hydrolysis of $CHCl_3$ in the presence of I^- is interrupted, there is recovered not only $CHCl_3$ but also $CHCl_2I$. (In the absence of base, $CHCl_3$ does not react with I^-.)

(e) In the presence of base, $CHCl_3$ reacts with acetone to give 1,1,1-trichloro-2-methyl-2-propanol.

28.13   Analysis of phenols

The most characteristic property of phenols is their particular degree of acidity. Most of them (Secs. 28.3 and 28.7) are stronger acids than water but weaker acids than carbonic acid. Thus, a water-insoluble compound that dissolves in aqueous sodium hydroxide but *not* in aqueous sodium bicarbonate is most likely a phenol.

Many (but not all) phenols form colored complexes (ranging from green through blue and violet to red) with ferric chloride. (This test is also given by *enols*.)

Phenols are often identified through bromination products and certain esters and ethers.

Problem 28.16   Phenols can often be identified as their aryloxyacetic acids, $ArOCH_2COOH$. Suggest a reagent and a procedure for the preparation of these derivatives. (*Hint*: See Sec. 28.8.) Aside from melting point, what other property of the aryloxyacetic acids would be useful in identifying phenols? (*Hint*: See Sec. 23.21.)

28.14   Spectroscopic analysis of phenols

Infrared   As can be seen in Fig. 28.3, phenols show a strong, broad band due to O—H stretching in the same region, 3200–3600 cm^{-1}, as alcohols.

O—H stretching, *strong, broad*

Phenols (or alcohols), 3200–3600 cm^{-1}

Figure 28.3   Infrared spectrum of *p*-cresol.

Phenols differ from alcohols, however, in the position of the C—O stretching band (compare Sec. 18.11).

C—O stretching, *strong, broad*

Phenols, about 1230 cm^{-1} Alcohols, 1050–1200 cm^{-1}

Phenolic ethers do not, of course, show the O—H band, but do show C—O stretching.

C—O stretching, *strong, broad*

Aryl and vinyl ethers, 1200–1275 cm^{-1}, and weaker, 1020–1075 cm^{-1}

Alkyl ethers, 1060–1150 cm^{-1}

(For a comparison of certain oxygen compounds, see Table 24.3, p. 890.)

NMR   Absorption by the O—H proton of a phenol, like that of an alcohol (Sec. 18.11), is affected by the degree of hydrogen bonding, and hence by the temperature, concentration, and nature of the solvent. The signal may appear anywhere in the range δ 4–7, or, if there is intramolecular hydrogen bonding, still lower: δ 6–12.

CMR   The OH of phenols exerts the usual effect of an electronegative substituent.

PROBLEMS

1. Write structural formulas for:

(a) 2,4-dinitrophenol
(b) *m*-cresol
(c) hydroquinone
(d) resorcinol

(e) 4-*n*-hexylresorcinol
(f) catechol
(g) picric acid
(h) phenyl acetate

(i) anisole
(j) salicylic acid
(k) ethyl salicylate

2. Give the reagents and any critical conditions necessary to prepare phenol from:

(a) aniline
(b) benzene

(c) chlorobenzene
(d) cumene (isopropylbenzene)

3. Outline the steps in a possible industrial synthesis of:

(a) catechol from *guaiacol*, *o*-CH$_3$OC$_6$H$_4$OH, found in beech-wood tar
(b) catechol from phenol
(c) resorcinol from benzene

(d) picric acid from chlorobenzene
(e) *veratrole*, *o*-C$_6$H$_4$(OCH$_3$)$_2$, from catechol

4. Outline a possible laboratory synthesis of each of the following compounds from benzene and/or toluene, using any needed aliphatic and inorganic reagents.

(a)–(c) the three cresols
(d) *p*-iodophenol
(e) *m*-bromophenol
(f) *o*-bromophenol
(g) 3-bromo-4-methylphenol
(h) 2-bromo-4-methylphenol
(i) 2-bromo-5-methylphenol

(j) 5-bromo-2-methylphenol
(k) 2,4-dinitrophenol
(l) *p*-isopropylphenol
(m) 2,6-dibromo-4-isopropylphenol
(n) 2-hydroxy-5-methylbenzaldehyde
(o) *o*-methoxybenzyl alcohol

5. Give structures and names of the principal organic products of the reaction (if any) of *o*-cresol with:

(a) aqueous NaOH

(b) aqueous $NaHCO_3$

(c) hot conc. HBr

(d) methyl sulfate, aqueous NaOH

(e) benzyl bromide, aqueous NaOH

(f) bromobenzene, aqueous NaOH

(g) 2,4-dinitrochlorobenzene, aqueous NaOH

(h) acetic acid, H_2SO_4

(i) acetic anhydride

(j) phthalic anhydride

(k) *p*-nitrobenzoyl chloride, pyridine

(l) benzenesulfonyl chloride, aqueous NaOH

(m) product (i) + $AlCl_3$

(n) thionyl chloride

(o) ferric chloride solution

(p) H_2, Ni, 200 °C, 20 atm.

(q) cold dilute HNO_3

(r) H_2SO_4, 15 °C

(s) H_2SO_4, 100 °C

(t) bromine water

(u) Br_2, CS_2

(v) $NaNO_2$, dilute H_2SO_4

(w) product (v) + HNO_3

(x) *p*-nitrobenzenediazonium chloride

(y) CO_2, NaOH, 125 °C, 5 atm.

(z) $CHCl_3$, aqueous NaOH, 70 °C

6. Answer Problem 5 for anisole.

7. Answer Problem 5, parts (a) through (o), for benzyl alcohol.

8. Without referring to tables, arrange the compounds of each set in order of acidity:

(a) benzenesulfonic acid, benzoic acid, benzyl alcohol, phenol

(b) carbonic acid, phenol, sulfuric acid, water

(c) *m*-bromophenol, *m*-cresol, *m*-nitrophenol, phenol

(d) *p*-chlorophenol, 2,4-dichlorophenol, 2,4,6-trichlorophenol

9. Describe simple chemical tests that would serve to distinguish between:

(a) phenol and *o*-xylene

(b) *p*-ethylphenol, *p*-methylanisole, and *p*-methylbenzyl alcohol

(c) 2,5-dimethylphenol, phenyl benzoate, *m*-toluic acid

(d) anisole and *o*-toluidine

(e) acetylsalicylic acid, ethyl acetylsalicylate, ethyl salicylate, and salicylic acid

(f) *m*-dinitrobenzene, *m*-nitroaniline, *m*-nitrobenzoic acid, and *m*-nitrophenol

Tell exactly what you would *do* and *see*.

10. Describe simple chemical methods for the separation of the compounds of Problem 9, parts (a), (c), (d), and (f), recovering each component in essentially pure form.

11. Outline all steps in a possible laboratory synthesis of each of the following compounds starting from the aromatic source given, and using any needed aliphatic and inorganic reagents:

(a) 2,4-diaminophenol (Amidol, used as a photographic developer) from chlorobenzene

(b) 4-amino-1,2-dimethoxybenzene from catechol

(c) 2-nitro-1,3-dihydroxybenzene from resorcinol (*Hint*: See Problem 14.7, p. 512.)

(d) 2,4,6-trimethylphenol from mesitylene

(e) *p-tert*-butylphenol from phenol

(f) 4-(*p*-hydroxyphenyl)-2,2,4-trimethylpentane from phenol

(g) 2-phenoxy-1-bromoethane from phenol (*Hint*: Together with $C_6H_5OCH_2CH_2OC_6H_5$.)

(h) phenyl vinyl ether from phenol

(i) What will phenyl vinyl ether give when heated with acid?

(j) 2,6-dinitro-4-*tert*-butyl-3-methylanisole (synthetic musk) from *m*-cresol

(k) 5-methyl-1,3-dihydroxybenzene (*orcinol*, the parent compound of the litmus dyes) from toluene

12. Outline a possible synthesis of each of the following from benzene, toluene, or any of the natural products shown in Sec. 28.4, using any other needed reagents.

(a) *caffeic acid*, from coffee beans

(b) *tyramine*, found in ergot (*Hint*: See Problem 25.21a, p. 919.)

(c) *noradrenaline*, an adrenal hormone

Caffeic acid   Tyramine   Noradrenaline

13. The reaction between benzyl chloride and sodium phenoxide follows second-order kinetics in a variety of solvents; the nature of the products, however, varies considerably. (a) In dimethylformamide, dioxane, or tetrahydrofuran, reaction yields only benzyl phenyl ether. Show in detail the mechanism of this reaction. To what general class does it belong? (b) In aqueous solution, the yield of ether is cut in half, and there is obtained, in addition, *o*- and *p*-benzylphenol. Show in detail the mechanism by which the latter products are formed. To what general class (or classes) does the reaction belong? (c) What is a possible explanation for the difference between (a) and (b)? (d) In methanol or ethanol, reaction occurs as in (a); in liquid phenol or 2,2,2-trifluoroethanol, reaction is as in (b). How can you account for these differences?

14. When *phloroglucinol*, 1,3,5-trihydroxybenzene, is dissolved in concentrated $HClO_4$, its NMR spectrum shows two peaks of equal area at δ 6.12 and δ 4.15. Similar solutions of 1,3,5-trimethoxybenzene and 1,3,5-triethoxybenzene show similar NMR peaks. On dilution, the original compounds are recovered unchanged. Solutions of these compounds in D_2SO_4 also show these peaks, but on standing the peaks gradually disappear.

How do you account for these observations? What is formed in the acidic solutions? What would you expect to recover from the solution of 1,3,5-trimethoxybenzene in D_2SO_4?

15. (a) When the terpene *citral* is allowed to react in the presence of dilute acid with *olivetol*, there is obtained a mixture of products containing I, the racemic form of one of the physiologically active components of hashish (marijuana). (C_5H_{11} is *n*-pentyl.) Show all steps in a likely mechanism for the formation of I.

$$(CH_3)_2C{=}CHCH_2CH_2C(CH_3){=}CHCHO \; +$$

Citral   Olivetol

I

Δ^1-3,4-*trans*-Tetrahydrocannabinol

(b) The olivetol used above was made from 3,5-dihydroxybenzoic acid. Outline all steps in such a synthesis.

16. Give structures of all compounds below:

(a) *p*-nitrophenol + C_2H_5Br + NaOH(aq) $\longrightarrow$ A ($C_8H_9O_3N$)
     A + Sn + HCl $\longrightarrow$ B ($C_8H_{11}ON$)
     B + $NaNO_2$ + HCl, then phenol $\longrightarrow$ C ($C_{14}H_{14}O_2N_2$)
     C + ethyl sulfate + NaOH(aq) $\longrightarrow$ D ($C_{16}H_{18}O_2N_2$)
     D + $SnCl_2$ $\longrightarrow$ E ($C_8H_{11}ON$)
     E + acetyl chloride $\longrightarrow$ *phenacetin* ($C_{10}H_{13}O_2N$), an analgesic ("pain-killer") and antipyretic ("fever-killer")

(b) β-(o-hydroxyphenyl)ethyl alcohol + HBr $\longrightarrow$ F (C_8H_9OBr)

  F + KOH $\longrightarrow$ *coumarane* (C_8H_8O), insoluble in NaOH

(c) phenol + $ClCH_2COOH$ + NaOH(aq), then HCl $\longrightarrow$ G ($C_8H_8O_3$)

  G + $SOCl_2$ $\longrightarrow$ H ($C_8H_7O_2Cl$)

  H + $AlCl_3$ $\longrightarrow$ 3-*cumaranone* ($C_8H_6O_2$)

(d) *p*-cymene (*p*-isopropyltoluene) + conc. H_2SO_4 $\longrightarrow$ I + J (both $C_{10}H_{14}O_3S$)

  I + KOH + heat, then H^+ $\longrightarrow$ *carvacrol* ($C_{10}H_{14}O$), found in some essential oils

  J + KOH + heat, then H^+ $\longrightarrow$ *thymol* ($C_{10}H_{14}O$), from oil of thyme

  I + HNO_3 $\longrightarrow$ K ($C_8H_8O_5S$)

  p-toluic acid + fuming sulfuric acid $\longrightarrow$ K

(e) anethole (p. 1003) + HBr $\longrightarrow$ L ($C_{10}H_{13}OBr$)

  L + Mg $\longrightarrow$ M ($C_{20}H_{26}O_2$)

  M + HBr, heat $\longrightarrow$ *hexestrol* ($C_{18}H_{22}O_2$), a synthetic estrogen (female sex hormone)

17. (a) The discovery of *crown ethers* was an "accident"—the kind of unplanned occurrence in the laboratory that has so often led observant and imaginative experimenters to important discoveries. The first crown ether made, unintentionally, by Pedersen (p. 711) was compound II, formed by reaction between catechol and di(2-chloroethyl) ether in the presence of NaOH.

II

Write equations for the reactions involved.

  (b) Pedersen obtained II as white crystals insoluble in hydroxylic solvents like methanol, but readily soluble upon addition of NaOH. At this point he thought II might be a phenol. Why was this? What is a likely structure for a phenol formed under these conditions?

  (c) The infrared and NMR spectra, however, showed the absence of —OH. Furthermore, Pedersen found that II was made soluble by the addition, not just of NaOH, but of *any* soluble sodium salt. How do you account for the effect of these salts? What was happening upon addition of, say, NaCl?

  (d) In Sec. 12.8 we learned that the usual technique for making large rings is to carry out the ring-closing reaction at high dilution. Why is this? Write equations to show why one might ordinarily expect this technique to be necessary here; that is, show what alternative course reaction might be expected to take.

  (e) Contrary to the expectations of (d), Pedersen found that he could obtain high yields of II at normal concentrations of reagents so long as Na^+ or, even better, K^+ was present *during the synthesis*. Can you suggest a way in which the presence of these cations could so dramatically affect the course of reaction? (*Hint*: See Fig. 19.1, p. 712.)

18. The adrenal hormone $(-)$-*adrenaline* was the first hormone isolated and the first synthesized. Its structure was proved by the following synthesis:

catechol + $ClCH_2COCl$ $\xrightarrow{POCl_3}$ N ($C_8H_7O_3Cl$)

N + CH_3NH_2 $\longrightarrow$ O ($C_9H_{11}O_3N$)

O + H_2, Pd $\longrightarrow$ $(\pm)$-adrenaline ($C_9H_{13}O_3N$)

N + NaOI, then H^+ $\longrightarrow$ 3,4-dihydroxybenzoic acid

What is the structure of adrenaline?

19. $(-)$-*Phellandral*, $C_{10}H_{16}O$, is a terpene found in eucalyptus oils. It is oxidized by Tollens' reagent to $(-)$-phellandric acid, $C_{10}H_{16}O_2$, which readily absorbs only one mole of hydrogen, yielding dihydrophellandric acid, $C_{10}H_{18}O_2$. $(\pm)$-Phellandral has been synthesized as follows:

isopropylbenzene + H_2SO_4 + SO_3 $\longrightarrow$ P ($C_9H_{12}O_3S$)
P + KOH, fuse $\longrightarrow$ Q ($C_9H_{12}O$)
Q + H_2, Ni $\longrightarrow$ R ($C_9H_{18}O$)
R + $K_2Cr_2O_7$, H_2SO_4 $\longrightarrow$ S ($C_9H_{16}O$)
S + KCN + H^+ $\longrightarrow$ T ($C_{10}H_{17}ON$)
T + acetic anhydride $\longrightarrow$ U ($C_{12}H_{19}O_2N$)
U + heat (600 °C) $\longrightarrow$ V ($C_{10}H_{15}N$) + CH_3COOH
V + H_2SO_4 + H_2O $\longrightarrow$ W ($C_{10}H_{16}O_2$)
W + $SOCl_2$ $\longrightarrow$ X ($C_{10}H_{15}OCl$)

X $\xrightarrow{\text{reduction}}$ ($\pm$)-phellandral

(a) What is the most likely structure of phellandral? (b) Why is synthetic phellandral optically inactive? At what stage in the synthesis does inactivity of this sort first appear? (c) Dihydrophellandric acid is actually a mixture of two optically inactive isomers. Give the structures of these isomers and account for their optical inactivity.

20. Compound Y, C_7H_8O, is insoluble in water, dilute HCl, and aqueous $NaHCO_3$; it dissolves in dilute NaOH. When Y is treated with bromine water it is converted rapidly into a compound of formula $C_7H_5OBr_3$. What is the structure of Y?

21. Two isomeric compounds, Z and AA, are isolated from oil of bay leaf; both are found to have the formula $C_{10}H_{12}O$. Both are insoluble in water, dilute acid, and dilute base. Both give positive tests with dilute $KMnO_4$ and Br_2/CCl_4. Upon vigorous oxidation, both yield anisic acid, p-$CH_3OC_6H_4COOH$.

(a) At this point what structures are possible for Z and AA?
(b) Catalytic hydrogenation converts Z and AA into the same compound, $C_{10}H_{14}O$. Now what structures are possible for Z and AA?
(c) Describe chemical procedures (other than synthesis) by which you could assign structures to Z and AA.
(d) Compound Z can be synthesized as follows:

p-bromoanisole + Mg + ether, then allyl bromide $\longrightarrow$ Z

What is the structure of Z?
(e) Z is converted into AA when heated strongly with concentrated base. What is the most likely structure for AA?
(f) Suggest a synthetic sequence starting with p-bromoanisole that would independently confirm the structure assigned to AA.

22. Compound BB ($C_{10}H_{12}O_3$) was insoluble in water, dilute HCl, and dilute aqueous $NaHCO_3$; it was soluble in dilute NaOH. A solution of BB in dilute NaOH was boiled, and the distillate was collected in a solution of NaOI, where a yellow precipitate formed.
The alkaline residue in the distillation flask was acidified with dilute H_2SO_4; a solid, CC, precipitated. When this mixture was boiled, CC steam-distilled and was collected. CC was found to have the formula $C_7H_6O_3$; it dissolved in aqueous $NaHCO_3$ with evolution of a gas.
(a) Give structures and names for BB and CC. (b) Write complete equations for all the above reactions.

23. *Chavibetol*, $C_{10}H_{12}O_2$, is found in betel-nut leaves. It is soluble in aqueous NaOH but not in aqueous $NaHCO_3$.
Treatment of chavibetol (a) with methyl sulfate and aqueous NaOH gives compound DD, $C_{11}H_{14}O_2$; (b) with hot hydriodic acid gives methyl iodide; (c) with hot concentrated base gives compound EE, $C_{10}H_{12}O_2$.
Compound DD is insoluble in aqueous NaOH, and readily decolorizes dilute $KMnO_4$ and Br_2/CCl_4. Treatment of DD with hot concentrated base gives FF, $C_{11}H_{14}O_2$.
Ozonolysis of EE gives a compound that is isomeric with vanillin (p. 1003).
Ozonolysis of FF gives a compound that is identical with the one obtained from the treatment of vanillin with methyl sulfate.
What is the structure of chavibetol?

24. *Piperine*, $C_{17}H_{19}O_3N$, is an alkaloid found in black pepper. It is insoluble in water, dilute acid, and dilute base. When heated with aqueous alkali, it yields *piperic acid*, $C_{12}H_{10}O_4$, and the cyclic secondary amine *piperidine* (see Sec. 35.12), $C_5H_{11}N$.

Piperic acid is insoluble in water, but soluble in aqueous NaOH and aqueous $NaHCO_3$. Titration gives an equivalent weight of 215 ± 6. It reacts readily with Br_2/CCl_4, without evolution of HBr, to yield a compound of formula $C_{12}H_{10}O_4Br_4$. Careful oxidation of piperic acid yields *piperonylic acid*, $C_8H_6O_4$, and *tartaric acid*, HOOCCHOHCHOHCOOH.

When piperonylic acid is heated with aqueous HCl at 200 °C it yields formaldehyde and *protocatechuic acid*, 3,4-dihydroxybenzoic acid.

(a) What kind of compound is piperine? (b) What is the structure of piperonylic acid? Of piperic acid? Of piperine?

(c) Does the following synthesis confirm your structure?

$$\text{catechol} + CHCl_3 + NaOH \longrightarrow GG\ (C_7H_6O_3)$$
$$GG + CH_2I_2 + NaOH \longrightarrow HH\ (C_8H_6O_3)$$
$$HH + CH_3CHO + NaOH \longrightarrow II\ (C_{10}H_8O_3)$$
$$II + \text{acetic anhydride} + \text{sodium acetate} \longrightarrow \text{piperic acid}\ (C_{12}H_{10}O_4)$$
$$\text{piperic acid} + PCl_5 \longrightarrow JJ\ (C_{12}H_9O_3Cl)$$
$$JJ + \text{piperidine} \longrightarrow \text{piperine}$$

25. *Hordinene*, $C_{10}H_{15}ON$, is an alkaloid found in germinating barley. It is soluble in dilute HCl and in dilute NaOH; it reprecipitates from the alkaline solution when CO_2 is bubbled in. It reacts with benzenesulfonyl chloride to yield a product **KK** that is soluble in dilute acids.

When hordinene is treated with methyl sulfate and base, a product, **LL**, is formed. When LL is oxidized by alkaline $KMnO_4$, there is obtained anisic acid, $p\text{-}CH_3OC_6H_4COOH$. When LL is heated strongly there is obtained *p*-methoxystyrene.

(a) What structure or structures are consistent with this evidence? (b) Outline a synthesis or syntheses that would prove the structure of hordinene.

26. The structure of the terpene *α-terpineol* (found in oils of cardamom and marjoram) was proved in part by the following synthesis:

$$p\text{-toluic acid} + \text{fuming sulfuric acid} \longrightarrow MM\ (C_8H_8O_5S)$$
$$MM + KOH \xrightarrow{\text{fusion}} NN\ (C_8H_8O_3)$$
$$NN + Na, \text{alcohol} \longrightarrow OO\ (C_8H_{14}O_3)$$
$$OO + HBr \longrightarrow PP\ (C_8H_{13}O_2Br)$$
$$PP + \text{base, heat} \longrightarrow QQ\ (C_8H_{12}O_2)$$
$$QQ + C_2H_5OH, HCl \longrightarrow RR\ (C_{10}H_{16}O_2)$$
$$RR + CH_3MgI, \text{then } H_2O \longrightarrow \alpha\text{-terpineol}\ (C_{10}H_{18}O)$$

What is the most likely structure for α-terpineol?

27. *Coniferyl alcohol*, $C_{10}H_{12}O_3$, is obtained from the sap of conifers. It is soluble in aqueous NaOH but not in aqueous $NaHCO_3$.

Treatment of coniferyl alcohol (a) with benzoyl chloride and pyridine gives compound **SS**, $C_{24}H_{20}O_5$; (b) with cold HBr gives $C_{10}H_{11}O_2Br$; (c) with hot hydriodic acid gives a volatile compound identified as methyl iodide; (d) with methyl iodide and aqueous base gives compound **TT**, $C_{11}H_{14}O_3$.

Both SS and TT are insoluble in dilute NaOH, and rapidly decolorize dilute $KMnO_4$ and Br_2/CCl_4.

Ozonolysis of coniferyl alcohol gives vanillin.

What is the structure of coniferyl alcohol?

Write equations for all the above reactions.

28. When α-(*p*-tolyloxy)isobutyric acid (prepared from *p*-cresol) is treated with Br_2, there is obtained UU.

UU

(a) To what class of compounds does UU belong? Suggest a mechanism for its formation.

(b) Give structural formulas for compounds VV, WW, and XX.

$$UU + AgNO_3, CH_3OH \longrightarrow VV\ (C_{12}H_{16}O_4)$$
$$VV + H_2, Rh \longrightarrow WW\ (C_{12}H_{20}O_4)$$
$$WW + H_2O, OH^- \longrightarrow XX\ (C_8H_{14}O)$$

(c) The reactions outlined in (b) can be varied. Of what general synthetic utility do you think this general process might be?

29. Compounds AAA–FFF are phenols or related compounds whose structures are given in Problem 18, p. 791, or Sec. 28.4. Assign a structure to each one on the basis of infrared and/or NMR spectra shown as follows.

> AAA, BBB, and CCC: infrared spectra in Fig. 28.4 (p. 1024)
> proton NMR spectra in Fig. 28.5 (p. 1025)
> DDD: proton NMR spectrum in Fig. 28.6 (p. 1026)
> EEE and FFF: infrared spectra in Fig. 28.7 (p. 1026)

(*Hint*: After you have worked out some of the structures, compare infrared spectra.)

Sadtler 15466 K

Sadtler 18061 K

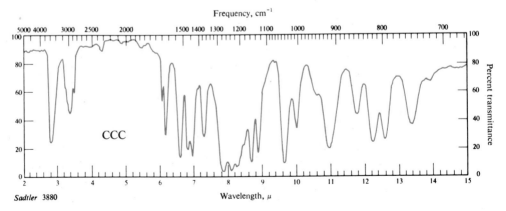

Sadtler 3880

Figure 28.4 Infrared spectra for Problem 29, p. 1023.

Figure 28.5   Proton NMR spectra for Problem 29, p. 1023.

Figure 28.6   Proton NMR spectrum for Problem 29, p. 1023.

Sadtler 8115 K

Sadtler 21073 K

Figure 28.7   Infrared spectra for Problem 29, p. 1023.

Part Two

Special Topics

29

Aryl Halides

Nucleophilic Aromatic Substitution

29.1 Structure

Aryl halides are compounds containing halogen attached directly to an aromatic ring. They have the general formula ArX, where Ar is phenyl, substituted phenyl, or one of the other aryl groups that we shall study (e.g., naphthyl, Chap. 34):

Bromobenzene   *m*-Chloronitrobenzene   *p*-Iodotoluene   *o*-Chlorobenzoic acid

An aryl halide is not just any halogen compound containing an aromatic ring. Benzyl chloride, for example, is not an aryl halide, for halogen is not attached to the aromatic ring; in structure and properties it is simply a substituted alkyl halide and was studied as such (Chap. 15).

We take up the aryl halides in a separate chapter because they differ so much from the alkyl halides in their preparation and properties. Aryl halides as a class are comparatively unreactive toward the nucleophilic substitution reactions so characteristic of the alkyl halides. The presence of certain other groups on the aromatic ring, however, greatly increases the reactivity of aryl halides; in the absence of such groups, reaction can still be brought about by very basic reagents or high temperatures. We shall find that **nucleophilic aromatic substitution** can follow two very different paths: the *bimolecular displacement mechanism*, for activated aryl halides; and the *elimination–addition mechanism*, which involves the remarkable intermediate called *benzyne*.

It will be useful to compare aryl halides with certain other halides that are not aromatic at all: *vinyl halides*, compounds in which halogen is attached directly to

$$-\overset{|}{C}=\overset{|}{C}-X$$

A vinyl halide

a doubly bonded carbon.

Vinyl halides, we have already seen, show an interesting parallel to aryl halides. Each kind of compound contains another functional group besides halogen: vinyl halides contain a carbon–carbon double bond, which undergoes electrophilic addition; aryl halides contain a ring, which undergoes electrophilic substitution. In each of these reactions, halogen exerts an anomalous influence on reactivity and orientation. In electrophilic addition, halogen deactivates, yet causes Markovnikov orientation (Sec. 10.15); in electrophilic substitution, halogen deactivates, yet directs *ortho,para* (Sec. 14.19). In both cases we attributed the influence of halogen to the working of opposing factors. Through its inductive effect, halogen withdraws electrons and deactivates the entire molecule toward electrophilic attack. Through its resonance effect, halogen releases electrons and tends to activate—but only toward attack *at certain positions*.

Problem 29.1 Drawing all pertinent structures, account in detail for the fact that: (a) addition of hydrogen iodide to vinyl chloride is slower than to ethylene, yet yields predominantly 1-chloro-1-iodoethane; (b) nitration of chlorobenzene is slower than that of benzene, yet occurs predominantly *ortho,para*.

The parallel between aryl and vinyl halides goes further: both are relatively unreactive toward nucleophilic substitution and, as we shall see, for basically the same reason. Moreover, this low reactivity is caused—partly, at least—by the same structural feature that is responsible for their anomalous influence on electrophilic attack: partial double-bond character of the carbon–halogen bond.

We must keep in mind that aryl halides are of "low reactivity" only with respect to certain sets of familiar reactions typical of the more widely studied alkyl halides. Before 1953, aryl halides appeared to undergo essentially only one reaction—and that one, rather poorly. It is becoming increasingly evident that aryl halides are actually capable of doing many different things; as with the "unreactive" alkanes (Sec. 3.18), it is only necessary to provide the proper conditions—and to have the ingenuity to *observe* what is going on. Of these reactions, we shall have time to take up only two. But we should be aware that there *are* others: free-radical reactions, for example, and what Joseph Bunnett (p. 277) has named the *base-catalyzed halogen dance* (Problem 23, p. 1057).

29.2 Physical properties

Unless modified by the presence of some other functional group, the physical properties of the aryl halides are much like those of the corresponding alkyl halides. Chlorobenzene and bromobenzene, for example, have boiling points very nearly the same as those of n-hexyl chloride and n-hexyl bromide; like the alkyl halides, the aryl halides are insoluble in water and soluble in organic solvents.

Table 29.1 ARYL HALIDES

	M.p., °C	B.p., °C	Ortho M.p., °C	Ortho B.p., °C	Meta M.p., °C	Meta B.p., °C	Para M.p., °C	Para B.p., °C
Fluorobenzene	− 45	85						
Chlorobenzene	− 45	132						
Bromobenzene	− 31	156						
Iodobenzene	− 31	189						
Fluorotoluene				115	−111	115		116
Chlorotoluene			− 34	159	− 48	162	8	162
Bromotoluene			− 26	182	− 40	184	28	185
Iodotoluene				206		211	35	211
Difluorobenzene			− 34	92	− 59	83	− 13	89
Dichlorobenzene			− 17	180	− 24	173	52	175
Dibromobenzene			6	221	− 7	217	87	219
Diiodobenzene			27	287	35	285	129	285
Nitrochlorobenzene			32	245	48	236	83	239
2,4-Dinitro-chlorobenzene	53	315						
2,4,6-Trinitro-chlorobenzene (picryl chloride)	83							
Vinyl chloride	−160	− 14						
Vinyl bromide	−138	16						

The physical constants listed in Table 29.1 illustrate very well a point previously made (Sec. 15.4) about the boiling points and melting points of *ortho, meta*, and *para* isomers. The isomeric dihalobenzenes, for example, have very nearly the same boiling points: between 173 °C and 180 °C for the dichlorobenzenes, 217 °C to 221 °C for the dibromobenzenes, and 285 °C to 287 °C for the diiodobenzenes. Yet the melting points of these same compounds show a considerable spread; in each case, the *para* isomer has a melting point that is some 70–100 °C higher than the *ortho* or *meta* isomer. The physical constants of the halotoluenes show a similar relationship.

Here again we see that, having the most symmetrical structure, the *para* isomer fits better into a crystalline lattice and has the highest melting point (Fig. 29.1, on the next page). We can see how it is that a reaction product containing both *ortho* and *para* isomers frequently deposits crystals of only the *para* isomer upon cooling. Because of the strong intracrystalline forces, the higher-melting *para* isomer also is less soluble in a given solvent than the *ortho* isomer, so that purification of the *para* isomer is often possible by recrystallization. The *ortho* isomer that remains in solution is generally heavily contaminated with the *para* isomer, and is difficult to purify.

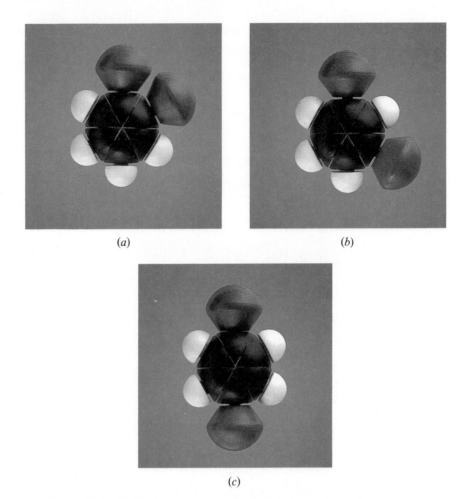

(a)   (b)

(c)

Figure 29.1 Molecular structure and physical properties: effect of symmetry. Models of the dichlorobenzenes: (a) *ortho*, m.p. −17 °C; (b) *meta*,
m.p. −24 °C; (c) *para*, m.p. 52 °C. The *para* isomer is most symmetrical, fits
into a crystal lattice best, and has the highest melting point and lowest
solubility.

29.3 Preparation

Aryl halides are most often prepared in the laboratory by the methods outlined
below, and on an industrial scale by adaptations of these methods.

These methods, we notice, differ considerably from the methods of preparing
alkyl halides. (a) Direct halogenation of the aromatic ring is more useful than direct
halogenation of alkanes; although mixtures may be obtained (e.g., *ortho* + *para*),
attack is not nearly so random as in the free-radical halogenation of aliphatic
hydrocarbons. (b) Alkyl halides are most often prepared from the corresponding
alcohols; aryl halides are not prepared from the phenols. Instead, aryl halides are
most commonly prepared by replacement of the nitrogen of a **diazonium salt**; as
the sequence above shows, this ultimately comes from a nitro group which was
itself introduced directly into the ring. *From the standpoint of synthesis, then, the
nitro compounds bear much the same relationship to aryl halides that alcohols do to*

PREPARATION OF ARYL HALIDES _____

   1. **From diazonium salts.** Discussed in Secs. 27.13 and 29.3.

$$Ar-H \xrightarrow[H_2SO_4]{HNO_3} Ar-NO_2 \xrightarrow{redn.} Ar-NH_2 \xrightarrow[0\ C]{HONO} Ar-N_2^+$$

Diazonium
salt

$$\xrightarrow{BF_4^-} Ar-F$$
$$\xrightarrow{CuCl} Ar-Cl$$
$$\xrightarrow{CuBr} Ar-Br$$
$$\xrightarrow{I^-} Ar-I$$

Example:

o-Toluenediazonium   *o*-Chlorotoluene
chloride

   2. **Halogenation.** Discussed in Secs. 14.11 and 15.13.

$$Ar-H + X_2 \xrightarrow{Lewis\ acid} Ar-X + HX$$

$$X_2 = Cl_2,\ Br_2$$

Lewis acid = $FeCl_3$, $AlCl_3$, $Tl(OAc)_3$, etc.

Examples:

Nitrobenzene   *m*-Chloronitrobenzene

Acetanilide   *p*-Bromoacetanilide
                                   Major product

alkyl halides. (These reactions of diazonium salts have been discussed in detail in Secs. 27.12–27.13.)

The preparation of aryl halides from diazonium salts is more important than direct halogenation for several reasons. First of all, fluorides and iodides, which can seldom be prepared by direct halogenation, can be obtained from the diazonium salts. Second, where direct halogenation yields a mixture of *ortho* and *para* isomers, the *ortho* isomer, at least, is difficult to obtain pure. On the other hand, the *ortho* and *para* isomers of the corresponding nitro compounds, from which the diazonium salts ultimately come, can often be separated by fractional distillation (Sec. 14.7).

For example, the *o*- and *p*-bromotoluenes boil only three degrees apart: 182 °C and 185 °C. The corresponding *o*- and *p*-nitrotoluenes, however, boil sixteen degrees apart: 222 °C and 238 °C.

29.4 Reactions

The typical reaction of alkyl halides, we have seen (Sec. 5.8), is nucleophilic substitution. Halogen is displaced as halide ion by such bases as OH^-, OR^-, NH_3, CN^-, etc., to yield alcohols, ethers, amines, nitriles, etc. Even Friedel–Crafts alkylation is, from the standpoint of the alkyl halide, nucleophilic substitution by the basic aromatic ring.

$$R:X + :Z \longrightarrow R:Z + :X^-$$

$$:Z = OH^-, OR^-, NH_3, CN^-, \text{etc.}$$

It is typical of **aryl halides** *that they undergo nucleophilic substitution only with extreme difficulty.* Except for certain industrial processes where very severe conditions are feasible, one does not ordinarily prepare phenols (ArOH), ethers (ArOR), amines ($ArNH_2$), or nitriles (ArCN) by nucleophilic attack on aryl halides. We cannot use aryl halides as we use alkyl halides in the Friedel–Crafts reaction.

However, aryl halides do undergo nucleophilic substitution readily if the aromatic ring contains, in addition to halogen, certain other properly placed groups: electron-withdrawing groups like $-NO_2$, $-NO$, or $-CN$, located *ortho* or *para* to halogen. For aryl halides having this special kind of structure, nucleophilic substitution proceeds readily and can be used for synthetic purposes.

The reactions of unactivated aryl halides with strong bases or at high temperatures, which proceed via *benzyne*, are finding increasing synthetic importance. The Dow process, which has been used for many years in the manufacture of phenol (Sec. 28.4), turns out to be what Bunnett (p. 277) calls "benzyne chemistry on the tonnage scale!"

The aromatic ring to which halogen is attached can, of course, undergo the typical electrophilic aromatic substitution reactions: nitration, sulfonation, halogenation, Friedel–Crafts alkylation. Like any substituent, halogen affects the reactivity and orientation in these reactions. As we have seen (Sec. 14.5), halogen is unusual in being deactivating, yet *ortho,para*-directing.

REACTIONS OF ARYL HALIDES

1. **Formation of Grignard reagent.** Limitations are discussed in Sec. 17.17.

$$Ar-Br + Mg \xrightarrow{\text{dry ether}} Ar-MgBr$$

$$Ar-Cl + Mg \xrightarrow{\text{tetrahydrofuran}} Ar-MgCl$$

2. **Substitution in the ring.** Electrophilic aromatic substitution. Discussed in Sec. 14.19.

X: Deactivates and directs *ortho,para*
in electrophilic aromatic substitution.

CONTINUED

———— CONTINUED ————

3. Nucleophilic aromatic substitution. Bimolecular displacement. Discussed in Secs. 29.7–29.13.

Ar:X + :Z $\longrightarrow$ Ar:Z + :X⁻

Ar must contain strongly electron-withdrawing groups ortho and/or para to —X.

Examples:

2,4-Dinitrochlorobenzene   2,4-Dinitrophenol

2,4-Dinitrochlorobenzene   2,4-Dinitroaniline

2,4-Dinitrochlorobenzene   Ethyl 2,4-dinitrophenyl ether

4. Nucleophilic aromatic substitution. Elimination–addition. Discussed in Sec. 29.14.

Ar:X + :Z $\longrightarrow$ Ar:Z + :X⁻
     Strong
     base

Ring not activated toward bimolecular displacement

Examples:

Fluorobenzene   Phenyllithium   Biphenyl

3-Bromo-4-methoxybiphenyl   2-Amino-4-methoxybiphenyl   ■

29.5   Low reactivity of aryl and vinyl halides

We have already discussed (Sec. 10.18) the extremely low reactivity toward nucleophilic substitution of vinylic halides. Similarly low reactivity is shown by aryl halides. Attempts to convert aryl or vinyl halides into phenols (or alcohols), ethers, amines, or nitriles by treatment with the usual nucleophilic reagents are also unsuccessful; aryl or vinyl halides cannot be used in place of alkyl halides in the Friedel–Crafts reaction.

How can the low reactivity of these halides be accounted for? To find possible answers, let us look at their structures.

29.6   Structure of aryl and vinyl halides

The low reactivity of aryl and vinyl halides toward displacement has, like the stabilities of alkenes and dienes (Secs. 10.22–10.24), been attributed to two different factors: (a) delocalization of electrons by resonance; and (b) differences in (σ) bond energies due to differences in hybridization of carbon.

Let us look first at the resonance interpretation.

Chlorobenzene is considered to be a hybrid of not only the two Kekulé structures, I and II, but also of three structures, III, IV, and V, in which chlorine

is joined to carbon by a double bond; in III, IV, and V chlorine bears a positive charge and the *ortho* and *para* positions of the ring bear a negative charge.

In a similar way, vinyl chloride is considered to be a hybrid of structure VI (the one we usually draw for it) and structure VII, in which chlorine is joined to carbon by a double bond; in VII chlorine bears a positive charge and C–2 bears a

negative charge. Other aryl and vinyl halides are considered to have structures exactly analogous to these.

Contribution from III, IV, and V, and from VII stabilizes the chlorobenzene and vinyl chloride molecules, and gives double-bond character to the carbon–chlorine bond. Carbon and chlorine are thus held together by something more than a single pair of electrons, and the carbon–chlorine bond is stronger than if it were a pure single bond. The low reactivity of these halides toward nucleophilic substitution is due (partly, at least) to resonance stabilization of the halides (by a factor that in this case does not stabilize the transition state to the same extent); this stabilization increases the E_{act} for displacement, and thus slows down reaction. For aryl halides, another factor—which may well be the most important one—is stabilization of the molecule by resonance involving the Kekulé structures.

The alternative interpretation is simple. In alkyl halides the carbon holding halogen is sp^3-hybridized. In aryl and vinyl halides, carbon is sp^2-hybridized; the bond to halogen is shorter and stronger, and the molecule is more stable (see Sec. 7.4).

What evidence is there to support either interpretation, other than the fact that it would account for *the low reactivity of aryl and vinyl halides*?

The carbon–halogen bonds of aryl and vinyl halides are unusually short. In chlorobenzene and vinyl chloride the C—Cl bond length is only 1.69 Å, as compared with a length of 1.77–1.80 Å in a large number of alkyl chlorides (Table 29.2). In bromobenzene and vinyl bromide the C—Br bond length is only 1.86 Å, as compared with a length of 1.91–1.92 Å in alkyl bromides.

Now, as we have seen (Sec. 7.2), a double bond is shorter than a single bond joining the same pair of atoms; if the carbon–halogen bond in aryl and vinyl halides has double-bond character, it should be shorter than the carbon–halogen bond in alkyl halides. Alternatively, a bond formed by overlap of an sp^2 orbital should be shorter than the corresponding bond involving an sp^3 orbital.

Dipole moments of aryl and vinyl halides are unusually small. Organic halogen compounds are polar molecules; displacement of electrons toward the more electronegative element makes halogen relatively negative and carbon relatively positive. Table 29.2 shows that the dipole moments of a number of alkyl chlorides and bromides range from 2.02 D to 2.15 D. The mobile π electrons of the benzene ring and of the carbon–carbon double bond should be particularly easy to displace; hence we might have expected aryl and vinyl halides to have even larger dipole moments than alkyl halides.

Table 29.2   BOND LENGTHS AND DIPOLE MOMENTS OF HALIDES

	Bond lengths, Å		Dipole moments, D	
	C—Cl	C—Br	R—Cl	R—Br
CH_3—X	1.77	1.91	—	—
C_2H_5—X	1.77	1.91	2.05	2.02
n-C_3H_7—X	—	—	2.10	2.15
n-C_4H_9—X	—	—	2.09	2.15
$(CH_3)_3C$—X	1.80	1.92	2.13	—
CH_2=CH—X	1.69	1.86	1.44	1.41
C_6H_5—X	1.69	1.86	1.73	1.71

However, we see that this is not the case. Chlorobenzene and bromobenzene have dipole moments of only 1.7 D, and vinyl chloride and vinyl bromide have dipole moments of only 1.4 D. This is consistent with the resonance picture of these molecules. In the structures that contain doubly bonded halogen (III, IV, V, and VII) there is a positive charge on halogen and a negative charge on carbon; to the extent that these structures contribute to the hybrids, they tend to oppose the usual displacement of electrons toward halogen. Although there is still a net displacement of electrons toward halogen in aryl halides and vinyl halides, it is less than in other organic halides.

Alternatively, sp^2-hybridized carbon is, in effect, a more electronegative atom than an sp^3-hybridized carbon (see Sec. 11.11), and is less willing to release electrons to chlorine.

As was discussed in Secs. 10.15, 14.19, and 29.1, contribution from structures in which halogen is doubly bonded and bears a positive charge accounts for *the way halogen affects the reactions of the carbon–carbon double bond or of the benzene ring to which it is joined.*

The counterargument is that this simply indicates that resonance of this kind can occur—but not how important it is in the halide molecules.

Finally, the *existence of cyclic halonium ions* (Secs. 8.14 and 9.6) certainly shows that halogen *can* share more than a pair of electrons.

It is hard to believe that the stability of these molecules is not affected by the particular kind of hybridization; on the other hand, it seems clear that there is resonance involving halogen and the π electrons. The question, once more, is one of their relative importance. As in the case of alkenes and dienes, it is probable that *both* are important.

As we shall see, in the rate-determining step of nucleophilic aromatic substitution a nucleophile attaches itself to the carbon bearing halogen; this carbon becomes tetrahedral, and the ring acquires a negative charge. Such a reaction is made more difficult by the fact that it destroys the aromaticity of the ring and disrupts the resonance between ring and halogen; and, if Dewar is correct (Sec. 10.24), because energy is required to change the hybridization of carbon from sp^2 to sp^3.

Problem 29.2   In Sec. 29.3 we learned that, unlike alkyl halides, aryl halides are not readily prepared from the corresponding hydroxy compounds. How might you account for this contrast between alcohols and phenols? (*Hint*: See Sec. 28.7.)

29.7   Nucleophilic aromatic substitution: bimolecular displacement

We have seen that the aryl halides are characterized by very low reactivity toward the nucleophilic reagents like OH^-, OR^-, NH_3, and CN^- that play such an important part in the chemistry of the alkyl halides. Consequently, nucleophilic aromatic substitution is much less important in synthesis than either nucleophilic aliphatic substitution or electrophilic aromatic substitution.

However, the presence of certain groups at certain positions of the ring markedly activates the halogen of aryl halides toward displacement. We shall have a look at some of these activation effects, and then try to account for them on the basis of the chemical principles we have learned. We shall find a remarkable parallel between the two kinds of aromatic substitution, electrophilic and nucleophilic, with respect both to mechanism and to the ways in which substituent groups affect reactivity and orientation.

Chlorobenzene is converted into phenol by aqueous sodium hydroxide only at temperatures over 300 °C. The presence of a nitro group *ortho* or *para* to the chlorine greatly increases its reactivity: *o*- or *p*-chloronitrobenzene is converted into the nitrophenol by treatment with aqueous sodium hydroxide at 160 °C. A nitro group *meta* to the chlorine, on the other hand, has practically no effect on reactivity. As the number of *ortho* and *para* nitro groups on the ring is increased, the reactivity increases: the phenol is obtained from 2,4-dinitrochlorobenzene by treatment with hot aqueous sodium carbonate, and from 2,4,6-trinitrochlorobenzene by simple treatment with water.

Chlorobenzene	Phenol

6–8% NaOH, 350 °C, 4500 lb/in.²

p-Chloronitrobenzene	*p*-Nitrophenol

15% NaOH, 160 °C

2,4-Dinitrochlorobenzene	2,4-Dinitrophenol

boiling aq. Na₂CO₃, 130 °C

2,4,6-Trinitrochlorobenzene	2,4,6-Trinitrophenol (Picric acid)

H₂O, warm

Similar effects are observed when other nucleophilic reagents are used. Ammonia or sodium methoxide, for example, reacts with chloro- or bromobenzene only under very vigorous conditions. For example:

Chlorobenzene	Aniline

NH₃, Cu₂O, 200 °C, 900 lb/in.²

Yet if the ring contains a nitro group—or preferably two or three of them—*ortho* or *para* to the halogen, reaction proceeds quite readily. For example:

2,4-Dinitrochlorobenzene   2,4-Dinitroaniline

2,4,6-Trinitrochlorobenzene   2,4,6-Trinitroanisole

Like $-NO_2$, certain other groups have been found to activate halogen located *ortho* or *para* to them: $-N(CH_3)_3{}^+$, $-CN$, $-SO_3H$, $-COOH$, $-CHO$, $-COR$. This is a familiar list. All these are electron-withdrawing groups, which are deactivating and *meta*-directing toward *electrophilic* substitution (see Table 14.3, p. 504).

Although our concern here is primarily with displacement of halogen, it is important to know that these electron-withdrawing substituents activate many groups other than halogen toward nucleophilic substitution. (Hydrogen is generally not displaced from the aromatic ring, since this would require the separation of the very strongly basic hydride ion, $:H^-$.)

Problem 29.3 When *p*-nitroso-*N*,*N*-dimethylaniline is heated with aqueous KOH, dimethylamine is evolved; this reaction is sometimes used to prepare pure dimethylamine, free from methylamine and trimethylamine. (a) What are the other products of the reaction? (b) To what class of organic reactions does this belong? (c) Upon what property of the nitroso group does this reaction depend? (d) Outline all steps in the preparation of pure diethylamine starting from nitrobenzene and ethyl alcohol.

Problem 29.4 How do you account for the following observations?

(a) Although most ethers are inert toward bases, 2,4-dinitroanisole is readily cleaved to methanol and 2,4-dinitrophenol when refluxed with dilute aqueous NaOH.

(b) Although amides can be hydrolyzed by either aqueous acid or aqueous alkali, hydrolysis of *p*-nitroacetanilide is best carried out in acidic solution.

(c) Treatment of *o*-chloronitrobenzene by aqueous sodium sulfite yields sodium *o*-nitrobenzenesulfonate. Give the structure of the reagent involved. How does this reagent compare with the one in ordinary sulfonations?

(d) Would you expect the method of (c) to be a general one for preparation of sulfonic acids? Could it be used, for example, to prepare benzenesulfonic acid?

(e) Washing crude *m*-dinitrobenzene with aqueous sodium sulfite removes contaminating *o*- and *p*-dinitrobenzene.

If electron-withdrawing groups activate toward nucleophilic substitution, we might expect electron-releasing groups to *deactivate*. This is found to be so. Furthermore, the degree of deactivation depends upon how strongly they release electrons: $-NH_2$ and $-OH$ deactivate strongly; $-OR$, moderately; and $-R$, weakly.

In nucleophilic as in electrophilic aromatic substitution, then, a substituent group affects reactivity by its ability to attract or release electrons; in nucleophilic as in electrophilic aromatic substitution, a substituent group exerts its effect chiefly at the positions *ortho* and *para* to it. The kind of effect that each group exerts, however, is exactly opposite to the kind of effect it exerts in electrophilic aromatic substitution. *In* **nucleophilic aromatic substitution** *electron withdrawal causes activation, and electron release causes deactivation.*

To account for these effects, we must look at the mechanism for the kind of nucleophilic aromatic substitution we have been talking about.

29.8 Bimolecular displacement mechanism for nucleophilic aromatic substitution

The bimolecular displacement mechanism for nucleophilic aromatic substitution (shown here for chlorobenzene) is:

(1) $C_6H_5-Cl + :Z \longrightarrow C_6H_5 \overset{\ominus}{\underset{Z}{\diagup}}^{Cl}$ *Slow*

**Bimolecular
displacement**

I

(2) $C_6H_5 \overset{\ominus}{\underset{Z}{\diagup}}^{Cl} \longrightarrow C_6H_5-Z + :Cl^-$ *Fast*

I

There are two essential steps: attack of a nucleophilic reagent upon the ring to form a carbanion (I), and the expulsion of halide ion from this carbanion to yield the product.

The intermediate carbanion (I) is a hybrid of II, III, and IV; this hybrid is sometimes represented by the single structure V:

II   III   IV   *equivalent to*   V

In nucleophilic aliphatic substitution (S_N2), the intermediate in which carbon is bonded to both the attacking group and the displaced group is considered to be a transition state; a structure (VI) containing carbon bonded to five atoms must be

**Intermediates in
nucleophilic substitution**

VI   V
Aliphatic S_N2 Aromatic
Pentavalent carbon,   *Tetrahedral carbon,*
transition state   *compound*

unstable and so corresponds to the top of an energy hill (Fig. 29.2). In nucleophilic aromatic substitution, on the other hand, the intermediate is an actual compound; a structure (V) containing tetrahedral carbon and having the negative charge distributed about the ring is comparatively stable, and corresponds to an energy valley (Fig. 29.3).

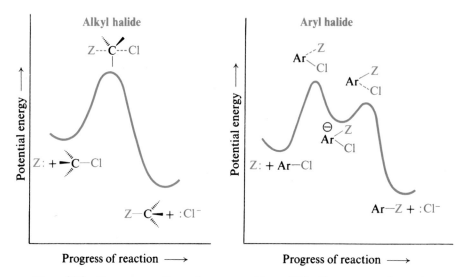

Figure 29.2 Energy curve for nucleophilic aliphatic (S_N2) substitution. One-step reaction: the intermediate is a transition state.

Figure 29.3 Energy curve for nucleophilic aromatic substitution. Two-step reaction: the intermediate is a compound.

29.9 Reactivity in nucleophilic aromatic substitution

For reactions involving an intermediate carbocation, we have seen that the overall rate depends only on the rate of formation of the carbocation. In nucleophilic aromatic substitution an analogous situation seems to exist: the first step, formation of the carbanion, largely determines the overall rate of reaction; once formed, the carbanion rapidly reacts to yield the final product.

For closely related reactions, we might expect a difference in rate of formation of carbanions to be largely determined by a difference in E_{act}, that is, by a difference in stability of the transition states. Factors that stabilize the carbanion by dispersing the charge should for the same reason stabilize the incipient carbanion of the transition state. Just as the more stable carbocation is formed more rapidly, so, we expect, the more stable carbanion should be formed more rapidly. We shall therefore concentrate our attention on the relative stabilities of the intermediate carbanions.

Transition state: *developing negative charge*

Carbanion: *full negative charge*

To compare the rates of substitution in chlorobenzene itself, a chlorobenzene containing an electron-withdrawing group, and a chlorobenzene containing an electron-releasing group, we compare the structures of carbanions I, II, and III.

A group that withdraws electrons (II) tends to neutralize the negative charge of the ring and so to become more negative itself; this dispersal of the charge stabilizes the carbanion. In the same way, electron withdrawal stabilizes the transition state with its developing negative charge, and thus speeds up reaction. A group that releases electrons (III) tends to intensify the negative charge, destabilizes the carbanion (and the transition state), and thus slows down reaction.

Nucleophilic aromatic substitution

G withdraws electrons:
stabilizes carbanion,
activates

$G = -N(CH_3)_3{}^+$
$-NO_2$
$-CN$
$-SO_3H$
$-COOH$
$-CHO$
$-COR$
$-X$

G releases electrons:
destabilizes carbanion,
deactivates

$G = -NH_2$
$-OH$
$-OR$
$-R$

It is clear, then, why a given substituent group affects nucleophilic and electrophilic aromatic substitution in opposite ways: it affects the stability of negatively and positively charged ions in opposite ways.

29.10 Orientation in nucleophilic aromatic substitution

To see why it is that a group activates the positions *ortho* and *para* to it most strongly, let us compare, for example, the carbanions formed from *p*-chloronitrobenzene and *m*-chloronitrobenzene. Each of these is a hybrid of three structures, I–III for *para* attack, IV–VI for *meta* attack. In one of these six structures, II, the negative charge is located on the carbon atom to which $-NO_2$ is attached. Although $-NO_2$ attracts electrons from all positions of the ring, it does so most from the carbon atom nearest it; consequently, structure II is a particularly stable one. Because of contribution from structure II, the hybrid carbanion resulting from attack on *p*-chloronitrobenzene is more stable than the carbanion resulting from attack on *m*-chloronitrobenzene. The *para* isomer therefore reacts faster than the *meta* isomer.

Para attack

I   II   III

Especially stable:
charge on carbon
bearing substituent

Meta attack

IV   V   VI

In the same way, it can be seen that attack on *o*-chloronitrobenzene (VII–IX) also yields a more stable carbanion, because of contribution from IX, than attack on *m*-chloronitrobenzene.

Ortho attack

VII   VIII   IX

Especially stable:
charge on carbon
bearing substituent

By considerations similar to those of Sec. 14.17, we can see that deactivation by an electron-releasing group should also be stongest when it is *ortho* or *para* to the halogen.

Nucleophilic and electrophilic aromatic substitutions are similar, then, in that a group exerts its strongest influence—whether activating or deactivating—at the positions *ortho* and *para* to it. This similarity is due to a similarity in the intermediate ions: in both cases the charge of the intermediate ion—whether negative or positive—is strongest at the positions *ortho* and *para* to the point of attack, and hence a group attached to one of these positions can exert the strongest influence.

29.11 Electron withdrawal by resonance

The activation by —NO_2 and other electron-attracting groups can be accounted for, as we have seen, simply on the basis of inductive effects. However, it is generally believed that certain of these groups withdraw electrons by resonance as well. Let us see what kind of structures are involved.

The intermediate carbanions formed by nucleophilic attack on *o*- and *p*-chloronitrobenzene are considered to be hybrids not only of structures with negative charges carried by carbons of the ring (as shown in the last section), but also of structures I and II in which the negative charge is carried by oxygen of the —NO$_2$ group. Being highly electronegative, oxygen readily accommodates a negative charge, and hence I and II should be especially stable structures. The carbanions to which these structures contribute are therefore much more stable

I

II

than the ones formed by attack on chlorobenzene itself or on *m*-chloronitrobenzene, for which structures like I and II are not possible. Thus resonance involving the —NO$_2$ group strengthens the activation toward nucleophilic substitution caused by the inductive effect.

The activating effect of a number of other electron-attracting groups is considered to arise, in part, from the contribution of similar structures (shown only for *para* isomers) to the intermediate carbanions.

—C≡N

—C—R

—N=O

Problem 29.5 There is evidence to suggest that the nitroso group, —N̈=Ö:, activates *ortho* and *para* positions toward *both* nucleophilic and electrophilic aromatic substitution; the group apparently can either withdraw or release electrons upon demand by the attacking reagent. Show how this might be accounted for. (*Hint*: See Sec. 14.18.)

29.12   Evidence for the two steps in bimolecular displacement

Our interpretation of reactivity and orientation in nucleophilic aromatic substitution has been based on one all-important assumption that we have not yet justified: *displacement involves two steps, of which the first one is much slower than the second.*

(1)　　　　　　　　　　$Ar-X + :Z \longrightarrow Ar \overset{\displaystyle \ominus}{\underset{\displaystyle Z}{\overset{\displaystyle X}{\diagup}}}$　　　　　　　　　　*Slow*

(2)　　　　　　　　　　$\underset{\displaystyle Z}{\overset{\displaystyle X}{Ar \diagup}} \ominus \longrightarrow Ar-Z + :X^-$　　　　　　　　　　*Fast*

The problem here reminds us of the analogous problem in electrophilic aromatic substitution (Sec. 14.14). There the answer was found in the absence of an isotope effect: although carbon–deuterium bonds are broken more slowly than carbon–hydrogen bonds, deuterium and hydrogen were found to be displaced at the same rate. Reactivity is determined by the rate of a reaction that does not involve the breaking of a carbon–hydrogen bond.

But in nucleophilic aromatic substitution, we are dealing with displacement, not of hydrogen, but of elements like the halogens; as was discussed in connection with dehydrohalogenation, any isotope effects would be small, and hard to measure.

The answer came from Joseph Bunnett (p. 277), who is responsible for much of what we understand about nucleophilic aromatic substitution. It was while studying this reaction that he first conceived the idea of *element effect* (Sec. 7.19), and showed how it gave evidence for the two-step mechanism.

In S_N1 and S_N2 displacement, we recall, the reactivity of alkyl halides follows the sequence

$$R-I > R-Br > R-Cl > R-F$$

The ease of breaking the carbon–halogen bond depends upon its strength, and the resulting differences in rate are quite large.

Yet, in nucleophilic *aromatic* substitution, there is often very little difference in reactivity among the various halides and, more often than not, the fluoride— containing the carbon–halogen bond hardest to break—is the *most* reactive. If reactivity is independent of the strength of the carbon–halogen bond, we can only conclude that the reaction *whose rate we are observing* does not involve breaking of the carbon–halogen bond. In nucleophilic aromatic substitution, as in electrophilic aromatic substitution, the rate of reaction is determined by the rate of attachment of the attacking particle to the ring (Fig. 29.4).

The *faster* reaction of aryl fluorides is attributed to the very strong inductive effect of fluorine; by withdrawing electrons it stabilizes the transition state of the first step of a reaction that will ultimately lead to its displacement.

Problem 29.6　When 2,4,6-trinitroanisole is treated with sodium ethoxide, a product of formula $C_9H_{10}O_8N_3^- Na^+$ is formed. A product of the same formula is formed by the treatment of trinitrophenetole with sodium methoxide. When treated with acid, both products give the same mixture of trinitroanisole and trinitrophenetole. What structure (or structures) would you assign to these products?

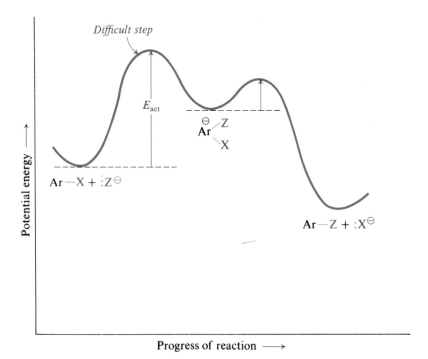

Figure 29.4   Potential energy changes during the course of reaction : nucleophilic aromatic substitution. Formation of the carbanion is the rate-controlling step; the strength of the C—X bond does not affect the overall rate.

29.13   Nucleophilic substitution: aliphatic and aromatic

We can see a regular progression in the three kinds of nucleophilic substitution that we have studied so far. The departing group leaves the molecule *before* the entering group becomes attached in an S_N1 reaction, *at the same time* in an S_N2 reaction, and *after* in nucleophilic aromatic substitution. A *positive charge* thus develops on carbon during an S_N1 reaction, *no particular charge* during an S_N2 reaction, and a *negative charge* during nucleophilic aromatic substitution. As a result, an S_N1 reaction is favored by *electron release*, an S_N2 reaction is relatively *insensitive to electronic factors*, and nucleophilic aromatic substitution is favored by *electron withdrawal*.

$$R-X \longrightarrow R^+ \xrightarrow{\;:Z\;} R-Z$$
$$+$$
$$X^-$$

S_N1
Positive charge develops on carbon

$$R-X \xrightarrow{\;:Z\;} \left[Z\text{---}R\text{---}X \right]^{\delta-} \longrightarrow R-Z + X^-$$

S_N2
Little charge develops on carbon

Nucleophilic aromatic
Negative charge develops on carbon

29.14 Elimination–addition mechanism for nucleophilic aromatic substitution. Benzyne

We have seen that electron-withdrawing groups activate aryl halides toward nucleophilic substitution. In the absence of such activation, substitution can be *made* to take place, by use of very strong bases, for example. But when this is done, substitution does not take place by the mechanism we have just discussed (the so-called *bimolecular mechanism*), but by an entirely different mechanism: the *benzyne* (or *elimination–addition*) *mechanism*. Let us first see what this mechanism is, and then examine some of the evidence for it.

When an aryl halide like chlorobenzene is treated with the very strong basic amide ion, NH_2^-, in liquid ammonia, it is converted into aniline. This is not the simple displacement that, on the surface, it appears to be. Instead, the reaction involves two stages: *elimination* and then *addition*. The intermediate is the molecule called *benzyne* (or *dehydrobenzene*).

Aryl halide   Benzyne   Aniline

Benzyne has the structure shown in Fig. 29.5, in which an additional bond is formed between two carbons (the one originally holding the halogen and the one

Figure 29.5 Benzyne molecule. The sideways overlap of sp^2 orbitals forms a π bond out of the plane of the aromatic π cloud.

originally holding the hydrogen) by sideways overlap of sp^2 orbitals. This new bond orbital lies along the side of the ring, and has little interaction with the π cloud lying above and below the ring. The sideways overlap is not very good, the new bond is a weak one, and benzyne is a highly reactive molecule.

The elimination stage, in which benzyne is formed, involves two steps: abstraction of a hydrogen ion (step 1) by the amide ion to form ammonia and carbanion I, which then loses halide ion (step 2) to form benzyne.

(1) $+ \; NH_2^- \longrightarrow$ $+ \; NH_3$

                                                I   **Elimination**

(2) $\longrightarrow$ $+ \; X^-$

                        I   Benzyne

The addition stage, in which benzyne is consumed, may also involve two steps: attachment of the amide ion (step 3) to form carbanion II, which then reacts with an acid, ammonia, to abstract a hydrogen ion (step 4). It may be that step (3) and step (4) are concerted, and addition involves a single step; if this is so, the transition state is probably one in which attachment of nitrogen has proceeded to a greater extent than attachment of hydrogen, so that it has considerable carbanion

(3)

Benzyne　　　　　　　　　II　　　　　　　　　　　　　　Addition

(4)

II　　　　　　　　　　　　Aniline

character. (This is analogous to hydroboration (Sec. 17.12), in which the transition state has considerable carbocation character.)

Let us look at the facts on which the above mechanism is based.

(**a**) *Fact*. Labeled chlorobenzene in which ^{14}C held the chlorine atom was allowed to react with amide ion. In *half* the aniline obtained the amino group was held by ^{14}C and in *half* it was held by an adjacent carbon.

(47%)　　　　　(53%)

Interpretation. In benzyne the labeled carbon and the ones next to it become equivalent, and NH_2^- adds randomly (except for a small isotope effect) to one or the other.

Although foreshadowed by certain earlier observations, this experiment, reported in 1953 by John D. Roberts of the California Institute of Technology, marks the real beginning of benzyne chemistry.

(b) *Fact.* Compounds containing two groups *ortho* to halogen, like 2-bromo-3-methylanisole, do not react at all.

Interpretation. With no *ortho* hydrogen to be lost, benzyne cannot form.

(c) *Fact.* When a 50:50 mixture of bromobenzene and *o*-deuteriobromobenzene is allowed to react with a limited amount of amide ion, recovered unreacted material contains more of the deuteriobromobenzene than bromobenzene; the deuterated compound is less reactive and is consumed more slowly.

o-Deuteriobromobenzene
Reacts more slowly:
more left unconsumed

Interpretation. This isotope effect (Sec. 7.17) shows not only that the *ortho* hydrogen is involved, but that it is involved in a rate-determining step. Deuterium is abstracted more slowly in the first step (equation 1, p. 1048), and the whole reaction sequence is slowed down.

Bond to D broken
more slowly

(d) *Fact. o*-Deuteriofluorobenzene is converted into aniline only very slowly, but loses its deuterium rapidly to yield ordinary fluorobenzene.

Interpretation. Abstraction of hydrogen (step 1) takes place, but before the very strong carbon–fluorine bond can break, the carbanion reacts with the acid—which is almost *all* NH_3 with only a trace of NH_2D—to regenerate fluorobenzene, but without its deuterium.

In the case of *o*-deuterio*bromo*benzene, on the other hand, breaking of the weaker carbon–bromine bond (step 2) is much faster than the protonation by ammonia (reverse of step 1): as fast as a carbanion is formed, it loses bromide ion. In this case, isotopic exchange is not important. (It may even be that here steps (1) and (2) are concerted.)

(1)

(2)

for X = F, $k_{-1} \gg k_2$

X = Br, $k_2 \gg k_{-1}$

(e) *Fact.* Both *m*-bromoanisole and *o*-bromoanisole yield the same product: *m*-anisidine (*m*-aminoanisole).

m-Bromoanisole   *m*-Anisidine   *o*-Bromoanisole

Interpretation. They yield the same product because they form the same intermediate benzyne.

Which benzyne is this, and how is it that it yields *m*-anisidine? To deal with orientation—both in the elimination stage and the addition stage—we must remember that a methoxyl group has an electron-withdrawing inductive effect. Since the electrons in carbanions like I and II (pp. 1048–1049) are out of the plane of the π cloud, there is no question of resonance interaction; only the inductive effect, working along the σ bonds (or perhaps through space), is operative.

o-Bromoanisole yields the benzyne shown (III, 2,3-dehydroanisole) because it has to. *m*-Bromoanisole yields III because, in the first step, the negative charge

o-Bromoanisole   III

appears preferentially on the carbon that can best accommodate it: the carbon next to the electron-withdrawing group. Whatever its source, III yields *m*-anisidine

for the same reason: addition of NH_2^- occurs in such a way that the negative charge appears on the carbon next to methoxyl.

More stable
carbanion

m-Anisidine

Actual product

Another common way to generate benzyne involves use of organolithium compounds. For example:

Here benzyne formation involves abstraction of a proton (reaction 5) by the base $C_6H_5^-$ to form a carbanion which loses fluoride ion (reaction 6) to give benzyne.

(5)

Stronger   Stronger   Weaker   Weaker
acid   base   base   acid

(6)

Problem 29.7 Account for the relative strengths of these acids and bases.

Addition of phenyllithium (reaction 7) to the benzyne gives the organolithium compound IV. From one point of view, this is the same reaction sequence observed for the amide ion–ammonia reaction (above), but it stops at the carbanion stage for want of strong acid. (Alternatively, the Lewis acid Li$^+$ has completed the sequence.) Addition of water—in this company, a very strong acid—yields (reaction 8) the final product. (The strong acid H$^+$ has displaced the weaker acid Li$^+$.)

(7)

IV

(8)

Organolithium compounds, RLi, resemble Grignard reagents, RMgX, in their reactions. As in Grignard reagents (Sec. 3.16), the carbon–metal bond can probably best be described as a highly polar covalent bond or, in another manner of speaking, as a bond with much *ionic character* (a resonance hybrid of R—M and R⁻M⁺). Because of the greater electropositivity of lithium, the carbon–lithium bond is even more ionic than the carbon–magnesium bond and, partly as a result of this, organolithium compounds are more reactive than Grignard reagents. As we have done with Grignard reagents, we shall for convenience focus our attention on the carbanion character of the organic group in discussing these reactions as acid–base chemistry. In the reactions involving $K^+NH_2^-$ we indicated free carbanions as intermediates, although even here the attractive forces—whatever they are—between carbon and potassium may be of great importance.

Problem 29.8   Account for the following facts: (a) treatment of the reaction mixture in reaction (8) with carbon dioxide instead of water gives V; (b) treatment of the

reaction mixture in reaction (8) with benzophenone gives VI; (c) benzyne can be generated by treatment of *o*-bromofluorobenzene with magnesium metal.

29.15   Analysis of aryl halides

Aryl halides show much the same response to characterization tests as the hydrocarbons from which they are derived: insolubility in cold concentrated sulfuric acid; inertness toward bromine in carbon tetrachloride and toward permanganate solutions; formation of orange to red colors when treated with chloroform and aluminum chloride; dissolution in cold fuming sulfuric acid, but at a slower rate than that of benzene.

Aryl halides are distinguished from aromatic hydrocarbons by the presence of halogen, as shown by elemental analysis. Aryl halides are distinguished from most alkyl halides by their inertness toward silver nitrate; in this respect they resemble vinyl halides (Sec. 29.5).

Any other functional groups that may be present in the molecule undergo their characteristic reactions.

Problem 29.9   Describe simple chemical tests (if any) that will distinguish between: (a) bromobenzene and *n*-hexyl bromide; (b) *p*-bromotoluene and benzyl bromide; (c) chlorobenzene and 1-chloro-1-hexene; (d) α-(*p*-bromophenyl)ethyl alcohol (p-$BrC_6H_4CHOHCH_3$) and *p*-bromo-*n*-hexylbenzene; (e) α-(*p*-chlorophenyl)ethyl alcohol and β-(*p*-chlorophenyl)ethyl alcohol (p-$ClC_6H_4CH_2CH_2OH$). Tell exactly what you would *do* and *see*.

Problem 29.10   Outline a procedure for distinguishing by chemical means (not necessarily simple tests) between: (a) *p*-bromoethylbenzene and 4-bromo-1,3-dimethylbenzene; (b) *o*-chloropropenylbenzene (o-ClC_6H_4CH=$CHCH_3$) and *o*-chloroallylbenzene (o-$ClC_6H_4CH_2CH$=CH_2).

PROBLEMS

1. Give structures and names of the principal organic products of the reaction (if any) of each of the following reagents with bromobenzene:

(a) Mg, ether
(b) boiling 10% aqueous NaOH
(c) boiling alcoholic KOH
(d) sodium acetylide
(e) sodium ethoxide
(f) NH$_3$, 100 °C
(g) boiling aqueous NaCN
(h) HNO$_3$, H$_2$SO$_4$

(i) fuming sulfuric acid
(j) Cl$_2$, Fe
(k) I$_2$, Fe
(l) C$_6$H$_6$, AlCl$_3$
(m) CH$_3$CH$_2$Cl, AlCl$_3$
(n) cold dilute KMnO$_4$
(o) hot KMnO$_4$

(\) no reaction)

2. Answer Problem 1 for *n*-butyl bromide.

3. Answer Problem 1, parts (b), (e), (f), and (g) for 2,4-dinitrobromobenzene.

4. Outline a laboratory method for the conversion of bromobenzene into each of the following, using any needed aliphatic and inorganic reagents.

(a) benzene
(b) *p*-bromonitrobenzene
(c) *p*-bromochlorobenzene
(d) *p*-bromobenzenesulfonic acid
(e) 1,2,4-tribromobenzene
(f) *p*-bromotoluene
(g) benzyl alcohol

(h) α-phenylethyl alcohol
(i) 2-phenyl-2-propanol
(j) 2,4-dinitrophenol
(k) allylbenzene
(l) benzoic acid
(m) aniline

5. Give the structure and name of the product expected when phenylmagnesium bromide is treated with each of the following compounds and then with water:

(a) H$_2$O
(b) HBr (dry)
(c) C$_2$H$_5$OH
(d) allyl bromide
(e) HCHO
(f) CH$_3$CHO
(g) C$_6$H$_5$CHO
(h) *p*-CH$_3$C$_6$H$_4$CHO

(i) CH$_3$COCH$_3$
(j) cyclohexanone
(k) 3,3-dimethylcyclohexanone
(l) C$_6$H$_5$COCH$_3$
(m) C$_6$H$_5$COC$_6$H$_5$
(n) (−)-C$_6$H$_5$COCH(CH$_3$)C$_2$H$_5$
(o) acetylene

Which products (if any) would be single compounds? Which (if any) would be racemic modifications? Which (if any) would be optically active as isolated?

6. Arrange the compounds in each set in order of reactivity toward the indicated reagent. Give the structure and name of the product expected from the compound you select as the most reactive in each set.

(a) NaOH: chlorobenzene, *m*-chloronitrobenzene, *o*-chloronitrobenzene, 2,4-dinitro-chlorobenzene, 2,4,6-trinitrochlorobenzene
(b) HNO$_3$/H$_2$SO$_4$: benzene, chlorobenzene, nitrobenzene, toluene
(c) alcoholic AgNO$_3$: 1-bromo-1-butene, 3-bromo-1-butene, 4-bromo-1-butene
(d) fuming sulfuric acid: bromobenzene, *p*-bromotoluene, *p*-dibromobenzene, toluene
(e) KCN: benzyl chloride, chlorobenzene, ethyl chloride
(f) alcoholic AgNO$_3$: 2-bromo-1-phenylethene, α-phenylethyl bromide, β-phenylethyl bromide

7. In the preparation of 2,4-dinitrochlorobenzene from chlorobenzene, the excess nitric acid and sulfuric acid must be washed from the product. Which would you select for this purpose: aqueous sodium hydroxide or aqueous sodium bicarbonate? Why?

8. Give structures and names of the principal organic products expected from each of the following reactions:

(a) 2,3-dibromopropene + NaOH(aq)
(b) *p*-bromobenzyl bromide + NH_3(aq)
(c) *p*-chlorotoluene + hot $KMnO_4$ *oxidize*
(d) *m*-bromostyrene + Br_2/CCl_4 *add to double bond*
(e) 3,4-dichloronitrobenzene + 1 mol $NaOCH_3$
(f) *p*-bromochlorobenzene + Mg, diethyl ether *Grignard*
(g) *p*-bromobenzyl alcohol + cold dilute $KMnO_4$ *oxidize*
(h) *p*-bromobenzyl alcohol + conc. HBr
(i) α-(*o*-chlorophenyl)ethyl bromide + KOH(alc) *makes a double bond*
(j) *p*-bromotoluene + 1 mol Cl_2, heat, light
(k) *o*-bromobenzotrifluoride + $NaNH_2$/NH_3
(l) *o*-bromoanisole + K^+ $^-NEt_2$/Et_2NH

9. Outline all steps in a possible laboratory synthesis of each of the following compounds from benzene and/or toluene, using any needed aliphatic or inorganic reagents:

(a) *m*-chloronitrobenzene
(b) *p*-chloronitrobenzene
(c) *m*-bromobenzoic acid
(d) *p*-bromobenzoic acid
(e) *m*-chlorobenzotrichloride
(f) 3,4-dibromonitrobenzene
(g) *p*-bromobenzal chloride

(h) 2,4-dinitroaniline
(i) *p*-bromostyrene
(j) 2,4-dibromobenzoic acid
(k) *m*-bromotoluene
(l) *p*-bromobenzenesulfonic acid
(m) *p*-chlorobenzyl alcohol
(n) 2-(*p*-tolyl)propane

10. Halogen located at the 2- or 4-position of the aromatic heterocyclic compound *pyridine* (Secs. 26.14 and 35.6) is fairly reactive toward nucleophilic displacement. For example:

4-Chloropyridine   4-Aminopyridine

How do you account for the reactivity of these compounds? (Check your answer in Sec. 35.10.)

11. The insecticide called DDT, 1,1,1-trichloro-2,2-bis-(*p*-chlorophenyl)ethane, (p-ClC_6H_4)$_2$$CHCCl_3$, can be manufactured by the reaction between chlorobenzene and trichloroacetaldehyde in the presence of sulfuric acid. Outline the series of steps by which this synthesis most probably takes place; make sure you show the function of the H_2SO_4. Label each step according to its fundamental reaction type.

12. In the Dow process for the manufacture of phenol, two by-products are diphenyl ether and *p*-phenylphenol. It has been suggested that these two compounds are formed via the same intermediate. How might this happen?

13. In KNH_2/NH_3, protium–deuterium exchange takes place at the following relative rates:

$$o\text{-}C_6H_4DF > m\text{-}C_6H_4DF > p\text{-}C_6H_4DF > C_6H_5D$$
4 000 000   4 000   200   1

How do you account for this sequence of reactivity?

14. Reduction of 2,6-dibromobenzenediazonium chloride, which would be expected to give *m*-dibromobenzene, actually yields chiefly *m*-bromochlorobenzene. How do you account for this?

15. The reaction of 2,4-dinitrofluorobenzene with *N*-methylaniline to give *N*-methyl-2,4-dinitrodiphenylamine is catalyzed by weak bases like acetate ion. The reaction of the corresponding bromo compound is faster, and is not catalyzed by bases. How do you account for these observations? (*Hint*: Examine in detail every step of the mechanism.)

16. (a) The labeled ether $2,4-(NO_2)_2C_6H_3{}^{18}OC_6H_5$ reacts more slowly than the unlabeled ether with the secondary amine piperidine (Sec. 35.12). How do you account for this?

(b) The isotope effect in part (a) becomes weaker as the piperidine concentration is raised. Account in detail for this observation. (*Hint*: See the preceding problem.)

17. The rate of reaction between *p*-fluoronitrobenzene and azide ion ($N_3{}^-$) is affected markedly by the nature of the solvent. How do you account for the following relative rates: in methanol, 1; in formamide, 5.6; in *N*-methylformamide, 15.7; in dimethylformamide, 2.4×10^4.

18. The dry diazonium salt I was subjected to a flash discharge, and an especially

I

adapted mass spectrometer scanned the spectrum of the products at rapid intervals after the flash. After about 50 microseconds there appeared simultaneously masses 28, 44, and 76. As time passed (about 250 microseconds) mass 76 gradually disappeared and a peak at mass 152 approached maximum intensity.

(a) What are the peaks at 28, 44, and 76 due to? What happens as time passes, and what is the substance of mass 152? (b) From what compound was the diazonium salt I prepared?

19. When a trace of KNH_2 is added to a solution of chlorobenzene and potassium triphenylmethide, $(C_6H_5)_3C^-K^+$, in liquid ammonia, a rapid reaction takes place to yield a product of formula $C_{25}H_{20}$. What is the product? What is the role of KNH_2, and why is it needed?

20. How do you account for each of the following observations?

(a) When *p*-iodotoluene is treated with aqueous NaOH at 340 °C, there is obtained a mixture of *p*-cresol (51%) and *m*-cresol (49%). At 250 °C, reaction is, of course, slower, and yields only *p*-cresol.

(b) When diazotized 4-nitro-2-aminobenzoic acid is heated in *tert*-butyl alcohol, there is obtained carbon dioxide, nitrogen, and a mixture of *m*- and *p*-nitrophenyl *tert*-butyl ethers.

(c) When *o*-chlorobenzoic acid is treated with $NaNH_2/NH_3$ in the presence of acetonitrile (CH_3CN) there is obtained a 70% yield of m-$HOOCC_6H_4CH_2CN$ and 10–20% of a 1:2 mixture of *o*- and *m*-aminobenzoic acids.

21. When either II or III is treated with $KN(C_2H_5)_2/HN(C_2H_5)_2$, there is obtained in

$CH_2CH_2NHCH_3$ / Cl	$CH_2CH_2NHCH_3$ / Cl
II	III

good yield the same product, of formula $C_9H_{11}N$. What is the product, and how is it formed?

22. An unknown compound is believed to be one of the following. Describe how you would go about finding out which of the possibilities the unknown actually is. Where possible use simple chemical tests; where necessary use more elaborate chemical methods like quantitative hydrogenation, cleavage, etc. Where necessary, make use of Table 23.1, p. 819.

(a) $C_6H_5CH=CHBr$ (b.p. 221 °C), o-$C_6H_4Br_2$ (b.p. 221 °C), $BrCH_2(CH_2)_3CH_2Br$ (b.p. 224 °C)

(b) o-$CH_3C_6H_4Br$ (b.p. 182 °C), m-$CH_3C_6H_4Br$ (b.p. 184 °C), p-$CH_3C_6H_4Br$ (b.p. 185 °C)

(c) o-$ClC_6H_4C_2H_5$ (b.p. 178 °C), $C_6H_5CH_2Cl$ (b.p. 179 °C), o-$C_6H_4Cl_2$ (b.p. 180 °C)

(d) $ClCH_2CH_2OH$ (b.p. 129 °C), 4-octyne (b.p. 131 °C), isopentyl alcohol (b.p. 132 °C), C_6H_5Cl (b.p. 132 °C), ethylcyclohexane (b.p. 132 °C), 1-chlorohexane (b.p. 134 °C)

(e) [structure: phenyl—CHBrCH₂Br] (m.p. 73 °C)   [structure: dimethyl-dibromobenzene with CH₃, Br, Br, CH₃] (m.p. 74 °C)   [structure: benzene with CH₂Br and CH₂Br] (m.p. 76 °C)

23. In studying the *base-catalyzed halogen dance*, Bunnett has made the following observations. When IV is treated with C_6H_5NHK/NH_3, it is isomerized to V. There is

[structures IV, V, VI, VII, VIII, IX — various bromo/iodo benzenes]

IV   V   VI   VII   VIII   IX

found, in addition, VI, m- and p-dibromobenzenes, and unconsumed IV. Similar treatment of VII gives chiefly VIII, along with IX, IV, and V. When IV labeled at the 1-position with radioactive bromine is allowed to react, the recovered IV had the label statistically distributed among all three positions.

(a) Bunnett first considered a mechanism involving intermediate benzynes. Show how you could account for the above observations on this basis.

(b) When IV is allowed to isomerize in the presence of much KI, no iodobromobenzenes are found. On this and other grounds, Bunnett rejected the benzyne mechanism. Explain.

(c) From the isomerization of IV, some unconsumed IV is *always* obtained. Yet the reaction of V gives IV *only if* there is present a small amount of VI to start with. (This is a *real* effect; highly purified materials give the same results.) In the presence of a little VI, the same mixture (about 50:50) of IV and V is formed whether one starts with IV or with V.

Suggest a complete mechanism for the base-catalyzed halogen dance, and show how it accounts for all the facts. It may help to go at the problem in this way. First, start with V and the base, in the presence of VI, and show how IV can be formed. Show how, under the same conditions, V can be formed from IV.

Next, start with *only* IV and base, and show how all the products are formed (V, VI, m- and p-dibromobenzenes), and account for the scrambling of the bromine label.

Finally, the hardest part: why must VI be added to bring about isomerization of V but not the isomerization of IV? (*Hint:* Simply write for V equations analogous to those you have written for IV, and keep in mind Problem 13, p. 1055.)

30

Carbanions II

Malonic Ester and Acetoacetic Ester Syntheses

$$CH_3-C-CH_2-COO^- \longrightarrow CH_3-C=CH_2 + CO_2$$

30.1 Carbanions in organic synthesis

We have already seen something of the importance to organic synthesis of the formation of carbon–carbon bonds: it enables us to make big molecules out of little ones. In this process a key role is played by negatively charged carbon. Such *nucleophilic carbon* attacks carbon holding a good leaving group—in alkyl halides or sulfonates, usually—or carbonyl or acyl carbon. Through nucleophilic substitution or nucleophilic addition, a new carbon–carbon bond is formed.

Nucleophilic carbon is of two general kinds. (a) There are the carbanion-like groups in organometallic compounds, usually generated through reaction of an organic halide with a metal: Grignard and organolithium reagents, for example. (b) There are the more nearly full-fledged carbanions generated through abstraction of α-hydrogens by base, as in the aldol and Claisen condensations and their relatives.

1059

The difference between these two kinds of carbon is one of degree, not kind. There is interaction—just how much depending on the metal and the solvent—even between electropositive ions like sodium or potassium or lithium and the anion from carbonyl compounds. These intermediates, too, could be called organometallic compounds; the bonding is simply more ionic than that in, say, a Grignard reagent.

In this chapter we shall continue with our study of carbanion chemistry, with emphasis on the attachment of alkyl groups to the α-carbons of carbonyl and acyl compounds. Such *alkylation* reactions owe their great importance to the special nature of the carbonyl group, and in two ways. First, the carbonyl group makes α-hydrogens acidic, so that alkylation can take place. Next, the products obtained still contain the carbonyl group and hence are highly reactive; they are ideal intermediates for *further* molecule-building.

Of the very many alkylation methods that have been developed, we can look at only a few: first, two classics of organic synthesis, the *malonic ester synthesis* and the *acetoacetic ester synthesis*; and then, several newer methods. In doing this we shall be concerned not only with learning a bit more about how to make new molecules from old ones, but also with seeing the variety of ways in which carbanion chemistry is involved.

30.2   Malonic ester synthesis of carboxylic acids

One of the most valuable methods of preparing carboxylic acids makes use of ethyl malonate (*malonic ester*), $CH_2(COOC_2H_5)_2$, and is called the **malonic ester synthesis**. This synthesis depends upon (a) the high acidity of the α-hydrogens of malonic ester, and (b) the extreme ease with which malonic acid and substituted malonic acids undergo decarboxylation. (As we shall see, this combination of properties is more than a happy accident, and can be traced to a single underlying cause.)

Like acetoacetic ester (Sec. 25.11), and for exactly the same reason, malonic ester contains α-hydrogens that are particularly acidic: they are *alpha* to *two* carbonyl groups. When treated with sodium ethoxide in absolute alcohol, malonic ester is converted largely into its salt, *sodiomalonic ester*:

$$H-CH(COOC_2H_5)_2 + Na^{+ -}OC_2H_5 \rightleftarrows CH(COOC_2H_5)_2^- Na^+ + H-OC_2H_5$$

Stronger acid   Sodiomalonic ester   Weaker acid

Reaction of this salt with an alkyl halide yields a substituted malonic ester, an *ethyl alkylmalonate*, often called an *alkylmalonic ester*:

$$CH(COOC_2H_5)_2^- Na^+ + R-X \longrightarrow R-CH(COOC_2H_5)_2 + Na^+X^-$$

Ethyl alkylmalonate
An alkylmalonic ester

This reaction involves nucleophilic attack on the alkyl halide by the carbanion, $CH(COOC_2H_5)_2^-$, and, as we might expect, gives highest yields with primary alkyl halides, lower yields with secondary alkyl halides, and is worthless for tertiary alkyl halides and for aryl halides.

The alkylmalonic ester still contains one ionizable hydrogen, and on treatment with sodium ethoxide it, too, can be converted into its salt; this salt can react with an alkyl halide—which may be the same as, or different from, the first alkyl halide—to yield a dialkylmalonic ester:

$$R—CH(COOC_2H_5)_2 + Na^+{}^-OC_2H_5 \rightleftarrows R—C(COOC_2H_5)_2{}^- Na^+ + C_2H_5OH$$

$$\downarrow R'X$$

$$\begin{array}{c} R' \\ | \\ R—C(COOC_2H_5)_2 + Na^+X^- \end{array}$$

A dialkylmalonic ester

The acidity of malonic ester thus permits the preparation of substituted malonic esters containing one or two alkyl groups. Now, how can these substituted malonic esters be used to make carboxylic acids? When heated above its melting point, malonic acid readily loses carbon dioxide to form acetic acid; in a similar way substituted malonic acids readily lose carbon dioxide to form substituted acetic acids. The monoalkyl- and dialkylmalonic esters we have prepared are readily converted into monocarboxylic acids by hydrolysis, acidification, and heat:

$$\begin{array}{ccc} COOC_2H_5 & COO^- & COOH \\ | & | & | \\ R—CH & \xrightarrow[\text{heat}]{H_2O,\ OH^-} R—CH & \xrightarrow{H^+} R—CH \\ | & | & | \\ COOC_2H_5 & COO^- & COOH \end{array}$$

A monoalkylmalonic ester

$$\downarrow \text{heat, 140 °C}$$

$$R—CH_2COOH + CO_2$$

A monosubstituted
acetic acid

$$\begin{array}{ccc} COOC_2H_5 & COO^- & COOH \\ | & | & | \\ R—C—R' & \xrightarrow[\text{heat}]{H_2O,\ OH^-} R—C—R' & \xrightarrow{H^+} R—C—R' \\ | & | & | \\ COOC_2H_5 & COO^- & COOH \end{array}$$

A dialkylmalonic ester

$$\downarrow \text{heat, 140 °C}$$

$$\begin{array}{c} R' \\ | \\ R—CHCOOH \end{array}$$

A disubstituted
acetic acid

A malonic ester synthesis yields an acetic acid in which one or two hydrogens have been replaced by alkyl groups.

In planning a malonic ester synthesis, our problem is to select the proper alkyl halide or halides; to do this, we have only to look at the structure of the acid we want. Isocaproic acid, for example, $(CH_3)_2CHCH_2CH_2COOH$, can be considered as acetic acid in which one hydrogen has been replaced by an isobutyl group. To prepare this acid by the malonic ester synthesis, we would have to use isobutyl bromide as the alkylating agent:

$$\underset{\substack{\text{Isocaproic acid}\\ \text{(4-Methylpentanoic acid)}}}{\overset{\displaystyle CH_3}{\underset{\displaystyle }{CH_3CHCH_2-CH_2COOH}}} \xleftarrow[-CO_2]{\text{heat,}} \underset{\substack{}}{\overset{\displaystyle COOH}{\underset{\displaystyle COOH}{CH_3CHCH_2-CH}}} \xleftarrow{H^+} \underset{}{\overset{\displaystyle COO^-}{\underset{\displaystyle COO^-}{CH_3CHCH_2-CH}}}$$

$\uparrow$ H_2O, OH^-, heat

$$\underset{\text{Isobutyl bromide}}{\overset{\displaystyle CH_3}{CH_3CHCH_2-Br}} + Na^+CH(COOC_2H_5)_2{}^- \longrightarrow \underset{\substack{\text{Ethyl isobutylmalonate}\\ \text{Isobutylmalonic ester}}}{\overset{\displaystyle CH_3 \quad COOC_2H_5}{\underset{\displaystyle COOC_2H_5}{CH_3CHCH_2-CH}}}$$

$\uparrow$ Na^+ $^-OC_2H_5$

$$CH_2(COOC_2H_5)_2$$
Malonic ester

An isomer of isocaproic acid, α-methylvaleric acid, $CH_3CH_2CH_2CH(CH_3)COOH$, can be considered as acetic acid in which one hydrogen has been replaced by a *n*-propyl group and a second hydrogen has been replaced by a methyl group; we must therefore use two alkyl halides, *n*-propyl bromide and methyl bromide.

$$\underset{\substack{\text{α-Methylvaleric acid}\\ \text{(2-Methylpentanoic acid)}}}{\overset{\displaystyle }{\underset{\displaystyle CH_3}{CH_3CH_2CH_2-CHCOOH}}} \xleftarrow[-CO_2]{\text{heat,}} \underset{\substack{}}{\overset{\displaystyle COOH}{\underset{\displaystyle CH_3}{CH_3CH_2CH_2-CCOOH}}} \xleftarrow{H^+} \underset{}{\overset{\displaystyle COO^-}{\underset{\displaystyle CH_3}{CH_3CH_2CH_2-CCOO^-}}}$$

$\uparrow$ H_2O, OH^-, heat

$$\underset{\text{Methyl bromide}}{CH_3-Br} + Na^+ CH_3CH_2CH_2-C(COOC_2H_5)_2{}^- \longrightarrow \underset{\substack{\text{Ethyl methyl-n-propylmalonate}\\ \text{Methyl-n-propylmalonic ester}}}{\overset{\displaystyle COOC_2H_5}{\underset{\displaystyle CH_3}{CH_3CH_2CH_2-CCOOC_2H_5}}}$$

$\uparrow$ Na^+ $^-OC_2H_5$

$$CH_3CH_2CH_2-CH(COOC_2H_5)_2$$

$\uparrow$

$$\underset{\text{n-Propyl bromide}}{CH_3CH_2CH_2-Br} + Na^+CH(COOC_2H_5)_2{}^-$$

$\uparrow$ Na^+ $^-OC_2H_5$

$$CH_2(COOC_2H_5)_2$$
Malonic ester

The basic malonic ester synthesis we have outlined can be modified. Often one can advantageously use: different bases as, for example, potassium *tert*-butoxide; alkyl sulfonates instead of halides; polar aprotic solvents like DMSO or DMF (Sec. 6.4).

In place of simple alkyl halides, certain other halogen-containing compounds may be used, in particular the readily available α-bromo esters (why can α-bromo *acids* not be used?), which yield substituted succinic acids by the malonic ester synthesis. For example:

$$\underset{\text{α-Methylsuccinic acid}}{\overset{\text{CH}_3}{\text{HOOCCH—CH}_2\text{COOH}}} \xleftarrow[\text{- CO}_2]{\text{heat,}} \underset{}{\overset{\text{CH}_3}{\text{HOOCCH—CH(COOH)}_2}} \xleftarrow{\text{H}^+} \overset{\text{CH}_3}{^-\text{OOCCH—CH(COO}^-)_2}$$

$$\Big\uparrow \text{H}_2\text{O, OH}^-, \text{heat}$$

$$\underset{\substack{\text{Br} \\ \text{Ethyl} \\ \text{α-bromopropionate}}}{\overset{\text{CH}_3\text{CHCOOC}_2\text{H}_5}{}} + \text{Na}^+\text{CH(COOC}_2\text{H}_5)_2^- \longrightarrow \overset{\text{CH}_3}{\text{C}_2\text{H}_5\text{OOCCH—CH(COOC}_2\text{H}_5)_2}$$

$$\Big\uparrow \text{Na}^+\,^-\text{OC}_2\text{H}_5$$

$$\underset{\text{Malonic ester}}{\text{CH}_2(\text{COOC}_2\text{H}_5)_2}$$

Problem 30.1 Outline the synthesis of each of the following compounds from malonic ester and alcohols of four carbons or fewer:

(a) the isomeric acids, *n*-valeric, isovaleric, and α-methylbutyric. (Why can the malonic ester synthesis not be used for the preparation of trimethylacetic acid?)
(b) *leucine* (α-aminoisocaproic acid)
(c) *isoleucine* (α-amino-β-methylvaleric acid)

Problem 30.2 *Adipic acid* is obtained from a malonic ester synthesis in which the first step is addition of one mole of ethylene bromide to a large excess of sodiomalonic ester in alcohol. *Cyclopropanecarboxylic acid* is the final product of a malonic ester synthesis in which the first step is addition of one mole of sodiomalonic ester to two moles of ethylene bromide followed by addition of one mole of sodium ethoxide.

HOOCCH₂CH₂CH₂CH₂COOH

Adipic acid

Cyclopropane-carboxylic acid

Cyclopentane-carboxylic acid

(a) Account for the difference in the products obtained in the two syntheses. (b) Tell exactly how you would go about synthesizing *cyclopentanecarboxylic acid*.

Problem 30.3 (a) Malonic ester reacts with benzaldehyde in the presence of piperidine (a secondary amine, Sec. 35.12) to yield a product of formula $C_{14}H_{16}O_4$. What is this compound, and how is it formed? (This is an example of the **Knoevenagel reaction**. Check your answer in Problem 25.22 (f), p. 920.) (b) What compound would be obtained if the product of (a) were subjected to the sequence of hydrolysis, acidification, and heating? (c) What is another way to synthesize the product of (b)?

Problem 30.4 (a) Cyclohexanone reacts with cyanoacetic ester (ethyl cyanoacetate, N≡CCH₂COOC₂H₅) in the presence of ammonium acetate to yield a product of formula $C_{11}H_{15}O_2N$. What is this compound, and how is it formed? (This is an example of the **Cope reaction**. Check your answer in Problem 25.22 (g), p. 920.) (b) What compound would be formed from the product of (a) by the sequence of hydrolysis, acidification, and heating?

Problem 30.5 In an example of the **Michael condensation**, malonic ester reacts with ethyl 2-butenoate in the presence of sodium ethoxide to yield A, of formula $C_{13}H_{22}O_6$. The sequence of hydrolysis, acidification, and heating converts A into 3-methyl-pentanedioic acid. What is A, and how is it formed? (*Hint*: See Sec. 10.26. Check your answer in Sec. 31.7.)

30.3 Acetoacetic ester synthesis of ketones

One of the most valuable methods of preparing ketones makes use of ethyl acetoacetate (*acetoacetic ester*), $CH_3COCH_2COOC_2H_5$, and is called the **aceto-acetic ester synthesis of ketones**. This synthesis closely parallels the malonic ester synthesis of carboxylic acids.

Acetoacetic ester is converted by sodium ethoxide into the sodioacetoacetic ester, which is then allowed to react with an alkyl halide to form an alkylacetoacetic ester (an ethyl alkylacetoacetate), $CH_3COCHRCOOC_2H_5$; if desired, the alkylation can be repeated to yield a dialkylacetoacetic ester, $CH_3COCRR'COOC_2H_5$. All alkylations are conducted in absolute alcohol.

When hydrolyzed by dilute aqueous alkali (or by acid), these monoalkyl- or dialkylacetoacetic esters yield the corresponding acids, $CH_3COCHRCOOH$ or $CH_3COCRR'COOH$, which undergo decarboxylation to form ketones, CH_3COCH_2R or $CH_3COCHRR'$. This loss of carbon dioxide occurs even more readily than from malonic acid, and may even take place before acidification of the hydrolysis mixture.

$$CH_3COCHCOOC_2H_5^- \xleftarrow{\ ^-OC_2H_5\ } CH_3COCH_2COOC_2H_5$$

Acetoacetic ester

$\downarrow RX$

$$CH_3COCHCOOC_2H_5 \xrightarrow{OH^-} CH_3COCHCOO^- \xrightarrow{H_2O\ or\ H^+} CH_3COCHCOOH$$
$$\qquad\quad |\qquad\qquad\qquad\qquad\quad |\qquad\qquad\qquad\qquad\qquad\qquad |$$
$$\qquad\quad R\qquad\qquad\qquad\qquad\quad R\qquad\qquad\qquad\qquad\qquad\qquad R$$

A monoalkylacetoacetic ester

$\downarrow\ ^-OC_2H_5$

$$CH_3COCCOOC_2H_5^-$$
$$\qquad\quad |$$
$$\qquad\quad R$$

$\downarrow -CO_2$

$$CH_3COCH_2\text{---}R$$
A monosubstituted
acetone

$\downarrow R'X$

$$CH_3COCCOOC_2H_5 \xrightarrow{OH^-} CH_3COCCOO^- \xrightarrow{H_2O\ or\ H^+} CH_3COCCOOH$$

with R' above and R below each central carbon.

A dialkylacetoacetic ester

$\downarrow -CO_2$

$$CH_3COCH\text{---}R$$
with R' above.
A disubstituted
acetone

The acetoacetic ester synthesis of ketones yields an acetone molecule in which one or two hydrogens have been replaced by alkyl groups.

In planning an acetoacetic ester synthesis, as in planning a malonic ester synthesis, our problem is to select the proper alkyl halide or halides. To do this, we have only to look at the structure of the ketone we want. For example, 5-methyl-2-hexanone can be considered as acetone in which one hydrogen has been replaced by an isobutyl group. In order to prepare this ketone by the acetoacetic ester synthesis, we would have to use isobutyl bromide as the alkylating agent:

$$
\underset{\substack{\text{CH}_3 \\ \text{CH}_3\text{CHCH}_2-\text{CH}_2\text{CCH}_3 \\ \parallel \\ \text{O}}}{} \xleftarrow{-\text{CO}_2}
\underset{\substack{\text{CH}_3 \qquad \text{COOH} \\ \text{CH}_3\text{CHCH}_2-\text{CHCCH}_3 \\ \parallel \\ \text{O}}}{} \xleftarrow{\text{H}_2\text{O or H}^+}
\underset{\substack{\text{CH}_3 \qquad \text{COO}^- \\ \text{CH}_3\text{CHCH}_2-\text{CHCCH}_3 \\ \parallel \\ \text{O}}}{}
$$

5-Methyl-2-hexanone

$\uparrow \text{OH}^-$

$$
\underset{\substack{\text{CH}_3 \\ \text{CH}_3\text{CHCH}_2-\text{Br} + \text{Na}^+\text{CH}_3\text{COCHCOOC}_2\text{H}_5^-}}{}
\longrightarrow
\underset{\substack{\text{CH}_3 \qquad \text{COOC}_2\text{H}_5 \\ \text{CH}_3\text{CHCH}_2-\text{CHCCH}_3 \\ \parallel \\ \text{O}}}{}
$$

Isobutyl bromide

$\uparrow \text{Na}^+ {}^- \text{OC}_2\text{H}_5$

$\text{CH}_3\text{COCH}_2\text{COOC}_2\text{H}_5$

Ethyl acetoacetate

Ethyl
α-isobutylacetoacetate

The isomeric ketone 3-methyl-2-hexanone can be considered as acetone in which one hydrogen has been replaced by a *n*-propyl group and a second hydrogen (on the same carbon) has been replaced by a methyl group; we must therefore use two alkyl halides, *n*-propyl bromide and methyl bromide:

$$
\underset{\substack{\text{CH}_3\text{CH}_2\text{CH}_2-\text{CH}-\text{CCH}_3 \\ | \quad \parallel \\ \text{CH}_3 \;\; \text{O}}}{} \xleftarrow{-\text{CO}_2}
\underset{\substack{\text{COOH} \\ \text{CH}_3\text{CH}_2\text{CH}_2-\text{C}-\text{CCH}_3 \\ | \quad \parallel \\ \text{CH}_3 \;\; \text{O}}}{} \xleftarrow{\text{H}_2\text{O or H}^+}
$$

3-Methyl-2-hexanone

$$
\underset{\substack{\text{COO}^- \\ \text{CH}_3\text{CH}_2\text{CH}_2-\text{C}-\text{CCH}_3 \\ | \quad \parallel \\ \text{CH}_3 \;\; \text{O}}}{}
$$

$\uparrow \text{OH}^-$

$$
\underset{\substack{\text{COOC}_2\text{H}_5 \\ \text{CH}_3-\text{Br} + \text{Na}^+\text{CH}_3\text{CH}_2\text{CH}_2-\text{CCOCH}_3^-}}{}
\longrightarrow
\underset{\substack{\text{COOC}_2\text{H}_5 \\ \text{CH}_3\text{CH}_2\text{CH}_2-\text{C}-\text{CCH}_3 \\ | \quad \parallel \\ \text{CH}_3 \;\; \text{O}}}{}
$$

Methyl bromide

$\uparrow \text{Na}^+ {}^- \text{OC}_2\text{H}_5$

Ethyl α-methyl-
α-*n*-propylacetoacetate

$$
\underset{\substack{\text{COOC}_2\text{H}_5 \\ \text{CH}_3\text{CH}_2\text{CH}_2-\text{CHCCH}_3 \\ \parallel \\ \text{O}}}{}
$$

$\uparrow$

$\text{CH}_3\text{CH}_2\text{CH}_2-\text{Br} + \text{Na}^+\text{CH}_2\text{COCHCOOC}_2\text{H}_5^-$

n-Propyl bromide

$\uparrow \text{Na}^+ {}^- \text{OC}_2\text{H}_5$

$\text{CH}_3\text{COCH}_2\text{COOC}_2\text{H}_5$

Ethyl acetoacetate

Like the malonic ester synthesis, this synthesis, too, can be modified by changes in the base, solvent, and alkylating agent.

Problem 30.6 To what general class does the reaction between sodioacetoacetic ester and an alkyl halide belong? Predict the relative yields using primary, secondary, and tertiary halides. Can aryl halides be used?

Problem 30.7 (a) Predict the product of the acetoacetic ester synthesis in which ethyl bromoacetate (why not bromoacetic *acid*?) is used as the halide. To what general class of compounds does this product belong? (b) Predict the product of the acetoacetic ester synthesis in which benzoyl chloride is used as the halide; in which chloroacetone is used as the halide. To what general classes of compounds do these products belong?

Problem 30.8 Outline the synthesis of each of the following compounds from acetoacetic ester, benzene, and alcohols of four carbons or fewer:

(a)–(c) the isomeric ketones:
      methyl *n*-butyl ketone (2-hexanone)
      methyl isobutyl ketone (4-methyl-2-pentanone)
      methyl *sec*-butyl ketone (3-methyl-2-pentanone)
(d) Why can the acetoacetic ester synthesis not be used for the preparation of methyl *tert*-butyl ketone?
(e) 2,4-pentanedione (*acetylacetone*)
(f) 2,5-hexanedione (*acetonylacetone*)
(g) 1-phenyl-1,4-pentanedione

Problem 30.9 The best general preparation of α-**keto acids** is illustrated by the sequence:

$$\text{ethyl propionate} + \text{ethyl oxalate} \xrightarrow{\text{NaOC}_2\text{H}_5} \text{A } (C_9H_{14}O_5)$$

$$\text{A} + \text{dil. H}_2\text{SO}_4 \xrightarrow{\text{boil}} \text{CO}_2 + 2C_2H_5OH + CH_3CH_2\underset{\underset{O}{\|}}{C}COOH \text{ (α-ketobutyric acid)}$$

What familiar reactions are involved? What is the structure of A?

Problem 30.10 Outline the synthesis from simple esters of:

(a) α-ketoisocaproic acid
(b) α-keto-β-phenylpropionic acid
(c) α-ketoglutaric acid
(d) *leucine* (α-aminoisocaproic acid) (*Hint*: See Sec. 26.11.)
(e) *glutamic acid* (α-aminoglutaric acid)

30.4 Decarboxylation of β-keto acids and malonic acids

The acetoacetic ester synthesis thus depends on (a) the high acidity of the α-hydrogens of β-keto esters, and (b) the extreme ease with which β-keto acids undergo decarboxylation. These properties are exactly parallel to those on which the malonic ester synthesis depends.

We have seen that the higher acidity of the α-hydrogens is due to the ability of the keto group to help accommodate the negative charge of the acetoacetic ester anion. The ease of decarboxylation is, in part, due to *exactly the same factor*. (So, too, is the occurrence of the Claisen condensation, by which the acetoacetic ester is made in the first place.)

Decarboxylation of β-keto acids involves both the free acid and the carboxylate ion. Loss of carbon dioxide from the anion yields the carbanion I. This carbanion

$$CH_3-\overset{\displaystyle \underset{\displaystyle O}{\|}}{C}-CH_2-COO^- \longrightarrow CH_3-\overset{\displaystyle \underset{\displaystyle O^{\ominus}}{\|}}{C}=CH_2 + CO_2$$

I

is formed faster than the simple carbanion $(R:^-)$ that would be formed from a simple carboxylate ion $(RCOO^-)$ because it is more stable. It is more stable, of course, due to the accommodation of the negative charge by the keto group.

Problem 30.11 Decarboxylation of malonic acid involves both the free acid and the monoanion, but not the doubly charged anion. (a) Account for the ease of decarboxylation of the monoanion. Which end loses carbon dioxide? (b) How do you account for the lack of reactivity of the doubly charged anion? (*Hint*: See Sec. 23.20.)

Problem 30.12 In contrast to most carboxylic acids (benzoic acid, say) 2,4,6-trinitrobenzoic acid is decarboxylated extremely easily: by simply boiling it in aqueous acid. How do you account for this?

Decarboxylation of free acetoacetic acid involves transfer of the acidic hydrogen to the keto group, either *prior to* (as shown here) or *simultaneously with*

$$CH_3-\overset{\displaystyle \underset{\displaystyle O}{\|}}{C}-CH_2-C\overset{\displaystyle O}{\underset{\displaystyle OH}{\diagdown}} \rightleftharpoons CH_3-\overset{\displaystyle \underset{\displaystyle OH^+}{\|}}{C}-CH_2-COO^- \longrightarrow CH_3-\overset{\displaystyle \underset{\displaystyle OH}{|}}{C}=CH_2 + CO_2$$

$$\downarrow$$

$$CH_3-\overset{\displaystyle \underset{\displaystyle O}{\|}}{C}-CH_3$$

loss of carbon dioxide. We are quite familiar with the function of protonation to reduce the basicity of a leaving group.

Problem 30.13 When dimethylacetoacetic acid is decarboxylated in the presence of iodine or bromine, there is obtained an iododimethylacetone or a bromodimethylacetone (3-halo-3-methyl-2-butanone), although under these conditions neither iodine nor bromine reacts significantly with the dimethylacetone. What bearing does this experiment have on the mechanism of decarboxylation?

Problem 30.14 Suggest a mechanism for the decarboxylation of free malonic acid.

Problem 30.15 Account for the comparative ease with which phenylpropiolic acid, $C_6H_5C{\equiv}CCOOH$, undergoes decarboxylation in alkaline solution.

30.5 Direct and indirect alkylation of esters and ketones

By the malonic ester and acetoacetic ester we make α-substituted acids and α-substituted ketones. But why not do the job *directly*? Why not convert simple acids (or esters) and ketones into their carbanions, and allow these to react with alkyl halides? There are a number of obstacles: (a) self-condensation—aldol condensation, for example, of ketones; (b) polyalkylation; and (c) for unsymmetrical ketones, alkylation at *both* α-carbons, or at the wrong one. Consider self-condensation. A carbanion can be generated from, say, a simple ketone; but, instead of attacking a molecule of alkyl halide, the carbanion may attack the carbonyl carbon of a second ketone molecule. What is needed is a base–solvent combination that can convert the ketone *rapidly* and essentially *completely* into the carbanion before appreciable self-condensation can occur. Steps toward solving this problem have been taken, and there are available methods—so far, of limited applicability—for the direct alkylation of acids and ketones.

A tremendous amount of work has gone into the development of alternatives to direct alkylation. Another group is introduced temporarily to do one or more of these things: increase the acidity of the α-hydrogens, prevent self-condensation, and direct alkylation to a specific position. The malonic ester and acetoacetic ester syntheses are, of course, typical of this approach. In the acetoacetic ester synthesis, for example, the carbethoxy group, —COOEt, enhances the acidity of α-hydrogens, but only those on one particular α-carbon, so that alkylation will take place there. Then, when alkylation is over, the carbethoxy group is easily removed by hydrolysis and decarboxylation.

In the biosynthesis of fats (Sec. 41.6), long-chain carboxylic acids are made via a series of what are basically malonic ester syntheses. Although in this case reactions are catalyzed by enzymes, the system still finds it worthwhile to consume carbon dioxide to make a malonyl compound, then form a new carbon–carbon bond, and finally eject the carbon dioxide.

To get some idea of the way problems like these are being approached, let us look at just a few of the other alternatives to direct alkylation.

30.6 Synthesis of acids and esters via 2-oxazolines

Reaction of a carboxylic acid with 2-amino-2-methyl-1-propanol yields a heterocyclic compound called a *2-oxazoline* (I). From this compound the acid can be regenerated, in the form of its ethyl ester, by ethanolysis.

2-Amino-2-methyl-1-propanol

A 2-alkyl-4,4-dimethyl-2-oxazoline

Using this way to protect the carboxyl group, A. I. Meyers (Colorado State University) has recently opened an elegant route to alkylated acetic acids—or, by modification, to β-hydroxy esters.

Treatment of the 2-oxazoline with the strong base, *n*-butyllithium, yields the lithio derivative II. This, like sodiomalonic ester, can be alkylated and, if desired, re-alkylated—up to a total of *two* substituents on the α-carbon. Ethanolysis of the new 2-oxazoline yields the substituted ester.

The synthesis depends upon: (a) the ease of formation and hydrolysis of 2-oxazolines; (b) the fact that the α-hydrogens retain their acidity in the oxazoline (*Why?*); and (c) the inertness of the 2-oxazoline ring toward the lithio derivative. (The ring is inert toward the Grignard reagent as well, and can be used to protect the carboxyl group in a wide variety of syntheses.)

Problem 30.16   Using the Meyers oxazoline method, outline all steps in the synthesis of: (a) *n*-butyric acid from acetic acid; (b) isobutyric acid from acetic acid; (c) isobutyric acid from propionic acid; (d) β-phenylpropionic acid from acetic acid.

Problem 30.17   (a) Give structural formulas of compounds A and B.

Oxazoline I (R = H) + *n*-BuLi, then $CH_3(CH_2)_5CHO$ $\longrightarrow$ A

A + EtOH, H_2SO_4 $\longrightarrow$ B $(C_{11}H_{22}O_3)$

    (b) Outline all steps in the synthesis of ethyl 3-(*n*-propyl)-3-hydroxyhexanoate. (c) Of ethyl 2-ethyl-3-phenyl-3-hydroxypropanoate.

Problem 30.18   (a) Give structural formulas of compounds C–E.

4-hydroxycyclohexanecarboxylic acid + $(CH_3)_2C(NH_2)CH_2OH$ $\longrightarrow$

$$C(C_{11}H_{19}O_2N)$$

C + CrO_3/pyridine $\longrightarrow$ D $(C_{11}H_{17}O_2N)$

D + C_6H_5MgBr, then C_2H_5OH, H_2SO_4 $\longrightarrow$ E $(C_{15}H_{18}O_2)$

(b) Using benzene, toluene, and any needed aliphatic and inorganic reagents, how would you make $C_6H_5COCH_2CH_2COOH$? (*Hint:* See Sec. 24.10.) (c) Now, how would you make $C_6H_5C(C_2H_5){=}CHCH_2COOH$? (d) Outline a possible synthesis of *p*-$CH_3CH_2CHOHC_6H_4COOC_2H_5$. (e) Of $C_6H_5CHOHC_6H_4COOC_2H_5$-*p*.

30.7   Organoborane synthesis of acids and ketones

Hydroboration of alkenes yields alkylboranes, and these, we have seen (Sec. 17.10), can be converted through oxidation into alcohols. But oxidation is only one of many reactions undergone by alkylboranes. Since the discovery of hydroboration in 1957, H. C. Brown and his co-workers (p. 648) have shown that alkylboranes are perhaps the most versatile class of organic reagents known.

In the presence of base, alkylboranes react with bromoacetone to yield alkylacetones, and with ethyl bromoacetate to yield ethyl alkylacetates.

$$R_3B + BrCH_2COCH_3 \xrightarrow{\text{base}} R{-}CH_2COCH_3$$
Bromoacetone   An alkylacetone

$$R_3B + BrCH_2COOEt \xrightarrow{\text{base}} R{-}CH_2COOEt$$
Ethyl bromoacetate   An ethyl alkylacetate

The following mechanism has been postulated, illustrated for reaction with bromoacetone. Base abstracts (1) a proton—one that is *alpha* both to the carbonyl group and to bromine—to give the carbanion I. Being a strong base, carbanion I

(1) $\qquad$ Base: + $CH_2BrCOCH_3$ $\rightleftharpoons$ $^-CHBrCOCH_3$ + Base:H
$\qquad\qquad\qquad\qquad\qquad\qquad\qquad\qquad\qquad\qquad$ I

(2) $\qquad$ R_3B + $^-CHBrCOCH_3$ $\longrightarrow$ $R_3\overset{\ominus}{B}$:$CHBrCOCH_3$
$\qquad\qquad\qquad\qquad$ I $\qquad\qquad\qquad\qquad\qquad$ II

(3) $\qquad$ $\overset{\ominus}{R}-B-CHCOCH_3$ $\longrightarrow$ $\overset{R}{R-B-CHCOCH_3}$ + Br^-
$\qquad\qquad\qquad\quad$ R $\quad Br$ $\qquad\qquad\qquad$ R
$\qquad\qquad\qquad\qquad$ II $\qquad\qquad\qquad\qquad\qquad$ III

(4) $\qquad$ $\overset{R}{R_2B-CHCOCH_3}$ + Base:H $\longrightarrow$ $R-CH_2COCH_3$ + R_2B:Base
$\qquad\qquad$ III

combines (2) with the (Lewis) acidic alkylborane to give II. Intermediate II now rearranges (3) with loss of halide ion to form III. Finally, III undergoes (4) protonolysis (a Lowry–Brønsted acid–base reaction this time) to yield the alkylated ketone.

The key step is (3), in which a new carbon–carbon bond is formed. In II, boron carries a negative charge. Made mobile by this negative charge, and attracted by the adjacent carbon holding a good leaving group, an alkyl group migrates to this carbon—taking its electrons along—and displaces the weakly basic halide ion.

We have, then, three acid–base reactions and a 1,2-alkyl shift: all familiar reaction types. Step (1) involves formation of a carbanion; step (3) involves intramolecular nucleophilic (S_N2) attack by a carbanion-like alkyl group; and step (4) involves attachment of a proton to a carbanion or a carbanion-like moiety.

Protonolysis of alkylboranes is much more difficult than protonolysis of, say, Grignard reagents. The course of reaction (4) is evidently not equilibrium-controlled, but rate-controlled: it is not the stronger base, $R:^-$, that gets the proton, but instead the resonance-stabilized (more abundant) carbanion $[RCHCOCH_3]^-$.

Problem 30.19 Trialkylboranes are inert to water, but are particularly prone to protonolysis by carboxylic acids. Can you suggest a specific mechanism for protonolysis of R_3B by a carboxylic acid?

As a synthetic route, this organoborane synthesis parallels the acetoacetic ester and malonic ester syntheses. An acetone unit is furnished by acetoacetic ester or, here, by bromoacetone; an acetic acid unit is furnished by malonic ester or, here, by bromoacetic ester. In these syntheses, bromine plays the same part that the —COOEt group did: by increasing the acidity of certain α-hydrogens, it determines *where* in the molecule reaction will take place; it is easily lost from the molecule when its job is done. Unlike the loss of —COOEt, the departure of —Br is an integral part of the alkylation process.

Consistently high yields depend on the proper selection of reagents. In general, the best base is the bulky potassium 2,6-di-*tert*-butylphenoxide. The best alkylating agent is *B*-alkyl-9-borabicyclo[3.3.1]nonane, or "*B*-alkyl-9-BBN", available via successive hydroborations of alkenes:

1,5-Cyclooctadiene   9-Borabicyclo[3.3.1]nonane   *B*-Alkyl-9-borabicyclo[3.3.1]nonane
                      (9-BBN)   (*B*-Alkyl-9-BBN)
                      As dimer

The overall synthesis thus amounts to the conversion of alkenes into ketones and esters. For example:

$(CH_3)_2C=CH_2$ $\xrightarrow{\text{9-BBN}}$ $\xrightarrow{BrCH_2COCH_3}$ $(CH_3)_2CHCH_2-CH_2COCH_3$
Isobutylene   5-Methyl-2-hexanone

B-Isobutyl-BBN

Cyclopentene $\xrightarrow{\text{9-BBN}}$ *B*-Cyclopentyl-9-BBN $\xrightarrow[\text{base}]{BrCH_2COOEt}$ CH_2COOEt Ethyl cyclopentylacetate

Besides bromoacetone, other bromomethyl ketones $(BrCH_2COR)$ can be used if they are available. Bromination is best carried out with cupric bromide as the reagent, and on ketones in which R contains no α-hydrogens to compete with those on methyl: acetophenone, for example, or methyl *tert*-butyl ketone.

Problem 30.20   Using 9-BBN plus any alkenes and unhalogenated acids or ketones, outline all steps in the synthesis of:

(a) 2-heptanone   (e) ethyl (*trans*-2-methylcyclopentyl)acetate
(b) 4-methylpentanoic acid   (f) 1-phenyl-4-methyl-1-pentanone
(c) 4-methyl-2-hexanone   (g) 1-cyclopentyl-3,3-dimethyl-2-butanone
(d) 1-cyclohexyl-2-propanone

30.8   Alkylation of carbonyl compounds via enamines

As we might expect, amines react with carbonyl compounds by nucleophilic addition. If the amine is *primary*, the initial addition product undergoes dehydration (compare Sec. 21.12) to form a compound containing a carbon–nitrogen double bond, an *imine*. Elimination occurs with this orientation even if the carbonyl

$$\text{>C=O} + \text{H}_2\text{NR}' \longrightarrow -\underset{|}{\overset{|}{\text{C}}}-\text{NHR}' \longrightarrow \text{>C=NR}'$$
$$\underset{\text{OH}}{}$$

A 1° amine   An imine

compound contains an α-hydrogen: that is, the preferred product is the imine rather than the *enamine* (*ene* for the carbon–carbon double bond, *amine* for the

$$-\underset{\overset{|}{\text{H}}}{\overset{|}{\text{C}}}-\underset{\overset{|}{\text{OH}}}{\overset{|}{\text{C}}}-\underset{\overset{|}{\text{H}}}{\text{N}}-\text{R}'$$

$$-\underset{\overset{|}{\text{H}}}{\text{C}}=\overset{|}{\text{C}}-\text{N}-\text{R}' \rightleftarrows -\overset{|}{\text{C}}-\underset{\overset{|}{\text{H}}}{\text{C}}=\text{N}-\text{R}' \qquad \text{Imine–enamine tautomerism}$$

Enamine   Imine
                      More stable form

amino group). If some enamine should be formed initially, it rapidly tautomerizes into the more stable imino form.

The system is strictly analogous to the keto–enol one (Secs. 11.10 and 25.4). The proton is acidic, and therefore separates fairly readily from the hybrid anion; it can return to either carbon or nitrogen, but when it returns to carbon, it tends to stay there. Equilibrium favors formation of the weaker acid.

Now, a secondary amine, too, can react with a carbonyl compound, and to yield the same kind of initial product. But here there is no hydrogen left on nitrogen; if dehydration is to occur, it must be in the other direction, to form a carbon–carbon double bond. A stable enamine is the product.

$$-\underset{\overset{|}{\text{H}}}{\overset{|}{\text{C}}}-\overset{|}{\text{C}}=\text{O} + \text{R}'_2\text{NH} \longrightarrow -\underset{\overset{|}{\text{H}}}{\overset{|}{\text{C}}}-\underset{\overset{|}{\text{OH}}}{\overset{|}{\text{C}}}-\overset{\overset{\text{R}'}{|}}{\text{N}}-\text{R}' \longrightarrow -\text{C}=\text{C}-\overset{\overset{\text{R}'}{|}}{\text{N}}-\text{R}'$$

A 2° amine   An enamine

In 1954 Gilbert Stork (of Columbia University) showed how enamines could be used in the alkylation and acylation of aldehydes and ketones, and in the years since then enamines have been intensively studied and used in organic synthesis in a wide variety of ways. All we can do here is to try to understand a little of the basic chemistry underlying the use of enamines.

The usefulness of enamines stems from the fact that they contain *nucleophilic carbon*. The electrons responsible for this nucleophilicity are, in the final analysis, the (formally) unshared pair on nitrogen; but they are available for nucleophilic attack by carbon of the enamine. Thus, in alkylation:

$$-\overset{|}{C}=\overset{|}{C}-\overset{..}{N}R'_2 \longrightarrow -\overset{|}{C}-\overset{|}{C}=\overset{\oplus}{N}R'_2 + X^-$$
$$X-R \qquad\qquad R$$

An iminium ion

The product of alkylation is an iminium ion, which is readily hydrolyzed to regenerate the carbonyl group. The overall process, then, is:

$$-\overset{|}{\underset{H}{C}}-\overset{|}{C}=O \xrightarrow{R'_2NH} -\overset{|}{C}=\overset{|}{C}-NR'_2 \xrightarrow{RX} -\overset{|}{\underset{R}{C}}-\overset{|}{C}=\overset{\oplus}{N}R'_2 \xrightarrow{H_2O,\ H^+} -\overset{|}{\underset{R}{C}}-\overset{|}{C}=O$$

Ketone   Enamine   Iminium ion   Alkylated ketone

(In enamines the nitrogen, too, is nucleophilic, but attack there, which yields quaternary *ammo*nium ions, is generally an unwanted side reaction. Heating often converts *N*-alkylated compounds into the desired *C*-alkylated products.)

Nitrogen in enamines plays the same role as it does in the chemistry of aromatic amines—not surprisingly, when we realize that enamines are, after all, *vinyl amines.* (Remember the similarities between vinyl and aryl halides.) For example, bromination of aniline involves, we say, electrophilic attack by bromine on the aromatic ring; but from the opposite, and equally valid, point of view, it involves nucleophilic attack on bromine by carbons of the ring—with nitrogen furnishing the electrons.

Commonly used secondary amines are the heterocyclic compounds *pyrrolidine* and *morpholine*:

Pyrrolidine   Morpholine

Best yields are obtained with reactive halides like benzyl and allyl halides, α-halo esters, and α-halo ketones. For example:

Cyclohexanone   Pyrrolidine   An enamine   An iminium ion

$$\downarrow H_2O,\ H^+$$

$$CH_2COOEt$$

β-Tetralone   Pyrrolidine   An enamine $BrCH_2COOEt$ $\dfrac{H_2O}{H^+}$

Problem 30.21 Outline all steps in the preparation of each of the following by the enamine synthesis:

(a) 2-benzylcyclohexanone
(b) 2,2-dimethyl-4-pentenal

(c)

(d)

(e) 2-(2,4-dinitrophenyl)cyclohexanone
(f) 2,2-dimethyl-3-oxobutanal, $CH_3COC(CH_3)_2CHO$

Problem 30.22 Give structural formulas of compounds A–F.

(a) cyclopentanone + morpholine, then TsOH $\longrightarrow$ A ($C_9H_{15}ON$)
     A + C_6H_5CHO, then H_2O, H^+ $\longrightarrow$ B ($C_{12}H_{12}O$)
(b) isobutyraldehyde + *tert*-butylamine $\longrightarrow$ C ($C_8H_{17}N$)
     C + C_2H_5MgBr $\longrightarrow$ D ($C_8H_{16}NMgBr$) + E
     D + $C_6H_5CH_2Cl$, then H_2O, H^+ $\longrightarrow$ F ($C_{11}H_{14}O$)

PROBLEMS

1. Outline the synthesis of each of the following from malonic ester and any other reagents:

(a) *n*-caproic acid
(b) isobutyric acid
(c) β-methylbutyric acid
(d) α,β-dimethylbutyric acid
(e) 2-ethylbutanoic acid

(f) dibenzylacetic acid
(g) α,β-dimethylsuccinic acid
(h) glutaric acid
(i) cyclobutanecarboxylic acid

2. Outline the synthesis of each of the following from acetoacetic ester and any other needed reagents. Do (j)–(m) after Problem 11, below.

(a) methyl ethyl ketone
(b) 3-ethyl-2-pentanone
(c) 3-ethyl-2-hexanone
(d) 5-methyl-2-heptanone
(e) 3,6-dimethyl-2-heptanone
(f) 4-oxo-2-methylpentanoic acid
(g) γ-hydroxy-*n*-valeric acid

(h) 3-methyl-2-hexanol
(i) 2,5-dimethylheptane
(j) β-methylcaproic acid
(k) β-methylbutyric acid
(l) methylsuccinic acid
(m) 2,5-hexanediol

3. What product would you expect from the hydrolysis by dilute alkali of 2-carbethoxycyclopentanone (see Problem 25.30, p. 924)? Suggest a method of synthesis of 2-methylcyclopentanone.

4. Give structures of compounds A through J:

(a) 1,3-dibromopropane + 2 mol sodiomalonic ester $\longrightarrow$ A $(C_{17}H_{28}O_8)$
    A + 2 mol sodium ethoxide, then CH_2I_2 $\longrightarrow$ B $(C_{18}H_{28}O_8)$
    B + OH^-, heat; then H^+; then heat $\longrightarrow$ C $(C_8H_{12}O_4)$
(b) ethylene bromide + 2 mol sodiomalonic ester $\longrightarrow$ D $(C_{16}H_{26}O_8)$
    D + 2 mol sodium ethoxide, then 1 mol ethylene bromide $\longrightarrow$ E $(C_{18}H_{28}O_8)$
    E + OH^-, heat; then H^+; then heat $\longrightarrow$ F $(C_8H_{12}O_4)$
(c) 2 mol sodiomalonic ester + I_2 $\longrightarrow$ G $(C_{14}H_{22}O_8)$ + 2NaI
    G + OH^-, heat; then H^+; then heat $\longrightarrow$ H $(C_4H_6O_4)$
(d) D + 2 mol sodium ethoxide, then I_2 $\longrightarrow$ I $(C_{16}H_{24}O_8)$
    I + OH^-, heat; then H^+; then heat $\longrightarrow$ J $(C_6H_8O_4)$
(e) Suggest a possible synthesis for 1,3-cyclopentanedicarboxylic acid;
    for 1,2-cyclopentanedicarboxylic acid; for 1,1-cyclopentanedicarboxylic acid.

5. Give structures of compounds K through O:

allyl bromide + Mg $\longrightarrow$ K (C_6H_{10})
K + HBr $\longrightarrow$ L $(C_6H_{12}Br_2)$
sodiomalonic ester + excess L $\longrightarrow$ M $(C_{13}H_{23}O_4Br)$
M + sodium ethoxide $\longrightarrow$ N $(C_{13}H_{22}O_4)$
N + OH^-, heat; then H^+; then heat $\longrightarrow$ O $(C_8H_{14}O_2)$

6. When sodium trichloroacetate is heated in diglyme solution with alkenes, there are formed 1,1-dichlorocyclopropanes. How do you account for this?

7. (a) How could you synthesize 2,7-octanedione? (*Hint*: See Problem 30.2, p. 1063.) (b) Actually, the expected ketone reacts further to give

$$\text{CH}_3$$

(structure of 1-methyl-2-acetylcyclopentene shown)

How does this last reaction occur? To what general types does it belong? (c) How could you synthesize 2,6-heptanedione? (d) What would happen to this ketone under the conditions of (b)?

8. Outline all steps in a possible synthesis of each of the following from simple esters:

(a) 1,2-cyclopentanedione (*Hint*: See Problem 25.33, p. 925.)
(b) $CH_3CH_2CH_2COCOOC_2H_5$ (*Hint*: See Problem 30.9, p. 1066.)

9. Outline the synthesis from readily available compounds of the following hypnotics (see Sec. 24.23):

(a) 5,5-diethylbarbituric acid (Barbital, Veronal; long-acting)
(b) 5-allyl-5-(2-pentyl)barbituric acid (Seconal; short-acting)
(c) 5-ethyl-5-isopentylbarbituric acid (Amytal; intermediate length of action)

10. (a) Contrast the structures of barbituric acid and Veronal (5,5-diethylbarbituric acid). (b) Account for the appreciable acidity $(K_a = 10^{-8})$ of Veronal.

11. When treated with *concentrated* alkali, acetoacetic ester is converted into two moles of sodium acetate. (a) Outline all steps in a likely mechanism for this reaction. (*Hint*: See Secs. 25.11 and 7.25.) (b) Substituted acetoacetic esters also undergo this reaction. Outline the steps in a general synthetic route from acetoacetic ester to carboxylic acids. (c) Outline the steps in the synthesis of 2-hexanone via acetoacetic ester. What acids will be formed as by-products? Outline a procedure for purification of the desired ketone. (Remember that the alkylation is carried out in alcohol; that NaBr is formed; that aqueous base is used for hydrolysis; and that ethyl alcohol is a product of the hydrolysis.)

12. (a) Suggest a mechanism for the alkaline cleavage of β-diketones, as, for example:

(b) Starting from cyclohexanone, and using any other needed reagents, outline all steps in a possible synthesis of 7-phenylheptanoic acid. (c) Of pentadecanedioic acid, $HOOC(CH_2)_{13}COOH$.

13. Give structures of compounds P through S:

heptanal (heptaldehyde) + ethyl bromoacetate + Zn, then H_2O $\longrightarrow$ P $(C_{11}H_{22}O_3)$
P + CrO_3 in glacial acetic acid $\longrightarrow$ Q $(C_{11}H_{20}O_3)$
Q + sodium ethoxide, then benzyl chloride $\longrightarrow$ R $(C_{18}H_{26}O_3)$
R + OH^-, heat; then H^+, warm $\longrightarrow$ S $(C_{15}H_{22}O)$

Useful information: In what is called the *Reformatsky* reaction, organozinc compounds act like (somewhat unreactive) Grignard reagents.

14. Treatment of 1,5-cyclooctadiene with diborane gives a material, T, which is oxidized by alkaline H_2O_2 to a mixture of 72% *cis*-1,5-cyclooctanediol and 28% *cis*-1,4-cyclooctanediol. If T is refluxed for an hour in THF solution (or simply distilled), there is obtained a white crystalline solid, U, which is oxidized to 99%-pure *cis*-1,5-cyclooctanediol.

(a) What is T? What is U? (b) Account for the conversion of T into U.

15. On treatment with concentrated KOH, 2,6-dichlorobenzaldehyde is converted into 1,3-dichlorobenzene and potassium formate. The kinetics shows that the aldehyde and two moles of hydroxide ion are in equilibrium with a reactive intermediate that (ultimately) yields product. (a) Outline a likely mechanism that is consistent with these facts. (*Hint:* See Sec. 21.14.) How do you account for the difference in behavior between this aldehyde and most aromatic aldehydes under these conditions?

16. Give structural formulas of compounds V and W, and tell *exactly* how each is formed:

γ-butyrolactone + CH_3ONa $\longrightarrow$ V $(C_8H_{10}O_3)$
V + conc. HCl $\longrightarrow$ W $(C_7H_{12}OCl_2)$
W + aq. NaOH $\longrightarrow$ dicyclopropyl ketone

17. The structure of *nerolidol*, $C_{15}H_{26}O$, a terpene found in oil of neroli, was established by the following synthesis:

geranyl chloride (RCl) + sodioacetoacetic ester $\longrightarrow$ X $(RC_6H_9O_3)$
X + $Ba(OH)_2$, then H^+, warm $\longrightarrow$ Y (RC_3H_5O)
Y + $NaC{\equiv}CH$, then H_2O $\longrightarrow$ Z (RC_5H_7O)
Z $\xrightarrow{\text{reduction}}$ AA (RC_5H_9O), nerolidol

(a) Give the structure of nerolidol, using R for the geranyl group.
(b) Referring to Problem 26, p. 692, what is the complete structure of nerolidol?

18. The structure of *menthone*, $C_{10}H_{18}O$, a terpene found in peppermint oil, was first established by synthesis in the following way:

ethyl β-methylpimelate + sodium ethoxide, then H_2O $\longrightarrow$ BB $(C_{10}H_{16}O_3)$
BB + sodium ethoxide, then isopropyl iodide $\longrightarrow$ CC $(C_{13}H_{22}O_3)$
CC + OH^-, heat; then H^+; then heat $\longrightarrow$ menthone

(a) What structures for menthone are consistent with this synthesis? (b) On the basis of the isoprene rule (Sec. 10.31) which structure is the more likely? (c) On vigorous reduction menthone yields *p-menthane*, 4-isopropyl-1-methylcyclohexane. Now what structure or structures are most likely for menthone?

19. The structure of *camphoronic acid* (a degradation product of the terpene camphor) was established by the following synthesis:

sodioacetoacetic ester + CH_3I $\longrightarrow$ DD $\xrightarrow{NaOC_2H_5}$ $\xrightarrow{CH_3I}$ EE ($C_8H_{14}O_3$)

EE + ethyl bromoacetate + Zn, then H_2O $\longrightarrow$ FF ($C_{12}H_{22}O_5$)

FF + PCl_5, then KCN $\longrightarrow$ GG ($C_{13}H_{21}O_4N$)

GG + H_2O, H^+, heat $\longrightarrow$ camphoronic acid ($C_9H_{14}O_6$)

What is the structure of camphoronic acid?

20. Two of the oxidation products of the terpene α-terpineol are *terebic acid* and *terpenylic acid*. Their structures were first established by the following synthesis:

ethyl chloroacetate + sodioacetoacetic ester $\longrightarrow$ HH ($C_{10}H_{16}O_5$)

HH + 1 mol CH_3MgI, then H_2O $\longrightarrow$ II ($C_{11}H_{20}O_5$)

II + OH^-, H_2O, heat, then H^+ $\longrightarrow$ [JJ ($C_7H_{12}O_5$)] $\longrightarrow$ terebic acid ($C_7H_{10}O_4$)

HH + sodium ethoxide, then ethyl chloroacetate $\longrightarrow$ KK ($C_{14}H_{22}O_7$)

KK + OH^-, then H^+, warm $\longrightarrow$ LL ($C_7H_{10}O_5$)

LL + ethyl alcohol, H^+ $\longrightarrow$ MM ($C_{11}H_{18}O_5$)

MM + 1 mol CH_3MgI, then H_2O $\longrightarrow$ NN ($C_{12}H_{22}O_5$)

NN + OH^-, H_2O, heat, then H^+ $\longrightarrow$ [OO ($C_8H_{14}O_5$)] $\longrightarrow$

terpenylic acid ($C_8H_{12}O_4$)

What is the structure of terebic acid? Of terpenylic acid?

21. Isopentenyl pyrophosphate, the precursor of isoprene units in nature (Sec. 10.31 and Problem 24, p. 413), is formed enzymatically from the pyrophosphate of *mevalonic acid* by the action of ATP (adenosine triphosphate) and Mn^{2+} ion.

CH_2-OPP	CH_2-OPP	CH_2-OPP
CH_2	CH_2	CH_2
$CH_3-\overset{\displaystyle }{\underset{\displaystyle }{C}}-OH$ $\xrightarrow[Mn^{2+}]{ATP}$	$CH_3-\overset{\displaystyle }{\underset{\displaystyle }{C}}-OPO_3H_2$ $\longrightarrow$	$\overset{\displaystyle }{\underset{\displaystyle }{C}}$
CH_2COOH	CH_2COOH	$CH_3 \quad CH_2$
Mevalonic acid 5-pyrophosphate	Mevalonic acid 5-pyrophosphate 3-phosphate	Isopentenyl pyrophosphate

It is believed that the function of ATP is to phosphorylate mevalonic acid pyrophosphate at the 3-position.

Just what happens in the last step of this conversion? Why should the 3-phosphate undergo this reaction more easily than the 3-hydroxy compound?

31

α, β-Unsaturated Carbonyl Compounds

Conjugate Addition

31.1 Structure and properties

In general, a compound that contains both a carbon–carbon double bond and a carbon–oxygen double bond has properties that are characteristic of both functional groups. At the carbon–carbon double bond an unsaturated ester or unsaturated ketone undergoes electrophilic addition of acids and halogens, hydrogenation, hydroxylation, and cleavage; at the carbonyl group it undergoes the nucleophilic substitution typical of an ester or the nucleophilic addition typical of a ketone.

Problem 31.1 What will be the products of the following reactions?

(a) $CH_3CH=CHCOOH + H_2 + Pt$
(b) $CH_3CH=CHCOOC_2H_5 + OH^- + H_2O + heat$
(c) $C_6H_5CH=CHCOCH_3 + I_2 + OH^-$
(d) $CH_3CH=CHCHO + C_6H_5NHNH_2 + acid\ catalyst$
(e) $CH_3CH=CHCHO + Ag(NH_3)_2{}^+$
(f) $C_6H_5CH=CHCOC_6H_5 + O_3$, followed by $Zn + H_2O$
(g) $CH_3CH=CHCHO + excess\ H_2 + Ni$, heat, pressure
(h) *trans*-$HOOCCH=CHCOOH + Br_2/CCl_4$
(i) *trans*-$HOOCCH=CHCOOH + cold\ alkaline\ KMnO_4$

Problem 31.2   What are A, B, and C, given the following facts?

(a) Cinnamaldehyde (C_6H_5CH=CHCHO) + H_2 + Ni, at low temperatures and pressures $\longrightarrow$ A

(b) Cinnamaldehyde + H_2 + Ni, at high temperatures and pressures $\longrightarrow$ B

(c) Cinnamaldehyde + 9-BBN, followed by $HOCH_2CH_2NH_2$ $\longrightarrow$ C

	A	B	C
$KMnO_4$ test	positive	negative	positive
Br_2/CCl_4 test	negative	negative	positive
Tollens' test	positive	negative	negative
2,4-$(NO_2)_2PhNHNH_2$	positive	negative	negative

In the α,β-unsaturated carbonyl compounds, the carbon–carbon double bond and the carbon–oxygen double bond are separated by just one carbon–carbon single bond; that is, the double bonds are *conjugated*. Because of this conjugation,

$$\overset{\beta}{-C}=\overset{\alpha}{C}-C=O$$

α,β-Unsaturated carbonyl compound

Conjugated system

such compounds possess not only the properties of the individual functional groups, but certain other properties besides. In this chapter we shall concentrate on the α,β-unsaturated compounds, and on the special reactions characteristic of the conjugated system.

Table 31.1   α,β-UNSATURATED CARBONYL COMPOUNDS

Name	Formula	M.p., °C	B.p., °C
Acrolein	CH_2=CHCHO	− 88	52
Crotonaldehyde	CH_3CH=CHCHO	− 69	104
Cinnamaldehyde	C_6H_5CH=CHCHO	− 7	254
Mesityl oxide	$(CH_3)_2C$=CHCOCH$_3$	42	131
Benzalacetone	C_6H_5CH=CHCOCH$_3$	42	261
Dibenzalacetone	C_6H_5CH=CHCOCH=CHC$_6H_5$	113	
Benzalacetophenone (Chalcone)	C_6H_5CH=CHCOC$_6H_5$	62	348
Dypnone	$C_6H_5C(CH_3)$=CHCOC$_6H_5$		150–5[1]
Acrylic acid	CH_2=CHCOOH	12	142
Crotonic acid	*trans*-CH_3CH=CHCOOH	72	189
Isocrotonic acid	*cis*-CH_3CH=CHCOOH	16	172*d*
Methacrylic acid	CH_2=C(CH$_3$)COOH	16	162
Sorbic acid	CH_3CH=CHCH=CHCOOH	134	
Cinnamic acid	*trans*-C_6H_5CH=CHCOOH	137	300
Maleic acid	*cis*-HOOCCH=CHCOOH	130.5	
Fumaric acid	*trans*-HOOCCH=CHCOOH	302	
Maleic anhydride		60	202
Methyl acrylate	CH_2=CHCOOCH$_3$		80
Methyl methacrylate	CH_2=C(CH$_3$)COOCH$_3$		101
Ethyl cinnamate	C_6H_5CH=CHCOOC$_2H_5$	12	271
Acrylonitrile	CH_2=CH−C≡N	− 82	79

Table 31.1 lists some of the more important of these compounds. Many have common names which the student must expect to encounter. For example:

$CH_2=CH-CHO$

Acrolein
Propenal

$CH_2=CH-COOH$

Acrylic acid
Propenoic acid

$CH_2=CH-C\equiv N$

Acrylonitrile
Propenenitrile

$\overset{\displaystyle CH_3}{\underset{\displaystyle |}{CH_2=C-COOH}}$

Methacrylic acid
2-Methylpropenoic acid

$CH_3CH=CHCHO$

Crotonaldehyde
2-Butenal

$C_6H_5CH=CHCHO$

Cinnamaldehyde
3-Phenylpropenal

$C_6H_5CH=CH\underset{\displaystyle O}{\overset{\displaystyle ||}{C}}CH_3$

Benzalacetone
4-Phenyl-
3-buten-2-one

$\overset{\displaystyle CH_3}{\underset{\displaystyle |}{CH_3}}C=CH\underset{\displaystyle O}{\overset{\displaystyle ||}{C}}CH_3$

Mesityl oxide
4-Methyl-
3-penten-2-one

Fumaric acid
trans-Butenedioic
acid

Maleic acid
cis-Butenedioic
acid

Maleic anhydride
cis-Butenedioic
anhydride

31.2 Preparation

There are several general ways to make compounds of this kind: the **aldol condensation**, to make unsaturated aldehydes and ketones; **dehydrohalogenation of α-halo acids** and the **Perkin condensation**, to make unsaturated acids. Besides these, there are certain methods useful only for making single compounds.

All these methods make use of chemistry with which we are already familiar: the fundamental chemistry of alkenes and carbonyl compounds.

Problem 31.3 Outline a possible synthesis of:

(a) crotonaldehyde from acetylene
(b) cinnamaldehyde from compounds of lower carbon number
(c) cinnamic acid from compounds of lower carbon number
(d) 4-methyl-2-pentenoic acid via a malonic ester synthesis

Problem 31.4 The following compounds are of great industrial importance for the manufacture of polymers: acrylonitrile (for Orlon), methyl acrylate (for Acryloid), methyl methacrylate (for Lucite and Plexiglas). Outline a possible industrial synthesis of: (a) acrylonitrile from ethylene; (b) methyl acrylate from ethylene; (c) methyl methacrylate from acetone and methanol.

(d) Polymerization of these compounds is similar to that of ethylene, vinyl chloride, etc. (Sec. 8.21). Draw a structural formula for each of the polymers.

Problem 31.5 Acrolein, CH₂=CHCHO, can be prepared by heating glycerol with sodium hydrogen sulfate, NaHSO₄. (a) Outline the likely steps in this synthesis, which involves acid-catalyzed dehydration and keto–enol tautomerization. (*Hint*: Which —OH is easier to eliminate, a primary or a secondary?) (b) How could acrolein be converted into acrylic acid?

31.3　Interaction of functional groups

We have seen (Sec. 8.12) that, toward electrophilic addition, a carbon–carbon double bond is activated by an electron-releasing substituent and deactivated by an electron-withdrawing substituent. The carbon–carbon double bond serves as a source of electrons for the electrophilic reagent; the availability of its electrons is determined by the groups attached to it. More specifically, an electron-releasing substituent stabilizes the transition state leading to the initial carbocation by dispersing the developing positive charge; an electron-withdrawing substituent destabilizes the transition state by intensifying the positive charge.

Electrophilic addition

$$-\overset{|}{\underset{|}{C}}=\overset{|}{\underset{|}{C}}-G + Y^+ \longrightarrow \left[\begin{array}{c} -\overset{|}{\underset{\vdots}{C}}\cdots\overset{|}{\underset{\delta+}{C}}-G \\ Y\delta. \end{array}\right] \longrightarrow -\overset{|}{\underset{|}{C}}-\overset{|}{\underset{\oplus}{C}}-G$$
$$\qquad\qquad\qquad\qquad\qquad\qquad Y$$

G *releases electrons: activates*
G *withdraws electrons: deactivates*

The C=O, —COOH, —COOR, and —CN groups are powerfully electron-withdrawing groups, and therefore would be expected to deactivate a carbon–carbon double bond toward electrophilic addition. This is found to be true: α,β-unsaturated ketones, acids, esters, and nitriles are in general less reactive than simple alkenes toward reagents like bromine and the hydrogen halides.

But this powerful electron withdrawal, which deactivates a carbon–carbon double bond toward reagents seeking electrons, at the same time *activates* toward reagents that are electron-rich. As a result, the carbon–carbon double bond of an α,β-unsaturated ketone, acid, ester, or nitrile is susceptible to nucleophilic attack, and undergoes a set of reactions, **nucleophilic addition**, that is uncommon for the simple alkenes. As we shall see (Sec. 31.5), this reactivity toward nucleophiles is primarily due, not to a simple inductive effect of these substituents, but rather to their *conjugation with* the carbon–carbon double bond.

31.4　Electrophilic addition

The presence of the carbonyl group not only lowers the **reactivity** of the carbon–carbon double bond toward electrophilic addition, but also controls the **orientation** of the addition.

In general, it is observed that addition of an unsymmetrical reagent to an α,β-unsaturated carbonyl compound takes place in such a way that hydrogen becomes attached to the α-carbon and the negative group becomes attached to the β-carbon. For example:

$$CH_2{=}CH{-}CHO + HCl(g) \xrightarrow{-10\,°C} \underset{\quad Cl \quad\; H}{CH_2{-}CH{-}CHO}$$

Acrolein

β-Chloropropionaldehyde

$$CH_2{=}CH{-}COOH + H_2O \xrightarrow{H_2SO_4,\,100\,°C} \underset{\quad OH \quad\; H}{CH_2{-}CH{-}COOH}$$

Acrylic acid

β-Hydroxypropionic acid

$$CH_3{-}CH{=}CH{-}COOH + HBr(g) \xrightarrow{20\,°C} \underset{\qquad\quad Br \quad\; H}{CH_3{-}CH{-}CH{-}COOH}$$

Crotonic acid

β-Bromobutyric acid

$$\underset{O}{\overset{CH_3}{CH_3{-}C{=}CH{-}C{-}CH_3}} + CH_3OH \xrightarrow{H_2SO_4} \underset{\quad CH_3O \quad H \quad\; O}{\overset{CH_3}{CH_3{-}C{-}CH{-}C{-}CH_3}}$$

Mesityl oxide

4-Methoxy-4-methyl-2-pentanone

Electrophilic addition to simple alkenes takes place in such a way as to form the most stable intermediate carbocation. Addition to α,β-unsaturated carbonyl compounds, too, is consistent with this principle; to see that this is so, however, we must look at the conjugated system as a whole. As in the case of conjugated dienes (Sec. 10.26), addition to an *end* of the conjugated system is preferred, since this yields (step 1) a resonance-stabilized carbocation. Addition of a proton to the carbonyl oxygen end would yield cation I; addition to the β-carbon end would yield cation II.

Of the two, I is the more stable, since the positive charge is carried by carbon atoms alone, rather than partly by the more highly electronegative oxygen atom.

In the second step of addition, a negative ion or basic molecule attaches itself either to the carbonyl carbon or to the β-carbon of the hybrid ion I.

(2)

$$-\overset{|}{\underset{\oplus}{C}}\overset{|}{\cdots}\overset{|}{C}\overset{|}{\cdots}\overset{|}{C}-OH + :Z$$

I

$$-\overset{|}{C}-\overset{|}{C}=\overset{|}{C}-O-H$$ *Actually formed*

$$\underset{Z}{|}$$

III

$$-\overset{|}{C}=\overset{|}{C}-\overset{|}{C}-O-H$$

$$\underset{Z}{|}$$

Unstable

Of the two possibilities, only addition to the β-carbon yields a stable product (III), which is simply the enol form of the saturated carbonyl compound. The enol form then undergoes tautomerization to the keto form to give the observed product (IV).

$$-\overset{|}{C}=\overset{|}{C}-\overset{|}{C}=O \underset{}{\overset{H^+}{\rightleftharpoons}} -\overset{|}{\underset{\oplus}{C}}\overset{|}{\cdots}\overset{|}{C}\overset{|}{\cdots}\overset{|}{C}-O-H \underset{}{\overset{:Z}{\rightleftharpoons}} -\overset{|}{C}-\overset{|}{C}=\overset{|}{C}-O-H$$

α,β-Unsaturated   I   III
compound   Carbocation   Enol form
                              $\underset{Z}{|}$

$$-\overset{|}{C}-\overset{|}{\underset{H}{C}}-\overset{|}{\underset{}{C}}=O$$

$$\underset{Z\ \ H}{}$$

IV

Keto form

31.5 Nucleophilic addition

Aqueous sodium cyanide converts α,β-unsaturated carbonyl compounds into β-cyano carbonyl compounds. The reaction amounts to addition of the elements of HCN to the carbon–carbon double bond. For example:

Benzalacetophenone → NaCN(aq) → 3-Cyano-1,3-diphenyl-1-propanone

$$CH_3-\overset{H}{\overset{|}{C}}=\overset{H}{\overset{|}{C}}-COOC_2H_5 \xrightarrow{NaCN(aq)} CH_3-\overset{H}{\overset{|}{C}}-\overset{H}{\overset{|}{C}}-COOC_2H_5$$

Ethyl crotonate $\underset{CN\ \ H}{}$

Ethyl β-cyanobutyrate

Ammonia or certain derivatives of ammonia (amines, hydroxylamine, phenyl-hydrazine, etc.) add to α,β-unsaturated carbonyl compounds to yield β-amino carbonyl compounds. For example:

$$CH_3-\underset{\underset{CH_3}{|}}{C}=\underset{\underset{H}{|}}{C}-\underset{\underset{O}{\|}}{C}-CH_3 + CH_3NH_2 \longrightarrow CH_3-\underset{\underset{CH_3NH}{|}}{\overset{\overset{CH_3}{|}}{C}}-\underset{\underset{H}{|}}{\overset{\overset{H}{|}}{C}}-\underset{\underset{O}{\|}}{C}-CH_3$$

Mesityl oxide   Methylamine   4-(N-Methylamino)-4-methyl-2-pentanone

$$trans\text{-}HOOCCH=CHCOOH + NH_3 \longrightarrow {}^-OOC-\underset{\underset{NH_3^+}{|}}{CH}-CH_2-COOH$$

Fumaric acid   Aminosuccinic acid
(Aspartic acid)

$$\text{Ph}-\underset{\underset{H}{|}}{C}=\underset{\underset{H}{|}}{C}-COOH + NH_2OH \longrightarrow \text{Ph}-\underset{\underset{NHOH}{|}}{\overset{\overset{H}{|}}{C}}-\underset{\underset{H}{|}}{\overset{\overset{H}{|}}{C}}-COOH$$

Cinnamic acid   Hydroxylamine   3-(N-Hydroxylamino)-3-phenylpropanoic acid

These reactions are believed to take place by the following mechanism:

(1)
$$-\underset{|}{C}=\underset{|}{C}-\underset{|}{C}=O + :Z \longrightarrow -\underset{\underset{Z}{|}}{C}-\underset{|}{C}{=\!\!=}\underset{\underset{\ominus}{}}{C}{=\!\!=}O$$
I

(2)
$$-\underset{\underset{Z}{|}}{C}-\underset{|}{C}{=\!\!=}\underset{\underset{\ominus}{}}{C}{=\!\!=}O + H^+$$
I

$$-\underset{\underset{Z}{|}}{C}-\underset{|}{C}=\underset{|}{C}-O-H$$
Enol

$$-\underset{\underset{Z}{|}}{C}-\underset{\underset{H}{|}}{C}-\underset{|}{C}=O$$
Keto

The nucleophilic reagent adds (step 1) to the carbon–carbon double bond to yield the hybrid anion I, which then accepts (step 2) a proton from the solvent to yield the final product. This proton can add either to the α-carbon or to oxygen, and thus yield either the keto or the enol form of the product; in either case the same equilibrium mixture, chiefly keto, is finally obtained.

In the examples we have just seen, the nucleophilic reagent, :Z, is either the strongly basic anion, $:CN^-$, or a neutral base like ammonia and its derivatives, $:NH_2-G$. These are the same reagents which, we have seen, add to the carbonyl group of simple aldehydes and ketones. (Indeed, nucleophilic reagents rarely add to the carbon–carbon double bond of α,β-unsaturated *aldehydes*, but rather to the highly reactive carbonyl group.)

These nucleophilic reagents add to the conjugated system in such a way as to form the most stable intermediate anion. The most stable anion is I, which is a hybrid of II and III.

As usual, initial addition occurs to an *end* of the conjugated system, and in this case to the particular end (β-carbon) *that enables the electronegative element oxygen to accommodate the negative charge.*

The tendency for α,β-unsaturated carbonyl compounds to undergo nucleophilic addition is thus due, not simply to the electron-withdrawing ability of the carbonyl group, but to the existence of the conjugated system that permits formation of the resonance-stabilized anion I. The importance in synthesis of α,β-unsaturated aldehydes, ketones, acids, esters, and nitriles is due to the fact that they provide such a conjugated system.

Problem 31.6   Draw structures of the anion expected from nucleophilic addition to each of the other positions in the conjugated system, and compare its stability with that of I.

Problem 31.7   Treatment of crotonic acid, $CH_3CH=CHCOOH$, with phenylhydrazine yields compound IV.

IV

To what simple class of compounds does IV belong? How can you account for its formation? (*Hint*: See Sec. 24.11.)

Problem 31.8   Treatment of acrylonitrile, $CH_2=CHCN$, with ammonia yields a mixture of two products: β-aminopropionitrile, $H_2NCH_2CH_2CN$, and di(β-cyanoethyl)amine, $NCCH_2CH_2NHCH_2CH_2CN$. How do you account for their formation?

Problem 31.9   Treatment of ethyl acrylate, $CH_2=CHCOOC_2H_5$, with methylamine yields $CH_3N(CH_2CH_2COOC_2H_5)_2$. How do you account for its formation?

31.6   Comparison of nucleophilic and electrophilic addition

We can see that nucleophilic addition is closely analogous to electrophilic addition: (a) addition proceeds in two steps; (b) the first and controlling step is the formation of an intermediate ion; (c) both orientation of addition and reactivity are determined by the stability of the intermediate ion, or, more exactly, by the stability of the transition state leading to its formation; (d) this stability depends upon dispersal of the charge.

The difference between nucleophilic and electrophilic addition is, of course, that the intermediate ions have opposite charges: negative in nucleophilic addition, positive in electrophilic addition. As a result, the effects of substituents are exactly opposite. Where an electron-withdrawing group deactivates a carbon–carbon double bond toward electrophilic addition, it activates toward nucleophilic addition. An electron-withdrawing group stabilizes the transition state leading to the formation of an intermediate anion in nucleophilic addition by helping to disperse the developing negative charge:

Nucleophilic addition

$$-\overset{|}{C}=\overset{|}{C}\text{--}G + :Z \longrightarrow \left[-\overset{|}{C}\text{=·=}\underset{\overset{\cdot}{Z}}{\overset{|}{C}}\text{--}G \right]_{\delta-} \longrightarrow -\overset{|}{\underset{Z}{C}}\text{--}\overset{|}{\underset{\ominus}{C}}\text{--}G$$

G *withdraws electrons; activates*

Addition to an α,β-unsaturated carbonyl compound can be understood best in terms of an attack on the entire conjugated system. To yield the most stable intermediate ion, this attack must occur at an end of the conjugated system. A nucleophilic reagent attacks at the β-carbon to form an ion in which the negative charge is partly accommodated by the electronegative atom oxygen; an electrophilic reagent attacks oxygen to form a carbocation in which the positive charge is accommodated by carbon.

$$-\overset{|}{C}=\overset{|}{C}\text{--}\overset{|}{C}\text{=O} \longrightarrow -\overset{|}{C}\text{=}\underbrace{\overset{|}{C}\text{=}\overset{|}{C}}\text{--O--H} \qquad \textit{Electrophilic attack}$$
$$\searrow H^+ \qquad\qquad \oplus$$

$$-\overset{|}{C}=\overset{|}{C}\text{--}\overset{|}{C}\text{=O} \longrightarrow -\overset{|}{C}\text{--}\underbrace{\overset{|}{C}\text{=·=}\overset{|}{C}\text{=O}} \qquad \textit{Nucleophilic attack}$$
$$Z: \qquad\qquad Z \qquad \ominus$$

31.7 The Michael addition

Of special importance in synthesis is the nucleophilic addition of carbanions to α,β-unsaturated carbonyl compounds known as the **Michael addition**. Like the reactions of carbanions that we studied in Chapter 30, it results in formation of carbon–carbon bonds. For example:

$$\underset{\text{Benzalacetophenone}}{C_6H_5-\overset{H}{\overset{|}{C}}=\overset{H}{\overset{|}{C}}-\underset{\overset{||}{O}}{C}-C_6H_5} + \underset{\text{Ethyl malonate}}{CH_2(COOC_2H_5)_2} \xrightarrow{\text{piperidine}} \underset{\underset{CH(COOC_2H_5)_2}{\overset{|}{}}}{C_6H_5-\overset{H}{\overset{|}{C}}-\overset{H}{\overset{|}{\underset{H}{C}}}-\underset{\overset{||}{O}}{C}-C_6H_5}$$

$$\underset{\text{Ethyl cinnamate}}{\overset{\overset{\displaystyle H \quad H}{|\quad\,|}}{C_6H_5-C=C-COOC_2H_5}} + \underset{\text{Ethyl malonate}}{CH_2(COOC_2H_5)_2} \xrightarrow{\;^-OC_2H_5\;} \underset{\substack{|\\ CH(COOC_2H_5)_2}}{\overset{\overset{\displaystyle H \quad H}{|\quad\,|}}{C_6H_5-C-C-COOC_2H_5}}$$

$$\underset{\text{Ethyl crotonate}}{\overset{\overset{\displaystyle H \quad H}{|\quad\,|}}{CH_3-C=C-COOC_2H_5}} + \underset{\text{Ethyl methylmalonate}}{CH_3-CH(COOC_2H_5)_2} \xrightarrow{\;^-OC_2H_5\;} CH_3-\overset{\overset{\displaystyle H}{|}}{\underset{\underset{\underset{CH_3}{|}}{C(COOC_2H_5)_2}}{C}}-\overset{\overset{\displaystyle H}{|}}{\underset{\underset{H}{|}}{C}}-COOC_2H_5$$

$$\underset{\text{Ethyl α-methylacrylate}}{\overset{\overset{\displaystyle H \quad CH_3}{|\quad\,|}}{H-C=C-COOC_2H_5}} + \underset{\underset{\underset{CN}{|}}{CH_2}}{\overset{COOC_2H_5}{|}} \xrightarrow{\;^-OC_2H_5\;} H-\overset{\overset{\displaystyle H}{|}}{\underset{\underset{\underset{CN}{|}}{CHCOOC_2H_5}}{C}}-\overset{\overset{\displaystyle CH_3}{|}}{\underset{\underset{H}{|}}{C}}-COOC_2H_5$$

$$\text{Ethyl cyanoacetate}$$

The Michael addition is believed to proceed by the following mechanism (shown for malonic ester):

(1) $H-CH(COOC_2H_5)_2 + \;:Base \longrightarrow H:Base^+ + CH(COOC_2H_5)_2^-$

(2) $\underset{\substack{ \\ \textit{Nucleophilic} \\ \textit{reagent}}}{-C=C-C=O} + CH(COOC_2H_5)_2^- \longrightarrow \underset{\substack{\ominus \\ CH(COOC_2H_5)_2}}{-C-C \cdots C \cdots O}$

(3) $\underset{\substack{\ominus \\ CH(COOC_2H_5)_2}}{-C-C \cdots C \cdots O} + H:Base^+ \longrightarrow \underset{\substack{H \\ CH(COOC_2H_5)_2}}{-C-C-C=O} + \;:Base$

The function of the base is to abstract (step 1) a proton from malonic ester and thus generate a carbanion which, acting as a nucleophilic reagent, then attacks (step 2) the conjugated system in the usual manner.

In general, the compound from which the carbanion is generated must be a fairly acidic substance, so that an appreciable concentration of the carbanion can be obtained. Such a compound is usually one that contains a $-CH_2-$ or $-CH-$ group flanked by two electron-withdrawing groups which can help accommodate the negative charge of the anion. In place of ethyl malonate, compounds like ethyl cyanoacetate and ethyl acetoacetate can be used.

Ethyl malonate

Ethyl cyanoacetate

Ethyl acetoacetate

Ammonia and primary and secondary amines are especially powerful catalysts for the Michael addition. They appear to play a specific role in this reaction: not just to abstract a proton from the reagent to generate a carbanion, but to react with the carbonyl group of the substrate to form an intermediate imine or iminium ion (Sec. 30.8) that is particularly reactive toward nucleophilic addition.

Problem 31.10 Predict the products of the following Michael additions:

(a) ethyl crotonate + malonic ester $\longrightarrow$ A $\xrightarrow{\text{OH}^-}$ $\xrightarrow{\text{H}^+}$ $\xrightarrow{\text{heat}}$ B

(b) ethyl acrylate + ethyl acetoacetate $\longrightarrow$ C $\xrightarrow{\text{H}_2\text{O,H}^+}$ D

(c) methyl vinyl ketone + malonic ester $\longrightarrow$ E

(d) benzalacetophenone + acetophenone $\longrightarrow$ F

(e) acrylonitrile + allyl cyanide $\longrightarrow$ G $\xrightarrow{\text{H}_2\text{O,H}^+}$ H + 2NH$_4^+$

(f) $C_2H_5OOC-C\equiv C-COOC_2H_5$ (1 mol) + ethyl acetoacetate (1 mol) $\longrightarrow$ I

(g) I $\xrightarrow{\text{strong OH}^-,\text{H}_2\text{O}}$ $\xrightarrow{\text{H}^+}$ J + CH$_3$COOH

Problem 31.11 Formaldehyde and malonic ester react in the presence of ethoxide ion to give K, $C_8H_{12}O_4$. (a) What is the structure of K? (*Hint*: See Problem 30.3, p. 1063.) (b) How can K be converted into L, $(C_2H_5OOC)_2CHCH_2CH(COOC_2H_5)_2$? (c) What would you get if L were subjected to hydrolysis, acidification, and heat?

Problem 31.12 Show how a Michael addition followed by an aldol condensation can transform a mixture of methyl vinyl ketone and cyclohexanone into $\Delta^{1,9}$-octalone.

$\Delta^{1,9}$-Octalone

Problem 31.13 When mesityl oxide, $(CH_3)_2C=CHCOCH_3$, is treated with ethyl malonate in the presence of sodium ethoxide, compound M is obtained. (a) Outline the steps in its formation. (b) How could M be turned into 5,5-dimethyl-1,3-cyclo-hexanedione?

$$
\begin{array}{c}
\text{O} \\
\| \\
\text{C} \\
\diagup \quad \diagdown \\
\text{H}_2\text{C} \qquad \text{CH}_2 \\
| \qquad\qquad | \\
(\text{CH}_3)_2\text{C} \qquad \text{C}=\text{O} \\
\diagdown\quad\diagup \\
\text{CH} \\
| \\
\text{COOC}_2\text{H}_5
\end{array}
$$

M

Problem 31.14 In the presence of piperidine (a secondary amine, Sec. 26.14), 1,3-cyclopentadiene and benzal-*p*-bromoacetophenone yield N. Outline the steps in its formation.

$$
\text{C}_6\text{H}_5{-}\text{CH}{-}\text{CH}_2{-}\text{C}{-}\text{C}_6\text{H}_4{-}\text{Br}
$$

N

Problem 31.15 (a) Using as your example the addition of ethyl malonate to benzal-acetophenone in the presence of dimethylamine, show how an iminium ion might be formed and act as an intermediate in this reaction.

(b) How do you account for the high reactivity toward nucleophilic addition of such an iminium ion?

(c) Why do tertiary amines not show specific catalytic action in the Michael addition?

31.8 The Diels–Alder reaction

$α,β$-Unsaturated carbonyl compounds undergo an exceedingly useful reaction with conjugated dienes, known as the **Diels–Alder reaction**. This is an addition reaction in which C–1 and C–4 of the conjugated diene system become attached to

Diene	Dienophile	Adduct
	(Greek: diene-loving)	*Six-membered ring*

the doubly bonded carbons of the unsaturated carbonyl compound to form a six-membered ring. A concerted, single-step mechanism is almost certainly involved; both new carbon–carbon bonds are partly formed in the same transition state, although not necessarily to the same extent. The Diels–Alder reaction is the most important example of *cycloaddition*, which is discussed further in Sec. 33.9. Since reaction involves a system of four π electrons (the diene) and a system of two π electrons (the dienophile), it is known as a [4 + 2] cycloaddition.

The Diels–Alder reaction is useful not only because a ring is generated, but also because it takes place so readily for a wide variety of reactants. Reaction is favored by electron-withdrawing substituents in the dienophile, but even simple alkenes can react. Reaction often takes place with the evolution of heat when the reactants are simply mixed together. A few examples of the Diels–Alder reaction are:

1,3-Butadiene

Maleic anhydride

benzene, 20 °C
quantitative

cis-1,2,3,6-Tetrahydrophthalic anhydride

1,3-Butadiene   Acrolein

100 °C
quantitative

1,2,3,6-Tetrahydrobenzaldehyde

1,3-Butadiene

p-Benzoquinone

benzene, 35 °C
quantitative

5,8,9,10-Tetrahydro-1,4-naphthoquinone

1,3-butadiene, 100 °C

1,4,5,8,11,12,13,14-Octahydro-9,10-anthraquinone

1,3-Cyclohexadiene

Maleic anhydride

benzene, warm
quantitative

Problem 31.16 From what reactants could each of the following compounds be synthesized?

Problem 31.17 (a) In one synthesis of the hormone *cortisone* (by Lewis Sarett of Merck, Sharp and Dohme), the initial step was the formation of I by a Diels–Alder reaction. What were the starting materials?

I

(b) In another synthesis of cortisone (by R. B. Woodward, p. 1040), the initial step was the formation of II by a Diels–Alder reaction. What were the starting materials?

II

31.9 Quinones

α,β-Unsaturated ketones of a rather special kind are given the name of **quinones**: these are cyclic diketones of such a structure that they are converted by reduction into hydroquinones, phenols containing two —OH groups. For example:

p-Benzoquinone   Hydroquinone
(Quinone)

Yellow

Because they are highly conjugated, quinones are colored: p-benzoquinone, for example, is yellow.

Also because they are highly conjugated, quinones are rather closely balanced, energetically, against the corresponding hydroquinones. The ready interconversion provides a convenient oxidation–reduction system that has been studied intensively. Many properties of quinones result from the tendency to form the aromatic hydroquinone system.

Quinones—some related to more complicated aromatic systems (Chap. 34)—have been isolated from biological sources (molds, fungi, higher plants). In many cases they seem to take part in oxidation–reduction cycles essential to the living organism.

Problem 31.18 When *p*-benzoquinone is treated with HCl, there is obtained 2-chlorohydroquinone. It has been suggested that this product arises via an initial 1,4-addition. Show how this might be so.

Problem 31.19 (a) Hydroquinone is used in photographic developers to aid in the conversion of silver ion into free silver. What property of hydroquinone is being taken advantage of here?

(b) *p*-Benzoquinone can be used to convert iodide ion into iodine. What property of the quinone is being taken advantage of here?

Problem 31.20 How do you account for the fact that the treatment of phenol with nitrous acid yields the mono-oxime of *p*-benzoquinone?

PROBLEMS

1. Outline all steps in a possible laboratory synthesis of each of the unsaturated carbonyl compounds in Table 31.1, p. 1079, using any readily available monofunctional compounds: simple alcohols, aldehydes, ketones, acids, esters, and hydrocarbons.

2. Give the structures of the organic products expected from the reaction of benzalacetone, $C_6H_5CH=CHCOCH_3$, with each of the following:

(a) H_2,Ni
(b) 9-BBN, then $HOCH_2CH_2NH_2$
(c) NaOI
(d) O_3, then Zn, H_2O
(e) Br_2
(f) HCl
(g) HBr
(h) H_2O, H^+
(i) CH_3OH, H^+
(j) NaCN (aq)
(k) CH_3NH_2
(l) aniline
(m) NH_3
(n) NH_2OH
(o) benzaldehyde, base
(p) ethyl malonate, base
(q) ethyl cyanoacetate, base
(r) ethyl methylmalonate, base
(s) ethyl acetoacetate, base
(t) 1,3-butadiene
(u) 1,3-cyclohexadiene
(v) 1,3-cyclopentadiene

3. In the presence of base the following pairs of reagents undergo Michael addition. Give the structures of the expected products.

(a) benzalacetophenone + ethyl cyanoacetate
(b) ethyl cinnamate + ethyl cyanoacetate
(c) ethyl fumarate + ethyl malonate
(d) ethyl acetylenedicarboxylate + ethyl malonate
(e) mesityl oxide + ethyl malonate
(f) mesityl oxide + ethyl acetoacetate
(g) ethyl crotonate + ethyl methylmalonate
(h) formaldehyde + 2 mol ethyl malonate
(i) acetaldehyde + 2 mol ethyl acetoacetate
(j) methyl acrylate + nitromethane
(k) 2 mol ethyl crotonate + nitromethane
(l) 3 mol acrylonitrile + nitromethane
(m) 1 mol acrylonitrile + $CHCl_3$

4. Give the structures of the compounds expected from the hydrolysis and decarboxylation of the products obtained in Problem 3, parts (a) through (i).

5. Depending upon reaction conditions, dibenzalacetone and ethyl malonate can be made to yield any of three products by Michael addition.

dibenzalacetone + 2 mol ethyl malonate ⟶ A (no unsaturation)
dibenzalacetone + 1 mol ethyl malonate ⟶ B (one carbon–carbon double bond)
dibenzalacetone + 1 mol ethyl malonate ⟶ C (no unsaturation)

What are A, B, and C?

6. Give the structure of the product of the Diels–Alder reaction between:

(a) maleic anhydride and isoprene
(b) maleic anhydride and 1,1′-bicyclohexenyl (I)
(c) maleic anhydride and 1-vinyl-1-cyclohexene
(d) 1,3-butadiene and methyl vinyl ketone
(e) 1,3-butadiene and crotonaldehyde
(f) 2 mol 1,3-butadiene and dibenzalacetone
(g) 1,3-butadiene and β-nitrostyrene ($C_6H_5CH{=}CHNO_2$)
(h) 1,3-butadiene and 1,4-naphthoquinone (II)
(i) p-benzoquinone and 1,3-cyclohexadiene
(j) p-benzoquinone and 1,1′-bicyclohexenyl (I)
(k) p-benzoquinone and 2 mol 1,3-cyclohexadiene
(l) p-benzoquinone and 2 mol 1,1′-bicyclohexenyl (I)
(m) 1,3-cyclopentadiene and acrylonitrile
(n) 1,3-cyclohexadiene and acrolein

I　　　　　　　　　　II

7. From what reactants could the following be synthesized by the Diels–Alder reaction?

8. The following observations illustrate one aspect of the stereochemistry of the Diels–Alder reaction:

maleic anhydride + 1,3-butadiene $\longrightarrow$ D ($C_8H_8O_3$)
D + H_2O, heat $\longrightarrow$ E ($C_8H_{10}O_4$)
E + H_2, Ni $\longrightarrow$ F ($C_8H_{12}O_4$), m.p. 192 °C
fumaryl chloride (*trans*-ClOCCH=CHCOCl) + 1,3-butadiene $\longrightarrow$ G ($C_8H_8O_2Cl_2$)
G + H_2O, heat $\longrightarrow$ H ($C_8H_{10}O_4$)
H + H_2, Ni $\longrightarrow$ I ($C_8H_{12}O_4$), m.p. 215 °C
I can be resolved; F cannot be resolved.

Does the Diels–Alder reaction involve a *syn*-addition or an *anti*-addition?

9. On the basis of your answer to Problem 8, give the stereochemical formulas of the products expected from each of the following reactions. Label *meso* compounds and racemic modifications.

(a) crotonaldehyde (*trans*-2-butenal) + 1,3-butadiene
(b) *p*-benzoquinone + 1,3-butadiene
(c) maleic anhydride + 1,3-butadiene, followed by cold alkaline $KMnO_4$
(d) maleic anhydride + 1,3-butadiene, followed by hot $KMnO_4$ $\longrightarrow$ $C_8H_{10}O_8$

10. Account for the following observations:

(a) Dehydration of 3-hydroxy-2,2-dimethylpropanoic acid yields 2-methyl-2-butenoic acid.

(b)

$C_2H_5OOC—COOC_2H_5$
Ethyl oxalate
$+$
$CH_3CH=CHCOOC_2H_5$
Ethyl crotonate

$\xrightarrow{OC_2H_5^-}$ $C_2H_5OOC—\underset{\underset{O}{\|}}{C}—CH_2CH=CHCOOC_2H_5$

(c) $CH_2=CH—\overset{+}{P}Ph_3\ Br^-$ + salicylaldehyde + a little base $\longrightarrow$

+ Ph_3PO

(d) $CH_3CH=CHCOOC_2H_5$ + $Ph_3P=CH_2$ $\longrightarrow$ $CH_3—CH—CH—COOC_2H_5$ + Ph_3P
                                                    $\underset{CH_2}{\diagdown\diagup}$

(e)

11. When *citral* (Problem 25, p. 793) is refluxed with aqueous potassium carbonate, acetaldehyde distills from the mixture and 6-methyl-5-hepten-2-one is obtained in high yield. Show all steps in a likely mechanism. (*Hint:* See Sec. 25.5.)

12. Outline all steps in each of the following syntheses:

(a) HOOC—CH=CH—CH=CH—COOH from adipic acid
(b) HC≡C—CHO from acrolein (*Hint:* See Problem 14(a) overleaf.)
(c) $CH_3COCH=CH_2$ from acetone and formaldehyde
(d) $CH_3COCH=CH_2$ from vinylacetylene
(e) β-phenylglutaric acid from benzaldehyde and aliphatic reagents
(f) phenylsuccinic acid from benzaldehyde and aliphatic reagents
(g) 4-phenyl-2,6-heptanedione from benzaldehyde and aliphatic reagents (*Hint:* See Problem 3(f), above.)

13. *Spermine*, $H_2NCH_2CH_2CH_2NHCH_2CH_2CH_2CH_2NHCH_2CH_2CH_2NH_2$, found in seminal fluid, has been synthesized from acrylonitrile and 1,4-diaminobutane (putrescine). Show how this was probably done.

14. Give structures of compounds J through CCC:

(a) glycerol + $NaHSO_4$, heat $\longrightarrow$ J (C_3H_4O)
J + ethyl alcohol + HCl $\longrightarrow$ K ($C_7H_{15}O_2Cl$)
K + NaOH, heat $\longrightarrow$ L ($C_7H_{14}O_2$)
L + cold neutral $KMnO_4$ $\longrightarrow$ M ($C_7H_{16}O_4$)
M + dilute H_2SO_4 $\longrightarrow$ N ($C_3H_6O_3$) + ethyl alcohol

(b) $C_2H_5OOC—C≡C—COOC_2H_5$ + sodiomalonic ester $\longrightarrow$ O ($C_{15}H_{22}O_8$)
O + OH^-, heat; then H^+; then heat $\longrightarrow$ P ($C_6H_6O_6$), *aconitic acid*, found in sugar cane and beetroot

(c) ethyl fumarate + sodiomalonic ester $\longrightarrow$ Q ($C_{15}H_{24}O_8$)
Q + OH^-, heat; then H^+; then heat $\longrightarrow$ R ($C_6H_8O_6$), *tricarballylic acid*

(d) benzil ($C_6H_5COCOC_6H_5$) + dibenzyl ketone ($C_6H_5CH_2COCH_2C_6H_5$) + base
$\longrightarrow$ S ($C_{29}H_{20}O$), "tetracyclone"
S + maleic anhydride $\longrightarrow$ T ($C_{33}H_{22}O_4$)
T + heat $\longrightarrow$ CO + H_2 + U ($C_{32}H_{20}O_3$)

(e) S + $C_6H_5C≡CH$ $\longrightarrow$ V ($C_{37}H_{26}O$)
V + heat $\longrightarrow$ CO + W ($C_{36}H_{26}$)

(f) acetone + $BrMgC≡COC_2H_5$, then H_2O $\longrightarrow$ X ($C_7H_{12}O_2$)
X + H_2, Pd/$CaCO_3$ $\longrightarrow$ Y ($C_7H_{14}O_2$)
Y + H^+, warm $\longrightarrow$ Z (C_5H_8O), β-methylcrotonaldehyde

(g) ethyl 3-methyl-2-butenoate + ethyl cyanoacetate + base $\longrightarrow$ AA ($C_{12}H_{19}O_4N$)
AA + OH^-, heat; then H^+; then heat $\longrightarrow$ BB ($C_7H_{12}O_4$)

(h) mesityl oxide + ethyl malonate + base $\longrightarrow$ CC ($C_{13}H_{22}O_5$)
CC + NaOBr, OH^-, heat; then H^+ $\longrightarrow$ $CHBr_3$ + BB ($C_7H_{12}O_4$)

(i) $CH_3C≡CNa$ + acetaldehyde $\longrightarrow$ DD (C_5H_8O)
DD + $K_2Cr_2O_7$, H_2SO_4 $\longrightarrow$ EE (C_5H_6O)

(j) 3-pentyn-2-one + H_2O, Hg^{2+}, H^+ $\longrightarrow$ FF ($C_5H_8O_2$)

(k) mesityl oxide + NaOCl, then H^+ $\longrightarrow$ GG ($C_5H_8O_2$)

(l) methallyl chloride (3-chloro-2-methylpropene) + HOCl $\longrightarrow$ HH ($C_4H_8OCl_2$)
HH + KCN $\longrightarrow$ II ($C_6H_8ON_2$)
II + H_2SO_4, H_2O, heat $\longrightarrow$ JJ ($C_6H_8O_4$)

(m) ethyl adipate + NaOEt $\longrightarrow$ KK ($C_8H_{12}O_3$)
KK + methyl vinyl ketone + base $\xrightarrow{\text{Michael}}$ LL ($C_{12}H_{18}O_4$)
LL + base $\xrightarrow{\text{aldol}}$ MM ($C_{12}H_{16}O_3$)

(n) hexachloro-1,3-cyclopentadiene + CH_3OH + KOH $\longrightarrow$ NN ($C_7H_6Cl_4O_2$)
NN + $CH_2=CH_2$, heat, pressure $\longrightarrow$ OO ($C_9H_{10}Cl_4O_2$)
OO + Na + t-BuOH $\longrightarrow$ PP ($C_9H_{14}O_2$)
PP + dilute acid $\longrightarrow$ QQ (C_7H_8O), *7-ketonorbornene*

(o) ethyl acetamidomalonate [$CH_3CONHCH(COOC_2H_5)_2$] + acrolein $\xrightarrow{\text{Michael}}$
RR ($C_{12}H_{19}O_6N$)
RR + KCN + acetic acid $\longrightarrow$ SS ($C_{13}H_{20}O_6N_2$)
SS + acid + heat $\longrightarrow$ TT ($C_{13}H_{18}O_5N_2$)
TT + H_2, catalyst, in acetic anhydride $\longrightarrow$ [UU ($C_{13}H_{24}O_5N_2$)]
UU $\xrightarrow{\text{acetic anhydride}}$ VV ($C_{15}H_{26}O_6N_2$)
VV + OH^-, heat; then H^+; then heat $\longrightarrow$ WW ($C_7H_{16}O_2N_2$)

(p) acrylonitrile + ethyl malonate $\xrightarrow{\text{Michael}}$ XX ($C_{10}H_{15}O_4N$)
XX + H_2, catalyst $\longrightarrow$ [YY ($C_{10}H_{19}O_4N$)] $\longrightarrow$ ZZ ($C_8H_{13}O_3N$)
ZZ + SO_2Cl_2 in $CHCl_3$ $\longrightarrow$ AAA ($C_8H_{12}O_3NCl$)
AAA + HCl, heat $\longrightarrow$ BBB ($C_5H_{10}O_2NCl$)
BBB $\xrightarrow{\text{base}}$ CCC ($C_5H_9O_2N$)

15. In connection with his new research problem, our naïve graduate student (Problem 16, p. 791, and Problem 20, p. 928) needed a quantity of the unsaturated alcohol $C_6H_5CH=CHC(OH)(CH_3)(C_2H_5)$. He added a slight excess of benzalacetone, $C_6H_5CH=CHCOCH_3$, to a solution of ethylmagnesium bromide, and, by use of a color test, found that the Grignard reagent had been consumed. He worked up the reaction mixture in the usual way with dilute acid. Having learned a little (but not much) from his earlier sad experiences, he tested the product with iodine and sodium hydroxide; when a copious precipitate of iodoform appeared, he concluded that he had simply recovered his starting material.

He threw his product into the waste crock, carefully and methodically destroyed his glassware, burned his laboratory coat, left school, and went into politics, where he did quite well; his career in Washington was marred only, in the opinion of some, by his blind antagonism toward all appropriations for scientific research and his frequent attacks—alternately vitriolic and caustic—on the French.

What had he thrown into the waste crock? How had it been formed?

16. Treatment of ethyl acetoacetate with acetaldehyde in the presence of the base piperidine was found to give a product of formula $C_{14}H_{22}O_6$. Controversy arose about its structure: did it have open-chain structure III or cyclic structure IV, each formed by combinations of aldol and Michael condensations?

(a) Show just how each possible product could have been formed.
(b) Then the NMR spectrum of the compound was found to be the following:

> *a* complex, δ 0.95–1.10, 3H
> *b* singlet, δ 1.28, 3H
> *c* triplet, centered at δ 1.28, 3H
> *d* triplet, centered at δ 1.32, 3H
> *e* singlet, δ 2.5, 2H
> *f* broad singlet, δ 3.5, 1H
> *g* complex, δ 2–4, total of 3H
> *h* quartet, δ 4.25, 2H
> *i* quartet, δ 4.30, 2H

Which structure is the correct one? Assign all peaks in the spectrum. Describe the spectrum you would expect from the other possibility.

17. Give the likely structures for GGG and HHH.

1,3-butadiene + propiolic acid ($HC\equiv CCOOH$) $\longrightarrow$ DDD ($C_7H_8O_2$)
DDD + 1 mol $LiAlH_4$ $\longrightarrow$ EEE ($C_7H_{10}O$)
EEE + methyl chlorocarbonate (CH_3OCOCl) $\longrightarrow$ FFF ($C_9H_{12}O_3$)
FFF + heat (short time) $\longrightarrow$ toluene + GGG (C_7H_8)
GGG + tetracyanoethylene $\longrightarrow$ HHH ($C_{13}H_8N_4$)

Compound GGG is not toluene or 1,3,5-cycloheptatriene; on standing at room temperature it is converted fairly rapidly into toluene. Compound GGG gives the following spectral data. Ultraviolet: λ_{max} 303 nm, ϵ_{max} 4400. Infrared: strong bands at 3020, 2900, 1595, 1400, 864, 692, and 645 cm^{-1}; medium bands at 2850, 1152, and 790 cm^{-1}.

18. Give structures of compounds III through KKK, and account for their formation:

cyclopentanone + pyrrolidine, then acid $\longrightarrow$ III ($C_9H_{15}N$)
III + $CH_2=CHCOOCH_3$ $\longrightarrow$ JJJ ($C_{13}H_{21}O_2N$)
JJJ + H_2O, H^+, heat $\longrightarrow$ KKK ($C_9H_{14}O_3$)

19. Irradiation by ultraviolet light of 2,2,4,4-tetramethyl-1,3-cyclobutanedione (V) produces tetramethylethylene and two moles of carbon monoxide. When the irradiation is carried out in furan (VI), there is obtained a product believed to have the structure VII.

V   VI   VII

(a) Chief support for structure VII comes from elemental analysis, mol. wt. determination, and NMR data:

  a singlet, δ 0.85, 6H
  b singlet, δ 1.25, 6H
  c singlet, δ 4.32, 2H
  d singlet, δ 6.32, 2H

Show how the NMR data support the proposed structure. Why should there be two singlets of 6H each instead of one peak of 12H?

(b) It is proposed that, in the formation of tetramethylethylene, one mole of carbon monoxide is lost at a time. Draw electronic structures to show all steps in such a two-stage mechanism. How does the formation of VII support such a mechanism?

20. In the reaction of benzaldehyde with semicarbazide to form the semicarbazone (Sec. 21.12) the anilinium ion is a *much* more effective catalyst than a carboxylic acid of the same acidity. How might you account for this?

21. β-Lactones cannot be made from β-hydroxy acids. The β-lactone VIII was obtained, however, by treatment of sodium maleate (or sodium fumarate) with bromine water.

VIII

This experiment, reported in 1937 by P. D. Bartlett and D. S. Tarbell (of Harvard University), was an important step in the establishment of the mechanism of addition of halogens to carbon–carbon double bonds. Why is this so? How do you account for the formation of the β-lactone?

22. When the sodium salt of diazocyclopentadiene-2-carboxylic acid (IX) is heated above 140 °C, N_2 and CO_2 are evolved. If IX is heated in solution with tetracyclone (X), CO

IX   X   XI

is evolved as well, and 4,5,6,7-tetraphenylindene (XI) is obtained. Show all steps in a likely mechanism for the formation of XI. (*Hint*: See Problem 10(e) above.) Of what special theoretical interest are these findings?

23. When ethyl methylmalonate, acetone, and α-chloroacrylonitrile (CH_2=CClCN) are allowed to react in the presence of base, there is obtained the epoxy compound XII. Show all steps in a likely mechanism for the formation of XII.

$$(EtOOC)_2\overset{\overset{\displaystyle CH_3}{\displaystyle |}}{C}-CH_2 \underset{\underset{\displaystyle NC}{}}{\overset{\overset{\displaystyle O}{}}{C}}-\underset{\underset{\displaystyle CH_3}{}}{\overset{\overset{\displaystyle CH_3}{}}{C}}$$

XII

32

Rearrangements

Neighboring carbon. Nonclassical ions

32.1 Rearrangements: intramolecular nucleophilic attack

The driving force for all reactions of carbocations, we have seen, is the need to provide electrons to the electron-deficient carbon. One way in which this need can be satisfied is through rearrangement (Sec. 5.23), in which the electrons are carried to where they are needed by a migrating organic group or hydrogen.

Now, other atoms besides carbon can be electron-deficient—in particular, nitrogen and oxygen—and they, too, can get electrons through rearrangement. The most important class of molecular rearrangements is that involving *1,2-shifts to electron-deficient atoms*. It is the kind of rearrangement that we shall deal with in this chapter.

In Sec. 20.3, we discussed rearrangement as *intramolecular nucleophilic attack*. A leaving group departs from the migration terminus, taking the bonding electrons with it. The migrating group, acting as a nucleophile, takes its place. Depending upon the timing of bond-breaking and bond-making, this nucleophilic attack can be either S_N1-like or S_N2-like. When it is S_N2-like, that is, when the migrating group helps to push out the leaving group, we say that it gives *anchimeric assistance*.

$$S\text{—}T \xrightarrow{\quad} :W + \overset{G}{S\text{—}T} \xrightarrow{\quad} \overset{G}{S\text{—}T} \qquad \text{S}_N1\text{-like}\atop\text{migration}$$

$$\overset{G}{S\text{—}T} \xrightarrow{\quad} \left[\overset{G}{S\text{—}T} \atop W \right] \xrightarrow{\quad} \overset{G}{S\text{—}T} + :W \qquad \text{S}_N2\text{-like}\atop\text{migration}$$

G = migrating group
S = migration source
T = migration terminus

So far, we have discussed migration only to electron-deficient carbon. Now let us see what happens in migrations to other electron-deficient atoms.

32.2 Hofmann rearrangement. Migration to electron-deficient nitrogen

Let us begin with a reaction that we encountered earlier (Sec. 26.12) as a method of synthesis of amines: the Hofmann degradation of amides. Whatever

$$R\text{—}C\!\!\overset{\displaystyle O}{\underset{\displaystyle NH_2}{\Big\langle}} \xrightarrow{\text{OBr}^-} R\text{—}NH_2 + CO_3^{2-}$$

An amide   A 1° amine

the mechanism of the reaction, it is clear that rearrangement occurs, since the group joined to carbonyl carbon in the amide is found joined to nitrogen in the product.

The reaction is believed to proceed by the following steps:

(1) $\quad R\text{—}C\!\!\overset{\displaystyle O}{\underset{\displaystyle \ddot{N}H_2}{\Big\langle}} + \text{OBr}^- \longrightarrow R\text{—}C\!\!\overset{\displaystyle O}{\underset{\displaystyle \underset{\displaystyle H}{\overset{\displaystyle |}{\ddot{N}\text{—}Br}}}{\Big\langle}} + \text{OH}^-$

(2) $\quad R\text{—}C\!\!\overset{\displaystyle O}{\underset{\displaystyle \underset{\displaystyle H}{\overset{\displaystyle |}{\ddot{N}\text{—}Br}}}{\Big\langle}} + \text{OH}^- \longrightarrow R\text{—}C\!\!\overset{\displaystyle O}{\underset{\displaystyle \overset{\displaystyle \ddot{N}\text{—}Br}{\ominus}}{\Big\langle}} + H_2O$

(3) $\quad R\text{—}C\!\!\overset{\displaystyle O}{\underset{\displaystyle \overset{\displaystyle \ddot{N}\text{—}Br}{\ominus}}{\Big\langle}} \longrightarrow R\text{—}C\!\!\overset{\displaystyle O}{\underset{\displaystyle \ddot{N}}{\Big\langle}} + \text{Br}^-$

(4) $\quad R\!\!\overset{\displaystyle \frown}{\text{—}C}\!\!\overset{\displaystyle O}{\underset{\displaystyle \ddot{N}}{\Big\langle}} \longrightarrow R\text{—}\ddot{N}{=}C{=}O$

Simultaneous

(5) $\qquad R-\overset{..}{N}=C=O + 2OH^- \xrightarrow{\text{H}_2\text{O}} R-\overset{..}{N}H_2 + CO_3^{2-}$

Step (1) is the halogenation of an amide. This is a known reaction, an *N*-haloamide being isolated if no base is present. Furthermore, if the *N*-haloamide isolated in this way is then treated with base, it is converted into the amine.

Step (2) is the abstraction of a hydrogen ion by hydroxide ion. This is reasonable behavior for hydroxide ion, especially since the presence of the electron-withdrawing bromine increases the acidity of the amide. Unstable salts have actually been isolated in certain of these reactions.

Step (3) involves the separation of a halide ion, which leaves behind an electron-deficient nitrogen atom.

In Step (4) the actual rearrangement occurs. Steps (3) and (4) are generally

(3,4) $\qquad\qquad$ $\qquad \longrightarrow \quad R-N=C=O + Br^-$

believed to occur simultaneously, the attachment of R to nitrogen helping to push out halide ion. That is, migration is S_N2-like, and provides anchimeric assistance.

Step (5) is the hydrolysis of an isocyanate ($R-N=C=O$) to form an amine and carbonate ion. This is a known reaction of isocyanates. If the Hofmann degradation is carried out in the absence of water, an isocyanate can actually be isolated.

Like the rearrangement of carbocations that we have already encountered (Sec. 5.23), the Hofmann rearrangement involves a 1,2-shift. In the rearrangement of carbocations a group migrates with its electrons to an electron-deficient carbon; in the present reaction the group migrates with its electrons to an electron-deficient *nitrogen*. We consider nitrogen to be electron-deficient even though it probably loses electrons—to bromide ion—*while* migration takes place, rather than before.

The strongest support for the mechanism just outlined is the fact that many of the proposed intermediates have been isolated, and that these intermediates have been shown to yield the products of the Hofmann degradation. The mechanism is also supported by the fact that analogous mechanisms account satisfactorily for observations made on a large number of related rearrangements. Furthermore, the actual rearrangement step fits the broad pattern of 1,2-shifts to electron-deficient atoms.

In addition to evidence indicating what the various steps in the Hofmann degradation are, there is also evidence that gives us a rather intimate view of just how the rearrangement step takes place. In following sections we shall see what some of that evidence is. We shall be interested in this not just for what it tells us about the Hofmann degradation, but because it will give us an idea of the kind of thing that can be done in studying rearrangements of many kinds.

Problem 32.1   Reaction of acid chlorides with sodium azide, NaN_3, yields *acyl azides*, $RCON_3$. When heated, these undergo the *Curtius rearrangement* to amines, RNH_2, or, in a non-hydroxylic solvent, to isocyanates, $RNCO$. Using the structure

$$
R-\overset{\displaystyle O}{\overset{\displaystyle \|}{C}}-\overset{\ominus}{N}-\overset{\oplus}{N}\equiv N
$$

for the azide, suggest a mechanism for the rearrangement. (*Hint*: Write balanced equations.)

32.3   Hofmann rearrangement. Intramolecular or intermolecular?

One of the first questions asked in the study of a rearrangement is this: Is the rearrangement *intra*molecular or *inter*molecular? That is, does the migrating group move from one atom to another atom within the same molecule, or does it move from one molecule to another?

In the mechanism outlined above, the Hofmann rearrangement is shown as intramolecular. How do we know that this is so? To answer this question, T. J. Prosser and E. L. Eliel (University of Notre Dame) carried out degradation of a mixture of *m*-deuteriobenzamide and benzamide-^{15}N. When they analyzed the product with the mass spectrometer, they found only *m*-deuterioaniline and aniline-^{15}N. There was *none* of the mixture of cross-products that would have been formed if a phenyl group from one molecule had become attached to the nitrogen of another. The results of this elegant double labeling experiment thus show beyond doubt that the Hofmann rearrangement is intramolecular.

32.4   Hofmann rearrangement. Stereochemistry at the migrating group

When optically active α-phenylpropionamide undergoes the Hofmann degradation, α-phenylethylamine of the same configuration and of essentially the same optical purity is obtained:

(+)-α-Phenylpropionamide   (−)-α-Phenylethylamine

Retention of configuration

Rearrangement proceeds *with complete retention of configuration* about the chiral center of the migrating group.

These results tell us two things. First, nitrogen takes the same relative position on the chiral carbon that was originally occupied by the carbonyl carbon. Second, the chiral carbon does not break away from the carbonyl carbon until it has started to attach itself to nitrogen. If the group were actually to become free during its migration, we would expect considerable loss of configuration and hence a partially racemic product. (If the group were to become free—*really* free—we would expect reaction to be, in part, intermolecular, also contrary to fact.)

We may picture the migrating group as moving from carbon to nitrogen via a transition state, I, in which carbon is pentavalent:

The migrating group *steps* from atom to atom; it does *not* jump.

There is much evidence to suggest that the stereochemistry of all 1,2-shifts has this common feature: *complete retention of configuration in the migrating group*.

Problem 32.2   Many years before the Hofmann degradation of optically active α-phenylpropionamide was studied, the following observations were made: when the cyclopentane derivative II, in which the —COOH and —CONH₂ groups are *cis* to each other, was treated with hypobromite, compound III was obtained; compound III could be converted by heat into the amide IV (called a *lactam*). What do these results show about the mechanism of the arrangement? (*Use models.*)

32.5   Hofmann rearrangement. Timing of the steps

We said that steps (3) and (4) of the mechanism are believed to be simultaneous, that is, that loss of bromide ion and migration occur in the same step:

One reason for believing this is simply the anticipated difficulty of forming a highly unstable intermediate in which an electronegative element like nitrogen has only a sextet of electrons. Such a particle should be even less stable than primary carbocations, and those, we know, are seldom formed; reaction takes the easier,

S_N2-like path. Another reason is the effect of structure on rate of reaction. Let us examine this second reason.

When the migrating group is aryl, the rate of the Hofmann degradation is increased by the presence of electron-releasing substituents in the aromatic ring; thus substituted benzamides show the following order of reactivity:

$$G: \ -OCH_3 > -CH_3 > -H > -Cl > -NO_2$$

Now, how could electron release speed up Hofmann degradation? One way could be through its effect on the rate of migration. Migration of an alkyl group must involve a transition state containing pentavalent carbon, like I in the preceding section. Migration of an aryl group, on the other hand, takes place via a structure like V. This structure is a familiar one; from the standpoint of the migrating aryl group, rearrangement is simply electrophilic aromatic substitution, with the electron-deficient atom—nitrogen, in this case—acting as the attacking reagent. In at least some rearrangements, as we shall see, there is evidence that

structures like V are actual intermediate compounds, as in the ordinary kind of electrophilic aromatic substitution (Sec. 14.14). Electron-releasing groups disperse the developing charge on the aromatic ring and thus speed up formation of V. Viewed in this way, substituents affect the rate of rearrangement—the *migratory aptitude*—of an aryl group in exactly the same way as they affect the rate of aromatic nitration, halogenation, or sulfonation. (In some cases, however, conformational effects can completely outweigh these electronic effects.)

There is another way in which electron release might be speeding up reaction: by speeding up formation of the electron-deficient species in equation (3). But the observed effect is a strong one, and more consistent with the development of the positive charge *in the ring itself*, as during rearrangement.

We should be clear about what the question is here. It is *not* whether some groups migrate faster than others—there is little doubt about that—but whether the rate of rearrangement affects the overall rate—the *measured* rate—of the Hofmann degradation.

It is likely, then, that electron-releasing substituents speed up Hofmann degradation by speeding up rearrangement. Now, under what conditions can this happen? Consider the sequence (3) and (4). Loss of bromide ion (3) could be fast and reversible, followed by slow rearrangement (4). Rearrangement would be rate-determining, as required, but in that case something else would not fit. The

(3)

$$R-C \overset{O}{\underset{\underset{\ominus}{N-Br}}{}} \longrightarrow R-C \overset{O}{\underset{N}{}} + Br^-$$

Simultaneous

(4)

$$R-C \overset{O}{\underset{N}{}} \longrightarrow R-\ddot{N}=C=O$$

reverse of (3) is combination of the particle ArCON with bromide ion; if this were taking place, so should combination of ArCON with the solvent, water—more abundant and more nucleophilic—to form the *hydroxamic acid* ArCONHOH. But hydroxamic acids are *not* formed in the Hofmann degradation.

If ArCON were indeed an intermediate, then, it would have to be undergoing rearrangement as fast as it is formed; that is, (4) would have to be fast compared with (3). But in that case, the overall rate would be independent of the rate of rearrangement, contrary to fact.

We are left with the concerted mechanism (3,4). Attachment of the migrating group helps to push out bromide ion, and overall rate *does* depend on the rate of rearrangement. As the amount of anchimeric assistance varies, so does the observed rate of reaction.

At the migrating group, we said, rearrangement amounts to electrophilic substitution. But at the electron-deficient nitrogen, rearrangement amounts to *nucleophilic* substitution: the migrating group (with its electrons) is a nucleophile, and bromide ion is the leaving group. The sequence (3) and (4) corresponds to an S_N1 mechanism; the concerted reaction (3,4) corresponds to a S_N2 mechanism. Dependence of overall rate on the nature of the nucleophile is consistent with the S_N2-like mechanism, but not with the S_N1-like mechanism.

32.6   Rearrangement of hydroperoxides. Migration to electron-deficient oxygen

In Sec. 28.4 we encountered the synthesis of phenol via cumene hydroperoxide.

$$C_6H_5CH(CH_3)_2 \xrightarrow{O_2} C_6H_5-\overset{CH_3}{\underset{CH_3}{C}}-O-OH \xrightarrow{H_2O, H^+} C_6H_5OH + CH_3COCH_3$$

Isopropylbenzene
(Cumene)

Cumene hydroperoxide

Phenol   Acetone

The phenyl group is joined to carbon in the hydroperoxide and to oxygen in phenol: clearly rearrangement takes place. This time, it involves a 1,2-shift to electron-deficient *oxygen*. Let us see how it is believed to take place.

(1)

$$CH_3-\overset{C_6H_5}{\underset{CH_3}{C}}-O-OH + H^+ \rightleftarrows CH_3-\overset{C_6H_5}{\underset{CH_3}{C}}-O-\overset{+}{O}H_2$$

Cumene hydroperoxide
I

(2) $\quad CH_3-\overset{\displaystyle C_6H_5}{\underset{\displaystyle CH_3}{C}}-O-\overset{+}{O}H_2 \longrightarrow CH_3-\overset{\displaystyle C_6H_5}{\underset{\displaystyle CH_3}{C}}-O^+ + H_2O$

(3) $\quad CH_3-\overset{\displaystyle C_6H_5}{\underset{\displaystyle CH_3}{C}}-O^+ \longrightarrow CH_3-\overset{+}{\underset{\displaystyle CH_3}{C}}=O-C_6H_5$

$\qquad\qquad\qquad\qquad\qquad\qquad\qquad\qquad\qquad\qquad\qquad$ II

$\left.\begin{array}{l} \\ \\ \\ \\ \\ \\ \end{array}\right\}$ *Simultaneous*

(4) $\quad CH_3-\underset{\displaystyle CH_3}{\overset{+}{C}}=O-C_6H_5 + H_2O \longrightarrow CH_3-\overset{\displaystyle ^+OH_2}{\underset{\displaystyle CH_3}{C}}-O-C_6H_5$

$\qquad\qquad$ II

$\qquad\qquad\qquad\qquad\qquad\qquad\qquad\qquad\qquad\qquad \updownarrow$

$\qquad\qquad\qquad\qquad\qquad\qquad\qquad CH_3-\overset{\displaystyle OH}{\underset{\displaystyle CH_3}{C}}-O-C_6H_5 + H^+$

$\qquad\qquad\qquad\qquad\qquad\qquad\qquad\qquad\qquad\qquad$ III

(5) $\quad CH_3-\overset{\displaystyle OH}{\underset{\displaystyle CH_3}{C}}-O-C_6H_5 \xrightarrow{H^+} CH_3-\underset{\displaystyle CH_3}{C}\overset{\displaystyle O}{\diagup} + HO-C_6H_5$

$\qquad\qquad$ III $\qquad\qquad\qquad\qquad\qquad$ Acetone $\qquad$ Phenol

    Acid converts (step 1) the peroxide I into the protonated peroxide, which loses (step 2) a molecule of water to form an intermediate in which oxygen bears only six electrons. A 1,2-shift of the phenyl group from carbon to electron-deficient oxygen yields (step 3) the "carbocation" II, which reacts with water to yield (step 4) the hydroxy compound III. Compound III is a hemiacetal (Sec. 21.13) which breaks down (step 5) to give phenol and acetone.

    Every step of the reaction involves chemistry with which we are already quite familiar: protonation of a hydroxy compound with subsequent dissociation to leave an electron-deficient particle; a 1,2-shift to an electron-deficient atom; reaction of a carbocation with water to yield a hydroxy compound; decomposition of a hemiacetal. In studying organic chemistry we encounter many new things; but much of what seems new is found to fit into old familiar patterns of behavior.

    It is very probable that steps (2) and (3) are simultaneous, the migrating phenyl group helping to push out (2,3) the molecule of water; that is to say, water is lost

(2,3) $\quad CH_3-\overset{\displaystyle \overset{\frown}{Ph}}{\underset{\displaystyle CH_3}{C}}-O-OH_2^+ \longrightarrow CH_3-\underset{\displaystyle CH_3}{\overset{+}{C}}=O-Ph + H_2O$

with anchimeric assistance. This concerted mechanism is supported by the same line of reasoning that we applied to the Hofmann rearrangement. (a) A highly unstable intermediate containing oxygen with only a sextet of electrons should be very difficult to form. (b) There is evidence that, if there *is* such an intermediate, it must undergo rearrangement as fast as it is formed; that is, if (2) and (3) are separate steps, (3) must be fast compared with (2). (c) The rate of overall reaction is speeded up by electron-releasing substituents in migrating aryl groups, and in a way that resembles, *quantitatively*, the effect of these groups on ordinary electrophilic aromatic substitution. Almost certainly, then, substituents affect the overall rate of reaction by affecting the rate of migration, and hence migration must take place in the rate-determining step. This rules out the possibility of a fast (3), and leaves us with the concerted reaction (2,3).

> **Problem 32.3**   When α-phenylethyl hydroperoxide, $C_6H_5CH(CH_3)O—OH$, undergoes acid-catalyzed rearrangement in $H_2{}^{18}O$, recovered unrearranged hydroperoxide is found to contain *no* oxygen-18. Taken with the other evidence, what does this finding tell us about the mechanism of reaction?

32.7   Rearrangement of hydroperoxides. Migratory aptitude

The rearrangement of hydroperoxides lets us see something that the Hofmann rearrangement could not: the preferential migration of one group rather than another. That is, we can observe the relative speeds of migration—the relative migratory aptitudes—of two groups, not as a difference in rate of reaction, but as a difference in the product obtained. In cumene hydroperoxide, for example, any one of three groups could migrate: phenyl and two methyls. If, instead of phenyl,

Acetophenone   Methanol
Not obtained

methyl were to migrate, reaction would be expected to yield methanol and acetophenone. Actually, phenol and acetone are formed quantitatively, showing that a phenyl group migrates much faster than a methyl.

It is generally true in 1,2-shifts that aryl groups have greater migratory aptitudes than alkyl groups. We can see why this should be so. Migration of an alkyl group must involve a transition state containing pentavalent carbon (IV). Migration of an aryl group, on the other hand, takes place via a structure of the

$$\left[\ \underset{S—T}{\overset{C}{\diagdown}}\ \right]^{+}$$

IV

Alkyl migration:
pentavalent carbon

$$\underset{S—T}{\left(+\right)}$$

V

Aryl migration:
benzenonium ion

benzenonium ion type (V); transition state or actual intermediate, V clearly offers an easier path for migration than does IV.

The hydroperoxide may contain several aryl groups and, if they are different, we can observe competition in migration between them, too. As was observed in

VI   VII

the Hofmann rearrangement, the relative migratory aptitude of an aryl group is raised by electron-releasing substituents, and lowered by electron-withdrawing substituents. For example, when *p*-nitrotriphenylmethyl hydroperoxide (VI) is treated with acid, it yields exclusively phenol and *p*-nitrobenzophenone (VII); as we would have expected, phenyl migrates in preference to *p*-nitrophenyl.

Problem 32.4 When *p*-methylbenzyl hydroperoxide, *p*-$CH_3C_6H_4CH_2O$—OH, is treated with acid, there are obtained *p*-methylbenzaldehyde (61%) and *p*-cresol (38%). (a) How do you account for the formation of each of these? What other products must have been formed? (b) What do the relative yields of the aromatic products show?

Problem 32.5 The treatment of aliphatic hydroperoxides, RCH_2O—OH and R_2CHO—OH, with aqueous acid yields aldehydes and ketones as the only organic products. What conclusion do you draw about migratory aptitudes?

32.8 Pinacol rearrangement. Migration to electron-deficient carbon

Upon treatment with mineral acids, 2,3-dimethyl-2,3-butanediol (often called *pinacol*) is converted into *tert*-butyl methyl ketone (often called *pinacolone*). The

Pinacol
2,3-Dimethyl-2,3-butanediol

Pinacolone
tert-Butyl methyl ketone
3,3-Dimethyl-2-butanone

1,2-diol undergoes dehydration, and in such a way that rearrangement of the carbon skeleton occurs. Other 1,2-diols undergo analogous reactions, which are known collectively as **pinacol rearrangements**.

The pinacol rearrangement is believed to involve two important steps: (1) loss of water from the protonated diol to form a carbocation; and (2) rearrangement of the carbocation by a 1,2-shift to yield the protonated ketone.

$$(2) \quad R-\overset{\overset{\displaystyle R}{|}}{\underset{\underset{\displaystyle OH}{|}}{C}}-\overset{\overset{\displaystyle R}{|}}{\underset{\oplus}{C}}-R \quad\longrightarrow\quad R-\overset{}{\underset{\underset{\displaystyle \oplus OH}{||}}{C}}-\overset{\overset{\displaystyle R}{|}}{\underset{\underset{\displaystyle R}{|}}{C}}-R \quad\rightleftharpoons\quad H^+ + R-\overset{}{\underset{\underset{\displaystyle O}{||}}{C}}-\overset{\overset{\displaystyle R}{|}}{\underset{\underset{\displaystyle R}{|}}{C}}-R$$

Both steps in this reaction are already familiar to us: formation of a carbocation from an alcohol under the influence of acid, followed by a 1,2-shift to the electron-deficient atom. The pattern is also familiar: rearrangement of a cation to a more stable cation, in this case to the protonated ketone. The driving force is the usual one behind carbocation reactions: the need to provide the electron-deficient carbon with electrons. The special feature of the pinacol rearrangement is the presence in the molecule of the second oxygen atom; it is this oxygen atom, with its unshared pairs, that ultimately provides the needed electrons.

Problem 32.6   Account for the products of the following reactions:

(a) 1-iodo-2-phenyl-2-propanol + Ag^+ $\longrightarrow$ benzyl methyl ketone
(b) 2-amino-1,1,2-triphenyl-1-propanol + HONO $\longrightarrow$ 1,1,2-triphenyl-1-propanone
    (*Hint*: See Problem 27.12, p. 972.)

When the groups attached to the carbon atoms bearing —OH differ from one another, the pinacol rearrangement can conceivably give rise to more than one compound. The product actually obtained is determined (a) by which —OH group is lost in step (1), and then (b) by which group migrates in step (2) to the electron-deficient carbon thus formed. For example, let us consider the rearrangement of 1-phenyl-1,2-propanediol. The structure of the product actually obtained, benzyl methyl ketone, indicates that the benzylic cation (I) is formed in preference to the secondary cation (II), and that —H migrates in preference to —CH$_3$.

H migrates →
$$C_6H_5-\overset{\overset{\displaystyle H}{|}}{\underset{\underset{\displaystyle H}{|}}{C}}-\overset{}{\underset{\underset{\displaystyle O}{||}}{C}}-CH_3$$
Benzyl methyl ketone
Actual product

$$\xrightarrow{-H_2O} \quad C_6H_5-\overset{\overset{\displaystyle H}{|}}{\underset{\oplus}{C}}-\overset{\overset{\displaystyle H}{|}}{\underset{\underset{\displaystyle OH}{|}}{C}}-CH_3$$
I

if CH$_3$ migrated →

if C$_6$H$_5$ migrated →
$$C_6H_5-\overset{\overset{\displaystyle H}{|}}{\underset{\underset{\displaystyle CH_3}{|}}{C}}-\overset{}{\underset{\underset{\displaystyle O}{||}}{C}}-H$$
α-Phenylpropionaldehyde

$$C_6H_5-\overset{\overset{\displaystyle H}{|}}{\underset{\underset{\displaystyle OH}{|}}{C}}-\overset{\overset{\displaystyle H}{|}}{\underset{\underset{\displaystyle OH}{|}}{C}}-CH_3 \xrightarrow{H^+}$$
1-Phenyl-1,2-propanediol

$$\xrightarrow{-H_2O} \quad C_6H_5-\overset{\overset{\displaystyle H}{|}}{\underset{\underset{\displaystyle OH}{|}}{C}}-\overset{\overset{\displaystyle H}{|}}{\underset{\oplus}{C}}-CH_3$$
II

if H migrated →
$$C_6H_5-\overset{}{\underset{\underset{\displaystyle O}{||}}{C}}-\overset{\overset{\displaystyle H}{|}}{\underset{\underset{\displaystyle H}{|}}{C}}-CH_3$$
Ethyl phenyl ketone

Study of a large number of pinacol rearrangements has shown that usually the product is the one expected if, first, ionization occurs to yield the more stable carbocation, and then, once the preferred ionization has taken place, migration takes place according to the sequence —Ar > —R. (We have already seen how it is that an aryl group migrates faster than an alkyl.) Hydrogen can migrate, too, but we cannot predict its relative migratory aptitude. Hydrogen may migrate in preference to —R or —Ar, but this is not always the case; indeed, it sometimes happens that with a given pinacol either —H or —R can migrate, depending upon experimental conditions.

Among aryl groups, relative migratory aptitude depends—other things being equal—on the ability of the ring to accommodate a positive charge. Although we cannot go into the matter here, we should be aware that strong stereochemical factors can operate in this reaction, and may outweigh these electronic factors.

Problem 32.7 For the rearrangement of each of the following 1,2-diols show which carbocation you would expect to be the more stable, and then the rearrangement that this carbocation would most likely undergo:

(a) 1,2-propanediol
(b) 2-methyl-1,2-propanediol
(c) 1-phenyl-1,2-ethanediol
(d) 1,1-diphenyl-1,2-ethanediol
(e) 1-phenyl-1,2-propanediol

(f) 1,1-diphenyl-2,2-dimethyl-1,2-ethanediol
(g) 1,1,2-triphenyl-2-methyl-1,2-ethanediol
(h) 3-ethyl-2-methyl-2,3-pentanediol
(i) 1,1-bis(p-methoxyphenyl)-2,2-diphenyl-1,2-ethanediol

We have depicted the pinacol rearrangement as a two-step process with an actual carbocation as intermediate. There is good evidence that this is so, at least when a tertiary or benzylic cation can be formed. Evidently the stability of the incipient cation in the transition state permits (S_N1-like) loss of water without anchimeric assistance from the migrating group. This is, we note, in contrast to what happens in migration to electron-deficient nitrogen or oxygen.

Problem 32.8 The following reactions have all been found to yield a mixture of pinacol and pinacolone, and *in the same proportions*: treatment of 3-amino-2,3-dimethyl-2-butanol with nitrous acid; treatment of 3-chloro-2,3-dimethyl-2-butanol with aqueous silver ion; and acid-catalyzed hydrolysis of the epoxide of 2,3-dimethyl-2-butene. What does this finding indicate about the mechanism of the pinacol rearrangement? (*Useful information*: Primary aliphatic amines, RNH_2, react with aqueous nitrous acid, HONO, to give mixtures of alcohols and alkenes, often with rearrangement.)

Problem 32.9 When pinacol was treated with acid in $H_2^{18}O$ solution, recovered unrearranged pinacol was found to contain oxygen-18. Studies showed that oxygen exchange took place two to three times as fast as rearrangement. What bearing does this fact have on the mechanism of rearrangement?

32.9 Neighboring group effects. Neighboring aryl

In Secs. 20.2–20.4, we studied neighboring group effects: intramolecular effects exerted on a reaction through direct participation by a group near the reaction center. We saw that these effects involved the same basic chemistry as rearrangements and that, in many cases, there *is* rearrangement—*hidden* rearrangement. What we see on the surface may be this:

$$
\begin{array}{c}
\overset{G}{\underset{W}{-C-C-}} \;+\; :Z \;\longrightarrow\; \overset{G}{\underset{Z}{-C-C-}} \;+\; :W
\end{array}
$$

But what is actually happening may be this:

$$
\underset{\text{I}}{\overset{G}{-C-C-}} \longrightarrow\; :W \;+\; \overset{G^{\oplus}}{-C-C-} \;\xrightarrow{:Z}\; \overset{G}{\underset{Z}{-C-C-}} \;+\; \overset{G}{-C-\;C-}
$$

This process, we saw, gives rise to symphoric effects: a special kind of stereochemistry or an unusual fast rate of reaction—or, very often, both.

In that earlier discussion, the neighboring groups that we were concerned with were those containing unshared pairs on atoms like sulfur, nitrogen, oxygen, or bromine. Now, we have just seen that aryl groups have a high migratory aptitude, and we have attributed this to the ability of the benzene ring to provide electrons. Is it possible that, using the π electrons of the ring instead of an unshared pair, an aryl group can exert neighboring group effects?

In 1949, at the University of California at Los Angeles, Donald J. Cram published the first of a series of papers on the effects of neighboring aryl groups, and set off a controversy that lasted twenty years before it was resolved. Let us look at just one example of the kind of thing he discovered.

Solvolysis of 3-phenyl-2-butyl tosylate in acetic acid yields the acetate. The tosylate contains two chiral centers, and exists as two racemic modifications; so,

$$
\underset{\substack{OTs}}{\overset{\substack{C_6H_5}}{CH_3-CH-CH-CH_3}} \xrightarrow[\text{KOAc}]{\text{CH}_3\text{COOH}} \underset{\substack{OAc}}{\overset{\substack{C_6H_5}}{CH_3-CH-CH-CH_3}}
$$

3-Phenyl-2-butyl tosylate	3-Phenyl-2-butyl acetate

too, does the acetate. Solvolysis is completely stereospecific and proceeds, it at first appears, with *retention* of configuration: racemic *erythro* tosylate gives only racemic *erythro* acetate,

Erythro
Racemic

Erythro
Racemic

and racemic *threo* tosylate gives only racemic *threo* acetate.

Threo		**Threo**
Racemic		*Racemic*

When, however, optically active *threo* tosylate is used, it is found to yield optically *inactive* product, racemic *threo* acetate. We see here the same pattern as in Sec. 20.2: retention at both carbons in half the molecules of the product, inversion at both carbons in the other half.

(+)-Threo	**Threo**
Optically active	*Racemic*

Cram interpreted these results in the following way. The neighboring phenyl group, with its π electrons, helps to push out (1) the tosylate anion. There is formed

(1)

A benzenonium ion

(2)

an intermediate *bridged* ion. This ion is of the kind we have encountered (Sec. 14.8) as the intermediate in electrophilic aromatic substitution: a *benzenonium ion*. The ring is bonded to both carbons by full-fledged σ bonds, to give a symmetrical structure. The ion owes its stability to the fact that the positive charge is distributed about the ring, being strongest at the positions *ortho* and *para* to the point of attachment.

In step (2) the benzenonium ion is attacked by the acetic acid at either of the two equivalent carbons to yield the product.

Problem 32.10   (a) Drawing structures of the kind in Fig. 20.3 (p. 736), show how Cram's mechanism accounts for the conversion of optically active *threo*-3-phenyl-2-butyl tosylate into racemic acetate. (b) In contrast, optically active *erythro* tosylate yields optically *active erythro* acetate. Show that this, too, fits Cram's interpretation of the reaction.

In the controversy that developed, the point under attack was not so much the existence of the intermediate bridged ion—although this was questioned, too—as its mode of formation. The 3-phenyl-2-butyl tosylates undergo solvolysis at much the same rate as does unsubstituted *sec*-butyl tosylate: formolysis a little faster, acetolysis a little slower. Yet, as depicted by Cram, phenyl gives anchimeric assistance to the reaction. Why, then, is there no rate acceleration?

Several alternatives were proposed: one, that participation by phenyl in expulsion of tosylate occurs, but is weak; another, that bridging occurs, not in the rate-determining step, but rapidly, following formation of an open cation. H. C. Brown (p. 648) suggested that—for unsubstituted phenyl, at least—the intermediate is not a bridged ion at all, but a pair of rapidly equilibrating open carbocations; phenyl, now on one carbon and now on the other, blocks back-side attack by the solvent and thus gives rise to the observed stereochemistry.

By 1971, a generally accepted picture of these reactions had begun to emerge, based on work by a number of investigators, prominent among them Paul Schleyer (University of Erlangen-Nürnberg). The big stumbling-block had been the widely held idea that secondary cations are formed, like tertiary cations, with little assistance from the solvent (Sec. 6.9). Using as standards certain special secondary substrates whose structure *prevents* solvent assistance, Schleyer showed that ordinary secondary substrates do indeed react with much solvent assistance.

Cram's original proposal seems to be essentially correct: aryl *can* give anchimeric assistance through formation of bridged ions. Competition is *not* between aryl-assisted solvolysis and unassisted solvolysis; competition is between aryl-assisted solvolysis and solvent-assisted solvolysis. Anchimeric assistance need not cause anchimeric acceleration. Formation of a bridged cation and an open cation may proceed at much the same rate, one with aryl assistance, the other with equally strong solvent assistance.

On the assumption of these two competing processes, successful quantitative correlations have been made among data of various kinds: rate of reaction, stereochemistry, scrambling of isotopic labels, and Hammett constants (Sec. 23.11) to represent the relative electronic effects of various substituents in aromatic rings. If neighboring aryl contains strongly electron-withdrawing substituents, reaction products are normal—chiefly alkenes plus inverted ester—and the rate of solvolysis is what one would expect for formation of an open cation slowed down by electron-withdrawing inductive effects. As substituents become increasingly electron-releasing (*p*-Cl, *m*-CH$_3$, *p*-CH$_3$, *p*-CH$_3$O) the rate increases *more* than expected if only inductive effects were operating; the amount of "extra" speed matches the amount of abnormal stereochemistry. Consider, for example, acetolysis of 3-aryl-2-butyl brosylates. One calculates from the rate data that *m*-tolyl assists in 73% of reaction; 68% of the product is found to have retained configuration. For *p*-methyl, calculated 87%, found 88%; for *p*-methoxyphenol, calculated 99%, found 100%.

How much anchimeric assistance there is, then, depends on how nucleophilic the neighboring group is. It also depends on how badly anchimeric assistance is *needed*. The more nucleophilic the solvent, the more assistance *it* gives, and the less the neighboring group participates. Or, if the open cation is a relatively stable one—tertiary or benzylic—it may need little assistance of any kind, either from the solvent or from the neighboring group.

In summary, an incipient cation can get electrons in three different ways: (a) from a substituent, through an inductive effect or resonance; (b) from the solvent; (c) from a neighboring group.

In all this, H. C. Brown played a role familiar to him: that of gad-fly—the organic chemist's conscience—forcing careful examination of ideas that had been accepted perhaps too readily because of their neatness. The turning point in this part of the great debate was marked by the joint publication of a paper by Brown and Schleyer setting forth essentially the interpretation we have just given.

In 1970, Olah (p. 194) prepared a molecule whose CMR spectrum was consistent with a bridged benzenonium ion, and *not* with a pair of equilibrating open cations.

CH₂CH₂Cl

β-Phenylethyl chloride   Bridged
                                 benzenonium ion

> **Problem 32.11** Quenching of Olah's solution with water gave a 3:1 mixture of β-phenylethyl alcohol and α-phenylethyl alcohol. The spectrum showed the presence not only of the bridged cation but, in lesser amounts, of an open cation. What is a likely structure for the open cation, and how is it formed?

In Secs. 20.2–20.4 we saw neighboring group effects in which electrons are provided by atoms with unshared pairs of electrons, atoms like sulfur or oxygen or halogen. We have just seen that carbon, too, can be involved in neighboring group effects; here, the electrons are provided by the π cloud on the aromatic ring. Now we are ready to follow this particular thread into a gray area, an area of continuing controversy, and see how neighboring group effects may even involve σ electrons of carbon and of hydrogen.

32.10   Neighboring group effects: neighboring saturated carbon. Nonclassical ions

The rearrangement of carbocations was first postulated, by Meerwein (p. 194) in 1922, to account for the conversion of camphene hydrochloride into isobornyl chloride. Oddly enough, this chemical landmark is the most poorly understood of all such rearrangements. With various modifications in structure, this bicyclic

Camphene hydrochloride   Isobornyl chloride

system has been for over 30 years the object of closer scrutiny than any other in organic chemistry.

We can see, in a general way, how this particular rearrangement could take place. Camphene hydrochloride loses chloride ion to form cation I, which rearranges by a 1,2-alkyl shift to form cation II. Using models, and keeping careful

Camphene hydrochloride   I   II

track of the various carbon atoms, we find that cation II need only combine with a chloride ion to yield isobornyl chloride.

II   II   Isobornyl chloride

We have accounted for the observed change in carbon skeleton, but we have not answered two questions that have plagued the organic chemist for more than a generation. Why is only the *exo* chloride, isobornyl chloride, obtained, and none of its *endo* isomer, bornyl chloride? Why does camphene hydrochloride undergo solvolysis thousands of times as fast as, say, *tert*-butyl chloride? To see the kind of answers that have been given, let us turn to a simpler but basically similar system.

In 1949, working at the same university where Cram was just then proposing aryl assistance, Saul Winstein (p. 244) reported these findings. On acetolysis, the diastereomeric *exo-* and *endo*-norbornyl brosylates both yield *exo*-norbornyl acetate:

and enantiomer   and enantiomer   OBs
                                                                     and enantiomer

exo-Norbornyl brosylate   *exo*-Norbornyl acetate   *endo*-Norbornyl brosylate

If the starting brosylate is optically active, the product is still the optically inactive racemic modification. For example:

exo-Norbornyl brosylate
Optically active

exo-Norbornyl acetate
Racemic

Finally, *exo*-norbornyl brosylate reacts 350 times as fast as the *endo* brosylate.

Winstein interpreted the behavior of these compounds in the following way (Fig. 32.1). Loss of brosylate anion yields (1) the bridged cation III, which undergoes nucleophilic attack by solvent (2) at either C–2 or C–1 to yield the product.

(1)

III   III

exo-Norbornyl brosylate
Optically active

(2)

III

exo-Norbornyl acetate
Racemic

Figure 32.1 Conversion of optically active *exo*-norbornyl brosylate into racemic *exo*-norbornyl acetate via nonclassical ion. Brosylate anion is lost with anchimeric assistance from C–6, to give the bridged cation III. Cation III undergoes back-side attack at either C–2 (path *a*) or C–1 (path *b*). Attacks *a* and *b* are equally likely, and give the racemic product.

Cation III is stabilized by resonance between two equivalent structures, IV and V, each corresponding to an open cation. The charge is divided between two carbons (C–1 and C–2) each of which—held in the proper position by the particular ring system—is bonded to C–6 by a half-bond. The bridging carbon (C–6) is thus pentavalent.

Reaction of the *exo* brosylate is S_N2-like, as shown in Fig. 32.1: back-side attack by C–6 on C–1 helps to push out brosylate, and yields the bridged ion in a single step. The geometry of the *endo* brosylate does not permit such back-side attack, and consequently it undergoes an S_N1-like reaction: slow formation of the open cation followed by rapid conversion into the bridged ion.

endo-Norbornyl brosylate   Open cation   Bridged cation

The two diastereomers yield the same product, racemic *exo* acetate, because they react via the same intermediate. But only the *exo* brosylate reacts with anchimeric assistance, and hence it reacts at the faster rate.

What Winstein was proposing was that *saturated* carbon using σ electrons could act as a neighboring group, to give anchimeric assistance to the expulsion of a leaving group, and to form an intermediate bridged cation containing pentavalent carbon. Bridged ions of this kind, with delocalized bonding σ electrons, have become known as *nonclassical ions*.

A nonclassical
bridged ion

Interpretation of the behavior of the norbornyl and many related systems on the basis of nonclassical ions seemed to be generally accepted until 1962, when H. C. Brown declared, "But the Emperor is naked!" Brown's point was not that the idea of nonclassical ions was necessarily wrong, but that it was *not necessarily right*. It had been accepted too readily, he thought, on the basis of too little evidence, and needed closer examination.

Brown suggested alternative interpretations. The norbornyl cation, for example, might not be a bridged ion but a pair of equilibrating open carbocations. That is to say, IV and V are not contributing structures to a resonance hybrid, but two distinct compounds in equilibrium with each other. Each ion can combine with solvent: IV at C–1, V at C–2. Substitution is exclusively *exo* because the *endo* face of each cation lies in a fold of the molecule, and is screened from attack. Differences in rate, too, are attributed to steric factors. It is not that the *exo* substrate reacts unusually fast, but that the *endo* substrate reacts unusually *slowly*, due to steric hindrance to the departure of the leaving group with its cluster of solvent molecules.

To test these alternative hypotheses, a tremendous amount of work has been done, by Brown and by others. For example, camphene hydrochloride is known to undergo ethanolysis 6000 times as fast as *tert*-butyl chloride, and this had been attributed to anchimeric assistance with formation of a bridged ion. Brown pointed out that the wrong standard for comparison had been chosen. He showed that a number of substituted (3°) cyclopentyl chlorides (examine the structure of camphene hydrochloride closely) *also* react much faster than *tert*-butyl chloride. He attributed these fast reactions—including that of camphene hydrochloride—to *relief of steric strain*. On ionization, chloride ion is lost and the methyl group on the sp^2-hybridized carbon moves into the plane of the ring: four non-bonded interactions thus disappear, two for chlorine and two for methyl. For certain systems at least, it became clear that one need not invoke a nonclassical ion to account for the facts.

In 1970, Olah reported that he had prepared a stable norbornyl cation in SbF_5–SO_2. From its NMR (both ^1H and ^{13}C) and Raman spectra, he concluded that it has, indeed, the nonclassical structure with delocalization of σ electrons. The 2-phenylnorbornyl cation, on the other hand, has the classical structure; this

Norbornyl cation	2-Methylnorbornyl cation	2-Phenylnorbornyl cation
Bridged ion	*Some bridging*	*Open ion*

benzylic cation, stabilized by electrons from the benzene ring, has no need of bridging. The tertiary 2-methylnorbornyl cation is intermediate in character: there is *partial* σ delocalization and hence bridging, but weaker than in the unsubstituted cation. (Interestingly enough, delocalization in the 2-methyl cation seems to come, not from the C(6)–C(1) bond, but from the C(6)–H bond; Olah pictures the back lobe of the carbon–hydrogen bond overlapping the *p* orbital of C(2).)

Thus, it seems, there *are* such things as nonclassical cations. What is still to be settled is just how much they are involved in the chemistry of ordinary solvolytic reactions.

Problem 32.12   (a) Show how a nonclassical ion intermediate could account for both the stereospecificity and the unusually fast rate (if it *is* unusually fast) of rearrangement of camphene hydrochloride into isobornyl chloride. (b) How do you account for the fact that optically active product is formed here, in contrast to what is obtained from solvolysis of norbornyl compounds?

PROBLEMS

1. Give a detailed interpretation of each of the following observations.

(a) $CH_3CH_2CH_2CD_2NH_2$ $\xrightarrow[H_2O]{HONO}$ 1-butanol + 2-butanol

The 2-butanol is $CH_3CH_2CHOHCHD_2$ + $CH_3CHOHCH_2CHD_2$
　　　　　　　　　　　　　 76.8%　　　　　　　　　　 23.2%

(b) $CH_3CH_2CD_2CH_2NH_2$ $\xrightarrow[H_2O]{HONO}$ $CH_3CH_2CD_2CH_2OH$ + $CH_3CH_2CDOHCH_2D$
　　　　　　　　　　　　　　　　　　　　　　　 68%　　　　　　　　　　 23.9%

　　　+ $CH_3CHOHCHDCH_2D$ + ($CH_3CHDCHOHCH_2D$ + $CH_3CDOHCH_2CH_2D$)
　　　　　 7.7%　　　　　　　　　　　　　　　　　 0.4%

2. Treatment of 1-methyl-1-cyclohexyl hydroperoxide with acid gives a product of formula $C_7H_{14}O_2$, which gives positive tests with CrO_3/H_2SO_4, 2,4-dinitrophenyl-hydrazine, and NaOI. What is a likely structure for this compound, and how is it formed?

3. (a) Describe simple chemical tests that would serve to distinguish among the possible products of rearrangement of 1-phenyl-1,2-propanediol shown on page 1111. Tell exactly what you would *do* and *see*. (b) Alternatively, you could use the proton NMR spectrum. Tell exactly what you would expect to see in the spectrum of each possible product.

4. In the presence of base, acyl derivatives of hydroxamic acids undergo the **Lossen rearrangement** to yield isocyanates or amines.

$$R-\overset{\overset{\displaystyle O}{\|}}{\underset{\underset{\displaystyle H}{|}}{C}}-N-O-\overset{\overset{\displaystyle O}{\|}}{C}-R' \xrightarrow{\text{base}} RNCO \xrightarrow{H_2O} RNH_2$$

　　(a) Write a detailed mechanism for the rearrangement.
　　(b) Study of a series of compounds in which R and R' were *m*- and *p*-substituted phenyl groups showed that reaction is speeded up by electron-releasing substituents in R and by electron-withdrawing substituents in R'. How do you account for these effects?

5. Benzophenone oxime, $C_{13}H_{11}ON$, m.p. 141 °C, like other oximes, is soluble in aqueous NaOH and gives a color with ferric chloride. When heated with acids it is transformed into a solid A, $C_{13}H_{11}ON$, m.p. 163 °C, which is insoluble in aqueous NaOH and in aqueous HCl.

After prolonged heating of A with aqueous NaOH, a liquid B separates and is collected by steam distillation. Acidification of the aqueous residue causes precipitation of a white solid C, m.p. 120–1 °C.

Compound B, b.p. 184 °C, is soluble in dilute HCl. When this acidic solution is chilled and then treated successively with $NaNO_2$ and β-naphthol, a red solid is formed. B reacts with acetic anhydride to give a compound that melts at 112.5–114 °C.

　　(a) What is the structure of A? (b) The transformation of benzophenone oxime into A illustrates a reaction to which the name **Beckmann** is attached. To what general class of reactions must this transformation belong? (c) Suggest a likely series of steps, each one basically familiar, for this transformation? (*Hint*: See Secs. 5.25 and 11.10.)

　　(d) Besides acids like sulfuric, other compounds "catalyze" this reaction. How might PCl_5 do the job? Tosyl chloride?

　　(e) What product or products corresponding to A would you expect from a similar transformation of acetone oxime; of acetophenone oxime; of *p*-nitrobenzophenone oxime; of methyl *n*-propyl ketoxime? (f) How would you go about identifying each of the products in (e)?

　　(g) Caprolactam (Problem 13, p. 1258) is made by the above reaction. With what ketone must the process start?

6. (a) Show all steps in the mechanisms probably involved in the following transformation. (*Hint*: Don't forget Sec. 25.5.)

3-Hydroperoxycyclohexene   Adipaldehyde   Cyclopentene-
    (6%)   1-carboxyaldehyde
    (39%)

(b) An important difference in migratory aptitude is illustrated here. What is it?

7. Urea is converted by hypohalites into nitrogen and carbonate. Given the fact that

$$H_2N-C-NH_2 \xrightarrow{Br_2, \ OH^-} N_2 + CO_3{}^{2-} + Br^-$$
(with C double-bonded to O)

hydrazine H_2N-NH_2, is oxidized to nitrogen by hypohalite, show that this reaction of urea is simply an example of the Hofmann degradation of amides.

8. Treatment of triarylmethanols, Ar_3COH, with acidic hydrogen peroxide yields a 50:50 mixture of ketone, $ArCOAr$, and phenol, $ArOH$. (a) Show all steps in a likely mechanism for this reaction. (b) Predict the major products obtained from *p*-methoxy-triphenylmethanol, $p\text{-}CH_3OC_6H_4(C_6H_5)_2COH$. From *p*-chlorotriphenylmethanol.

9. (a) Upon treatment with acid I ($R = C_2H_5$) yields II and III. Show all steps in these transformations.

I   II   III

(b) Account for the fact that when $R = C_6H_5$, I yields only II.
(c) Show the most likely steps in the following transformation:

(d) Predict the products of the pinacol rearrangement of 2,3-diphenyl-2,3-butanediol; of 3-phenyl-1,2-propanediol. Describe a simple chemical test that would show whether your prediction was correct or incorrect.

10. When dissolved in $HSO_3F\text{-}SbF_5\text{-}SO_2$, the glycol 1,3-propanediol is rapidly converted into propionaldehyde. Write all steps in a likely mechanism for this reaction.

11. Spectroscopic and thin-layer chromatographic analysis has shown that, even when not found in the final product, epoxides are present during the reaction of such pinacols as 1,1,2,2-tetraphenyl-1,2-ethanediol. It seems most likely that epoxides represent a blind alley down which many molecules stray before pinacolone is finally formed. (a) How are these epoxides probably formed? (b) What probably happens to them in the reaction medium?

12. In the oxidation stage of hydroboration–oxidation, alkylboranes are converted into alkyl borates, which are hydrolyzed to alcohols. It has been suggested that the formation of the borates involves the reagent HOO^-.

$$H_2O_2 + OH^- \underset{}{\overset{(1)}{\rightleftharpoons}} HOO^- + H_2O$$

$$R_3B + 3HOO^- \xrightarrow{(2)} (RO)_3B \xrightarrow{(3)} 3ROH$$
$$\text{Trialkylborane} \qquad\qquad \text{Alkyl borate}$$

(a) Show all steps in a possible mechanism for step (2), the formation of the borate.

(b) What did you conclude (Problem 17.9, p. 648) was the likely stereochemistry of the oxidation stage of hydroboration–oxidation? Is your mechanism in (a) consistent with this stereochemistry?

13. Labeled $ArCH_2{}^{14}CH_2NH_2$ was treated with HONO, and the $ArCH_2CH_2OH$ obtained was oxidized to $ArCOOH$. The fraction of the original radioactivity found in the $ArCOOH$ depended on the nature of Ar: $p\text{-}NO_2C_6H_4$ 8%, C_6H_5 27%, $p\text{-}CH_3OC_6H_4$ 45%. How do you account for these findings?

14. Clair Collins (Oak Ridge National Laboratory) prepared 1,1,2-triphenylethyl acetate triply labeled with ^{14}C (indicated as C*) and studied reactions (1)–(3) in ordinary

(1) $Ph_2CH-\underset{\underset{OAc}{|}}{C}{*}HPh \xrightarrow{\text{HOAc}} Ph_2CH-\underset{\underset{OAc}{|}}{C}{*}HPh + Ph_2\underset{\underset{OAc}{|}}{C}{*}H-CHPh$
                                                         50%   *50%*

(2) $Ph_2CH-\underset{\underset{OAc}{|}}{C}HPh{*} \xrightarrow{\text{HOAc}} Ph_2CH-\underset{\underset{OAc}{|}}{C}HPh{*} + Ph{*}PhCH-\underset{\underset{OAc}{|}}{C}HPh$
                                                         33%   *67%*

(3) $Ph_2CH-\underset{\underset{OAc{*}}{|}}{C}HPh \xrightarrow{\text{HOAc}} Ph_2CH-\underset{\underset{OAc}{|}}{C}HPh$

acetic acid. The equilibrium product of (1) and (2) had the indicated distribution of labels. The rate of acetate exchange (3) was found to be *identical* with the rates of (1) and (2).

Collins concluded that bridged ions are *not* involved in this particular system.

(a) Explain in detail how his conclusion is justified. Show just what probably *does* happen. (Among other things: what results would be expected if a bridged ion *were* involved?)

(b) Why might this system be expected to differ from, say, the 3-phenyl-2-butyl one?

15. Account in detail for each of the following sets of observations.

(a) Compound IV reacts with acetic acid 1200 times as fast as does ethyl tosylate,

$(CH_3)_2C{=}CHCH_2CH_2OTs \xrightarrow{\text{HOAc}}$
    IV

$(CH_3)_2C{=}CHCH_2CH_2OAc +$ $\underset{H_2C}{\overset{H_2C}{\diagdown}}\hspace{-0.3em}\underset{}{\bigg\rangle}CHC(CH_3){=}CH_2$
                    V   VI

and yields not only V but also VI. When the labeled compound IVa is used, product V consists of equal amounts of Va and Vb.

$$(CH_3)_2C=CHCH_2CD_2OTs \qquad\qquad (CH_3)_2C=CHCH_2CD_2OAc$$
IVa
Va

$$(CH_3)_2C=CHCD_2CH_2OAc$$
Vb

(b) The cyclopentene derivative VII (ONs = *p*-nitrobenzenesulfonate) undergoes solvolysis in acetic acid 95 times as fast as the analogous saturated compound (VIII), and gives *exo*-norbornyl acetate (IX).

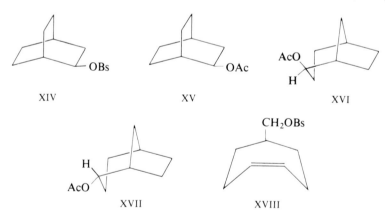

(c) *anti*-7-Norbornylene tosylate (X) reacts with acetic acid 10^{11} times as fast

X   XI   XII   XIII

as the saturated analog, and yields *anti*-7-norbornylene acetate (XI) with *retention* of configuration. Solvolysis of X in the presence of $NaBH_4$ gives XII and XIII.

16. (a) We saw (Sec. 32.11) that optically active *exo*-norbornyl brosylate reacts with acetic acid to give optically inactive *exo*-norbornyl acetate. The related brosylate XIV similarly reacts to give XV; yet in this case optically active brosylate yields optically *active*

XIV   XV   XVI

XVII   XVIII

acetate. Oddly enough, the complete racemization in the norbornyl reaction and the complete retention here are taken as evidence of the same fundamental behavior. On what common basis can you account for all of the above observations? (*Hint*: See also part (b).)

(b) Brosylate XIV also yields XVI, but no XVII. When XIV is optically active, so is the XVI that is obtained. How do these facts fit into your answer to (a)?

(c) Brosylate XVIII reacts with acetic acid 30 times as fast as the corresponding saturated compound does, and yields (optically inactive) XVII but no XVI. How do you account for these observations?

17. Treatment of XIX with $NaOCH_3$ gives product XX; treatment of XIX with R_2NH gives the corresponding product XXI. (a) Show all steps in the most likely mechanism for these rearrangements.

XIX

XX

XXI

XXII

(b) From the reaction of **XIX** with R_2NH, there is also obtained **XXII**. How is **XXII** probably formed? Of what general significance is its isolation?

33

Molecular Orbitals.
Orbital Symmetry

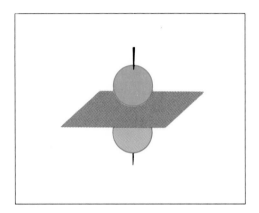

33.1 Molecular orbital theory

The structure of molecules is best understood through quantum mechanics. Exact quantum mechanical calculations are enormously complicated, and so various methods of approximation have been worked out to simplify the mathematics. The method that is often the most useful for the organic chemist is based on the concept of *molecular orbitals*: orbitals that are centered, not about individual nuclei, but about all the nuclei in the molecule.

What are the various molecular orbitals of a molecule like? What is their order of stability? How are electrons distributed among them? These are things we must know if we are to understand the relative stability of molecules: why certain molecules are aromatic, for example. These are things we must know if we are to understand the course of many chemical reactions: their stereochemistry, for example, and how easy or difficult they are to bring about; indeed, whether or not they will occur at all.

We cannot learn here how to make quantum mechanical calculations, but we can see what the results of some of these calculations are, and learn a little about how to use them.

In this chapter, then, we shall learn what is meant by the *phase* of an orbital, and what *bonding* and *antibonding* orbitals are. We shall see, in a non-mathematical way, what lies behind the Hückel $4n + 2$ rule for aromaticity. And finally, we shall take a brief look at a recent—and absolutely fundamental—development in chemical theory: the application of the concept of *orbital symmetry* to the understanding of organic reactions.

33.2   Wave equations. Phase

In our first description of atomic and molecular structure, we said that electrons show properties not only of particles but also of waves. We must examine a little more closely the wave character of electrons, and see how this is involved in chemical bonding. First, let us look at some properties of waves in general.

Let us consider the *standing waves* (or *stationary waves*) generated by the vibration of a string secured at both ends: the wave generated by, say, the plucking of a guitar string (Fig. 33.1). As we proceed horizontally along the string from left

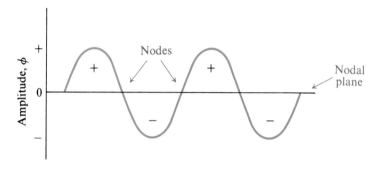

Figure 33.1   Standing waves. Plus and minus signs show relative phases.

to right, we find that the vertical displacement—the *amplitude* of the wave—increases in one direction, passes through a maximum, decreases to zero, and then increases in the opposite direction. The places where the amplitude is zero are called *nodes*. In Fig. 33.1 they lie in a plane—the *nodal plane*—perpendicular to the plane of the paper. Displacement upward and displacement downward correspond to opposite *phases* of the wave. To distinguish between phases, we arbitrarily assign algebraic signs to the amplitude: plus for, say, displacement upward, and minus for displacement downward. If we were to superimpose similar waves on one another exactly *out of phase*—that is, with the crests of one lined up with the troughs of the other—they would cancel each other; that is to say, the sum of their amplitudes, $+$ and $-$, would be zero.

The differential equation that describes the wave is a *wave equation*. Solution of this equation gives the amplitude, ϕ, as a function, $f(x)$, of the distance, x, along the wave. Such a function is a *wave function*.

Now, electron waves are described by a wave equation of the same general form as that for string waves. The wave functions that are acceptable solutions to this equation again give the amplitude, ϕ, this time as a function, not of a single coordinate, but of the three coordinates necessary to describe motion in three dimensions. It is these electron wave functions that we call *orbitals*.

Any wave equation has a *set* of solutions—an infinity of them, actually—each corresponding to a different energy level. The *quantum* thus comes naturally out of the mathematics.

Like a string wave, an electron wave can have nodes, where the amplitude is zero. On opposite sides of a node the amplitude has opposite signs, that is, the wave is of opposite phases. Of special interest to us is the fact that between the two lobes of a *p* orbital lies a nodal plane, perpendicular to the axis of the orbital (Fig. 33.2). The two lobes are of opposite phase, and this is often indicated by + and − signs.

As used here, the signs do not have anything to do with charge. They simply indicate that the amplitude is of opposite algebraic sign in the two lobes. To avoid confusion, we shall show lobes as blue and green. Two blue lobes are of the same phase, both plus or both minus—it does not matter which. Similarly, two green lobes are of the same phase; a blue lobe and a green lobe are of opposite phase.

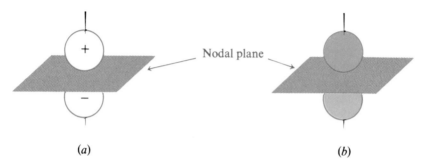

(a)   (b)

Figure 33.2   The *p* orbital. The two lobes are of opposite phase, indicated either (*a*) by plus and minus signs or (*b*) by color.

The amplitude or wave function, ϕ, *is* the orbital. As is generally true for waves, however, it is the square of the amplitude, ϕ^2, that has physical meaning. For electron waves, ϕ^2 represents the probability of finding an electron at any particular place. The fuzzy balls or simple spheres we draw to show the "shapes" of orbitals are crude representations of the space within which ϕ^2 has a particular value—the space within which the electron spends, say, 95% of its time. Whether ϕ is positive or negative, ϕ^2 is of course positive; this makes sense, since probability cannot be negative. The usual practice is to draw the lobes of a *p* orbital to represent ϕ^2; if + or − signs are added, or one lobe is shaded and the other unshaded, this is to show the relative signs of ϕ.

33.3   Molecular orbitals. LCAO method

As chemists, we picture molecules as collections of atoms held together by bonds. We consider the bonds to arise from the overlap of an atomic orbital of one atom with an atomic orbital of another atom. A new orbital is formed, which is occupied by a pair of electrons of opposite spin. Each electron is attracted by both positive nuclei, and the increase in electrostatic attraction gives the bond its strength, that is, stabilizes the molecule relative to the isolated atoms.

This highly successful qualitative model parallels the most convenient quantum mechanical approach to molecular orbitals: **the method of linear combination of atomic orbitals (LCAO)**. We have assumed that the shapes and dispositions of bond orbitals are related in a simple way to the shapes and dispositions of atomic orbitals. The LCAO method makes the same assumption *mathematically*: to calculate an approximate molecular orbital, ψ, one uses a *linear combination* (that is, a combination through addition or subtraction) of atomic orbitals.

$$\psi = \phi_A + \phi_B$$

where
ψ is the molecular orbital
ϕ_A is atomic orbital A
ϕ_B is atomic orbital B

The rationale for this assumption is simple: when the electron is near atom A, ψ resembles ϕ_A; when the electron is near atom B, ψ resembles ϕ_B.

Now this combination is *effective*—that is, the molecular orbital is appreciably more stable than the atomic orbitals—only if the atomic orbitals ϕ_A and ϕ_B:

(a) overlap to a considerable extent;
(b) are of comparable energy; and
(c) have the same symmetry about the bond axis.

These requirements can be justified mathematically. Qualitatively, we can say this: if there is not considerable overlap, the energy of ψ is equal to either that of ϕ_A or that of ϕ_B; if the energy of ϕ_A and ϕ_B are quite different, the energy of ψ is essentially that of the more stable atomic orbital. In either case, there is no significant stabilization, and no bond formation.

When we speak of the symmetry of orbitals, we are referring to the relative phases of lobes, and their disposition in space. To see what is meant by requirement (c), that the overlapping orbitals have the same symmetry, let us look at one example: hydrogen fluoride. This molecule can be pictured as resulting from overlap of the s orbital of hydrogen with a p orbital of fluorine. In Fig. 33.3a, we use the $2p_x$ orbital, where the x coordinate is taken as the H—F axis. The blue s orbital overlaps the blue lobe of the p orbital, and a bond forms. If, however, we were to use the $2p_z$ (or $2p_y$) orbital as in Fig. 33.3b, overlap of *both* lobes—plus and minus—would occur and cancel each other. That is, the positive overlap integral would be exactly canceled by the negative overlap integral; the net effect would be *no overlap*, and no bond formation. The dependence of overlap on phase is fundamental to chemical bonding.

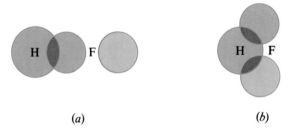

(a)   (b)

Figure 33.3 The hydrogen fluoride molecule: dependence of overlap on orbital symmetry. (*a*) Overlap of lobes of the same phase leads to bonding. (*b*) Positive overlap and negative overlap cancel each other.

33.4 Bonding and antibonding orbitals

Quantum mechanics shows that linear combination of two functions gives, not one, but *two* combinations and hence *two* molecular orbitals: a *bonding* orbital, more stable than the component atomic orbitals; and an *antibonding* orbital, less stable than the component orbitals.

$$\psi_+ = \phi_A + \phi_B$$ Bonding orbital:
stabilizes molecule

$$\psi_- = \phi_A - \phi_B$$ Antibonding orbital:
destabilizes molecule

Two *s* orbitals, for example, can be added,

A   +   B   ⟶   A   B

s   *s*   σ orbital

Bonding

A   −   B   ⟶   A   B

s   *s*   σ* orbital

Antibonding

We can see, in a general way, why there must be two combinations. There can be as many as two electrons in each component atomic orbital, making a total of four electrons; two molecular orbitals are required to accommodate them.

Figure 33.4 (p. 1132) shows schematically the shapes of the molecular orbitals, bonding and antibonding, that result from overlap of various kinds of atomic orbitals. We recognize the bonding orbitals, σ and π, although until now we have not shown the two lobes of a π orbital as being of opposite phase. An antibonding orbital, we see, has a nodal plane perpendicular to the bond axis, and cutting between the atomic nuclei. The antibonding sigma orbital, σ*, thus consists of two lobes, of opposite phase. The antibonding pi orbital, π*, consists of four lobes.

In a bonding orbital, electrons are concentrated in the region between the nuclei, where they can be attracted by both nuclei. The increase in electrostatic attraction lowers the energy of the system. In an antibonding orbital, by contrast, electrons are *not* concentrated between the nuclei; electron charge is zero in the nodal plane. Electrons spend most of their time farther from a nucleus than in the separated atoms. There is a decrease in electrostatic attraction, and an increase in repulsion between the nuclei. The energy of the system is higher than that of the separated atoms. *Where electrons in a bonding orbital tend to hold the atoms together, electrons in an antibonding orbital tend to force the atoms apart.*

It may at first seem strange that electrons in certain orbitals can actually weaken the bonding. Should not *any* electrostatic attraction, even if less than optimum, be better than none? We must remember that it is the bond dissociation energy we are concerned with. We are not comparing the electrostatic attraction in an antibonding orbital with no electrostatic attraction; we are comparing it with the stronger electrostatic attraction in the separated atoms.

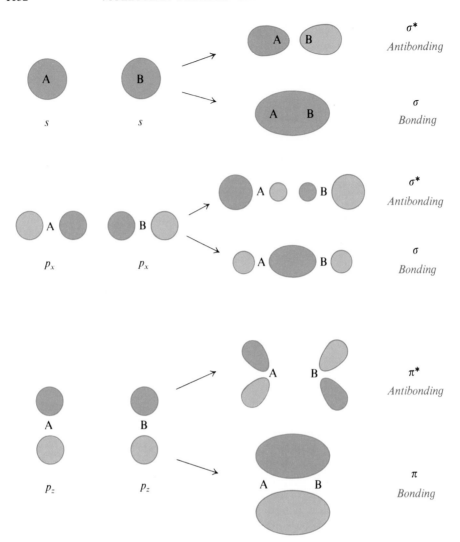

Figure 33.4   Bonding and antibonding orbitals.

There are, in addition, orbitals of a third kind, *non-bonding orbitals*. As the name indicates, electrons in these orbitals—unshared pairs, for example—neither strengthen nor weaken the bonding between atoms.

33.5   Electronic configurations of some molecules

Let us look at the electronic configurations of some familiar molecules. The shapes and relative stabilities of the various molecular orbitals are calculated by quantum mechanics, and we shall simply use the results of these calculations. We picture the nuclei in place, with the molecular orbitals mapped out about them, and we feed electrons into the orbitals. In doing this we follow the same rules that we followed in arriving at the electronic configurations of atoms. There can be only two electrons—and of opposite spin—in each orbital, with orbitals of lower energy being filled up first. If there are orbitals of equal energy, each gets an electron before any one of them gets a pair of electrons. We shall limit our attention to orbitals containing π electrons, since these electrons will be the ones of chief interest to us.

For the π electrons of ethylene (Fig. 33.5), there are two molecular orbitals since there are two linear combinations of the two component p orbitals. The broken line in the figure indicates the non-bonding energy level; below it lies the bonding orbital, π, and above it lies the antibonding orbital, π^*.

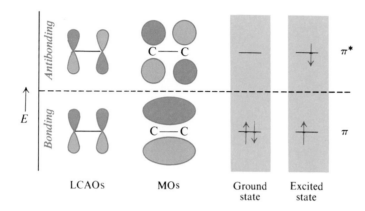

Figure 33.5   Ethylene. Configuration of π electrons in the ground state and the excited state.

Normally, a molecule exists in the state of lowest energy, the *ground state*. But, as we have seen (Sec. 16.5), absorption of light of the right frequency (in the ultraviolet region) raises a molecule to an *excited state*, a state of higher energy. In the ground state of ethylene, we see, both π electrons are in the π orbital; this configuration is specified as π^2, where the superscript tells the number of electrons in that orbital. In the excited state one electron is in the π orbital and the other—still of opposite spin—is in the π^* orbital; this configuration, $\pi\pi^*$, is naturally the less stable since only one electron helps to hold the atoms together, while the other tends to force them apart.

For 1,3-butadiene, with four component p orbitals, there are four molecular orbitals for π electrons (Fig. 33.6, p. 1134). The ground state has the configuration $\psi_1{}^2\psi_2{}^2$; that is, there are two electrons in each of the bonding orbitals, ψ_1 and ψ_2. The higher of these, ψ_2, resembles two isolated π orbitals, although it is of somewhat lower energy. Orbital ψ_1 encompasses all four carbons; this delocalization provides the net stabilization of the conjugated system. Absorption of light of the right frequency raises one electron to ψ_3.

$$\psi_1{}^2\psi_2{}^2 \xrightarrow{\ h\nu\ } \psi_1{}^2\psi_2\psi_3$$
Ground   Lowest
state   excited state

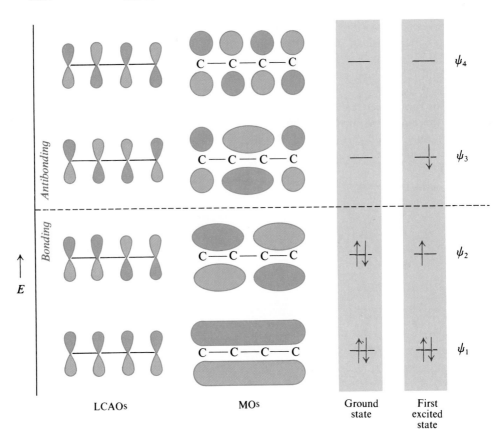

Figure 33.6　1,3-Butadiene. Configuration of π electrons in the ground state and the first excited state.

Next, let us look at the allyl system: cation, free radical, and anion. Regardless

$$\underbrace{CH_2\text{---}CH\text{---}CH_2}_{\oplus}\qquad \underbrace{CH_2\text{---}CH\text{---}CH_2}_{\cdot}\qquad \underbrace{CH_2\text{---}CH\text{---}CH_2}_{\ominus}$$

Allyl cation　　　　Allyl free radical　　　Allyl anion

of the number of π electrons, there are three component p orbitals, one on each carbon, and they give rise to three molecular orbitals, ψ_1, ψ_2, and ψ_3. As shown in Fig. 33.7, ψ_1 is bonding and ψ_3 is antibonding. Orbital ψ_2 encompasses only the end carbons (there is a node at the middle carbon) and is of the same energy as an isolated p orbital; it is therefore non-bonding.

The allyl cation has π electrons only in the bonding orbital. The free radical has one electron in the non-bonding orbital as well, and the anion has two in the non-bonding orbital. The bonding orbital ψ_1 encompasses all three carbons, and is more stable than a localized π orbital involving only two carbons; it is this delocalization that gives allylic particles their special stability. We see the symmetry

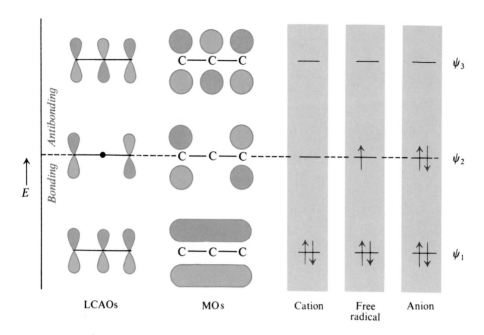

Figure 33.7   Allyl system. Configuration of π electrons in the cation, the free radical, and the anion.

we have attributed to allylic particles on the basis of the resonance theory; the two ends of each of these molecules are equivalent.

Finally, let us look at benzene. There are six combinations of the six component *p* orbitals, and hence six molecular orbitals. Of these, we shall consider only three combinations, which correspond to the three most stable molecular orbitals, all bonding orbitals (Fig. 33.8, p. 1136). Each contains a pair of electrons. The lowest orbital, ψ_1, encompasses all six carbons. Orbitals ψ_2 and ψ_3 are of different shape, but equal energy; together they provide—as does ψ_1—equal electron density

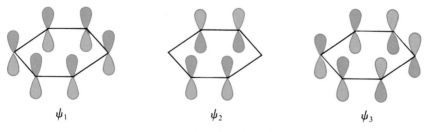

Benzene: first three LCAOs

at all six carbons. The net result, then, is a highly symmetrical molecule with considerable delocalization of π electrons. But this is only part of the story; in the next section we shall look more closely at just what makes benzene such a special kind of molecule.

Figure 33.8 Benzene. Configuration of π electrons in the ground state.

33.6 Aromatic character. The Hückel 4n + 2 rule

In Chapter 13 we discussed the structure of aromatic compounds. An aromatic molecule is flat, with cyclic clouds of delocalized π electrons above and below the plane of the molecule. We have just seen, for benzene, the molecular orbitals that permit this delocalization. But delocalization alone is not enough. For that special degree of stability we call *aromaticity*, the number of π electrons must conform to **Hückel's rule**: *there must be a total of* **(4n + 2) π electrons**.

In Sec. 13.10, we saw evidence of special stability associated with the "magic" numbers of two, six, and ten π electrons, that is, with systems where n is 0, 1, and 2 respectively. Problem 6 (p. 623) described the NMR spectrum of cyclooctadeca-nonaene, which contains eighteen π electrons (n is 4). Twelve protons lie outside the ring, are deshielded, and absorb downfield; but, because of the particular geometry of the large flat molecule, six protons lie *inside* the ring, are shielded (see

Cyclooctadecanonaene

Eighteen π electrons

Aromatic

Fig. 16.4, p. 584), and absorb upfield. The spectrum is unusual, but exactly what we would expect if this molecule were aromatic.

Hückel (p. 488) was a pioneer in the field of molecular orbital theory. He developed the LCAO method in its simplest form, yet "Hückel molecular orbitals" have proved enormously successful in dealing with organic molecules. Hückel proposed the $4n + 2$ rule in 1931. It has been tested in many ways since then, and it *works*. Now, what is the theoretical basis for this rule?

Let us begin with the cyclopentadienyl system. Five sp^2-hybridized carbons have five component p orbitals, which give rise to five molecular orbitals (Fig. 33.9). At the lowest energy level there is a single molecular orbital. Above this, the orbitals appear as *degenerate* pairs, that is, pairs of orbitals of equal energy. The lowest degenerate pair are bonding, the higher ones are antibonding.

The cyclopentadienyl cation has four electrons. Two of these go into the lower orbital. Of the other two electrons, one goes into each orbital of the lower degenerate pair. The cyclopentadienyl free radical has one more electron, which fills one orbital of the pair. The anion has still another electron, and with this we fill the remaining orbital of the pair. The six π electrons of the cyclopentadienyl anion are *just enough to fill all the bonding orbitals*. Fewer than six leaves bonding orbitals unfilled; more than six, and electrons would have to go into antibonding orbitals. Six π electrons gives maximum bonding and hence maximum stability.

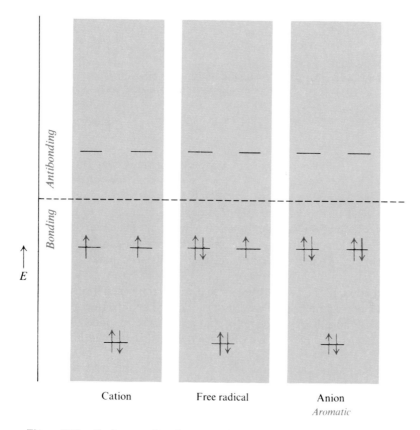

Figure 33.9 Cyclopentadienyl system. Configuration of π electrons in the cation, the free radical, and the anion.

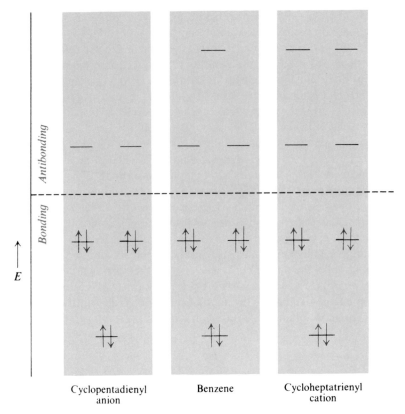

Figure 33.10 Aromatic compounds with six π electrons. Configuration of π electrons in the cyclopentadienyl anion, benzene, and the cycloheptatrienyl cation.

Figure 33.10 shows the molecular orbitals for rings containing five, six, and seven sp^2-hybridized carbons. We see the same pattern for all of them: a single orbital at the lowest level, and above it a series of degenerate pairs. It takes $(4n + 2)$ π electrons to *fill* a set of these bonding orbitals: two electrons for the lowest orbital, and four for each of n degenerate pairs. Such an electron configuration has been likened to the rare-gas configuration of an atom, with its closed shell. It is the filling of these molecular orbital shells that makes these molecules aromatic.

In Problem 13.6 (p. 491) we saw that the cyclopropenyl cation is unusually stable: 20 kcal/mol more stable even than the allyl cation. In contrast, the cyclopropenyl free radical and anion are *not* unusually stable; indeed, the anion seems to be particularly *unstable*. The cation has the Hückel number of two π electrons

Cyclopropenyl   Cyclopropenyl   Cyclopropenyl
cation   free radical   anion
Two π electrons   *Three π electrons*   *Four π electrons*
Aromatic

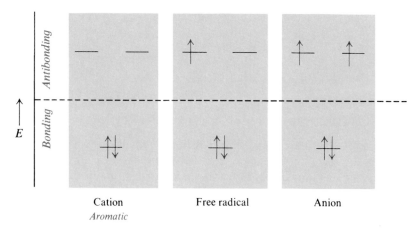

Figure 33.11 Cyclopropenyl system. Configuration of π electrons in the cation, the free radical, and the anion.

(n is zero) and is aromatic. Here, too, aromaticity results from the filling up of a molecular orbital shell (Fig. 33.11).

In the allyl system (Fig. 33.7) the third and fourth electrons go into a non-bonding orbital, whereas here they go into antibonding orbitals. As a result, the cyclopropenyl free radical and anion are less stable than their open-chain counterparts. For the cyclopropenyl anion in particular, with two electrons in antibonding orbitals, simple calculations indicate no net stabilization due to delocalization, that is, zero resonance energy. Some calculations indicate that the molecule is actually less stable than if there were no conjugation at all. Such cyclic molecules, in which delocalization actually leads to *de*stabilization, are not just non-aromatic; they are *anti*aromatic.

Problem 33.1 When 3,4-dichloro-1,2,3,4-tetramethylcyclobutene was dissolved at $-78\,^{\circ}\text{C}$ in SbF_5—SO_2, the solution obtained gave three NMR peaks, at $\delta\,2.07$, $\delta\,2.20$, and $\delta\,2.68$, in the ratio $1:1:2$. As the solution stood, these peaks slowly disappeared and were replaced by a single peak at $\delta\,3.68$. What compound is each spectrum probably due to? Of what theoretical significance are these findings?

Problem 33.2 (a) Cyclopropenones (I) have been made, and found to have rather unusual properties.

R = phenyl or *n*-propyl

They have very high dipole moments: about 5 D, compared with about 3 D for benzophenone or acetone. They are highly basic for ketones, reacting with perchloric acid to yield salts of formula $(R_2C_3OH)^+ClO_4^-$. What factor may be responsible for these unusual properties?

(b) Diphenylcyclopropenone was allowed to react with phenylmagnesium bromide, and the reaction mixture was hydrolyzed with perchloric acid. There was obtained, not a tertiary alcohol, but a salt of formula $[(C_6H_5)_3C_3]^+ClO_4^-$. Account for the formation of this salt.

(c) The synthesis of the cyclopropenones involved the addition to alkynes of CCl_2, which was generated from $Cl_3CCOONa$. Show all steps in the most likely mechanism for the formation of CCl_2. (*Hint*: See Sec. 12.17.)

33.7 Orbital symmetry and the chemical reaction

A chemical reaction involves the crossing of an energy barrier. In crossing this barrier, the reacting molecules seek the easiest path: a low path, to avoid climbing any higher than is necessary; and a broad path, to avoid undue restrictions on the arrangement of atoms. As reaction proceeds, there is a change in bonding among the atoms, from the bonding in the reactants to the bonding in the products. Bonding is a stabilizing factor; the stronger the bonding, the more stable the system. If a reaction is to follow the easiest path, it must take place in the way that *maintains maximum bonding during the reaction process*. Now, bonding, as we visualize it, results from overlap of orbitals. Overlap requires that portions of different orbitals occupy the same space, and that they be *of the same phase*.

This line of reasoning seems perfectly straightforward. Yet the central idea, that the course of reaction can be controlled by orbital symmetry, was a revolutionary one, and represents one of the really giant steps forward in chemical theory. A number of people took part in the development of this concept: K. Fukui in Japan, H. C. Longuet-Higgins in England. But organic chemists became aware of the power of this approach chiefly through a series of papers published in 1965 by R. B. Woodward and Roald Hoffmann working at Harvard University. (For their work, Woodward, Hoffmann, and Fukui have received Nobel Prizes.)

Very often in organic chemistry, theory lags behind experiment; many facts are accumulated, and a theory is proposed to account for them. This is a perfectly respectable process, and extremely valuable. But with orbital symmetry, just the reverse has been true. The theory lay in the mathematics, and what was needed was the spark of genius to see the applicability to chemical reactions. Facts were sparse, and Woodward and Hoffmann made *predictions*, which have since been borne out by experiment. All this is the more convincing because these predictions were of the kind called "risky": that is, the events predicted seemed unlikely on any grounds other than the theory being tested.

Orbital symmetry effects are observed in *concerted* reactions, that is, in reactions where several bonds are being made or broken simultaneously. Woodward and Hoffmann formulated "rules", and described certain reaction paths as *symmetry-allowed* and others as *symmetry-forbidden*. *All of this applies only to concerted reactions*, and refers to the relative ease with which they take place. A "symmetry-forbidden" reaction is simply one for which the concerted mechanism is very difficult, so difficult that, if reaction is to occur at all, it will probably do so in a different way: by a different concerted path that is symmetry-allowed; or, if there is none, by a stepwise, non-concerted mechanism. In the following brief discussion, and in the problems based on it, we have not the space to give the evidence indicating that each reaction is indeed concerted; but there must *be* such evidence, and gathering it is often the hardest job the investigator has to do.

Nor have we space here for a full, rigorous treatment of concerted reactions, which considers the correlation of symmetry between all the molecular orbitals of the products. We shall focus our attention on certain key orbitals, which contain the "valence" electrons of the molecules. Even this simplified approach, we shall find, is tremendously powerful; it is highly graphic, and in some cases gives information that the more detailed treatment does not.

33.8   Electrocyclic reactions

Under the influence of heat or light, a conjugated polyene can undergo isomerization to form a cyclic compound with a single bond between the terminal carbons of the original conjugated system; one double bond disappears, and the remaining double bonds shift their positions. For example, 1,3,5-hexatrienes yield 1,3-cyclohexadienes:

A 1,3,5-hexatriene   A 1,3-cyclohexadiene

The reverse process can also take place: a single bond is broken and a cyclic compound yields an open-chain polyene. Cyclobutenes, for example, are converted into butadienes:

A cyclobutene   A 1,3-butadiene

Such interconversions are called **electrocyclic reactions**.

It is the stereochemistry of electrocyclic reactions that is of chief interest to us. To observe this, we must have suitably substituted molecules. Let us consider first the interconversion of 3,4-dimethylcyclobutene and 2,4-hexadiene (Fig. 33.12). The cyclobutene exists as *cis* and *trans* isomers. The hexadiene exists in three forms: *cis,cis*; *cis,trans*; and *trans,trans*. As we can see, the *cis*-cyclobutene yields only one of the three isomeric dienes; the *trans*-cyclobutene yields a

cis-3,4-Dimethylcyclobutene   cis,trans-2,4-Hexadiene

trans-3,4-Dimethylcyclobutene   trans,trans-2,4-Hexadiene

Figure 33.12   Interconversions of 3,4-dimethylcyclobutenes and 2,4-hexadienes.

different isomer. Reaction is thus *completely stereoselective* and *completely stereospecific*. Furthermore, photochemical cyclization of the *trans,trans* diene gives a different cyclobutene from the one from which the diene is formed by the thermal (heat-promoted) ring-opening.

The interconversions of the corresponding dimethylcyclohexadienes and the 2,4,6-octatrienes are also stereoselective and stereospecific (Fig. 33.13). Here, too, thermal and photochemical reactions differ in stereochemistry. If we examine the

trans,cis,trans-2,4,6-Octatriene   *cis*-5,6-Dimethyl-1,3-cyclohexadiene

trans,cis,cis-2,4,6-Octatriene   *trans*-5,6-Dimethyl-1,3-cyclohexadiene

Figure 33.13 Interconversions of 2,4,6-octatrienes and 5,6-dimethyl-1,3-cyclohexadienes.

structures closely, we see something else: the stereochemistry of the triene–cyclohexadiene interconversions is *opposite to* that of the diene–cyclobutene interconversions. For the thermal reactions, for example, *cis* methyl groups in the cyclobutene become *cis* and *trans* in the diene; *cis* methyl groups in the cyclohexadiene are *trans* and *trans* in the related triene.

Electrocyclic reactions, then, are completely stereoselective and stereospecific. The exact stereochemistry depends upon two things: (a) the number of double bonds in the polyene, and (b) whether reaction is thermal or photochemical. It is one of the triumphs of the orbital symmetry approach that it can account for all these facts; indeed, most of the examples known today were *predicted* by Woodward and Hoffmann before the facts were known.

It is easier to examine these interconversions from the standpoint of cyclization; according to the principle of microscopic reversibility, whatever applies to this reaction applies equally well to the reverse process, ring-opening. In cyclization, two π electrons of the polyene form the new σ bond of the cycloalkene. But which two electrons? We focus our attention on the *highest occupied molecular orbital (HOMO) of the polyene*. Electrons in this orbital are the "valence" electrons of the molecule; they are the least tightly held, and the most easily pushed about during reaction.

Let us begin with the thermal cyclization of a disubstituted butadiene, $RCH=CH-CH=CHR$. As we have already seen (Fig. 33.6, p. 1134), the highest occupied molecular orbital of a conjugated diene is ψ_2. It is the electrons in this orbital that will form the bond that closes the ring. Bond formation requires overlap, in this case overlap of lobes on C–1 and C–4 of the diene: the front

ψ_2 HOMO of
                  ground state

Conjugated diene

carbons in Fig. 33.14. We see that to bring these lobes into position for overlap, there must be rotation about two bonds, C(1)–C(2) and C(3)–C(4). This rotation can take place in two different ways: there can be **conrotatory** motion, in which the bonds rotate in the same direction,

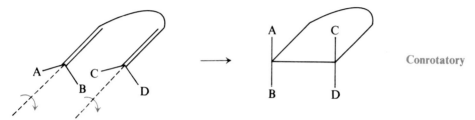

Conrotatory

or there can be **disrotatory** motion, in which the bonds rotate in opposite directions.

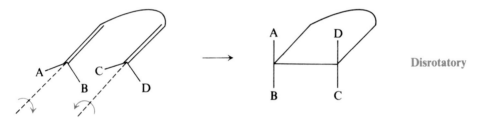

Disrotatory

Now, in this case, as we see in Fig. 33.14, conrotatory motion brings together lobes of the *same phase*; overlap occurs and a bond forms. Disrotatory motion, on

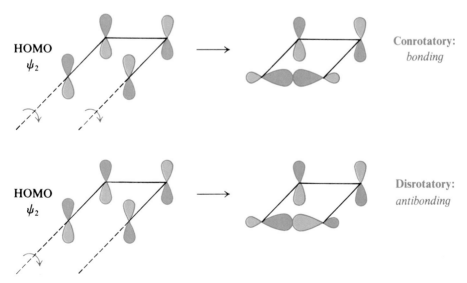

HOMO
ψ_2

Conrotatory:
bonding

HOMO
ψ_2

Disrotatory:
antibonding

Figure 33.14   Thermal cyclization of a 1,3-butadiene to a cyclobutene. Conrotatory motion leads to bonding. Disrotatory motion leads to antibonding.

the other hand, brings together lobes of *opposite phase*; here interaction is antibonding, and repulsive. As Fig. 33.15 shows, it is conrotatory motion that produces the stereochemistry actually observed.

cis,trans-2,4-Hexadiene   cis-3,4-Dimethylcyclobutene

trans,trans-2,4-Hexadiene   trans-3,4-Dimethylcyclobutene

Figure 33.15 Thermal cyclization of substituted butadienes. Observed stereochemistry indicates conrotatory motion.

How are we to account for the opposite stereochemistry in the photochemical reaction? On absorption of light, butadiene is converted into the excited state shown in Fig. 33.6, in which one electron from ψ_2 has been raised to ψ_3. Now the highest occupied orbital is ψ_3, and it is the electron here that we are concerned

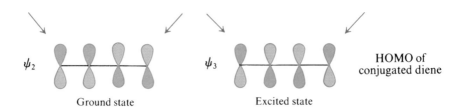

ψ_2 ψ_3 HOMO of conjugated diene

Ground state   Excited state

with. But in ψ_3 the relative symmetry of the terminal carbons is opposite to that in ψ_2. Now it is the *disrotatory* motion that brings together lobes of the same phase, and the stereochemistry is reversed (Fig. 33.16).

Next, let us look at the thermal cyclization of a disubstituted hexatriene, $RCH{=}CH{-}CH{=}CH{-}CH{=}CHR$, whose electronic configuration is shown in

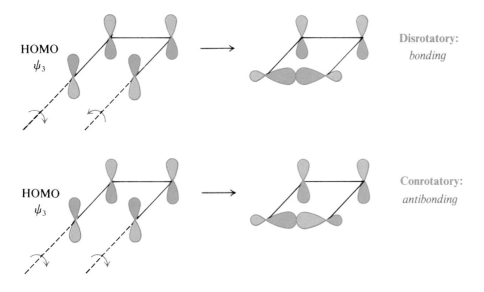

HOMO
ψ_3

Disrotatory:
bonding

HOMO
ψ_3

Conrotatory:
antibonding

Figure 33.16   Photochemical cyclization of a 1,3-butadiene to a cyclobutene. Disrotatory motion leads to bonding. Conrotatory motion leads to antibonding.

Fig. 33.17 (p. 1146). The HOMO for the ground state of the hexatriene is ψ_3. If we compare this with the HOMO for the ground state of butadiene (ψ_2 in Fig. 33.6, p. 1134), we see that the relative symmetry about the terminal carbons is opposite in

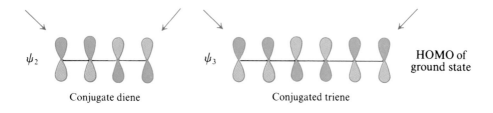

ψ_2 ψ_3 HOMO of
                                                                                   ground state

Conjugate diene   Conjugated triene

the two cases. For ground-state hexatriene it is disrotatory motion that leads to bonding and, as shown in Fig. 33.18 (p. 1147), gives rise to the observed stereochemistry.

In the excited state of hexatriene, ψ_4 is the HOMO, and once again we see a reversal of symmetry: here, conrotatory motion is the favored process.

What we see here is part of a regular pattern (Table 33.1, p. 1146) that emerges from the quantum mechanics. As the number of pairs of π electrons in the polyene increases, the relative symmetry about the terminal carbons in the HOMO alternates regularly. Furthermore, symmetry in the HOMO of the first excited state is always opposite to that in the ground state.

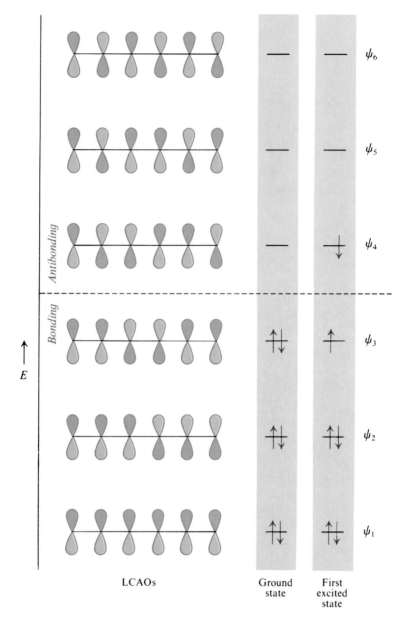

Figure 33.17   A 1,3,5-hexatriene. Configuration of π electrons in the ground state and the first excited state.

Table 33.1   WOODWARD–HOFFMANN RULES FOR ELECTROCYCLIC REACTIONS

Number of π electrons	Reaction	Motion
$4n$	thermal	conrotatory
$4n$	photochemical	disrotatory
$4n + 2$	thermal	disrotatory
$4n + 2$	photochemical	conrotatory

trans,cis,trans-2,4,6-Octatriene   5,6-*cis*-Dimethyl-1,3-cyclohexadiene

trans,cis,cis-2,4,6-Octatriene   5,6-*trans*-Dimethyl-1,3-cyclohexadiene

Figure 33.18   Thermal cyclization of substituted hexatrienes. Observed stereochemistry indicates disrotatory motion.

Problem 33.3   Thermal ring closure of three stereoisomeric 2,4,6,8-decatetraenes (I, II, and III) has been found to be in agreement with the Woodward–Hoffmann

rules. Two of these stereoisomers give one dimethylcyclooctatriene, and the third stereoisomer gives a different dimethylcyclooctatriene. (a) Which decatetraenes give which cyclooctatrienes? (b) Predict the product of photochemical ring closure of each.

Problem 33.4   The commonly observed conversion of cyclopropyl cations into allyl cations is considered to be an example of an electrocyclic reaction. (a) What is the

Cyclopropyl cation   Allyl cation

HOMO of the allyl cation? How many π electrons has it? (b) Where does this reaction fit in Table 33.1? Would you expect conrotatory or disrotatory motion? (c) What prediction would you make about interconversion of allyl and cyclopropyl *anions*? (d) About the interconversion of pentadienyl cations and cyclopentenyl cations?

Pentadienyl cation   Cyclopentenyl cation

Problem 33.5   Each of the following reactions involves one or more concerted steps that take place in accordance with the Woodward–Hoffmann rules. In each case, show exactly what is happening.

(a)

7-Cyano-7-trifluoromethyl-
norcaradiene

7-Cyano-7-trifluoromethyl-
tropylidene

(b)   25 °C

(c)

cis,cis,cis-Cyclo-
nona-1,3,5-triene

100 °C

cis-Bicyclo[4.3.0]-
nona-2,4-diene

hν
−20 °C

cis,cis,trans-Cyclo-
nona-1,3,5-triene

25 °C

trans-Bicyclo[4.3.0]-
nona-2,4-diene

(d)

cis-Bicyclo[6.2.0]-
deca-2,9-diene

140 °C

trans,cis,cis-Cyclo-
deca-1,3,5-triene

220 °C

trans-Bicyclo[4.4.0]-
deca-2,4-diene

(e)

Cl   Br

+   Ag⁺

H₂O

+   AgCl

endo-2-Chloro-exo-2-bromo-
bicyclo[3.1.0]hexane

2-Bromo-2-cyclohexenol

(f)

H₃PO₄

Problem 33.6 Stereoisomers IV and V are easily interconverted by heating. After 51 days at 124 °C—during which time, it was calculated, 2.6×10^6 interconversions took

place—*only* IV and V were found to be present; there was *none* of their stereoisomers VI and VII. Propose a mechanism for the interconversion that would account for this remarkable stereoselectivity.

33.9 Cycloaddition reactions

In Sec. 31.8, we encountered the Diels–Alder reaction, in which a conjugated diene and a substituted alkene—the dienophile—react to form a cyclohexene.

Diels–Alder reaction:

a [4 + 2] cycloaddition

Diene Dienophile Adduct

This is an example of **cycloaddition**, a reaction in which two unsaturated molecules combine to form a cyclic compound, with π electrons being used to form two new σ bonds. The Diels–Alder reaction is a [4 + 2] cycloaddition, since it involves a system of four π electrons and a system of two π electrons.

Reaction takes place very easily, often spontaneously, and at most requires moderate application of heat.

There are several aspects to the stereochemistry of the Diels–Alder reaction.

(a) First, we have taken for granted—correctly—that the diene must be in the conformation (*s-cis*) that permits the ends of the conjugated system to *reach* the doubly bonded carbons of the dienophile.

s-cis

*Required for
Diels–Alder reaction*

s-trans

(b) Next, with respect to the alkene (dienophile) addition is clear-cut *syn* (Problem 8, p. 1095); this stereoselectivity is part of the evidence that the Diels–

Diels–Alder reaction:
syn-addition

Alder reaction is, indeed, a concerted one, that is, that both new bonds are formed in the same transition state.

(c) Finally, the Diels–Alder reaction takes place in the *endo*, rather than *exo*, sense. That is to say, any other unsaturated groups in the dienophile (for example, —CO—O—CO— in maleic anhydride) tend to lie *near* the developing double bond in the diene moiety (Fig. 33.19). For the *endo* preference to be *seen*, of course, the diene must be suitably substituted.

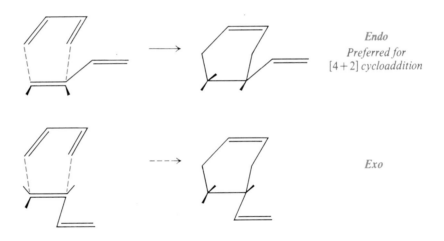

Endo
Preferred for
[4 + 2] cycloaddition

Exo

Figure 33.19   Stereochemistry of the Diels–Alder reaction, illustrated for the reaction between two moles of 1,3-butadiene.

Now are there such reactions as [2 + 2] cycloadditions? Can, say, two molecules of ethylene combine to form cyclobutane? The answer is: yes, but not easily under thermal conditions. Under vigorous conditions cycloaddition may occur, but stepwise—via diradicals—and not in a concerted fashion. Photochemical [2 + 2] cycloadditions, on the other hand, are very common. (Although some of these, too, may be stepwise reactions, many are clearly concerted.)

Of thermal cycloadditions, then, [4 + 2] is easy and [2 + 2] is difficult. Of [2 + 2] cycloadditions, the thermal reaction is difficult and the photochemical reaction is easy. How are we to account for these contrasts?

In cycloaddition, two new σ bonds are formed by use of π electrons of the reactants. The concerted reaction results from overlap of orbitals of one molecule

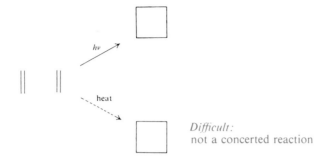

Difficult:
not a concerted reaction

with orbitals of the other. As before, it is on electrons in the HOMO that we focus attention. But which orbital does the HOMO overlap? It picks an orbital *into which its electrons can flow*: an unoccupied orbital. And of unoccupied orbitals it picks the most stable, the *lowest unoccupied molecular orbital (LUMO)*. In the transition state of cycloaddition, then, *stabilization comes chiefly from overlap between the HOMO of one reactant and the LUMO of the other.*

On this basis, let us examine the [4 + 2] cycloaddition of 1,3-butadiene and ethylene, the simplest example of the Diels–Alder reaction. The electronic configurations of these compounds—and of dienes and alkenes in general—have been given in Fig. 33.5 (p. 1133) and Fig. 33.6 (p. 1134). There are two combinations: overlap of the HOMO of butadiene (ψ_2) with the LUMO of ethylene (π^*); and overlap of the HOMO of ethylene (π) with the LUMO of butadiene (ψ_3). In either case, as Fig. 33.20 shows, overlap brings together lobes of the same phase. There is a flow of electrons from HOMO to LUMO, and bonding occurs.

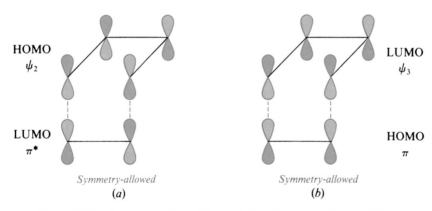

HOMO
ψ_2

LUMO
ψ_3

LUMO
π^*

HOMO
π

Symmetry-allowed
(*a*)

Symmetry-allowed
(*b*)

Figure 33.20 Symmetry-allowed thermal [4 + 2] cycloaddition: 1,3-butadiene and ethylene. Overlap of (*a*) the HOMO of 1,3-butadiene and the LUMO of ethylene, and (*b*) the HOMO of ethylene and the LUMO of 1,3-butadiene.

Now, consider a thermal [2 + 2] cyclization, dimerization of ethylene. This would involve overlap of the HOMO, π, of one molecule with the LUMO, π^*, of the other. But π and π^* are of opposite symmetry, and, as Fig. 33.21 (p. 1152) shows, lobes of opposite phase would approach each other. Interaction is antibonding and repulsive, and concerted reaction does not occur.

π HOMO

π^* LUMO

Symmetry-forbidden

Figure 33.21   Symmetry-forbidden thermal [2 + 2] cycloaddition: two molecules of ethylene. Interaction is antibonding.

Photochemical [2 + 2] cycloadditions are symmetry-allowed. Here we have (Fig. 33.22) overlap of the HOMO (π^*) of an excited molecule with the LUMO (also π^*) of a ground-state molecule.

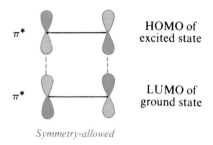

π^* HOMO of excited state

π^* LUMO of ground state

Symmetry-allowed

Figure 33.22   Symmetry-allowed photochemical [2 + 2] cycloaddition: two molecules of ethylene, one excited and one in the ground state. Interaction is bonding.

If, in a concerted reaction of this kind, both bonds to a component are being formed (or broken) on the same face, the process is said to be *suprafacial*. If the bonds are being formed (or broken) on opposite faces, the process is *antarafacial*.

Suprafacial   Antarafacial

These terms resemble the familiar ones *syn* and *anti*, but with this difference. *Syn* and *anti* describe the net stereochemistry of a reaction. We have seen *anti* addition, for example, as the overall result of a two-step mechanism. *Suprafacial* and *antarafacial*, in contrast, refer to actual processes: the simultaneous making (or breaking) of two bonds on the same face or opposite faces of a component.

So far, our discussion of cycloaddition has assumed that reaction is suprafacial with respect to both components. For [4 + 2] cycloadditions, the stereochemistry shows that this is indeed the case. Now, as far as orbital symmetry is concerned, thermal [2 + 2] cycloaddition *could* occur if it were suprafacial with respect to one component and antarafacial with respect to the other (Fig. 33.23). Almost

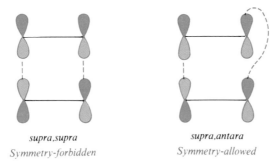

supra,supra   *supra,antara*

Symmetry-forbidden   *Symmetry-allowed*

Figure 33.23   [2 + 2] Cycloaddition. *Supra,supra*: geometrically possible, but symmetry-forbidden. *Supra,antara*: symmetry-allowed, but geometrically difficult.

certainly, such a *supra,antara* process is impossible here on geometric grounds. But if the ring being formed is big enough, both *supra,supra* and *supra,antara* processes are geometrically possible; in that case orbital symmetry determines, not *whether* cycloaddition occurs, but *how* it occurs (Table 33.2).

Table 33.2   Woodward–Hoffmann Rules for [i + j] Cycloadditions

$i + j$	Thermal	Photochemical
$4n$	supra,antara antara,supra	supra,supra antara,antara
$4n + 2$	supra,supra antara,antara	supra,antara antara,supra

Cycloadditions are reversible. These *cycloreversions* (for example, the *retro-* Diels–Alder reaction) follow the same symmetry rules as cycloadditions—as they must, of course, since they occur via the same transition states.

Problem 33.7   Give structural formulas for the products expected from each of the following reactions. Tell *why* you expect the particular products.

(a) *trans,trans*-2,4-hexadiene + ethylene

(b) *trans*-1,3-pentadiene + maleic anhydride

(c) *trans,trans*-1,4-diphenyl-1,3-butadiene + maleic anhydride

(d) *cis*-2-butene $\xrightarrow{h\nu}$ A + B

(e) *trans*-2-butene $\xrightarrow{h\nu}$ A + C

(f) *cis*-2-butene + *trans*-2-butene $\xrightarrow{h\nu}$ A + B + C + D

Problem 33.8 On standing, cyclopentadiene spontaneously forms *dicyclopentadiene* (I), from which it can be regenerated by heating under a fractionating column.

I   II

Dicyclopentadiene

(a) What reaction has taken place in the formation of dicyclopentadiene? In the regeneration of cyclopentadiene? (b) On what basis could you have predicted that dicyclopentadiene would have the structure I rather than the structure II?

Problem 33.9 Each of the following reactions is believed to be concerted. Tell what kind of reaction is involved in each case, and what significance it bears on orbital symmetry theory.

(a)

(b)

(c)

(d)

(e)

33.10   Sigmatropic reactions

A concerted reaction of the type,

$$
\begin{array}{ccc}
\overset{\displaystyle G}{\underset{\displaystyle |}{C}}-(C{=}C)_n & \longrightarrow & \overset{\displaystyle G}{\underset{\displaystyle |}{(C{=}C)_n{-}C}}
\end{array}
$$

in which a group migrates with its σ bond within a π framework—an ene or a polyene—is called a **sigmatropic reaction**.

The migration is accompanied by a shift in π bonds. For example:

|1,3| Sigmatropic reaction

|1,5| Sigmatropic reaction

In the designations [1,3] and [1,5] the "3" and "5" refer to the number of the carbon to which group G is migrating (the migration terminus). The "1" does *not* refer to the migration source; instead, it specifies that in both reactant and product bonding is to the same atom (number 1) in the migrating group. The important *Cope rearrangment* of hexa-1,5-dienes, for example, is a [3,3] sigmatropic reaction,

A 1,5-hexadiene

|3,3| Sigmatropic reaction

in which there is a change in position of attachment in G as well as in the π framework—indeed, G itself is a π framework.

In the transition state of a sigmatropic reaction, the migrating group is bonded to both the migration source and the migration terminus; it is the nature of this transition state that we are concerned with. In Sec. 1.8, for convenience we considered bonding in the H_2 molecule to arise from overlap between orbitals on two hydrogen atoms. In the same way, and simply *for convenience*, we consider bonding in the transition state for sigmatropic reactions to arise from overlap between an orbital of an atom or free radical (G) and an orbital of an allylic free radical (the π framework).

This does *not* mean that rearrangement actually involves the separation and reattachment of a free radical. Such a stepwise reaction would not be a concerted one, and hence is not the kind of reaction we are dealing with here. Indeed, a stepwise reaction would be a (high-energy) alternative open to a system if a (concerted) sigmatropic rearrangement were symmetry-forbidden.

In the transition state, there is overlap between the HOMO of one component and the HOMO of the other. Each HOMO is singly occupied, and together they provide a pair of electrons.

The HOMO of an allylic radical depends on the number of carbons in the π framework. The migrating group is passed from one end of the allylic radical to the other, and so it is the end carbons that we are concerned with. We see that the

HOMO of allylic radicals

symmetry at these end carbons alternates regularly as we pass from C_3 to C_5 to C_7, and so on. The HOMO of the migrating group depends, as we shall see, on the nature of the group.

Let us consider first the simplest case: **migration of hydrogen**. Stereochemically, this shift can be suprafacial or antarafacial:

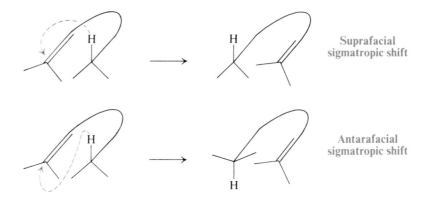

In the transition state, a *three-center bond* is required, and this must involve overlap between the *s* orbital of the hydrogen and lobes of *p* orbitals of the two terminal carbons. Whether a suprafacial or antarafacial shift is allowed depends upon the symmetry of these terminal orbitals:

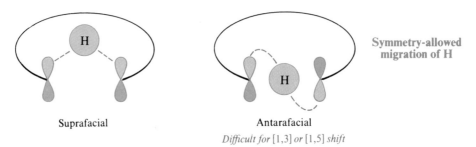

Suprafacial

Antarafacial

Difficult for [1,3] *or* [1,5] *shift*

Symmetry-allowed migration of H

Whether a sigmatropic rearrangement actually takes place, though, depends not only on the symmetry requirements but also on the *geometry* of the system. In particular, [1,3] and [1,5] *antara* shifts should be extremely difficult, since they would require the π framework to be twisted far from the planarity that it requires for delocalization of electrons.

Practically, then, [1,3] and [1,5] sigmatropic reactions seem to be limited to *supra* shifts. A [1,3] *supra* shift of hydrogen is symmetry-forbidden; since the s orbital of hydrogen would have to overlap p lobes of opposite phase, hydrogen cannot be bonded simultaneously to both carbons. A [1,5] *supra* shift of hydrogen, on the other hand, is symmetry-allowed.

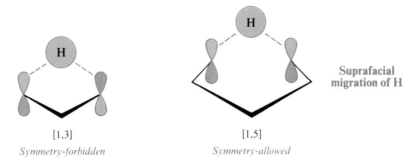

Suprafacial migration of H

[1,3]   [1,5]

Symmetry-forbidden   *Symmetry-allowed*

For larger π frameworks, both *supra* and *antara* shifts should be possible on geometric grounds, and here we would expect the stereochemistry to depend simply on orbital symmetry. A [1,7]-H shift, for example, should be *antara*, a [1,9]-H shift, *supra*, and so on. For photochemical reactions, as before, predictions are exactly reversed.

The facts agree with the above predictions: [1,3] sigmatropic shifts of hydrogen are not known, whereas [1,5] shifts are well known. For example:

$$\underset{=CD_2}{\overset{-CH_3}{\Big\langle}} \quad \xrightarrow[\text{heat}]{[1,5]\text{-H}} \quad \underset{-CHD_2}{\overset{=CH_2}{\Big\langle}}$$

The preference for [1,5]-H shifts over [1,3]-H shifts has been demonstrated many times. For example, the heating of 3-deuterioindene (I) causes scrambling of the label to *all three* non-aromatic positions. Let us examine this reaction.

I   II   III

We cannot account for the formation of II on the basis of [1,3] shifts: migration of D would regenerate I;

I   I

migration of H would yield only III.

But if we include the *p* orbitals of the benzene ring, and count along the edge of this ring, we see that a [1,5] shift of D would yield the unstable non-aromatic

intermediate IVa. This, in turn, can transfer H or D by [1,5] shifts to yield all the observed products (see Fig. 33.24).

Figure 33.24   Deuterium scrambling in indene via unstable intermediates IVa and IVb: a series of [1,5] hydrogen shifts.

So far we have discussed only migration of hydrogen, which is necessarily limited to the overlap of an *s* orbital. Now let us turn to **migration of carbon**. Here, we have two possible kinds of bonding to the migrating group. One of these is similar to what we have just described for migration of hydrogen: bonding of both ends of the π framework to the same lobe on carbon. Depending on the symmetry of the π framework, the symmetry-allowed migration may be suprafacial or antarafacial.

With carbon, a new aspect appears: the stereochemistry in the migrating group. Bonding through the same lobe on carbon means attachment to the same face of the atom, that is to say, *retention of configuration in the migrating group*.

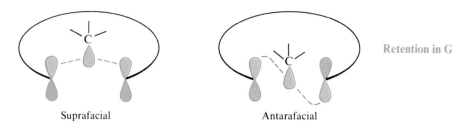

Suprafacial   Antarafacial

Retention in G

But there is a second possibility for carbon: bonding to the two ends of the π framework through different lobes of a *p* orbital. These lobes are on opposite faces of carbon—exactly as in an S_N2 reaction—and there is *inversion of configuration in the migrating group.*

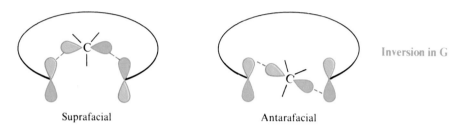

Suprafacial   Antarafacial

Inversion in G

For [1,3] and [1,5] shifts, the geometry pretty effectively prevents antarafacial migration. Limiting ourselves, then, to suprafacial migrations, we make these predictions: [1,3] migration with inversion; [1,5] migration with retention. *These predictions have been borne out by experiment.*

In 1968, Jerome Berson (of Yale University) reported that the deuterium-labeled bicyclo[3.2.0]heptene V is converted stereospecifically into the

$$\xrightarrow{300\ °C}$$

V   VI

exo-norbornene VI. As Fig. 33.25 shows, this reaction proceeds by a [1,3] migration and with *complete inversion* of configuration in the migrating group.

V   VI

Figure 33.25   The deuterium-labeled bicyclo[3.2.0]heptene V rearranges via a [1,3]-C shift to the norbornene VI. There is *inversion of configuration* at C–7: from *R* to *S*. (Or, using C–6 as our standard, we see that H eclipses OAc in V, and D eclipses OAc in VI.)

In 1970, H. Kloosterziel (of the University of Technology, Eindhoven, The Netherlands) reported a study of the rearrangement of the diastereomeric 6,9-dimethylspiro[4.4]nona-1,3-dienes (*cis*-VII and *trans*-VII) to the dimethyl-bicyclo-[4.3.0]nonadienes VIII, IX, and X. These reactions are completely stereoselective and stereospecific.

As Fig. 33.26 shows, they proceed by [1,5] migrations and with *complete retention* of configuration in the migrating group.

To predict a different stereochemistry between [1,3] and [1,5] migrations, and in particular to predict *inversion* in the [1,3] shift—certainly not the easier path on geometric grounds—is certainly "risky". The fulfillment of such predictions demonstrates both the validity and the power of the underlying theory.

Figure 33.26 Rearrangement of *cis*-6,9-dimethylspiro[4.4]nona-1,3-diene. Migration of C–6 from C–5 to C–4 is a [1,5]-C shift. (We count 5, 1, 2, 3, 4.) Configuration at C–6 is *retained*, as shown by its relationship to configuration at C–9. Successive [1,5]-H shifts then yield the other products.

Problem 33.10 In each of the following, the high stereoselectivity or regioselectivity provides confirmation of predictions based on orbital symmetry. Show how this is so. (*Use models.*)

(a)

(b) When 1,3,5-cyclooctatriene labeled with deuterium at the 7 and 8 positions was heated, it gave products labeled only at the 3, 4, 7, and 8 positions.

(c)

PROBLEMS

1. Tropolone (I, $C_7H_7O_2$) has a flat molecule with all carbon–carbon bonds of the same

Tropolone

I

length (1.40 Å). The measured heat of combustion is 20 kcal lower than that calculated by the method of Problem 13.2 (p. 483). Its dipole moment is 3.71 D; that of 5-bromotropolone is 2.07 D.

Tropolone undergoes the Reimer–Tiemann reaction, couples with diazonium ions, and is nitrated by dilute nitric acid. It gives a green color with ferric chloride, and does not react with 2,4-dinitrophenylhydrazine. Tropolone is both acidic ($K_a = 10^{-7}$) and weakly basic, forming a hydrochloride in ether.

(a) What class of compounds does tropolone resemble? Is it adequately represented by formula I? (b) Using both valence-bond and orbital structures, account for the properties of tropolone.

(c) In what direction is the dipole moment of tropolone? Is this consistent with the structure you have proposed?

(d) The infrared spectrum of tropolone shows a broad band at about 3150 cm^{-1} that changes only slightly upon dilution. What does this tell you about the structure of tropolone?

2. Each transformation shown below is believed to involve a concerted reaction. In each case show just what is happening.

(a)

$\xrightarrow{110\ °C}$ $\xrightarrow{250\ °C}$

cis-Bicyclo[4.2.0]-octa-7-ene   *cis,trans*-Cyclo-octa-1,3-diene   *cis,cis*-Cyclo-octa-1,3-diene

(b)

$\xrightarrow{\text{heat}}$ $\longrightarrow$

cis-Bicyclo[6.2.0]-deca-2,4,6,9-tetraene   *cis,cis,cis,cis,trans*-Deca-1,3,5,7,9-pentaene   *trans*-9,10-Dihydronaphthalene

(c)

$\xrightarrow{h\nu}$ +

$\updownarrow h\nu$

(d)

cis-9,10-Dihydronaphthalene

(e)   + CH$_2$=C(CH$_3$)CH$_2$I + AgOOCCCl$_3$ $\longrightarrow$

(f)

(g)

3. Each of the following transformations is believed to proceed by the indicated sequence of concerted reactions. Show just what each step involves, and give structures of compounds A–J.

(a) Electrocyclic closure; electrocyclic closure.

(b) [1,5]-H shift; electrocyclic opening.

(c) Electrocyclic opening; electrocyclic closure. Final products are not interconvertible at 170 °C; be sure you account for *both* of them.

(d) Three electrocyclic closures.

cis,cis-Cyclonona-1,3-diene

$$\text{E} \xleftarrow{\ hv\ } \text{?} \xleftarrow{175\ °C} \text{F} \xrightarrow{\ hv\ } \text{G}$$

(e) A series of *supra* H shifts.

4. Account for the difference in conditions required to bring about the following transformations:

5. Give stereochemical structures of K and L, and tell exactly what process is taking place in each reaction.

cis,cis,cis-Cycloocta-1,3,5-triene $\xrightarrow{80–100\ °C}$ K (C_8H_{10})

K + $CH_3OOCC\equiv CCOOCH_3$ $\longrightarrow$ L ($C_{14}H_{16}O_4$)

L $\xrightarrow{\text{heat}}$ cyclobutene + dimethyl phthalate

6. (a) The familiar rearrangement of a carbocation by a 1,2-alkyl shift is, as we have described it (Sec. 5.23), a concerted reaction. Its ease certainly suggests that it is symmetry-allowed. Discuss the reaction from the standpoint of orbital symmetry. What stereochemistry would you predict in the migrating group?

(b) There is evidence that concerted 1,4-alkyl shifts of the kind

can occur. What stereochemistry would you predict in the migrating group?

7. Discuss the direct, concerted, non-catalytic addition of H_2 to an alkene from the standpoint of orbital symmetry.

8. The deuterium scrambling between II and III has been accounted for on the basis of intramolecular Diels–Alder and *retro*-Diels–Alder reactions. Show how this might occur.

(*Hint*: Look for an intermediate that is symmetrical except for the presence of deuterium.)

9. Suggest an explanation for each of the following facts.
(a) When the diazonium salt IV is treated with *trans,trans*-2,4-hexadiene, N_2 and CO_2 are evolved, and there is obtained stereochemically pure V. (*Hint*: See Problem 18, p. 1056.)

(b) In contrast, decomposition of IV in either *cis*- or *trans*-1,2-dichloroethene yields a mixture of *cis*- and *trans*-VI.

10. For each of the following reactions suggest an intermediate that would account for the formation of the product. *Show exact stereochemistry*. (For a hint, see Fig. 33.24, p. 1158.)

11. (a) The diastereomeric 6,9-dimethylspiro[4.4]nona-1,3-dienes (p. 1160) were synthesized by reaction of cyclopentadiene with diastereomeric 2,5-dibromohexanes in the presence of sodium amide. Which 2,5-dibromohexane would you expect to yield each spirane?
(b) The stereochemistry of the spiranes obtained was shown by comparison of their NMR spectra, specifically, of the peaks due to the olefinic hydrogens. Explain.

12. (a) Berson synthesized the stereospecifically labeled compound V (p. 1159) by the following sequence. Give structures for compounds M–R.

+ B_2D_6, then H_2O_2 ⟶   M + N (both C_7H_9DO)

M + N $\xrightarrow{\text{oxidation}}$ O + P (both C_7H_7DO), *separated*

O + LiAl(OBu-*t*)$_3$H, then $(CH_3CO)_2O$ ⟶ Q + R (both $C_8H_{11}O_2$), *separated*

Q is compound V on p. 1159.

(b) Berson's study of the rearrangement of V to VI (p. 1159) was complicated by the tendency of VI, once formed, to decompose into cyclopentadiene and vinyl acetate. What kind of reaction is this decomposition?

13. (a) Woodward and Hoffmann have suggested that the *endo* preference in Diels–Alder reactions is a "secondary" effect of orbital symmetry, and there is experimental evidence to support this suggestion. Using the dimerization of butadiene (Fig. 33.19, p. 1150) as your example, show how these secondary effects could arise. (*Hint*: Draw the orbitals involved and examine the structures closely.)

(b) In contrast, [6 + 4] cycloaddition was predicted to take place in the *exo* sense. This has been confirmed by experiment. Using the reaction of *cis*-1,3,5-hexatriene with 1,3-butadiene as example, show how this prediction could have been made.

14. (a) The sex attractant of the male boll weevil has been synthesized by the following sequence. Give stereochemical structures for compounds S–Y.

ethylene + 3-methyl-2-cyclohexenone $\xrightarrow{hv}$ S ($C_9H_{14}O$)

S $\xrightarrow{\text{bromination}}$ T ($C_9H_{13}OBr$)

T + CO_3^{2-} ⟶ U ($C_9H_{12}O$)

U + CH_3Li ⟶ V ($C_{10}H_{16}O$), *a single stereoisomer* (*Hint*: Examine structure of U.)

V + IO_4^-/OsO_4 ⟶ W ($C_9H_{14}O_3$), *a carboxylic acid*

W + excess $Ph_3P{=}CH_2$ ⟶ X ($C_{10}H_{16}O_2$) $\xrightarrow{\text{NaAl(OR)}_2H_2}$ Y ($C_{10}H_{18}O$),

the sex attractant

(b) The stereochemistry of the sex attractant was confirmed by the following reaction. Give a stereochemical formula for Z, and show how this confirms the stereochemistry.

Y + $Hg(OAc)_2$, then $NaBH_4$ ⟶ Z ($C_{10}H_{18}O$)

15. (a) Although "Dewar benzene", VII, is less stable by 60 kcal than its isomer benzene, its conversion into benzene is surprisingly slow, with an E_{act} of about 37 kcal. It has a half-life at room temperature of two days; at 90 °C complete conversion into benzene takes half an hour.

The high E_{act} for conversion of VII into benzene is attributed to the fact that the reaction is symmetry-forbidden. Explain.

VII   VIII

(b) In Problem 17, p. 1097, we outlined the synthesis of VIII. Although much less stable than its aromatic isomer toluene, this compound is surprisingly long-lived. Here, too, the conversion is considered to be symmetry-forbidden. Explain.

16. (a) In the skin of animals exposed to sunlight, 7-dehydrocholesterol is converted

$C_8H_{17} = -\overset{\underset{\displaystyle |}{CH_3}}{CH}CH_2CH_2CH_2CH\overset{\displaystyle CH_3}{\underset{\displaystyle CH_3}{}}$

7-Dehydrocholesterol

into the hormone *cholecalciferol*, the so-called "vitamin" D_3 that plays a vital role in the development of bones. In the laboratory, the following sequence was observed:

7-Dehydrocholesterol   Pre-cholecalciferol   Cholecalciferol

What processes are actually taking place in these two reactions? Show details.

(b) An exactly analogous reaction sequence is used to convert the plant steroid ergosterol (p. 660) into *ergocalciferol*, the "vitamin" D_2 that is added to milk:

$$ergosterol \xrightarrow{hv} pre\text{-}ergocalciferol \xrightarrow{warm} ergocalciferol$$

What is the structure of pre-ergocalciferol? Of ergocalciferol?

(c) On heating at 190 °C, pre-ergocalciferol is converted into IX and X, stereoisomers of ergosterol. What reaction is taking place, and what are the structures of IX and X?

(d) Still another stereoisomer of ergosterol, XI, can be converted by ultraviolet light into pre-ergocalciferol. What must XI be?

17. On photolysis at room temperature, *trans*-XII was converted into *cis*-XII. When *trans*-XII was photolyzed at − 190 °C, however, no *cis*-XII could be detected in the reaction

trans-XII   *cis*-XII

mixture. When *trans*-XII was photolyzed at − 190 °C, allowed to warm to room temperature, and then cooled again to − 190 °C, *cis*-XII was obtained. If, instead, the low-temperature photolysis mixture was reduced at − 190 °C, cyclodecane was formed; reduction of the room-temperature photolysis mixture gave only a trace of cyclodecane.

On the basis of these and other facts, E. E. van Tamelen (of Stanford University) proposed a two-step mechanism, consistent with orbital symmetry theory, for the conversion of *trans*-XII into *cis*-XII.

(a) Suggest a mechanism for the transformation. Show how it accounts for the facts.

(b) The intermediate proposed by van Tamelen—never isolated and never before identified—is of considerable theoretical interest. Why? What conclusion do you draw about this compound from the facts?

34

Polynuclear Aromatic Compounds

34.1 Fused-ring aromatic compounds

Two aromatic rings that share a pair of carbon atoms are said to be *fused*. In this chapter we shall study the chemistry of the simplest and most important of the fused-ring hydrocarbons, **naphthalene,** $C_{10}H_8$, and look briefly at two others of formula $C_{14}H_{10}$, **anthracene** and **phenanthrene**.

Naphthalene   Anthracene   or   Phenanthrene

All three of these hydrocarbons are obtained from coal tar, naphthalene being the most abundant (5%) of all constituents of coal tar.

If diamond (p. 438) is the ultimate polycyclic aliphatic system, then the other allotropic form of elemental carbon, *graphite*, might be considered the ultimate in fused-ring aromatic systems. X-ray analysis shows that the carbon atoms are arranged in layers. Each layer is a continuous network of planar, hexagonal rings; the carbon atoms within a layer are held together by strong, covalent bonds 1.42 Å long (only slightly longer than those in benzene, 1.39 Å). The different layers, 3.4 Å apart, are held to each other by comparatively weak

1167

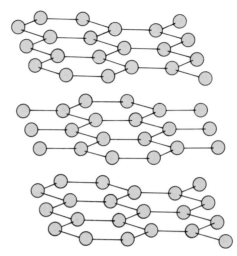

forces. The lubricating properties of graphite (its "greasy" feel) may be due to slipping of layers (with adsorbed gas molecules between) over one another.

Table 34.1 POLYNUCLEAR AROMATIC COMPOUNDS

Name	M.p., °C	B.p., °C	Name	M.p., °C	B.p., °C
Naphthalene	80	218	1-Naphthalenesulfonic acid	90	
1,4-Dihydronaphthalene	25	212	2-Naphthalenesulfonic acid	91	
Tetralin	− 30	208	1-Naphthol	96	280
cis-Decalin	− 43	194	2-Naphthol	122	286
trans-Decalin	− 31	185	1,4-Naphthoquinone	125	
1-Methylnaphthalene	− 22	241			
2-Methylnaphthalene	38	240	Anthracene	217	354
1-Bromonaphthalene	6	281	9,10-Anthraquinone	286	380
2-Bromonaphthalene	59	281	Phenanthrene	101	340
1-Chloronaphthalene		263	9,10-Phenanthrenequinone	207	
2-Chloronaphthalene	46	265	Chrysene	255	
1-Nitronaphthalene	62	304	Pyrene	150	
2-Nitronaphthalene	79		1,2-Benzanthracene	160	
1-Naphthylamine	50	301	Dibenz[a,h]anthracene	262	
2-Naphthylamine	113	294	3-Methylcholanthrene	180	

NAPHTHALENE

34.2 Nomenclature of naphthalene derivatives

Positions in the naphthalene ring system are designated as in I. Two isomeric

I

monosubstituted naphthalenes are differentiated by the prefixes 1- and 2-, or α- and β-. The arrangement of groups in more highly substituted naphthalenes is indicated by numbers. For example:

1,5-Dinitronaphthalene

6-Amino-2-naphthalenesulfonic acid

2-Naphthol
β-Naphthol

2,4-Dinitro-1-naphthylamine

Problem 34.1   How many different mononitronaphthalenes are possible? Dinitro-naphthalenes? Nitronaphthylamines?

34.3   Structure of naphthalene

Naphthalene is classified as aromatic because its properties resemble those of benzene (see Sec. 13.10). Its molecular formula, $C_{10}H_8$, might lead one to expect a high degree of unsaturation; yet naphthalene is resistant (although less so than benzene) to the addition reactions characteristic of unsaturated compounds. Instead, the typical reactions of naphthalene are electrophilic substitution reactions, in which hydrogen is displaced as hydrogen ion and the naphthalene ring system is preserved. Like benzene, naphthalene is unusually stable: its heat of combustion is 61 kcal lower than that calculated on the assumption that it is aliphatic (see Problem 13.2, p. 483).

From the experimental standpoint, then, naphthalene is classified as aromatic on the basis of its properties. From a theoretical standpoint, naphthalene has the structure required of an aromatic compound: it contains flat six-membered rings, and consideration of atomic orbitals shows that the structure can provide π clouds containing six electrons—the *aromatic sextet* (Fig. 34.1). Ten carbons lie at the

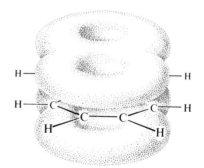

Figure 34.1   Naphthalene molecule. π clouds above and below the plane of the rings.

corners of two fused hexagons. Each carbon is attached to three other atoms by σ bonds; since these σ bonds result from the overlap of trigonal sp^2 orbitals, all carbon and hydrogen atoms lie in a single plane. Above and below this plane there is a cloud of π electrons formed by the overlap of p orbitals and shaped like a figure 8. We can consider this cloud as two partially overlapping sextets that have a pair of π electrons in common. (See Fig. 34.2.)

Figure 34.2   Electronic configuration and molecular shape. Model of the naphthalene molecule: two views.

In terms of valence bonds, naphthalene is considered to be a resonance hybrid of the three structures I, II, and III. Its resonance energy, as shown by the heat of combustion, is 61 kcal/mol.

I   II   III

X-ray analysis shows that, in contrast to benzene, all carbon–carbon bonds in naphthalene are not the same; in particular, the C(1)–C(2) bond is considerably shorter (1.365 Å) than the C(2)–C(3) bond (1.404 Å). Examination of structures I, II, and III shows us that this difference in bond lengths is to be expected. The C(1)–C(2) bond is double in two structures and single in only one; the C(2)–C(3) bond is single in two structures and double in only one. We would therefore expect the C(1)–C(2) bond to have more double-bond character than single, and the C(2)–C(3) bond to have more single-bond character than double.

For convenience, we shall represent naphthalene as the single structure IV, in which the circles stand for partially overlapping aromatic sextets.

IV

Although representation IV suggests a greater symmetry for naphthalene than exists, it has the advantage of emphasizing the aromatic nature of the system.

34.4   Reactions of naphthalene

Like benzene, naphthalene typically undergoes electrophilic substitution; this is one of the properties that entitle it to the designation of "aromatic". An electrophilic reagent finds the π cloud a source of available electrons, and attaches itself to the ring to form an intermediate carbocation; to restore the stable aromatic system, the carbocation then gives up a proton.

Naphthalene undergoes oxidation or reduction more readily than benzene, but only to the stage where a substituted benzene is formed; further oxidation or reduction requires more vigorous conditions. Naphthalene is stabilized by resonance to the extent of 61 kcal/mol; benzene is stabilized to the extent of 36 kcal/mol. When the aromatic character of one ring of naphthalene is destroyed, only 25 kcal of resonance energy is sacrificed; in the next stage, 36 kcal has to be sacrificed.

REACTIONS OF NAPHTHALENE

1. Oxidation. Discussed in Sec. 34.5.

CrO$_3$, HOAc, 25 °C

1,4-Naphthoquinone
α-Naphthoquinone
(40% yield)

Naphthalene

O$_2$, V$_2$O$_5$, 460–480 °C

Phthalic anhydride
(76% yield)

2. Reduction. Discussed in Sec. 34.6.

Na, C$_2$H$_5$OH, reflux

1,4-Dihydronaphthalene

Naphthalene

Na, C$_5$H$_{11}$OH, reflux

1,2,3,4-Tetrahydronaphthalene
Tetralin

H$_2$, catalyst

Decahydronaphthalene
Decalin

CONTINUED

CONTINUED

3. **Electrophilic substitution.** Discussed in Secs. 34.8–34.13.

  (a) **Nitration.** Discussed in Sec. 34.8.

Naphthalene $\xrightarrow{HNO_3, H_2SO_4, 50-60\ °C}$ 1-Nitronaphthalene
α-Nitronaphthalene
(90–95% yield)

  (b) **Halogenation.** Discussed in Sec. 34.8.

Naphthalene $\xrightarrow[\substack{No\ Lewis\ acid \\ needed}]{Br_2, CCl_4, reflux}$ 1-Bromonaphthalene
α-Bromonaphthalene
(75% yield)

  (c) **Sulfonation.** Discussed in Sec. 34.11.

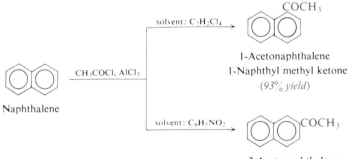

Naphthalene

conc. H_2SO_4, 80 °C →

1-Naphthalenesulfonic acid
α-Naphthalenesulfonic acid

conc. H_2SO_4, 160 °C

conc. H_2SO_4, 160 °C →

2-Naphthalenesulfonic acid
β-Naphthalenesulfonic acid

  (d) **Friedel–Crafts acylation.** Discussed in Sec. 34.10.

Naphthalene $\xrightarrow{CH_3COCl, AlCl_3}$

solvent: $C_2H_2Cl_4$ →

1-Acetonaphthalene
1-Naphthyl methyl ketone
(93% yield)

solvent: $C_6H_5NO_2$ →

2-Acetonaphthalene
2-Naphthyl methyl ketone
(90% yield)

34.5   Oxidation of naphthalene

Oxidation of naphthalene by oxygen in the presence of vanadium pentoxide destroys one ring and yields phthalic anhydride. Because of the availability of naphthalene from coal tar, and the large demand for phthalic anhydride (for example, see Secs. 34.18 and 36.7), this is an important industrial process.

Oxidation of certain naphthalene derivatives destroys the aromatic character of one ring in a somewhat different way, and yields diketo compounds known as *quinones* (Sec. 31.9). For example:

2-Methylnaphthalene

2-Methyl-1,4-naphthoquinone
(*70% yield*)

Because of this tendency to form quinones, it is not always feasible to prepare naphthalenecarboxylic acids as we do benzoic acids, by oxidation of methyl side chains.

Problem 34.2 Show how 1- and 2-naphthalenecarboxylic acids (α- and *β-naphthoic acids*) can be obtained from naphthalene by way of the corresponding acetonaphthalenes.

34.6   Reduction of naphthalene

In contrast to benzene, naphthalene can be reduced by chemical reducing agents. It is converted by sodium and ethanol into 1,4-dihydronaphthalene, and by sodium and isopentyl alcohol into 1,2,3,4-tetrahydronaphthalene (*tetralin*). The temperature at which each of these sodium reductions is carried out is the boiling point of the alcohol used; at the higher temperature permitted by isopentyl alcohol (b.p. 132 °C), reduction proceeds further than with the lower-boiling ethyl alcohol (b.p. 78 °C).

Naphthalene

1,4-Dihydronaphthalene

1,2,3,4-Tetrahydronaphthalene
Tetralin

The tetrahydronaphthalene is simply a dialkyl derivative of benzene. As with other benzene derivatives, the aromatic ring that remains is reduced only by vigorous catalytic hydrogenation.

Tetralin $\xrightarrow{\text{H}_2, \text{Pt or Ni}}$ Decalin

Problem 34.3 *Decalin* exists in two stereoisomeric forms, *cis*-decalin (b.p. 194 °C) and *trans*-decalin (b.p. 185 °C).

cis-Decalin

trans-Decalin

(a) Build models of these compounds and see that they differ from one another. Locate in the models the pair of hydrogen atoms, attached to the fused carbons, that are *cis* or *trans* to each other.

(b) In *trans*-decalin is one ring attached to the other by two equatorial bonds, by two axial bonds, or by one axial bond and one equatorial bond? In *cis*-decalin? Remembering (Sec. 12.12) that an equatorial position gives more room than an axial position for a bulky group, predict which should be the more stable isomer, *cis*- or *trans*-decalin.

(c) Account for the following facts: rapid hydrogenation of tetralin over a platinum black catalyst at low temperatures yields *cis*-decalin, while slow hydrogenation of tetralin over nickel at high temperatures yields *trans*-decalin. Compare this with 1,2- and 1,4-addition to conjugated dienes (Sec. 10.27), Friedel–Crafts alkylation of toluene (Sec. 15.12), sulfonation of phenol (Problem 28.12, p. 1013), and sulfonation of naphthalene (Sec. 34.11).

34.7 Dehydrogenation of hydroaromatic compounds. Aromatization

Compounds like 1,4-dihydronaphthalene, tetralin, and decalin, which contain the carbon skeleton of an aromatic system but too many hydrogen atoms for aromaticity, are called *hydroaromatic compounds*. They are sometimes prepared, as we have seen, by partial or complete hydrogenation of an aromatic system.

More commonly, however, the process is reversed, and hydroaromatic compounds are converted into aromatic compounds. Such a process is called **aromatization**.

One of the best methods of aromatization is **catalytic dehydrogenation**, accomplished by heating the hydroaromatic compound with a catalyst like platinum, palladium, or nickel. We recognize these as the catalysts used for hydrogenation;

since they lower the energy barrier between hydrogenated and dehydrogenated compounds, they speed up reaction in *both* directions (see Sec. 8.3). The position of the equilibrium is determined by other factors: hydrogenation is favored by an excess of hydrogen under pressure; dehydrogenation is favored by sweeping away the hydrogen in a stream of inert gas. For example:

Tetralin   Naphthalene

In an elegant modification of dehydrogenation, hydrogen is *transferred* from the hydroaromatic compound to a compound that readily accepts hydrogen. For example:

1-(α-Naphthyl)-   Chloranil   1-Phenylnaphthalene   Tetrachloro-
cyclohexene   Tetrachloro-   hydroquinone
                     benzoquinone

The tendency to form the stable aromatic system is so strong that, when necessary, groups can be eliminated: for example, a methyl group located at the point of fusion between two rings, a so-called *angular methyl group* (Sec. 17.18).

Abietic acid   1-Methyl-7-isopropylphenanthrene
(in rosin)

Aromatization has also been accomplished by heating hydroaromatic compounds with selenium, sulfur, or organic disulfides, RSSR. Here hydrogen is eliminated as H_2Se, H_2S, or RSH.

Problem 34.4   In a convenient laboratory preparation of dry hydrogen bromide, Br_2 is dripped into boiling tetralin; the vapors react to form naphthalene and four moles of hydrogen bromide. Account, step by step, for the formation of these products. What familiar reactions are involved in this aromatization?

Aromatization is important in both *synthesis* and *analysis*. Many polynuclear aromatic compounds are made from open-chain compounds by ring closure; the last step in such a synthesis is aromatization (see, for example, Secs. 34.14, 34.19, and 35.13). Many naturally occurring substances are hydroaromatic; conversion into identifiable aromatic compounds gives important information about their structures. For example:

Cholesterol: a steroid
Occurs in all animal tissues

3′-Methyl-1,2-cyclopentenophenanthrene
(Diels' hydrocarbon)

Problem 34.5 *Cadinene*, $C_{15}H_{24}$, is found in oil of cubebs. Dehydrogenation with sulfur converts cadinene into *cadalene*, $C_{15}H_{18}$, which can be synthesized from *carvone* by the following sequence:

$$+ \quad BrCH_2COOC_2H_5 \quad + \quad Zn \quad \longrightarrow \quad A\ (C_{14}H_{22}O_3)$$

Carvone

A + acid $\xrightarrow{\text{heat}}$ [B] $\xrightarrow{\text{isomerization}}$ C $(C_{12}H_{16}O_2)$, *a benzene derivative*

C + C_2H_5OH + H_2SO_4 $\longrightarrow$ D $(C_{14}H_{20}O_2)$

D + Na + alcohol $\longrightarrow$ E $(C_{12}H_{18}O)$ $\xrightarrow{\text{HBr}}$ F $(C_{12}H_{17}Br)$

F + $CH_3C(COOC_2H_5)_2{}^-Na^+$ $\longrightarrow$ G $(C_{20}H_{30}O_4)$

G + H_2SO_4 $\xrightarrow{\text{heat}}$ H $(C_{15}H_{22}O_2)$ $\xrightarrow{SOCl_2}$ I $(C_{15}H_{21}OCl)$

I + $AlCl_3$ $\longrightarrow$ J $(C_{15}H_{20}O)$ $\xrightarrow{H_2,\ Ni}$ K $(C_{15}H_{22}O)$

K + sulfur $\xrightarrow{\text{strong heating}}$ cadalene

(a) What is the structure and systematic name of cadalene? (b) What is a likely carbon skeleton for cadinene? (*Useful information*: In what is called the *Reformatsky* reaction, organozinc compounds act like (somewhat unreactive) Grignard reagents.)

34.8 Nitration and halogenation of naphthalene

Nitration and halogenation of naphthalene occur almost exclusively in the 1-position. Chlorination or bromination takes place so readily that a Lewis acid is not required for catalysis.

As we would expect, introduction of these groups opens the way to the preparation of a series of *alpha*-substituted naphthalenes: from 1-nitronaphthalene via the amine and diazonium salts, and from 1-bromonaphthalene via the Grignard reagent.

Synthesis of α-substituted naphthalenes

Problem 34.6   Starting with 1-nitronaphthalene, and using any inorganic or aliphatic reagents, prepare:

(a) 1-naphthylamine
(b) α-iodonaphthalene
(c) α-naphthonitrile
(d) α-naphthoic acid
    (1-naphthalenecarboxylic acid)
(e) α-naphthoyl chloride
(f) 1-naphthyl ethyl ketone

(g) 1-(aminomethyl)naphthalene, $C_{10}H_7CH_2NH_2$
(h) 1-(n-propyl)naphthalene
(i) α-naphthaldehyde
(j) (1-naphthyl)methanol
(k) 1-chloromethylnaphthalene
(l) (1-naphthyl)acetic acid
(m) N-(1-naphthyl)acetamide

Problem 34.7   Starting with 1-bromonaphthalene, and using any inorganic or aliphatic reagents, prepare:

(a) 1-naphthylmagnesium bromide
(b) α-naphthoic acid
    (1-naphthalenecarboxylic acid)
(c) 2-(1-naphthyl)-2-propanol

(d) 1-isopropylnaphthalene
(e) (1-naphthyl)methanol
(f) 1-(1-naphthyl)ethanol
(g) 2-(1-naphthyl)ethanol

Problem 34.8   (a) When 1-chloronaphthalene is treated with sodium amide, $Na^+NH_2^-$, in the secondary amine *piperidine* (Sec. 26.14), there is obtained not only I but also II,

in the ratio of 1:2. Similar treatment of 1-bromo- or 1-iodonaphthalene yields the same products *and in the same* 1:2 *ratio*. Show all steps in a mechanism that accounts for these observations. Can you suggest possible factors that might tend to favor II over I?

(b) Under the conditions of part (a), 1-fluoro-2-methylnaphthalene reacts to yield III. By what mechanism must this reaction proceed?

III

(c) Under the conditions of part (a), 1-fluoronaphthalene yields I and II, but in the ratio of 3:2. How do you account for this different ratio of products? What two factors make the fluoronaphthalene behave differently from the other halonaphthalenes?

34.9 Orientation of electrophilic substitution in naphthalene

Nitration and halogenation of naphthalene take place almost exclusively in the α-position. Is this orientation of substitution what we might have expected?

In our study of electrophilic substitution in the benzene ring (Chap. 14), we found that we could account for the observed orientation on the following basis: (a) the controlling step is the attachment of an electrophilic reagent to the aromatic ring to form an intermediate carbocation; and (b) this attachment takes place in such a way as to yield the most stable intermediate carbocation. Let us see if this approach can be applied to the nitration of naphthalene.

Attack by nitronium ion at the α-position of naphthalene yields an intermediate carbocation that is a hybrid of structures I and II in which the positive charge is accommodated by the ring under attack, and several structures like III in which the charge is accommodated by the other ring.

I	II	III	Alpha attack
More stable: *aromatic sextet* *preserved*	*More stable:* *aromatic sextet* *preserved*	*Less stable:* *aromatic sextet* *disrupted*	

Attack at the β-position yields an intermediate carbocation that is a hybrid of IV and V in which the positive charge is accommodated by the ring under attack, and several structures like VI in which the positive charge is accommodated by the other ring.

IV	V	VI	Beta attack
More stable: *aromatic sextet* *preserved*	*Less stable:* *aromatic sextet* *disrupted*	*Less stable:* *aromatic sextet* *disrupted*	

In structures I, II, and IV, the aromatic sextet is preserved in the ring that is not under attack; these structures thus retain the full resonance stabilization of one benzene ring (36 kcal/mol). In structures like III, V, and VI, on the other hand, the aromatic sextet is disrupted in both rings, with a large sacrifice of resonance stabilization. Clearly, structures like I, II, and IV are much the more stable.

But there are two of these stable contributing structures (I and II) for attack at the α-position and only one (IV) for attack at the β-position. On this basis we would expect the carbocation resulting from attack at the α-position (and also the transition state leading to that ion) to be much more stable than the carbocation (and the corresponding transition state) resulting from attack at the β-position, and that nitration would therefore occur much more rapidly at the α-position.

Throughout our study of polynuclear hydrocarbons, we shall find that the matter of orientation is generally understandable on the basis of this principle: of the large number of structures contributing to the intermediate carbocation, the important ones are those that require the smallest sacrifice of resonance stabilization. Indeed, we shall find that this principle accounts for orientation not only in electrophilic substitution but also in oxidation, reduction, and addition.

34.10   Friedel–Crafts acylation of naphthalene

Naphthalene can be acetylated by acetyl chloride in the presence of aluminum chloride. The orientation of substitution is determined by the particular solvent used: predominantly *alpha* in carbon disulfide or solvents like tetrachloroethane, predominantly *beta* in nitrobenzene. (The effect of nitrobenzene has been attributed to its forming a complex with the acid chloride and aluminum chloride which, because of its bulkiness, attacks the roomier *beta* position.)

solvent: $C_2H_2Cl_4$

1-Acetonaphthalene
Methyl α-naphthyl ketone

Naphthalene

$CH_3COCl, AlCl_3$

solvent: $C_6H_5NO_2$

2-Acetonaphthalene
Methyl β-naphthyl ketone

Thus acetylation (as well as sulfonation, Sec. 34.11) affords access to the *beta* series of naphthalene derivatives. Treatment of 2-acetonaphthalene with hypohalite, for example, provides the best route to β-naphthoic acid.

NaOCl, 60–70 °C

2-Acetonaphthalene
Methyl β-naphthyl ketone

β-Naphthoic acid
(88% yield)

$+ \ CHCl_3$

Acylation of naphthalene by succinic anhydride yields a mixture of *alpha* and *beta* products. These are separable, however, and both are of importance in the synthesis of higher ring systems (see Sec. 34.19).

COCH₂CH₂COOH

Naphthalene + Succinic anhydride → (AlCl₃, C₆H₅NO₂)

4-(1-Naphthyl)-4-oxobutanoic acid
β-(1-Naphthoyl)propionic acid

COCH₂CH₂COOH

4-(2-Naphthyl)-4-oxobutanoic acid
β-(2-Naphthoyl)propionic acid

Friedel–Crafts alkylation of naphthalene is of little use, probably for a combination of reasons: the high reactivity of naphthalene which causes side reactions and polyalkylations, and the availability of alkylnaphthalenes via acylation or ring closure (Sec. 34.14).

Problem 34.9 The position of the —COOH in β-naphthoic acid was shown by vigorous oxidation and identification of the product. What was this product? What product would have been obtained from α-naphthoic acid?

Problem 34.10 Outline the synthesis of the following compounds via an initial acylation:

- (a) 2-ethylnaphthalene
- (b) 2-(2-naphthyl)-2-butanol
- (c) 2-(*sec*-butyl)naphthalene
- (d) 1-(2-naphthyl)ethanol
- (e) γ-(2-naphthyl)butyric acid
- (f) 4-(2-naphthyl)-1-butanol
- (g) 5-(2-naphthyl)-2-methyl-2-pentanol
- (h) 2-isohexylnaphthalene
- (i) 1-amino-1-(2-naphthyl)ethane
- (j) β-vinylnaphthalene

34.11 Sulfonation of naphthalene

Sulfonation of naphthalene at 80 °C yields chiefly 1-naphthalenesulfonic acid; sulfonation at 160 °C or higher yields chiefly 2-naphthalenesulfonic acid. When 1-naphthalenesulfonic acid is heated in sulfuric acid at 160 °C, it is largely converted into the 2-isomer. These facts become understandable when we recall that sulfonation is readily reversible (Sec. 14.12).

SO₃H

conc. H₂SO₄, 80 °C

1-Naphthalenesulfonic acid
α-Naphthalenesulfonic acid

Naphthalene

conc. H₂SO₄, 160 °C

conc. H₂SO₄, 160 °C

SO₃H

2-Naphthalenesulfonic acid
β-Naphthalenesulfonic acid

Sulfonation, like nitration and halogenation, occurs more rapidly at the α-position, since this involves the more stable intermediate carbocation. But, for the same reason, attack by a proton, with subsequent desulfonation, also occurs more readily at the α-position. Sulfonation at the β-position occurs more slowly but, once formed, the β-sulfonic acid tends to resist desulfonation. At low temperatures desulfonation is slow and we isolate the product that is formed faster, the *alpha* naphthalenesulfonic acid. At higher temperatures, desulfonation becomes important, equilibrium is more readily established, and we isolate the product that is more stable, the *beta* naphthalenesulfonic acid.

α-Isomer	β-Isomer
Formed rapidly;	*Formed slowly;*
desulfonated rapidly	*desulfonated slowly*

We see here a situation exactly analogous to one we have encountered several times before: in 1,2- and 1,4-addition to conjugated dienes (Sec. 10.27), in Friedel–Crafts alkylation of toluene (Sec. 15.12), and in sulfonation of phenols (Problem 28.12, p. 1013). At low temperatures the controlling factor is *rate of reaction*, at high temperatures, *position of equilibrium.*

Sulfonation is of special importance in the chemistry of naphthalene because it gives access to the *beta*-substituted naphthalenes, as shown in the next section.

Problem 34.11   (a) Show all steps in the sulfonation and desulfonation of naphthalene. (b) Draw a potential energy curve for the reactions involved. (Compare your answer with Fig. 10.7, p. 404.)

34.12   Naphthols

Like the phenols we have already studied, naphthols can be prepared from the corresponding sulfonic acids by fusion with alkali. Naphthols can also be made

Sodium	Sodium	2-Naphthol
2-naphthalenesulfonate	2-naphthoxide	β-Naphthol

from the naphthylamines by direct hydrolysis under acidic conditions. (This reaction, which does not work in the benzene series, is superior to hydrolysis of diazonium salts.)

1-Naphthylamine	1-Naphthol
	α-Naphthol
	(95% yield)

The α-substituted naphthalenes, like substituted benzenes, are most commonly prepared by a sequence of reactions that ultimately goes back to a nitro compound (Sec. 34.8). Preparation of β-substituted naphthalenes, on the other hand, cannot start with the nitro compound, since nitration does not take place in the β-position. The route to β-naphthylamine, and through it to the versatile diazonium salts, lies through β-naphthol. β-Naphthol is made from the β-sulfonic acid; it is converted

Synthesis of β-substituted naphthalenes

Naphtha-   2-Naphthalene-   2-Naphthol   2-Naphthylamine
lene   sulfonic acid

Halides,
nitriles, azo
compounds, etc.
(See Chap. 27)

2-Naphthalenediazonium
salt

into β-naphthylamine when heated under pressure with ammonia and ammonium sulfite (the **Bucherer reaction**, not useful in the benzene series except in rare cases).

Naphthols undergo the usual reactions of phenols. Coupling with diazonium salts is particularly important in dye manufacture (see Sec. 27.18); the orientation of this substitution is discussed in the following section.

Problem 34.12   Starting from naphthalene, and using any readily available reagents, prepare the following compounds:

    (a) 2-bromonaphthalene   (d) β-naphthoic acid
    (b) 2-fluoronaphthalene   (e) β-naphthaldehyde
    (c) β-naphthonitrile   (f) 3-(2-naphthyl)propenoic acid

Problem 34.13   Diazonium salts can be converted into nitro compounds by treatment with sodium nitrite, usually in the presence of a catalyst. Suggest a method for preparing 2-nitronaphthalene.

34.13   Orientation of electrophilic substitution in naphthalene derivatives

We have seen that naphthalene undergoes nitration and halogenation chiefly at the α-position, and sulfonation and Friedel–Crafts acylation at either the α- or β-position, depending upon conditions. Now, to what position will a *second* substituent attach itself, and how is the orientation influenced by the group already present?

Orientation of substitution in the naphthalene series is more complicated than in the benzene series. An entering group may attach itself either to the ring that already carries the first substituent, or to the other ring; there are seven different positions open to attack, in contrast to only three positions in a monosubstituted benzene.

The major products of further substitution in a monosubstituted naphthalene can usually be predicted by the following rules. As we shall see, these rules are reasonable ones in light of structural theory and our understanding of electrophilic aromatic substitution.

(a) An activating group (electron-releasing group) tends to direct further substitution into the same ring. An activating group in position 1 directs further substitution to position 4 (and, to a lesser extent, to position 2). An activating group in position 2 directs further substitution to position 1.

(b) A deactivating group (electron-withdrawing group) tends to direct further substitution into the other ring: at an α-position in nitration or halogenation, or at an α- or β-position (depending upon temperature) in sulfonation.

For example:

OH
(1-Naphthol) + C₆H₅N₂⁺Cl⁻ —NaOH, 0–10 °C→ 4-Phenylazo-1-naphthol (OH, N=N—C₆H₅)

OH
(1-Naphthol) + HNO₃ —H₂SO₄, 20 °C→ 2,4-Dinitro-1-naphthol (OH, NO₂, NO₂)

OH
(2-Naphthol) + C₆H₅N₂⁺Cl⁻ —NaOH, 0–5 °C→ 1-Phenylazo-2-naphthol (N=N—C₆H₅, OH)

NO₂
(1-Nitronaphthalene) + HNO₃ —H₂SO₄, 0 °C→ 1,5-Dinitronaphthalene (NO₂, NO₂) + 1,8-Dinitronaphthalene (NO₂ NO₂) *Chief product*

CH₃
(2-Methylnaphthalene) + Br₂ —dark→ 1-Bromo-2-methylnaphthalene (Br, CH₃)

These rules do not always hold in sulfonation, because the reaction is reversible and at high temperatures tends to take place at a β-position. However, the observed products can usually be accounted for if this feature of sulfonation is kept in mind.

Problem 34.14 Predict the orientation in each of the following reactions, giving structural formulas and names for the predicted products:

(a) 1-methylnaphthalene + Br_2
(b) 1-methylnaphthalene + HNO_3 + H_2SO_4
(c) 1-methylnaphthalene + CH_3COCl + $AlCl_3$
(d) the same as (a), (b), and (c) for 2-methylnaphthalene
(e) 2-nitronaphthalene + Br_2
(f) 2-methoxynaphthalene + Br_2

Problem 34.15 How do you account for the following observed orientations?

(a) 2-methoxynaphthalene + CH_3COCl + $AlCl_3$ + CS_2 $\longrightarrow$ 1-aceto compound
(b) 2-methoxynaphthalene + CH_3COCl + $AlCl_3$ + $C_6H_5NO_2$ $\longrightarrow$ 6-aceto compound
(c) 2-methylnaphthalene + H_2SO_4 above 100 °C $\longrightarrow$ 6-sulfonic acid
(d) 2,6-dimethylnaphthalene + H_2SO_4 at 40 °C $\longrightarrow$ 8-sulfonic acid
(e) 2,6-dimethylnaphthalene + H_2SO_4 at 140 °C $\longrightarrow$ 3-sulfonic acid
(f) 2-naphthalenesulfonic acid + HNO_3 + H_2SO_4 $\longrightarrow$ 5-nitro and 8-nitro compounds

Problem 34.16 Give the steps for the synthesis of each of the following from naphthalene and any needed reagents:

(a) 4-nitro-1-naphthylamine
(b) 1,4-dinitronaphthalene
  (*Hint*: See Problem 34.13, p. 1182.)
(c) 2,4-dinitro-1-naphthylamine
(d) 1,3-dinitronaphthalene
(e) 1,2-dinitronaphthalene

(f) 4-amino-1-naphthalenesulfonic acid
  (*naphthionic acid*)
(g) 8-amino-1-naphthalenesulfonic acid
(h) 5-amino-2-naphthalenesulfonic acid
(i) 8-amino-2-naphthalenesulfonic acid

We have seen (Sec. 34.9) that orientation in naphthalene can be accounted for on the same basis as orientation in substituted benzenes: formation of the more stable intermediate carbocation. In judging the relative stabilities of these naphthalene carbocations, we have considered that those in which an aromatic sextet is preserved are by far the more stable and hence the more important. Let us see if we can account for orientation in substituted naphthalenes in the same way.

The structures preserving an aromatic sextet are those in which the positive charge is carried by the ring under attack; it is in this ring, therefore, that the charge chiefly develops. Consequently, attack occurs most readily on whichever ring can best accommodate the positive charge: the ring that carries an electron-releasing (activating) group or the ring that does *not* carry an electron-withdrawing (deactivating) group. (We have arrived at the quite reasonable conclusion that a substituent exerts its greatest effect—activating or deactivating—on the ring to which it is attached.)

G *is electron-releasing:*
activating,
attack in same ring

G *is electron-withdrawing:*
deactivating,
attack in other ring

An electron-releasing group located at position 1 can best help accommodate the positive charge if attack occurs at position 4 (or position 2), through the contribution of structures like I and II.

G *is electron-releasing:*
when on position 1,
it directs attack to
positions 4 or 2

This is true whether the group releases electrons by an inductive effect or by a resonance effect. For example:

An electron-releasing group located at position 2 could help accommodate the positive charge if attack occurred at position 1 (through structures like III), or if attack occurred at position 3 (through structures like IV).

III

IV

G *is electron-releasing:*
when on position 2,
it directs attack to
position 1

More stable:
aromatic sextet
preserved

Less stable:
aromatic sextet
disrupted

However, we can see that only the structures like III preserve an aromatic sextet; these are much more stable than the structures like IV, and are the important ones. It is not surprising, therefore, that substitution occurs almost entirely at position 1.

34.14 Synthesis of naphthalene derivatives by ring closure. The Haworth synthesis

Derivatives of benzene, we have seen, are almost always prepared from a compound that already contains the benzene ring: benzene itself or some simple substituted benzene. One seldom generates the benzene ring in the course of a synthesis.

While compounds containing other aromatic ring systems, too, are often prepared from the parent hydrocarbon, there are important exceptions: syntheses in which the ring system, or part of it, is actually generated. Such syntheses usually involve two stages: **ring closure** (or **cyclization**) and **aromatization**.

As an example, let us look at just one method used to make certain naphthalene derivatives: the **Haworth synthesis** (developed by R. D. Haworth at the University of Durham, England). Figure 34.3 shows the basic scheme, which would yield naphthalene itself (not, of course, actually prepared in this way).

All the steps are familiar ones. The reaction in which the second ring is formed is simply Friedel–Crafts acylation that happens to involve two parts of the same molecule. Like many methods of ring closure, this one does not involve a new reaction, but merely an adaptation of an old one.

Problem 34.17 Why is ring closure possible after the first Clemmensen reduction but not before?

To obtain substituted naphthalenes, the basic scheme can be modified in any or all of the following ways:

(a) A substituted benzene can be used in place of benzene and a β-substituted naphthalene obtained. Toluene or anisole or bromobenzene, for example, undergoes the initial Friedel–Crafts reaction chiefly at the *para* position; when the ring is closed, the substituent originally on the benzene ring must occupy a β-position in naphthalene.

$G = -R, -X, -OCH_3$ β-Substituted naphthalene

(b) The intermediate cyclic ketone (an α-tetralone) can be treated with a Grignard reagent, and an alkyl (or aryl) group introduced into an α-position.

α-Tetralone α-Alkylnaphthalene

(c) The original keto acid (in the form of its ester) can be treated with a Grignard reagent, and an alkyl (or aryl) group introduced into an α-position. The success of this reaction depends upon the fact that a ketone reacts much faster than an ester with a Grignard reagent. (See p. 1188.)

Figure 34.3　Haworth synthesis of naphthalene derivatives.

Keto acid

(1) dehydration
(2) hydrolysis
(3) hydrogenation

1,6-Disubstituted naphthalene

By proper combinations of these modifications, a wide variety of substituted naphthalenes can be prepared.

Problem 34.18   Outline all steps in the synthesis of the following compounds, starting from benzene and using any necessary aliphatic and inorganic reagents.

(a) 2-methylnaphthalene
(b) 1-methylnaphthalene
(c) 1,4-dimethylnaphthalene
(d) 1,7-dimethylnaphthalene
(e) 1,6-dimethylnaphthalene

(f) 1,4,6-trimethylnaphthalene
(g) 1-ethyl-4-methylnaphthalene
(h) 7-bromo-1-ethylnaphthalene
(i) 1-phenylnaphthalene

Problem 34.19   Outline the Haworth sequence of reactions, starting with naphthalene and succinic anhydride. What is the final hydrocarbon or hydrocarbons? (Remember the orientation rules for naphthalene.) Check your answer in Sec. 34.19.

ANTHRACENE AND PHENANTHRENE

34.15   Nomenclature of anthracene and phenanthrene derivatives

The positions in anthracene and phenanthrene are designated by numbers as shown:

Anthracene

Phenanthrene   or

Examples are found in Secs. 34.17–34.19.

34.16   Structure of anthracene and phenanthrene

Like naphthalene, anthracene and phenanthrene are classified as aromatic on the basis of their properties. Consideration of atomic orbitals follows the same pattern as for naphthalene, and leads to the same kind of picture: a flat structure with partially overlapping π clouds lying above and below the plane of the molecule. (See Fig. 34.4.)

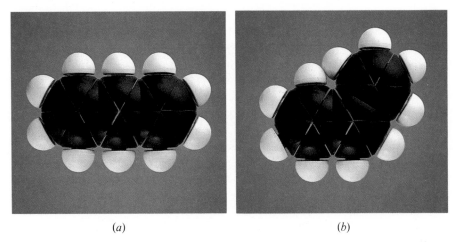

(a)   (b)

Figure 34.4 Electronic configuration and molecular shape. Models of (a) anthracene, (b) phenanthrene.

In terms of valence bonds, anthracene is considered to be a hybrid of structures I–IV,

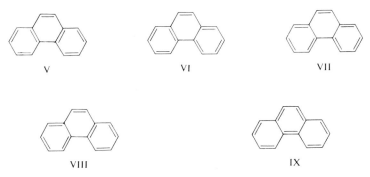

anthracene has a resonance energy of 84 kcal/mol, and that phenanthrene has a resonance energy of 92 kcal/mol.

For convenience we shall represent anthracene as the single structure X, and phenanthrene as XI, in which the circles can be thought of as representing partially overlapping aromatic sextets.

X
Anthracene

XI
Phenanthrene

34.17 Reactions of anthracene and phenanthrene

Anthracene and phenanthrene are even less resistant toward oxidation or reduction than naphthalene. Both hydrocarbons are oxidized to the 9,10-quinones and reduced to the 9,10-dihydro compounds. Both the orientation of these reactions and the comparative ease with which they take place are understandable on the basis of the structures involved. Attack at the 9- and 10-positions leaves two benzene rings intact; thus there is a sacrifice of only 12 kcal of resonance energy $(84 - 2 \times 36)$ for anthracene, and 20 kcal $(92 - 2 \times 36)$ for phenanthrene. (In the

Anthracene

$K_2Cr_2O_7, H_2SO_4$

9,10-Anthraquinone

Na, C_2H_5OH, reflux

9,10-Dihydroanthracene

Phenanthrene

$K_2Cr_2O_7, H_2SO_4$

9,10-Phenanthrenequinone

Na, $C_5H_{11}OH$, reflux

9,10-Dihydrophenanthrene

case of phenanthrene, the two remaining rings are conjugated; to the extent that this conjugation stabilizes the product—estimated at anywhere from 0 to 8 kcal/mol—the sacrifice is even less than 20 kcal.)

Problem 34.20   How much resonance energy would be sacrificed by oxidation or reduction of one of the outer rings of anthracene? Of phenanthrene?

Both anthracene and phenanthrene undergo electrophilic substitution. With a few exceptions, however, these reactions are of little value in synthesis because of the formation of mixtures and polysubstitution products. Derivatives of these two hydrocarbons are usually obtained in other ways: by electrophilic substitution in 9,10-anthraquinone or 9,10-dihydrophenanthrene, for example, or by ring-closure methods (Secs. 34.18 and 34.19).

Bromination of anthracene or phenanthrene takes place at the 9-position. (9-Bromophenanthrene is a useful intermediate for the preparation of certain 9-substituted phenanthrenes.) In both cases, especially for anthracene, there is a tendency for addition to take place with the formation of the 9,10-dibromo-9,10-dihydro derivatives.

9-Bromophenanthrene

Phenanthrene   + Br$_2$

9,10-Dibromo-9,10-dihydrophenanthrene

Anthracene   9,10-Dibromo-9,10-dihydroanthracene   9-Bromoanthracene

This reactivity of the 9- and 10-positions toward electrophilic attack is understandable, whether reaction eventually leads to substitution or addition. The carbocation initially formed is the most stable one, I or II, in which aromatic sextets are preserved in two of the three rings. This carbocation can then either (a) give up a proton to yield the substitution product, or (b) accept a base to yield the addition product. The tendency for these compounds to undergo addition is undoubtedly due to the comparatively small sacrifice in resonance energy that this entails (12 kcal/mol for anthracene, 20 kcal/mol or less for phenanthrene).

Problem 34.21 Nitric acid converts anthracene into any of a number of products, III–VI, depending upon the exact conditions. How could each be accounted for?

(a) Nitric acid and acetic acid yields III   (c) Excess nitric acid yields V
(b) Nitric acid and ethyl alcohol yields IV   (d) Nitric acid and acetic anhydride
                                                   yields 9-nitroanthracene (VI)

III   IV   V   VI

Problem 34.22 Account for the following observations: (a) Upon treatment with hydrogen and nickel, 9,10-dihydroanthracene yields 1,2,3,4-tetrahydroanthracene. (b) In contrast to bromination, sulfonation of anthracene yields the 1-sulfonic acid.

34.18 Preparation of anthracene derivatives by ring closure. Anthraquinones

Derivatives of anthracene are seldom prepared from anthracene itself, but rather by ring-closure methods. As in the case of naphthalene, the most important method of ring closure involves adaptation of Friedel–Crafts acylation. The products initially obtained are **anthraquinones**, which can be converted into corresponding anthracenes by reduction with zinc and alkali. This last step is seldom carried out, since the quinones are by far the more important class of compounds.

The following reaction sequence shows the basic scheme. (Large amounts of anthraquinones are manufactured for the dye industry in this way.)

Phthalic anhydride   *o*-Benzoylbenzoic acid   9,10-Anthraquinone

The basic scheme can be modified in a number of ways.

(a) A monosubstituted benzene can be used in place of benzene, and a 2-substituted anthraquinone obtained. (The initial acylation goes chiefly *para*. If the *para* position is blocked, *ortho* acylation is possible.) For example:

Phthalic anhydride   *o*-(*p*-Toluyl)benzoic acid   2-Methyl-9,10-anthraquinone

(b) A polynuclear compound can be used in place of benzene, and a product having more than three rings obtained. For example:

Phthalic anhydride   *o*-(2-Naphthoyl)benzoic acid   1,2-Benz-9,10-anthraquinone

(c) The intermediate *o*-aroylbenzoic acid can be reduced before ring closure, and 9-substituted anthracenes obtained via Grignard reactions.

o-Benzoylbenzoic acid   *o*-Benzylbenzoic acid   Anthrone

9-Alkylanthracene

Anthraquinoid dyes are of enormous technological importance, and much work has been done in devising syntheses of large ring systems embodying the quinone structure. Several examples of anthraquinoid dyes are:

Alizarin

Indanthrene Golden Yellow GK

Indanthrone

Problem 34.23   Outline the synthesis of the following, starting from compounds having fewer rings:

(a) 1,4-dimethylanthraquinone   (d) 2,9-dimethylanthracene
(b) 1,2-dimethylanthraquinone   (e) 9-methyl-1,2-benzanthracene (a potent
(c) 1,3-dimethylanthraquinone   cancer-producing hydrocarbon)

Problem 34.24   What anthraquinone or anthraquinones would be expected from a sequence starting with 3-nitrophthalic anhydride and (a) benzene, (b) toluene?

34.19   Preparation of phenanthrene derivatives by ring closure

Starting from naphthalene instead of benzene, the Haworth succinic anhydride synthesis (Sec. 34.14) provides an excellent route to substituted phenanthrenes.

The basic scheme is outlined in Fig. 34.5. Naphthalene is acylated by succinic anhydride at both the 1- and 2-positions; the two products are separable, and either can be converted into phenanthrene. We notice that γ-(2-naphthyl)butyric acid undergoes ring closure at the 1-position to yield phenanthrene rather than at the 3-position to yield anthracene; the electron-releasing side chain at the 2-position directs further substitution to the 1-position (Sec. 34.13).

Substituted phenanthrenes are obtained by modifying the basic scheme in the ways already described for the Haworth method (Sec. 34.14).

Problem 34.25   Apply the Haworth method to the synthesis of the following, starting from naphthalene or a monosubstituted naphthalene:

(a) 9-methylphenanthrene   (f) 1,4-dimethylphenanthrene
(b) 4-methylphenanthrene   (g) 1,4,9-trimethylphenanthrene
(c) 1-methylphenanthrene   (h) 2-methoxyphenanthrene
(d) 1,9-dimethylphenanthrene   (*Hint*: See Problem 34.15, p. 1184.)
(e) 4,9-dimethylphenanthrene

Figure 34.5   Haworth synthesis of phenanthrene derivatives.

Problem 34.26 Give structural formulas for all intermediates in the following synthesis of 2-methylphenanthrene. Tell what kind of reaction each step involves.

$$\text{naphthalene} + CH_3CH_2COCl + AlCl_3 \xrightarrow{C_6H_5NO_2} \text{A } (C_{13}H_{12}O)$$

$$\text{A} + Br_2 \longrightarrow \text{B } (C_{13}H_{11}OBr)$$

$$\text{B} + CH(COOC_2H_5)_2{}^- Na^+ \longrightarrow \text{C } (C_{20}H_{22}O_5)$$

$$\text{C} \xrightarrow{\text{aq. KOH, heat}} \text{D} \xrightarrow{HCl} \text{E} \xrightarrow{\text{heat}} \text{F } (C_{15}H_{14}O_3) + CO_2$$

$$\text{F} + Zn(Hg) + HCl \longrightarrow \text{G } (C_{15}H_{16}O_2)$$

$$\text{G} \xrightarrow{\text{polyphosphoric acid}} \text{H } (C_{15}H_{14}O)$$

$$\text{H} + Zn(Hg) + HCl \longrightarrow \text{I } (C_{15}H_{16})$$

$$\text{I} \xrightarrow{\text{Pd, heat}} \text{2-methylphenanthrene}$$

Problem 34.27 Follow the instructions for Problem 34.26 for the following synthesis of phenanthrene (the **Bogert–Cook synthesis**):

$$\beta\text{-phenylethyl bromide} + Mg \longrightarrow \text{A } (C_8H_9MgBr)$$

$$\text{A} + \text{cyclohexanone} \longrightarrow \text{B} \xrightarrow{H_2O} \text{C } (C_{14}H_{20}O)$$

$$\text{C} \xrightarrow{H_2SO_4} \text{D } (C_{14}H_{18})$$

$$\text{D} \xrightarrow{H_2SO_4} \text{E } (C_{14}H_{18})$$

$$\text{E} \xrightarrow{\text{Se, heat}} \text{phenanthrene}$$

How could β-phenylethyl bromide be made from benzene?

Problem 34.28 Follow the instructions for Problem 34.26 for the following synthesis of phenanthrene (the **Bardhan–Sengupta synthesis**):

$$\text{potassium} + \text{ethyl 2-keto-1-cyclohexanecarboxylate} \longrightarrow \text{A } (C_9H_{13}O_3K)$$

$$\text{A} + \beta\text{-phenylethyl bromide} \longrightarrow \text{B } (C_{17}H_{22}O_3)$$

$$\text{B} \xrightarrow{\text{aq. KOH, heat}} \text{C} \xrightarrow{HCl} \text{D } (C_{14}H_{18}O)$$

$$\text{D} + Na + \text{moist ether} \longrightarrow \text{E } (C_{14}H_{20}O)$$

$$\text{E} \xrightarrow{P_2O_5} [\text{F } (C_{14}H_{18})] \xrightarrow{P_2O_5} \text{G } (C_{14}H_{18})$$

$$\text{G} \xrightarrow{\text{Se, heat}} \text{phenanthrene}$$

Problem 34.29 Follow the instructions for Problem 34.26 for the following synthesis of *pyrene*:

4-keto-1,2,3,4-tetrahydrophenanthrene ($C_{14}H_{12}O$)

$$+ BrCH_2COOC_2H_5 + Zn \xrightarrow{\text{ether}} \text{A} \xrightarrow{H_2O, H^+} \text{B } (C_{18}H_{20}O_3)$$

$$\text{B} + \text{acid} + \text{heat} \longrightarrow \text{C } (C_{18}H_{18}O_2)$$

$$\text{C} + \text{aq. NaOH} + \text{heat} \longrightarrow \text{D} \xrightarrow{HCl} \text{E } (C_{16}H_{14}O_2)$$

$$\text{E} \xrightarrow{HF} \text{F } (C_{16}H_{12}O)$$

$$\text{F} + Zn(Hg) + HCl \longrightarrow \text{G } (C_{16}H_{14})$$

$$\text{G} \xrightarrow{\text{Pd, heat}} \text{pyrene } (C_{16}H_{10})$$

How could you make the starting material?

Problem 34.30   Outline a possible synthesis of *chrysene* by the Bogert–Cook method (Problem 34.27, p. 1196), starting from naphthalene and using any aliphatic or inorganic reagents. (*Hint*: See Problem 34.7(g), p. 1177.)

Chrysene

Problem 34.31   Outline an alternative synthesis of chrysene by the Bogert–Cook method, starting from benzene and using any aliphatic or inorganic reagents.

34.20   Carcinogenic hydrocarbons. Arene oxides

In recent years a rapidly increasing number of compounds have been found to be *carcinogenic*, that is, cancer-producing. Indeed, it has been suggested that cancer is primarily an environmental disease; just as the draining of swamps and the elimination of mosquitoes can be used to control malaria and yellow fever, so the elimination from our environment of carcinogens could reduce the incidence of cancer drastically.

Among the most potent carcinogens are certain polynuclear hydrocarbons. For example:

7,12-Dimethyl-
benz[*a*]anthracene

Dibenz[*a*,*h*]anthracene

Benzo[*a*]pyrene

3-Methylcholanthrene

Since such hydrocarbons are the products of incomplete combustion of organic material—coal, petroleum, tobacco—they are widespread, and may well be an important cause of human cancer. As a result, much of the research into the mode of action of carcinogens has centered about polynuclear hydrocarbons.

When a foreign substance enters an organism, the organism tries to eliminate it. If the intruder is of low water solubility, this elimination generally takes the form of conversion into more water-soluble substances, which are readily excreted. Polynuclear hydrocarbons, it now seems clear, are converted first into *arene oxides*:

epoxides in which the aromaticity of one ring has necessarily been destroyed. For example:

Benzo[a]pyrene

Benzo[a]pyrene-7,8-oxide   Benzo[a]pyrene-7,8-dihydro-7,8-diol

An epoxide, we know (Secs. 19.13–19.14) typically undergoes *nucleophilic substitution*, a reaction that opens the epoxide ring to give a product containing two functional groups:

In hydrolysis, the nucleophile is water, and the product is a 1,2-diol. In an organism, hydrolysis (enzyme-catalyzed, of course) is a principal reaction of arene oxides; the resulting diols undergo further reaction and the material is eventually excreted.

But some of these diols undergo *further* epoxidation—in a regioselective and stereoselective way—to yield dihydroxy epoxides. Benzo[a]pyrene, for example, is converted into the diol epoxide I:

I

It is these dihydroxy epoxides that are believed to be the actual carcinogens formed by metabolism of polynuclear hydrocarbons.

Now, how do these epoxides cause cancer? When diol epoxide I is allowed to react in the test tube with the *nucleic acid* DNA (Sec. 41.8), and the product is degraded, there is obtained compound II.

I   II

A perfectly straightforward, familiar kind of reaction has taken place: *nucleophilic attack on an epoxide*. The —NH$_2$ of the nucleoside base guanine has attacked C–10 of the epoxide and opened the ring with stereochemical inversion to yield the *trans* product II.

This same compound, II, is the principal product of the action of the parent hydrocarbon benzo[*a*]pyrene on DNA and RNA in human cells. The observed damage to the DNA that results from this reaction is easy to understand: attachment of this large hydrocarbon group to guanine clearly prevents it from fitting into the double helix of DNA and from hydrogen bonding to a cytosine in the opposite strand (Sec. 41.8). This damage leads to mutations and, with mutations, an increased likelihood of carcinogenesis.

Many carcinogenic substances are believed to work in much the same way: they are electrophilic and suffer nucleophilic attack by a nucleoside base of DNA. The nitrosamines (Sec. 27.11) formed by the action of nitrites on the proteins of meat are believed to exert their carcinogenic effects through their alkylating power. Methylating agents are in general carcinogenic: even the attachment of the tiny methyl groups is enough to interfere with base-pairing in the double helix.

PROBLEMS

1. Give the structures and names of the principal products of the reaction (if any) of naphthalene with:

(a) CrO$_3$, CH$_3$COOH

(b) O$_2$, V$_2$O$_5$

(c) Na, C$_2$H$_5$OH

(d) Na, C$_5$H$_{11}$OH

(e) H$_2$, Ni

(f) HNO$_3$, H$_2$SO$_4$

(g) Br$_2$

(h) conc. H$_2$SO$_4$, 80 °C

(i) conc. H$_2$SO$_4$, 160 °C

(j) CH$_3$COCl, AlCl$_3$, CS$_2$

(k) CH$_3$COCl, AlCl$_3$, C$_6$H$_5$NO$_2$

(l) succinic anhydride, AlCl$_3$, C$_6$H$_5$NO$_2$

2. Give the structure and names of the principal products of the reaction of HNO$_3$/H$_2$SO$_4$ with:

(a) 1-methylnaphthalene

(b) 2-methylnaphthalene

(c) 1-nitronaphthalene

(d) 2-nitronaphthalene

(e) 1-naphthalenesulfonic acid

(f) 2-naphthalenesulfonic acid

(g) *N*-(1-naphthyl)acetamide

(h) *N*-(2-naphthyl)acetamide

(i) α-naphthol

(j) β-naphthol

(k) anthracene

3. When 2-methylnaphthalene is nitrated, three isomeric mononitro derivatives are obtained. Upon vigorous oxidation one of these yields 3-nitro-1,2,4-benzenetricarboxylic acid, and the other two both yield 3-nitrophthalic acid. Give the names and structures of the original three isomeric nitro compounds.

4. Outline all steps in a possible synthesis of each of the following from naphthalene, using any needed organic and inorganic reagents:

(a) α-naphthol
(b) β-naphthol
(c) α-naphthylamine
(d) β-naphthylamine
(e) 1-iodonaphthalene
(f) 2-iodonaphthalene
(g) 1-nitronaphthalene
(h) 2-nitronaphthalene
(i) α-naphthoic acid
(j) β-naphthoic acid
(k) 4-(1-naphthyl)butanoic acid
(l) α-naphthaldehyde
(m) β-naphthaldehyde
(n) 1-phenylazo-2-naphthol

(o) 1-amino-2-naphthol (*Hint*: Use product of (n).)
(p) 4-amino-1-naphthol
(q) 1-bromo-2-methoxynaphthalene
(r) 1,5-diaminonaphthalene
(s) 4,8-dibromo-1,5-diiodonaphthalene
(t) 5-nitro-2-naphthalenesulfonic acid
(u) 1,2-diaminonaphthalene
(v) 1,3-diaminonaphthalene
(w) *o*-aminobenzoic acid
(x) phenanthrene
(y) 9,10-anthraquinone
(z) anthracene

5. Naphthalene was transformed into another hydrocarbon by the following sequence of reactions:

naphthalene + Na, $C_5H_{11}OH \longrightarrow$ A $(C_{10}H_{12})$
A + succinic anhydride, $AlCl_3 \longrightarrow$ B $(C_{14}H_{16}O_3)$
B + Zn(Hg) + HCl $\longrightarrow$ C $(C_{14}H_{18}O_2)$
C + anhydrous HF $\longrightarrow$ D $(C_{14}H_{16}O)$
D + Zn(Hg) + HCl $\longrightarrow$ E $(C_{14}H_{18})$
E + Pd/C + heat $\longrightarrow$ F $(C_{14}H_{10}$, m.p. 100–101 °C$) + 4H_2$

What was F? What were A–E?

6. Outline all steps in a possible synthesis of each of the following from hydrocarbons containing fewer rings:

(a) 6-methoxy-4-phenyl-1-methylnaphthalene
(b) 1,2-benzanthracene

(c) 9-phenylanthracene
(d) 1-phenylphenanthrene
(e) 1,9-diphenylphenanthrene

7. Acylation of phenanthrene by succinic anhydride takes place at the 2- and 3-positions. The sequence of reduction, ring closure, and aromatization converts the 2-isomer into G and H, and converts the 3-isomer into G.

What is the structure and name of G? Of H?

8. When 4-phenyl-3-butenoic acid is refluxed there is formed a product, $C_{10}H_8O$, which is soluble in aqueous NaOH but not in aqueous $NaHCO_3$, and which reacts with benzenediazonium chloride to yield a red-orange solid. What is the product, and by what series of steps is it probably formed?

9. Anthracene reacts readily with maleic anhydride to give I, $C_{18}H_{12}O_3$, which can be hydrolyzed to J, a dicarboxylic acid of formula $C_{18}H_{14}O_4$. (a) What reaction do you think is involved in the formation of I? (b) What is the most probable structure of I? Of J?

Anthracene reacts with methyl fumarate to give a product that on hydrolysis yields K, a dicarboxylic acid of formula $C_{18}H_{14}O_4$. (c) Compare the structures of J and K. (*Hint*: See Problem 8, p. 1095.)

Anthracene reacts with *p*-benzoquinone to yield L, $C_{20}H_{14}O_2$. In acid, L undergoes rearrangement to a hydroquinone M, $C_{20}H_{14}O_2$. Oxidation of M gives a new quinone N, $C_{20}H_{12}O_2$. Reductive amination of N gives a diamine O, $C_{20}H_{16}N_2$. Deamination of O by the usual method gives the hydrocarbon *triptycene*, $C_{20}H_{14}$. (d) What is a likely structure for triptycene? For L–O?

10. Reduction of aromatic rings by the action of Li metal in ammonia generally gives 1,4-addition and yields a dihydro compound. Thus from naphthalene, $C_{10}H_8$, one can obtain $C_{10}H_{10}$. (a) Draw the structure of this dihydro compound.

Similar reduction is possible for 2-methoxynaphthalene (methyl 2-naphthyl ether). (b) Draw the structure of this dihydro compound. (c) If this dihydro ether is cleaved by acid, what is the structure of the initial product? (d) What further change will this initial product undoubtedly undergo, and what will be the final product?

11. Reduction of naphthalene by Li metal in $C_2H_5NH_2$ gives a 52% yield of 1,2,3,4,5,6,7,8-octahydronaphthalene. (a) What will this compound yield upon ozonolysis?

Treatment of the ozonolysis product ($C_{10}H_{16}O_2$) with base yields an unsaturated ketone ($C_{10}H_{14}O$). (b) What is its structure? (c) Show how this ketone can be transformed into *azulene*, $C_{10}H_8$, a blue hydrocarbon that is isomeric with naphthalene.

I

Azulene

12. (a) Azulene (preceding problem) is a planar molecule, and has a heat of combustion about 40 kcal/mol lower than that calculated by the method of Problem 13.2 (p. 483). It couples with diazonium salts and undergoes nitration and Friedel–Crafts acylation. Using both valence-bond and orbital structures, account for these properties of azulene. What might be a better representation of azulene than the formula I?

(b) The dipole moment of azulene is 1.08 D; that of 1-chloroazulene is 2.69 D. What is the direction of the dipole of azulene? Is this consistent with the structure you arrived at in (a)?

13. (a) In CF_3COOH solution, azulene gives the following NMR spectrum:

a	singlet,	δ 4.4, 2H
b	doublet,	δ 7.8, 1H
c	doublet,	δ 8.1, 1H
d	multiplet,	δ 9, 5H

and in CF_3COOD solution, the following spectrum:

a	singlet,	δ 8.1, 1H
b	multiplet,	δ 9, 5H

What compound gives rise to the spectrum in CF_3COOH? In CF_3COOD? Identify all NMR signals.

(b) In light of your structure for azulene (preceding problem), how do you account for what happens in CF_3COOH solution? What would you expect to obtain on neutralization of this solution?

(c) Show in detail just how the compound giving rise to the spectrum observed in CF_3COOD must have been formed. What would you expect to obtain on neutralization of this solution?

(d) At which position or positions in azulene would you expect nitration, Friedel–Crafts acylation, and diazonium coupling to occur?

14. Azulene reacts with *n*-butyllithium to yield, after hydrolysis and dehydrogenation, a *n*-butylazulene, and similarly with sodamide to yield an aminoazulene. To what class of reactions do these substitutions belong? In which ring would you expect such substitution to have occurred? At which position?

15. The structure of *eudalene*, $C_{14}H_{16}$, a degradation product of eudesmol (a terpene found in eucalyptus oil), was first established by the following synthesis:

p-isopropylbenzaldehyde + ethyl bromoacetate, Zn; then H_2O $\longrightarrow$ P ($C_{14}H_{20}O_3$)
P + acid, heat $\longrightarrow$ Q ($C_{14}H_{18}O_2$)
Q + Na, ethyl alcohol $\longrightarrow$ R ($C_{12}H_{18}O$)
R $\xrightarrow{\text{HBr}}$ $\xrightarrow{\text{KCN}}$ $\xrightarrow{H_2O, H^+}$ $\xrightarrow{\text{SOCl}_2}$ S ($C_{13}H_{17}OCl$)
S + $AlCl_3$, warm $\longrightarrow$ T ($C_{13}H_{16}O$)
T + CH_3MgBr, then H_2O $\longrightarrow$ U ($C_{14}H_{20}O$)
U + acid, heat $\longrightarrow$ V ($C_{14}H_{18}$)
V + sulfur, heat $\longrightarrow$ eudalene ($C_{14}H_{16}$)

What is the structure and systematic name of eudalene? (*Useful information*: See note at end of Problem 34.5, p. 1176.)

16. Many polynuclear aromatic compounds do not contain fused ring systems, e.g., biphenyl and triphenylmethane. Give structures and names of compounds W through II, formed in the following syntheses of such polynuclear compounds.

(a) 1,2,4,5-tetrachlorobenzene + H_2O, heat $\longrightarrow$ W ($C_6H_3OCl_3$)

   W + HCHO + H_2SO_4 $\longrightarrow$ [X, $C_7H_5O_2Cl_3$] $\longrightarrow$ Y ($C_{13}H_6O_2Cl_6$),

     "Hexachlorophene," soluble in base

(b) *m*-bromotoluene + Mg, ether $\longrightarrow$ Z (C_7H_7MgBr)

   Z + 4-methylcyclohexanone, then H_2O $\longrightarrow$ AA ($C_{14}H_{20}O$)

   AA + H^+, heat $\longrightarrow$ BB ($C_{14}H_{18}$)

   BB + Pd/C, heat $\longrightarrow$ CC ($C_{14}H_{14}$)

(c) ethyl benzoate + C_6H_5MgBr, then H_2O $\longrightarrow$ DD ($C_{19}H_{16}O$)

   DD + conc. HBr $\longrightarrow$ EE ($C_{19}H_{15}Br$)

   EE + Ag $\longrightarrow$ FF ($C_{38}H_{30}$)

(d) $(C_6H_5)_3COH + C_6H_5NH_2$ + acid $\longrightarrow$ GG ($C_{25}H_{21}N$)

   GG + $NaNO_2$ + HCl; then H_3PO_2 $\longrightarrow$ HH ($C_{25}H_{20}$)

(e) $C_6H_5COCH_3$ + acid + heat $\longrightarrow$ II ($C_{24}H_{18}$)

   (*Hint*: Acids catalyze aldol condensations.)

17. When 1-nitro-2-aminonaphthalene is treated with sodium nitrite and HCl, and then with warm water, there is obtained not only 1-nitro-2-naphthol, but also 1-chloro-2-naphthol. How do you account for the formation of the chloronaphthol? Consider carefully the stage at which chlorine is introduced into the molecule.

18. Treatment of phenanthrene with diazomethane yields a product JJ for which mass spectrometry indicates a molecular weight of 192. The infrared spectrum of JJ resembles that of 9,10-dihydrophenanthrene; its NMR spectrum shows two signals of one proton each at $\delta -0.12$ and $\delta 1.48$.

(a) What is a likely structure for JJ, and how is it probably formed? How do you account for the formation of JJ rather than one of its isomers?

(b) When a solution of JJ in *n*-pentane was irradiated with ultraviolet light, there were obtained phenanthrene, 2-methylpentane, 3-methylpentane, and *n*-hexane; the alkanes were obtained in the ratio 34:17:49. What happened in this reaction? What is the driving force?

(c) The irradiation of JJ in cyclohexene gave four products of formula C_7H_{12}. What would you expect these products to be?

(d) What would you expect to obtain from the irradiation of JJ in *cis*-4-methyl-2-pentene? In *trans*-4-methyl-2-pentene?

19. When *dihydropentalene* is treated with a little more than two moles of *n*-butyllithium,

Dihydropentalene

a stable white crystalline material KK is obtained. In contrast to the rather complicated NMR spectrum of dihydropentalene, the NMR spectrum of KK is simple:

     a doublet, $\delta 4.98, J = 3$ Hz

     b triplet, $\delta 5.73, J = 3$ Hz

     peak area ratio $a:b = 2:1$

What is a likely structure for KK? Of what theoretical significance is its formation and stability?

20. (a) When *either* 1-chloronaphthalene *or* 2-chloronaphthalene is treated with lithium piperidide and piperidine (Sec. 26.14) dissolved in ether, the *same* mixture of products is obtained: I and II of Problem 34.8 (p. 1177) in the ratio 31:69. Show all steps in a mechanism that accounts for these observations. In particular, show why 2-chloronaphthalene yields the same mixture as 1-chloronaphthalene.

(b) Under the conditions of (a), 1-bromonaphthalene and 1-iodonaphthalene give I and II in the same ratio as 1-chloronaphthalene does. With 1-fluoronaphthalene, however, the ratio of products depends on the concentration of piperidine. At high piperidine concentration, I makes up as much as 84% of the product; at low piperidine concentrations the product ratio levels off at the 31:69 value.

Account in detail for these facts. Tell what is happening to change the product ratio, why the ratio is affected by piperidine concentration, and why the fluoride should behave differently from the other halides.

21. Give structural formulas for LL through UU. Account in detail for the properties of compound UU.

3,5-dibromo-4-methylanisole + CuCN $\longrightarrow$ LL ($C_{10}H_8ON_2$)

LL + KOH, then CH_3OH, H^+ $\longrightarrow$ MM ($C_{12}H_{14}O_5$)

MM + LiAlH$_4$ $\longrightarrow$ NN ($C_{10}H_{14}O_3$)

NN + PBr$_3$ $\longrightarrow$ OO ($C_{10}H_{12}OBr_2$)

OO + Na $\longrightarrow$ PP ($C_{20}H_{24}O_2$)

PP + CrO$_3$ $\longrightarrow$ QQ ($C_{18}H_{18}O_2$), a pale yellow solid

 (*Hint*: A carbon–carbon bond is formed)

QQ + 2NaOH $\longrightarrow$ RR ($C_{18}H_{16}O_2Na_2$), soluble in water

RR + O$_2$ $\longrightarrow$ SS ($C_{18}H_{14}O_2$), a yellow solid

SS + LiAlH$_4$ $\longrightarrow$ TT ($C_{18}H_{18}$)

TT + 2,3-dichloro-5,6-dicyanoquinone ("DDQ") $\longrightarrow$ UU ($C_{18}H_{16}$)

Compound UU undergoes nitration, bromination, and Friedel–Crafts acylation. X-ray analysis shows that (except for the two methyl groups) UU is flat or nearly flat. Ten carbon–carbon bonds are between 1.386 Å and 1.401 Å long. The NMR spectrum shows peaks for 10H *downfield*, and for 6H *far upfield*:

> *a* singlet, $\delta - 4.25$ (τ 14.25), 6H
> *b* triplet, δ 8.11, 2H
> *c* doublet, δ 8.62, 4H
> *d* singlet, δ 8.67, 4H

35

Heterocyclic Compounds

35.1 Heterocyclic systems

A **heterocyclic compound** is one that contains a ring made up of more than one kind of atom.

In many of the cyclic compounds that we have studied so far—benzene, naphthalene, cyclohexanol, cyclopentadiene—the rings are made up only of carbon atoms; such compounds are called *homocyclic* compounds. But there are also rings containing, in addition to carbon, other kinds of atoms, most commonly nitrogen, oxygen, or sulfur. For example:

Pyrrole   Furan   Thiophene   Imidazole   Oxazole   Thiazole

Pyrazole   3-Pyrroline   Pyrrolidine   Pyridine   Pyrimidine   Purine

Quinoline   Isoquinoline   Carbazole

We notice that, in the numbering of ring positions, hetero atoms are generally given the lowest possible numbers.

Table 35.1 HETEROCYCLIC COMPOUNDS

Name	M.p., °C	B.p., °C	Name	M.p., °C	B.p., °C
Furan	− 30	32	Pyridine	− 42	115
Tetrahydrofuran	−108	66	α-Picoline	− 64	128
Furfuryl alcohol		171	β-Picoline		143
Furfural	− 36	162	γ-Picoline		144
Furoic acid	134		Piperidine	− 9	106
Pyrrole		130	Picolinic acid	137	
Pyrrolidine		88	Nicotinic acid	237	
Thiophene	− 40	84	Isonicotinic acid	317	
			Indole	53	254
			Quinoline	− 19	238
			Isoquinoline	23	243

Actually, of course, we have already dealt with numerous heterocyclic compounds: *cyclic anhydrides* (Sec. 24.9) and *cyclic imides* (Sec. 24.14), for example; *lactones* (Sec. 24.15) and *lactams* (Problem 32.2, p. 1105); *cyclic acetals* of dihydroxy alcohols (Problem 22, p. 792); the solvents *dioxane* and *tetrahydrofuran* (Sec. 19.9). In all these, the chemistry is essentially that of their open-chain analogs.

Crown ethers (Sec. 19.10) are, of course, heterocyclic compounds, and with them we found an ordinary property of ethers—the ability to solvate cations—taking on a special importance because these molecules are rings, and rings of a particular size. In Sec. 26.14 we looked very briefly at a few nitrogen heterocycles, but only for the property they share with other amines: basicity.

We have encountered three-membered heterocyclic rings which, because of ring strain, are highly reactive: *epoxides* (Secs. 19.11–19.16) and *aziridines* (Sec. 26.6); the fleeting but important intermediates, cyclic *halonium ions* (Secs. 8.14 and 20.2) and cyclic *sulfonium ions* (Sec. 20.4).

Heterocyclic intermediates are being used more and more in synthesis as *protecting groups*, readily generated and, when their job is done, readily removed. We have seen two examples of this: the temporary incorporation of the carboxyl group into a *2-oxazoline* ring (Sec. 30.6), and the temporary formation of *tetrahydropyranyl* (*THP*) *ethers* and *esters*, resistant toward alkali but extremely easily cleaved by acid (Sec. 19.9 and Problem 16, p. 892).

In the biological world, as we shall see in the final chapters of this book, heterocyclic compounds are everywhere. Carbohydrates are heterocyclic; so are chlorophyll and heme, which make leaves green and blood red and bring life to plants and animals. Heterocycles form the sites of reaction in many enzymes and coenzymes. Heredity comes down, ultimately, to the particular sequence of attachment of a half-dozen heterocyclic rings to the long chains of nucleic acids.

In this chapter we can take up only a very few of the many different heterocyclic systems, and look only briefly at them. Among the most important and most interesting heterocycles are the ones that possess aromatic properties; we shall focus our attention on a few of these, and in particular upon their aromatic properties.

We can get some idea of the importance—as well as complexity—of heterocyclic systems from the following examples. Some others are *heme* (p. 1367), *nicotinamide adenine dinucleotide* (p. 1367), and *oxytocin* (p. 1357).

Penicillin G
Antibiotic

Thiamine
Vitamin B$_1$
Anti-beriberi factor

Reserpine
A tranquilizing drug

Nicotine
A tobacco alkaloid

Copper phthalocyanine
A blue pigment

Chlorophyll a
Green plant pigment:
catalyst for photosynthesis

FIVE-MEMBERED RINGS

35.2 Structure of pyrrole, furan, and thiophene

The simplest of the five-membered heterocyclic compounds are **pyrrole**, **furan**, and **thiophene**, each of which contains a single hetero atom.

I	II	III
Pyrrole	Furan	Thiophene

Judging from the commonly used structures I, II, and III, we might expect each of these compounds to have the properties of a conjugated diene and of an amine, an ether, or a sulfide (thioether). Except for a certain tendency to undergo addition reactions, however, these heterocycles do not have the expected properties: thiophene does not undergo the oxidation typical of a sulfide, for example; pyrrole does not possess the basic properties typical of amines.

Instead, these heterocycles and their derivatives most commonly undergo electrophilic substitution: nitration, sulfonation, halogenation, Friedel–Crafts acylation, even the Reimer–Tiemann reaction and coupling with diazonium salts. Heats of combustion indicate resonance stabilization to the extent of 22–28 kcal/mol; somewhat less than the resonance energy of benzene (36 kcal/mol), but much greater than that of most conjugated dienes (about 3 kcal/mol). On the basis of these properties, pyrrole, furan, and thiophene must be considered *aromatic*. Clearly, formulas I, II, and III do not adequately represent the structures of these compounds.

Let us look at the orbital picture of one of these molecules, pyrrole. Each atom of the ring, whether carbon or nitrogen, is held by a σ bond to three other atoms. In forming these bonds, the atom uses three sp^2 orbitals, which lie in a plane and are 120° apart. After contributing one electron to each σ bond, each carbon atom of the ring has left *one* electron and the nitrogen atom has left *two* electrons; these electrons occupy p orbitals. Overlap of the p orbitals gives rise to π clouds, one above and one below the plane of the ring; the π clouds contain a total of six electrons, the *aromatic sextet* (Fig. 35.1).

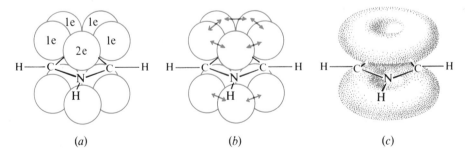

(a)   (b)   (c)

Figure 35.1 Pyrrole molecule. (*a*) Two electrons in the *p* orbital of nitrogen; one electron in the *p* orbital of each carbon. (*b*) Overlap of the *p* orbitals to form π bonds. (*c*) Clouds above and below the plane of the ring; a total of six π electrons, the aromatic sextet.

Delocalization of the π electrons stabilizes the ring. As a result, pyrrole has an abnormally low heat of combustion; it tends to undergo reactions in which the stabilized ring is retained, that is, to undergo substitution.

Nitrogen's extra pair of electrons, which is responsible for the usual basicity of nitrogen compounds, is involved in the π cloud, and is not available for sharing with acids. In contrast to most amines, therefore, pyrrole is an extremely weak base ($K_b \sim 2.5 \times 10^{-14}$). By the same token, there is a high electron density in the ring, which causes pyrrole to be extremely reactive toward electrophilic substitution: it undergoes reactions like nitrosation and coupling with diazonium salts which are characteristic of only the most reactive benzene derivatives, phenols and amines.

It thus appears that pyrrole is better represented by IV,

IV

Pyrrole

in which the circle represents the aromatic sextet.

What does IV mean in terms of conventional valence-bond structures? Pyrrole can be considered a hybrid of structures V–IX. Donation of electrons to the ring by nitrogen is

V   VI   VII   VIII   IX

Pyrrole

indicated by the ionic structures in which nitrogen bears a positive charge and the carbon atoms of the ring bear a negative charge.

Furan and thiophene have structures that are analogous to the structure of pyrrole. Where nitrogen in pyrrole carries a hydrogen atom, the oxygen or sulfur carries an unshared pair of electrons in an sp^2 orbital. Like nitrogen, the oxygen

Furan   Thiophene

or sulfur atom provides two electrons for the π cloud; as a result these compounds, too, behave like extremely reactive benzene derivatives.

35.3   Source of pyrrole, furan, and thiophene

Pyrrole and thiophene are found in small amounts in coal tar. During the fractional distillation of coal tar, thiophene (b.p. 84 °C) is collected along with the benzene (b.p. 80 °C); as a result ordinary benzene contains about 0.5% of thiophene, and must be specially treated if *thiophene-free benzene* is desired.

Thiophene can be synthesized on an industrial scale by the high-temperature reaction between *n*-butane and sulfur.

$$CH_3CH_2CH_2CH_3 + S \xrightarrow{560\ °C} \underset{\substack{S}}{\bigcirc} + H_2S$$

n-Butane   Thiophene

Pyrrole can be synthesized in a number of ways. For example:

$$HC{\equiv}CH + 2HCHO \xrightarrow{Cu_2C_2} HOCH_2C{\equiv}CCH_2OH \xrightarrow{NH_3,\ pressure} \underset{\substack{N\\H}}{\bigcirc}$$

1,4-Butynediol   Pyrrole

The pyrrole ring is the basic unit of the *porphyrin* system, which occurs, for example, in chlorophyll (p. 1207) and in hemoglobin (p. 1367).

Furan is most readily prepared by decarbonylation (elimination of carbon monoxide) of **furfural** (furfuraldehyde), which in turn is made by the treatment of oat hulls, corncobs, or rice hulls with hot hydrochloric acid. In the latter reaction pentosans (polypentosides) are hydrolyzed to pentoses, which then undergo dehydration and cyclization to form furfural.

$$(C_5H_8O_4)_n \xrightarrow{H_2O,\ H^+} \underset{\substack{CH_2OH}}{\overset{\substack{CHO}}{(CHOH)_3}} \xrightarrow{-3H_2O} \underset{\substack{O}}{\bigcirc}CHO \xrightarrow[\substack{steam,\ 400\ °C}]{oxide\ catalyst,} \underset{\substack{O}}{\bigcirc}$$

Pentosan   Pentose   Furfural   Furan
(2-Furancarboxaldehyde)

Certain substituted pyrroles, furans, and thiophenes can be prepared from the parent heterocycles by substitution (see Sec. 35.4); most, however, are prepared from open-chain compounds by ring closure. For example:

$\xrightarrow{P_2O_5,\ heat}$ CH$_3$—◯—CH$_3$ (O)

2,5-Dimethylfuran

$\underset{\substack{O\ \ \ \ O}}{\overset{\substack{H_2C-CH_2}}{H_3C-C\ \ \ \ \ C-CH_3}}$ $\xrightarrow{(NH_4)_2CO_3,\ 100\ °C}$ CH$_3$—◯—CH$_3$ (N,H)

Acetonylacetone   2,5-Dimethylpyrrole
(2,5-Hexanedione)
A 1,4-diketone

$\xrightarrow{P_2S_5,\ heat}$ CH$_3$—◯—CH$_3$ (S)

2,5-Dimethylthiophene

Problem 35.1   Give structural formulas for all intermediates in the following synthesis of acetonylacetone (2,5-hexanedione):

ethyl acetoacetate + NaOC$_2$H$_5$ $\longrightarrow$ A (C$_6$H$_9$O$_3$Na)
A + I$_2$ $\longrightarrow$ B (C$_{12}$H$_{18}$O$_6$) + NaI
B + dilute acid + heat $\longrightarrow$ 2,5-hexanedione + carbon dioxide + ethanol

Problem 35.2   Outline a synthesis of 2,5-diphenylfuran, starting from ethyl benzoate and ethyl acetate.

35.4   Electrophilic substitution in pyrrole, furan, and thiophene. Reactivity and orientation

Like other aromatic compounds, these five-membered heterocycles undergo nitration, halogenation, sulfonation, and Friedel–Crafts acylation. They are much more reactive than benzene, and resemble the most reactive benzene derivatives (amines and phenols) in undergoing such reactions as the Reimer–Tiemann reaction, nitrosation, and coupling with diazonium salts.

Reaction takes place predominantly at the 2-position. For example:

Furan   2-Furansulfonic acid

Furan   Boron trifluoride   2-Acetylfuran
               etherate

Thiophene   2-Benzoylthiophene

Pyrrole   2-(Phenylazo)pyrrole

Pyrrole   2-Pyrrolecarboxaldehyde
                               (*Low yield*)

In some of the examples we notice modifications in the usual electrophilic reagents. The high reactivity of these rings makes it possible to use milder reagents in many cases, as, for example, the weak Lewis acid stannic chloride in the Friedel–

Crafts acylation of thiophene. The sensitivity to protic acids of furan (which undergoes ring opening) and pyrrole (which undergoes polymerization) makes it necessary to modify the usual sulfonating agent.

Problem 35.3 Furan undergoes ring opening upon treatment with sulfuric acid; it reacts almost explosively with halogens. Account for the fact that 2-furoic acid, however, can be sulfonated (in the 5-position) by treatment with fuming sulfuric acid, and brominated (in the 5-position) by treatment with bromine at 100 °C.

2-Furoic acid

Problem 35.4 Upon treatment with formaldehyde and acid, ethyl 2,4-dimethyl-3-pyrrolecarboxylate is converted into a compound of formula $C_{19}H_{26}O_4N_2$. What is the most likely structure for this product? How is it formed? (*Hint*: See Sec. 36.7.)

Problem 35.5 Predict the products from the treatment of furfural (2-furancarboxaldehyde) with concentrated aqueous NaOH.

Problem 35.6 Sulfur trioxide dissolves in the tertiary amine pyridine to form a salt:

Pyridine

Show all steps in the most likely mechanism for the sulfonation of an aromatic compound by this reagent.

In our study of electrophilic aromatic substitution (Sec. 14.17 and Sec. 34.9), we found that we could account for orientation on the following basis: the controlling step is the attachment of the electrophilic reagent to the aromatic ring,

More stable ion

which takes place in such a way as to yield the most stable intermediate carbocation. Let us apply this approach to the reactions of pyrrole.

Attack at position 3 yields a carbocation that is a hybrid of structures I and II. Attack at position 2 yields a carbocation that is a hybrid not only of structures III and IV (analogous to I and II) but also of structure V; the extra stabilization conferred by V makes this ion the more stable one.

Viewed differently, attack at position 2 is faster because the developing positive charge is accommodated by *three* atoms of the ring instead of by only two.

Pyrrole is highly reactive, compared with benzene, because of contribution from the relatively stable structure III. In III *every atom has an octet of electrons*; nitrogen accommodates the positive charge simply by *sharing* four pairs of electrons. It is no accident that pyrrole resembles aniline in reactivity: both owe their high reactivity to the ability of nitrogen to share four pairs of electrons.

Orientation of substitution in furan and thiophene, as well as their high reactivity, can be accounted for in a similar way.

Problem 35.7 The heterocycle *indole*, commonly represented as formula VI, is found in coal tar and in orange blossoms.

VI

Indole

It undergoes electrophilic substitution, chiefly at position 3. Account (a) for the aromatic properties of indole, and (b) for the orientation in electrophilic substitution. (*Hint*: See Sec. 34.9.)

35.5 Saturated five-membered heterocycles

Catalytic hydrogenation converts pyrrole and furan into the corresponding saturated heterocycles, *pyrrolidine* and *tetrahydrofuran*. Since thiophene poisons

Pyrrole
($K_b \sim 10^{-14}$)

Pyrrolidine
($K_b \sim 10^{-3}$)

Furan

Tetrahydrofuran

$$BrCH_2CH_2CH_2CH_2Br + Na_2S \xrightarrow{\text{heat}}$$

Tetrahydrothiophene

most catalysts, *tetrahydrothiophene* is synthesized instead from open-chain compounds.

Saturation of these rings destroys the aromatic structure and, with it, the aromatic properties. Each of the saturated heterocycles has the properties we would expect of it: the properties of a secondary aliphatic amine, an aliphatic ether, or an aliphatic sulfide. With nitrogen's extra pair of electrons now available for sharing with acids, pyrrolidine ($K_b \sim 10^{-3}$) has the normal basicity of an aliphatic amine. Hydrogenation of pyrrole increases the base strength by a factor of 10^{11} (100 billion); clearly a fundamental change in structure has taken place. (See Fig. 35.2.)

(a)   *(b)*   *(c)*

Figure 35.2 Electronic configuration and molecular shape: *(a)* and *(b)* pyrrole, aromatic; *(c)* pyrrolidine, aliphatic.

The fundamental difference in structure is reflected by the striking difference in shape between the two molecules. As we see, pyrrole has the characteristic aromatic shape: flat, like benzene—or, closer yet, like the cyclopentadienyl anion (Fig. 13.7, p. 490), with which it is isoelectronic. Pyrrolidine, on the other hand, is clearly aliphatic, and closely resembles cyclopentane (Fig. 12.8, p. 451), with an unshared pair of electrons taking the place of one hydrogen atom.

Tetrahydrofuran is an important solvent, used, for example, in reductions with lithium aluminum hydride, in the preparation of arylmagnesium chlorides (Sec. 29.4), and in hydroborations. Oxidation of tetrahydrothiophene yields *tetramethylene sulfone* (or *sulfolane*), also used as an aprotic solvent (Sec. 6.4).

Tetramethylene sulfone
(Sulfolane)

We have encountered pyrrolidine as a secondary amine commonly used in making enamines (Sec. 30.8). The pyrrolidine ring occurs naturally in a number of alkaloids (Sec. 4.27), providing the basicity that gives these compounds their name (*alkali-like*).

Problem 35.8 An older process for the synthesis of both the adipic acid and the hexamethylenediamine needed in the manufacture of nylon-6,6 (Sec. 36.7) started with tetrahydrofuran. Using only familiar chemical reactions, suggest possible steps in their synthesis.

Problem 35.9 Predict the products of the treatment of pyrrolidine with:

(a) aqueous HCl
(b) aqueous NaOH
(c) acetic anhydride
(d) benzenesulfonyl chloride + aqueous NaOH
(e) methyl iodide, followed by aqueous NaOH
(f) repeated treatment with methyl iodide, followed by Ag_2O and then strong heating

Problem 35.10 The alkaloid *hygrine* is found in the coca plant. Suggest a structure for it on the basis of the following evidence:

Hygrine ($C_8H_{15}ON$) is insoluble in aqueous NaOH but soluble in aqueous HCl. It does not react with benzenesulfonyl chloride. It reacts with phenylhydrazine to yield a phenylhydrazone. It reacts with NaOI to yield a yellow precipitate and a carboxylic acid ($C_7H_{13}O_2N$). Vigorous oxidation by CrO_3 converts hygrine into *hygrinic acid* ($C_6H_{11}O_2N$).

Hygrinic acid can be synthesized as follows:

$BrCH_2CH_2CH_2Br + CH(COOC_2H_5)_2{}^-Na^+ \longrightarrow A\ (C_{10}H_{17}O_4Br)$

$A + Br_2 \longrightarrow B\ (C_{10}H_{16}O_4Br_2)$

$B + CH_3NH_2 \longrightarrow C\ (C_{11}H_{19}O_4N)$

$C + aq.\ Ba(OH)_2 + heat \longrightarrow D \xrightarrow{HCl} E \xrightarrow{heat} hygrinic\ acid + CO_2$

SIX-MEMBERED RINGS

35.6 Structure of pyridine

Of the six-membered aromatic heterocycles, we shall take up only one, **pyridine.**

Pyridine is classified as aromatic on the basis of its properties. It is flat, with bond angles of 120°; the four carbon–carbon bonds are of the same length, and so are the two carbon–nitrogen bonds. It resists addition and undergoes electrophilic substitution. Its heat of combustion indicates a resonance energy of 23 kcal/mol.

Pyridine can be considered a hybrid of the Kekulé structures I and II. We shall represent it as structure III, in which the circle represents the aromatic sextet.

In electronic configuration, the nitrogen of pyridine is considerably different from the nitrogen of pyrrole. In pyridine the nitrogen atom, like each of the carbon atoms, is bonded to other members of the ring by the use of sp^2 orbitals, and provides one electron for the π cloud. The third sp^2 orbital of each carbon atom is used to form a bond to hydrogen; the third sp^2 orbital of nitrogen simply contains a pair of electrons, which are available for sharing with acids (Fig. 35.3).

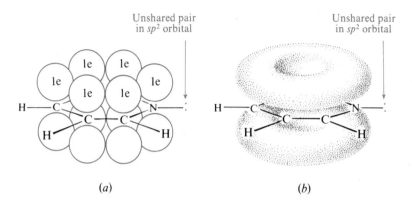

Figure 35.3 Pyridine molecule. (*a*) One electron in each *p* orbital; two electrons in an sp^2 orbital of nitrogen. (*b*) The *p* orbitals overlap to form π clouds above and below the plane of the ring; two unshared electrons are still in an sp^2 orbital of nitrogen.

Because of this electronic configuration, the nitrogen atom makes pyridine a much stronger base than pyrrole, and affects the reactivity of the ring in a quite different way, as we shall see.

35.7   Source of pyridine compounds

Pyridine is found in coal tar. Along with it are found a number of methyl-pyridines, the most important of which are the monomethyl compounds, known as *picolines*.

Oxidation of the picolines yields the pyridinecarboxylic acids.

Picoline   Pyridinecarboxylic acid
(2-, 3-, or 4-)   (2-, 3-, or 4-)

The 3-isomer (*nicotinic acid* or *niacin*) is a vitamin. The 4-isomer (*isonicotinic acid*) has been used, in the form of its hydrazide, in the treatment of tuberculosis.

Nicotinic acid
Niacin
3-Pyridinecarboxylic acid
Anti-pellagra factor

Isonicotinic acid hydrazide
(Isoniazid)

The increasing demand for certain pyridine derivatives has led to the development of syntheses involving ring closure. For example:

$$2CH_2{=}CH{-}CHO + NH_3 \longrightarrow$$

Acrolein

3-Methylpyridine
β-Picoline

Nicotinic acid

35.8   Reactions of pyridine

The chemical properties of pyridine are those we would expect on the basis of its structure. The ring undergoes the substitution, both electrophilic and nucleophilic, typical of aromatic rings; our interest will lie chiefly in the way the nitrogen atom affects these reactions.

There is another set of reactions in which pyridine acts as a base or nucleophile; these reactions involve nitrogen directly and are due to its unshared pair of electrons.

35.9   Electrophilic substitution in pyridine

Toward electrophilic substitution pyridine resembles a highly deactivated benzene derivative. It undergoes nitration, sulfonation, and halogenation only under very vigorous conditions, and does not undergo the Friedel–Crafts reaction at all.

Substitution occurs chiefly at the 3- (or β-) position.

$KNO_3, H_2SO_4, 300\ ^\circ C$

3-Nitropyridine

$H_2SO_4, 350\ ^\circ C$

3-Pyridinesulfonic acid

Pyridine

$Br_2, 300\ ^\circ C$

3-Bromo-
pyridine

3,5-Dibromo-
pyridine

RX or $RCOX$, $AlCl_3$ no reaction

Let us see if we can account for the reactivity and orientation on our usual basis of stability of the intermediate carbocation. Attack at the 4-position yields a carbocation that is a hybrid of structures I, II, and III.

Electrophilic
attack at
4-position

I   II   III

Especially unstable:
nitrogen has sextet

Attack at the 3-position yields an ion that is a hybrid of structures IV, V, and VI.

Electrophilic
attack at
3-position

IV   V   VI

(Attack at the 2-position resembles attack at the 4-position just as *ortho* attack resembles *para* attack in the benzene series.)

All these structures are less stable than the corresponding ones for attack on benzene, because of electron withdrawal by the nitrogen atom. As a result, pyridine undergoes substitution more slowly than benzene.

Of these structures, III is *especially* unstable, since in it the electronegative nitrogen atom has only a sextet of electrons. As a result, attack at the 4-position (or 2-position) is especially slow, and substitution occurs predominantly at the 3-position.

It is important to see the difference between substitution in pyridine and substitution in pyrrole. In the case of pyrrole, a structure in which nitrogen bears a positive charge (see Sec. 35.4) is especially stable since every atom has an octet of electrons; nitrogen accommodates the positive charge simply by sharing four pairs of electrons. In the case of pyridine, a structure in which nitrogen bears a positive charge (III) is especially unstable since nitrogen has only a sextet of electrons; nitrogen *shares* electrons readily, but as an electronegative atom it resists the *removal* of electrons.

Problem 35.11   2-Aminopyridine can be nitrated or sulfonated under much milder conditions than pyridine itself; substitution occurs chiefly at the 5-position. Account for these facts.

Problem 35.12   Because of the difficulty of nitrating pyridine, 3-aminopyridine is most conveniently made via nicotinic acid. Outline the synthesis of 3-aminopyridine from β-picoline.

35.10   Nucleophilic substitution in pyridine

Here, as in electrophilic substitution, the pyridine ring resembles a benzene ring that contains strongly electron-withdrawing groups. Nucleophilic substitution takes place readily, particularly at the 2- and 4-positions. For example:

2-Bromopyridine   2-Aminopyridine

4-Chloropyridine   4-Aminopyridine

The reactivity of pyridine toward nucleophilic substitution is so great that even the powerfully basic hydride ion, :H⁻, can be displaced. Two important examples of this reaction are amination by sodium amide (**Chichibabin reaction**), and alkylation or arylation by organolithium compounds.

Pyridine   Sodium amide   2-Aminopyridine

Sodium salt of
2-aminopyridine

Pyridine   Phenyllithium   2-Phenylpyridine

As we have seen (Sec. 29.8), nucleophilic aromatic substitution can take place by a mechanism that is quite analogous to the mechanism for electrophilic substitution. Reaction proceeds by two steps; the rate of the first step, formation of a charged particle, determines the rate of the overall reaction. In electrophilic substitution, the intermediate is positively charged; in nucleophilic substitution, the intermediate is negatively charged. The ability of the ring to accommodate the charge determines the stability of the intermediate and of the transition state leading to it, and hence determines the rate of the reaction.

Nucleophilic attack at the 4-position yields a carbanion that is a hybrid of structures I, II, and III:

Nucleophilic attack at 4-position

I   II   III

Especially stable:
negative charge
on nitrogen

Attack at the 3-position yields a carbanion that is a hybrid of structures IV, V, and VI:

IV   V   VI

Nucleophilic attack at 3-position

(As before, attack at the 2-position resembles attack at the 4-position.)

All these structures are more stable than the corresponding ones for attack on a benzene derivative, because of electron withdrawal by the nitrogen atom. Structure III is *especially* stable, since the negative charge is located on the atom that can best accommodate it, the electronegative nitrogen atom. It is reasonable, therefore, that nucleophilic substitution occurs more rapidly on the pyridine ring than on the benzene ring, and more rapidly at the 2- and 4-positions than at the 3-position.

The same electronegativity of nitrogen that makes pyridine unreactive toward electrophilic substitution makes pyridine highly reactive toward nucleophilic substitution.

35.11 Basicity of pyridine

Pyridine is a base with $K_b = 2.3 \times 10^{-9}$. It is thus much stronger than pyrrole ($K_b \sim 2.5 \times 10^{-14}$) but much weaker than aliphatic amines ($K_b \sim 10^{-4}$).

Pyridine has a pair of electrons (in an sp^2 orbital) that is available for sharing with acids; pyrrole has not, and can accept an acid only at the expense of the aromatic character of the ring.

The fact that pyridine is a weaker base than aliphatic amines is more difficult to account for, but at least it fits into a pattern. Let us turn for a moment to the basicity of the carbon analogs of amines, the carbanions, and use the approach of Sec. 11.11.

Benzene is a stronger acid than an alkane, as shown by its ability to displace an alkane from its salts; this, of course, means that the phenyl anion, $C_6H_5^-$, is a weaker base than an an alkyl anion, R^-.

$$R:^- Na^+ \ + \ C_6H_5:H \ \rightleftharpoons \ R:H \ + \ C_6H_5:^- Na^+$$

Stronger base   Stronger acid   Weaker acid   Weaker base

In the same way, acetylene is a stronger acid than benzene, and the acetylide ion is a weaker base than the phenyl anion.

$$C_6H_5:^- Na^+ \ + \ HC{\equiv}C:H \ \rightleftharpoons \ C_6H_5:H \ + \ HC{\equiv}C:^- Na^+$$

Stronger base   Stronger acid   Weaker acid   Weaker base

Thus we have the following sequences of acidity of hydrocarbons and basicity of their anions:

Acidity $HC{\equiv}C:H > C_6H_5:H > R:H$

Basicity $HC{\equiv}C:^- < C_6H_5:^- < R:^-$

A possible explanation for these sequences can be found in the electronic configuration of the carbanions. In the alkyl, phenyl, and acetylide anions, the unshared pair of electrons occupies respectively an sp^3, an sp^2, and an sp orbital. The availability of this pair for sharing with acids determines the basicity of the particular anion. As we proceed along the series sp^3, sp^2, sp, the p character of the orbital decreases and the s character increases. Now, an electron in a p orbital is at some distance from the nucleus and is held relatively loosely; an electron in an s orbital, on the other hand, is close to the nucleus and is held more tightly. Of the three anions, the alkyl ion is the strongest base since its pair of electrons is held most loosely, in an sp^3 orbital. The acetylide ion is the weakest base since its pair of electrons is held most tightly, in an sp orbital.

Pyridine bears the same relationship to an aliphatic amine as the phenyl anion bears to an alkyl anion. The pair of electrons that gives pyridine its basicity occupies an sp^2 orbital; it is held more tightly and is less available for sharing with acids than the pair of electrons of an aliphatic amine, which occupies an sp^3 orbital.

Problem 35.13 Predict the relative basicities of amines (RCH_2NH_2), imines ($RCH{=}NH$), and nitriles ($RC{\equiv}N$).

Pyridine is widely used in organic chemistry as a water-soluble base, as, for example, in the Schotten–Baumann acylation procedure (Sec. 24.8).

Problem 35.14 Ethyl bromosuccinate is converted into the unsaturated ester ethyl fumarate by the action of pyridine. What is the function of the pyridine? What advantage does it have here over the usual alcoholic KOH?

Like other amines, pyridine has nucleophilic properties, and reacts with alkyl halides to form quaternary ammonium salts.

Pyridine

N-Methylpyridinium iodide
(Pyridine methiodide)

Problem 35.15 Like any other tertiary amine, pyridine can be converted (by peroxybenzoic acid) into its *N*-oxide. In contrast to pyridine itself, pyridine *N*-oxide readily undergoes nitration, chiefly in the 4-position. How do you account for this reactivity and orientation?

Pyridine *N*-oxide

Problem 35.16 Pyridine *N*-oxides not only are reactive toward electrophilic substitution, but also seem to be reactive toward nucleophilic substitution, particularly at the 2- and 4-positions. For example, treatment of 4-nitropyridine *N*-oxide with hydrobromic acid gives 4-bromopyridine *N*-oxide. How do you account for this reactivity and orientation?

Problem 35.17 The oxygen of pyridine *N*-oxide is readily removed by treatment with PCl_3. Suggest a practical route to 4-nitropyridine. To 4-bromopyridine.

35.12 Reduction of pyridine

Catalytic hydrogenation of pyridine yields the aliphatic heterocyclic compound **piperidine**, $C_5H_{11}N$.

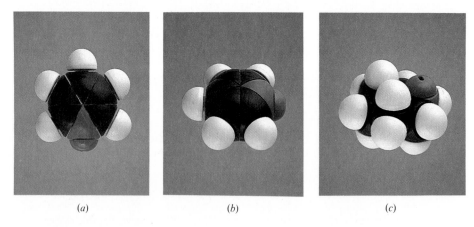

Pyridine
$(K_b = 2.3 \times 10^{-9})$
Piperidine
$(K_b = 2 \times 10^{-3})$

H₂, Pt, HCl, 25 °C, 3 atm.

Piperidine ($K_b = 2 \times 10^{-3}$) has the usual basicity of a secondary aliphatic amine, a million times greater than that of pyridine; again, clearly, a fundamental change in structure has taken place (see Fig. 35.4). Like pyridine, piperidine is often used as a basic catalyst in such reactions as the Knoevenagel reaction (Problem 25.22(f), p. 920) or Michael addition (Sec. 31.7).

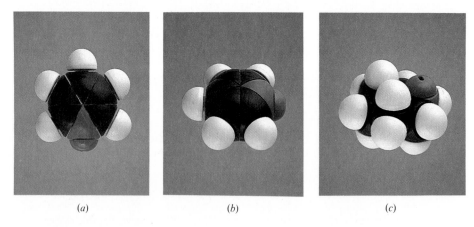

(a)   (b)   (c)

Figure 35.4 Electronic configuration and molecular shape: (*a*) and (*b*) pyridine, aromatic; (*c*) piperidine, aliphatic.

Here again we see the contrast between aromatic and aliphatic structures reflected in a contrast in molecular shape. Pyridine has the shape of benzene (Fig. 13.5, p. 487), with an unshared pair of electrons taking the place of one hydrogen. Pyrimidine has the familiar shape of chair cyclohexane (Fig. 12.5, p. 448), with an unshared pair occupying an equatorial—or, in another conformation, an axial—position.

Like the pyrrolidine ring, the piperidine and pyridine rings are found in a number of alkaloids, including *nicotine, strychnine, cocaine,* and *reserpine* (see p. 1207).

Problem 35.18   Why can piperidine not be used in place of pyridine in the Schotten–Baumann procedure?

FUSED RINGS

35.13   Quinoline. The Skraup synthesis

Quinoline, C_9H_7N, contains a benzene ring and a pyridine ring fused as shown in I.

I

Quinoline

$(K_b = 3 \times 10^{-10})$

In general, its properties are the ones we would expect from what we have learned about pyridine and naphthalene.

Problem 35.19   Account for the following properties of quinoline:
(a) Treatment with nitric and sulfuric acids gives 5- and 8-nitroquinolines; treatment with fuming sulfuric acid gives 5- and 8-quinolinesulfonic acids.
(b) Oxidation by $KMnO_4$ gives 2,3-pyridinedicarboxylic acid (*quinolinic acid*).
(c) Treatment with sodamide gives 2-aminoquinoline; treatment with alkyllithium compounds gives 2-alkylquinolines.

Problem 35.20   *8-Hydroxyquinoline* (8-quinolinol) is a reagent in inorganic analysis. Suggest a method of synthesizing it.

Quinoline is found in coal tar. Although certain derivatives of quinoline can be made from quinoline itself by substitution, most are prepared from benzene derivatives by ring closure.

Perhaps the most generally useful method for preparing substituted quinolines is the **Skraup synthesis**. In the simplest example, quinoline itself is obtained from the reaction of aniline with glycerol, concentrated sulfuric acid, nitrobenzene, and ferrous sulfate.

The following steps seem to be involved:

(1) Dehydration of glycerol by hot sulfuric acid to yield the unsaturated aldehyde acrolein:

(2) Nucleophilic addition of aniline to acrolein to yield β-(phenylamino)-propionaldehyde:

(3) Electrophilic attack on the aromatic ring by the electron-deficient carbonyl carbon of the protonated aldehyde (this is the actual ring-closing step):

(4) Oxidation by nitrobenzene resulting in the aromatization of the newly formed ring:

Ferrous sulfate in some way moderates the otherwise very vigorous reaction.

Thus we see that what at first appears to be a complicated reaction is actually a sequence of simple steps involving familiar, fundamental types of reactions:

acid-catalyzed dehydration, nucleophilic addition to an α,β-unsaturated carbonyl compound, electrophilic aromatic substitution, and oxidation.

The components of the basis synthesis can be modified to yield a wide variety of quinoline derivatives. For example:

aniline + crotonaldehyde $\longrightarrow$ 2-methylquinoline (quinaldine)

3-nitro-4-aminoanisole + glycerol $\longrightarrow$ 6-methoxy-8-nitroquinoline

2-aminonaphthalene + glycerol $\longrightarrow$

5,6-Benzoquinoline
(1-Azaphenanthrene)

Nitrobenzene is often replaced as oxidizing agent by arsenic acid, H_3AsO_4, which usually gives a less violent reaction; vanadium pentoxide is sometimes added as a catalyst. Sulfuric acid can be replaced by phosphoric acid or other acids.

Problem 35.21   Show all steps in the Skraup syntheses mentioned above.

Problem 35.22   The dehydration of glycerol to yield acrolein involves acid-catalyzed dehydration and keto–enol tautomerization. Outline the possible steps in the dehydration. (*Hint*: Which —OH is easier to eliminate, a primary or a secondary?)

Problem 35.23   What is the product of the application of the Skraup synthesis to (a) *o*-nitroaniline, (b) *o*-aminophenol, (c) *o*-phenylenediamine, (d) *m*-phenylenediamine, (e) *p*-toluidine?

Problem 35.24   Outline the synthesis of 6-bromoquinoline. Of 8-methylquinoline.

Problem 35.25   In the **Doebner–von Miller** modification of the Skraup synthesis, aldehydes, ketones, or mixtures of aldehydes and ketones replace the glycerol. If acetaldehyde is used, for example, the product from aniline is 2-methylquinoline (*quinaldine*). (a) Account for its formation. (b) Predict the product if methyl vinyl ketone were used. (c) If a mixture of benzaldehyde and pyruvic acid, $CH_3COCOOH$, were used.

Problem 35.26   Account for the formation of 2,4-dimethylquinoline from aniline and acetylacetone (2,4-pentanedione) by the Doebner–von Miller synthesis. (*Hint*: See Problem 23, p. 928.)

35.14   Isoquinoline. The Bischler–Napieralski synthesis

Isoquinoline, C_9H_7N, contains a benzene ring and a pyridine ring fused as shown in I. Isoquinoline, like quinoline, has the properties we would expect from what we know about pyridine and naphthalene.

I

Isoquinoline
($K_b = 1.1 \times 10^{-9}$)

Problem 35.27   Account for the following properties of isoquinoline. (*Hint*: Review orientation in β-substituted naphthalenes, Sec. 34.13.)

(a) Nitration gives 5-nitroisoquinoline.

(b) Treatment with potassium amide, KNH₂, gives 1-aminoisoquinoline, and treatment with alkyllithium compounds gives 1-alkylisoquinoline; the 3-substituted products are not obtained.

(c) 1-Methylisoquinoline reacts with benzaldehyde to yield compound II, whereas 3-methylisoquinoline undergoes no reaction. (*Hint*: See Problem 25.22 (c), p. 920.)

CH=CHC₆H₅

II

An important method for making derivatives of isoquinoline is the **Bischler–Napieralski synthesis**. Acyl derivatives of β-phenylethylamine are cyclized by treatment with acids (often P₂O₅) to yield dihydroisoquinolines, which can then be aromatized.

N-(2-phenylethyl)acetamide   1-Methyl-3,4-dihydroisoquinoline   1-Methylisoquinoline

Problem 35.28   To what general class of reactions does the ring closure belong? What is the function of the acid? (Check your answers in Sec. 36.7.)

Problem 35.29   Outline the synthesis of *N*-(2-phenylethyl)acetamide from toluene and aliphatic and inorganic reagents.

PROBLEMS

1. Give structures and names of the principal products from the reaction (if any) of pyridine with:

(a) Br₂, 300 °C
(b) H₂SO₄, 350 °C
(c) acetyl chloride, AlCl₃
(d) KNO₃, H₂SO₄, 300 °C
(e) NaNH₂, heat
(f) C₆H₅Li
(g) dilute HCl
(h) dilute NaOH

(i) acetic anhydride
(j) benzenesulfonyl chloride
(k) ethyl bromide
(l) benzyl chloride
(m) peroxybenzoic acid
(n) peroxybenzoic acid, then HNO₃, H₂SO₄
(o) H₂, Pt

2. Give structures and names of the principal products from each of the following reactions:

(a) thiophene + conc. H_2SO_4
(b) thiophene + acetic anhydride, $ZnCl_2$
(c) thiophene + acetyl chloride, $TiCl_4$
(d) thiophene + fuming nitric acid in acetic anhydride
(e) product of (d) + Sn, HCl
(f) thiophene + 1 mol Br_2
(g) product of (f) + Mg; then CO_2; then H^+
(h) pyrrole + pyridine: SO_3
(i) pyrrole + diazotized sulfanilic acid
(j) product of (i) + $SnCl_2$
(k) pyrrole + H_2, Ni $\longrightarrow$ C_4H_9N
(l) furfural + acetone + base
(m) quinoline + HNO_3/H_2SO_4
(n) quinoline N-oxide + HNO_3/H_2SO_4
(o) isoquinoline + n-butyllithium

3. Pyrrole can be reduced by zinc and acetic acid to a *pyrroline*, C_4H_7N. (a) What structures are possible for this pyrroline?
(b) On the basis of the following evidence which structure must the pyrroline have?

pyrroline + O_3; then H_2O; then H_2O_2 $\longrightarrow$ A $(C_4H_7O_4N)$
chloroacetic acid + NH_3 $\longrightarrow$ B $(C_2H_5O_2N)$
B + chloroacetic acid $\longrightarrow$ A

4. Furan and its derivatives are sensitive to protic acids. The following reactions illustrate what happens.

2,5-dimethylfuran + dilute H_2SO_4 $\longrightarrow$ C $(C_6H_{10}O_2)$
C + NaOI $\longrightarrow$ succinic acid

(a) What is C? (b) Outline a likely series of steps for its formation from 2,5-dimethylfuran.

5. Pyrrole reacts with formaldehyde in hot pyridine to yield a mixture of products from which there can be isolated a small amount of a compound of formula $(C_5H_5N)_4$. Suggest a possible structure for this compound. (*Hint:* See Sec. 36.7 and p. 1207.)

6. There are three isomeric pyridinecarboxylic acids, $(C_5H_4N)COOH$: D, m.p. 137 °C; E, m.p. 234–237 °C; and F, m.p. 317 °C. Their structures were proved as follows:

quinoline + $KMnO_4$, OH^- $\longrightarrow$ a diacid $(C_7H_5O_4N)$ $\xrightarrow{\text{heat}}$ E, m.p. 234–237 °C
isoquinoline + $KMnO_4$, OH^- $\longrightarrow$ a diacid $(C_7H_5O_4N)$ $\xrightarrow{\text{heat}}$ E, m.p. 234–237 °C
and F, m.p. 317 °C

What structures should be assigned to D, E, and F?

7. (a) What structures are possible for G?

m-toluidine + glycerol $\xrightarrow{\text{Skraup}}$ G $(C_{10}H_9N)$

(b) On the basis of the following evidence which structures must G actually have?

2,3-diaminotoluene + glycerol $\xrightarrow{\text{Skraup}}$ H $(C_{10}H_{10}N_2)$
H + $NaNO_2$, HCl; then H_3PO_2 $\longrightarrow$ G

8. Outline all steps in a possible synthesis of each of the following from benzene, toluene, and any needed aliphatic and inorganic reagents:

(a) 1-phenylisoquinoline
(b) 1-benzylisoquinoline
(c) 1,5-dimethylisoquinoline
(d) 6-nitroquinoline

(e) 2-methyl-6-quinolinecarboxylic acid
(f) 1,8-diazaphenanthrene (*Hint:* Use the Skraup synthesis twice.)

1,8-Diazaphenanthrene

9. Outline all steps in each of the following syntheses, using any other needed reagents:

(a) β-cyanopyridine from β-picoline
(b) 2-methylpiperidine from pyridine
(c) 5-aminoquinoline from quinoline
(d) ethyl 5-nitro-2-furoate from furfural

(e) furylacrylic acid, [furan ring]—CH CHCOOH, from furfural

(f) 1,2,5-trichloropentane from furfural
(g) 3-indolecarboxaldehyde from indole

10. Give the structures of compounds I through JJ formed in the following syntheses of heterocyclic systems.

(a) ethyl malonate + urea, base, heat $\longrightarrow$ I ($C_4H_4O_3N_2$), a *pyrimidine* (1,3-diazine)
(b) 2,5-hexanedione + H_2N—NH_2 $\longrightarrow$ J ($C_6H_{10}N_2$)
$\quad$ J + air $\longrightarrow$ K ($C_6H_8N_2$), a *pyridazine* (1,2-diazine)
(c) 2,4-pentanedione + H_2N—NH_2 $\longrightarrow$ L ($C_5H_8N_2$), a *pyrazole*
(d) 2,3-butanedione + o-$C_6H_4(NH_2)_2$ $\longrightarrow$ M ($C_{10}H_{10}N_2$), a *quinoxaline*
(e) ethylene glycol + phosgene $\longrightarrow$ N ($C_3H_4O_3$), a *1,3-dioxolanone*
(f) o-aminobenzoic acid + chloroacetic acid $\longrightarrow$ O ($C_9H_9O_4N$)
$\quad$ O + base, strong heat $\longrightarrow$ P (C_8H_7ON), *indoxyl*, an intermediate in the synthesis
$\qquad$ of indigo
(g) aminoacetone $\longrightarrow$ Q ($C_6H_{10}N_2$)
$\quad$ Q + air $\longrightarrow$ R ($C_6H_8N_2$), a *pyrazine* (1,4-diazine)
(h) ethylenediamine + ethyl carbonate $\longrightarrow$ S ($C_3H_6ON_2$), an *imidazolidone*
(i) o-$C_6H_4(NH_2)_2$ + acetic acid, strong heat $\longrightarrow$ T ($C_8H_8N_2$), a *benzimidazole*
(j) ethyl o-aminobenzoate + malonic ester $\longrightarrow$ U ($C_{14}H_{17}O_5N$), insoluble in dilute
$\qquad$ acid

$\quad$ U $\xrightarrow{\text{NaOC}_2\text{H}_5}$ V ($C_{12}H_{11}O_4N$)
$\quad$ V + acid, warm $\longrightarrow$ W ($C_9H_7O_2N$), a *quinoline*
(k) repeat (j) starting with ethyl 3-amino-2-pyridinecarboxylate $\longrightarrow$
$\qquad$ a *1,5-diazanaphthalene*
(l) benzalacetophenone + KCN + acetic acid $\longrightarrow$ X ($C_{16}H_{13}ON$)
$\quad$ X + CH_3OH, H^+, H_2O $\longrightarrow$ Y ($C_{17}H_{16}O_3$) + NH_4^+
$\quad$ Y + phenylhydrazine $\longrightarrow$ Z ($C_{22}H_{18}ON_2$), a *dihydro-1,2-diazine*
(m) acrylic acid + H_2N—NH_2 $\longrightarrow$ AA ($C_3H_8O_2N_2$) $\longrightarrow$ BB ($C_3H_6ON_2$),
$\qquad$ a *pyrazolidone*
(n) o-$C_6H_4(NH_2)_2$ + glycerol $\xrightarrow{\text{Skraup}}$ CC ($C_{12}H_8N_2$), a *4,5-diazaphenanthrene*
(o) di(o-nitrophenyl)acetylene + Br_2 $\longrightarrow$ DD ($C_{14}H_8O_4N_2Br_2$)
$\quad$ DD + Sn, HCl $\longrightarrow$ EE ($C_{14}H_{12}N_2Br_2$)

$\quad$ EE $\xrightarrow{\text{warm}}$ [FF ($C_{14}H_{11}N_2Br$)] $\longrightarrow$ GG ($C_{14}H_{10}N_2$), which contains four
$\qquad$ fused aromatic rings
(p) m-$ClC_6H_4CH_2CH_2CH_2NHCH_3$ + C_6H_5Li $\longrightarrow$ HH ($C_{10}H_{13}N$), a
$\qquad$ *tetrahydroquinoline*
(q) o-$ClC_6H_4NHCOC_6H_5$ + KNH_2/NH_3 $\longrightarrow$ II ($C_{13}H_9ON$), a *benzoxazole*
(r) *trans*-I + base $\longrightarrow$ JJ ($C_{13}H_{15}ON$), an *oxazoline*

[structure of I: cyclohexane ring with OTs and NHCOC₆H₅ substituents]

OTs

NHCOC$_6$H$_5$

I

(s) How do you account for the fact that *cis*-I undergoes reaction (r) much more slowly than *trans*-I?

11. The structure of *papaverine*, $C_{20}H_{21}O_4N$, one of the opium alkaloids, has been established by the following synthesis:

3,4-dimethoxybenzyl chloride + KCN $\longrightarrow$ KK ($C_{10}H_{11}O_2N$)
KK + hydrogen, Ni $\longrightarrow$ LL ($C_{10}H_{15}O_2N$)
KK + aqueous acid, heat $\longrightarrow$ MM $\xrightarrow{PCl_5}$ NN ($C_{10}H_{11}O_3Cl$)
LL + NN $\longrightarrow$ OO ($C_{20}H_{25}O_5N$)
OO + P_2O_5, heat $\longrightarrow$ PP ($C_{20}H_{23}O_4N$)
PP + Pd, 200 °C $\longrightarrow$ papaverine

12. *Plasmochin* (also called *Pamaquine*), a drug effective against malaria, has been synthesized as follows:

ethylene oxide + diethylamine $\longrightarrow$ QQ ($C_6H_{15}ON$)
QQ + $SOCl_2$ $\longrightarrow$ RR ($C_6H_{14}NCl$)
RR + sodioacetoacetic ester $\longrightarrow$ SS ($C_{12}H_{23}O_3N$)
SS + dilute H_2SO_4, warm $\longrightarrow$ TT ($C_9H_{19}ON$) + CO_2 + C_2H_5OH
TT + H_2, Ni $\longrightarrow$ UU ($C_9H_{21}ON$)
UU + conc. HBr $\longrightarrow$ VV ($C_9H_{20}NBr$)

4-amino-3-nitroanisole + glycerol $\xrightarrow{Skraup}$ WW ($C_{10}H_8O_3N_2$)
WW + Sn + HCl $\longrightarrow$ XX ($C_{10}H_{10}ON_2$)
VV + XX $\longrightarrow$ Plasmochin ($C_{19}H_{29}ON_3$)

What is the most likely structure of Plasmochin?

13. (−)-*Nicotine*, the alkaloid in tobacco, can be synthesized in the following way:

nicotinic acid + $SOCl_2$, heat $\longrightarrow$ nicotinoyl chloride (C_6H_4ONCl)
nicotinoyl chloride + $C_2H_5OCH_2CH_2CH_2CdCl$ $\longrightarrow$ YY ($C_{11}H_{15}O_2N$), a ketone
YY + NH_3, H_2, catalyst $\longrightarrow$ ZZ ($C_{11}H_{18}ON_2$)
ZZ + HBr + strong heat $\longrightarrow$ AAA ($C_9H_{12}N_2$) + ethyl bromide
AAA + CH_3I, NaOH $\longrightarrow$ ($\pm$)-nicotine ($C_{10}H_{14}N_2$)
($\pm$)-nicotine + (+)-tartaric acid $\longrightarrow$ BBB and CCC (both $C_{14}H_{20}O_6N_2$)
BBB + NaOH $\longrightarrow$ (−)-nicotine + sodium tartrate

What is the structure of ($\pm$)-nicotine? Write equations for all the above reactions.

14. The red and blue colors of many flowers and fruits are due to the *anthocyanins*, glycosides of pyrylium salts. The parent structure of the pyrylium salts is *flavylium chloride*, which can be synthesized as follows:

salicylaldehyde + acetophenone $\xrightarrow{aldol}$ DDD ($C_{15}H_{12}O_2$)
DDD + HCl $\longrightarrow$ flavylium chloride, a salt containing three aromatic rings

Flavylium chloride

(a) What is the structure of DDD? (b) Outline a likely series of steps leading from DDD to flavylium chloride. (c) Account for the aromatic character of the fused ring system.

15. (a) Account for the aromatic properties of the imidazole ring.
(b) Arrange the nitrogen atoms of *histamine* (the substance responsible for many allergenic reactions) in order of their expected basicity, and account for your answer.

Histamine

(c) Account for the particular dipolar structure given for the amino acid *histidine* in Table 40.1, p. 1346.
(d) Account for the particular point of attachment to the *guanine* residue in compound II, p. 1198.

16. *Tropinic acid*, $C_8H_{13}O_4N$, is a degradation product of atropine, an alkaloid of the deadly nightshade, *Atropa belladonna*. It has a neutralization equivalent of 94 ± 1. It does not react with benzenesulfonyl chloride, cold dilute $KMnO_4$, or Br_2/CCl_4. Exhaustive methylation gives the following results:

tropinic acid + CH_3I $\longrightarrow$ EEE ($C_9H_{16}O_4NI$)
EEE + Ag_2O, then strong heat $\longrightarrow$ FFF ($C_9H_{15}O_4N$)
FFF + CH_3I $\longrightarrow$ GGG ($C_{10}H_{18}O_4NI$)
GGG + Ag_2O, then strong heat $\longrightarrow$ HHH ($C_7H_8O_4$) + $(CH_3)_3N$ + H_2O
HHH + H_2, Ni $\longrightarrow$ heptanedioic acid (pimelic acid)

(a) What structures are likely for tropinic acid?
(b) Tropinic acid is formed by oxidation with CrO_3 of *tropinone*, whose structure has been shown by synthesis to be

Tropinone

Now what is the most likely structure for tropinic acid?

17. *Tropilidene*, 1,3,5-cycloheptatriene, has been made from tropinone (Problem 16). Show how this might have been done. (*Hint*: See Problem 24, p. 990.)

18. Reduction of tropinone (Problem 16) gives *tropine* and *pseudotropine*, both $C_8H_{15}ON$. When heated with base, tropine is converted into pseudotropine. Give likely structures for tropine and pseudotropine, and explain your answer.

19. *Arecaidine*, $C_7H_{11}O_2N$, an alkaloid of betel nut, has been synthesized in the following way:

ethyl acrylate + NH_3 $\xrightarrow{\text{Michael}}$ III ($C_5H_{11}O_2N$)

III + ethyl acrylate $\xrightarrow{\text{Michael}}$ JJJ ($C_{10}H_{19}O_4N$)

JJJ + sodium ethoxide $\xrightarrow{\text{Dieckmann}}$ KKK ($C_8H_{13}O_3N$)

KKK + benzoyl chloride $\longrightarrow$ LLL ($C_{15}H_{17}O_4N$)

LLL + H_2, Ni $\longrightarrow$ MMM ($C_{15}H_{19}O_4N$)
MMM + acid, heat $\longrightarrow$ NNN ($C_6H_9O_2N$), *guvacine*, another betel nut alkaloid +
$\qquad\qquad\qquad\qquad\qquad\qquad\qquad\qquad\qquad C_6H_5COOH + C_2H_5OH$

NNN + CH_3I $\longrightarrow$ arecaidine ($C_7H_{11}O_2N$)

(a) What is the most likely structure of arecaidine? Of guvacine?
(b) What will guvacine give upon dehydrogenation?

20. Give the structures of compounds OOO through UUU. (*Hint*: Sec. 36.7.)

thiophene + 3-hexanone + H_2SO_4 $\longrightarrow$ OOO ($C_{14}H_{18}S_2$)
OOO + $(CH_3CO)_2O$ + $HClO_4$ $\longrightarrow$ PPP ($C_{16}H_{20}OS_2$)
PPP + N_2H_4 + KOH + heat $\longrightarrow$ QQQ ($C_{16}H_{22}S_2$)
QQQ + $C_6H_5N(CH_3)CHO$ $\longrightarrow$ RRR ($C_{17}H_{22}OS_2$), an aldehyde
RRR + Ag_2O $\longrightarrow$ SSS ($C_{17}H_{22}O_2S_2$)
SSS *was resolved*
(+)-SSS + Cu, quinoline, heat $\longrightarrow$ CO_2 + (+)-TTT ($C_{16}H_{22}S_2$)
(+)-TTT + H_2/Ni $\longrightarrow$ UUU ($C_{16}H_{34}$), *optically inactive*

What is the significance of the optical inactivity of UUU?

21. When heated in solution, 2-pyridinecarboxylic acid (II) loses carbon dioxide and forms pyridine. The rate of this decarboxylation is slowed down by addition of either acid or base. When decarboxylation is carried out in the presence of the ketone, R_2CO, there is obtained not only pyridine but also the tertiary alcohol III. The N-methyl derivative (IV) is decarboxylated much faster than II.

II	III	IV	V

(a) Show all steps in the most likely mechanism for decarboxylation of II. Show how this mechanism is consistent with each of the above facts.

(b) In the decarboxylation of the isomeric pyridinecarboxylic acids (II and its isomers), the order of reactivity is:

$$2 > 3 > 4$$

In the decarboxylation of the isomeric pyridineacetic acids (V and its isomers), on the other hand, the order of reactivity is:

$$2 \text{ or } 4 > 3$$

How do you account for each order of reactivity? Why is there a difference between the two sets of acids? (The same mechanism seems to be involved in both cases.)

36

Macromolecules.
Polymers and Polymerization

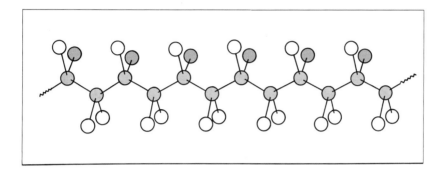

36.1 Macromolecules

So far, our study of organic chemistry has dealt mainly with rather small molecules, containing perhaps as many as 50 to 75 atoms. But there also exist enormous molecules called *macromolecules*, which contain hundreds of thousands of atoms. Some of these are naturally occurring, and make up classes of compounds that are, quite literally, vital: the *polysaccharides* starch and cellulose, which provide us with food, clothing, and shelter (Chap. 39); *proteins*, which constitute much of the animal body, hold it together, and run it (Chap. 40); and *nucleic acids*, which control heredity on the molecular level (Chap. 41).

Macromolecules can be man-made, too. The first syntheses were aimed at making substitutes for the natural macromolecules, rubber and silk; but a vast technology has grown up that now produces hundreds of substances that have no natural counterparts. Synthetic macromolecular compounds include: **elastomers**, which have the particular kind of elasticity characteristic of rubber; **fibers**, long, thin, and threadlike, with the great strength *along the fiber* that characterizes cotton, wool, and silk; and **plastics**, which can be extruded as sheets or pipes, painted on surfaces, or molded to form countless objects. We wear these man-made materials, eat and drink from them, sleep between them, sit and stand on them; turn knobs, pull switches, and grasp handles made of them; with their help we hear sounds and see sights remote from us in time and space; we live in houses and move about in vehicles that are increasingly made of them.

We sometimes deplore the resistance to the elements of these seemingly all too immortal materials, and fear that civilization may some day be buried beneath a pile of plastic debris—plastic cigar tips have been found floating in the Sargasso Sea—but with them we can do things never before possible. By use of plastics, blind people can be made to see, and cripples to walk; heart valves can be repaired and arteries patched; damaged tracheas, larynxes, ureters, and even entire hearts can be replaced. These materials protect us against heat and cold, electric shock and fire, rust and decay. As tailor-made solvents, they may soon be used to extract fresh water from the sea. Surely the ingenuity that has produced these substances can devise ways of disposing of the waste they create: the problem is not one of technology, but of sociology and, ultimately, of politics.

In this chapter, we shall be first—and chiefly—concerned with the chemical reactions by which macromolecules are formed, and the structures that these reactions produce. We shall look briefly at how these structures lead to the properties on which the use of the macromolecules depends: why rubber is elastic, for example, and why nylon is a strong fiber. Then, in the chapters on biomolecules, we shall take up the natural macromolecules—polysaccharides, proteins, and nucleic acids—and study them in much the same way.

In all this, we must remember that what makes macromolecules special is, of course, their great size. This great size permits a certain complexity of structure, not just on the molecular level, but on a *secondary* level that involves the disposition of molecules with respect to each other. Are the molecules stretched out neatly alongside one another, or coiled up independently? What forces act between different molecules? What happens to a collection of giant molecules when it is heated, or cooled, or stretched? As we shall see, the answers to questions like these are found ultimately in structure as we have known it: the nature of functional groups and substituents, their sequence in the molecule, and their arrangement in space.

36.2   Polymers and polymerization

Macromolecules, both natural and man-made, owe their great size to the fact that they are *polymers* (Greek: many parts); that is, each one is made up of a great many simpler units—identical to each other or at least chemically similar—joined together in a regular way. They are formed by a process we touched on earlier: **polymerization**, the *joining together of many small molecules to form very large molecules*. The simple compounds from which polymers are made are called *monomers*.

Polymers are formed in two general ways.

(a) In **chain-reaction polymerization**, there is a series of reactions each of which consumes a reactive particle and produces another, similar particle; each individual reaction thus depends upon the previous one. The reactive particles can be free radicals, cations, or anions. A typical example is the polymerization of ethylene (Sec. 8.21). Here the chain-carrying particles are free radicals, each of which adds to a monomer molecule to form a new, bigger free radical.

$$\text{Rad}\cdot \ + \ CH_2{=}CH_2 \ \longrightarrow \ RadCH_2CH_2\cdot \ \xrightarrow{\ CH_2{=}CH_2\ }$$

$$RadCH_2CH_2CH_2CH_2\cdot \ \longrightarrow \ \textit{etc.}$$

(b) In **step-reaction polymerization**, there is a series of reactions each of which is essentially independent of the preceding one; a polymer is formed simply because the monomer happens to undergo reaction at more than one functional group. A diol, for example, reacts with a dicarboxylic acid to form an ester; but each moiety of the simple ester still contains a group that can react to generate another ester linkage and hence a larger molecule, which itself can react further, and so on.

$$HOCH_2CH_2OH \ + \ HOOC \bigcirc COOH$$

1,2-Ethanediol   Terephthalic acid

$$HOCH_2CH_2-O-\underset{O}{\overset{|}{C}}-\bigcirc-\underset{O}{\overset{|}{C}}-OH$$

$$\downarrow \quad HOCH_2CH_2OH$$

$$HOCH_2CH_2-O-\underset{O}{\overset{|}{C}}-\bigcirc-\underset{O}{\overset{|}{C}}-O-CH_2CH_2OH$$

$$\downarrow \quad p\text{-}C_6H_4(COOH)_2$$

$$HO-\underset{O}{\overset{|}{C}}-\bigcirc-\underset{O}{\overset{|}{C}}-O-CH_2CH_2-O-\underset{O}{\overset{|}{C}}-\bigcirc-\underset{O}{\overset{|}{C}}-O-CH_2CH_2OH$$

There is an alternative, somewhat less meaningful system of classification: *addition polymerization*, in which molecules of monomer are simply added together; and *condensation polymerization*, in which monomer molecules combine with loss of some simple molecules like water. As it happens, the two systems almost exactly coincide; nearly all cases of chain-reaction polymerization involve addition polymerization; nearly all cases of step-reaction polymerization involve condensation polymerization. Indeed, some chemists use the term "addition polymerization" to *mean* polymerization via chain reactions.

Let us look first at chain-reaction polymerization, starting with the kind that involves free radicals.

Problem 36.1 Examine the structure of the following synthetic polymers. Tell what class of compound it belongs to and give structures of the most likely monomers.

(a) nylon-6,6 (fibers),

$$\text{\raise1ex\hbox{$\sim$}}C(CH_2)_4CNH(CH_2)_6NHC(CH_2)_4CNH(CH_2)_6NH\text{\raise1ex\hbox{$\sim$}}$$
$$\quad \overset{||}{O} \quad \overset{||}{O} \qquad\qquad \overset{||}{O} \quad \overset{||}{O}$$

(b) nylon-6 (fibers),

$$\sim CCH_2(CH_2)_4NHCCH_2(CH_2)_4NH\sim$$
$$\quad \| \qquad\qquad\qquad \| $$
$$\quad O \qquad\qquad\qquad O$$

(c) Carbowax (water-soluble wax),

$$\sim OCH_2CH_2OCH_2CH_2OCH_2CH_2\sim$$

(d) Neoprene (oil-resistant elastomer),

$$\sim CH_2C{=}CHCH_2CH_2C{=}CHCH_2\sim$$
$$\qquad\quad | \qquad\qquad\qquad\quad |$$
$$\qquad\quad Cl \qquad\qquad\qquad\quad Cl$$

(e) Saran (packaging film, seat covers),

$$\sim CH_2CCl_2CH_2CCl_2\sim$$

Problem 36.2   Answer the questions of Problem 36.1 for each of the following kinds of natural macromolecules: (a) a protein, p. 1365; (b) a nucleic acid, p. 1394; (c) starch (amylose), p. 1332; (d) cellulose, p. 1339.

36.3   Free-radical vinyl polymerization

In Sec. 8.21 we discussed briefly the polymerization of ethylene and substituted ethylenes under conditions where free radicals are generated—typically in the presence of small amounts of an initiator, such as a peroxide. Reaction occurs

$$n CH_2{=}CH \xrightarrow{\text{initiator}} \sim CH_2CHCH_2CHCH_2CHCH_2CH\sim$$
$$\qquad\quad | \qquad\qquad\qquad\quad | \quad\quad | \quad\quad | \quad\quad |$$
$$\qquad\quad G \qquad\qquad\qquad\quad G \quad\quad G \quad\quad G \quad\quad G$$

Vinyl monomer

$$\text{or} \ ({-}CH_2CH{-})_n$$
$$\qquad\qquad |$$
$$\qquad\qquad G$$

Polymer

at the doubly bonded carbons—the vinyl groups—and is called *vinyl polymerization*. A wide variety of unsaturated monomers may be used, to yield polymers with different *pendant groups* (G) attached to the polymer backbone. For example:

$$CH_2{=}CH \longrightarrow \sim CH_2{-}CH{-}CH_2{-}CH{-}CH_2{-}CH\sim$$
$$\qquad\quad | \qquad\qquad\qquad\quad | \qquad\quad | \qquad\quad |$$
$$\qquad\quad Cl \qquad\qquad\qquad\quad Cl \qquad\quad Cl \qquad\quad Cl$$

Vinyl chloride   Poly(vinyl chloride)
                                              (PVC)

$$CH_2{=}CH \longrightarrow \sim CH_2{-}CH{-}CH_2{-}CH{-}CH_2{-}CH\sim$$
$$\qquad\quad | \qquad\qquad\qquad\quad | \qquad\quad | \qquad\quad |$$
$$\qquad\quad CN \qquad\qquad\qquad\quad CN \qquad\quad CN \qquad\quad CN$$

Acrylonitrile   Polyacrylonitrile
                                          (Orlon)

$$CH_2=CH \longrightarrow \sim CH_2-CH-CH_2-CH-CH_2-CH\sim$$
$$\quad\;\; | \qquad\qquad\qquad\qquad | \qquad\quad | \qquad\quad |$$
$$\quad C_6H_5 \qquad\qquad\qquad\; C_6H_5 \quad\; C_6H_5 \quad\; C_6H_5$$

Styrene   Polystyrene

$$\qquad\quad CH_3 \qquad\qquad\qquad\qquad CH_3 \quad\;\; CH_3 \quad\;\; CH_3$$
$$\qquad\quad | \qquad\qquad\qquad\qquad\qquad | \qquad\quad | \qquad\quad |$$
$$CH_2=C \longrightarrow \sim CH_2-C-CH_2-C-CH_2-C\sim$$
$$\qquad\quad | \qquad\qquad\qquad\qquad\qquad | \qquad\quad | \qquad\quad |$$
$$\qquad COOCH_3 \qquad\qquad\qquad COOCH_3 \; COOCH_3 \; COOCH_3$$

Methyl methacrylate   Poly(methyl methacrylate)
                                     (Plexiglas, Lucite)

Polymerization involves addition of free radicals to the double bond of the monomer: addition, first, of the free radical generated from the initiator, and then of the growing polymer molecule. This is, of course, an example of chain-reaction polymerization.

(1)   Peroxide $\longrightarrow$ Rad$\cdot$

(2)   Rad$\cdot$ + CH$_2$=CH $\longrightarrow$ RadCH$_2$—CH$\cdot$
                            | |
                            G   G

} Chain-initiating steps

(3)   RadCH$_2$—CH$\cdot$ + CH$_2$=CH $\longrightarrow$ RadCH$_2$—CH—CH$_2$—CH$\cdot$
              | | | |
              G   G   G   G

} Chain-propagating steps

then steps like (3) *repeated, until finally*:

(4)   2Rad(CH$_2$CH)$_n$CH$_2$CH$\cdot$
                            | |
                            G   G

| combination
$\downarrow$

Rad(CH$_2$CH)$_n$CH$_2$CH—CHCH$_2$(CHCH$_2$)$_n$Rad
            | | | |
            G   G   G   G

or

(5)   2Rad(CH$_2$CH)$_n$CH$_2$CH$\cdot$
                            | |
                            G   G

| disproportionation
$\downarrow$

Rad(CH$_2$CH)$_n$CH$_2$CH$_2$ + Rad(CH$_2$CH)$_n$CH=CH
            | | | |
            G   G   G   G

} Chain-terminating steps

In each step the consumption of a free radical is accompanied by the formation of a new, bigger free radical. Eventually, the reaction chain is terminated by steps that consume but do not form free radicals: *combination* or *disproportionation* of two free radicals.

Problem 36.3   Free-radical polymerization of 1,3-butadiene gives molecules containing both the following units,

$$\sim CH_2CH\!=\!CHCH_2\sim \quad and \quad \sim CH_2CH\sim$$
$$| $$
$$CH\!=\!CH_2$$

the exact proportions depending on the temperature. Account in detail for the formation of the two different units.

Problem 36.4   Polystyrene formed with isotopically labeled AIBN as initiator was found to contain *two* initiator fragments per molecule. What termination reaction is indicated by this finding?

$$(CH_3)_2C\!-\!N\!=\!N\!-\!C(CH_3)_2 \quad \longrightarrow \quad N_2 \; + \; 2(CH_3)_2C\cdot$$

with CN, CN and CN substituents as shown

Azoisobutyronitrile
(AIBN)

Added compounds can modify the polymerization process drastically. For example, in the presence of carbon tetrachloride, styrene undergoes polymerization at the same rate as in its absence, but the polystyrene obtained has a lower average molecular weight; furthermore, it contains small amounts of chlorine. This is an example of **chain transfer**, the termination of one polymerization chain (7) with the simultaneous initiation of another (8).

(6) $\sim CH_2\!-\!CH\cdot \; + \; CH_2\!=\!CH \xrightarrow{\text{polymerization}} \sim CH_2\!-\!CH\!-\!CH_2\!-\!CH\cdot$
            | | | |
            Ph   Ph   Ph   Ph

(7) $\sim CH_2\!-\!CH\cdot \; + \; CCl_4 \xrightarrow{\text{chain-transfer}} \sim CH_2\!-\!CH\!-\!Cl \; + \; \cdot CCl_3$
            | |
            Ph   Ph

(8) $Cl_3C\cdot \; + \; CH_2\!=\!CH \longrightarrow Cl_3C\!-\!CH_2\!-\!CH\cdot \xrightarrow{\text{styrene}} polymer$
                       | |
                       Ph   Ph

Ordinarily a growing polystyrene radical adds (6) to styrene monomer to continue the reaction chain. Every so often, however, it abstracts an atom from the chain-transfer agent (7) to end the original polymerization chain and generate a new particle ($CCl_3\cdot$ in this case) that initiates a new polymerization chain (8). Since one reaction chain is replaced by another, the rate of polymerization is unaffected. Since the average number of chain-propagating steps in each reaction chain is reduced, the average molecular weight of the polymer is lowered. A transfer agent thus competes with the monomer for the growing radicals. The ratio of rate constants for (7) and (6), $k_{transfer}/k_{polymerization}$, is called the *transfer constant*; it is a measure of how effective the transfer agent is at lowering the molecular weight of the polymer.

Problem 36.5 For polymerization of styrene at 60 °C, the following chain-transfer constants have been measured. Account for the relative effectiveness of the members of each sequence.

(a) benzene, 0.018; *tert*-butylbenzene, 0.04; toluene, 0.125; ethylbenzene, 0.67; iso-propylbenzene, 0.86

(b) *n*-heptane, 0.42; 2-heptene, 2.7

(c) CCl_4, 90; CBr_4, 13 600

An added compound may react with the growing free radical to generate a new free radical that is not reactive enough to add to monomer; a reaction chain is terminated but no new one is begun. Such a compound is, of course, an **inhibitor** (Sec. 2.14). Many amines, phenols, and quinones act as inhibitors. Although their exact mode of action is not understood, it seems clear that they are converted into free radicals that do not add to monomer; instead, they may combine or dispro-portionate, or combine with another growing radical to halt a second reaction chain.

$$\text{CH}_2-\text{CH} \cdot \; + \; \text{Inhibitor} \; \longrightarrow \; \text{CH}_2-\text{CH}_2 \; + \; \text{Inhibitor} \cdot$$
$$\underset{\text{G}}{|} \qquad\qquad\qquad\qquad \underset{\text{G}}{|} \qquad\qquad \textit{Unreactive:}$$
$$\textit{cannot initiate}$$
$$\textit{new chain}$$

Since even traces of certain impurities, acting as chain-transfer agents or inhibitors, can interfere with the polymerization process, the monomers used are among the purest organic chemicals produced.

In an extreme case—if the alkene is of low reactivity and the transfer agent of high reactivity—chain transfer is so effective that there is *no* polymerization. Then we observe simply addition of the "transfer agent" to the double bond, a reaction we encountered in Sec. 8.20. For example:

$$n\text{-}C_6H_{13}CH{=}CH_2 \; + \; CBr_4 \; \xrightarrow{\text{peroxides}} \; n\text{-}C_6H_{13}\underset{\underset{Br}{|}}{C}HCH_2CBr_3$$

Problem 36.6 (a) Chain transfer can cause *branching* of a polymer molecule. Show how this could happen. What is the chain-transfer agent? (b) Rather short branches (four or five carbons) are attributed to "back-biting". What do you think is meant by this term? Show the chemical reactions probably involved.

36.4 Copolymerization

So far, we have discussed only polymerization of a single monomeric compound to form a *homopolymer*, a polymer made up—except, of course, at the two ends of the long molecule—of identical units.

Now, if a mixture of two (or more) monomers is allowed to undergo polymerization, there is obtained a **copolymer**: a polymer that contains two (or more) kinds of monomeric units in the same molecule. For example:

$$
\begin{array}{ccc}
& & CH_3 \\
& & | \\
CH_2{=}CH \ + \ CH_2{=}C & \xrightarrow{\text{initiator}} & \sim\sim CH_2{-}CH{-}CH_2{-}C\sim\sim \\
| & & | \\
Ph & COOCH_3 & Ph \qquad COOCH_3
\end{array}
$$

Styrene   Methyl   Polystyrene(co-methyl methacrylate)
                 methacrylate

Through copolymerization there can be made materials with different properties than those of either homopolymer, and thus another dimension is added to the technology. Consider, for example, styrene. Polymerized alone, it gives a good electric insulator that is molded into parts for radios, television sets, and automobiles. Copolymerization with butadiene (30%) adds toughness; with acrylonitrile (20–30%) increases resistance to impact and to hydrocarbons; with maleic anhydride yields a material that, on hydrolysis, is water-soluble, and is used as a dispersant and sizing agent. The copolymer in which butadiene predominates (75% butadiene, 25% styrene) is an elastomer, and since World War II has been the principal rubber substitute manufactured in the United States.

Let us look more closely at the copolymerization process. Consider free-radical vinyl polymerization of two monomers, M_1 and M_2. In each step the growing free radical can react with either monomer to continue the reaction chain. What are the factors that determine *which* monomer it preferentially reacts with?

$$
M_1M_2M_2M_1M_2M_1M_1\cdot \ - \left[\begin{array}{l} \xrightarrow{M_1} \quad M_1M_2M_2M_1M_2M_1M_1{-}M_1\cdot \\[2ex] \xrightarrow{M_2} \quad M_1M_2M_2M_1M_2M_1M_1{-}M_2\cdot \end{array}\right.
$$

First, of course, there are the relative *concentrations* of the two monomers; the higher the concentration of a particular monomer, the greater its chance of being incorporated into the chain, and the more abundant its units are in the final product.

Next, there are the relative *reactivities* of the monomers toward free-radical addition; in general, the more reactive the monomer, the greater its chance of being incorporated into the polymer. We know that the reactivity of a carbon–carbon double bond toward free-radical addition is affected by the stability of the new free radical being formed; factors that tend to stabilize the free-radical product tend to stabilize the incipient free radical in the transition state, so that the more stable free radical tends to be formed faster. Now, stability of a free radical depends upon accommodation of the odd electron. The group G stabilizes the radical by delocalization: the phenyl group in styrene, through formation of a benzylic radical; the vinyl group of 1,3-butadiene, through formation of an allylic radical; the $-COOCH_3$ group of methyl methacrylate, through formation of a radical in which acyl oxygen helps carry the odd electron. (*Problem*: Draw resonance structures to show how this last effect could arise.)

$$\sim M\cdot \;+\; CH_2 \!=\! \underset{\underset{G}{\big|}}{CH} \;\longrightarrow\; \left[\sim \overset{\delta\cdot}{M} \text{----} CH_2 \text{---} \underset{\underset{G}{\big|}}{\overset{\delta\cdot}{CH}} \right] \;\longrightarrow\; \sim M\!-\!CH_2\!-\!\underset{\underset{G}{\big|}}{CH}\cdot$$

(We notice that the above discussion does not take into account the nature of the attacking radical, and hence would predict the same relative reactivities for a pair of alkenes toward all free radicals. We shall return to this point later.)

Now, let us see what kind of copolymer we would expect to get on the basis of what we have said so far. In the copolymerization of styrene (M_1) and butadiene (M_2), for example, reaction can proceed via either of two growing radicals: one ending in a styrene unit ($\sim M_1 \cdot$), or one ending in a butadiene unit ($\sim M_2 \cdot$). Either radical can add to either monomer, to form a copolymer with styrene and butadiene units distributed *randomly* along the molecule:

$$\sim M_1 M_2 M_2 M_1 M_2 M_1 M_1 \sim \qquad \text{Random copolymer}$$

With these particular monomers, copolymerization is in fact random. Now, toward either free-radical type, it happens, butadiene is about 1.4 times as reactive as styrene, so that, if monomer concentrations were equal, butadiene units would tend to predominate in the product. Furthermore, since butadiene is consumed faster, the relative concentrations of monomers would change as reaction goes on, and so would the composition of the polymer being produced. These effects can be compensated for by adjusting the ratio of monomers fed into the reaction vessel; indeed, by control of the feed ratio, copolymers of any desired composition can be made.

Random copolymerization, of the kind observed for styrene and butadiene, is actually rather rare. In general, copolymerization shows, to a greater or lesser extent, a tendency to *alternation* of monomer units. An extreme case is that of

$$\sim M_1 M_2 M_1 M_2 M_1 M_2 M_1 M_2 \sim \qquad \text{Alternating copolymer}$$

stilbene (1,2-diphenylethene) and maleic anhydride, which copolymerize with absolutely regular alternation of units; regardless of the feed ratio, a 50:50 copolymer is obtained.

How are we to account for this tendency toward alternation? It must mean that a growing radical ending in one unit tends to add to the *opposite* monomer. Clearly, the relative reactivity of a monomer *does* depend upon the nature of the radical that is attacking it. Maleic anhydride is much more reactive than stilbene toward radicals ending in a stilbene unit, and stilbene is much more reactive than maleic anhydride toward the other kind of radical. (Indeed, these two compounds, individually, undergo self-polymerization only with extreme difficulty.) A more modest—and more typical—tendency toward alternation is shown by styrene and methyl methacrylate. Here, toward either radical ($\sim M_1 \cdot$) the "opposite" monomer (M_2) is about twice as reactive as the "same" monomer (M_1).

$$\sim M_1 \cdot \; \xrightarrow{M_2} \; \sim M_1 M_2 \cdot \; \xrightarrow{M_1} \; \sim M_1 M_2 M_1 \cdot \; \xrightarrow{M_2} \; etc.$$

The alternating tendency in copolymerization was established on a quantitative basis by Frank R. Mayo (of the Stanford Research Institute) and Cheves

Walling (of the University of Utah) while working in the laboratories of the U.S. Rubber Company. Their work was fundamental to the development of free-radical chemistry: it showed clearly for the first time the dependence of reactivity on the nature of the attacking free radical, and led directly to the concept of *polar factors*, working not only in copolymerization and other additions of free radicals, but in free-radical reactions of all kinds (compare Sec. 8.19).

Basically, Mayo and Walling's interpretation was the following. Although free radicals are neutral, they have certain tendencies to gain or lose electrons, and hence they partake of the character of electrophilic or nucleophilic reagents. The transition states for their reactions can be polar, with the radical moiety acquiring a partial negative or positive charge at the expense of the substrate—the alkene, in the case of addition. In copolymerization, a substituent generally exerts the same polar effect—electron-withdrawing or electron-releasing—on a free radical as on the alkene (monomer) from which the free radical was derived. Electron withdrawal makes a free radical electrophilic, but makes an alkene less able to supply the electrons which that radical is seeking. An electrophilic radical will, then, preferentially add to a monomer containing an electron-releasing group. In a similar way, a nucleophilic radical, containing an electron-releasing substituent, will seek out a monomer containing an electron-withdrawing substituent.

Styrene and methyl methacrylate tend to alternate because their substituents are of opposite polarity: in methacrylate the —COOCH$_3$ group tends to withdraw electrons; in styrene the phenyl group tends (via resonance) to release electrons. The transition states for addition to the opposite monomers are thus stabilized:

$$\sim CH_2-\underset{\underset{Ph}{|}}{CH}\cdot \; + \; CH_2=\underset{\underset{COOCH_3}{|}}{\overset{\overset{CH_3}{|}}{C}} \longrightarrow \left[\sim CH_2-\underset{\underset{Ph}{|}}{\overset{\overset{\delta+}{}}{CH}}\text{----}CH_2\!=\!=\!\underset{\underset{COOCH_3}{|}}{\overset{\overset{CH_3}{|}}{C}}{}^{\delta-} \right]$$

$$\sim CH_2-\underset{\underset{COOCH_3}{|}}{\overset{\overset{CH_3}{|}}{C}}\cdot \; + \; CH_2=\underset{\underset{Ph}{|}}{CH} \longrightarrow \left[\sim CH_2-\underset{\underset{COOCH_3}{|}}{\overset{\overset{CH_3}{|}}{\overset{\delta-}{C}}}\text{----}CH_2\!=\!=\!\underset{\underset{Ph}{|}}{\overset{\overset{\delta+}{}}{CH}} \right]$$

Perhaps the most convincing evidence for the play of polar forces comes from copolymerization of a series of ring-substituted styrenes; here relative reactivities toward a variety of monomers not only fall into a pattern consistent with the familiar electronic effects of the substituents, but show the same quantitative relationships (the Hammett sigma–rho relationship, Sec. 23.11) as do *ionic* reactions: dissociation of carboxylic acids, for example, or hydrolysis of esters.

The concept of polar transition states in free-radical reactions has been questioned, at least for reactions in which hydrogen is abstracted—halogenation, for example. Here, it has been suggested, electron-withdrawing or electron-releasing groups affect reactivity simply by strengthening or weakening the bonds holding hydrogen in the substrate.

If we define polar effects on free-radical reactions as effects due to electron withdrawal or electron release—rather than to accommodation of the odd electron—then there is no doubt about their existence; it is the *interpretation* of such effects that is open to question.

We must realize that polar effects are *superimposed on* effects due to delocalization of the odd electron. Styrene and butadiene, for example, are highly reactive toward any radical since the transition state contains an incipient benzylic or allylic free radical. This high reactivity is modified—enhanced or lowered—by the demands of the particular attacking radical.

Problem 36.7   (a) Draw structures to account for the strong alternating tendency in copolymerization of butadiene (M_1) and acrylonitrile (M_2). (b) Toward ~M_1 · acrylonitrile is 2.5 times as reactive as butadiene, but toward ~M_2 · butadiene is *20 times as reactive* as acrylonitrile. How do you account for this contrast?

Copolymers can be made not just from two different monomers but from three, four, or even more. They can be made not only by free-radical chain reactions, but by any of the polymerization methods we shall take up: ionic, coordination, or step-reaction. The monomer units may be distributed in various ways, depending on the technique used. As we have seen, they may alternate along a chain, either randomly or with varying degrees of regularity. In *block copolymers*, sections made up of one monomer alternate with sections of another:

$$\sim\!M_1M_1M_1M_1M_1\!-\!M_2M_2M_2M_2\!\sim \qquad\qquad \text{Block copolymer}$$

In *graft copolymers*, a branch of one kind is grafted to a chain of another kind:

$$\sim\!M_1M_1M_1M_1M_1M_1M_1\!\sim$$
$$\qquad\qquad\;\; | \qquad\qquad\qquad\qquad \text{Graft copolymer}$$
$$\qquad\; M_2M_2M_2M_2$$

Problem 36.8   Graft copolymers can be made by each of the following processes. Show the chemistry most likely involved, and the structure of the product. (a) Poly-butadiene is treated with styrene in the presence of a free-radical initiator. (b) Poly-(vinyl chloride) is treated with methyl methacrylate in the presence of benzoyl peroxide, $(C_6H_5COO)_2$.

36.5   Ionic polymerization. Living polymers

Chain-reaction polymerization can proceed with ions instead of free radicals as the chain-carrying particles: either cations or anions, depending on the kind of initiator that is used.

Cationic polymerization

$$Y \quad CH_2\!\!=\!\!CH \longrightarrow Y\!:\!CH_2\!-\!\overset{\oplus}{C}H$$

An acid
$$\qquad\qquad |\qquad\qquad\qquad\qquad |$$
$$\qquad\qquad G \qquad\qquad\qquad\qquad G$$

A carbocation

$$Y\!:\!CH_2\!-\!\overset{\oplus}{C}H \quad CH_2\!\!=\!\!CH \longrightarrow Y\!:\!CH_2\!-\!CH\!-\!CH_2\!-\!\overset{\oplus}{C}H \longrightarrow etc.$$
$$\qquad\; |\qquad\qquad |\qquad\qquad\qquad\qquad\;\; |\qquad\qquad\;\; |$$
$$\qquad\; G\qquad\qquad G\qquad\qquad\qquad\qquad G\qquad\qquad G$$

Anionic polymerization

$$Z: \quad CH_2\!=\!CH \quad \longrightarrow \quad Z\!:\!CH_2\!-\!\overset{\ominus}{C}H\!:$$
$$\underset{\text{A base}}{} \qquad \underset{G}{} \qquad\qquad \underset{G}{}$$

A carbanion

$$Z\!:\!CH_2\!-\!\overset{\ominus}{C}H\!: \quad CH_2\!=\!CH \quad \longrightarrow \quad Z\!:\!CH_2\!-\!CH\!-\!CH_2\!-\!\overset{\ominus}{C}H\!: \quad \longrightarrow \quad etc.$$
$$\underset{G}{} \qquad \underset{G}{} \qquad\qquad \underset{G}{} \qquad \underset{G}{}$$

Cationic polymerization is initiated by *acids*. Isobutylene, for example, undergoes cationic polymerization to a tacky material used in adhesives. Copolymerization with a little isoprene gives *butyl rubber*, used to make automobile innertubes and tire liners. A variety of acids can be used: sulfuric acid; $AlCl_3$ or BF_3 plus a trace of water. We recognize this process as an extension of the dimerization discussed in Sec. 8.16.

$$H^+ \quad CH_2\!=\!\overset{CH_3}{\underset{CH_3}{C}} \quad CH_2\!=\!\overset{CH_3}{\underset{CH_3}{C}} \quad CH_2\!=\!\overset{CH_3}{\underset{CH_3}{C}} \quad \longrightarrow \quad CH_3\!-\!\overset{CH_3}{\underset{CH_3}{C}}\!-\!CH_2\!-\!\overset{CH_3}{\underset{CH_3}{C}}\!-\!CH_2\!-\!\overset{CH_3}{\underset{CH_3}{C}}\!\sim$$

Anionic polymerization, as we might expect, is initiated by *bases*: $Li^+NH_2{}^-$, for example, or organometallic compounds like *n*-butyllithium. For example:

$$NH_2{}^- K^+ + CH_2\!=\!\underset{Ph}{CH_2} \quad \longrightarrow \quad NH_2\!-\!CH_2\!-\!\underset{Ph}{CH^-} K^+ \quad \xrightarrow{CH_2=CHPh}$$

Styrene

$$NH_2\!-\!CH_2\!-\!\underset{Ph}{CH}\!-\!CH_2\!-\!\underset{Ph}{CH^-} K^+ \quad \longrightarrow \quad etc.$$

$$\textit{n}\text{-BuLi} + CH_2\!=\!\overset{CH_3}{\underset{COOCH_3}{C}} \quad \longrightarrow \quad \textit{n}\text{-Bu}\!-\!CH_2\!-\!\overset{CH_3}{\underset{COOCH_3}{C^-}} Li^+ \quad \longrightarrow \quad etc.$$

Methyl
methacrylate

Active metals like Na or Li can be used; here the initiation becomes a little more complicated, as in the polymerization of styrene by the action of sodium metal and naphthalene. A sodium atom transfers an electron (1) to naphthalene to form a

(1) $$Na + naphthalene \quad \rightleftharpoons \quad Na^+ \; naphthalene\overline{\cdot}$$

Naphthalene radical–anion

(2) $$naphthalene\overline{\cdot} + CH\!=\!\underset{Ph}{CH_2} \quad \longrightarrow \quad naphthalene + \underset{Ph}{CHCH_2\overline{\cdot}}$$

Styrene radical–anion
Formed by one-electron transfer

(3) $2CHCH_2{}^{\cdot}{}^{-} \longrightarrow {}^{-}CH-CH_2-CH_2-CH^{-}$
               | | |
               Ph   Ph   Ph

                                    A dianion

radical–anion, which then donates the electron to styrene (2) to form the styrene radical–anion. Like many other free radicals, these dimerize (3). The resulting dianion is the true initiator, and begins to grow *at both ends*:

                    ${}^{-}CHCH_2-CH_2CH^{-}$
                         | |
                         Ph   Ph

                              | $CH_2=CHPh$
                              ↓

${}^{-}CH-CH_2{\sim\!\!\sim}CH-CH_2-CH-CH_2-CH_2-CH-CH_2-CH{\sim\!\!\sim}CH_2-CH^{-}$
     | | | | | |
     Ph   Ph   Ph   Ph   Ph   Ph

Anionic polymerization is not limited to the vinyl kind, involving addition to carbon–carbon double bonds. Ethylene oxide, for example, is converted by a small amount of base into a high-molecular-weight polyether.

$CH_3O^{-} + CH_2{-}CH_2 \longrightarrow CH_3O-CH_2-CH_2O^{-}$
                      \O/

$CH_3O-CH_2-CH_2O^{-} + CH_2{-}CH_2 \longrightarrow CH_3O-CH_2-CH_2-O-CH_2-CH_2O^{-}$
                                  \O/

> **Problem 36.9**   The presence of methanol during the polymerization of ethylene oxide by sodium methoxide tends to lower the molecular weight of the product. (a) How do you think it does this? What process is this an example of? (b) What product will be obtained in the presence of *much* methanol?

In 1956, Michael Szwarc (of the State University of New York at Syracuse) reported the following observations. When a sample of styrene was treated with a little sodium naphthalene initiator, rapid polymerization took place and was complete within a few seconds. When more styrene was added, it, too, underwent polymerization; viscosity measurements showed that the molecular weight of the polystyrene was now *higher than before*. If, instead of the second batch of styrene, there was added butadiene, polymerization again took place; the product was found to contain no (homo)polystyrene—all styrene units were now part of a (block) copolymer with butadiene. How are these results to be interpreted?

So far, we have not mentioned chain-terminating steps for ionic polymerization. Such steps do exist. In cationic polymerization, for example, the growing carbocation can undergo either of two familiar reactions: ejection of a proton to form an alkene, or combination with an anion. But, particularly in the case of anionic polymerization, termination often involves reaction with an impurity or some other molecule not a necessary part of the polymerization system. Under carefully controlled conditions, there are *no* termination steps—or at least none that happen very fast. Reaction stops when all monomer is consumed, but the reaction mixture contains what Szwarc has named *living* polymer molecules. When these are "fed" additional monomer—either styrene or butadiene, for example— they continue to grow. They are not immortal, however, but can be "killed" by addition of a compound that reacts with carbanions: water, say. The generation of living polymers is of immense practical importance; it provides the best route to block copolymers, and permits the introduction of a variety of terminal groups.

Problem 36.10   Draw the structure of the product expected from the killing of living polystyrene by each of the following reagents: (a) water; (b) carbon dioxide, then water; (c) a small amount of ethylene oxide, then water; (d) a large amount of ethylene oxide, then water.

36.6   Coordination polymerization

Until 1953, almost all vinyl polymerization of commercial importance was of the free-radical type. Since that time, however, a new kind of polymerization, *coordination polymerization*, has revolutionized the field. Following discoveries by Karl Ziegler (Max Planck Institute for Coal Research) and by Giulio Natta (Polytechnic Institute of Milan)—who jointly received the Nobel Prize in 1963 for this work—catalysts have been developed that permit control of the polymerization process to a degree never before possible.

In Chapter 20 we saw examples of the remarkable power of transition metal complexes to bring about and control organic reactions. Here, in these polymerization catalysts, we see another.

Ziegler–Natta catalysts are made up of a transition metal salt—typically titanium trichloride—and a metal alkyl like triethylaluminum. These react to form the active catalyst: a titanium complex holding an ethyl group.

Now the alkene—ethylene, say—is introduced. According to the generally accepted mechanism, the alkene attaches itself to titanium by a π bond: the π cloud of the alkene overlaps an empty orbital of the metal (Sec. 20.5). Next, with

ethyl and the alkene both held by the metal, the first of many similar steps takes place. The ethylene unit *inserts itself* between metal and the ethyl group. In place of ethyl there is now a *n*-butyl group attached to titanium. The bonding site where ethylene was held is vacant again, and the catalyst is ready to work again. Another ethylene becomes π bonded to the metal, and then inserts itself between the metal and alkyl to form, this time, a *n*-hexyl group. And so the process continues over and over again, with the alkyl group growing by two carbons in each cycle. Finally,

$$\text{>Ti}-CH_2CH_2-CH_2CH_3 \quad \xrightarrow{CH_2=CH_2} \quad \text{>Ti}\underset{CH_2CH_2-CH_2CH_3}{\overset{H_2C\diagup CH_2}{\big\langle}}$$

$$\downarrow$$

$$\text{>Ti}-CH_2CH_2-CH_2CH_2-CH_2CH_3$$

perhaps through the insertion of hydrogen, the long chain separates from the metal and a molecule of polyethylene has been formed.

We see here another example of *symphoria*—the bringing together of molecules for a useful purpose. The basic similarity of this mechanism to the ones for homogeneous hydrogenation (Secs. 20.5–20.7) and the oxo process (Sec. 20.8) is striking. Titanium holds an alkyl group where rhodium or cobalt held hydrogen. In each case there is a vacant bonding site—an empty orbital—on the metal, through which the alkene can become π bonded before it inserts itself into a bond: between titanium and alkyl, between rhodium and hydrogen, or between cobalt and hydrogen. Here, as in those other reactions, the net process is *addition*: the insertion amounts to the addition of metal and alkyl across the double bond.

Polymerization with Ziegler–Natta catalysts has two important advantages over free-radical polymerization: (a) it gives *linear* polymer molecules; and (b) it permits *stereochemical control*.

Polyethylene made by the free-radical process has highly branched chains due to chain transfer of a special kind, in which the chain-transfer agent *is a polymer molecule*. At the high temperatures required for this particular polymerization, the growing free radicals not only *add* to the double bond of a monomer but also *abstract* hydrogen from a chain already formed. This abstraction generates a free-radical center from which a branch can now grow. These highly branched polyethylene molecules fit together poorly and in a random way; the compound is said to have low *crystallinity*. It has a low melting point and is mechanically weak.

$$\underset{\sim CH_2\overset{H}{\underset{|}{C}}HCH_2CH_2\sim}{} \quad \xrightarrow{\sim CH_2CH_2\cdot} \quad \sim CH_2CH_2-H \; + \; \sim CH_2\overset{\cdot}{C}HCH_2CH_2\sim \quad \xrightarrow{CH_2=CH_2}$$

$$\underset{\sim CH_2\overset{\displaystyle CH_2}{\underset{\displaystyle |}{\overset{\displaystyle |}{\underset{\displaystyle CH_2}{C}}}}HCH_2CH_2\sim}{} \quad \longrightarrow \quad etc.$$

In contrast, polyethylene made by the coordination process is virtually unbranched. These unbranched molecules fit together well, and the polymer has a high degree of crystallinity. It has a higher melting point and higher density than the older (*low density*) polyethylene, and is mechanically much stronger. (We shall look at the crystallinity of polymers and its effect on their properties in Sec. 36.8.)

A second, far-reaching development in coordination polymerization is *stereochemical control*. Propylene, for example, could polymerize to any of three different arrangements (Fig. 36.1): *isotactic*, with all methyl groups on one side of an extended chain; *syndiotactic*, with methyl groups alternating regularly from side to side; and *atactic*, with methyl groups distributed at random.

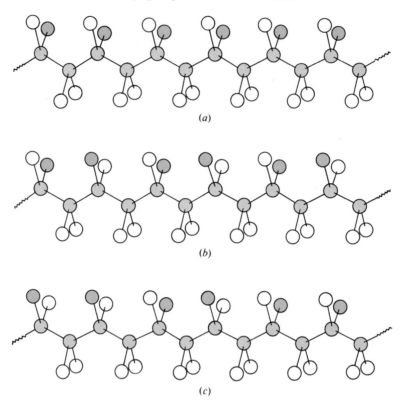

Figure 36.1   Polypropylene. (*a*) Isotactic. (*b*) Syndiotactic. (*c*) Atactic.

By proper choice of experimental conditions—catalyst, temperature, solvent—each of these stereoisomeric polymers has been made. Atactic polypropylene is a soft, elastic, rubbery material. Both isotactic and syndiotactic polypropylenes are highly crystalline: regularity of structure permits their molecules to fit together well. Over three billion pounds of isotactic polypropylene is produced every year, to be molded or extruded as sheets, pipes, and filaments; it is one of the principal synthetic fibers.

Coordination catalysts also permit stereochemical control about the carbon–carbon double bond. By their use, isoprene has been polymerized to a material virtually identical with natural rubber: *cis*-1,4-polyisoprene. (See Sec. 10.30.) This, like formation of isotactic polypropylene—and like hydrogenation with Wilkinson's catalyst—we recognize as an example of stereoselective synthesis (Sec. 9.2).

The Ziegler–Natta polymerization of ethylene can be adapted to make molecules of only modest size (C_6–C_{20}) and containing certain functional groups. If, for example, the metal alkyls initially obtained are heated (in the presence of ethylene and a nickel catalyst), the hydrocarbon groups are displaced as straight-chain 1-alkenes of even carbon number. Large quantities of such alkenes in the

$$M-(CH_2CH_2)_nCH_2CH_3 \xrightarrow[\text{heat}]{CH_2=CH_2,\ Ni} CH_2=CH-(CH_2CH_2)_{n-1}CH_2CH_3$$

C_{12}–C_{20} range are consumed in the manufacture of detergents (Sec. 37.5). Alternatively, the metal alkyls can be oxidized by air to give straight-chain primary alcohols:

$$M-(CH_2CH_2)_nCH_2CH_3 \xrightarrow[30-95\,°C]{air} M-O(CH_2CH_2)_nCH_2CH_3$$

$$\xrightarrow[40\,°C]{H_2O,\ H_2SO_4} HO(CH_2CH_2)_nCH_2CH_3$$

"A chemist setting out to build a giant molecule is in the same position as an architect designing a building. He has a number of building blocks of certain shapes and sizes, and his task is to put them together in a structure to serve a particular purpose. . . . What makes high polymer chemistry still more exciting just now is that almost overnight, within the last few years, there have come discoveries of new ways to put the building blocks together—discoveries which promise a great harvest of materials that have never existed on the earth."—Giulio Natta, *Scientific American*, September, 1957, p. 98.

36.7　Step-reaction polymerization

Carboxylic acids react with amines to yield amides, and with alcohols to form esters. When an acid that contains more than one —COOH group reacts with an amine that contains more than one —NH_2 group, or with an alcohol that contains more than one —OH group, then the products are *polyamides* and *polyesters*. For example:

$$HOOC(CH_2)_4COOH + H_2N(CH_2)_6NH_2 \longrightarrow salt$$

　　　Adipic acid　　　Hexamethylenediamine

$$\downarrow heat,\ -H_2O$$

Nylon-6,6
A polyamide

Methyl terephthalate　　Ethylene glycol

Dacron
A polyester

Phthalic
anhydride　　　　Glycerol

$\xrightarrow{-H_2O}$ Glyptal (an alkyd resin)
A polyester

These are examples of *step-reaction polymerization* (Sec. 36.2). In contrast to chain-reaction polymerization, reaction here does not depend on chain-carrying free radicals or ions. Instead, the steps are essentially independent of each other; they just happen to involve more than one functional group in a monomer molecule.

If each monomer molecule contains just two functional groups, growth can occur in only two directions, and a *linear* polymer is obtained, as in Dacron. But if reaction can occur at more than two positions in a monomer, there is formed a highly cross-linked *space network* polymer, as in Glyptal, an *alkyd resin*. Dacron and Glyptal are both polyesters, but their structures are quite different and, as we shall see, so are their uses.

Problem 36.11   Work out a possible structure for an alkyd resin formed from phthalic anhydride and glycerol, considering the following points: (a) In the first stage a linear polyester is formed. (Which hydroxyl groups are esterified more rapidly, primary or secondary?) (b) In the second stage these linear molecules are cross-linked to give a rather rigid network.

Step-reaction polymerization can involve a wide variety of functional groups and a wide variety of reaction types. Among the oldest of the synthetic polymers, and still extremely important, are those resulting from reaction between phenols and formaldehyde: the *phenol–formaldehyde resins* (Bakelite and related polymers). When phenol is treated with formaldehyde in the presence of alkali or acid, there is obtained a high-molecular-weight substance in which many phenol rings are held together by $-CH_2-$ groups:

o-Hydroxymethylphenol

The stages involved in the formation of the polymer seem to be the following. First, phenol reacts with formaldehyde to form *o*- or *p*-hydroxymethylphenol. Hydroxymethylphenol then reacts with another molecule of phenol, with the loss of water, to form a compound in which two rings are joined by a $-CH_2-$ link. This process then continues, to yield a product of high molecular weight. Since three positions in each phenol molecule are susceptible to attack, the final product contains many cross-links and hence has a rigid three-dimensional structure. It is thus a space-network polymer, and its properties reflect this (Sec. 36.8).

The first stage can be viewed as both electrophilic substitution on the ring by the electron-deficient carbon of formaldehyde, and nucleophilic addition of the aromatic ring to the carbonyl group. Base catalyzes reaction by converting phenol into the more reactive (more nucleophilic) phenoxide ion. Acid catalyzes reaction by protonating formaldehyde and increasing the electron deficiency of the carbonyl carbon.

Nucleophilic   Electrophilic
reagent   reagent

Basic catalysis

Nucleophilic   Electrophilic
reagent   reagent

Acidic catalysis

As a *di*amide, urea is capable of forming polymers; it reacts with formaldehyde to form the *urea–formaldehyde resins*, highly important in molded plastics. Here, too, a space-network polymer is formed.

Formaldehyde   Urea   Methylolurea

Dimethylolurea

Organic *isocyanates*, RNCO, undergo reactions of the following kinds (compare Sec. 24.23), all of which are used, in one way or another, in the synthesis of polymers. Reaction of *di*hydroxy alcohols with *di*isocyanates gives the important polyurethanes.

$$RN{=}C{=}O + R'OH \longrightarrow RNH{-}\underset{\underset{O}{\|}}{C}{-}OR'$$

An isocyanate

A urethane

$$RN{=}C{=}O + R'NH_2 \longrightarrow RNH{-}\underset{\underset{O}{\|}}{C}{-}NHR'$$

A substituted urea

$$RN{=}C{=}O + H_2O \longrightarrow \left[RNH{-}\underset{\underset{O}{\|}}{C}{-}OH\right] \longrightarrow RNH_2 + CO_2$$

A carbamic acid

Unstable

$$\xrightarrow{RNCO} RNH{-}\underset{\underset{O}{\|}}{C}{-}NHR$$

Problem 36.12 Give the structure of the polymer expected from the reaction of 1,2-ethanediol and 2,4-tolylene diisocyanate, 2,4-$(OCN)_2C_6H_3CH_3$.

36.8 Structure and properties of macromolecules

The characteristic thing about macromolecules, we have said, is their great size. This size has little effect on chemical properties. A functional group reacts much as we would expect, whether it is in a big or little molecule: an ester is hydrolyzed, an epoxide undergoes ring-opening, an allylic hydrogen is susceptible to abstraction by free radicals.

Problem 36.13 Describe reagents and conditions—if any—that would be expected to cleave the natural polymers of Problem 36.2 (p. 1236) into monomers.

Problem 36.14 When poly(vinyl acetate) is treated with methanol (b.p. 65 °C) in the presence of a little sulfuric acid, a substance of b.p. 57 °C distills from the mixture, and a new polymer is left behind. (a) What reaction has taken place? What is the structure of the new polymer? Why must it be prepared in this indirect manner? (b) When this new polymer is treated with *n*-butyraldehyde in the presence of a little phosphoric acid, a third polymer is formed, Butvar, which is used in making safety glass. What reaction has taken place here, and what is the structure of Butvar?

It is in their physical properties that macromolecules differ from ordinary molecules, and it is on these that their special functions depend. To begin with, let us look at the property of *crystallinity*. In a crystalline solid, we know, the structural units—molecules, in the case of a non-ionic compound—are arranged in a very regular, symmetrical way, with a geometric pattern repeated over and over. If a long molecule is to fit into such a pattern, it cannot be looped and coiled into a random conformation, but must be extended in a regular zig-zag (see Fig. 36.2).

(a)

(b)

Figure 36.2   Long chain (a) in a random conformation, and (b) extended.

This lack of randomness corresponds to an unfavorable entropy for the system (Secs. 2.23 and 23.11). On the other hand, the regularity and close fitting of the molecules in a crystal permits operation of strong intermolecular forces—hydrogen bonding, dipole–dipole attractions, van der Waals forces—which result in a favorable enthalpy (heat content). As we shall see, this tug-of-war between entropy and enthalpy is a key factor in determining the use to which a macromolecule can be put.

Now, in general, a high polymer does not exist entirely in crystalline form—not even a polymer whose regularity of molecular structure might be expected to permit this. The problem is the size of the molecule. As solidification begins, the viscosity of the material rises and the polymer molecules find it difficult to move about and arrange their long chains in the regular pattern needed for crystal formation. Chains become entangled; a change in shape of a chain must involve rotation about single bonds, and this becomes difficult because of hindrance to the swinging about of pendant groups. Polymers, then, form solids made up of regions of crystallinity, called *crystallites*, embedded in amorphous material. We speak of the *degree of crystallinity* of a polymer to mean the extent to which it is composed of crystallites.

Problem 36.15   Although both polymers are prepared by free-radical processes, poly(vinyl chloride) is amorphous and poly(vinylidene chloride) (Saran) is highly crystalline. How do you account for the difference? (Vinylidine chloride is 1,1-dichloroethene.)

Let us examine the various uses of polymers, and see how these depend on their structure—molecular and intermolecular.

Fibers are long, thin, threadlike bits of material that are characterized by great tensile (pulling) strength *in the direction of the fiber*. The natural fibers—cotton, wool, silk—are typical. Fibers are twisted into threads, which can then be woven into cloth, or embedded in plastic material to impart strength. The tensile strength can be enormous, some synthetic fibers rivaling—on a weight basis—steel.

The gross characteristics of fibers are reflected on the molecular level—the molecules, too, are long, thin, and threadlike. Furthermore, and most essential, they lie stretched out alongside each other, *lined up in the direction of the fiber*. The strength of the fiber resides, ultimately, in the strength of the chemical bonds of the polymer chains. The lining-up is brought about by *drawing*—stretching—the polymeric material. Once lined up, the molecules stay that way; the tendency to return to random looping and coiling is overcome by strong intermolecular attractions. In a fiber, enthalpy wins out over entropy. This high degree of molecular orientation is usually—although not always—accompanied by appreciable crystallinity.

The key requirements of a fiber are, then, a molecular shape—linear—that permits side-by-side alignment, and strong intermolecular forces to maintain this alignment. In addition, the intermolecular forces prevent "slipping" of one molecule past another. Now, what are these intermolecular forces?

The principal synthetic fibers are polyamides (the nylons), polyesters (Dacron, Terylene, Vycron), polyacrylonitrile ("acrylic fibers", Orlon, Acrilan), polyurethanes (Spandex, Vycra), and isotactic polypropylene. In nylon and polyurethanes, molecular chains are held to each other by hydrogen bonds (Fig. 36.3). In polyesters and polyacrylonitriles, the polar carbonyl and cyano groups lead to powerful dipole–dipole attractions. The stereoregular chains of isotactic polypropylene fit together so well that van der Waals forces are strong enough to maintain alignment.

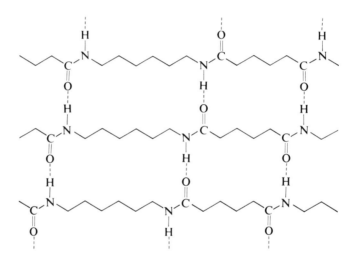

Figure 36.3   Hydrogen bonding in crystallites of nylon-6,6.

An **elastomer** possesses the high degree of elasticity that is characteristic of rubber: it can be greatly deformed—stretched to eight times its original length, for example—and yet return to its original shape. Here, as in fibers, the molecules are long and thin; as in fibers; they become lined up when the material is stretched. The big difference is this: when the stretching force is removed, the molecular chains of an elastomer do not remain extended and aligned but return to their original random conformations favored by entropy. They do not remain aligned because the intermolecular forces necessary to hold them that way are weaker than in a fiber. In general, elastomers do not contain highly polar groups or sites for hydrogen bonding; the extended chains do not fit together well enough for van der Waals forces to do the job. In an elastomer entropy beats enthalpy.

One further requirement: the long chains of an elastomer must be connected to each other by occasional cross-links: enough of them to prevent slipping of molecules past one another; not so many as to deprive the chains of the flexibility that is needed for ready extension and return to randomness.

Natural rubber illustrates these structural requirements of an elastomer: long, flexible chains; weak intermolecular forces; and occasional cross-linking. Rubber is *cis*-1,4-polyisoprene. With no highly polar substituents, intermolecular attraction is largely limited to van der Waals forces. But these are weak because of the all-*cis*

Natural rubber
All-cis configuration

configuration about the double bond. Figure 36.4 compares the extended chains of rubber with those of its *trans* stereoisomer. As we can see, the *trans* configuration permits highly regular zig-zags that fit together well; the *cis* configuration does not. The all-*trans* stereoisomer occurs naturally as *gutta percha*; it is highly crystalline and non-elastic.

(*a*)

(*b*)

Figure 36.4   Extended chains of (*a*) natural rubber, *cis*-1,4-polyisoprene, and of (*b*) gutta percha, its *trans* stereoisomer.

Cross-linking in rubber, as we have seen (Sec. 10.30), is brought about by *vulcanizing*—heating with sulfur—which causes formation of sulfur bridges between molecules. This reaction involves reactive allylic positions, and thus depends on the double bond in the polymer.

Chief among the synthetic elastomers is SBR, a copolymer of butadiene (75%) and styrene (25%) produced under free-radical conditions; it competes with natural rubber in the main use of elastomers, the making of automobile tires. All-*cis* polybutadiene and polyisoprene can be made by Ziegler–Natta polymerization.

An elastomer that is entirely or mostly polydiene is, of course, highly unsaturated. All that is required of an elastomer, however, is enough unsaturation to permit cross-linking. In making butyl rubber (Sec. 36.5), for example, only 5% of isoprene is copolymerized with isobutylene.

Problem 36.16   (a) A versatile elastomer is obtained by Ziegler–Natta copolymerization of ethylene and propylene in the presence of a little diene, followed by vulcanization. How does the use of ethylene *and* propylene—instead of just one or the other—help to give the polymer elasticity?

(b) A similar copolymer can be made without the diene. This is cured by heating, not with sulfur, but with benzoyl peroxide. Why is this? What is the nature of the cross-links generated here?

Although enormous quantities of man-made fibers and elastomers are produced each year, the major consumption of synthetic polymers is as **plastics**, materials used in the form of sheets, pipes, films, and, most important of all, molded objects: toys and bottles; knobs, handles, and switches; dishes, fountain pens, toothbrushes; valves, gears, bearings; cases for radios and television sets; boats, automobile bodies, and even houses.

The molecular structure of plastics is of two general kinds: long molecules, either linear or branched; and space-network molecules.

The *linear* and *branched* polymers may be more or less crystalline, and include some of the materials also used as fibers: nylon, for example. They include the various polyalkenes we have mentioned: polyethylene, poly(vinyl chloride), polystyrene, etc. On heating, these polymers soften, and for this reason are called *thermoplastic*. It is in this softened state that they can be molded or extruded.

Space-network polymers (or *resins*) are highly cross-linked to form a rigid but irregular three-dimensional structure, as in phenol–formaldehyde or urea–formaldehyde resins. A sample of such material is essentially one gigantic molecule; heating does not soften it, since softening would require breaking of covalent bonds. Indeed, heating may cause formation of additional cross-links and thus make the material harder; for this reason, these polymers are called *thermosetting* polymers. This continuation of the polymerization process through heating is often coupled with the shaping of the product.

Certain linear, thermoplastic polymers are, like the space-network polymers, amorphous—and for basically the same reason. On cooling, their molecules form a rigid but irregular three-dimensional structure; they are held there, not by covalent cross-links, but by powerful dipole–dipole forces which lock the molecules into position before they can shake down into the regular arrangement required of a crystal. These materials are called *glasses*; poly(methyl methacrylate)—Plexiglas, Lucite—is the commonest one. Like ordinary (inorganic) glass, they lack crystalline planes for reflecting light, and are transparent. Like ordinary glass—and like the space-network polymers—they are brittle; when struck, these molecules cannot "give" with the blow through the sliding of crystalline planes over one another; they either resist—or break.

The rest of this book is devoted to organic compounds of biological importance. Many of these are macromolecules. We shall find that, just as the technological function of a macromolecule—fiber, elastomer, plastic—depends upon its structure, so does the biological function: to hold the organism together, to nourish it, to control it, to allow it to reproduce itself.

PROBLEMS

1. Account for the fact that, whatever the mechanism—free-radical, cationic, anionic—vinyl polymerization gives products with almost exclusively "head-to-tail" arrangement of units.

2. Like other oxygen-containing compounds, alcohols dissolve in cold concentrated H_2SO_4 (Sec. 8.24). In the case of some secondary and tertiary alcohols, dissolution is followed by the gradual separation of an insoluble liquid of high boiling point. How do you account for this behavior?

3. Isobutylene does not give the kinds of stereoisomeric polymers (isotactic, etc.) that propylene does. Why not? What can you say about 1-butene?

4. Formaldehyde is polymerized by the action of a strong base like sodium methoxide. Suggest a mechanism for the process, and a structure for the polymer. To what general class of organic reactions does this polymerization belong?

5. A simple process for recycling polyurethanes has been developed by the Ford Motor Company. Can you suggest a way to accomplish this? What products would you expect to obtain?

6. Suggest an explanation for the following order of reactivity toward the addition of $BrCCl_3$ in the presence of peroxides: $C_6H_5CH{=}CH_2$, over 100; 1-octene, 1.0; $C_6H_5CH_2CH{=}CH_2$, 0.7; $ClCH_2CH{=}CH_2$, 0.5; $Cl_3CCH_2CH{=}CH_2$, 0.3.

7. Treatment of benzyl alcohol ($C_6H_5CH_2OH$) with cold concentrated H_2SO_4 yields a high-boiling resinous material. What is a likely structure for this material, and how is it probably formed?

8. Account for each of the following observations. (a) In the presence of peroxides, CCl_4 reacts with 1-octene, $RCH{=}CH_2$, to give not only the 1:1 adduct, $RCHClCH_2CCl_3$, but also the 2:1 adduct, $RCHClCH_2CH(R)CH_2CCl_3$. (b) In contrast, CBr_4 adds to the 1-octene to give *only* the 1:1 product. (c) Styrene reacts with peroxides in the presence of CCl_4 to give only polymer.

9. Outline all steps in a possible synthesis from non-polymeric starting materials of each of the following polymers.

(a) Elastic fibers, used in girdles and bathing suits (Spandex, Lycra).

$$\text{—}CH_2CH_2O(CH_2CH_2O)_nCH_2CH_2O\text{—}CO\text{—}NH\!\!\underset{NH\text{—}CO\text{—}OCH_2CH_2\text{—}}{\overset{CH_3}{\bigcirc}}$$

(b) A polyester resin, used in making pipe, boats, automobile bodies, etc.

$$\text{—}OCH_2CH_2O\text{—}CO\text{—}CH\text{—}CH_2\text{—}CO\text{—}OCH_2CH_2O\text{—}CO\text{—}$$
$$|$$
$$CH_2$$
$$|$$
$$CHC_6H_5$$
$$\}$$
$$CH_2$$
$$|$$
$$CHC_6H_5$$
$$|$$
$$\text{—}OCH_2CH_2O\text{—}CO\text{—}CH\text{—}CH_2\text{—}CO\text{—}OCH_2CH_2O\text{—}CO\text{—}$$

(c) A surface-active polymer.

$$HO(CH_3)CHCH_2[O(CH_3)CHCH_2]_n \qquad\qquad [CH_2CH(CH_3)O]_nCH_2CH(CH_3)OH$$
$$\underset{NCH_2CH_2N}{\searrow\qquad\qquad\swarrow}$$
$$HO(CH_3)CHCH_2[O(CH_3)CHCH_2]_n \qquad\qquad [CH_2CH(CH_3)O]_nCH_2CH(CH_3)OH$$

(d)

$\sim$OCH$_2$CH$_2$(OCH$_2$CH$_2$)$_n$OCH$_2$CH$_2$CHCH$_2$(CHCH$_2$)$_n$(CH$_2$CH)$_n$CH$_2$CHOCH$_2$CH$_2\sim$

Ph   Ph   Ph   Ph

(e)

CH$_3$CCOOCH$_3$

CH$_2$

CH$_3$CCOOCH$_3$

CH$_2$

CH$_3$CCOOCH$_3$

CH$_2$

$\sim$CH$_2$CHCH$_2$CHCH$_2$CCH$_2$CHCH$_2$CHCH$_2$CH$\sim$

OAc   Cl   OAc Cl   Cl   OAc

10. Treatment of β-propiolactone with base gives a polymer. Give a likely structure for this polymer, and show a likely mechanism for the process. Is this an example of chain-reaction or step-reaction polymerization?

11. When styrene is treated with KNH$_2$ in liquid ammonia, the product is a dead polymer that contains one —NH$_2$ group per molecule and no unsaturation. Suggest a termination step for the process.

12. When poly(vinyl acetate) was hydrolyzed, and the product treated with periodic acid and then re-acetylated, there was obtained poly(vinyl acetate) of lower molecular weight than the starting material. What does this indicate about the structure of the original polymer? About the polymerization process?

13. (a) What is the structure of nylon-6, made by alkaline polymerization of caprolactam?

Caprolactam

(b) Suggest a mechanism for the process. Is polymerization of the chain-reaction or step-reaction type?

14. In the *Beckmann rearrangement* (Problem 5, p. 1121) oximes are converted into amides by the action of acids. For example:

$$(C_6H_5)_2C{=}NOH \xrightarrow{\text{acid}} C_6H_5C\overset{O}{\underset{NHC_6H_5}{}}$$

Benzophenone oxime   Benzanilide

Caprolactam (preceding problem) can be made by the Beckmann rearrangement. With what ketone must the process start?

15. Fibers of very high tensile strength ("high-modulus fibers") have been made by reactions like the one between terephthalic acid and p-phenylenediamine, p-C$_6$H$_4$(NH$_2$)$_2$. Of key importance is the isomer composition of the monomers: the more exclusively *para*, the higher the melting point and the lower the solubility of the polymer, and the stronger the fibers. How do you account for this effect?

16. Evidence of many kinds shows that the metal–carbon bond in compounds like *n*-butyllithium is covalent, although highly polar. Yet living polystyrene solutions, which are colored, have virtually identical spectra whether the metal involved is sodium, potassium, cesium, or lithium. Can you suggest an explanation for this?

17. (a) When the alkane, 2,4,6,8-tetramethylnonane was synthesized by an unambiguous method (Problem 12(l), p. 690), there was obtained a product which was separated by gas chromatography into two components, A and B. The two components had identical mol. wt. and elemental composition, but different m.p., b.p., and infrared and NMR spectra. Looking at the structure of the expected product, what are these two components?

(b) When the same synthesis was carried out starting with an optically active reactant, compound B was obtained in optically active form, but A was still inactive. What is the structure of A? Of B?

(c) The NMR and infrared spectra of A and B were compared with the spectra of isotactic and syndiotactic polypropylenes (Fig. 36.1, p. 1248). With regard to their spectra, A showed a marked resemblance to one of the polymers, and B showed a marked resemblance to the other. It was concluded that the results "confirm the structures originally assigned [by Natta, p. 1246] for the two crystalline polymers of propylene". Which polymer did A resemble? Which polymer did B resemble?

18. Material similar to foam rubber can be made by the following sequence:

adipic acid + excess 1,2-ethanediol $\longrightarrow$ C

C + excess *p*-OCN—C$_6$H$_4$—C$_6$H$_4$—NCO-*p* $\longrightarrow$ D

D + limited H$_2$O $\longrightarrow$ E

Write equations for all steps, and show structures for C, D, and E. Be sure to account for the cross-linking in the final polymer, and its *foamy* character. (*Remember*: A foam is a dispersion of a gas in a solid.)

19. In the presence of benzoyl peroxide, allyl acetate gives poor yields of polymer of low molecular weight. The deuterium-labeled ester, CH$_2$=CHCD$_2$OAc, polymerizes 2 to 3 times as fast as the ordinary ester, and gives polymer of about twice the molecular weight. How do you account for these facts?

20. Linseed oil and tung oil, important constituents of paints, are esters (Sec. 37.6) derived from acids that contain two or three double bonds per molecule: 9,12-octadecadienoic acid, for example. On exposure to air, paint forms a tough protective film; oddly enough, after the initial rapid evaporation of solvent, this "drying" of paint is accompanied by a *gain* in weight. What kind of process do you think is involved? Be as specific as you can be.

21. To use an epoxy cement, one mixes the fluid "cement" with the "hardener", applies the mixture to the surfaces being glued together, brings them into contact, and waits for hardening to occur. The fluid cement is a low-molecular-weight polymer prepared by the following reaction:

$$\text{HO}\!\!\left\langle\bigcirc\right\rangle\!\!\overset{\overset{\displaystyle CH_3}{|}}{\underset{\underset{\displaystyle CH_3}{|}}{C}}\!\!\left\langle\bigcirc\right\rangle\!\!\text{OH} \quad + \quad \text{CH}_2\!\!-\!\!\text{CHCH}_2\text{Cl} + \text{OH}^- \quad \longrightarrow \quad \text{"cement"}$$

$$\underset{O}{}$$

2,2-Bis(*p*-hydroxyphenyl)propane   Epichlorohydrin   *Contains no chlorine*
("Bisphenol A")   *Excess*

The hardener can be any of a number of things: NH$_2$CH$_2$CH$_2$NHCH$_2$CH$_2$NH$_2$, diethylene-triamine, for example.

(a) What is the structure of the fluid cement, and how is it formed? What is the purpose of using *excess* epichlorohydrin? (b) What happens during hardening? What is the structure of the final epoxy resin? (c) Suggest a method of making bisphenol A, starting from phenol.

22. Poly(methyl methacrylate) was prepared in two different ways: polymer F, with initiation by benzoyl peroxide at 100 °C; polymer G, with initiation by *n*-butyllithium at −62 °C. Their NMR spectra were, with considerable simplification, as follows:

F   *a* singlet, δ 1.10
      b singlet, δ 2.0
      c singlet, δ 3.58
      approximate area ratios, $a:b:c = 3:2:3$

G   *a* singlet, δ 1.33
      b doublet, δ 1.7
      c doublet, δ 2.4
      d singlet, δ 3.58
      approximate area ratios, $a:b:c:d = 3:1:1:3$

Account in detail for the difference in spectra. What, essentially, is polymer F? Polymer G?

Part Three

Biomolecules

37

Fats

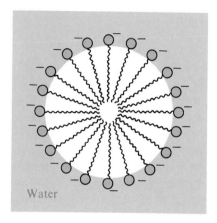

Water

37.1 The organic chemistry of biomolecules

The study of biology at the molecular level is called biochemistry. It is a branch of biology, but it is equally a branch of organic chemistry. In general the molecules involved, the *biomolecules*, are bigger and more complicated than most of the ones we have studied so far, and their environment—a living organism—is a far cry from the stark simplicity of the reaction mixture of the organic chemist. But the physical and chemical properties of these compounds depend on molecular structure in exactly the same way as do the properties of other organic compounds.

The detailed chemistry of biological processes is vast and complicated, and is beyond the scope of this book; indeed, the study of biochemistry must be *built upon* a study of the fundamentals of organic chemistry. We can, however, attempt to close the gap between the subject "organic chemistry" and the subject "biochemistry".

In the remaining chapters of this book, we shall take up the principal classes of biomolecules: fats, carbohydrates, proteins, and nucleic acids. Our chief concern will be with their structures—since structure is fundamental to everything else—and with the methods used to determine these structures. Because most biomolecules are big ones—*macromolecules* (Chap. 36)—we shall encounter structure on several levels: first, of course, the *sequence of functional groups* and the *configuration* at any chiral centers or double bonds; then, *conformation*, with loops, coils, and zig-zags on a grander scale than anything we have seen yet; finally, the arrangement of *collections of molecules*, and even of collections of these collections. We shall see

1263

remarkable effects due to our familiar intermolecular forces: operating between biomolecules; between biomolecules—or *parts* of them—and the solvent; between different parts of the same biomolecule. In all this we shall see, as we did for the man-made macromolecules (Sec. 36.8), how the functions of these giant molecules depend upon their structure at all levels.

We shall study the chemical properties of these compounds observed in the test tube, since these properties must lie behind the reactions they undergo in living organisms. In doing this, we shall reinforce our knowledge of basic organic chemistry by applying it to these more complex substances. Finally, we shall look— very briefly—at a few biochemical processes, just to catch a glimpse of the ways in which molecular structure determines biological behavior.

37.2 Occurrence and composition of fats

Biochemists have found it convenient to define one set of biomolecules, the *lipids*, as substances, insoluble in water, that can be extracted from cells by organic solvents of low polarity like ether or chloroform. This is a catch-all sort of definition, and lipids include compounds of many different kinds: steroids (Sec. 17.18), for example, and terpenes (Sec. 10.31). Of the lipids, we shall take up only the *fats* and certain closely related compounds. These are not the only important lipids— indeed, every compound in an organism seems to play an important role, if only as an unavoidable waste product of metabolism—but they are the most abundant.

Fats are the main constituents of the storage fat cells in animals and plants, and are one of the important food reserves of the organism. We can extract these animal and vegetable fats—liquid fats are often referred to as *oils*—and obtain such substances as corn oil, coconut oil, cottonseed oil, palm oil, tallow, bacon grease, and butter.

Chemically, fats are carboxylic esters derived from the single alcohol, glycerol, $HOCH_2CHOHCH_2OH$, and are known as *glycerides*. More specifically, they are *triacylglycerols*. As Table 37.1 shows, each fat is made up of glycerides derived

$$\begin{array}{c}
CH_2\!-\!O\!-\!\underset{\underset{O}{\|}}{C}\!-\!R \\[2.5ex]
CH\!-\!O\!-\!\underset{\underset{O}{\|}}{C}\!-\!R' \\[2.5ex]
CH_2\!-\!O\!-\!\underset{\underset{O}{\|}}{C}\!-\!R''
\end{array}$$

A triacylglycerol
(A glyceride)

from many different carboxylic acids. The proportions of the various acids vary from fat to fat; each fat has its characteristic composition, which does not differ very much from sample to sample.

With only a few exceptions, the fatty acids are all straight-chain compounds, ranging from three to eighteen carbons; except for the C_3 and C_5 compounds, only acids containing an even number of carbons are present in substantial amounts. As we shall see in Sec. 41.7, these even numbers are a natural result of the biosynthesis of fats: the molecules are built up two carbons at a time from acetate

Table 37.1 Fatty Acid Composition of Fats and Oils

Fat or oil	Saturated acids, %							Unsaturated acids %						
								Enoic				Dienoic	Trienoic
	C_8	C_{10}	C_{12}	C_{14}	C_{16}	C_{18}	$>C_{18}$	$<C_{16}$	C_{16}	C_{18}	$>C_{18}$	C_{18}	C_{18}
Beef tallow			0.2	2–3	25–30	21–26	0.4–1	0.5	2–3	39–42	0.3	2	
Butter	1–2[a]	2–3	1–4	8–13	25–32	8–13	0.4–2	1–2	2–5	22–29	0.2–1.5	3	
Coconut	5–9	4–10	44–51	13–18	7–10	1–4				5–8	0–1	1–3	
Corn				0–2	8–10	1–4			1–2	30–50	0–2	34–56	
Cottonseed				0–3	17–23	1–3				23–44	0–1	34–55	
Lard				1	25–30	12–16		0.2	2–5	41–51	2–3	3–8	
Olive			0–1	0–2	7–20	1–3	0–1		1–3	53–86	0–3	4–22	
Palm				1–6	32–47	1–6				40–52		2–11	
Palm kernel	2–4	3–7	45–52	14–19	6–9	1–3	1–2		0–1	10–18		1–2	
Peanut				0.5	6–11	3–6	5–10		1–2	39–66		17–38	
Soybean				0.3	7–11	2–5	1–3		0–1	22–34		50–60	2–10
Cod liver				2–6	7–14	0–1		0–2	10–20	25–31	C_{20} 25–32; $>C_{20}$ 10–20	8–29[b]	
Linseed				0.2	5–9	4–7				9–29		8–15	45–67[c]
Tung					5–9	4–7	0.5–1			4–13			78–82[d]

[a] 3–4% C_4, 1–2% C_6.
[b] Linoleic acid, cis,cis-9,12-octadecadienoic acid.
[c] Linolenic acid, cis,cis,cis-9,12,15-octadecatrienoic acid.
[d] Eleostearic acid, cis,trans,trans-9,11,13-octadecatrienoic acid, and 3–6% saturated acids.

units, in steps that closely resemble the malonic ester synthesis of the laboratory (Sec. 30.2).

Problem 37.1 *n*-Heptadecane is the principal *n*-alkane found both in a 50 million-year-old shale and in the blue-green algae, primitive organisms still existing. When blue-green algae were grown on a medium containing [18-^{14}C]stearic acid, essentially all the radioactivity that was not left in unconsumed stearic acid was found in *n*-heptadecane. By what kind of chemical reaction is the hydrocarbon evidently produced? Of what geological significance is this finding?

Problem 37.2 (a) Acetate is not the only building block for the long chains of lipids. From a 50 million-year-old shale (see Problem 37.1)—as well as from modern organisms—there has been isolated 3,7,11,15-tetramethylhexadecanoic acid.

3,7,11,15-Tetramethylhexadecanoic acid

What familiar structural unit occurs here?
   (b) The long side chain of chlorophyll (p. 1207) is derived from the alcohol *phytol*, which is *cis*-7(*R*),11(*R*)-3,7,11,15-tetramethyl-2-hexadecen-1-ol. The acid in (a) was

cis-7(*R*),11(*R*)-3,7,11,15-Tetramethyl-2-hexadecen-1-ol
Phytol

found to be a mixture of two diastereomers: the 3(*S*),7(*R*),11(*R*) and 3(*R*),7(*R*),11(*R*). Of what biogenetic significance is this finding?

Besides saturated acids, there are unsaturated acids containing one or more double bonds per molecule. The most common of these acids are:

$$CH_3(CH_2)_7CH=CH(CH_2)_7COOH \qquad CH_3(CH_2)_4CH=CHCH_2CH=CH(CH_2)_7COOH$$

Oleic acid   Linoleic acid
(*cis* isomer)   (*cis,cis* isomer)

$$CH_3CH_2CH=CHCH_2CH=CHCH_2CH=CH(CH_2)_7COOH$$

Linolenic acid
(*cis,cis,cis* isomer)

The configuration about these double bonds is almost invariably *cis*, rather than the more stable *trans*.

Unsaturation *with this particular stereochemistry* has an effect that is seemingly trivial but is actually (Sec. 37.8) of vital biological significance: it lowers the melting point. In the solid phase, the molecules of a fat fit together as best they can; the closer they fit, the stronger the intermolecular forces, and the higher the melting point. Saturated acid chains are extended in a linear fashion—with, of course, the zig-zag due to the tetrahedral bond angles—and fit together rather well. *trans*-Unsaturated acid chains can be similarly extended to linear conformations that match saturated chains rather well (Fig. 37.1). But *cis*-unsaturated acid chains have a *bend* at the double bond, and fit each other—and saturated chains—badly. The net result is that *cis* unsaturation lowers the melting point of fat.

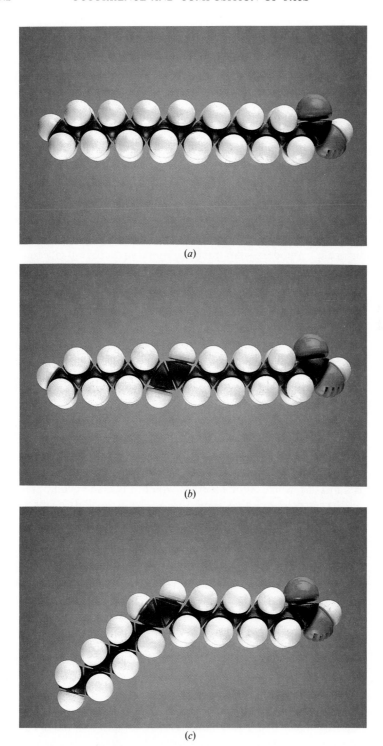

(a)

(b)

(c)

Figure 37.1   Molecular structure and physical properties. Extended chains of fatty acids: (a) hexadecanoic acid, (b) *trans*-9-hexadecenoic acid, (c) *cis*-9-hexadecenoic acid. The saturated acid (a) and the *trans*-unsaturated acid (b) are straight. The *cis*-unsaturated acid (c) has a bend at the double bond; it fits poorly into a crystalline lattice and hence has a lower melting point.

While we synthesize fats in our own bodies, we also eat fats synthesized in plants and other animals; they are one of the three main classes of foods, the others being carbohydrates (Chap. 39) and proteins (Chap. 40). Fats are used in enormous amounts as raw materials for many industrial processes; let us look at some of these before we turn our attention to some close relatives of the fats.

37.3   Hydrolysis of fats. Soap. Micelles

The making of soap is one of the oldest of chemical syntheses. (It is not nearly so old, of course, as the production of ethyl alcohol; man's desire for cleanliness is much newer than his desire for intoxication.) When the German tribesmen of Caesar's time boiled goat tallow with potash leached from the ashes of wood fires, they were carrying out the same chemical reaction as the one carried out on a tremendous scale by modern soap manufacturers: *hydrolysis of glycerides*. Hydrolysis yields salts of the carboxylic acids, and glycerol, $CH_2OHCHOHCH_2OH$.

$$
\begin{array}{c}
CH_2\!-\!O\!-\!\underset{\underset{O}{\|}}{C}\!-\!R \\[2.2em]
CH\!-\!O\!-\!\underset{\underset{O}{\|}}{C}\!-\!R' \\[2.2em]
CH_2\!-\!O\!-\!\underset{\underset{O}{\|}}{C}\!-\!R''
\end{array}
\xrightarrow{\ \text{NaOH}\ }
\begin{array}{c}
CH_2OH \\[0.5em]
CHOH \\[0.5em]
CH_2OH \\[0.5em]
\text{Glycerol}
\end{array}
\ + \
\left\{
\begin{array}{c}
RCOO^-\,Na^+ \\[0.5em]
R'COO^-\,Na^+ \\[0.5em]
R''COO^-\,Na^+
\end{array}
\right\}
$$

A glyceride
(A fat)   Soap

Ordinary soap today is simply a mixture of sodium salts of long-chain fatty acids. It is a mixture because the fat from which it is made is a mixture, and for washing our hands or our clothes a mixture is just as good as a single pure salt. Soap may vary in composition and method of processing: if made from olive oil, it is *Castile soap*; alcohol can be added to make it transparent; air can be beaten in to make it float; perfumes, dyes, and germicides can be added; if a potassium salt (instead of sodium salt), it is *soft soap*. Chemically, however, soap remains pretty much the same, and does its job in the same way.

We might at first expect these salts to be water-soluble—and, indeed, one can prepare what are called "soap solutions". But these are not true solutions, in which solute molecules swim about, separately and on their own. Instead, soap is dispersed in spherical clusters called **micelles**, each of which may contain hundreds of soap molecules. A soap molecule has a polar end, $-COO^-Na^+$, and a non-polar end, the long carbon chain of 12 to 18 carbons. The polar end is water-soluble, and is thus *hydrophilic*. The non-polar end is water-insoluble, and is thus *hydrophobic* (or *lipophilic*, Sec. 6.3); it is, of course, soluble in non-polar solvents. Molecules like these are called *amphipathic*: they have both polar and non-polar ends and, in addition, are big enough for each end to display its own solubility behavior. In line with the rule of "like dissolves like", each non-polar end seeks a non-polar environment; in this situation, the only such environment about is the non-polar ends of other soap molecules, which therefore huddle together in the center of the micelle (Fig. 37.2). The polar ends project outward into the polar solvent, water.

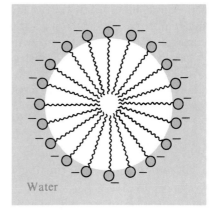

Figure 37.2 Soap micelle. The non-polar hydrocarbon chains "dissolve" in each other. The polar —COO⁻ groups dissolve in water. Similarly charged micelles repel each other.

Water

Negatively charged carboxylate groups stud the surface of the micelle, and it is surrounded by an ionic atmosphere. Repulsion between similar charges keeps the micelles dispersed.

Now, how does a soap clean? The problem in cleansing is the fat and grease that make up and contain the dirt. Water alone cannot dissolve these hydrophobic substances; oil droplets in contact with water tend to coalesce so that there is a water layer and an oil layer. But the presence of soap changes this. The non-polar ends of soap molecules dissolve in the oil droplet, leaving the carboxylate ends projecting into the surrounding water layer. Repulsion between similar charges keeps the oil droplets from coalescing; a stable emulsion of oil and water forms, and can be removed from the surface being cleaned. As we shall see, this emulsifying, and hence cleansing, property is not limited to carboxylate salts, but is possessed by other amphipathic molecules (Sec. 37.5).

Hard water contains calcium and magnesium salts, which react with soap to form insoluble calcium and magnesium carboxylates (the "ring" in the bathtub).

37.4   Fats as sources of pure acids and alcohols

Treatment of the sodium soaps with mineral acid (or hydrolysis of fats under acidic conditions) liberates a mixture of the free carboxylic acids. In recent years, fractional distillation of these mixtures has been developed on a commercial scale to furnish individual carboxylic acids of over 90% purity.

Fats are sometimes converted by transesterification into the methyl esters of carboxylic acids; the glycerides are allowed to react with methanol in the presence of a basic or acidic catalyst. The mixture of methyl esters can be separated by

$$
\begin{array}{c}
CH_2-O-C-R \\
\quad\quad\;\; \| \\
\quad\quad\;\; O \\
| \\
CH-O-C-R' \;\; + \;\; CH_3OH \;\; \xrightarrow{\;base\;} \;\; CHOH \;\; + \;\; \left\{ \begin{array}{l} RCOOCH_3 \\ R'COOCH_3 \\ R''COOCH_3 \end{array} \right\} \\
\end{array}
$$

CH₂—O—C—R
        ‖
        O

CH—O—C—R′   +   CH₃OH   —base→   CH₂OH   ⎧ RCOOCH₃ ⎫
        ‖ | ⎨ R′COOCH₃ ⎬
        O   CHOH   +   ⎩ R″COOCH₃ ⎭
            Methanol |
CH₂—O—C—R″   CH₂OH
        ‖   Glycerol   Mixture of
        O   methyl esters

A glyceride

fractional distillation into individual esters, which can then be hydrolyzed to individual carboxylic acids of high purity. Fats are thus the source of straight-chain acids of even carbon number ranging from six to eighteen carbons.

Alternatively, these methyl esters, either pure or as mixtures, can be catalytically reduced to straight-chain primary alcohols of even carbon number, and from these can be derived a host of compounds (as in Problem 23.10, p. 844). Fats thus provide us with long straight-chain units to use in organic synthesis.

37.5   Detergents

Of the straight-chain primary alcohols obtained from fats—or in other ways (Sec. 36.6)—the C_8 and C_{10} members are used in the production of high-boiling esters used as *plasticizers* (e.g., octyl phthalate). The C_{12} to C_{18} alcohols are used in enormous quantities in the manufacture of *detergents* (cleansing agents).

Although the synthetic detergents vary considerably in their chemical structure, the molecules of all of them have one common feature, a feature they share with ordinary soap: they are amphipathic, and have a large non-polar hydrocarbon end that is oil-soluble, and a polar end that is water-soluble. The C_{12} to C_{18} alcohols are converted into the salts of alkyl hydrogen sulfates. For example:

$$n\text{-}C_{11}H_{23}CH_2H \xrightarrow{H_2SO_4} n\text{-}C_{11}H_{23}CH_2OSO_3H \xrightarrow{NaOH} n\text{-}C_{11}H_{23}CH_2OSO_3^-Na^+$$

Lauryl alcohol   Lauryl hydrogen sulfate   Sodium lauryl sulfate

For these, the non-polar end is the long chain, and the polar end is the $-OSO_3^-Na^+$.

Treatment of alcohols with ethylene oxide (Sec. 19.14) yields a *non-ionic* detergent:

$$CH_3(CH_2)_{10}CH_2OH + 8CH_2\overset{}{-}CH_2 \xrightarrow{base} CH_3(CH_2)_{10}CH_2(OCH_2CH_2)_8OH$$

Lauryl alcohol O An ethoxylate

Ethylene oxide

Hydrogen bonding to the numerous oxygen atoms makes the polyether end of the molecule water-soluble. Alternatively, the ethoxylates can be converted into sulfates and used in the form of the sodium salts.

Perhaps the most widely used detergents are sodium salts of alkylbenzenesulfonic acids. A long-chain alkyl group is attached to a benzene ring by the action

$$SO_3^-Na^+$$

of a Friedel–Crafts catalyst and an alkyl halide, an alkene, or an alcohol. Sulfonation and neutralization yields the detergent.

Formerly, polypropylene was commonly used in the synthesis of these alkylbenzenesulfonates; but the highly branched side chain it yields blocks the rapid biological degradation of the detergent residues in sewage discharge and septic tanks. Since about 1965 in the United States, such "hard" detergents have been replaced by "soft" (biodegradable) detergents: alkyl sulfates, ethoxylates and their sulfates; and alkylbenzenesulfonates in which the phenyl group is randomly attached to the various secondary positions of a long straight chain (C_{12}–C_{18} range). (See Problem 15, p. 566.) The side chains of these "linear" alkylbenzenesulfonates are derived from straight-chain l-alkenes (Sec. 36.6), or chlorinated straight-chain alkanes separated (by use of molecular sieves) from kerosene.

These detergents act in essentially the same way as soap does. They are used because they have certain advantages. For example, the sulfates and sulfonates retain their efficiency in hard water, since the corresponding calcium and magnesium salts are soluble. Being salts of strong acids, they yield neutral solutions, in contrast to the soaps, which, being salts of weak acids, yield slightly alkaline solutions (Sec. 23.10).

37.6   Unsaturated fats. Hardening of oils. Drying oils

We have seen that fats contain, in varying proportions, glycerides of unsaturated carboxylic acids. We have also seen that, other things being equal, unsaturation in a fat tends to lower its melting point and thus tends to make it a liquid at room temperature. In the United States the long-established use of lard and butter for cooking purposes has led to a prejudice against the use of the cheaper, equally nutritious oils. Hydrogenation of some of the double bonds in such cheap fats as cottonseed oil, corn oil, and soybean oil converts these liquids into solids having a consistency comparable to that of lard or butter. This *hardening* of oils is the basis of an important industry that produces cooking fats (for example, Crisco, Spry) and oleomargarine. Hydrogenation of the carbon–carbon double bonds takes place under such mild conditions (Ni catalyst, 175–190 °C, 20–40 lb/in.^{2}) that hydrogenolysis of the ester linkage does not occur.

Hydrogenation not only changes the physical properties of a fat, but also—and this is even more important—changes the chemical properties: a hydrogenated fat becomes *rancid* much less readily than does a non-hydrogenated fat. Rancidity is due to the presence of volatile, bad-smelling acids and aldehydes. These compounds result (in part, at least) from attack by oxygen at reactive allylic positions in the fat molecules; hydrogenation slows down the development of rancidity presumably by decreasing the number of double bonds and hence the number of allylic positions.

(In the presence of hydrogenation catalysts, unsaturated compounds undergo not only hydrogenation but also isomerization—shift of double bonds, or stereochemical transformations—which also affects physical and chemical properties.)

Linseed oil and tung oil have special importance because of their high content of glycerides derived from acids that contain two or three double bonds. They are known as **drying oils** and are important constituents of paints and varnishes. The "drying" of paint does not involve merely evaporation of a solvent (turpentine, etc.), but rather a chemical reaction in which a tough organic film is formed. Aside from the color due to the pigments present, protection of a surface by this film is the chief purpose of paint. The film is formed by a polymerization of the unsaturated oils that is brought about by oxygen. The polymerization process and the structure of the polymer are extremely complicated and are not well understood. The process seems to involve, in part, free-radical attack at reactive allylic hydrogens, free-radical chain-reaction polymerization similar to that previously described (Secs. 8.21 and 36.6), and cross-linking by oxygen analogous to that by sulfur in vulcanized rubber (Sec. 10.30).

37.7 Phosphoglycerides. Phosphate esters

So far, we have talked only about glycerides in which all three ester linkages are to acyl groups, that is, triacylglycerols. There also occur lipids of another kind, phosphoglycerides, which contain only two acyl groups and, in place of the third, a *phosphate* group. The parent structure is *diacylglycerol phosphate*, or *phosphatidic acid*.

$$
\begin{array}{c}
\overset{C_{17}H_{33}}{R'-\overset{\displaystyle\|}{\underset{\displaystyle O}{C}}-O-CH_2} \\[2pt]
R''-\overset{\displaystyle\|}{\underset{\displaystyle O}{C}}-O-CH \qquad \overset{\displaystyle O}{} \\[2pt]
\Big\{\qquad\qquad CH_2-O-\overset{\displaystyle\|}{\underset{\displaystyle OH}{P}}-OH \\[2pt]
C_{17}H_{33}
\end{array}
$$

Phosphatidic acid
(A phosphoglyceride)

Phosphoglycerides are, then, not only carboxylate esters but phosphate esters as well. Just what are phosphate esters like? It will be well for us to learn something about them since we shall be encountering them again and again: phospholipids make up the membranes of cells (Sec. 37.8); adenosine triphosphate lies at the heart of the energy system of organisms, and it does its job by converting hosts of other compounds into phosphate esters (Sec. 41.3); nucleic acids, which control heredity, are polyesters of phosphoric acid.

To begin with, phosphates come in various kinds. Phosphoric acid contains three hydroxy groups and can form esters in which one, two, or three of these have been replaced by alkoxy groups. Phosphoric acid is highly acidic, and so are the

$$
\overset{\displaystyle O}{\underset{\displaystyle OH}{HO-\overset{\displaystyle|}{\underset{\displaystyle|}{P}}-OH}} \qquad
\overset{\displaystyle O}{\underset{\displaystyle OR}{HO-\overset{\displaystyle|}{\underset{\displaystyle|}{P}}-OH}} \qquad
\overset{\displaystyle O}{\underset{\displaystyle OR}{RO-\overset{\displaystyle|}{\underset{\displaystyle|}{P}}-OH}} \qquad
\overset{\displaystyle O}{\underset{\displaystyle OR}{RO-\overset{\displaystyle|}{\underset{\displaystyle|}{P}}-OR}}
$$

Phosphoric acid   Phosphate esters

monoalkyl and dialkyl esters; in aqueous solution they tend to exist as anions, the exact extent of ionization depending, of course, upon the acidity of the medium. For example:

$$
\overset{\displaystyle O}{\underset{\displaystyle OH}{RO-\overset{\displaystyle|}{\underset{\displaystyle|}{P}}-OH}}
\;\rightleftharpoons\;
\overset{H^+}{\underset{+}{}}\;\;
\overset{\displaystyle O^-}{\underset{\displaystyle OH}{RO-\overset{\displaystyle|}{\underset{\displaystyle|}{P}}-O}}
\;\rightleftharpoons\;
\overset{H^+}{\underset{+}{}}\;\;
\left.\overset{\displaystyle O}{\underset{\displaystyle O}{RO-\overset{\displaystyle|}{\underset{\displaystyle|}{P}}-O}}\right\}^{2-}
$$

Like other esters, phosphates undergo hydrolysis to the parent acid and alcohol. Here, the acidity of —OH attached to phosphorus has several effects. In the first place, since acidic phosphate esters can undergo ionization, there may be many species present in the hydrolysis solution. A monoalkyl ester, for example, could exist as dianion, monoanion, neutral ester, and protonated ester; any or all of these could conceivably be undergoing hydrolysis. Actually, the situation is not quite that complicated. From the dissociation constants of these acidic esters, one can calculate the fraction of ester in each form in a given solution. The dependence of rate on acidity of the solution often shows which species is the principal reactant.

In carboxylates, we remember, attack generally occurs at acyl carbon, and in sulfonates, at alkyl carbon, with a resulting difference in point of cleavage. In

hydrolytic behavior, phosphates are intermediate between carboxylates and sulfonates. Cleavage can occur at either position, depending on the nature of the alcohol group.

C—O cleavage   P—O cleavage

Here again the acidity of phosphoric acids comes in. Cleavage of the alkyl–oxygen bond in carboxylates is difficult because the carboxylate anion is strongly basic and a poor leaving group; in sulfonates such cleavage is favored because the weakly basic sulfonate anion is a very good leaving group. Phosphoric acid is intermediate in acidity between carboxylic and sulfonic acid; as a result, the phosphate anion is a better leaving group than carboxylate but a poorer one than sulfonate. In these esters, phosphorus is bonded to four groups; but it can accept more—witness stable pentacovalent compounds like PCl_5—and nucleophilic attack at phosphorus competes with attack at alkyl carbon.

In acidic solution, phosphate esters are readily cleaved to phosphoric acid. In alkaline solution, however, only trialkyl phosphates, $(RO)_3PO$, are hydrolyzed, and only one alkoxy group is removed. Monoalkyl and dialkyl esters, $ROPO(OH)_2$ and $(RO)_2PO(OH)$, are inert to alkali, even on long treatment. This may seem unusual behavior, but it has a perfectly rational explanation. The monoalkyl and dialkyl esters contain acidic —OH groups on phosphorus, and in alkaline solution exist as anions; repulsion between like charges prevents attack on these anions by hydroxide ion.

In most phospholipids, phosphate is of the kind

in which G is the glyceryl group—with its two carboxylates—and R is derived from some other alcohol, ROH, most often *ethanolamine*, $HOCH_2CH_2NH_2$, or *choline*, $HOCH_2CH_2N(CH_3)_3{}^+$. Since the remaining —OH on phosphorus is highly

$$
\begin{array}{cc}
\text{R}'\text{—C—O—CH}_2 & \text{R}'\text{—C—O—CH}_2 \\
\quad\ \ \text{O} & \quad\ \ \text{O} \\
\text{R}''\text{—C—O—CH} & \text{R}''\text{—C—O—CH} \\
\quad\ \ \text{O} & \quad\ \ \text{O} \\
\end{array}
$$

Phosphatidyl ethanolamine   Phosphatidyl choline
(Ethanolamine phosphoglyceride)   (Choline phosphoglyceride)

acidic, the ester exists mostly in the ionic form. Furthermore, since the alcohol ROH usually contains an amino group, the phosphate unit carries both positive and negative charges, and the phospholipid is—at this end—a *dipolar ion*. On hydrolysis, these phosphates generally undergo cleavage between phosphorus and oxygen, P⫶O—R.

> **Problem 37.4** Consider hydrolysis of $(RO)_2PO(OH)$ by aqueous hydroxide, and grant that for electrostatic reasons attack by OH^- cannot occur. Even so, why does not attack by the nucleophile water lead to hydrolysis? After all, water *is* the successful nucleophile in acidic hydrolysis. (*Hint*: See Sec. 24.18.)

37.8   Phospholipids and cell membranes

The fats are found, we said, in storage fat cells of plants and animals. Their function rests on their chemical properties: through oxidation, they are consumed to help provide energy for the life processes.

The phospholipids, on the other hand, are found in the membranes of cells—all cells—and are a basic structural element of living organisms. This vital function depends, in a fascinating way, on their physical properties.

Phosphoglyceride molecules are amphipathic, and in this respect differ from fats—but resemble soaps and detergents. The lipophilic part is, again, the long fatty acid chains. The hydrophilic part is the dipolar ionic end: the substituted phosphate group with its positive and negative charges. In aqueous solution, as we would expect, phosphoglycerides form micelles. In certain situations, however—at an aperture between two aqueous solutions, for example—they tend to form bilayers: two rows of molecules are lined up, back to back, with their polar ends projecting into water on the two surfaces of the bilayer (Fig. 37.3). Although the polar groups are needed to hold molecules in position, the bulk of the bilayer is made up of the fatty acid chains. Non-polar molecules can therefore dissolve in this mostly hydrocarbon wall and pass through it, but it is an effective barrier to polar molecules and ions.

Fig. 37.3 A phospholipid bilayer. The lipophilic fatty chains are held together by van der Waals forces. The hydrophilic ends dissolve in water.

It is in the form of bilayers that phosphoglycerides are believed to exist in cell membranes. They constitute walls that not only enclose the cell but also very selectively control the passage, in and out, of the various substances—nutrients, waste products, hormones, etc.—even from a solution of low concentration to a solution of high concentration. Now, many of these substances that enter and leave the cells are highly polar molecules like carbohydrates and amino acids, or ions like sodium and potassium. How can these molecules pass through cell membranes when they cannot pass through simple bilayers? And how can permeability be so highly selective?

The answer to both these questions seems to involve the proteins that are also found in cell membranes: embedded in the bilayer, and even extending clear through it. Proteins, as we shall see in Chapter 40, are very long-chain amides, polymers of twenty-odd different amino acids. Protein chains can be looped and coiled in a variety of ways; the conformation that is favored for a particular protein molecule depends on the exact sequence of amino acids along its chain.

It has been suggested that transport through membranes happens in the following way. A protein molecule, coiled up to turn its lipophilic parts outward, is dissolved in the bilayer, forming a part of the cell wall. A molecule approaches: a potassium ion, say. If the particular protein is the one designed to handle potassium ion, it receives the ion into its polar interior. Hidden in this lipophilic wrapping, the ion is smuggled through the bilayer and released on the other side.

This mechanism for ionic transport is exactly the one we gave earlier to account for the action of an antibiotic like Nonactin (Sec. 19.10). Here, it is necessary for normal cell function; there, it upset the ionic balance and disrupted the cell function. In both cases we are seeing a *host–guest* relationship of basically the same kind as that between a crown ether and a cation: there is the same kind of bonding between host and guest, and the function is the same one—to carry a cation into a non-polar medium.

Now, if the transport protein is to do its job, it must be free to move within the membrane. The molecules of the bilayer, while necessarily aligned, must not be locked into a rigid crystalline lattice—as they would be if all the fatty acid chains were saturated. Actually, some of the chains in the membrane phospholipids are unsaturated and these, with their *cis* stereochemistry and the accompanying bend (Fig. 37.1), disrupt the alignment enough to make the membrane semiliquid at physiological temperatures.

Here, we have had a glimpse of just one complex biological process. Yet we can begin to see how the understanding of biology rests on basic chemical concepts: van der Waals forces and ion–dipole bonds; polarity and solubility; melting point and molecular shape; configuration and conformation; and, ultimately, the sequence of atoms in molecular chains.

Problem 37.5 The degree of unsaturation of the membrane lipids in the legs of reindeer is higher in cells near the hooves than in cells near the body. What survival value does this unsaturation gradient have?

PROBLEMS

1. From saponification of cerebrosides, lipids found in the membranes of brain and nerve cells, there is obtained *nervonic acid*. This acid rapidly decolorizes dilute $KMnO_4$ and Br_2/CCl_4 solutions. Hydrogenation in the presence of nickel yields tetracosanoic acid, $n\text{-}C_{23}H_{47}COOH$. Vigorous oxidation of nervonic acid yields one acid of neutralization equivalent 156 ± 3 and another acid of neutralization equivalent 137 ± 2. What structure or structures are possible for nervonic acid?

2. When peanut oil is heated very briefly with a little sodium methoxide, its properties are changed dramatically—it becomes so viscous it can hardly be poured—yet saponification yields the same mixture of fatty acids as did the untreated oil. What has probably happened? What is the function of the sodium methoxide?

3. On oxidation with O_2, methyl oleate (methyl 9-*cis*-octadecenoate) was found to yield a mixture of hydroperoxides of formula $C_{19}H_{36}O_4$. In these, the —OOH group was found attached not only to C–8 and C–11 but also to C–9 and C–10. What is the probable structure of these last two hydroperoxides? How did they arise? Show all steps in a likely mechanism for the reaction.

4. Although alkaline hydrolysis of monoalkyl or monoaryl phosphates is ordinarily very difficult, 2,4-dinitrophenyl phosphate, $2,4\text{-}(NO_2)_2C_6H_3OPO_3H_2$, does react with aqueous base, and with cleavage at the phosphorus–oxygen bond. Suggest an explanation for this.

5. *Spermaceti* (a wax from the head of the sperm whale) resembles high-molecular-weight hydrocarbons in physical properties and inertness toward Br_2/CCl_4 and $KMnO_4$; on qualitative analysis it gives positive tests only for carbon and hydrogen. However, its infrared spectrum shows the presence of an ester group, and quantitative analysis gives the empirical formula $C_{16}H_{32}O$.

A solution of the wax and KOH in ethanol is refluxed for a long time. Titration of an aliquot shows that one equivalent of base has been consumed for every 475 ± 10 grams of wax. Water and ether are added to the cooled reaction mixture, and the aqueous and ethereal layers are separated. Acidification of the aqueous layer yields a solid A, m.p. 62–63 °C, neutralization equivalent 260 ± 5. Evaporation of the ether layer yields a solid B, m.p. 48–49 °C. (a) What is a likely structure of spermaceti? (b) Reduction by $LiAlH_4$ of either spermaceti or A gives B as the only product. Does this confirm the structure you gave in (a)?

6. As the acidity of the solution is increased, the rate of hydrolysis of monoalkyl phosphates, $ROPO(OH)_2$, rises from essentially zero in alkaline solution, and *passes through a maximum* at the point (moderate acidity, pH about 4) where the concentration of monoanion, $ROPO(OH)(O^-)$, is greatest. Cleavage is at the phosphorus–oxygen bond.

(a) Can you suggest a mechanism or mechanisms that might account for the fact that this species is more reactive than either the dianion, $ROPO(O^-)_2$, or the neutral ester?

(b) At still higher acidity, the rate rises again and continues to rise. To what is the high reactivity now due?

7. On the basis of the following synthesis, give the structure of *vaccenic acid*.

n-hexyl chloride + sodium acetylide $\longrightarrow$ C (C_8H_{14})
C + Na, NH_3; then $I(CH_2)_9Cl$ $\longrightarrow$ D ($C_{17}H_{31}Cl$)
D + KCN $\longrightarrow$ E ($C_{18}H_{31}N$)
E + OH^-, heat; then H^+ $\longrightarrow$ F ($C_{18}H_{32}O_2$)
F + H_2, Pd $\longrightarrow$ vaccenic acid ($C_{18}H_{34}O_2$)

8. From the lipids of *Corynebacterium diphtherium* there is obtained *corynomycolenic acid*. Its structure was confirmed by the following synthesis.

n-$C_{13}H_{27}CH_2Br$ + sodiomalonic ester $\longrightarrow$ G ($C_{21}H_{40}O_4$)
G + exactly 1 mol alc. KOH $\longrightarrow$ H ($C_{19}H_{36}O_4$)
H + dihydropyran (Problem 16, p. 892) $\longrightarrow$ I ($C_{24}H_{44}O_5$)
cis-9-hexadecenoic acid + $SOCl_2$ $\longrightarrow$ J ($C_{16}H_{29}OCl$)
I + Na, then J $\longrightarrow$ K ($C_{40}H_{72}O_6$)
K + dilute acid $\longrightarrow$ L ($C_{34}H_{64}O_3$)
L + $NaBH_4$ $\longrightarrow$ M ($C_{35}H_{66}O_3$)
M + OH^-, heat; then H^+ $\longrightarrow$ ($\pm$)-corynomycolenic acid ($C_{32}H_{62}O_3$)

What is the structure of corynomycolenic acid?

9. From saponification of the fatty capsule of the tubercle bacillus, there is obtained *tuberculostearic acid*. Its structure was established by the following synthesis.

2-decanol + PBr_3 $\longrightarrow$ N ($C_{10}H_{21}Br$)
N + sodiomalonic ester; then OH^-, heat; then H^+; then heat $\longrightarrow$ O ($C_{12}H_{24}O_2$)

O + $SOCl_2$ $\longrightarrow$ P $\xrightarrow{C_2H_5OH}$ Q ($C_{14}H_{28}O_2$)

Q + $LiAlH_4$ $\longrightarrow$ R ($C_{12}H_{26}O$) $\xrightarrow{PBr_3}$ S ($C_{12}H_{25}Br$)

S + Mg; $CdCl_2$; then $C_2H_5OOC(CH_2)_5COCl$ $\longrightarrow$ T ($C_{21}H_{40}O_3$), a ketone
(Compare Sec. 21.6.)
T + Zn, HCl $\longrightarrow$ U ($C_{21}H_{42}O_2$)
U + OH^-, heat; then H^+ $\longrightarrow$ tuberculostearic acid ($C_{19}H_{38}O_2$)

What is the structure of tuberculostearic acid?

10. Besides tuberculostearic acid (preceding problem), the capsule of the tubercle bacillus yields C_{27}-*phthienoic acid*, which on injection into animals causes the lesions typical of tuberculosis. On the basis of the following data, assign a structure to this acid.

C_{27}-phthienoic acid ($C_{27}H_{52}O_2$) + O_3; then Zn, H_2O $\longrightarrow$ $CH_3COCOOH$
C_{27}-phthienoic acid + $KMnO_4$ $\longrightarrow$ acid V ($C_{24}H_{48}O_2$)
methyl ester of V + $2C_6H_5MgBr$; then H_2O $\longrightarrow$ W ($C_{36}H_{58}O$)
W + H^+, heat $\longrightarrow$ X ($C_{36}H_{56}$)
X + CrO_3 $\longrightarrow$ (C_6H_5)$_2CO$ + ketone Y ($C_{23}H_{46}O$)
Y + I_2, NaOH $\longrightarrow$ CHI_3

V + Br_2, P $\longrightarrow$ Z $\xrightarrow{\text{alc. KOH}}$ acid AA ($C_{24}H_{46}O_2$)
AA + $KMnO_4$ $\longrightarrow$ among other things, BB ($C_{20}H_{40}O$)
Compound BB was shown to be identical with a sample of $CH_3(CH_2)_{17}COCH_3$.

Caution: $KMnO_4$ is a vigorous reagent, and not all the cleavage occurs at the double bond. Compare the number of carbons in AA and BB.

11. On the basis of the following NMR spectra, assign likely structures to the isomeric fatty acids, CC and DD, of formula $C_{17}H_{35}COOH$.

Isomer CC   *a* triplet, δ 0.8, 3H
                 b broad band, δ 1.35, 30H
                 c triplet, δ 2.3, 2H
                 d singlet, δ 12.0, 1H

Isomer DD   *a* triplet, δ 0.8, 3H
                 b doublet, δ 1.15, 3H
                 c broad band, δ 1.35, 28H
                 d multiplet, δ 2.2, 1H
                 e singlet, δ 12.05, 1H

12. *Juvenile hormones* take part in the delicate balance of hormonal activity that controls development of insects. Applied artificially, they prevent maturing, and thus offer a highly specific way to control insect population.

The structure of the juvenile hormone of the moth *Hyalophora cecropia* was confirmed by the following synthesis. (At each stage where geometric isomers were obtained, these were separated and the desired one—(Z) or (E)—was selected on the basis of its NMR spectrum.)

2-butanone + $[(CH_3O)_2P(O)CHCOOCH_3]^- Na^+$ (See Sec. 30.2)

$$\longrightarrow [EE (C_9H_{18}O_6PNa)]$$

$[EE] \longrightarrow (CH_3)_2PO_4^- + FF ((Z)\text{-}C_7H_{12}O_2)$ (See Sec 25.10)

$FF + LiAlH_4 \longrightarrow GG (C_6H_{12}O) \xrightarrow{PBr_3} HH (C_6H_{11}Br)$

$HH + [CH_3CH_2COCHCOOC_2H_5]^- Na^+ \longrightarrow II (C_{13}H_{22}O_3)$

$II + OH^-$, heat; then H^+; then heat $\longrightarrow JJ (C_{10}H_{18}O)$

$JJ + [(CH_3O)_2P(O)CHCOOCH_3]^- Na^+ \longrightarrow [KK] \longrightarrow LL ((E)\text{-}C_{13}H_{22}O_2)$

$LL + LiAlH_4 \longrightarrow MM \xrightarrow{PBr_3} NN (C_{12}H_{21}Br)$

$NN + $ sodioacetoacetic ester $\longrightarrow OO (C_{18}H_{30}O_3)$

$OO + OH^-$, heat; then H^+; then heat $\longrightarrow PP (C_{15}H_{26}O)$

$PP + [(CH_3O)_2P(O)CHCOOCH_3]^- Na^+ \longrightarrow [QQ] \longrightarrow RR ((E)\text{-}C_{18}H_{30}O_2)$

$RR + m\text{-}ClC_6H_4CO_2OH \longrightarrow SS (racemic\text{-}C_{18}H_{30}O_3)$

SS was a mixture of positional isomers, corresponding to attack by perbenzoic acid at various double bonds in RR. Of these, one isomer (a *racemic* modification) was found to be identical, in physical and biological properties, to the natural juvenile hormone. This isomer was the one resulting from reaction at the double bond first introduced into the molecule.

What is the structure of the juvenile hormone of *Hyalophora cecropia*? Account for the fact that the synthesis yields a racemic modification.

38

Carbohydrates I. Monosaccharides

38.1 Introduction

In the leaf of a plant, the simple compounds carbon dioxide and water are combined to form the sugar **(+)-glucose**. This process, known as *photosynthesis*, requires catalysis by the green coloring matter *chlorophyll*, and requires energy in the form of light. Thousands of (+)-glucose molecules can then be combined to form the much larger molecules of **cellulose**, which constitutes the supporting framework of the plant. (+)-Glucose molecules can also be combined, in a somewhat different way, to form the large molecules of **starch**, which is then stored in the seeds to serve as food for a new, growing plant.

When eaten by an animal, the starch—and in the case of certain animals also the cellulose—is broken down into the original (+)-glucose units. These can be carried by the bloodstream to the liver to be recombined into **glycogen**, or animal starch; when the need arises, the glycogen can be broken down once more into (+)-glucose. (+)-Glucose is carried by the bloodstream to the tissues, where it is oxidized, ultimately to carbon dioxide and water, with the release of the energy originally supplied as sunlight. Some of the (+)-glucose is converted into fats; some reacts with nitrogen-containing compounds to form amino acids, which in turn are combined to form the proteins that make up a large part of the animal body.

1279

(+)-Glucose, cellulose, starch, and glycogen all belong to the class of organic compounds known as **carbohydrates**. Carbohydrates are the ultimate source of most of our food: we eat starch-containing grain, or feed it to animals to be converted into meat and fat which we then eat. We clothe ourselves with cellulose in the form of cotton and linen, rayon and cellulose acetate. We build houses and furniture from cellulose in the form of wood. Thus carbohydrates quite literally provide us with the necessities of life: food, clothing, and shelter.

Basic necessities aside, our present civilization depends to a surprising degree upon cellulose, particularly as *paper*: the books and newspapers we read, the letters we write, the bills we pay and the money and checks with which we pay them; marriage licenses, drivers' licenses, birth certificates, mortgages; paper in the form of bags and boxes, sheets and rolls.

The study of carbohydrates is one of the most exciting fields of organic chemistry. It extends from the tremendously complicated problem of under-standing the process of photosynthesis to the equally difficult problem of unravel-ing the tangled steps in the enzyme-catalyzed reconversion of (+)-glucose into carbon dioxide and water. Between these two biochemical problems there lie the more traditional problems of the organic chemist: determination of the structure and properties of the carbohydrates, and the study of their conversion into other organic compounds.

In this book we shall learn something of the fundamental chemical properties of the carbohydrates, knowledge that is basic to any further study of these compounds.

38.2   Definition and classification

Carbohydrates are polyhydroxy aldehydes, polyhydroxy ketones, or com-pounds that can be hydrolyzed to them. A carbohydrate that cannot be hydrolyzed to simpler compounds is called a **monosaccharide**. A carbohydrate that can be hydrolyzed to two monosaccharide molecules is called a **disaccharide**. A carbo-hydrate that can be hydrolyzed to many monosaccharide molecules is called a **polysaccharide**.

A monosaccharide may be further classified. If it contains an aldehyde group, it is known as an **aldose**; if it contains a keto group, it is known as a **ketose**. Depending upon the number of carbon atoms it contains, a monosaccharide is known as a **triose, tetrose, pentose, hexose**, and so on. An **aldohexose**, for example, is a six-carbon monosaccharide containing an aldehyde group; a **ketopentose** is a five-carbon monosaccharide containing a keto group. Most naturally occurring monosaccharides are pentoses or hexoses.

Carbohydrates that reduce Fehling's (or Benedict's) or Tollens' reagent (p. 1285) are known as **reducing sugars**. All monosaccharides, whether aldose or ketose, are reducing sugars. Most disaccharides are reducing sugars; sucrose (common table sugar) is a notable exception, for it is a non-reducing sugar.

38.3   (+)-Glucose: an aldohexose

Because it is the unit of which starch, cellulose, and glycogen are made up, and because of its special role in biological processes, (+)-**glucose** is by far the most abundant monosaccharide—there are probably more (+)-glucose units in nature than any other organic group—and by far the most important monosaccharide.

Most of what we need to know about monosaccharides we can learn from the study of just this one compound, and indeed from the study of just one aspect: its structure, and how that structure was arrived at. In learning about the structure of

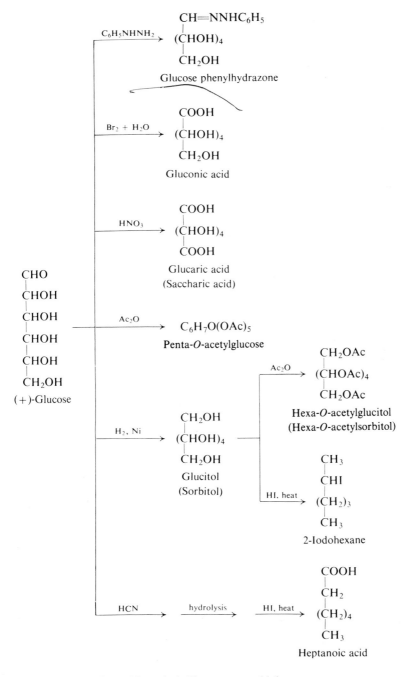

Figure 38.1   (+)-Glucose as an aldohexose.

(+)-glucose, we shall at the same time learn about its properties, since it is from these properties that the structure has been deduced. (+)-Glucose is a typical monosaccharide, so that in learning about its structure and properties, we shall be learning about the structure and properties of the other members of this family.

(+)-Glucose has the molecular formula $C_6H_{12}O_6$, as shown by elemental analysis and molecular weight determination. In Fig. 38.1 is summarized other evidence about its structure: evidence consistent with the idea that (+)-glucose is a six-carbon, straight-chain, pentahydroxy aldehyde, that is, that (+)-glucose is an aldohexose. But this is only the beginning. There are, as we shall see, 16 possible aldohexoses, all stereoisomers of each other, and we want to know which one (+)-glucose is. Beyond this, there is the fact that (+)-glucose exists in *alpha* and *beta* forms, indicating still further stereochemical possibilities that are not accommodated by the simple picture of a pentahydroxy aldehyde. Finally, we must pinpoint the predominant conformation in which the compound exists. All this is the structure of (+)-glucose and, when we have arrived at it, we shall see the features that make it the very special molecule that it is.

Problem 38.1 Assume that you start knowing only the molecular formula of (+)-glucose. You carry out each of the reactions of Fig. 38.1, and study each of the products obtained: characterize the product as to family; determine its molecular weight and, if any, its neutralization equivalent. You identify 2-iodohexane and heptanoic acid by comparison with authentic samples.

(a) For each product, tell what you would actually observe. (b) Take each piece of evidence in turn, and tell what it shows about the structure of (+)-glucose.

38.4 (−)-Fructose: a 2-ketohexose

The most important ketose is **(−)-fructose**, which occurs widely in fruits and, combined with glucose, in the disaccharide *sucrose* (common table sugar).

The following sequence shows that (−)-fructose is a ketone rather than an aldehyde, and gives the position of the keto group in the chain:

CH_2OH		CH_2OH		CH_2OH		CH_3
C=O		C(OH)CN		C(OH)COOH		CHCOOH
CHOH	HCN	CHOH	hydrolysis	CHOH	HI, heat	CH_2
CHOH	→	CHOH	→	CHOH	→	CH_2
CHOH		CHOH		CHOH		CH_2
CH_2OH		CH_2OH		CH_2OH		CH_3
Fructose		Cyanohydrin		Hydroxy acid		α-Methylcaproic acid
		(two diastereomers)		(two diastereomers)		(racemic modification)

Fructose is thus a 2-ketohexose.

38.5 Stereoisomers of (+)-glucose. Nomenclature of aldose derivatives

If we examine the structural formula we have drawn for glucose, we see that it contains four chiral centers (marked by asterisks):

```
1   CHO
    |
2  *CHOH
    |
3  *CHOH
    |
4  *CHOH
    |
5  *CHOH
    |
6   CH₂OH
```

Each of the possible stereoisomers is commonly represented by a cross formula, as, for example, in I. As always in formulas of this kind, it is understood that

```
        CHO                                    CHO
         |                                      ‖
    H ——|—— OH                            H — C — OH
         |                                      |
   HO ——|—— H          stands for        HO — C — H
         |                                      |
    H ——|—— OH                            H — C — OH
         |                                      |
    H ——|—— OH                            H — C — OH
         |                                      ‖
       CH₂OH                                  CH₂OH

         I
```

horizontal lines represent bonds coming *toward us* out of the plane of the paper, and *vertical* lines represent bonds going *away from us* behind the plane of the paper. (See Fig. 38.2, on the next page.)

The dissimilarity of the two ends of an aldohexose molecule prevents the existence of *meso* compounds (Sec. 4.18), and hence we expect that there should be 2^4 or 16 stereoisomers—eight pairs of enantiomers. All 16 of these possible stereoisomers are now known, through either synthesis in the laboratory or isolation from natural sources; only three—(+)-glucose, (+)-mannose, (+)-galactose—are found in abundance.

Problem 38.2 Draw a cross formula of one enantiomer of each of these eight pairs, placing —CHO at the top, —CH₂OH at the bottom, and —OH on the right on the lowest chiral center (C–5).

Of these 16 isomers, only one is the (+)-glucose that we have described as the most abundant monosaccharide. A second isomer is (−)-glucose, the enantiomer of the naturally occurring compound. The other 14 isomers are all diastereomers of (+)-glucose, and are given names of their own, for example, *mannose, galactose, gulose,* etc. As we might expect, these other aldohexoses undergo the same set of reactions that we have described for glucose. Although as diastereomers they undergo these reactions at different rates and yield different individual compounds, the chemistry is essentially the same.

The products obtained from these other aldohexoses are generally given names that correspond to the names of the products obtained from glucose. This principle is illustrated in Table 38.1 for the aldohexose (+)-mannose, which occurs naturally in many plants (the name is derived from the Biblical word *manna*).

(a)

(b)

(c)

Figure 38.2 Meaning of cross formulas of carbohydrates. (For simplicity, red balls are used for hydroxyl groups as well as aldehyde oxygen.)

Only molecular models can show us what is really meant by formulas like I. A correct model of one of these stereoisomers is difficult to build unless we follow certain rules first clearly stated by the great carbohydrate chemist Emil Fischer:

(1) Construct a chain of carbon atoms with a —CHO group at one end, and a —CH₂OH group at the other.

(2) Hold the —CHO group in one hand (a) and let the rest of the chain hang down.

(3) Take the —CH₂OH at the bottom end in the other hand and bring it up *behind* the chain until it touches the —CHO group.

(4) Now one hand can hold both groups firmly (b) and the rest of the chain will form a rather rigid ring projecting *toward you.* (This is the object of the whole operation up to this point: to impart rigidity to an otherwise flexible chain.) By this procedure you have —CHO above —CH₂OH as in formula I, and both these groups directed *away from you.*

(5) Finally, still holding the ring as described above, look in turn at each carbon atom, and attach the —OH or —H to the right or to the left just as it appears in the cross formula (c). In each case, these groups will be directed *toward you.*

Table 38.1   NAMES OF ALDOSE DERIVATIVES

Type of compound	Type name	Examples of specific names	
Monosaccharide $HOCH_2(CHOH)_nCHO$	Aldose	*Glucose*	*Mannose*
Monocarboxylic acid $HOCH_2(CHOH)_nCOOH$	Aldonic acid	*Gluconic acid*	*Mannonic acid*
Dicarboxylic acid $HOOC(CHOH)_nCOOH$	Aldaric acid	*Glucaric acid* (*Saccharic acid*)	*Mannaric acid* (*Mannosaccharic acid*)
Polyhydroxy alcohol $HOCH_2(CHOH)_nCH_2OH$	Alditol	*Glucitol* (*Sorbitol*)	*Mannitol*
Aldehydo acid $HOOC(CHOH)_nCHO$	Uronic acid	*Glucuronic acid*	*Mannuronic acid*

The structural formula we have drawn to represent (+)-glucose so far could actually represent any of the 16 aldohexoses. Only when we have specified the configuration about each of the chiral centers will we have the structural formula that applies only to (+)-glucose itself. Before we can discuss the brilliant way in which the configuration of (+)-glucose was worked out, we must first learn a little more about the chemistry of monosaccharides.

Problem 38.3   (a) How many chiral centers are there in (−)-fructose? (b) How many stereoisomeric 2-ketohexoses should there be? (c) Draw a cross formula of one enantiomer of each pair, placing C=O near the top, and —OH on the right on the lowest chiral center (C–5).

38.6   Oxidation. Effect of alkali

Aldoses can be oxidized in four important ways: (a) by Fehling's or Tollens' reagent; (b) by bromine water; (c) by nitric acid; and (d) by periodic acid, HIO_4.

Aldoses reduce **Tollens' reagent**, as we would expect aldehydes to do. They also reduce **Fehling's solution**, an alkaline solution of cupric ion complexed with tartrate ion (or **Benedict's solution**, in which complexing is with citrate ion); the deep-blue color of the solution is discharged, and red cuprous oxide precipitates. These reactions are less useful, however, than we might at first have expected.

In the first place, they cannot be used to differentiate aldoses from ketoses. Ketoses, too, reduce Fehling's and Tollens' reagents; this behavior is characteristic of α-hydroxy ketones.

In the second place, oxidation by Fehling's or Tollens' reagent cannot be used for the preparation of aldonic acids (monocarboxylic acids) from aldoses. Both Fehling's and Tollens' reagents are alkaline reagents, and the treatment of sugars with alkali can cause extensive isomerization and even decomposition of the chain. Alkali exerts this effect, in part at least, by establishing an equilibrium between the monosaccharide and an enediol structure.

$$
\begin{array}{ccc}
\underset{\text{Aldose}}{
\begin{array}{c}
\text{CHO} \\
\text{H--C--OH} \\
\text{H--C--OH} \\
\end{array}}
&\rightleftharpoons&
\underset{\text{Enediol}}{
\begin{array}{c}
\text{CHOH} \\
\text{C--OH} \\
\text{H--C--OH} \\
\end{array}}
&\rightleftharpoons&
\underset{\text{Aldose}}{
\begin{array}{c}
\text{CHO} \\
\text{HO--C--H} \\
\text{H--C--OH} \\
\end{array}}
\end{array}
$$

$$\Updownarrow$$

$$
\underset{\text{Ketose}}{
\begin{array}{c}
\text{CH}_2\text{OH} \\
\text{C=O} \\
\text{H--C--OH} \\
\end{array}}
\rightleftharpoons
\underset{\text{Enediol}}{
\begin{array}{c}
\text{CH}_2\text{OH} \\
\text{C--OH} \\
\text{C--OH} \\
\end{array}}
\rightleftharpoons etc.
$$

Bromine water oxidizes aldoses, but not ketoses; as an acidic reagent it does not cause isomerization of the molecule. It can therefore be used to differentiate an aldose from a ketose, and is the reagent chosen to synthesize the *aldonic acid* (monocarboxylic acid) from an aldose.

$$
\underset{\text{Aldose}}{
\begin{array}{c}
\text{CHO} \\
(\text{CHOH})_n \\
\text{CH}_2\text{OH}
\end{array}}
\quad
\begin{array}{l}
\xrightarrow{\text{Br}_2\ +\ \text{H}_2\text{O}}
\underset{\text{Aldonic acid}}{
\begin{array}{c}
\text{COOH} \\
(\text{CHOH})_n \\
\text{CH}_2\text{OH}
\end{array}} \\
\\
\xrightarrow{\text{HNO}_3}
\underset{\text{Aldaric acid}}{
\begin{array}{c}
\text{COOH} \\
(\text{CHOH})_n \\
\text{COOH}
\end{array}}
\end{array}
$$

Treatment of an aldose with the more vigorous oxidizing agent **nitric acid** brings about oxidation not only of the —CHO group but also of the —CH$_2$OH group, and leads to the formation of the *aldaric acid* (dicarboxylic acid).

Like other compounds that contain two or more —OH or =O groups on *adjacent* carbon atoms, carbohydrates undergo oxidative cleavage by **periodic acid**, HIO$_4$ (Sec. 18.10). This reaction, introduced in 1928 by L. Malaprade (at the University of Nancy, France), is one of the most useful tools in modern research on carbohydrate structure.

Problem 38.4 Treatment of (+)-glucose with HIO$_4$ gives results that confirm its aldohexose structure. What products should be formed, and how much HIO$_4$ should be consumed?

Problem 38.5 Identify each of the following glucose derivatives:

$$
\begin{array}{l}
\text{A} + 4\text{HIO}_4 \longrightarrow 3\text{HCOOH} + \text{HCHO} + \text{OHC--COOH} \\
\text{B} + 5\text{HIO}_4 \longrightarrow 4\text{HCOOH} + 2\text{HCHO} \\
\text{C} + 3\text{HIO}_4 \longrightarrow 2\text{HCOOH} + 2\text{OHC--COOH} \\
\text{D} + 4\text{HIO}_4 \longrightarrow 4\text{HCOOH} + \text{OHC--COOH}
\end{array}
$$

38.7   Osazone formation. Epimers

As aldehydes, aldoses react with phenylhydrazine to form phenylhydrazones. If an excess of phenylhydrazine is used, the reaction proceeds further to yield products known as **osazones**, which contain two phenylhydrazine residues per molecule; a third molecule of the reagent is turned into aniline and ammonia. (Just how the —OH group is oxidized is not quite clear.)

$$
\begin{array}{c}
\text{CHO} \\
|\\
\text{CHOH} \\
\vdots
\end{array}
\quad\xrightarrow{\;3C_6H_5NHNH_2\;}\quad
\begin{array}{c}
\text{CH}{=}\text{NNHC}_6\text{H}_5 \\
|\\
\text{C}{=}\text{NNHC}_6\text{H}_5 \\
\vdots
\end{array}
+\ C_6H_5NH_2\ +\ NH_3
$$

Aldose   Osazone

Osazone formation is not limited to carbohydrates, but is typical of α-hydroxy aldehydes and α-hydroxy ketones in general (e.g., *benzoin*, $C_6H_5CHOHCOC_6H_5$).

Removal of the phenylhydrazine groups yields dicarbonyl compounds known as **osones**. For example:

$$
\begin{array}{c}
\text{CH}{=}\text{NNHC}_6\text{H}_5 \\
|\\
\text{C}{=}\text{NNHC}_6\text{H}_5 \\
\vdots
\end{array}
\quad\xrightarrow{\;C_6H_5CHO,\ H^+\;}\quad
\begin{array}{c}
\text{CHO} \\
|\\
\text{C}{=}\text{O} \\
\vdots
\end{array}
\ +\ \ 2C_6H_5CH{=}\text{NNHC}_6\text{H}_5
$$

Osazone   Osone   Benzaldehyde phenylhydrazone

Problem 38.6   Aldehydes are more easily reduced than ketones. On this basis what product would you expect from the reduction of glucosone by zinc and acetic acid? Outline a sequence of reactions by which an aldose can be turned into a 2-ketose.

In 1858 Peter Griess (in time taken from his duties in an English brewery) discovered diazonium salts. In 1875 Emil Fischer (at the University of Munich) found that reduction of benzenediazonium chloride by sulfur dioxide yields phenylhydrazine. Nine years later, in 1884, Fischer reported that the phenyl-hydrazine he had discovered could be used as a powerful tool in the study of carbohydrates.

One of the difficulties of working with carbohydrates is their tendency to form sirups; these are fine for pouring on pancakes at breakfast, but hard to work with in the laboratory. Treatment with phenylhydrazine converts carbohydrates into solid osazones, which are readily isolated and purified, and can be identified by their characteristic crystalline forms.

Fischer found osazone formation to be useful not only in identifying carbo-hydrates, but also—and this was much more important—in determining their configurations. For example, the two diastereomeric aldohexoses (+)-glucose and (+)-mannose yield the same osazone. Osazone formation destroys the configuration about C–2 of an aldose, but does not affect the configuration of the rest of the molecule. It therefore follows that (+)-glucose and (+)-mannose differ only in configuration about C–2, and have the same configuration about C–3, C–4, and C–5. We can see that whenever the configuration of either of these compounds is established, the configuration of the other is immediately known through this osazone relationship.

$$
\begin{array}{ccc}
\begin{array}{l}
1 \quad \text{CHO} \\
2 \quad \text{H}\!-\!\text{OH} \\
3 \\
4 \\
5 \\
6 \quad \text{CH}_2\text{OH}
\end{array}
&
\xrightarrow{\;3\text{C}_6\text{H}_5\text{NHNH}_2\;}
\quad
\begin{array}{l}
1 \quad \text{HC}\!=\!\text{NNHC}_6\text{H}_5 \\
2 \quad \text{C}\!=\!\text{NNHC}_6\text{H}_5 \\
3 \\
4 \\
5 \\
6 \quad \text{CH}_2\text{OH}
\end{array}
\quad
\xleftarrow{\;3\text{C}_6\text{H}_5\text{NHNH}_2\;}
&
\begin{array}{l}
1 \quad \text{CHO} \\
2 \quad \text{HO}\!-\!\text{H} \\
3 \\
4 \\
5 \\
6 \quad \text{CH}_2\text{OH}
\end{array}
\end{array}
$$

Epimers give the same osazone

A pair of diastereomeric aldoses that differ only in configuration about C–2 are called **epimers**. One way in which a pair of aldoses can be identified as epimers is through the formation of the same osazone.

> **Problem 38.7**   When the ketohexose (–)-fructose is treated with phenylhydrazine, it yields an osazone that is identical with the one prepared from either (+)-glucose or (+)-mannose. How is the configuration of (–)-fructose related to those of (+)-glucose and (+)-mannose?

38.8   Lengthening the carbon chain of aldoses. The Kiliani–Fischer synthesis

In the next few sections we shall examine some of the ways in which an aldose can be converted into a different aldose. These conversions can be used not only to synthesize new carbohydrates, but also, as we shall see, to help determine their configurations.

First, let us look at a method for converting an aldose into another aldose containing one more carbon atom, that is, at a method for lengthening the carbon chain. In 1886, Heinrich Kiliani (at the Technische Hochschule in Munich) showed that an aldose can be converted into two aldonic acids of the next higher carbon number by addition of HCN and hydrolysis of the resulting cyanohydrins. In 1890, Fischer reported that reduction of an aldonic acid (in the form of its lactone, Sec. 24.15) can be controlled to yield the corresponding aldose. In Fig. 38.3, the entire **Kiliani–Fischer synthesis** is illustrated for the conversion of an aldopentose into two aldohexoses.

Addition of cyanide to the aldopentose generates a new chiral center, about which there are two possible configurations. As a result, two diastereomeric cyanohydrins are obtained, which yield diastereomeric carboxylic acids (aldonic acids) and finally diastereomeric aldoses.

The situation is strictly analogous to that in Sec. 4.26. Using models, we can see that the particular configuration obtained here depends upon which face of the carbonyl group is attacked by cyanide ion. Since the aldehyde is already chiral, attack at the two faces is not equally likely. Both possible diastereomeric products are formed, and in unequal amounts.

Since a six-carbon aldonic acid contains —OH groups in the γ- and δ-positions, we would expect it to form a lactone under acidic conditions (Sec. 24.15). This occurs, the γ-lactone generally being the more stable product. It is the lactone that is actually reduced to an aldose in the last step of a Kiliani–Fischer synthesis.

The pair of aldoses obtained from the sequence differ only in configuration

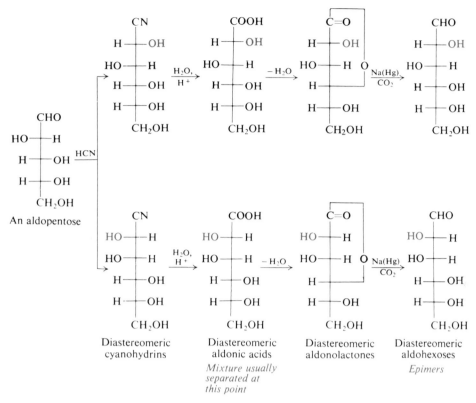

Figure 38.3   An example of the Kiliani–Fischer synthesis.

about C–2, and hence are epimers. A pair of aldoses can be recognized as epimers not only by their conversion into the same osazone (Sec. 38.7), but also by their formation in the same Kiliani–Fischer synthesis.

Like other diastereomers, these epimers differ in physical properties and therefore are separable. However, since carbohydrates are difficult to purify, it is usually more convenient to separate the diastereomeric products at the acid stage, where crystalline salts are easily formed, so that a single pure lactone can be reduced to a single pure aldose.

Problem 38.8   As reducing agent, Fischer used sodium amalgam and acid. Today, lactones are reduced to aldoses by the addition of $NaBH_4$ to an aqueous solution of lactone. If, however, lactone is added to the $NaBH_4$, another product, not the aldose, is obtained. What do you think this other product is? Why is the order of mixing of reagents crucial?

Problem 38.9   (a) Using cross formulas to show configuration, outline all steps in a Kiliani–Fischer synthesis, starting with the aldotriose R-(+)-glyceraldehyde, $CH_2OHCHOHCHO$. How many aldotetroses would be expected? (b) Give configurations of the aldopentoses expected from each of these aldotetroses by a Kiliani–Fischer synthesis; of the aldohexoses expected from each of these aldopentoses.

(c) Make a "family tree" showing configurations of these aldoses hypothetically descended from R-(+)-glyceraldehyde. If the —CHO is placed at the top in each case, what configurational feature is the same in all these formulas? Why?

Problem 38.10 (a) Give the configuration of the dicarboxylic acid (aldaric acid) that would be obtained from each of the tetroses in Problem 38.9 by nitric acid oxidation. (b) Assume that you have actually carried out the chemistry in part (a). In what simple way could you assign configuration to each of your tetroses?

38.9 Shortening the carbon chain of aldoses. The Ruff degradation

There are a number of ways in which an aldose can be converted into another aldose of one less carbon atom. One of these methods for shortening the carbon chain is the **Ruff degradation**. An aldose is oxidized by bromine water to the aldonic acid; oxidation of the calcium salt of this acid by hydrogen peroxide in the presence of ferric salts yields carbonate ion and an aldose of one less carbon atom (see Fig. 38.4).

$$
\begin{array}{ccc}
\text{CHO} & \text{COOH} & (\text{COO}^-)_2\text{Ca}^{2+} \\
\text{H}-\text{OH} & \text{H}-\text{OH} & \text{H}-\text{OH} \\
\text{HO}-\text{H} & \text{HO}-\text{H} & \text{HO}-\text{H} \\
\text{H}-\text{OH} & \text{H}-\text{OH} & \text{H}-\text{OH} \\
\text{H}-\text{OH} & \text{H}-\text{OH} & \text{H}-\text{OH} \\
\text{CH}_2\text{OH} & \text{CH}_2\text{OH} & \text{CH}_2\text{OH}
\end{array}
$$

An aldohexose $\xrightarrow{Br_2 + H_2O}$ An aldonic acid $\xrightarrow{CaCO_3}$ A calcium aldonate $\xrightarrow{H_2O_2,\ Fe^{3+}}$

$$
\begin{array}{c}
\text{CHO} \\
\text{HO}-\text{H} + \text{CO}_3^{2-} \\
\text{H}-\text{OH} \\
\text{H}-\text{OH} \\
\text{CH}_2\text{OH}
\end{array}
$$

An aldopentose

Figure 38.4 An example of the Ruff degradation.

38.10 Conversion of an aldose into its epimer

In the presence of a tertiary amine, in particular pyridine (Sec. 35.6), an equilibrium is established between an aldonic acid and its epimer. This reaction is the basis of the best method for converting an aldose into its epimer, since the only configuration affected is that at C–2. The aldose is oxidized by bromine water to the aldonic acid, which is then treated with pyridine. From the equilibrium mixture thus formed, the epimeric aldonic acid is separated, and reduced (in the form of its lactone) to the epimeric aldose. See, for example, Fig. 38.5.

CHO   COOH   COOH

H——OH   H——OH   HO——H

HO——H   HO——H   HO——H

H——OH   H——OH   H——OH

H——OH   H——OH   H——OH

CH₂OH   CH₂OH   CH₂OH

Reagents: Br₂, H₂O ; pyridine ; H₂O

An aldohexose   Epimeric aldonic acids

An aldonolactone   Na(Hg), CO₂   Epimeric aldohexose

Figure 38.5   Conversion of an aldose into its epimer.

38.11 Configuration of (+)-glucose. The Fischer proof

Let us turn back to the year 1888. Only a few monosaccharides were known, among them (+)-glucose, (−)-fructose, (+)-arabinose. (+)-Mannose had just been synthesized. It was known that (+)-glucose was an aldohexose and that (+)-arabinose was an aldopentose. Emil Fischer had discovered (1884) that phenylhydrazine could convert carbohydrates into osazones. The Kiliani cyanohydrin method for lengthening the chain was just two years old.

It was known that aldoses could be reduced to alditols, and could be oxidized to the monocarboxylic aldonic acids and to the dicarboxylic aldaric acids. A theory of stereoisomerism and optical activity had been proposed (1874) by van't Hoff and Le Bel. Methods for separating stereoisomers were known and optical activity could be measured. The concepts of racemic modifications, *meso* compounds, and epimers were well established.

(+)-Glucose was known to be an aldohexose; but as an aldohexose it could have any one of 16 possible configurations. The question was: *which* configuration did it have? In 1888, Emil Fischer (at the University of Würzburg) set out to find the answer to that question, and in 1891 announced the completion of a most remarkable piece of chemical research, for which he received the Nobel Prize in 1902. Let us follow Fischer's steps to the configuration of (+)-glucose. Although somewhat modified, the following arguments are essentially those of Fischer.

The 16 possible configurations consist of eight pairs of enantiomers. Since methods of determining absolute configuration were not then available, Fischer realized that he could at best limit the configuration of (+)-glucose to a pair of enantiomeric configurations; he would not be able to tell which one of the pair was the correct absolute configuration.

To simplify the problem, Fischer therefore rejected eight of the possible configurations, arbitrarily retaining only those (I–VIII) in which C–5 carried the —OH on the right (with the understanding that —H and —OH project toward the observer). He realized that any argument that led to the selection of one of these formulas applied with equal force to the mirror image of that formula. (As it turned out, his arbitrary choice of an —OH on the right of C–5 in (+)-glucose was the correct one.)

	I	II	III	IV	
1	CHO	CHO	CHO	CHO	1
2	H—OH	HO—H	H—OH	HO—H	2
3	H—OH	H—OH	HO—H	HO—H	3
4	H—OH	H—OH	H—OH	H—OH	4
5	H—OH	H—OH	H—OH	H—OH	5
6	CH_2OH	CH_2OH	CH_2OH	CH_2OH	6

	V	VI	VII	VIII	
1	CHO	CHO	CHO	CHO	1
2	H—OH	HO—H	H—OH	HO—H	2
3	H—OH	H—OH	HO—H	HO—H	3
4	HO—H	HO—H	HO—H	HO—H	4
5	H—OH	H—OH	H—OH	H—OH	5
6	CH_2OH	CH_2OH	CH_2OH	CH_2OH	6

Since his proof depended in part on the relationship between (+)-glucose and the aldopentose (−)-arabinose, Fischer also had to consider the configurations of the five-carbon aldoses. Of the eight possible configurations, he retained only four, IX–XII, again those on which the bottom chiral center carried the —OH on the right.

IX	X	XI	XII
CHO	CHO	CHO	CHO
H—OH	HO—H	H—OH	HO—H
H—OH	H—OH	HO—H	HO—H
H—OH	H—OH	H—OH	H—OH
CH_2OH	CH_2OH	CH_2OH	CH_2OH
IX	X	XI	XII

The line of argument is as follows:

(1) Upon oxidation by nitric acid, (−)-arabinose yields an optically active dicarboxylic acid. Since the —OH on the lowest chiral center is arbitrarily placed on the right, this fact means that the —OH on the uppermost chiral center is on the left (as in X or XII),

for if it were on the right (as in IX or XI), the diacid would necessarily be an inactive *meso* acid.

(2) (−)-Arabinose is converted by the Kiliani–Fischer synthesis into (+)-glucose and (+)-mannose. (+)-Glucose and (+)-mannose therefore are epimers, differing only in configuration about C-2, and have the same configuration about C–3, C–4, and C–5 as does (−)-arabinose. (+)-Glucose and (+)-mannose must be III and IV, or VII and VIII.

(3) Upon oxidation by nitric acid, both (+)-glucose and (+)-mannose yield dicarboxylic acids that are optically active. This means that the —OH on C–4 is on the right, as in III and IV,

for if it were on the left, as in VII and VIII, *one* of the aldaric acids would necessarily be an inactive *meso* acid.

(−)-Arabinose must also have that same —OH on the right, and hence has configuration X.

(+)-Glucose and (+)-mannose have configurations III and IV, but one question remains: which compound has which configuration? One more step is needed—the most elegant step in this elegant sequence.

(4) Oxidation of another hexose, (+)-gulose, yields the same dicarboxylic acid, (+)-glucaric acid, as does oxidation of (+)-glucose. (The gulose was synthesized for this purpose by Fischer.) If we examine the two possible configurations for (+)-glucaric acid, IIIa and IVa, we see that only IIIa can be derived from two different hexoses: from III and the enantiomer of V.

```
  1        CHO                          COOH                        CH₂OH   6
  2     H ──── OH                   H ──── OH                     H ──── OH   5
  3    HO ──── H                   HO ──── H                     HO ──── H    4
              HNO₃                          HNO₃
  4     H ──── OH                   H ──── OH                     H ──── OH   3
  5     H ──── OH                   H ──── OH                     H ──── OH   2
  6        CH₂OH                       COOH                          CHO     1
           III                          IIIa               Enantiomer of V
```

The acid IVa can be derived from just one hexose: from IV.

```
  1        CHO                          COOH                        CH₂OH   6
  2    HO ──── H                   HO ──── H                     HO ──── H    5
  3    HO ──── H                   HO ──── H                     HO ──── H    4
              HNO₃                          HNO₃
  4     H ──── OH                   H ──── OH                     H ──── OH   3
  5     H ──── OH                   H ──── OH                     H ──── OH   2
  6        CH₂OH                       COOH                          CHO     1
           IV                           IVa              IV (rotated 180°)
```

It follows that (+)-glucaric acid has configuration IIIa, and therefore that (+)-glucose has configuration III.

```
  1        CHO
  2     H ──── OH
  3    HO ──── H
  4     H ──── OH
  5     H ──── OH
  6        CH₂OH
           III
       (+)-Glucose
```

(+)-Mannose, of course, has configuration IV, and (−)-gulose (the enantiomer of the one used by Fischer) has configuration V.

```
        CHO                          CHO
   HO ——— H                   H ——— OH
   HO ——— H                   H ——— OH
   H ——— OH                  HO ——— H
   H ——— OH                   H ——— OH
       CH₂OH                       CH₂OH
         IV                          V
    (+)-Mannose                 (−)-Gulose
```

38.12   Configurations of aldoses

Today all possible aldoses (and ketoses) of six carbons or fewer, and many of more than six carbons, are known; most of these do not occur naturally and have been synthesized. The configurations of all these have been determined by application of the same principles that Fischer used to establish the configuration of (+)-glucose; indeed, 12 of the 16 aldohexoses were worked out by Fischer and his students.

So far in our discussion, we have seen how configurations III, IV, V, and X of the previous section were assigned to (+)-glucose, (+)-mannose, (−)-gulose, and (−)-arabinose, respectively. Let us see how configurations have been assigned to some other monosaccharides.

The aldopentose **(−)-ribose** forms the same osazone as (−)-arabinose. Since (−)-arabinose was shown to have configuration X, (−)-ribose must have configuration IX. This configuration is confirmed by the reduction of (−)-ribose to the optically inactive (*meso*) pentahydroxy compound *ribitol*.

```
      CHO                      HC=NNHC₆H₅                   CHO
  HO ——— H                   C=NNHC₆H₅                 H ——— OH
                    C₆H₅NHNH₂                    C₆H₅NHNH₂
  H ——— OH      ————————>   H ——— OH      <————————    H ——— OH
  H ——— OH                   H ——— OH                  H ——— OH
     CH₂OH                      CH₂OH                     CH₂OH
       X                                                    IX
 (−)-Arabinose                 Osazone                  (−)-Ribose
                                                          │ H₂, Ni
                                                          ↓
                                                         CH₂OH
                                                     H ——— OH
                                                     H ——— OH
                                                     H ——— OH
                                                         CH₂OH
                                                         Ribitol
                                                    A meso compound
                                                       Inactive
```

The two remaining aldopentoses, **(+)-xylose** and **(−)-lyxose**, must have the configurations XI and XII. Oxidation by nitric acid converts (+)-xylose into an

```
   CHO              COOH              CHO              COOH
H——OH            H——OH           HO——H            HO——H
HO——H    HNO₃→   HO——H           HO——H   would give HO——H
H——OH            H——OH           H——OH            H——OH
  CH₂OH            COOH            CH₂OH              COOH
   XI          Xylaric acid         XII
(+)-Xylose    A meso compound   (−)-Lyxose          Active
                 Inactive
```

optically inactive (*meso*) aldaric acid. (+)-Xylose must therefore be XI, and (−)-lyxose must be XII.

Degradation of (−)-arabinose yields the tetrose **(−)-erythrose**, which therefore has configuration XIII. In agreement with this configuration, (−)-erythrose is found to yield *meso*tartaric acid upon oxidation by nitric acid.

```
   CHO
HO——H                          CHO              COOH
H——OH    Ruff degradation→  H——OH   HNO₃→   H——OH
H——OH                         H——OH            H——OH
  CH₂OH                         CH₂OH            COOH
(−)-Arabinose                   XIII         Mesotartaric acid
                             (−)-Erythrose      Inactive
```

Degradation of (+)-xylose by the Ruff method yields the tetrose **(−)-threose**, which must therefore have configuration XIV. This is confirmed by oxidation of (−)-threose to optically active (−)-tartaric acid.

```
   CHO
H——OH                          CHO              COOH
HO——H    Ruff degradation→  HO——H   HNO₃→   HO——H
H——OH                         H——OH            H——OH
  CH₂OH                         CH₂OH            COOH
                                 XIV
(+)-Xylose                   (−)-Threose    (−)-Tartaric acid
                                                 Active
```

Problem 38.11   Assign a name to I, II, VI, VII, and VIII (p. 1292) on the basis of the following evidence and the configurations already assigned:

(a) The aldohexoses **(+)-galactose** and **(+)-talose** yield the same osazone. Degradation of (+)-galactose yields (−)-lyxose. Oxidation of (+)-galactose by nitric acid yields an inactive *meso* acid, *galactaric acid* (also called *mucic acid*).

(b) (−)-Ribose is converted by the Kiliani–Fischer synthesis into the two aldo-hexoses (+)-**allose** and (+)-**altrose**. Oxidation of (+)-altrose yields optically active (+)-*altraric acid*. Reduction of (+)-allose to a hexahydroxy alcohol yields optically inactive *allitol*.

(c) The aldohexose (−)-**idose** yields the same osazone as (−)-gulose.

Problem 38.12   Go back to the "family tree" you constructed in Problem 38.9, p. 1289, and assign names to all structures.

Problem 38.13   What is the configuration of the 2-ketohexose (−)-fructose? (See Problem 38.7, p. 1288.)

Problem 38.14   Give the configurations of (−)-glucose, (−)-mannose, and (+)-fructose.

38.13   Optical families. D and L

Before we can explore further the structure of (+)-glucose and its relatives, we must examine a topic of stereochemistry we have not yet touched on: use of the prefixes D and L.

Most applications of stereochemistry, as we have already seen, are based upon the *relative* configurations of different compounds, not upon their absolute config-urations. We are chiefly interested in whether the configurations of a reactant and its product are the same or different, not in what either configuration actually is. In the days before any absolute configurations had been determined, there was the problem not only of determining the relative configurations of various optically active compounds, but also of indicating these relationships once they had been established. This was a particularly pressing problem with the carbohydrates.

The compound **glyceraldehyde**, CH$_2$OHCHOHCHO, was selected as a stan-dard of reference, because it is the simplest carbohydrate—an aldotriose—capable of optical isomerism. Its configuration could be related to those of the carbo-hydrates, and because of its highly reactive functional groups, it could be converted into, and thus related to, many other kinds of organic compounds. (+)-Glycer-aldehyde was arbitrarily assigned configuration I, and was designated D-glycer-aldehyde; (−)-glyceraldehyde was assigned configuration II and was designated L-glyceraldehyde. Configurations were assigned to the glyceraldehydes purely for

CHO   CHO

H — C — OH   HO — C — H

CH$_2$OH   CH$_2$OH

I   II

D-Glyceraldehyde   L-Glyceraldehyde

convenience; the particular assignment had a 50:50 chance of being correct, and, as it has turned out, the configuration chosen actually is the correct abso-lute configuration.

Other compounds could be related configurationally to one or the other of the glyceraldehydes by means of reactions that did not involve breaking bonds to a chiral center (Sec. 4.24). On the basis of the *assumed* configuration of the glyceraldehyde, these related compounds could be assigned configurations, too. As it has turned out, these configurations are the correct absolute ones; in any case, for many years they served as a convenient way of indicating structural relationships. See, for example, Fig. 38.6.

Figure 38.6   Relating configurations to glyceraldehyde.

To indicate the relationship thus established, compounds related to D-glyceraldehyde are given the designation D, and compounds related to L-glyceraldehyde are given the designation L. The symbols D and L (pronounced "dee" and "ell") thus refer to configuration, not to sign of rotation, so that we have, for example, D-(−)-glyceric acid and L-(+)-lactic acid. (One sometimes encounters the prefixes *d* and *l*, pronounced "dextro" and "levo", but their meaning is not always clear. Today they usually refer to direction of rotation; in some of the older literature they refer to optical family. It was because of this confusion that D and L were introduced.)

Unfortunately, the use of the designations D and L is not unambiguous. In relating glyceraldehyde to lactic acid, for example, we might envision carrying out a sequence of steps in which the —CH₂OH rather than the —CHO group is converted into the —COOH group:

By this series of reactions, (+)-glyceraldehyde would yield (+)-lactic acid; by the previous sequence, (+)-glyceraldehyde yields (−)-lactic acid. It would appear that, depending upon the particular sequence used, we could designate either of the lactic acids as D-lactic acid; the first sequence is the more direct, and by convention is the accepted one. We should notice that, whatever the ambiguity associated with the use of D and L, there is no ambiguity about the configurational relationship; we arrive at the proper configurations for (+)- and (−)-lactic acids whichever route we use.

The prefixes *R* and *S* enable us to specify unambiguously the absolute configuration of a compound, because their use does not depend on a relationship to any other compound. But, by the same token, the letters *R* and *S* do not immediately reveal configurational relationships between two compounds; we have to work out and compare the configurations in each case.

The designations D and L, on the other hand, tell us nothing of the configuration of the compound unless we know the route by which the configurational relationship has been established. However, in the case of the carbohydrates (and the amino acids, Chapter 40), there are certain conventions about this which make these designations extremely useful.

Problem 38.15 Which specification, *R* or *S*, would you give to the following? (a) D-(+)-glyceraldehyde; (b) D-(−)-glyceric acid; (c) D-(−)-3-bromo-2-hydroxy-propionic acid; (d) D-(−)-lactic acid.

Problem 38.16 The transformation of L-(+)-lactic acid into (+)-2-butanol was accomplished by the following sequence of reactions:

L-(+)-lactic acid $\xrightarrow{\text{C}_2\text{H}_5\text{OH, H}_2\text{SO}_4}$ A $\xrightarrow{\text{Na, C}_2\text{H}_5\text{OH}}$ B $\xrightarrow{\text{HBr}}$ C

C $\xrightarrow{\text{KCN}}$ D $\xrightarrow{\text{H}_2\text{O, HCl, heat}}$ E $\xrightarrow{\text{CH}_3\text{OH, HCl}}$ F $\xrightarrow{\text{Na,CH}_3\text{COOH}}$ G

G $\xrightarrow{\text{HI}}$ H $\xrightarrow{\text{H}_2,\ \text{Pd}}$ (+)-2-butanol

What is the absolute configuration of (+)-2-butanol?

38.14 Tartaric acid

Tartaric acid, HOOCCHOHCHOHCOOH, has played a key role in the development of stereochemistry, and particularly the stereochemistry of the carbohydrates. In 1848 Louis Pasteur, using a hand lens and a pair of tweezers, laboriously separated a quantity of the sodium ammonium salt of racemic tartaric acid into two piles of mirror-image crystals and, in thus carrying out the first resolution of a racemic modification, was led to the discovery of enantiomerism. Almost exactly 100 years later, in 1951, Bijvoet, using x-ray diffraction—and also laboriously—determined the actual arrangement in space of the atoms of the sodium rubidium salt of (+)-tartaric acid, and thus made the first determination of the absolute configuration of an optically active substance.

(+)-Tartaric acid

As we shall see in the next section, tartaric acid is the stereochemical link between the carbohydrates and our standard of reference, glyceraldehyde. In 1917, the configurational relationship between glyceraldehyde and tartaric acid was worked out. When the reaction sequence outlined in Fig. 38.7 was carried out starting with D-glyceraldehyde, two products were obtained, one inactive and one which rotated the plane of polarized light to the left. The inactive product was, of course, mesotartaric acid, III. The active (−)-tartaric acid thus obtained was assigned configuration IV; since it is related to D-glyceraldehyde, we designate it D-(−)-tartaric acid.

$$
\begin{array}{ccc}
\begin{array}{c}
\text{CN} \\
\text{H}\!-\!\!-\text{OH} \\
\text{H}\!-\!\!-\text{OH} \\
\text{CH}_2\text{OH}
\end{array}
&
\xrightarrow{\text{Ba(OH)}_2}
\begin{array}{c}
\text{COOH} \\
\text{H}\!-\!\!-\text{OH} \\
\text{H}\!-\!\!-\text{OH} \\
\text{CH}_2\text{OH}
\end{array}
&
\xrightarrow{\text{HNO}_3}
\begin{array}{c}
\text{COOH} \\
\text{H}\!-\!\!-\text{OH} \\
\text{H}\!-\!\!-\text{OH} \\
\text{COOH} \\
\text{III} \\
meso\text{-Tartaric acid}
\end{array}
\end{array}
$$

$$
\begin{array}{c}
\text{CHO} \\
\text{H}\!-\!\!-\text{OH} \\
\text{CH}_2\text{OH}
\end{array}
\xrightarrow{\text{HCN}}
$$

D-(+)-Glyceraldehyde

$$
\begin{array}{ccc}
\begin{array}{c}
\text{CN} \\
\text{HO}\!-\!\!-\text{H} \\
\text{H}\!-\!\!-\text{OH} \\
\text{CH}_2\text{OH}
\end{array}
&
\xrightarrow{\text{Ba(OH)}_2}
\begin{array}{c}
\text{COOH} \\
\text{HO}\!-\!\!-\text{H} \\
\text{H}\!-\!\!-\text{OH} \\
\text{CH}_2\text{OH}
\end{array}
&
\xrightarrow{\text{HNO}_3}
\begin{array}{c}
\text{COOH} \\
\text{HO}\!-\!\!-\text{H} \\
\text{H}\!-\!\!-\text{OH} \\
\text{COOH} \\
\text{IV} \\
\text{D-(−)-Tartaric acid}
\end{array}
\end{array}
$$

Figure 38.7 Configurational relationship between glyceraldehyde and tartaric acid.

On the basis of the assumed configuration of D-(+)-glyceraldehyde, then, L-(+)-tartaric acid, the enantiomer of D-(−)-tartaric acid, would have configuration V, the mirror image of IV. When Bijvoet determined the absolute configura-

$$
\begin{array}{cc}
\begin{array}{c}
\text{COOH} \\
\text{HO}-\text{C}-\text{H} \\
\text{H}-\text{C}-\text{OH} \\
\text{COOH}
\end{array}
&
\begin{array}{c}
\text{COOH} \\
\text{H}-\text{C}-\text{OH} \\
\text{HO}-\text{C}-\text{H} \\
\text{COOH}
\end{array}
\end{array}
$$

$$
\begin{array}{cc}
\begin{array}{c}
\text{COOH} \\
\text{HO}\!-\!\!-\text{H} \\
\text{H}\!-\!\!-\text{OH} \\
\text{COOH} \\
\text{IV} \\
\text{D-(−)-Tartaric acid}
\end{array}
&
\begin{array}{c}
\text{COOH} \\
\text{H}\!-\!\!-\text{OH} \\
\text{HO}\!-\!\!-\text{H} \\
\text{COOH} \\
\text{V} \\
\text{L-(+)-Tartaric acid}
\end{array}
\end{array}
$$

tion of (+)-tartaric acid, he found that it actually has the configuration that had been previously assumed. The assumed configurations of the glyceraldehydes, and hence the assumed configurations of all compounds related to them, were indeed the correct ones.

The designation of even the tartaric acids is subject to ambiguity. In this book, we have treated the tartaric acids as one does carbohydrates: by considering —CHO of glyceraldehyde as the position from which the chain is lengthened, via the cyanohydrin reaction. Some chemists, on the other hand, view the tartaric acids as one does the amino acids (Sec. 40.5) and, considering —COOH to be derived from —CHO of glyceraldehyde, designate (−)-tartaric acid as L, and (+)-tartaric acid as D.

Regardless of which convention one follows, this fact remains: (−)- and (+)-tartaric acid—and (+)- and (−)-glyceraldehyde—have the absolute configurations shown above and on p. 1298.

Problem 38.17 Give the specification by the R/S system of: (a) $(-)$-tartaric acid; (b) $(+)$-tartaric acid; (c) mesotartaric acid.

Problem 38.18 (a) From the sequence of Fig. 38.7 the ratio of products III:IV is about 1:3. Why would you have expected to obtain III and IV in unequal amounts?

(b) Outline the same sequence starting from L-$(-)$-glyceraldehyde. Label each product with its name, showing its rotation and D/L designation. In what ratio will these products be obtained?

(c) Outline the same sequence starting from racemic $(\pm)$-glyceraldehyde. How do you account for the fact that only inactive material is obtained in spite of the unequal amounts of diastereomeric products formed from each of the enantiomeric glyceraldehydes?

38.15 Families of aldoses. Absolute configuration

The evidence on which Fischer assigned a configuration to $(+)$-glucose leads to either of the enantiomeric structures I and II. Fischer, we have seen, arbitrarily selected I, in which the lowest chiral center carries —OH on the right.

I
D-$(+)$-Glucose

II
L-$(+)$-Glucose

We recognize I as the enantiomer that would hypothetically be derived from D-$(+)$-glyceraldehyde by a series of Kiliani–Fischer syntheses, the chiral center of $(+)$-glyceraldehyde being retained as the *lowest* chiral center of the aldoses derived from it. (See Problem 38.9, p. 1289.) That $(+)$-glucose is related to D-$(+)$-glyceraldehyde has been established by a number of reaction sequences, one of which is shown in Fig. 38.8. On this basis, then, structure I becomes D-$(+)$-glucose, and structure II becomes L-$(-)$-glucose.

In 1906 the American chemist M. A. Rosanoff (then an instructor at New York University) proposed glyceraldehyde as the standard to which the configurations of carbohydrates should be related. Eleven years later experiment showed that it is the *dextrorotatory* $(+)$-glyceraldehyde that is related to $(+)$-glucose. On that basis, $(+)$-glyceraldehyde was then given the designation D and was assigned a configuration to conform with the one arbitrarily assigned to $(+)$-glucose by Fischer. Although rejected by Fischer, the Rosanoff convention became universally accepted.

Regardless of the direction in which they rotate polarized light, all monosaccharides are designated as D or L on the basis of the configuration about the lowest chiral center, the carbonyl group being at the top: D if the —OH is on the right, L if the —OH is on the left. (As always, it is understood that —H and —OH project toward us from the plane of the paper.) $(+)$-Mannose and $(-)$-arabinose, for example, are both assigned to the D-family on the basis of their relationship to D-(+)-glucose, and, through it, to D-(+)-glyceraldehyde.

CHO
H—OH
HO—H
H—OH
H—OH
CH₂OH
D-(+)-Glucose

$\xrightarrow{\text{HNO}_3}$

COOH
H—OH
HO—H
H—OH
H—OH
COOH
(+)-Glucaric acid

$\xleftarrow{\text{HNO}_3}$

CH₂OH
H—OH
HO—H
H—OH
H—OH
CHO
(+)-Gulose

CHO
H—OH
HO—H
H—OH
CH₂OH
(+)-Xylose

$\xleftarrow[\text{dation}]{\text{Ruff degra-}}$

CHO
H—OH
H—OH
HO—H
H—OH
CH₂OH
(−)-Gulose

Ruff degradation ↓

CHO
H—OH
CH₂OH
D-(+)-Glyceraldehyde

$\xrightarrow{\text{Fig. 38.7}}$

COOH
HO—H
H—OH
COOH
(−)-Tartaric acid

$\xleftarrow{\text{HNO}_3}$

CHO
HO—H
H—OH
CH₂OH
(−)-Threose

Figure 38.8   Relating (+)-glucose to D-(+)-glyceraldehyde.

Until 1951, these configurations were accepted on a purely empirical basis; they were a convenient way to show configurational relationships among the various carbohydrates, and between them and other organic compounds. But so far as anyone knew, the configurations of these compounds might actually have been the mirror images of those assigned; the lowest chiral center in the D-series of monosaccharides might have carried —OH on the left. As we have seen, however, when Bijvoet determined the absolute configuration of (+)-tartaric acid by x-ray analysis in 1951, he found that it actually has the configuration that had been up to then merely assumed. The arbitrary choice that Emil Fischer made in 1891 was the correct one; the configuration he assigned to (+)-glucose—and, through it, to every carbohydrate—is the correct absolute configuration.

Problem 38.19   The (+)-gulose that played such an important part in the proof of configuration of D-(+)-glucose was synthesized by Fischer via the following sequence:

D-(+)-glucose $\xrightarrow{\text{HNO}_3}$ (+)-glucaric acid $\xrightarrow{-\text{H}_2\text{O}}$ A and B (lactones, separated)

A $\xrightarrow{\text{Na(Hg)}}$ C (aldonic acid) $\xrightarrow{-\text{H}_2\text{O}}$ D (lactone) $\xrightarrow{\text{Na(Hg), acid}}$ D-(+)-glucose

B $\xrightarrow{\text{Na(Hg)}}$ E (aldonic acid) $\xrightarrow{-\text{H}_2\text{O}}$ F (lactone) $\xrightarrow{\text{Na(Hg), acid}}$ (+)-gulose

Give the structures of A through F. What is the configuration of (+)-gulose? Is it a member of the D-family or of the L-family? Why?

38.16   Cyclic structure of D-(+)-glucose. Formation of glucosides

We have seen evidence indicating that D-(+)-glucose is a pentahydroxy aldehyde. We have seen how its configuration has been established. It might seem, therefore, that D-(+)-glucose had been definitely proved to have structure I.

$$
\begin{array}{c}
\text{CHO} \\
\text{H}\!-\!\!-\!\!-\!\text{OH} \\
\text{HO}\!-\!\!-\!\!-\!\text{H} \\
\text{H}\!-\!\!-\!\!-\!\text{OH} \\
\text{H}\!-\!\!-\!\!-\!\text{OH} \\
\text{CH}_2\text{OH}
\end{array}
$$

I
D-(+)-Glucose

But during the time that much of the work we have just described was going on, certain facts were accumulating that were inconsistent with this structure of D-(+)-glucose. By 1895 it had become clear that the picture of D-(+)-glucose as a pentahydroxy aldehyde had to be modified.

Among the facts that had still to be accounted for were the following:

(a) **D-(+)-Glucose fails to undergo certain reactions typical of aldehydes.** Although it is readily oxidized, it gives a negative Schiff test and does not form a bisulfite addition product.

(b) **D-(+)-Glucose exists in two isomeric forms which undergo mutarotation.** When crystals of ordinary D-(+)-glucose of m.p. 146 °C are dissolved in water, the specific rotation gradually drops from an initial + 112° to + 52.7°. On the other hand, when crystals of D-(+)-glucose of m.p. 150 °C (obtained by crystallization at temperatures above 98 °C) are dissolved in water, the specific rotation gradually rises from an initial + 19° to + 52.7°. The form with the higher positive rotation is called **α-D-(+)-glucose** and that with lower rotation **β-D-(+)-glucose**. The change in rotation of each of these to the equilibrium value is called **mutarotation**.

(c) **D-(+)-Glucose forms two isomeric methyl D-glucosides.** Aldehydes, we remember, react with alcohols in the presence of anhydrous HCl to form acetals (Sec. 21.13). If the alcohol is, say, methanol, the acetal contains two methyl groups:

$$\underset{\text{Aldehyde}}{-\underset{\underset{\text{H}}{|}}{\overset{\overset{\text{H}}{|}}{C}}=O} \quad \underset{\longleftarrow}{\overset{\text{CH}_3\text{OH, H}^+}{\longrightarrow}} \quad \underset{\text{Hemiacetal}}{-\underset{\underset{\text{OH}}{|}}{\overset{\overset{\text{H}}{|}}{C}}-OCH_3} \quad \underset{\longleftarrow}{\overset{\text{CH}_3\text{OH, H}^+}{\longrightarrow}} \quad \underset{\text{Acetal}}{-\underset{\underset{\text{OCH}_3}{|}}{\overset{\overset{\text{H}}{|}}{C}}-OCH_3}$$

When D-(+)-glucose is treated with methanol and HCl, the product, **methyl D-glucoside**, contains only one —CH$_3$ group; yet it has properties resembling those of a full acetal. It does not spontaneously revert to aldehyde and alcohol on contact with water, but requires hydrolysis by aqueous acids.

Furthermore, not just one but two of these monomethyl derivatives of D-(+)-glucose are known, one with m.p. 165 °C and specific rotation + 158°, and the other with m.p. 107 °C and specific rotation − 33°. The isomer of higher positive rotation is called **methyl α-D-glucoside**, and the other is called **methyl β-D-glucoside**. These glucosides do not undergo mutarotation, and do not reduce Tollens' or Fehling's reagent.

To fit facts like these, ideas about the structure of D-(+)-glucose had to be changed. In 1895, as a result of work by many chemists, including Tollens, Fischer, and Tanret, there emerged a picture of D-(+)-glucose as a *cyclic* structure. In 1926 the ring size was corrected, and in recent years the preferred conformation has been elucidated.

D-(+)-Glucose has the cyclic structure represented crudely by IIa and IIIa, more accurately by IIb and IIIb, and best of all by IIc and IIIc (Fig. 38.9).

Glucose anomers: Hemiacetals
Reducing sugars
Mutarotate

Figure 38.9   Cyclic structures of D-(+)-glucose.

D-(+)-Glucose is the hemiacetal corresponding to reaction between the aldehyde group and the C–5 hydroxyl group of the open-chain structure (I). It has a cyclic structure simply because aldehyde and alcohol are part of the same molecule.

There are two isomeric forms of D-(+)-glucose because this cyclic structure has one more chiral center than Fischer's original open-chain structure (I). α-D-(+)-Glucose and β-D-(+)-glucose are diastereomers, differing in configuration about C–1. Such a pair of diastereomers are called **anomers**.

As hemiacetals, α- and β-D-(+)-glucose are readily hydrolyzed by water. In aqueous solution either anomer is converted—via the open-chain form—into an equilibrium mixture containing both cyclic isomers. This mutarotation results from the ready opening and closing of the hemiacetal ring (Fig. 38.10).

Mutarotation

α-D-Aldohexose   β-D-Aldohexose

Open-chain form

Figure 38.10 Mutarotation.

The typical aldehyde reactions of D-(+)-glucose—osazone formation, and perhaps reduction of Tollens' and Fehling's reagents—are presumably due to a small amount of open-chain compound, which is replenished as fast as it is consumed. The concentration of this open-chain structure is, however, too low (less than 0.5%) for certain easily reversible aldehyde reactions like bisulfite addition and the Schiff test.

The isomeric forms of methyl D-glucoside are anomers and have the cyclic structures IV and V (Fig. 38.11).

Although formed from only one mole of methanol, they are nevertheless full acetals, the other mole of alcohol being D-(+)-glucose itself through the C–5 hydroxyl group. The glucosides do not undergo mutarotation since, being acetals, they are fairly stable in aqueous solution. On being heated with aqueous acids, they undergo hydrolysis to yield the original hemiacetals (II and III). Toward bases glycosides, like acetals generally, are stable. Since they are not readily hydrolyzed to the open-chain aldehyde by the alkali in Tollens' or Fehling's reagent, glucosides are non-reducing sugars.

Like D-(+)-glucose, other monosaccharides exist in anomeric forms capable of mutarotation, and react with alcohols to yield anomeric **glycosides**.

Glucoside anomers: Acetals

Non-reducing sugars

Do not mutarotate

Methyl α-D-glucoside (m.p. 165 °C, [α] = +158°)

Methyl β-D-glucoside (m.p. 107 °C, [α] = −33°)

Figure 38.11   Cyclic structures of methyl D-glucosides.

We have represented the cyclic structures of D-glucose and methyl D-glucoside in several different ways: β-D-glucose, for example, by IIIa, IIIb, and IIIc. At this point we should convince ourselves that all three representations correspond to the same structure, and that the configurations about C–2, C–3, C–4, and C–5 are the same as in the open-chain structure worked out by Fischer. Once again, only molecular models can show us what these relationships really are (Fig. 38.12).

Problem 38.20   (a) From the values for the specific rotations of aqueous solutions of pure α- and β-D-(+)-glucose, and for the solution after mutarotation, calculate the relative amounts of α- and of β-forms at equilibrium (assuming a negligible amount of open-chain form).

(b) From examination of structures IIc and IIIc, suggest a reason for the greater proportion of one isomer. (*Hint*: See Sec. 12.14.)

Problem 38.21   From what you learned in Secs. 21.7 and 21.13, suggest a mechanism for the acid-catalyzed mutarotation of D-(+)-glucose.

Problem 38.22   (+)-Glucose reacts with acetic anhydride to give two isomeric pentaacetyl derivatives neither of which reduces Fehling's or Tollens' reagent. Account for these facts.

(*a*)

(*b*) (*c*)

Figure 38.12 Conversion of an open-chain model to cyclic models: (*a*) open-chain D-glucose, (*b*) α-D-glucose, (*c*) β-D-glucose. (For simplicity, red balls are used for hydroxyl groups as well as aldehyde oxygen.)

To convert (*a*) into (*b*) or (*c*), we join oxygen of the C–5 hydroxyl to the aldehyde carbon C–1. Depending upon which face of the flat carbonyl group we join the C–5 oxygen to, we end up with either the α-structure (*b*) or the β-structure (*c*).

Like (*a*) and (*b*), formulas IIb and IIIb represent the ring lying on its side, so that groups that were on the right in the open-chain model (held as in Fig. 38.2(*c*), p. 1284) are directed downward, and groups that were on the left in the open-chain model are directed upward. (Note particularly that the —CH₂OH group points *upward*.)

In the more accurate representations IIc and IIIc, the disposition of these groups is modified by puckering of the six-membered ring; this puckering can be seen in (*b*) and (*c*), and will be discussed further in Sec. 38.20.

the pyranose ring should be quite similar to the cyclohexane ring (Sec. 12.14). It should be puckered and, to minimize torsional and van der Waals strain, should exist in chair conformations in preference to twist-boat conformations. X-ray analysis shows this reasoning to be correct.

But there are *two* chair conformations possible for a D-(+)-glucopyranose anomer: I and II for β-D-(+)-glucopyranose, for example.

I

More stable:
all bulky groups equatorial

II

Less stable:
all bulky groups axial

β-D-(+)-Glucopyranose

Which of these is the more stable one, the one in which the molecules spend most of the time? For β-D-(+)-glucopyranose, the answer seems clear: I, in which all bulky substituents (—CH_2OH and —OH) occupy roomy equatorial positions, should certainly be much more stable than II, in which all bulky groups are crowded into axial positions. Again, x-ray analysis shows this reasoning to be correct. (See Fig. 38.12(c), p. 1308.)

What can we say about α-D-(+)-glucose and the other aldohexoses? This problem has been largely worked out by R. E. Reeves (then at the U.S. Southern Regional Research Laboratory) through study of copper complexes.

In general, the more stable conformation is the one in which the bulkiest group, —CH_2OH, occupies an equatorial position. For example:

III

α-D-Glucopyranose
Stable conformation

IV

β-D-Mannopyranose
Stable conformation

V

α-D-Galactopyranose
Stable conformation

In an extreme case, to permit many —OH groups to take up equatorial positions, the —CH_2OH group may be forced into an axial position. For example:

More stable:	*Less stable:*
4 equatorial OH groups	4 axial OH groups
1 axial —CH_2OH	1 equatorial —CH_2OH

α-D-Idopyranose

We notice that *of all* D-*aldohexoses it is* β-D-(+)-*glucose that can assume a conformation in which every bulky group occupies an equatorial position.* It is hardly accidental that β-D-(+)-glucose is the most widely occurring organic group in nature.

In drawing structural formulas or making models for the aldohexoses, a convenient point of reference is β-D-(+)-glucose. We draw the ring as shown in I, with C–1 down, C–4 up, and oxygen at the right-hand back corner, and place all —OH groups and the —CH_2OH group in equatorial positions. We draw the structures of other D-family aldohexoses merely by taking into account their differences from I. Thus α-D-(+)-glucose (III) differs in configuration at C–1; β-D-mannose (IV) differs in configuration at C–2; α-D-galactose (V) differs at C–1 and C–4. Compounds of the L-family are, of course, mirror images of these.

In methylated and acetylated pyranoses, too, bulky groups tend to occupy equatorial positions, with one general exception: a methoxy or acetoxy group on C–1 tends to be axial. This *anomeric effect* is attributed to repulsion between the dipoles associated with the C–1 oxygen and the oxygen of the ring.

Anomeric effect

More stable

As we would expect for dipole–dipole interactions, the anomeric effect weakens as the polarity of the solvent increases (Sec. 12.10). For free sugars dissolved in water, the anomeric effect is usually outweighed by other factors; D-glucose, for example, exists predominantly as the β-anomer, with the —OH on C–1 equatorial.

Problem 38.28 Draw the conformation you predict to be the most stable for:

(a) β-D-allopyranose
(b) β-D-gulopyranose
(c) β-D-xylopyranose
(d) α-D-arabinopyranose
(e) β-L-(−)-glucopyranose
(f) β-D-(−)-fructopyranose

PROBLEMS

1. Give structures and, where possible, names of the principal products of the reaction (if any) of D-(+)-galactose with:

(a) hydroxylamine $=N$
(b) phenylhydrazine
(c) bromine water
(d) HNO_3
(e) HIO_4
(f) acetic anhydride
(g) benzoyl chloride, pyridine

(h) CH_3OH, HCl
(i) CH_3OH, HCl; then $(CH_3)_2SO_4$, NaOH
(j) reagents of (i), then dilute HCl
(k) reagents of (i) and (j), then vigorous oxidation
(l) H_2, Ni
(m) $NaBH_4$
(n) CN^-, H^+; then hydrolysis; then one mole $NaBH_4$
(o) H_2, Ni; then oxidation to monocarboxylic acid
(p) $Br_2(aq)$; then pyridine; then H^+; then Na(Hg), CO_2
(q) phenylhydrazine; then benzaldehyde; then H^+
(r) reagents of (q), then reduction to monocarbonyl compound
(s) $Br_2(aq)$; then $CaCO_3$; then H_2O_2, Fe^{3+}
(t) reagents of (i), then NaOH
(u) CH_3OH, HCl; then HIO_4
(v) reagents of (u); then $Br_2(aq)$; then dilute HCl

2. Write equations to show how D-(+)-glucose could be converted into:

(a) methyl β-D-glucoside
(b) methyl β-2,3,4,6-tetra-*O*-methyl-D-glucoside
(c) 2,3,4,6-tetra-*O*-methyl-D-glucose
(d) D-mannose
(e) L-gulose
(f) D-arabinose
(g) mesotartaric acid
(h) hexa-*O*-acetyl-D-glucitol
(i) D-fructose

(j)

3. Besides D-fructose, there are three D-2-ketohexoses: D-*psicose*, D-*sorbose*, and D-*tagatose*. (a) Draw the possible configurations for these three ketoses. (b) Given the configurations of all aldohexoses, tell how you could assign definite configurations to the ketoses.

4. Draw stereochemical formulas for products A through O, and tell what aldoses E, E', F, H, I, I', N, and O are related to.

(a) $ClCH_2CHO + BrMgC{\equiv}CMgBr + OHCCH_2Cl \longrightarrow$ A $(C_6H_8O_2Cl_2)$, mainly *meso*
  meso-A + KOH $\longrightarrow$ B $(C_6H_6O_2)$, a diepoxide
  B + H_2O, $OH^- \longrightarrow$ C $(C_6H_{10}O_4)$
  C + H_2, $Pd/CaCO_3 \longrightarrow$ D $(C_6H_{12}O_4)$
  D + cold dilute $KMnO_4 \longrightarrow$ E and E' (both $C_6H_{14}O_6$)
  D + peroxyformic acid $\longrightarrow$ F $(C_6H_{14}O_6)$
  C + Na, $NH_3 \longrightarrow$ G $(C_6H_{12}O_4)$
  G + cold dilute $KMnO_4 \longrightarrow$ H $(C_6H_{14}O_6)$
  G + peroxyformic acid $\longrightarrow$ I and I' (both $C_6H_{14}O_6$)

(b) *trans*-2-penten-4-yn-1-ol + $HCO_2OH \longrightarrow$ J $(C_5H_8O_3)$, 4-pentyn-1,2,3-triol
  J + acetic anhydride, then $Pd/CaCO_3 + H_2 \longrightarrow$ K $(C_{11}H_{16}O_6)$
  K + HOBr $\longrightarrow$ L and M (both $C_{11}H_{17}O_7Br$)
  L + hydrolysis $\longrightarrow$ N $(C_5H_{12}O_5)$
  M + hydrolysis $\longrightarrow$ O $(C_5H_{12}O_5)$, a racemic modification

(c) Starting with 2-butyn-1,4-diol, outline a synthesis of erythritol; of DL-threitol.

(d) 2-Butyn-1,4-diol (above) is made by reaction under pressure of acetylene with formaldehyde. What kind of reaction is this?

5. When borneol (ROH) is fed to a dog, this toxic substance is excreted as compound P, $C_6H_9O_6$—OR, where R stands for the bornyl group. Compound P does not reduce Benedict's solution. It reacts with aqueous $NaHCO_3$ with the liberation of a gas. Treatment of P with aqueous acid yields borneol (ROH) and D-glucuronic acid (Table 38.1), which is oxidized by bromine water to D-glucaric acid.

(a) What is the structure of P?

(b) Hydrolysis of the polysaccharide *pectin* (from fruits and berries) gives chiefly D-galacturonic acid; hydrolysis of the polysaccharide *algin* (from seaweed) yields D-mannuronic acid. Give the structures of these uronic acids.

(c) There are two uronic acids related to D-fructose. Draw their structures. Give the name and family of the aldonic acids formed from each "fructuronic acid" by reduction of the carbonyl group.

(d) What compound would you expect from the treatment of D-glucosone with bromine water?

6. The rate of oxidation of reducing sugars by cupric ion is found to be proportional to sugar and $[OH^-]$, and to be independent of $[Cu^{2+}]$. What does the kinetics suggest about the mechanism of oxidation?

7. Upon oxidation by HIO_4 the methyl glycoside Q yields the same product (shown on p. 1310) as that obtained from methyl α-glycosides of the D-aldohexoses; however, it consumes only one mole of HIO_4 and yields *no* formic acid.

(a) How many carbon atoms are there in Q, and what is the ring size? (b) For which carbon atoms do you know the configuration? (c) When Q is methylated, hydrolyzed, and then vigorously oxidized, the dicarboxylic acid obtained is the di-*O*-methyl ether of (−)-tartaric acid. What is the complete structure and configuration of Q?

8. *Salicin*, $C_{13}H_{18}O_7$, found in willow (*Salix*, whence the name *salicylic*), is hydrolyzed by emulsin to D-glucose and saligenin, $C_7H_8O_2$. Salicin does not reduce Tollens' reagent. Oxidation of salicin by nitric acid yields a compound that can be hydrolyzed to D-glucose and salicylaldehyde.

Methylation of salicin gives pentamethylsalicin, which on hydrolysis gives 2,3,4,6-tetra-*O*-methyl-D-glucose.

What is the structure of salicin?

9. The optically inactive carbohydrate *bio-inonose*, $C_6H_{10}O_6$, reduces Benedict's solution, but does not react with bromine water. It is reduced to R and S, of formula $C_6H_{12}O_6$. Compounds R and S are oxidized by HIO_4 to six moles of HCOOH, and react with acetic anhydride to yield products of formula $C_{18}H_{24}O_{12}$. Vigorous oxidation of bio-inonose yields DL-idaric acid (the dicarboxylic acid from idose) as the only six-carbon fragment.

What is the structure of bio-inonose? Of R and S?

10. Much of what is known about photosynthesis has been learned by determining the fate of radioactive carbon dioxide, $^{14}CO_2$. The ^{14}C was found in many products, including glucose, fructose, and sucrose. To measure the radioactivity of each carbon atom in a particular molecule, degradations to one-carbon fragments were carried out.

Tell which position or positions in the molecule each of the following one-carbon products came from.

Show how the activity of the carbon atom in every position could be figured out.

(a) glucose $\xrightarrow{\text{Ruff degradation}}$ CO_2 + arabinose $\xrightarrow{\text{Ruff degradation}}$ CO_2

glucose + HIO_4 $\longrightarrow$ HCHO

glucose + CH_3OH, HCl; then HIO_4 $\longrightarrow$ HCOOH

glucose $\xrightarrow{\text{Lactobacillus casei}}$ 2 lactic acid (carboxyls are C–3 and C–4)

$$\downarrow \text{KMnO}_4$$

CO_2 + CH_3CHO $\xrightarrow{\text{NaOI}}$ CHI_3 + HCOOH

(b) ribulose (a 2-ketopentose) + HIO_4 $\longrightarrow$ $HOCH_2COOH$ + $2HCOOH$ + HCHO

ribulose + H_2, Pt; then HIO_4 $\longrightarrow$ 2HCHO + 3HCOOH

ribulose + $C_6H_5NHNH_2$ $\longrightarrow$ ribosazone

$$\text{ribosazone} + HIO_4 \longrightarrow HCHO + HCOOH + \begin{array}{c} HC\!=\!NNHC_6H_5 \\ | \\ C\!=\!NNHC_6H_5 \\ | \\ CHO \end{array}$$

11. *Nucleic acids*, the compounds that control heredity on the molecular level, are polymers composed of nucleotide units. The structures of nucleotides have been determined in the following way, as illustrated for *adenylic acid*, a nucleotide isolated from yeast cells.

Hydrolysis of adenylic acid yields one molecule each of a heterocyclic base, a sugar T, and phosphoric acid. The base is called *adenine*, and will be represented as R_2NH. Adenylic acid has the formula $R_2N-C_5H_8O_3-OPO_3H_2$.

The sugar T is levorotatory and has the formula $C_5H_{10}O_5$; it reduces Tollens' reagent and Benedict's solution. T is oxidized by bromine water to optically active $C_5H_{10}O_6$, and by nitric acid to optically inactive $C_5H_8O_7$. T forms an osazone that is identical with the osazone obtained from another pentose, $(-)$-U. Degradation of $(-)$-U, followed by oxidation by nitric acid, yields optically inactive $C_4H_6O_6$.

(a) What is T?

Careful acidic hydrolysis of adenylic acid yields adenine and a phosphate of T, $C_5H_9O_4-$OPO$_3H_2$. Reduction of the phosphate with H_2/Pt yields optically inactive V, $C_5H_{11}O_4-$OPO$_3H_2$. Hydrolysis of V yields optically inactive W, $C_5H_{12}O_5$, which reacts with acetic anhydride to yield optically inactive X, $C_{15}H_{22}O_{10}$.

(b) What is the structure of the phosphate of T?

Adenylic acid does not reduce Tollens' reagent or Benedict's solution. When hydrolyzed by aqueous ammonia, adenylic acid yields phosphoric acid and the nucleoside *adenosine*. Treatment of adenosine with methyl sulfate and NaOH, followed by acidic hydrolysis, yields Y, a methylation product of T. Compound Y has the formula $C_8H_{16}O_5$. Vigorous oxidation of Y yields 2,3-di-O-methylmesotartaric acid and no larger fragments.

Synthesis of adenosine shows that a nitrogen atom of adenine is joined to a carbon atom in T; synthesis also shows that T has the β-configuration.

(c) Give the structure of adenylic acid, using R_2NH for the adenine unit.

(Check your answers in Fig. 41.5, p. 1395.)

12. Give structural formulas for compounds Z through II. Tell what each piece of information—(a), (b), (c), etc.—shows about the structures of Z and AA.

(a) D-glucose + CH_3COCH_3, H_2SO_4 $\longrightarrow$ Z ($C_{12}H_{20}O_6$) + AA ($C_9H_{16}O_6$)

Z or AA + H_2O, OH^- $\longrightarrow$ no reaction

Z $\xrightarrow{H_2O, H^+}$ AA $\xrightarrow{H_2O, H^+}$ D-glucose + CH_3COCH_3

To what class of compounds do Z and AA belong?

(b) Z or AA + Benedict's solution $\longrightarrow$ no reaction

(c) Z + $(CH_3)_2SO_4$, NaOH $\longrightarrow$ BB ($C_{13}H_{22}O_6$)

BB + H_2O, H^+ $\longrightarrow$ CC ($C_7H_{14}O_6$)

CC + $C_6H_5NHNH_2$ $\longrightarrow$ DD (an osazone)

(d) $AA + (CH_3)_2SO_4, NaOH \longrightarrow EE (C_{12}H_{22}O_6)$
$EE + H_2O, H^+ \longrightarrow FF (C_9H_{18}O_6)$
$FF + C_6H_5NHNH_2 \longrightarrow GG$ (an osazone)

(e) $FF + (CH_3)_2SO_4, NaOH \longrightarrow$ 2,3,5,6-tetra-O-methyl-D-glucofuranose

(f) $CC + HNO_3 \longrightarrow HH (C_6H_{10}O_7)$

(g) $CC + HCN$, then $H_2O, H^+ \longrightarrow$ II (a δ-lactone)

13. When either D-glyceraldehyde or dihydroxyacetone, $HOCH_2COCH_2OH$, is treated with base, there is obtained a mixture of the following compounds:

CH_2OH	CH_2OH	CH_2OH
$C=O$	$C=O$	$C=O$
HO——H	H——OH	H——OH
H——OH	HO——H	$HOCH_2$——OH
H——OH	H——OH	CH_2OH
CH_2OH	CH_2OH	and enantiomer
D-Fructose	D-Sorbose	DL-Dendroketose

Suggest a possible mechanism for this reaction. (*Hints*: See Sec. 38.6. Count the carbons in reactants and products, and consider the reagent used.)

14. In dilute acid, hydrolysis of D-glucose-1-phosphate differs from ordinary alkyl esters of its type $(ROPO_3H_2)$ in two ways: it is abnormally fast, and it takes place with cleavage of the carbon–oxygen bond. Can you suggest an explanation for its unusual behavior?

15. In Chapter 16, we learned about certain relationships between NMR spectra and the conformations of six-membered rings: in Problems 10 and 11 (p. 623), that a given proton absorbs farther downfield when in an equatorial position than when in an axial position; in Sec. 16.11, that the coupling constant, J, between *anti* protons (axial,axial) is bigger than between *gauche* protons (axial,equatorial or equatorial,equatorial). It was in the study of carbohydrates that those relationships were first recognized, chiefly by R. U. Lemieux (p. 1330).

(a) In the NMR spectra of aldopyranoses and their derivatives, the signal from one proton is found at lower fields than any of the others. Which proton is this, and why?

(b) In the NMR spectra of the two anomers of D-tetra-O-acetylxylopyranose the downfield peak appears as follows:

Anomer JJ: doublet, δ 5.39, $J = 6$ Hz
Anomer KK: doublet, δ 6.03, $J = 3$ Hz

Identify JJ and KK; that is, tell which is the α-anomer, and which is the β-anomer. Explain your answer.

(c) Answer (b) for the anomers of D-tetra-O-acetylribopyranose:

Anomer LL: doublet, δ 5.72, $J = 5$ Hz
Anomer MM: doublet, δ 5.82, $J = 2$ Hz

(d) Consider two pairs of anomers: NN and OO, and PP and QQ. One pair are the D-penta-O-acetylglucopyranoses, and the other pair are the D-penta-O-acetylmannopyranoses.

Anomer NN: doublet, δ 5.97, $J = 3$ Hz
Anomer OO: doublet, δ 5.68, $J = 3$ Hz
Anomer PP: doublet, δ 5.54, $J = 8$ Hz
Anomer QQ: doublet, δ 5.99, $J = 3$ Hz

Identify NN, OO, PP, and QQ. Explain your answer.

16. The rare sugar $(-)$-*mycarose* occurs as part of the molecules of several antibiotics. Using the following evidence, work out the structure and configuration of mycarose.

(i) lactone of $CH_3CH(OH)CH=C(CH_3)CH_2COOH$ $\xrightarrow{syn\text{-hydroxylation}}$ RR $(C_7H_{12}O_4)$
   RR + KBH_4 $\longrightarrow$ $(\pm)$-mycarose

(ii) In the NMR spectrum of $(-)$-mycarose and several derivatives, the coupling constant between the C–4 proton and the C–5 proton is 9.5–9.7 Hz.

(iii) methyl mycaroside + HIO_4 $\longrightarrow$ SS $(C_8H_{14}O_4)$
   SS + cold $KMnO_4$ $\longrightarrow$ TT $(C_8H_{14}O_5)$
   TT $\xrightarrow{\text{hydrolysis}}$ L-lactic acid

(a) Disregarding stereochemistry, what is the structure of mycarose?

(b) What are the relative configurations about C–3 and C–4? About C–4 and C–5?

(c) What is the absolute configuration at C–5?

(d) What is the absolute configuration of $(-)$-mycarose? To which family, D or L, does it belong? In what conformation does it preferentially exist?

(e) $(-)$-Mycarose can be converted into two methyl mycarosides. In the NMR spectrum of one of these, the downfield peak appears as a triplet with $J = 2.4$ Hz. Which anomer, α or β, is this one likely to be? What would you expect to see in the NMR spectrum of the other anomer?

(f) In the NMR spectrum of free $(-)$-mycarose, the downfield peak (1H) appears as two doublets with $J = 9.5$ and 2.5 Hz. Which anomer of mycarose, α or β, does this appear to be?

17. How do you account for the following facts? (a) In an equilibrium mixture of methyl α-D-glucoside and methyl β-D-glucoside, the α-anomer predominates. (b) In the more stable conformation of *trans*-2,5-dichloro-1,4-dioxane, both chlorines occupy axial positions.

18. From study of the NMR spectra of many compounds, Lemieux (p. 1330) found that the protons of axial acetoxy groups ($-OOCCH_3$) generally absorb at lower field than those of equatorial acetoxy groups.

(a) Draw the two chair conformations of tetra-*O*-acetyl-β-L-arabinopyranose. On steric grounds, which would you expect to be the more stable? Taking into account the anomeric effect, which would you expect to be the more stable?

(b) In the NMR spectrum of this compound, absorption by the acetoxy protons appears upfield as two equal peaks, at δ 1.92 and δ 2.04. How do you account for the equal sizes of these peaks? What, if anything, does this tell about the relative abundances of the two anomers?

(c) When the acetoxy group on C–1 is replaced by the deuteriated group $-OOCCD_3$, the total area of the upfield peaks is decreased, of course, from 12H to 9H. The ratio of peak areas δ 1.92 : δ 2.04 is now 1.46 : 1.00. Which conformation predominates, and by how much? Is the predominant conformation the one you predicted to be the more stable?

39

Carbohydrates II. Disaccharides and Polysaccharides

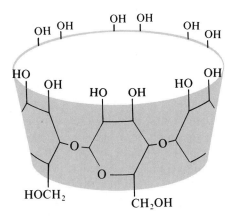

39.1 Disaccharides

Disaccharides are carbohydrates that are made up of two monosaccharide units. On hydrolysis a molecule of disaccharide yields two molecules of monosaccharide.

We shall study four disaccharides: **(+)-maltose** (malt sugar), **(+)-cellobiose**, **(+)-lactose** (milk sugar), and **(+)-sucrose** (cane or beet sugar). As with the monosaccharides, we shall focus our attention on the structure of these molecules: on which monosaccharides make up the disaccharide, and how they are attached to each other. In doing this, we shall also learn something about the properties of these disaccharides.

39.2 (+)-Maltose

(+)-Maltose can be obtained, among other products, by partial hydrolysis of starch in aqueous acid. (+)-Maltose is also formed in one stage of the fermentation of starch to ethyl alcohol; here hydrolysis is catalyzed by the enzyme *diastase*, which is present in malt (sprouted barley).

Let us look at some of the facts from which the structure of (+)-maltose has been deduced.

(+)-Maltose has the molecular formula $C_{12}H_{22}O_{11}$. It reduces Tollens' and Fehling's reagents and hence is a reducing sugar. It reacts with phenylhydrazine to yield an osazone, $C_{12}H_{20}O_9(=NNHC_6H_5)_2$. It is oxidized by bromine water to a monocarboxylic acid, $(C_{11}H_{21}O_{10})COOH$, *maltobionic acid.* (+)-Maltose exists in *alpha* ([α] = + 168°) and *beta* ([α] = + 112°) forms which undergo mutarotation in solution (equilibrium [α] = + 136°).

All these facts indicate the same thing: (+)-maltose contains a carbonyl group that exists in the reactive hemiacetal form as in the monosaccharides we have studied. It contains only one such "free" carbonyl group, however, since (a) the osazone contains only two phenylhydrazine residues, and (b) oxidation by bromine water yields only a *mono*carboxylic acid.

When hydrolyzed in aqueous acid, or when treated with the enzyme *maltase* (from yeast), (+)-maltose is completely converted into D-(+)-glucose. This indicates that (+)-maltose ($C_{12}H_{22}O_{11}$) is made up of two D-(+)-glucose units joined together in some manner with the loss of one molecule of water:

$$2C_6H_{12}O_6 - H_2O = C_{12}H_{22}O_{11}$$

Hydrolysis by acid to give a new reducing group (two reducing D-(+)-glucose molecules in place of one (+)-maltose molecule) is characteristic of glycosides; hydrolysis by the enzyme maltase is characteristic of *alpha* glucosides. A glycoside is an acetal formed by interaction of an alcohol with a carbonyl group of a carbohydrate (Sec. 38.16); in this case the alcohol concerned can only be a second molecule of D-(+)-glucose. We conclude that (+)-maltose contains two D-(+)-glucose units, joined by an *alpha*-glucoside linkage between the carbonyl group of one D-(+)-glucose unit and an —OH group of the other.

Two questions remain: which —OH group is involved, and what are the sizes of the rings in the two D-(+)-glucose units? Answers to both these questions are given by the sequence of oxidation, methylation, and hydrolysis shown in Fig. 39.1.

Oxidation by bromine water converts (+)-maltose into the monocarboxylic acid D-maltobionic acid. Treatment of this acid with methyl sulfate and sodium hydroxide yields octa-*O*-methyl-D-maltobionic acid. Upon hydrolysis in acidic solution, the methylated acid yields two products, 2,3,5,6-tetra-*O*-methyl-D-gluconic acid and 2,3,4,6-tetra-*O*-methyl-D-glucose.

These facts indicate that (+)-maltose has structure I, as shown in Fig. 39.1; this is given the name 4-*O*-(α-D-glucopyranosyl)-D-glucopyranose. It is the —OH group on C–4 that serves as the alcohol in the glucoside formation; both halves of the molecule contain the six-membered, pyranose ring.

Let us see how we arrive at structure I from the experimental facts.

First of all, the initial oxidation labels (with a —COOH group) the D-glucose unit that contains the "free" aldehyde group. Next, methylation labels (as —OCH$_3$) every free —OH group. Finally, upon hydrolysis, the absence of a methoxyl group shows which —OH groups were *not* free.

The oxidized product, 2,3,5,6-tetra-*O*-methyl-D-gluconic acid, must have arisen from the reducing (oxidizable) D-glucose unit. The presence of a free —OH group at C–4 shows that this position was not available for methylation at the maltobionic acid stage; hence it is the —OH on C–4 that is tied up in the glucoside linkage of maltobionic acid and of (+)-maltose itself. This leaves only the —OH group on C–5 to be involved in the ring of the reducing (oxidizable)

Figure 39.1   Sequence of oxidation, methylation, and hydrolysis shows that (+)-maltose is 4-*O*-(α-D-glucopyranosyl)-D-glucopyranose.

I

(+)-Maltose (α-anomer)

4-O-(α-D-Glucopyranosyl)-D-glucopyranose

unit in the original disaccharide. On the basis of these facts, therefore, we designate one D-(+)-glucose unit as a 4-O-substituted-D-glucopyranose.

The unoxidized product, 2,3,4,6-tetra-O-methyl-D-glucose, must have arisen from the non-reducing (non-oxidizable) D-glucose unit. The presence of the free —OH group at C–5 indicates that this position escaped methylation at the maltobionic acid stage; hence it is the —OH on C–5 that is tied up as a ring in maltobionic acid and in (+)-maltose itself. On the basis of these facts, therefore, we designate the second D-(+)-glucose unit as an α-D-glucopyranosyl group.

Problem 39.1   Formula I shows the structure of only the α-form of (+)-maltose. What is the structure of the β-(+)-maltose that in solution is in equilibrium with I?

Problem 39.2   The position of the free —OH group in 2,3,4,6-tetra-O-methyl-D-glucose was shown by the products of oxidative cleavage, as described in Sec. 38.19. What products would be expected from oxidative cleavage of 2,3,5,6-tetra-O-methyl-D-gluconic acid?

Problem 39.3   What products would be obtained if (+)-maltose itself were subjected to methylation and hydrolysis? What would this tell us about the structure of (+)-maltose? What uncertainty would remain in the (+)-maltose structure? Why was it necessary to oxidize (+)-maltose first before methylation?

Problem 39.4   When (+)-maltose is subjected to two successive one-carbon degradations, there is obtained a disaccharide that reduces Tollens' and Fehling's reagents but does not form an osazone. What products would be expected from the acidic hydrolysis of this disaccharide? What would these facts indicate about the structure of (+)-maltose?

39.3   (+)-Cellobiose

When cellulose (cotton fibers) is treated for several days with sulfuric acid and acetic anhydride, a combination of acetylation and hydrolysis takes place; there is obtained the octaacetate of (+)-cellobiose. Alkaline hydrolysis of the octaacetate yields (+)-cellobiose itself.

Like (+)-maltose, (+)-cellobiose has the molecular formula $C_{12}H_{22}O_{11}$, is a reducing sugar, forms an osazone, exists in *alpha* and *beta* forms that undergo mutarotation, and can be hydrolyzed to two molecules of D-(+)-glucose. The sequence of oxidation, methylation, and hydrolysis (as described for (+)-maltose) shows that (+)-cellobiose contains two pyranose rings and a glucoside linkage to an —OH group on C–4.

(+)-Cellobiose differs from (+)-maltose in one respect: it is hydrolyzed by the enzyme *emulsin* (from bitter almonds), not by maltase. Since emulsin is known to hydrolyze only β-glucoside linkages, we can conclude that the structure of (+)-cellobiose differs from that of (+)-maltose in only one respect: the D-glucose units are joined by a *beta* linkage rather than by an *alpha* linkage. (+)-Cellobiose is therefore 4-*O*-(β-D-glucopyranosyl)-D-glucopyranose.

(+)-Cellobiose (β-anomer)
4-*O*-(β-D-Glucopyranosyl)-D-glucopyranose

Although the D-glucose unit on the right in the formula of (+)-cellobiose may look different from the D-glucose unit on the left, this is only because it has been turned over to permit a reasonable bond angle at the glycosidic oxygen atom.

Problem 39.5 Why is *alkaline* hydrolysis of cellobiose octaacetate (better named octa-*O*-acetylcellobiose) to (+)-cellobiose preferred over acidic hydrolysis?

Problem 39.6 Write equations for the sequence of oxidation, methylation, and hydrolysis as applied to (+)-cellobiose.

39.4   (+)-Lactose

(+)-Lactose makes up about 5% of human milk and of cows' milk. It is obtained commercially as a by-product of cheese manufacture, being found in the *whey*, the aqueous solution that remains after the milk proteins have been coagulated. Milk *sours* when lactose is converted into lactic acid (sour, like all acids) by bacterial action (e.g., by *Lactobacillus bulgaricus*).

(+)-Lactose has the molecular formula $C_{12}H_{22}O_{11}$, is a reducing sugar, forms an osazone, and exists in *alpha* and *beta* forms which undergo mutarotation. Acidic hydrolysis or treatment with emulsin (which splits β-linkages only) converts (+)-lactose into equal amounts of D-(+)-glucose and D-(+)-galactose. (+)-Lactose is evidently a β-glycoside formed by the union of a molecule of D-(+)-glucose and a molecule of D-(+)-galactose.

The question next arises: which is the reducing monosaccharide unit and which the non-reducing unit? Is (+)-lactose a glucoside or a galactoside? Hydrolysis of lactosazone yields D-(+)-galactose and D-glucosazone; hydrolysis of *lactobionic acid* (monocarboxylic acid) yields D-gluconic acid and D-(+)-galactose (see Fig. 39.2, on the following page). Clearly, it is the D-(+)-glucose unit that contains the "free" aldehyde group and undergoes osazone formation or oxidation to the acid. (+)-Lactose is thus a substituted D-glucose in which a D-galactosyl unit is attached to one of the oxygens; it is a galactoside, not a glucoside.

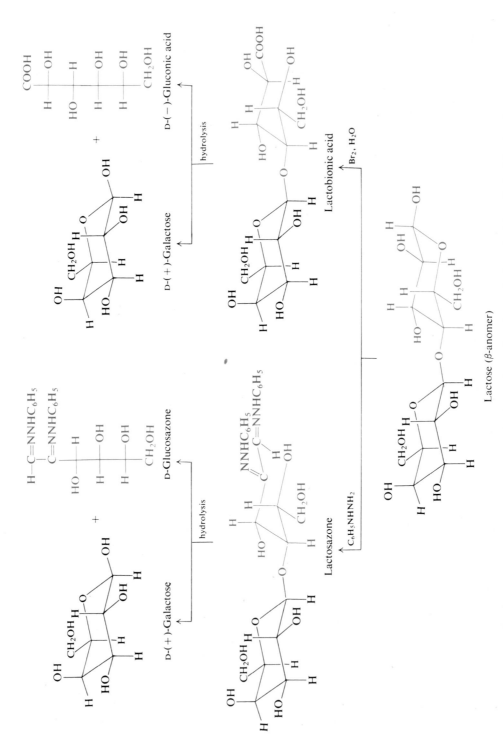

Figure 39.2   Hydrolysis of (+)-lactose derivatives: shows that glucose is the reducing unit. (+)-Lactose is 4-*O*-(β-D-galactopyranosyl)-D-glucopyranose.

The sequence of oxidation, methylation, and hydrolysis gives results analogous to those obtained with (+)-maltose and (+)-cellobiose: the glycoside linkage involves an —OH group on C–4, and both units exist in the six-membered, pyranose form. (+)-Lactose is therefore 4-*O*-(β-D-galactopyranosyl)-D-glucopyranose.

Problem 39.7 (a) Write equations for the sequence of oxidation, methylation, and hydrolysis as applied to (+)-lactose.

(b) What compounds would be expected from oxidative cleavage of the final products of (a)?

Problem 39.8 What products would be expected if (+)-lactose were subjected to two successive one-carbon degradations followed by acidic hydrolysis?

39.5 (+)-Sucrose

(+)-Sucrose is our common table sugar, obtained from sugar cane and sugar beets. Of organic chemicals, it is the one produced in the largest amount in pure form.

(+)-Sucrose has the molecular formula $C_{12}H_{22}O_{11}$. It does not reduce Tollens' or Fehling's reagent. It is a non-reducing sugar, and in this respect it differs from the other disaccharides we have studied. Moreover, (+)-sucrose does not form an osazone, does not exist in anomeric forms, and does not show mutarotation in solution. All these facts indicate that (+)-sucrose does not contain a "free" aldehyde or ketone group.

When (+)-sucrose is hydrolyzed by dilute aqueous acid, or by the action of the enzyme *invertase* (from yeast), it yields equal amounts of D-(+)-glucose and D-(−)-fructose. This hydrolysis is accompanied by a change in the sign of rotation from positive to negative; it is therefore often called the *inversion* of (+)-sucrose, and the levorotatory mixture of D-(+)-glucose and D-(−)-fructose obtained has been called *invert sugar*. (Honey is mostly invert sugar; the bees supply the invertase.) While (+)-sucrose has a specific rotation of + 66.5° and D-(+)-glucose has a specific rotation of + 52.7°, D-(−)-fructose has a large negative specific rotation of − 92.4°, giving a net negative value for the specific rotation of the mixture. (Because of their opposite rotations and their importance as components of (+)-sucrose, D-(+)-glucose and D-(−)-fructose are commonly called **dextrose** and **levulose**.)

Problem 39.9 How do you account for the experimentally observed $[\alpha] = -19.9°$ for invert sugar?

(+)-Sucrose is made up of a D-glucose unit and a D-fructose unit; since there is no "free" carbonyl group, it must be both a D-glucoside and a D-fructoside. The two hexose units are evidently joined by a glycoside linkage between C–1 of glucose and C–2 of fructose, for only in this way can the single link between the two units effectively block *both* carbonyl functions.

Problem 39.10 What would be the molecular formula of (+)-sucrose if C–1 of glucose were attached to, say, C–4 of fructose, and C–2 of fructose were joined to C–4 of glucose? Would this be a reducing or non-reducing sugar?

Determination of the stereochemistry of the D-glucoside and D-fructoside linkages is complicated by the fact that both linkages are hydrolyzed at the same time. The weight of evidence, including the results of x-ray studies and finally the synthesis of (+)-sucrose (1953), leads to the conclusion that (+)-sucrose is a *beta* D-fructoside and an *alpha* D-glucoside. (The synthesis of sucrose, by R. U. Lemieux of the University of Alberta, has been described as "the Mount Everest of organic chemistry".)

(+)-Sucrose
α-D-Glucopyranosyl β-D-fructofuranoside
β-D-Fructofuranosyl α-D-glucopyranoside
(no anomers: *non-mutarotating*)

Problem 39.11 When (+)-sucrose is hydrolyzed enzymatically, the D-glucose initially obtained mutarotates *downward* to +52.7°. What does this fact indicate about the structure of (+)-sucrose?

Methylation and hydrolysis show that (+)-sucrose contains a D-glucopyranose unit and a D-fructofuranose unit. (The unexpected occurrence of the relatively rare five-membered, furanose ring caused no end of difficulties in both structure proof and synthesis of (+)-sucrose.) (+)-Sucrose is named equally well as either α-D-glucopyranosyl β-D-fructofuranoside or β-D-fructofuranosyl α-D-glucopyranoside.

Problem 39.12 (a) Write equations for the sequence of methylation and hydrolysis as applied to (+)-sucrose.
(b) What compounds would be expected from oxidative cleavage of the final products of (a)?

39.6 Polysaccharides

Polysaccharides are compounds made up of many—hundreds or even thousands—monosaccharide units per molecule. As in disaccharides, these units are held together by glycoside linkages, which can be broken by hydrolysis.

Polysaccharides are naturally occurring polymers, which can be considered as derived from aldoses or ketoses by polymerization with loss of water. A polysaccharide derived from hexoses, for example, has the general formula $(C_6H_{10}O_5)_n$. This formula, of course, tells us very little about the structure of the polysaccharide. We need to know what the monosaccharide units are and how many there are in each molecule; how they are joined to each other; and whether the huge molecules thus formed are straight-chained or branched, looped or coiled.

By far the most important polysaccharides are **cellulose** and **starch**. Both are produced in plants from carbon dioxide and water by the process of photosynthesis, and both, as it happens, are made up of D-(+)-glucose units. Cellulose is the chief structural material of plants, giving the plants rigidity and form. It is probably the most widespread organic material known. Starch makes up the reserve food supply of plants and occurs chiefly in seeds. It is more water-soluble than cellulose, more easily hydrolyzed, and hence more readily digested.

Both cellulose and starch are, of course, enormously important to us. Generally speaking, we use them in very much the same way as the plant does. We use cellulose for its structural properties: as wood for houses, as cotton or rayon for clothing, as paper for communication and packaging. We use starch as a food: potatoes, corn, wheat, rice, cassava, etc.

39.7   Starch

Starch occurs as granules whose size and shape are characteristic of the plant from which the starch is obtained. When intact, starch granules are insoluble in cold water; if the outer membrane has been broken by grinding, the granules swell in cold water and form a gel. When the intact granule is treated with warm water, a soluble portion of the starch diffuses through the granule wall; in hot water the granules swell to such an extent that they burst.

In general, starch contains about 20% of a water-soluble fraction called **amylose**, and 80% of a water-insoluble fraction called **amylopectin**. These two fractions appear to correspond to different carbohydrates of high molecular weight and formula $(C_6H_{10}O_5)_n$. Upon treatment with acid or under the influence of enzymes, the components of starch are hydrolyzed progressively to dextrin (a mixture of low-molecular-weight polysaccharides), (+)-maltose, and finally D-(+)-glucose. (A mixture of all these is found in corn sirup, for example.) Both amylose and amylopectin are made up of D-(+)-glucose units, but differ in molecular size and shape.

39.8   Structure of amylose. End group analysis

(+)-Maltose is the only disaccharide that is obtained by hydrolysis of amylose, and D-(+)-glucose is the only monosaccharide. To account for this, it has been proposed that amylose is made up of chains of many D-(+)-glucose units, each unit joined by an *alpha*-glycoside linkage to C-4 of the next one.

Amylose
(chair conformations assumed)

We could conceive of a structure for amylose in which α- and β-linkages regularly alternate. However, a compound of such a structure would be expected to yield (+)-cellobiose as well as (+)-maltose unless hydrolysis of the β-linkages occurred much faster than hydrolysis of the α-linkages. Since hydrolysis of the β-linkage in (+)-cellobiose is actually slower than hydrolysis of the α-linkage in (+)-maltose, such a structure seems unlikely.

How many of these α-D-(+)-glucose units are there per molecule of amylose, and what are the shapes of these large molecules? These are difficult questions, and attempts to find the answers have made use of chemical and enzymatic methods, and of physical methods like x-ray analysis, electron microscopy, osmotic pressure and viscosity measurements, and behavior in an ultracentrifuge.

Valuable information about molecular size and shape has been obtained by the combination of methylation and hydrolysis that was so effective in studying the structures of disaccharides. D-(+)-Glucose, a monosaccharide, contains five free —OH groups and forms a pentamethyl derivative, methyl tetra-O-methyl-

Amylose

$\xrightarrow{\text{Me}_2\text{SO}_4, \text{NaOH}}$

Methylated amylose

$\downarrow$ HCl

2,3,6-Tri-O-methyl-D-glucose
(α-anomer)

D-glucopyranoside. When two D-(+)-glucose units are joined together, as in (+)-maltose, each unit contains four free —OH groups; an octamethyl derivative is formed. If each D-(+)-glucose unit in amylose is joined to two others, it contains only three free —OH groups; methylation of amylose should therefore yield a compound containing only three —OCH$_3$ groups per glucose unit. What are the facts?

When amylose is methylated and hydrolyzed there is obtained, as expected, 2,3,6-tri-O-methyl-D-glucose. But there is also obtained a little bit of 2,3,4,6-tetra-O-methyl-D-glucose, amounting to about 0.2–0.4% of the total product. Con-

2,3,4,6-Tetra-O-methyl-D-glucose
(α-anomer)

sideration of the structure of amylose shows that this, too, is to be expected, and an important principle emerges: that of **end group analysis** (Fig. 39.3, on the next page).

Each D-glucose unit in amylose is attached to two other D-glucose units, one through C–1 and the other through C–4, with C–5 in every unit tied up in the pyranose ring. As a result, free —OH groups at C–2, C–3, and C–6 are available for methylation. But this is not the case for *every* D-glucose unit. Unless the amylose chain is cyclic, it must have two ends. At one end there should be a D-glucose unit that contains a "free" aldehyde group. At the other end there should be a D-glucose unit that has a free —OH on C–4. This last D-glucose unit should undergo methylation at *four* —OH groups, and on hydrolysis should give a molecule of 2,3,4,6-tetra-O-methyl-D-glucose.

Thus each molecule of completely methylated amylose that is hydrolyzed should yield one molecule of 2,3,4,6-tetra-O-methyl-D-glucose; from the number of molecules of tri-O-methyl-D-glucose formed *along with* each molecule of the tetramethyl compound, we can calculate the length of the amylose chain.

Here we see an example of the use of end group analysis to determine chain length. A methylation that yields 0.25% of tetra-O-methyl-D-glucose shows that for every end group (with a free —OH on C–4) there are about 400 chain units.

But physical methods suggest that the chains are even longer than this. Molecular weights range from 150 000 to 600 000, indicating 1000 to 4000 glucose units per molecule. Evidently some degradation of the chain occurs during the methylation step; hydrolysis of only a few glycoside linkages in the alkaline medium would break the chain into much shorter fragments.

Problem 39.13 Consider an amylose chain of 4000 glucose units. At how many places must cleavage occur to lower the average length to 2000 units? To 1000? To 400? What percentage of the total number of glycoside links are hydrolyzed in each case?

Amylose

Me$_2$SO$_4$, NaOH

Methylated amylose

HCl

2,3,4-Tetra-*O*-methyl-D-glucose
0.3% yield

and

2,3,6-Tri-*O*-methyl-D-glucose
(*n* + 1) molecules

Figure 39.3   End group analysis. Hydrolysis of methylated amylose. End unit of long molecule gives 2,3,4,6-tetra-*O*-methyl-D-glucose; other units give 2,3,6-tri-*O*-methyl-D-glucose.

Amylose, then, is believed to be made up of long chains, each containing 1000 or more D-glucose units joined together by α-linkages as in (+)-maltose; there is little or no branching of the chain.

Amylose is the fraction of starch that gives the intense blue color with iodine. X-ray analysis shows that the chains are coiled in the form of a helix (like a spiral staircase), inside which is just enough space to accommodate an iodine molecule; the blue color is due to entrapped iodine molecules.

Problem 39.14 On the basis of certain evidence, it has been suggested that the rings of amylose have a twist-boat conformation, rather than the usual chair conformation. (a) What feature would tend to make any chair conformation unstable? (b) Suggest a twist-boat conformation that would avoid this difficulty. (*Hint*: What are the largest groups attached to a ring in amylose?)

Problem 39.15 When one mole of a disaccharide like (+)-maltose is treated with periodic acid (under conditions that minimize hydrolysis of the glycoside link), three moles of formic acid (and one of formaldehyde) are obtained.

H
CH₂OH H
HO
HO
H   OH
H
HCOOH

HCHO

CH₂OH H
O
H H
HO
H   OH
H   OH
HCOOH

HCOOH

(a) Show what would happen to amylose (see formula on p. 1332) when treated with HIO₄. (b) How could this reaction be used to determine chain length? (c) Oxidation by HIO₄ of 540 mg of amylose (from the sago plant) yielded 0.0102 millimoles of HCOOH. What is the chain length of this amylose?

39.9 Structure of amylopectin

Amylopectin is hydrolyzed to the single disaccharide (+)-maltose; the sequence of methylation and hydrolysis yields chiefly 2,3,6-tri-*O*-methyl-D-glucose. Like amylose, amylopectin is made up of chains of D-glucose units, each unit joined by an *alpha*-glycoside linkage to C–4 of the next one. However, its structure is more complex than that of amylose.

Molecular weights determined by physical methods show that there are up to a million D-glucose units per molecule. Yet hydrolysis of methylated amylopectin gives as high as 5% of 2,3,4,6-tetra-*O*-methyl-D-glucose, indicating only 20 units per chain. How can these facts be reconciled by the same structure?

The answer is found in the following fact: along with the trimethyl and tetramethyl compounds, hydrolysis yields 2,3-di-*O*-methyl-D-glucose and in an amount nearly equal to that of the tetramethyl derivative.

Methylated amylopectin

2,3,6-Tri-O-methyl-D-glucose
~90%

+

2,3,4,6-Tetra-O-methyl-D-glucose
~5%

+

2,3-Di-O-methyl-D-glucose
~5%

Amylopectin has a highly branched structure consisting of several hundred short chains of about 20–25 D-glucose units each. One end of each of these chains is joined through C–1 to a C–6 on the next chain.

Amylopectin
(chair conformations assumed)

Schematically the amylopectin molecule is believed to be something like this:

CHO

Amylopectin

Glycogen, the form in which carbohydrate is stored in animals to be released upon metabolic demand, has a structure very similar to that of amylopectin, except that the molecules appear to be more highly branched, and to have shorter chains (12–18 D-glucose units each).

Problem 39.16 Polysaccharides known as *dextrans* have been used as substitutes for blood plasma in transfusions; they are made by the action of certain bacteria on (+)-sucrose. Interpret the following properties of a dextran: Complete hydrolysis by acid yields only D-(+)-glucose. Partial hydrolysis yields only one disaccharide and only one trisaccharide, which contain only α-glycoside linkages. Upon methylation and hydrolysis, there is obtained chiefly 2,3,4-tri-*O*-methyl-D-glucose, together with smaller amounts of 2,4-di-*O*-methyl-D-glucose and 2,3,4,6-tetra-*O*-methyl-D-glucose.

Problem 39.17 Polysaccharides called *xylans* are found along with cellulose in wood and straw. Interpret the following properties of a sample of xylan: Its large negative rotation suggests β-linkages. Complete hydrolysis by acids yields only D-(+)-xylose. Upon methylation and hydrolysis, there is obtained chiefly 2,3-di-*O*-methyl-D-xylose, together with smaller amounts of 2,3,4-tri-*O*-methyl-D-xylose and 2-*O*-methyl-D-xylose.

39.10 Cyclodextrins

When starch is treated with a particular enzyme (the amylase of *Bacillus macerans*), there is formed a mixture of *cyclodextrins*: polysaccharides of low molecular weight belonging to the general class called *oligosaccharides* (*oligo* = few).

A cyclodextrin consists of six, seven, eight, or more D-glucose units joined through 1,4-*alpha* linkages in such a way as to form a ring—a chain bracelet each link of which is a pyranose hexagon. These rings are doughnut-shaped, much as crown ethers are (Sec. 19.10), but with a number of important differences. The smallest of them, α-cyclodextrin, has a diameter about twice that of 18-crown-6, and its hole (4.5 Å across) is about twice as broad.

This hole is tapered slightly, so that the molecule is shaped like a tiny pail with the bottom knocked out (see Fig. 39.4, on the next page). Making up the sides is a loop of six or more hexagons, each one lying roughly in the plane of the sides; the depth of the pail is thus the width of the pyranose ring. Outside the pail, around the "upper", larger rim lie the secondary —OH groups of C–2 and C–3; around the "lower", smaller rim lie the primary —OH groups of C–6, that is, the —CH$_2$OH groups. The inside of the pail consists of three bands, one on top of another: two bands of C—H's and, in between, a band of glycosidic O's.

Like a crown ether, a cyclodextrin can act as a host to guest molecules. Indeed, it was in connection with this property of cyclodextrins that the phenomenon now

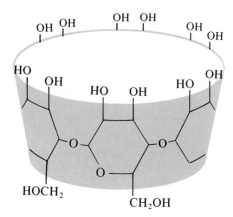

Figure 39.4   A schematic representation of α-cyclodextrin. The secondary —OH groups face outward about the "upper" rim; the —CH₂OH groups face outward about the "lower" rim. The cavity is lined with C—H's and glycosidic O's in three bands lying one above another.

known as the host–guest relationship was first recognized. But, in contrast to a crown ether, a cyclodextrin has a polar, hydrophilic outside and a relatively non-polar lipophilic inside. This leads naturally to two important results: (a) into its lipophilic interior a cyclodextrin typically takes as a guest, not an ion, but a non-polar organic molecule or the non-polar end of an organic molecule; and (b) its hydrophilic exterior confers water solubility on the resulting complex. How well a guest molecule is accommodated depends upon its size and polarity, and the size of the particular cyclodextrin.

Cyclodextrins can be used: to catalyze organic reactions, often with regio-selectivity and a degree of stereoselectivity; and, most important, as comparatively simple models by which to study the action of enzymes.

The effects of cyclodextrins on chemical reactions can arise in a number of ways.

(a) They can simply hide certain parts of a guest molecule and expose other parts.

(b) They can change the conformation of the guest.

(c) Their lipophilic lining provides a non-polar medium for the guest—but within a polar solvent.

(d) Their —OH groups can participate in the reaction: either directly—as bases and nucleophiles or as hydrogen-bonding sites—or via transient inter-mediates (esters, for example) formed by reaction with the host or with the attacking reagent.

The particular usefulness of cyclodextrins as enzyme models comes from the fact that, like enzymes (see, for example, Sec. 41.2), they first *bind* the substrate and then, through substituent groups, *act upon it*: clearly, an example of symphoria.

Problem 39.18   The structure of cyclodextrins is shown, not only by x-ray analysis, but also by evidence of the kind we have already dealt with. Predict in detail the response expected from cyclodextrins to each of the following reagents or analyses: (a) Fehling's solution; (b) acidic hydrolysis; (c) methylation followed by acidic hydrolysis; (d) periodic acid; (e) molecular weight determination.

Problem 39.19 When sodium benzenesulfonate is held by α-cyclodextrin, one end of the molecule is believed to protrude. Which end would you expect this to be, and why?

Problem 39.20 A mixture of α-, β-, and γ-cyclodextrins (which contain, respectively, six, seven, and eight glucose units) can be separated by the selective precipitation of each component upon the successive addition of three compounds: cyclohexane, fluorobenzene, and anthracene (Sec. 34.15). Which compound precipitates which cyclodextrin, and why?

Problem 39.21 Cyclodextrins can be used to separate a mixture of o-, m-, and p-cymenes (isopropyltoluenes). Can you suggest how this might be done?

39.11 Structure of cellulose

Cellulose is the chief component of wood and plant fibers; cotton, for instance, is nearly pure cellulose. It is insoluble in water and tasteless; it is a non-reducing carbohydrate. These properties, in part at least, are due to its extremely high molecular weight.

Cellulose has the formula $(C_6H_{10}O_5)_n$. Complete hydrolysis by acid yields D-(+)-glucose as the only monosaccharide. Hydrolysis of completely methylated cellulose gives a high yield of 2,3,6-tri-O-methyl-D-glucose. Like starch, therefore,

Cellulose

cellulose is made up of chains of D-glucose units, each unit joined by a glycoside linkage to C–4 of the next.

Cellulose differs from starch, however, in the configuration of the glycoside linkage. Upon treatment with acetic anhydride and sulfuric acid, cellulose yields octa-O-acetylcellobiose; there is evidence that all glycoside linkages in cellulose, like the one in (+)-cellobiose, are *beta* linkages.

Physical methods give molecular weights for cellulose ranging from 250 000 to 1 000 000 or more; it seems likely that there are at least 1500 glucose units per molecule. End group analysis by both methylation and periodic acid oxidation gives a chain length of 1000 glucose units or more. X-ray analysis and electron microscopy indicate that these long chains lie side by side in bundles, undoubtedly held together by hydrogen bonds between the numerous neighboring —OH groups. These bundles are twisted together to form rope-like structures, which themselves are grouped to form the fibers we can see. In wood these cellulose "ropes" are embedded in lignin to give a structure that has been likened to reinforced concrete.

39.12 Reactions of cellulose

We have seen that the glycoside linkages of cellulose are broken by the action of acid, each cellulose molecule yielding many molecules of D-(+)-glucose. Now let us look briefly at reactions of cellulose in which the chain remains essentially intact. Each glucose unit in cellulose contains three free —OH groups; these are the positions at which reaction occurs.

These reactions of cellulose, carried out to modify the properties of a cheap, available, ready-made polymer, are of tremendous industrial importance.

39.13   Cellulose nitrate

Like any alcohol, cellulose forms esters. Treatment with a mixture of nitric and sulfuric acid converts cellulose into *cellulose nitrate*. The properties and uses of the product depend upon the extent of nitration.

Guncotton, which is used in making smokeless powder, is very nearly completely nitrated cellulose, and is often called *cellulose trinitrate* (three nitrate groups per glucose unit).

Pyroxylin is less highly nitrated material containing between two and three nitrate groups per glucose unit. It is used in the manufacture of plastics like celluloid and collodion, in photographic film, and in lacquers. It has the disadvantage of being flammable, and forms highly toxic nitrogen oxides upon burning.

39.14   Cellulose acetate

In the presence of acetic anhydride, acetic acid, and a little sulfuric acid, cellulose is converted into the triacetate. Partial hydrolysis removes some of the acetate groups, degrades the chains to smaller fragments (of 200–300 units each), and yields the vastly important commercial *cellulose acetate* (roughly a *di*acetate).

Cellulose acetate is less flammable than cellulose nitrate and has replaced the nitrate in many of its applications, in safety-type photographic film, for example. When a solution of cellulose acetate in acetone is forced through the fine holes of a spinnerette, the solvent evaporates and leaves solid filaments. Threads from these filaments make up the material known as *acetate rayon*.

39.15   Rayon. Cellophane

When an alcohol is treated with carbon disulfide and aqueous sodium hydroxide, there is obtained a compound called a *xanthate*. Treatment of the xanthate with aqueous acid regenerates the starting materials.

$$RONa + S{=}C{=}S \longrightarrow RO{-}\underset{\underset{S}{\|}}{C}{-}SNa$$

$$A \text{ xanthate}$$

$$\xrightarrow{\;H^+\;} ROH + CS_2$$

Cellulose undergoes an analogous reaction to form *cellulose xanthate*, which dissolves in the alkali to form a viscous colloidal dispersion called *viscose*.

When viscose is forced through a spinnerette into an acid bath, cellulose is regenerated in the form of fine filaments which yield threads of the material known as *rayon*. There are other processes for making rayon, but the viscose process is still the principal one used in the United States.

If viscose is forced through a narrow slit, cellulose is regenerated as thin sheets which, when softened by glycerol, are used for protective films (Cellophane).

Although rayon and Cellophane are often spoken of as "regenerated cellulose", they are made up of much shorter chains than the original cellulose because of degradation by the alkali treatment.

39.16 Cellulose ethers

Industrially, cellulose is alkylated by the action of alkyl chlorides (cheaper than sulfates) in the presence of alkali. Considerable degradation of the long chains is unavoidable in these reactions.

Methyl, ethyl, and benzyl ethers of cellulose are important in the production of textiles, films, and various plastic objects.

PROBLEMS

1. (+)-*Gentiobiose*, $C_{12}H_{22}O_{11}$, is found in the roots of gentians. It is a reducing sugar, forms an osazone, undergoes mutarotation, and is hydrolyzed by aqueous acid or by emulsin to D-glucose. Methylation of (+)-gentiobiose, followed by hydrolysis, gives 2,3,4,6-tetra-*O*-methyl-D-glucose and 2,3,4-tri-*O*-methyl-D-glucose. What is the structure and systematic name of (+)-gentiobiose?

2. (a) (+)-*Trehalose*, $C_{12}H_{22}O_{11}$, a non-reducing sugar found in young mushrooms, gives only D-glucose when hydrolyzed by aqueous acid or by maltase. Methylation gives an octa-*O*-methyl derivative that, upon hydrolysis, yields only 2,3,4,6-tetra-*O*-methyl-D-glucose. What is the structure and systematic name for (+)-trehalose?

(b) (−)-*Isotrehalose* and (+)-*neotrehalose* resemble trehalose in most respects. However, isotrehalose is hydrolyzed by either emulsin or maltase, and neotrehalose is hydrolyzed only by emulsin. What are the structures and systematic names for these two carbohydrates?

3. *Ruberythric acid*, $C_{25}H_{26}O_{13}$, a non-reducing glycoside, is obtained from madder root. Complete hydrolysis gives *alizarin* ($C_{14}H_8O_4$), D-glucose, and D-xylose; graded

Alizarin

hydrolysis gives alizarin and *primeverose*, $C_{11}H_{20}O_{10}$. Oxidation of primeverose with bromine water, followed by hydrolysis, gives D-gluconic acid and D-xylose. Methylation of primeverose, followed by hydrolysis, gives 2,3,4-tri-*O*-methyl-D-xylose and 2,3,4-tri-*O*-methyl-D-glucose.

What structure or structures are possible for ruberythric acid? How can any uncertainties be cleared up?

4. (+)-*Raffinose*, a non-reducing sugar found in beet molasses, has the formula $C_{18}H_{32}O_{16}$. Hydrolysis by acid gives D-fructose, D-galactose, and D-glucose; hydrolysis by the enzyme α-galactosidase gives D-galactose and sucrose; hydrolysis by invertase (a sucrose-splitting enzyme) gives D-fructose and the disaccharide *melibiose*.

Methylation of raffinose, followed by hydrolysis, gives 1,3,4,6-tetra-*O*-methyl-D-fructose, 2,3,4,6-tetra-*O*-methyl-D-galactose, and 2,3,4-tri-*O*-methyl-D-glucose.

What is the structure of raffinose? Of melibiose?

5. (+)-*Melezitose*, a non-reducing sugar found in honey, has the formula $C_{18}H_{32}O_{16}$. Hydrolysis by acid gives D-fructose and two moles of D-glucose; partial hydrolysis gives D-glucose and the disaccharide *turanose*. Hydrolysis by maltase gives D-glucose and D-fructose; hydrolysis by another enzyme gives sucrose.

Methylation of melezitose, followed by hydrolysis, gives 1,4,6-tri-*O*-methyl-D-fructose and two moles of 2,3,4,6-tetra-*O*-methyl-D-glucose.

(a) What structure of melezitose is consistent with these facts? What is the structure of turanose?

Melezitose reacts with four moles of HIO_4 to give two moles of formic acid but no formaldehyde.

(b) Show that the absence of formaldehyde means either a furanose or pyranose structure for the fructose unit, and either a pyranose or septanose (seven-membered ring) structure for the glucose units.

(c) How many moles of HIO_4 would be consumed and how many moles of formic acid would be produced if the two glucose units had septanose rings? (d) Answer (c) for one septanose ring and one pyranose ring. (e) Answer (c) for two pyranose rings. (f) What can you say about the size of the rings in the glucose units?

(g) Answer (c) for a pyranose ring in the fructose unit; for a furanose ring.

(h) What can you say about the size of the ring in the fructose unit?

(i) Are the oxidation data consistent with the structure of melezitose you gave in (a)?

6. The sugar, (+)-*panose*, was first isolated by S. C. Pan and co-workers (at Joseph E. Seagram and Sons, Inc.) from a culture of *Aspergillus niger* on maltose. Panose has a mol. wt. of approximately 475–500. Hydrolysis gives glucose, maltose, and an isomer of maltose called isomaltose. Methylation and hydrolysis of panose gives 2,3,4-tri-, 2,3,6-tri-, and 2,3,4,6-tetra-*O*-methyl-D-glucose in essentially equimolar amounts. The high positive rotation of panose is considered to exclude the possibility of any β-linkages.

(a) How many monosaccharide units make up a molecule of panose? In how many ways might these be arranged?

(b) Oxidation of panose to the aldonic acid, followed by hydrolysis, gives *no* maltose; reduction of panose to panitol, followed by hydrolysis, gives glucitol and maltitol (the reduction product of maltose). Can you now draw a single structure for panose? What must be the structure of isomaltose?

(c) Panose and isomaltose can be isolated from the partial hydrolysis products of amylopectin. What bearing does this have on the structure of amylopectin?

7. Cellulose can be oxidized by N_2O_4 to $[(C_5H_7O_4)COOH]_n$. (a) What is the structure of this product? (b) What will it give on hydrolysis of the chain? What is the name of this hydrolysis product?

(c) The oxidation product in (a) is readily decarboxylated to $(C_5H_8O_4)_n$. What will this give on hydrolysis of the chain? What is the name of this hydrolysis product? Is it a D or L compound?

8. Suggest structural formulas for the following polysaccharides, neglecting the stereochemistry of the glycoside linkages:

(a) An *araban* from peanut hulls yields only L-arabinose on hydrolysis. Methylation, followed by hydrolysis, yields equimolar amounts of 2,3,5-tri-*O*-methyl-L-arabinose, 2,3-di-*O*-methyl-L-arabinose, and 3-*O*-methyl-L-arabinose.

(b) A *mannan* from yeast yields only D-mannose on hydrolysis. Methylation, followed by hydrolysis, yields 2,3,4,6-tetra-*O*-methyl-D-mannose, 2,4,6-tri-*O*-methyl-D-mannose, 3,4,6-tri-*O*-methyl-D-mannose, and 3,4-di-*O*-methyl-D-mannose in a molecular ratio of 2:1:1:2, together with small amounts of 2,3,4-tri-*O*-methyl-D-mannose.

9. When a *xylan* (see Problem 39.17, p. 1337) is boiled with dilute hydrochloric acid, a pleasant-smelling liquid, *furfural*, $C_5H_4O_2$, steam-distills. Furfural gives positive tests with Tollens' and Schiff's reagents; it forms an oxime and a phenylhydrazone but not an osazone. Furfural can be oxidized by $KMnO_4$ to A, $C_5H_4O_3$, which is soluble in aqueous $NaHCO_3$.

Compound A can be readily decarboxylated to B, C_4H_4O, which can be hydrogenated to C, C_4H_8O. C gives no tests for functional groups except solubility in cold concentrated H_2SO_4; it gives negative tests for unsaturation with dilute $KMnO_4$ or Br_2/CCl_4.

Prolonged treatment of C with HCl gives D, $C_4H_8Cl_2$, which on treatment with KCN gives E, $C_6H_8N_2$. E can be hydrolyzed to F, $C_6H_{10}O_4$, identifiable as adipic acid.

What is the structure of furfural? Of compounds A through E?

10. Give a likely structure for each of the following polysaccharides:

(a) *Alginic acid*, from sea weed, is used as a thickening agent in ice cream and other foods. Hydrolysis yields only D-mannuronic acid. Methylation, followed by hydrolysis, yields 2,3-di-*O*-methyl-D-mannuronic acid. (Mannuronic acid is $HOOC(CHOH)_4CHO$.) The glycoside linkages in alginic acid are thought to be *beta*.

(b) *Pectic acid* is the main constituent of the *pectin* responsible for the formation of jellies from fruits and berries. Methylation of pectic acid, followed by hydrolysis, gives only 2,3-di-*O*-methyl-D-galacturonic acid. The glycoside linkages in pectic acid are thought to be *alpha*.

(c) *Agar*, from sea weed, is used in the growing of microorganisms. Hydrolysis yields a 9:1:1 molar ratio of D-galactose, L-galactose, and sulfuric acid. Methylation, followed by hydrolysis, yields 2,4,6-tri-*O*-methyl-D-galactose, 2,3-di-*O*-methyl-L-galactose, and sulfuric acid in the same 9:1:1 ratio. What uncertainties are there in your proposed structure?

11. The main constituent of the capsule surrounding the Type III pneumonococcus, and the substance responsible for the specificity of its antigen–antibody reactions, is a polysaccharide (mol. wt. about 150 000). Hydrolysis yields equimolar amounts of D-glucose and D-glucuronic acid, $HOOC(CHOH)_4CHO$; careful hydrolysis gives cellobiuronic acid (the uronic acid related to cellobiose). Methylation, followed by hydrolysis, gives equimolar amounts of 2,3,6-tri-*O*-methyl-D-glucose and 2,4-di-*O*-methyl-D-glucuronic acid.

What is a likely structure for the polysaccharide?

12. Draw structures of compounds G through J:

amylose + HIO_4 $\longrightarrow$ G + a little HCOOH and HCHO
G + bromine water $\longrightarrow$ H
H + H_2O, H^+ $\longrightarrow$ I $(C_4H_8O_5)$ + J $(C_2H_2O_3)$

13. (a) Show what would happen to cellulose when treated with HIO_4. (b) How could this reaction be used to determine chain length? (c) If oxidation by HIO_4 of 203 mg of a sample of cellulose yields 0.0027 millimoles of HCOOH, what is the chain length of the cellulose?

14. Aromatic chlorination can be brought about not only by hypochlorous acid, HOCl (Problem 14.5, p. 511), but also by alkyl hypohalites, ROCl, formed by the reaction between alcohols and HOCl.

(a) Outline all steps in a likely mechanism for the acid-catalyzed chlorination of anisole by *tert*-butyl hypochlorite, *t*-BuOCl.

(b) Chlorination of anisole by HOCl or *t*-BuOCl gives a mixture of *o*- and *p*-chloro-anisoles. In the presence of α-cyclodextrin, however, chlorination by HOCl gives almost exclusively the *para* product, and takes place faster than in the absence of the cyclodextrin. How might you account for both the regioselectivity and the enhancement of rate?

(c) An α-cyclodextrin methylated at all C–2 and C–6 positions exerts an effect comparable to that of the unmethylated cyclodextrin. Can you now be more specific in your answer to part (b)?

40

Amino Acids and Proteins

40.1 Introduction

The name **protein** is taken from the Greek *proteios*, which means *first*. This name is well chosen. Of all chemical compounds, proteins must almost certainly be ranked first, for they are the substance of life.

Proteins make up a large part of the animal body, they hold it together, and they run it. They are found in all living cells. They are the principal material of skin, muscle, tendons, nerves, and blood; of enzymes, antibodies, and many hormones.

(Only the nucleic acids, which control heredity, can challenge the position of proteins; and the nucleic acids are important because they direct the synthesis of proteins.)

Chemically, proteins are high polymers. They are polyamides, and the monomers from which they are derived are the α-amino carboxylic acids. A single protein molecule contains hundreds or even thousands of amino acid units; these units can be of twenty-odd different kinds. The number of different combinations, that is, the number of different protein molecules that are possible, is almost infinite. It is likely that tens of thousands of different proteins are required to make up and run an animal body; and this set of proteins is not identical with the set required by an animal of a different kind.

In this chapter we shall look first at the chemistry of the amino acids, and then briefly at the proteins that they make up. Our chief purpose will be to see the ways in which the structures of these enormously complicated molecules are being worked out, and how, in the last analysis, all this work rests on the basic principles of organic structural theory: on the concepts of bond angle and bond length, group size and shape, hydrogen bonding, resonance, acidity and basicity, optical activity, configuration and conformation.

40.2 Structure of amino acids

Table 40.1 gives the structures and names of 23 amino acids that have been found in proteins. Certain of these (marked e) are the *essential* amino acids, which

Table 40.1 NATURAL AMINO ACIDS

Name	Abbreviation	Formula
(+)-Alanine	Ala A	$CH_3-\underset{\underset{^+NH_3}{\vert}}{C}HCOO^-$
(+)-Arginine^e	Arg R	$H_2N\underset{\underset{^+NH_2}{\Vert}}{C}NHCH_2CH_2CH_2-\underset{\underset{NH_2}{\vert}}{C}HCOO^-$
(−)-Asparagine	Asn N	$H_2NCOCH_2-\underset{\underset{^+NH_3}{\vert}}{C}HCOO^-$
(+)-Aspartic acid	Asp D	$HOOCCH_2-\underset{\underset{^+NH_3}{\vert}}{C}HCOO^-$
(−)-Cysteine	Cys C	$HSCH_2-\underset{\underset{^+NH_3}{\vert}}{C}HCOO^-$
(−)-Cystine	Cys—Cys	$^-OOC\underset{\underset{^+NH_3}{\vert}}{C}H-CH_2S-SCH_2-\underset{\underset{^+NH_3}{\vert}}{C}HCOO^-$
(+)-Glutamic acid	Glu E	$HOOCCH_2CH_2-\underset{\underset{^+NH_3}{\vert}}{C}HCOO^-$
(+)-Glutamine	Gln Q	$H_2NCOCH_2CH_2-\underset{\underset{^+NH_3}{\vert}}{C}HCOO^-$
Glycine	Gly G	$\underset{\underset{^+NH_3}{\vert}}{C}H_2COO^-$
(−)-Histidine^e	His H	(imidazole ring)$CH_2-\underset{\underset{^+NH_3}{\vert}}{C}HCOO^-$

^{e} Essential amino acid

Table 40.1   NATURAL AMINO ACIDS (*continued*)

Name	Abbreviation	Formula
(−)-Hydroxylysine	Hyl	$^+H_3NCH_2CHCH_2CH_2-CHCOO^-$ $\quad\quad\quad\ \ OH \quad\quad\quad\quad NH_2$
(−)-Hydroxyproline	Hyp	
(+)-Isoleucine[e]	Ile I	$CH_3CH_2CH(CH_3)-CHCOO^-$ $\quad\quad\quad\quad\quad\quad\quad ^+NH_3$
(−)-Leucine[e]	Leu L	$(CH_3)_2CHCH_2-CHCOO^-$ $\quad\quad\quad\quad\quad\quad ^+NH_3$
(+)-Lysine[e]	Lys K	$^+H_3NCH_2CH_2CH_2CH_2-CHCOO^-$ $\quad\quad\quad\quad\quad\quad\quad\quad NH_2$
(−)-Methionine[e]	Met M	$CH_3SCH_2CH_2-CHCOO^-$ $\quad\quad\quad\quad\quad\quad ^+NH_3$
(−)-Phenylalanine[e]	Phe F	
(−)-Proline	Pro P	
(−)-Serine	Ser S	$HOCH_2-CHCOO^-$ $\quad\quad\quad\quad ^+NH_3$
(−)-Threonine[e]	Thr T	$CH_3CHOH-CHCOO^-$ $\quad\quad\quad\quad\quad ^+NH_3$
(−)-Tryptophane[e]	Trp W	
(−)-Tyrosine	Tyr Y	
(+)-Valine[e]	Val V	$(CH_3)_2CH-CHCOO^-$ $\quad\quad\quad\quad\quad ^+NH_3$

[e] Essential amino acid

must be fed to young animals if proper growth is to take place; these particular amino acids evidently cannot be synthesized by the animal from the other materials in its diet.

We see that all are *alpha*-amino carboxylic acids; in two cases (proline and hydroxyproline) the amino group forms part of a pyrrolidine ring. This common feature gives the amino acids a common set of chemical properties, one of which is the ability to form the long polyamide chains that make up proteins. It is on these common chemical properties that we shall concentrate.

In other respects, the structures of these compounds vary rather widely. In addition to the carboxyl group and the amino group *alpha* to it, some amino acids contain a second carboxyl group (e.g., aspartic acid or glutamic acid), or a potential carboxyl group in the form of a carboxamide (e.g., asparagine); these are called *acidic amino acids*. Some contain a second basic group, which may be an amino group (e.g., lysine), a guanidino group (arginine), or the imidazole ring (histidine); these are called *basic amino acids*. Some of the amino acids contain benzene or heterocyclic ring systems, phenolic or alcoholic hydroxyl groups, halogen or sulfur atoms. Each of these ring systems or functional groups undergoes its own typical set of reactions. (See Fig. 40.1.)

40.3   Amino acids as dipolar ions

Although the amino acids are commonly shown as containing an amino group and a carboxyl group, $H_2NCHRCOOH$, certain properties, both physical and chemical, are not consistent with this structure:

(a) In contrast to amines and carboxylic acids, the amino acids are non-volatile crystalline solids which melt with decomposition at fairly high temperatures.

(b) They are insoluble in non-polar solvents like petroleum ether, benzene, or ether, and are appreciably soluble in water.

(c) Their aqueous solutions behave like solutions of substances of high dipole moment.

(d) Acidity and basicity constants are ridiculously low for —COOH and —NH$_2$ groups. Glycine, for example, has $K_a = 1.6 \times 10^{-10}$ and $K_b = 2.5 \times 10^{-12}$, whereas most carboxylic acids have K_a values of about 10^{-5} and most aliphatic amines have K_b values of about 10^{-4}.

All these properties are quite consistent with a dipolar ion structure for the amino acids (I).

$$^+H_3N—CHR—COO^-$$

I

Amino acids: *dipolar ions*

The physical properties—melting point, solubility, high dipole moment—are just what would be expected of such a salt. The acid–base properties also become understandable when it is realized that the measured K_a actually refers to the

Figure 40.1 Models of some amino acids, $^+H_3N—CH—COO^-$. (*a*) Glycine, (*b*) alanine, (*c*) methionine, (*d*) phenylalanine, (*e*) histidine. All contain the same core structure, but the side chain R varies widely. Proteins are made up of amino acid units, and it is the particular sequence of side chains that gives each protein its characteristic set of properties.

acidity of an ammonium ion, $RNH_3{}^+$,

$$^+H_3NCHRCOO^- + H_2O \rightleftarrows H_3O^+ + H_2NCHRCOO^-$$
Acid

$$K_a = \frac{[H_3O^+][H_2NCHRCOO^-]}{[^+H_3NCHRCOO^-]}$$

and K_b actually refers to the basicity of a carboxylate ion, $RCOO^-$.

$$^+H_3NCHRCOO^- + H_2O \rightleftarrows {}^+H_3NCHRCOOH + OH^-$$
Base

$$K_b = \frac{[^+H_3NCHRCOOH][OH^-]}{[^+H_3NCHRCOO^-]}$$

In aqueous solution, the acidity and basicity of an acid and its conjugate base (CH_3COOH and CH_3COO^-, or $CH_3NH_3{}^+$ and CH_3NH_2, for example) are related by the expression $K_a \times K_b = 10^{-14}$. From this it can be calculated that a K_a of 1.6×10^{-10} for the $-NH_3{}^+$ of glycine means $K_b = 6.3 \times 10^{-5}$ for $-NH_2$: a quite reasonable value for an aliphatic amine. In the same way, a K_b of 2.5×10^{-12} for the $-COO^-$ of glycine means $K_a = 4 \times 10^{-3}$ for $-COOH$: a quite reasonable value for a carboxylic acid containing the strongly electron-withdrawing (acid-strengthening) $-NH_3{}^+$ group.

When the solution of an amino acid is made alkaline, the dipolar ion I is converted into the anion II; the stronger base, hydroxide ion, removes a proton from the ammonium ion and displaces the weaker base, the amine.

$$^+H_3NCHRCOO^- + OH^- \rightleftarrows H_2NCHRCOO^- + H_2O$$

I		II	
Stronger acid	Stronger base	Weaker base	Weaker acid

When the solution of an amino acid is made acidic, the dipolar ion I is converted into the cation III; the stronger acid, H_3O^+, gives up a proton to the carboxylate ion, and displaces the weaker carboxylic acid.

$$^+H_3NCHRCOO^- + H_3O^+ \rightleftarrows {}^+H_3NCHRCOOH + H_2O$$

I		III	
Stronger base	Stronger acid	Weaker acid	Weaker base

In summary, the acidic group of a simple amino acid like glycine is $-NH_3{}^+$ not $-COOH$, and the basic group is $-COO^-$ not $-NH_2$.

Problem 40.1 In quite alkaline solution, an amino acid contains two basic groups, $-NH_2$ and $-COO^-$. Which is the more basic? To which group will a proton preferentially go as acid is added to the solution? What will the product be?

Problem 40.2 In quite acidic solution, an amino acid contains two acidic groups, $-NH_3{}^+$ and $-COOH$. Which is the more acidic? Which group will more readily give up a proton as base is added to the solution? What will the product be?

Problem 40.3   Account for the fact that *p*-aminobenzoic acid or *o*-aminobenzoic acid does not exist appreciably as the dipolar ion, but *p*-aminobenzenesulfonic acid (*sulfanilic acid*) does. (*Hint*: What is K_b for most aromatic amines?)

We must keep in mind that ions II and III, which contain a free —NH_2 or —COOH group, are in equilibrium with dipolar ion I; consequently, amino acids undergo reactions characteristic of amines and carboxylic acids. As ion II is removed, by reaction with benzoyl chloride, for example, the equilibrium shifts to supply more of ion II so that eventually the amino acid is completely benzoylated.

$$H_2NCHRCOO^- \underset{OH^-}{\overset{H^+}{\rightleftharpoons}} \ ^+H_3NCHRCOO^- \underset{OH^-}{\overset{H^+}{\rightleftharpoons}} \ ^+H_3NCHRCOOH$$

$$\text{II} \qquad\qquad\qquad \text{I} \qquad\qquad\qquad \text{III}$$

Where feasible we can speed up a desired reaction by adjusting the acidity or basicity of the solution in such a way as to increase the concentration of the reactive species.

Problem 40.4   Suggest a way to speed up (a) esterification of an amino acid; (b) acylation of an amino acid.

40.4   Isoelectric point of amino acids

What happens when a solution of an amino acid is placed in an electric field depends upon the acidity or basicity of the solution. In quite alkaline solution,

$$H_2NCHRCOO^- \underset{OH^-}{\overset{H^+}{\rightleftharpoons}} \ ^+H_3NCHRCOO^- \underset{OH^-}{\overset{H^+}{\rightleftharpoons}} \ ^+H_3NCHRCOOH$$

$$\text{II} \qquad\qquad\qquad \text{I} \qquad\qquad\qquad \text{III}$$

anions II exceed cations III, and there is a net migration of amino acid toward the anode. In quite acidic solution, cations III are in excess, and there is a net migration of amino acid toward the cathode. If II and III are exactly balanced, there is no net migration; under such conditions any one molecule exists as a positive ion and as a negative ion for exactly the same amount of time, and any small movement in the direction of one electrode is subsequently canceled by an equal movement back toward the other electrode. The hydrogen ion concentration of the solution in which a particular amino acid does not migrate under the influence of an electric field is called the **isoelectric point** of that amino acid.

A monoamino monocarboxylic acid, $^+H_3NCHRCOO^-$, is somewhat more acidic than basic (for example, glycine: $K_a = 1.6 \times 10^{-10}$ and $K_b = 2.5 \times 10^{-12}$). If crystals of such an amino acid are added to water, the resulting solution contains more of the anion II, $H_2NCHRCOO^-$, than of the cation III, $^+H_3NCHRCOOH$. This "excess" ionization of ammonium ion to amine ($\text{I} \rightleftharpoons \text{II} + \text{H}^+$) must be repressed, by addition of acid, to reach the isoelectric point, which therefore lies somewhat on the acid side of neutrality (pH 7). For glycine, for example, the isoelectric point is at pH 6.1.

Problem 40.5 (a) Will the isoelectric point be on the acid or alkaline side of pH 7 (neutrality) for a monoamino dicarboxylic acid? (b) For a diamino monocarboxylic acid? (c) Compare each of these isoelectric points with that for glycine.

An amino acid usually shows its lowest solubility in a solution at the isoelectric point, since here there is the highest concentration of the dipolar ion. As the solution is made more alkaline or more acidic, the concentration of one of the more soluble ions, II or III, increases.

Problem 40.6 Account for the fact that sulfanilic acid dissolves in alkalies but not in acids.

Problem 40.7 Suggest a way to separate a mixture of amino acids into three fractions: monoamino monocarboxylic acids, monoamino dicarboxylic acids (the acidic amino acids), and diamino monocarboxylic acids (the basic amino acids).

40.5 Configuration of natural amino acids

From the structures in Table 40.1, we can see that every amino acid except glycine contains at least one chiral center. As obtained by acidic or enzymatic hydrolysis of proteins, every amino acid except glycine has been found optically active. Stereochemical studies of these naturally occurring amino acids have shown that all have the same configuration about the carbon atom carrying the *alpha*-amino group, and that this configuration is the same as that in L-(-)-glyceraldehyde. Since the group R nearly always happens to have a lower Cahn–Ingold–Prelog priority than COOH, most of these have the S configuration (Sec. 4.16).

$$
\begin{array}{ccc}
& \text{COO}^- & & \text{CHO} \\
& | & & | \\
\text{H}_3\overset{+}{\text{N}} - & \text{C} - \text{H} & \text{HO} - & \text{C} - \text{H} \\
& | & & | \\
& \text{R} & & \text{CH}_2\text{OH} \\
& \text{L-Amino acid} & & \text{L-Glyceraldehyde}
\end{array}
$$

Problem 40.8 There are two amino acids in Table 40.1 that, although L like the others, happen to have the R configuration. Which ones are these, and why is their specification different?

Problem 40.9 Draw all possible stereoisomeric formulas for the amino acid threonine. Naturally occurring threonine gets its name from its relationship to the tetrose *threose*; on this basis which is the correct configuration for natural threonine?

Problem 40.10 Besides threonine, there are four amino acids in Table 40.1 that can exist in more than two stereoisomeric forms. (a) What are they? (b) How many isomers are possible in each case? Indicate enantiomers, diastereomers, any *meso* compounds.

40.6   Preparation of amino acids

Of the many methods that have been developed for synthesizing amino acids, we shall take up only one: **amination of α-halo acids**. Considered in its various modifications, this method is probably the most generally useful, although, like any of the methods, it cannot be applied to the synthesis of all the amino acids.

Sometimes an α-chloro or α-bromo acid is subjected to **direct ammonolysis** with a large excess (*Why?*) of concentrated aqueous ammonia. For example:

$$CH_3CH_2COOH \xrightarrow{Br_2, P} CH_3CHCOOH \xrightarrow{NH_3 \text{ (excess)}} CH_3CHCOO$$

Propionic acid   α-Bromopropionic acid (Br)   Alanine (NH$_3^+$)
                                                     70% yield

The necessary α-halo acids or esters can be prepared by the Hell–Volhard–Zelinsky halogenation of the unsubstituted acids (Sec. 23.19), or by a modification of the **malonic ester synthesis**, the usual route to the unsubstituted acids. For example:

$$Na^+ \begin{bmatrix} COOC_2H_5 \\ CH \\ COOC_2H_5 \end{bmatrix} \xrightarrow{C_6H_5CH_2Cl} HC-CH_2C_6H_5 \xrightarrow[\text{heat}]{KOH} \xrightarrow{HCl} H-C-CH_2C_6H_5$$

Sodiomalonic ester   Ethyl benzylmalonate (COOC$_2$H$_5$)   Benzylmalonic acid (COOH)

$\downarrow$ Br$_2$, ether, reflux

$$C_6H_5CH_2CHCOO^- \xleftarrow{NH_3 \text{ (excess)}} C_6H_5CH_2CHCOOH \xleftarrow{heat} Br-C-CH_2C_6H_5$$

(NH$_3^+$)   (Br)   (COOH / COOH)

Phenylalanine
35% overall yield

Better yields are generally obtained by the **Gabriel phthalimide synthesis** (Problem 11, p. 950); the α-halo esters are used instead of α-halo acids (*Why?*). A further modification, the **phthalimidomalonic ester method**, is a combined malonic ester–Gabriel synthesis.

Potassium phthalimide   +   Cl—CH$_2$COOC$_2$H$_5$ →   (phthalimide)N—CH$_2$COOC$_2$H$_5$

Ethyl chloroacetate

$\downarrow$ HCl, H$_2$O

$$Cl^- \, {}^+H_3N-CH_2COOH + \text{phthalic acid}$$

Glycine hydrochloride
89% overall yield

Potassium
phthalimide

+

Ethyl
bromomalonate

Phthalimidomalonic
ester

excess
ClCH$_2$COOC$_2$H$_5$,
heat

$$\text{HOOCCH}_2\text{—CHCOO}^- \xleftarrow[\text{heat}]{\text{base} \quad \text{conc. HCl}}$$

Aspartic acid
43% overall yield

These synthetic amino acids are, of course, optically inactive, and must be resolved if the active materials are desired for comparison with the naturally occurring acids or for synthesis of peptides (Sec. 40.10). There is growing interest in enantiotopic syntheses, which yield directly optically active amino acids; such preparation must, of course, be carried out in a chiral medium. We have already seen a promising example of such syntheses in Sec. 20.7.

Problem 40.11   Various amino acids have been made in the following ways:

Direct ammonolysis: glycine, alanine, valine, leucine, aspartic acid
Gabriel synthesis: glycine, leucine
Malonic ester synthesis: valine, isoleucine
Phthalimidomalonic ester method: serine, glutamic acid, aspartic acid

List the necessary starting materials in each case, and outline the entire sequence for one example from each group.

Problem 40.12   Acetaldehyde reacts with a mixture of KCN and NH$_4$Cl (**Strecker synthesis**) to give a product, C$_3$H$_6$N$_2$ (What is its structure?), which upon hydrolysis yields alanine. Show how the Strecker synthesis can be applied to the synthesis of glycine, leucine, isoleucine, valine, and serine (start with C$_2$H$_5$OCH$_2$CH$_2$OH). Make all required carbonyl compounds from readily available materials.

Problem 40.13   (a) Synthesis of amino acids by **reductive amination** (Sec. 26.11) is illustrated by the following synthesis of leucine:

$$\text{ethyl isovalerate + ethyl oxalate} \xrightarrow{\text{NaOC}_2\text{H}_5} \text{A (C}_{11}\text{H}_{18}\text{O}_5)$$

$$\text{A + 10\% H}_2\text{SO}_4 \xrightarrow{\text{boil}} \text{B (C}_6\text{H}_{10}\text{O}_3) + \text{CO}_2 + \text{C}_2\text{H}_5\text{OH}$$

$$\text{B + NH}_3 + \text{H}_2 \xrightarrow{\text{Pd. heat}} \text{leucine}$$

(b) Outline the synthesis by this method of alanine. Of glutamic acid.

40.7   Reactions of amino acids

The reactions of amino acids are in general the ones we would expect of compounds containing amino and carboxyl groups. In addition, any other groups that may be present undergo their own characteristic reactions.

Problem 40.14   Predict the products of the treatment of glycine with:

(a) aqueous NaOH   (d) acetic anhydride
(b) aqueous HCl   (e) $NaNO_2$ + HCl
(c) benzoyl chloride + aqueous NaOH   (f) C_2H_5OH + H_2SO_4
    (g) benzyl chlorocarbonate (carbobenzoxy chloride), $C_6H_5CH_2OCOCl$

Problem 40.15   Predict the products of the following reactions:

(a) *N*-benzoylglycine (*hippuric acid*) + $SOCl_2$ (g) proline + methyl iodide
(b) product of (a) + NH_3 (h) tyrosine + methyl sulfate, OH^-
(c) product of (a) + alanine   (i) glutamic acid + one mole
(d) product of (a) + C_2H_5OH $NaHCO_3$
(e) tyrosine + Br_2(aq)   (j) glutamic acid + excess ethyl
(f) asparagine + hot aqueous NaOH   alcohol + H_2SO_4 + heat

Problem 40.16   The reaction of primary aliphatic amines with nitrous acid gives a quantitative yield of nitrogen gas, and is the basis of the **Van Slyke determination of amino nitrogen**. What volume of nitrogen gas at S.T.P. would be liberated from 0.001 mole of: (a) leucine, (b) lysine, (c) proline?

Problem 40.17   When a solution of 9.36 mg of an unknown amino acid was treated with excess nitrous acid, there was obtained 2.01 mL of nitrogen at 748 mm and 20 °C. What is the minimum molecular weight for this compound? Can it be one of the amino acids found in proteins? If so, which one?

40.8   Peptides. Geometry of the peptide linkage

Peptides are amides formed by interaction between amino groups and carboxyl groups of amino acids. The amino group, —NHCO—, in such compounds is often referred to as the *peptide linkage*.

Depending upon the number of amino acid residues per molecule, they are known as *dipeptides*, *tripeptides*, and so on, and finally *polypeptides*. (By convention, peptides of molecular weight up to 10 000 are known as polypeptides and above that as proteins.) For example:

$$^+H_3NCH_2\overset{\overset{O}{\|}}{C}-NHCH_2COO^-$$

Gly-Gly
Glycylglycine
A dipeptide

$$^+H_3NCH_2\overset{\overset{O}{\|}}{C}-NH\overset{\underset{CH_3}{|}}{C}H\overset{\overset{O}{\|}}{C}-NH\overset{\underset{CH_2C_6H_5}{|}}{C}HCOO^-$$

Gly-Ala-Phe
Glycylalanylphenylalanine
A tripeptide

$$^+H_3N\overset{\underset{R}{|}}{C}H\overset{\overset{O}{\|}}{C}-(NH\overset{\underset{R}{|}}{C}H\overset{\overset{O}{\|}}{C}-)_n NH\overset{\underset{R}{|}}{C}HCOO^-$$

A polypeptide

A convenient way of representing peptide structures by use of standard abbreviations (see Table 40.1) is illustrated here. According to convention, the **N-terminal amino acid residue** (having the free amino group) is written at the left end, and the **C-terminal amino acid residue** (having the free carboxyl group) at the right end.

X-ray studies of amino acids and dipeptides indicate that the entire amide group is flat: carbonyl carbon, nitrogen, and the four atoms attached to them all lie in a plane. The short carbon–nitrogen distance (1.32 Å as compared with 1.47 Å for the usual carbon–nitrogen single bond) indicates that the carbon–nitrogen bond has considerable double-bond character (about 50%); as a result the angles of the bonds to nitrogen are similar to the angles about the trigonal carbon atom (Fig. 40.2).

Figure 40.2   Geometry of the peptide link. The carbon–nitrogen bond has much double-bond character. Carbonyl carbon, nitrogen, and the atoms attached to them lie in a plane.

Problem 40.18   (a) What contributing structure(s) would account for the double-bond character of the carbon–nitrogen bond? (b) What does this resonance mean in terms of orbitals?

Problem 40.19   At room temperature, N,N-dimethylformamide gives the following NMR spectrum:

   a singlet, δ 2.88, 3H b singlet, δ 2.97, 3H c singlet, δ 8.02, 1H

As the temperature is raised, signals a and b broaden and coalesce; finally, at 170 °C, they are merged into one sharp singlet. (a) How do you account for these observations? (b) What bearing do they have on the structure of the peptide linkage? (*Hint*: See Sec. 16.13.)

Peptides have been studied chiefly as a step toward the understanding of the much more complicated substances, the proteins. However, peptides are extremely important compounds in their own right: the tripeptide *glutathione*, for example, is found in most living cells; the nonapeptide *oxytocin* is a posterior pituitary hormone concerned with contraction of the uterus; α-*corticotropin*, made up of 39 amino acid residues, is one component of the adrenocorticotropic hormone ACTH.

We shall look at two aspects of the chemistry of peptides: how their structures are determined, and how they can be synthesized in the laboratory.

$$^+H_3NCHCH_2CH_2\overset{\overset{\displaystyle O}{\|}}{C}-NHCH\overset{\overset{\displaystyle O}{\|}}{C}-NHCH_2COOH \quad \textit{or} \quad \text{Glu-Cys-Gly}$$

COO⁻ ... CH₂SH

Glutathione
(Glutamylcysteinylglycine)

or

```
Ile-Tyr-Cys
 |       |
Gln-Asn-Cys-Pro-Leu-Gly(NH₂)
```

Oxytocin

```
Ser-Tyr-Ser-Met-Glu-His-Phe-Arg-Trp-Gly-Lys-Pro-Val-┐
 └─Gly-Lys-Lys-Arg-Arg-Pro-Val-Lys-Val-Tyr-Pro-Ala-Gly-┐
   └─Glu-Asp-Asp-Glu-Ala-Ser-Glu-Ala-Phe-Pro-Leu-Glu-Phe
```

α-Corticotropin (sheep)

40.9 Determination of structure of peptides. Terminal residue analysis. Partial hydrolysis

To assign a structure to a particular peptide, one must know (a) what amino acid residues make up the molecule and how many of each there are, and (b) the sequence in which they follow one another along the chain.

To determine the composition of a peptide, one hydrolyzes the peptide (in acidic solution, since alkali causes racemization) and determines the amount of each amino acid thus formed. One of the best ways of analyzing a mixture of amino acids is to separate the mixture into its components by chromatography—most commonly by ion-exchange chromatography, but sometimes, after conversion into the methyl esters (*Why?*), by gas chromatography.

From the weight of each amino acid obtained, one can calculate the number of moles of each amino acid, and in this way know the relative numbers of the various amino acid residues in the peptide. At this stage one knows what might be called the "empirical formula" of the peptide: the relative abundance of each amino acid residue in the peptide.

Problem 40.20 An analysis of the hydrolysis products of *salmine*, a polypeptide from salmon sperm, gave the following results:

	g/100 g salmine
Isoleucine	1.28
Alanine	0.89
Valine	3.68
Glycine	3.01
Serine	7.29
Proline	6.90
Arginine	86.40

What are the relative numbers of the various amino acid residues in salmine; that is, what is its empirical formula? (Why do the weights add up to more than 100 g?)

To calculate the "molecular formula" of the peptide—the actual number of each kind of residue in each peptide molecule—one needs to know the molecular weight. Molecular weights can be determined by chemical methods and by various physical methods: behavior in an ultracentrifuge, electrophoresis (Sec. 40.14), chromatography with molecular sieves.

Problem 40.21 The molecular weight of salmine (see the preceding problem) is about 10 000. What are the actual numbers of the various amino acid residues in salmine; that is, what is its molecular formula?

Problem 40.22 A protein was found to contain 0.29% tryptophane (mol. wt. 204). What is the minimum molecular weight of the protein?

Problem 40.23 (a) Horse hemoglobin contains 0.335% Fe. What is the minimum molecular weight of the protein? (b) Osmotic pressure measurements give a molecular weight of about 67 000. How many iron atoms are there per molecule?

There remains the most difficult job of all: to determine the sequence in which these amino acid residues are arranged along the peptide chain, that is, the structural formula of the peptide. This is accomplished by a combination of terminal residue analysis and partial hydrolysis.

Terminal residue analysis is the identifying of the amino acid residues at the ends of the peptide chain. The procedures used depend upon the fact that the residues at the two ends are different from all the other residues and from each other: one, the *N-terminal residue*, contains a free *alpha* amino group and the other, the *C-terminal residue*, contains a free carboxyl group *alpha* to a peptide linkage.

A very successful method of identifying the N-terminal residue (introduced in 1945 by Frederick Sanger of Cambridge University) makes use of 2,4-dinitrofluorobenzene (DNFB), which undergoes nucleophilic substitution by the free amino group to give an N-dinitrophenyl (DNP) derivative. The substituted peptide is hydrolyzed to the component amino acids, and the *N*-terminal residue, *labeled by the 2,4-dinitrophenyl group,* is separated and identified.

$$O_2N\langle\bigcirc\rangle F + H_2NCHC\!-\!NHCHC\!\sim \xrightarrow{\text{alkaline medium}} O_2N\langle\bigcirc\rangle NHCHC\!-\!NHCHC\!\sim$$

2,4-Dinitrofluorobenzene (DNFB)	Peptide	Labeled peptide

aq HCl, heat

$$O_2N\langle\bigcirc\rangle NHCHCOOH + {}^+H_3NCHCOOH, \text{etc.}$$

N-(2,4-Dinitrophenyl)amino acid   Unlabeled amino acids
(DNP-AA)

In its various modifications, however, the most widely used method of *N*-terminal residue analysis is one introduced in 1950 by Pehr Edman (Max Planck Institute of Biochemistry, Munich). This is based upon the reaction between an amino group and phenyl isothiocyanate to form a substituted thiourea (compare Sec. 36.7). Mild hydrolysis with hydrochloric acid selectively removes the *N*-terminal residue as the phenylthiohydantoin, which is then identified. The great

$$C_6H_5NCS + H_2NCHC\!-\!NHCHC\!\sim \xrightarrow{\text{alkaline medium}} C_6H_5N\!-\!C\!-\!NHCHC\!-\!NHCHC\!\sim$$

Phenyl isothiocyanate	Peptide	Labeled peptide

H₂O, HCl

$$C_6H_5N\quad NH + H_2NCHC\!\sim$$

A phenylthiohydantoin	Degraded peptide

One less residue

advantage of this method is that it leaves the rest of the peptide chain intact, so that the analysis can be repeated and the *new* terminal group of the shortened peptide identified. In 1967, Edman reported that this analysis could be carried out *automatically* in his "protein sequenator", which is now available in commercial form; with all operations controlled by a computer and the results displayed continuously on a recorder, residue after residue is identified. In practice it is not feasible to extend this analysis beyond about 20 residues, since by that point there is interference from the accumulation of amino acids formed by (slow) hydrolysis during the acid treatment.

Problem 40.24 Edman has also devised the highly sensitive "dansyl" method in which a peptide is treated with 5-dimethylaminonaphthalenesulfonyl chloride, followed by acidic hydrolysis. A derivative of the *N*-terminal residue is obtained which can be followed during its analysis by virtue of its characteristic fluorescence. What is the derivative? Why does it survive the acid treatment that cleaves the peptide bonds?

One successful method of determining the *C*-terminal residue has been enzymatic rather than chemical. The *C*-terminal residue is removed selectively by the enzyme *carboxypeptidase* (obtained from the pancreas), which cleaves only peptide linkages adjacent to *free alpha*-carboxyl groups in polypeptide chains. The analysis can be repeated on the shortened peptide and the *new C*-terminal residue identified, and so on.

Problem 40.25 The use of carboxypeptidase has an inevitable disadvantage. What would you expect this to be, and how could you allow for it in interpreting the analytical results?

Problem 40.26 There are a number of chemical methods for determining the *C*-terminal residue. For each of the following write equations to show what is happening and how it gives the identity of this residue: (a) treatment of the peptide with LiBH$_4$, followed by acidic hydrolysis and analysis; (b) treatment with hydrazine, NH$_2$NH$_2$, and analysis of the products. (*Hint*: What basic properties would you expect hydrazine to have?)

In practice it is not feasible to determine the sequence of all the residues in a long peptide chain by the stepwise removal of terminal residues. Instead, the chain is subjected to partial hydrolysis (acidic or enzymatic), and the fragments formed— dipeptides, tripeptides, and so on—are identified, with the aid of terminal residue analysis. When enough of these small fragments have been identified, it is possible to work out the sequence of residues in the entire chain.

To take an extremely simple example, there are six possible ways in which the three amino acids making up glutathione could be arranged; partial hydrolysis to the dipeptides glutamylcysteine (Glu-Cys) and cysteinylglycine (Cys-Gly) makes it clear that the cysteine is in the middle and that the sequence Glu-Cys-Gly is the correct one.

It was by the use of the approach just outlined that structures of such peptides as oxytocin and α-corticotropin (see p. 1357) were worked out. A milestone in protein chemistry was the determination of the entire amino acid sequence in the insulin molecule by a Cambridge University group headed by Frederick Sanger, who received the Nobel Prize in 1958 for this work. (See Problem 11, p. 1376.) Since then the number—and complexity—of completely mapped proteins has grown rapidly: the four chains of hemoglobin, for example, each containing 140-odd amino acid residues; chymotrypsinogen, with a single chain 246 units long; an immunoglobulin (*gamma*-globulin) with two chains of 446 units each and two chains of 214 units each—a total of 1320 amino acid residues.

As usual, final confirmation of the structure assigned to a peptide lies in its synthesis by a method that must unambiguously give a compound of the assigned structure. This problem is discussed in the following section.

Problem 40.27 Work out the sequence of amino acid residues in the following peptides:

(a) Asp,Glu,His,Phe,Val (commas indicate unknown sequence) *gives*
    Val-Asp + Glu-His + Phe-Val + Asp-Glu.
(b) Cys,Gly,His$_2$,Leu$_2$,Ser *gives* Cys-Gly-Ser + His-Leu-Cys + Ser-His-Leu.
(c) Arg,Cys,Glu,Gly$_2$,Leu,Phe$_2$,Tyr,Val *gives*
    Val-Cys-Gly + Gly-Phe-Phe + Glu-Arg-Gly + Tyr-Leu-Val + Gly-Glu-Arg.

40.10 Synthesis of peptides

Methods have been developed by which a single amino acid (or sometimes a di- or tripeptide) can be polymerized to yield polypeptides of high molecular weight. These products have been extremely useful as model compounds: to show, for example, what kind of x-ray pattern or infrared spectrum is given by a peptide of known, comparatively simple structure.

Most work on peptide synthesis, however, has had as its aim the preparation of compounds identical with naturally occurring ones. For this purpose a method must permit the joining together of optically active amino acids to form chains of predetermined length and with a predetermined sequence of residues. Syntheses of this sort not only have confirmed some of the particular structures assigned to natural peptides, but also—and this is more fundamental—have proved that peptides and proteins are indeed polyamides.

It was Emil Fischer who first prepared peptides (ultimately one containing 18 amino acid residues) and thus offered support for his proposal that proteins contain the amide link. It is evidence of his extraordinary genius that Fischer played the same role in laying the foundations of peptide and protein chemistry as he did in carbohydrate chemistry.

The basic problem of peptide synthesis is one of *protecting the amino group*. In bringing about interaction between the carboxyl group of one amino acid and the amino group of a different amino acid, one must prevent interaction between the carboxyl group and the amino group of the same amino acid. In preparing glycylalanine, for example, one must prevent the simultaneous formation of glycylglycine. Reaction can be forced to take place in the desired way by attaching to one amino acid a group that renders the —NH$_2$ unreactive. There are many such protecting groups; the problem is to find one that can be removed later without destruction of any peptide linkages that may have been built up.

$$^+H_3NCHCOO^- \longrightarrow Z-NHCHCOOH \longrightarrow Z-NHCHCOCl$$

|
R

Protection of amino group

$$Z-NHCHCOCl + {}^+H_3NCHCOO^- \longrightarrow Z-NHCHC-NHCHCOOH$$

Formation of peptide linkage

$$Z-NHCHC-NHCHCOOH \longrightarrow {}^+H_3NCHC-NHCHCOO^-$$

Removal of the protecting group

Peptide

We could, for example, benzoylate glycine ($Z = C_6H_5CO$), convert this into the acid chloride, allow the acid chloride to react with alanine, and thus obtain benzoylglycylalanine. But if we attempted to remove the benzoyl group by hydrolysis, we would simultaneously hydrolyze the other amide linkage (the peptide linkage) and thus destroy the peptide we were trying to make.

Of the numerous methods developed to protect an amino group, we shall look at just one: **acylation by benzyl chlorocarbonate**, also called **carbobenzoxy chloride**. (This method was introduced in 1932 by Max Bergmann and Leonidas Zervas of the University of Berlin, later of the Rockefeller Institute.) The reagent, $C_6H_5CH_2OCOCl$, is both an ester and an acid chloride of carbonic acid, $HOCOOH$; it is readily made by reaction between benzyl alcohol and phosgene (carbonyl chloride), $COCl_2$. (In what order should the alcohol and phosgene be mixed?)

$$CO + Cl_2 \xrightarrow{\text{active carbon, 200 °C}} Cl-\underset{O}{\overset{\|}{C}}-Cl \xrightarrow{C_6H_5CH_2OH} C_6H_5CH_2O-\underset{O}{\overset{\|}{C}}-Cl$$

Phosgene
(Carbonyl chloride)

Carbobenzoxy chloride
(Benzyl chlorocarbonate)

Like any acid chloride, the reagent can convert an amine into an amide, in this case, a carbamate (Sec. 24.23):

$$C_6H_5CH_2O-\underset{O}{\overset{\|}{C}}-Cl + H_2NR \longrightarrow C_6H_5CH_2O-\underset{O}{\overset{\|}{C}}-NHR$$

Amine

An amide
(A carbamate)

Such amides, $C_6H_5CH_2OCONHR$, differ from most amides, however, in one feature that is significant for peptide synthesis. The carbobenzoxy group can be

$$C_6H_5CH_2O-\underset{O}{\overset{\|}{C}}-NHR \begin{cases} \xrightarrow{H_2,\ Pd} C_6H_5CH_3 + \left[HO-\underset{O}{\overset{\|}{C}}-NHR\right] \longrightarrow CO_2 + RNH_2 \\ \\ \xrightarrow[\substack{cold \\ HOAc}]{HBr,} C_6H_5CH_2Br + \left[HO-\underset{O}{\overset{\|}{C}}-NHR\right] \longrightarrow CO_2 + RNH_2 \end{cases}$$

A carbamic acid
Unstable

cleaved by reagents that do not disturb peptide linkages: catalytic hydrogenation or hydrolysis with hydrogen bromide in cold acetic acid.

The carbobenzoxy method is illustrated by the synthesis of glycylalanine (Gly-Ala):

$$C_6H_5CH_2OCOCl + {}^+H_3NCH_2COO^- \longrightarrow C_6H_5CH_2OCO—NHCH_2COOH$$

Carbobenzoxy   Glycine   Carbobenzoxyglycine
chloride

$$\downarrow SOCl_2$$

$$C_6H_5CH_2OCO—NHCH_2COCl$$
Acid chloride of carbobenzoxyglycine

$$C_6H_5CH_2OCO—NHCH_2COCl + {}^+H_3NCHCOO^-$$
$$\underset{\underset{\text{Alanine}}{CH_3}}{|}$$

$$C_6H_5CH_2OCO—NHCH_2CO—NHCHCOOH$$
$$\underset{CH_3}{|}$$
Carbobenzoxyglycylalanine

$$C_6H_5CH_2OCO—NHCH_2CO—NHCHCOOH \xrightarrow{H_2,\ Pd} {}^+H_3NCH_2CO—NHCHCOO^-$$
$$\underset{CH_3}{|} \qquad\qquad\qquad\qquad\qquad \underset{CH_3}{|}$$

Glycylalanine
Gly-Ala

$$+ C_6H_5CH_3 + CO_2$$

Problem 40.28 (a) How could the preceding synthesis be extended to the tripeptide glycylalanylphenylalanine (Gly-Ala-Phe)?
(b) How could the carbobenzoxy method be used to prepare alanylglycine (Ala-Gly)?

Methods like this can be repeated over and over with the addition of a new unit each time. In this way the hormone oxytocin (p. 1357) was synthesized by Vincent du Vigneaud of Cornell Medical College, who received the Nobel Prize in 1955 for this and other work. In 1963, the total synthesis of the insulin molecule— with the 51 amino acid residues in the sequence mapped out by Sanger—was reported.

But the bottle-neck in such syntheses is the need to isolate and purify the new peptide made in each cycle; the time required is enormous, and the yield of product steadily dwindles. A major breakthrough came with the development of *solid-phase* peptide synthesis by R. Bruce Merrifield at Rockefeller University. Synthesis is carried out with the growing peptide *attached* chemically to polystyrene beads; as each new unit is added, the reagents and by-products are simply washed away, leaving the growing peptide behind, ready for another cycle. The method was automated, and in 1969 Merrifield announced that, using his "protein-making machine", he had synthesized—in *six weeks*—the enzyme ribonuclease, made up of 124 amino acid residues. In 1984 Merrifield received the Nobel Prize.

Problem 40.29 Give formulas for compounds A–G, and tell what is happening in each reaction.

polystyrene + CH_3OCH_2Cl $\xrightarrow{SnCl_4}$ A + CH_3OH

A + $C_6H_5CH_2OCONHCH_2COO^{-+}NHEt_3$ $\longrightarrow$ B + Et_3NHCl

B + dil. HBr $\longrightarrow$ C + $C_6H_5CH_2Br$ + CO_2

C + carbobenzoxyalanyl chloride $\longrightarrow$ D

D + dil. HBr $\longrightarrow$ E + $C_6H_5CH_2Br$ + CO_2

E + HBr $\xrightarrow{CF_3COOH}$ F $(C_5H_{10}O_3N_2)$ + G

40.11 Proteins. Classification and function. Denaturation

Proteins are divided into two broad classes: **fibrous proteins**, which are insoluble in water, and **globular proteins**, which are soluble in water or aqueous solutions of acids, bases, or salts. (Because of the large size of protein molecules, these solutions are colloidal.) The difference in solubility between the two classes is related to a difference in molecular shape, which is indicated in a rough way by their names.

Molecules of fibrous proteins are long and thread-like, and tend to lie side by side to form fibers; in some cases they are held together at many points by hydrogen bonds. (See, for example, Fig. 40.4, p. 1370, or Fig. 40.5, p. 1371.) As a result, the intermolecular forces that must be overcome by a solvent are very strong.

Molecules of globular proteins are folded into compact units that often approach spheroidal shapes. (See, for example, Fig. 41.1, p. 1381.) The folding takes place in such a way that the lipophilic parts are turned inward, toward each other, and away from water; hydrophilic parts—charged groups, for example—tend to stud the surface where they are near water. Hydrogen bonding is chiefly intramolecular. Areas of contact between molecules are small, and intermolecular forces are comparatively weak.

Molecular and intermolecular structure determines not only the solubility of a protein but also the general kind of function it performs.

Fibrous proteins serve as the chief structural materials of animal tissues, a function to which their insolubility and fiber-forming tendency suit them. They make up: *keratin*, in skin, hair, nails, wool, horn, and feathers; *collagen*, in tendons; *myosin*, in muscle; *fibroin*, in silk.

Globular proteins serve a variety of functions related to the maintenance and regulation of the life process, functions that require mobility and hence solubility. They make up: all enzymes; many hormones, as, for example, *insulin* (from the pancreas), *thyroglobulin* (from the thyroid gland), *ACTH* (from the pituitary gland); antibodies, responsible for allergies and for defense against foreign organisms; *albumin* in eggs; *hemoglobin*, which transports oxygen from the lungs to the tissues; *fibrinogen*, which is converted into the insoluble, fibrous protein *fibrin*, and thus causes the clotting of blood.

Within the two broad classes, proteins are subdivided on the basis of physical properties, especially solubility: for example, albumins (soluble in water, coagulated by heat), globulins (insoluble in water, soluble in dilute salt solutions), etc.

Irreversible precipitation of proteins, called **denaturation**, is caused by heat, strong acids or bases, or various other agents. Coagulation of egg white by heat, for example, is denaturation of the protein egg albumin. The extreme ease with which many proteins are denatured makes their study difficult. Denaturation causes a fundamental change in a protein, in particular destroying any physiological activity. (Denaturation appears to involve changes in the secondary structure of proteins, Sec. 40.16.)

Only one other class of compounds, the *nucleic acids* (Sec. 41.8), shows the phenomenon of denaturation. Although closely related to the proteins, polypeptides do not undergo denaturation, presumably because their molecules are smaller and less complex.

40.12   Structure of proteins

We can look at the structure of proteins on a number of levels. At the lowest level, there is the *primary* structure: the way in which the atoms of protein molecules are joined to one another by covalent bonds to form chains. Next, there is the *secondary* structure: the way in which these chains are arranged in space to form coils, sheets, or compact spheroids, with hydrogen bonds holding together different chains or different parts of the same chain. Even higher levels of structure are gradually becoming understood: the weaving together of coiled chains to form ropes, for example, or the clumping together of individual molecules to form larger aggregates. Let us look first at the primary structure of proteins.

40.13   Peptide chain

Proteins are made up of peptide chains, that is, of amino acid residues joined by amide linkages. They differ from polypeptides in having higher molecular

$$
\overset{\displaystyle H}{\underset{}{\sim N}}-\overset{}{\underset{\displaystyle O}{C}}-\overset{\displaystyle H}{\underset{}{N}}-\overset{}{\underset{\displaystyle O}{C}}-\overset{}{\underset{}{C}}-\overset{\displaystyle H}{\underset{}{N}}-\overset{}{\underset{\displaystyle O}{C}}-C\sim
$$

weights (by convention over 10 000) and more complex structures.

The peptide structure of proteins is indicated by many lines of evidence: hydrolysis of proteins by acids, bases, or enzymes yields peptides and finally amino acids; there are bands in their infrared spectra characteristic of the amide group; secondary structures based on the peptide linkage can be devised that exactly fit x-ray data.

40.14   Side chains. Isoelectric point. Electrophoresis

To every third atom of the peptide chain is attached a side chain. Its structure depends upon the particular amino acid residue involved: $-H$ for glycine, $-CH_3$ for alanine, $-CH(CH_3)_2$ for valine, $-CH_2C_6H_5$ for phenylalanine, etc.

$$
\overset{\displaystyle H}{\underset{\displaystyle R}{\sim N}}-\overset{}{\underset{}{CH}}-\overset{}{\underset{\displaystyle O}{C}}-\overset{\displaystyle H}{\underset{\displaystyle R'}{N}}-\overset{}{\underset{}{CH}}-\overset{}{\underset{\displaystyle O}{C}}-\overset{\displaystyle H}{\underset{\displaystyle R''}{N}}-\overset{}{\underset{}{CH}}-\overset{}{\underset{\displaystyle O}{C}}\sim
$$

Some of these side chains contain basic groups: $-NH_2$ in lysine, or the imidazole ring in histidine. Some side chains contain acidic groups: $-COOH$ in aspartic acid or glutamic acid. Because of these acidic and basic side chains, there are positively and negatively charged groups along the peptide chain. The behavior

$$\begin{array}{ccccccc}
& H & & O & & H & & O \\
& | & & \| & & | & & \| \\
\sim\!N\!-\!\!&CH\!-\!\!&C\!\!&\sim\!\!\sim\!\!\sim\!\!\sim\!N\!-\!\!&CH\!-\!\!&C\!\!\sim \\
& | & & & & | \\
& CH_2 & & & & (CH_2)_4 \\
& | & & & & | \\
& COO^- & & & & {}^+NH_3
\end{array}$$

of a protein in an electric field is determined by the relative numbers of these positive and negative charges, which in turn are affected by the acidity of the solution. At the isoelectric point, the positive and negative charges are exactly balanced and the protein shows no net migration; as with amino acids, solubility is usually at a minimum here. On the acid side of the isoelectric point, positive charges exceed negative charges and the protein moves to the cathode; on the basic side of the isoelectric point, negative charges exceed positive charges and the protein moves to the anode.

While all proteins contain the peptide backbone, each protein has its own characteristic sequence of side chains, which gives it its characteristic properties. Different proteins have different proportions of acidic and basic side chains, and hence have different isoelectric points. In a solution of a particular hydrogen ion concentration, some proteins move toward a cathode and others toward an anode; depending upon the size of the charge as well as upon molecular size and shape, different proteins move at different speeds. This difference in behavior in an electric field is the basis of one method of separation and analysis of protein mixtures: **electrophoresis**.

Side chains affect the properties of proteins not only by their acidity or basicity, but also by their other chemical properties and even by their sizes and shapes. Hydroxyl and sulfhydryl (—SH) groups can form esters; amino nitrogen is not only basic but nucleophilic. It seems likely that the "permanent" waving of hair depends upon changes in disulfide (—S—S—) cross-linkages provided by cysteine side chains; that much of the difference between silk and wool is related to the small side chains, —H and —CH_3, that predominate in silk fibroin; that the toughness of tendon is due to the flatness of the pyrrolidine ring and the ability of the —OH group of hydroxyproline to form hydrogen bonds. Replacement of *one* glutamic acid side chain in the hemoglobin molecule (300 side chains in all) by a valine unit is the cause of the fatal sickle-cell anemia.

The sequence of amino acids in hemoglobin has been used to study evolution, in the science called *chemical paleogenetics*. In the *beta*-chain of hemoglobin, for example, the horse differs from man at 26 of the 146 sites; a pig, at 10 sites; and the gorilla at just *one* site. It has been estimated that, on the average, it takes roughly ten million years for one successful amino acid substitution to occur—that is, a substitution that improves the chances of survival. (Such a change is due to a change in the base sequence in a molecule of nucleic acid, Sec. 41.9.)

40.15   Conjugated proteins. Prosthetic groups. Coenzymes

Some protein molecules contain a non-peptide portion called a **prosthetic group**; such proteins are called *conjugated proteins*. The prosthetic group is intimately concerned with the specific biological action of the protein.

The prosthetic group of hemoglobin, for example, is *heme*. As we see, heme

Heme

contains iron bound to the pyrrole system known as *porphin* (compare with the structure of chlorophyll, p. 1207). It is the formation of a reversible oxygen–heme complex that enables hemoglobin to carry oxygen from the lungs to the tissues. Carbon monoxide forms a similar, but more stable, complex; it thus ties up hemoglobin, prevents oxygen transport, and causes death. Heme is held to the peptide portion (*globin*) of the protein by a combination of forces: coordination of iron by histidine nitrogen of the protein, hydrogen bonding, and van der Waals forces between hydrophobic parts of the two molecules.

Many enzymes require *cofactors* if they are to exert their catalytic effects: metal ions, for example. The peptide portion of such an enzyme—the protein without the coenzyme—is called an **apoenzyme**. An organic cofactor is called a **coenzyme** and, if it is covalently bonded to the apoenzyme, it too is a prosthetic group.

The coenzyme *nicotinamide adenine dinucleotide* (NAD), for example, is associated with a number of dehydrogenation enzymes. (We have already seen it at work in the enzymatic oxidation of ethanol, Sec. 22.2.) This coenzyme, we see, is

Nicotinamide adenine dinucleotide (NAD)
(Diphosphopyridinenucleotide)

made up of two molecules of D-ribose linked as phosphate esters, the fused heterocyclic system known as *adenine*, and nicotinamide in the form of a quaternary ammonium salt. In some systems one encounters *nicotinamide adenine dinucleotide phosphate* (NADP), in which the —OH on C–2 of the left-hand ribose unit of NAD has been phosphorylated. The characteristic biological function of these dehydrogenation enzymes (see, for example, Sec. 41.6) involves conversion of the nicotinamide portion of NAD or NADP into the dihydro structure.

NAD   Reduced NAD

Like nicotinamide, many molecules making up coenzymes are **vitamins**, that is, substances that must be supplied in the diet to permit proper growth or maintenance of structure. Undoubtedly it is for their coenzyme activity that these substances are needed.

40.16   Secondary structure of proteins

It seems clear that proteins are made up of polypeptide chains. How are these chains arranged in space and in relationship to each other? Are they stretched out side by side, looped and coiled about one another, or folded into independent spheroids?

Much of our understanding of the secondary structure of proteins is the result of x-ray analysis. For many proteins the x-ray diffraction pattern indicates a regular repetition of certain structural units. For example, there are *repeat distances* of 7.0 Å in silk fibroin, and of 1.5 Å and 5.1 Å in α-keratin of unstretched wool.

The problem is to devise structures that account for the characteristic x-ray diffraction patterns, and are at the same time consistent with what is known about the primary structure: bond lengths and bond angles, planarity of the amide group, similarity of configuration about chiral centers (all L-family), size and sequence of side chains. Of key importance in this problem has been recognition of the stabilizing effect of hydrogen bonds (5–10 kcal per mole per hydrogen bond), and the principle that the most stable structure is one that permits formation of the maximum number of hydrogen bonds. On the basis of the study of simpler compounds, it has been further assumed that the N—H—O bond is very nearly linear, hydrogen lying on, or within 20° of, the line between nitrogen and oxygen. In all this work the simultaneous study of simpler, synthetic polypeptides containing only a single kind of amino acid residue has been of great help.

The progress made on a problem of this size and difficulty has necessarily been the work of many people. Among them is Linus Pauling, of the California Institute of Technology, who received the Nobel Prize in 1954. In 1951 Pauling wrote: "Fourteen years ago Professor Robert B. Corey and I, after we had made a vigorous but unsuccessful attack on the problem of formulating satisfactory configurations of polypeptide chains in proteins, decided to attempt to solve the problem by an indirect method—the method of investigating with great thoroughness crystals of amino acids, simple peptides, and related substances, in order to obtain completely reliable and detailed information about the structural characteristics of substances of this sort, and ultimately to permit the confident prediction of precisely described configurations of polypeptide chains in proteins." (*Record Chem. Prog.*, 1951, **12**, 156–7.) This work on simple substances, carried on for more than 14 years, gave information about the geometry of the amide group that eventually led Pauling and his co-workers to propose what may well be the most important secondary structure in protein chemistry: the α-helix.

Let us look at some of the secondary structures that have been proposed.

As a point of departure, it is convenient to consider a structure (perhaps hypothetical) in which peptide chains are fully extended to form flat zig-zags:

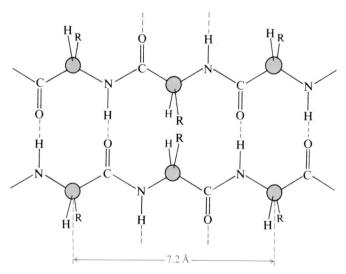

Extended peptide chain

These chains lie side by side to form a *flat sheet*. Each chain is held by hydrogen bonds to the two neighboring chains (Fig. 40.3). This structure has a repeat distance of 7.2 Å, the distance between *alternate* amino acid residues. (Notice that alternate side chains lie on the same side of the sheet.) However, crowding between side chains makes this idealized flat structure impossible, except perhaps for synthetic polyglycine.

Figure 40.3 Hypothetical flat sheet structure for a protein. The chains are fully extended; adjacent chains head in opposite directions; the hydrogen bonding is between adjacent chains. The side chains (R) are crowded.

Room can be made for small or medium-sized side chains by a slight contraction of the peptide chains:

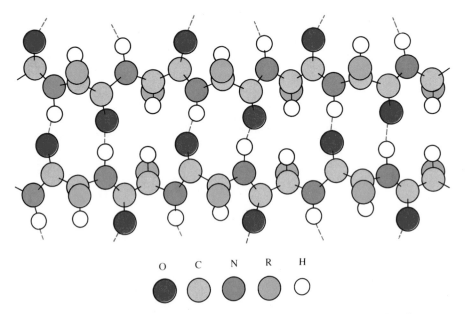

Contracted peptide chain

The chains still lie side by side, held to each other by hydrogen bonds. The contraction results in a *pleated sheet*, with a somewhat shorter distance between alternate amino acid residues (see Fig. 40.4). Such a structure, called the **beta** arrangement, has been proposed for silk fibroin, which has a repeat distance of 7.0 Å and most closely approaches the fully extended, flat-sheet structure. It is significant that, although 15 kinds of amino acid residue are found in silk fibroin, 46% of the residues are glycine, which has no side chain, and another 38% are alanine and serine with the small side chains $-CH_3$ and $-CH_2OH$. (See Fig. 40.5.)

O C N R H

Figure 40.4 Pleated sheet structure (*beta arrangement*) proposed by Pauling for silk fibroin. The chains are contracted to make room for small side chains. Adjacent chains head in opposite directions; the hydrogen bonding is between adjacent chains.

Figure 40.5   Model of poly(L-alanine) in the pleated sheet structure. As we look along a chain from left to right, we can see the methyl side chains projecting alternately up and (most visibly in the front row) down. The pleats run from front to back, and the repeat distance is that between the ridges created by the projecting methyl groups.

When the side chains are quite large, they are best accommodated by a quite different kind of structure. Each chain is coiled to form a *helix* (like a spiral staircase). Hydrogen bonding occurs between different parts of the *same* chain,

A helix
(right-handed)

and holds the helix together. For α-keratin (unstretched wool, hair, horn, nails) Pauling has proposed a helix in which there are 3.6 amino acid residues per turn (Fig. 40.6). Models show that this 3.6-helix provides room for the side chains and allows all possible hydrogen bonds to form. It accounts for the repeat distance of 1.5 Å, which is the distance between amino acid residues measured along the axis of the helix. To fit into this helix, all the amino acid residues must be of the same configuration, as, of course, they are; furthermore, their L-configuration requires

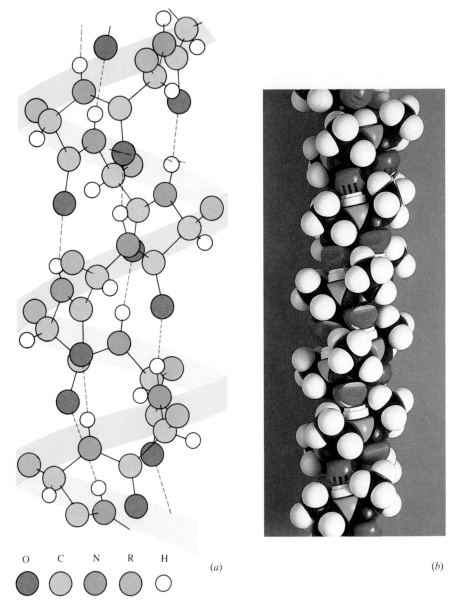

O   C   N   R   H   (a)   (b)

Figure 40.6 Alpha helix structure proposed by Pauling for α-keratin. (a) Schematic representation, with nine residues. (b) CPK model of poly(L-alanine), with twenty-two residues; all side chains are methyl.

This arrangement makes room for large side chains. It is a right-handed helix with 3.6 residues per turn; the hydrogen bonding is within a chain.

the helix to be *right-handed*, as shown. The **alpha helix**, as it is called, is of fundamental importance in the chemistry of proteins.

(To account for the second repeat distance of 5.1 Å for α-keratin, we must go to what is properly the *tertiary structure*. Pauling has suggested that each helix can itself be coiled into a superhelix which has one turn for every 35 turns of the *alpha* helix. Six of these superhelixes are woven about a seventh, straight helix to form a seven-strand cable.)

When wool is stretched, α-keratin is converted into β-keratin, with a change in the x-ray diffraction pattern. It is believed that the helixes are uncoiled and the chains stretched side by side to give a sheet structure of the *beta* type. The hydrogen bonds within the helical chain are broken, and are replaced by hydrogen bonds between adjacent chains. (Compare the effect (Sec. 36.8) of drawing a synthetic fiber—nylon, for example, also a polyamide.) Because of the larger side chains, the peptide chains are less extended (repeat distance 6.4 Å) than in silk fibroin (repeat distance 7.0 Å).

Besides the x-ray diffraction patterns characteristic of the *alpha*- and *beta*-type proteins, there is a third kind: that of *collagen*, the protein of tendon and skin. On the primary level, collagen is characterized by a high proportion of proline and hydroxyproline residues, and by frequent repetitions of the sequence Gly-Pro-Hyp. The pyrrolidine ring of proline and hydroxyproline can affect the secondary

Proline residue   Hydroxyproline residue

structure in several ways. The amido nitrogen carries no hydrogen for hydrogen bonding. The flatness of the five-membered ring, in conjunction with the flatness of the amide group, prevents extension of the peptide chain as in the *beta* arrangement, and interferes with the compact coiling of the *alpha* helix.

The structure of collagen combines the helical nature of the *alpha*-type proteins with the inter-chain hydrogen bonding of the *beta*-type proteins. Three peptide chains—each in the form of a left-handed helix—are twisted about one another to form a three-strand right-handed superhelix. A small glycine residue at every third position of each chain makes room for the bulky pyrrolidine rings on the other two chains. The three chains are held strongly to each other by hydrogen bonding between glycine residues and between the —OH groups of hydroxyproline.

When collagen is boiled with water, it is converted into the familiar water-soluble protein *gelatin*; when cooled, the solution does not revert to collagen but sets to a gel. Gelatin has a molecular weight one-third that of collagen. Evidently the treatment separates the strands of the helix, breaking inter-chain hydrogen bonds and replacing them with hydrogen bonds to water molecules.

Turning from the insoluble, fibrous proteins to the soluble, globular proteins (e.g., hemoglobin, insulin, *gamma*-globulin, egg albumin), we find that the matter of secondary structure can be even more complex. Evidence is accumulating that here, too, the *alpha* helix often plays a key role. These long peptide chains are not uniform: certain segments may be coiled into helixes or folded into sheets; other segments are looped and coiled into complicated, irregular arrangements. Look, for example, at α-chymotrypsin in Fig. 41.1 (p. 1381).

This looping and coiling appears to be random, but it definitely is *not*. The sequence of amino acids is determined genetically (Sec. 41.8) but, once formed, the chain *naturally* falls into the arrangement that is *most stable* for that particular sequence.

We find all our kinds of "intermolecular" forces at work here—but acting between different parts of the same molecule: van der Waals forces, hydrogen bonds, interionic attraction (or repulsion) between charged groups. There is chemical cross-linking by disulfide bonds. The characteristic feature of these globular proteins is that lipophilic parts are turned inward, toward each other and away from water—like the lipophilic tails in a soap micelle.

In their physiological functions, proteins are highly specific. We have encountered, for example, an enzyme that will cleave α-glucosides but not β-glucosides, and an enzyme that will cleave only C-terminal amino acid residues in polypeptides. In Secs. 22.4–22.7 we saw how the enzyme alcohol dehydrogenase discriminates between enantiotopic hydrogens of ethanol and between enantiotopic faces of acetaldehyde, and (Problem 3, p. 814) how a different oxidation–reduction enzyme also discriminates, but *in the opposite manner*.

It seems clear that the biological activity of a protein depends not only upon its prosthetic group (if any) and its particular amino acid sequence, but also upon its molecular shape. As Emil Fischer said in 1894: ". . . enzyme and glucoside must fit together like a lock and key. . . ." In Sec. 41.2 we shall see how one enzyme is believed to exert its effect, and how that effect depends, in a very definite and specific way, on the shape of the enzyme molecule.

Denaturation uncoils the protein, destroys the characteristic shape, and with it the characteristic biological activity.

In 1962, M. F. Perutz and J. C. Kendrew of Cambridge University were awarded the Nobel Prize in chemistry for the elucidation of the structure of hemoglobin and the closely related oxygen-storing molecule, myoglobin. Using x-ray analysis, and knowing the amino acid sequence (p. 1361), they determined the shape—in three dimensions—of these enormously complicated molecules: precisely for myoglobin, and very nearly so for hemoglobin. They can say, for example, that the molecule is coiled in an *alpha* helix for sixteen residues from the *N*-terminal unit, and then turns through a right angle. They can even say *why*: at the corner there is an aspartic acid residue; its carboxyl group interferes with the hydrogen bonding required to continue the helix, and the chain changes its course. The four folded chains of hemoglobin fit together to make a spheroidal molecule, 64 Å × 55 Å × 50 Å. Four flat heme groups, each of which contains an iron atom that can bind an oxygen molecule, fit into separate pockets in this sphere. When oxygen is being carried, the chains move to make the pockets slightly smaller; Perutz has described hemoglobin as "a breathing molecule". These pockets are lined with the hydrocarbon portions of the amino acids; such a non-polar environment prevents electron transfer between oxygen and ferrous iron, and permits the complexing necessary for oxygen transport.

PROBLEMS

1. Outline all steps in the synthesis of phenylalanine from toluene and any needed aliphatic and inorganic reagents by each of the following methods:

(a) direct ammonolysis   (d) phthalimidomalonic ester method
(b) Gabriel synthesis   (e) Strecker synthesis
(c) malonic ester synthesis   (f) reductive amination

2. (a) Give structures of all intermediates in the following synthesis of proline:

potassium phthalimide + bromomalonic ester $\longrightarrow$ A

$A + Br(CH_2)_3Br \xrightarrow{NaOC_2H_5} B\ (C_{18}H_{20}O_6NBr)$
B + potassium acetate $\longrightarrow$ $C\ (C_{20}H_{23}O_8N)$
$C + NaOH$, heat; then H^+, heat $\longrightarrow$ $D\ (C_5H_{11}O_3N)$
$D + HCl \longrightarrow [E\ (C_5H_{10}O_2NCl)] \longrightarrow$ proline

(b) Outline a possible synthesis of lysine by the phthalimidomalonic ester method.

3. Using the behavior of hydroxy acids (Sec. 24.15) as a pattern, predict structures for the products obtained when the following amino acids are heated:

(a) the α-amino acid, glycine $\longrightarrow$ $C_4H_6O_2N_2$ (*diketopiperazine*)
(b) the β-amino acid, $CH_3CH(NH_2)CH_2COOH \longrightarrow C_4H_6O_2$
(c) the γ-amino acid, $CH_3CH(NH_2)CH_2CH_2COOH \longrightarrow C_5H_9ON$ (*a lactam*)
(d) the δ-amino acid, $H_2NCH_2CH_2CH_2CH_2COOH \longrightarrow C_5H_9ON$ (*a lactam*)

4. (a) Draw the two possible dipolar structures for lysine. Justify the choice of structure given in Table 40.1. (b) Answer (a) for aspartic acid. (c) Answer (a) for arginine. (*Hint*: See Problem 24.23, p. 887.) (d) Answer (a) for tyrosine.

5. *Betaine*, $C_5H_{11}O_2N$, occurs in beet sugar molasses. It is a water-soluble solid that melts with decomposition at 300 °C. It is unaffected by base but reacts with hydrochloric acid to form a crystalline product, $C_5H_{12}O_2NCl$. It can be made in either of two ways: treatment of glycine with methyl iodide, or treatment of chloroacetic acid with trimethylamine.

Draw a structure for betaine that accounts for its properties.

6. Addition of ethanol or other organic solvents to an aqueous "solution" of a globular protein brings about denaturation. Such treatment also tends to break up micelles of, say soap (Sec. 37.3). What basic process is at work in both cases?

7. An amino group can be protected by acylation with phthalic anhydride to form an *N*-substituted phthalimide. The protecting group can be removed by treatment with hydrazine, H_2N—NH_2, without disturbing any peptide linkages. Write equations to show how this procedure (exploited by John C. Sheehan of the Massachusetts Institute of Technology) could be applied to the synthesis of glycylalanine (Gly-Ala) and alanylglycine (Ala-Gly).

8. An elemental analysis of *cytochrome c*, an enzyme involved in oxidation–reduction processes, gave 0.43% Fe and 1.48% S. What is the minimum molecular weight of the enzyme? What is the minimum number of iron atoms per molecule? Of sulfur atoms?

9. A protein, *β-lactoglobulin*, from cheese whey, has a molecular weight of 42 020 $\pm$ 105. When a 100-mg sample was hydrolyzed by acid and the mixture was made alkaline, 1.31 mg of ammonia was evolved. (a) Where did the ammonia come from, and approximately how many such groups are there in the protein?

Complete hydrolysis of a 100-mg sample of the protein used up approximately 17 mg of water. (b) How many amide linkages per molecule were cleaved?

(c) Combining the results of (a) and (b), and adding the fact that there are four *N*-terminal groups (four peptide chains in the molecule), how many amino acid residues are there in the protein?

10. The complete structure of *gramicidin S*, a polypeptide with antibiotic properties, has been worked out as follows:

(a) Analysis of the hydrolysis products gave an empirical formula of Leu,Orn,Phe,Pro,Val. (*Ornithine*, Orn, is a rare amino acid of formula $^+H_3NCH_2CH_2CH_2CH(NH_2)COO^-$.) It is interesting that the phenylalanine has the unusual D-configuration.

Measurement of the molecular weight gave an approximate value of 1300. On this basis, what is the molecular formula of gramicidin S?

(b) Analysis for the *C*-terminal residue was negative: analysis for the *N*-terminal residue using DNFB yielded only $DNP{-}NHCH_2CH_2CH_2CH(\overset{+}{N}H_3)COO^-$. What structural feature must the peptide chain possess?

(c) Partial hydrolysis of gramicidin S gave the following di- and tripeptides:

Leu-Phe	Phe-Pro	Phe-Pro-Val	Val-Orn-Leu
Orn-Leu	Val-Orn	Pro-Val-Orn	

What is the structure of gramicidin S?

11. The structure of beef insulin was determined by Sanger (see Sec. 40.9) on the basis of the following information. Work out for yourself the sequence of amino acid residues in the protein.

Beef insulin appears to have a molecular weight of about 6000 and to consist of two polypeptide chains linked by disulfide bridges of cystine residues. The chains can be separated by oxidation, which changes any Cys—Cys or Cys residues to sulfonic acids (CySO$_3$H).

One chain, A, of 21 amino acid residues, is acidic and has the empirical formula

$$GlyAlaVal_2Leu_2IleCys_4Asp_2Glu_4Ser_2Tyr_2$$

The other chain, B, of 30 amino acid residues, is basic and has the empirical formula

$$Gly_3Ala_2Val_3Leu_4ProPhe_3Cys_2ArgHis_2LysAspGlu_3SerThrTyr_2$$

(Chain A has four simple side-chain amide groups, and chain B has two, but these will be ignored for the time being.)

Treatment of chain B with 2,4-dinitrofluorobenzene (DNFB) followed by hydrolysis gave DNP-Phe and DNP-Phe-Val; chain B lost alanine (Ala) when treated with carboxypeptidase.

Acidic hydrolysis of chain B gave the following tripeptides:

Glu-His-Leu	Leu-Val-Cys	Tyr-Leu-Val
Gly-Glu-Arg	Leu-Val-Glu	Val-Asp-Glu
His-Leu-Cys	Phe-Val-Asp	Val-Cys-Gly
Leu-Cys-Gly	Pro-Lys-Ala	Val-Glu-Ala
	Ser-His-Leu	

Many dipeptides were isolated and identified; two important ones were Arg-Gly and Thr-Pro.

(a) At this point construct as much of the B chain as the data will allow.

Among the numerous tetrapeptides and pentapeptides from chain B were found:

His-Leu-Val-Glu	Tyr-Leu-Val-Cys
Ser-His-Leu-Val	Phe-Val-Asp-Glu-His

(b) How much more of the chain can you reconstruct now? What amino acid residues are still missing?

Enzymatic hydrolysis of chain B gave the necessary final pieces:

Val-Glu-Ala-Leu	His-Leu-Cys-Gly-Ser-His-Leu
Tyr-Thr-Pro-Lys-Ala	Tyr-Leu-Val-Cys-Gly-Glu-Arg-Gly-Phe-Phe

(c) What is the complete sequence in the B chain of beef insulin?

Treatment of chain A with DNFB followed by hydrolysis gave DNP-Gly; the *C*-terminal group was shown to be aspartic acid (Asp).

Acidic hydrolysis of chain A gave the following tripeptides:

Cys-Cys-Ala	Glu-Leu-Glu
Glu-Asp-Tyr	Leu-Tyr-Glu
Glu-Cys-Cys	Ser-Leu-Tyr
Glu-Glu-Cys	Ser-Val-Cys

Among other peptides isolated from acidic hydrolysis of chain A were:

Cys-Asp   Tyr-Cys   Gly-Ile-Val-Glu-Glu

(d) Construct as much of chain A as the data will allow. Are there any amino acid residues missing?

Up to this point it is possible to arrive at the sequences of four parts of chain A, but it is still uncertain which of the two center fragments, Ser-Val-Cys or Ser-Leu-Tyr, etc., comes first. This was settled by digestion of chain A with pepsin, which gave a peptide that contained no aspartic acid (Asp) or tyrosine (Tyr). Hydrolysis of this peptide gave Ser-Val-Cys and Ser-Leu.

(e) Now what is the complete structure of chain A of beef insulin?

In insulin the cysteine units (Cys) are involved in cystine disulfide links (Cys—Cys). Residue 7 of chain A (numbering from the N-terminal residue) is linked to residue 7 of chain B, residue 20 of chain A to residue 19 of chain B, and there is a link between residues 6 and 11 of chain A.

There are amide groups on residues 5, 15, 18, and 21 of chain A, and on residues 3 and 4 of chain B.

(f) Draw a structure of the complete insulin molecule. (*Note*: The disulfide loop in chain A is a 20-atom, pentapeptide ring, of the same size as the one in oxytocin.)

In the analysis for the N-terminal group in chain B of insulin, equal amounts of *two* different DNP derivatives of single amino acids actually were found. One was DNP-Phe; what could the other have been?

(g) What would have been obtained if that second amino acid had been N-terminal?

41

Biochemical Processes

Molecular Biology

41.1 Biochemistry, molecular biology, and organic chemistry

In the past four chapters, we have learned something about fats, carbohydrates, and proteins: their structures and how these are determined, and the kind of reactions they undergo in the test tube. These, we said, are biomolecules: they are participants in the chemical process we call life. But just what do they *do*? What reactions do they undergo, not in the test tube, but in a living organism?

Even a vastly simplified answer to that question would—and does—fill a book as big as this one. Having come this far, though, we cannot help being curious. And so, in this chapter, we shall take a brief glance at the answer—or, rather, at the kind of thing the answer entails.

We shall look at just a few examples of biochemical processes: how one enzyme—of the thousands in our bodies—may work; the simple chemical transformation that is the basis of vision; what happens in one of the dozens of reactions by which carbohydrates are oxidized to furnish energy; how one kind of chemical compound—fatty acids—is synthesized. Finally, we shall learn a little about another class of biomolecules, the nucleic acids, and how they are involved in the most fascinating biochemical process of all—heredity.

The study of nucleic acids has become known as "molecular biology". Actually, of course, all of these processes are a part of molecular biology—biology on the molecular level—and they are, in the final analysis, organic chemistry. And it is as organic chemistry that we shall treat them. We shall see how all these vital processes—even the mysterious powers of enzymes—come down to a matter of molecular structure as we know it: to molecular size and shape; to intermolecular and intramolecular forces; to the chemistry of functional groups; to acidity and basicity, oxidation and reduction; to energy changes and rate of reaction; to the host–guest relationship and symphoria.

Since catalysis by enzymes is fundamental to everything else, let us begin there.

41.2 Mechanism of enzyme action. Chymotrypsin

Enzymes, we have said, are proteins that act as enormously effective catalysts for biological reactions. To get some idea of how they work, let us examine the action of just one: *chymotrypsin*, a digestive enzyme whose job is to promote hydrolysis of certain peptide links in proteins. The sequence of the 241 amino acid residues in chymotrypsin has been determined and, through x-ray analysis, the conformation of the molecule is known (Fig. 41.1). It is, like all enzymes, a soluble globular protein coiled in the way that turns its lipophilic parts inward, toward each other and away from water, and that permits maximum intramolecular hydrogen bonding.

The action of chymotrypsin has been more widely explored than that of any other enzyme. In crystalline form, it is available for studies in the test tube under a variety of conditions. It catalyzes hydrolysis not only of proteins but of ordinary amides and esters, and much has been learned by use of these simpler substrates. Compounds modeled after portions of the chymotrypsin molecule have been made, and their catalytic effects measured.

To begin with, it seems very likely that chymotrypsin acts in two stages. In the first stage, acting as an alcohol, it breaks the peptide chain. We recognize this as alcoholysis of a substituted amide: nucleophilic acyl substitution. The products are an amine—the liberated portion of the substrate molecule—and, as we shall see

$$\text{(Stage 1)} \quad \overset{\displaystyle O}{\overset{\displaystyle \|}{\sim\!\!\text{C}}}\!-\!\text{NH}\!\sim \;+\; \text{E}\!-\!\text{O}\!-\!\text{H} \;\longrightarrow\; \overset{\displaystyle O}{\overset{\displaystyle \|}{\sim\!\!\text{C}}}\!-\!\text{O}\!-\!\text{E} \;+\; \text{NH}_2\!\sim$$

Protein	Enzyme	Acyl enzyme	Part of protein chain
Amide	*Alcohol*	*Ester*	*Amine*

$$\text{(Stage 2)} \quad \overset{\displaystyle O}{\overset{\displaystyle \|}{\sim\!\!\text{C}}}\!-\!\text{O}\!-\!\text{E} \;+\; \text{H}_2\text{O} \;\longrightarrow\; \sim\!\!\text{COOH} \;+\; \text{E}\!-\!\text{O}\!-\!\text{H}$$

Acyl enzyme		Rest of protein chain	Regenerated enzyme
Ester		*Carboxylic acid*	*Alcohol*

shortly, an ester of the enzyme. In the second stage, the enzyme ester is hydrolyzed. This yields a carboxylic acid—the other portion of the substrate molecule—and the regenerated enzyme, ready to go to work again.

Figure 41.1   Three-dimensional structure of α-chymotrypsin. Residues are numbered from 1 to 245 as in its precursor, chymotrypsinogen (p. 1361), but residues 14–15 and 147–148 have been lost. The three chains (with the *N*-terminal units given first) are: A, 1–13; B, 16–146; C, 149–245.

Histidine-57, serine-195, and isoleucine-16 are key units. The lipophilic pocket lies to the right of histidine-57 and serine-195, and is indicated by red dots; it is bounded by residues 184–191 and 214–227.

We can see one short segment of α-helix at residues 234–245; another (mostly hidden) lies at 164–170. There is a hint of a twisted sheet beginning with residues 91–86 and 103–108, and extending to their right.

What is the structure of this intermediate ester formed from the enzyme? The answer has been found by use of simple esters as substrates, *p*-nitrophenyl acetate, for example. An appreciable steady-state concentration of the intermediate ester builds up and, by quenching of the reaction mixture in acid, it can be isolated.

Sequence analysis of the enzyme ester showed that the acetyl group from the substrate was linked to *serine-195*. It is, then, at the —OH group of this particular amino acid residue that the enzyme reacts.

$$HOCH_2\underset{\underset{^+NH_3}{|}}{C}HCOO^-$$

Serine

But evidence shows that certain other amino acid residues are also vital to enzyme activity. The rate of enzyme-catalyzed hydrolysis changes as the acidity of the reaction medium is changed. If one plots the rate of hydrolysis against the pH of the solution, one gets a bell-shaped curve: as the pH is increased, the rate rises to a maximum and then falls off. The rate is fastest at about pH 7.4 (fittingly, the physiological pH) and slower in either more acidic or more basic solution. Analysis of the data shows the following. Hydrolysis requires the presence of a free base, of K_b about 10^{-7}, and a protonated base, of K_b about 3×10^{-5}. At low pH (acid solution), both bases are protonated; at high pH (alkaline solution), both bases are free. Hydrolysis is fastest at the intermediate pH where the weaker base is mostly free and the stronger base is mostly protonated.

The K_b of the weaker base fits that of the imidazole ring of histidine, and there is additional evidence indicating that this is indeed the base: studies involving

Histidine   Protonated histidine
Base   *Acid*

catalysis by imidazole itself, for example. Now, examination of the conformation of chymotrypsin (Fig. 41.1) shows that very close to serine-195 there *is* a histidine residue. This is *histidine-57*, and it is believed to be the one involved in enzyme activity.

What about the stronger base which, according to the kinetics, is involved in its protonated form? Its K_b fits the α-amino group of most amino acids—an α-amino group, that is, which is not tied up in a peptide link. But all the (free) amino groups in chymotrypsin—*except one*—may be acetylated without complete loss of activity. The exception is *isoleucine-16*, the *N*-terminal unit of chain B.

$$CH_3CH_2CH(CH_3)\underset{\underset{^+NH_3}{|}}{C}HCOO^-$$

Isoleucine

Presumably, then, this amino group cannot be acetylated, but must be free to be protonated and do its part of the job.

Now, what is the job of each of these key units in the enzyme molecule? It is clear what serine-195 does: it provides the —OH for ester formation. What does isoleucine-16 do? The descending leg of the bell-shaped rate curve was attributed to protonation of this unit. But something else happens as the pH is raised above

7.4: the optical activity of the solution decreases—evidently due to a change in conformation of the enzyme molecule—and in a way that parallels the decrease in rate of hydrolysis. It is believed that the $-NH_3^+$ of isoleucine-16 is attracted by the $-COO^-$ of aspartic acid-194; this ion pairing helps hold the enzyme chain in the proper shape for it to act as a catalyst: to keep histidine-57 near serine-195, among other things. At higher pH the $-NH_3^+$ is converted into $-NH_2$, and the chain changes its shape; with the change in shape goes loss of catalytic power and a change in optical rotation.

Next, we come to the question: what is the role of histidine-57? We are observing an example of *general acid–base catalysis*: catalysis not just by hydroxide ions and oxonium ions, but by all the bases and conjugate acids that are present, each contributing according to its concentration and its acid or base strength.

Let us look at this concept first with a simple example: hydrolysis of an ester catalyzed by the simple heterocyclic base, imidazole. Catalysis by hydroxide ions

$$RCOOR' + H_2O \xrightarrow{\text{imidazole}} RCOOH + R'OH$$

we understand: these highly nucleophilic ions are more effective than water at attacking acyl carbon. Imidazole generates some hydroxide ions by reaction with water, but these are already taken into account. We are talking now about hydrolysis that is directly proportional to the concentration of the base itself: imidazole. What seems to be involved in such reactions is something like the following. In step (1), water adds to acyl carbon with *simultaneous loss of a proton to the base*; reaction is

(1)

(2)

fast because, in effect, the attacking nucleophile is not just water, but an incipient hydroxide ion. In step (2), transfer of the proton from the protonated base is simultaneous with loss of the ethoxy group; again reaction is fast, this time because the leaving group is not the strongly basic ethoxide ion, but an incipient alcohol molecule.

Reactions like (1) and (2) need not involve unlikely three-body collisions among the reactive molecules. Instead, there is prior hydrogen bonding between the base and water or between the protonated base and ester; it is these double molecules that collide with the third reagent and undergo reaction, with the dipole–dipole attraction of the hydrogen bonding being replaced by a covalent bond.

Figure 41.2 depicts the action of chymotrypsin, with the imidazole group of histidine-57 playing the same role of general base as that just described—and with protonated imidazole necessarily acting as general acid. There is general acid–base catalysis of both reactions involved: first, in the formation of the acyl enzyme, and then in its hydrolysis.

Chymotrypsin is not, as enzymes go, very specific in its action; it hydrolyzes proteins, peptides, simple amides, and esters alike. There is one structural requirement, nevertheless; a relatively non-polar group in the acyl moiety of the substrate, typically an aromatic ring. Now, turning once more to Fig. 41.1, we find that at the reactive site in the enzyme there is a pocket; this pocket is lined with lipophilic substituents to receive the non-polar group of the substrate and thus hold the molecule in position for hydrolysis. It is the size of this pocket and the nature of its lining that gives the enzyme its specificity; here we find, in a very real sense, Emil Fischer's lock into which the substrate key must fit.

We see here clearly the symphoria that gives enzymes their catalytic powers. The substrate is bound to a particular site in the enzyme, where the necessary functional groups are gathered: here, hydroxyl of serine and imidazole of histidine. In most cases, there are other functional groups as well, in molecules of cofactors— reagents, really—bound by the enzyme near the reactive site. In the enzyme– substrate complex, these functional groups are *parts of the same molecule*, and in their reactions enjoy all the advantages we listed (Sec. 20.1) for such groups. *They*

Figure 41.2 Catalysis by the enzyme chymotrypsin of the cleavage of one peptide bond in a protein: a proposed mechanism. Histidine and protonated histidine act as general base and acid in two successive nucleophilic substitution reactions: (*a*) cleavage of protein with formation of acyl enzyme and liberation of one protein fragment; (*b*) hydrolysis of acyl enzyme with regeneration of the enzyme and liberation of the other protein fragment.

are there, poised in just the right position for attack on the substrate. They need not wait for the lucky accident of a molecular collision; in effect, concentration of reagents is very high. Orientation of reacting groups is exactly right. There are no clinging solvent molecules to be stripped away as reaction occurs.

And there may be other factors at work here: it has been suggested, for example, that the pocket in which reaction occurs fits the transition state better than it fits the reactants, so that relief of strain or an increase in van der Waals attractions provides a driving force.

41.3 The organic chemistry of vision

To see, in perhaps the most graphic way possible, the part that molecular shape plays in determining biological action, let us look very briefly at the chemistry of *vision*—or, rather, at just one aspect of that chemistry. Vision, in the final analysis, comes down to the *detection of light*: light strikes the eye, and the brain receives a signal that something is there. The recognition of just *what* is there—the size, shape, brightness, and distance of the object seen—is a matter of the physics of the eye and the biology of the brain. But all this depends upon one initial event: light does something in the eye—something which starts off the entire process and without which there would be no vision. That "something", it turns out, is a simple, purely chemical transformation; it is that rare occurrence in biology, an organic reaction that does not require catalysis by an enzyme. It is so direct and uncomplicated—so *elegant*—that it has been adopted as the basis of vision in every form of animal life.

In the rod cells of the retina of a mammal there is a conjugated protein called *rhodopsin*. The prosthetic group of this protein is *11-cis-retinal*: an unsaturated aldehyde derived from vitamin A, which in turn is derived from β-carotene, the

β-Carotene

Vitamin A₁
(Retinol)

11-*cis*-Retinal

pigment that makes carrots yellow. Retinal is not only bonded covalently to the protein—the carbonyl group reacts with an amino group to form an imine (Sec. 26.11)—but is held in a lipophilic pocket.

When light strikes rhodopsin it does just one thing, and then plays no further part: it transforms the 11-*cis*-retinal into 11-*trans*-retinal. *It is this transformation, this change of one geometric isomer into another, that is the beginning of the visual process*; it is the link between the impingement of light and the series of chemical reactions that generates the nerve impulses that let us *see*.

11-*cis*-Retinal   11-*trans*-Retinal

Light brings energy to the rhodopsin, energy that causes a $\pi \rightarrow \pi^*$ transformation in the retinal moiety (Sec. 16.5); in effect, it opens carbon–carbon double bonds and permits the rotation that is necessary for *cis–trans* isomerization. This isomerization changes the *shape* of the retinal; the bend is removed and the molecule straightens out. (This difference in shape between *cis* and *trans* isomers is the same as what we saw for rubber and gutta percha (Sec. 36.8), and for the unsaturated carboxylic acids of fats (Sec. 37.2).) With the change in shape of the retinal moiety there is a change in shape of the entire rhodopsin molecule; the protein portion must adjust its conformation to accommodate this altered guest. This, it is believed, affects the permeability of certain membranes, and permits the passage of Ca^{2+} ions that trigger off nerve impulses to the brain. The entire process is amazingly efficient: the human eye can detect the absorption of as few as *five* photons of light by five rod cells!

A great deal more then happens: a series of enzyme-catalyzed reactions that supply the energy needed to convert the *trans*-retinal back into the less stable *cis* isomer, so that the process can start all over again.

What we have described is the absorption of light by the rod cells of a mammal. Animals of very different kinds—arthropods, mollusks—have very different optical systems. But, regardless of differences in anatomy, the process of seeing always begins with the same simple organic reaction: the transformation of 11-*cis*-retinal into its geometric isomer.

41.4   The source of biological energy. The role of ATP

In petroleum we have a fuel reserve on which we can draw for energy—as long as it lasts. We burn it, and either use the heat produced directly to warm ourselves or convert it into other kinds of energy: mechanical energy to move things about; electrical energy, which is itself transformed—at a more convenient place than where the original burning happened—into light, or mechanical energy, or back into heat.

In the same way, the energy our bodies need to keep warm, move about, and build new tissue comes from a food reserve: carbohydrates, chiefly in the form of starch. (We eat other animals, too, but ultimately the chain goes back to a carbohydrate-eater.) In the final analysis, we get energy from food just as we do from petroleum: we oxidize it to carbon dioxide and water.

This food reserve is not, however, a limited one that we steadily deplete. Our store of carbohydrates—and the oxygen to go with it—is constantly replenished by the recombining, in plants, of carbon dioxide and water. The energy for recombination comes, of course, from the sun.

We speak of both petroleum and carbohydrates as sources of energy; we could speak of them as "energy-rich molecules". But the oxygen that is also consumed in oxidation is equally a source of energy. What we really mean is that the energy content of carbohydrates (or petroleum) plus oxygen is greater than that of carbon dioxide plus water. (In total, the bonds that are to be broken are weaker—contain more energy—than the bonds that are to be formed.) These reactants are, of course, energy-rich only in relation to the particular products we want to convert them into. But this is quite sufficient; in our particular kind of world, these *are* our sources of energy.

The body takes in carbohydrates and oxygen, then, and eventually gives off carbon dioxide and water. In the process considerable energy is generated. But in what form? And how is it used to move muscles, transport solutes, and build new molecules? Certainly each of our cells does not contain a tiny fire in which carbohydrates burn merrily, running a tiny steam engine, and over which a tiny organic chemist stews up reaction mixtures. Nor do we contain a central power plant where, again, carbohydrates are burned, and the energy sent about in little steam pipes or electric cables to run muscle-machines and protein-and-fat factories.

In a living organism, virtually the whole energy system is a chemical one. Energy is generated, transported, and consumed by way of chemical reactions and chemical compounds. Instead of a single reaction with a long plunge from the energy level of carbohydrates and oxygen to that of carbon dioxide and water—as in the burning of a log, say—there are long series of chemical reactions in which the energy level descends in gentle cascades. Energy resides, ultimately, in the molecules involved; as they move through the organism, they carry energy with them.

Constantly appearing in these reactions is one compound, *adenosine triphosphate* (ATP). It is called by biochemists an "energy-rich" molecule, but there is

Adenosine triphosphate
ATP

nothing magical about this. ATP does not carry about a little bag of energy which it sprinkles on molecules to make them react. Nor does it undergo hydrolysis alongside other molecules and in some mystical way make this energy available to them. ATP simply undergoes reactions—only one reaction, really. It *phosphorylates*, that is, transfers a phosphoryl group, $-PO_3H_2$, to some other molecule. For example:

$$ATP \quad + \quad R-O-H \quad \longrightarrow \quad ADP \quad + \quad R-O-PO_3H_2$$

| Adenosine triphosphate | An alcohol | Adenosine diphosphate | A phosphate ester |

ATP is called a "high-energy phosphate" compound, but this simply means that it is a fairly reactive phosphorylating agent. It is exactly as though we were to call acetic anhydride "high-energy acetate" because it is a better acetylating agent than acetic acid. And, indeed, there is a true parallel here: ATP is an anhydride, too, an anhydride of a substituted phosphoric acid, and it is a good phosphorylating agent for much the same reasons that acetic anhydride is a good acetylating agent.

When ATP loses a phosphoryl group to another molecule, it is converted into ADP, *adenosine diphosphate*. If ATP is to be regenerated, ADP must itself be phosphorylated, and it is: by certain other compounds that are good enough phosphorylating agents to do this. The important thing in all this is not really the energy level of these various phosphorylating agents—so long as they are reactive enough to do the job they must—but the fact *that the energy level of the carbohydrates and their oxidation products is gradually sinking to the level of carbon dioxide and water*. These compounds—and oxygen—are where the energy is, and ATP is simply a chemical reagent that helps to make it available.

We have seen that very often factors that stabilize products also stabilize the transition state leading to those products, that is, that often there is a parallel between ΔH and E_{act}. To that extent, the energy level of the various phosphorylating agents may enter in, too: less stable phosphorylating agents—less stable, let us say, relative to phosphate anion—may in general tend to transfer phosphate to more stable phosphorylating agents. In addition, of course, if any of the phosphate transfers should be too highly endothermic, this would require a prohibitively high E_{act} for reaction (see Sec. 2.17).

In following sections, we shall see some of the specific reactions in which ATP is involved.

41.5 Biological oxidation of carbohydrates

Next, let us take a look at the overall picture of the biological oxidation of carbohydrates. We start with glycogen ("sugar-former"), the form in which carbohydrates are stored in the animal body. This, we have seen (Sec. 39.9), is a starch-like polymer of D-glucose.

The trip from glycogen to carbon dioxide and water is a long one. It is made up of dozens of reactions, each of which is catalyzed by its own enzyme system. Each of these reactions must, in turn, take place in several steps, most of them unknown. (Consider what is involved in the "reaction" catalyzed by chymotrypsin.) We can divide the trip into three stages. (a) First, glycogen is broken down into its component D-glucose molecules. (b) Then, in *glycolysis* ("sugar-splitting"), D-glucose is itself broken down, into three-carbon compounds. (c) These, in *respiration*, are converted into carbon dioxide and water. Oxygen appears in only the third stage; the first two are anaerobic ("without-air") processes.

The first stage, **cleavage of glycogen**, is simply the hydrolytic cleavage of acetal linkages (Sec. 38.16), this time enzyme-catalyzed.

$$(C_6H_{10}O_5)_n + nH_2O \xrightarrow{enzyme} nC_6H_{12}O_6$$

Glycogen   D-Glucose

The second stage, **glycolysis**, takes eleven reactions and eleven enzymes. The sum of these reactions is:

$$\text{D-glucose} + 2HPO_4^{2-} + 2ADP^{3-} \longrightarrow 2CH_3CHOHCOO^- + 2H_2O + 2ATP^{4-}$$

Phosphate   Lactate

No oxygen is consumed, and we move only a little way down the energy hill toward carbon dioxide and water. What is important is that a start has been made in breaking the five carbon–carbon bonds of glucose, and that two molecules of ADP are converted into ATP. (ATP is required for some of the steps of glycolysis, but there is a *net* production of two molecules of ATP for each molecule of glucose consumed.)

The third stage, **respiration**, is a complex system of reactions in which molecules provided by glycolysis are oxidized. Oxygen is consumed, carbon dioxide and water are formed, and energy is produced.

Let us look at the linking-up between glycolysis and respiration. Ordinarily, the energy needs of working muscles are met by respiration. But, during short periods of vigorous exercise, the blood cannot supply oxygen fast enough for respiration to carry the entire load; when this happens, glycolysis is called upon to supply the energy difference. The end-product of glycolysis, lactic acid, collects in the muscle, and the muscle feels tired. The lactic acid is removed by the blood and rebuilt into glycogen, which is ready for glycolysis again.

The last step of glycolysis is reduction of pyruvic acid to lactic acid. (The reducing agent is, incidentally, an old aquaintance, reduced nicotinamide adenine

$$CH_3COCOO^- + NADH + H^+ \longrightarrow CH_3CHOHCOO^- + NAD^+$$

| Pyruvate | Reduced nicotinamide adenine dinucleotide | | Lactate | Nicotinamide adenine dinucleotide |

dinucleotide, Secs. 22.2–22.4 and 40.15.) Most of the time, however, glycolysis does not proceed to the very end. Instead, pyruvic acid is diverted, and oxidized to acetic acid in the form of a thiol ester, $CH_3CO–S–CoA$, derived from *coenzyme A* and called "acetyl CoA".

Coenzyme A
CoA

It is as acetyl CoA that the products of glycolysis are fed into the respiration cycle.

The acetyl CoA that is fuel for respiration comes not only from carbohydrates but also from the breakdown of amino acids and fats. It is thus the common link between all three kinds of food and the energy-producing process. (Acetyl CoA is even more than that: as we shall see, it is the building block from which the long chains of fatty acids are synthesized.)

Thiols are sulfur analogs of alcohols. They contain the sulfhydryl group, —SH, which plays many parts in the chemistry of biomolecules. Easily oxidized, two —SH groups are converted into disulfide links, —S—S—, which hold together different peptide chains or different parts of the same chain. (See, for example, oxytocin on p. 1357.) Thiols form the same kinds of derivatives as alcohols: *thio*ethers, *thio*acetals, *thiol* esters. Thiol ester groups show the chemical behavior we would expect—they undergo nucleophilic acyl substitution and they make α-hydrogens acidic—this last more effectively than their oxygen counterparts.

41.6   Mechanism of a biological oxidation

Now let us take just one of the many steps in carbohydrate oxidation and look at it in some detail.

Although there is no *net* oxidation in glycolysis, certain individual reactions do involve oxidation and reduction. About mid-way in the eleven steps we arrive

$$H_2O_3P-O-CH_2CHOHCHO \qquad\qquad H_2O_3P-O-CH_2CHOHCOOH$$

ᴅ-Glyceraldehyde-3-phosphate   3-Phosphoglyceric acid

at ᴅ-glyceraldehyde-3-phosphate and its oxidation to 3-phosphoglyceric acid. In the course of this conversion, a phosphate ion becomes attached to ADP to generate a molecule of ATP.

Two reactions are actually involved. First, ᴅ-glyceraldehyde-3-phosphate is oxidized, but not directly to the corresponding acid, 3-phosphoglyceric acid. In-

$$^{2-}O_3POCH_2CHOH-CHO \; + \; NAD^+ \; + \; HPO_4{}^{2-} \; \rightleftharpoons$$

ᴅ-Glyceraldehyde-3-phosphate   Phosphate

$$^{2-}O_3POCH_2CHOH-\underset{\underset{O}{\|}}{C}-O-PO_3{}^{2-} \; + \; NADH \; + \; H^+$$

1,3-Diphosphoglycerate

$$^{2-}O_3POCH_2CHOH-\underset{\underset{O}{\|}}{C}-O-PO_3{}^{2-} \; + \; ADP^{3-} \; \rightleftharpoons$$

$$^{2-}O_3POCH_2CHOH-COO^- \; + \; ATP^{4-}$$

1,3-Diphosphoglycerate   3-Phosphoglycerate

stead, a phosphate ion is picked up to give the mixed anhydride, 1,3-diphospho-glycerate. This is a highly reactive phosphorylating agent and, in the second reaction, transfers a phosphoryl group to ADP to form ATP.

Now, how does all this happen? The enzyme required for the first reaction is *glyceraldehyde-3-phosphate dehydrogenase* ("enzyme-that-dehydrogenates-glycer-aldehyde-3-phosphate"). Its action is by no means as well understood as that of chymotrypsin, but let us look at the kind of thing that is believed to happen. A sulfhydryl group (—SH) of the enzyme adds to the carbonyl group of glyceraldehyde-3-phosphate. Thiols are sulfur analogs of alcohols, and the product is a hemiacetal:

$$E-S-H + RCHO \longrightarrow E-S-\underset{\underset{OH}{|}}{\overset{\overset{H}{|}}{C}}-R$$

Enzyme   Aldehyde   Hemithioacetal

more precisely, a hemi*thio*acetal. Like other acetals, this is both an ether (a *thio* ether) and an alcohol. Such an alcohol group is especially easily oxidized to a carbonyl group.

The oxidizing agent is a compound that, like ATP, constantly appears in these reactions: our old acquaintance nicotinamide adenine dinucleotide (NAD). The functional group here, we remember (Sec. 40.15), is the pyridine ring, which can accept a hydride ion to form NADH. Like the hemiacetal moiety, NAD is bound to the enzyme, and in a position for easy reaction (Fig. 41.3).

$$
\begin{array}{c}
\left|\!-NAD^+ \right. \\
\left|\quad\right. \quad + \text{ RCHO} \rightleftharpoons \\
\left|\!-S\!-\!H\right.
\end{array}
\quad
\begin{array}{c}
\left|\!-NAD^+\right. \\
\left|\quad H\right. \\
\left|\quad |\right. \\
\left|\!-S\!-\!C\!-\!R\right. \\
\left|\quad |\right. \\
\left|\quad OH\right.
\end{array}
\xrightarrow[\;]{\text{base}}
\begin{array}{c}
\left|\!-NAD^+\right. \\
\left|\quad(H\right. \\
\left|\quad |\right. \\
\left|\!-S\!-\!C\!-\!R\right. \\
\left|\quad O^-\right.
\end{array}
\xrightleftharpoons[\text{transfer}]{\text{hydride}}
\begin{array}{c}
\left|\!-NAD\!-\!H\right. \\
\left|\quad\right. \\
\left|\!-S\!-\!C\!-\!R\right. \\
\left|\quad \|\right. \\
\left|\quad O\right.
\end{array}
$$

Enzyme–   1,3-Diphospho-
coenzyme   glyceraldehyde

NAD

NAD—H
+

$$
\begin{array}{c}
\left|\!-NAD^+\right. \\
\left|\quad\right. \quad + \text{ R}\!-\!\underset{\|}{\underset{O}{C}}\!-\!O\!-\!PO_3{}^{2-} \\
\left|\!-S\!-\!H\right.
\end{array}
\xrightleftharpoons[\;]{HPO_4{}^{2-}}
\begin{array}{c}
\left|\!-NAD^+\right. \\
\left|\quad\right. \\
\left|\!-S\!-\!\underset{\|}{\underset{O}{C}}\!-\!R\right.
\end{array}
$$

Enzyme–   1,3-Diphospho-
coenzyme   glycerate

Figure 41.3 Enzymatic conversion of glyceraldehyde-3-phosphate into 1,3-diphosphoglycerate.

Oxidation converts the hemithioacetal into a thiol ester—an acyl enzyme. Like other esters, this one is prone to nucleophilic acyl substitution. It is cleaved, with phosphate ion as nucleophile, to regenerate the sulfhydryl group in the enzyme. The other product is 1,3-diphosphoglycerate. The molecule is (still) a phosphate ester at the 3-position, and has become a mixed anhydride at the 1-position.

The anhydride phosphoryl group is easily transferred; in another enzyme-catalyzed reaction, 1,3-diphosphoglycerate reacts with ADP to yield 3-phospho-glycerate and ATP. The 3-phosphoglycerate goes on in the glycolysis process.

The ATP is available to act as a phosphorylating agent: to convert a molecule of D-glucose into D-glucose-6-phosphate, for example, and help start another molecule through glycolysis; to assist in the synthesis of fatty acids; to change the cross-linking between molecules of *actin* and *myosin*, and thus cause muscular contraction.

The NADH produced is also available to do *its* job, that of reducing agent. It may, for example, reduce pyruvate to lactate in the last step of glycolysis. The extra electrons that make it a reducing agent are passed along, and ultimately are accepted by molecular oxygen.

We are in a strange, complex chemical environment here, but in it we recognize familiar kinds of compounds—hemiacetals, esters, anhydrides, carboxylic acids—and familiar kinds of reactions—nucleophilic carbonyl addition, hydride transfer, nucleophilic acyl substitution. And underlying it all, symphoria: the bringing together of molecules so that they can react rapidly and with selectivity and specificity.

41.7 Biosynthesis of fatty acids

When an animal eats more carbohydrate than it uses up, it stores the excess: some as the polysaccharide glycogen (Sec. 39.9), but most of it as fats. Fats, we know (Sec. 37.2), are triacylglycerols, esters derived (in most cases) from long straight-chain carboxylic acids containing an *even number* of carbon atoms. These even numbers, we said, are a natural consequence of the way fats are synthesized in biological systems.

There are even numbers of carbons in fatty acids because the acids are built up, two carbons at a time, from acetic acid units. These units come from acetyl CoA: the thiol ester derived from acetic acid and coenzyme A (Sec. 41.5). The acetyl CoA itself is formed either in glycolysis, as we have seen, or by oxidation of fatty acids.

Let us see how fatty acids are formed from acetyl CoA units. As before, we must realize that every reaction is catalyzed by a specific enzyme and proceeds by several steps—steps that in some direct, honest-to-goodness chemical way, involved the enzyme.

First, acetyl CoA takes up carbon dioxide (1) to form malonyl CoA. (To illustrate the point made above: this does not happen directly; carbon dioxide

(1) $CH_3CO-S-CoA + CO_2 + ATP \rightleftharpoons$
Acetyl CoA

$HOOC-CH_2CO-S-CoA + ADP + phosphate$
Malonyl CoA

combines with the prosthetic group of the enzyme—*acetyl CoA carboxylase*—and is then transferred to acetyl CoA.) Just as in the carbonation of a Grignard reagent, the *carbanion character* of the α-carbon of acetyl CoA must in some way be involved.

In the remaining steps, acetic and malonic acids react, not as CoA esters, but as thiol esters of *acyl carrier protein* (ACP), a small protein with a prosthetic group quite similar to CoA. These esters are formed by (2) and (3), which we recognize as examples of transesterification.

(2) $CH_3CO-S-CoA + ACP-SH \rightleftharpoons CH_3CO-S-ACP + CoA-SH$
Acetyl-S-ACP

(3) $HOOCCH_2CO-S-CoA + ACP-SH \rightleftharpoons HOOCCH_2CO-S-ACP + CoA-SH$
Malonyl-S-ACP

Now starts the first of many similar cycles. Acetyl—S—ACP condenses (4) with malonyl—S—ACP to give a four-carbon chain.

(4) $CH_3CO-S-ACP + HOOCCH_2CO-S-ACP \rightleftharpoons$

$CH_3CO-CH_2CO-S-ACP + CO_2 + ACP-SH$
Acetoacetyl-S-ACP

At this point we see a strong parallel to the malonic ester synthesis (Sec. 30.2). The carbon dioxide taken up in reaction (1) is lost here; its function was to generate malonate, with its highly acidic α-hydrogens, its carbanion-like α-carbon. Here, as in test tube syntheses, the formation of carbon–carbon bonds is all-important;

here, as in test tube syntheses (Sec. 30.1), carbanion-like carbon plays a key role. In the malonic ester synthesis, decarboxylation follows the condensation step; here, it seems, the steps are concerted, with loss of carbon dioxide providing driving force for the reaction.

The next steps are exact counterparts of what we would do in the laboratory: reduction to an alcohol (5), dehydration (6), and hydrogenation (7). The reducing agent for both (5) and (7) is reduced nicotinamide adenine dinucleotide phosphate, NADPH (Sec. 40.15).

(5) $CH_3CO-CH_2CO-S-ACP + NADPH + H^+ \rightleftharpoons$

$$D\text{-}CH_3CHOH-CH_2CO-S-ACP + NADP^+$$
D-β-Hydroxybutyryl–S–ACP

(6) D-$CH_3CHOH-CH_2CO-S-ACP \rightleftharpoons$ $trans\text{-}CH_3CH{=}CHCO-S-ACP + H_2O$
Crotonyl–S–ACP

(7) $trans\text{-}CH_3CH{=}CHCO-S-ACP + NADPH + H^+ \rightleftharpoons$

$$CH_3CH_2-CH_2CO-S-ACP + NADP^+$$
n-Butyryl–S–ACP

We now have a straight-chain saturated fatty acid, and with this the cycle begins again: reaction of it with malonyl—S—ACP, decarboxylation, reduction, dehydration, hydrogenation. After seven such cycles we arrive at the 16-carbon acid, palmitic acid—and here, for some reason, the process stops. Additional carbons can be added, but by a different process. Double bonds can be introduced, to produce unsaturated acids. Finally, glycerol esters are formed: triacylglycerols, to be stored and, when needed, oxidized to provide energy; and phosphoglycerides (Sec. 37.8), to help make up cell walls.

Enzymes are marvelous catalysts. Yet, even aided by powerful symphoric effects, these biological reactions seek the easiest path. In doing this, they take advantage of the same structural effects that the organic chemist does: the acidity of α-hydrogens, the leaving ability of a particular group, the ease of decarboxylation of β-keto acids.

41.8 Nucleoproteins and nucleic acids

In every living cell there are found **nucleoproteins**: substances made up of proteins combined with natural polymers of another kind, the **nucleic acids**. Of all fields of chemistry, the study of the nucleic acids is perhaps the most exciting, for these compounds are the substance of heredity. Let us look very briefly at the structure of nucleic acids and, then, in the next section, see how this structure may be related to their literally vital role in heredity.

Although chemically quite different, nucleic acids resemble proteins in a fundamental way: there is a long chain—a backbone—that is the same (except for length) in all nucleic acid molecules; and attached to this backbone are various groups, which by their nature and sequence characterize each individual nucleic acid.

Where the backbone of the protein molecule is a polyamide chain (a polypeptide chain), the backbone of the nucleic acid molecule is a polyester chain (called a *polynucleotide* chain). The ester is derived from phosphoric acid (the acid portion) and a sugar (the alcohol portion).

Polynucleotide chain

The sugar is D-ribose (p. 1296) in the group of nucleic acids known as ribonucleic acids (RNA), and D-2-deoxyribose in the group known as deoxyribonucleic acids (DNA). (The prefix *2-deoxy* simply indicates the lack of an —OH group at the 2-position.) The sugar units are in the furanose form, and are joined to phosphate through the C–3 and C–5 hydroxyl groups (Fig. 41.4).

Figure 41.4   Deoxyribonucleic acid (DNA) and ribonucleic acid (RNA).

Attached to C–1 of each sugar, through a β-linkage, is one of a number of heterocyclic bases. A base–sugar unit is called a *nucleoside*; a base–sugar–phosphoric acid unit is called a *nucleotide*. An example of a nucleotide is shown in Fig. 41.5.

Figure 41.5 A nucleotide: an adenylic acid unit of RNA. Here, the nucleoside is adenosine, and the heterocyclic base is adenine.

Four principal bases are found in DNA: *adenine* (A) and *guanine* (G), which contain the purine ring system, and *cytosine* (C) and *thymine* (T), which contain the pyrimidine ring system. RNA contains adenine, guanine, cytosine, and *uracil* (U). (See Fig. 41.6, on p. 1396.)

The proportions of these bases and the sequence in which they follow each other along the polynucleotide chain differ from one kind of nucleic acid to another. Study of this primary structure is approached in the same general way as in the case of proteins: by degradation and identification of fragments. The enormous length of a DNA molecule makes this job a formidable one; in 1968 it was predicted that the sequence of bases in even the shortest DNA could hardly be determined before the 21st century. But only nine years later, in 1977, Sanger (p. 1360) reported the complete sequence of the DNA of the bacteriophage ϕX174, a virus that infects *E. coli*. This DNA molecule is looped to form a giant ring made up of 5386 nucleotide residues! (In 1980, Sanger received a second Nobel Prize for this work.)

Now, what is the secondary structure of nucleic acids? In the autumn of 1951, J. D. Watson (now of Cold Spring Harbor Laboratory) and F. H. C. Crick (now of Cambridge University) began work together on the structure of DNA. They approached the problem along the path that Pauling had laid out in his study of proteins (Sec. 40.16). They had to devise a structure which would account for the chemical and x-ray evidence, and at the same time be consistent with all the structural features of the units involved: molecular size and shape, bond angles and bond lengths, configurations and conformations. Of the chemical evidence the most puzzling piece—and, of course, the most valuable clue—was this: although the proportions of bases vary from one DNA to another, it is always found that A = T and G = C.

Working with molecular models, Watson and Crick assembled a structure in which all the building blocks fitted together without crowding and, of prime importance, which permitted the greatest stabilization by hydrogen bonds: not only many hydrogen bonds, but hydrogen bonds of the kind that Pauling had shown to be the strongest, those with a linear disposition of N---H---N or N---H---O. In April 1953 Watson and Crick reported the structure they had arrived at, the now-famous *double helix*, and in 1962 they received the Nobel Prize. Figure 41.7, on page 1397, shows a model of a tiny portion of the double helix; on page 1379 is a computer-generated representation of DNA as viewed looking *along* the double helix.

Adenine

Guanine

Cytosine

Thymine

Uracil

Figure 41.6　The heterocyclic bases of DNA and RNA. Each base is bonded to the sugar through the lower —NH in each formula; that is, the —H is replaced by C–1 of the sugar.

Figure 41.7   The double helix of DNA. Two-and-a-half turns are shown.

DNA is made up of two polynucleotide chains wound about each other to form a double helix 20 Å in diameter (shown schematically in Fig. 41.8). Each

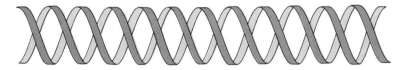

Figure 41.8   Schematic representation of the double helix structure proposed for DNA. Both helixes are right-handed and head in opposite directions; ten residues per turn. There is hydrogen bonding between the helixes.

helix is right-handed and has ten nucleotide units for each complete turn, which occurs every 34 Å along the axis. The two chains head in opposite directions; that is, the deoxyribose units are oriented in opposite ways, so that the sequence is C–3, C–5 in one chain and C–5, C–3 in the other.

The chains are held together at intervals by hydrogen bonds. These are linear hydrogen bonds between adenine and thymine and between guanine and cytosine. Quite simply, A = T and G = C because A is always *bonded to* T and G is always *bonded to* C. Hydrogen bonding between other pairs of bases would not allow them to fit into the double helical structure. The two strands are thus not identical but complementary: opposite every A of one chain is a T in the other, and opposite every G is a C. (See Fig. 41.9, on the next page.)

(a)

(b)

Figure 41.9 Hydrogen bonding in DNA bases; (a) adenine–thymine; (b) guanine–cytosine. The bases fit together exactly right for the formation of two hydrogen bonds between adenine and thymine, and three hydrogen bonds between guanine and cytosine. (The other potential hydrogen-bond donor seen here is the point of attachment to deoxyribose.)

In the secondary structure of RNA helixes are again involved, but this time nearly always single-strand helixes. These molecules vary a good deal in size: some are very large, like DNA molecules; others much smaller and containing fewer than a hundred residues.

So far we have discussed only the secondary structure of nucleic acids. At the tertiary—and higher—level one deals with the way in which they are bound to proteins, and how these nucleoproteins are coiled and folded to make up the chromosome—how, for example, *four meters* of DNA can be fitted into a single cell only two ten-thousandths of a meter across!

But at the heart of all this lies the double helix, which not only meets all the standards that Watson and Crick had set but also, with a simplicity and beauty that could not have been anticipated, accounts for the ability of DNA to play its dual role: as the repository of hereditary information and the director of protein synthesis.

41.9   Chemistry and heredity. The genetic code

Just how is the structure of nucleic acids related to their function in heredity? Nucleic acids control heredity *on the molecular level.* The double helix of DNA is the repository of the hereditary information of the organism. The information is stored as the sequence of bases along the polynucleotide chain; it is a message "written" in a language that has only four letters, A, G, T, C (adenine, guanine, thymine, cytosine).

DNA must both *preserve* this information and *use* it. It does these things through two properties: (a) DNA molecules can duplicate themselves, that is, can bring about the synthesis of other DNA molecules identical with the originals; this process is called *replication.* (b) DNA molecules can control the synthesis, in an exact and specific way, of the proteins that are characteristic of each kind of organism.

(All this is a reciprocal affair, a tightly interwoven system of give-and-take. Every activity of DNA requires catalysis by an enzyme: replication, for example, needs DNA polymerase. Yet all these enzymes are proteins, and exist only because they were originally made—with enzyme catalysis—at the direction of DNA.)

First, there is the matter of replication. The sequences of bases in one chain of the double helix controls the sequence in the other chain. The two chains fit together, as Crick puts it, like a hand and a glove. They separate, and about the hand is formed a new glove, and inside the glove is formed a new hand. Thus, the pattern is preserved, to be handed down to the next generation.

Next, there is the matter of guiding the synthesis of proteins. A particular sequence of bases along a polynucleotide chain leads to a particular sequence of amino acid residues along a polypeptide chain. A protein has been likened to a long sentence written in a language of 20 letters: the 20 different amino acid residues. But the hereditary message is written in a language of only four letters; it is written in a *code*, with each word standing for a particular amino acid.

The genetic code has been broken, but this is only a beginning; research is now aimed at, among other things, tracking down the lines of communication. DNA serves as a template on which molecules of RNA are formed in the process called *transcription.* The double helix of DNA partially uncoils, and about one of the separated strands is formed a chain of RNA; the process thus resembles replication of DNA, except that this newly formed chain contains ribose instead of deoxyribose and corresponds to only a segment of the DNA chain. The base sequence along the RNA chain is different from that along the DNA template, but is determined *by it*: opposite each adenine of DNA, there appears on RNA a uracil; opposite guanine, cytosine; opposite thymine, adenine; opposite cytosine, guanine. Thus, AATCAGTT on DNA becomes UUAGUCAA on RNA.

One kind of RNA—called, fittingly, *messenger RNA*—carries a message to the ribosome, where protein synthesis actually takes place. At the ribosome, messenger RNA calls up a series of *transport RNA* molecules, each of which is loaded with a particular amino acid. The order in which the transport RNA molecules are called up—the sequence in which the amino acids are built into the protein chain—depends upon the sequence of bases along the messenger RNA chain. Thus, GAU is the code for aspartic acid; UUU, phenylalanine; GUG, valine. There are 64 three-letter code words (*codons*) and only 20-odd amino acids, so that more than one codon can call up the same amino acids: CUU and CUC, leucine; GAA and GAG, glutamic acid.

A difference of a single base in the DNA molecule, or a single error in the "reading" of the code, can cause a change in the amino acid sequence. The tiny defect in the hemoglobin molecule that results in sickle-cell anemia (p. 1366) has been traced to a single gene—a segment of the DNA chain—where, perhaps, the codon CAC appears instead of CTC. There is evidence that some antibiotics, by altering the ribosome, cause misreading of the code and, with this, the production of defective proteins and death to the organism.

When the nature of the base is changed by a chemical reaction—oxidation, for example, or alkylation—its size and hydrogen-bonding ability are altered, and base-pairing between strands is impaired. This damage can lead to *mutations*— changes in the sequence of bases—and, with mutations, an increased likelihood of the development of cancerous cells. Carcinogenic compounds exert their effects in this way, many of them by a familiar reaction: nucleophilic substitution, with attack by a basic nitrogen of one of these purine or pyrimidine rings on an electrophilic substrate—an epoxide, for example (Sec. 34.20).

Thus, the structure of nucleic acid molecules determines the structure of protein molecules. The structure of protein molecules, we have seen, determines the way in which they control living processes. Biology is becoming more and more a matter of shapes and sizes of molecules.

For these molecules to do the kinds of things they must—the kinds of things we have seen in this chapter—they must be *big* ones. Only big molecules can offer the infinite variety of shapes that are needed to carry on the myriad different activities that constitute life. Of all the elements only carbon can form the framework of such big molecules. Thus, it would seem, biomolecules are inevitably organic molecules, and the chemistry of life is organic chemistry.

PROBLEMS

1. Carbon dioxide is required for the conversion of acetyl CoA into fatty acids. Yet when carbon dioxide labeled with ^{14}C is used, none of the labeled carbon appears in the fatty acids that are formed. How do you account for these facts?

2. Taken together, what do the following two facts show about chymotrypsin action? (a) The two esters, p-nitrophenyl acetate and p-nitrophenyl thiolacetate, $p\text{-}NO_2C_6H_4SCOCH_3$, undergo chymotrypsin-catalyzed hydrolysis at the same rate and with the same pH dependence of rate, despite the fact that —SR is a much better leaving group than —OR. (b) There is no oxygen exchange (Sec. 24.17) in chymotrypsin-catalyzed hydrolysis of an ester RCOOR'.

3. For each enzyme-catalyzed reaction shown in the following equations, tell what fundamental organic chemistry is involved.

(a) So that acetyl CoA can get through the membrane from the mitochondria where it is formed to the cytoplasm where fatty acids are made, it is converted into citric acid.

$$CH_3CO\text{—}S\text{—}CoA + HOOCCOCH_2COOH \rightleftharpoons \underset{\underset{\displaystyle COOH}{|}}{HOOCCH_2\overset{\overset{\displaystyle OH}{|}}{C}CH_2COOH} + CoA\text{—}SH$$

Oxaloacetic acid   Citric acid

(b) Cholesterol is made up of isoprene units derived from isopentenyl pyrophosphate (Sec. 10.31), which is, in turn, formed from mevalonic acid.

$$CH_3CO-S-CoA + CH_3COCH_2CO-S-CoA \rightleftharpoons$$

$$\underset{\underset{OH}{|}}{HOOCCH_2\overset{\overset{CH_3}{|}}{C}CH_2CO-S-CoA} + CoA-SH$$

$$\underset{\underset{OH}{|}}{HOOCCH_2\overset{\overset{CH_3}{|}}{C}CH_2CO-S-CoA} + 2NADPH + 2H^+ \rightleftharpoons$$

$$\underset{\underset{OH}{|}}{HOOCCH_2\overset{\overset{CH_3}{|}}{C}CH_2CH_2OH} + 2NADP^+ + CoA-SH$$

Mevalonic acid

4. In 1904, Franz Knoop outlined a scheme for the biological oxidation of fatty acids that was shown—50 years later—to be correct. In his key experiments, he fed rabbits fatty acids of formula $C_6H_5(CH_2)_nCOOH$. When the side chain ($n + 1$) contained an even number of carbons, a derivative of phenylacetic acid, $C_6H_5CH_2COOH$, was excreted in the urine; an odd number, and a derivative of benzoic acid was excreted. What general hypothesis can you formulate from these results?

5. In the actual cleavage reaction of glycolysis, D-fructose-1,6-diphosphate is converted into D-glyceraldehyde-3-phosphate and dihydroxyacetone, $CH_2OHCOCH_2OH$. What kind of reaction is this, basically? Sketch out a possible mechanism, neglecting, of course, the all-important role of the enzyme. (*Hints*: The enzyme required is called *aldolase*. See Problem 25.14, p. 915.)

6. The particular wavelengths of electromagnetic radiation that are absorbed by rhodopsin are, by definition, "visible light" (Sec. 16.3). Imagine a creature whose vision depended upon the *cis–trans* isomerization, not of a compound like retinal, but of a simple, unconjugated alkene. What wavelengths of radiation would constitute "visible light" to such a creature?

7. When RNA is hydrolyzed there is *no* relationship among the quantities of the four bases obtained similar to that observed for the bases obtained from DNA. What does this fact suggest about the structure of RNA?

8. When DNA partially uncoils in the process of transcription, only one of the separated strands serves as a template for RNA synthesis. What disadvantage would there be if *both* separated strands were to act as templates?

Suggested Readings

General

Wheland, G. W. *Advanced Organic Chemistry,* 3rd ed.; Wiley: New York, 1960.

Hine, J. *Physical Organic Chemistry,* 2nd ed.; McGraw-Hill: New York, 1962.

Ingold, C. K. *Structure and Mechanism in Organic Chemistry,* 2nd ed.; Cornell University: Ithaca, 1969.

March, J. *Advanced Organic Chemistry,* 3rd ed.; McGraw-Hill: New York, 1985.

Hammett, L. P. *Physical Organic Chemistry,* 2nd ed.; McGraw-Hill: New York, 1970.

Carey, F.; Sundberg, R. *Advanced Organic Chemistry.* 2nd ed.; Plenum: New York, 1983; 2 vols.

Gould, E. S. *Mechanism and Structure in Organic Chemistry;* Holt: New York, 1959.

Sykes, P. *A Guide to Mechanism in Organic Chemistry,* 5th ed.; Longman: New York, 1981.

Breslow, R. *Organic Reaction Mechanisms,* 2nd ed.; W. A. Benjamin: New York, 1969.

Jones, R. A. Y. *Physical and Mechanistic Organic Chemistry,* 2nd ed.; Cambridge University: New York, 1984.

Bender, M. L. *Mechanisms of Homogeneous Catalysis from Protons to Proteins;* Wiley: New York, 1971.

Carruthers, W. *Some Modern Methods of Organic Synthesis,* 2nd ed.; Cambridge University: New York, 1978.

House, H. O. *Modern Synthetic Reactions,* 2nd ed.; W. A. Benjamin: Menlo Park, CA, 1972.

Advances in Physical Organic Chemistry; Gold, V.; Bethell, D., Eds.; Academic: New York; a series starting in 1963.

Progress in Physical Organic Chemistry; Taft, R. W., Ed.; Wiley: New York; a series starting in 1963.

Topics in Stereochemistry; Allinger, N. L.; Eliel, E. L., Eds.; Wiley: New York; a series starting in 1967.

The Chemistry of Functional Groups; Patai, S., Ed.; Wiley-Interscience: New York; a series starting in 1964.

Chemistry of Carbon Compounds; Rodd, E. H., Ed.; Elsevier: New York; a series starting in 1951, 2nd ed. starting in 1964, with supplements in 1974.

Comprehensive Organic Chemistry; Barton, D.; Ollis, W. D., Eds.; Pergamon: New York, 1979; 6 vols.

Organic Reactions; Wiley: New York; a series starting in 1942. Each chapter discusses one
    reaction ("The Clemmensen Reduction," "Periodic Acid Oxidation," etc.) with par-
    ticular emphasis on its application to synthesis.
Note: Some of the above books will be referred to later by abbreviated names, e.g., *O. R.;*
    III-2 for *Organic Reactions;* Vol. III, Chapter 2.

Molecular Structure and Intermolecular Forces

Wheland, G. W. *Adv. Org. Chem.;* Chapters 1 and 3.
Ingold, C. K. *Struct. and Mech.;* Chapters I, II, and IV.
Pauling, L. *The Nature of the Chemical Bond,* 3rd ed.; Cornell University: Ithaca, 1960.
Benson, S. W. "Bond Energies"; *J. Chem. Educ.* **1965,** *42,* 502.
Wheland, G. W. *Resonance in Organic Chemistry;* Wiley: New York, 1955.
Coulson, C. A. "The Meaning of Resonance in Quantum Chemistry"; *Endeavour* **1947,**
    6, 42.
Dewar, M. J. S. *Hyperconjugation;* Ronald: New York, 1962.
Dewar, M. J. S.; Dougherty, R. C. *The PMO Theory of Organic Chemistry;* Plenum: New
    York, 1975.
Verkade, P. E. "August Kekulé"; *Proc. Chem. Soc.* **1958,** 205.
Baker, W. "The Widening Outlook in Aromatic Chemistry"; *Chemistry in Britain* **1965,** *1,*
    191, 250.
Breslow, R. "The Nature of Aromatic Molecules"; *Sci. American,* **1972,** August, p 28.
Fowles, G. W. A. "Lone Pair Electrons"; *J. Chem. Educ.* **1957,** *34,* 187.
Cartmell, E.; Fowles, G. W. A. *Valency and Molecular Structure,* 4th ed.; Butterworths:
    London, 1977.
Orchin, M.; Jaffé, H. H. *The Importance of Antibonding Orbitals;* Houghton-Mifflin:
    Boston, 1967.
Woodward, R. B.; Hoffmann, R. *The Conservation of Orbital Symmetry;* Academic: New
    York, 1970.
Vollmer, J. J.; Service, K. L. "Woodward-Hoffmann Rules: Electrocyclic Reactions"; *J.
    Chem. Educ.* **1968,** *45,* 214; "Woodward-Hoffmann Rules: Cycloaddition Reac-
    tions"; *J. Chem. Educ.* **1970,** *47,* 491.
Pearson, R. G. "Molecular Orbital Symmetry Rules"; *Chem. Eng. News* **1970,** Sept. 28,
    p 66.
Gilchrist, T. L.; Storr, R. C. *Organic Reactions and Orbital Symmetry,* 2nd ed.; Cambridge
    University: New York, 1979.
Fukui, K. *Theory of Orientation and Stereoselection;* Springer-Verlag: New York, 1975.
Fleming, I. *Frontier Orbitals and Organic Chemical Reactions;* Wiley: New York, 1976.

Isomerism and Stereochemistry

Eliel, E. L. *Stereochemistry of Carbon Compounds;* McGraw-Hill: New York, 1962.
Eliel, E. L. *Elements of Stereochemistry;* Wiley: New York, 1969.
Wheland, G. W. *Adv. Org. Chem.;* Chapters 2, 6–9.
Mislow, K. *Introduction to Stereochemistry;* W. A. Benjamin: New York, 1965.
Kagan, H. *Organic Stereochemistry;* Wiley: New York, 1979.
Eliel, E. L., et al. *Conformational Analysis;* Wiley-Interscience: New York, 1965.
Bassendale, A. *The Third Dimension in Organic Chemistry;* Wiley: New York, 1984.
Eliel, E. L. "Stereochemical Nonequivalence of Ligands and Faces"; *J. Chem. Educ.* **1980,**
    57, 52.
Cahn, R. S. "An Introduction to the Sequence Rule"; *J. Chem. Educ.* **1964,** *41,* 116.
Mowery, Jr., D. F. "The Cause of Optical Inactivity"; *J. Chem. Educ.* **1952,** *29,* 138.
Bijvoet, J. M. "Determination of the Absolute Configuration of Optical Antipodes"; *En-
    deavour* **1955,** *14,* 71.
Prelog, V. "Chirality in Chemistry (Nobel Lecture)"; *Science* **1976,** *193,* 17.
Mislow, K.; Raban, M. "Stereochemical Relationships of Groups in Molecules"; *Topics in
    Stereochem.* **1967,** *1,* 1.

Arigoni, D.; Eliel, E. L. "Chirality Due to the Presence of Hydrogen at Nonequivalent Positions"; *Topics in Stereochem.* **1969,** *4,* 127.

Klyne, W.; Buckingham, J. *Atlas of Stereochemistry,* 2nd ed.; Chapman and Hall: London, 1978.

Tollenaere, J. P., et al. *Atlas of the Three-Dimensional Structure of Drugs;* Elsevier: Amsterdam, 1979.

Loewus, F. A.; Westheimer, F. H.; Vennesland, B. "Enzymatic Syntheses of Enantiomorphs of Ethanol-1-*d*"; *J. Am. Chem. Soc.* **1953,** *75,* 5018.

Morrison, J. D.; Mosher, H. S. *Asymmetric Organic Reactions;* American Chemical Society: Washington, DC, 1976.

ApSimon, J. W.; Seguin, R. P. "Recent Advances in Asymmetric Synthesis"; *Tetrahedron* **1979,** *35,* 2797.

Boyle, P. H. "Methods of Optical Resolution"; *Quart. Revs.* (London) **1971,** *25,* 323.

Applequist, J. "Optical Activity: Biot's Legacy"; *American Scientist* **1987,** *75,* 58.

March, J. *Adv. Org. Chem.;* pp 82–109, or Chapter 4.

Acids and Bases

Lewis, G. N. "Acids and Bases"; *J. Franklin Inst.* **1938,** *226,* 293.

Seaborg, G. T. "Acids and Bases. The Research Style of G. N. Lewis"; *J. Chem. Educ.* **1984,** *61,* 93.

Jensen, W. B. *The Lewis Acid–Base Concepts;* Wiley-Interscience: New York, 1979.

Wheland, G. W. *Adv. Org. Chem.;* Chapter 5.

VanderWerf, C. A. *Acids, Bases, and the Chemistry of the Covalent Bond;* Reinhold: New York, 1961.

Bell, R. P. *The Proton in Chemistry,* 2nd ed.; Cornell University: Ithaca, 1973.

Bender, M. L. *Mech. Homog. Catal.;* Chapters 1–5.

Hine, J. *Phys. Org. Chem.;* Chapter 2, "Acids and Bases."

March, J. *Adv. Org. Chem.;* Chapter 8.

Ingold, C. K. *Struct. and Mech.;* Chapter XIV.

Jones, J. R. "Acidities of Carbon Acids"; *Quart. Revs.* (London) **1971,** *25,* 365.

Ho, T-L *Hard and Soft Acid and Base Principle in Organic Chemistry;* Academic: New York, 1977.

Olah, G. A., et al. *Superacids;* Wiley: New York, 1985.

Nomenclature

Cahn, R. S.; Dermer, O. C. *Introduction to Chemical Nomenclature,* 5th ed.; Butterworths: Boston, 1979.

Orchin, M.; Kaplan, F.; Macomber, R. S.; Wilson, R. M.; Zimmer, H. *The Vocabulary of Organic Chemistry;* Wiley: New York, 1980.

IUPAC Nomenclature of Organic Chemistry; Rigaudy, J.; Klesney, S. P., Eds.; Pergamon: New York, 1979.

Ring Systems Handbook; American Chemical Society: Washington, DC, 1984; with supplements.

Free Radicals

Gomberg, M. "An Instance of Trivalent Carbon: Triphenylmethyl"; *J. Amer. Chem. Soc.* **1900,** *22,* 757.

Wheland, G. W. *Adv. Org. Chem.;* Chapter 15.

Hine, J. *Phys. Org. Chem.;* Chapters 18–23.

Nonhebel, D. C.; Tedder, J. M.; Walton, J. C. *Radicals;* Cambridge University: Cambridge, 1979.

Pryor, W. A. *Free Radicals;* McGraw-Hill: New York, 1965.

Pryor, W. A. *Introduction to Free Radical Chemistry;* Prentice-Hall: Englewood Cliffs, NJ, 1965.
Huyser, E. S. *Free-Radical Chain Reactions;* Wiley: New York, 1970.
Walling, C. *Free Radicals in Solution;* Wiley: New York, 1957.
Symons, M. C. R. "The Identification of Organic Free Radicals by Electron Spin Resonance," *Adv. Phys. Org. Chem.* **1965,** *3,* 284.
Wertz, J. E.; Bohton, J. R. *ESR: Elementary Theory and Practical Application;* McGraw-Hill: New York, 1972.

Carbocations

Whitmore, F. C. "Alkylation and Related Processes of Modern Petroleum Practice"; *Chem. Eng. News* **1948,** *26,* 668.
Ingold, C. K. *Struct. and Mech.;* Chapter VII.
Bethell, D.; Gold, V. *Carbonium Ions, An Introduction;* Academic: New York, 1967.
Brown, H. C. *The Nonclassical Ion Problem;* Plenum: New York, 1977.
Stang, P. J.; Rappaport, Z.; Hanack, M.; Subramanian, L. R. *Vinyl Cations;* Academic: New York, 1979.
Dewar, M. J. S., et al. *Topics in Current Chemistry No. 80. In Memory of H. L. Meerwein;* Springer-Verlag: New York, 1979.
Bartlett, P. D. *Nonclassical Ions;* W. A. Benjamin: New York, 1965.
Carbonium Ions; Olah, G. A.; Schleyer, P. v. R., Eds. Wiley: New York; in 4 vols.; Vol. I, 1968.
Olah, G. A. *Carbocations and Electrophilic Reactions;* Wiley: New York, 1974.

Carbanions and Tautomerism

Hine, J. *Phys. Org. Chem.;* Chapter 10.
Ingold, C. K. *Struct. and Mech.;* Chapter X.
Wheland, G. W. *Adv. Org. Chem.;* Chapter 14.
Buncel, E.; Durst, T. *Comprehensive Carbanion Chemistry;* Elsevier: New York, 1980.
Cram, D. J. *Fundamentals of Carbanion Chemistry;* Academic: New York, 1965.
Stowell, J. C. *Carbanions in Organic Synthesis;* Wiley: New York, 1979.
Jones, J. R. "Acidities of Carbon Acids"; *Quart. Revs.* (London) **1971,** *25,* 365.
Saltzman, M. "Arthur Lapworth: The Genesis of Reaction Mechanism"; *J. Chem. Educ.* **1972,** *49,* 750.

Nucleophilic Aliphatic Substitution

Hine, J. *Phys. Org. Chem.;* Chapters 6–7.
Ingold, C. K. *Struct. and Mech.;* Chapter 7.
March, J. *Adv. Org. Chem.;* Chapter 10.
Bunton, C. A. *Nucleophilic Substitution at a Saturated Carbon Atom;* Elsevier: New York, 1963.
Streitwieser, Jr., A. *Solvolytic Displacement Reactions;* McGraw-Hill: New York, 1962.
Thornton, E. R. *Solvolysis Mechanisms;* Ronald: New York, 1964.
Arnett, E. M.; Petro, C.; Schleyer, P. v. R. "Carbonium Ions in Solution. 9. The Relationship between Carbocations in Superacid and Solvolysis Transition States"; *J. Am. Chem. Soc.* **1979,** *101,* 522.
Harris, J. M. "Solvolytic Substitution in Simple Aklyl Systems"; *Progr. Phys. Org.* **1974,** *11,* 89.
Bentley, T. W.; Schleyer, P. v. R. "Medium Effects on the Rates and Mechanisms of Solvolytic Reactions"; *Adv. Phys. Org.* **1977,** *14,* 1.
Hartshorn, S. R. *Aliphatic Nucleophilic Substitution;* Cambridge University: New York, 1973.
Brown, H. C.; Kim, C. J.; Lancelot, C. J.; Schleyer, P. v. R. "Product-Rate Correlation in

Acetolysis of *threo*-3-Aryl-2-butyl Brosylates. Supporting Evidence for the Existence of Two Discrete Pathways"; *J. Amer. Chem. Soc.* **1970,** *92,* 5244.

Pedersen, C. J. "Macrocyclic Polyethers for Complexing Metals"; *Aldrichimica acta* **1971,** *4,* 1.

Weber, W. P.; Gokel, G. W. *Phase Transfer Catalysis in Organic Synthesis;* Springer-Verlag: New York, 1977.

Starks, C. M. "Phase-Transfer Catalysis I"; *J. Amer. Chem. Soc.* **1971,** *93,* 195; Starks, C. M.; Owens, R. M. "Phase-Transfer Catalysis II"; *J. Amer. Chem. Soc.* **1973,** *95,* 3613.

Cram, D. J.; Cram, J. M. "Host-Guest Chemistry"; *Science* **1974,** *183,* 803.

Gokel, G. W.; Weber, W. P. "Phase Transfer Catalysis"; *J. Chem. Educ.* **1978,** *55,* 350, 429.

Electrophilic Aromatic Substitution

Hine, J. *Phys. Org. Chem.;* Chapter 16.

Ingold, C. K. *Struct. and Mech.;* Chapter VI.

March, J. *Adv. Org. Chem.;* Chapter 11.

Norman, R. O. C.; Taylor, R. *Electrophilic Substitution in Benzenoid Compounds;* Elsevier: New York, 1965.

Ridd, J. H. "Mechanism of Nitration"; *Accts. Chem. Res.* **1971,** *4,* 248.

Roberts, R. M. "Friedel-Crafts Chemistry"; *Chem. Eng. News* **1965,** Jan. 25, p 96.

Olah, G. A. *Friedel-Crafts Chemistry;* Wiley: New York, 1973.

Nucleophilic Aromatic Substitution

Hine, J. *Phys. Org. Chem.:* Chapter 17.

March, J. *Adv. Org. Chem.;* Chapter 13.

Bunnett, J. F. "Mechanism and Reactivity in Aromatic Nucleophilic Substitution Reactions"; *Quart. Revs.* (London) **1958,** *12,* 1.

Bunnett, J. F. "The Chemistry of Benzyne"; *J. Chem. Educ.* **1961,** *38,* 278.

Miller, J. *Aromatic Nucleophilic Substitution;* Elsevier: New York, 1968.

Hoffmann, R. W. *Dehydrobenzene and Cycloalkynes;* Academic: New York, 1967.

Pietra, F. "Mechanisms for Nucleophilic and Photonucleophilic Aromatic Substitution"; *Quart. Revs.* (London) **1969,** *23,* 504.

Bunnett, J. F. "The Base-Catalyzed Halogen Dance, and Other Reactions of Aryl Halides"; *Accts. Chem. Res.* **1972,** *5,* 139.

Addition to Carbon–Carbon Multiple Bonds

Hine, J. *Phys. Org. Chem.;* Chapter 9, "Polar Addition to Carbon–Carbon Multiple Bonds."

Ingold, C. K. *Struct. and Mech.;* Chapter XIII, "Additions and Their Retrogression."

March, J. *Adv. Org. Chem.;* Chapter 15.

de la Mare, P. B. D.; Bolton, R. *Electrophilic Additions to Unsaturated Systems;* Elsevier: New York, 1966.

O. R.; XIII-3, Walling, C.; Huyser, E. S. "Free Radical Additions to Olefins and Acetylenes to Form Carbon–Carbon Bonds"; XIII-1, Zweifel, G.; Brown, H. C. "Hydration of Olefins, Dienes, and Acetylenes via Hydroboration"; XIII-2, Parham, W. E.; Schweizer, E. E. "Halocyclopropanes from Halocarbenes"; XX-1, Simmons, H. E., et al. "Cyclopropanes."

Jones, Jr., M. "Carbenes"; *Sci. American* **1976,** Feb., p 101.

Brown, H. C. *Boranes in Organic Chemistry;* Cornell University: Ithaca, 1972.

Melloni, G., et al. "Relative Reactivities of Carbon–Carbon Double Bonds and Carbon–Carbon Triple Bonds toward Electrophiles"; *Accts. Chem. Res.* **1981,** *14,* 227.

Orchin, M. "HCo(CO)$_4$ the Quintessential Catalyst"; *Accts. Chem. Res.* **1981,** *14,* 259.

Elimination

Hine, J. *Phys. Org. Chem.;* Chapter 8, "Polar Elimination Reactions."
Ingold, C. K. *Struct. and Mech.;* Chapter IX, "Olefin-Forming Eliminations."
March, J. *Adv. Org. Chem.;* Chapter 17.
Saunders, Jr., W. H. *Mechanisms of Elimination Reactions;* Wiley: New York, 1973.
Banthorpe, D. V. *Elimination Reactions;* Elsevier: New York, 1963.
O. R.; XI-5, Cope, A. C.; Trumbull, E. R. "Hofmann Elimination."
Sicher, J. "The *Syn* and *anti* Course in Bimolecular Olefin-Forming Eliminations"; *Angew. Chem. Int. Ed.* **1972,** *11,* 20.
Saunders, W. H.; Cockerill, A. F. *Mechanisms of Elimination Reactions;* Wiley: New York, 1973.

Oxidation

Carruthers, W. *Mod. Meth. Synth.;* Chapter 6.
House, H. O. *Mod. Syn. Reactions;* Chapters 5–7.
March, J. *Adv. Org. Chem.;* Chapter 19.
Rinehart, K. L. *Oxidation and Reduction of Organic Compounds;* Prentice-Hall: Englewood Cliffs, NJ, 1973.
Stewart, R. *Oxidation Mechanisms;* W. A. Benjamin: New York, 1964.
Oxidation in Organic Chemistry; K. B. Wiberg, Ed.; Academic: New York, 1965; Part A.
O. R.; VII-7, Swern, D. "Epoxidation and Hydroxylation of Ethylenic Compounds with Organic Peracids"; II-8, Jackson, E. L. "Periodic Acid Oxidation."
Haines, A. H. *Methods for the Oxidation of Organic Compounds;* Academic: Orlando, FL, 1985; in the series called *Best Synthetic Methods.*

Reduction and Hydrogenation

Carruthers, W. *Mod. Meth. Synth.;* Chapter 7.
House, H. O. *Mod. Syn. Reactions;* Chapters 1–4.
March, J. *Adv. Org. Chem.;* Chapter 19.
Rinehart, K. L. *Oxidation and Reduction of Organic Compounds;* Prentice-Hall: Englewood Cliffs, NJ, 1973.
Rylander, P. N. *Hydrogenation Methods;* Academic: Orlando, FL, 1985.
O. R.; I-7, Martin, E. L. "The Clemmensen Reduction"; XXII-3, Vedejs, E. "Clemmensen Reduction of Ketones in Anhydrous Organic Solvents"; IV-8, Todd, D. "The Wolff-Kishner Reduction"; II-5, Wilds, A. L. "Reduction with Aluminum Alkoxides (the Meerwein-Ponndorf-Verley Reduction)"; VI-10, Brown, W. G. "Reductions by Lithium Aluminum Hydride"; XXIII-1, Caine, D. "Reduction and Related Reactions of α,β-Unsaturated Compounds with Metals in Liquid Ammonia"; XXIV-1, Birch, A. J.; Williamson, D. H. "Catalytic Homogeneous Hydrogenation"; XXXIV-1, Malek, J. "Reduction by Metal Alkoxyaluminum Hydrides."
Walker, E. R. H. "The Functional Group Selectivity of Complex Hydride Reducing Agents"; *Chem. Soc. Revs.* **1976,** *5,* 23.
Reucroft, J.; Sammes, P. G. "Stereoselective and Stereospecific Olefin Synthesis"; *Quart. Revs.* (London) **1971,** *25,* 135.
Fryzuk, M.; Bosnich, B. "Preparation of Chiral Methyllactic Acid by Catalytic Asymmetric Hydrogenation"; *J. Amer. Chem. Soc.* **1979,** *101,* 3043.

Rearrangements

Whitmore, F. C. "The Common Basis of Intramolecular Rearrangements"; *J. Amer. Chem. Soc.* **1932,** *54,* 3274.
Wheland, G. W. *Adv. Org. Chem.;* Chapters 12–13.
Hine, J. *Phys. Org. Chem.;* Chapters 14, 15, and 23.
Ingold, C. K. *Struct. and Mech.;* Chapters X–XII.

March, J. *Adv. Org. Chem.;* Chapter 18.

Molecular Rearrangements; de Mayo, P., Ed.; Interscience: New York, 1963.

O. R.; III-7, Wallis, E. S.; Lane, J. F. "The Hofmann Reaction"; III-8, Wolff, H. "The Schmidt Reaction"; III-9, Smith, P. A. S. "The Curtius Reaction."

Collins, C. J. "The Pinacol Rearrangement"; *Quart. Revs.* (London) **1960,** *14,* 357.

Collins, C. J. "Reactions of Primary Aliphatic Amines with Nitrous Acid"; *Accts. Chem. Res.* **1971,** *4,* 315.

Acyl Compounds

Ingold, C. K. *Struct. and Mech.;* Chapter XXV.

Hine, J. *Phys. Org. Chem.;* Chapters 12–13.

Carruthers, W. *Mod. Meth. Synth.;* Chapter 1.

House, H. O. *Mod. Syn. Reactions;* Chapter 11, "Acylation at Carbon."

O. R.; I-9, Hauser, C. R.; Hudson, Jr., B. E. "The Acetoacetic Ester Condensation and Related Reactions"; IV-4, McElvain, S. M. "The Acyloins"; II-4, Johnson, W. S. "The Formation of Cyclic Ketones by Intramolecular Acylation"; VIII-2, Shirley, D. A. "The Synthesis of Ketones from Acid Chlorides and Organometallic Compounds of Magnesium, Zinc, and Cadmium."

Satchell, D. P. N. "An Outline of Acylation"; *Quart. Revs.* (London) **1963,** *17,* 160.

Davies, A. G.; Kenyon, J. "Alkyl-Oxygen Heterolysis in Carboxylic Esters and Related Compounds"; *Quart. Revs.* (London) **1955,** *9,* 203.

Bender, M. L. "Mechanisms of Catalysis of Nucleophilic Reactions of Carboxylic Acid Derivatives"; *Chem. Revs.* **1960,** *60,* 53.

Carbonyl Compounds

Ingold, C. K. *Struct. and Mech.;* pp 994–1015.

Hine, J. *Phys. Org. Chem.;* Chapter 11.

Carruthers, W. *Mod. Meth. Synth.;* Chapters 1 and 5.

House, H. O. *Mod. Syn. Reactions;* Chapter 9, "The Alkylation of Active Methylene Compounds"; Chapter 10, "The Aldol Condensation and Related Reactions."

Gutsche, C. D. *The Chemistry of Carbonyl Compounds;* Prentice-Hall: Englewood Cliffs, NJ, 1967.

O. R.; XVIII-1, Jorgenson M. J. "Preparation of Ketones from the Reaction of Organolithium Reagents with Carboxylic Acids"; XVI (the entire volume), Nielsen, A. J.; Houlihan, W. J. "The Aldol Condensation"; IV-5, Ide, W. S.; Buck, J. S. "The Synthesis of Benzoins"; II-3, Geissman, T. A. "The Cannizzaro Reaction"; V-6, Crounse, N. N. "The Gattermann-Koch Reaction"; I-10, Blicke, F. F. "The Mannich Reaction"; XV-1, Schaefer, J. P.; Bloomfield, J. J. "The Dieckmann Condensation"; XV-2, Jones, G. "The Knoevenagel Reaction"; XXII-4, Rathke, M. W. "The Reformatsky Reaction"; XIV-3, Maercker, A. "The Wittig Reaction"; XXII-2, Posner, G. H. "Substitution Reactions Using Organocopper Compounds"; XXVIII-3, Mukaiyama, T. "The Directed Aldol Reaction."

α, β-Unsaturated Carbonyl Compounds

Ingold, C. K. *Struct. and Mech.;* pp 1015–1037.

House, H. O. *Mod. Syn. Reactions;* Chapter 9, "The Alkylation of Active Methylene Compounds."

Carruthers, W. *Mod. Meth. Synth.;* Chapters 1 and 3.

O. R.; IV-1, Kloetzel, M. C. "The Diels-Alder Reaction with Maleic Anhydride"; V-2, Bruson, H. A. "Cyanoethylation"; X-3, Bergmann, E. D.; Ginsburg, D.; Pappo, E. "The Michael Reactions"; XXV-3, Nagata, W.; Yoshioka, M. "Hydrocyanation of Conjugated Carbonyl Compounds"; XXXII-1, Ciganek, E. "The Intramolecular Diels-Alder Reaction."

Nitrogen Compounds

Zollinger, H. *Azo and Diazo Chemistry;* Interscience: New York, 1961.
Zollinger, H. "Reactivity and Stability of Arenediazonium Ions"; *Accts. Chem. Res.* **1973,** *6,* 335.
Ridd, J. H. "Nitrosation, Diazotization, and Deamination"; *Quart. Revs.* (London) **1961,** *15,* 418.
Paquette, L. A. *Principles of Modern Heterocyclic Chemistry;* W. A. Benjamin: New York, 1968.
Katritzky, A. R. *Handbook of Heterocyclic Chemistry;* Pergamon: Elmsford, NY, 1968.
Meyers, A. I. *Heterocycles in Organic Synthesis;* Wiley: New York, 1974.
Cook, A. G. *Enamines: Synthesis, Structure, and Reactions;* Dekker: New York, 1969.
Adams, E. "Barbiturates"; *Sci. American* **1958,** Jan., p 60.
Miller, I. T.; Springall, H. D. *Sidgwick's Organic Chemistry of Nitrogen,* 3rd ed.; Clarendon: Oxford, 1966.
Guida, W. C.; Raber, D. J. "The Chemistry of Color Photography"; *J. Chem. Educ.* **1975,** *52,* 622.
Armarego, W. L. F. *Stereochemistry of Heterocyclic Compounds, Part I, Nitrogen Heterocycles;* Wiley: New York, 1977.
Riddell, F. G. *The Conformational Analysis of Heterocyclic Compounds;* Academic: New York, 1980.

Synthesis

Carruthers, W. *Some Modern Methods of Organic Synthesis;* Cambridge University: New York, 1978.
House, H. O. *Modern Synthetic Reactions,* 2nd ed.; W. A. Benjamin: New York, 1972.
Norman, R. O. C. *Principles of Organic Synthesis,* 2nd ed.; Chapman and Hall: London, 1978.
Warren, S. *Organic Synthesis, The Disconnection Approach;* Wiley: New York, 1982.
Warren, S. *Designing Organic Syntheses. A Programmed Introduction to the Synthon Approach;* Wiley: New York, 1978.
Greene, T. W. *Protective Groups in Organic Synthesis;* Wiley: New York, 1981.
Houghton, R. P. *Metal Complexes in Organic Chemistry;* Cambridge University: New York, 1979.
Brown, H. C. *Organic Syntheses via Boranes;* Wiley: New York, 1975.
Colquhoun, H. M., et al. *New Pathways for Organic Synthesis: Practical Application of Transition Metals;* Plenum: New York, 1984.
Hanessian, S. *Total Synthesis of Natural Products: The "Chiron" Approach;* Pergamon: Oxford, 1983.
Merrifield, B. "Solid Phase Synthesis (Nobel Lecture)"; *Science* **1986,** *232,* 341.
Brown, H. C. *Boranes in Organic Chemistry;* Cornell University: Ithaca, 1972.
Henrick, C. A. "The Synthesis of Insect Sex Pheromones," *Tetrahedron,* 33, 1848 (1977).

Polymers and Polymerization

Treloar, L. R. G. *Introduction to Polymer Science;* Springer-Verlag: New York, 1970.
Mandelkern, L. *Introduction to Macromolecules;* Springer-Verlag: New York, 1972.
Seymour, R. B.; Carraher, Jr., C. E. *Polymer Chemistry: An Introduction;* M. Dekker: New York, 1981.
Bawn, C. E. H. "New Kinds of Macromolecules"; *Endeavour* **1956,** *15,* 137.
Natta, G. "How Giant Molecules Are Made"; *Sci. American* **1957,** Sept., p 98.
Natta, G. "Precisely Constructed Polymers"; *Sci. American* **1961,** Aug., p 33.
Bawn, C. E. H.; Ledwith, A. "Stereoregular Addition Polymerization"; *Quart. Revs.* (London) **1962,** *16,* 361.
"Symposium on 'State of the Art: Polymer Chemistry' "; *J. Chem. Educ.* **1981,** *58,* 837–955.

Szwarc, M. *Carbanions, Living Polymers, and Electron Transfer Processes;* Wiley: New York, 1968.
Quirk, R. P. "Stereochemistry and Macromolecules"; *J. Chem. Educ.* **1981,** *58,* 540.

Natural Products

Hudson, C. S. "Emil Fischer's Discovery of the Configuration of Glucose"; *J. Chem. Educ.* **1941,** *18,* 353. Fiftieth Anniversary Lecture.
Sharon, N. "Carbohydrates"; *Sci. American,* **1980,** Nov., p 90.
Guthrie, R. D.; Honeyman, J. *An Introduction to the Chemistry of Carbohydrates,* 3rd ed.; Clarendon: Oxford, 1968.
Bender, M. L.; Komiyama, M. *Cyclodextrin Chemistry;* Springer-Verlag: New York, 1978.
Breslow, R. "Biomimetic Chemistry"; *Chem. Soc. Revs.* **1972,** *1,* 574–580.
Tabushi, I. "Cyclodextrin Catalysis as a Model for Enzyme Action"; *Accts. Chem. Res.* **1982,** *15,* 66.
Finar, I. L. *Organic Chemistry;* Longmans, Green: New York, 1959; Vol. II, 2nd ed.
Fieser, L. F.; Fieser, M. *Steroids;* Reinhold: New York, 1959.
Klyne, W. *The Chemistry of the Steroids;* Wiley: New York, 1957.
Mulheirn, L. J.; Ramm, P. J. "The Biosynthesis of Sterols"; *Chem. Soc. Revs.* **1972,** *1,* 259.
Simonsen, J. *The Terpenes;* Cambridge University; Vols. I–III, 2nd ed. 1947; Vols. IV–V, with Ross, W. C. J., 1957.
Erman, W. F. *Chemistry of the Monoterpenes;* M. Dekker: New York, 1985; Parts A and B.
Hendrickson, J. B. *The Molecules of Nature;* W. A. Benjamin: New York, 1965.
Nakanishi, K., et al. *Natural Products Chemistry,* Vols. 1–2, Academic: New York, 1974–1975; Vol. 3, University Science Books: Mill Valley, CA, 1983.
Wood, W. F. "Chemical Ecology: Chemical Communication in Nature"; *J. Chem. Educ.* **1983,** *60,* 531.

Amino Acids and Proteins

Dickerson, R. E.; Geis, I. *The Structure and Action of Proteins;* Harper and Row: New York, 1969.
Schroeder, W. A. *Primary Structures of Proteins;* Harper and Row: New York, 1968.
Schulz, G. E.; Schirmer, R. H. *Principles of Protein Structure;* Springer-Verlag: New York, 1979.
Kopple, K. D. *Peptides and Amino Acids;* W. A. Benjamin: New York, 1966.
Bodansky, M.; Klausner, Y. S.; Ondetti, M. A. *Peptide Synthesis,* 2nd ed.; Wiley: New York, 1976.
Pauling, L. "The Configuration of Polypeptide Chains in Proteins"; *Rec. Chem. Progr.* (Kresge-Hooker Sci. Lib.) **1951,** *12,* 155.
Doty, P. "Proteins"; *Sci. American* **1957,** Sept., p 173.
Pauling, L.; Corey, R. B.; Hayward, R. "Structure of Protein Molecules"; *Sci. American* **1954,** Oct., p 54.
Sanger, F.; Smith, L. F. "The Structure of Insulin"; *Endeavour* **1957,** *16,* 48.
Atlas of Protein Sequence and Structure; National Biomedical Research Foundation; Silver Spring, MD, beginning in 1965.
Sanger, F. "The Chemistry of Insulin (Nobel Lecture)"; *Chemistry and Industry* **1959,** 104.
Li, C. H. "The ACTH Molecule"; *Sci. American* **1963,** July, p 46.
Koshland, Jr., D. E. "Protein Shape and Biological Control"; *Sci. American* **1973,** Oct., p 52.
Dickerson, R. E. "The Structure and Ancient History of an Ancient Protein"; *Sci. American* **1972,** April, p 58.
Calvin, M. "Chemical Evolution"; *American Scientist* **1975,** *63,* 169.
Dayhoff, M. O. "Computer Analysis of Protein Evolution"; *Sci. American* **1969,** July, p 86.
Lehninger, A. L. *Principles of Biochemistry;* Worth: New York, 1982.
Stryer, L. *Biochemistry,* 2nd ed.; Freeman: San Francisco, CA, 1981.

Bohinski, R. C., et al. *Modern Concepts in Biochemistry,* 5th ed.; Allyn and Bacon: Boston, 1987.

Wood, W. B.; Wilson, J. H.; Benbow, R. M.; Hood, L. E. *Biochemistry: A Problems Approach,* 2nd ed.; Benjamin/Cummings: Menlo Park, CA, 1981.

Chemistry of Biological Processes

Wold, F. *Macromolecules: Structure and Function;* Prentice-Hall: Englewood Cliffs, NJ, 1971.

The Molecular Basis of Life: Readings from Scientific American; Freeman: San Francisco, CA, 1968.

Organic Chemistry of Life; Calvin, M.; Pryor, W. A., Eds.; Freeman: San Francisco, CA, 1973.

The Molecules of Life: Readings from Scientific American; Freeman: Salt Lake City, UT, 1986; or the entire October 1985 issue of *Scientific American.*

Watson, J. D. *Molecular Biology of the Gene,* 3rd ed.; W. A. Benjamin: New York, 1976.

Nirenberg, M. W. "The Genetic Code II"; *Sci. American* **1963,** *208*(3), 80.

Crick, F. H. C. "The Genetic Code III"; *Sci. American* **1966,** *215*(4), 55.

Clayton, R. B. "Biosynthesis of Sterols, Steroids, and Terpenoids"; *Quart. Revs.* (London) **1965,** *19,* 168, 201.

Lehninger, A. L. *Bioenergetics,* 2nd ed.; W. A. Benjamin: New York, 1971.

Dugas, H.; Penney, C. *Bioorganic Chemistry;* Springer-Verlag: New York, 1981.

Bender, M. L. *Mech. Homog. Catal.;* Chapters 12–19.

Jencks, W. P. *Catalysis in Chemistry and Enzymology;* McGraw-Hill: New York, 1969.

Bruice, T. C.; Bentkovic, S. *Bioorganic Mechanisms;* W. A. Benjamin: New York, 1966; 2 vols.

Levy, H. R.; Talalay, P.; Vennesland, B. "The Steric Course of Enzymatic Reactions at Meso Carbon Atoms: Application of Hydrogen Isotopes"; in *Progress in Stereochemistry* **1962,** *3,* 299; de la Mare, P. B. D.; Klyne, W., Eds.

Allworth, W. *Stereochemistry and Its Application in Biochemistry;* Wiley: New York, 1972.

Cornforth, J. W. "Enzymes and Stereochemistry"; *Tetrahedron* **1974,** *30,* 1515.

Cornforth, J. W. "Asymmetry and Enzyme Action (Nobel Lecture)"; *Science* **1976,** *193,* 121.

Westheimer, F. H. "Enzyme Action"; Vol. 21, pp 1–35; in *Advances in Physical Organic Chemistry;* Academic: New York, 1985.

Knowles, J. R.; Albery, W. J. "Perfection in Enzyme Catalysis"; *Accts. Chem. Res.* **1977,** *10,* 105.

Dickerson, R. E. "Cytochrome C and the Evolution of Energy Metabolism"; *Sci. American* **1980,** Mar., p 136.

Cohen, C. "The Protein Switch of Muscle Contraction"; *Sci. American* **1975,** Nov., p 36.

"The Chemistry of Vision (special issue)"; *Accts. Chem. Res.* **1975,** *8,* 81–112.

Kunan, W-H. "Chemistry and Biochemistry of Unsaturated Fatty Acids"; *Angew. Chem. Int. Ed.* **1976,** *15,* 61.

Hathway, D. E.; Kolar, G. F. "Mechanisms of Reaction between Ultimate Chemical Carcinogens and Nucleic Acid"; *Chem. Soc. Revs.* **1980,** *9,* 241.

Eigen, M., et al. "Origin of Genetic Information"; *Sci. American* **1981,** Apr., p 88.

Perutz, M. "Hemoglobin Structure and Respiratory Transport"; *Sci. American* **1978,** Dec., p 92.

Polycyclic Hydrocarbons and Carcinogenesis; Harvey, R. G., Ed.; American Chemical Society: Washington, DC, 1985.

Shapiro, R. *Origins: A Skeptic's Guide to the Creation of Life;* Summit Books: New York, 1986.

Cairns-Smith, A. G. *Seven Clues to the Origin of Life;* Cambridge University: Cambridge, 1985.

Use of Isotopes

Semenow, D. A.; Roberts, J. D. "Uses of Isotopes in Organic Chemistry"; *J. Chem. Educ.* **1956,** *33,* 2.

Burr, J. G. *Tracer Applications for the Study of Organic Reactions;* Interscience: New York, 1957.

Collins, C. J. "Isotopes and Organic Reaction Mechanisms"; Vol. 2, p 3; in *Adv. in Phys. Org. Chem.*

Isotopes in Molecular Rearrangements; Buncel, E.; and Lee, C. C., Eds.; Elsevier: New York, 1975.

Zollinger, H. "Hydrogen Isotope Effects in Aromatic Substitution Reactions"; Vol. 2, p 163; in *Adv. in Phys. Org. Chem.*

Melander, L.; Saunders, Jr., W. H. *Reaction Rates of Isotopic Molecules;* Wiley: New York, 1979.

Wightman, R. H., et al. "Enzyme-Mediated Reactions I. Stereochemical Course of Elimination"; *J. Chem. Soc.,* Perkin Transactions **1972,** *1,* 2355.

Banthorpe, D. V.; Baxendale, D. "Terpene Biosynthesis III"; *J. Chem. Soc.,* Sect. C **1970,** 2694.

Bassham, J. A.; Benson, A. A.; Calvin, M. "Isotope Studies in Photosynthesis"; *J. Chem. Educ.* **1953,** *30,* 274.

Bassham, J. A. "The Path of Carbon in Photosynthesis"; *Sci. American* **1962,** June, p 88.

Levine, R. P. "The Mechanism of Photosynthesis"; *Sci. American* **1969,** Dec., p 58.

Libby, W. F. *Radiocarbon Dating,* 2nd ed.; University of Chicago: Chicago, 1955.

Libby, W. F. *Radiocarbon and Tritium;* Geo Science Analytical: Santa Monica, CA, 1981.

Synthesis and Applications of Isotopically Labeled Compounds; Duncan, W. P.; Susan, A. B., Eds. Elsevier: Amsterdam, 1983.

Analysis

Shriner, R. L.; Fuson, R. C.; Curtin, D. Y.; Morrill, T. C. *The Systematic Identification of Organic Compounds,* 6th ed.; Wiley: New York, 1980.

Pasto, D. J.; Johnson, C. R. *Organic Structure Determination;* Prentice-Hall; Englewood Cliffs, NJ, 1969.

Silverstein, R. M.; Bassler, G. C.; Morrill, T. C. *Spectrometric Identification of Organic Compounds,* 4th ed.; Wiley: New York, 1981.

Sternhell, S.; Kalman, J. R. *Organic Structures from Spectra;* Wiley: New York, 1986.

Levy, G. C.; Lichter, R. L.; Nelson, G. L. *Carbon-13 NMR,* 2nd ed.; Wiley-Interscience: New York, 1980.

Levy, G. C.; Lichter, R. L. ^{15}N *NMR Spectroscopy;* Wiley: New York, 1979.

Tonelli, A. E.; Schilling, F. C. "^{13}C NMR Chemical Shifts and the Microstructure of Polymers"; *Accts. Chem. Res.* **1981,** *14,* 233.

McLafferty, F. W. *Interpretation of Mass Spectroscopy,* 2nd ed.; W. A. Benjamin: Menlo Park, CA, 1973.

Budzikiewicz, H.; Djerassi, C.; Williams, D. H. *Mass Spectrometry of Organic Compounds;* Holden-Day: San Francisco, CA, 1967.

Johnson, L. F.; Jankowski, W. C. *Carbon-13 NMR Spectra;* Wiley-Interscience: New York, 1972.

Breitmaier, E., et al. *Atlas of C-13 NMR Data;* Heyden: Philadelphia, 1978; Vols. 1–3.

Industrial Methods

Wittcoff, H. A. *Industrial Organic Chemicals in Perspective;* Wiley: New York, 1980; 2 vols.

Wiseman, P. *An Introduction to Industrial Organic Chemistry,* 2nd ed.; Applied Science: London, 1979.

Falk, J.; Bahrmann, H. "Homogeneous Catalysis—Industrial Applications"; *J. Chem. Educ.* **1984,** *61,* 961.

Encyclopedia of Chemical Technology, 3rd ed.; Kirk, R. E.; Othmer, D. F., Eds.; Interscience: New York, 1978–1984; 25 vols., with supplements. Also in a one-volume version: Kirk, R. E. *Kirk-Othmer Concise Encyclopedia of Chemical Technology;* Wiley: New York, 1985.

Answers to Problems

Chapter 1

1.1 Ionic: a, e, f. **1.4** All tetrahedral (sp^3). **1.5** Structure (a), not (b). **1.6** Linear. **1.7** (a) Expect zero; (b) expect $NF_3 > NH_3$. **1.8** d, e. **1.9** (a) $CH_3OH > CH_3NH_2$; (b) $CH_3SH > CH_3OH$; (c) $H_3O^+ > NH_4^+$. **1.10** (a) H_3O^+; (b) NH_4^+; (c) H_2S; (d) H_2O. **1.11** (a) $CH_3^- > NH_2^- > OH^- > F^-$; (b) $NH_3 > H_2O > HF$; (c) $SH^- > Cl^-$; (d) $F^- > Cl^- > Br^- > I^-$; (e) $OH^- > SH^- > SeH^-$. **1.12** $CH_3NH_2 > CH_3OH > CH_3F$. **1.13** (a) $OH^- > H_2O > H_3O^+$; (b) $NH_2^- > NH_3$; (c) $S^{2-} > HS^- > H_2S$.

1. Ionic: a, d, e, g. **3.** a, c: trigonal. Others: tetrahedral. **4.** Octahedral. **7.** Li compound: ionic. Be compound: non-ionic, covalent. **10.** (a) H_3O^+; (b) HCl; (c) HCl in benzene.

Chapter 2

2.1 (a) -8 kcal; (b) $+13$ kcal; (c) -102 kcal. **2.2** (a) $+46$, $+16$, -24 kcal; (b) $+36$, $+33$, -20 kcal; (c) $+38$, -32, -70 kcal. **2.5** Cation: sp^2, trigonal, flat. Anion: sp^3, pyramidal. **2.7** (a) ($\%C + \%H$) < 100\%; (b) 34.8\%. **2.8** (a) 69.6\% Cl; (b) 70.3\% Cl; (c) 24.84 mg; (d) 26.51 mg; (e) 27.43 mg. **2.9** (a) CH_3; (b) $C_3H_6Cl_2$. **2.10** C_6H_6. **2.11** $C_4H_8O_2$.

1. X, 93.9\% C, 6.3\% H; Y, 64.0\% C, 4.5\% H, 31.4\% Cl; Z, 62.0\% C, 10.3\% H, 27.7\% O. **2.** (a) 45.9\% C, 8.9\% H, 45.1\% Cl; (b) 52.1\% C, 13.1\% H, 34.8\% O; (c) 54.5\% C, 9.2\% H, 36.3\% O; (d) 41.8\% C, 4.7\% H, 18.6\% O, 16.3\% N, 18.6\% S; (e) 20.0\% C, 6.7\% H, 26.6\% O, 46.7\% N; (f) 55.6\% C, 6.2\% H, 10.8\% O, 27.4\% Cl. **3.** (a) CH_2; (b) CH; (c) CH_2O; (d) C_2H_5OCl; (e) $C_3H_{10}N_2$; (f) $C_3H_4O_2Cl_2$. **4.** $C_{20}H_{21}O_4N$. **5.** $C_{14}H_{14}O_3N_3SNa$. **6.** (a) 85.8\% C, 14.3\% H; (b) CH_2; (c) C_6H_{12}. **7.** $C_2H_4O_2$. **8.** CH_2O. **9.** $C_{16}H_{10}O_2N_2$. **10.** (a) 942; (b) 6. **11.** (a) -130; (b) -44; (c) -26; (d) -2; (e) -13; (f) -8; (g) -1; (h) 1st step $+46$; 2nd steps $+10$, -3, 0; 3rd steps -23, -5, -1. **12.** (b) Highly improbable, since E_{act} for reaction with Cl_2 is much smaller. **14.** (b) Chain-carrying $E_{act} \geq 33$ kcal.

Chapter 3

3.2 No. **3.3** Van der Waals repulsion between "large" methyls. **3.9** (a) and (b) C_3H_8; (c) $CH_3CH_2CH_2D$ and CH_3CHDCH_3. **3.10** (a) 3; (b) 4; (c) 2; (d) 1. **3.11** (b) R′X should be 1°. **3.13** (a) 44% 1-Cl, 56% 2-Cl; (b) 64% 1°, 36% 3°; (c) 55% 1°, 45% 3°; (d) 21% 1-Cl, 53% 2-Cl, 26% 3-Cl; (e) 28% 1-Cl-2-Me, 23% 2-Cl-2-Me, 35% 2-Cl-3-Me, 14% 1-Cl-3-Me; (f) 45% 1-Cl-2,2,3-triMe, 25% 3-Cl-2,2,3-triMe, 30% 1-Cl-2,3,3-triMe; (g) 33% 1-Cl-2,2,4-triMe, 28% 3-Cl-2,2,4-triMe, 18% 4-Cl-2,2,4-triMe, 22% 1-Cl-2,4,4-triMe. **3.14** (a) 4% 1-Br, 96% 2-Br; (b) 0.6% 1°, 99.4% 3°; (c) 0.3% 1°, 99.7% 3°; (d) 1% 1-Br, 66% 2-Br, 33% 3-Br; (e) 0.3% 1-Br-2-Me, 90% 2-Br-2-Me, 9% 2-Br-3-Me, 0.2% 1-Br-3-Me; (f) 0.6% 1-Br-2,2,3-triMe, 99% 3-Br-2,2,3-triMe, 0.4% 1-Br-2,3,3-triMe; (g) 0.5% 1-Br-2,2,4-triMe, 9% 3-Br-2,2,4-triMe, 90% 4-Br-2,2,4-triMe, 0.3% 1-Br-2,4,4-triMe. **3.15** 40:1. **3.16** 1.15:1. **3.21** 2,2-Dimethylhexane.

5. (e) 6. **6.** One monochloro, three dichloro, four trichloro. **7.** c, b, e, a, d. **10.** (a) 1-, 2-, and 3-chlorohexane; (b) 1-, 2-, and 3-chloro-2-methylpentane, and 1- and 2-chloro-4-methylpentane; (c) 1-, 3-, and 4-chloro-2,2,4-trimethylpentane, and 1-chloro-2,4,4-trimethylpentane; (d) 1- and 3-chloro-2,2-dimethylbutane, and 1-chloro-3,3-dimethylbutane. **11.** Order of isomers as in Problem 10: (a) 16, 42, 42%; (b) 21, 17, 26, 26, 10%; (c) 33, 28, 18, 22%; (d) 46, 39, 15%. **16.** (a) 2650 g; (b) 8710 kcal; (c) 169 g. **17.** Carius: mono, 45.3% Cl; di, 62.8% Cl. Mol.wt.: mono, 78.5; di, 113. **19.** (a) Methane gas; 1.49 mg CH_3OH; (b) 59, *n*-propyl or isopropyl alcohol; (c) 3; $CH_2OHCHOHCH_2OH$.

Chapter 4

4.1 2 (mirror images). **4.2** (a) 3; (b) 2; (c) 3 (2 are mirror images); (d) 1. **4.3** (a) −39.0°; (b) −2.4°; (c) −0.6°. **4.4** Use a shorter or longer tube, measure rotation. **4.5** Chiral: b, d, f, g, h. **4.6** (b) 3 of 5 are chiral. **4.7** (d) Mirror images: a, b. **4.9** 3°, 2°, 1°, Me. **4.15** (b) Neither active: one is achiral, other is a racemic modification. **4.17** (a) 4; (c) none. **4.19** c, d, e, g. **4.21** −1.01°. **4.22** (a) 5 fractions, two inactive, others active; (b) 5, all inactive; (c) 6, all inactive. **4.23** Rapidly inverting pyramid.

3. Equal but opposite specific rotations; opposite R/S specifications: all other properties the same. **4.** (a) Screw, scissors, spool of thread; (b) glove, shoe, coat sweater, tied scarf; (c) helix, double helix; (d) football (laced), tennis racket (looped trim), golf club, rifle barrel; (e) hand, foot, ear, nose, yourself. **5.** (a) Sawing; (b) opening a soft-drink can; (c) throwing a ball. **7.** (a) and (b) 3-Methylhexane and 2,3-dimethylpentane. **8.** a, b, e, k, 2 pairs of enantiomers; c, d, h, 1 pair of enantiomers + 1 *meso*; f, 4 pairs of enantiomers; g, 1 pair of enantiomers + 2 *meso*; i, 2 diastereomers; j, 1 pair of enantiomers. **9.** A, $CH_3CCl_2CH_3$; B, $ClCH_2CH_2CH_2Cl$; C, $CH_3CHClCH_2Cl$, chiral; D, $CH_3CH_2CHCl_2$; (d) active, $CH_3CHClCHCl_2$. **11.** Attractive dipole–dipole interaction. **12.** (c) 12% *gauche* (as non-resolvable racemic modification), 88% *anti*. **13.** Any optical activity: "accident". Optical inactivity: "murder". **14.** (a) 3; (b) 5; (c) 7 (5 active); (d) 7 (6 active); (e) 1; (f) 2 (1 active); (g) 2. **15.** E, (S,S); F, (R,S); G, (S,S); H (S,S); I, ($2R,3S$)-4-bromo-1,2,3-butanetriol; J, (R,R); K, (R,S).

Chapter 5

5.1 1°: four; 2°: three; 3°: one. **5.6** Starting bromide 63% optical purity; expect alcohol of −6.5°, same purity. **5.7** Optical purity of final product 13% that of reactant: 13% inversion, 87% racemization; or 43.5% front-side attack, 56.5% back-side attack. **5.10** (a) 1.9%; (b) 16.4%; (c) 66.2%; (d) 95.1%; (e) 99.0%. **5.11** Free-radical chlorination of neopentane.

12. (a) Retention; (b) inversion in last step. **15.** (a) E1; (b) E2; (c) change from E1 to E2; (d) S_N1 in (a), S_N2 in (b). (See Secs. 7.13–7.14 for E1 and E2.)

Chapter 6

6.1 Aprotic: b, c, e, g, j, k, l. **6.4** (a) Transition state more stabilized; (b) reactants more stabilized; (c) reactants more stabilized.

5. Anions little solvated; sequence as in gas phase (p. 237). **10.** Much hydrogen bonding.

Chapter 7

7.5 (g) None. **7.8** (a) 2.05; (b) 1.02 mol HCl:1 mol DCl. **7.9** (g) None. **7.10** *i*-Bu > *n*-Pr > Et(≫ neopentyl). **7.13** Anion unsolvated, highly basic. **7.16** Unsolvated F⁻ very basic. **7.17** Principal base is *t*-BuOH. **7.19** Step (3), p. 290, is slower than reverse of (2). **7.21** RNH₃⁺ is deprotonated by OH⁻.

3. b, d, g, h, i, k (3 isomers). **4.** (b) 4 show geometric isomerism. **5.** Differ in all except (h); (l) dipole moment would tell. **9.** 3° > 2° > 1°. **13.** Unsolvated *tert*-butoxide ion very basic. **14.** $(CH_3)_2C{=}CHCH_3$ (major product) and $CH_2{=}C(CH_3)C_2H_5$.

Chapter 8

8.1 (c) 1-Butene 649.8, *cis*-2-butene 648.1, *trans*-2-butene 647.1; (d) 1-pentene 806.9, *cis*-2-pentene 805.3, *trans*-2-pentene 804.3. **8.2** (a) H_3O^+; HBr; (b) HBr; (c) HBr. **8.4** (a) Nucleophilic substitution by water; (b) S_N2; (c) S_N1. **8.5** (a) $Et^+ < i\text{-}Pr^+ < t\text{-}Bu^+$; (b) $i\text{-}Pr^+ - Et^+ = 19.8$ kcal, $t\text{-}Bu^+ - Et^+ = 32.9$ kcal. **8.15** A, alkane; B, 2° alcohol; C, alkyl halide; D, alkene; E, 3° alcohol.

4. 3° radical more stable than 2° radical, forms faster. **8.** (d) Steps (2) and (4) are too difficult with HCl. **10.** Polar effect of —Br substituent. **11.** (b) Formation of carbocation rate-determining. **12.** (e) Cyclopentene (see p. 436). **14.** 3-Hexene.

Chapter 9

9.1 (a) Racemic, *meso*; (b) *syn*; (c) *anti*. **9.2** Enantiomeric, a, c, d; identical, b. **9.5** *Anti*.

1. *Syn* (see Figs. 20.6 and 20.7, pp. 750–751). **3.** (a) *Anti*. **5.** (a) Heterolytic addition via cyclic iodonium ion, favored by polar solvents; (b) free-radical chain addition. **6.** (a) One (inactive, racemic); (b), (c), (e), (f) two (one active, one *meso*); (d) two (both inactive, racemic). **8.** (a) (*E*)-2-pentene + Cl_2; (b) (*E*)-3-hexene + HCO_2H; (c) (*Z*)-3-hexene + $KMnO_4$; (d) (*Z*)-2-butene + $Cl_2(aq)$; (e) (*E*)-2-butene + D_2 + Wilkinson's catalyst (Sec. 20.6). **10.** See Sec. 20.2.

Chapter 10

10.1 React with relatively scarce HCl, with a minimum E_{act} of 15 kcal. **10.2** 1-Chloro-2-butene and 3-chloro-1-butene. **10.6** Attachment of Br to an allylic radical. **10.7** "One-and-a-half" bonds effectively prevent rotation between configurations. **10.12** Steric hindrance by Me to S_N2 attack. **10.13** Vinylic cation is intermediate. **10.14** (a) 56–60 kcal. **10.15** 1,3-Hexadiene. **10.17** (c) Position of equilibrium. **10.21** Head-to-tail (1,4) polymer of isoprene.

13. (a) Two CH_2 planes perpendicular to each other. **19.** Cyclohexene. **20.** 1,3,5-Hexatriene. **21.** (b) Myrcene, $(CH_3)_2C{=}CHCH_2CH_2C({=}CH_2)CH{=}CH_2$. **22.** (a) Dihydromyrcene, $(CH_3)_2C{=}CHCH_2CH_2C(CH_3){=}CHCH_3$; (b) 1,4-addition. **23.** (c) Two farnesyl units, head to head, form squalene skeleton.

Chapter 11

11.5 H goes to terminal C. **11.6** Calcium acetylide.

10. Muscalure, (*Z*)-9-tricosene.

Chapter 12

12.4 *trans* is resolvable. **12.7** (a) 0 kcal; (b) 2.7 kcal; (c) 5.4 kcal (3.6 from methyl–methyl interaction); (d) 0 kcal; (e) 0 kcal; (f) 3.6 kcal. **12.8** (b) 3.6 kcal. **12.9** (a) *cis* > *trans*; (b) *trans* > *cis*; (c) 1.8 kcal/mol in each case. **12.10** More than: (a) 3.2 kcal; (b) 6.8 kcal; (c) 2.3 kcal. **12.11** Resolvable: b, d. *Meso*: c (e and f do not contain chiral centers). **12.12** (a) e; (b) a; (c) c, f; (d) d; (e) b; (f) none. **12.13** Pairs of enantiomers: a, b, c, d. No *meso* compounds. None are non-resolvable racemic modifications. **12.20** (e) For the same degree of unsaturation, there are two fewer hydrogens for each ring. **12.21** All are C_6H_{12}; no information about ring size. **12.22** Two, two, one, none.

4. (a) 4; (b) 6; (c) 7; (d) 9; (e) 5; (f) 2; (g) all-equatorial. **5.** A, *cis*-dimethyl; B, *trans*-dimethyl. **8.** (d) In the *trans* isomer, both large substituents (the other ring) are equatorial; (e) high energy barrier (E_{act}) between decalins since bond must be broken. **14.** (d) Keto–enol tautomerization. **17.** *cis* Isomer: E2; *trans* isomer: E1. **21.** (a) 2; (b) 4; (c) 1; (d) 3; (e) 4. **22.** (a) 1; (b) 2; (c) 1; (d) 3; (e) 3; (f) 5; (g) 4.

23. (b)　　　　**24.** (c)　　　　　　　　　　**25.** (b)

Limonene　　*p*-Menthane　　　α-Terpinene

26. Ring closure: addition of carbocation to an alkene.

Chapter 13

13.1 (a) +5.6 kcal; (b) −26.8 kcal. **13.2** (a) 824.1 kcal; (b) 35.0 kcal greater. **13.8** *Ortho*, +6 °C; *meta*, −7 °C; *para*, +87 °C. **13.10** 26.0%. **13.11** 22.8%. **13.12** 18.5%. **13.13** 25.9%, 22.9%, 18.6%.

2. (a) 3; (b) 3; (c) 3; (d) 6; (e) 10; (f) 6. **3.** (a) 2, 3, 3, 1, 2; (b) 5, 5, 5, 2, 4 (neglecting stereoisomers); (c) none. **4.** (a) 2; (b) 3; (c) 1; (d) 4; (e) 4; (f) 2; (g) 4; (h) 4; (i) 2; (j) 1; (k) 3; (l) 2. **5.** (a) 1; (b) 1; (c) 2; (d) 1; (e) 2; (f) 3; (g) 2. **6.** Yes. **7.** (c) No, the *ortho* isomer would be chiral, and enantiomers would be possible. **8.** *Ortho*, 104 °C; *meta*, 63 °C; *para*, 142 °C. **9.** (a) Those with 3, 5, 7, 9 double bonds; actually, poor geometry for 5, 7; (b) $C_9H_9^-$. **11.** (a) $C_6H_6Cl_6$; (e)–(f) 9 stereoisomers (2 are enantiomers). **13.** (a) $10e^-$, a "magic" number; (b) see Sec. 34.3.

Chapter 14

14.3 (d) Carbocation mechanism. **14.6** Large size of complex. **14.8** (a) $RC{\equiv}O^+$;

(b) $Ar'N_2^+$; (c) NO^+. **14.9** (b) D^+. **14.10**

$$CH_3$$

$$H_3C \quad \overset{+}{\bigcirc} \quad CH_3 \qquad BF_4^-$$

$$H \qquad H$$

1. Activated (faster): a, c, d, g, h, k. Deactivated (slower): b, e, f, i, j. **8.** Via $H_2\overset{+}{O}NO_2$. **12.** (a) Expect lower rate with C_6D_6; (b) expect more C_6H_5Y; (c) expect higher *ortho*:*para* ratio; (d) expect more $C_6H_2D_3Y$ than $C_6H_3D_2Y$. **13.** See Sec. 35.4. **14.** See Sec. 34.9.

Chapter 15

15.8 (a) Similar to Fig. 2.4, with $E_{act} = 19$ kcal, and $\Delta H = +11$ kcal; (b) 8 kcal; (c) steric hindrance to combination. **15.17** See Sec. 10.24. **15.19.** See Fig. 10.8, p. 407.

6. $\sim\!CH_2C_6H_4CH_2C_6H_4CH_2C_6H_4\!\sim$.   **15.** 2-, 3-, 4-, 5-, and 6-phenyldodecanes.

22.

Indene   Indane

23. A and B, racemic and *meso*-$C_6H_5CH(CH_3)\!-\!CH(CH_3)C_6H_5$; C, $[C_6H_5C(CH_3)_2\!-\!]_2$.

Chapter 16

16.1 (a) $(CH_3)_3C^+$; $CH_2\!=\!CH\!-\!CH_2{}^+$; $CH_3CH_2{}^+$; $CH_2\!=\!CH^+$.

16.2

β-Carotene

16.3 (a) C, 1,4-pentadiene; D and E, *cis*-

and *trans*-1,3-pentadiene. **16.4** (a) 2, 1; (b) 1, 2, 3, 4 (for 1,2-dibromopropane); (c) 3, 2; (d) 2, 4, 3; (e) 3, 1; (f) 2, 4, 3, 5; (g) 2, 4; (h) 3, 1, 5. **16.6** 1 signal. **16.7** Electron release by methyl groups. **16.9** (a) Neopentylbenzene; (b) isobutylene bromide, $(CH_3)_2CBrCH_2Br$; (c) benzyl alcohol, $C_6H_5CH_2OH$. **16.12** (a) Ethylbenzene; (b) 1,3-dibromopropane; (c) *n*-propyl bromide. **16.16** (a) 2-Chloro-2-methylbutane; (b) 1-chloro-3-methylbutane; (c) 1-bromopentane. **16.22** A, *trans*; B, *cis*. **16.24** (a) $CH_3\cdot$; (b) $CH_3\dot{C}HCH_3$, $CH_3CH_2\dot{C}HCH_3$; (c) $Ph_3C\cdot$. **16.25** Cyclohexane.

1. (a) $CHCl_2CHClCHCl_2$;   (b) $CH_2ClCCl_2CH_3$;   (c) $(CH_3)_2CHCH_2Br$;

(d) $C_6H_5C(CH_3)_3$; (e) $C_6H_5CH_2CH(CH_3)_2$; (f) indane, ; (g) $C_6H_5CH_2CCl(CH_3)_2$

(actually) or $C_6H_5C(CH_3)_2CH_2Cl$; (h) 1-methyl-1-phenylcyclopropane; (i) $C_6H_5CH_2$-CH_2CH_2Br; (j) $CH_2ClCF_2CH_3$. **2.** (a) $CH_2ClCHClCH_2Cl$; (b) $(CH_3)_2CHCH_2Br$; (c) $CH_3CHClCH_2Cl$; (d) $CH_2\!=\!CHCH_2Br$; (e) cyclohexene; (f) $CH_3CH_2CHBrCH_2Br$. **3.** X, *trans*-1,3-dibromo-*trans*-1,3-dimethylcyclobutane; Y, the *cis,cis* isomer. **4.** See answer

to Problem 14.10. **5.** 1,2-Dimethylcyclopropene. **6.** See Sec. 33.6. **8.** B, C, ;

9. (a) eeeeee, eeeaaa; (b) eeeeea; (c) eeeeaa, eeaeea; (d) eeeeee, no change; eeeaaa, split into two peaks of equal area. **10.** (a) H on C–1; (b) equatorial H downfield from axial H. **11.** 82% equatorial $-$Br (axial H on C–1). **13.** (b) $(CH_3)_3C^+$, $(CH_3)_2CH^+$. **15.** (a)

1-Methylcyclopropene; (b) cyclopropene. **16.** D, ;   **17.** (a) Isopropylbenzene;

(b) isobutylene; (c) phenylacetylene. **18.** (a) Isobutylbenzene; (b) *tert*-butylbenzene; (c) *p*-isopropyltoluene. **19.** (a) Indane; (b) 2,2,4-trimethylpentane; (c) *p-tert*-butyltoluene. **20.** (a) α-Phenylethyl bromide, $C_6H_5CHBrCH_3$; (b) *tert*-pentylbenzene; (c) *sec*-butyl bromide. **21.** (a) 1,2,4-Trimethylbenzene; (b) mesitylene (1,3,5-trimethylbenzene); (c) isopropylbenzene. **22.** (a) 3,3-Dimethyl-1-butene; (b) methylcyclopentane; (c) *trans*-4-octene. **23.** 2,4,4-Trimethyl-2-pentene. **24.** E, α-methylstyrene, $C_6H_5C(CH_3)\!=\!CH_2$.

Chapter 17

17.1 Intramolecular H-bond in *cis* isomer (see Sec. 28.2). **17.5** (a) Leucine → isopentyl alcohol; isoleucine → active amyl alcohol. **17.7** $CH_3CH(OCH_3)CH_3$. **17.8** *syn*-Hydration with anti-Markovnikov orientation. **17.9** Retention.

2. d (highest), e, a, c, b. **4.** (a) *p*-Cresol; (b) and (c) propionic acid. **7.** Intramolecular H-bond between —OH and —G. **8.** (a) Coprostane-3β,6β-diol, by *syn*-hydration at more hindered "top" face of molecule. (b) *syn*-Hydration from beneath gives *alpha* —OH at C–11. **9.** (b) e,e; (c) a,a. **10.** Twist-boat. **11.** Allyllithium: considerable ionic character. Anion: 4 equivalent Hs. **14.** *syn*-Addition: Rh—C and C—H bonds formed on same face. **15.** *anti*-Elimination.

Chapter 18

18.1 Electron-withdrawing groups increase acid strength. **18.4** Complete inversion. **18.6** 1HIO$_4$, a, b, c, e; 4HIO$_4$, f, g; no reaction, d. **18.7** A, $(CH_3)_2C(OH)CH_2OH$; B, 1,2-cyclohexanediol; C, 2-hydroxycyclohexanone; D, HOOCCHOHCHOHCOOH; E, HOCH$_2$CHOHCHOHCH$_2$OH; F, HOCH$_2$CHOHCOCHO; G, HOCH$_2$(CHOH)$_4$CHO. **18.8** Change the concentration.

1. (a) Two give iodoform; (c) one gives negative test. **10.** (a) *anti*-Elimination. **12.** B, HOCH$_2$CH$_2$OH; C, ClCH$_2$COOH; D, HOCH$_2$COOH; E, 1,2-cyclohexanediol; F, CH$_3$(CH$_2$)$_7$CH=CH(CH$_2$)$_7$COOH; J, CH$_2$=CHCOOH; M, HOCH$_2$C≡CH; O, CH$_3$COCH$_3$; S, CH$_3$COONa; U, diacetate of *cis*-1,2-cyclohexanediol; W, triacetate of glycerol; AA, C$_6$H$_5$CO(CH$_2$)$_4$COOH; GG, active 2,4,6,8-tetramethylnonane; HH, *meso*-2,4,6,8-tetramethylnonane. **14.** Salt, R$^+$HSO$_4^-$, formed. **15.** Douglas-fir tussock moth pheromone, (*Z*)-6-henicosen-11-one. **16.** (a) R$_3$C$^+$, stabilized by overlap of empty *p* orbital with π clouds of rings. (b) Methyls located unsymmetrically; plane of methyls and trigonal carbon perpendicular to and bisecting ring. **18.** MM, 1,2,2-triphenylethanol; NN, 1,1,2-triphenylethanol. Use CrO$_3$/H$_2$SO$_4$ test. **19.** (a) *sec*-Butyl alcohol; (b) isobutyl alcohol; (c) diethyl ether. **20.** (a) α-Phenylethyl alcohol; (b) β-phenylethyl alcohol; (c) benzyl methyl ether. **21.** (a) Isopentyl alcohol; (b) 2-ethyl-1-butanol; (c) 4-methyl-2-pentanol. **22.** OO, 2-methyl-2-propen-1-ol; PP, isobutyl alcohol. **23.** QQ, 3,3-dimethyl-2-butanol. **24.** RR, 2-butyn-1-ol. **25.** Geraniol, $(CH_3)_2C$=CHCH$_2$CH$_2$C(CH$_3$)=CHCH$_2$OH. **26.** (a) Same as Problem 25; (b) geometric isomers; (c) in geraniol, —H and —CH$_3$ are *trans*. **27.** Same hybrid allylic cation; gives same bromide.

Chapter 19

19.7 (a) Configuration of (−)-ether same as (−)-alcohol; (b) maximum rotation is −19.5°. **19.8** (a) Complete inversion. **19.9** Phase-transfer catalysis. **19.11** Trifluoroacetate is weaker base, weaker nucleophile, does not compete with alcohol. **19.20** (f) None.

6. Polyisobutylene. **7.** (a) *t*-BuOH. **13.** C, $(CH_2$=CH)$_2$O; D, ClCH$_2$-CHOHCH$_2$OCH$_3$; E, CH$_3$OCH$_2$COOH; F, CH$_3$OCH$_2$CH—CH$_2$; G, CH$_2$—CH$_2$; H;

; I, racemic *trans*-2-chlorocyclohexanol; J, racemic 1-methyl-*trans*-1,2-cyclohexane-

diol; K, racemic and *meso*-HOCH$_2$CHOHCHOHCH$_2$OH; L, racemic 2,3-butanediol; M, *meso*-2,3-butanediol. **14.** Grape berry moth pheromone, (*Z*)-9-dodecen-1-yl acetate. **15.** One component of gossyplure, (7*Z*,11*Z*)-7,11-hexadecadien-1-yl acetate. **16.** *m*-Methylanisole, *m*-CH$_3$C$_6$H$_4$OCH$_3$. **17.** BB, *p*-methoxybenzyl alcohol, *p*-CH$_3$OC$_6$H$_4$CH$_2$OH. **18.** (a) *tert*-Butyl ethyl ether; (b) di-*n*-propyl ether; (c) diisopropyl ether. **19.** (a) 2-Ethoxyethanol; (b) 3-methoxy-1-butanol; (c) 2,5-dihydrofuran, . **20.** CC, *p*-ethoxytoluene; DD, benzyl ethyl ether; EE, 3-phenyl-1-propanol.

Chapter 20

20.2 Intermediate is an α-lactone. **20.4** Neighboring *trans*-Br and *trans*-I give anchimeric assistance.

4. Both *cis* and *trans* isomers give the same intermediate bromonium ion. Back-side attack on cyclic bromonium ion gives *trans* dibromide. **7.** Disparlure, (7*R*,8*S*)-7,8-epoxy-2-methyloctadecane.

Chapter 21

21.2 (a) Acetic, propionic, and *n*-butyric acids; (b) adipic acid. **21.3** (a) 1; (b) 1; (c) 1; (d) 2 (both active); (e) 2; (f) no change. **21.4** Semicarbazone formation reversible: rate control *vs.* equilibrium control. **21.6** (a) Williamson synthesis of ethers; (b) acetals (cyclic). **21.13** Internal "crossed" Cannizzaro reaction.

5. (a) Cannizzaro; (b) "crossed" Cannizzaro.
10.

A, B, C, PhC(CH$_3$)$_2$CH$_2$CH$_2$OH.

11. See Fig. 38.7, Sec. 38.14. **12.** Cyclic ketal. **16.** Hydride transfer from Ph$_2$CHO$^-$ to excess PhCHO. **19.** Protonated aldehyde is electrophile, double bond is nucleophile. **20.** Chair: in N, all —CCl$_3$ groups equatorial; in O, two equatorial, one axial. **22.** (b) *trans* Isomer: intramolecular hydrogen bonding between —OH and ring oxygen. **25.** (CH$_3$)$_2$C=CHCH$_2$CH$_2$C(CH$_3$)=CHCHO, citral *a* (H and CH$_3$ *trans*), citral *b* (H and CH$_3$ *cis*). **26.** Carvotanacetone, 5-isopropyl-2-methyl-2-cyclohexen-1-one. **27.** (a) 2-Butanone; (b) isobutyraldehyde; (c) 3-buten-2-ol. **28.** (a) 4-Heptanone; (b) 3-heptanone; (c) 2-heptanone. **29.** (a) 2-Pentanone; (b) methyl isopropyl ketone; (c) methyl ethyl ketone. **30.** P, *p*-methoxybenzaldehyde; Q, *p*-methoxyacetophenone; R, isobutyrophenone. **31.** S, citronellal, 3,7-dimethyl-6-octenal, (CH$_3$)$_2$C=CHCH$_2$CH$_2$CH(CH$_3$)CH$_2$CHO.

Chapter 22

22.1 Two identical pairs: b. Three pairs: k. Two pairs: d, f, i. One pair: e, g, h, j. None: a, c, l (all *R*). **22.2** Two pairs: f. One pair: a, c, d, e. None: b. **22.3** Enantiotopic: a, d, f, g. Diastereotopic: c, h. None: b. e.

1. Enantiotopic ligands: three pairs, a, d; two pairs, b, g; one pair, c, h. Enantiotopic faces: one pair, c, j. Diastereotopic ligands: four pairs, a, g; one pair, d, j. Diastereotopic faces: one pair, f, h. None of these: e, i. **2.** (a) Enantiomer of II. (b) NADD. **4.** (a) V → (*S*)-amino acid. (c) H adds to *Re* face. **6.** (c) Yes, it is prochiral. **7.** Yes: a, b, d, e, f. No: c.

Chapter 23

23.1 91 at 110 °C, 71 at 156 °C; association occurs even in the vapor phase, decreasing as temperature increases. **23.2** (b) 2-Methyldecanoic acid; (c) 2,2-dimethyldodecanoic acid; (d) ethyl *n*-octylmalonate, *n*-C$_8$H$_{17}$CH(COOEt)$_2$. **23.3** (b) 2-Methylbutanoic acid. **23.4** (a) *p*-Bromobenzoic acid; (b) *p*-bromophenylacetic acid. **23.6** See Sec. 25.1. **23.7** (a) F > Cl > Br > I; (b) electron-withdrawing. **23.12** To prevent generation of HCN. **23.17** (a) Cyclic diester; (b) cyclic anhydride; (c) see Sec. 20.24. **23.19** *o*-Chlorobenzoic acid. **23.20** (a) 103; (b) ethoxyacetic acid. **23.21** (a) Two, 83; (b) N.E. = mol.wt./number acidic H per molecule; (c) 70, 57. **23.22** Sodium carbonate.

7. No reaction, f, h, l, n, o, p. **19.** A and B, *erythro-* and *threo*-2,3-dibromobutanoic acid; C, *meso*-HOOCCHOHCHOHCOOH; F, *cis*-HOOCCH₂CH—CHCH₂COOH.

$$\underset{\underset{CH_2}{\diagdown}}{}$$

20. G, $HC{\equiv}CMgBr$; J, $OHCCH_2COOH$. **26.** N.E. 165; *o*-nitrobenzoic acid. **27.** Q, *m*-ethylbenzoic acid; U, 3,5-dimethylbenzoic acid. **28.** Tropic acid, C_6H_5CH-$(CH_2OH)COOH$; atropic acid, $C_6H_5C({=}CH_2)COOH$; hydratropic acid, $C_6H_5CH(CH_3)$-$COOH$. **29.** (a) $CH_3CHClCOOH$; (b) $ClCH_2COOCH_3$; (c) $BrCH_2COOCH_2$-CH_3; (d) $CH_3CH_2CHBrCOOH$; (e) $CH_3CH_2OCH_2COOH$. **30.** (a) Crotonic acid; (b) mandelic acid; (c) *p*-nitrobenzoic acid.

Chapter 24

24.3 (a) *cis* Acid: the only one that can form a cyclic anhydride. **24.4** G, naphthalene. See Fig. 34.3, p. 1187. **24.5** Final product is 1-phenylnaphthalene. **24.6** 9,10-Anthraquinone. See Sec. 34.18. **24.7** *o*-(*p*-toluyl)benzoic acid (p. 1193). **24.8** The product, *o*-$HOOCC_6H_4COOR$, has an acidic "handle". **24.12** (a) Cyclic double ester; (b) linear polyester by step-reaction polymerization (Sec. 36.7). **24.15** Basicity of leaving group: $Cl^- < RCOO^- < OR^- < NH_2^-$. **24.16** Structure II in Sec. 24.17. **24.20** (a) Formic acid. **24.21** 1-Octadecanol and 1-butanol. **24.24** Linear: *sp*-carbon. **24.25** Urea, $CaCO_3$, NH_3. **24.26** (b) Nucleophilic addition. **24.27** (a) $RCOCl$; (b) $RCOO^-NH_4^+$, $RCONH_2$, RCN, amides of low mol.wt. amines; (c) $RCOO^-NH_4^+$; (d) $(RCO)_2O$; (e) $RCOOR'$. **24.28** (a) 102; (c) 4; (d) no. **24.29** (a) Two, 97; (b) S.E. = mol.wt./number ester groups per molecule; (c) 297.

1. (b) Contain aromatic ring; (c) diesters. **2.** No reaction: f, k. **3.** No reaction: f, k. **10.** (a) $HOCH_2CH_2CH_2CONH_2$; (b) $HOCH_2CH_2CH_2CH_2OH$; (c) $HOCH_2$-CH_2CH_2COOEt. **11.** Second step is S_N2 attack by benzoate anion. **17.** A, *meso*; B, racemic. **18.** Phase-transfer catalysis speeds reaction (Sec. 6.7). **19.** C, CO_3^{2-}; D, $C_2H_5OCONH_2$; K, 1-indanone (compare Problem 10a, Chap. 21); M, indene (see Prob-

lem 22, Chap. 15); O, *trans*-2-methylcyclohexanol. **20.** Progesterone,

21. AA, 1,3-propanediol; BB, 1,2-propanediol; CC, 2-methoxyethanol; DD, dimethoxymethane (di-*O*-methylacetal of formaldehyde); EE, α-hydroxypropionaldehyde; FF, hydroxyacetone; GG, β-hydroxypropionaldehyde; HH, propionic acid; II, ethyl formate; JJ, methyl acetate; KK, *cis*-1,2-cyclopropanediol; LL, ; MM, CH_2—CH—CH_2OH.

22. (a) Methyls are *trans* in NN, PP; *cis* in OO, QQ, RR; (b) NN is resolvable. **23.** See p. 1297. **25.** (a) Ethyl acetate; (b) methacrylic acid; (c) phenylacetamide. **26.** (a) *n*-Propyl formate; (b) methyl propionate; (c) ethyl acetate. **27.** Ethyl *p*-methoxybenzoate. **28.** (a) Isopropyl acetate; (b) γ-butyrolactone. **29.** (a) *n*-Butyl methacrylate; (b) cyclohexyl acetate; (c) diethyl fumarate. **30.** SS, benzyl acetate; TT, methyl phenylacetate; UU, hydrocinnamic acid, $PhCH_2CH_2COOH$. **31.** VV, vinyl acetate. **32.** (a) Ethyl adipate; (b) ethyl ethylphenylmalonate; ethyl acetamidomalonate.

Chapter 25

25.1 III, in which the negative charge resides on oxygen, the atom that can best accommodate it. **25.3** Order of decreasing delocalization of the negative charge of the anion. **25.5** Formation of carbanion is rate-determining in: bromination, racemization, H/D exchange. **25.6** (b) Hard to generate *second* negative charge. **25.7** Expect rate of racemization to be twice as fast as exchange. **25.8** (a) Both reactions go through the same slow step (2), formation of the enol. **25.9** (a) HSO_4^-; (b) H_2O or D_2O. **25.11** Gives a mixture of aldol products. **25.12** Electrophile is protonated aldehyde; nucleophile is enol. **25.14** *Retro* (reverse) aldol condensation. **25.15** π-Orbital overlap of C=C and C=O. (Compare Fig. 10.5, p. 397.) **25.18** 1,5-Cyclooctadiene. **25.20** (a) γ-Hydrogen will be acidic. **25.23** Elimination → 1- and 2-butenes. **25.26** A, Ph_3P=CHOPh; B, $C_2H_5C(CH_3)$=CHOPh; C, $C_2H_5CH(CH_3)CHO$; a general route to aldehydes. **25.27** D,

1-phenylcyclopentene; E, Ph_3P=CHCH$_2$CH=PPh$_3$; F, **25.30** (a) Intra-

molecular Claisen condensation leading to cyclization; (b) 2-carbethoxycyclohexanone; (c) ethyl 2,5-dioxocyclohexane-1,4-dicarboxylate. **25.32** (b) 2,4-Hexanedione; (c) 1,3-diphenyl-1,3-propanedione (dibenzoylmethane); (d) 2-(EtOOCCO)cyclohexanone. **25.33** (a) PhCOOEt and PhCH$_2$COOEt; (b) EtOOCCOOEt and ethyl glutarate; (c) ethyl phthalate and CH_3COOEt. **25.36** C, citric acid, $(HOOCCH_2)_2C(OH)COOH$.

1. (e) Allylbenzene. **2.** (e) Methylenecyclohexane. **3.** (a) No reaction; (m) PhCH=CHCH=CH$_2$; (n) PhCH=CHOPh; (o) PhCH$_2$CHO. **6.** All Claisen condensations. In (e) and (i): two successive condensations. **7.** (b) No: poor yield contaminated by others. **12.** From acetone via mesityl oxide. **13.** (b) Iodoform test. **15.** Triple aldol condensation, followed by crossed Cannizzaro reaction. **18.** Dehydrocitral, $(CH_3)_2C$=CHCH=CHC(CH$_3$)=CHCHO, formed by aldol condensation on γ-carbon of α,β-unsaturated aldehyde. **19.** By Wittig reaction at two points in the molecule. **20.** $CH_3COCH_2COOEt + CH_3MgI \rightarrow CH_4\uparrow + (CH_3COCHCOOEt)^- Mg^{2+}I^-$. **21.** (a) Pheromone is (9Z,11E)-9,11-tetradecadien-1-yl acetate. (b) C is a mixture of Z and E diastereomers. **22.** Bombykol, (10E,12Z)-10,12-hexadecadien-1-ol. **23.** (b) C=C conjugated with second C=O; (c) intramolecular H bonding.

24. L, M, =O, a triketone. **25.** (a) *a*, enol —CH$_3$; *b*, keto —CH$_3$;

c, keto —CH$_2$—; *d*, enol —CH=; *e*, enol —OH. Ratios *a*:*b* and 2*d*:*c* are equal (5.5 and 5.6) and show 85% enol. (b) All enol; conjugation with ring.

Chapter 26

26.2 R:$^-$ undergoes rapid inversion.

6. (a) Putrescine, 1,4-diaminobutane; (b) cadaverine, 1,5-diaminopentane. **8.** NH$_3$; OBr$^-$; H$^+$, **9.** Pair of enantiomers, a, c, e, f; one inactive compound, b; inactive *cis–trans* pair, d. **11.** C, $CH_3CH_2CH_2NH_2$. Gabriel synthesis gives 1° amines free from 2° and 3°.

Chapter 27

27.2 $(CH_3)_3\overset{+}{N}:\overset{-}{B}F_3$. **27.4** 1,3-Pentadiene (from thermal isomerization of 1,4-pentadiene); 2-methyl-1,3-butadiene (isoprene). **27.9** Attack at acyl carbon less hindered than at sulfur; sulfonate better leaving group than carboxylate. **27.10** Free amine is much more reactive. **27.12** (a) *n*-Butyl cation. **27.13** (b) 2-Methyl-2-butene, 2-methyl-1-butene, *tert*-pentyl alcohol. **27.14** Leaving groups Cl$^-$ > H$_2$O > OH$^-$. **27.19** (a) Electron withdrawal makes diazonium ion more electrophilic. **27.22** (a) 2'-Bromo-4-hydroxy-3,4'-dimethylazobenzene. **27.23** Reduction of azo compound formed by coupling N,N-dimethylaniline with some diazonium salt (usually $^-O_3SC_6H_4N_2{}^+$ from sulfanilic acid). **27.25** (a) That unknown is 2°; (b) separate, acidify aqueous solution.

13. (a) See Sec. 32.7. (b) Acidic hydrolysis of amide linkages. **15.** Poor leaving group (OH^-) converted into a good leaving group (OTs^-). **16.** Reaction of $PhN_2{}^+$ is S_N1-like; reaction of p-$O_2NC_6H_4N_2{}^+$ is S_N2-like. **21.** Choline, $HOCH_2CH_2N(CH_3)_3{}^+OH^-$; acetylcholine, $CH_3COOCH_2CH_2N(CH_3)_3{}^+OH^-$. **22.** Novocaine, p-$H_2NC_6H_4COOCH_2CH_2$-$N(C_2H_5)_2$. **23.** D, N-methyl-N-phenyl-p-toluamide. **24.** Q, 1,3,5,7-cyclooctatetraene. **25.** Pantothenic acid, $HOCH_2C(CH_3)_2CHOHCONHCH_2CH_2COOH$. **26.** W, Ph-CONHPh; X, $PhNH_2$; Y, $PhCOOH$. **27.** (a) n-Butylamine; (b) N-methylformamide; (c) m-anisidine. **28.** (a) α-Phenylethylamine; (b) β-phenylethylamine; (c) p-toluidine. **29.** (a) Cyclohexylamine; (b) 4-methylpiperidine (see Sec. 26.14); (c) 4-ethylpyridine (see Sec. 35.1). **30.** X, p-ethoxyaniline; Y, N-ethylbenzylamine; Z, Michler's ketone, p,p'-bis(dimethylamino)benzophenone.

Chapter 28

28.1 Intramolecular hydrogen bond in *ortho* isomer unaffected by dilution. **28.4** Benzene, propylene, HF. **28.8** p-Bromophenyl benzoate, p-$BrC_6H_4OOCC_6H_5$. **28.11** (a) The $-SO_3H$ group is displaced by electrophilic reagents, in this case by nitronium ion. **28.12** Sulfonation is reversible: rate *vs.* equilibrium control. **28.13** Phenol, HONO, 7–8 °C; HNO_3. **28.16** N.E.

5. No reaction: b, c, f, n. **6.** Reaction only with: c, p, r, s, t, u. **7.** Reaction only with: c, h, i, j, k, l, n. **13.** (a) Nucleophilic aliphatic substitution; (b) electrophilic aromatic substitution. **16.** Phenacetin, p-$CH_3CONHC_6H_4OC_2H_5$; coumarane, ; 3-cu-

maranone, ; carvacrol, 5-isopropyl-2-methylphenol; thymol, 2-isopropyl-5-methylphenol; hexestrol, 3,4-bis(p-hydroxyphenyl)hexane. **18.** Adrenaline, 1-(3,4-dihydroxyphenyl)-2-(N-methylamino)ethanol. **19.** Phellandral, 4-isopropyl-3,4,5,6-tetrahydrobenzaldehyde. **20.** Y, m-cresol. **21.** Z, p-allylanisole; AA, p-propenylanisole. **22.** BB, isopropyl salicylate. **23.** Chavibetol, 2-methoxy-5-allylphenol.

24. Piperine,

25. Hordinene, p-$HOC_6H_4CH_2CH_2N(CH_3)_2$ or p-$HOC_6H_4CH(CH_3)N(CH_3)_2$ (actually the former). **26.** α-Terpineol, 2-(4-methyl-3-cyclohexenyl)-2-propanol. **27.** Coniferyl alcohol, 3-(4-hydroxy-3-methoxyphenyl)-2-propen-1-ol. **28.** (a) UU, a ketal and lactone. **29.** AAA, piperonal; BBB, vanillin; CCC, eugenol; DDD, thymol; EEE, isoeugenol; FFF, safrole.

Chapter 29

29.1 (a) See Sec. 10.15; (b) see Sec. 14.19. **29.3** (b) Nucleophilic aromatic substitution; (c) electron withdrawal.

1. No reaction: b, c, d, e, f, g, k, l, n, o. **2.** No reaction: h, i, j, k, m, n, o. **5.** (o) $C_6H_6 + HC\equiv CMgBr$. Racemic modifications: f, h, k. Optically active: n. **13.** Inductive effect, $o \gg m > p$. **14.** $-N_2{}^+$ activates molecule toward nucleophilic substitution.

15. **18.** (a) 28, N_2;

44, CO_2; 76, benzyne, C_6H_4; 152, biphenylene. (b) Anthranilic acid.

Biphenylene

19. Tetraphenylmethane. 21. 23. $Ar^\ominus + Ar'—Br \rightleftarrows Ar—Br + Ar'^\ominus$. Only

carbanions with negative charge *ortho* to halogen are involved.

Chapter 30

30.3 (a) Ethyl benzalmalonate, $PhCH= C(COOEt)_2$. **30.4** (b) Cyclohexylideneacetic acid. **30.6** Nucleophilic substitution (S_N2); $1° > 2° \gg 3°$ (or none); aryl halides not used. **30.7** (a) $CH_3COCH_2CH_2COOH$, a γ-keto acid; (b) $PhCOCH_2COCH_3$, $CH_3COCH_2CH_2COCH_3$, both diketones. **30.9** A, $EtOOCCOCH(CH_3)COOEt$. **30.11** (a) Charged end loses CO_2. **30.12** Gives relatively stable anion, $2,4,6\text{-}(NO_2)_3C_6H_2:^-$. **30.15** Gives relatively stable anion, $PhC\equiv C:^-$. **30.17** B, ethyl 3-hydroxynonanoate.

30.18 E, PhCOOEt. **30.22** B, 2-benzalcyclopentanone; F, 3-phenyl-2,2-dimethyl-

propanal.

3. Cyclopentanone. **4.** C, 1,3-cyclohexanedicarboxylic acid; F, 1,4-cyclohexane-dicarboxylic acid; H, succinic acid; J, 1,2-cyclobutanedicarboxylic acid. **5.** K, 1,5-hexa-diene; O, 2,5-dimethylcyclopentanecarboxylic acid. **7.** (b) Intramolecular aldol condensa-tion; (d) gives 3-methyl-2-cyclohexen-1-one. **11.** (a) *Retro* (reverse) Claisen condensation.

13. S, 1-phenyl-3-nonanone. **14.** U is 9-BBN. **16.** V, ;

W, $ClCH_2CH_2CH_2COCH_2CH_2CH_2Cl$. **17.** Nerolidol, $RCH_2C(CH_3)(OH)CH= CH_2$. **18.** Menthone, 2-isopropyl-5-methylcyclohexanone. **19.** Camphoronic acid, $HOOCCH_2C(CH_3)(COOH)C(CH_3)_2COOH$.

20. Terebic acid, Terpenylic acid,

21. Dihydrogenphosphate ion, $H_2PO_4^-$, a better leaving group than OH^-.

Chapter 31

31.2 A, $PhCH_2CH_2CHO$; B, $PhCH_2CH_2CH_2OH$; C, $PhCH= CHCH_2OH$.

31.4 (d)

$\sim\sim CH_2CH\sim\sim$, $\sim\sim CH_2CH\sim\sim$, $\sim\sim CH_2C(CH_3)\sim\sim$
   | | |
  CN   COOMe   COOMe

Orlon   Acryloid   Lucite, Plexiglas

31.6 All less stable than I. **31.7** An amide. **31.8** Two successive nucleophilic additions. **31.9** Two successive nucleophilic additions. **31.10** B, $CH_3CH(CH_2COOH)_2$; D, δ-keto-caproic acid; E, $CH_3COCH_2CH_2CH(COOEt)_2$; F, $PhCH(CH_2COPh)_2$; H,

H_2C=CHCH(COOH)CH$_2$CH$_2$COOH; I, EtOOCCH=C(COOEt)CH(COOEt)COCH$_3$; J, HOOCCH=C(COOH)CH$_2$COOH. **31.11** (a) K, H_2C=C(COOEt)$_2$; (c) glutaric acid. **31.15** (c) Cannot form iminium ions. **31.16** 1,4-Diphenyl-1,3-butadiene + maleic anhydride; 1,3-butadiene + 2-cyclopentenone; 1,3-butadiene (2 mol). **31.17** (a) 3-Ethoxy-1,3-pentadiene + *p*-benzoquinone; (b) 5-methoxy-2-methyl-1,4-benzoquinone + 1,3-butadiene. **31.18** This is one case in which "enol" is more stable than "keto". **31.19** (a) Ease of oxidation; (b) ease of reduction. **31.20** *p*-Nitrosophenol undergoes keto–enol tautomerization to give the mono-oxime.

3. (a) C$_6$H$_5$COCH$_2$CH(C$_6$H$_5$)CH(CN)COOC$_2$H$_5$; (f) CH$_3$COCH$_2$C(CH$_3$)$_2$CH-(COOEt)COCH$_3$; (h) (EtOOC)$_2$CHCH$_2$CH(COOEt)$_2$; (j) O$_2$NCH$_2$CH$_2$CH$_2$COOMe; (l) O$_2$NC(CH$_2$CH$_2$CN)$_3$; (m) Cl$_3$CCH$_2$CH$_2$CN. **5.** A, (EtOOC)$_2$CHCHPhCH$_2$COCH$_2$-CHPhCH(COOEt)$_2$; B, (EtOOC)$_2$CHCHPhCH$_2$COCH=CHPh; C, 4,4-dicarbethoxy-3,5-diphenylcyclohexanone. **6.** (d) 4-Acetylcyclohexene; (g) 5-nitro-4-phenylcyclohexene. **7.** (a) 1,3,5-Hexatriene + maleic anhydride; (b) 1,4-dimethyl-1,3-cyclohexadiene + maleic anhydride; (c) 1,3-butadiene + benzalacetone; (d) 1,3-butadiene + acetylenedicarboxylic acid; (e) 1,3-cyclopentadiene + *p*-benzoquinone; (f) 1,1′-bicyclohexenyl (see Problem 6(b)) + 1,4-naphthoquinone (see Problem 6(h)); (g) 1,3-cyclopentadiene + crotonaldehyde; (h) 1,3-cyclohexadiene + methyl vinyl ketone; (i) 1,3-cyclopentadiene (2 mol). **8.** *syn*-Addition. **9.** (a) Racemic modification; (b) *meso*; (c) 2 *meso*; (d) *meso*. **11.** Conjugate addition of H$_2$O, then *retro* aldol condensation. **14.** N, glyceraldehyde; P, aconitic acid, HOOCCH=C(COOH)CH$_2$COOH; R, tricarballylic acid, HOOCCH(CH$_2$COOH)$_2$; S, "tetracyclone", tetraphenylcyclopentadienone; U, tetraphenylphthalic anhydride; W, pentaphenylbenzene; BB, (CH$_3$)$_2$C(CH$_2$COOH)$_2$; DD, CH$_3$CHOHC≡CCH$_3$; EE, CH$_3$COC≡CCH$_3$; FF, acetylacetone; GG, (CH$_3$)$_2$C=CHCOOH (isoprene skeleton);

JJ, HOOCCH=C(CH$_3$)CH$_2$COOH;   MM,   QQ,

RR, CH$_3$CONHC(COOC$_2$H$_5$)$_2$CH$_2$CH$_2$CHO; VV, CH$_3$CONHC(COOC$_2$H$_5$)$_2$CH$_2$-(CH$_2$)$_2$CH$_2$NHCOCH$_3$; XX, NCCH$_2$CH$_2$CH(COOC$_2$H$_5$)$_2$; BBB, $^+$H$_3$NCH$_2$(CH$_2$)$_2$-CHClCOO$^-$. **15.** C$_6$H$_5$CH(C$_2$H$_5$)CH$_2$COCH$_3$, 4-phenyl-2-hexanone. **16.** IV is correct.

17. GGG,   **18.** KKK,   **19.** (b)

is intermediate. **22.** Intermediate aryne: dehydrocyclopentadienyl anion. **23.** Michael addition; then carbonyl addition; finally intramolecular nucleophilic substitution.

Chapter 32

32.1 N$_2$ is leaving group. **32.2** Goes with retention, since only *cis* amino acid can form lactam. **32.3** If reaction (2), Sec. 32.6, occurs, it is not reversible; in view of substituent effect, then, (2) and (3) are concerted. **32.4** (a) *p*-Methylbenzaldehyde formed by migration of H; *p*-cresol (and formaldehyde), by migration of *p*-tolyl; (b) H migrates somewhat faster than *p*-tolyl. **32.5** H migrates much faster than alkyl. **32.6** Carbocation undergoes pinacol-like rearrangement. **32.8** Competition between solvent attack and rearrangement independent of leaving group; hence reaction is S$_N$1-like, with intermediate cation. **32.9** Intermediate is a carbocation, which recombines with water faster than it rearranges. **32.11** α-Phenylethyl cation, by H shift.

1. Successive H shifts occur. **2.** CH$_3$CO(CH$_2$)$_4$CH$_2$OH, formed by migration of ring carbon. **4.** (a) Analogous to Hofmann rearrangement, with R′COO$^-$ leaving group instead of X$^-$. **5.** A, PhCONHPh; B, PhNH$_2$; C, PhCOOH; (g) cyclohexanone. **6.** Vinyl migrates predominantly, to give adipaldehyde, most of which undergoes aldol condensation to cyclopentene-1-carboxaldehyde. **8.** (b) *p*-Methoxyphenol and benzophenone; phenol and *p*-chlorobenzophenone. **10.** Two successive H shifts. **11.** (a) Neighboring—OH; (b) epoxide hydrolyzed to carbocation, which rearranges to pinacolone. **12.** R group undergoes 1,2-shift, with retention of configuration, from boron to oxygen in intermediate R$_3$B—OOH, with displacement of OH$^-$. **13.** With *p*-CH$_3$OPh, nearly all reaction via (symmetrical) bridged ion; with *p*-NO$_2$Ph, most reaction via open cation; with Ph, about 50:50. **15.** Assistance by π electrons to give the following intermediates (in (b), may be nonclassical ion):

(a)   (b)   (c)

16. Start with XIV and write equations similar to those in Fig. 32.1, page 1118, for *exo*-norbornyl chloride. **17.** Nucleophilic attack on acyl carbon of XIX by Z to give *tetrahedral intermediate*:

Chapter 33

33.1 First, monocation; then aromatic dication with two π electrons. **33.2** (a) Aromatic, with two π electrons:

33.3 (a) *Con* closure; I or III → *trans*; II → *cis*; (b) *dis* closure; I or III → *cis*; II → *trans*. **33.4** (a) ψ_1; two π electrons; (b) $4n + 2$; *dis* (thermal); (c) $4n$, *con* (thermal); (d) cation, $4n$, *con* (thermal). **33.5** (a) *Dis* opening; (b) *dis* closure; (c) *dis* closure; *con* opening; *dis* closure; (d) *con* opening (4 e); *dis* closure (6 e); (e) *dis* opening of cation (2 e), then combination with water; (f) protonated ketone like a pentadienyl cation, with four π electrons; *con* closure. **33.6** Via the cyclobutene, with *con* closures and openings. **33.7** (a) *cis*-3,6-Dimethylcyclohexene; [4 + 2]; (c) phenyls are *cis* to each other (*syn*-addition) and *cis* to anhydride bridge (*endo* reaction); (d), (e), (f) all are tetramethylcyclobutanes; in D, one methyl is *trans* to other three. **33.8** (a) Diels–Alder; *retro* Diels–Alder; (b) *endo* not *exo*. **33.9** (a) [4 + 2], not [6 + 2]; (b) photochemical (intramolecular) *supra,supra* [2 + 2]; (c) *supra,supra* [6 + 4]; (d) *supra,supra* [8 + 2]; (e) *supra,antara* [14 + 2]. **33.10** (a) *supra* [1,5]-H to either face of trigonal carbon; (b) [1,5]-D, not [1,3]-D or [1,7]-D; (c) [1,3]-C (*supra*) with inversion at migrating C.

1. (a) Phenols; no; (b) dipolar structure is aromatic with six π electrons (compare answer to Problem 33.2); (d) intramolecular H-bond. **2.** (a) *Con* opening (4 e); [1,5]-H *supra*; (b) *con* opening (4 e); *dis* closure (6 e); (c) [1,7]-C *supra* and *dis* closure (4 e); [1,7]-H *supra*; (d) [4 + 4] *supra,supra*; *retro* [4 + 2] *supra,supra* (presumably thermal); (e) allylic cation (two π electrons) undergoes [4 + 2] cycloaddition, followed by loss of proton; (f) bridge walks around the ring in a series of *supra* [1,5]-C shifts; (g) intramolecular *syn* [4 + 2] cycloaddition. **3.** (a) A, *trans*-7,8-dialkyl-*cis,cis,cis*-cycloocta-1,3,5-triene; (b) C,

$(CH_3)_2C=C(CH_3)C(=CH_2)C(CH_3)=CH_2$; (c) D, 9-ethyl-9-methyl-*trans,cis,cis,cis*-cyclo-nona-1,3,5,7-tetraene; the *dis* closure takes place with both possible rotations; (d) E, *cis*-bicyclo[5.2.0]nona-8-ene; F, *cis,trans*-cyclonona-1,3-diene; G, *trans*-bicyclo[5.2.0]nona-8-ene. **4.** Symmetry-allowed *con* opening impossible on geometric grounds for bicyclo compound; reaction is probably not concerted. **5.** K, *cis*-bicyclo[4.2.0]-octa-2,4-diene; L, Diels–Alder adduct which undergoes *retro* Diels–Alder. **6.** (a) [1,2] *supra* sigmatropic shift; π framework is a vinyl radical cation; HOMO is π; predict retention in migrating group; (b) π framework is diene radical cation; HOMO is ψ_2; predict inversion in migrating group.

7. Symmetry-forbidden. **8.**

9. (a) [4 + 2] cycloaddition of benzyne and diene; (b) [2 + 2] thermal cycloaddition symmetry-forbidden; reaction non-concerted, probably via diradicals. **10.**

11. (a) *Meso* dibromide gives *cis*-VII (Fig. 33.26); racemic dibromide gives *trans*-VII; *cis*-VII contains four non-equivalent olefinic hydrogens; *trans*-VII, two equivalent pairs. **12.** (a) M and N, position isomers, both from *syn exo* addition; O and P, position isomers; (b) *retro* Diels–Alder. **13.** (a) (numbering from left to right in Fig. 33.19) Overlap between lobe of C–3 of diene and C–3 of ene, carbons to which bonds are not being formed; (b) lobes corresponding to those in (a) are of opposite phase.

14. (a)

(b) intramolecular solvomercuration possible only for *cis* isomer. **15.** (a) Allowed thermal *con* opening (4 e) would give impossibly strained *cis,cis,trans*-cyclohexa-1,3,5-triene; (b) allowed *antara* [1,3]-H impossible on geometric grounds. **16.** (a) *Con* opening (6 e); [1,7]-H *antara*; (c) *dis* closure (6 e); (d) *con* opening (6 e). **17.** (a) Via *cis,cis,cis,cis,cis*-cyclo-deca-1,3,5,7,9-pentaene; (b) 10 π electrons fits Hückel rule, but evidently not very stable for steric reasons.

Chapter 34

34.1 2; 10; 14. **34.3** (b) *trans*-Decalin more stable; both large groups (the other ring) on each ring are equatorial; (c) *syn*-addition, rate control; *anti*-addition, equilibrium control. **34.4** Benzylic substitution; elimination of HBr to give conjugated alkenylbenzene; benzylic–allylic substitution; elimination to give aromatic ring. **34.5** (a) Cadalene, 4-iso-propyl-1,6-dimethylnaphthalene; (b) cadinene has same carbon skeleton as cadalene, follows isoprene rule. **34.8** (a) Via aryne; (b) direct displacement of —F by amine; (c) both direct displacement and elimination–addition occur. **34.9** 1,2,4-Benzenetricarboxylic acid; 1,2,3-benzenetricarboxylic acid. **34.17** Deactivating acyl group transformed into activating alkyl group. **34.19** Phenanthrene (see Sec. 34.19, and Fig. 34.5, p. 1195). **34.20** 23 kcal/mol; 31 kcal/mol. **34.22** (a) Most stable tetrahydro product; (b) reversible sulfonation yields more stable product. **34.24** (a) 1-Nitro-9,10-anthraquinone; (b) 5-nitro-2-methyl-9,10-anthra-quinone (with some 8-nitro isomer).

34.29 Pyrene,

3. 1-, 5-, and 8-nitro-2-methylnaphthalene. **5.** F, phenanthrene. **7.** G, 1,2-benzanthracene; H, chrysene. **8.** α-Naphthol. **9.** (a) Diels–Alder; (c) J, *meso*; K, racemic modification. **10.** (d) β-Tetralone (2-oxo-1,2,3,4-tetrahydronaphthalene).   **11.** (a) 1,6-Cyclodec-

anedione; (b) bicyclic unsaturated ketone, one 7-ring and one 5-ring. **12.** (a)

Azulene

six π electrons in each ring. (b) From 7-ring toward 5-ring; augmented by C—Cl dipole.

13. (a) [structures] (b) Aromaticity of 7-ring preserved. Protona-

tion at C–1; azulene upon neutralization. (c) Deuteration via electrophilic substitution at C–1 and C–3, and deuteration again at C–1 comparable to the protonation in (b); expect 1,3-dideuterioazulene upon neutralization; (d) at C–1. **14.** Nucleophilic substitution in the

7-ring, at C–4, [structure] Aromaticity of 5-ring preserved, conjugation in 7-ring.

15. Eudalene, 7-isopropyl-1-methylnaphthalene. **16.** Y, 2,2′,3,3′,5,5′-hexachloro-6,6′-di-hydroxydiphenylmethane; CC, 3,4′-dimethylbiphenyl; FF, compound I, p. 555; HH, tetra-phenylmethane; II, 1,3,5-triphenylbenzene. **17.** —N$_2^+$ activates molecule toward nucleo-philic aromatic substitution. **18.** (a) JJ, methylene bridge between 9- and 10-positions of phenanthrene; (b) random insertion of methylene into *n*-pentane; (c) three insertion products

and one addition product. **19.** KK, [structure] Each ring contains six π electrons. **20.** (a)

Via an aryne; (b) direct displacement accompanies elimination–addition. Fluoride least reactive toward benzyne formation (p. 1050), most reactive toward direct displacement (Sec. 29.12). Piperidine shifts equilibrium (1) toward left, tends to inhibit benzyne formation. **21.** UU is aromatic, with 14 π electrons. Methyl protons are *inside* aromatic ring; compare Fig. 16.4, p. 583.

[structure with CH₃ groups]

UU

Chapter 35

35.1 B, [—CH(COOEt)COCH$_3$]$_2$. **35.3** —COOH deactivates ring. **35.4** Two units of starting material linked at the 5-positions through a —CH$_2$— group. **35.5** Sodium furoate and furfuryl alcohol (Cannizzaro reaction). **35.10** Hygrine, 2-acetonyl-*N*-methyl-pyrrolidine; hygrinic acid, *N*-methyl-2-pyrrolidinecarboxylic acid. **35.11** Orientation ("*para*") controlled by activating —NH$_2$ group.   **35.13** Amine > imine > nitrile; $sp^3 > sp^2 > sp$.   **35.18** Piperidine, a 2° amine, would itself be acylated.   **35.23** (a) 8-Nitroquinoline; (b) 8-hydroxyquinoline (8-quinolinol); (c) 4,5-diazaphenanthrene; (d) 1,5-diazaphenanthrene; (e) 6-methylquinoline.   **35.28** Electrophilic aromatic substitution or acid-catalyzed nucleophilic carbonyl addition, depending upon viewpoint.

1. No reaction: c, h, i, j.   **3.** Pyrroline has double bond between C–3 and C–4.   **4.** C, 2,6-hexanedione (acetonylacetone).   **5.** Porphin, with same ring skeleton as in heme, p. 1367.   **6.** D, 2-COOH; E, 3-COOH; F, 4-COOH.   **7.** (a) 5- or 7-methylquinoline; (b) G, 7-methylquinoline.   **9.** (e) Perkin reaction; (g) Reimer–Tiemann reaction.   **10.** (See below for parent ring systems.) I, 2,4,6-trihydroxy-1,3-diazine; K, 3,6-dimethyl-1,2-diazine; L, 3,5-dimethyl-1,2-diazole; M, 2,3-dimethyl-1,4-diazanaphthalene; N, 1,3-dioxolan-2-one (ethylene carbonate); P, 3-indolol; R, 2,5-dimethyl-1,4-diazine; S, 1,3-diazolid-2-one (2-imidazolidone, ethyleneurea); T, 4,5-benzo-2-methyl-1,3-diazole (2-methylbenzimidazole); W, 2,4-dihydroxyquinoline; BB, 1,2-diazolid-3-one (3-pyrazolidone); CC, 4,5-diazaphenanthrene; GG, two indole units fused 2,3 to 3′,2′; HH, *N*-methyl-1,2,3,4-tetrahydroquinoline; II, 2-phenylbenzoxazole; JJ, the benzene ring of II completely hydrogenated.

1,3-Diazine   1,2-Diazine   1,2-Diazole   1,4-Diazanaphthalene
(Pyrimidine)   (Pyridazine)   (Pyrazole)   (Quinoxaline)

1,3-Dioxolane   Indole   1,4-Diazine   1,3-Diazole
                   (Benzopyrrole)   (Pyrazine)   (Imidazole)

Benzo-1,3-diazole 4,5-Diazaphenanthrene   Benzoxazole
(Benzimidazole)   (4,5-Phenanthroline)

11. LL, 3,4-(CH$_3$O)$_2$C$_6$H$_3$CH$_2$CH$_2$NH$_2$; NN, 3,4-(CH$_3$O)$_2$C$_6$H$_3$CH$_2$COCl; OO, amide; PP, a 1-substituted 7,8-dimethoxy-3,4-dihydroisoquinoline; papaverine, the corresponding substituted isoquinoline.   **12.** VV, (C$_2$H$_5$)$_2$NCH$_2$CH$_2$CH$_2$CHBrCH$_3$; XX, 8-amino-6-methoxyquinoline; Plasmochin, 8-amino group of XX alkylated by VV.   **13.** Nicotine, 2-(3-pyridyl)-*N*-methylpyrrolidine.   **14.** DDD, *o*-hydroxybenzalacetophenone; (c) oxygen contributes a pair of electrons to complete an aromatic sextet.   **15.** Aliphatic NH$_2$ > "pyridine" N > "pyrrole" NH.   **16.** Tropinic acid, 2-COOH-5-CH$_2$COOH-*N*-methylpyrrolidine.   **18.** Pseudotropine has equatorial—OH, is more stable.   **19.** (a) Guvacine, 1,2,5,6-tetrahydro-3-pyridinecarboxylic acid; arecaidine, *N*-methylguvacine; (b) nicotinic acid.   **20.** UUU, one enantiomer of ethyl-*n*-propyl-*n*-butyl-*n*-hexylmethane; chirality does not necessarily lead to measurable optical activity (see Sec. 4.13).   **21.** Dipolar ion loses CO$_2$.

Chapter 36

36.1 (a) Amide; see Sec. 36.7; (b) amide; 6-aminohexanoic acid; (c) ether; ethylene oxide; (d) chloroalkene; 2-chloro-1,3-butadiene; (e) chloroalkane; 1,1-dichloroethene.

36.2 (a) Amide; (b) ester; (c) acetal; (d) acetal. **36.3** 1,2- and 1,4-addition. **36.4** Combination. **36.6** Polymer is transfer agent. **36.9** (a) Chain transfer.

2. Dehydration, polymerization. **4.** Nucleophilic carbonyl addition. **5.** Hydrolysis gives amine, alcohol, and carbon dioxide. **10.** $\sim\sim OCH_2CH_2COOCH_2CH_2COO\sim\sim$; chain reaction. **11.** Growing anion abstracts proton from solvent. **12.** Some head-to-head polymerization. **13.** (a) $\sim\sim NHCH_2(CH_2)_4CO\sim\sim$; (b) chain reaction. **14.** Cyclohexanone. **16.** Compounds are ionic, due to stability of benzylic anions. **17.** A, *meso*, resembles isotactic; B, racemic, resembles syndiotactic. **19.** Monomer acts as chain-transfer agent. **20.** Cross-linking by oxygen between allylic positions. **22.** F, syndiotactic; G, isotactic.

Chapter 37

37.1 Decarboxylation. Fatty acids could be precursors of petroleum hydrocarbons. **37.2** (a) Isoprene unit. (b) Likely that petroleum comes from green plants. **37.3** Tung oil is high in eleostearic acid (3 double bonds). **37.4** Alkoxide is a poor leaving group. **37.5** Preserves semiliquidity of membranes in colder part of body.

1. Nervonic acid, *cis-* or *trans-*$CH_3(CH_2)_7CH=CH(CH_2)_{13}COOH$ (actually, *trans*). **2.** Transesterification to more random distribution of acyl groups among glyceride molecules. **3.** Hybrid (allylic) free radical is intermediate. **4.** $2,4\text{-}(NO_2)_2C_6H_3O^-$ is a good leaving group. **5.** Spermaceti, *n*-hexadecyl *n*-hexadecanoate. **6.** Cleavage of monoanion as dipolar ion (or with simultaneous transfer of proton) is easiest because of (a) protonation of alkoxy group and (b) double negative charge on other oxygens:

$$ R\overset{+}{\underset{H}{-O}}{-}PO_3{}^{2-} \xrightarrow{\ H_2O\ } ROH + H_2PO_4{}^- $$

7. Vaccenic acid, *cis*-$CH_3(CH_2)_5CH=CH(CH_2)_9COOH$. **8.** Corynomycolenic acid, *cis*-n-$C_{13}H_{27}CH_2CH(COOH)CHOH(CH_2)_7CH=CHC_6H_{13}$-$n$. **9.** Tuberculostearic acid, 10-methyloctadecanoic acid. **10.** C_{27}-phthienoic acid, $CH_3(CH_2)_{17}CH(CH_3)CH_2CH\text{-}(CH_3)CH=C(CH_3)COOH$. **11.** CC, octadecanoic acid; DD, 2-methyloctadecanoic acid. **12.** Juvenile hormone,

Chapter 38

38.2 Formulas I–VIII, p. 1292. **38.3** (a) 3; (b) 8. **38.4** Glucose + $5HIO_4 \rightarrow 5HCOOH + HCHO$. **38.5** A, gluconic acid; B, glucitol; C, glucaric acid; D, glucuronic acid. **38.6** Fructose. Aldose → osazone → osone → 2-ketose. **38.7** Identical in configuration at C–3, C–4, and C–5. **38.8** Alditol. **38.9** (a) 2 tetroses; (b) 4 pentoses, 8 hexoses (see Problem 38.2); (c) lowest chiral C has OH on right. **38.10** One product (S,S) would be optically active, one product (*meso*) optically inactive. **38.11** I, (+)-allose; II, (+)-altrose; VI, (−)-idose; VII, (+)-galactose; VIII, (+)-talose. **38.15** (a) R; (b) R; (c) S; (d) R. **38.16** (S)-(+)-2-butanol. **38.17** (a) S,S-; (b) R,R-; (c) R,S-. **38.18** (b) 1:3; (c) the isomer favored in the L-series will be the mirror image of the isomer favored in the D-series. **38.19** L-(+)-Gulose. **38.20** (a) 36.2% α, 63.8% β. **38.22** Acetylation occurs at C–1 to give diastereomers (anomers). **38.23** (a) CH_3OH, HOOCCHO, and D-glyceric acid. **38.24** HCHO instead of HCOOH. **38.25** (a) Six-membered ring; (b) HCOOH, OHC—CHO, and HOCH_2CHO. **38.26** (a) Six-membered ring; (b) enantiomer. **38.27** (a) Five-membered ring; (b) optically active, L-family; (c) enantiomer.

4. E and E', allitol and galactitol; F, glucitol (or gulitol); H, glucitol (or gulitol); I and I', allitol and galactitol; N, ribitol; O, arabitol (or lyxitol). **5.** (a) P, glycoside of glucuronic acid; (d) $HOCH_2(CHOH)_3COCOOH$. **6.** Rate-determining step involves OH^- before reaction with Cu^{2+}: probably abstraction of proton leading to formation of enediol. **7.** (a) 5 carbons, five-ring; (b) C–1 and C–4; (c) Q, methyl α-D-arabinofuranoside. **8.** Salicin, o-(hydroxymethyl)phenyl β-D-glucopyranoside. **9.** Bio-inonose, the pentahydroxycyclo-hexanone in which successive —OH groups are trans to each other. **11.** (a) T, D-ribose; U, D-arabinose; (b) 3-phosphate. **12.** Z and AA are ketals: Z, furanose with acetone bridging C–1 to C–2 and C–5 to C–6; AA, furanose, with acetone bridging C–1 to C–2. **14.** S_N1-like, with separation of relatively stable oxonium ion (see Sec. 21.13). **15.** (a) Proton on C–1 most deshielded by two oxygens. (b) JJ, β-anomer; KK, α-anomer; (c) LL, β-anomer; MM, α-anomer; (d) NN, α-mannose; OO, β-mannose; PP, β-glucose; QQ, α-glucose.

16. L-(−)-Mycarose,

(e) α-glycoside; (f) β-anomer.

17. (a) Anomeric effect (Sec. 38.20) stabilizes the α-anomer; (b) anomeric effect stabilizes diaxial chlorines. **18.** (a) On steric grounds, neither; anomeric effect would favor axial OAc on C–1. (b) Tells nothing: in either conformation two OAc are equatorial, two are axial. (c) The e:a peak area ratio would be 2:1 if C–1 OAc were all axial, 1:1 if half axial, 0.5:1 if none axial. Ratio of 1.46:1.00 shows C–1 OAc is axial in 78% of molecules.

Chapter 39

39.1 Differ at C–1 of reducible glucose unit only. **39.2** Methoxyacetic acid and di-O-methyl-D-glyceric acid. **39.3** 2,3,4,6-Tetra- and 2,3,6-tri-O-methyl-D-glucose. **39.4** D-Glucose and D-erythrose; indicates attachment to other ring is at C–4. **39.6** Same as in Fig. 39.1 except for β-linkage in first three formulas. **39.7** 2,3,4,6-Tetra-O-methyl-D-galactose and 2,3,5,6-tetra-O-methyl-D-gluconic acid. **39.8** D-Galactose and D-erythrose. **39.9** $(-92.4° + 52.7°)/2 = -19.9°$. **39.10** $C_{12}H_{20}O_{10}$, non-reducing. **39.11** Sucrose is an α-glucoside. **39.12** Di-O-methyl-L- and D-tartaric acids. **39.13** 1 (0.025%); 3 (0.075%); 9 (0.225%). **39.14** (a) A large group in an axial position. **39.15** (a) 3 molecules of HCOOH per molecule of amylose; (b) moles HCOOH/3 = moles amylose; wt. amylose/moles amylose = mol.wt. amylose; mol.wt. amylose/wt. (of 162) per glucose unit = glucose units per molecule of amylose; (c) 980. **39.16** A poly-α-D-glucopyranoside; chain-forming unit, attachment at C–1 and C–6; chain-linking unit, attachment at C–1, C–3, and C–6; chain-terminating unit, attachment at C–1. **39.17** A poly-β-D-xylopyranoside; chain-forming unit, attachment at C–1 and C–4; chain-linking unit, attachment at C–1, C–3, and C–4; chain-terminating unit, attachment at C–1. **39.19** The ionic sulfonate end. **39.20** α: cyclohexane; β: PhF; γ: anthracene. **39.21** The para is smallest; the meta largest.

1. Gentiobiose, 6-O-(β-D-glucopyranosyl)-D-glucopyranose. **2.** (a) Trehalose, α-D-glucopyranosyl α-D-glucopyranoside; (b) isotrehalose, α-D-glucopyranosyl β-D-glucopyranoside; neotrehalose, β-D-glucopyranosyl β-D-glucopyranoside. **4.** Raffinose, α-D-galactosyl unit attached at C–6 of glucose unit of sucrose; melibiose, 6-O-(α-D-galactopyranosyl)-D-glucopyranose. **5.** (a) Melezitose, α-D-glucopyranosyl unit attached at C–3 of fructose unit of sucrose; turanose, 3-O-(α-D-glucopyranosyl)-D-fructofuranose. **6.** Panose, α-D-gluco-pyranosyl unit attached at C–6 of non-reducing moiety of maltose; isomaltose, 6-O-(α-D-glucopyranosyl)-D-glucopyranose. **7.** (b) D-Glucuronic acid; (c) D-xylose. **9.** B, furan (p. 1209); C, tetrahydrofuran (p. 1213 and Sec. 19.9); E, N≡C(CH₂)₄C≡N. Furfural (p. 1210). **12.** I, D-CH₂OHCHOHCHOHCOOH, D-erythronic acid; J, HOOCCHO, glyoxylic acid. **13.** (a) 3 molecules of HCOOH per molecule of cellulose; (c) 1390 glucose units.

Chapter 40

40.1 $-NH_2 > -COO^-$; proton goes to $-NH_2$ to form $^+H_3NCHRCOO^-$.
40.2 $-COOH > -NH_3^+$; $-COOH$ gives up proton to form $^+H_3NCHRCOO^-$. **40.5** (a)
On acid side; (b) on basic side; (c) more acidic and more basic than for glycine. **40.8** $(-)$-
Cysteine and $(-)$-cystine. S takes priority over O of carbonyl group. **40.9** 4 isomers.
40.10 Cys–Cys, Hyl, Hyp, Ile. **40.12** Intermediate for Ala is $CH_3CH(NH_2)CN$. **40.13** A,
$(CH_3)_2CHCH(COOEt)COCOOEt$; B, $(CH_3)_2CHCH_2COCOOEt$. **40.16** (a) 22.4 mL;
(b) 44.8 mL; (c) no N_2. **40.17** Minimum mol.wt. = 114; could be valine. **40.20** Salmine,
$AlaArg_{50}Gly_4IlePro_6Ser_7Val_3$. **40.21** Same as empirical formula (preceding problem).
40.22 70 300. **40.23** (a) 16 700; (b) 4. **40.24** A sulfonamide, which is more resistant
to hydrolysis than carboxamides (see Sec. 27.7). **40.26** (a) $-COOH \rightarrow -CH_2OH$;

(b) $-COOH \rightarrow -C\overset{\displaystyle O}{\underset{\displaystyle NHNH_2}{\big\langle}}$ a hydrazide. **40.27** (a) Phe-Val-Asp-Glu-His; (b) His-Leu-

Cys-Gly-Ser-His-Leu; (c) Tyr-Leu-Val-Cys-Gly-Glu-Arg-Gly-Phe-Phe. **40.28** (a) Cbz-Gly-
Ala, $SOCl_2$; Phe; H_2, Pd. (b) $PhCH_2OCOCl$, Ala; $SOCl_2$; Gly; H_2, Pd. **40.29** In A,
polystyrene has $-CH_2Cl$ groups attached to rings; in G, $-CH_2Br$ groups.

2. D, $HOCH_2CH_2CH_2CH(NH_3^+)COO^-$. **3.** (a) Diketopiperazine, cyclic diamide;
(b) unsaturated acid; (c) γ-lactam, 5-ring amide; (d) δ-lactam, 6-ring amide. **5.** Betaine,
$^+(CH_3)_3NCH_2COO^-$. **6.** Polarity of solvent lowered; lipophilic parts of organic molecules
come out of their huddle. **8.** Minimum mol.wt. = 13 000; minimum of one Fe atom and
six S atoms. **9.** (a) Approx. 32 $-CONH_2$ groups; (b) 395–398 peptide links plus $-CONH_2$
groups; (c) 367–370 amino acid residues.

Val-Orn-Leu-Phe

10.

Gramicidin S
Cyclic decapeptide

11. Beef insulin:

Chain A:

Chain B:

(g) $DNP-NH(CH_2)_4CH(NH_3^+)COO^-$ from ε-amino group of Lys. If Lys had been
N-terminal, would have got a double DNP derivative of it, and no DNP-Phe.

Chapter 41

1. CO_2 becomes the $-COOH$ of malonyl CoA in reaction (1), Sec. 41.7; this is the
carbon lost in reaction (4). **2.** Slow (rate-determining) formation of a tetrahedral interme-

diate (see Sec. 24.17) followed by fast loss of OR or SR.   **3.** (a) Aldol-like condensation between ester and keto group of oxaloacetate; (b) aldol-like condensation between ester and keto group of acetoacetyl CoA; reduction of ester to 1° alcohol by hydride transfer. **4.** Biological oxidation of fatty acids removes 2 carbons at a time, starting at the carboxyl end: "*beta*-oxidation".   **5.** *Retro* (reverse) aldol condensation.   **6.** Far ultraviolet. **7.** Single-strand helix.   **8.** Two different RNAs would be generated, and hence two different sets of amino acids.

Index

M

PERIODIC TABLE OF THE ELEMENTS

Group	I	II										III	IV	V	VI	VII	VIII
Period																	
1	H 1																He 2
2	Li 3	Be 4										B 5	C 6	N 7	O 8	F 9	Ne 10
3	Na 11	Mg 12				Transition elements						Al 13	Si 14	P 15	S 16	Cl 17	Ar 18
4	K 19	Ca 20	Sc 21	Ti 22	V 23	Cr 24	Mn 25	Fe 26	Co 27	Ni 28	Cu 29	Zn 30	Ga 31	Ge 32	As 33	Se 34	Br 35	Kr 36
5	Rb 37	Sr 38	Y 39	Zr 40	Nb 41	Mo 42	Tc 43	Ru 44	Rh 45	Pd 46	Ag 47	Cd 48	In 49	Sn 50	Sb 51	Te 52	I 53	Xe 54
6	Cs 55	Ba 56	* 57–71	Hf 72	Ta 73	W 74	Re 75	Os 76	Ir 77	Pt 78	Au 79	Hg 80	Tl 81	Pb 82	Bi 83	Po 84	At 85	Rn 86
7	Fr 87	Ra 88	† 89–103	Unq 104	Unp 105	Unh 106												

*Lanthanide series	La 57	Ce 58	Pr 59	Nd 60	Pm 61	Sm 62	Eu 63	Gd 64	Tb 65	Dy 66	Ho 67	Er 68	Tm 69	Yb 70	Lu 71
†Actinide series	Ac 89	Th 90	Pa 91	U 92	Np 93	Pu 94	Am 95	Cm 96	Bk 97	Cf 98	Es 99	Fm 100	Md 101	No 102	Lr 103